Lecture Notes in Civil Enginee

Volume 171

Lecture Notes in Civil Engineering (LNCE) publishes the latest developments in Civil Engineering—quickly, informally and in top quality. Though original research reported in proceedings and post-proceedings represents the core of LNCE, edited volumes of exceptionally high quality and interest may also be considered for publication. Volumes published in LNCE embrace all aspects and subfields of, as well as new challenges in, Civil Engineering. Topics in the series include:

- Construction and Structural Mechanics
- Building Materials
- Concrete, Steel and Timber Structures
- Geotechnical Engineering
- Earthquake Engineering
- Coastal Engineering
- Ocean and Offshore Engineering; Ships and Floating Structures
- Hydraulics, Hydrology and Water Resources Engineering
- Environmental Engineering and Sustainability
- Structural Health and Monitoring
- Surveying and Geographical Information Systems
- Indoor Environments
- Transportation and Traffic
- Risk Analysis
- Safety and Security

To submit a proposal or request further information, please contact the appropriate Springer Editor:

- Pierpaolo Riva at pierpaolo.riva@springer.com (Europe and Americas);
- Swati Meherishi at swati.meherishi@springer.com (Asia - except China, and Australia, New Zealand);
- Wayne Hu at wayne.hu@springer.com (China).

All books in the series now indexed by Scopus and EI Compendex database!

More information about this series at http://www.springer.com/series/15087

Giuseppe Carlo Marano · Samit Ray Chaudhuri ·
G. Unni Kartha · P. E. Kavitha ·
Reshma Prasad · Rinu J. Achison
Editors

Proceedings of SECON'21

Structural Engineering and Construction Management

Editors
Giuseppe Carlo Marano
Department of Structural, Geotechnical and Building Engineering
Politecnico di Torino
Turin, Italy

G. Unni Kartha
Department of Civil Engineering
Federal Institute of Science and Technology (FISAT)
Angamaly, Kerala, India

Reshma Prasad
Department of Civil Engineering
Federal Institute of Science and Technology (FISAT)
Angamaly, Kerala, India

Samit Ray Chaudhuri
Department of Civil Engineering
Indian Institute of Technology Kanpur
Kanpur, Uttar Pradesh, India

P. E. Kavitha
Department of Civil Engineering
Federal Institute of Science and Technology (FISAT)
Angamaly, Kerala, India

Rinu J. Achison
Department of Civil Engineering
Federal Institute of Science and Technology (FISAT)
Angamaly, Kerala, India

ISSN 2366-2557 ISSN 2366-2565 (electronic)
Lecture Notes in Civil Engineering
ISBN 978-3-030-80314-8 ISBN 978-3-030-80312-4 (eBook)
https://doi.org/10.1007/978-3-030-80312-4

This Springer imprint is published by the registered company Springer Nature Switzerland AG
The registered company address is: Gewerbestrasse 11, 6330 Cham, Switzerland

Contents

Utilization of Paper Sludge Ash in Lime Based Geopolymer Concrete: An Experimental Study

P. L. Meyyappan, R. Sutharsan, and M. Jemimah Carmichael

Abstract In past researches, it is indicated that the usage of industrial waste materials is the better approach to make the sustainable concrete industry. Cement manufacturing industry is one of biggest industry which emits large amount of Co_2 that's roughly estimated as 7% of green house to gas to the earth's atmosphere. Some of the challenges related mechanical and durability properties of conventional concrete are addressed by the geopolymer concrete. This geopolymer concrete consists of a type of inorganic polymer which has silica content and low calcium oxide content. Also the usage of lime on the concrete enhances the strength property of the geopolymer concrete. In this aspect, the experimental work is conducted with the utilization of paper sludge ash in the proportions of 0, 25, 50, 75 and 100% in the lime based geopolymer concrete. The specimens of size 150 mm × 150 mm × 150 mm were casted, cured in the ambient temperature against 7, 14 and 28 days. These specimens are tested to arrive the compressive strength, split tensile strength and water absorption properties. Appropriate discussions were made based on the experimental test results in the representation of tables and graphs. It is concluded that, 25% of paper sludge ash in the lime based geopolymer concrete is found to be optimum in enhancing those mechanical and durability properties.

Keywords Paper sludge ash · Lime · Geopolymer concrete · Flyash · Strength · Durability

---

P. L. Meyyappan (✉) · R. Sutharsan
Kalasalingam Academy of Research and Education, Krishnankoil, India
e-mail: meyyappan@klu.ac.in

M. Jemimah Carmichael
Vignan's LARA Institute of Technology and Science, Guntur, India

G. C. Marano et al. (eds.), *Proceedings of SECON'21*, Lecture Notes in Civil Engineering 171, https://doi.org/10.1007/978-3-030-80312-4_1

1 Introduction

Due to the tremendous increasing rate of construction activities in the urban areas demanding the production of Portland cement to the larger extent. This leads to several environmental threat to the environment in terms of Co_2 emission and green house effect. This can be lowered by limiting the usage of cement without neglecting the requirements. Several studies [1, 2] indicating the alternative building materials in terms of partial/full replacement proportions of the sustainable waste materials such as flyash, GGBS, silica fume, metakaolin, rice husk ash, granite dust, marble dust, quarry dust etc. Due to advantage of full replacement of cement is possible in concept of geopolymer concrete. The binder material of flyash or GGBS or metakaolin with the prepared alkaline solution (NaoH and Na_2SiO_3) to make a concrete material like conventional cement concrete. The role of alkaline activator is act as a catalyst to activate the geopolymeric source materials containing Si and Al such as flyash or GGBS as described in Kong et al. [3], Ganesan et al. [4]. As mentioned in the past researches such as Wangsa et al. [5] and Hakea and Damgirb [6], the presence of lime in the geopolymer concrete will activate the heat generation requirement for the completion of geopolymer reaction. Moreover it is indicated that, the addition of lime will not required temperature curing and ensure the strength properties. The main intention of this work is to arrive the compression strength, split tension strength and volume of pores of the lime based geopolymer concrete with and without the utilization of paper sludge ash in the proportions of 0, 25, 50, 75 and 100% in replacement for flyash content. The optimized utilization of these sustainable materials in the geopolymer concrete is arrived for achieving the enhancement in the strength and durability characteristics.

2 Materials Used

(i) ***Fine aggregate***
The river sand is taken as fine aggregate material. For this study, material retaining in 2.36 sieve is considered. It has specific gravity 2.78, fineness modulus 3.24 and confirming to zone II category as per the norms of IS 382-2016.

(ii) ***Coarse aggregate***
Crushed coarse aggregate obtained from the nearby quarry is taken for this study. Retaining in the sieve 4.75 mm is considered. It has specific gravity 2.782 as per the standards of IS 383-1970.

(iii) ***Fly ash***
Class C category of flyash is acquired from Tuticorin thermal power plant, Tamil Nadu. The collected flyash is sieved from 90 micron sieve and the passing material is considered for this study.

(iv) ***Lime***
A white caustic alkaline substance consisting of CaO which is obtained by heating of limestone is taken for this study.

(v) ***Paper sludge ash***
The paper sludge is collected from one of the leading paper industry in Tamil Nadu. It is dried and is burnt into ashes which are having specific gravity 2.63 and fineness of 10% is considered. The chemical composition of sludge ash found and it is having CaO—64.94%, SiO_2—17.35%, SO_3—4.68%, Al_2O_3—4.12%, MgO—3.69%, Fe_2O_3—1.15% and minor composition of K_2O and TiO_2.

(vi) ***Alkaline liquids***
For preparing alkaline activator solution, the solution of sodium hydroxide and sodium silicate are mixed in the ratio of 1(NaOH): 2.5 (Na_2SO_3).

(vii) ***Admixture***
The admixture La-Hypercrete S-30 (based on type of Modified Carboxylic acid) is in this study.

3 Experimental Investigations

Activator solution is prepared in the ratio of 1 parts of sodium hydroxide solution (NaOH) and 2.5 parts of sodium silicate solution (Na_2SiO_3) just before 24 h of casting of the specimens for achieving 8 molarities. Binders (90% Flyash and 10% of lime), fillers (both aggregates) are homogeneously mixed in the condition of dry state and essential care has to be ensured for avoiding some chunk or lumps. By mixing together of activator and admixture solution, the geopolymer concrete mix is prepared into the dry mixture of binding and filler materials in the rotated mixing machine to make sure the homogeneity conditions as per the mix proportions stated in Table 1. The mix design is arrived based on the mix design procedure for geopolymer concrete as referred by Pavithira et al. [7]. Once the geopolymer concrete mix, is prepared, it has cast in to the cube mould of having size 150 mm × 150 mm × 150 mm and cylinder of size 150 mm × 300 mm. Table 2 indicates the number of cube and cylinder specimens to be casted for this experimental programme. For each proportion of paper sludge ash (0, 25, 50, 75 and 100%) 12 cubes were casted for 7 days and 28 age of curing for compression and water absorption test. Similarly 6 cylinder specimens were casted against each

Table 1 Specimen details

Fly ash (kg/m^3)	Lime (kg/m^3)	Fine aggregate (kg/m^3)	Coarse aggregate (kg/m^3)	NaOH (kg/m^3)	Na_2SiO_3 (kg/m^3)	Super plasticizer (kg/m^3)
360	40	542	1281	60	150	5.5

Table 2 Specimen details

S. No	Paper sludge ash proportions (%)	No. of cube specimens		No. of cylinder specimens
		Compression test	Water absorption test	Split tensile test
1	0	6	6	6
2	25	6	6	6
3	50	6	6	6
4	75	6	6	6
5	100	6	6	6
Total specimens		60		30

proportion. Totally 60 cubes and 30 cubes were casted this study against various paper sludge ash proportions and age of curing. All the 90 specimens were kept under ambient curing. The compression tests and split tension test were conducted in the compression testing machine of 1000 kN capacity and tests were carried out at a unfailing standardized stress rate of 14 N/mm^2/min, once the specimen is kept in the machine. The failure load is noted and compressive strength is obtained by dividing contact area of the specimen.

Water absorption test were conducted on the cube specimens to estimate the volume of pores. The specimens were permitted to keep in the hot air oven in the typical temperature of range of 70–90°. The specimen in the dry conditions, weights to be taken before and after keeping in the hot air oven which is considered as W1 and W2. The differences in the weights were taken for the estimating the volume of pores in terms of percentage.

4 Results and Discussions

Table 3 shows the experimental test results of compression strength, split tension strength and volume of pores on the lime based geopolymer concrete having without and with replacement for flyash by paper sludge ash in the proportions of 0, 25, 50, 75 and 100%. It is observed that, the 7 days curing, the compression strength of this geopolymer concrete for the replacement proportions of 0%, 25%, 50%, 75% and 100% are 16.81 N/mm^2, 20.16 N/mm^2, 18.23 N/mm^2, 11.49 N/mm^2 and 6.92 N/mm^2 respectively similarly for the 28 days curing, the compressive strength is 26.37 N/mm^2, 31.52 N/mm^2, 27.71 N/mm^2, 18.93 N/mm^2 and 10.13 N/mm^2 respectively. It is evident that the age of curing accelerates the compressive strength as seen from Fig. 1. It is visible that there is an enhancement in the compressive strength by 16.34%, if 25% paper sludge ash replacement when compared with 0% replacement proportions (Control specimens). But if the percentage of replacement of paper sludge ash is increased to 50, 75 and 100% to the

Table 3 Strength and water absorption test results

S. No	Paper sludge ash proportions (%)	Compressive strength (N/mm^2)	Split tensile strength (N/mm^2)	Water absorption test (%)			
		7 days	28 days	7 days	28 days	7 days	28 days
1	0	16.81	26.37	1.42	2.51	9.84	6.56
2	25	20.16	31.52	1.56	2.79	7.76	5.18
3	50	16.23	26.71	1.37	2.44	7.84	5.23
4	75	11.49	18.93	1.01	1.63	10.29	7.21
5	100	6.92	10.13	0.61	0.92	14.34	10.74

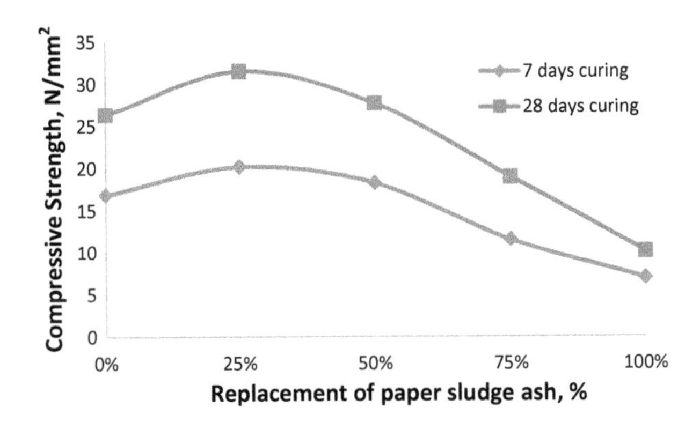

Fig. 1 Compressive strength versus paper sludge ash replacements

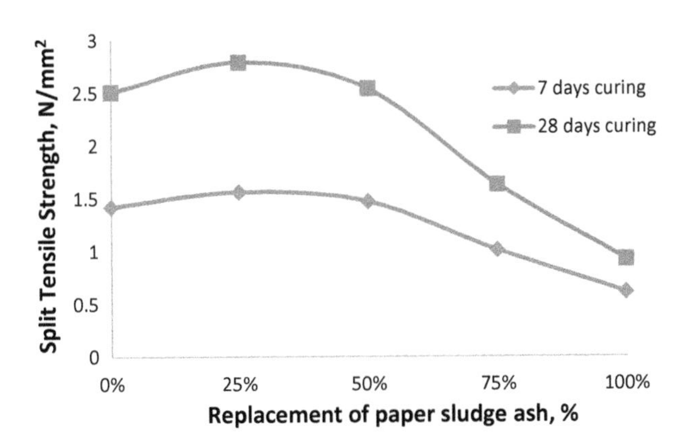

Fig. 2 Split tensile strength versus paper sludge ash replacements

flyash content, the compressive strength is decreased to 15.26, 40.95 and 67.86% as compared with 25% replacement of paper sludge ash. It is seen that, the reduction level of compressive strength is gradual from 25 to 50% and drastic reduction from 50 to 100%. The 50% replacement of paper sludge ash has compressive strength which is almost same for 0% replacement levels. In a lower replacement percentage, the paper sludge ash along with the flyash content seems to be advantage for the enhancement of compressive strength (Figs. 1, 2).

The split tensile strength of this geopolymer concrete for 7 days age of curing is found as 1.42 N/mm^2, 1.56 N/mm^2, 1.37 N/mm^2, 1.01 N/mm^2 and 0.61 N/mm^2 for 0%, 25%, 50%, 75% and 100% replacement proportions respectively. For 28 days age of curing, the split tensile strength for above replacement proportions are 2.51 N/mm^2, 2.79 N/mm^2, 2.44 N/mm^2, 1.63 N/mm^2 and 0.92 N/mm^2 respectively. A slight increase (8.89%) is in 25% replacement of paper sludge ash as compared with 0% replacement. If the paper sludge ash replacement is increased beyond 25%, the split tensile strength is getting decreased as seen in Fig. 2. For the replacement of paper sludge 50%, 75% and 100%, the split tensile strength is decreased to 12.54%, 41.57% and 67.02% respectively. As like compressive strength, the 25% replacement of paper sludge ash into the flyash seems to be optimum in terms of increase in the split tensile strength.

The water absorption test results for the specimens having 7 days age of curing is found as 9.84%, 7.76%, 7.84%, 10.29% and 14.34% for the replacement proportions of 0%, 25%, 50%, 75% and 100% respectively. For specimens of 28 days age of curing, percentage of pores are 6.56%, 5.18%, 5.23%, 7.21% and 10.74% respectively. It is observed that, 7 days age of curing has more percentage of voids since the formation of bonding between the aggregates and filler materials is not sufficient. As the age of curing days increased, the pores have been reduced and thereby the bonding between these constituent materials binder and fillers are adequate. It is also observed that, if the replacement of paper sludge ash upto 25% there is some reduction of 21.03% in the total volume of pores which indicating that

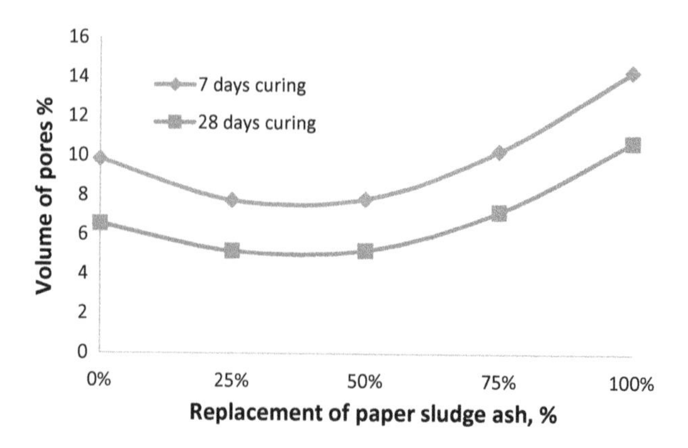

Fig. 3 Volume of pores versus paper sludge ash replacements

the fine particles of paper sludge ash are occupying the pores of the concrete. From Fig. 3, it is seen that, the percentage of voids is almost same for both 25% and 50% replacement of paper sludge ash. Further if the replacement percentage is increased to 75% and 100%, the presences of pores are increased to 28.15% and 51.76% respectively since the binding was not effective due to over saturation of paper sludge ashes. The higher amount of paper sludge ash leads a path which can able to percolates more water into it due to non binding effect within the structure. The higher water absorption rate is due to less binding effect. This effect is witnessed during the strength test of the sample, that failure crack is seen only on the interface of the aggregates within a short interval.

5 Conclusions

The following concluding remarks have been arrived based on this experimental work.

(i) A sustainable attempt is made by utilizing the waste paper sludge ash in the various replacement proportions such as 0%, 25%, 50%, 75% and 100% to the flyash content in making lime based geopolymer concrete and their behaviour is studied through the strength and durability tests.
(ii) The optimum utilization of the paper sludge ash is found to be 25% based on the enhancement of strength and durability properties of this concrete.
(iii) The waste paper sludge ash is found effective upto 25% replacement in terms of increasing 16.34% and 10.03% of compressive strength and split tensile strength respectively. Maximum pores reduction is also seen as 21.04%.
(iv) If the replacement percentage is increased beyond 25%, there is a reduction trend which is seen and this is found common in both strength and durability properties.

References

1. Singh B, Ishwarya G, Gupta M, Bhattacharyya SK (2015) Geopolymer concrete: a review of some recent developments. Constr Build Mater 85
2. Meyyappan PL, Kumaran K, Gopalakrishnan M, Harikrishnan E (2018) Effect of glass fibers, flyash and quarry dust on strength and durability aspects of concrete—an experimental study. IOP Conf Ser Mater Sci Eng. https://doi.org/10.1088/1757-899X/396/1/012001
3. Kong DLY, Sanjayan JG (2010) Effect of elevated temperatures on geopolymer paste, mortar and concrete. Cement Concr Res 40
4. Ganesan N, Ramesh Babu C, Meyyappan PL (2019) Influence of alkaline activator ratio on compressive strength of GGBS based geopolymer concrete. In: IOP Conference Series, Material Science and Engineering, p 012083

5. Wangsa FA, Tjaronge MW, Djamaluddin AR, Muhiddin AB (2017) Effect of hydrated lime on compressive strength mortar of fly ash laterite soil geopolymer mortar. IOP Conf Ser, Mater Sci Eng:012068
6. Hakea L, Damgirb RM (2019) Investigate mixing ingredient for lime added geopolymer concrete. Glob J Civil Eng 1
7. Pavithra P, Srinivasula Reddy M, Pasla Dinakar, Hanumantha Rao B, Satpathy BK, Mohanty AN (2016) A mix design procedure for geopolymer concrete with flyash. J Clean Prod 133(1):117–125

Seismic Resisting Performance and Strengthening of Single and Double Corrugated Steel Plate Shear Wall

Abraham Varughese Lalu and Aswathy Ann Mathew

Abstract Corrugated steel plate shear walls (CSPSWs) are most commonly used as a lateral load resisting mechanism in earthquake prone areas. They are used in modular building structures (MBS) and steel structures for its high ductility, and strength. Double corrugated steel plate shear wall (DCSPSW) contains two trapezoidal corrugated steel plates that are connected together either by weld or bolts. It is most often used as an alternative for the conventional steel plate shear walls. Since openings like window and door are unavoidable, the performance of the system with and without openings must be studied. The performance and change of strength of the walls on addition of stiffeners around the openings are to be evaluated. In this paper, the seismic performance of DCSPSW with and without, openings and stiffeners are investigated and compared with that of the ordinary (single) CSPSW.

Keywords Corrugated Steel Plate Shear Wall (CSPSW) · Double-Corrugated Steel Plate Shear Wall (DCPSW) · Seismic performance · Modular building structures

1 Introduction

Steel structures are commonly used in seismic hazard area, for its high strength and malleability. CSPSW are generally used as seismic resisting member principally in MBS and in steel structures in those seismic zones. It includes infill plate (stiffened or unstiffened), with and without openings, with vertical and horizontal structural components. From similar researches, the CSPSW is found to be a cost-effective economical methodology for high rise buildings than typical strategies.

Numerical studies concerning the SPSW was carried out in the past was on flat plates being used as infill plates. The corrugated steel plates are used as replacement

A. V. Lalu · A. A. Mathew (✉)
Department of Civil Engineering, Saintgits College of Engineering, Saintgits Group of Institutions, Kuttukulam Hills, Pathamuttom, Kottayam, Kerala 686532, India

G. C. Marano et al. (eds.), *Proceedings of SECON'21*, Lecture Notes in Civil Engineering 171, https://doi.org/10.1007/978-3-030-80312-4_2

for their high out-of- plane geometric stability. Due to the high stiffness of corrugated plates despite of its lower thickness than flat plates have been helpful for the construction of light girders [1–3].

The strength of the RBS shear wall could be calculated by a proposed equation, verified on comparison with the value obtained by FE analysis by providing RBS on the beam section to ensure the plastic hinges form on the beam rather on the beam span or on columns [4].

By experimental and numerical means the seismic performance of low and midrise buildings with CSPSW with slits indicated that the shear walls with perforation provided desirable ductility and strength than shear walls without perforation [5]. On application of monotonic loading on CSPSW with and without opening the parameters such as ductility, stiffness, strength and buckling stability were studied to understand the seismic performance by preparing FE models [6].

Predominantly the factor to be considered in the corrugated plates is their low stiffness is the direction perpendicular to the corrugation of the plates and the high strength to arrest the in-plane forces along the direction of the corrugation. The CSPSW in the MBS are commonly a part of the walls incorporated with openings such as doors and windows. The connection of the CSPSW in the normal and MBS make them distinct [7].

In regular structures they are connected on the edges whereas in modular structures they are connected in corners. Since the modules in MBS are connected in corners it helps in the load transfer vertically from column to column. Due to the high stiffness in MBS the CSPSW act as seismic resisting mechanism. On the studies that have been done by other researchers to understand the behavior of the CSPSW, the addition of the openings to the walls impairs the strength and performance of the element [8].

On comparison studies done by the application of pushover and cyclic loading on CSPSW and SPSW on numerous models [9]. Experimental studies on CSPSW with and without openings was carried out and addition of constructional column around the openings to arrest the buckling of the infill plate. The results showed that the initial stiffness of the models with openings are reduced when compared to the model without opening [10].

Since the slit and perforation on the walls are inevitable, small steel strips are used as reinforcement in CSPSW. They are connected, welded in perpendicular to the peak of each corrugation. Since the reinforcement are provided, they enhance the out-of-plane stiffness of the walls. They also enhance the ductility of the plates and also limit the deformation by energy dissipation betwixt the corrugations [11].

A DCSPSW was recently planned by the authors, consist of two corrugated steel plates connected together either by weld or by bolts. The general dimension of the DCSPSW as shown in Fig. 1.

An analytical formula was proposed to predict the ultimate shear strength of DSCSW. Proposal for calculation of the shear yield and the local and global shear elastic buckling were given as three analytical formulas and accuracy was defined

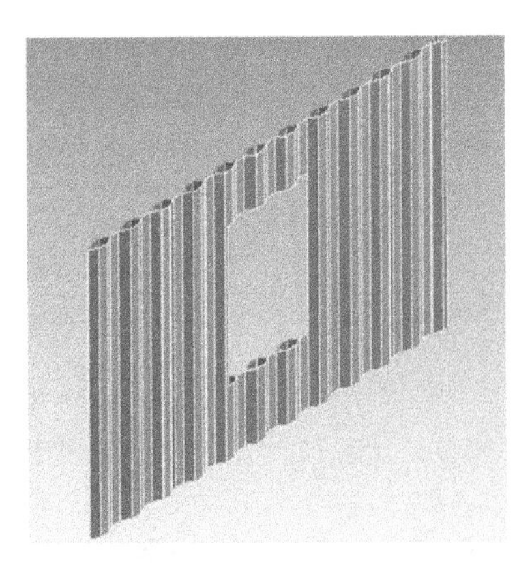

Fig. 1 Geometrical model of double corrugated plate

by experimental means [12]. By the application of monotonic shear loads on finite element models the shear resisting behaviors of the double corrugated walls were investigated [13].

2 Methodology

SCSPSWs and DCSPSWs were modelled and analyzed using ANSYS 16.1 finite element software. Fixed support was provided at the bottom of the columns as a boundary condition for the wall. Monotonic loading was provided in transverse direction.

The load was provided at the top edge of the column in displacement control and in incremental manner. the elements used around the openings for this study is considered from Korean Standard (KS) profiles.

The element type used for the analysis is 20-node solid 186 element, which is higher order 3D element which exhibits quadratic displacement behavior. The 20 nodes of the element have 3 degrees of freedom i.e., translation in nodal x, y, and z direction. The element supports plasticity, hyper elasticity, creep, stress stiffening, large deflection, and large strain capabilities. The meshing used in the model is combined tetra and hexa mesh. The meshing near the opening is not refined.

For the current study, a corrugated wall of a single-story residential building of height 3.1m and length of 4.5m from center to center was considered. The dimensional description of the beams and columns used in the study are provided in Table 1 and material properties of plates, beam and column are shown in Table 2. The material properties of the boundary elements are provided in Table 3 and cross-sectional details are shown in Fig. 2.

CSPSW for the analysis was modelled by changing the corrugation angle with 0°, 45°, 90°. Angle of inclination changed with respect to X axis. The main purpose of changing angle of corrugation for selecting best angle which carrying higher ultimate strength. Some models with different openings, alignment and stiffeners

Table 1 Cross sectional dimensions [4]

	Specimen	Dimension (mm)
Beam	Width of flange (b_f)	398
	Depth	394
	Thickness of web (t_w)	11
	Thickness of flange (t_f)	18
Column	Width of flange (b_f)	432
	Depth	498
	Thickness of web (t_w)	45
	Thickness of flange (t_f)	70

Table 2 Material model behaviour [4]

Type	Young's modulus (MPa)	f_y (MPa)	f_u (MPa)	F_u/F_y
Plate	210,000	341	341	1
Columns and beams	210,000	390	480	1.23

Table 3 Material model behaviour [10]

Type	Young's modulus (MPa)	f_y (MPa)	f_u (MPa)	F_u/F_y
Stiffener elements	192	441	544	1.23

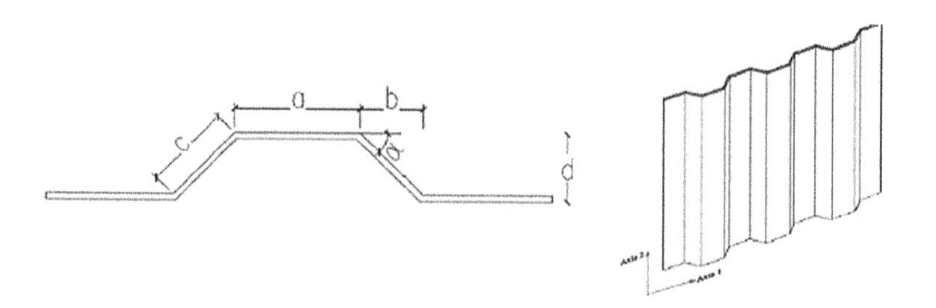

Fig. 2 Geometry of the model [4]

are shown in Fig. 3. The dimensional details of the panel element are given in Table 4.

Study was carried out on the journal of "Alireza Farzampour (Analysis and design recommendations for corrugated steel plate shear wall reduced beam section)" and the results obtained on comparison is less than 0.5% error. Table 5 shows the dimensions of openings used for the study.

Deformation was mostly affected around the door and the window opening provided, therefore to arrest these deformations a small thickness steel element was provided around the opening and thereby improving the strength of the load carrying capacity. The size of the element provided around the opening were

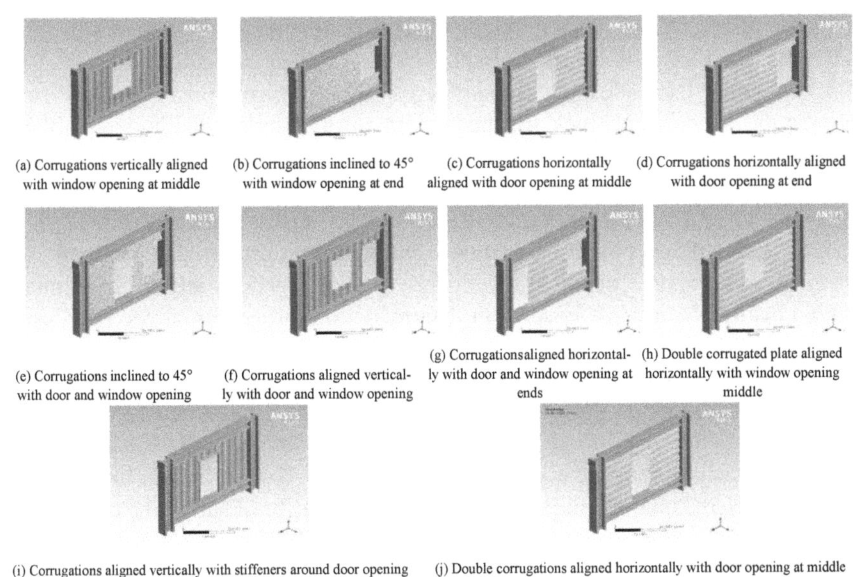

(a) Corrugations vertically aligned with window opening at middle
(b) Corrugations inclined to 45° with window opening at end
(c) Corrugations horizontally aligned with door opening at middle
(d) Corrugations horizontally aligned with door opening at end
(e) Corrugations inclined to 45° with door and window opening
(f) Corrugations aligned vertically with door and window opening
(g) Corrugationsaligned horizontally with door and window opening at ends
(h) Double corrugated plate aligned horizontally with window opening middle
(i) Corrugations aligned vertically with stiffeners around door opening
(j) Double corrugations aligned horizontally with door opening at middle

Fig. 3 Models of CSPSW and DCSPSW

Table 4 Corrugated panel geometry [4]

Specimen	t	a	d	a
Dimension (mm)	1.5	100	50	30

Table 5 Dimensions of opening [10]

Function of opening	Size l × h (mm)
Door	1000 × 2290
Window	1000 × 1600

120 × 60 × 4mm [10]. Steel plate shear walls with boundary elements were one in all of the advanced models.

Therefore, getting more accurate results, frame elements and infill plates were meshed separately by using different element size. The mesh size was kept 150 mm for the frame column and beam and constructional column and 100 mm for the infill plate.

3 Results

In this part, analysis and discussion on the performance of the CSPSW and DCSPSW considering both provision for perforation and without perforations, with different alignment of corrugations were carried out. The ultimate load carrying capacity of each model was different. The ultimate strength of different angle of CSPSW was shown in Tables 6, 7 and 8. The pushover curve and stiffness of the CSPSW under lateral loading are shown in Fig. 4. On incorporation of stiffeners around the window openings, the model with corrugation aligned horizontally was found to be more effective by having better load carrying capacity than other models. But when the door opening was provided with stiffeners the maximum load carrying capacity was found for the model with corrugations aligned vertically. For models with stiffeners for combined door and window opening, the maximum load carrying capacity was obtained for models with corrugations aligned horizontally with a negligible increase in strength than model of corrugation aligned vertically.

For DCSPSW, load carrying capacity of models with corrugations aligned vertically has augmented by 5% to 13% than CSPSW. Likewise, for models with corrugations aligned at 45°, the strength has augmented by 6% to 11% and with corrugations intensified horizontally the strength has increased by 7% to 9%.

As the results signify, the strength of the CSPSW have increased by providing stiffeners around the opening and additionally it can be made more effective by connecting the constructional column from top beam to bottom beam. The strength of the CSPSW have increased much more by making single corrugated plate shear wall to double corrugated shear wall, than the models with stiffeners provided around openings. Among the three-corrugation alignment the maximum load carrying capacity for the DCSPSW was obtained for the model with corrugation aligned vertically.

The addition of stiffeners adds auxiliary strength to the model as for CSPSW, since the openings on the model in real life are unavoidable, the addition of stiffeners to the models are applicable and hence it provides more strength to DCSPSW with openings. The position of the openings, i.e., providing the perforation at the center of the wall and at the end of the walls also affect the strength and stiffness of the models.

For CSPSW models, with corrugations aligned vertically, at 45° and horizontally, with the door and window openings when provided separately, the maximum loads were carried when the placement of the openings were provided at the ends. From the results obtained the maximum loads were carried when the corrugations were aligned horizontally and on the placement of the opening at the end.

On providing both the openings on the shear wall the loads were significantly reduced, and the maximum load were carried on the positioning of the openings on the ends of the plate. On providing both the openings on the same wall, the maximum load was found to be carried by the models with corrugations aligned vertically. Among the models of three configurations, on providing window and door opening separately, the plates aligned horizontally was found to be effective and for models with combined door and window opening the models with plates aligned vertically was found to be more effective under lateral loads.

For DCSPSW models, with corrugations aligned vertically, at 45° and horizontally, with the door and window openings when provided separately, the maximum loads were carried when the placement of the openings were provided at the ends.

Table 6 Results of CSPSW and DCSPSW with corrugation vertically aligned

Opening		Yield deformation (mm)	Yield load (kN)	Stiffness (kN/mm)	Ultimate deformation (mm)	Ultimate load (kN)	Ultimate stiffness (kN/mm)	Ductility	Percentage of strength
CSPSW with corrugations vertically aligned									
–	–	146.350	4267.900	29.162	377.770	6209.400	16.437	2.581	1.000
Window	Middle	150.330	4064.700	27.039	341.000	5222.300	15.315	2.268	1.000
	End	140.690	3992.200	28.376	337.500	5300.500	15.705	2.399	1.000
Door	Middle	150.200	4083.900	27.190	412.630	5123.000	12.415	2.747	1.000
	End	137.540	3848.200	27.979	335.760	5214.500	15.530	2.441	1.000
Window and door	Window at end	114.180	3943.000	34.533	380.200	5026.300	13.220	3.330	1.000
	Door at end	143.610	3939.200	27.430	384.030	5033.800	13.108	2.674	1.000
	Window and door at ends	140.450	3681.400	26.211	513.220	5099.200	9.936	3.654	1.000
CSPSW with corrugations vertically aligned with stiffeners									
Window	Middle	138.340	4474.600	32.345	355.840	5494.400	15.441	2.572	5.210
	End	165.650	4991.900	30.135	377.610	5464.200	14.470	2.280	3.088
Door	Middle	120.290	4032.600	33.524	362.200	5313.500	14.670	3.011	3.719
	End	182.530	5007.300	27.433	364.700	5362.400	14.704	1.998	2.836
Window and door	Window at end	123.820	4354.800	35.170	380.750	5350.800	14.053	3.075	6.456
	Door at end	130.310	4412.200	33.859	373.190	5305.600	14.217	2.864	5.399
	Window and door at ends	175.670	4889.000	27.831	399.820	5316.500	13.297	2.276	4.261

(continued)

Table 6 (continued)

Opening		Yield deformation (mm)	Yield load (kN)	Stiffness (kN/mm)	Ultimate deformation (mm)	Ultimate load (kN)	Ultimate stiffness (kN/mm)	Ductility	Percentage of strength
DCSPSW with corrugations vertically aligned									
Window	Middle	84.252	4656.200	55.265	305.240	5788.900	18.965	3.623	10.850
	End	115.940	5021.700	43.313	325.210	5924.300	18.217	2.805	11.769
Door	Middle	98.903	4870.900	49.249	293.680	5695.000	19.392	2.969	11.165
	End	102.670	5001.900	48.718	325.290	5893.100	18.116	3.168	13.014
Window and door	Window at end	106.180	4781.900	45.036	341.740	5426.400	15.879	3.218	7.960
	Door at end	96.908	4792.900	49.458	323.990	5426.000	16.747	3.343	7.791
	Window and door at ends	72.544	4124.700	56.858	462.340	5505.000	11.907	6.373	7.958
DCSPSW with corrugations vertically aligned with stiffeners									
Window	Middle	98.058	5180.600	52.832	306.790	6112.100	19.923	3.129	17.038
	End	103.480	4567.000	44.134	346.620	6171.000	17.803	3.350	16.423
Door	Middle	93.369	5354.500	57.348	321.510	6075.300	18.896	3.443	18.589
	End	90.628	4431.600	48.899	322.390	6071.100	18.832	3.557	16.427
Window and door	Window at end	93.259	5005.500	53.673	300.300	5955.600	19.832	3.220	18.489
	Door at end	94.901	4958.800	52.252	324.260	5897.800	18.188	3.417	17.164
	Window and door at ends	99.636	5520.000	55.402	406.410	5931.300	14.594	4.079	16.318

Table 7 Results of CSPSW and DCSPSW with corrugation aligned to 45°

Opening		Yield deformation (mm)	Yield load (kN)	Stiffness (kN/mm)	Ultimate deformation (mm)	Ultimate load (kN)	Ultimate stiffness (kN/mm)	Ductility	Percentage of strength
CSPSW with corrugations inclined 45°									
-	-	79.587	4527.100	56.882	330.460	5483.300	16.593	4.152	1.000
Window	Middle	110.920	4388.600	39.565	360.660	5110.000	14.168	3.252	1.000
	End	89.846	4089.300	45.515	359.800	5193.300	14.434	4.005	1.000
Door	Middle	157.440	4050.800	25.729	404.180	5072.700	12.551	2.567	1.000
	End	81.176	4098.400	50.488	366.960	5170.300	14.090	4.521	1.000
Window and door	Window at end	97.561	3927.000	40.252	371.420	4976.700	13.399	3.807	1.000
	Door at end	81.690	3889.900	47.618	366.200	4955.600	13.532	4.483	1.000
	Window and door at ends	164.450	3984.200	24.227	467.330	5062.400	10.833	2.842	1.000
CSPSW with corrugations inclined 45° with stiffeners									
Window	Middle	87.024	4452.200	51.161	353.030	5378.300	15.235	4.057	5.250
	End	112.450	4723.000	42.001	349.640	5359.400	15.328	3.109	3.198
Door	Middle	98.407	4967.900	50.483	397.250	5332.300	13.423	4.037	5.118
	End	132.950	4814.800	36.215	351.580	5284.900	15.032	2.644	2.217
Window and door	Window at end	152.250	4757.200	31.246	331.050	5227.500	15.791	2.174	5.039
	Door at end	119.270	4586.700	38.456	368.940	5245.500	14.218	3.093	5.850
	Window and door at ends	77.475	4088.700	52.774	376.520	5288.200	14.045	4.860	4.460

(continued)

Table 7 (continued)

Opening		Yield deformation (mm)	Yield load (kN)	Stiffness (kN/mm)	Ultimate deformation (mm)	Ultimate load (kN)	Ultimate stiffness (kN/mm)	Ductility	Percentage of strength
DCSPSW with corrugations inclined 45°									
Window	Middle	98.624	5023.200	50.933	314.110	5523.300	17.584	3.185	8.088
	End	88.578	4980.900	56.232	288.410	5768.600	20.001	3.256	11.078
Door	Middle	113.050	4967.900	43.944	283.330	5519.100	19.479	2.506	8.800
	End	83.066	4840.000	58.267	244.700	5677.300	23.201	2.946	9.806
Window and door	Window at end	101.450	4751.400	46.835	360.990	5310.800	14.712	3.558	6.713
	Door at end	99.225	4674.200	47.107	343.630	5283.500	15.376	3.463	6.617
	Window and door at ends	85.778	4590.200	53.513	397.910	5421.900	13.626	4.639	7.101
DCSPSW with corrugations inclined 45° with stiffeners									
Window	Middle	174.110	5749.600	33.023	354.390	6071.000	17.131	2.035	18.806
	End	64.888	4537.700	69.931	335.810	6012.800	17.905	5.175	15.780
Door	Middle	60.076	4183.200	69.632	365.590	5980.700	16.359	6.085	17.900
	End	71.325	4636.500	65.005	337.940	5835.800	17.269	4.738	12.872
Window and door	Window at end	94.832	4954.500	52.245	353.320	5803.400	16.425	3.726	16.611
	Door at end	180.750	5471.100	30.269	372.240	5824.600	15.647	2.059	17.536
	Window and door at ends	66.394	4271.500	64.336	378.950	5835.400	15.399	5.708	15.269

Table 8 Results of CSPSW and DCSPSW with corrugation aligned horizontally

Opening		Yield deformation (mm)	Yield load (kN)	Stiffness	Ultimate deformation (mm)	Ultimate load (kN)	Ultimate stiffness	Ductility	Percentage of strength
CSPSW with corrugations horizontally aligned									
–	–	249.950	5209.500	20.842	383.670	6216.800	16.204	1.535	1.000
Window	Middle	198.910	3906.700	19.641	476.600	5198.100	10.907	2.396	1.000
	End	247.600	4234.800	17.103	498.000	5307.200	10.657	2.011	1.000
Door	Middle	217.860	3971.900	18.231	427.630	5144.600	12.411	1.963	1.000
	End	261.740	4081.900	15.595	442.070	5271.200	11.924	1.689	1.000
Window and door	Window at end	255.780	4062.200	15.882	392.880	5008.200	12.747	1.536	1.000
	Door at end	180.180	3995.800	22.177	415.050	5009.600	12.070	2.304	1.000
	Window and door at ends	229.030	3743.200	16.344	453.140	5035.800	11.113	1.979	1.000
CSPSW with corrugations horizontally aligned with stiffeners									
Window	Middle	138.320	3903.600	28.222	409.400	5421.000	13.241	2.960	4.288
	End	111.100	4009.700	36.091	374.990	5507.300	14.687	3.375	3.770
Door	Middle	192.450	4460.900	23.180	472.840	5441.600	11.508	2.457	5.773
	End	163.260	3818.100	23.387	404.760	5355.200	13.231	2.479	1.594
Window and door	Window at end	211.000	4934.000	23.384	392.320	5360.700	13.664	1.859	7.038
	Door at end	92.435	3745.400	40.519	341.000	5290.300	15.514	3.689	5.603
	Window and door at ends	75.714	3381.100	44.656	365.350	5288.500	14.475	4.825	5.018

(continued)

Table 8 (continued)

Opening		Yield deformation (mm)	Yield load (kN)	Stiffness	Ultimate deformation (mm)	Ultimate load (kN)	Ultimate stiffness	Ductility	Percentage of strength
DCSPSW with corrugations horizontally aligned									
Window	Middle	99.069	4790.000	48.350	334.490	5669.200	16.949	3.376	9.063
	End	197.690	5179.300	26.199	397.120	5821.900	14.660	2.009	9.698
Door	Middle	126.230	4584.400	36.318	401.680	5591.000	14.494	3.182	8.677
	End	220.240	5087.200	23.098	338.530	5640.800	16.663	1.537	7.012
Window and door	Window at end	124.790	4734.100	37.937	391.880	5441.300	13.885	3.140	8.648
	Door at end	127.050	4764.300	37.499	372.120	5317.100	14.289	2.929	6.138
	Window and door at ends	121.620	4404.700	36.217	482.590	5422.900	11.237	3.968	7.687
DCSPSW with corrugations horizontally aligned with stiffeners									
Window	Middle	74.532	4721.100	63.343	349.210	6112.900	17.505	4.685	17.599
	End	77.054	5121.000	66.459	298.100	6170.800	20.700	3.869	16.272
Door	Middle	73.662	4209.500	57.146	320.460	5921.900	18.479	4.350	15.109
	End	72.181	4294.800	59.500	309.310	6096.500	19.710	4.285	15.657
Window and door	Window at end	49.021	3818.900	77.903	305.230	5891.900	19.303	6.227	17.645
	Door at end	51.657	3860.000	74.723	321.570	5918.400	18.405	6.225	18.141
	Window and door at ends	60.257	4320.000	71.631	343.930	5893.900	17.137	5.708	17.040

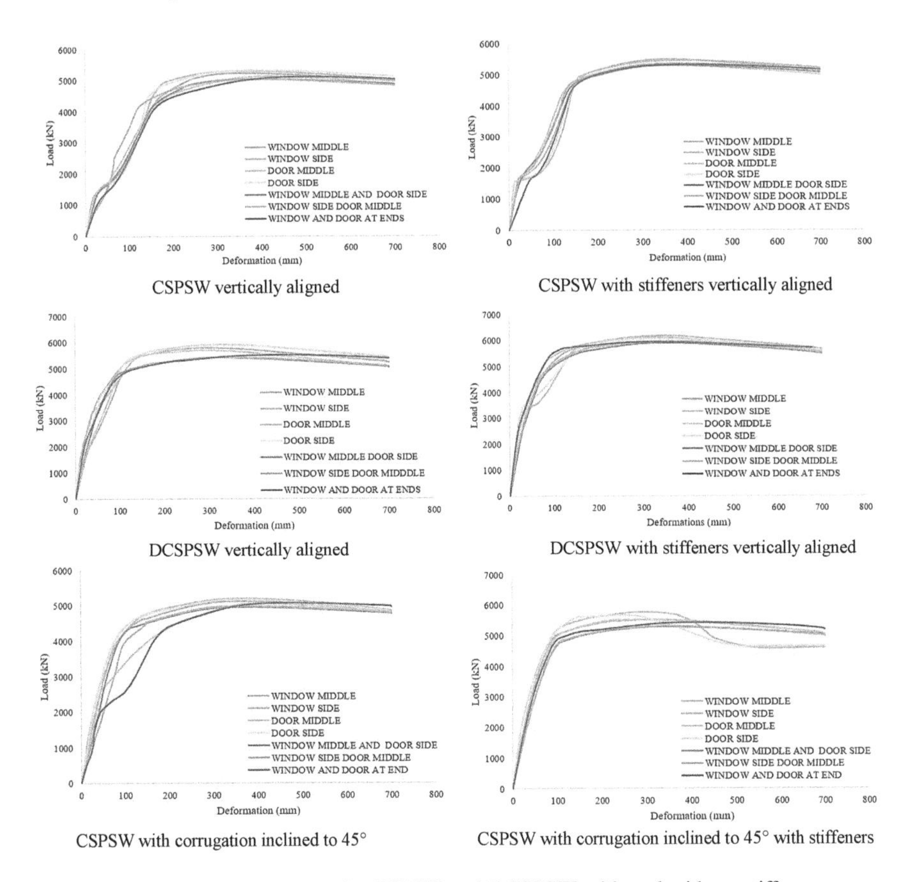

CSPSW vertically aligned

CSPSW with stiffeners vertically aligned

DCSPSW vertically aligned

DCSPSW with stiffeners vertically aligned

CSPSW with corrugation inclined to 45°

CSPSW with corrugation inclined to 45° with stiffeners

Fig. 4 Load–displacement curves for CSPSW and DCSPSW with and without stiffeners

From the results obtained the maximum loads were carried when the corrugations were aligned vertically and on the placement of the opening at the end. On providing both the openings on the shear wall the loads were significantly reduced as similar to CSPSW, and the maximum load were carried on the positioning of the openings at the ends of the plate. On providing both the openings on the same wall, the maximum load was found to be carried by the models with corrugations aligned vertically.

Among the models of three configurations, on providing window and door opening separately and for models with combined door and window opening the shear wall models with plates aligned vertically was found to be more effective under lateral loads. Addition of stiffeners to the model increases strength to shear wall as in CSPSW.

On providing stiffeners around the opening and throughout the top beam to bottom beam, the lateral out plane bucking of the plates can be arrested. On the model with window opening the maximum load was carried on the positioning of

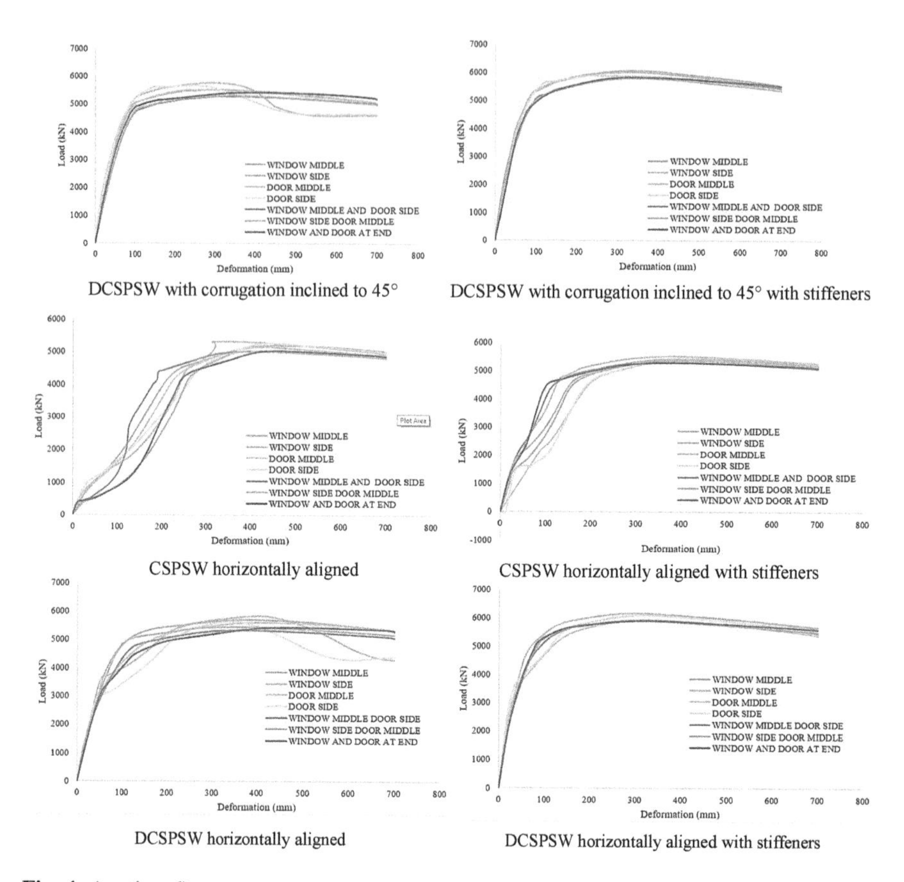

DCSPSW with corrugation inclined to 45°

DCSPSW with corrugation inclined to 45° with stiffeners

CSPSW horizontally aligned

CSPSW horizontally aligned with stiffeners

DCSPSW horizontally aligned

DCSPSW horizontally aligned with stiffeners

Fig. 4 (continued)

the opening on the end of plate and on models with door opening maximum load was carried by models with openings provided at the ends.

Based on the change in the alignment the maximum load was carried by the models with corrugations vertically aligned and model with combined door and window openings the maximum load was carried by the models with the positioning of the door at the middle and window at the end with the corrugations aligned vertically.

4 Conclusion

By inferring from the results, the study was carried out to identify the performance and change in strength of the models with openings, by changing the position of the opening, i.e., by providing the opening at the end and middle and by the addition of stiffeners around the opening and also providing two infill corrugated plates in CSPSW making it DCSPSW. Similarly, study on different corrugation alignment (horizontal, vertical and aligned 45°) was carried out to identify the best alignment for the shear wall to carry the maximum lateral load.

Addition of the opening to the models reduce the strength of the model and addition of the stiffeners around the opening increase the strength of the model. DCSPSW provides better strength when compared to that of CSPSW without openings and addition of stiffeners to the openings in DCSPSW increases the strength of the model to 15% to 18% than CSPSW with openings. Based on the study of change in the alignment of the corrugations the maximum load was carried by the models with corrugations aligned horizontally for models with door and window opening separately. And for models with combined door and window opening the maximum loads was carried by the models with corrugations aligned vertically.

For DCSPSW models with and without stiffeners the maximum load was carried on the arrangement of corrugations in vertical direction. On the models with door and window opening separately, and on models without stiffeners, the maximum load was carried by the models with perforations at the end of the plate of wall. On the models with stiffeners maximum load was carried by the models with corrugations vertically aligned and the model with combined door and window openings the maximum load was carried by the models with the positioning of the door at the middle and window at the end with the corrugations aligned vertically.

Acknowledgements I take this opportunity to convey my deep sense of gratitude to all the staff members and students of Department of Civil Engineering, Saintgits College of Engineering, Kerala, for their cooperation and encouragement during the course of this study.

References

1. Emami F, Mofid M, Vafai A (2013) Experimental study on cyclic behavior of trapezoidally corrugated steel shear walls. Eng Struct 48:750–762
2. Emami F, Mofid M (2012) On the hysteretic behavior of trapezoidally corrugated steel shear walls. Struct Des Tall Spec Build 23(2):94–104
3. Nilsson P, Al-Emrani M, Atashipour SR (2017) Transverse shear stiffness of corrugated core steel sandwich panels with dual weld lines. Thin-Walled Struct 117:98–112
4. Farzampour A, Mansouri I, Lee C, Sim H, Hu JW (2018) Analysis and design recommendations for corrugated steel plate shear walls with a reduced beam section. Thin-Walled Struct 132:658–666
5. Zhang W, Yu C, Mahdavian M (2019) Seismic performance of cold-formed steel shear walls using corrugated sheathing with slits. J Struct Eng 145(4):04019014
6. Bahrebar M, Kabir MZ, Hajsadeghi M, Zirakian T, Lim JB (2016) Structural performance of steel plate shear walls with trapezoidal corrugations and centrally-placed square perforations. Int J Steel Struct 16(3):845–855
7. Liew RJ, Dai Z, Chau YS (2018) Steel concrete composite systems for modular construction of high-rise buildings. In: Proceedings 12th international conference on advances in steel-concrete composite structures—ASCCS 2018

8. Farzampour A, Laman JA, Mofid M (2015) Behavior prediction of corrugated steel plate shear walls with openings. J Constr Steel Res 114:258–268
9. Zhao Q, Sun J, Li Y, Li Z (2017) Cyclic analyses of corrugated steel plate shear walls. Struct Des Tall Spec Build 26(16)
10. Ding Y, Deng E-F, Zong L, Dai X-M, Lou N, Chen Y (2017) Cyclic tests on corrugated steel plate shear walls with openings in modularized- constructions. J Constr Steel Res 138:675–691
11. Dai X-M, Ding Y, Zong L, Deng E-F, Lou N, Chen Y (2018) Experimental study on seismic behavior of steel strip reinforced CSPSWs in MBS. J Constr Steel Res 151:228–237
12. Tong J-Z, Guo Y-L, Zuo J-Q (2018) Elastic buckling and load-resistant behaviors of double-corrugated-plate shear walls under pure in-plane shear loads. Thin-Walled Struct 130:593–612
13. Labibzadeh M, Hamidi R (2019) A design formula for lateral load resistance of concrete filled double-steel-plate walls with small height-to-length ratio. KSCE J Civ Eng 23(8):3493–3508

Use of Parametric Software for Selecting Building Materials Based on Embodied Energy

Vighnesh Rameshkumar

Abstract Major amount of Green House Gases discharges are contributed from the building and automobile industries. The two vital energy parameters of a structure are Operational (OE) and Embodied Energy (EE). Several studies have dealt with operational energy since it is a main contributor to the entire energy of the building lifespan. Lately, work on embodied energy and embracing sustainable construction has increased as it is important for the future. This paper looks into process of selecting materials for components like walls, roofs floors and windows based on total embodied energy. Parametric software like grasshopper was used in defining several choices for each component existing in the codal provisions. Plugins like archsim and climate studio was used for defining the zonal loads, conditioning and construction. Lastly, energy plus and galapagos were used for finding the best selection of materials based on the total embodied energy for the construction. The study selected components for the construction, which provided the least EE out of several options given to the software.

Keywords Generative design · Rhino and grasshopper · Selection · Building components

1 Introduction

Constructions foster economic growth and it has increased manifolds over the past years. Construction industry constitute 45% of worldwide energy, with India recognized as one of the topmost 10 nations in the world to consume energy [1]. Similar reports from the world energy outlook shows that buildings occupied the

V. Rameshkumar (✉)
Department of Civil Engineering, Sri Krishna College of Engineering and Technology, Coimbatore, Tamilnadu 641008, India
e-mail: vighneshr@skcet.ac.in

V. Rameshkumar
Department of Civil Engineering, Amrita University, Coimbatore, Tamilnadu, India

G. C. Marano et al. (eds.), *Proceedings of SECON'21*, Lecture Notes in Civil Engineering 171, https://doi.org/10.1007/978-3-030-80312-4_3

second position in the consumption of energy [2] since it utilizes large quantities of material. The total energy in a building will be a summation of Embodied Energy (EE) and Operational energy (OE). In conventional buildings, OE constitute 80% of the total energy since it expanded over its lifetime [3]. Embodied energy is the total energy necessary to produce a particular product, which includes the energy for extracting, manufacturing and transporting the product [4–7].

Shadram et al. [8] used a BIM based methodology to establish a trade-off between the two energy parameters, embodied and operational energy, in the initial planning stages. The study also evaluated the life cycle energy using a combination of parametric software like grasshopper and dynamo. Raja et al. [9] reduced the embodied energy consumption involved in transportation of elements by using BIM based platform. The author also explores the possibilities of reducing the energy in material production and construction process as a whole. Gabriele et al. [10] investigated the concept of Net Zero Energy Building (NZEB) in Nordic climate conditions. Multi objective optimization of OE and EE was done using a model based on electricity consumption. Nicola et al. [11] used a parametric analysis tool (PAT) to evaluate the EE and OE in the initial stages of the material production. Farshid et al. [8] developed a framework to reducing the EE in the building design. The inclusion of Environmental product declarations (EPDS) and Extract transform load (ETL) into the software was done, which reduced the overall carbon footprint of the building. Van Den et al. [12] examined the relation between the heat demand and EE to evaluate the construction materials, reducing the EE of the Swedish buildings.

The literature study shows that research done on EE has great advantages over the years and should be a part of the planning process. This can be done using simulations to get information about the materials of less EE for the overall construction.

2 Methodology

The motivation of this research was to minimize the EE of the construction based on the material selection. Floors, walls and roofs were explored with several combinations and the embodied energy for a span of 50 years was calculated.

2.1 Selection of Materials for Components

Components are made up of materials, which are essential in evaluating the EE. Materials were selected conforming to "Energy conservation Building Code for Residential Buildings" standards. OE, which is the other energy parameter, was not considered in the study. The various components for walls, roofs and floors are given in Tables 1, 2, 3.

Table 1 Wall components

Sl. No	Wall components
1	1.25 cm cement plaster + 20 cm brick + 1.25 cm cement plaster
2	1.25 cm cement plaster + 22.5 mm brick + 12.5 mm cement plaster
3	1.25 cm cement plaster + 11.25 cm brick + 5 cm air gap + 11.25 cm brick + 1.25 cm cement plaster
4	1.25 cm gypsum plaster + 7.5 cm brick + 5 cm air gap + 7.5 cm brick + 1.25 cm gypsum plaster
5	1.25 cm gypsum plaster + 11.25 cm brick + 5 cm air gap + 11.25 mm brick + 1.25 cm gypsum plaster
6	1.25 cm cement plaster + 11.25 cm flyash brick + 5 cm air gap + 11.25 cm flyash brick + 1.25 cm cement plaster
7	1.25 cm cement plaster + 7.5 cm brick + 5 cm hardboard + 7.5 cm brick + 1.25 cm cement plaster
8	1.25 cm gypsum plaster + 7.5 cm brick + 5 cm extended polystyrene + 7.5 cm brick + 1.25 cm gypsum plaster
9	1.25 cm cement plaster + 20 cm AAC block + 1.25 cm cement plaster
10	23 cm concrete wall

Table 2 Roof components

Sl. No	Roof components
1	10 cm reinforced concrete + 7.5 cm lime concrete
2	10 cm reinforced concrete + 10 cm lime concrete
3	10 cm reinforced concrete + 5 cm cinder concrete + 5 cm brick tile
4	10 cm reinforced concrete + 5 cm foam concrete
5	10 cm reinforced concrete + 7.5 cm cinder concrete + 5 cm brick tile
6	11.5 cm reinforced concrete + 5 cm mud phuska + 5 cm brick tile
7	11.5 cm reinforced concrete + 7.5 cm mud phuska + 5 cm brick tile
8	10 cm reinforced concrete + 4 cm extended polystyrene (36 kg/m^3)
9	10 cm reinforced concrete + 4 cm extended polystyrene (24 kg/m^3)

Table 3 Floor components

Sl. No	Floor components
1	10 cm reinforced concrete + 7.5 cm lime concrete
2	10 cm lightweight concrete + 7.5 cm + 5 cm brick tile
3	10 cm reinforced concrete + 7.5 cm gap + 5 cm acoustic tile
4	10 cm reinforced concrete + 7.5 cm gap + 5 cm brick tile

Table 4 Material embodied energy and embodied carbon

Sl. No	Name of the material	Embodied energy (MJ/kg)	Embodied carbon (kg CO_2 e/kg)
1	Cement plaster	1.5	0.3
2	Clay brick	2.98	0.2
3	Flyash brick	0.6	0.04
4	Gypsum plaster	2	0.1
5	Hardboard	15	1.9
6	Expanded polystyrene	89	5.3
7	AAC block	3.57	0.3
8	Concrete	0.8	0.1
9	RCC	1.05	0.07
10	Lime concrete	1.3	0.2
11	Cinder concrete	1	0.1
12	Brick tile	6.6	0.5
13	Foam concrete	1.5	0.2
14	Mud phuska	0.5	0.02
15	Acoustic tile	12	0.8
16	Lightweight concrete	1.3	0.22

Several factors like density, conductivity, heat capacity, embodied carbon and embodied energy were defined using Inventory of Carbon and Energy (ICE) drafted by the University of Bath [12] and ASHRAE [14, 15]. The values for embodied energy and embodied carbon are listed in Table 4.

2.2 *Model Generation*

A shoebox model was simulated (Fig. 3) using Rhino3D, a parametric software. The dimensions for the models is tabulated in Table 5. This simulation aids in estimating the total EE using Climate studio and archsim plugins [13].

Table 5 Shoebox dimensions

Dimension (meters)	
Width	6
Length	9
Height	3.25

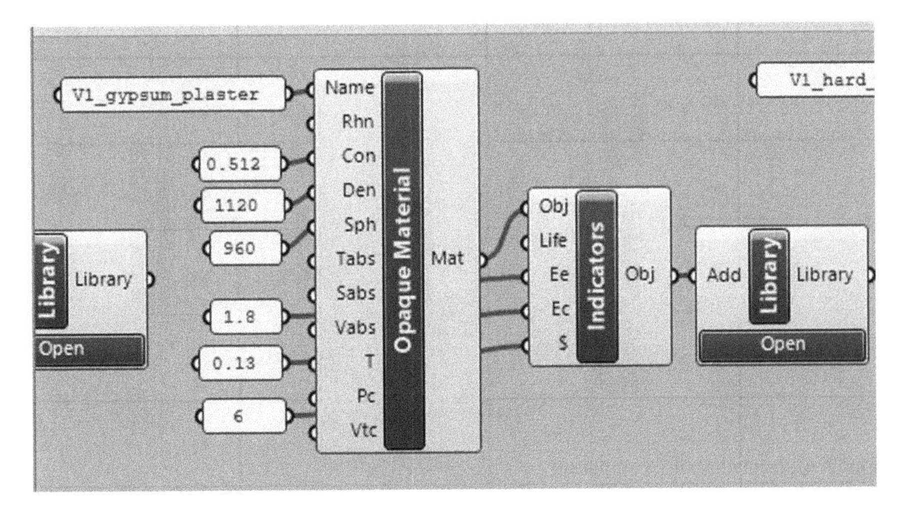

Fig. 1 Definition of materials using opaque material command

2.2.1 Material Definition

Materials form components and their applicability in the real world scenario depends on the accuracy of its definition in the simulation. Commands in archsim and climate studio were used for defining the materials. The defined materials can be used in the study after they were added to the climate studio library (Fig. 1).

2.2.2 Component Construction

Construction of the components was done using layer and merge commands. Thickness of each layer was also assigned in the definition. Any number of layers can be defined and merged to make a single construction as shown in Fig. 2. The constructions need to be included in the library to use it in the subsequent simulations.

3 Simulation and Discussions

3.1 Model Generation

A shoebox model was created using rhinocerous 3D using the basic commands as shown in Fig. 3. The thermal zone settings requires inputs for various parameters like conditioning of the zone, ventilation required, zonal constructions of components and numerous loads applied (Fig. 4).

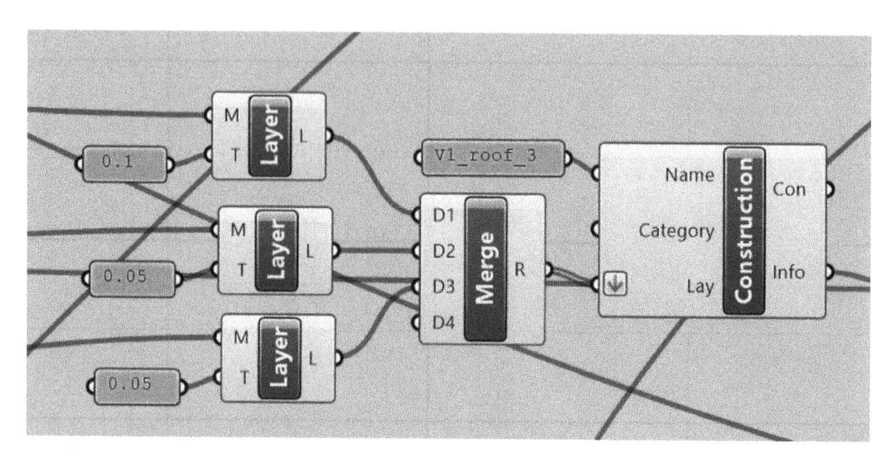

Fig. 2 Construction of a component

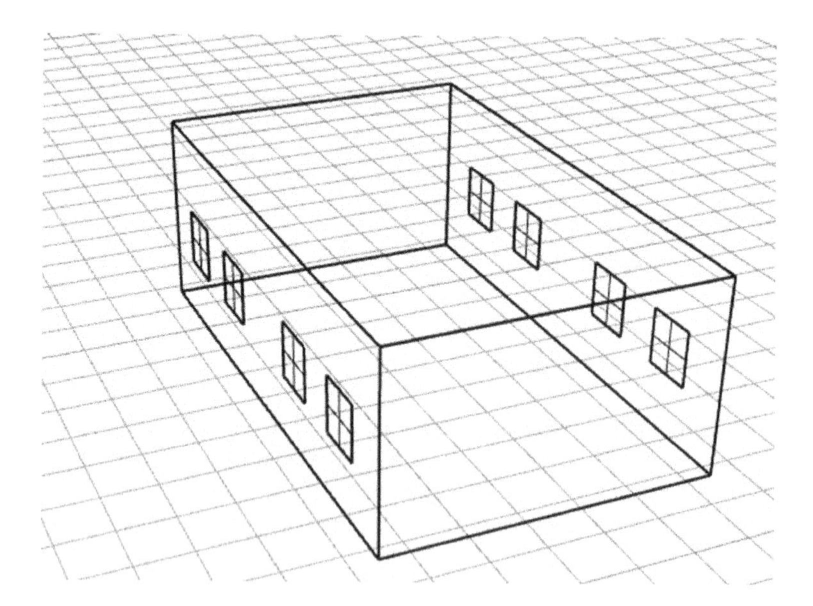

Fig. 3 Shoebox model created in Rhinoceros3D

Cooling set point was set to 26 °C and heating was not assigned since the microclimate does not require it. Mechanical ventilation, humidity was all set to a constant value in the energy parameter study. Ventilation setting such as scheduled ventilation, air changes per hour were defaulted to the software settings.

As the shoebox model was a simple simulation and assumed to house lesser people, occupancy loads was set to 0.18 p/m^2. The power and light density were inputted as 8 W/m^2 and 6 W/m^2 respectively.

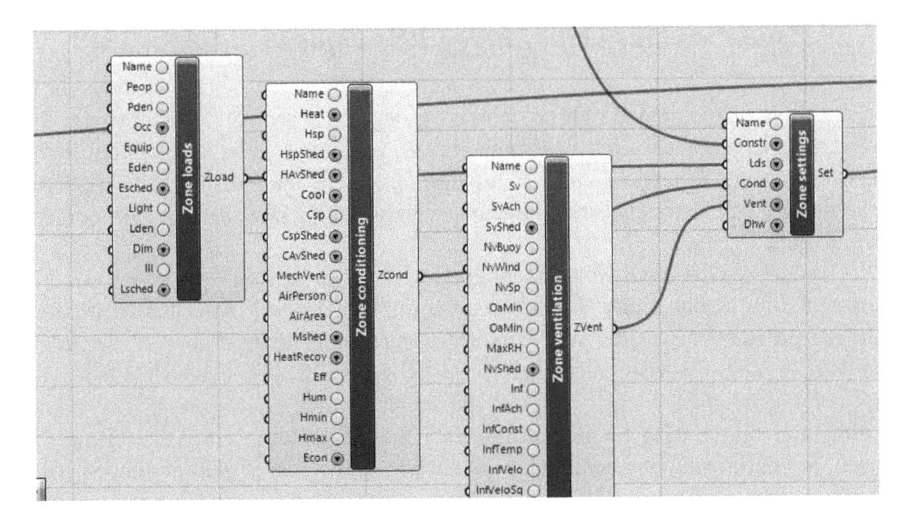

Fig. 4 Thermal zone and the settings used

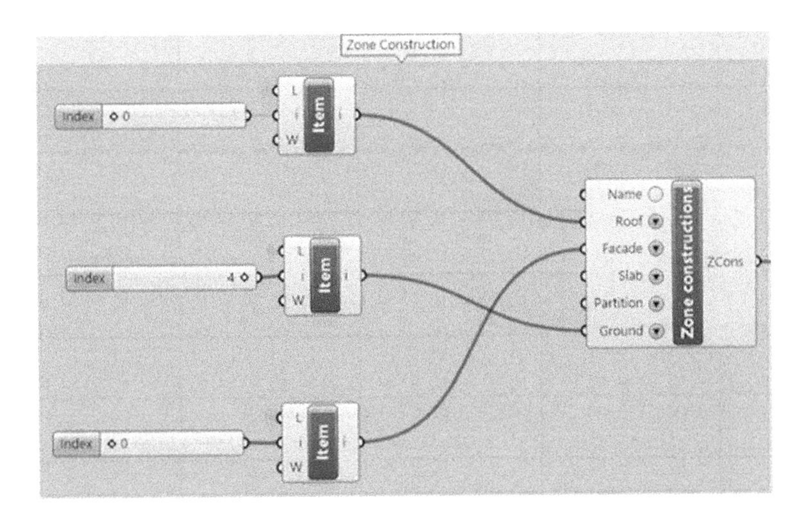

Fig. 5 Zone construction using sliders

Construction of the zone requires the inputs for all the components. A list was made for each component and it can be accessed using sliders as shown in Fig. 5b.

3.2 Zone Networker and Energy Plus Inputs

The networker command (Fig. 6) combines the zone geometry (breps) with the zone settings and makes it into a single model, which was given as the input to the E+ simulation. The panel attached to the zone networker displays the selected parameters like settings for ventilation, conditioning, construction etc. for the simulation.

The energy plus simulation was inputted with the model and energy plus weather (epw) file for Chennai city was selected for the locality [16]. After the simulation, the embodied energy results was extracted using "Integrate indicators" as displayed in Fig. 7. The EE was calculated for a period of 50 years, which means the recurring embodied energy will also be included in the simulation. The above simulation can be used for finding the EE for a single construction but cannot be used for optimising the materials. Galapagos will be used for minimising the embodied energy based on its genome and fitness selections. Fitness was the function to be minimised (EE in this study) and the genome were the variables that influenced the calculation of the fitness function (Components).

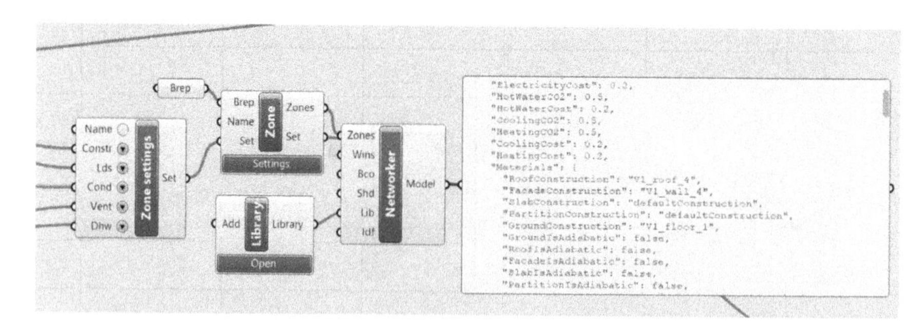

Fig. 6 Zone networker and panel showing the selected constructions

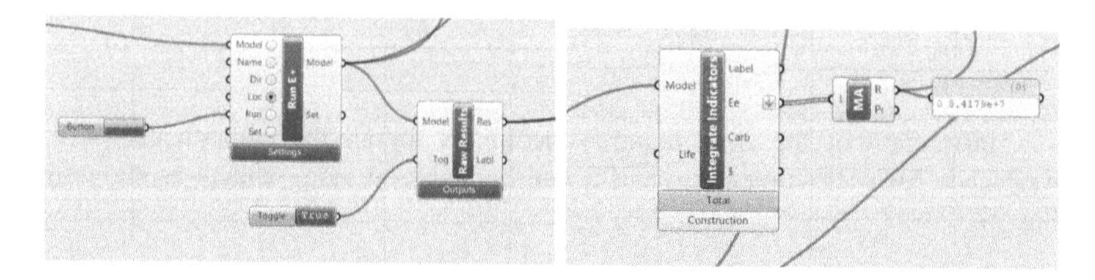

Fig. 7 Energy plus simulation and extracting the EE (Joules) from "Integrate indicators" command

3.3 Component Selection Based on the Minimal Embodied Energy

Sliders indicating the corresponding component selection were connected to the genome of the galapagos command. The calculated embodied energy, extracted from "Integrate indicators" was the fitness function to be reduced as indicated in Fig. 8.

Galapagos was set to minimize the total EE and 50 generations (iterations) were done to select the construction in which the total EE is less (Fig. 9). The selected components from the study is listed in Table 6 and gives the least EE of all the combinations that can be done. These energy simulation will be helpful for determining the best combination of materials and components for a low EE construction.

The selection of materials provided by the software is really important on the context of energy efficient buildings. The usages of locally available natural materials will be of very low EE but there are limitations of the usage of such materials in terms of the performance. This study has taken a list of materials available in the market according to "Energy conservation Building Code for Residential Buildings". Simulations done using a parametric software takes the various options for the building construction components and selects the construction with the least EE.

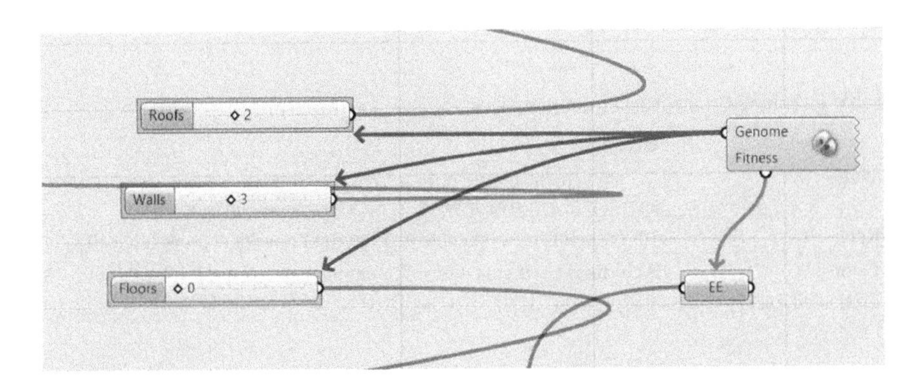

Fig. 8 Sliders indicating the components connected to galapagos

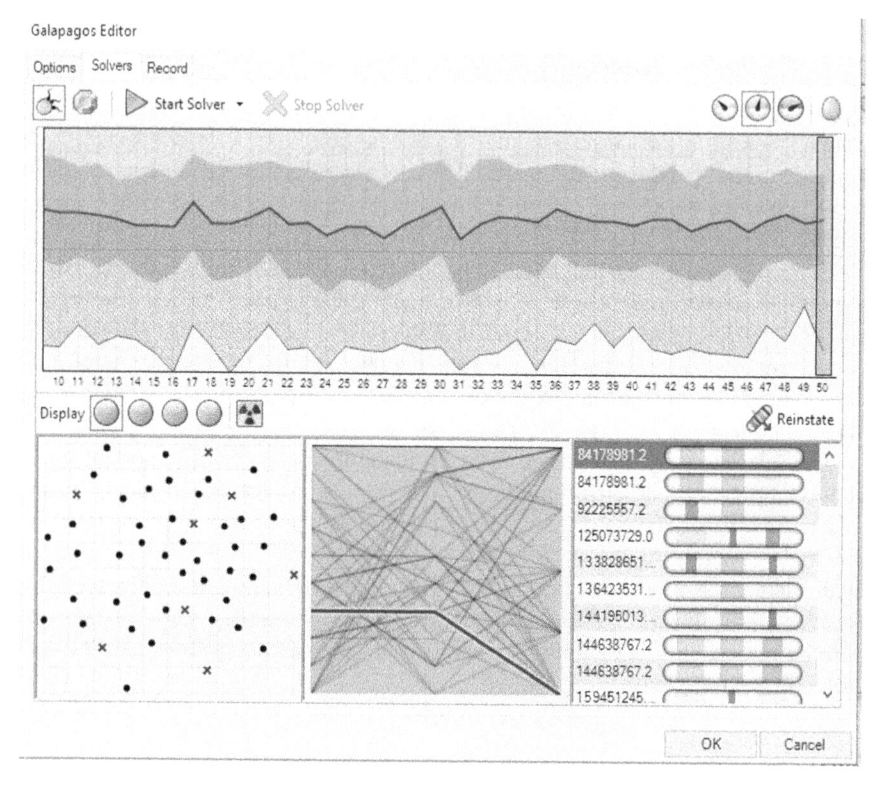

Fig. 9 Selection of the components based on the total EE

Table 6 Selected components from galapagos

Component	Selection
Wall	1.25 cm cement plaster + 11.25 cm flyash brick + 5 cm air gap + 11.25 cm flyash brick + 1.25 cm cement plaster
Roof	11.5 cm reinforced concrete + 5 cm mud phuska + 5 cm brick tile
Floor	10 cm reinforced concrete + 7.5 cm air gap + 5 cm brick tile
Embodied energy (MJ)	84.17

4 Conclusions

Components for the structure was selected based on the total EE for a period of 50 years. The following pointers were the main takeaways from the study.

(a) The wall selection was an arrangement of plaster, flyash brick and air gap acting as the separation. The selection was influenced by reduced EE of flyash brick and the air gap, which has zero EE. The selection of the floor consisted of RC and brick tile with an air gap in the middle. The same reason as the wall selection can be attributed to this.

(b) The roof selection was reinforced concrete combined with mud phuska and brick tile on top. The construction consisted of natural materials, which will not have high-embodied energy.

(c) Though natural materials have low embodied energy, multi storey constructions cannot be constructed only with such materials. A trade-off between the OE and EE can be looked into, as both the parameters are critical.

5 Conflict of Interest

There is no Conflict of Interest for this manuscript.

References

1. Nejat P, Jomehzadeh F, Taheri MM, Gohari M, Majid MZA (2015) A global review of energy consumption, CO_2 emissions and policy in the residential sector (with an overview of the top ten CO_2 emitting countries). Renew Sustain Energy Rev 43:843–862
2. India Energy Outlook (2015) World energy outlook special report-2015. International Energy Agency
3. Praseeda KI, Reddy BV, Mani M. Embodied energy assessment of building materials in India using process and input–output analysis. Energy Build. https://doi.org/10.1016/j.enbuild.2014.10.042
4. Zhang Z, Wu X, Yang X, Zhu Y. BEPAS—a life cycle building environmental performance assessment model. Build Environ https://doi.org/10.1016/j.buildenv.2005.02.028
5. Dixit MK, Culp CH, Fernandez-Solis JL. Embodied energy of construction materials: integrating human and capital energy into an IO-based hybrid model. Energy Build. https://doi.org/10.1021/es503896
6. Rauf, Crawford RH. Building service life and its effect on the life cycle embodied energy of buildings. Energy. https://doi.org/10.1016/j.energy.2014.10.093
7. Sartori, Hestnes AG. Energy use in the life cycle of conventional and low-energy buildings: a review article. Energy Build. https://doi.org/10.1016/j.enbuild.2006.07.001
8. Shadram F, Johansson TD, Lu W, Schade J, Olofsson T. An integrated BIM-based framework for minimizing embodied energy during building design. Energy Build. https://doi.org/10.1016/j.enbuild.2016.07.007
9. Nizam RS, Zhang C, Tian L. A BIM based tool for assessing embodied energy for buildings. Energy Build. https://doi.org/10.1016/j.enbuild.2018.03.067
10. Lobaccaro G, Wiberg AH, Ceci G, Manni M, Lolli N, Berardi U. Parametric design to minimize the embodied GHG emissions in a ZEB. Energy Build. https://doi.org/10.1016/j.enbuild.2018.02.025
11. Lolli N, Fufa SM, Inman M. A parametric tool for the assessment of operational energy use, embodied energy and embodied material emissions in building. Energy Procedia. https://doi.org/10.1016/j.egypro.2017.03.004

12. Energy Conservation Building Code for Residential Buildings (Part I: Building Envelope Design)
13. Grasshopper. https://www.rhino3d.com/6/new/grasshopper
14. ASHRAE (2004) Standard 55–2004: Thermal environmental conditions for human occupancy. American Society of Heating, Refrigerating and Air-Conditioning Engineers, Atlanta, GA
15. Hammond G, Jones C (2012) Inventory of carbon and energy (ICE) version 2.0. Sustainable Energy Research Team (SERT), Department of Mechanical Engineering, University of Bath, U.K
16. Energy Plus. https://energyplus.net/

Application of Generalised Beam Theory in Predicting Lateral Torsional Buckling of Thin-Walled Steel Sections

Shashi Kant Sharma, K. V. Praveen Kumar, M. Abdul Akbar, and Dadi Rambabu

Abstract Thin-walled structures are extensively used in buildings and automobile industries, and majority of the thin-walled structures are of lightweight cold-formed steel or aluminum. Structural elements such as cold-rolled profiles often fail in instability before even achieving their yield capacity/stress. Lateral Torsional Buckling (LTB) is a common instability failure in both thin-walled columns and beams. It is important to assess the structural response of this instability failure. The aim of this study is to derive cross-sectional properties and warping functions of a channel section required to predict the pure Lateral Torsional Buckling strength of light gauge steel channel sections using first order Generalised Beam Theory (GBT). Further, a first order GBT formulation and numerical analysis of cold-formed steel lipped channel beams (C80 × 40 × 10 × 1, C90 × 40 × 10 × 1, C100 × 40 × 10 × 1, C80 × 40 × 10 × 1.6, C90 × 40 × 10 × 1.6 and C100 × 40 × 10 × 1.6) of two different thicknesses (1 and 1.6 mm) and three different depths (80, 90, and 100 mm) subjected to uniform moment are undertaken to predict pure LTB and the critical buckling strength. The results show that theoretical buckling load values are 0.3% more than GBT buckling values and critical buckling values obtained from GBT shows little variation when distortions are involved in the final failure mode.

Keywords Lateral torsional buckling · Generalised Beam Theory · Lipped channel beams · Light gauge steel channel sections

S. K. Sharma · K. V.P. Kumar · M. A. Akbar · D. Rambabu (✉)
Department of Civil Engineering, Dr. B. R. Ambedkar National Institute of Technology, Jalandhar 144011, India
e-mail: dadir.ce.19@nitj.ac.in

G. C. Marano et al. (eds.), *Proceedings of SECON'21*, Lecture Notes in Civil Engineering 171, https://doi.org/10.1007/978-3-030-80312-4_4

1 Introduction

Majority of thin-walled structural members used in industrial and residential buildings includes cold-formed steel made by rolling or pressing thin gauges of steel. Earlier, cold form structural elements were mainly used as secondary member systems such as roof purlins, steel frameworks, floor decking etc. However recent applications in the use of thin-walled steel members have been concerned with developing large scale systems like steel portals and space frames build entirely from cold-formed steel members. The growth of its application in recent times is mainly due to the advantages such as high stiffness and strength to weight ratio and easy erection and installation compared to thicker hot-rolled steel members [1]. As there is a wide range of engineering applications which seek efficiency in strength and cost by minimizing material, the relevance of thin-walled structural elements is quite high. But in the construction industry, the thin-walled frame elements with very slender open cross-sections and low torsional stiffness must withstand a complex loading condition where axial, bending, shear and torsion are present. Hence these often fail in instability even before the yield capacity is reached [2]. The simplest and major example of instability is the buckling of column member under uniform axial compression. Instability condition arises when a structure starts to lose its stiffness. This can be characterized in such a way that when the structural deformation corresponding to any load factor tends to infinite values for even smallest of the load increments. One of the most common instability failures associated with thin-walled structures is Lateral Torsional Buckling (LTB). It can be defined as the form of buckling in which the beam deflects laterally and twists out of plane of loading when the beam bends in the plane of greatest structural rigidity. Members, which have low lateral bending, and torsional stiffness compared with its stiffness in plane of loading are more prone to this type of buckling [3]. Lateral torsional buckling comes under the category of global buckling whereas there is other two buckling modes namely local buckling and distortional buckling known as cross-sectional buckling modes. As the name implies these, occur as local failure modes at the cross-sections. In general, the deformed shape of any structural section is normally contributed by cross-section buckling (local and distortional buckling) and rigid body displacements, which leads to a lower critical stress value than the lowest individual critical stress value. Incorporating the local–global interaction in the analysis still is a challenging task, mainly because of the difficulties in modeling the simultaneous short and long wave structural response. Additionally, the results yielded by the numerical analyses are often difficult to process and interpret [4–6]. However, some methods which have been followed by the researchers are Finite Element Method, Finite Strip Method, Direct Strength Method, Codal Methods (AS/NZS4600, BS5950 PART-5, EUROCODE3-1.3, AISI method etc.) and Generalised Beam Theory etc.

In this study, six cold formed lipped channel sections are used corresponding to the commonly used industrial buildings for a bay spacing of 6 m.

2 Generalised Beam Theory Formulation

One of the most important features associated with GBT is that it can take care of the significance of individual as well as any selected combinations of the buckling modes which will allow the designer to understand modes which are important for a particular analysis [7, 8]. As per conventional beam theory, there are four fundamental modes of deformation, namely extension, bending about two principal axes and torsion. These are called rigid body modes since distortions of cross-section are not involved. GBT unifies the analytical treatment of these rigid body modes and further extends it to include higher order deformation modes that involve cross-section distortion. In the first order GBT, all these modes are orthogonal which means that they are uncoupled, and their effects can be combined by simple superposition. A unifying feature of GBT is the introduction of 'warping functions' whereby each mode (K) of deformation is related with a distribution of axial strain ${}^{k}\tilde{u}$. Accordingly, for the four rigid body modes, the first mode of deformation is the axial extension, and which has a constant axial strain distribution throughout the cross-section. Hence at any point, the warping function ${}^{k}\tilde{u} = -1$ for this mode. The second and third modes are bending about major and minor axis respectively and the warping functions linked to these modes are linear distribution of strain about the two principal axes. The final rigid body mode is torsion, and here warping function is the sectorial coordinate, which reflects the distribution of axial strain due to a bimoment. For a lipped C section, there are 6 orthogonal modes. The first four initial modes are rigid body modes and the higher modes ($k > 4$) are distortional modes. For the sake of considering the effect of local buckling of flat elements between the fold lines or to increase the distortional modes, intermediate nodal points are to be added in-between the fold lines [9–11].

2.1 Assumptions of GBT Theory

GBT theory is based on the following assumptions:

- In the first order GBT, all the modes are orthogonal which means that they are uncoupled and their effects can be combined by simple superposition.
- All the nodes can warp independently and therefore the section has six orthogonal dimensional modes.
- The warping functions associated with each mode have six degrees of freedom and the warping functions are linear between the nodes.
- The loading condition is a constant bending moment about the major axis throughout the beam.

The basic equation of GBT is:

$$E^kC^kV''' - G^kD^kV'' + B^kV^k = q^k \quad (1)$$

where the forward superscript is used to represent the mode number.

E = Young's modulus
G = Shear modulus
C^k = Stiffness regarding direct stress.
D^k = Stiffness regarding shear stress
B^k = Stiffness regarding transverse bending stress
V^k = Generalized deformation in mode k
q^k = Distributed load applicable to mode k

The solution for the fundamental beam equation gives the stresses and deformations of a member under specified loading and support condition. The equation can be solved in three different steps. The first step comprises of finding out the warping function ${}^k\tilde{u}$ and section properties ${}^kC$, ${}^kD$ and ${}^kB$ corresponding to every single mode k by only considering the cross-section. These basic section properties found out are made use in the next step to obtain the solutions for fundamental GBT equation for each mode taking account of the relevant loading and boundary conditions. Once the cross-sectional properties are found out, material properties such as Elastic modulus (E) and Shear modulus (G) along with the support conditions, load quantities and length of the member are incorporated into the GBT basic equation. Hence an eigen value problem is formed by a system of ordinary differential equations (ODEs) corresponding to each mode of deflection. The solution of this problem yields the member bifurcation stress resultants (eigen values) and corresponding buckling mode shapes (eigen functions). The third step then involves combining the results of step 2 to calculate the required stresses and deflections. In terms of GBT, the stress resultant is determined as,

$$W^k = -E^kC^kV''' = \int \sigma^k u dA \quad (2)$$

where, W^k is the stress resultant for mode k (e.g. bending moment for modes 2 and 3, bi moment for mode 4, etc.)

In this case, the load is applied exclusively in the second mode of buckling. Therefore, the distributed load for all other modes is zero. However, if the loading condition were different such as a concentrated load or a uniformly distributed load, the load should be apportioned between various nodes. It is done using the principle of virtual work. In the fundamental beam equation, the load terms represent the virtual work of loads acting on the modal displacements. Therefore, by considering the virtual work of horizontal and vertical components of the nodal loads acting on the modal displacements, it is possible to obtain the distributed load applicable to each mode. The virtual work of nodal loads can be expressed as,

$$^{k}q = \sum_{r=1}^{n+1} q_{y,r}{}^{k}v_r + q_{x,r}{}^{k}w_r$$

where '${}^{k}v_r$ and ${}^{k}w_r$' are the horizontal and vertical unit displacements for mode k and at node r.

$q_{y,r}$ and $q_{x,r}$ are the distributed load in the corresponding mode in Y and X directions, *'r'* is the plate number and *'n'* refers to the node number

3 Analysis Using GBTUL

3.1 General

GBTUL is a code developed by Bebiano et al. [11] to solve the buckling and vibration problems of thin-walled open cross-sections. It constitutes the implementation of GBT formulation so that to avoid the complex manual solving procedures of higher order differential equations involving several combinations of deformation modes. As already discussed, GBT can be considered as an elegant approach to solve several structural problems involving thin-walled sections and can also be used to obtain the contributions of each deformation mode (modal participation) in the final failure mode. GBTUL as a code can only handle elastic buckling and vibration analyses of different cross-sections made of isotropic/special orthotropic materials. GBT analysis procedure is given in Fig. 1 based on which the GBTUL interface structure is made.

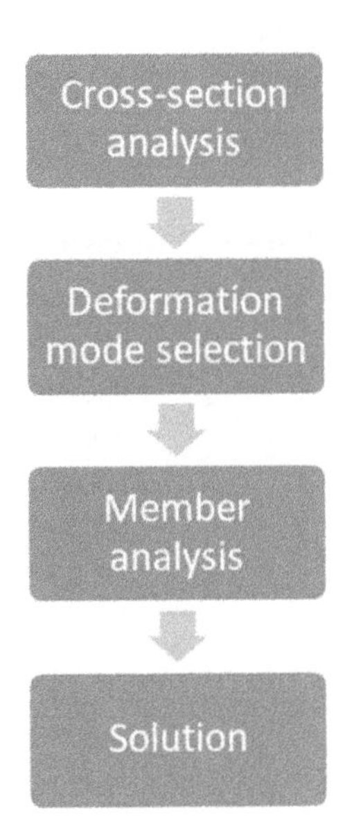

Fig. 1 GBT analysis procedure

3.2 Theoretical Analysis

Theoretical buckling load for a beam under compression is found out using Euler's buckling load formula.

For section C80 × 40 × 10 × 1.6, Euler's Critical buckling load, $P_{cr} = \frac{\pi^2 EI}{L^2}$.
Modulus of elasticity, E = 2 × 10^5 N/mm^2.
Moment of inertia about minor axis, I = 68337.58 mm^4.
Effective length, L = 2500 mm.

$$(P_{cr})_{theoretical} = (\pi^2 \times 200000 \times 68337.58)/2500^2 = 21561\,N = 21.561\,kN.$$

3.3 Validation

This can be done in a similar way by doing the buckling analysis with a unit axial load and checking the results with theoretical values. When the channel sections are axially loaded, all the sections underwent flexural buckling about minor axis in the second eigen mode and hence 2nd eigen mode represents the critical buckling load. The comparison with the theoretical values obtained from Euler's buckling theory are tabulated in Table 1 for a 2.5 m long beam.

The buckling loads from the GBTUL models and theoretical buckling loads are comparable. The difference between the GBTUL and theoretical values is less than 0.3% in all the cases. This indicates that, the geometry, material property and boundary conditions have been modelled properly in GBTUL and hence the results obtained can be used for the comparative study.

Table 1 Comparison of GBTUL critical buckling loads with theoretical buckling values

Model no	Designation	GBTUL (kN)	Theoretical buckling value (kN)
1	80CS40 × 10 × 1	13.457	13.5
2	90CS40 × 10 × 1	13.979	14.001
3	100CS40 × 10 × 1	14.443	14.486
4	80CS40 × 10 × 1.6	21.562	21.561
5	90CS40 × 10 × 1.6	22.409	22.412
6	100CS40 × 10 × 1.6	23.168	23.178

3.4 Limitations

The study carried out has the following limitations:

- In this study, local buckling modes has not been considered in analysis.
- The study is carried out for lipped channel sections, and the accuracy of the predicted results for other types of including round lipped channel sections needs to be studied.
- The current study is based on linear analysis and the applicability of the results when the boundary goes to non-linear range (increasing post-buckling effects) is not captured in this study.

4 Results and Discussion

In this study, six cold-formed steel channel sections were analysed to find out the critical buckling load. Since the study deals with a stability failure problem and follows eigen value analysis the comparison of first three eigen modes is given here. The participation of each individual mode contributing to the failure corresponding to each eigen mode is also obtained from GBTUL analysis. By checking the modal participation and deflected shape of the beam the type of failure can be understood whether it is Lateral Torsional Buckling or distortional. All the sections are of 2.5 m length and analysed under a uniform/constant moment case. Table 2 gives the comparison of critical buckling loads obtained from GBTUL for channels C80 × 40 × 10 × 1, C90 × 40 × 10 × 1, C100 × 40 × 10 × 1, C80 × 40 × 10 × 1.6, C90 × 40 × 10 × 1.6 and C100 × 40 × 10 × 1.6. Table 3 gives the modal participation (in percentage) for the same sections.

From Table 2, it can be observed that for all the sections first mode of failure is contributed by 50% bending minor (buckling) and 50% torsion which represents the global buckling failure mode LTB (lateral torsional buckling). And from the second eigen mode, distortion also contributes to the failure. The lowest buckling load (first eigen value) in all the cases corresponds to LTB Mode. The critical buckling load obtained for the first failure mode from GBT analyses largely differ

Table 2 Critical buckling load values from GBTUL analysis (1 and 1.6 mm thick sections)

Sl. No	Channel section (h × b) mm	Critical buckling moment (k Nm)					
		Eigen value 1		Eigen value 2		Eigen value 3	
		1 mm	1.6 mm	1 mm	1.6 mm	1 mm	1.6 mm
1	80CS40 × 10	0.556	1.02	1.91	3.34	1.99	5.47
2	90CS40 × 10	0.63	1.14	2.18	3.82	2.23	6.01
3	100CS40 × 10	0.706	1.26	2.37	4.33	2.43	6.56

Table 3 Modal participation for corresponding Eigen modes (1 and 1.6 mm thick sections)

Sl. No	Channel section	Eigen mode no	Modal participation (%)											
			Axial		Bending major		Bending minor		Torsion		Symmetric distortion		Anti-symmetric distortion	
			1	1.6	1	1.6	1	1.6	1	1.6	1	1.6	1	1.6
1	C80 × 40 × 10	EM 1	0	0	0	0	50	50	50	50	0	0	0	0
		EM 2	0	0	0	0	43	50	50	47	7	3	0	0
		EM 3	0	0	0	0	0	0	3	3	50	50	47	47
2	C90 × 40 × 10	EM 1	0	0	0	0	50	50	50	50	0	0	0	0
		EM 2	0	0	0	0	0	48	2	50	50	2	48	0
		EM 3	0	0	0	0	0	0	1.5	2	50	48	48.5	50
3	C100 × 40 × 10	EM 1	0	0	0	0	50	50	50	50	0	0	0	0
		EM 2	0	0	0	0	0	48	1	50	50	2	49	0
		EM 3	0	0	0	0	0	0	1	1	50	50	49	49

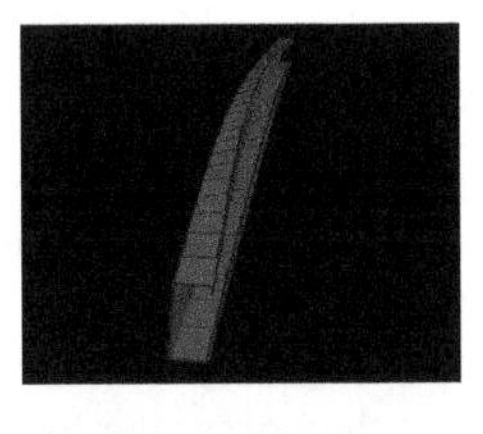

Fig. 2 Deflected shape in first eigen mode

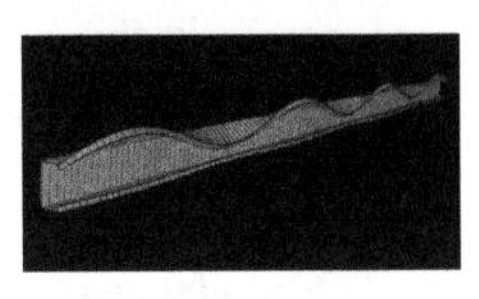

Fig. 3 Deflected shape in third eigen mode

with each other. From the next failure mode onwards, critical loads obtained from both the analyses were comparable with a maximum percentage difference of 3%.

Deflected shapes of C80 × 40 × 10 × 1 channel section corresponding to first eigen mode are in Fig. 2. From the modal participation values shown in Table 3 and the deflection diagrams obtained from GBTUL, the first mode of buckling corresponds to Lateral Torsional Buckling. Even if the critical buckling load obtained varies from each other, the deflected diagrams confirm that the first mode of buckling in thin-walled steel sections under a uniform/constant moment is LTB. The lower buckling load obtained from GBT can be probably due to the limited modelling options available in GBTUL which is a code developed by researchers to solve the buckling and vibration problems.

The support condition chosen for the analysis was simply supported but the options to simulate support conditions with an additional restrained/free degree of freedom are not available in GBTUL as of now. Similarly, the comparison for the remaining three sections (C80 × 40 × 10 × 1.6, C90 × 40 × 10 × 1.6 and C100 × 40 × 10 × 1.6) are discussed as follows. Unlike sections with 1 mm thickness (slenderer) the buckling load values for these sections show considerable difference up to second Eigen mode (Table 3). This can be due to the modelling difference as mentioned earlier. But like the previous sections this disparity in values is present only when the failure mode is contributed solely by the rigid body deformations (Table 3). When distortions are involving, the values obtaining are the same with less than 3% difference (Table 3 and Fig. 3).

5 Conclusion

This study covers buckling analysis of thin-walled lipped channel sections using GBT for a constant/uniform moment load case. The results are compared with each other to arrive at certain conclusions and the same models under unit axial load condition are compared with the Euler's buckling values for validation of models. Based on the study following general conclusions have been made:

- Generalized beam theory can satisfactorily explain the lateral modes of lipped cold form channel sections. However, more studies need to be carried to conform the accuracy of the prediction of GBT for round lipped channel sections and non-linear loading conditions.
- Critical buckling values obtained from GBT shows little variation when distortions are involved in the final failure mode. For the sections with 1 mm thickness the distortions are involved from second failure mode onwards whereas in the case of sections with 1.6 mm thickness the distortions are found to be involved from third failure mode onwards.
- GBT provides a better interpretation of the final failure mode with the help of modal participation values.
- The modal participation values (from GBTUL) shows that the failure is solely contributed by torsion (50%) and buckling (50%) and the deflection diagrams obtained from the analyses confirm the occurrence of LTB.
- When the distortions (symmetric and anti-symmetric) are majorly involved in the final failure mode, the critical buckling values are found to be comparable with less than 3% difference.
- The buckling analysis of channel sections under axial load using GBT techniques give critical buckling values marginally close to the Euler's buckling value with a maximum percentage difference of 0.3%.

Further the linear GBT analysis can be extended to non-linear analysis to perform post buckling analysis. Moreover, the GBT analysis can consider various types of loading and presence of arbitrary initial imperfections.

References

1. Adany S (2019) Modal identification of thin-walled members by using the constrained finite element method. Thin-Walled Struct 140:31–42. https://doi.org/10.1016/j.tws.2019.03.029
2. Schardt R (1994) Generalized beam theory-an adequate method for coupled stability problems. Thin-Walled Struct 19(2–4):161–180
3. Nguyen CT, Moon J, Le VN, Lee HE (2010) Lateral-torsional buckling of I-girders with discrete torsional bracings. J Constr Steel Res 66(2):170–177
4. Gonçalves R, Dinis PB, Camotim D (2009) GBT formulation to analyse the first-order and buckling behaviour of thin-walled members with arbitrary cross-sections. Thin-Walled Struct 47(5):583–600
5. Gonçalves R, Camotim D (2007) Thin-walled member plastic bifurcation analysis using generalised beam theory. Adv Eng Softw 38(8–9):637–646
6. Gonçalves R, Camotim D (2012) Geometrically non-linear generalised beam theory for elastoplastic thin-walled metal members. Thin-Walled Struct 51:121–129
7. Yuan W, Nan S (2019) An analytical solution of local—global interaction buckling of cold—formed steel channel—section columns. Int J Steel Struct 19(5):1578–1591
8. Davies JM, Leach P (1992) Some applications of generalized beam theory, pp 479–501
9. Taig G, Ranzi G (2014) Generalised beam theory (GBT) for stiffened sections. Int J Steel Struct 14(2):381–397

10. Davies JM, Leach P (1994) First-order generalised beam theory. J Constr Steel Res 31(2–3):187–220
11. Davies JM, Leach P, Heinz D (1994) Second-order generalised beam theory. J Constr Steel Res 31(2–3):221–241

Design of Standard Questionnaire for Occupational Risk Perception Caused Due to Air Pollution

Abinaya Sekar, George K. Varghese, and Ravi Mundakkara Kovilakam Varma

Abstract India tops the list of countries in the number of premature deaths caused due to industrialization and urbanization. In the year 2017, around 0.8 million deaths were due to occupational exposure to hazardous chemicals globally. In India, Section 91-A of the Factories Act deals with safety and occupational health surveys and such information is also provided in the Mines act of 1952. Despite the strict legislation, 90% of the working sectors are still unorganized. The occupational health nursing concept is very new to India and it is non-existent in unorganized sectors. Considering the severe effect of such exposure, periodic occupational exposure survey (POES) will be mandatory in India to assess the exposure caused due to air pollution and also to determine the health cost, economic cost and cost of suffering. This study aimed at designing a standard questionnaire for knowing the perception of occupational risk caused due to air pollution. The questionnaire should be adequately structured with appropriate question type, wording and sequence of the questions. The questionnaire structure may include demographics, exposure information, epidemiology, perception and precautions taken. A conceptual risk perception model was developed based on the factors collected, which can be used for any occupational exposure perception studies. The questionnaire designed is never a diagnostic tool to substitute clinical assessment but will serve as a source for epidemiological information.

Keywords Air pollution · Occupational exposure survey · Questionnaire · Risk perception · Conceptual model

---

A. Sekar · G. K. Varghese (✉)
Department of Civil Engineering, National Institute of Technology Calicut, Kozhikode, India
e-mail: gkv@nitc.ac.in

R. M. K. Varma
Department of Physics, National Institute of Technology Calicut, Kozhikode, India

G. C. Marano et al. (eds.), *Proceedings of SECON'21*, Lecture Notes in Civil Engineering 171, https://doi.org/10.1007/978-3-030-80312-4_5

1 Introduction

Indian economy is one of the fastest-growing economies with development in infrastructure for industrialization [1]. Among the total workforce in India, only 8–10% work in the organized sector, whereas 90–92% work in the unorganized sector [2]. Occupational health care is mostly missing in the unorganized segments [3]. It is discussed in several studies that the workplace environment has a direct relationship to the productivity, safety, health and comfort of the worker [4]. The most important physical factors that should be considered in the work environment include the design of the building, age of the building, quality of the equipment and their design, furniture, space and layout of the workplace. Factors involved under Indoor environmental quality (IEQ) include temperature, ventilation of the building, lighting, acoustics and vibrations, hazardous radiation and air quality. Among the factors, air quality is one of the most important factors that should be considered for a good workplace environment as it is directly linked to severe health effects [5–8]. Exposure to air pollutants, either for an individual or a population, can be determined by two means (1) qualitatively (For example, questionnaire survey and number of people visiting hospitals) (2) quantitatively (by measuring the concentration of the pollutant) [9]. Several researchers have used the qualitative method to get quick, easy, cheap and efficient means to collect information about exposure from a larger group of people [10]. In the questionnaire survey method, the essential part of the research is survey questions [11, 12]. The questions should be designed to correctly measure the opinions, experiences, and behavior of the respondents. In earlier days, the design of survey questions was considered just an art, but over the past thirty years, there is much science involved in crafting a good questionnaire [8]. The primary design method used in such surveys are (1) the Cross-sectional design method and (2) the Panel or longitudinal design method. Out of the two, the Cross-sectional method is the most commonly used one in public opinion research [13]. Panel design is mainly used for social research that surveys the same people over a long period [14].

Considering the severe effects caused due to exposure to air pollutants, a periodic occupational exposure survey (POES) is very important in the Indian scenario. Such a survey will help estimate the health cost, economic cost and cost of suffering. This study aims to identify and explain the major components of a standard questionnaire for obtaining the perception of workers and general public on the occupational risk due to air pollution. Efforts have been taken to compile the most important factors considered in every occupational exposure study. A conceptual questionnaire model has been developed based on the identified factors that can be used for similar studies among any set of the targeted population.

2 Methodology

The present study comes under descriptive research, where the source of information is the secondary data obtained from several other studies conducted across the world. Based on the literature, factors influencing the perception of occupational exposure to air pollutants have been shortlisted and used to develop a conceptual model.

3 Discussions

3.1 Elements of a Research Survey

The basic structure of every survey-based research is shown in Fig. 1 as compiled and modified [10, 20, 21]. In Fig. 1, steps 4–7 primarily affect the quality and the credibility of the research.

Hypothesis Formation and Testing of Variable

Most of the research proceeds with an observation, formulation of hypothesis and testing of hypothesis. There are four essential criteria that every hypothesis must meet [15].

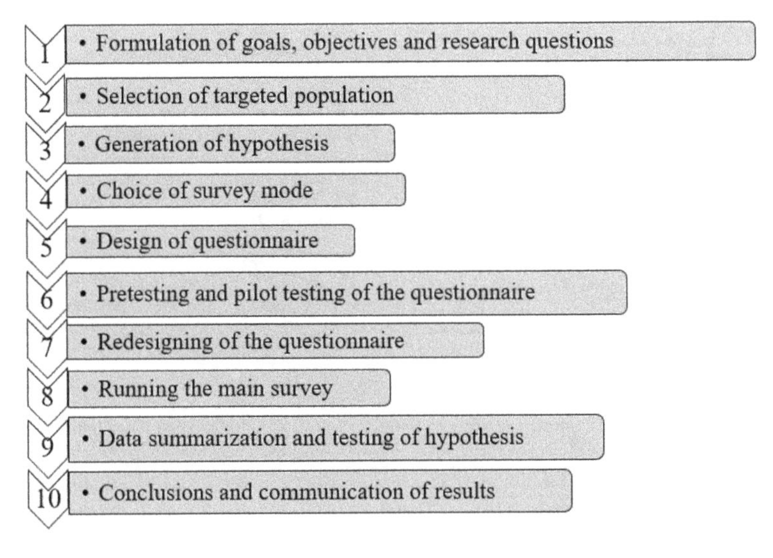

Fig. 1 General steps involved in questionnaire survey research

1. Must define the expected relationship between the variables
2. Must be testable and falsifiable, i.e. it should enable one to test whether the hypothesis is true or false
3. Must be consistent with the existing knowledge
4. Must be concise and straightforward

For example.

> "Workers who use personal protective gears are less exposed to pollutants in their occupational environment than who do not use."
>
> "Indoor workers face fewer health issues when compared with outdoor workers."
>
> "Perception on occupational exposure is largely related to the actual exposure to pollution."

By defining the hypothesis, the researcher can narrow down the focus to question formation and data collection [16]. Testing the hypothesis can be done with statistical tools, with the type of testing depending on the nature of the outcome required and the predictor variable. The most commonly used tests are provided in Fig. 2. Suitable alternative tests may also be employed in testing the hypothesis.

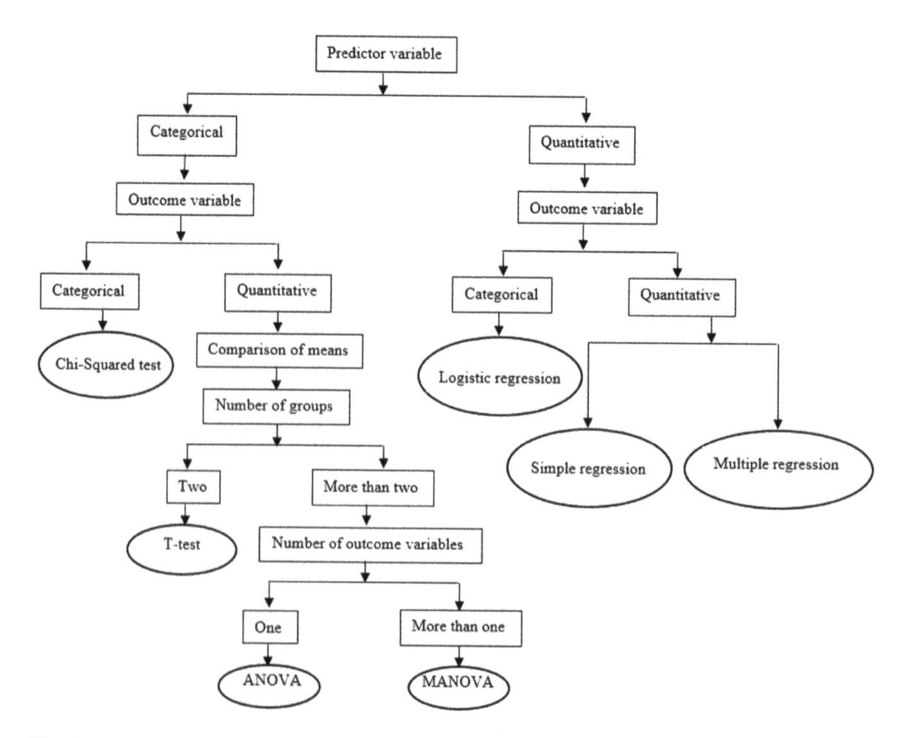

Fig. 2 Type of test to be used based on the type of variable (modified and adopted from [17–19])

Choice of Survey Mode

Efforts should be put in wisely choosing the survey mode based on the target population. The most commonly used survey modes include (1) Face to face interview, (2) Telephonic interview, (3) Mail questions and (4) Internet questions [22]. Face-to-face interviews and telephonic interviews are considered the best survey mode and recommended in the case of an occupational health assessment survey for a small group of targeted pollution. This is because it provides clarity and a more profound understanding to the respondents about the questions posed to them [23, 24]. However, in recent days due to enormous computer and Internet technology development, online surveys are mostly recommended for surveys involving a larger population. Online surveys are effective in terms of questionnaire preparation, data collection, data storage, data visualization and communication of results [25].

The significant advantage of online surveys is that it permits automated collection data such as date and time of filling the questionnaire, geo-spatial location of the participant, and IP address of the gadget the respondent uses for filling. In earlier days, the major drawback of online surveys was a lack of control on respondent selection. However, the advancement in collecting automated data helps controlling respondents with regards to the geographical location and multiple submissions from the same IP address [25]. Another important drawback of online surveys is the poor response rate. Online survey platforms offer automatic storage of the responses in the central server of the website and the researcher can download the data at any time and the stored data will not be accessible for others. In addition to this, those survey platforms provide tools for data visualization in the form of charts and graphs. However, statistical analysis cannot be performed on most of the online survey platforms [26, 27]. In recent days, this deficiency is overcome by software packages like Epi-info, Epi-data and Epi-collect [25].

Design of Questionnaire

Types of Questionnaire

There are three main types of questionnaire (1) structured questionnaire, (2) unstructured questionnaire, and (3) quasi-structured questionnaire. A structured questionnaire has pre-coded questions and options with well-defined patterns to follow the sequence of the questions. Most of the research which uses quantitative data employs structured questionnaire. Data obtained from a structured questionnaire are readily amenable for further analysis [26]. Unstructured questionnaires include primarily open-ended questions and questions with vague opinions [28]. In practice, it is not possible to include pre-coded questions with all possible alternatives for the respondent to answer and hence in most of the standard questionnaires, there will be an option called "Others-Please specify," which allows the respondent to have his own choice of answering the questions. Such a type of questionnaire is called quasi-structured [29]. In the case of occupational health risk assessment surveys, it is suggested to employ a quasi-structured questionnaire as it permits the respondents to share their views other than what the predefined options provide.

Type of Questions

One of the decisive factors in obtaining quality responses is the type of question asked. In general, there are two types of questions, open-ended questions, and closed-ended questions. In the case of open-ended, the respondent will have the freedom to express his response in his own words, whereas in the case of closed-ended, there will be multiple pre-fixed options and the respondent has to select from the options provided [10]. There are seven types of closed-ended questions and they are listed in Table 1 with explanation and example. The third category of question is called 'mixed category', which includes both open and closed categories. Studies have shown significant difference in responses when the same question is asked in both open and closed-ended format for the same targeted population [30, 31]. To avoid such biased data, it is necessary to conduct a pretesting/pilot testing with open-ended questions and after knowing the possible responses, these can be included in the closed-ended questions in the main survey. It is vital to ensure that the provided options are mutually exclusive [32]. It is essential to avoid "double-barreled" questions in the health assessment surveys [33]. For example, "Do you feel that exposure to air pollution cause both mental and physical illness?". Here the respondents might find it challenging to interpret and so question about a mental and physical illness should be asked separately.

Wordings and Order of the Questions

Wordings used in the questions also play an essential role in the responses [34]. For example. "Do you favour or oppose the closure of industries emitting hazardous chemicals?" and "Do you favour or oppose the closure of industries emitting hazardous chemicals even if it meant that it affects the employees of the industry and the productivity of the country?" There may be a difference in response if the same question is asked differently. A similar observation was presented in the study conducted by Pew research center survey in the year 2003. When the question was "favour or oppose taking military action in Iraq to end Saddam Hussein's rule," 68% favoured military action and 25% opposed military action. When the same question was asked in a different manner "favour or oppose taking military action in Iraq to end Saddam Hussein's rule *even if it meant that U.S. forces might suffer thousands of casualties,*" the responses were found to be dramatically different; around 48% opposed the military action [30]. Hence proper attention should be given to crafting the questions with appropriate wordings. The order of the questions also plays a vital role in a survey. If a closed-ended question is placed before an open-ended question, the respondent is more likely to answer based on the concepts provided in the previous closed-ended questions. Two types of order effects are to be observed in closed-ended questions (1) contrast effect- in this effect,

Table 1 Types of closed-ended questions

Type of question	Explanation	Example
Leading question	It is forcing the respondent to opt for a specific answer and makes every response similar	What do you feel about the horrible effect caused due to occupational exposure to air pollutants?
Importance questions	Asking the respondents to rate the importance of any specific matter on a scale of 1–5	How do you rate the usefulness of a mask in preventing exposure to air pollutants? 1-Extremely helpful 2-very helpful 3-Somewhat helpful 4-Not very helpful 5-Not at all helpful
Likert scale questions	The degree to which the respondents agree to a specific statement can be obtained using Likert scale questions	How often you wear a mask in your workplace? Never, rarely, sometimes, often, always
Dichotomous questions	Make the respondents answer either yes or no. Nevertheless, a middle perception is not available in this type of question	Are you exposed to air pollutants in your workplace? Yes No
Bipolar questions	The questions will have two different levels of extremities written at the opposite ends of a scale	Rate your occupational exposure Minimum [] [] [] [] [] [] Maximum
Rating scale questions	In this type of question, the respondent will be asked to rate a specific matter on a scale of 1 to 10 or Good to Very poor	How do you rate the occupational health support service in your organization? 1-Good 2-Fair 3-Poor 4-Very poor
Risk propensity questions	These questions aim to assess the respondent's future intentions and determine the propensity towards the health risk	Do you think occupational exposure to air pollutants can have a severe effect on your health? (Unlikely/rarely/possible/highly likely/invariably)

order results in significant differences in the opinion and (2) assimilation effect- in this effect, the responses become more similar due to the order of the questionnaire. Therefore proper attention should be taken while arranging the order of the questionnaire [16, 25].

3.2 Factors Influencing Perception on Occupational Exposure to Pollution

Demographics

The most commonly collected demographics (population characteristics) include gender, age, income, education level, ownership of the house, marital status, size of the family and other investigation-based factors [35]. Demographics play a significant role in every research because it is crucial to determine whether the respondents represent the targeted population [36]. These demographic variables always act as independent variables in a research design [37]. The demographic information can also identify a subset of the entire population and characterize their responses [38]. In most occupational exposure surveys conducted earlier, demographic information on gender, age, residence, smoking habits, educational qualifications, designation, income, nature of activity and years of experience in the current job were collected [35, 39, 40]. Other than this, job-specific information were also collected.

Exposure Information

In case of any exposure assessment, the variables considered include concentration level of pollutant, exposure time in hours/day, exposure frequency days/year, exposure duration years and the averaging time in days [41]. Similarly, in the indirect exposure assessment method, details on working hours, shift details, work location (indoor or outdoor), ambient pollution level and levels of particulate and gaseous pollutants in the work environment should be collected to obtain a clear workplace environmental inventory.

Epidemiology

The term epidemiology means "study of distribution and determinants of health-related states and events specified for a particular group of the population" [42]. In occupational exposure assessments, the commonly asked questions include family health history, personal past medical history, symptoms/health problems, frequency of occurrence of the disease, details on hospitalization, treatments and medications [43, 44].

Perception

Questions on risk perception can be on the visibility of exposure, choice of exposure, type of risk, the familiarity of the risk, catastrophic potential, individual control, origin, risk management ability, knowledge about risks, uncertainty, manifestation, damage, fair distribution of damage, damage visibility, victims and social or scientific status [45]. The information collected on the perception can be subjected to any statistical analysis for finding out the difference in opinion among the different subset of the targeted population [46].

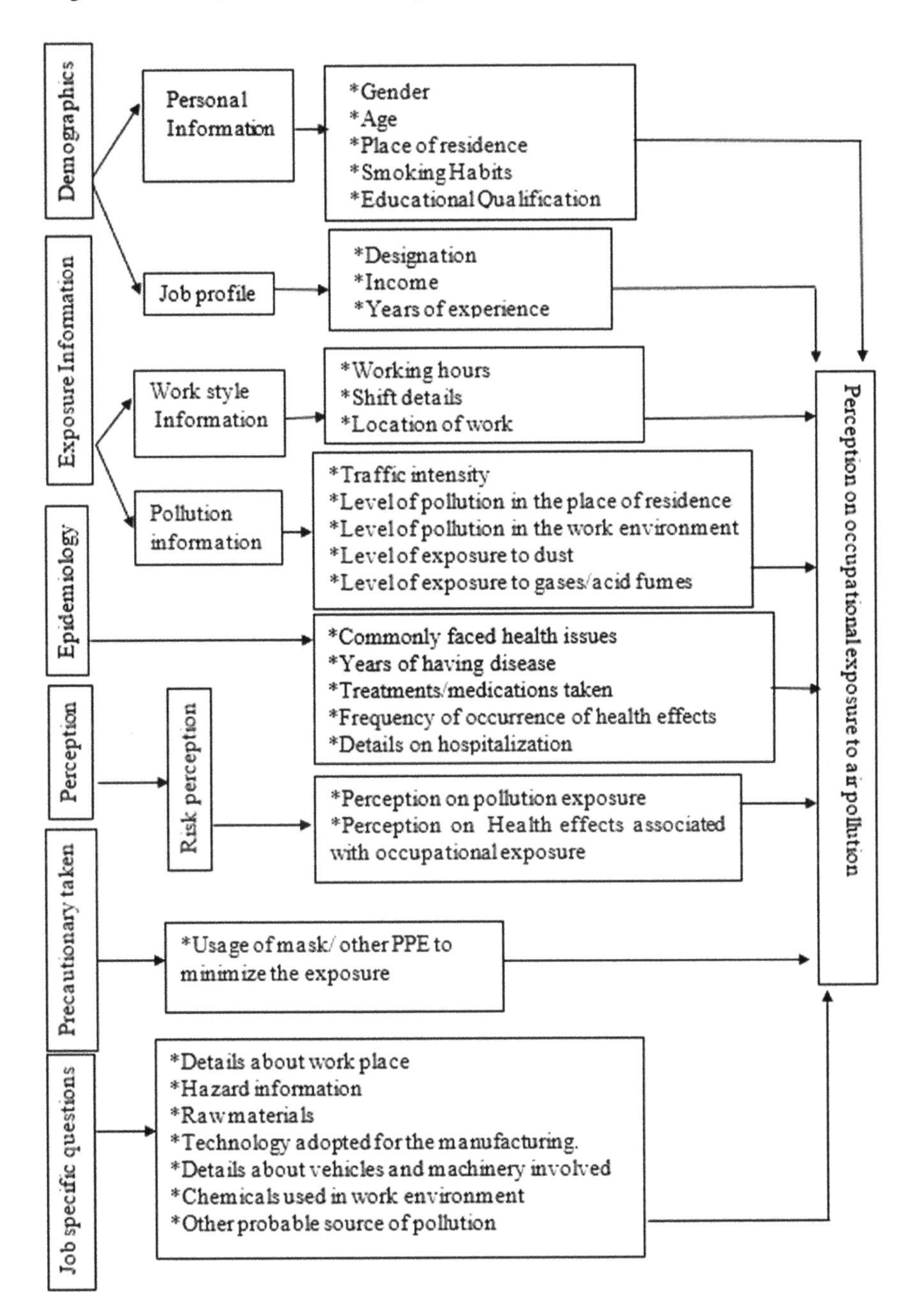

Fig. 3 A conceptual model developed for perception on occupational based exposure to air pollution

Precautionary Measures Taken

Use of personal protective equipment (PPE) of any sort in the workplace greatly minimizes the exposure to pollutants and hence details on the usage, frequency of usage, type of PPE, days of reusing the PPE, fit of PPE and changes made in the PPE to make it fit can be collected [43, 44].

Other than the above-said factors, other job-specific questions may be needed to determine the exposure and perception. For example. "How many liters of toluene you handle every day for the extraction process?".

Based on the factors considered, a conceptual model was developed as in Fig. 3, which can determine the factors influencing perception on occupational exposure for any set of a given population.

4 Conclusion

A questionnaire is one of the most effective tools used in occupational exposure research studies. This article provides information on factors to be considered before designing a standard questionnaire. It is recommended to conduct a pilot study before conducting the main study so that the suggestions obtained in the pilot study can be adopted in the main study. A properly designed questionnaire will help us to get reliable information. A conceptual model was developed considering various factors used in the earlier studies to assess occupational exposure. Such compilation will be helpful to future researchers for the determination of perception on occupational exposure to air pollution.

References

1. Tiwari RR, Sharma A, Zodpey SP, Khandare SM (2014) Does occupational health nursing exist in India? Indian J Occup Environ Med 18:113–117. https://doi.org/10.4103/0019-5278.146907
2. Sakthivel S, Joddar P (2006) Unorganised sector workforce in India trends, patterns and social security coverage. Econ Polit Wkly 41:2107–2114
3. Nag A, Vyas H, Nag P (2016) Occupational health scenario of Indian informal sector. Ind Health 54:377–385
4. Sekar A, Mohan P, Varghese GK, Varma MR (2020) Exposure to particulate matter in classrooms and laboratories of a university building. In: Lecture notes in civil engineering indoor environmental quality. https://doi.org/10.1007/978-981-15-1334-3_12
5. Al Y, Arif M, Katafygiotou M, Mazroei A, Kaushik A, Elsarrag E (2016) Impact of indoor environmental quality on occupant well-being and comfort : a review of the literature. Int J Sustain Built Environ 5:1–11. https://doi.org/10.1016/j.ijsbe.2016.03.006
6. Al Y, Arif M, Kaushik A, Mazroei A (2016) Occupant productivity and office indoor environment quality : a review of the literature. Build Environ 105:369–389. https://doi.org/10.1016/j.buildenv.2016.06.001

7. Kubba S (2014) Indoor environmental quality. In: Handbook of green building design and construction, pp 353–412. https://doi.org/10.1016/B978-0-12-810433-0.00007-1
8. Danielsson CB (2014) Indoor environmental quality-the office architecture: a contextual experience with influences at the individual and group level. In: Encyclopedia of toxicology
9. Khafaie MA (2016) Methodological approach in Air pollution Health effects. J Air Pollut Heal 1:219–226
10. Nieuwenhuijsen MJ (2005) Design of exposure questionnaires for epidemiological studies. Occup Env Med 62:272–280. https://doi.org/10.1136/oem.2004.015206
11. Preston V, Production C, Systems M, Manage W (2009) Questionnaire survey. In: International encyclopedia of human geography
12. Kelley K, Clark B, Brown V, Sitzia J (2003) Good practice in the conduct and reporting of survey research. Int J Qual Health Care 15:261–266
13. Lavrakas PJ (2008) Cross-sectional survey design. In: Encyclopedia of survey research methods. Sage Publications. https://doi.org/10.4135/9781412963947
14. Salkind NJ (2010) Panel design. In: Encyclopedia of research design, vols 1–0. Sage Publications. https://doi.org/10.4135/9781412961288
15. Allen M (2017) The sage encyclopedia of communication research methods, vols 1–4. SAGE Publications, Inc., Thousand Oaks, CA. https://doi.org/10.4135/9781483381411
16. Lavrakas PJ (2008) Encyclopedia of survey research methods, vols 1–0. Sage Publications, Inc., Thousand Oaks, CA. https://doi.org/10.4135/9781412963947
17. Nayak BK, Hazra A (2011) How to choose the right statistical test? Indian J Ophthalmol 59:85–86. https://doi.org/10.4103/0301-4738.77005
18. Mishra P, Pandey C, Singh U, Keshri A, Sabaretnam M (2019) Selection of appropriate statistical methods for data analysis. Ann Card Anaesth 22:297–301. https://doi.org/10.4103/aca.ACA_248_18
19. Bevans R Statistical tests: which one should you use?. https://www.scribbr.com/statistics/statistical-tests/
20. OECD (2012) Good practices in survey design step-by-step. In: Measuring regulatory performance a practitioner's guide to perception surveys, pp 31–44
21. Chirk N (2006) Designing a questionnaire. Malaysian Fam. Physician 1:32–35
22. Roopa S (2012) Questionnaire designing for a survey. J Indian Orthod Soc 46:273–277
23. Darcy-Jones C, Harriss A Telephone health assessments: good practice in occupational health. https://www.personneltoday.com/hr/telephone-health-assessments-good-practice-occupational-health/
24. International Labour Organisation (2017) Survey on how occupational safety experts carry out their roles and responsibilities
25. Durga MS, Nayak P, Narayan KA (2019) Strengths and weakness of online surveys. IOSR J Humanit Soc Sci. https://doi.org/10.9790/0837-2405053138
26. Duffy B, Smith K, Terhanian G, Bremer J, Interactive H (2017) Comparing data from online and face-to- face surveys. Int J Mark Res 47:1–12
27. Wright KB (2006) Researching internet-based populations: advantages and disadvantages of online survey research, Online questionnaire authoring software packages, and web survey services. J Comput Mediat Commun
28. Given LM (2008) Unstructured interview. In: The SAGE encyclopedia of qualitative research methods, vols 1–0. SAGE Publications, Thousand Oaks, CA. https://doi.org/10.4135/9781412963909
29. Acharya B (2010) Questionnaire design
30. Pew research center: questionnaire design. https://www.pewresearch.org/methods/u-s-survey-research/questionnaire-design/
31. Desai SC (2019) Comparing the use of open and closed questions for Web-based measures of the continued-influence effect. Behav Res Methods 51:1426–1440
32. Hu S (2014) Pretesting. In: Michalos AC (ed) Encyclopedia of quality of life and well-being research. Springer, Dordrecht. https://doi.org/10.1007/978-94-007-0753-5_2256

33. Lavrakas PJ (2008) Double-barreled question. In: Encyclopedia of survey research methods, vols 1–0. SAGE Publications, Inc., Thousand Oaks, CA. https://doi.org/10.4135/9781412963947
34. Madson L (2014) Teaching of psychology. Teach Psychol. https://doi.org/10.1207/s15328023top3201
35. Pelissier C, Dutertre V, Fournel P, Gendre I, Vergnon JM, Kalecinski J, Tinquaut F, Fontana L, Chauvin F (2016) Design and validation of a self-administered questionnaire as an aid to detection of occupational exposure to lung carcinogens. Public Health 3:3–10. https://doi.org/10.1016/j.puhe.2016.10.026
36. Hammer CS (2020) The importance of participant demographics. Am J Speech-Language Pathol 0360:9–11. https://doi.org/10.1044/1058-0360(2011/ed-04)
37. Rughiniş C, Humă B (2015) Who theorizes age ? the variables device and age-period-cohort analysis in the rhetoric of survey who theorizes age? the " socio—demographic variables" device and age—period—cohort. J Aging Stud 35:144–159
38. Connelly LM (2013) Demographic data in research studies. MedSurg Nurs 22:269+
39. Susitaival P, Flyvholm MA, Meding B, Kanerva L, Lindberg M, Svensson A, O'Lafsson JH (2003) Nordic occupational skin questionnaire (NOSQ-2002): a new tool for surveying occupational skin diseases and exposure. Contact Dermat 49:70–76
40. Hobson AJ, Sterling DA, Emo B, Evanoff BA, Callen S, Good L, Seixas N, Checkoway H, Racette BA (2009) Validity and reliability of an occupational exposure questionnaire for Parkinsonism in welders. J Occup Env Hyg 2009(6):324–331. https://doi.org/10.1080/15459620902836856.Validity
41. US EPA (2020) Exposure assessment tools by routes—inhalation
42. Center for disease control and prevention: Epidemiology, https://www.cdc.gov/careerpaths/k12teacherroadmap/epidemiology.html
43. Occupational Safety and Health Administration (1996) Occupational safety and health standards-toxic and hazardous substances-medical questionnaires, Non-mandatory
44. Occupational Safety and Health Administration (1986) Safety and health regulations for construction-toxic and hazardous substances-medical questionnaires; Mandatory
45. European Agency for Safety and Health at Work (2012) Risk perception and risk communication with regard to nanomaterials in the workplace. https://doi.org/10.2802/93075
46. Chaswa EN, Bobby I, Kosamu M, Kumwenda S, Utembe W (2020) Risk perception and its influencing factors among construction workers in Malawi. Safety 6:1–12

Studies on Performance of Isolated Footing with Load Transfer Platform

Shilpa K. Cherian and T. Soosan George

Abstract Foundations transfer the load of superstructure to the foundation soil, so the stability of a structure can be ensured by constructing foundation over good foundation soil. But due to increased developments, situations arise to construct structures on soft soil. Various methods of ground improvement techniques are adopted to construct a stable structure over soft soil. Studies show that the use of load transfer platform under structure foundation can homogenize settlement by establishing a resistant base layer. In this project compacted fiber reinforced load transfer platform (FRLTP) are used between isolated footing and foundation soil. In this study polypropylene fiber was mixed with clay, the fiber content (0%, 0.2%, 0.4%, 0.6% and 0.8% by weight of clay) in the clay was a varying parameter. The clay containing 0.4% of fiber is used as compacted FRLTP based on UCC strength. Plate load test was conducted on "Foundation soil + Compacted FRLTP" of varying thickness (0B, 0.25B, 0.5B and 1.0B, where B is the width of footing). For studying the effect of soil reinforcement in Compacted FRLTP, plate load test was conducted on "Foundation soil + Compacted Non-reinforced soil" of varying thickness (0B, 0.25B, 0.5B and 1.0B). Results show that, when Compacted FRLTP or Compacted Non-reinforced soil is used above foundation soil, the bearing capacity increases. The effect of reinforcement on soil was studied by comparing the results of plate load test of "Foundation soil + Compacted FRLTP or Compacted Non-reinforced soil". Results show that, with the use of Compacted FRLTP instead of Compacted Non-reinforced soil above foundation soil, the bearing capacity increased by 47.2%, 39.5% and 35.3% for thickness of 0.25B, 0.5B and 1.0B respectively. By plotting a graph between percentage increase in bearing capacity versus Thickness of Compacted Non-reinforced soil or Compacted FRLTP, an optimum thickness of 0.35B is taken for both Compacted FRLTP or Compacted Non-reinforced soil above foundation soil.

S. K. Cherian (✉)
PG Student, Geomechanics and Structures, Department of Civil Engineering, Saintgits College of Engineering, Kottayam, Kerala, India

T. S. George
Professor and Dean, Department of Civil Engineering, Saintgits College of Engineering, Kottayam, Kerala, India

G. C. Marano et al. (eds.), *Proceedings of SECON'21*, Lecture Notes in Civil Engineering 171, https://doi.org/10.1007/978-3-030-80312-4_6

Keywords Load transfer platform · Treated soil · Footing · Soil reinforcement · Bearing capacity

1 Introduction

Footings transfer the load of superstructure to the foundation soil. The stability of a structure depends on the property of soil in which the foundation rest. As far as possible the foundations should be constructed over good foundation soil so that the stability of structure can be ensured. But due to increased developments, situations arise to construct structures on soft soil. Settlement of the foundation soil and the stability of structure is a major concern while constructing structures on soft soil. Various methods of ground improvement techniques such as soil stabilization and soil reinforcement are adopted to construct a stable structure over soft soil. This project studies the performance of isolated footing with load transfer platform.

Load transfer platform are treated soil used under structure foundations in order to homogenize settlement and establish a resistant base layer. In this study fiber reinforced load transfer platform (FRLTP) are used between the isolated footing and foundation soil as shown in Fig. 1. By providing a strong top layer over weak foundation soil causes an increase in bearing pressure.

1.1 Literature Survey

- Wei et al. [1] study the effect of four kinds of fiber. Wheat straw, rice straw, jute and polypropylene fiber was added to the soil and lime-soil. Triaxial compressive test of fiber-lime-soil was carried out to study the shear strength, deviatoric stress–strain properties and sample failure pattern. Reinforcement significantly increased the cohesion and lightly improved the internal friction angle. From the test results polypropylene fiber is selected as the best fiber for soil reinforcement from the four kinds of fiber. The study show that fiber content and fiber length are the important parameters associated with the mechanical propertiesof soil. The study use different percentage of fiber to determine the optimum fiber content based on previous study.

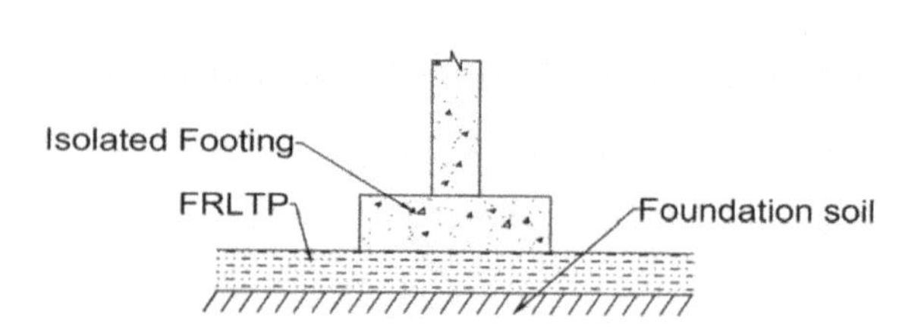

Fig. 1 FRLTP between isolated footing and foundation soil

- Dang et al. [2] conducted numerical investigation in the performance of NFRLTP. Results show that natural fiber-lime reinforced soil helps load transfer process from the embankment to pile. It prevents the total and differential settlement of the embankment. The natural fiber-lime reinforced load transfer platform has comparable engineering characteristics as geogrid reinforced LTP.
- Cai et al. [3] conducted their study with different percentages of fiber and lime for stabilization. Fiber content, lime content and curing duration has significant influence on engineering properties of soil. The unconfined compressive strength, cohesion and friction angles increase while increasing the length of curing.
- Oliveira et al. [4] study the effect of geotextile reinforcements on embankment behavior. The laboratory model test results indicated the effect of reinforcement is considerable if the reinforcement is stiff and strong. To achieve a substantial improvement on embankment behavior, stiffer and stronger reinforcements should be used.
- Arora and Kumar [5] study the bearing capacity of a circular footing resting on randomly distributed fiber-reinforced pond ash overlying soft clay was investigated using a model footing. Thickness of pond ash layer was a varying parameter. Also the reinforcement percentage and length was a varying parameter. Results show that the ultimate bearing capacity increased by 80% when compacted pond ash layer of 1.5D was used. The increase in thickness of the top strong layer leads to a decrease in pressure over the weak lower layer, thus resulting in increased bearing pressure.

2 Materials and Methods

2.1 Material Collection

For conducting the experimental study clay was collected from a construction site at Kuttanad region in Alapuzha district. The clay was collected from a depth of 1–2 m for excluding the organic matter. The clay samples were transported to the laboratory after being sealed in plastic bags and then wrapped in gunny bags for minimum moisture change during transportation. The polypropylene fiber was bought from Buddha Building Technology, Maharashtra.

2.2 Basic Properties of Clay

Field density of clay was determined by inserting PVC pipe in to the field. The mass of soil filled in the pipe divided by the volume of the pipe gives the field density. Laboratory tests such as Natural moisture content (as per IS 2720 part 2),

Atterberg's limits (as per IS 2720 part 5 and IS 2720 part 6), Specific gravity (as per IS 2720 part 3) and Grain size analysis (as per IS 2720 part 4) of clay was conducted.

2.3 Clay-Fiber Mix for Compacted FRLTP

The clay sample collected from the field was air dried and powdered. The percentage of polypropylene fiber in the clay was a varying parameter, five different percentage of fiber content was used in the clay soil, i.e. 0, 0.2, 0.4, 0.6 and 0.8% by weight of the clay soil. The powdered clay was mixed with polypropylene fiber. Compaction test were conducted (as per IS 2720 part 7) on different mix to determine the maximum dry density (MDD) and optimum moisture content (OMC). Unconfined compressive strength test (as per IS 2720 part 10) were conducted on clay-fiber mix at maximum dry density (MDD) and optimum moisture content (OMC) for determining the optimum mix which can be used as compacted FRLTP. The polypropylene fiber used in the study is shown in Fig. 2.

2.4 Model Plate Load Test

A square plate of 100 mm width and 6 mm thick was used as a footing in a tank of 500 mm × 500 mm × 600 mm size. Footing is made of cast iron to have perfect rigidity. The tank is made of sufficient thickness to withstand lateral expansion under loading. The dimensions of the tank are kept more than five times the width of footing tested so that it should not include boundary effect. Figure 3 shows the

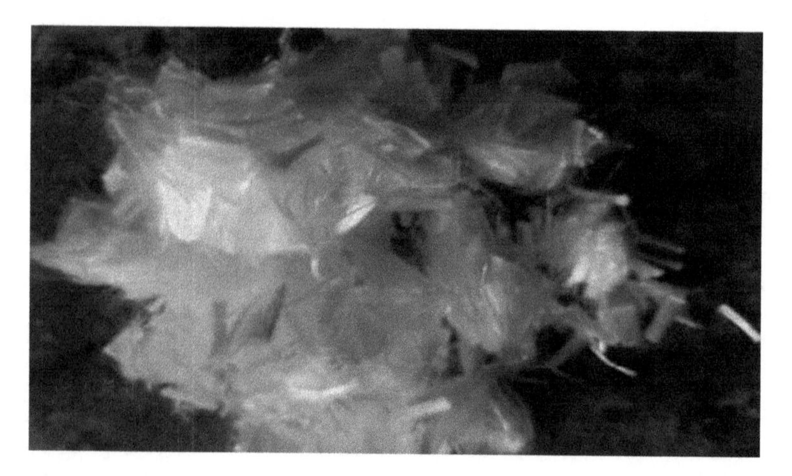

Fig. 2 Polypropylene fiber

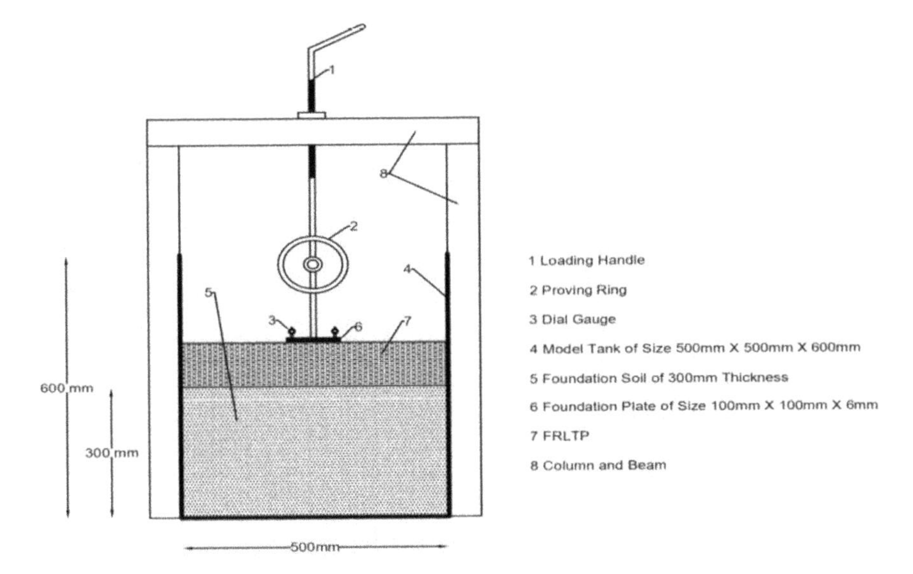

Fig. 3 Schematic diagram of experimental set up

schematic diagram of experimental set up. The tank was filled with clay collected from Kuttanad region at field condition to a depth of 3B, where B is the width of footing. The clay filled in the tank at field condition represents the foundation soil. The bearing capacity of foundation soil was find out from the load-settlement curve obtained by conducting plate load test. In order to find the effect of soil reinforcement on bearing capacity, plate load test were conducted over foundation soil + compacted non-reinforced and foundation soil + compacted FRLTP (optimum mix). The depth of compacted non-reinforced soil and compacted FRLTP was a varying parameter. A depth of 0B (0 mm), 0.25B (25 mm), 0.5B (50 mm) and 1.0B (100 mm) of compacted nonreinforced and compacted FRLTP are placed over foundation soil. The footing was loaded by pushing it towards the soil. Dial gauge measurements and loading on the proving ring are recorded.

3 Results and Discussion

3.1 Properties of Clay

The particle size distribution curve of clay is sown in Fig. 4 and the properties of clay obtained by various laboratory investigation is shown in Table 1.

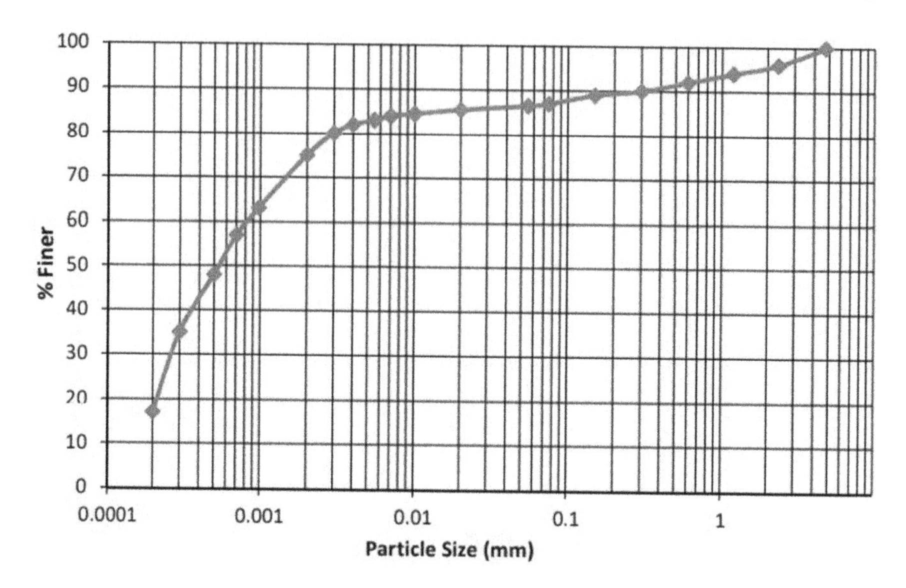

Fig. 4 Particle size distribution of clay

Table 1 Properties of clay

Properties	Value
Natural moisture content (%)	101.2
Field density (g/cm^3)	1.67
Liquid limit (%)	150
Plastic limit (%)	49
Shrinkage limit (%)	32
Specific gravity	2.42
Clay (%)	75
Silt (%)	10.5

3.2 Properties of Polypropylene Fiber

The properties of polypropylene fiber provided by Buddha Building Technology is shown in Table 2.

Table 2 Properties of polypropylene fiber

Properties	Value
Length (mm)	12
Diameter (μ)	24
Specific gravity	0.92
Thermal and electrical conductivity	Low
Alkaline resistance	Very high
Acid and salt resistance	Very high

3.3 Compaction and UCC Test Results on Clay-Fiber Mix

The compaction curve for different mix is shown in Fig. 5. Table 3 shows the MDD, OMC and UCC strength for different mix. From the results we can see that as the fiber content increases the maximum dry density (MDD) decreases and optimum moisture content (OMC) increases. The decrease in maximum dry density in the mix is due to the lower specific gravity of polypropylene fiber used in the clay-fiber mix. From the test results as the fiber content increases from 0 to 0.4% the unconfined compressive strength of the clay-fiber mix increases. But when the fiber content is greater than 0.4% the UCC strength decreases. As the fiber content increases by 0.4% the unconfined compressive strength increases by 54.1%.But when the fiber content is greater than 0.4% the unconfined compressive strength decreases. This decrease in unconfined compressive strength is due to the fact that with higher fiber content the quantity of soil matrix available for holding the fiber is insufficient to develop effective bond between fiber and soil. The clay-fiber mix with 0.4% of fiber is taken as the optimum mix due to its high unconfined compressive strength. This is used as the compacted FRLTP.

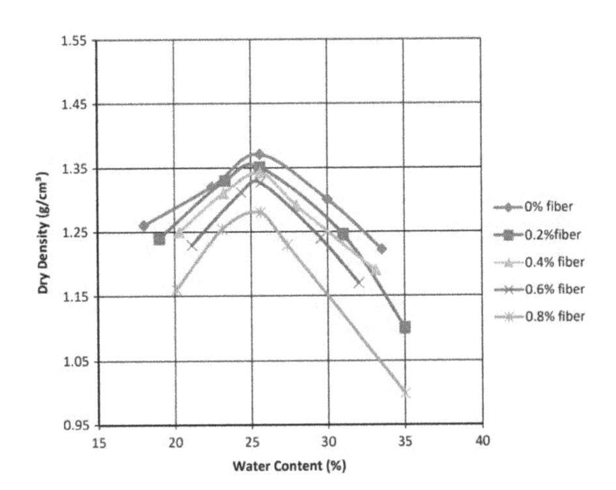

Fig. 5 Compaction curve of different clay-fiber mix

Table 3 Variation of MDD, OMC and UCC strength for different mix

Fiber (%)	MDD (g/cm^3)	OMC (%)	UCC strength (kPa)
0	1.371	25.58	19.6
0.2	1.351	25.59	32.8
0.4	1.342	26.61	42.7
0.6	1.327	25.62	40.6
0.8	1.28	25.63	37.6

3.4 Bearing Capacity for Different Conditions

The bearing capacity for different conditions are find out from the load settlement curve obtained by conducting plate load test. Figure 6 shows the load settlement curve for different conditions. The bearing capacities and percentage increase in bearing capacities for different conditions of plate load test are shown in Table 4. From the results as the thickness of compacted FRLTP or compacted non-reinforced soil increases above foundation soil the bearing capacity also increases. Because by increasing the thickness of top strong layer decreases the pressure over weak lower layer. This results in increase in bearing capacity.

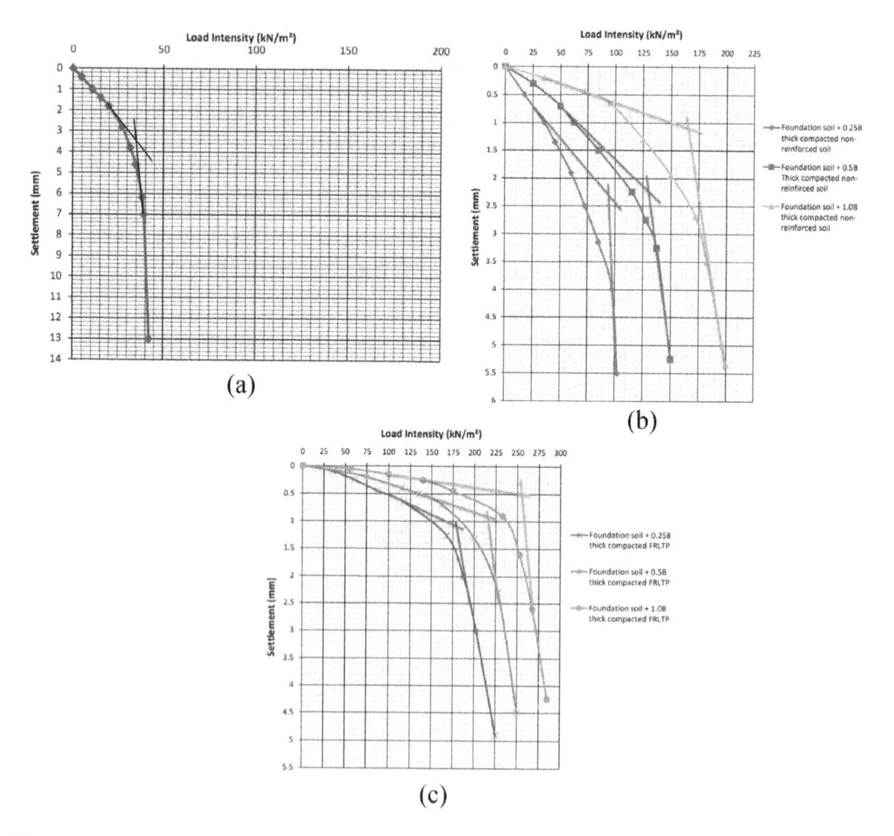

Fig. 6 Load settlement curve for (**a**) Foundation soil, (**b**) Foundation soil + compacted non-reinforced soil of different thickness and (**c**) Foundation soil + compacted FRLTP of different thickness

Table 4 Bearing capacities and percentage increase in bearing capacities for different conditions of plate load test

Condition of plate load test	Bearing capacity (kN/m^2)	Percentage increase in bearing capacity (%)
Foundation soil	35	0
Foundation soil + 0.25B thick compacted non-reinforced soil	95	63.1
Foundation soil + 0.5B thick compacted non-reinforced soil	130	73
Foundation soil + 1.0B thick compacted non-reinforced soil	165	78.7
Foundation soil + 0.25B thick compacted reinforced soil	180	80.5
Foundation soil + 0.5B thick compacted reinforced soil	215	83.7
Foundation soil + 1.0B thick compacted reinforced soil	255	86.2

3.4.1 Effect of Soil Reinforcement on the Bearing Capacity of Compacted FRLTP

The effect of soil reinforcement on compacted FRLTP can be determined by comparing the load settlement results of "Foundation soil + compacted non-reinforced soil" with load settlement results of "Foundation soil + compacted FRLTP". The load settlement curve foundation soil + compacted non-reinforced soil or compacted FRLTP of different thickness is shown in Fig. 7. The results show that when compacted FRLTP was used instead of compacted non-reinforced soil of same thickness above foundation soil there is an increase in bearing capacity. Because the top strong layer decreases the pressure over weak lower layer. When compacted FRLTP is used instead of compacted non-reinforced soil above foundation soil, the bearing capacity increases, indicating that the fiber addition on the top compacted soil makes it stronger.

For 0.25B thick compacted non-reinforced soil above foundation soil has a bearing capacity of 95 kN/m^2 and for 0.25B thick FRLTP above foundation soil has a bearing capacity of 180 kN/m^2, i.e., the bearing capacity increased by 47.2%. For 0.5B thick compacted non-reinforced soil above foundation soil has a bearing capacity of 130 kN/m^2 and for 0.5B thick FRLTP above foundation soil has a bearing capacity of 215 kN/m^2, i.e., the bearing capacity increased by 39.5%. For 1.0B thick compacted non-reinforced soil above foundation soil has a bearing capacity of 165 kN/m^2 and for 1.0B thick FRLTP above foundation soil has a bearing capacity of 255 kN/m^2, i.e., the bearing capacity increased by 35.3%.

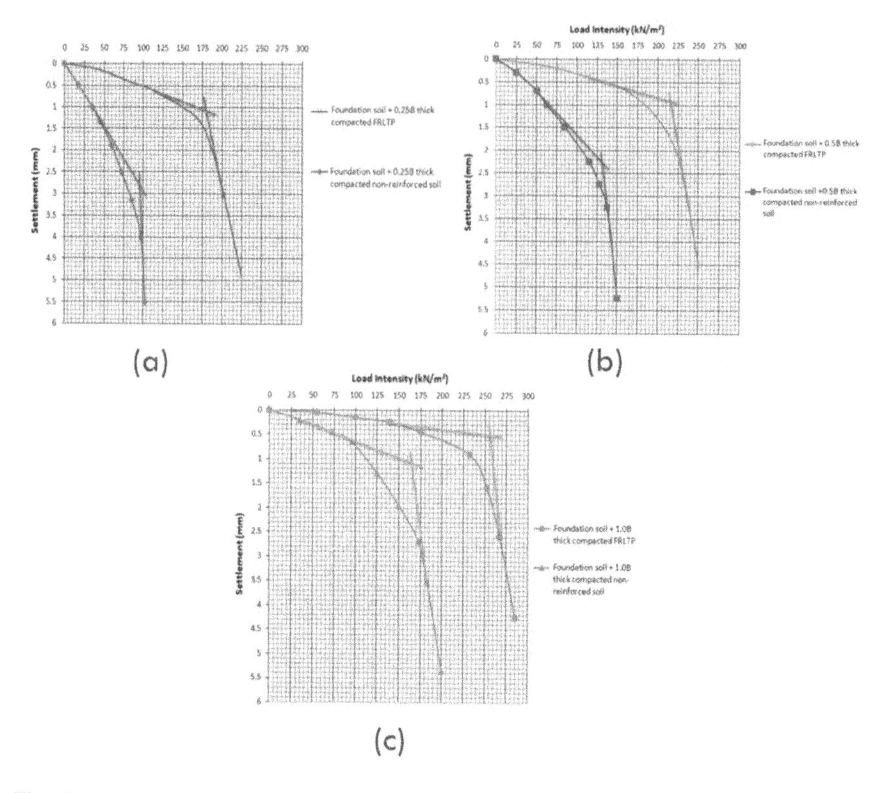

Fig. 7 The load settlement curve of foundation soil + compacted non-reinforced soil or compacted FRLTP of (**a**) 0.25B Thickness, (**b**) 0.5B Thickness and (**c**) 1.0B Thickness

3.4.2 Percentage Increase in Bearing Capacity and Optimum Thickness

As the thickness of compacted non-reinforced soil above foundation soil increases from 0B to 1.0B the bearing capacity increases by 78.7%. As the thickness of compacted FRLTP above foundation soil increases from 0B to 1.0B the bearing capacity increases by 86.2%. By increasing the thickness of top strong layer decreases the pressure over weak lower layer. This results in increase in bearing capacity. The percentage increase in bearing capacity for foundation soil + compacted non-reinforced soil or compacted FRLTP of different thickness is shown in Fig. 8. In both cases the percentage increase in bearing capacity is maximum between 0.25B and 0.5B and the percentage increase in bearing capacity between 0.5B and 1.0B is negligible. So for both compacted non-reinforced soil and compacted FRLTP an optimum thickness of 0.35B above the foundation soil can be taken.

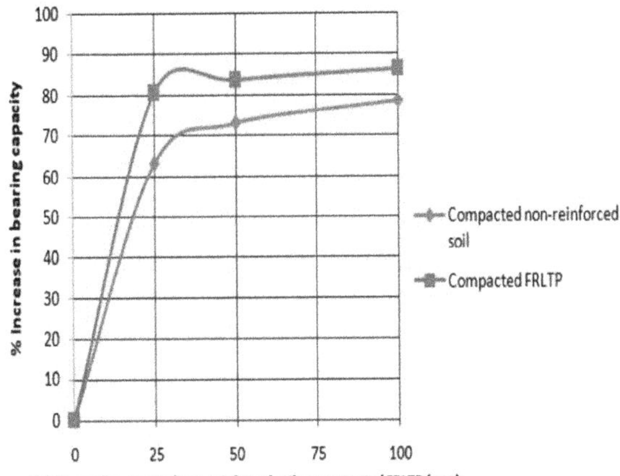

Fig. 8 Percentage increase in bearing capacity for foundation soil + compacted non-reinforced soil or compacted FRLTP of different thickness

4 Conclusions

This study determines the possibility of using compacted fiber reinforced load transfer platform (FRLTP) between isolated footing and foundation soil. The results show that the bearing capacity is increased by placing compacted FRLTP between isolated footing and foundation soil. To study the effect of soil reinforcement in FRLTP plate load test was conducted by placing compacted non-reinforced soil layer above foundation soil. The result shows that the bearing capacity obtained by placing compacted FRLTP is greater than the bearing capacity obtained by placing compacted non-reinforced soil layer above foundation soil. The thickness of compacted FRLTP and compacted non-reinforced soil above foundation sol was a varying parameter (0B, 0.25B, 0.5B and 1.0B) while conducting plate load test. Results show that as the thickness of compacted FRLTP and compacted non-reinforced soil above foundation soil increases the bearing capacity also increases. The results obtained from the experimental study are as follows,

- Fiber addition in the clay decreases the maximum dry density (MDD), due to the low specific gravity of polypropylene fiber
- Fiber addition up to 0.4%in clay increases the unconfined compressive strength by 54.1%. Fiber content greater than 0.4% in clay decreases the unconfined compressive strength value. So clay-fiber mix with 0.4% of fiber is taken as the optimum mix for compacted FRLTP
- As the thickness of compacted non-reinforced soil increases from 0B to 1.0B above the foundation soil, the bearing capacity increases by 78.7%
- As the thickness of compacted FRLTP increases from 0B to 1.0B above the foundation soil, the bearing capacity increases by 86.2%
- The bearing capacity of foundation soil + compacted FRLTP of 0.25B thickness is 47.2% greater than the bearing capacity of foundation soil + compacted non-reinforced soil of 0.25B thickness

- The bearing capacity of foundation soil + compacted FRLTP of 0.5B thickness is 39.5% greater than the bearing capacity of foundation soil + compacted non-reinforced soil of 0.5B thickness
- The bearing capacity of foundation soil + compacted FRLTP of 1.0B thickness is 35.3% greater than the bearing capacity of foundation soil + compacted non-reinforced soil of 1.0B thickness
- For compacted non-reinforced soil and compacted FRLTP above foundation soil the percentage increase in bearing capacity is maximum between 0.25B and 0.5B and the percentage increase in bearing capacity between 0.5B and 1.0B is negligible. So for both compacted non-reinforced soil and compacted FRLTP an optimum thickness of 0.35B above the foundation soil can be taken

References

1. Wei L, Chai S, Zhang H, Shi Q (2018) Mechanical properties of soil reinforced with both lime and four kinds of fiber. Constr Build Mater 172:300–308
2. Dang L, Dang C, Khabbaz H, Fatahi B (2016) Numerical assessment of fiber inclusion in a load transfer platform for pile supported embankment over soft soil. In: Geo-China, pp148–155
3. Cai Y, Shi B, Ng C, Tang C (2006) Effect of polypropylene fibre and lime admixture on engineering properties of clayey soil. Eng Geol 87(3–4):230–240
4. Oliveira PJV, Pinheiro JLP, Correia AAS (2011) Numerical analysis of an embankment built on soft soil reinforced with deep mixing columns: parametric study. Comput Geotechnics 38 (4):566–576
5. Arora S, Kumar A (2019) Bearing capacity of strip footing resting on fibre-reinforced pond ash overlying soft clay. J Hazard Toxic Radioactive Waste 4(1)

Experimental Investigation on Artificial Light Weight Fly Ash Aggregates in Concrete

K. Rhishi Raj and R. Vasudev

Abstract Construction sector has a significant influence on environment. Massive amounts of natural resources are used by these industries. A desirable alternative to disposal has been the recycling or reuse of industrial by-products. The usage of artificial aggregates plays a crucial role in this developing infrastructural era, sustainable development and energy efficient building construction. This paper studies the effect of artificial light weight fly ash aggregate in concrete and its mechanical properties. Artificial light weight fly ash aggregates can be produced by a process called pelletization by nodulizing the fly ash with the correct amount of water and binder in a pelletizer and further hardened by cold bonding method. The cold bonding technique is more conservative than other methods of hardening due to the use of minimum energy. Natural aggregate replacement adopted for artificial cold bonded lightweight aggregate are 0, 25, 50, 75 and 100% and metakaolin is added by 10% of weight of cement to concrete for achieving more strength [1, 2]. The concrete formed is light in nature and has the additional benefit of reducing overall cost savings, especially in transport and placement, it has good sound absorption, fire resistance and high thermal insulation properties. In this paper an experimental study has been conducted to know about the mechanical properties of artificial light weight fly ash aggregates through partial and complete replacement of coarse aggregates and find optimum percentage of replacement and also its effects on the properties of concrete by compressive strength, flexural strength and split tensile strength.

Keywords Artificial light weight aggregate · Cold bonding · Fly ash · Fly ash aggregate · Pelletization

---

K. R. Raj (✉) · R. Vasudev
Department of Civil Engineering, Toc H Institute of Science and Technology, Ernakulam, Kerala, India

G. C. Marano et al. (eds.), *Proceedings of SECON'21*, Lecture Notes in Civil Engineering 171, https://doi.org/10.1007/978-3-030-80312-4_7

1 Introduction

In the emerging infrastructural era, concrete has some of the most used construction materials. The constant overuse of environmental raw materials for the production of concrete generates many problems to eco system. The problem of the disposal of by-products from industries such as fly ash, silica fume, waste sludge etc. is increasing with each day. The use of artificial aggregates plays a crucial role in sustainable development [3–6]. Artificial light weight aggregates (ALWA) can be produced by process called pelletization, through which nodulizing the fly ash (FA) with the right amount of water and binder in a pelletizer, and additional hardening by sintering, cold bonding or autoclaving [7–9]. The cold bonding process is more energy efficient than sintering because it uses less energy [10, 11]. Weight of aggregates in conventional concrete is the factors to be compared with the fly ash aggregate concrete. As compared to the less self-weight of fly ash aggregate concrete, this high self-weight results in a construction material that is not cost-effective. In order to produce concrete of the preferred density to satisfy the mandated application, the self-weight of structural and non-structural members should be decreased. Although lightweight concrete (LWC) cannot always replace normal concrete (NC) with its strength capable, it has its own returns, such as decreased dead load, cost effective structure and improved seismic resistance, high sound absorption and good resistance to fire [12–14]. Rounded aggregates support concrete workability, while natural gravel of angular nature offers an improved bonding property, but needs extra cement mortar for good workability. This can be reached by lowering the void ratio, thereby reducing the amount of paste needed to fill the voids. Well graded aggregate reduces the consumption of cement paste, thus reducing concrete costs. This can be achieved by lowering the void ratio, thus reducing the quantity of paste needed to fill the voids. One of the reasons to compare in normal concrete with weight of fly ash aggregate concrete is the weight of the concrete. Compared to the low self-weight of fly ash aggregate concrete, the heavy self-weight of NC creates an uneconomic structural material. The self-weight of the structural and non-structural members should remain lowered to manufacture concrete of the wanted density to uniform the necessary application. As a result, the construction of supporting structural members is made more cost-effective, resulting in the advancement of light-weight concrete.

2 Pelletization and Cold Bonding

Artificial light weight fly ash aggregates can be manufactured by a process called pelletization [15–20]. An artificial lightweight aggregate is crushed or agglomerated to achieve the desired grain size distribution. The disc type pelletizer is shown in

Fig. 1. Agglomeration is a method of manufacturing aggregate pellets due to rotational motion or tumbling force emerged during the pelletization process without any external force. Production involves adding dry materials into the disc pellatizer and spraying the determined water during the action. Fly ash (90% of the total weight) and cement (10% of the weight) were the materials used in the production. During the manufacturing process, maximum pelletization efficiency and fresh aggregate stability were achieved, resulting in the formation of uniform spherical shaped pellets in 20 min. The incorporation of binder in fly ash evoked the initial development of pellets, and quick hardening of fly ash balls without collapse was provided by the activator addition[21]. This has provided a stable fly ash pellet formation. Initially, in the disc pelletizer, the fly ash and cement powder were properly mixed and then 25% of the water content was sprayed during processing. The pellets were air-dried after production. Pellets formed after pelletization are shown in Fig. 1. Pelletization is followed by a process called cold bonding which is one of the techniques for the hardening of the pellets and the aggregates were permitted to be cured for 1, 3 and 7 days in the cold bonding process to achieve green strength. The method of cold bonding allows the pelletized aggregates are treated at room temperature or be stored with steam in an enfolded space. Curing until its strengths are appropriate for concrete. This technique requires medium technical ability with least energy consumption and follows the fly ash agglomeration with Portland cement at Ambient temperature. Structural lightweight concretes with a compressive strength of up to 50 MPa have been produced by the use of these aggregates.

Fig. 1 Pellets formed from pelletization

3 Materials

The different materials used for the study are portland pozzolana cement, M-sand, coarse aggregates, fly ash, and metakaolin.

3.1 Cement

The cement collected is PPC (Portland Pozzolana Cement) with 5% fines and specific gravity 3.209.

3.2 Fine Aggregates and Coarse Aggregates

M sand of specific gravity 2.632 and water absorption 1.23% is used as fine aggregate in this study. Coarse aggregates of size 20 mm, specific gravity 2.65 and water absorption 0.72% is used for casting.

3.3 Fly Ash

Fly ash is an industrial waste product which is accepted as an environmental pollutant, generated during the combustion of coal for energy production. The fly ash of specific gravity 2.6 and Class-F was collected from 'Marymatha Construction Company', Ambalamugal, Ernakulam. These fly ashes are used for producing fly ash aggregate in the pelletizer.

3.4 Metakaolin

Metakaolin of specific gravity 2.6 is used for this study and collected from Chennai.

3.5 Superplasticizer

Conflo LN is a chloride free, high range water reducing and waterproofing admixture designed as an aid to high quality concrete production.

4 Experimental Programme

The experimental programme involves of compressive strength, flexural strength and split tensile strength of concrete.

4.1 *Mix Design*

Experimental study is conducted for M40 grade concrete. Different proportions of materials are shown in Table 1.

4.2 *Procedures*

After the proper mix design procedure, Fly ash aggregates can be produced by mixing fly ash, cement (90:10) and proportionate quantity of water in a pelletizer. The time required for forming fly pellets are 20 min. After forming the green pellets, it is allowed to cure in a room temperature for 28 days to achieve required strength. Fly ash aggregate concrete was mixed in the same way as traditional concrete. Cement, metakaolin (10% by weight of cement), fine aggregate, water was mixed first and is called as mortar then coarse aggregate that means normal coarse aggregate and fly ash aggregate were added with water, fly ash aggregate is added by different percentage of total amount of coarse aggregate required and these ingredients were mixed thoroughly and was then ready for casting. The different percentage of replacements of natural aggregate by fly ash aggregated used are 0, 25, 50, 75 and 100%.

4.3 *Casting and Curing of Specimen*

Specimens were cast to find out the mechanical properties of concrete. Compressive strength was determined using 150 × 150 × 150 mm cubes, flexural strength was determined using 100 × 100 × 500 mm beams, and split tensile strength was determined using 300 mm length and 150 mm diameter cylinders. In this study the

Table 1 Mix design

Materials	Cement	Fine aggregate	Coarse aggregate	Water	Super plasticizer
Quantity (kg/m^3)	371	702.46	1194.89	148	4.12
Ratio	1	1.893	3.22	0.36	0.02

specimens are prepared by replacing natural aggregate with fly ash aggregate by different percentage i.e., 0% (fly ash aggregate 0%, natural aggregate 100%), 25% (fly ash aggregate 25%, natural aggregate 75%), 50% (fly ash aggregate 50%, natural aggregate 50%), 75% (fly ash aggregate 25%, natural aggregate 75%), and 100% (fly ash aggregate 100%, natural aggregate 0%). After casting the specimens with varying percentage of fly ash aggregates, the specimens were allowed to air cure for 24 h at room temperature. After 24 h of casting, the specimens were demoulded and allowed to cure for 28 days. The specimens were then taken out of the water and allowed to dry in the shade before being examined. The following is a list of the various mixes. FA-0, FA-25, FA-50, FA-75, and FA-100.FA represents pelletized cold bonded fly ash light weight aggregate.

4.4 Compressive Strength

Compressive strength test on cubes of M40 grade equivalent cold bonded light-weight aggregate concrete was conducted. The compressive strength of concrete was determined by crushing the test cubes until they were destroyed using a compression testing machine in accordance with IS 516:1959. (Reaffirmed 2004). Figure 2 shows the compressive strength testing of the specimen.

Fig. 2 Compressive strength

$$\text{Compressive strength} = \frac{\text{load applied}}{\text{cross sectional area}}$$

4.5 Flexural Strength

The flexural test was performed on reinforced beams 700 mm long and 150 × 150 mm in cross section in compliance with IS 516:1959 (reaffirmed in 2004). The tests were carried out on a Universal Testing Machine (UTM) with a two-point load configuration. The load increased continuously without shock until the specimen failed and the maximum load was recorded as P. Figure 3 shows the flexural strength testing of the specimen.

The flexural strength is given by,

$$\text{Flexural strength} = \text{Pl}/\left(\text{bd}^2\right)$$

where

l = length of specimen.
b = width of specimen.
d = depth of specimen.

Fig. 3 Flexural strength

4.6 Split Tensile Strength

The split tensile test was performed in accordance with IS 5816-1999 by positioning the cylinder specimen horizontally between the loading surfaces of the compression testing system and applying load until the cylinder failed along the vertical diameter. This is an indirect method for determining the tensile strength of concrete cylinder of diameter 150 mm and height 300 mm were used for this test. Figure 4 shows the split tensile strength testing of the specimen.

The split tensile strength was calculated by,

$$\text{Tensile strength} = 2P/\pi DL$$

where

P = applied load.
D = diameter of the cylinder.
L = length of the cylinder.

Fig. 4 Split tensile strength

5 Results and Discussion

The overall result of mechanical strength of light weight aggregate concrete at twenty eighth day is shown in Table 2.

5.1 Compressive Strength

The compressive strength and percentage of fly ash aggregate replacing natural aggregate at 7th- and 28th-day curing is shown in Fig. 5. and observed that with the addition of FA aggregate the cube compressive strength decreases continuously up to 100% replacement of FA aggregate. More than the target mean strength of M40 concrete i.e., 48.25 N/mm^2 has been achieved even when the natural aggregate is replaced with 50% of FA aggregate.

Table 2 Result of mechanical strength of Light weight aggregate concrete (28th day)

Mix	Compressive strength (MPa)	Flexural strength (MPa)	Split tensile strength (MPa)
FA-0	49.2	4.47	4.8
FA-25	52.25	5	4.98
FA-50	49.6	4.2	4.83
FA-75	46.8	4	4.73
FA-100	43.23	3.65	4.5

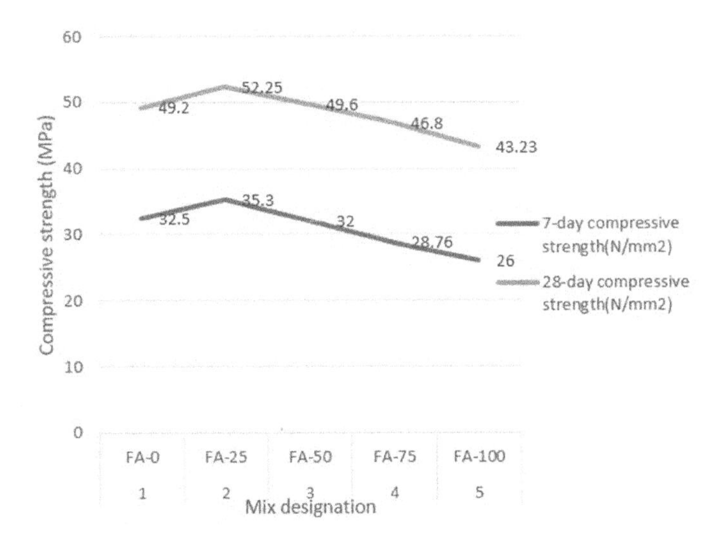

Fig. 5 Compressive strength of LWAC

5.2 Split Tensile Strength

Due to its brittle nature, concrete is very poor under tensile force and is not projected to withstand direct tension. When tensile forces are applied to concrete, cracks appear. As a result, the tensile strength of concrete must be determined in order to determine the load at which the concrete members will crack. Figure 6 shows the split tensile strength of LWAC.

5.3 Flexural Strength

Figure 7 gives the flexural strength values of LWAC. The graph between flexural strength versus percentage of pelletized fly ash aggregate replacing natural aggregate at 7- and 28-days curing periods are shown. With increase in percentage of replacement of natural aggregate by FA aggregate, the flexural strength values are decrease continuously up to 100% replacement.

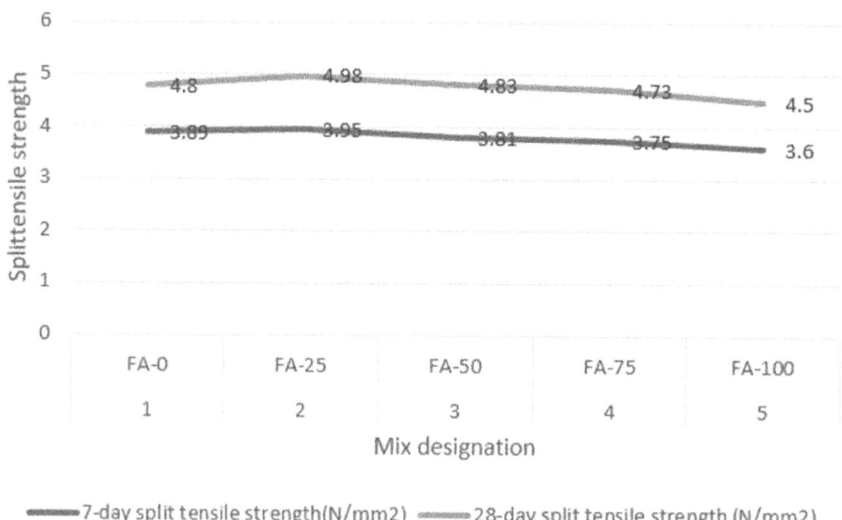

Fig. 6 Split tensile strength of LWAC

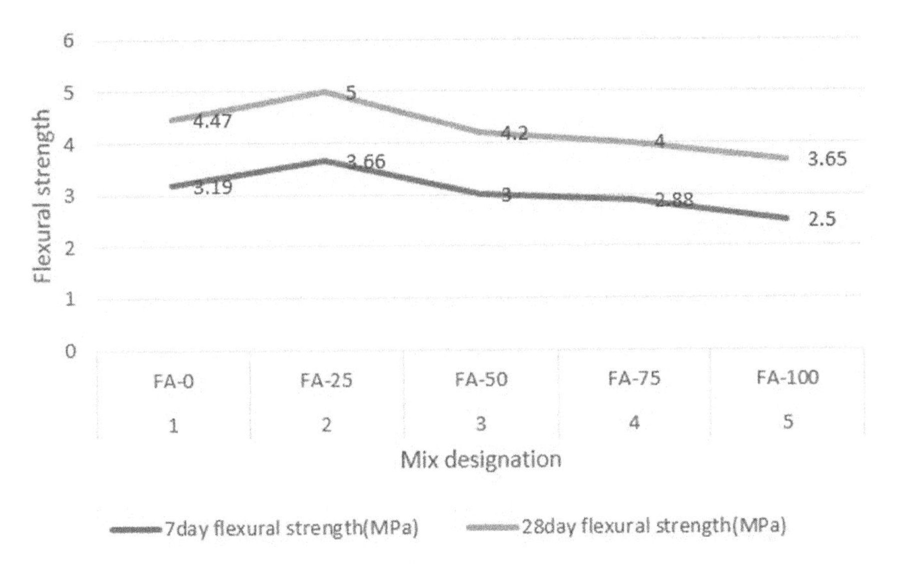

Fig. 7 Flexural strength of LWAC

6 Conclusion

The advancement of the latest waste control systems is one of the most important fields of research in modern times. This is due to the need to alter the materials to stay away from depleting ordinary resources that are used sufficiently for people to develop. Light weight aggregate density is lower when compared to natural aggregate. It provides greater insulation and can be used to produce lightweight concrete due to its lower density. During the pelletizing process, Production of fly ash aggregates in the pelletizer was found to be successful according to the addition of a binder and an alkali activator, which projected stable aggregate formation as well as enhanced strength properties after sufficient curing. From this study it is concluded that FA-25 and FA-50 shows better strength characteristics (compressive strength, flexural strength and split tensile strength) than control mix. It is clear from the study that, FA-25 attain the maximum strength (compressive, split and flexural strength) than all other replacements. The requirement for lightweight structural design is evolving today because of the effect of natural disasters such as earthquake pressure all over the world, as it reduces the structure's self-weight.

References

1. Karahan O, Khandaker MA, Hossain Erdogan O, Lachemi M, Sancak E (2012) Effect of metakaolin content on the properties self-consolidating lightweight concrete. Constr Build Mater 31:320–325

2. Muduli R, Bhusan B, Mukharjee (2019) Performance assessment of concrete incorporating recycled coarse aggregates and metakaolin: a systematic approach. Constr Build Mater 21:233–254
3. Debeib F, Kenai S (2008) The use of course and fine crushed bricks as aggregate in 546 concrete. Constr Build Mater 22:886–893
4. Nadesan MS, Dinakar O (2017) Structural concrete using sintered fly ash lightweight aggregate: a review. Constr Build Mater 154:928–944
5. Saraber A, Overhof R, Green T, Pels J (2012) Artificial lightweight aggregates as utilization for future ashes—a case study. Waste Manage 32:144–152
6. Tay JH, Hong SY, Show KY (2015) Reuse of industrial sludge as pelletized aggregate for concrete. J Environ Eng 126:279–287
7. Harikrishnan KI, Ramamurthy K (2006) Influence of pelletization process on the properties of fly ash aggregates. Waste Manage 26:846–852
8. Kockal NU, Ozturan T (2010) Effects of lightweight fly ash aggregate properties on the behavior of lightweight concretes. J Hazard Mater 179:954–965
9. Kotapati S, Brahmini A (2017) Concrete made by using fly ash pellets as coarse aggregate. Int J Inno Res Sci Eng Technol 6:9790–9796
10. Bui LA, Hwang CI, Chen C, Lin K, Hsieh M (2012) Manufacture and performance of cold bonded lightweight aggregate using alkaline activators for high performance concrete. Constr Build Mater 35:1056–1062
11. Narattha C, Chaipanich A (2018) Phase characterizations, physical properties and strength of environment-friendly cold-bonded fly ash lightweight aggregates. J Clean Prod 171:1094–1100
12. Desai BV, Sathyam A (2014) A study on partial replacement of natural granite aggregate with pelletized fly ash aggregate. Int J Comput Eng Res 4:31–40
13. Gesoglu M, Ozturan T, Guneyisi E (2006) Effects of cold-bonded fly ash aggregate properties on the shrinkage cracking of lightweight concretes. Cement Concr Compos 28(7):598–605
14. Gomathi P, Sivakumar A (2015) Accelerated curing effects on the mechanical performance of cold bonded and sintered fly ash aggregate concrete. Constr Build Mater 77:276–287
15. Frankovi A, Bosiljkov VB, Ducman V (2017) Lightweight aggregates made from fly ash using the cold-bond process and their use in lightweight concrete. J Mater Technol 51:267–274
16. Gesoglu M, Gooneys E, Oz OE (2012) Properties of light weight aggregate produced with cold–bonding pelletization of fly ash and ground granulated blast furnace slag. Mater Struct 45:1535–1546
17. Gesoglu M, Ozturan T, Guneyisi E (2007) Effects of fly ash properties on characteristics of cold-bonded fly ash lightweight aggregates. Constr Build Mater 21(9):69–78
18. Kockal NU, Ozturan T (2011) Durability of lightweight concretes with lightweight fly ash aggregates. Constr Build Mater 25:1430–1438
19. Mohan AB, Vasudev R (2018) Artificial lightweight aggregate through cold bonding pelletization of fly ash: a review. Int Res J Eng Technol 5:778–783
20. Sathyam D, Srikanth K, Desai BV (2017) Brief study on concrete modified with artificial cold bonded pelletized light weight fly ash aggregates. IOSR J Eng 4:31–40
21. Jagadeeswari K, Phanindra HR, Ramadevi G (2017) Replacement of natural aggregate with fly ash aggregate with alkali blinders. Int J Tech Inno Mod Eng Sci 4:300–304

Experimental Investigation on the Effect of Fly Ash on Plastic Shrinkage Cracking of Self-Compacting Concrete

Anju George and C. Abin Thomas

Abstract The plastic shrinkage may sometimes cause early age cracking before the setting on the young concrete surface. One of the serious problems in concrete elements especially in structures with large surface area/volume ratios is the shrinkage cracking and it affects the durability of the structure. This paper quantifies the effectiveness of fly ash on reducing plastic shrinkage cracking of self-compacting concrete. The aim of this experimental work was to study the plastic shrinkage evolution in self-compacting concrete (SCC) in which the cement was replaced by fly ash at various percentages of 20, 25, 30, 35, 40 and 45% by weight of cement. The results show that replacing cement by fly ash at higher percentages exhibited a delayed formation in the crack in the concrete and the crack width was reduced at higher dosage of fly ash. The mould based on ASTM C1579 was used to carry out the test. At early ages compressive strength of the SCC mixes was similar to that of the reference mix and the maximum strength was observed for 25% replacement by Fly ash. The test result of normal concrete, SCC with various percentages of fly ash was compared.

Keywords Plastic concrete · Shrinkage crack · Shrinkage · Fly ash · Self-compacting concrete

1 Introduction

Plastic shrinkages are those which cause early age cracking on the young concrete surface before the setting. It can be mainly seen in concrete members with large surface to volume ratio, like pavements and slabs. The cracks start its formation at the concrete surface when the fluidity of the concrete has been lost but has not yet achieved any appreciable cohesion and it often propagate inwards the structure facilitating the ingress of harmful materials and eventually can affect the aesthetics,

A. George (✉) · C. A. Thomas
Department of Civil Engineering, Federal Institute of Science and Technology (FISAT), Ernakulam 68377, India

G. C. Marano et al. (eds.), *Proceedings of SECON'21*, Lecture Notes in Civil Engineering 171, https://doi.org/10.1007/978-3-030-80312-4_8

durability and serviceability of the structure. This study focuses on investigating and quantifying the effectiveness of using fly ash on reducing early shrinkage in self compacting concrete.

When the evaporation rate becomes higher than the bleeding rate, the thin water layer that is commonly formed on the surface of the fresh concrete disappears. Following the surface evaporation, evaporation takes place inside the pore system, in which the adhesive force and surface tension of water form menisci. These menisci developed cause a negative pressure in the concrete pore system, as result of which a tensile force is applied on the solid particles. Consequently, the concrete tends to contract as the inter-particle distances decrease. The shrinkage thus developed leads to a tensile strain accumulation, at the concrete surface. If the tensile strain is greater than the very low early age tensile strain capacity, the young concrete may start to crack [1, 2].

Plastic shrinkage is greater the greater the rate of evaporation of water, which in turn depends on the air temperature, the concrete temperature, the relative humidity of the air, and wind speed. The plastic shrinkage is greater when larger is the cement content of the mix, or lower when larger the volumetric aggregate content [3, 4].

Turcry and Loukili [5] in the paper describe an experimental investigation of plastic shrinkage cracking of self-consolidating concrete (SCC). Five SCC mixtures were compared to five ordinary concrete (OC) mixtures. For restrained plastic tests, the authors used a mould very similar to ASTM C 1579 mould. Restrained shrinkage tests show that SCC tend to have less wider cracks than OC. This indicates that SCC could be more vulnerable to shrinkage cracking, especially during setting.

Haque and Kayali [6] studied the properties of the high strength concrete using fine fly ash. The author arrived at the result that 10% replacement level can reduce the w/c ratio by 35% and this gives higher strength for concrete. 20% replacement is considered as the optimum percentage since it gives maximum 28-day compressive strength for concrete.

McCarthy and Dhir [7] conducted research regarding maximising the use of fly ash as a binder. 30% cement volume can be considered as a replacement level in concrete but his studies revealed that beyond a particular percentage replacement level early strength problem can be encountered.

Kristiawan et al. [8] studied long term shrinkage on SCC by replacing cement by fly ash in percentages of 35, 55 and 65%. For each mix proportion of SCC, six cylindrical specimens (75 mm × 275 mm) were cast for shrinkage measurements as per the RILEM Recommendation. Three of them were used to measure the drying shrinkage and the other three for autogenous shrinkage. The results show, a higher cement replacement by fly ash tends to decrease both the drying and autogenous shrinkage. The temperature and humidity were in the range of 25–32 °C and 70–80%, respectively.

2 Experimental Investigation

2.1 Materials

Ordinary Portland cement (OPC) of 53-grade (IS: 12269, 1987) and commercially available fly ash were used. The specific gravity of cement is 3.17. For improving workability, a poly-carboxylate ether based superplasticiser with a specific gravity of 1.1 is used at a dosage of 0.3–0.8% by weight of cement. The fly ash dry dense form conforming to IS 3812: 1981 is used as a mineral admixture. It is brown to gray in colour with its average particle size of 22.2 μm. The physical and the chemical properties of fly ash are shown in Table 1.

As per IS 383:1970 the grading for fine aggregate and different zones are provided and the fine aggregate used is Manufactured sand and belongs to zone 2 [3].

2.2 Mix Proportion of Normal and SCC Mix

Initially normal M30 mix and control mix for SCC was prepared. Then SCC mixes with varying percentages of fly ash at 20, 25, 30, 35, 40 and 45% was prepared. The mix proportion was arrived by conducting various trials according to IS 10262: 2019 guidelines. A control mix of characteristic compressive strength of 30 N/mm^2 is developed. Mix proportions with varying fly ash dosage are tabulated in Table 2.

Table 1 Physical and chemical properties of fly ash

Physical properties		Chemical compositions	
Specific gravity	2.2	SiO_2	63.8%
Specific surface area	649.9 m^2/kg	Al_2O_3	21.29%
Average particle size	22.2 μm	CaO	2.9%
Form	Powder	Fe_2O_3	0.39%

Table 2 Mix proportions of various mixes

Mix designation	W/P	Fly ash	cement	FA	CA	Water	S. P
		Kg/m^3	Kg/m^3	Kg/m^3	Kg/m^3	Kg/m^3	Kg/m^3
Mn	0.4	–	396	840	1046	178.8	1.584
M_{F20}	0.43	81.397	325	1200	631	175	3.256
M_{F25}	0.43	101.741	305.238	1200	631	175	3.256
M_{F30}	0.43	122.1	284.9	1200	631	175	3.256
M_{F35}	0.43	98.851	264.54	1200	631	175	3.256
M_{F40}	0.43	112.943	244.191	1200	631	175	3.256
M_{F45}	0.43	183.141	223.837	1200	631	175	3.256

2.3 Assessment of the Rheological Properties of the Mix

To ensure the rheological properties of SCC such as flow ability, passing ability and segregation resistance, tests were performed on control SCC mix according to IS 10262:2019 guidelines. The V-funnel test for the viscosity and segregation resistance, Slump flow and T500 slump flow time tests for filling ability, L-box test to assess the flow of concrete and passing ability where performed to study the workability of the mix. In case of the normal mix (Mn) to determine the workability of the mix slump test is performed.

2.4 Plastic Shrinkage Tests

The concrete mixes were poured into the plastic shrinkage mold and no external vibration or compaction. Figure 1 shows the plastic shrinkage measurement mould designed and fabricated based on ASTM C1579 with some modifications [9–11]. The dimension of the steel mould is 600 × 200 × 100 mm size with three stress risers. The central riser is used to provide maximum stress concentration on the central area and promote cracking with a height of 63.5 mm. The other two risers provided with a height 32 mm each are used to provide restrain in concrete. To increase the restrain, 10 numbers of bolts with 5 mm diameter are provided at both ends as shown in Fig. 3. The mould is made of cast iron and a transparent acrylic sheet is provided on one side of the mould to make cracking visible along the depth of the mould. The parameters like the time of occurrence of initial centre line crack, the time for propagation of the crack on the concrete surface and the crack width

Fig. 1 Plastic shrinkage mould

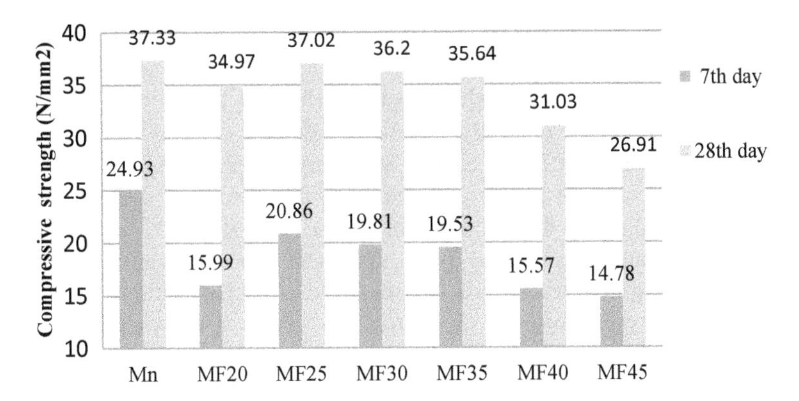

Fig. 2 Compressive strength of mixes with various fly ash percentage

Fig. 3 Side view of crack evolution in specimen

was recorded. Wind condition at 3 m/s air flow condition was adopted during the test. The temperature and humidity parameters were noted, respectively, in the range of 26–35 °C and 50–70%.

2.5 *Compressive Strength*

Three cubes of size 150 × 150 × 150 mm were cast and tested for each mix to determine the compressive strength according to IS: 516-2018. Figure 2 shows the results obtained from the test.

3 Experimental Results and Discussions

3.1 *Rheological Properties of the Mix*

For ordinary concrete mix (Mn) slump test is performed. A slump of 100 mm was obtained for the mix. The mix falls to medium workable category (75–100 mm) as per IS 456:2000. Except M_{F45}, all the SCC mixes developed satisfied the SF2 category. The rheological property of the mix with maximum strength i.e. M_{F25} shows that it has a slump flow of 678 mm with a T500 of 3.7 s. The viscosity and segregation resistance were tested by V funnel and showed a value of 6.5 s. The passing ability and flow tested by L box depicted a blocking ratio of 0.9 which is within the range prescribed by IS 10262:2019. Table 3 shows the rheological properties of the SCC mixes.

3.2 *Compressive Strength*

From Fig. 2, 25% the weight of cement used is found to be the optimum replacement percentage of fly ash to be used. Compared to normal mix, the 7-day strength is less for the SCC mixes. The 28-day strength of mixes with fly ash was nearly similar to that of the normal mix (Mn). An increase in percentage addition of above 25% shows a decreasing trend in compressive strength of mix. This may be due to the fact that at higher dosages, the extra added fly ash may simply stay as a filler material and no CSH gel is formed hence the reduction in strength due to pozzolanic action.

3.3 *Plastic Shrinkage*

The plastic shrinkage behaviour of the concrete indicates the time of occurrence of the initial centre line crack and the time of crack propagation. The time of propagation of crack is the time difference between the time of occurrence of center line

Table 3 Rheological property of SCC mix

Mix designation	% of Fly ash	Slump flow (mm)	T_{500} t (sec)
M_{F20}	20	655	4.2
M_{F25}	25	678	3.7
M_{F30}	30	684	3.8
M_{F35}	35	676	4.11
M_{F40}	40	662	4.34
M_{F45}	45	607	4.4

full length crack and centre line initial crack. [3] The full-length crack can be observed with naked eye. The side view of crack developed and the full-length crack of the specimen of concrete formed in mould during testing is shown in Figs. 3 and 4.

Plastic shrinkage behaviour and slump flow of various concrete mixes is tabulated in Table 5. The Table 5 shows that the normal Mn concrete mix without fly ash showed a faster initial centerline crack. The Results in case of mix of SCC with fly ash showed a delayed initial centre line crack but the crack propagation was fast. It can be observed that as the percentage replacement increases the delay in crack formation also increases. This shows the positive influence of fly ash on reducing plastic shrinkage cracks [12]. The plastic shrinkage test performed in the windy condition revealed that SCC tends to have less cracking than OC. SCC is thought to have a greater strain capacity than ordinary concrete (OC) in fresh state because of its fluid consistency. Figure 5 shows the time for full length crack propagation.

A 14.5% increase in time for initial crack development in M_{F25} is observed compared to Mn. The delayed crack occurrence can be considered as a positive effect of supplementary cementitious materials on the plastic shrinkage of concrete. For M_{F25}, M_{F30}, M_{F35}, M_{F40} and M_{F45} a delay in the time for initial cracking is 1.21%, 5.09%, 7.08%, 13.1% and 4.61% respectively compared to M_{F20} (Fig. 5).

Fig. 4 Full length crack evolution

Table 5 Plastic shrinkage behaviour and slump flow of various concrete mixes

Mix designation	% of Fly ash	Time for initial crack (min)	Time for full length crack (min)	Time for crack propagation (min)	Crack width (mm)
M_n	–	94	127	33	0.4
M_{F20}	20	149	182	33	0.7
M_{F25}	25	154	188	34	0.5
M_{F30}	30	170	200	30	0.4
M_{F35}	35	178	209	31	0.3
M_{F40}	40	203	225	22	0.3
M_{F45}	45	168	188	20	0.2

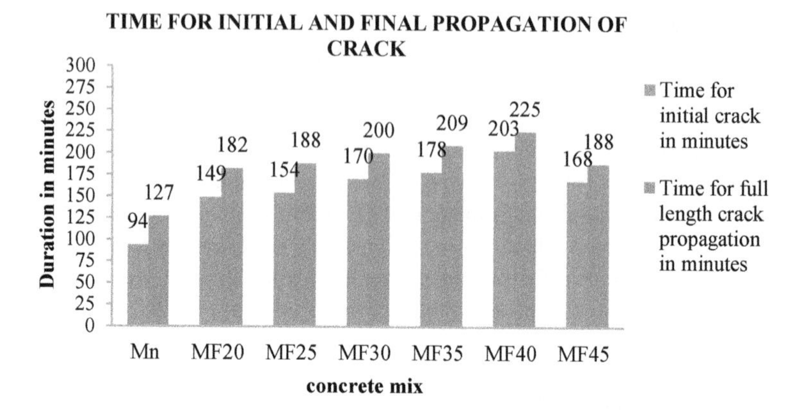

Fig. 5 Time for initial and final propagation of cracks

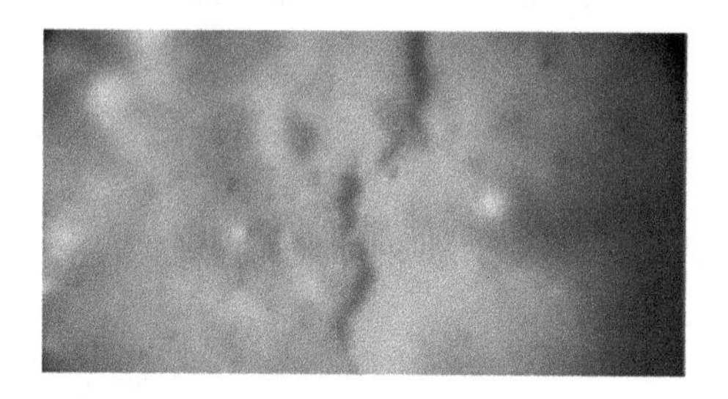

Fig. 6 Crack width measured by hand held microscope

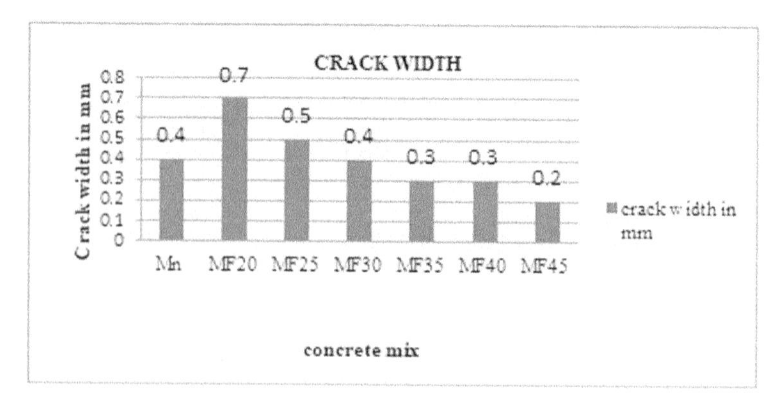

Fig. 7 Crack width of different mixes

A hand held microscope is used to measure the crack width. Figure 6 shows the crack width measured. The least count of the microscope is 0.02 mm. Figure 7 shows the crack width measured for different mixes. It is observed that with the increased percentages of fly ash the width of the crack reduced. The least width was observed for M_{F45}. The crack width of ordinary concrete Mn is less compared to that of $M_{F20.}$ This may be due to the fact that normal concrete Mn contains higher amount of course aggregate which may tent to act as crack arrester and hence have less wide crack.

4 Conclusions

Following conclusion can be drawn from this current experimental study.

- SCC with higher fly ash content i.e. upto 40% as cement replacement tends to reduce the plastic shrinkage. The crack width was considerably reduced and there was a delayed initial centre line crack. But the crack propagation was faster. When replacement was to 45% the duration for crack propagation slightly decreased.
- The delay on the beginning of plastic shrinkage registered in SCC samples, with regard to standard concrete is a result of the delay of SCC bleeding and several factors that contribute to this reduction include the effect of water content, cement content and pore size refinement. The filling effect of fly ash turns big pores in the structure of concrete into small voids lowering the permeability of concrete and hence it retains the moisture in the concrete which reduce the shrinkage.
- The fluid consistency of SCC makes it a concrete with a greater strain capacity compared to ordinary concrete. However, the mix with a cement replacement by fly ash at 25% shows the maximum compressive strength out of all the mixes.

References

1. IS: 4031 (1996) Method of physical tests for hydraulic cement—specification, Bureau of Indian Standards, New Delhi (March)
2. Brooks (2015) Shrinkage of concrete. Concr Masonry Movements 22(6):137–185. https://doi.org/10.1016/B978-0-12-801525-4.00006-6
3. George A, Abin Thomas CA (2021) Influence of alkali resistant glass fiber on the reduction of plastic shrinkage cracking of self compacting concrete. In: Dasgupta K, Sudheesh TK, Praseeda KI, Unni Kartha G, Kavitha PE, Jawahar Saud S (eds) Proceedings of SECON 2020. SECON 2020. Lecture Notes in Civil Engineering, vol 97. Springer, Cham. https://doi.org/10.1007/978-3-030-55115-5_31
4. Linmei Wu, Farzadnia N, Shi C, Zhang Z, Wang H (2017) Autogenous shrinkage of high performance concrete: a review. Constr Build Mater 149:62–75
5. Turcry P, Loukili A (2006) Evaluation of plastic shrinkage cracking of self-consolidating concrete. ACI Mater J 103-M30:272–279
6. Haque MN, Kayali O (1998) Properties of high strength concrete using fine fly ash. Cem Concr Res 28:1445–1452

7. McCarthy MJ, Dhir RK (1999) Towards maximising the use of fly ash as a binder. Fuel 78:121–132
8. Kristiawan SA, Taib M, Aditya M (2015) Effect of high-volume fly ash on shrinkage of self-compacting concrete. Procedia Eng 125(2015):705–712
9. Combrinch R, Ateyl L, Boshoff WP (2018) Interaction between settlement and shrinkage cracking in plastic concrete. Constr Build Mater 185:1–11
10. Sayahi F, Emborg M, Hedlund H (2017) Plastic shrinkage cracking in concrete—influence of test methods. In: 2nd International RILEM/COST conference on early age cracking and serviceability in cement-based materials and structures—EAC2 12–14 Sept 2017, ULB-VUB, Brussels, Belgium
11. Mora-Ruacho J, Gettu R, Aguado A (2009) Influence of shrinkage-reducing admixtures on the reduction of plastic shrinkage cracking in concrete. Cem Concr Res 39(2009):141–146
12. Altoubat S, Junaid MT, Leblouba M, Badran D (2017) Effectiveness of fly ash on the restrained shrinkage cracking resistance of self-compacting concrete. Cement Concr Compos 79 (2017):9–20

Development of a Spreadsheet Solver for Regional Bus Timetabling

Shazia Mariam and M. V. L. R. Anjaneyulu

Abstract One of the vital processes in bus operation is timetable development. Timetable generation influences the whole operational structure of a bus network. Custom tailoring of commercially available software packages remains a challenge for small and medium bus organisations/companies operating in cities in India. This study aims at developing a tool for bus timetabling in a city that can be run on a widely available spreadsheet software. The objective is to produce timetables with maximum synchronizations to facilitate passenger transfers and avoid bus bunching along the network. The scheduling tool enables the users to input passenger data and analyse characteristics of each route in the network to study the operations of the bus service. The proposed solution algorithm coded as a Multi-start iterated local search algorithm was tested on different real-life examples. The total number of synchronizations was found to depend on the headway and the specified time limits mainly. From the test results, it can be inferred that the solver can be used to obtain efficient timetables with maximum synchronizations between routes.

Keywords Bus scheduling · Bus timetable · Synchronization · Spreadsheet

1 Introduction

The rapid urbanization of cities lead to an unending cycle of increase in number of vehicles and the associated congestion. Public transportation modes are promoted as it will help in alleviating traffic congestion and, on the whole, reduce vehicular emissions in urban areas. Buses prove to be the most popular and most commonly used ones because of their inherent flexibility, adaptability to changing employment and residential patterns, and low capital costs [4].

The bus network planning process consists of several steps such as network design, timetable generation, vehicle scheduling and crew scheduling [3].

S. Mariam (✉) · M. V. L. R. Anjaneyulu
Department of Civil Engineering, National Institute of Technology Calicut, Kozhikode, Kerala, India

G. C. Marano et al. (eds.), *Proceedings of SECON'21*, Lecture Notes in Civil Engineering 171, https://doi.org/10.1007/978-3-030-80312-4_9

Timetable development is crucial as it influences the subsequent planning steps especially the total number of buses on the road. Reliability of bus timetable also encourages transportation mode-shift from cars to buses.

Although the benefits of bus network planning are many, local bus operators are deterred from adopting them due to the added cost and lack of technical knowledge. Instead, bus operators' perceptions and field experience are used to estimate passenger demand and schedule buses [1]. Therefore, a need arises to solve bus timetabling problems on a familiar work interface that would enable local bus services to improve their operational effectiveness.

2 Problem Description

Synchronization of arrival times of different routes is chosen as the objective of timetable development. This is because of its ability to reduce passenger waiting times and avoid bunching of buses which is characteristic of large bus networks [6]. Since instant transfers are not feasible due to uncertainties in the transit system, two trips are considered to be synchronized if the difference between arrival times at a node is within a given time window. Passenger activity varies during the day and therefore each day is split into several planning periods (peak hours, morning, night etc.) to achieve a more accurate deterministic passenger demand [5]. Thus, the objective of this model is to maximize the number of synchronizations along the network during a planning period to reduce bus bunching and to optimize passenger transfer.

3 Model Formulation

The bus timetabling problem is formulated as a Mixed-Integer Linear Programing problem. The formulation is similar to that given by Ibarra-Rojas and Rios-Solis [5] with the addition of holding times at nodes. This formulation is an extension to the models proposed by Ceder et al. [2] and Eranki [4]. It has been already been proved that the bus timetabling optimization problem belongs to the NP Hard class [5]. Hence a heuristic is proposed to solve the Transit network timetabling problem (TNTP).

3.1 Multi-start Iterated Local Search Algorithm (MILS)

A Multi start approach to the iterated local search algorithm is used to achieve diverse results by resuming the search from a new solution once a region has been thoroughly investigated. This procedure is carried out in order to prevent the local search method to be stuck in a small area of the solution space and eliminating the possibility of finding a global optimum. The basic outline of the MILS algorithm is based on that developed by Ibarra-Rojas and Rios-Solis [5]. The following are changes that were incorporated to the solution algorithm:

- Holding time, λ_b
  Holding time at each node has been considered in order to obtain a more realistic representation of the problem. Buses were assumed to depart immediately as they arrived in the previous literature.
- Iterations of the Initial solution before improvement process
  The random departure times initially generated are iterated for a given number of times and the best solution is chosen for further improvement. This is done in order to improve the chances of obtaining better results with fewer iterations.
- Changes to Hill climbing local search
  After the departure times are generated in the improvement procedure, they are further iterated within the available departure window until an improvement to the incumbent solution is obtained. This ensures that the departure time chosen improves the result.

The proposed solution algorithm for the Bus network timetabling problem [5] is as shown by Algorithm below. Here, the solution 's' refers to the total number synchronizations in departure times. Initially, a number of main iterations (step 2) are defined. In each iteration, the following are done. At first, an initial solution is constructed. Next, a hill climbing local search algorithm attempts to improve the current solution (step 3). Then, the iterations of the iterated local search are defined. In each one, the perturbation phase (step 6) that searches for trips that can be forced to synchronize, followed by the local search to improve the perturbed solution (step 7) are applied, and the new solution is only accepted if there is an improvement in the total number of synchronizations (step 8).

Termination criteria Ideally the process runs until all the specified number of iterations are completed. However, an option to limit the processing time is provided so that process maybe stopped and answers maybe retrieved if the programs runs for an extended period of time.

Pseudo code of algorithm—Multi-start Iterated Local Search (MILS)

Input: Bus network timetabling problem (BNTP) details
Output: Solution's' for BNTP

1: **for** (it = 1 to it = $iter_{MAIN}$) **do**
2: s_o = *GenerateInitialSolution*
3: s* = *Improve* (s_o)
4: **for** (iter = 1 to $iter_{ILS}$) **do**
5: **for** (each pair of lines i and j that can synchronize) **do**
6: s ′ = *Resynchronize* (s*, i, j)
7: s*′ = *Improve* (s′)
8: s = max{s*, s*'}
9: **end for**
10: **end for**
11: **end for**

4 How to Use the Spreadsheet Solver

The solver is designed in Microsoft Excel using VBA, thereby providing a familiar interface to work on and the flexibility to modify codes according to specific needs. The program is built upon the three components of scheduling process: data input, scheduling process and timetable development [1]. The scheduling tool can be used to create timetables for both a single bus route and for that of a network. Several transit route characteristics such as average running time, average passenger volume, etc., can also be calculated for each route.

Timetables for a single bus route is generated on the basis of the peak load factor concept with even headways between successive departures. The timetables for the network are designed with the objectives of maximum synchronization by encoding the MILS algorithm using VBA. Both timetables are intended for a single planning period such as morning period, afternoon period or peak hour etc., as desired by the planner. The steps involved in scheduling process of the solver are as shown in Fig. 1. All components of the scheduling process are automated with user intervention required only for data entry, data verification and result generation. A brief summary of the sheets along with information if it is user defined is provided in Tables 1 and 2.

5 Experimental Results

MILS was coded in Excel using VBA and executed on a HP laptop with an Intel core i7, 2.7 GHz processor and 16 GB of RAM. Four example problems on bus networks are tested with two of them being real life examples using the spreadsheet solver (Table 3). These problems are from Ceder et al. [2] and Eranki [4] has introduced waiting times at each node.

Studies were conducted to understand the optimum number of iterations, processing time required and the effect of time limits. The number of synchronizations obtained for each network were also compared with that of the models (Fig. 2) suggested by Ibarra-Rojas and Rios-Solis [5], Eranki [4] and Ceder [2]. Calculations for the Ceder [2] model were found setting waiting time limits to zero.

The following inferences were made from the experimental studies:

- The number of iterations required to obtain the best result was less for the proposed model in comparison to that required by Ibarra-Rojas and Rios-Solis [5]. This can be attributed to changes incorporated such as using a good initial solution and the improvements made to the local search algorithm.
- The number of iterations required to obtain the best solution varies for each sample network due to random factor involved in calculating the departure time. More the number of iterations, better the chances of obtaining an optimal solution.

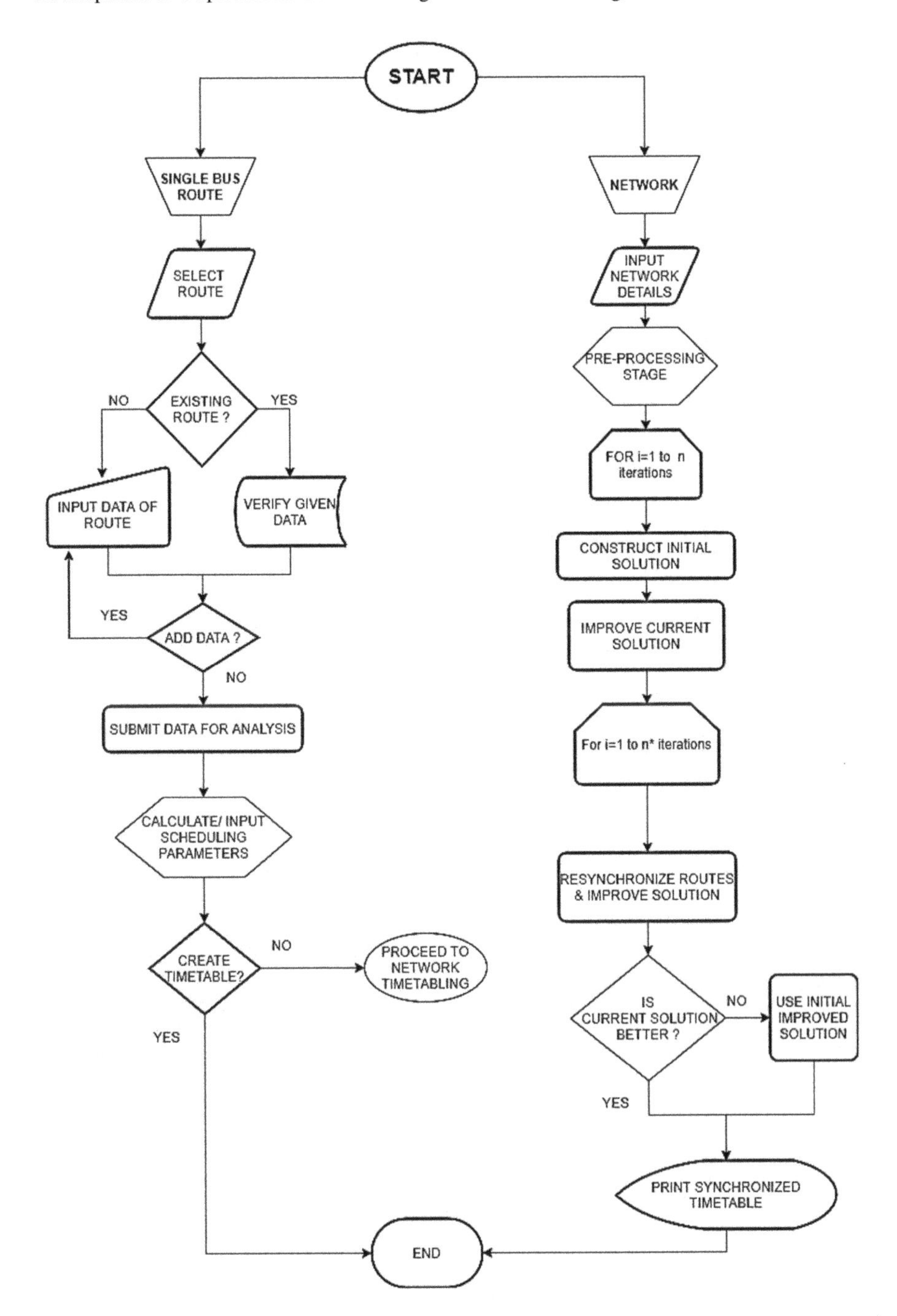

Fig. 1 Flowchart of Scheduling process of solver

Table 1 Summary of sheets of single route scheduling process

Sheet No	Name of sheet	Function	User defined (Yes/No/partially)
1	Add route	Choose desired route for scheduling	Yes
2	Data entry and results	Analyse data using Ride check data and develop timetables	Partially
3	Route name	Stores the bus stop names of the route	No
4	Route distance	Stores distance to each stop	No

Table 2 Summary of sheets for scheduling of bus network

Sheet No	Name of sheet	Function	User defined (Yes/No/partially)
1	Network information sheet	Provide information on frequency, headway, number of routes and nodes etc	Yes
2	Node input	Stipulate waiting time limits and maximum holding time at each node	Yes
3	Travel time	Construct Travel time matrix of the network	Yes
4	Dashboard	To state number of iterations of MILS and display summary of results	Partially
5	Result	Displays node wise number of synchronization events	No
6	Departure times	Displays departure times of proposed timetable	No
7	Arrival times	Displays arrivals times at each node of the route	No
8	Synchronization window	Displays synchronization event according to the trip number of each route	No

Table 3 Input details of Example problems

Example network	Number of routes	Number of nodes	Planning horizon	Headway times	Waiting times	Holding times
			(In minutes)			
1	2	2	60	[5, 20]	[4, 13]	0
2	4	4	45	[8, 20]	[9, 20]	0
3	6	3	240	[14, 20]	[5, 10]	0
4	5	9	120	[15, 30]	[5, 10]	0

- The processing time required for the computation is directly proportional to the input data including the network details (such number of routes and nodes and their respective frequencies etc.) and the number of iterations of chosen for the process.

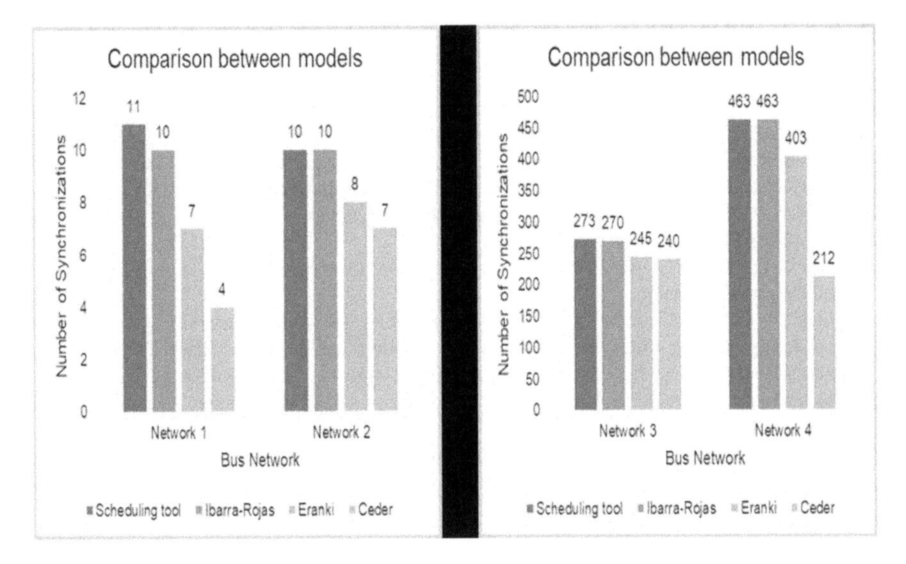

Fig. 2 Comparison between models on obtained synchronizations

- Timetables were obtained in less than a minute for all the four sample networks. Thus fast results can be obtained using the scheduling tool.
- As noted by Eranki [4], synchronizations are directly proportional to the waiting time limit range.
- The number of synchronizations obtained by the scheduling model used in this study is the highest among the models compared. Thus, it can be used to obtain efficient timetables with maximum synchronizations between routes.

6 Conclusion

The regional bus timetabling problem can be solved using commonly used spreadsheet software in reasonable processing time. Synchronization of routes of a bus network while developing timetables is essential as it favors passenger transfer with minimum waiting time and avoids bus bunching at bus stops. The approach of including holding time of the bus at stops helps simulate real-world instances.

On comparing the timetables of example networks generated by the solver with models developed previously, it was found that either better or equally good results can be obtained. However, the iterated local search algorithm can be enhanced in the future by improving the performance of its modules. The spreadsheet solver may be incorporated with other steps of the transit planning process, such as vehicle scheduling and the network routing problem to provide a complete bus network planning experience.

References

1. Abad RBP et al (2014) Developing a public transit scheduling tool for metro manila bus operators. In: 6th International conference humanoid, nanotechnology, information technology communication and control, environment and management (HNICEM), The Institute of Electrical and Electronics Engineers Inc. (IEEE)—Philippine Section, 12–14 Nov, 2014
2. Ceder A, Golany B, Tal O (2001) Creating bus timetables with maximal synchronization. Transp Res Part A: Policy Pract 35(10):913–928
3. Ceder A (2007) Public transit planning and operation: theory modeling and practice. Elsevier, Butterworth-Heinemann
4. Eranki A (2004) A model to create bus timetables to attain maximum synchronization considering waiting times at transfer stops. Unpublished Master's thesis, Department of Industrial and Management Systems Engineering, University of South Florida, Tampa, FL
5. Ibarra-Rojas OJ, Rios-Solis YA (2012) Synchronization of bus timetabling. Transp Res Part B: Methodol 46(5):599–614
6. Liu Z, Shen J, Wang H, Yang W (2007) Regional bus timetabling model with synchronization. J Transp Syst Eng Inform Tech 7(2):109–112

Influence of Two Discrete Heaters on the Laminar Natural Convection Inside a Closed Room

J. Jeseema Nisrin, R. Velkennedy, and K. Kalidasan

Abstract This numerical investigation aimed to analyze laminar natural convection inside a closed room containing two discrete heaters using Computational Fluid Dynamics (CFD). The Navier–Stokes equation was adopted to study the fluid flow inside the room. The fluid inside the room was standard air with the Prandtl number of 0.71. The top wall, bottom wall, and center portion of the right walls were considered adiabatic, while the left-side wall remains cold. Two discrete heaters were positioned near the top-and-bottom- side of the right wall, and the heater sizes were 20 and 30% of the room's length. Rayleigh number (Ra) was considered within the laminar range that, they were varied from 10^3 to 10^6. Natural convection is induced between the walls due to the difference in temperature inside the room. The rate of heat transfer, velocity profile, streamlines, and isotherms were measured. The results show a strong influence of the heater's size and the Rayleigh number on how the fluid in the room heated up. The heat transfer was influenced by conduction when Ra = 10^3. As the Ra increases, buoyancy force increases and shows the dominance of natural convection at Ra = 10^6. The findings suggest that by varying the heater size and the Rayleigh number, optimal thermal comfort can be achieved in the living place without going for any external mechanical forces. This understanding could help engineers prepare an energy-efficient structural design for buildings and provide excellent thermal comfort to the occupants.

Keywords Discrete heaters · Natural convection · Navier–Stokes equation · Energy-efficient design · Thermal comfort

Nomenclature

α	Thermal diffusion of fluid
β	Thermal expansion coefficient

J. Jeseema Nisrin (✉) · R. Velkennedy
Department of Civil Engineering, Thiagarajar College of Engineering, Madurai, Tamilnadu 625015, India

K. Kalidasan
Department of Civil Engineering, Arulmigu Palaniandavar Polytechnic College, Palani, Tamilnadu 624601, India

G. C. Marano et al. (eds.), *Proceedings of SECON'21*, Lecture Notes in Civil Engineering 171, https://doi.org/10.1007/978-3-030-80312-4_10

γ	Fluid kinematic viscosity
μ	Dynamic viscosity
c_p	Specific heat
k	Thermal conductivity
g	Acceleration due to gravity
u	Dimensionless horizontal velocity
v	Dimensionless vertical velocity
T	Dimensionless temperature
t	Dimensionless time
w	Dimensionless vorticity
Pr	Prandtl number
Ra	Rayleigh number
Nu	Local Nusselt number
$\overline{\text{Nu}}$	Average of Local Nusselt number
CFD	Computation Fluid Dynamics
FDM	Finite Difference Method

1 Introduction

The world's constructed floor area is expected to reach a massive range of about 400 billion square meters by the end of 2050 [1]. The energy consumption of these build-up areas will also be proportionately increasing due to the emerging economies and buying power. This puts huge pressure on the growing countries in terms of green gas emissions, depleting natural resources like oil and coal, investment in other sources of energy, etc. Due to this scenario, many countries have started implementing strict energy efficiency norms on commercial and residential buildings. More than 62 of 192 countries in the world have formulated their building energy efficiency code in the past decade. According to the International Energy Agency (IEA), residential usage is the third-largest power-consuming sector in the world and more than 50% of the end energy consumption in residential usage accounts for space heating and cooling [1]. The amount of energy used for thermal comforting will be even higher by including the commercial buildings.

In space conditioning, energy loss due to infiltration is considered a major source of loss of energy. Building enveloping is suggested as one of the effective methods for combating the effect of infiltration by many building codes. Hence, it becomes necessary to understand the flow of fluid in a closed envelop due to the positioning of the heater and cooler. Computational fluid dynamics is considered as an effective tool in studying air circulation and thermal modeling in many fields. It is a proved and successfully used tool to simulate and study the cavity flow. Owing to the diversity of application of cavity flow, the fluid flow and heat transfer inside the cavity were studied by many researchers in the past few decades. A pioneering study in buoyancy-driven natural convection inside a differently heated cavity was done by de Vahl Davis [2]. His work is used as the benchmark result for validating the computer simulation codes by many researchers. A detailed review of natural convection inside an enclosure was done by Ostrach [3]. In his study, he has reviewed analytical, numerical, and experimental works of natural convection inside different geometries of the enclosure.

A steady-state laminar natural convection in a rectangular cavity with passive cooling from the top ceiling having a small localized heater at the bottom was investigated by Ben Nasr et al. [4]. The study depicts the passive cooling of a room from the ceiling provided with a solar radiation transmission through the window. In a numerical study done by Rahman et al. [5], the effect of a flush mount heater on the vertical right wall and a cool left wall was investigated. They presented the relationship between the length, location of the heater, heat-generating parameters, and the rate of heat transfer by means of Nusselt number. The distribution and dilution of buoyancy-driven pollutants in a suburban skyline were studied by Mei et al. [6] using CFD. Hu et al. [7] have analyzed the moisture convection due to buoyancy difference inside a closed building enclosure at a range of Ra 10^3–10^7. Saha et al. [8] investigated the heat flow pattern inside an attic of a building. In this study, they have considered a closed triangle enclosure having sinusoidal temperature variations on the sloping sides and an ambient temperature on the bottom side. The influence of aspect ratio and Rayleigh's number on thermal distribution was also determined. The thermal comfort of a office room was analyzed subjected to cooling was analyzed using CFD by Ozsagiroglu et al. [9] recently.

This present investigation aims to find the influence of two heaters on the laminar natural convection inside the enclosed room, which is cooled from one of its vertical sides. The heating source acts as the heater installed inside the room. The floor and the roof of the room is considered as adiabatic. Two heaters are installed to study its efficiency in heating the entire room. As only the top heater will not allow the heat to transfer to the bottom of the room due to buoyancy created by convection. This space heaters will not change the absolute humidity inside the room, because the room is assumed to be completely enveloped. But the relative humidity of the room will go down because of the rising temperature which makes the air feels dry. This also aids in the space comforting.

2 Problem Formulation

The schematic representation of the formulated problem is shown in Fig. 1. The problem is considered two-dimensional. It consists of a square room with an adiabatic top and bottom wall. The room is considered to be cooled from the left wall. Two heaters are installed on the top and bottom corners of the right wall. These two heaters are placed inside the room to study the heat flow due to this configuration to the entire room. The capacity of the heater is considered constant with dimensionless temperature T = 1 for the entire study. The center portion of the right wall

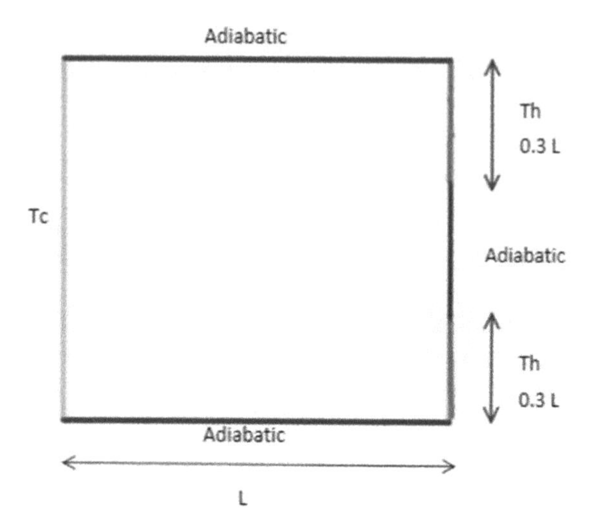

Fig. 1 Schematic representation of the problem

is considered adiabatic. The length of the heater is varied between 20 and 30% of the length of the room. The fluid inside the room is considered as normal air with Prandtl number equals to 0.71 and the flow is varied within the laminar field by changing the Rayleigh's number from 10^3 to 10^6. The fluid flow and heat transfer due to the temperature difference created between these two walls is studied.

3 Governing Differential Equation

The airflow inside the room is governed by the Navier Stokes equation. The flow is assumed to obey the Boussinesq approximation that the fluid flow inside the room is assumed to be unsteady, laminar, and incompressible. The continuity, momentum, and energy equations are solved to evaluate the flow. The stream function vorticity formulation is used. The equations are adopted as used by [5, 10]. The dimensionless form of governing equations is derived by adopting the dimensionless parameters as follows

$$u = \frac{\overline{\mathrm{u}}\mathrm{L}}{\alpha};\ v = \frac{\overline{\mathrm{v}}\mathrm{L}}{\alpha};\ x = \frac{\overline{\mathrm{x}}}{\mathrm{L}};\ y = \frac{\overline{\mathrm{y}}}{\mathrm{L}};\ w = \frac{\overline{w}}{\overline{\mathrm{u}}/\mathrm{L}};$$
$$\mathrm{s} = \frac{\overline{\mathrm{s}}}{\mathrm{L}\overline{\mathrm{u}}};\ \mathrm{t} = \frac{\overline{\mathrm{t}}}{\mathrm{L}/\overline{\mathrm{u}}};\ \mathrm{T} = \frac{(\overline{\mathrm{T}} - \mathrm{T_c})}{(\mathrm{T_h} - \mathrm{T_c})} \tag{1}$$

The dimensionless form of continuity, vorticity transport, and energy equations are

$$\frac{\partial u}{\partial x} + \frac{\partial v}{\partial y} = 0 \tag{2}$$

$$\frac{\partial w}{\partial t} + u\frac{\partial w}{\partial x} + v\frac{\partial w}{\partial y} = \mathrm{Pr}\left(\frac{\partial^2 w}{\partial x^2} + \frac{\partial^2 w}{\partial y^2}\right) + \mathrm{Ra}\,\mathrm{Pr}\frac{\partial T}{\partial x} \tag{3}$$

$$\frac{\partial T}{\partial t} + u\frac{\partial T}{\partial x} + v\frac{\partial T}{\partial y} = \left(\frac{\partial^2 T}{\partial x^2} + \frac{\partial^2 T}{\partial y^2}\right) \tag{4}$$

where Ra is the dimensionless Rayleigh number which gives the relation between viscosity and buoyancy in the fluid and Prandtl number relates the momentum and thermal diffusivity of the fluid. They are given as

$$\mathrm{Ra} = \frac{g\beta\Delta T L^3}{\gamma\alpha} \tag{5}$$

$$\mathrm{Pr} = \frac{c_p \mu}{k} \tag{6}$$

u velocity, v velocity, vorticity, and stream function are related as follows

$$\mathrm{w} = \frac{\partial v}{\partial x} - \frac{\partial u}{\partial y} \tag{7}$$

$$u = \frac{\partial \mathrm{s}}{\partial y};\, v = -\frac{\partial \mathrm{s}}{\partial x} \tag{8}$$

The following Poisson equation gives the relationship between stream function and vorticity

$$\frac{\partial^2 \mathrm{s}}{\partial x^2} + \frac{\partial^2 \mathrm{s}}{\partial y^2} = -w \tag{9}$$

Nusselt number (Nu) quantifies the heat transfer rate at the cavity boundaries. The local Nusselt number at the wall is calculated as follows

$$\mathrm{Nu} = -\frac{\partial \mathrm{T}}{\partial n} \tag{10}$$

whereas the average Nusselt number of the respective wall the room is calculated as follows

$$\overline{\mathrm{Nu}} = \int_0^1 \frac{\partial \mathrm{T}}{\partial n} dn \tag{11}$$

The convergence criteria for the defined problem is given by

$$\sum_{i,j=1}^{i\max,j\max}\left(\emptyset_{i,j}^{n+1}-\emptyset_{i,j}^{n}\right)\leq 10^{-5} \tag{12}$$

where $\emptyset$ implies stream function, vorticity, or temperature.

4 Mathematical Scheme and Grid Allocation

A code using the finite-difference formulation was written on MATLAB to solve the above-stated Navier Stokes equation. The entire region is divided into an equal number of grids and the grid-independent study is carried out to fix the number of grids. The partial differential equations are first discretized using Taylor's series expansion. The boundary conditions are imposed on the boundary nodes of the room and the internal nodes were assumed with the initial guess values for the iteration. The internal nodes are solved by adopting Alternate Direct Implicit (ADI) scheme. At first, the temperature equation is solved in the x-direction sweep and the resulting values are utilized for the y-direction sweep. The vorticity transport equation is solved in the same way by using the previous resulting values. Then the temperature and vorticity values are used to calculate the stream function and u, v velocities. The same procedure is followed till the convergence is reached, as shown in Eq. (12). A time step of 0.001 is used throughout the process. For fixing the grid numbers, the above procedure is repeated on a different number of grids until the difference value of Nu for consecutive grids falls below 0.1 percentage. The value of the grid-independent study is given in Table 1. As seen in Table 1, the value of Nu_Avg falls below 0.1% at 141 × 141, the number of grids for further computation is fixed as 141 equally spaced grid in both the direction.

5 Validation of the Formulated Code

To validate the usage of this code for the computation of the stated problem, it is validated with the lid-driven cavity problem by Ghia et al. [11] and Indukuri and Maniyeri [12] as shown in Fig. 2. The same code is validated with the results of de Vahl Davis [2] and Kalidasan et al. [13] for the natural convection in differently

Table 1 Grid independent study

Grid size	Nu_Avg	Difference	% Difference
81 × 81	1.9284		
111 × 111	1.9208	0.00758	0.39
141 × 141	1.9196	0.00122	0.06

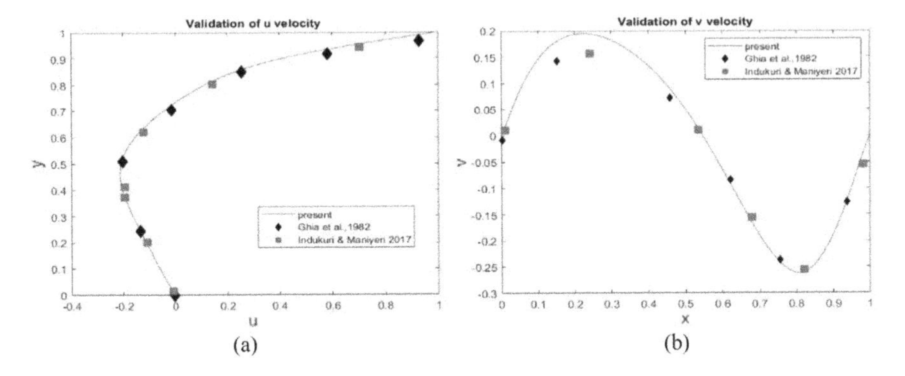

Fig. 2 Validation of lid-driven cavity with the results of [11, 12]

Table 2 Validation of differently heated cavity with the results of [2, 13]

Ra		10^3	10^4	10^5
$\overline{\text{Nu}}$	de Vahl Davis [2]	1.116	2.4519	4.519
	Kalidasan et al. [13]	1.086	2.214	4.447
	Present	1.124	2.293	4.508

heated cavity problem and the results of difference in average Nusselt number is tabulated in Table 2. The results are in good agreement with the published values.

6 Boundary Conditions

The boundary conditions on the enclosed room wall are considered as the following dimensionless values.

At x = 0; u = 0; v = 0; $T = 0$;

At x = 1; u = 0; v = 0; T = 1 at heater position and $\frac{\partial T}{\partial x} = 0$ at the center part.

At y = 0; u = 0; v = 0; $\frac{\partial T}{\partial x} = 0$;

At y = 1; u = 0; v = 0; $\frac{\partial T}{\partial x} = 0$;

The vorticity at the boundaries was calculated using the following formula used by Das and Kanna [14].

$$w_w = \frac{2(s_w - s_{w+1})}{\delta n^2} \tag{13}$$

7 Results

The influence of two discrete heaters inside a room of aspect ratio 1 is studied. The fluid inside the room is considered as air with Pr 0.71. The Rayleigh number is varied from 1×10^3 to 1×10^6. The heater size is varied as 0.2 and 0.3 times the length of the heater. The thermal distribution, the flow pattern of the streamlines, the horizontal and vertical components of velocity, and heat transfer rate inside the enclosure are investigated. The findings are discussed in the subsequent sections.

7.1 Streamline

The path traced by the massless fluid particle inside the enclosed space is visualized by the streamline contour. It can be seen from Fig. 3 that the concentration of the core is more intense in the room with 30% heater than the 20% heater for all the Ra considered. This is due to the amount of heat produced by the 30% heater, which is more than the heat produced by the 20% heater.

This difference in heat intensifies the flow of particles inside the room. At Ra 10^3, both the enclosed rooms show a conductive mode of transfer since both the core are circular in form and all the particles trace a circular streamline around the core. Convection slightly begins at Ra 10^4, but it is not predominant. This can be observed from the oval-shaped core of the streamline. Due to the increase in the heater size, the convection inside the room increases as seen in streamline figure (d) and this is higher than the figure (c), but this convection force is not strong enough to break the inertial force of the fluid. Natural convection accrues energy to break the inertial force at Ra 10^5. Hence, the fluid at the core starts to elongate, as seen in Fig. 3e, f. Due to the increase in heater size, core splits and forms a bipolar vortex. Ra 10^6 intensifies the buoyancy force as seen in Fig. 3g, h, and the fluid particle with higher stream function moves away from the core. When the heater is 30%, the core is occupied by the fluid with lower stream function.

7.2 Isotherms

The temperature distribution inside the room due to the heaters place at the right wall is shown in Fig. 4. At a lower Rayleigh number (Ra 10^3), the thermal contours are formed relatively parallel to the cold wall and a higher temperature is seen close to the heater location. The increase in heater size leads to the higher heat transfer, as seen by the position of 0.7 and 0.8 contours in Fig. 4a, b. Here the mode of heat transfer is by conduction. Natural convection begins at Ra 10^4, as seen by the low-temperature fluids moving to the bottom. Due to the higher intensity of heating by 0.3 size heater, the temperature at the base is high in figure (d) than (c).

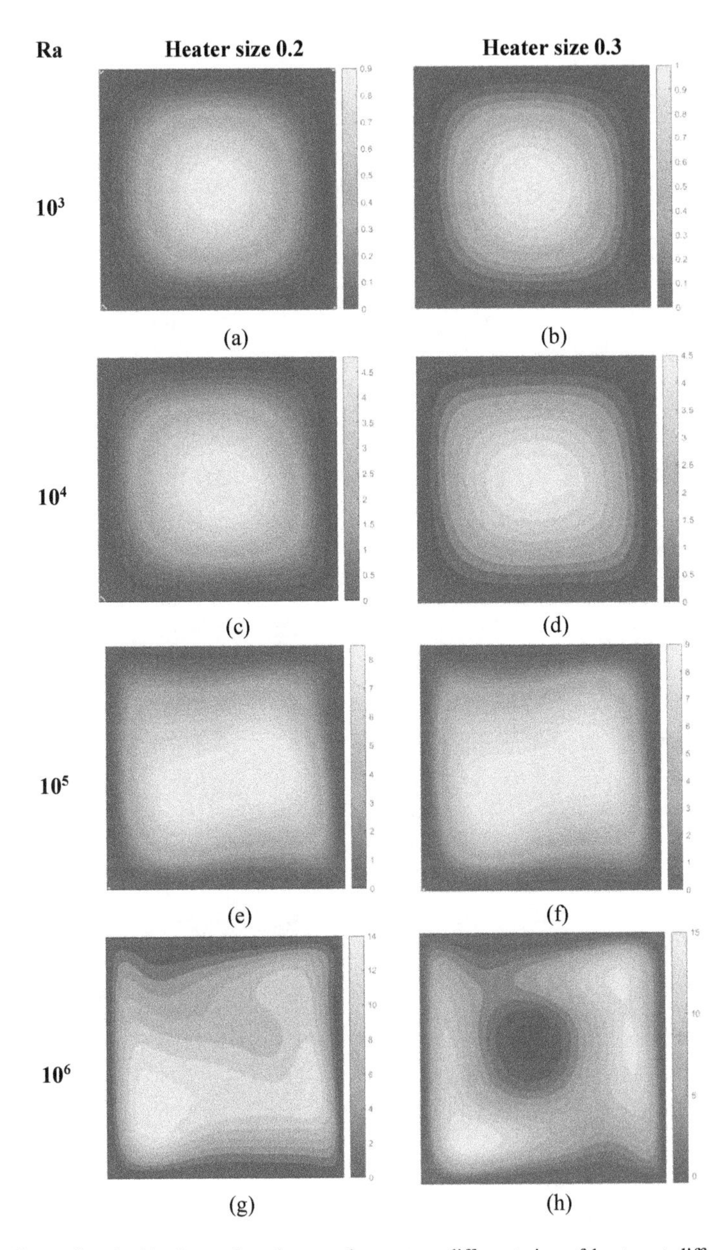

Fig. 3 Streamline inside the enclosed room due to two different size of heater at different Ra

At Ra 10^5, the buoyancy increases; hence the hot fluid air tends to move upward, leaving the bottom portion of the room cooler. The effect of increasing the heater size can only be seen at the top. The buoyance force dominates when the Ra reaches 10^6. This force pushes the hot air to the top, leaving a heavy, dense cool air at the

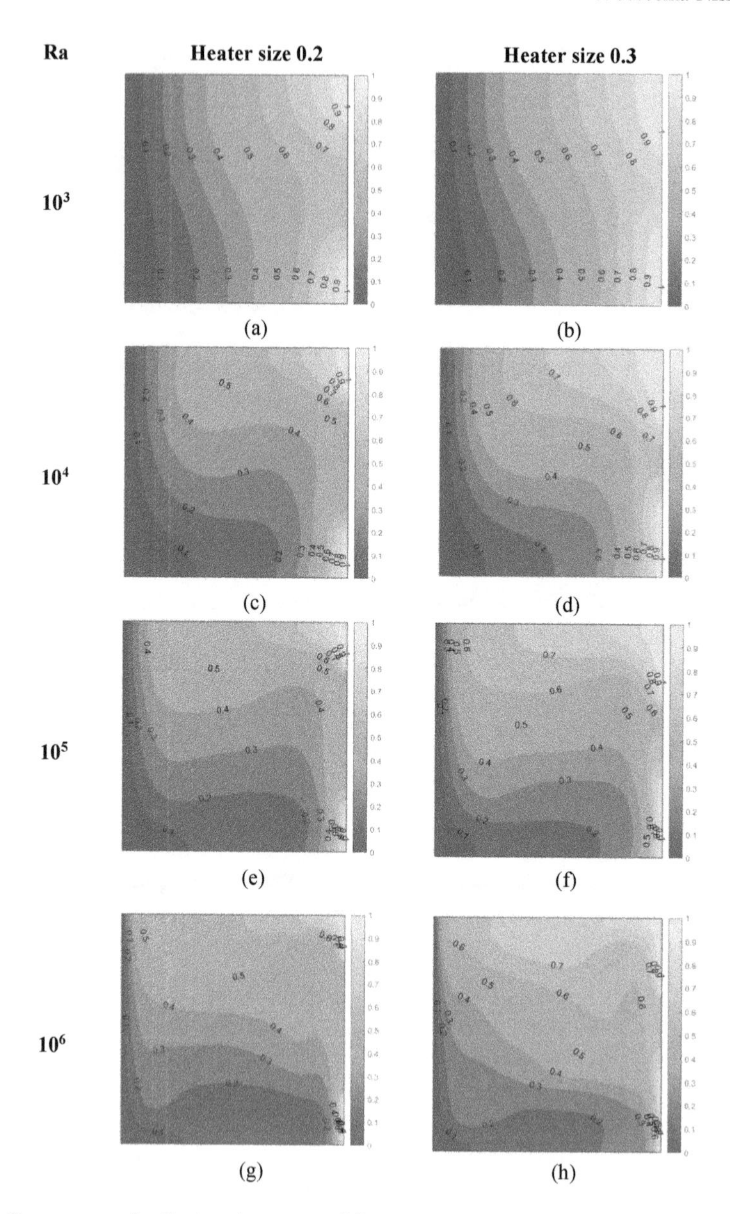

Fig. 4 Temperature distribution due to two different heaters at different Ra

bottom. Due to the increase in the heater size, the lower contour is slightly pushed away from the bottom heater, and the diagonal upper half of the room holds warm air.

7.3 *Centerline U and V Velocity*

The horizontal and the vertical component of the velocity along the centerline of the y-direction are plotted in Fig. 5, respectively. When the Ra is 10^3, the fluid acquires a very little velocity of the negligible amount. This evident that, the heat transfer takes place by conduction, as discussed in the previous sections. Until Ra $\leq 10^5$, the fluid at the bottom of the room moves clockwise, known from the positive u and v velocity value, and the fluid at the top moves anti-clockwise. At the higher Ra (Ra = 10^6), the fluid at the core gets a significant opposite velocity due to buoyancy. When the heater size is increased from 0.2 to 0.3 times the length of the room, the above said phenomena pile on, making the fluid at the core and outer fluid move in the opposite direction. This is the reason for the formation of the low-intensity stream function core seen in Fig. 3h.

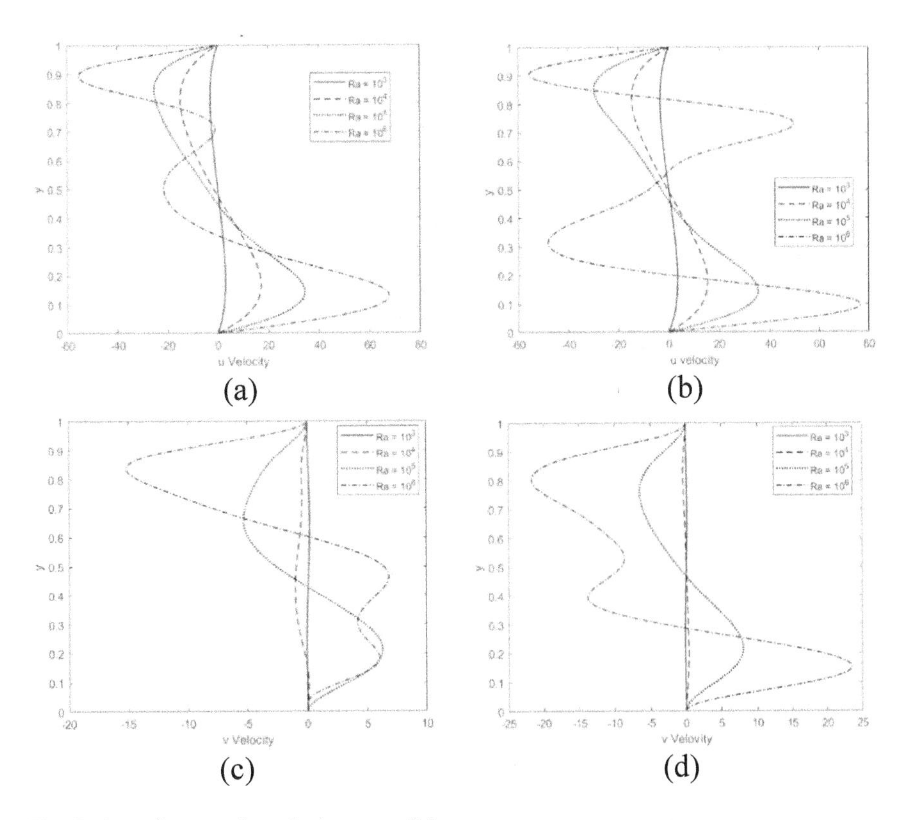

Fig. 5 Centerline u and v velocity at x = 0.5

7.4 *Rate of Heat Transfer*

The rate of heat transfer is measured by calculating the Nusselt number as given in Eq. (11). The Nusselt number is calculated for the left cold wall. It can be seen from Fig. 6 that the rate of heat transfer to the left wall is low when Ra is less than 10^4. When Ra increase to 10^5, Nu increases due to the effect of buoyancy. It gets intensified when Ra further to 10^6. The higher Ra pushes a large amount of heat to the wall and the rate of heat transferred to the left wall increases. The pattern of the line traced by the Nusselt number is the same irrespective of the heater size. The heater size affects the intensity of heat transfer, as seen from Fig. 6a, b.

8 Conclusion

Space heating has become the essence of these days. It consumes more than 50% of residential energy usage. This numerical study has been carried out to find the effect of two heaters placed inside a close room in space heating. The other side of the room (left wall) is assumed to be exposed to cool temperature. All other walls are considered adiabatic. The streamline, isotherm, centerline velocity, and rate of heat transfer are analyzed. The intensity of heat flow is varied by changing the Rayleigh number. Air is considered as the fluid inside the room with Prandtl number 0.71. The results show that the buoyancy increase with the increase in Ra. The mode of heat transfer at lower Ra is by conduction. Natural convection influences the flow when Ra $\geq 10^5$. The fluid-particle acquires a thermal boundary when Ra = 10^6. Increase in the size of the heater from 0.2 to 0.3, the intensity of temperature, velocity, and Nu increases. The effect of the heater at the bottom position is low

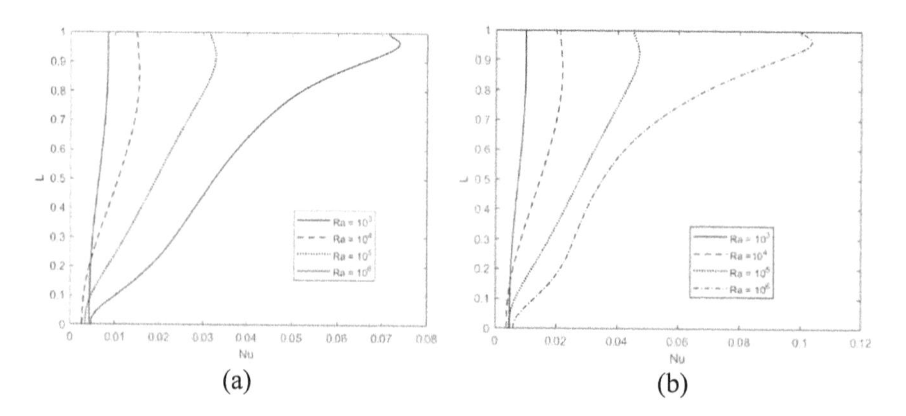

Fig. 6 Average Nusselt Number along the left side wall of the room **a** 20% heater **b** 30% heater

compared to the top one since most of the hot air is pushed to the top due to buoyancy. Further study has to be carried out to find the optimal position and size of the heater for effective space heating by natural convection.

References

1. Energy efficiency indicators—analysis—IEA, International Energy Agency (2020). Available https://www.iea.org/reports/energy-efficiency-indicators
2. De Vahl Davis G (1983) Natural convection of air in a square cavity: a bench mark numerical solution. Int J Num Methods Fluids 3(3):249–264
3. Ostrach S (1972) Natural convection in enclosures. Adv Heat Transf 8:161–227
4. Ben Nasr K, Chouikh R, Kerkeni C, Guizani A (2006) Numerical study of the natural convection in cavity heated from the lower corner and cooled from the ceiling. Appl Therm Eng 26(7):772–775
5. Rahman MM, Mamun MAH, Billah MM, Saidur R (2010) Natural convection flow in a square cavity with internal heat generation and a flush mounted heater on a side wall. J Nav Archit Mar Eng 8235(July)
6. Mei SJ, Hu JT, Liu D, Zhao FY, Li Y, Wang HQ (2019) Airborne pollutant dilution inside the deep street canyons subjecting to thermal buoyancy driven flows: effects of representative urban skylines. Build Environ 149:592–606
7. Hu JT, Mei SJ, Liu D, Zhao FY, Wang HQ (2018) Buoyancy driven double diffusive moisture convection inside the fluid-porous-solid sandwiched building enclosure containing internal heating sources. Int J Heat Mass Transf 123:600–615
8. Saha SC, Patterson JC, Lei C (2010) Natural convection and heat transfer in attics subject to periodic thermal forcing. Int J Therm Sci 49(10):1899–1910
9. Ozsagiroglu S, Camci M, Taner T, Acikgoz O, Dalkilic AS, Wongwises S (2021) CFD analyses on the thermal comfort conditions of a cooled room: a case study. J Therm Anal Calorim 1–25
10. Kalidasan K, Velkennedy R, Rajesh Kanna P (2014) Buoyancy enhanced natural convection inside the ventilated square enclosure with a partition and an overhanging transverse baffle. Int Commun Heat Mass Transf 56:121–132
11. Ghia, U, Ghia KN, Shina CT (1982) High-resolutions for incompressible flow using Navier-Stokes equation and a multigrid method. J Comput Phys 20(3):387–411
12. Indukuri JV, Maniyeri R (2018) Numerical simulation of oscillating lid driven square cavity. Alexandria Eng J 57(4)"2609–2625
13. Kalidasan K, Velkennedy R, Rajesh Kanna P (2014) Numerical investigation on natural convection inside the side ventilated square enclosure with vertical mid-partition. Num Heat Transf Part A Appl 66(12):1389–1418
14. Das MK, Kanna PR (2007) Application of an ADI scheme for steady and periodic solutions in a lid-driven cavity problem. Int J Num Methods Heat Fluid Flow 17(8):799–822

Salinity Mapping of Coastal Aquifers Using GIS and GALDIT

Subin Abraham Jacob, Robin Alexander, Kevin Varghese, Juval Joy, and Rinu J. Achison

Abstract Aquifers are vulnerable to saltwater intrusion, making the groundwater pumped from these aquifers unfit for consumption. The vulnerability of aquifers to saltwater intrusion can be assessed by GALDIT, an index based method. GALDIT measures the vulnerability in terms of six parameters Groundwater occurrence, Aquifer conductivity, Level above mean sea level, Distance from the shore, Existing impact, and Thickness of the aquifer. This study aims to find the vulnerability of aquifers in the coastal areas of Ernakulam district in Kerala to saltwater intrusion. The study area includes the taluks of Kanayannur and Kochi. GALDIT method is applied to each aquifer to get the vulnerability in terms of low, moderate, and highly vulnerable. This classification can be used for assessing the vulnerable areas for timely intervention measures. The data for the aquifers was collected from IWRIS (Indian Water Resource Information System) website and CGWB (Central Ground Water Board). Then using the data compiled from the GALDIT method, vulnerability maps were prepared in ArcGIS. Kriging Interpolation method was used to prepare maps for the influence of the six parameters of GALDIT on saltwater intrusion and then weighted overlay analysis was used to classify the vulnerable zones of the study area into low, moderate, and highly vulnerable.

Keywords Coastal aquifer · Saltwater intrusion · GALDIT · GIS

1 Introduction

Freshwater is the driving need of human civilization. The rise in the level of pollution in terrestrial freshwaters has led to an increased usage of groundwater. The major source of freshwater is said to be the groundwater which is under severe threat to saltwater intrusion, caused by natural as well as man-made causes. The

S. A. Jacob (✉) · R. Alexander · K. Varghese · J. Joy · R. J. Achison
APJ Abdul Kalam Technological University, Thiruvananthapuram, Kerala, India
e-mail: rinuachison@fisat.ac.in

G. C. Marano et al. (eds.), *Proceedings of SECON'21*, Lecture Notes in Civil Engineering 171, https://doi.org/10.1007/978-3-030-80312-4_11

increased need and use of water by the public sector and by individuals have resulted in the decline of quality and quantity of freshwater and has led to increased saltwater intrusion.

The freshwater storage mainly in arid and semi-arid regions consists of coastal aquifers. Due to the demographic growth, there is increased usage of water especially in the agriculture sector and aquifers are also intensively exploited despite their extreme vulnerability to quality degradation by seawater intrusion. This situation is also expected to exacerbate by climate change and associated sea level rise. The groundwater vulnerability assessment to seawater intrusion is a major technique used to assist the development of groundwater protection strategies. These methods involve mapping of coastline area which are highly vulnerable. An appropriate method for vulnerability mapping is chosen based on the purpose, scope, data availability, time, and end-user requirement.

The coastal parts of Ernakulam district are under threat of saltwater intrusion being close to the sea and being densely populated. In recent years, environmental assessments of groundwater resources have resulted in the development of models that help identify the vulnerable zones such as GALDIT. The GALDIT model is developed to determine the vulnerability of coastal aquifers to saltwater intrusion. There are several approaches for assessing the vulnerability of coastal aquifers to saltwater intrusion, numerical simulations with the variable density flow-transport model, the sharp interface approach, and the overlay and index method. Recently assessments using overlay and index techniques based on spatial mapping have been developed and improved to intuitively show the intrinsic vulnerability of aquifers. One of the common methods used for classification and mapping of groundwater vulnerability zones is GALDIT. The GALDIT method was specifically created for coastal aquifers as it considers the saltwater intrusion is mainly directed from sea water. The index-based GALDIT numerical ranking method is advantageous because it is relatively independent of the boundary condition problem.

1.1 Study Area

The present study focuses on the coastal plains of Ernakulam district (Fig. 1) namely the taluks of Kochi and Kanayannur. The study area covers a total area of 432.44 Km^2, is bound on the western side by the Arabian sea. The study area has a wet monsoon type climate with average annual precipitation of 3233 mm and a temperature of 31.4–26 °C. The Lithology of the study area consists of Younger Alluvium soil of clay, silt, sand, and calcareous rocks and the aquifers are usually unconfined in their occurrence. The area primarily depends on water from municipal corporation sources to meet daily water requirements, though most of the households have either a well/ bore well since there are periodic occurrences of water shortages.

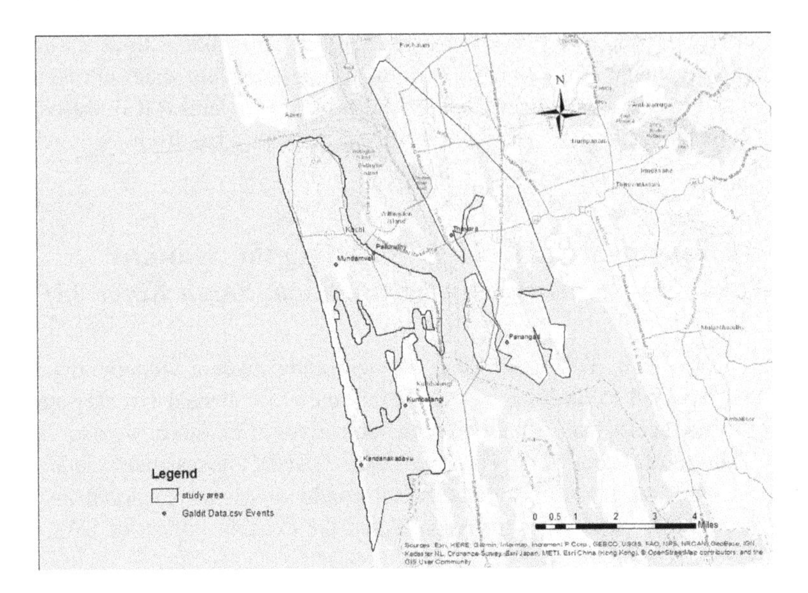

Fig. 1 Study area

2 Literature Review

2.1 *Sea Water Intrusion Vulnerability Mapping of Aquifers Using GALDIT Method [1]*

Lobo-Ferreira and Chaachadi has studied the effect of increase in mean sea level on the ingress of salt water using GALDIT method. The aquifer of Monte Gordo was chosen, it was unconfined in its occurrence and the rest of the parameters of GALDIT were rated based on its characteristics. Distance from the aquifer was calculated from 3 perpendicular distances from coastline and banks of rivers in the north, southeast of the aquifer. The Ratio of $Cl^-/(HCO_3^- + CO_3^{2-})$ was determined from wells. The maps were developed using the point values of vulnerability which were obtained from equations and then contoured in surfer software to get the vulnerability score map.

2.2 *Mapping of Salinity Ingress Using GALDIT Model for Sirkali Coastal Region [2]*

V. Satishkumar, S. Sankaran, Taufique Warsi, B. Amarender and Ratnakar Dhakate developed a model which is traced to evaluate and enumerate the significance of vulnerability to seawater intrusion due to excessive groundwater withdrawals and some anthropogenic activities at coastal aquifers. So taking these issues into account few thematic maps which were influencing the saline water intrusion were prepared and overlaid using Geographical Information Systems (GIS). Based on the GALDIT method, groundwater vulnerability cartography has been assessed.

2.3 Application of GALDIT in Assessing the Seawater Intrusion Vulnerability of Jeju Island, South Korea [3]

Chang, Chung, Kim, Tolera and Koh, in their study made a vulnerability assessment of Jeju Island to SWI based on several years of collected groundwater level data and hydro geological values where the objectives of the study were to visualize the distribution of recent SWI. To improve the GALDIT assessment method to fit the Jeju model, the possibility of electrical conductivity was explored instead of standard GALDIT parameters that represented the existing impact of SWI.

2.4 Groundwater Vulnerability Assessment to Seawater Intrusion Through GIS-Based GALDIT Method. Case Study: Atalanti Coastal Aquifer, Central Greece [4]

Lappas, Kallioras, Pliakas, and Rondogianni, In their study has used GIS-based GALDIT index method to characterize the coastal hydrogeological settings and obtain a potential vulnerability distribution map of Atalanti coastal aquifer involving measurable weighted and rated parameters (numerical ranking system). Each factor was evaluated with respect to assigning a relative weight depending on local field conditions. The spatial maps were prepared using the kriging geostatistical interpolation technique.

3 Methodology

The study was carried out by using data obtained from exploratory and monitoring wells of CGWB. Water quality data was obtained from CGWB groundwater information booklet and IWRIS, Hydrological data of aquifers were obtained from website of IWRIS.

To evaluate the vulnerability of aquifers GALDIT method [1] was used which evaluates the vulnerability based on six parameters—Groundwater occurrence, Aquifer conductivity, Level above mean sea level, Distance from the shore, Impact of existing intrusion and Thickness of aquifer. Each parameter is assigned a different weightage depending upon their influence on saltwater intrusion [1]. Each aquifer was selected and scored/rated on the basis of its characteristics. The rating for each parameter was multiplied by their respective weights which is then summed and divided by the sum of all the weights to get the final GALDIT score which classifies the aquifer as low, moderate or highly vulnerable.

3.1 GALDIT

GALDIT is an indicator based model which assesses and quantifies the significance of vulnerability to seawater intrusion due to excessive groundwater withdrawals. The system contains three significant parts: weights, ranges, and ratings. Each GALDIT factor has been evaluated with respect to the other to determine the relative importance of each factor [1]. Each of the six indicators has a predetermined fixed weight that reflects its relative importance to seawater intrusion.

Groundwater occurrence (G): Groundwater occurrence is the parameter that shows which type of aquifer is under consideration. The type of aquifer has a significant role in vulnerability to saltwater intrusion. This parameter is given a weightage of 1 [1].
Aquifer Hydraulic conductivity (A): Hydraulic conductivity governs the magnitude of intrusion into an aquifer. The more the soil is hydraulically conductive, the more vulnerable it is to saltwater intrusion. This parameter is given a weightage of 3 [1].
Height of the groundwater level above sea level (L): The height of the aquifer also plays a significant role in SWI. When the freshwater aquifer is located high above sea level, the chances of it getting intruded by seawater is on the lower side. The aquifer located at a lower elevation compared to sea-level is highly vulnerable to saltwater intrusion. This parameter is given a weightage of 4 [1].
Distance inland perpendicular from the shoreline (D): SWI decreases as the distance from the shoreline increases. Saltwater has to penetrate more if the aquifer under consideration is far from the coastline. This parameter is given a weightage of 4 [1].
Impact of existing seawater intrusion in the area (I): This parameter takes into account the amount of SWI that has already taken place. The molar ratio of chloride ions to bicarbonate ions is generally used, but electrolytic conductivity [3] of water can also be used as a measure of existing impact, it is controlled by the concentration of dissolved ions. This Parameter is given a weightage of 1 [1].
Thickness of the aquifer (T): Thicker an aquifer may be, the more vulnerable it is to salt water intrusion. Thinner aquifers are less vulnerable to SWI. This parameter is given a weightage of 2 [1].

Each of the above six parameters were given rating based on their respective values, which is done for each and every aquifer. The different weights as well as the rating given for different values of the aquifer is summarized in Table 1.

The GALDIT Index is then obtained by computing the individual indicator scores and summing them as per Eq. (1).

$$\text{GALDIT - Index} = \sum[(Wi \times Ri)] \div \sum Wi \quad (1)$$

where W_i is the weight of the ith indicator and R_i is the importance rating of the ith indicator. The GALDIT index of a particular aquifer is obtained by multiplying the ratings of each parameter with their respective weights and then dividing the sum of all the products with the sum of all weights which is 15. Based on the GALDIT Index the area can be classified as low vulnerability, moderate vulnerability and high vulnerability as given in Table 2.

3.2 *GIS*

In the present study ArcGIS version 10.7.1 was used for developing raster maps. Spatial changes can be evaluated more easily and quickly using GIS. The shape file for the study area was made in ArcGIS using the polygon reshape tool, and the x–y coordinates of the monitoring wells of CGWB were used as the sampling points for the analyses.

In this study, GIS has been used to collect diverse spatial data to represent spatially variable phenomena by applying a series of overlay analysis of data layers that are in spatial register. The GIS is also utilized in the map classification

Table 1 GALDIT rates and weights

Parameters	Groundwater occurrence (G)	Aquifer hydraulic conductivity (A) (m/day)	Height (L) (m)	Distance from shore (D) (m)	Impact of existing status of seawater intrusion (I) (μs/m)	Aquifer thickness (T) (m)
	Weights					
Rates	1	3	4	4	1	2
10	Confined aquifer	>40	<0	<500	>3000EC	>10
7.5	Unconfined aquifer	10–40	0–1.0	500–750	2000–3000	7.5–10
5	Leaky confined aquifer	5–10	1.0–2.0	750–1000	1000–2000	5–7.5
2.5	Bounded aquifer	<5	>2.0	>1000	<1000	<5

Table 2 Vulnerability classification

Sl. No	GALDIT index	Vulnerability classification
1	>7.5	Highly vulnerable
2	5–7.5	Moderately vulnerable
3	<5	Low vulnerability

according to the rates and weights of each parameter and to develop the final vulnerability map.

4 Results and Discussions

4.1 Data Collection

The parameters of GALDIT were found using 6 monitoring wells of CGWB in Panangad, Kumbalangy, Kandanakadavu, Thevara, Mundamveli, and Palluruthy. The aquifer in the study area was found to be made of young alluvium soil which occurred in the unconfined state and was given a rating of 7.5.

Aquifer conductivity was found from the transmissivity and thickness data of six monitoring wells, Using the relation between transmissivity, conductivity and thickness of the aquifer as shown in Eq. (2).

$$K = T/d \tag{2}$$

where K is the hydraulic conductivity, T hydraulic transmissivity and d aquifer thickness. The conductivity values were found to be ranging from 0.994 to 17.8 m/day. The aquifers in Mundamveli, Panangad and Kandanakadavu had a conductivity value in the range 10–40 m/day and were given a rating of 7.5. The conductivity values of Thevara and Palluruthy were less than 5 and was given a rating of 2.5. The aquifer in Kumbalangy was given a rating 5.

The water level above mean sea level of aquifers were found out by subtracting the depth to water level from the reduced level of the surface. The reduced level of the surface as obtained from Google earth and depth to water level was found out from the data provided for monitoring wells of CGWB by IWRIS. The water level in aquifers of Kumbalangy and Panangad were above 2 m and was given a rating of 2, while the water level in rest of the aquifers were less than 1 m and was given a maximum rating of 10. Aquifers in Palluruthy and Mundamveli, the water level was below the mean sea level indicating high probability of seawater intrusion.

Distance from shore was found by measuring the perpendicular distance from the location coordinates of the monitoring well's to the sea [1]. This was done with help of google maps. Panangad, Palluruthy and Kumbalangy were the farthest from the sea coast and were given a minimum rating of 2.5, Mundamveli was moderately far from the sea and was given a rating of 5, while Kandanakadvu and Thevara were very close to sea less than 500 m and was given a maximum rating of 10.

Impact of Existing Intrusion, was rated based on the electrical conductivity of water in the monitoring wells. Instead of evaluating the existing impact by the ratio of chloride to carbonate and bicarbonate ions [1] the factor range was modified to electrical conductivity [3]. The EC values ranged from 3280 to 660 μs/cm. The values were rated based on the modified ranges for electrical conductivity.

Thickness of the aquifer was found out from IWRIS. Thickness of the aquifers ranged from less than 10 to greater than 50 m. Most of the study area had a thickness in the range of 10–40 m and was given a rating of 7.5. For a thickness less than 10 m a rating of 5 was given and for a thickness greater than 50 m a maximum rating of 10 was given.

After the values for different parameters were obtained, substituting these values in Eq. (1) the GALDIT index was determined for the six aquifer systems, which has been summarized in Table 3.

4.2 Saltwater Intrusion Vulnerability Map

Saltwater intrusion vulnerability maps were prepared in ArcGIS. The sampling points used were the monitoring wells of CGWB located in the study area. The study area was made from the shape file of Kanyanoor and Kochi taluks and polygon reshape tool in ArcGIS. Then spatial variability maps for each of the six parameters of GALDIT were prepared using Kriging interpolation model [4] as shown in "Fig. 2." The raster maps were then overlaid on top of each other using the weighted overlay tool in ArcGIS to develop the final saltwater intrusion vulnerability map as shown in "Fig. 3".

5 Limitations

The major drawback of this method is the unawareness of the pumping effect on the seawater intrusion process [5]. This method is also limited by the lack of availability of hydrogeological parameters of GALDIT. Some parameters, like groundwater level and distance from the shore, are the ones with higher magnitude; while the existing state of the aquifer resulted in a weaker index. The overexploitation of groundwater resources throughout the summer could represent the main problem even if an optimal management plan will be adopted. More analysis could be applied in the area like hydro-geochemical analyses in the porous aquifer as well as in the granodiorite fissured rock aquifer and an evaluation of all the possible impacts induced by anthropic activities in order to perform a better management plan.

Table 3 GALDIT parameters and their ratings

No.	Place	Groundwater occurrence (G) (1)		Aquifer hydraulic conductivity (A) (3)		Height (L) (4)		Distance from shore (D) (4)		Impact (I) (1)		Aquifer thickness (T) (2)		GALDIT index
		Value	R	V	R	V	R	V	R	V	R	V	R	
1	Panangad	UC*	7.5	11.06	7.5	2.4	2.5	6800	2.5	5500	10	17.5	10	5.3
2	Kumbalangi	UC*	7.5	5.2	5	2.17	2.5	2000	2.5	660	2.5	27.5	10	4.3
3	Kandanakadavu	UC*	7.5	16.36	7.5	0.76	10	188	10	3248.2	10	50	10	9.3
4	Mundanveli	UC*	7.5	17.8	7.5	1.21	7.5	979.7	5	1560	5	10	7.5	6.7
5	Thevara	UC*	7.5	1.604	2.5	−2.5	10	310.2	10	1500	5	17.5	10	8
6	Palluruthy	UC*	7.5	0.994	2.5	−1.76	10	2630	2.5	3280	10	10	7.5	6

*Unconfined aquifer

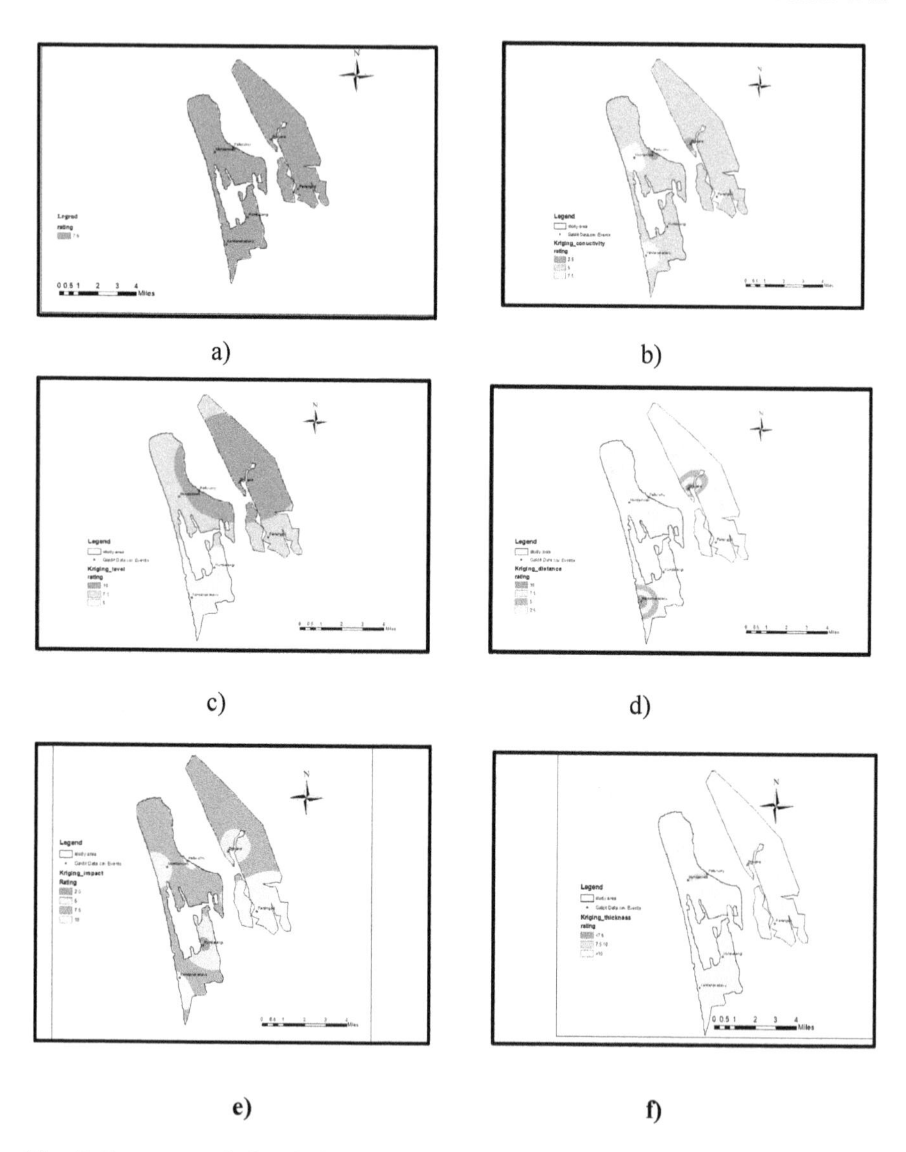

Fig. 2 Parameters of the GALDIT method: **a** Aquifer occurrence; **b** aquifer hydraulic conductivity (A); **c** height of groundwater level above sea level (L); **d** distance from shore; **e** impact of existing status of seawater intrusion in the area (I), **f** thickness of aquifer (T)

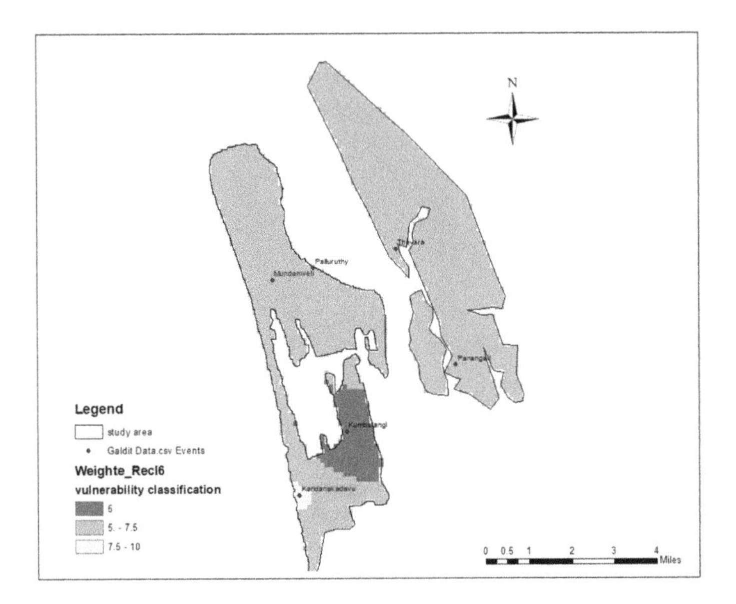

Fig. 3 Saltwater Intrusion vulnerability map

6 Conclusion

The GALDIT scores of Kandanakadavu and Thevara were 9.33 and 8 respectively indicating high vulnerability to saltwater intrusion. The index values of Pananagad, Mundamveli, and Palluruthy, all lay within the range of 5–7.5 and were hence classified as moderately vulnerable, Kumbalangy had an index of 4.33 which is less than 5, and was classified as a low vulnerability area. The ratings for highly vulnerable zones are to the parameters of GALDIT method, distance from the shore, level above mean sea level, and aquifer conductivity which had a rating of 4, 4, and 3 respectively. This is because the major factors which influences an aquifer's vulnerability to saltwater intrusion is its conductivity, its water level, and distance of the aquifer from the sea. The GALDIT index values of highly vulnerable zones were a result of high rating on these three parameters.

The vulnerability map is prepared to classify the study area into different zones of vulnerability. The map can be used to assess the saltwater vulnerable zones of Kochi and implement suitable preventive strategies such as artificial recharge, extraction barrier, injection barrier, subsurface barrier. The map shows that most of the study area falls under the category of moderately vulnerable area except small region surrounding Kandanakadavu which lie in the high vulnerability zone as it is surrounded by water on both sides. In moderately vulnerable zones preventive measures are required to prevent further intrusion of saltwater, by artificial recharge during the summer season. High vulnerability zones require water treatment before it can be used for everyday activities.

GALDIT method can be considered as an effective tool for assessing groundwater vulnerability to seawater intrusion. The resulting vulnerability map is a useful management tool that can be used to control and manage seawater intrusion in different types of coastal aquifers, the map illustrates the parts of Kochi where the aquifer is more vulnerable and should be monitored more intensively.

References

1. Chachadi AG, Lobo-Ferreira JP (2001) Sea water intrusion vulnerability mapping of aquifers using GALDIT method. Coastin 4:7–9
2. Satishkumar V et al (2016) Mapping of salinity ingress using Galdit model for Sirkali coastal region: a case study. J Geogr Inf Syst 8(4):526–536
3. Chang SW et al (2019) Application of GALDIT in assessing the seawater intrusion vulnerability of Jeju Island, South Korea. Water 11(9):1824
4. Lappas I, Kallioras A, Pliakas F, Rondogianni T (2016) Groundwater vulnerability assessment to seawater intrusion through GIS-based Galdit method. Case study: Atalanti coastal aquifer, central Greece. Bull Geol Soc Greece 50(2):798–807
5. Trabelsi N et al (2016) Aquifer vulnerability and seawater intrusion risk using GALDIT, GQI SWI and GIS: case of a coastal aquifer in Tunisia. Environ Earth Sci 75(8):669

Design Aids for Special Shaped Reinforced Concrete Columns

Sangeeta Gadve and Rangesh Jajodia

Abstract A column is an important component of a structure. Given the potential for catastrophic failure and the relatively low ratio of cost to additional load bearing capacity, it has been found that to make the column safe, much larger dimensions than required, are provided to account for all the loading conditions. However, if the cross section of the column is properly adjusted according to the requirement, then much economy may be achieved. The use of Special Shaped reinforced concrete columns (SSCs) with L-, T-, and + -shaped cross-sections is a breakthrough in this context. This paper presents a procedure that allows the construction of simplified axial load—bending moment interaction diagrams for the SSCs for practical design applications. In the proposed methodology, the analysis of Special Shaped Columns is carried out based on principles of equilibrium and strain compatibility equivalent to that of conventional RC columns. The method determines the interaction diagrams for the Special Shaped Reinforced Concrete (RC) short columns under axial load combined with uniaxial and biaxial bending. The reliability of the obtained charts is verified by the use of a computer analysis program.

Keywords Special shaped reinforced concrete columns (SSC's) · Interaction diagram · Strain compatibility · Uniaxial and biaxial bending

1 Introduction

Columns are vertical elements in reinforced concrete structures used to transfer loads to the foundation. According to IS 456: 2000, column is defined as a compression member whose effective length exceeds three times the least lateral dimension. Most commonly used shapes of column cross section are square, rectangle and circle. The choice of shape of a column depends upon various factors

S. Gadve · R. Jajodia (✉)
Department of Applied Mechanics, Visvesvaraya National Institute of Technology, Nagpur, Maharashtra 440010, India

G. C. Marano et al. (eds.), *Proceedings of SECON'21*, Lecture Notes in Civil Engineering 171, https://doi.org/10.1007/978-3-030-80312-4_12

such as the loading condition, adequacy for space, types of forces and architectural requirements. However, use of standard shapes leads to wastage of space due to protruding corners of columns. This problem can be overcome using special shaped cross sections (SSC) such as T (Tee), L (Ell) and + (Plus). T (Tee) shaped columns can be used as interior/intermediate column without disturbing the free space out of the wall. Similarly, + (Plus) column can be used as meeting/intermediate columns and L (ELL) column can be used as corner columns. When compared with conventional square or rectangular cross-sections, columns with T (Tee) and + (Plus) shaped cross-sections have the advantage of not projecting outside the walls. Figure 1 shows the layout of the different sections used at different locations of a building.

In order to check the adequacy of special shaped columns, comparison between columns of different shape viz. rectangular, T and Plus was done. The comparison was carried out in two ways. Firstly, the cross-sectional area of conventional and special shaped columns was kept same and change in the moment of inertia was observed. Secondly, the moment of inertia was kept constant and reduction in area of special shaped column was witnessed. The comparison of these two cases is tabulated in Table 1. The moment of inertia of special shaped column increases up to 30% as compared to rectangular column of same cross-sectional area. Upsurge of M.I. leads to better moment carrying capacity of the member. Contrastingly, when moment of inertia is kept constant there is significant reduction in the cross-sectional area of special shaped column, leading to increase in economy.

Liu et al. [1] performed studies to find the seismic capacity of the specially shaped column-frame structure system. A comparative study on the seismic behaviour between specially shaped column frame structure and rectangular column frame structures was performed by Yang et al. [2]. A regular six-story RC frame with specially shaped columns and three RC frames with conventional rectangular columns were designed with their columns respectively based on the criterion of the same section area, same moment of inertia and same initial stiffness of the specially shaped frame structure. It was observed that the maximum concrete compressive strain of the specially shaped column frame structure is much higher than the rectangular columns, but less than the concrete ultimate compressive strain.

Yu-Ye and Bo [3] performed experiments to study the fire resistance characteristics of special shaped columns. Four L-shaped, four T-shaped, three +-shaped and one square-shaped RC columns were experimentally investigated. Later, Li et al. [4] conducted tests to overcome the problem that steel bars are put too close in a frame joint with special-shaped beam and column. Experimental comparison was conducted between special-shaped (L, T and +) column and normal beams. Wang et al. [5] investigated the effect of axial compression ratios and stirrup spacing on failure mode, bearing capacity, displacement, and curvature ductility of the L-shaped columns.

Zhou et al. [6] made a detailed investigation of specially shaped concrete filled steel tubes. Experimental investigation was carried out to study the performance of column subjected to axial compression. The FEM model was also proposed and

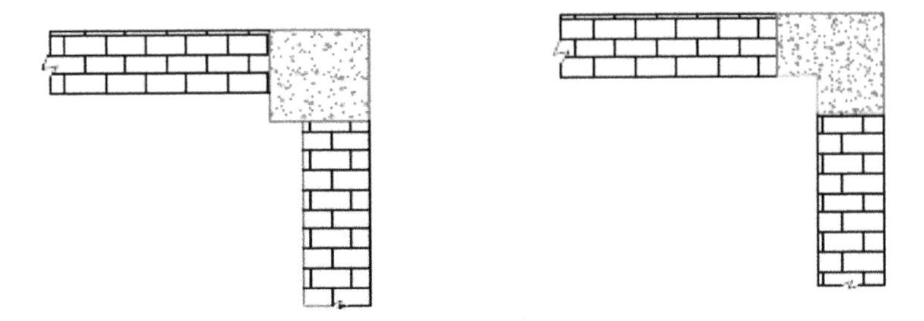

(a) Rectangular Column replaced by L shaped column at exterior corners

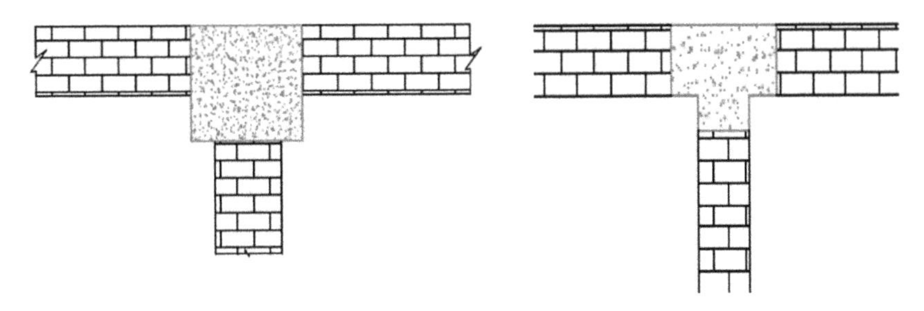

(b) Rectangular Column replaced by T shaped column at exterior edges

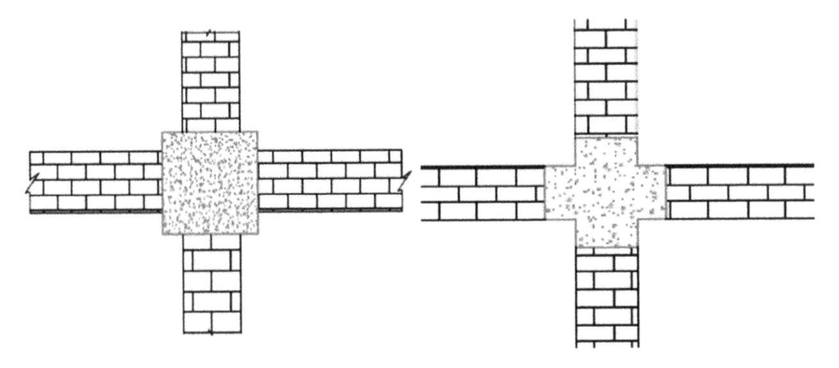

(c) Rectangular Column replaced by + shaped column at interior corners

Fig. 1 Layout of columns with different sections

influence of various parameters was studied. Later, Zheng et al. [7] studied the behaviour of L-shaped and T-shaped concrete filled steel tabular. They investigated the design procedure with the FEM models and suggested new design models for them.

Table 1 Comparison between conventional rectangular columns and equivalent SSCs

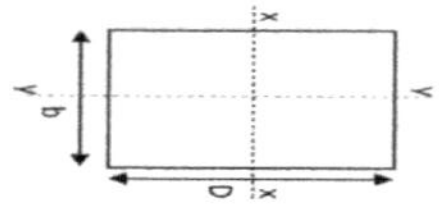

Square Section
b = 400 mm
D = 400 mm
Ag = 160000 mm ^{2}
M. I. = 21.33 × 108 mm ^{4}

Shape	Constant area (Ag = 160,000 mm^2)						Constant M. I. (21.33 × 10^8 mm^4)			
	Size (mm)	b/D	M. I. (10^8 mm^4)		% increase in M. I		Size (mm)	b/D	Area (mm^2)	% decrease in area
			I_{xx}	I_{yy}						
+ 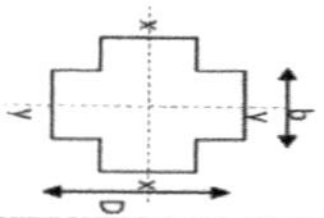	150 × 605	0.25	29.13		27		150 × 525	0.28	135,000	16
	175 × 545	0.32	28.5		25		180 × 500	0.36	147,500	8
	200 × 500	0.4	22.8		6		200 × 475	0.42	150,000	6
L 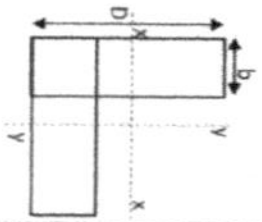	150 × 605	0.25	49.12		57		150 × 475	0	120,000	25
	175 × 545	0.32	38.4		44		200 × 450	0.4	140,000	13
	200 × 500	0.4	31.2		32		250 × 430	0.5	152,500	5
T 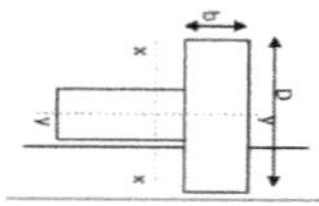	150 × 605	0.25	49.12	29.13	57	27	150 × 475	0.3	120,000	25
	175 × 545	0.32	38.4	28.5	44	25	200 × 450	0.4	140,000	13
	200 × 500	0.4	31.2	22.8	32	6	250 × 430	0.5	152,500	5

In the standard guidelines, design charts are available only for conventional shaped sections. The present study proposes the interaction chart for design of special shaped columns subjected to uniaxial bending moments. The reliability of the obtained charts would be verified using a computer analysis program.

2 Mathematical Analysis

2.1 Interaction Curve Diagrams

The analysis and design of conventional shaped reinforced column subjected to combined action of axial load and uniaxial bending moment is simple, as the neutral axis is parallel to the moment axis. However, in case of asymmetric members subjected to axial load and uniaxial bending moments, the neutral axis may not remain parallel to moment axis. The analysis in this case is much tedious and involves determination of correct position of neutral axis, which requires several trials involving lengthy calculations. Therefore, a simplified approach has been evolved for the design using moment-load interaction curve.

An interaction curve is a graphical summary of the ultimate bending capacity of a range of reinforced columns with different dimensions and areas of longitudinal reinforcement. Each point on the interaction diagram corresponds to the column capacity at a specific load eccentricity. The points also represent the combination of axial forces and bending moment corresponding to the resistance of a column cross-section. The plot in Fig. 2 shows the outside envelope of typical interaction diagram.

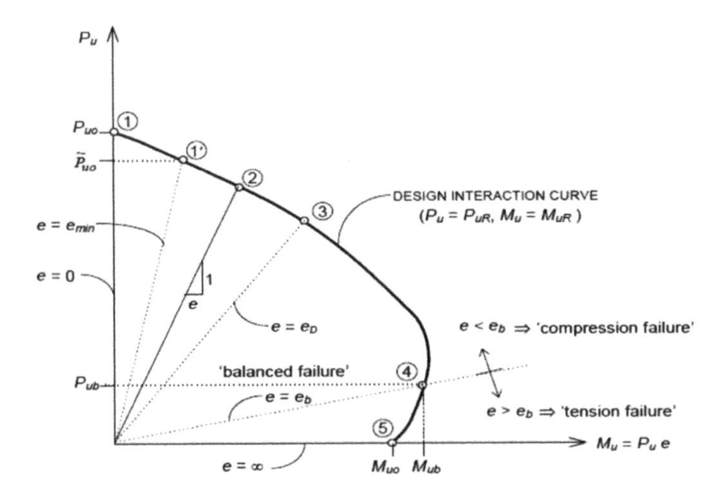

Fig. 2 Typical P_u-M_u interaction diagram

The interaction diagram for a column is constructed using a series of values for P_u and M_u. The coordinates of the design interaction curve, viz. M_u (on the x-axis) and P_u (on the y-axis), can be determined for any arbitrary neutral axis depth x_u. The salient points indicated in the typical interaction curve can be observed in the strain profile under ultimate limit state. These points can be established by considering the different locations of neutral axis.

The starting value of x_u corresponding to $P_u = 0$, must be determined by trial-and-error. Having located (approximately) x_u/D, the coordinates of the design interaction curve can be obtained and tabulated for incremental values. The coordinates (M_u, P_u) of the design interaction curve can then be tabulated and plotted. By creating interaction diagrams at various angles of neutral axis, a three-dimensional plot of P_{uR}—M_{ux}—M_{uy} is obtained as shown below.

2.2 Analysis of Column

For the analysis of special shaped columns, the Limit State Method (LSM) as given in IS 456-2000 is adopted. The assumptions used in the analysis are as given in clause 38 and 39 of IS 456-2000. The stress and stain block diagrams for T-shape and +-shape column are used to calculate the design load of each element of column. The ultimate axial load (P_u) and bending moment (M_u) is given as

$$P_u = P_{uc} + P_{us}$$

$$M_u = M_{uc} + M_{us}$$

where,

P_{uc} is the axial load in concrete.
P_{us} is the axial load in steel.
M_{uc} is the bending moment in concrete.
M_{us} is the bending moment in steel.

2.3 Analysis of T Shaped Section

In flange portion

$$P_{uc} = 0.36 f_{ck} * b * x_u + 0.447 f_{ck} * x_u * (D - b)$$

$$P_{us} = \frac{\rho}{f_{ck}} * \rho_i * (f_{si} - f_{ci})$$

$$M_{uc} = P_{uc} * (y - 0.416x_u)$$

$$M_{us} = P_{us} * y_i$$

In web portion

$$P_{uc} = 0.36f_{ck} * b * x_u + 0.447f_{ck} * b * (D - b)$$

$$P_{us} = \frac{\rho}{f_{ck}} * \rho_i * (f_{si} - f_{ci})$$

$$M_{uc} = P_{uc} * (y' - 0.416x_u)$$

$$M_{us} = P_{us} * y_i$$

See Fig. 3.

2.4 *Analysis of + Shaped Section*

In top web portion

$$P_{uc} = 0.36f_{ck} * b * x_u$$

$$P_{us} = \frac{\rho}{f_{ck}} * \rho_i * (f_{si} - f_{ci})$$

$$M_{uc} = 0.36f_{ck} * b * x_u * (y' - 0.416x_u)$$

$$M_{us} = P_{us} * y_i$$

In flange portion

$$P_{uc} = 0.36f_{ck} * b * x_u + 0.447f_{ck} * \left(x_u - \frac{(D-b)}{2}\right) * (D - b)$$

$$P_{us} = \frac{\rho}{f_{ck}} * \rho_i * (f_{si} - f_{ci})$$

$$M_{uc} = 0.36f_{ck} * b * x_u * (y' - 0.416x_u) + 0.447f_{ck} * \left(x_u - \frac{(D-b)}{2}\right) * (D - b) * \left(y' * \left(x_u - \frac{(D-b)}{2}\right)/2\right)$$

$$M_{us} = P_{us} * y_i$$

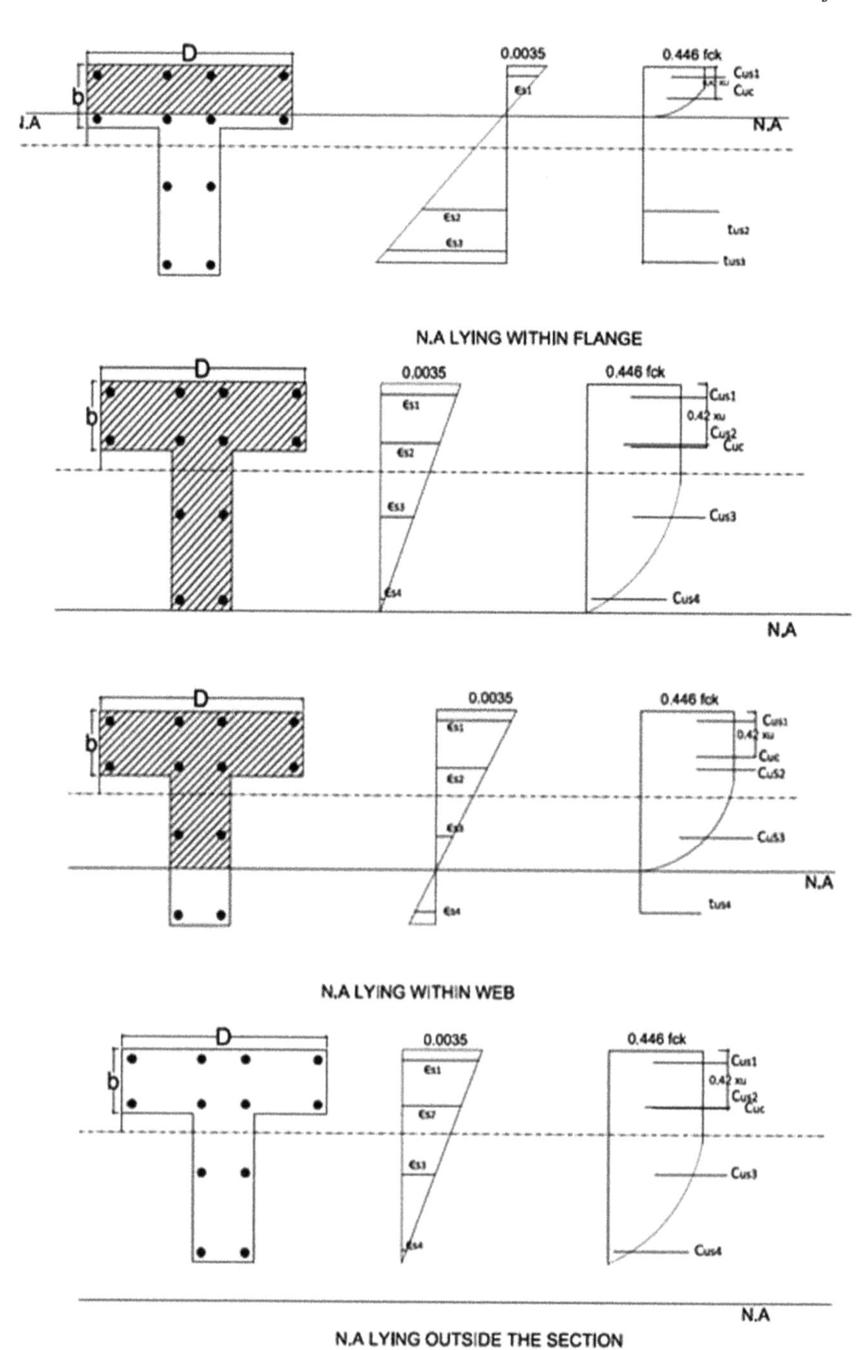

Fig. 3 Analysis of design strength for T section

In bottom web portion

$$P_{uc} = 0.36f_{ck} * b * x_u + 0.447f_{ck} * b * (D - b)$$

$$P_{us} = \frac{\rho}{f_{ck}} * \rho_i * (f_{si} - f_{ci})$$

$$M_{uc} = 0.36f_{ck} * b * x_u * (y' - 0.416x_u)$$

$$M_{us} = P_{us} * y_i$$

See Fig. 4.

3 Design Methodology

3.1 Steps for Designing Column

The design of specially shaped column involves analyses of numerous design parameters governing design equations, and interaction ratio formulae. For the designing of reinforced concrete columns the Indian code (IS 456-2000) is adopted. It suggests the use of interaction curves for design of RC columns subjected to uniaxial or biaxial moments. The use of design charts is explained below in four steps while designing columns.

Step 1: Verification of Eccentricity

The foremost step in design of short column is to check for minimum eccentricity along both major and minor axis. The clause 25 of IS 456-2000 suggests the equations for minimum eccentricity due to axial loads. In case of column subjected to uniaxial or biaxial bending moment, the eccentricity occurring due to bending moments should also be evaluated.

Step 2: Selection of the Interaction Diagram

After the assessment of minimum eccentricity, the designer has to assume some trial section for designing considering the known load and bending moments. Based on the value of *b*/*D* and considering the uniform distribution of longitudinal steel bars in the column the interaction diagram is selected.

Step 3: Selection of the Particular Curve

For the above selected interaction diagram there will be many curve for different ρ/f_{ck} value. Hence, the design parameters,

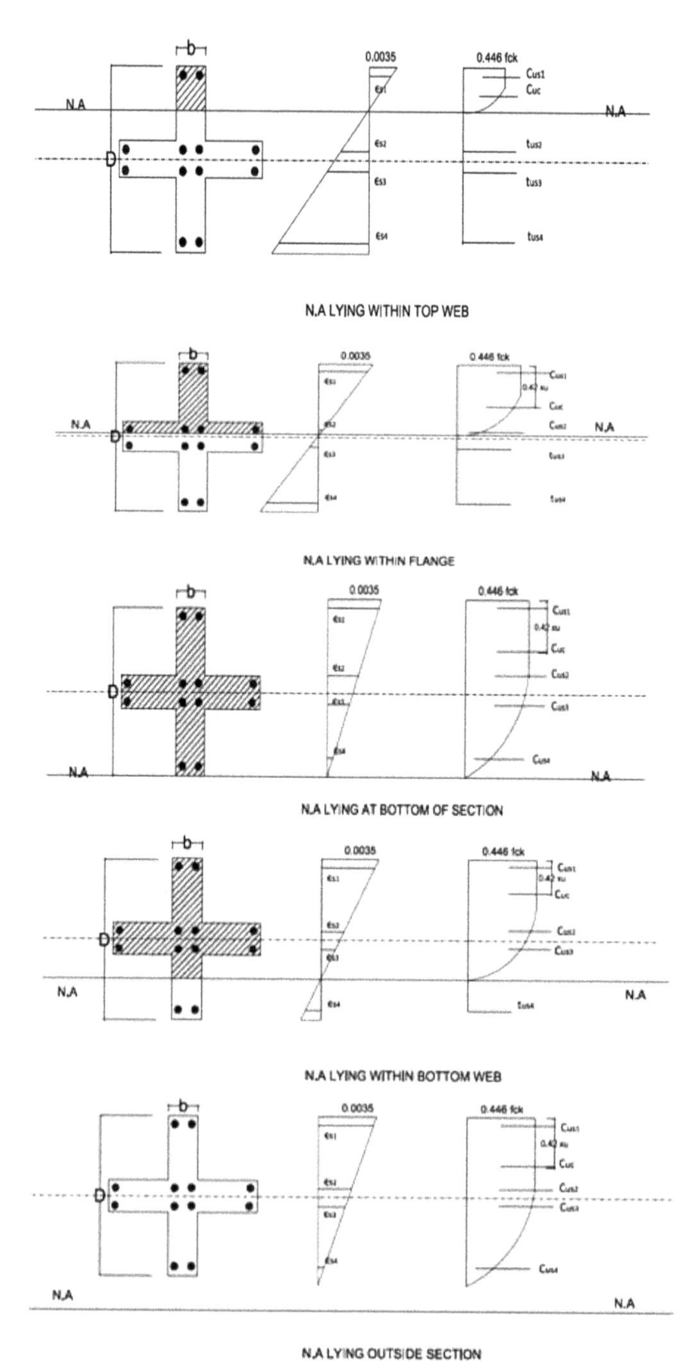

Fig. 4 Analysis of design strength for + section

$$\frac{P_u}{f_{ck}A_g}$$

$$\text{and } \frac{M_u}{f_{ck}A_gD}$$

are assessed and the curve corresponding to design parameters is further used for designing. If the point lies on the standard curve, the corresponding ρ/f_{ck} value can be directly obtained. Otherwise, liner interpolation is to be done.

Step 4: Determination of the Percentage of Longitudinal Steel

From the above selected curve, the percentage of longitudinal reinforcement required will be calculated. The percentage of steel obtained will be used for designing the reinforcement.

The design spreadsheet is prepared for design of specially shaped columns by incorporating the design methodology of IS 456-2000 as discussed above. Table A of SP 16: 1980 is also incorporated in MS Excel program for determining stress from strain of cold-worked bars.

4 Design Curves

A family of non-dimensional design interaction curves is thus generated for a given arrangement of bars. In practice, *b*/*D* usually varies in the range of 0.2–0.75, and $\rho f_y/f_{ck}$ in the range of 10–130. The design curves developed are added in the Appendix 1.

4.1 Design Charts

The L shaped are invariably subjected to biaxial bending and in this paper our primary focus is on member subjected to uniaxial bending. The design charts (non-dimensional interaction curves) are obtained for the following three cases of symmetrically arranged reinforcement:

i. Plus (+) sections with uniaxial eccentric compression
ii. T sections with uniaxial bending over x-x-axis
iii. T sections with uniaxial bending over y-y-axis

In each case, the bars are arranged at the corners and then if required, at positions according to the maximum allowable spacing criteria. Corresponding to each of the above three cases, there are as many as six charts with three values of *b*/*D* i.e. 0.2, 0.4, 0.6.

For intermediate values of b/D, linear interpolation may be done. Each of the charts covers a family of non-dimensional design interaction curves with $\rho f_y/f_{ck}$ values ranging from 10 to 130. It may be noted that there are other types of symmetrical reinforcement arrangements possible for these sections, which are not covered in this study. In such cases, the designer may make judicious approximations.

The non-dimensional interaction curves can be used to handle all types of design and analysis problems. In the design problem, the desired value of percentage reinforcement ρ can be easily obtained from the family of interaction curves for a given loading conditions. This is demonstrated in example in the following chapter. The generated design charts are shown below. The arrangement of steel is indicated in the inset shown at the top or bottom of each chart.

4.2 *Validation of the Design Curves Using CSICOL 8.4*

CSICOL is a comprehensive software package for analysing and designing concrete, reinforced concrete, and reinforced concrete composite columns. The program provides several predefined parametric shapes, including a variety of solid, hollow, and flanged shapes, in addition to a large collection of Standard Steel Database Shapes that can be used in composite columns. The results are compared with the mathematical results.

4.3 *Sample Problem*

An example from the journal of Ramamurthy is considered below to demonstrate the method of design using the interaction curves. The same problem is then redesigned using the software and the results are compared.

Design a T-shaped column to carry an axial load of 1200 kN and moment of 120 kN-m about the X-axis. Cube strength of concrete = 4000 psi (27.6 N/mm^2); while yield stress of steel = 40,000 psi (276 N/mm^2).

4.3.1 Design Using Interaction Curves

$$\mathrm{b} = 200\,\mathrm{mm},\ \mathrm{D} = 500\,\mathrm{mm}$$

$$P_u = 1200\,\mathrm{kN},\ M_{ux} = 120\ \mathrm{kN.m}$$

$$f_{ck} = 30\,\mathrm{N/mm^2},$$

$$f_y = 250\,\text{N/mm}^2$$

$$A_g = 160,000\,\text{mm}^2$$

$$b/D = 0.4$$

$$\frac{P_u}{f_{ck}A_g} = 0.25$$

$$\frac{M_{ux}}{f_{ck}A_gD} = 0.05$$

By interpolation from the developed chart number 9

$$\frac{\rho f_y}{f_{ck}} = 10$$

$$\rho = 1.20\%$$

Using CSiCoL the longitudinal reinforcement obtained is $\rho = 1.25\%$

From the comparison of both the results, it can be seen that the results are satisfactorily matching. A number of such columns of all the cross sections (+ and T) were similarly designed using both methods. The results are given in Table 2.

From the observed agreement between the theoretical values and the test results, it is concluded that the design curves are reliable and can be safely used for the analysis and design of special shaped columns.

5 Summary

In present work,

1. The analysis and design of reinforced concrete special shaped cross sections using conventional method is tedious and time consuming.
2. It is observed that for same cross-sectional area of rectangular column and SSC, the increased moment inertia can be obtained. This will result in increased stability or resistance of a structure against the externally applied moment. Additionally, for some designated loading condition the reduced dimension of specially shaped columns is obtained leading to economical design of column.
3. A comprehensive set of 18 charts were proposed for columns with T (Tee), and + (Plus) sections subjected to uniaxial eccentric compression. The design charts can be conveniently used for designing of special shaped columns.

Table 2 Comparison of results from design charts and software for columns under uniaxial bending

Type	Loads			Design charts													CSiCOL	% deviation
				Assumed dimensions						$\frac{P_u}{f_{ck}A_g}$	$\frac{M_{ux}}{f_{ck}A_gD}$	$\frac{M_{uy}}{f_{ck}A_gD}$	Chart No	A_{st}				
	P	M_{xx}	M_{yy}	b	D	b/D	Ag	f_{ck}	f_y					$\rho f_y/f_{ck}$	ρ		ρ	
+	1000	120	0	200	500	0.4	160,000	30	250	0.2	0.05	0.00	3	10	1.20		1.30	8
	800	0	200	250	500	0.5	187,500	35	415	0.1	0.00	0.06	3 and 4	17.5	1.48		1.55	5
	1200	0	1000	350	1500	0.2	927,500	30	415	0.0	0.00	0.02	2	15	1.08		1.21	10
	1500	700	0	300	900	0.3	450,000	30	250	0.1	0.06	0.00	1 and 3	18	2.16		2.17	1
	2000	1000	0	450	800	0.6	517,500	40	500	0.1	0.06	0.00	6	18	1.44		1.56	8
T	1200	120	0	200	500	0.4	160,000	30	250	0.3	0.05	0.00	9	10	1.20		1.25	4
	1000	500	0	200	1000	0.2	360,000	35	415	0.1	0.04	0.00	7	15	1.27		1.30	3
	1200	0	1000	350	1500	0.2	927,500	30	415	0.0	0.00	0.02	2	15	1.08		1.21	9
	2000	3000	0	450	1500	0.3	1,147,500	30	500	0.1	0.06	0.00	8 and 10	20	1.20		1.30	8

4. The results obtained using design charts and the software, CSiCoL, shows good degree of accuracy; an average ratio of 1.1 with a deviation of 6–7% has been achieved.

Even though, using charts as design aids for columns is limited to very simple geometries when only a few loading cases are to be handled, these results prove that the design charts are a good design tool when use of computer program is not available.

Appendix 1

Interaction Curves for + Shaped Columns Subjected to Compression with Uniaxial Bending

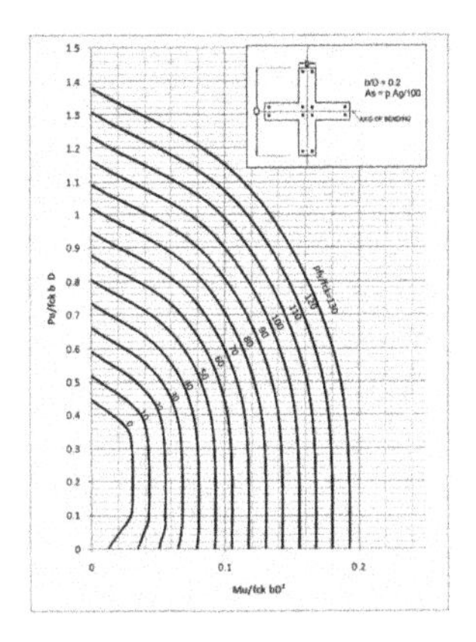

Chart 1 Design curves for + section ($0.2 \leq b/D < 0.4$) (with bars at corners)

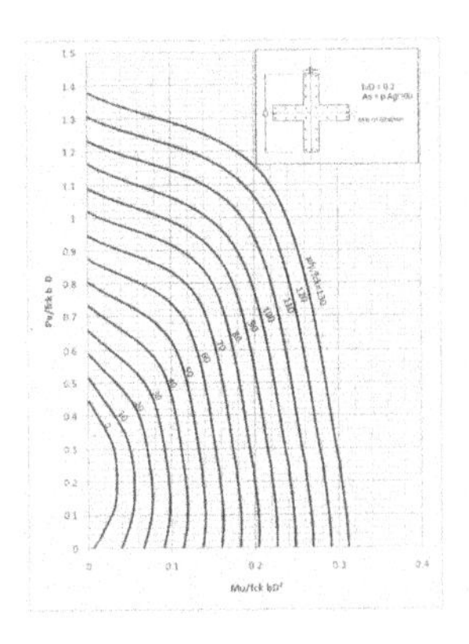

Chart 2 Design curves for + section ($0.2 \leq b/D < 0.4$) (with bars along edges)

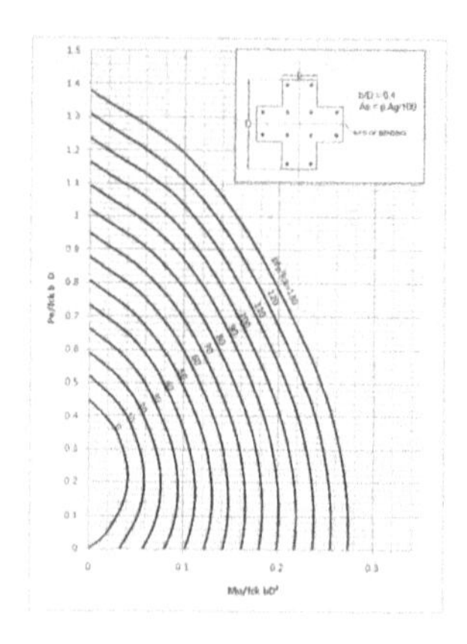

Chart 3 Design curves for + section (0.4 $\leq$ *b/D* < 0.6) (with bars at corners)

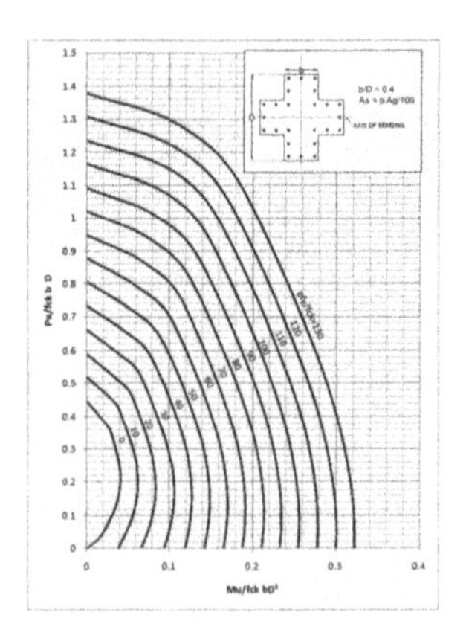

Chart 4 Design curves for + section (0.4 $\leq$ *b/D* < 0.6) (with bars along edges)

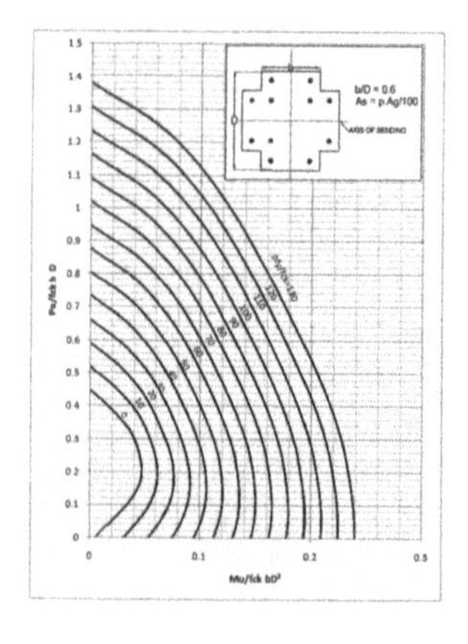

Chart 5 Design curves for + section (0.6 $\leq$ *b/D* < 0.75) (with bars at corners)

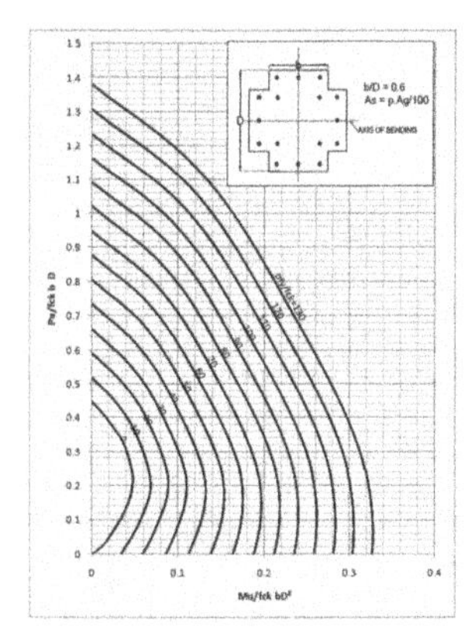

Chart 6 Design curves for + section (0.6 $\leq$ *b/D* < 0.75) (with bars along edges)

Interaction Curves for T Shaped Columns Subjected to Compression with Uniaxial Bending—Moment Along X-Axis

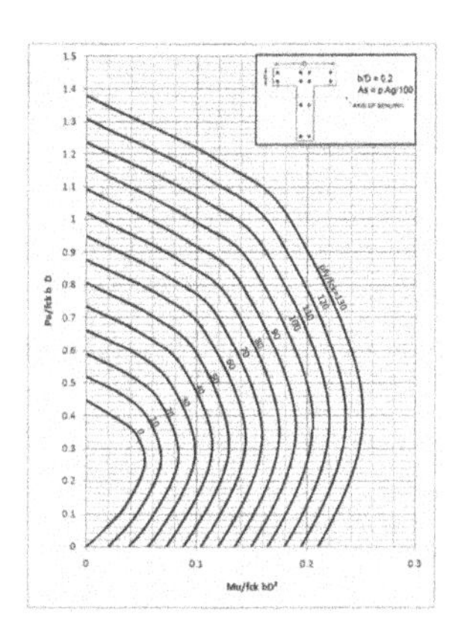

Chart 7 Design curves for T section (0.2 $\leq$ *b/D* < 0.4) (with bars at corners)

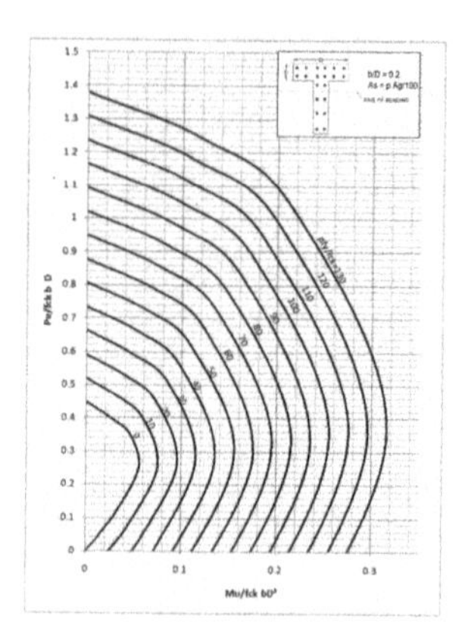

Chart 8 Design curves for T section (0.2 $\leq$ *b/D* < 0.4) (with bars along edges)

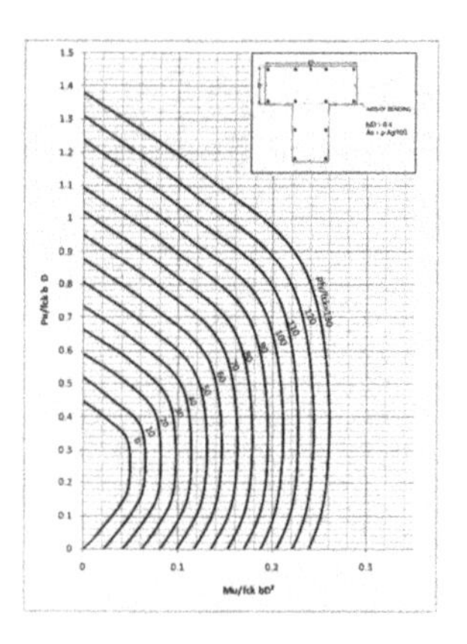

Chart 9 Design curves for T section (0.4 $\leq$ *b/D* < 0.6) (with bars at corners)

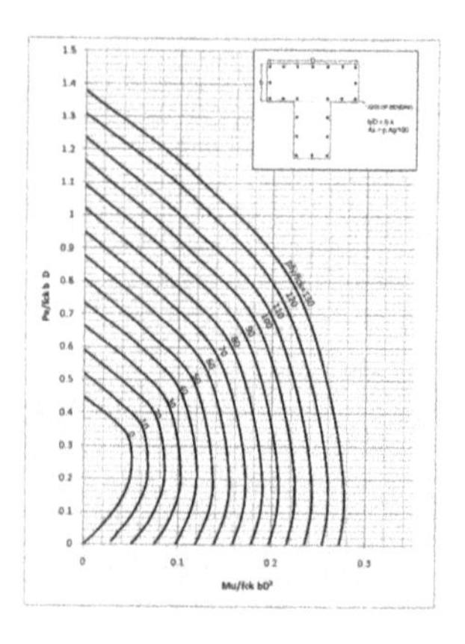

Chart 10 Design curves for T section (0.4 $\leq$ *b/D* < 0.6) (with bars along edges)

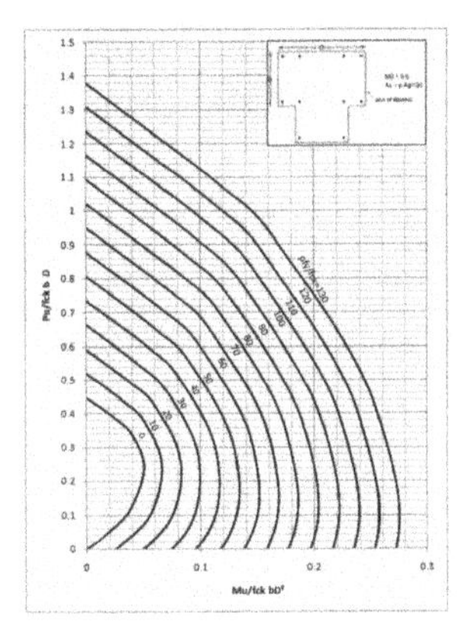

Chart 11 Design curves for T section (0.6 $\leq$ *b/D* < 0.75) (with bars at corners)

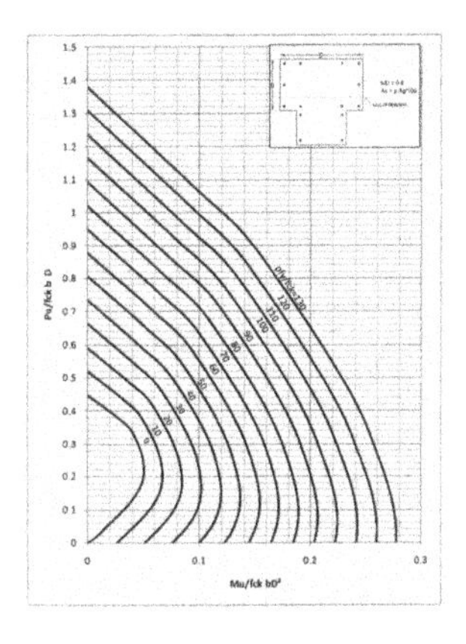

Chart 12 Design curves for T section (0.6 $\leq$ *b/D* < 0.75) (with bars along edges)

Interaction Curves for T Shaped Columns Subjected to Compression with Uniaxial Bending—Moment Along Y-Axis

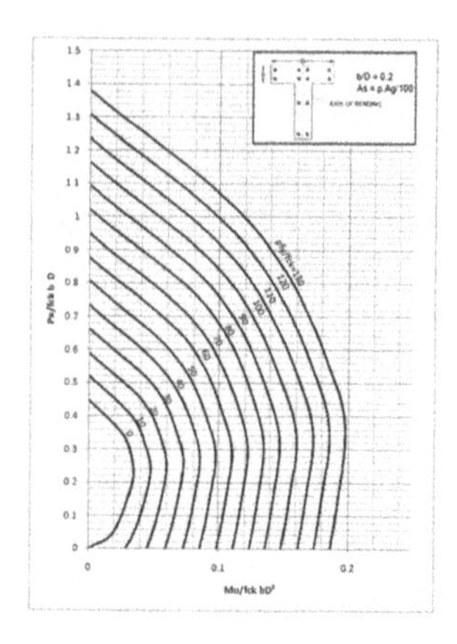

Chart 13 Design curves for T section ($0.2 \leq b/D < 0.4$) (with bars at corners)

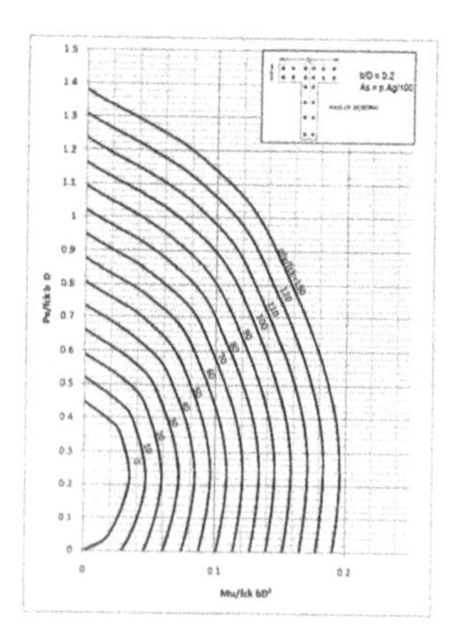

Chart 14 Design curves for T section ($0.2 \leq b/D < 0.4$) (with bars along edges)

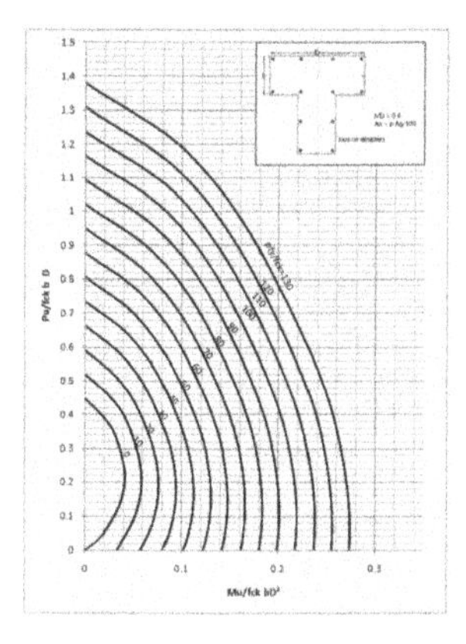

Chart 15 Design curves for T section (0.4 $\leq$ b/D < 0.6) (with bars at corners)

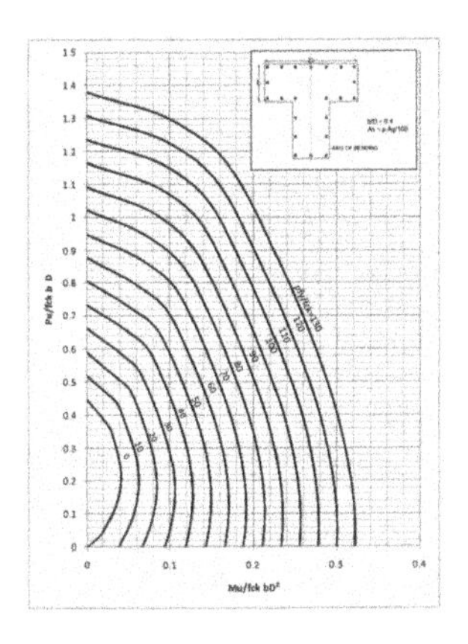

Chart 16 Design curves for T section (0.4 $\leq$ b/D < 0.6) (with bars along edges)

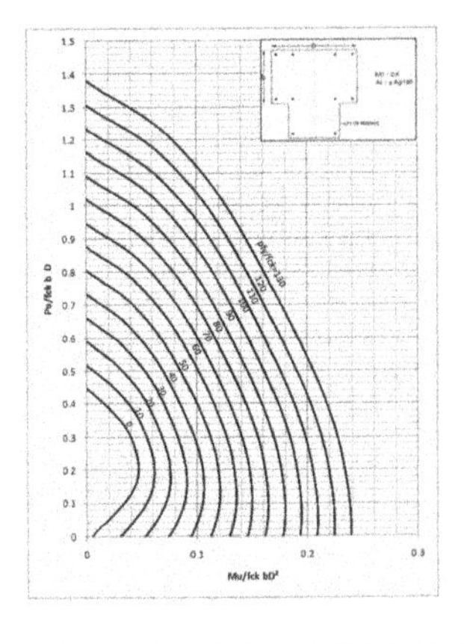

Chart 17 Design curves for T section (0.6 $\leq$ b/D < 0.75) (with bars at corners)

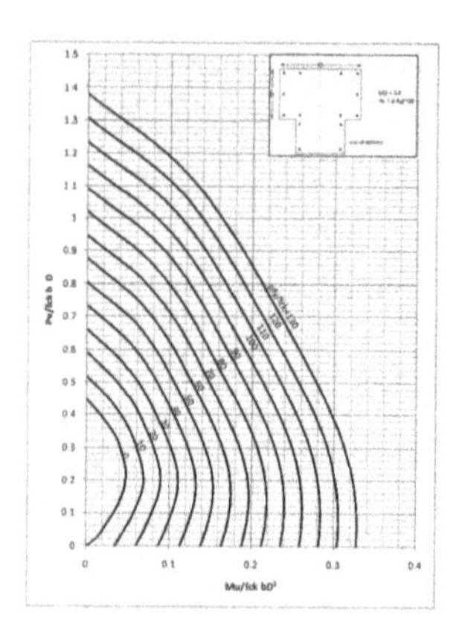

Chart 18 Design curves for T section ($0.6 \leq b/D < 0.75$) (with bars along edges)

References

1. Liu J, Li Z, Feng J (2002) Study on shaking table test of a 9 storey RC frame model with specially shaped columns and transfer storey. J Build Struct
2. Yang P, Liu H, Huang Z (2008) A comparison of seismic behaviour between specially shaped column frame structure and rectangular column frame structures. In: The 14th world conference on earthquake engineering, 12–17 Oct, Beijing, China
3. Yu-Ye Xu, Bo Wu (2009) Fire resistance of reinforced concrete columns with L-, T-, +-shaped cross-sections. Fire Saf J 44(6):869–880
4. Li S, Diao Bo, Youpo Su (2009) Seismic behaviour experimental study of frame joints with special-shaped column and dispersed steel bar beam. Front Arch Civil Eng China 3(4):378–383
5. Wang T, Liu X, Zhao H (2014) Experimental study of the seismic performance of L-shaped columns with 500MPa steel bars. Sci World J 105826
6. Zhou T, Minyang Xu, Wang X, Chen Z, Qin Y (2015) Experimental study and parameter analysis of L-shaped composite column under axial loading. Int J Steel Struct 15(4):797–807
7. Zheng Y, Zeng S (2020) Design of L-shaped and T-shaped concrete-filled steel tubular stub columns under axial compression. Eng Struct 207:110262

Disruption of Construction Industry During COVID-19 Pandemic—A Case Study from Ernakulam, Kerala, India

Niya Thomas and S. Jawahar Saud

Abstract The wide spread of COVID-19 has generated unforeseen obstacles as well as unpredictable financial consequences. The construction industry was far different from other industries which require an onsite involvement of all the project members. It was important to know how the building sector would deal with this unexpected condition. The building sector makes a significant contribution to Gross domestic product, the slowdown triggered a global recession and economic downturn. The COVID-19 created huge impact on construction sectors causing most of them to stop their ongoing projects, but the work from home switch made the design engineers to strive through it. The aim of this research was to evaluate the effect of COVID-19 on the building industry at Ernakulam district of Kerala. For this, a Delphi study involving all main stakeholders within the construction industry involving project managers, design engineers, site execution engineers, planning engineers, field workers, and supervisors were conducted. The statistical study aims to understand how the COVID-19 pandemic affected construction industries and civil engineers also to study the effect of Work from Home system. The impact was calculated using the Relative importance index, and the study allowed for the identification of relevant conditions based on expert responses. The five point Likert scale range from 1–5 was adopted and transformed to relative importance indices.

Keywords Covid-19 · Construction industry · Delphi study · Relative importance index

1 Introduction

The World has been undergoing a rare disaster of Coronavirus pandemic (COVID-19) since December 2019. Most of the countries in the World have been instigating quarantines and social distancing practices to contain the pandemic and

N. Thomas · S. J. Saud (✉)
Department of Civil Engineering, FISAT, Angamaly, Kerala, India

G. C. Marano et al. (eds.), *Proceedings of SECON'21*, Lecture Notes in Civil Engineering 171, https://doi.org/10.1007/978-3-030-80312-4_13

implemented lockdown. The COVID-19 pandemic found as the biggest fitness disaster in decades. With exception of death and health issues the pandemic has ended in fiscal slowdowns, considerable enterprise disruptions, and significant sufferings [1–3]. The construction sector was one of the sectors that had to bear the worst brunt of COVID-19, as it was already battling a liquidity crunch as fallout of the non-banking financial institutions in the financial sector for more than a year. The Indian construction industry was a key indicator of progress because it generates investment opportunities in a variety of related industries. The construction industry here plays a pivotal role in making or breaking that economy's substructure largely contributing to maintain that trade link within the national boundaries as well as outside of it. Latterly Past few months, since the outbreak of coronavirus, when no industry vertical has remained untouched from the havoc it created; the construction sector is bound to face the upshots of the pandemic. Construction activities in the state are largely dependent on the availability of the guest labour force and since the lockdown, the majority of the labourers are either confined to their camps or continue to remain blocked in their states. Several workers who were forcefully trying to migrate to their hometown were stopped by the authorities and quarantined at the state. Most notably, a big quantity of construction workers have been confirmed to have tested positive for COVID-19. Many projects remained in the unfinished stages because of the lack of funds. Those were finished, remained unsold, because of changing buyer preferences [2]. In the construction industry, all the workers and technical engineers need to nearly work on-site either to perform activities or to monitor the work. The construction industry is far different from other industries which typically requires on-site involvement of all the project members. Hence, it is crucial to appreciate how the construction industry addresses this unforeseen situation. Following the open out of virus, many countries began to take steps to restrict people's movement, which has hampered construction because it necessitates on-site work, and every team participant must be available to work, search, and track all work activities [3].

Due to pandemic conditions, delays were the main issue in construction projects that affected all the parties in the construction sector. It lead to many negative effects such as lawsuits between owners and contractors, increased costs, loss of productivity and revenue, and contract termination [4]. So it was important to study the effects of the pandemic on the construction industry of Kerala. This study intended to explore the effect of this pandemic on civil engineers, and their perception of long-term economic and legal consequences on their sector and to identify the other major causes of delays on construction projects in Ernakulam during the COVID-19 Pandemic. Impact and delay were measured based on Relative importance index. To determine the impacts of Covid on workers, civil engineers, the Delphi process was chosen as the research tool to conduct the investigation.

1.1 Study Area

The study, in vast engrossed on analyzing the impact of the COVID-19 in the construction Industry of Ernakulam. Ernakulam is located in the state's central region, covers over 3000 square kilometres (1200 square miles), and is home to about 9% of Kerala's population and it is located at 9.98° N 76.28° E. The headquarters are located at Kakkanad. The district includes Kochi, also known as the IT and commercial capital of Kerala. This district Ernakulam is the main revenue resource and consisting most of the industries. It is the third famous district in Kerala which procure 60% of annual revenue of the state. Kerala has 3.5 to 4 million inter-state migrant workers [5] and most of them depend on the construction industries in Ernakulam district.

In India, Kerala was the first state affected by COVID-19, and the first coronavirus case was confirmed in Thrissur district on 30 January 2020. Due to a large number of immigrants from other countries and unrestricted contacts the state soon had the highest number of active cases in India by early March. As of 17 November 2020, there have been 533,500 confirmed cases with 461,394 (86.48%) recoveries and 1915 deaths in the state. The largest single-day spike (11,755 cases) was reported on 10 October and Kerala now has the 6th highest number of confirmed cases in India. As of October 2020, more than 90% of known cases were due to community spread. Ernakulam (10.9%) is one of the worst-hit districts following Thiruvananthapuram (12.6%), Malappuram (11.7%), and Kozhikode (11.3%) [6]. The high numbers of migrant labourers coming to Ernakulam every year the construction sector is still reeling under labour shortage due to the current pandemic situations. Due to the presence of a high number of migrant workers they are facing several safety issues, most of the migrant workers returning to the home town will make a huge impact on the construction sector in Ernakulam [7]. Therefore respondents are selected from the Ernakulam, those who are working in different construction companies.

2 Methodology

2.1 Introduction to Research Methodology

COVID 19 has impacted every field of life and results in the disruption of all major businesses around the globe. The study, in particular, was focused on analyzing the impact of the COVID-19 on the construction Industry of Ernakulam. The study was not only focused on the economic issues, but also on the major problems faced by the labour and construction-based companies in Ernakulam, for identifying the impacts. This was accomplished through a desk study review of limited (due to the pandemic's recent emergence) literature and a field study gathering primary data from construction professionals in architecture, building engineering, civil/

structural engineering, construction/project management, workers, and quantity surveying. The introductory part of the methodology focused on the literature to define and identify the scope, problem, and objectives of the study. The second part involved survey designs based on literature review then the third section was survey distribution. To achieve the objectives, a two-round Delphi method was conducted for collecting the needed information.

2.2 Preparation of Questionnaire

Questions were generated in such a way to collect the maximum data considering Respondents knowledge, experiences, and thoughts on the effect of the global pandemic in the construction industry. Two sets of questionnaires were prepared for the two rounds of the Delphi process through literature study. Five-point Likert scale questions were used in the survey. In the first round questionnaire survey, Likert scale 5, 4, 3, 2 and 1 represented response of very frequently, frequently, occasionally, rarely and never respectively. In the second round questionnaire survey, Likert scale 5, 4, 3, 2 and 1 represented response of very high impact, high impact, moderate impact, low impact and very low impact. The questionnaire was divided into five sections. The first section was the demographics section, which collected the age, gender, field of work, and the effect of COVID-19 on office working civil engineers, the effect of COVID-19 on field working civil Engineers and fourth section inquired about their perception of long term economic and the legal implication of COVID-19 pandemic on civil engineering and the construction industry and the last section was a questionnaire related to the delay.

2.3 Data Collection

The data collections were carried out by the administration of questionnaire survey. Questionnaire surveys were done by telephonic interviews, face-to-face interviews, and also Google forms. Personnel with direct relation with the construction industry like site/design engineers, architects, contractors, project managers, material distributor, and workers were selected as the target population.

2.4 Delphi Method

The Delphi Process includes sending questions to a panel of experts with the aim of finding answers to particular questions. It all begins with the Delphi panel experts' recognition, placement, and confirmation [8, 9]. The approach then moves on to creating the first questionnaire, disseminating it, and evaluating the results. The

Delphi Method comes to an end when consensus are achieved and the conclusions are issued. If no consensus is made, a new round must be completed, and a new questionnaire must be produced, it was used in variety to achieve expert consensus by intense questionnaire rounds [10, 11]. The Delphi methodology has been used in a variety of research fields for decades, including strategic planning, health, and social science fields [12]. However, its use in the building industry has only recently been considered. The method usually entails the appointment of appropriate experts, the development of an appropriate questionnaire, and the review of the answers in multiple rounds [13]. The Delphi method was used in this analysis, and then a questionnaire was developed. For this, 44 Ernakulam-based construction industry professionals, including employees, were chosen as panelists. The questionnaire was launched in 18th of January 2021 to 20th of March 2021. A five-point Likert scale ranging from 1 to 5 was used to measure Relative Importance Index. For data recording and analysis, Microsoft Excel 2016 (Microsoft Corp., Redmond, WA, USA) was used.

2.5 *Relative Importance Index (RII)*

The Relative Importance Index (RII) rating system is used to calculate the relative importance of the various delay causes. Since RII best suited well for the function of this study, it was used for this research. It would classify the most important delay factors in the construction industry based on the rankings given to each cause of delays. The RII has been used to test comparative effectiveness in a variety of fields and to assess the value of one object to another. RII used to identify the important variable out of many. The relative importance index for all of the causes was calculated using the equation below [14].

$$\mathrm{RII} = \frac{\sum \mathrm{w}}{\mathrm{A} \times \mathrm{N}} \tag{1}$$

where, W = weight given to each factor (ranging from 1 to 5) by the participants, A = the highest weight (i.e. 5 in this research study), and N = the total number of participants. The higher the RII value means that it is more important than the others.

3 Results

3.1 *Survey Responses*

The survey responses obtained through two rounds of the Delphi method; 44 participants completed the entire two rounds of the Delphi study. The distribution of respondents based on their profession is shown in Fig. 1. And the demographics of the survey personals are as in Table 1. Out of the 44 personals, 47.3% (21) of them worked in the office while 52.7% (23) were working in the field job. Main objective of this study was to determine the impact of COVID-19 it had construction sectors therefore profession of the respondents are in the field of construction sector as shown in Fig. 1.

3.2 *The Impact of the COVID-19 Pandemic on Office-Based Civil Engineers*

From the investigation, the impact of COVID-19 on office working civil engineers are inconsiderable because they adopted work from home model and the productivity was found to be increasing shown as in Table 2.

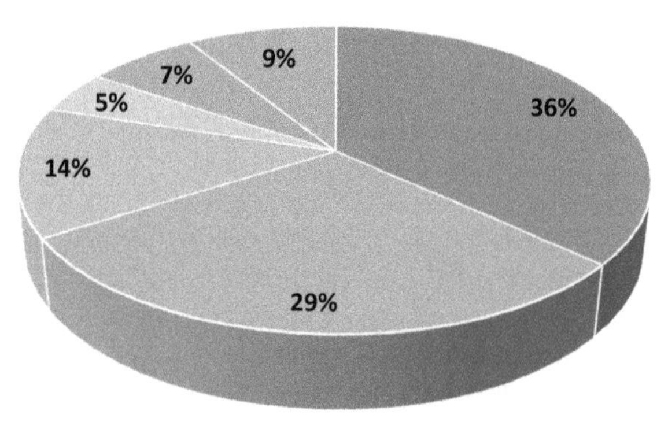

Fig. 1 Respondent's profession

Table 1 Demographic data of survey responders

Characteristic		Value
Age		23–50
Gender	Male	63.6%
	Female	36.4%
Work field	Office job	47.3%
	Field job	52.27%

Table 2 The impact of the COVID-19 pandemic on office-based civil engineers

Sl. No	Features	RII (%)
1	During the lock-down, their company decided to work from home	62.32
2	Their productivity during the pandemic increased	62.79
3	working from home would be a good replacement for office work	65.11

3.3 *The Impact of the COVID-19 Pandemic on Field-Based Civil Engineers and Workers*

After examining the survey results, the impact of COVID-19 pandemic may in huge amount because applicability of social distancing and provision for personal protective equipment were understudied. In order to prevent labour shortages in coming months, proper safety measures must be added at work site like proper usage of gloves, provision of additional hand washing stations, staggered working conditions and also adopt work plan for reducing access to site and to avoiding close contacts.one of the main issue that impact the construction industry were the unavailability of migrant workers shown as in Table 3.

3.4 *The Impact of the COVID-19 Pandemic on Construction Industries*

One of the important impacts of COVID-19 pandemic on construction industries were, they were not aware of the economic assistance available for the construction projects followed by the financial crisis due to lockdown and this may lead to disruption of construction works and most of the construction companies faces legal consequences due to delay of construction works followed by the lockdown as shown in Table 4.

Table 3 The impact of the COVID-19 pandemic on field-based civil engineers and workers

Sl. No	Characteristic	RII %
1	PPE for COVID-19 infection prevention will be completely accessible at their worksite	29
2	Applicability to thoroughly implicate social distancing at their work place	46.36
3	concerned about safety of fieldwork during the pandemic	50.90
4	Migrant workers are replaced with local workers	37.73
5	Migrant workers returns to their home town due to COVID-19 Pandemic	72.72
6	Local employees are well-trained and experienced enough to fill in for migrant workers	26.82

Table 4 The Impact of the COVID-19 pandemic on construction industries

Sl.No	Features	RII %
1	Legal consequences for delay in ongoing and planned projects	57.67
2	Unpredicted incidents were listed as an appropriate cause of delays in their company's contracts as a section	53.48
3	The lockdown would have a detrimental financial impact on construction industry	60
4	The firm was aware of the economic assistance available for construction projects that had been devastated by the COVID-19 pandemic	47
5	As a result of the COVID-19 pandemic, some engineers might lose their jobs	59

3.5 Outcome of First Round Delphi Analysis

As a result of the pandemic and to improve safety, civil engineers who work in offices Shift the working model from an office to a home-based one. According to Respondents there were no huge impact on office working civil engineers because 65.11% of respondents believe that working at home will substitute for office work and the productivity found to be increasing. But in case of field working civil engineers, due to onsite construction there may be some safety issues because applicability of social distancing in their workplace was not carried in better way. 59% of respondents mentioned that some of the engineers lost their job as the consequence of COVID-19 pandemic, significant portion of the migrant workers were worried about their family and their health therefore they returned to their home town during COVID-19 Pandemic, this lead to disruption of construction works. Therefore second round of Delphi carried to determine the central reason for delay and proper category of delay then detailed delay analysis were carried out.

3.6 The Principal Reason of Disruption in Construction Industry

The survey's findings revealed that the majority of companies in Ernakulum were experiencing delays in their building projects. Employees were unable to function at any time due to COVID-19 pandemic, which was the frequent sources of project delays as shown in Fig. 2.

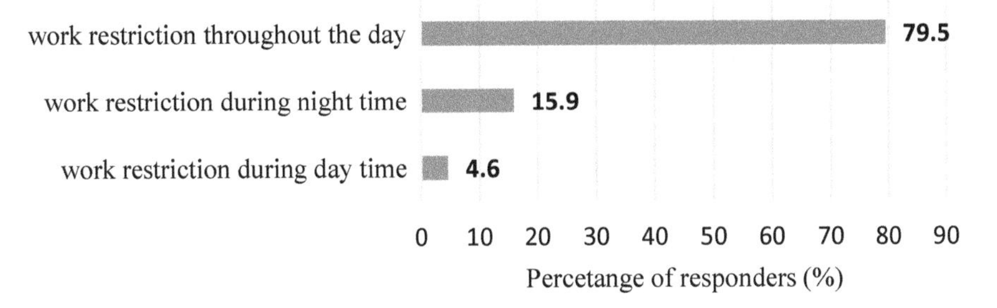

Fig. 2 The principal reason of disruption in construction industry

3.7 The Proper Category of Delay in Construction Projects

Non-excusable delay, serial delay, independent delay, critical delay, excusable delay, and concurrent delay are the six types of delays described in building projects [15, 16]. A critical delay is a postponement of the project's completion date, when a contractor fails to finish the job within the agreed time [17, 18]. Better part of the respondents mentioned that critical delay and excusable delays are the main reason for delays in construction projects. The rest of the delays are comparatively in small amount as shown in Fig. 3.

3.8 Delay Analysis

In construction, the delay could be defined as the time overrun either beyond the completion date specified in a contract or beyond the date that the parties agreed upon for delivery of a project. It was a typical issue in building projects where a work deviates from the original timeline [18, 19]. The findings revealed that COVID-19 pandemic had caused delays in building works in Ernakulam. Then the detailed delay analysis were carried out and is represented as in Table 5.

Fig. 3 The proper category of delay in construction project

Table 5 Delay analysis

Sl. No	Features	RII%	Rank
1	Non-payment to contractors	75	7
2	Lag in approval of finished works	69	12
3	Delay from consultants in providing instructions	65	13
4	Delay in delivery	82.7	1
5	Delay from subcontractors	73.6	9
6	Delay from the main contractor	73.6	9
7	Delay in revising and approving documents	75	7
8	Unsafe working conditions	74.54	3
9	Imperfect site monitoring and administration	72.27	10
10	Poor scheduling and planning of project	86.8	2
11	Poor communication with other parties	81.8	15
12	Low productivity of workers	69.54	11
13	Lack of construction materials	80.45	5
14	Lack of skilled labour	80.45	5
15	Non availability of construction equipment	65	16
16	Inflation of material prices and operational cost	77.72	6

4 Discussions

The COVID-19 have made huge impact on construction sectors causing most of them to stop their ongoing projects, but the work from home switch made the design engineers to strive through it [20]. From the questionnaire survey it was clear that unavailability of migrant workers, material shortages and escalations of prices were the main issues faced by the construction sectors followed by the lockdown, All these made negative impact on productivity of the construction sector Delay was described as a condition in which the contractor, consultant, and customer all contributed to the project not being completed during the initial or stipulated or negotiated contract time [17]. From the questionnaire survey results, 79.5% of participants mentioned that disruption of work was due to the result of workers' inability to function at any given time. 57.67% of respondents mentioned that most of the construction companies faces legal consequences due to delay of construction works followed by the lockdown. Better part of the respondents mentioned that critical delay and excusable delays are important reason for delays in construction projects.Most of the respondents mentioned that one of the important impacts of COVID-19 pandemic on construction industries are they are, not aware of the economic assistance available for the construction projects followed by the financial crisis due to lockdown.According to the findings of a detailed delay evaluation, unhealthy working condition and delay in delivery are some of the main reasons for building project delays which were ranked with one and three. Unavailability of material, delay in delivery, labour shortages that affected the planning and scheduling of the project and this lead to disruption of construction works and also 77.72% of respondents mentioned that inflation of material and operational cost followed by the lockdown also impacted the building construction.

5 Conclusions

This was the first research to look at COVID-19's effect on civil engineering and the building industries in Ernakulam. The constructing enterprises made good contribution to the economy. Another consequence of the lockdown was a labor shortage as migrant workers returned home, causing construction projects to be postponed and associated prices to be increased. The pandemic had an impact on the majority of engineer's work. The financial consequences of conducting the shutdown were shocking, despite the fact that the shutdown was necessary for managing the propagation of COVID-19, limiting the lot of critical cases and, ultimately decreasing the number of cases that result in death. This highlights the significance of implementing collaborative and cohesive strategy for post lockdown period by considering all sectors including building sector. The acceptability of social distancing, enhancing sanitation facilities and providing personal protective equipment (PPE) for workers and engineers at some stage during the pandemic were questionable, because 29% of respondents mentioned that workers, civil engineers are unaware of personal protective equipment, social distancing and safety of field work during pandemic. One of the main issues that impact the construction industry were the unavailability of migrant workers because 72.72% of respondents mentioned that Migrant workers returned to their home town due to COVID-19 Pandemic this effected the productivity of the both engineers, workers and construction industry because local workers are not available as substitute for migrant workers. According to 26.82% of respondents, local workers are not well trained to replacing migrant workers which caused delay in building projects. More than half of the respondent mentioned that construction companies undergo legal consequences followed by financial crisis due to lockdown. After completing second round Delphi questionnaire, respondents reached to a common reason that critical and excusable delays were the main reason for disruption of construction projects because there would be a work restriction during the lockdown period.

Results of detailed delay analysis indicates that unsafe working conditions and delay in delivery were the main reasons for the delay of construction because it was ranked with one and three respectively. Unavailability of material, delay in delivery, labour shortages that affect planning and scheduling this lead to disruption in construction. 77.72% of respondents mentioned that inflation of material and operational cost followed by the lockdown also impacted the building construction. This research reflects not just the importance of scheduling and planning from the begin, along with this emphasis the necessity of proper communication between the contractors, distributers and stakeholders within the construction industry throughout the project and the need to keep people aware of major issues when they occur, such as material shortages and changes affecting the construction. It is in everyone's best interests to minimize disruptions as far as possible. The owner faces the financial repercussions of the delay and the likely effects of the lack of service, but the provider will be responsible for new and unpredicted labor and machinery costs. The COVID-19 pandemic may have unforeseeable socio-economic effects, affecting the global building industry. In conclusion, impact of COVID-19 during lockdown period in Ernakulam on various construction sectors were severe because from the results, shortages of migrant workers and delay delivery are the main issues. According to the District Labour Office, 46,000 migrant labours in Ernakulam out of 86,000 have returned to their native places [21]. Due to the pandemic situation only the migrant workers who have guaranteed jobs were prioritized by the Construction contractors to return back to Kerala. Kerala's construction industry was largely reliant on skilled migrant labour. A possibility of the pandemic outbreak in rural areas in the coming months, leads labour shortages in Ernakulam [21], to resolve this situation, create a proper work plan for the post-construction and pre-construction periods, as well as a future plan for the construction industry to manage the unpredictable situations and economic consequences.

References

1. Alsharef A et al (2021) Early impacts of the COVID-19 pandemic on the United States construction industry. Int J Environ Res Public Health 18(4):1559
2. Gamil Y (2020) The impact of pandemic crisis on the survival of construction industry: a case of COVID-19. Mediterr J Soc Sci 11(4):122
3. Hao F et al (2020) COVID-19 and China's hotel industry: impacts, a disaster management framework, and post-pandemic agenda. Int J Hosp Manage
4. Allan-Blitz L-T, Turner I, Hertlein F, Klausner JD (2020) High frequency and prevalence of community-based asymptomatic SARS-CoV-2 infection. MedRxiv 2020
5. Narayana D, Venkiteswaran CS, Joseph MP (2013) Domestic migrant labour in Kerala. Gulati Institute of Finance and Taxation, Thiruvananthapuram (2013)
6. https://en.wikipedia.org/wiki/Timeline_of_the_COVID-19_pandemic_in_Kerala
7. Choudhari R (2020) COVID 19 pandemic: mental health challenges of internal migrant workers of India. Asian J Psychiatry 1–4
8. Ameyaw EE et al (2014) Application of Delphi method in construction engineering and management research: a quantitative perspective. J Civil Eng Manage 22(8):991–1000
9. Hallowell MR et al (2011) Interrelationships among highly effective construction injury prevention strategies. J Constr Eng Manage (2011): 985–999
10. Chan APC (2016) Strategies for improving safety and health of ethnic minority construction workers. American Society of Civil Engineers
11. Foster CJ (2020) A Delphi study of human factors methods for the evaluation of adaptation in safety-related organisations. Safety Sci (2020)
12. Hallowell MR et al (2010) Qualitative research: application of the Delphi method to CEM research. J Constr Eng Manage 99–107
13. Tymvios N (2016) Direction for generating interest for design for construction worker safety —a Delphi study. Am Soc Civil Eng 1 (2016)

14. Tikote RH (2017) Delay analysis in industrial projects by using relative importance index method. Int Res J Eng Technol 7(2017)
15. Desai M, Bhatt R (2013) Critical causes of delay in residential construction projects: case study of central Gujarat region of India. Int J Eng Trends Technol 4(4):762–768
16. Alenezi TAN (2020) Covid-19 causes of delays on construction projects in Kuwait. 8(4), July–August, 2020
17. Tosniwal RB et al (2018) Delay analysis and its effects in construction industry. 5:121–127
18. Udasi AP (2013) Analysis of causes and effects of delays in construction projects. 5 (2013):1382–1390
19. Paray WA et al (2020) Delay analysis in construction projects. Int Res J Eng Technol 7
20. Bsisu KAD (2020) The impact of COVID-19 pandemic on Jordanian civil engineers and construction industry. Int J Eng Res Tech 13(5):828–830
21. https://www.newindianexpress.com/cities/kochi/2020/jun/17/labour-shortage-retards-construction-sector-2157447.html

3D-FE Analysis of RC Tunnel with GFRP Shielding Under Internal Blast Loading

Anshul Kaushik, Gyanesh Patnaik, Abhishek Rajput, and Guru Prakash

Abstract Due to high risk of terrorist attacks inside underground subway system, the blast resistant design has become very crucial. The present study deals with the non-linear analysis of underground RC tunnels with a layer of GFRP shielding subjected to internal blast of 10 kg TNT at the center point of tunnel using FE software ABAQUS/EXPLICIT [1]. The explosion inside tunnel has been modeled using CONWEP tool. The soil neighboring the RC tunnel has been modeled using Mohr–Coulomb plasticity model, concrete part of RC tunnel is modeled using CDP (Concrete Damaged Plasticity) model, reinforcement part is modeled using JC (Johnson Cook) model, while GFRP shielding is modeled using Hashin model. The performance of tunnel and soil is evaluated in terms of stress and displacement values at the crown of tunnel and the mid-point of top surface of soil. Two different soil conditions are also considered in order to check the effect of surrounding soil on the response of RC tunnel in case of explosion. The results obtained indicate that the use of GFRP shielding effectively reduces the displacement and stress values in soil and RC tunnel. Thus, it is concluded that this method can be utilized for blast resistant design of underground RC tunnels.

Keywords Subway tunnel · Blast resistant design · Conwep · Finite element method · Strain rate · Saturated and unsaturated soil

1 Introduction

Subsurface tunnels are significantly used for transportation, water pipelines, and other utilities. However, in the recent time, transportation tunnels especially underground subway systems have been highly prone to terrorist attacks including the bombings in Saint Petersburg (2017), Belgium (2016), Istanbul (2015), Minsk (2011), Moscow (2010 and 2004). Explosion inside a tunnel maybe very detri-

A. Kaushik (✉) · G. Patnaik · A. Rajput · G. Prakash
IIT Indore, Madhya Pradesh, Simrol 452020, India
e-mail: phd1901204004@iiti.ac.in

G. C. Marano et al. (eds.), *Proceedings of SECON'21*, Lecture Notes in Civil Engineering 171, https://doi.org/10.1007/978-3-030-80312-4_14

mental in terms of loss of lives, infrastructure, and wealth. Thus, it has become very necessary to explore for potential methods which could be used for safeguarding the tunnels from damage and collapse caused by internal blast.

The use of experimental techniques for studying the behavior of underground RC tunnel against blast loading is not feasible from socio-political as well as economic considerations. Due to this, no experimental studies have been reported in this domain. Also, the analytical methodologies advanced till now are based on simplified assumptions which may not be fully applicable for practical scenarios. Due to the high accuracy of numerical methods for non-linear dynamic problems, it is considered a good platform for carrying out the analysis of underground RC tunnels subjected to explosion and have been used by many researchers [2–9]. Also, few researches aimed at improving the blast performance of underground tunnels using different lining materials such as steel, steel fiber reinforced concrete, sandwich panels, and foam materials have been performed [6, 8]. It is found that the use of sandwich panels and foam materials effectively reduces the damage of tunnels. Also, the use of box shaped tunnels must be avoided since they go through the highest amount of damage while, circular tunnels are the most effective in dissipating blast energy [6, 8, 9].

The use of FRP's such as CFRP and GFRP as external strengthening material has been effectively carried out for several structural members such as beams, columns, and slabs due to its high tensile strength and excellent corrosion resistance properties [10–12]. Also, they have been used with structural members against blast loading [13, 14]. However, the effectiveness of using FRP layers as shielding over RC tunnel against internal blast has not been checked. Due to the high cost of CFRP material, it is not suitable to be used where larger spans demand shielding. GFRP has a comparatively lower cost due to which this material can be effectively utilized as a shielding over RC tunnels.

In this study, the efficacy of utilization of 20 mm thick GFRP shield over RC tunnel enclosed by two different soil conditions (saturated and unsaturated) has been checked against 10 kg TNT explosion. The analysis has been done using ABAQUS/EXPLICIT software [1]. The performance of tunnel enclosed by soil medium is evaluated in terms of displacement and von-Mises stress values at crucial points: (1) crown of RC tunnel, and (2) mid-point of top of soil surface.

2 Finite Element Modeling

2.1 Modeling of the Subway System

The 3D FE model of the subway system is prepared using ABAQUS/EXPLICIT [1]. A 20 m long underground RC tunnel which is surrounded by soil domain measuring 20 m × 26 m × 26 m is considered. The tunnel has an internal diameter of 5.4 m with 300 mm thick RC lining. The mid-axis of tunnel is considered to

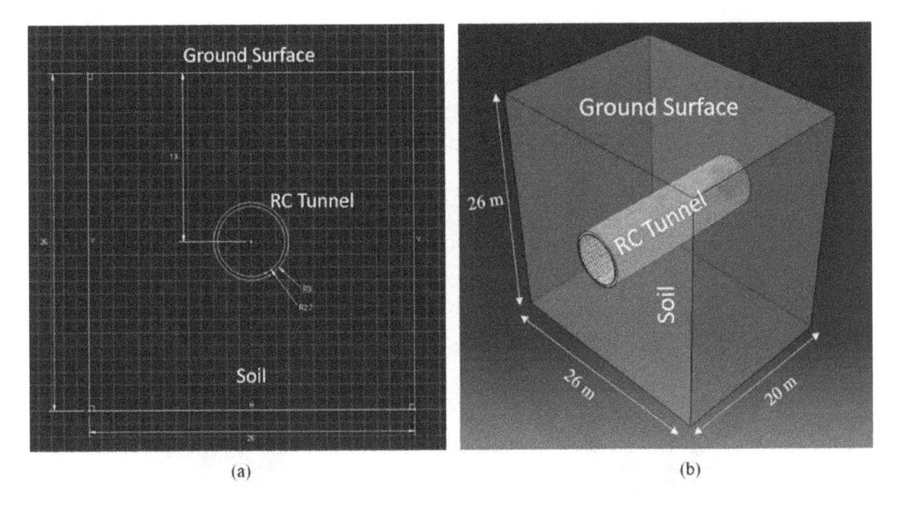

Fig. 1 FE model details **a** 2D geometry, **b** complete model (ABAQUS/EXPLICIT [1])

be at a depth of 13 m below the ground surface. The complete geometric details of the FE model are shown in Fig. 1. 20 mm thick GFRP shielding is considered in this study. Steel bars of 10 mm diameter having yield strength of 415 MPa are used as reinforcement. The transverse reinforcement rings are provided at spacing of 300 mm. At each cross section, two rings are provided with thickness of 120 mm in between them. 20 longitudinal bars are provided at the perimeter of each ring. Thus, a total number of 40 longitudinal bars are provided. The reinforcing details used in this study have been taken from [7] which is based on the specifications of Indian Metro Rail Corporation. Since, the geometric dimension of RC tunnel is much smaller compared to that of soil domain, no reflection of blast waves will occur.

The details of FE mesh as well as the geometry of individual components of the FE model are shown in Fig. 2. The soil and concrete part of the FE model are developed using 3D deformable solid part option with reduced integration eight node brick elements (C3D8R). In order to prevent the excessive distortion of elements under blast loading, enhanced hourglass stiffness option with distortion control is also utilized. Steel bars are developed using 3D deformable wires with two node truss elements (T3D2) and embedded inside concrete tunnel using Embedded constraints option. General contact option is used in which hard contact is defined in normal direction and penalty 0.25 for frictional behavior in tangential direction is adopted for interaction between concrete surface and soil domain. GFRP shield is developed using shell part and meshed using continuum hexahedral elements (SC8R). Surface to surface contact option is utilized for defining the cohesive interaction between GFRP and internal surface of concrete tunnel. The adhesive properties are given in Table 1. The bottom surface of soil is fixed in all directions using ENCASTRE option, while in the outer vertical surfaces of soil domain, displacements normal to the surface are constrained.

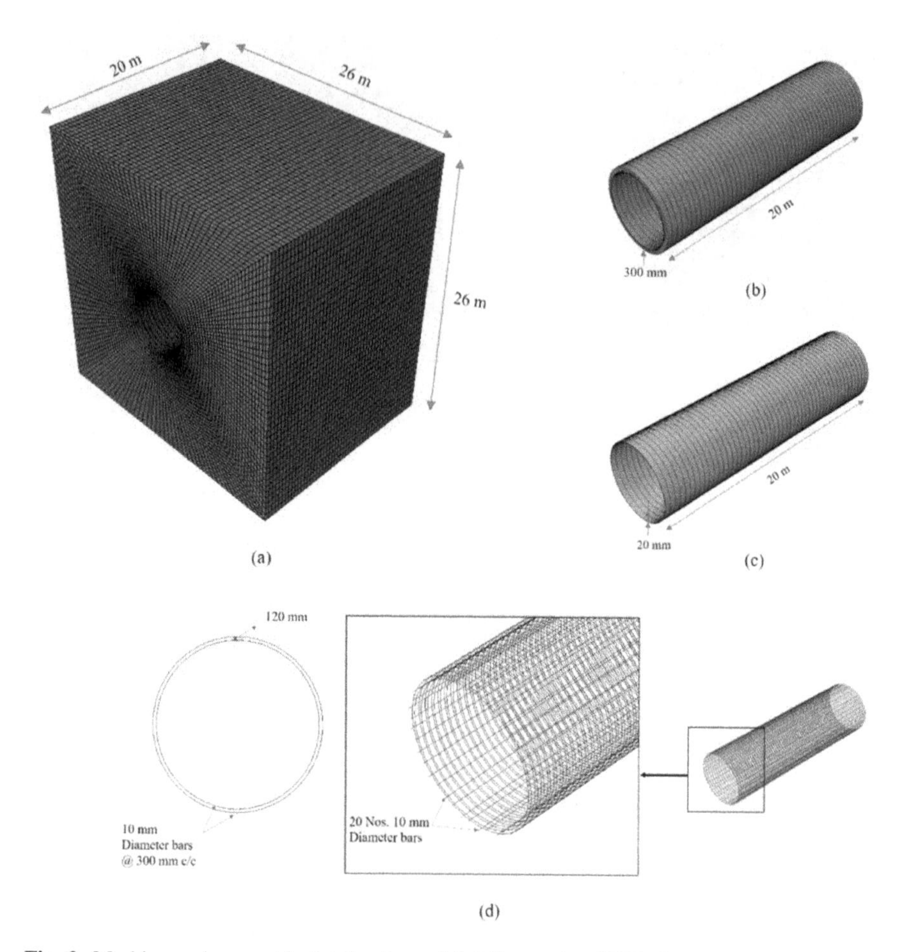

Fig. 2 Meshing and geometric details of **a** soil **b** RC tunnel **c** GFRP **d** complete reinforcement details (ABAQUS/EXPLICIT [1])

Table 1 Properties of adhesive utilized for cohesive interaction between GFRP blanket and RC lining

Adhesive Property	Value
Young's modulus	12.7 GPa [15]
Shear modulus	0.665 GPa [16]
Thickness (t)	0.1 mm (Assumed)
Stiffness coefficients	$K_{nn} = 1.724 \times 10^{14}$ N/m^3 $K_{ss} = 6.65 \times 10^{12}$ N/m^3 $K_{tt} = 6.65 \times 10^{12}$ N/m^3
Cohesive strength	6 MPa [17]
Shear strength	2.84 MPa [16]
Fracture energy	900 N/m (Estimated)

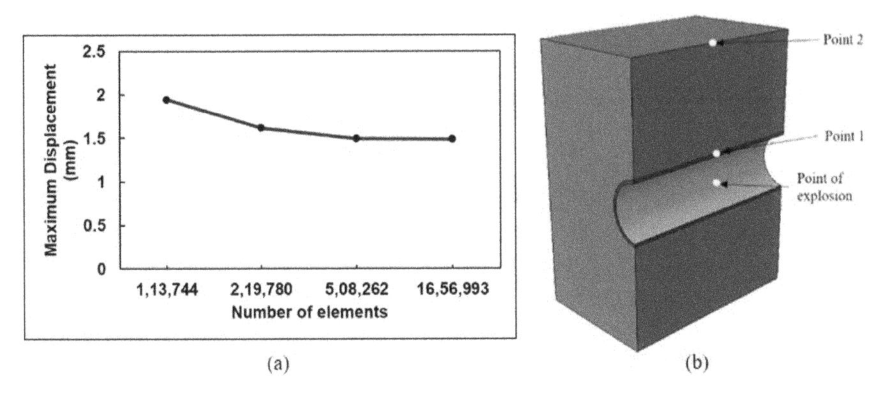

Fig. 3 **a** Mesh convergence study, **b** key points in the FE model (ABAQUS/EXPLICIT [1])

Proper mesh size and number of elements for the FE model are finalized based on the mesh convergence study. For this, the values of peak displacement obtained at the tunnel crown are considered. The results of mesh convergence study are shown in Fig. 3a. Finer mesh is provided in the periphery of point of explosion including the GFRP layer, RC tunnel, and soil. The mesh is made coarser away from the point of explosion in order to make the FE computations efficient. The crucial points where the outputs of displacement, velocity and stress values are obtained are shown in Fig. 3b.

2.2 *Modeling of Explosion*

An explosive of 10 kg TNT charge is considered at the central point of tunnel as shown in Fig. 3b. The assumption of 10 kg explosive charge is based on the previous research carried out by Chaudhary et al. [8] in which it has been considered that this is the maximum weight of explosive which could be carried by the terrorists in a backpack. The blast condition is developed using CONWEP tool. The total pressure ($\bar{P}(t)$) developed using CONWEP as given in ABAQUS user manual [1] is given as:

$$\bar{P}(t) = P_{incident}(t) + \left[1 + \cos\theta - 2\cos^2\theta\right] + P_{reflect}(t)\cos^2\theta \quad for \ \cos\theta \geq 0 \quad (1)$$

$$\bar{P}(t) = P_{incident}(t) \quad for \ \cos\theta < 0 \quad (2)$$

where, $P_{incident}(t)$ and $P_{reflect}(t)$ represent the incident and reflected pressure waves respectively, and θ is the angle between normal to the surface of loading and the line connecting that point on surface to the point of explosion.

2.3 Material Properties

2.3.1 Material Properties of Concrete

The M50 grade concrete of RC tunnel is modeled using CDP model. The yield function used in CDP model is developed by Lubliner et al. [18] for quasi-static loading and later modified by Lee and Fenves [19] for dynamic loading. The two main modes of failure of concrete in CDP model are cracking failure in tension and crushing failure in compression. The values of parameters of CDP model can be adjusted so as to model different types of concrete such as plain concrete, fibre-reinforced concrete at different strain rates. In CDP model, the tensile and compressive stress–strain relationships are as follows:

$$f_t = (1 - D_t)E_o : \left(\varepsilon_t - \varepsilon_t^{pl}\right) \tag{3}$$

$$\sigma_c = (1 - D_c)E_o : \left(\varepsilon_c - \varepsilon_c^{pl}\right) \tag{4}$$

where, tension and compression are denoted as t and c respectively, ε_t^{pl} and ε_c^{pl} represent the equivalent plastic strain values, D_t and D_c represent the tensile and compressive damage variables, E_o is the initial Young's modulus of elasticity.

In this model, the potential function is given as:

$$G = \sqrt{(\varepsilon\sigma_{t0}\tan\varphi)^2 + \bar{q}^2} - \bar{p}\tan\varphi \tag{5}$$

where, φ is the dilation angle, σ_{t0} is the failure stress value in tension, ε is the eccentricity parameter.

The elastic properties of concrete are given in Table 2. The parameters of CDP model and the stress–strain model of concrete have been adopted from data for M50 grade concrete available in [20]. The stress–strain model of concrete is shown in Fig. 4. In order to consider the effects of strain rate, DIF values of 2 and 6 are adopted for compression and tension respectively [21].

2.3.2 Material Properties of Steel

The response of steel is modeled using Johnson–Cook (JC) model [22]. The stress–strain relationship in this model is given as:

Table 2 Elastic properties of concrete

Material	Density (kg/m^3)	Elastic modulus (GPa)	Poisson's ratio
M50 Concrete	2400	35.36	0.19

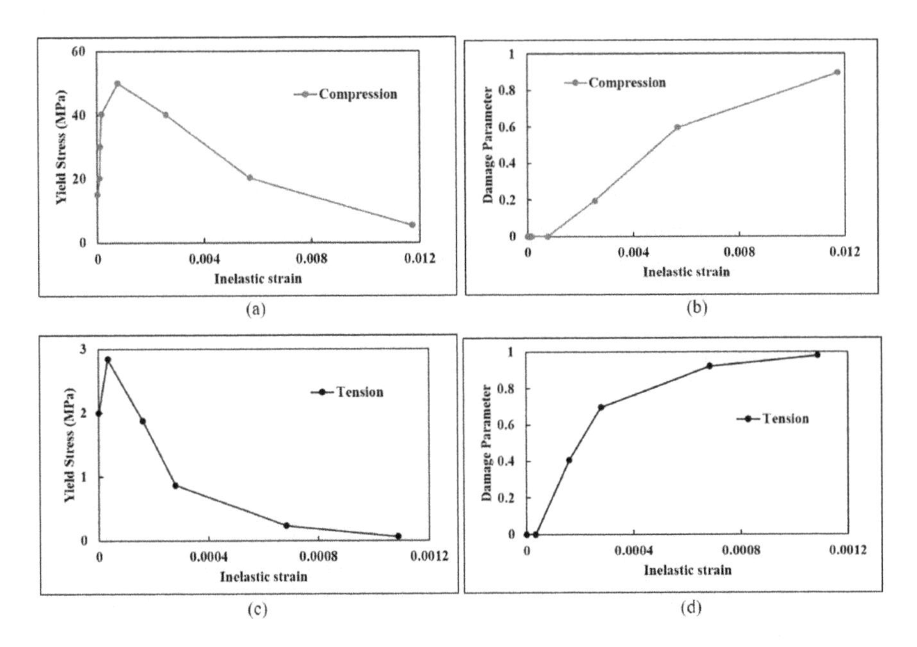

Fig. 4 Stress–strain model of concrete in **a**, **b** compression and **c**, **d** tension

$$\sigma = (A + B\varepsilon^n)(1 + C\log_e \varepsilon^*)(1 - T^{*m}) \tag{6}$$

where, ε^* is the dimensionless strain in plastic condition; $\varepsilon^* = \dot{\varepsilon}/\dot{\varepsilon}_0$, $\dot{\varepsilon}_0$ represents the reference strain rate (1/s), $\dot{\varepsilon}$ represents the strain rate in plastic condition, and T^* represents the homologous temperature. A, B, C, m, and n are constants. The elastic properties for FE415 steel are given in Table 3. The values of constants are taken as A = 360 MPa, B = 635 MPa, n = 0.114, C = 0.075 from [23]. Temperature effects are neglected.

2.3.3 Material Properties of Soil

The soil neighboring the RC tunnel is modeled using Mohr–Coulomb plasticity model. The equation for this model is given as:

$$s + \sigma_m \sin\emptyset - c\cos\emptyset = 0 \tag{7}$$

Table 3 Elastic properties of steel

Material	Density (kg/m^3)	Elastic modulus (GPa)	Poisson's ratio
Steel Fe415	7800	200	0.3

Table 4 Mohr–Coulomb plasticity parameters for unsaturated and saturated soil (clay)

Soil type	Elastic modulus (MPa)	Poisson's ratio	Cohesion (kPa)	Friction angle (°)	Dilation angle (°)
Saturated	40	0.49	50	0	0
Unsaturated	50	0.3	90	20	0

where, s is the maximal shear stress value, σ_m is the mean of maximum and minimum principal stress values, c is the material cohesion, and $\varnothing$ is the friction angle.

In this study, the values of the parameters of Mohr–Coulomb model have been taken from [24] and [10]. The properties for saturated and unsaturated soil are given in Table 4.

2.3.4 Material Properties of GFRP

GFRP shield used in this study has been modeled using Hashin damage model [25, 26]. The properties of GFRP are given in Table 5.

3 Validation of FE Analysis

In order to validate the FE scheme used in this study, the results are compared with the (1) Experimental and numerical results of blast loading on RC slabs reported by Zhao and Chen [28], and (2) Numerical simulation results of explosion inside RC tunnel reported by Onal [10].

3.1 Validation with Results of Zhao and Chen [28]

Zhao and Chen performed experimental and numerical investigations on RC slabs of size 1 m × 1 m × 0.4 m subjected to TNT explosions of 0.2 and 0.31, and 0.46 kg at 0.4 m standoff distance. In this validation study, the RC slab is modeled with similar geometric as well as material properties and subjected to explosions of same TNT charge. The results obtained are compared as shown in Table 6. It is observed that the results from present study match well with the results reported in [28].

Table 5 Material properties of GFRP [27]

Material properties		Value
Elastic modulus	E_{11} (GPa)	138
	$E_{22} = E_{33}$ (GPa)	9
Shear modulus	$E_{12} = E_{13}$ (GPa)	5.7
	E_{23} (GPa)	3
Poisson's ratio	$\upsilon_{12} = \upsilon_{13}$	0.3
	υ_{23}	0.42
Tensile strength	X^T (MPa)	2004
	$Y^T = Z^T$ (MPa)	1197
Compressive strength	X^C (MPa)	53
	$Y^C = Z^C$ (MPa)	204
Shear strength	S^L (MPa)	137
	S^T (MPa)	42
Fiber volume		62%

Table 6 Comparison of results obtained from present study and the results of Zhao and Chen [28]

Explosive charge (kg)	Displacement (mm) at the center of slab			Error (%)	
	Zhao and Chen [28]		Present study	w.r.t. experimental analysis	w.r.t. numerical analysis
	Experimental analysis	Numerical analysis	Numerical analysis		
0.20	10	8.8	8.232	17.68	6.45
0.31	15	12.7	13.72	8.53	8.03
0.46	35	29	28.06	19.83	3.24

3.2 Validation with Results of Goel et al. [10]

In order to validate the FE model for problems involving explosion inside a tunnel, the results from present FE model are compared with the results reported in [10]. Gupta et al. [10] carried out simulations of 100 kg explosive inside RC tunnel of length 20 m and internal radius of 2.7 m. The thickness of RC tunnel is 300 mm. Two different soil conditions are considered: saturated and unsaturated. The geometric as well as the material specifications utilized in the present numerical study is similar to that of [10]. The comparison of time histories of displacement at crown of tunnel (point 1) and at mid-point of top surface of soil (point 2) are shown in Fig. 5. Also, the von-Mises stress values obtained at these two points are also compared and are shown in Table 7.

As shown in Fig. 5, the displacement time histories at tunnel crown and middle point of soil surface are in good agreement, also, the peak stress values match well with the numerical simulation results of Goel et al. [10]. Hence, the present FE modeling approach is considered to be validated.

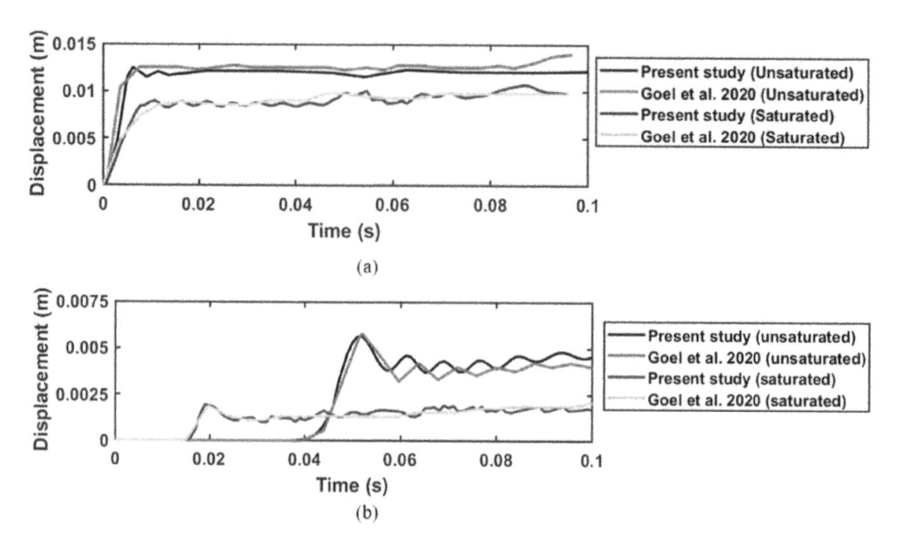

Fig. 5 Comparison of displacement values at **a** RC tunnel crown, **b** mid-point of soil top surface

Table 7 Comparison of peak stress values obtained from present study and Goel et al. [10]

Type of soil	Maximum von-Mises stress value (Pa)					
	Crown of RC tunnel			Soil top surface		
	Goel et al. [10]	Present study	Error (%)	Goel et al. [10]	Present study	Error (%)
Unsaturated	5,910,449	6,020,000	−1.85	71,502.4	73,114	−2.26
Saturated	4,119,404	4,010,000	2.66	9862.4	8989.09	8.85

4 Results of FE Analysis

In this study, the FE analysis of underground subway RC tunnels with and without GFRP shielding, surrounded by two different soil mass conditions, and exposed to internal explosion of 10 kg TNT are carried out. The analysis is performed for an explicit time of 100 ms.

The time histories of displacement and velocity values obtained at the crown of RC tunnel are shown in Fig. 6. As the TNT charge explodes, the velocity and displacement values at crown of tunnel increase suddenly and attain their peak values at 1 ms. This is followed by vibrations of tunnel lining as observed from the displacement time histories. It is observed that the use of GFRP shielding effectively minimizes the displacement values and nullifies the vibrations induced in tunnel lining due to internal explosion. The enhancement in the performance of RC tunnel is observed due to dissipation of blast energy by GFRP shield. Also, similar to observations made by Goel et al. [10] it is observed that lower magnitude of displacement and velocity are obtained in case of saturated soil due to its higher

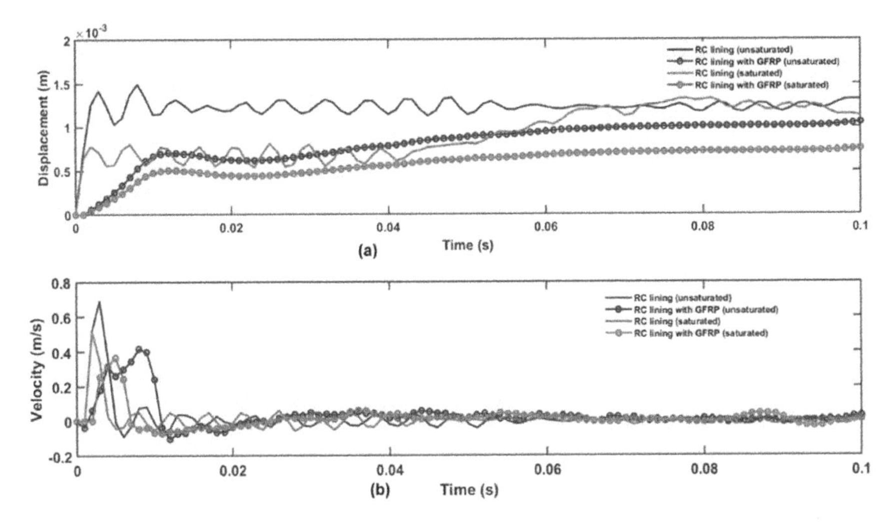

Fig. 6 Response of RC tunnel under 10 kg internal explosion **a** displacement at crown of RC tunnel, **b** velocity at crown of RC tunnel

density compared to unsaturated soil. The RC tunnel surrounded by saturated soil undergoes 46% lesser displacement compared to that in unsaturated soil.

The performance of RC tunnels is further evaluated by analyzing the displacement and von-Mises stress values obtained at the mid-point of top surface of soil (point 2) as shown in Fig. 7. The presence of GFRP shield effectively dissipates a higher amount of blast energy as compared to unshielded RC tunnel due to which stress waves propagating through soil medium are reduced. Consequently, displacement and stress values obtained at the soil top surface are also reduced. It is also observed that lower values of displacement and stresses are obtained at soil top surface for saturated soil. However, the peak values are achieved earlier in case of saturated soil, this is due to faster propagation of blast induced pressure waves in saturated soil due to its higher density. The time lag between the peak values of displacement at point 1 and point 2 exhibits the time taken by blast induced waves to propagate through the soil domain.

5 Effectiveness of Utilization of GFRP as Shielding Material Under Internal Explosion

In order to evaluate the effectiveness of provision of GFRP layer as shielding material over RC tunnel in case of blast loading, outputs at crucial points i.e., crown of tunnel lining and mid-point of soil top surface are analyzed. The peak displacement and von-Mises stress values obtained at crown of RC tunnel (point 1) are shown in Fig. 8. Due to the dissipation of blast energy by GFRP shielding, the peak

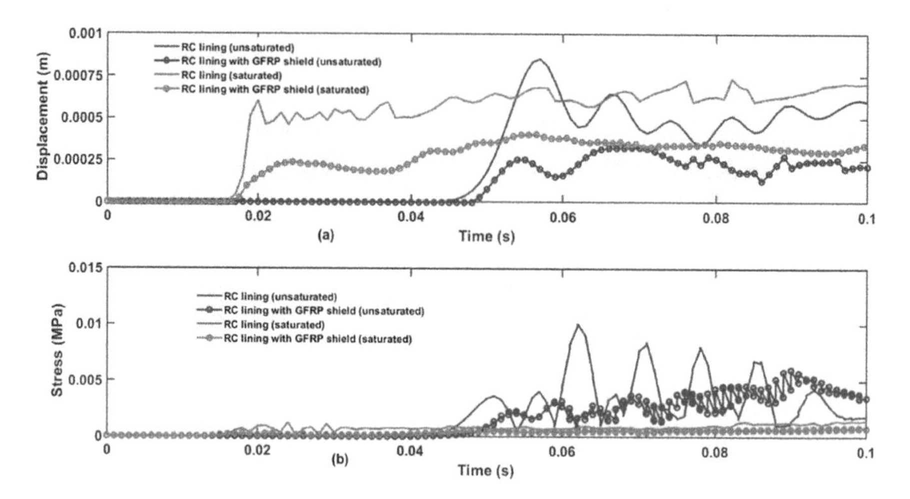

Fig. 7 **a** Displacement and **b** stress time histories obtained at the mid-point of soil top surface (point 2)

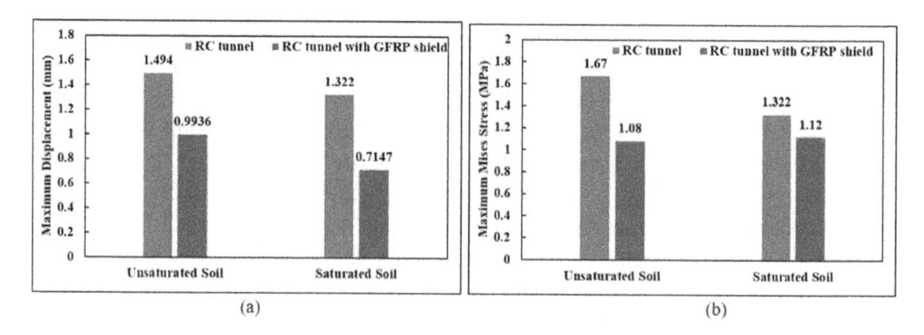

Fig. 8 Maximum values of **a** displacement and **b** von-Mises stress at crown of RC tunnel

displacement values are reduced by 34% and 46% respectively for unsaturated and saturated soil, while, the peak von-Mises stress values are reduced by 36% and 15% respectively for unsaturated and saturated soil.

In order to further check the performance of GFRP shielding, the response of neighboring soil is also analyzed. The peak displacement and von-Mises stress values obtained at the mid-point of soil top surface (point 2) are shown in Fig. 9. GFRP shielding depletes a portion of blast energy due to its excellent tensile resistance before conveying it to the RC tunnel and soil medium, consequently resulting in reduced displacement and stress values at soil surface. The presence of GFRP shielding reduces the peak displacement value by 31% and 28% respectively for unsaturated and saturated soil, while the peak von-Mises stress values are reduced by 30% and 15% respectively for unsaturated and saturated soil.

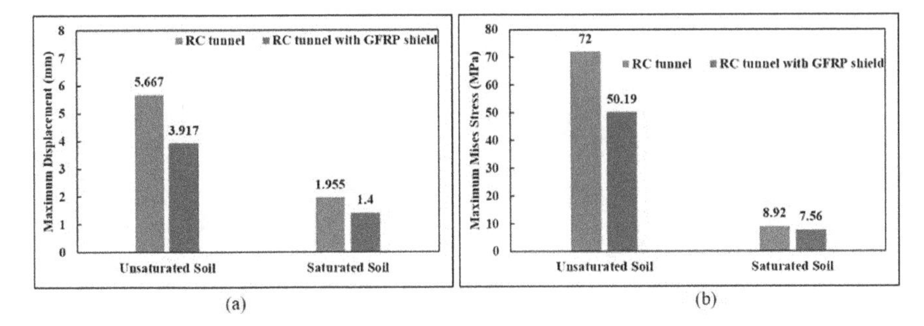

Fig. 9 Maximum values of **a** displacement **b** von-Mises stress at mid-point of soil top surface

The damage evaluation of RC tunnels under blast loading is carried out using Equivalent plastic strain in tension (PEEQT) parameter. This parameter represents the tensile plastic deformation and its value higher than zero denotes that the material has already yielded in tension. The PEEQT fringes of RC tunnel near the detonation point with and without GFRP shield for different time steps is shown in Fig. 10. The internal blast inside RC tunnel results in longitudinal dispersion of stress waves originating near point of explosion. Also, it is observed that the use of GFRP shield reduces the strain values developed in RC tunnel. Thus, the provision of GFRP shield effectively reduces the damage and improves the blast resistant of RC tunnel.

6 Conclusions

In the present study, the efficacy of utilization of GFRP shielding over RC tunnel under internal blast loading has been checked using 3D-FE analysis. Considering a typical context of explosion inside subway system, a 10 kg explosive is considered in the present study which could be easily carried inside subway system by the terrorists. Two different soil conditions are considered. CONWEP tool is used for simulating the internal explosion inside RC tunnel. Based on the present study, following conclusions are obtained:

1. The use of GFRP shield enhances the performance of RC tunnel in case of internal explosion. The peak displacement values at crown of RC tunnel are reduced by approximately 40%, while the peak stress values are reduced by approximately 35%. Also, there is a reduction in the induced vibrations in tunnel lining due to internal explosion.
2. The provision of GFRP shield depletes a portion of blast energy before transmitting it to RC tunnel due to which lower magnitude of pressure waves propagate through soil and hence displacement and stresses at the soil top surface are reduced.

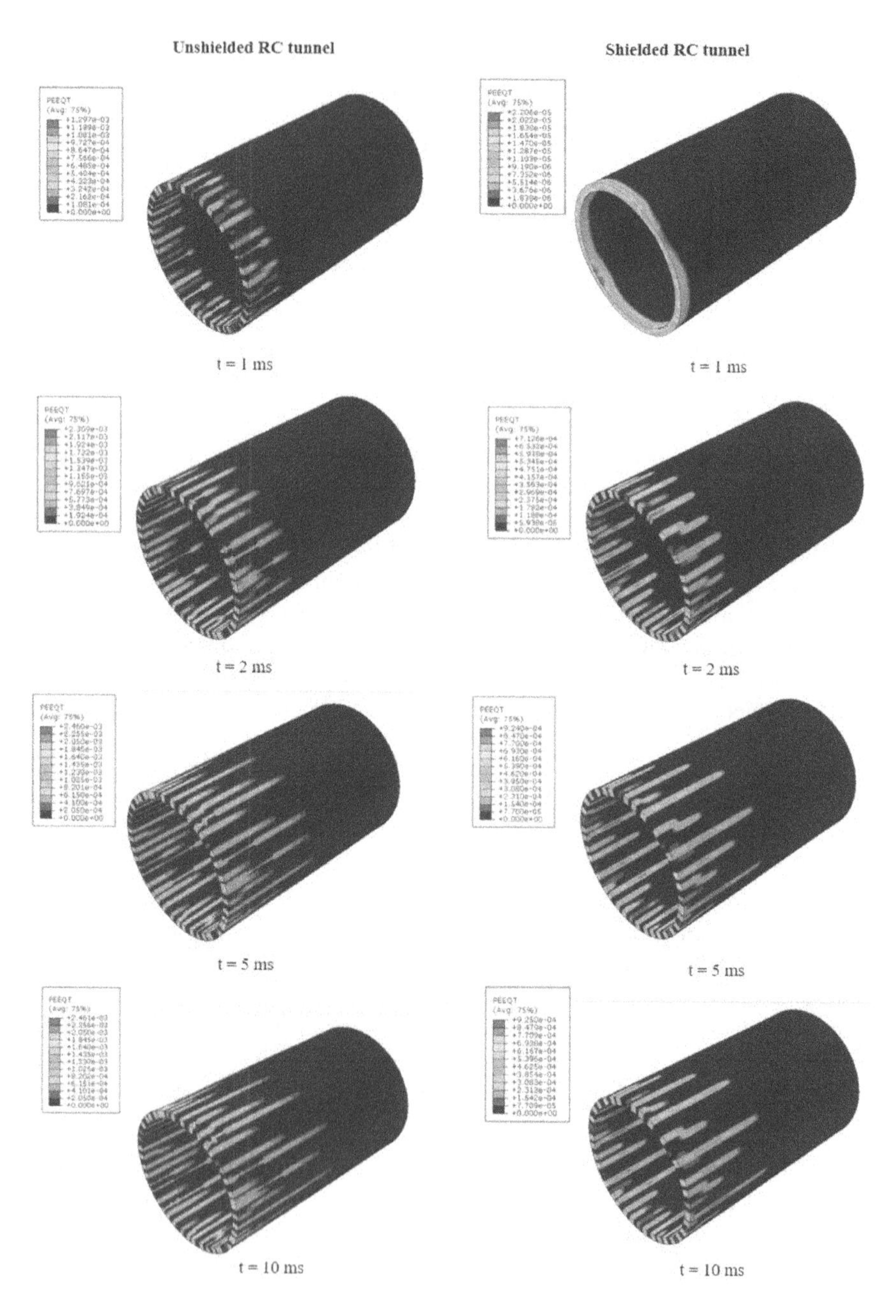

Fig. 10 Comparison of PEEQT for RC tunnel subjected to 10 kg explosion without GFRP and with GFRP layer at different time instances (Abaqus CAE 2017)

3. The provision of GFRP layer effectively reduces the damage and plastic strain values in RC tunnel lining.
4. The neighboring soil substantially effects the performance of RC tunnel in case of internal explosion. In case of saturated soil, the values of displacement and stresses at RC lining and soil top surface are reduced.

The use of GFRP shield provides excellent strengthening effect due to which this method can be effectively utilized for improving the blast resistance of RC tunnels.

References

1. Manual, ABAQUS User (2017) DS Simulia (2017)
2. Gui MW, Chien MC (2006) Blast-resistant analysis for a tunnel passing beneath Taipei Shongsan airport—a parametric study. Geotech Geol Eng 24:227–248
3. Liu H (2009) Dynamic analysis of subway structures under blast loading. Geotech Geol Eng 27(6):699–711
4. Liu H (2012) Soil-structure interaction and failure of cast-iron subway tunnels subjected to medium internal blast loading. J Perform Constr Fac ASCE 26(5):691–701
5. Chakraborty T, Larcher M, Gebbeken N (2014) Performance of tunnel lining materials under internal blast loading. Int J Protective Struct 5(1):83–96. https://doi.org/10.1260/2041-4196.5.1.83
6. Prasanna R, Boominathan A (2015) Numerical simulation on behaviour of concrete tunnels in internal blast loading. In: Computer methods and recent advances in geomechanics: proceedings of the 14th international conference of international association for computer methods and recent advances in geomechanics. Taylor & Francis Books Ltd., pp 1907–1911
7. Tiwari R, Chakraborty T, Matsagar V (2017) Dynamic analysis of tunnel in soil subjected to internal blast loading. Geotech Geol Eng 35(4):1491–1512. https://doi.org/10.1007/s40098-016-01
8. Chaudhary RK, Mishra S, Chakraborty T, Matsagar V (2019) Vulnerability analysis of tunnel linings under blast loading. Int J Protective Struct 10(1):73–94
9. Goel MD, Verma S, Panchal S (2020) Effect of internal blast on tunnel lining and surrounding soil. Indian Geotech J 25:1
10. Önal MM (2014) Strengthening reinforced concrete beams with CFRP and GFRP. Adv Mater Sci Eng 1–8
11. Asrani NP, Murali G, Parthiban K, Surya K, Prakash A, Rathika K, Chandru U (2019) A feasibility of enhancing the impact resistance of hybrid fibrous geopolymer composites: experiments and modelling. Constr Build Mater 203:56–68. https://doi.org/10.1016/j.conbuildmat.2019.01.072
12. Murali G, Asrani NP, Ramkumar VR, Siva A, Haridharan MK (2019) Impact resistance and strength reliability of novel two-stage fibre-reinforced concrete. Arab J Sci Eng 44:4477–4490. https://doi.org/10.1007/s13369-018-3466-x
13. Buchan PA, Chen JF (2007) Blast resistance of FRP composites and polymer strengthened concrete and masonry structures–a state-of-the-art review. Compos B Eng 38(5–6):509–522
14. Shooshtari M, Gomar H (2017) Retrofit of reinforced concrete structures by CFRP/GFRP sheets against blast load. Math Models Eng 3(1):49–57
15. Elarbi AM (2011) Durability performance of frp strenghtened concrete beams and columns exposed to hygrothermal environment. Wayne State University Dissertations, 307. https://digitalcommons.wayne.edu/oa_dissertations/307
16. Obaidat YT, Heyden S, Dahlblom O (2010) The effect of CFRP and CFRP/concrete interface models when modelling retrofitted RC beams with FEM. Compos Struct 92(6):1391–1398

17. Mikami M, Toki S, Endo M (2015) Comparison of CRISPR/Cas9 expression constructs for efficient targeted mutagenesis in rice. Plant Mol Biol 88(6):561–572
18. Lubliner J, Oliver J, Oller S, Oñate E (1989) A plastic-damage model for concrete. Int J Solids Struct 25(3):299–326
19. Lee J, Fenves GL (1998) Plastic-damage model for cyclic loading of concrete structures. J Eng Mech 124(8):892–900
20. Jankowiak T, Lodygowski T (2005) Identification of parameters of concrete damage plasticity constitutive model. Found Civil Environ Eng 6(1):53–69
21. UFC 3-340-02 (2008) Structures to resist the effects of accidental explosions. Unified Facilities Criteria, US Departments of Army and Navy and Air Force, USA
22. Johnson GR, Cook WH (1983) A constitutive model and data for metals subjected to large strains, high strain rates and high temperatures. In: Proceedings of 7th international symposium on ballistics, Hague, Netherlands, pp 541–547
23. Goel MD, Matsagar VA, Gupta AK (2011) Dynamic response of stiffened plates under air blast. Int J Prot Struct 2(1):139–155. https://doi.org/10.1260/2041-4196.2.1.139
24. Amli A, Sabah A, Al-Ansari N, Laue J (2019) Study numerical simulation of stress-strain behavior of reinforced concrete bar in soil using theoretical models. Civil Eng J 11(5): 2349–2358
25. Hashin Z, Rotem A (1973) A fatigue failure criterion for fiber reinforced materials. J Compos Mater 4:448–464
26. Hashin Z (1980) Failure criteria for unidirectional fiber composites. J Appl Mech 47(2): 329–334
27. Naderi M, Maligno AR (2012) Fatigue life prediction of carbon/epoxy laminates by stochastic numerical simulation. Compos Struct 94(3):1052–1059
28. Zhao CF, Chen JY (2013) Damage mechanism and mode of square reinforced concrete slab subjected to blast loading. Theoret Appl Fract Mech 63:54–62

Seismic Performance of Floor Diaphragm Openings in Multi Storey Building

Sreelakshmi Ranjit and Nincy Jose

Abstract Diaphragms, which transmit lateral forces to the vertical elements of the lateral force resisting system, are to be built as part of the seismic force resisting system of the building. Opening in the floor diaphragms are common for many purposes such as stair cases, escalators, lifts, architectural lighting, etc. These diaphragm openings cause stresses to discontinuous joints with building elements. Diaphragm openings, if unaccounted for may potentially lead to structural instability or possible diaphragm failure. Correct positioning of openings will give the structure effective strength and serviceability. This study focuses on the general effects of floor diaphragm openings on seismic response of RC multistoried building having diaphragm openings of various shapes and aims to find out the optimum shape. The effect of opening shape in the diaphragm is studied using model properties such as maximum storey displacement, storey drift, and storey shear of the building by time history analysis using ETABS 2018 software, and the optimum opening shape is finalized. Effects of diaphragm openings in asymmetric buildings with plan irregularities are also considered.

Keywords Diaphragm openings · Optimum shape · Time history analysis · ETABS 2018 · Asymmetric buildings

1 Introduction

Earthquake usually starts in multi-storeyed framed structures in the lateral load resisting frames at places with structural weaknesses. During strong earthquakes, the action of multi-story framed buildings is dependent on the distribution of mass, stiffness, and intensity in each of the longitudinal and transverse directions of the building. These vulnerabilities are also created in a few cases by changes in the diaphragm's stiffness, strength or mass. Buildings with normal shape and uniformly distributed mass and stiffness in plan and elevation sustain far less damage than irregular configurations.

S. Ranjit (✉) · N. Jose
Angamaly, Ernakulam, India

G. C. Marano et al. (eds.), *Proceedings of SECON'21*, Lecture Notes in Civil Engineering 171, https://doi.org/10.1007/978-3-030-80312-4_15

According to IS 1893-2002 part 1;clause 4.8, diaphragm is a horizontal, or nearly horizontal system, which transmits lateral forces to the vertical resisting elements, for example, reinforced concrete floors and horizontal bracing systems. In structural engineering, diaphragm is a structural system used to transfer lateral loads to shear walls or frames. Wind and earthquake loads are the most common lateral loads, but diaphragm action can also resist other lateral loads such as lateral earth pressure or hydrostatic pressure. Horizontal diaphragms cause the walls or frames to act as a group in resisting lateral forces. If they were not connected together, the vertical structural elements that resist lateral forces would be a disorganised mass trying to work alone. The presence of openings makes the behavior of floor diaphragms significantly more complicated and unpredictable [1]. Openings reduce diaphragm stiffness and can reduce the load carrying capacity of the member [2].

Vas et al. [3] performed a parametric study on a set of models by changing the percentage of opening. Response spectrum and Time history analysis were carried out. The results showed that the least value of maximum storey displacement and drift was observed in 20% opening [3]. Vinod et al. [4] conducted analysis on G + 4 and G + 8 buildings and the study concluded that 20% openings can be used in the different higher story buildings so that it decreases the seismic forces in any earthquake affected areas [4]. The influence of openings on the in-plane structural behavior of reinforced concrete floor slabs was investigated by Khajehdehi et al. [5]. The inelastic behaviour of RC floor diaphragms with openings was investigated using the finite element (FE) approach. Since the yielding of bars at opening corners appeared to have a major impact on slab action, it was discovered that the failure mechanism of FE slabs with openings differs significantly from that of FE solid slabs [5]. Khajehdehi et al. [6] conducted studies to show how the presence of openings reduces the in-plane load carrying capacity of floor panels. The research used nonlinear 3-D finite element (FE) modeling. The study concluded that by reinforcing both the top and bottom of the openings, in-plane load carrying capacity can be increased [6]. As the diaphragm discontinuity is closer to the building's center of mass, the building's behavior improves. Slab openings in the center were found to be more effective in resisting lateral forces from the Maximum Storey drift and Base shear views [7]. When comparing flexible and rigid diaphragms in terms of story drift in tall buildings, it is clear that the structure with the flexible diaphragm has higher drift values in the upper portion of the structure. The story drifts were almost identical in the middle and lower portions of the system, at least in terms of mean values. A rigid diaphragm assumption in tall building will produce "safer" structure, with only the roof acceleration under predicted [8].

1.1 Diaphragm Openings

The distribution of load is often disrupted by openings in the slab. However, by placing these openings in suitable positions, the structure's susceptibility to damage can be reduced. While several publications have dealt with the actions and design of diaphragms, it is understood that there are many problems that have not yet been resolved. Openings are provided in reinforced concrete slab in buildings to provide way for lifts, cables, or other instruments to pass through one floor to other floor such as escalators. Nowadays, the use of escalators and lifts in high rise buildings has increased for easy movement from one floor to another. The use of floor openings is also increasing, introducing a new aspect design in architectural way. Diaphragm openings could potentially lead to structural instability or possible diaphragm failure if unaccounted for.

This study aims to compare the seismic performance of RC structures with and without diaphragm openings of different shapes using ETABS and to analyse the effect of optimum shape of diaphragm openings in asymmetric building plans. All the analyses will be carried out in RC structures with and without diaphragm openings of different shapes. Only parameters such as storey displacement, storey drift and base shear of the building are considered.

2 Methodology Adopted

3D models of symmetric and irregular plan configurations with no opening and 20% openings of square, circular and diamond shapes were created in ETABS 2018. The seismic behavior of structures with floor diaphragm openings is studied using time history analysis. To arrive at conclusions, the results based on the parameters maximum storey displacement, storey drift, and base shear are evaluated. For the study, data from the El Centro California time history feature was entered (magnitude 6.9). These structures were examined in accordance with the Indian Standard Code IS 1893-2002 (Part I). At the foundation of the building, fixed joints are provided. It is assumed that all structural sections are rectangular.

3 Modeling of Structure

For the study, G + 10 Storey building having symmetric, rectangular, L shaped and H-shaped plan models are considered. Each one having 4 models for no opening, 20% of different openings of square, circular and diamond shapes. Storey height is taken as 3 and 3.5 m for the ground storey. The structural members are made of M30 concrete and Fe 415 steel and the slab thickness is 150 mm. The columns measure 550 × 550 mm and the beams are 350 × 450 mm. Models with the same building configuration but different floor diaphragm openings are compared to diaphragms with no openings to find the best opening shape. Figure 1 shows the plan view of all the configurations with no floor diaphragm openings. Figure 2 shows the different shapes of opening.

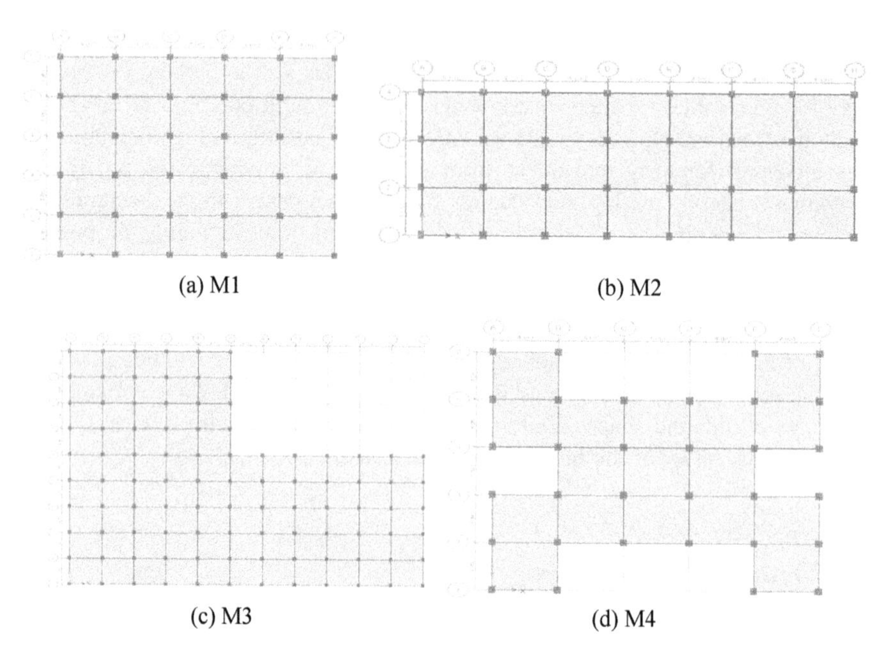

Fig. 1 Plan view of models

3.1 Loading

According to IS: 875 (Part II) 1987, a live load of 3 kN/m^2 is applied, and the dead load is software allocated. 1.5 kN/m^2 floor finish and 2 kN/m^2 roof load is also applied. According to IS 1893-2002, lateral loads are applied as seismic loads in the X and Y directions (Part 1). The design earthquake load is calculated using a 0.16 zone factor, medium soil, importance factor of 1 and a 5 response reduction factor (IS: 1893 (Part-I), 2016). Fixed support is provided.

4 Analysis Results

Table 1 shows the analysis results of all models. Maximum storey displacement, storey drifts and base shear are determined by using time history analysis, by providing different shapes. Fig. 3 shows the maximum displacement and storey drift results of model M1.

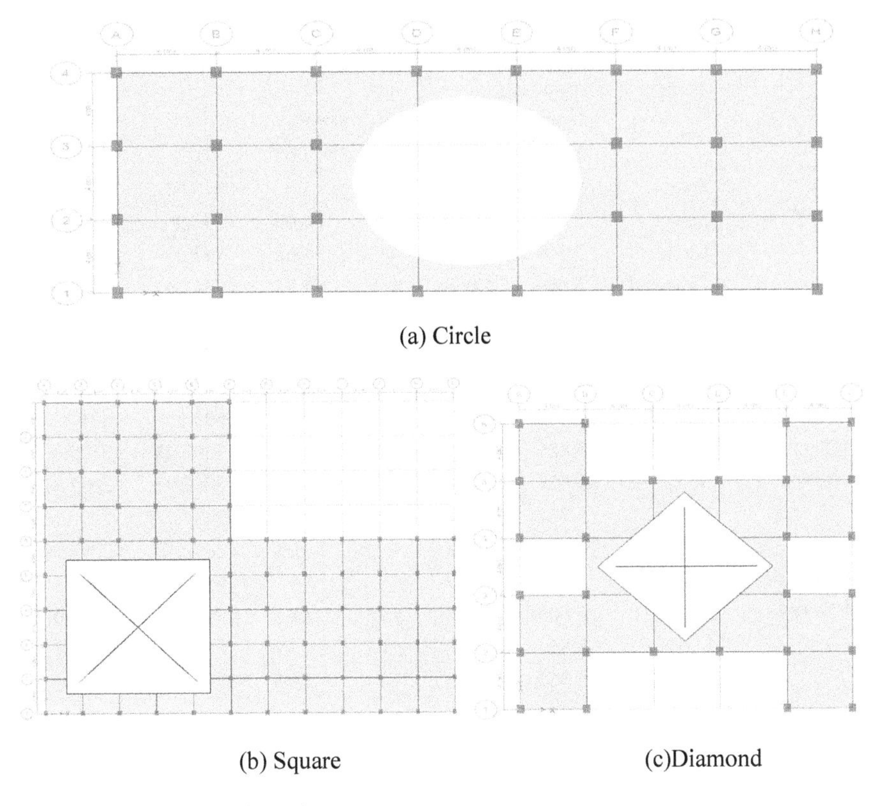

(a) Circle

(b) Square (c)Diamond

Fig. 2 Different shapes of opening

4.1 M1-Symmetric Model Results

Figure 4 shows the analysis results comparison of model M1.Models with square and circular shaped opening had almost identical responses for all values. For symmetric G + 10 buildings M1, in case of 20% diamond shaped openings maximum storey displacement is 13% less than that of solid slab models and storey drift increases by 6%. In all different opening cases maximum storey shear is less than that for solid slabs.

4.2 M2-Rectangular Modal Results

Figure 5 shows the analysis results comparison of model M2. Maximum storey displacement is least for the case of square shaped opening and the value of diamond

Table 1 Analysis results

G + 10 models		Max storey displacement (mm)		Max storey drift (10^{-4} mm)		Base shear (kN)	
		THX	THY	THX	THY	THX	THY
(M1) Symmetric	Solid	5.731	5.731	4.6	4.6	870.03	870.03
	Square	5.569	5.570	4.64	4.64	746.90	746.90
	Circle	5.565	5.565	4.63	4.63	747.53	747.54
	Diamond	5.069	5.068	4.9	4.9	744.36	744.43
(M2) Rectangular	Solid	5.642	5.337	4.89	4.72	671.81	769.28
	Square	5.614	4.973	4.95	5.22	598.05	622.14
	Circle	5.626	4.968	4.94	5.23	599.72	622.03
	Diamond	5.642	4.968	4.93	5.26	603.68	621.42
(M3) L-Shaped	Solid	5.816	5.715	4.68	4.96	2502.21	2525.88
	Square	5.538	5.414	4.65	4.91	2063.95	2047.11
	Circle	5.761	5.593	4.65	4.89	2077.82	2095.14
	Diamond	5.588	5.386	4.75	5.02	2090.47	2093.99
(M4) H-shaped	Solid	5.694	5.689	5.1	5.1	657.81	657.43
	Square	5.600	5.584	5.15	4.92	595.44	573.40
	Circle	5.550	5.502	5.08	4.85	586.90	568.77
	Diamond	5.564	5.526	5.1	4.87	589.72	566.49

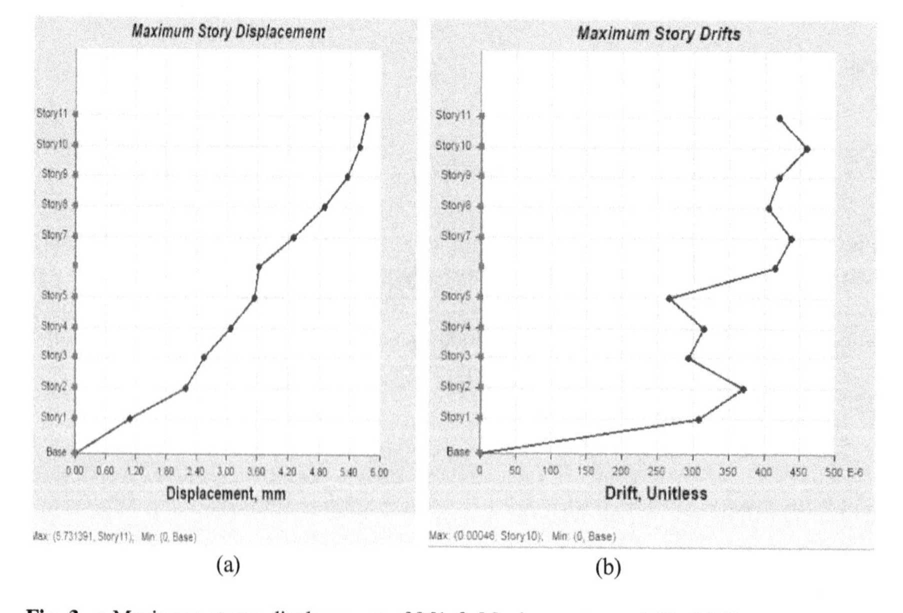

(a) (b)

Fig. 3 **a** Maximum storey displacement of M1. **b** Maximum storey drift of M1

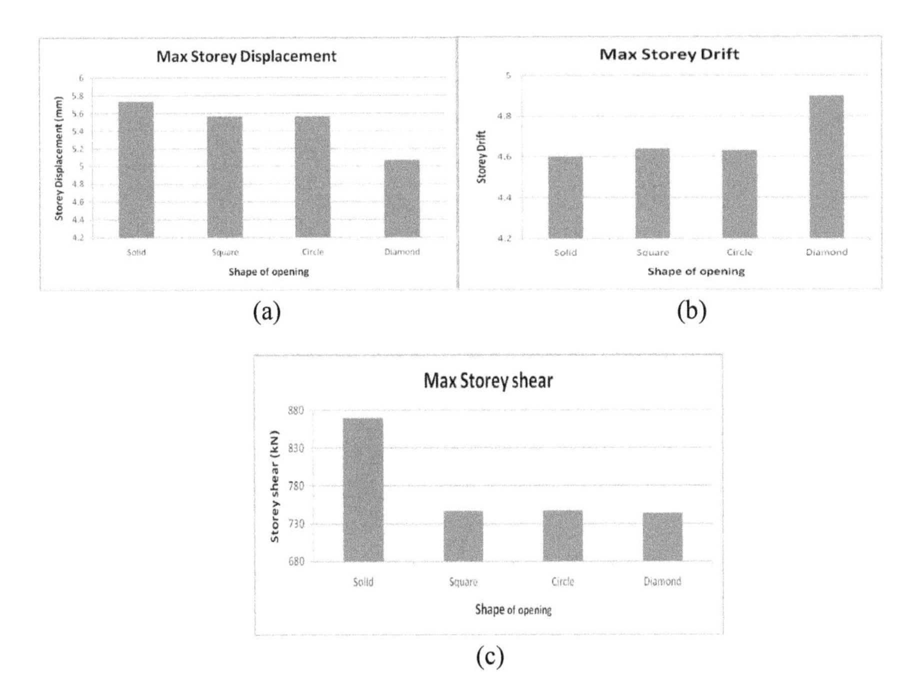

Fig. 4 **a** Maximum storey displacement. **b** Maximum storey drift. **c** Maximum storey shear

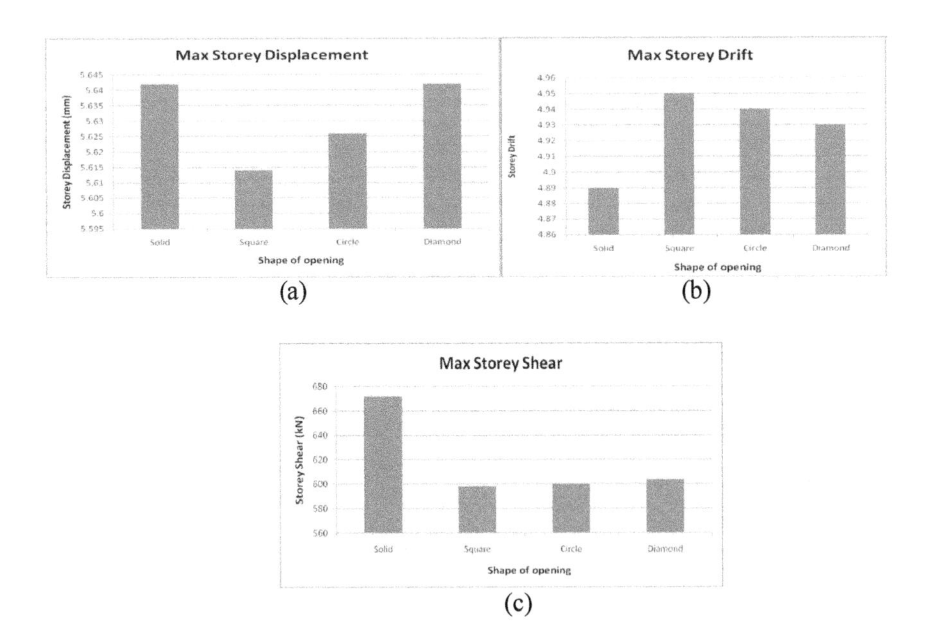

Fig. 5 **a** Maximum storey displacement. **b** Maximum storey drift. **c** Maximum storey shear

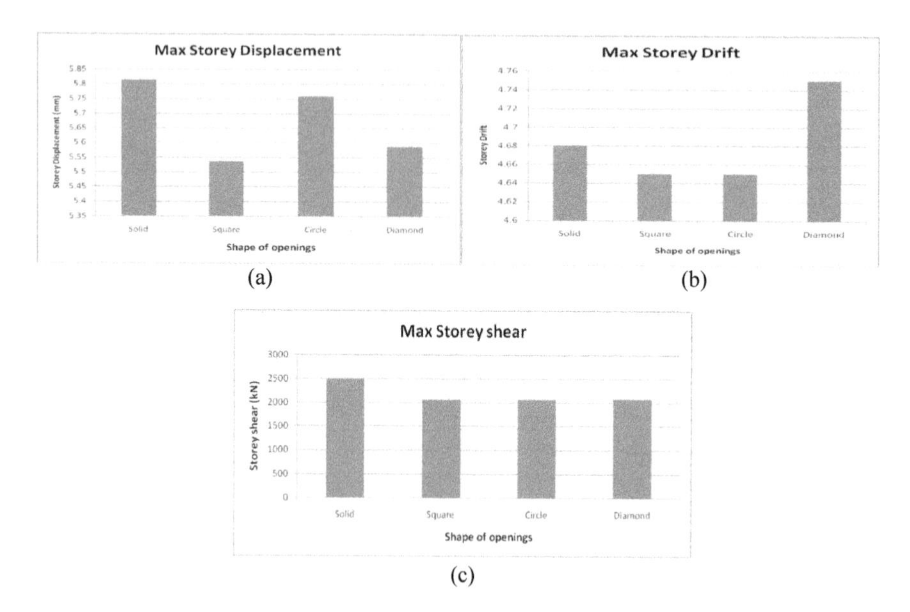

Fig. 6 **a** Maximum storey displacement. **b** Maximum storey drift. **c** Maximum storey shear

shaped opening is equal to that of solid slab opening. In case of square shaped opening, maximum storey shear reduces by 12.3% than that of solid slab opening.

4.3 M3-L-shaped Modal Results

Figure 6 shows the analysis results comparison of model M3.In case of L-shaped building models M3, with 20% square opening, maximum storey displacement is 5% less than that of building with solid slabs. Square and circular shaped opening have equal drift values. Base shear is reduced by 21% by proving square shaped openings than that of solid slabs.

4.4 M4-H-shaped Modal Results

Figure 7 shows the analysis results comparison of model M4. Maximum storey disp7lacement and maximum storey drift is least for circular shaped openings and storey shear reduces by 12% than that of solid slab case.

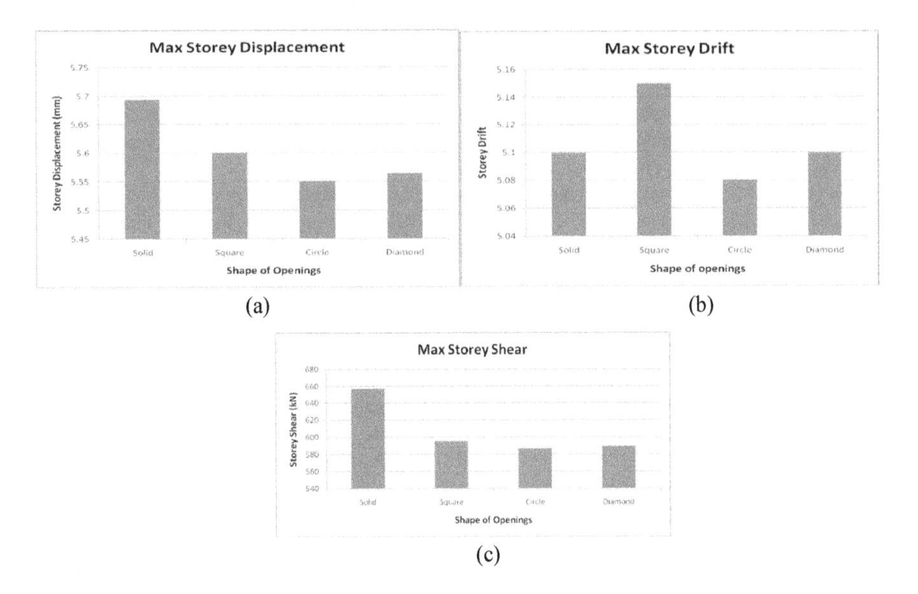

Fig. 7 **a** Maximum storey displacement. **b** Maximum storey drift. **c** Maximum storey shear

5 Conclusions

- The performance of different opening shapes in building varies according to the plan configurations. The behavior depends on the type of building plan.
- The presence of different diaphragm opening shapes changes the seismic behavior of structures. Models with square and circular shaped opening had almost identical responses for all parameters for buildings symmetrical in both directions, while models with symmetry changes had different responses.
- The earthquake resistance of the structure can be improved by providing optimum shape of openings. In case of L-shaped model using square shaped opening can reduce the seismic vulnerability of the structure. Maximum storey displacement is 5% less than that of building with solid slabs and base shear reduces by 21% by opting square shaped openings.
- For H-shaped building models, circular shaped opening is suitable. Maximum storey displacement is 2% less than that of building with solid slabs and base shear reduces by 12% by opting circular shaped openings.
- The effect of different opening shapes is more significant in irregular building plans than symmetrical configurations.

References

1. Al Harash MT, Panahshahi N (2012) Inelastic seismic response of reinforced concrete buildings with symmetric and unsymmetric floor diaphragm openings. In: Proceeding, 15th world conference on earthquake engineering
2. Al Harash M, Rathore A, Panahshahi N (2010) Inelastic seismic response of rectangular RC buildings with plan aspect ratio of 3: 1 with floor diaphragm openings. In: Structures Congress, ASCE
3. Vas VM, Nagaraja P, Venkataramana K (2021) Effect of diaphragm discontinuity on the seismic response of an RC building. Trends in civil engineering and challenges for sustainability. Lecture Notes in Civil Engineering 99, Springer Nature Singapore Pvt Ltd.
4. Vinod V, Pramod Kumar HV (2017, June) Influence of stiffness discontinuous diaphragm characteristics on the seismic behavior of Rc structure. Int J Sci Dev Res (IJSDR) 2(6)
5. Khajehdehi R, Panahshahi N (2016) Effect of openings on in-plane structural behavior of reinforced concrete floor slabs. J Build Eng 7 (Elsevier)
6. Khajehdehi R, Panahshahi N, Ghaffari R (2018) In-plane flexibility of reinforced concrete floor diaphragm with openings. Structure Congress, ASCE
7. Vinod Kumar P, Gundakalle VD (2015) Effect of diaphragm openings in multi-storeyed RC framed buildings using Pushover analysis (IRJET)
8. Bhuiyan MT, Leon RT (2013) Effect of diaphragm flexibility on tall building responses. Structures Congress, ASCE

Numerical Investigation and Parametric Analysis of Hybrid Shear Wall with Energy Dissipating Reinforcements

Sneha Benoy and Asha Joseph

Abstract Experiences from past earthquakes show that incorporating shear walls is an effective way in improving a buildings resilience to tolerate seismic activity. To overcome the limitations of conventional RC shear wall, a new resilient hybrid shear wall system with post tensioned tendons and external energy dissipating reinforcements (EEDR) is utilized. Previous research works on shear wall with EDR is limited and this research is intended to fill this gap. The study focuses on evaluating the effect of aspect ratio, post-tensioning force and tendon location on the behaviour of post-tensioned hybrid shear wall with external energy dissipating reinforcements. The scope of effort includes simulating twelve hybrid post-tensioned shear wall with varying aspect ratio, four specimens with varying post-tensioning force and two specimens with varying tendon profile.

Keywords Hybrid shear wall · Self-centering · Energy dissipating reinforcements · Post-tensioning · Aspect ratio

1 Introduction

Conventional RC shear wall has been widely used as a lateral force-resisting system in a number of buildings to provide primary resistance against seismic excitations. One of the key drawbacks of using traditional shear wall construction is that the building is greatly weakened when a powerful earthquake hits, posing a danger to lives as well. With traditional constructions, substantial damage is likely to occur with large residual lateral displacements and wide residual cracks; thus, the cost and effects of damage during an earthquake may be significant to the occupant of the building.

In response for the need to improve seismic performance of buildings, there has been a growing body of experimental research on self-centering structural systems

S. Benoy (✉) · A. Joseph
Department of Civil Engineering, Federal Institute of Science and Technology, Angamaly 683577, India

G. C. Marano et al. (eds.), *Proceedings of SECON'21*, Lecture Notes in Civil Engineering 171, https://doi.org/10.1007/978-3-030-80312-4_16

that can sustain severe earthquake shaking with minimal damage or limited residual deformations. Since the early 1990s, researchers have been investigating the behavior and design of precast concrete structures subjected to seismic loading. To reduce residual drift of conventional structural systems after earthquakes, Priestley and Tao [1] first proposed the idea of utilizing unbonded post-tensioned (PT) tendons in precast RC frame structures to provide the primary lateral force resistance and restoring force. The findings of the study conducted by Erkmen and Schultz [2] show that self-centering behaviour can be achieved even when the PT forces fully dies out by proper design of the end anchorages of PT tendons. The use of post-tensioned (PT) tendons proves to be an efficient solution to resolve the shortcomings of conventional RC shear wall during an earthquake such as flexural and shear cracks, toe crushing, and rebar failure because of their superior self-centering potential and higher energy dissipation capacity [3–10]. Further, it is found that the PT shear walls require lesser steel than conventional shear walls, which makes them an economical choice as well [11].

Earlier studies primarily focused only on fully post-tensioned walls with no energy dissipating (ED) steel reinforcement [12–15]. The biggest limitation in using fully post-tensioned walls in seismic regions is that the energy dissipation is very small; and thus, the lateral displacements during an earthquake can be considerably larger. Kurama and Shen [4] proposed the concept of a self-centering shear wall using unbonded PT tendons and formulated the corresponding design method. A number of researchers have investigated the use of supplemental energy dissipators, such as friction dampers, fluid dampers, and yielding dampers, to reduce the seismic displacements of fully post-tensioned precast walls [1, 10, 13, 16, 17].

Instead of resorting to complicated configurations, Zhu and Guo [18] developed hybrid PT wall system which utilized internal energy dissipating reinforcements and PT bars. PT tendons allows the shear wall to undergo larger nonlinear displacement without yielding of PT bars [7, 18]. The restoring force provided by post-tensioning steel results in higher self-centering capacity upon unloading which eliminates residual deformation after the seismic action. Consequently, the shear wall can experience larger lateral deformation without failure [19]. The study carried out by Smith et al. [5, 6] shows that hybrid shear walls with continuous mild steel bars showed superior restoring power, energy dissipation and ductile behavior over larger lateral displacements.

While the performance of hybrid shear walls with internal EDR is satisfactory, post-yielding replacement is a major concern. To overcome this challenge, Taori et al. [9] developed a new model which utilises external energy dissipating reinforcements to meet the needs of easy placement and replacement of EDR and convenient rehabilitation of the wall post any seismic activity. This study aims at investigating the behaviour of hybrid shear wall under various governing parameters like effect of aspect ratio, effect of post-tensioning force and tendon location and the results may be used for further research.

2 Model Configuration

In the present study, the hybrid PT shear wall developed by Taori et al. [9] having a height (h_w) of 2830 mm, length (l_w) of 1000 mm and thickness (t_w) of 100 mm is considered for validation. The term "hybrid" reflects that a combination of mild reinforcing steel and high-strength unbounded post-tensioning (PT) steel is used. The foundation is having a dimension of 1200 mm × 300 mm × 100 mm and the section is modelled as a rigid beam to simulate strong floor beam. The length of energy dissipating reinforcement (EDR) is kept as $25d_b$ as per ACI-ITG 5.2 (ACI 2009) [20] and diameter is 9 mm at the middle portion and 10 mm at both the ends. EDR spans externally between angle and channel sections. The angle section (ISA 75 × 75 × 10) is attached to the shear wall and the channel section (ISMC 100) is embedded inside the foundation beam. 2 # 12.7 mm ø PT tendons are used. The structure is modelled and analyzed in ANSYS v16 since it is a sophisticated software for FE analysis.

The concrete wall, foundation beam, PT tendons and loading plates are modelled using higher order three-dimensional, twenty-node solid element that exhibits quadratic displacement behavior (SOLID186) available in ANSYS. Each node of these elements has three translational degrees of freedom. The finite element model of hybrid shear wall is shown in Fig. 1. The reinforcing elements, including longitudinal reinforcement, transverse reinforcement, and confining reinforcement, are modelled using three-dimensional two-node beam elements (BEAM188) with six DOFs in each node. The longitudinal and lateral reinforcements are embedded inside the concrete shear wall. The EDR, angle, and channel sections are modelled using the SOLID186 element.

3 Material Property

The behavior of concrete is described using concrete damaged plasticity model and reinforcement is modelled using a bilinear curve with strain-hardening with reference to the yield and ultimate stress and strain values. Material properties of various reinforcements, concrete and PT tendon used in the present study are given in Tables 1 and 2 respectively.

4 Boundary Conditions, Meshing and Loading

The bottom end of the foundation is considered as a fixed base. The PT shear walls are subjected to a displacement controlled incremental loading to simulate its behavior under seismic loads. Prior to that, gravity load and post-tensioning of the tendons is done. A design tensile force of 73.9 kN is induced in each PT tendon. In

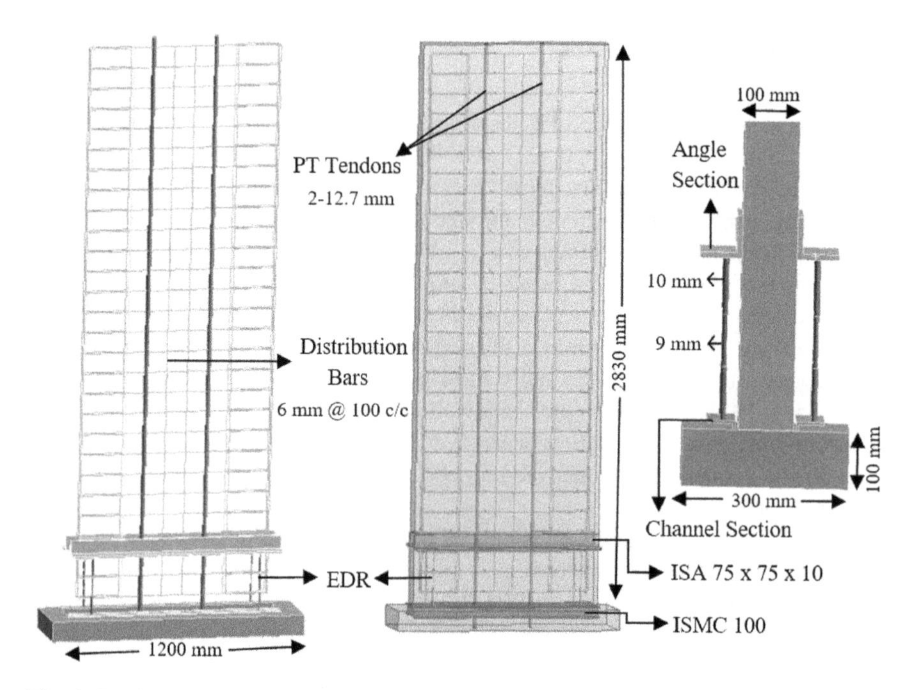

Fig. 1 Details of hybrid shear wall **a** Front view. **b** Side view

Table 1 Material properties of reinforcements [9]

Reinforcement type	Diameter (mm)	f_y (MPa)	f_u (MPa)	ε_y	ε_u	E (GPa)
Distribution steel	6	528	622	0.0025	0.15	210
EDR	9	550	650	0.0027	0.14	206
Tie rod	12	502	563	0.0024	0.14	208

Table 2 Material properties of concrete and PT tendons [9]

Material type	Modulus of elasticity (MPa)	Compressive strength (MPa)	Poisson's ratio	Yield strength (MPa)	Ultimate strength (MPa)
Concrete	31,622	40	0.15	–	–
PT tendons	2×10^5	–	0.3	1650	1850

the next step, displacement-controlled loading is applied at a height of 2.3 m from the base of the wall as shown in Fig. 2. The meshing of the concrete shear wall foundation beam and all other elements is done using medium size mesh with a total number of nodes and elements of 79,947 and 15,447 respectively. The finite element model of the shear wall is shown in Fig. 3.

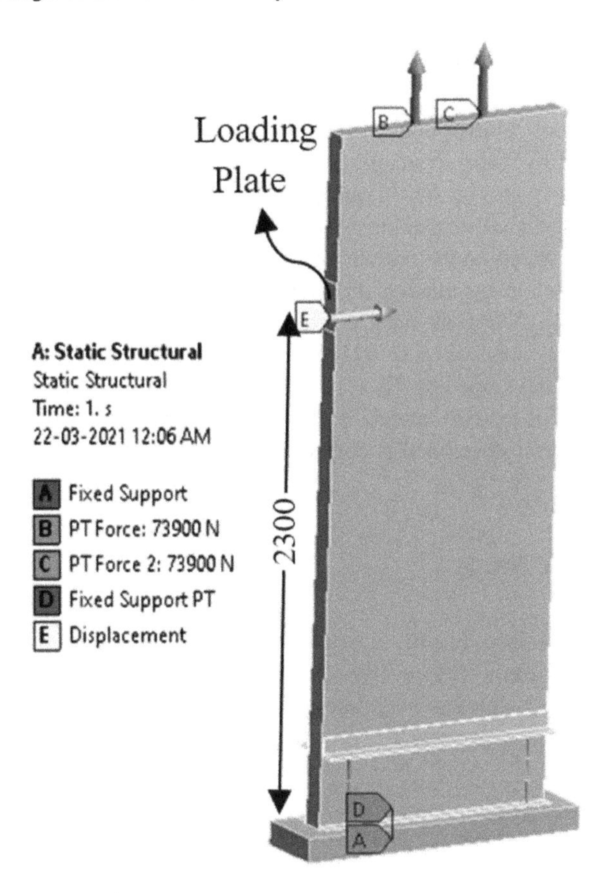

Fig. 2 Loading and boundary conditions

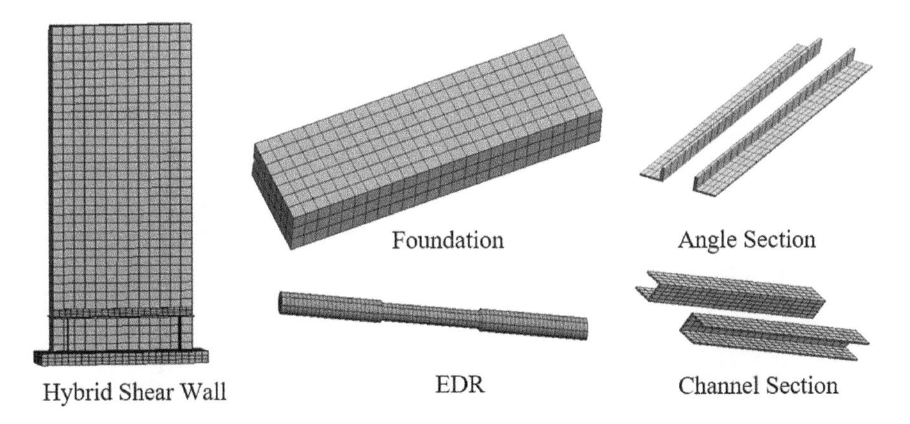

Fig. 3 Finite element model of hybrid shear wall

5 Verification of Finite Element Model

The comparison of results obtained from the verification model and the numerical study conducted by Taori et al. [9] is made in terms of base shear—lateral drift response, as shown in Fig. 4b. It was observed that the wall experienced rocking behavior with stress concentrating on the EDR portion. The maximum stress in EDR was accumulated on the middle portion as the weak zone of EDR was shifted to the middle portion by making the diameter of the end portion larger than the middle portion [9]. The base shear corresponding to 2.4% drift obtained from the present numerical analysis was 921 kN whereas the value obtained from the numerical model developed by Taori et al. [9] is 870 kN. The maximum difference in the base shear of the two models is about 5.86% observed at 2.4% drift. Hence, the numerical model developed is considered suitable for further studies.

6 Parametric Study

Based on the obtained results, a parametric analysis is carried out to further investigate the feasibility of the hybrid shear wall. Finite element study of shear wall with various design variables like aspect ratio (h_w/L_w), post-tensioning force and tendon location is investigated.

6.1 Effect of Aspect Ratio

It has been noted that aspect ratio (height to length ratio) greatly influences the performance of shear walls [21–24]. To understand the influence better, finite element model of 12 post tensioned shear walls with EDR models with varying

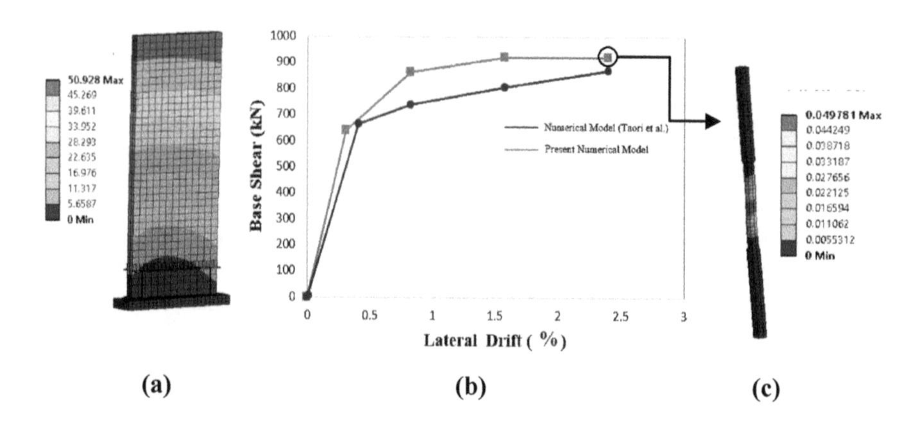

Fig. 4 **a** Total deformation. **b** FE model validation chart. **c** Plastic strain in EDR

aspect ratio is conducted. The aspect ratio for shear walls was varied between 2 and 4, which classifies them as medium-rise walls according to CSA and ACI codes [21].

For aspect ratio ranging from 2 to 3, the specimens were of height (h_w) 3600 mm and thickness (t_w) 100 mm. For aspect ratio varying from 3 to 4, height of the specimen was kept at 4370 and 100 mm thick (t_w). Table 3 shows the description of wall specimens for each aspect ratio. The concrete compressive strength for all specimens is 40 MPa. The specimens are subjected to displacement controlled incremental loading and the response is compared in terms of base shear obtained.

Figure 5 gives the base shear versus displacement response of the wall system under monotonic loading for varying aspect ratio. For specimens with aspect ratio between 2 and 3, it is observed that all the specimens reached their ultimate strength capacity. The uplift of the wall has been delayed in SW1 compared to SW6. The yielding of energy-dissipating reinforcement started early in specimens with lower aspect ratio. The stiffness degradation of EDR in SW1 and SW2 is gradual compared to the other three specimens. It is further noted that as the aspect ratio increases, the base shear value decreases. For SW1-SW6, base shear value has been reduced by an amount of 29% and for SW7-SW12, the value has reduced by 19%.

Compared to Fig. 5a, the specimens of aspect ratio varying from 3 to 4 (Fig. 5b) doesn't have a smooth curve. Specimens SW9 and SW10 failed before reaching the ultimate capacity. The region between yielding and failure is small compared to specimens SW1-SW6. It is observed that, the lateral load response of SW1-SW6 is mostly linear after the decompression state while it is not that profound in SW7-SW12.

Table 3 Details of wall specimen

Wall	Aspect ratio	Height (mm)	L_w (mm)	No. of PT tendons	No. of EDR
SW1	2	3600	1800	4	5
SW2	2.2	3600	1636	4	5
SW3	2.4	3600	1500	4	4
SW4	2.6	3600	1384	4	4
SW5	2.8	3600	1285	3	4
SW6	3	3600	1200	3	2
SW7	3	4370	1456	4	4
SW8	3.2	4370	1365	3	4
SW9	3.4	4370	1285	3	4
SW10	3.6	4370	1213	3	2
SW11	3.8	4370	1150	3	3
SW12	4	4370	1092	3	2

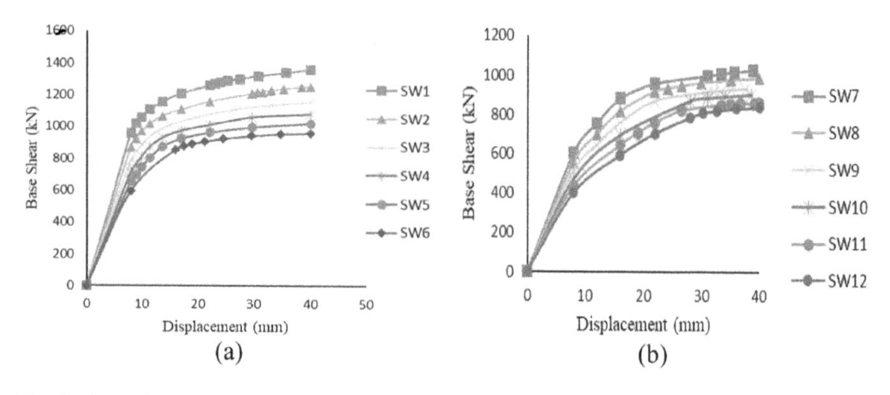

Fig. 5 Base shear versus displacement **a** Aspect ratio—2 to 3. **b** Aspect ratio—3 to 4

6.2 *Effect of Post-tensioning Force*

To analyze the effect of post-tensioning force on the behavior of hybrid shear wall, 4 different PT forces are applied to the tendons and the structure is analyzed under monotonic loading. As per ACI-ITG 5.2 (ACI 2009), effective stress in PT tendons less than 0.3 f_{pu} is not recommended. The stress value is selected such that the tensile stress in prestressing should not exceed 0.94 f_{py} but not greater than 0.80 f_{pu}, where f_{py} and f_{pu} are yield stress and ultimate stress of PT steel respectively [7].

For the present numerical study, an effective stress value of PT tendon is varied in the range of 0.3 f_{pu}, 0.4 f_{pu}, 0.5 f_{pu} and 0.6 f_{pu}. The height and length of the shear wall is 3600 mm and 1200 mm respectively. The thickness is kept the same as 100 mm. The cross-section of the prototype shear wall is shown in Fig. 6.

Figure 7 shows the base shear vs displacement for all the four specimens with varying prestressing force. Table 4 gives the response quantities like decompression at the base of the wall (V_{dec}), yielding state (V_y) and compression of concrete walls (V_c) and their corresponding lateral displacement (Δ).

From Table 4, it is noted that as the effective prestressing force f_{pe} is increased, the base shear values (i.e., V_{dec}, V_y and V_c) has been improved significantly. V_{dec} depends upon the prestressing force and it is observed that V_{dec} decreases as f_{pe}

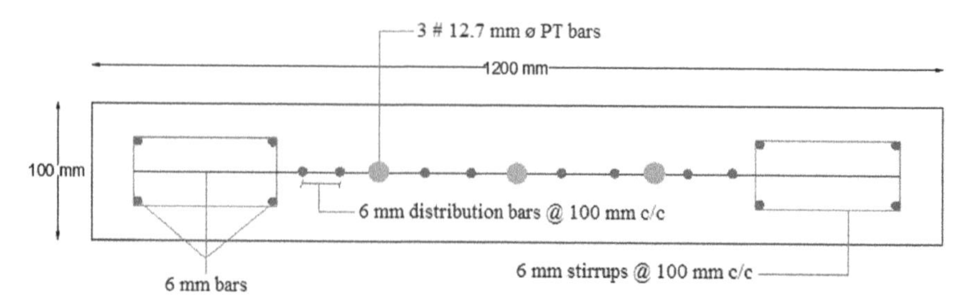

Fig. 6 Cross-sectional details of shear wall

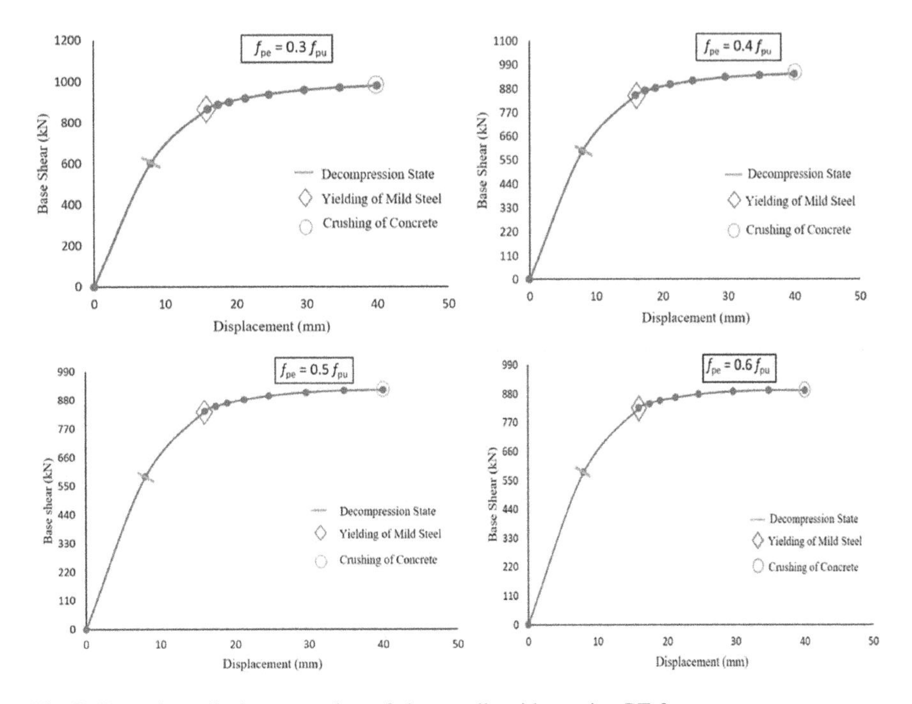

Fig. 7 Base shear-displacement plots of shear walls with varying PT force

increases. The maximum base shear corresponding to yielding of mild bar occurred at 0.3 f_{pu}. 8.7% decrease in V_c has been achieved when f_{pe} has been increased from 0.3 f_{pu} to 0.6 f_{pu}. The effect of f_{pe} on Δ_{dec} and Δ_c is negligible. However, the values of lateral displacement corresponding to yielding was dropped by increasing the tendon forces.

6.3 *Effect of Tendon Location*

To identify the effect of tendon location, reference specimen with four tendons and two other models with varying tendon profile were analyzed. All the wall specimens were having a cross-section of 3600 mm × 1200 mm × 100 mm. In the first model (centered tendon), the outermost two tendons were removed, and the area of the centered tendons was doubled. In the second model (exterior tendon), two tendons closest to the wall center were removed, and the area of two outermost tendons was tripled.

The cross-sectional model for reference specimen, centered tendon specimen and exterior tendon specimen is shown in Fig. 8. A prestressing force of 0.45 f_{pu} is induced in all the cases and the walls are analysed under monotonic incremental

Table 4 Response Quantities of Wall Specimen

PT Force (f_{pe})	Response Quantity					
	V_{dec} (kN)	Δ_{dec} (mm)	V_y (kN)	Δ_y (mm)	V_c (kN)	Δ_c (mm)
0.3 f_{pu}	599.51	8	916.89	20	977.13	40
0.4 f_{pu}	592.53	8	884.03	19	947.84	40
0.5 f_{pu}	585.58	8	868.19	18.85	919.53	40
0.6 f_{pu}	578.67	8	852.52	18.67	892.28	40

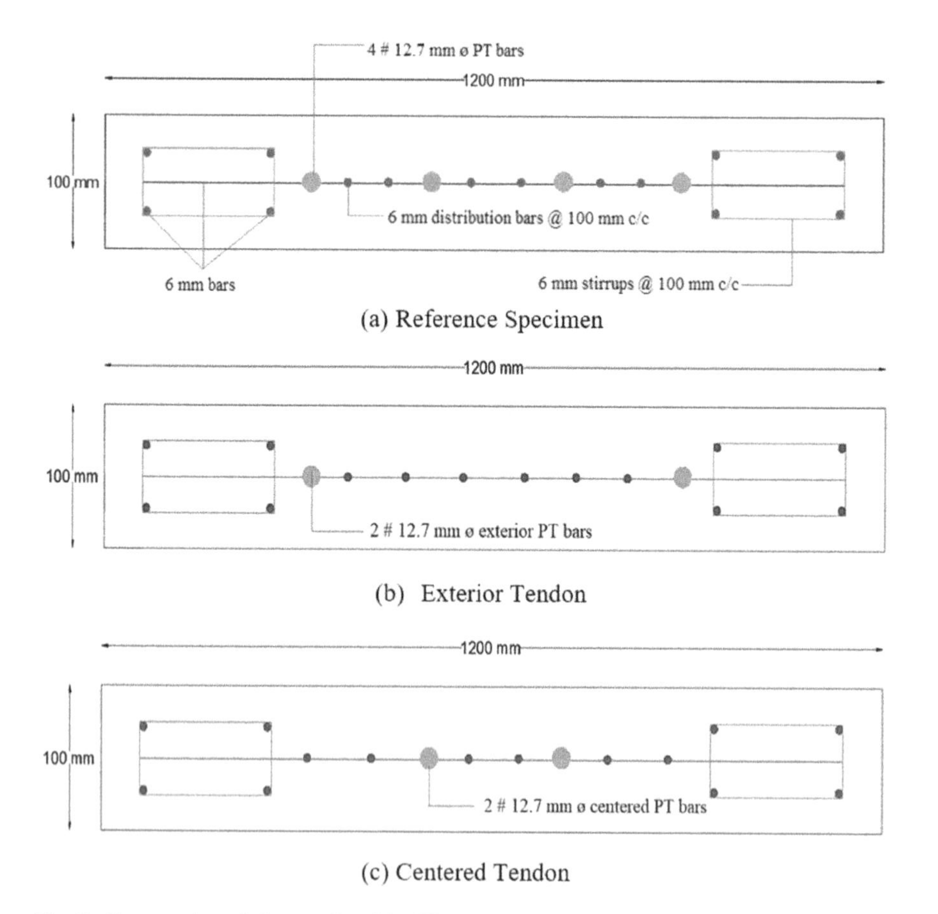

Fig. 8 Cross-section of shear walls with different tendon profile

loading. The lateral load v/s drift relationships for the reference specimen, centered tendon and exterior tendon models of the shear wall are presented in Fig. 9.

Compared to reference specimen, the base shear value is 8.59% less in shear wall with exterior tendons and it is 2.53% higher in centered tendons. The effective stiffness (prior to system yield) of both the models are almost the same. The

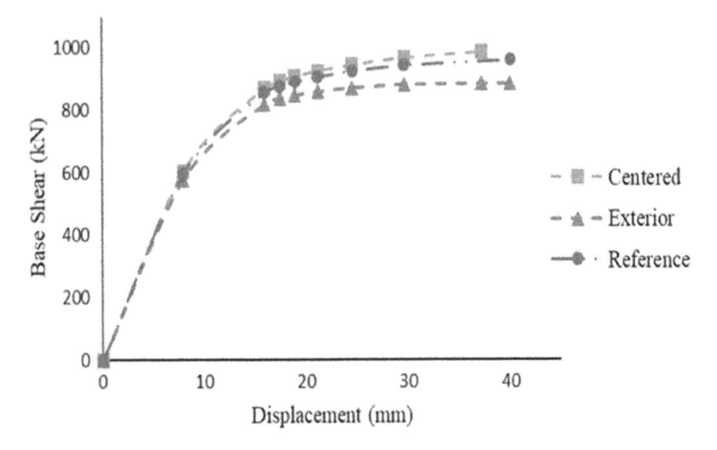

Fig. 9 Comparison of reference, centered and exterior tendons

difference is found in the lateral load values as exterior tendons experienced a lower base shear value compared to centered tendon. It is believed that the variation caused is due to the cross-sectional difference of PT tendons in both the models.

From Fig. 9, it has been observed that changing the tendons to the exterior has desirable characteristics than placing the tendons centrally. The base shear is found to reduce by 10.24% in externally placed tendons compared to centered tendons. Nevertheless, the self-centering property is least effected by changing the tendon location.

7 Conclusions

In the present study, concrete shear wall with both post-tensioned tendons and energy dissipating reinforcements in the form of mild steel bars is considered. This hybrid shear wall with varying governing parameters like aspect ratio, post-tensioning force and tendon location were analyzed. The shear wall was subjected to monotonic loading and the conclusive outcomes drawn from the analysis carried out is presented below.

Aspect ratio plays an important role in the overall structural performance of hybrid shear wall. To understand the effect of aspect ratio, 12 specimens with aspect ratio varying from 2 to 4 were analysed. It has been noted that:

- For aspect ratio ranging from 2 to 3, the base shear has been reduced by an amount of 29% while for shear wall with aspect ratio 3–4, the value has only been reduced by 19%.
- Increasing the aspect ratio can reduce the base shear but can negatively affect the stiffness of shear walls.

- Aspect ratio affects the uplift, yielding and stiffness degradation. Aspect ratio of 2.4–3.4 has desirable properties and can be adopted for practical purposes.
- The yielding of energy-dissipating reinforcement started early in specimens with lower aspect ratio.

Prestressing force can be considered as a governing factor for the response of shear wall under seismic forces. From the present study, it has been observed that:

- Effective prestressing force can increase a structures capacity to resist lateral deformations.
- A minimum prestressing force of 0.4 f_{pu} is required to acquire the desired seismic performance and enhance the lateral load carrying capacity.
- Base shear corresponding to V_c has been reduced by an amount of 8.7% when f_{pe} has been increased from 0.3 f_{pu} to 0.6 f_{pu}.
- Yielding of mild reinforcement bars can be reduced by increasing the tendon force.

To understand the effect of tendon location, two models with centered tendons and exterior tendons was studied. The base shear vs displacement plot for both the specimens showed that:

- The base shear is found to reduce by 10.24% in externally placed tendons compared to centered tendons.
- The effect of tendon profile on the self-centering behavior of hybrid shear wall is negligible.

References

1. Priestley MJN, Tao JR (1993) Seismic response of precast prestressed concrete frames with partially debonded tendons. PCI J 38(1):58–69
2. Erkmen B, Schultz AE (2009) Self-centering behavior of unbonded precast concrete shear walls. Earthq Resistant Eng Struct 13(7):1047–1064
3. Holden T, Restrepo J, Mander JB (2003) Seismic performance of precast reinforced and prestressed concrete walls. J Struct Eng 129(3):286–296
4. Kurama YC, Shen Q (2004) Post Tensioned hybrid coupled walls under lateral loads. J Struct Eng 130(2):297–309
5. Smith BJ, Kurama YC, McGinnis MJ (2011) Design and measured behavior of a hybrid precast concrete wall specimen for seismic regions. J Struct Eng 137(10):1052–1062
6. Smith BJ, Kurama YC, McGinnis MJ (2013) Behavior of precast concrete shear walls for seismic regions: comparison of hybrid and emulative specimens. J Struct Eng 139(11):1917–1927
7. Shatnawi A, Abdallah SG, Tarawneh B (2019) Seismic behavior of hybrid post-tensioned cast in place concrete shear walls. Arab J Sci Eng 44(5):4095–4109
8. Li Y, Liu Ye (2019) A study on coupled wall system with post-tensioned steel truss coupling beams: Concept and performance evaluation. KSCE J Civ Eng 23(8):3560–3570
9. Taori P, Dash SR, Mondal G (2020) Seismic response of post tensioned hybrid shear walls with external energy dissipating reinforcement (EEDR). J Earthq Eng 1–16

10. Zareian MS, Esfahani MR, Hosseini A (2020) Experimental evaluation of self-centering hybrid coupled wall subassemblies with friction dampers. Eng Struct 214:110644
11. Chavan R, Kannan RM (2017) Analysis and design of vertical post-tensioned precast shear wall. Int J Recent Innov Trends Comput Commun 5(4):50–54
12. Kurama Y et al (1999) Lateral load behavior and seismic design of unbonded post-tensioned precast concrete walls. Struct J 96(4):622–632
13. Kurama YC et al (2002) Seismic response evaluation of unbonded post-tensioned precast walls. Struct J 99(5):641–651
14. Rahman AM, Restrepo-Posada JI (2000) Earthquake resistant precast concrete buildings: seismic performance of cantilever walls prestressed using unbonded tendons
15. Perez FJ, Sause R, Pessiki S (2007) Analytical and experimental lateral load behavior of unbonded posttensioned precast concrete walls. J Struct Eng 133(11):1531–1540
16. Perez FJ, Pessiki S, Sause R (2004) Seismic design of unbonded post-tensioned precast concrete walls with vertical joint connectors. PCI J 49(1):58–79
17. Ajrab JJ, Pekcan G, Mander JB (2004) Rocking wall–frame structures with supplemental tendon systems. J Struct Eng 13(6):895–903
18. Zhu Z, Guo Z (2017) Experimental study on emulative hybrid precast concrete shear walls. KSCE J Civ Eng 21(1):329–338
19. Li Y, Liu Y, Chen Z (2020) Seismic response assessment of a hybrid coupled wall structure with novel self-centering steel truss coupling beams. Bull Earthq Eng 1–24
20. ACI (2009) Requirements for design of a special unbonded post-tensioned precast shear wall satisfying ACI-ITG-5.1 and commentary. ACI-ITG-5.1-09, ACI Committee 93, American Concrete Institute
21. Mohamed N et al (2014) Experimental investigation of concrete shear walls reinforced with glass fiber–reinforced bars under lateral cyclic loading. J Compos Constr 18(3):A4014001
22. Wallace JW, Moehle JP (1992) Ductility and detailing requirements of bearing wall buildings. J Struct Eng 118(6):1625–1644
23. Jiang H, Kurama YC (2010) Analytical modeling of medium-rise reinforced concrete shear walls. ACI Struct J 107(4)
24. Mostofinejad D, Anaei MM (2012) Effect of confining of boundary elements of slender RC shear wall by FRP composites and stirrups. Eng Struct 41:1–13

Estimating Landslide Risk Management Index for Nilambur Taluk, Kerala Using Analytic Hierarchy Process—A Case Study

P. K. Ajmal and S. Jawahar Saud

Abstract The Western Ghats of Kerala are extremely vulnerable, with landslides posing a persistent danger. In the last few years, Kerala has faced several landslides. As many as 145 people have died in landslides over the last two years in Kerala. The occurrence of natural events is uncontrollable, landslides can be mitigated by good risk control. For reducing the risk of disasters in a particular area, they need to estimate the extent of risk and the various types of vulnerabilities correlated with it. No specific indicators exist in India for the assessment of landslide risk management. The deficiency of a formal conceptual structure of disaster risk that could enable a multidisciplinary assessment and intervention has hampered successful disaster risk management. Cardona et al. (Disaster risk and risk management benchmarking: a methodology based on indicators at national level. IDB-IDEA Program on Indicators for Disaster Risk Management, Universitad National De Colombian, Manizales, 2004) [12] put forward the Risk Management Index (RMI) to satisfy these requirements. Four public policies were used to calculate the RMI, and each of them is calculated by the weighted values of its indicators. The Analytic Hierarchy Process (AHP) is used to distribute the weights. The study focuses on developing a replicable methodology to determine RMI at the local level. The procedure for estimating this index is centered on surveys of professional personnel, deciders, and stakeholders engaged in risk mitigation activities at all times. The area selected for this study is Nilambur Taluk of Malappuram district and its majority falls under the landslide susceptible area.

Keywords Landslides · Disaster risk management · Analytic hierarchy process · Risk management index · Nilambur

P. K. Ajmal (✉) · S. J. Saud
Department of Civil Engineering, FISAT, Angamaly, Kerala, India

G. C. Marano et al. (eds.), *Proceedings of SECON'21*, Lecture Notes in Civil Engineering 171, https://doi.org/10.1007/978-3-030-80312-4_17

1 Introduction

A landslide is a displacement of earth, rock, and organic materials caused by gravity. The movement may be triggered by weathering or by external mechanisms. Heavy rainfall is the primary cause of landslides [1]. Rainwater percolates and raises hydraulic pressure, which reaches the elastic limit of the soil or rocks. As a result, the pressure builds up, causing the soil and rocks to loosen their elastic strengths, resulting in landslides [2]. The raging flood, groundwater level fluctuations, and variation in the level of water along shorelines, reservoir banks, waterways, earth dams, wetlands, and rivers are all examples of this influence. Falls, slides, topples, flows, and spreads are all examples of how landslides can manifest themselves. Geomorphological and geographical conditions are the primary determinants of landslides [3] and precipitation, volcanic activity, seismic activity, fluctuations in groundwater, and anthropogenic activities are all potential triggers [4]. Every year, landslides in India pose a serious threat to human life, resulting in an annual loss of $400 million [5]. Economic losses from landslides have been increasing over recent decades [6], rising construction and investment in landslide-prone areas are mostly to blame [7]. Landslides are a major threat to human life, buildings, resources, and natural ecosystems in most mountainous areas of the world. According to data from the Centre for Research on Disaster Epidemiology (CRED) (Brussels, Belgium) shows that landslides are to blame for at least 17% of all fatalities from natural disasters all over the world. Kerala is especially prone to natural disasters and shifting climatic dynamics due to its geographical position along the sea coast and steep gradient along the Western Ghats slopes. Kerala is one of India's most heavily populated districts (860 persons per km^2) [8]. As a result, it is much more susceptible to damages and losses as a consequence of disasters. Kerala is one of India's most vulnerable states when it comes to mass movements [9, 10]. Slope instability has a long history in Kerala dating back to 1341 AD [11]. In Kerala, landslides are common during monsoon periods, especially in hilly tracts of Idukki, Malappuram, Palakkad, Kozhikode, and Wayanad districts.

The occurrence of natural events is uncontrollable, proper risk management can reduce the consequences of landslides. For reducing the risk of disasters in a particular area, they need to assess the level of risk and the different types of vulnerabilities associated with it. No specific indicators exist in India for the assessment of landslide risk management. The absence of a formal conceptual definition of catastrophe risk that could enable a multidisciplinary assessment and response has hampered successful disaster risk management. The Risk Management Index (RMI) put forward by Cardona et al. [12] meets these needs. These performance assessment methods for landslide risk management assist in understanding the current issues, but they can also focus decision-makers on the strategies and measures that will be taken in the future. This work's key goals have been defined as; (1) to perform a perception survey on the landslide risk management (2) to analyze the survey data using the AHP technique (3) to identify the Risk Management Index (RMI) for Nilambur Taluk, Malappuram (4) to learn about the current state of risk reduction programs in the research area.

2 Selection of Study Area

Nilambur taluk (Latitude 11° 16′ 37″ and Longitude 76° 13′ 33″) in Malappuram district, Kerala, is selected as the study area. It is located close to the Nilgiris range of the Western Ghats on the banks of the Chaliyar river. Nilambur taluk is one of the 5 most landslide-prone taluk of the state. About 198.6 sq. km area of Malappuram district falls under landslide high hazard zone and 267.6 sq. km under low hazard zone. From the landslide susceptibility map of Malappuram (Fig. 1), it is clear that a major portion of Nilambur taluk is under a landslide high hazard zone. On the 8th of August 2019, massive losses were reported in Kavalappara in the Puthukkad area of Malappuram district, with 81 lives lost and extensive property damage. Over 39 homes and a pedestrian bridge were completely destroyed. From a field study of landslide-affected areas in 2019, 18 locations under Nilambur taluk were found to be unfit for human settlement. Nilambur taluk has a total population of 574,059 as per the Census 2011. Out of which 273,909 are males while 300,150 are females. In 2011 total of 120,552 families were residing in Nilambur taluk. Out of the total population, 8.1% of people live in urban areas while the major population of 91.9% lives in the rural areas. Out of 171,955 workers engaged in main workers, 8,862 were cultivators (owner or co-owner) while 20,027 were agricultural laborers. As per the report in 2015 more than 3600 people were directly affected by landslides in Nilambur taluk and around 300 houses were damaged.

2.1 History of Risk Management Index (RMI)

Cardona et al. in 2004 [12] proposed the RMI method, Cardona et al. [13] later updated and used this definition in 2005 and Carreño et al. [14] to evaluate the viability of natural disaster risk reduction and management in Latin American and Caribbean countries (LAC). This index was one of the outputs of the Inter-American Development Bank/Institute of Environmental Studies of the National University of Colombia (IADB/IDEA) [15] disaster impact assessment metrics initiative. For 17 countries in the LAC region, the index has been updated to 2008 [16]. To perform a comparative analysis, Chiu [17] used the RMI method for the study of landslide risk management perceptions in Norway and Hong Kong. In early 2015, a study was undertaken in Norway to analyse public views of landslide risk management for different periods such as 2004, 2014, and 2024 at the county and national levels.

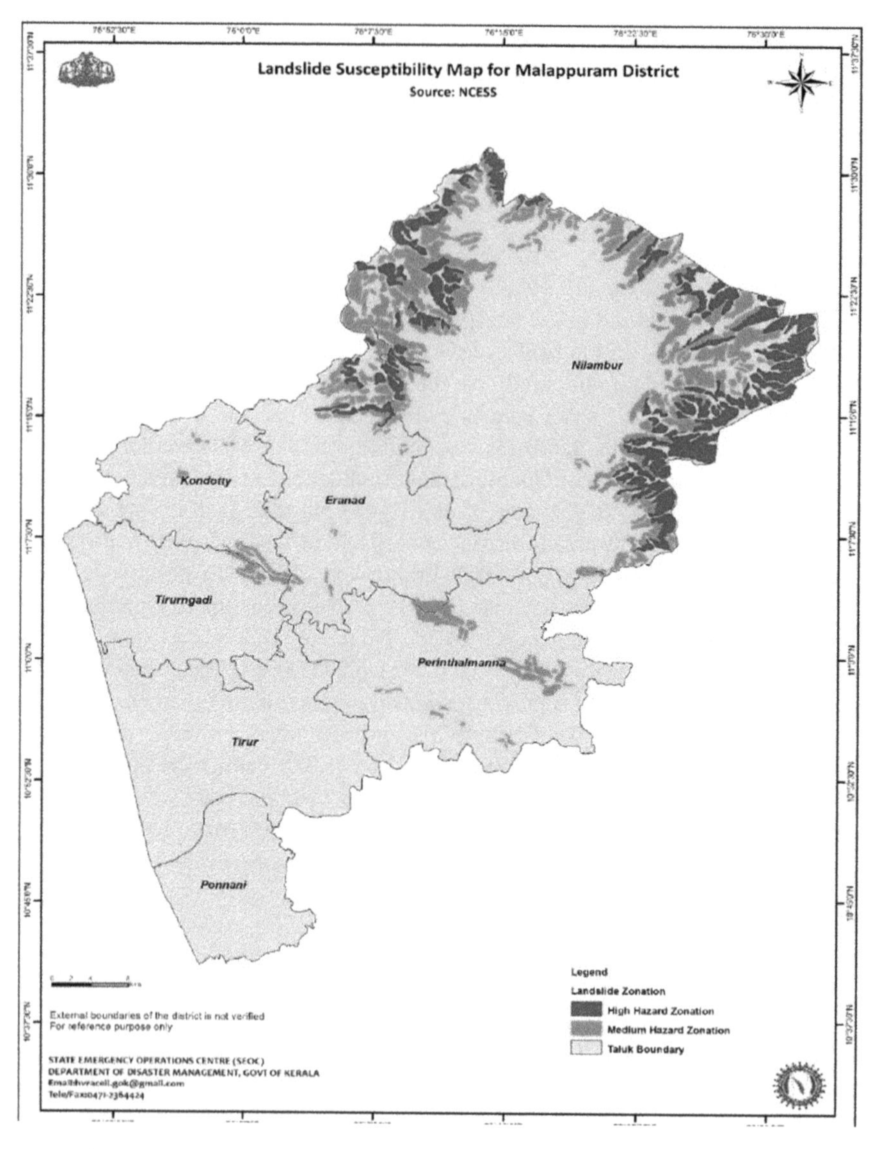

Fig. 1 The landslide susceptibility map of the Malappuram district. *Source* District disaster management plan, Malappuram 2015

3 Methodology

The indices proposed by Cardona et al. were modified in this paper [12, 16] with the aim of creating sub-indices that can be used on a local level. The Analytic Hierarchy Process (AHP) [18, 19] is used to calculate the indicator's weights (w_i)

within the four composite indices, just like the method developed by Cardona et. al. [13]. The RMI, on the other hand, is calculated as mentioned below, using simple statistical analysis. The modified subindices used in this project is as follows:

3.1 Risk Identification (RMI_{RI})

1. RI_1. Systematic disaster and loss databases
2. RI_2. Hazard surveillance and prediction
3. RI_3. Hazard assessment and mapping
4. RI_4. Risk and vulnerability evaluation
5. RI_5. Citizen's participation and community information
6. RI_6. Risk management awareness and training

3.2 Risk Reduction (RMI_{RR})

1. RR_1. Zoning and land use planning
2. RR_2. Hydrographic basin intervention and environmental protection
3. RR_3. Execution of hazard-event control and protection techniques
4. RR_4. Housing improvement and relocation from prone-areas
5. RR_5. Updating of safety standards and construction codes
6. RR_6. Reinforcement and retrofitting of public and private assets

3.3 Disaster Management (RMI_{DM})

1. DM_1. Emergency preparedness and continuity planning
2. DM_2. Information and warning systems
3. DM_3. Emergency response
4. DM_4. Community preparedness and training
5. DM_5. Rehabilitation and reconstruction planning

3.4 Governance and Financial Protection (Loss Transfer) (RMI_{FP})

1. FP_1. Inter-institutional organization and strengthening
2. FP_2. Budget allocation and mobilization
3. FP_3. Insurance and disaster funds

The major stakeholders in the community evaluate all of these sub-indicators based on five performance levels: low, incipient, significant, outstanding, and optimal. For data analysis, the Likert scale is used to denote the performance levels in values ranging from low = 1 to optimal = 5.

3.5 Step 1: Data Gathering

Data is gathered from local authorities in charge of disaster mitigation, prevention, and response in the evaluated community. The first questionnaire asks for ratings of each indicator's performance level. The questionnaire developed by Chiu et al. [17] is taken as a reference for the first questionnaire. It is modified according to the conditions in Kerala and simplified for the use of the local level. Each indicator's value is assigned to one of five performance levels. A brief description of each indicator is also provided to have an overview of the scope of each indicator. If appropriate, participants can select 'not relevant' or 'unable to answer'. It's worth mentioning that the technique used by Cardona et al. [13] does not account for the 'not relevant' or 'unable to answer' options. A form for distributing relative importance between pairs of measures based on the AHP is included in the second questionnaire.

3.6 Step 2: Calculation of Weights (w_i) for Each Sub-indicator Using AHP

The AHP method is used to weigh each sub-indicator as in Cardona et al. [13]. Centered on mathematical and psychological principles, the AHP is a formal methodology for arranging and evaluating complex decisions. It represents and quantifies the components of a decision-making process, links them to general priorities, and evaluates possible alternatives, providing a systematic and realistic basis for structuring a decision-making process. The importance of one indicator over the other one is rated with numerical values ranging from 1 to 9. Here 1 means both measures are equally important, while degree 9 indicates that one measure is 9 times more significant than the other. Each comparison's results are tabulated to create a comparison matrix. With the help of eigenvalues and eigenvector, the values for relative weights are calculated. The final step is to quantify a Consistency Ratio (CR) to see how accurate the decisions were compared to vast samples with completely random judgments.

$$Consistency\,Index(CI) = \frac{\lambda_{max} - n}{n - 1} \tag{1}$$

$$Consistency\,Ratio(CR) = \frac{CI}{RI} \tag{2}$$

Here λ_{max} is the largest eigenvalue and 'n' is the number of criteria for comparison. And *RI* denotes the Random Index and it will vary concerning the value of 'n'. Saaty [18] suggested that if the value of CR larger than 0.1 the results of pair vice comparison have to be re-examined.

Local officials are being asked to compare all RMI sub-indicators by comparing in pair vice and evaluating their effect on disaster avoidance, reduction, and recovery planning in our situation. The survey responders are asked to use specific statistics and information about the sub-indicators, and they are advised to do so. However, usually, they use their assumptions about the relative significance and meaning of these sub-indicators. As a result, the purpose of using AHP is to accommodate both individual decisions as well as the underlying data used to measure weights (w_i) for each sub-indicator.

3.7 Step 3: Data Processing

The weighted average of each sub-indicator is used to measure the four components of the RMI, resulting in a user-friendly and reliable formula for use at the local scale. The following equations can be used to calculate each risk management indicator.

$$RMI_{RI} = \frac{\sum_{i=1}^{6} w_i RI_i}{\sum_{i=1}^{6} w_i} \tag{3}$$

$$RMI_{RR} = \frac{\sum_{i=1}^{6} w_i RR_i}{\sum_{i=1}^{6} w_i} \tag{4}$$

$$RMI_{DM} = \frac{\sum_{i=1}^{5} w_i DM_i}{\sum_{i=1}^{5} w_i} \tag{5}$$

$$RMI_{FP} = \frac{\sum_{i=1}^{3} w_i FP_i}{\sum_{i=1}^{3} w_i} \tag{6}$$

where w_i is the weight obtained using the AHP and RI_i, RR_i, DM_i, and, FP_i represents given values to each sub-indicator between 1 and 5 during the first questionnaire. Finally, the average of the four combined sub-indicators is used to measure the cumulative RMI., as in Cardona et al. [13]:

$$RMI = \frac{(RMI_{RI} + RMI_{RR} + RMI_{DM} + RMI_{FP})}{4} \tag{7}$$

The following is a list of the major differences between the methodology used in this paper and that developed by Cardona et al. [13]

1. Changes were made to the sub-indicator scheme to make it even more applicable and to accommodate for problems at the local level (instead of the country level).
2. The questionnaires were modified according to the local conditions. 'Not relevant' or 'unable to answer' options are introduced.
3. Performance levels are graded on a scale of 1–5, with 1 being the lowest and 5 being the highest (instead of 0–100).
4. Statistical analysis is used to determine the qualification of each descriptor (instead of fuzzy sets).
5. The weight for each index is calculated by expert views from local inhabitants, rather than using the same weights measured at the country level.

Each society can define and categorize the most relevant factors for disaster management, reduction, and prevention using this approach, and the weights can be determined accordingly.

4 Results and Discussions

4.1 Survey Response

A total of 25 personals including Malappuram district sub-collector, Additional District Magistrate (ADM), officials from District Disaster Management Authority (DDMA), officials from the local administrative level, etc. were answered the first questionnaire. But the second questionnaire is answered by only 15 of them. Figures 2 and 3 shows the details about the survey responders of first and second questionnaires.

4.2 Numerical Weights (w_i) for the Sub-indicators from AHP

The AHP weights are calculated as the average of every set of AHP weights received from different respondents. The AHP weights obtained from the survey is as shown in the graphs.

Under each policy, relative weights differ between metrics, as shown in Figs. 4 and 5. Furthermore, one measure has a slightly higher relative weight in each strategy than the others, such as

1. RI_6-Risk management awareness and training
2. RR_3-Execution of hazard event control and protection techniques
3. DM_1-Emergency preparedness and continuity planning
4. FP_2-Budget allocation and mobilization

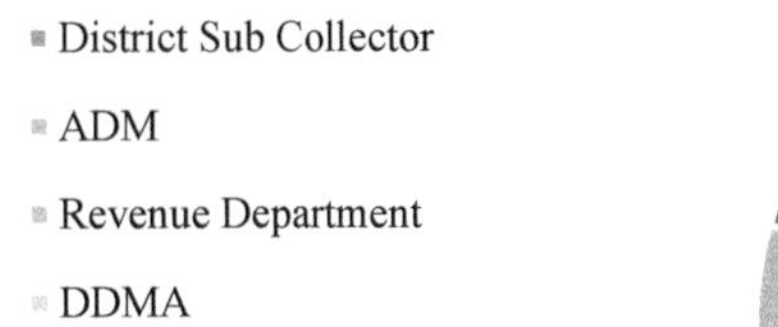

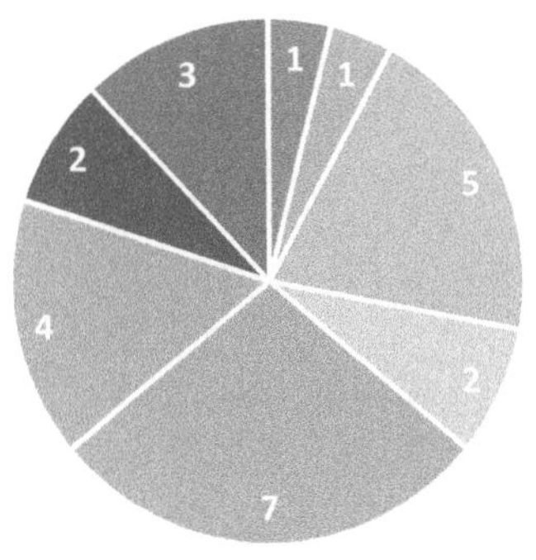

Fig. 2 Survey responders for the first questionnaire

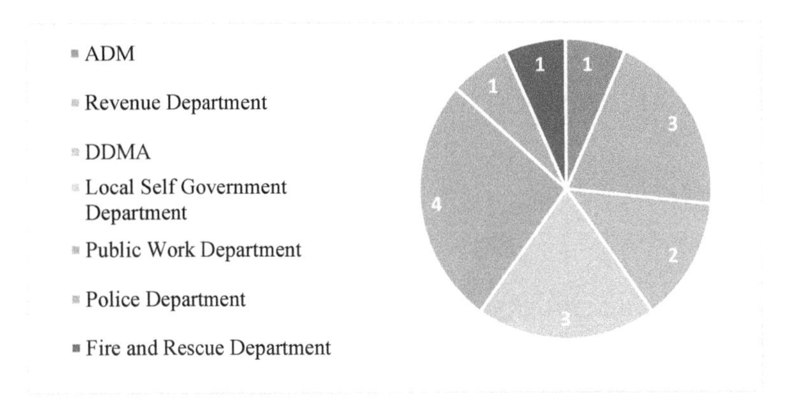

Fig. 3 Survey responders for the second questionnaire

4.3 Performance Level

From the responses of the first questionnaire performance level of each sub-indicators is found out. The average values of each sub-indicator are obtained as shown in Fig. 6.

The performance level of the sub-indicators ranges from a level of 1.84 to 4.44. If we consider 4 (outstanding) as a desired limit only 2 of the 20 indicators reached the desired performance level, those are:

1. RI_1. Systematic disaster and loss databases
2. RR_1. Land use and urban planning

It is really important to note that the major portions of the indicators are below performance level 3 (significant) (11 out of 20).

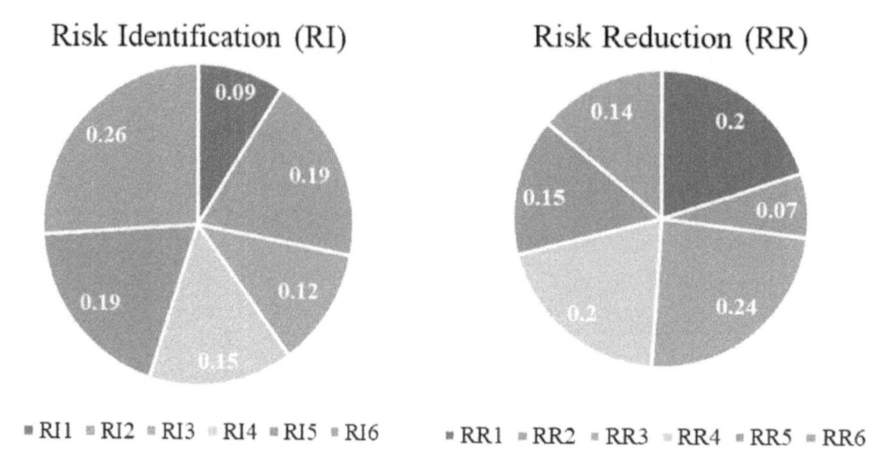

Fig. 4 AHP weight distribution of Risk identification and Risk reduction

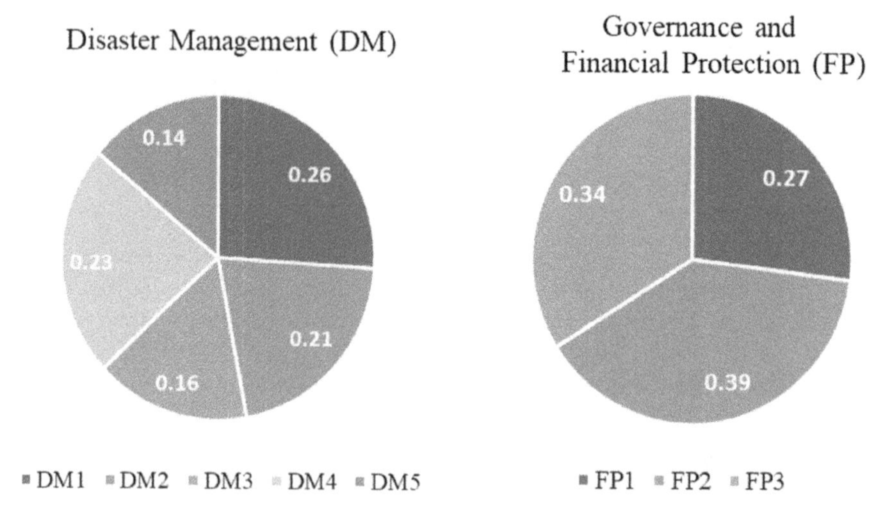

Fig. 5 AHP weight distribution of Disaster management and Governance & Financial protection

All of the performance levels of the sub-indicators of disaster management and governance and financial protection except DM_1 and FP_2 fall under the significant level. From the AHP weightage, we saw that RI_6 has the highest weightage among the sub-indicators of risk identification. But in contrast in the case of performance level RI_6 is the least one under risk identification. The sub-indicator FP_3 (Insurance and disaster funds) under governance and financial protection the lowest one among all the performance levels, which even didn't attain the incipient level.

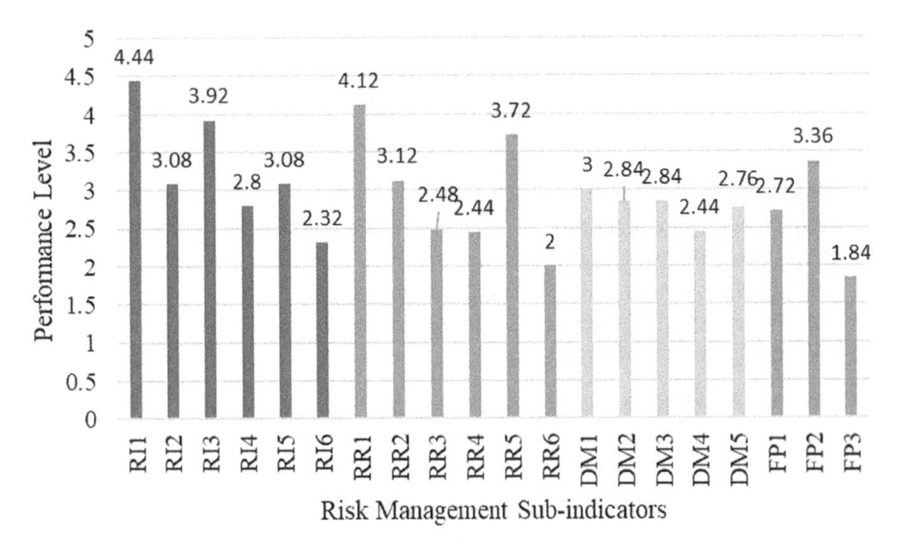

Fig. 6 The average performance level of risk management sub-indicators

4.4 Landslide Risk Management Index

Taking into account all available output data and AHP weights, the landslide risk management indices RMI_{RI}, RMI_{RR}, RMI_{DM}, RMI_{FP}, and RMI, are calculated and summarized in Table 1.

As per RMI study findings, it is clear that, to date, the Nilambur taluk has implemented a significant action for risk identification. When opposed to the other two indexes, activities relating to emergency relief and financial security performed poorly. It is important to note that only one of the four indices attained at least a significant level. From the results, it is clear that, Risk Identification Index is the highest one, and the sub-indicators of RMI_{RI} show that the Nilambur taluk is good at keeping systematic disaster and loss databases. The performance level for the community training is low which has given a major AHP weightage. 'Insurance and disaster fund' are the sub-indicator with the lowest performance level. Also, retrofitting works of the principal public and private buildings is showed very low value compared to other sub-indicators.

The landslide risk management index (RMI) for Nilambur taluk is obtained as 2.87 which denotes the incipient level in overall performance. The performance levels for the sub-indicators clearly shows that they have to improve in many factors, especially in;

Table 1 Landslide risk management indices

Indices	RMI_{RI}	RMI_{RR}	RMI_{DM}	RMI_{FP}	RMI
	3.06	2.96	2.78	2.67	2.87

1. RI_4. Risk and vulnerability evaluation
2. RI_6. Risk management awareness and training
3. RR_3. Implementation of hazard-event control and protection techniques
4. RR_4. Housing improvement and relocation from prone-areas
5. RR_6. Reinforcement and retrofitting of public and private assets
6. DM_2. Information and warning systems
7. DM_3. Emergency response
8. DM_4. Community preparedness and training
9. DM_5. Rehabilitation and reconstruction planning
10. FP_1. Inter-institutional organization and strengthening
11. FP_3. Insurance and disaster funds

These results indicate that projects and programs for catastrophe relief must be implemented, prevention and management including housing improvement, retrofitting of structures, monitoring, and warning systems, community training, and introducing emergency response training. It is also critical to improve its financial security efficiency to achieve further outstanding and optimal standards in these sub-indicators. The poor risk management performance indicates that it must establish short- and medium-term initiatives and services to assess the current extent of danger (hazard and vulnerability) to which its inhabitants are exposed and to put in place the appropriate disaster avoidance, reduction, and response management measures.

5 Conclusions

Landslides are one of the most common natural incidents, affecting 15% of the landmass [20]. It results in the loss of life and property [21]. 13 out of 14 districts of Kerala are vulnerable to landslides. Nilambur taluk is one of the 5 most landslide-prone taluks of the state, and which faced several major landslides in the past few years and lead to the loss of hundreds of people's lives. To substantially lower the damage caused by landslides, it is important to take a constructive approach to risk management. Practitioners' perspectives on the extent of landslide risk management will help to provide insight into how to boost landslide risk management practices. They also could be used as a guide for prioritizing risk management plans in the future. The risk management metrics represent operational, growth, capability, and structural measures taken to minimize vulnerability and disruptions in a geographic region, plan for a crisis, and bounce back rapidly from disasters.

The framework of indexes, combined with this methodology, provides the necessary evaluation tools to get a better understanding of the issue to guide local decision-making. The findings can be used to develop landslide risk management innovations in Nilambur taluk, and they could be considered as a help full reference for upcoming works. However, extreme caution should be exercised when interpreting the findings, since they were fraught with unknown variables and the number of responses was very few, the respondent's minimal experience and the subjective essence of experiences. The system's main benefit is it's capacity to break down data and recognize causes that need to be prioritized in risk management works at the local level and also assessing their effectiveness. This methodology's most important contribution is that it allows local decision-makers to begin a structured process of assessing and documenting their risk assessment success and needs.

Based on the survey findings, it is possible to infer that certain elements of Nilambur's landslide risk management can be strengthened. For example, vulnerability assessment, training for risk management, and hazard-event control and protection techniques have to be prioritized in Nilambur. Landslide risk avoidance techniques can also involve asset upgrades, retrofitting, and restoration. Besides, for coping with landslides, more emphasis should be placed on inter-institutional coordination, similarly in the allocation and usage of financial capital. Community training for disaster management, insurance coverage for private and public assets, and the introduction of early warning systems need immediate considerations.

The present approach can be modified and extended to other forms of natural disasters in Kerala, such as floods. However, according to the disaster situation and the local scenario, the survey questions should be modified or updated. It is also recommended that the current approach of surveying landslide risk management expectations be extended to include public perceptions, as comparing expert and public opinions will provide a more comprehensive measure of landslide risk management effectiveness.

References

1. Biansoongnern S, Plungkang B, Susuk S (2016) Development of low-cost vibration sensor network for early warning system of landslides. Energy Procedia 89:417–420
2. Pawar P, Patil A, Rathod H, Hadale R, Kharche S (2019) IoT Based Landslide Detection and Monitoring
3. Van Westen CJ et al (2009) Multi-hazard risk assessment: distance education course guide book. United Nations University—ITC School on Disaster Geoinformation Management: Tokyo, Japan
4. Highland L, Bobrowsky PT (2008) The landslide handbook: a guide to understanding landslides. US Geological Survey, Reston
5. Ramesh MV (2009) Real-time wireless sensor network for landslide detection. In: 2009 third international conference on sensor technologies and applications, IEEE
6. Guha-Sapir D et al (2011) Annual disaster statistical review 2010. Centre Res Epidemiol Disasters
7. Perera ENC et al (2018) Direct impacts of landslides on socio-economic systems: a case study from Aranayake, Sri Lanka. Geoenviron Disasters 5(1):1–12
8. Floods, Kerala Post Disaster Needs Assessment, and Landslides August (2018) Government of Kerala. Thiruvananthapuram, October
9. Sreekumar S (2009) Techniques for slope stability analysis: Site specific studies from Idukki district, Kerala. J Geol Soc India 73(6):813–820

10. Vasudevan N, Ramanathan K (2016) Geological factors contributing to landslides: case studies of a few landslides in different regions of India. IOP Conf Ser Earth Environ Sci 30(1). https://doi.org/10.1088/1755-1315/30/1/012011
11. Kuriakose SL, Sankar G, Muraleedharan C (2009) History of landslide susceptibility and a chorology of landslide-prone areas in the Western Ghats of Kerala, India. Environ Geol 57 (7):1553–1568
12. Cardona OD et al (2004)Disaster risk and risk management benchmarking: a methodology based on indicators at national level. In: IDB-IDEA program on indicators for disaster risk management, Universitad National De Colombian, Manizales
13. Cardona OD, Hurtado JE, Duque G, Moreno A, Chardon AC, Velásquez LS, Prieto SD (2005) System of indicators for disaster risk management: program for Latin America and the Caribbean: main technical report. Universidad Nacional de Colombia, Manizales, IDB/IDEA Program on Indicators for Disaster Risk Management
14. Carreño ML, Cardona OD, Barbat AH (2007) A disaster risk management performance index. Nat Hazards 41:1–20
15. Inter-American Development Bank (IADB)-Universidad Nacional de Colombia, Manizales, Instituto de Estudios Ambientales (UNC/IDEA) (2005) Indicators of disaster risk and risk management. Summary report, program for Latin America and the Caribbean. http://idea.unalmzl.edu.co. Accessed 11 June 2014
16. Cardona, O.D., and Carreño, M.L. (2011) Updating the Indicators of Disaster Risk and Risk Management for the Americas. Journal of Integrated Disaster Risk Management
17. Chiu JKY (2015) Landslide risk management perceptions in territories—comparative case studies of Hong Kong and Norway. Master's thesis, Department of Geosciences, University of Oslo, Norway
18. Saaty TL, Vargas LG (1991) Prediction, projection, and forecasting: applications of the analytical hierarchy process in economics, finance, politics, games, and sports. Kluwer Academic Publishers, Boston, p 251
19. Yadollahi M, Rosli MZ (2011) Development of the Analytical Hierarchy Process (AHP) method for rehabilitation project ranking before disasters. WIT Trans Built Environ 119:209–220
20. Huang R, Li W (2011) Formation, distribution and risk control of landslides in China. J Rock Mech Geotech Eng 3(2):97–116
21. Pilgrim NK (1999) Landslides, risk and decision-making in Kinnaur district: bridging the gap between science and public opinion. Disasters 23(1):45–65. https://doi.org/10.1111/1467-7717.00104

Seismic Performance of Multi-storey RC Building with Coupled Shear Walls

N. Y. Anjali and B. R. Beena

Abstract Earthquakes are one of the most dangerous disasters which cause severe threats to life and property. Lack of land availability and rising demands for better housing facilities have made vertical construction to gain significant importance. Nowadays, multi-storey buildings are being constructed even in high earthquake prone regions. Conventionally, buildings are made seismic resistant by constructing shear walls or by installing passive energy dissipating devices like dampers. The use of coupled shear wall (CSW) systems instead of conventional shear walls is an innovative technique that aims at enhancing the performance of buildings which are located even in high seismic zones. In the present study, a multi-storey RC building which is symmetrical in plan and having coupled shear wall system was modelled using ETABS software and it was assumed to be located in a region of high seismic activity. Response spectrum analysis was carried out on the building model and the results were compared with seismic responses of the same building having conventional shear walls instead of coupled shear walls. Therefore, improved performance of buildings having CSW systems and the use of such systems for better seismic resistance was understood.

Keywords Earthquake · Seismic resistance · Coupled shear walls · Response spectrum analysis · ETABS

1 Introduction

In the present era, construction of multi-storey structures have become widespread and highly important due to the increase in population and lack of availability of sufficient land. Nowadays, high-rise buildings are being constructed even in highly earthquake prone regions. Such practices can lead to severe damages during an

---

N. Y. Anjali (✉) · B. R. Beena
Federal Institute of Science and Technology, Angamaly, Kerala, India

G. C. Marano et al. (eds.), *Proceedings of SECON'21*, Lecture Notes in Civil Engineering 171, https://doi.org/10.1007/978-3-030-80312-4_18

earthquake scenario as the lateral seismic loads cause sway in the buildings which increases with increase in number of storey and due to this, it is necessary to make them seismic resistant. Thus it is important to counteract the lateral loads and to increase the strength and stiffness of the buildings so as to avoid huge damages to life and properties. Dynamic analysis methods are commonly adopted to conduct detailed studies on the behavior of multi-storey buildings during earthquakes in order to determine their maximum responses to various base excitations [1, 2].

The major concerns to be dealt with during the seismic analysis, design and construction of multi-storey buildings are to have their lateral storey drift, displacement and base shear values within the accepted limits as preferred by the codes, which is quite difficult in the case of seismic activities of high intensity. Conventionally, these requirements are satisfied by constructing shear walls in buildings for improving their lateral stiffness [3]. Nowadays, shear walls are extensively constructed in multi-storey structures as lateral force resisting members in the aim of reducing the responses of buildings to very high seismic forces. They are even used for retrofitting existing structures which does not comply with updating of codes or design practices [4].

Coupled shear walls are modifications of shear walls in which the wall piers are connected by coupling beams. In such systems, overturning moments are resisted by axial compression-tension couples due to accumulation of shear in the coupling beams, instead of lateral load resistance by individual flexural action in shear walls. They show excellent energy dissipation capacities and the dissipation of input energy is distributed along the coupling beams instead of concentrating at the bottom of wall piers [5, 6]. Under the action of large seismic loads, the coupling beams undergo degradation in strength and stiffness and their performance evolves from that of a coupled shear wall system to a system of linked wall piers which further resisted the seismic forces. This indicated the presence of reserve strength and stiffness in the wall piers even after the degradation in the capacity of coupling beams [7].

Nowadays, the concept of design codes have changed from strength-based to performance-based design and so, many researches are being carried out to determine the behavior and performance of coupled shear walls under severe earthquakes. Researches have shown that these structural systems not only provide large strength and stiffness to structures but they also control the horizontal deformation of buildings under large seismic loads [8, 9]. Hence, understanding the behavior of coupled shear wall systems on buildings that encounter high seismic forces are of great relevance for further upgradations in construction techniques as it will aid engineers in carrying out safer design and construction of building structures. This paper deals with understanding the seismic performance of a multi-storey structure which uses coupled shear walls over conventional shear walls as lateral load resisting members.

2 Methodology

A Model of multi-storey RC building was created in ETABS 2018 with shear walls which were provided along two of its bays in the X-direction, throughout the entire height. Another model of the same building was created in which the shear walls were coupled together by RC coupling beams at all storey heights, thereby forming a system of coupled shear walls. Design check was carried out for the models to ensure proper designing of all structural members. Linear dynamic analysis by response spectrum analysis was conducted to study the seismic responses of the structure under the action of both shear walls and couples shear wall systems when the building was located at a zone of high seismic activity. Analysis results were obtained in terms of storey displacement, storey drift and storey shear, and the results were evaluated to arrive at conclusions regarding response of the structure to high seismic forces.

3 Model Description

For the study, a G + 15 storey RC residential building having square plan was modelled. The total height of the building is 45 m with a typical storey height of 3 m. It has 5 bays with 5 m spacing in each direction and a typical floor plan of 625 m^2 at all storey levels. Dimensions of columns were 500 mm × 500 mm and that of beams were 250 mm × 600 mm. The thickness of slabs were considered to be 150 mm. M30 grade concrete and Fe500 grade steel were used for the structural members, whereas Fe415 grade steel was used for the lateral ties in beams and columns.

In the first model, shear walls were provided in two bays along the x-direction and the wall piers have a thickness of 250 mm. In the second model, the shear walls were coupled together using coupling beams. The thickness of wall piers of the CSW system is 250 mm and the coupling beams were modelled as spandrels having size of 250 mm × 1000 mm. The floors were considered to be semi-rigid diaphragms. Figure 1a, b represent the plan view and 3D view of the model having shear walls. Figure 2a, b represents the plan view and 3D view of the model having coupled shear walls. The building was modeled in ETABS 2018.

3.1 Loads

Live load of 2 kN/m^2 was applied as per IS 875 (Part II): 1987 and dead loads were assigned by the software. Wall loads were assigned to the beam elements as superimposed dead loads of 11.04 kN/m and 4.6 kN/m for parapets. Floor finish loads were taken to be 1 kN/m^2. Support conditions of the building were assumed to be fixed.

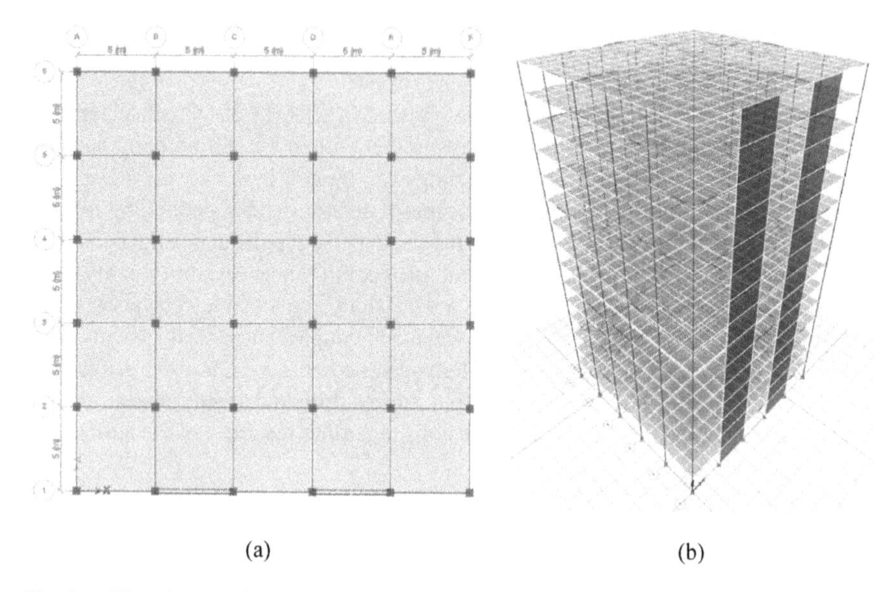

(a) (b)

Fig. 1 **a** Plan view and **b** 3D view of building with shear walls

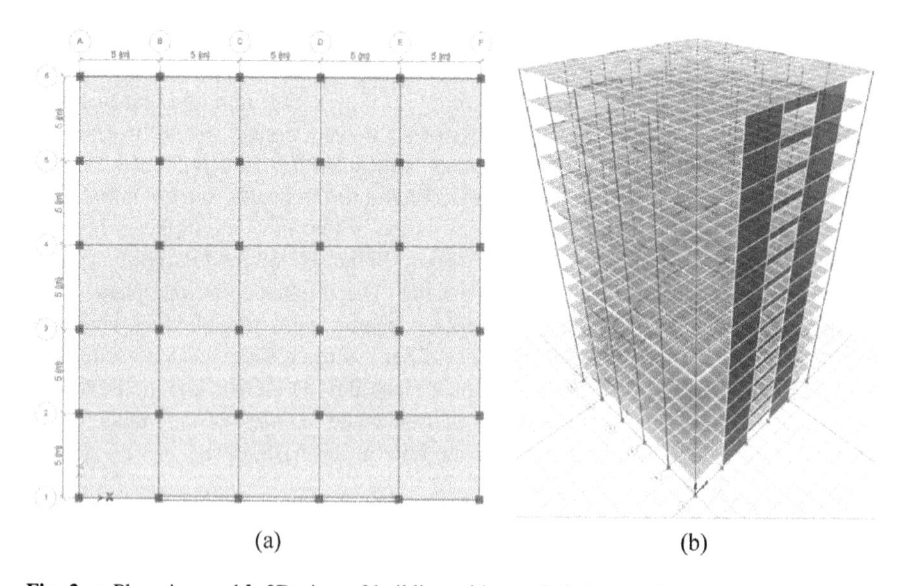

(a) (b)

Fig. 2 **a** Plan view and **b** 3D view of building with coupled shear walls

4 Response Spectrum Analysis

Response spectrum analysis is a linear-dynamic statistical analysis method that helps in the determination of response of a structure to seismic events. In this study, response spectrum function was set as per IS 1893: 2016 and was defined for seismic zone factor of 0.36, Medium (Type II) soil, importance factor of 1.5, response reduction factor of 5 and 5% damping.

4.1 Design Check

Design checks were conducted on the structure that was modelled to ensure that the beams and columns were not overstressed or had not undergone failure. Generally, after a design check, overstressed beams and columns will be indicated by red color and all the structural members that have passed the check will be indicated by pink color. In this study, all the structural members had passed the design check, indicating that they were sufficient to safely carry the assigned loads. Figure 3a, b show the plan view of design check result of models having shear walls and coupled shear walls respectively.

It can be seen that all members in both models passed the design check, indicating that they were safe to carry the assigned loads. Response spectrum analysis was carried out on the two models and results were obtained in terms of maximum storey displacements, storey drifts and storey shears. The seismic analysis results obtained from both models were compared to reach valid conclusions.

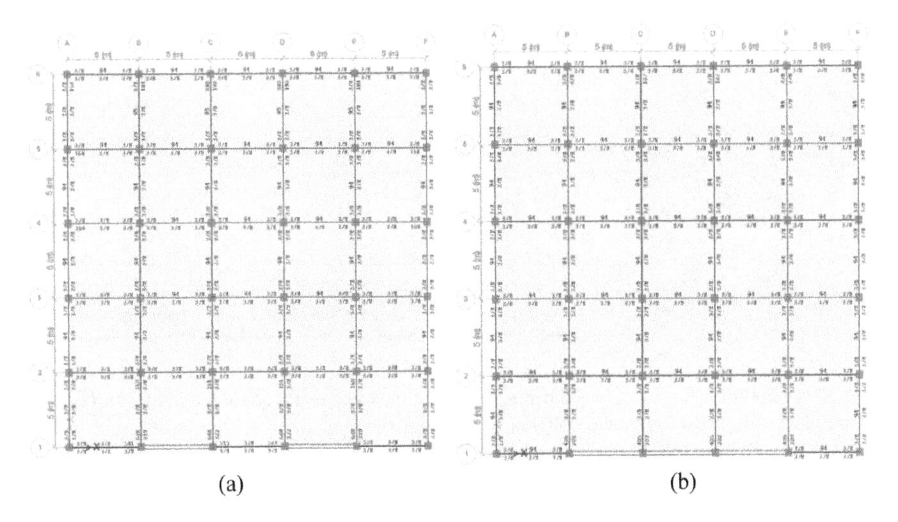

(a) (b)

Fig. 3 Design check of model with **a** shear walls and **b** coupled shear walls

4.2 Storey Displacement

Figure 4a shows maximum storey displacements in the X-direction versus elevation for buildings having shear walls and coupled shear walls. It was observed that the displacements increased with elevation of the building for both models and the displacement was maximum at the top storey. It was also observed that the model having coupled shear walls showed reduction in storey displacements at all storey heights when compared to the building having conventional shear walls.

4.3 Storey Drift

Figure 4b shows the maximum storey drifts observed in the response spectrum analysis conducted for the buildings having shear walls and coupled shear walls. The storey drift values were maximum at the third storey for both models and the drift versus elevation curve conformed to the general storey drift pattern observed for multi-storey buildings. Storey drifts were lesser for the building having coupled shear walls when compared to the same building when it had conventional shear walls.

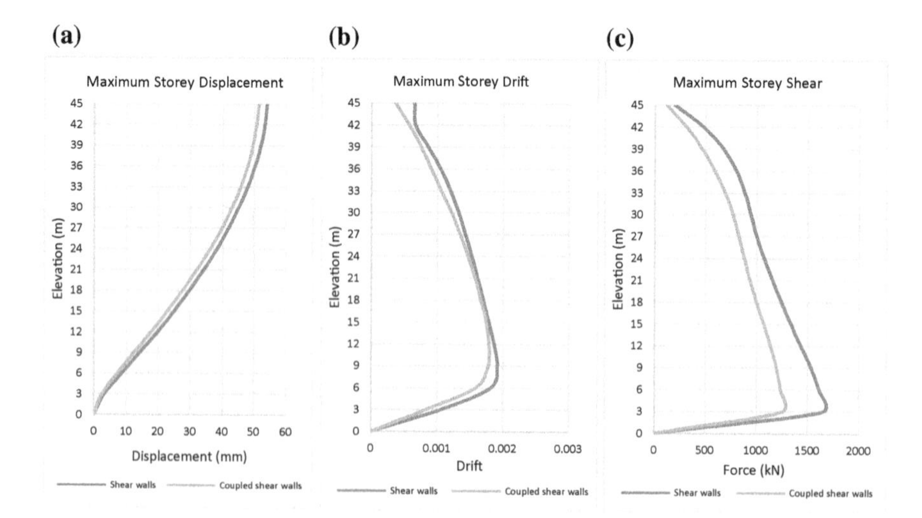

Fig. 4 **a** Maximum storey displacements, **b** maximum storey drifts and **c** maximum storey shear of building with shear walls and coupled shear walls

Table 1 Results of response spectrum analysis

Model	Building with conventional shear walls	Building with coupled shear walls	Inference
Maximum storey displacement (mm)	54.054	51.479	Decrease of 4.76%
Maximum storey drift	0.001916	0.001788	Decrease of 6.68%
Storey shear (kN/m)	1648.366	1252.785	Decrease of 24%

4.4 Storey Shear

Figure 4c shows maximum shear forces in the x-direction versus elevation of the building with conventional shear walls and coupled shear walls. Storey shear forces had a more pronounced effect at the bottom storeys of the building. The forces were maximum at the first storey for both models but the building with coupled shear walls showed considerable reduction in shear forces at all storey levels when compared to the same building with conventional shear walls.

5 Results and Discussions

The effect of coupled shear walls on the seismic performance of a 15-storey RC building was studied and the results were compared with that of the same building having conventional shear walls. Response spectrum analysis was conducted and the results are summarized in Table 1. The views on the results obtained are discussed in the conclusions.

From the response spectrum analysis results, maximum storey displacement of 54.054 mm, maximum storey drift of 0.001916 and storey shear forces of 1648.366 kN/m were observed for buildings having conventional shear walls and the corresponding values were reduced to 51.479 mm, 0.001788 and 1252.785 kN/m respectively, when the conventional shear walls were replaced by the coupled shear wall systems. Therefore, the building with coupled shear walls showed an overall reduction of 4.76% in storey displacement, 6.68% in storey drift and considerable reduction of 24% in storey shear forces, when compared to a similar conventional shear wall building, proving them to provide better seismic resistance to structures.

6 Conclusions

A detailed response spectrum analysis was conducted on two models of a G + 15 storey RC building having conventional shear walls and coupled shear walls in ETABS 2018 software based on IS 1893: 2016. Based on the seismic performance of the models, following conclusions were drawn:

- Coupled shear walls reduced the seismic responses by providing increased stiffness to the structure, when compared with the building having conventional shear walls.
- There was a reduction of 4.76% in maximum storey displacement and 6.68% in maximum storey drift values for coupled shear wall building when compared to the conventional shear wall building, owing to the increased stiffness of coupled shear walls and coupling action of the coupling beams.
- Coupled shear wall building showed considerable reduction in storey shear values by 24% when compared to the same building having conventional shear walls which indicates the improved seismic performance of Coupled shear walls.

Therefore, conventional shear walls can be adequately replaced by coupled shear walls in the present era of construction, which will help in improving the resistance of the structures to even very high seismic forces. This reduces the susceptibility of such buildings to severe damages induced by earthquake scenarios, further reducing possibilities of loss of lives and also preventing damages to adjacent buildings and hence proving them to be better lateral load resisting structural systems.

References

1. Ahamad SA, Pratap KV (2020) Dynamic analysis of G + 20 multi storeyed building by using shear walls in various locations for different seismic zones by using Etabs. Mater Today: Proc (Elsevier)
2. Khy K, Chintanapakdee C, Warnitchai P, Wijeyewickrema AC (2019) Modified response spectrum analysis to compute shear force in tall RC shear wall buildings. Eng Struct 180:295–309 (Elsevier)
3. Dahesh MA, Tuken A, Siddiqui NA (2015) Controlling the earthquake-induced lateral displacement of RC buildings using shear walls: parametric study. Arab J Geosci (Springer)
4. Cavdar O, Cavdar A, Bayraktar E (2017) Earthquake performance of reinforced-concrete shear-wall structure using nonlinear methods. J Perform Constr Facil (ASCE)
5. Lu X, Chen L (2005) Modeling of coupled shear walls and its experimental verification. J Struct Eng (ASCE)
6. Harries KA, Moulton JD, Clemenson RL (2004) Parametric study of coupled wall behavior—implications for the design of coupling beams. J Struct Eng 130(3) (ASCE)
7. Eljadei AA, Harries KA (2014) Design of coupled wall structures as evolving structural systems. Eng Struct 73:100–113 (Elsevier)
8. Asghari A, Zarnagh BA (2017) A new study of seismic behavior of perforated coupled shear walls. Int J Civ Eng 15: 775–789 (Springer)
9. Afefy HM (2020) Seismic retrofitting of reinforced-concrete coupled shear walls: a review. Pract Periodical Struct Des Constr 25(3) (ASCE)

Study on Seismic Performance of Braced Single Storey Framed Structure Using X-Shaped Pipe Damper

P. S. Thapasya and Neeraja Nair

Abstract X-shaped pipe damper (XPD) is a recently proposed metallic-yielding steel damper which is fabricated from commonly used pipes. This damper is assembled through welding the two oppositely positioned pipe halves to form an x-shaped core and connecting the x-shaped core to side plates with fillet and circumferential welds. The x-shaped pipe damper provides energy dissipation and resistance behaviours through tension brace mechanism. This study examines the effects of x-shaped pipe damper in a single storey frame equipped with XPD of various geometrical properties on chevron brace by static non-linear pushover analysis using ANSYS 19.0 software. The more effective x-shaped pipe damper numbers and its dimensions are determined by comparing with the developed models.

Keywords X-shaped pipe dampers (XPD) · ANSYS · Geometrical properties · Non-linear pushover analysis · Braced single storey steel frame

1 Introduction

Isolators or dampers are often used to cut the input of earthquake. Buckling-restrained braces and shear-panel dampers are the metallic dampers that utilize low yield point steel [1]. Due to many advantages like stable performance, lower cost, simple mechanism and the convenience in adjusting strength and stiffness, the metallic passive dampers are more popular [2]. Geometrical elasticity of circular steel pipes helps for the easy installation and maintenance so they can be utilized as dampers [3]. There are further developments made in the design of pipe dampers in order to improve the damper performance. Pipe dampers are used for energy dissipation through the flexural yielding or mild steel pipes deformation. Bare steel pipes show an excellent hysteretic behavior than the pipes filled with concrete [4]. This study mainly focuses on the steel pipes behaviour as a possible

P. S. Thapasya (✉) · N. Nair
Federal Institute of Science and Technology, Angamaly, Ernakulam, India

G. C. Marano et al. (eds.), *Proceedings of SECON'21*, Lecture Notes in Civil Engineering 171, https://doi.org/10.1007/978-3-030-80312-4_19

hysteretic damper. The application of pipe dampers in structures reduces the seismic demand [5, 6]. It can easily be installed, replaced and it is economical. The modifications on pipe dampers, thus developed are Dual pipe dampers (DPD) in 2012 [7], Infilled-pipe dampers (IPD) in 2014 [8], X-shape pipe dampers (XPD) in 2020 [9].

1.1 X-Shaped Pipe Damper (XPD)

X-shaped pipe damper (XPD) [9], is a new type of steel damper. X-shaped pipe damper (XPD) is designed from the developments made in the most effective parts of dual pipe damper (DPD). In x-shaped pipe damper (XPD), the most effective parts of the dual pipe damper has been considered as in Fig. 1. The energy dissipation in x-shaped pipe damper takes place by the tension brace mechanism of pipe halves. Through the flexural bending of the pipe plates tensile stretching at composite pipe halves, the x-shaped pipe damper provides resistance laterally and dissipation energy.

2 Summary of Literature Review

The pipe dampers are very cost effective, showed excellent hysteretic behavior, high energy dissipation and can be used anywhere on the structure depending on the needs.

Comparative studies on seismic performance of cost effective pipe dampers conclude that metallic steel dampers were more on compared with other types of dampers appears most popular due to low manufacturing cost and their better performance [2]. Several modifications are made on pipe dampers to improve their performance in an economical manner include Dual pipe dampers (DPD) (2013) [7], Infilled pipe dampers (IPD) (2014) [8] and X-shaped pipe dampers (XPD) (2020) [9]. From the results it was concluded that pipes in this manner can be used as a metallic yielding damper that provides energy dissipation in structures

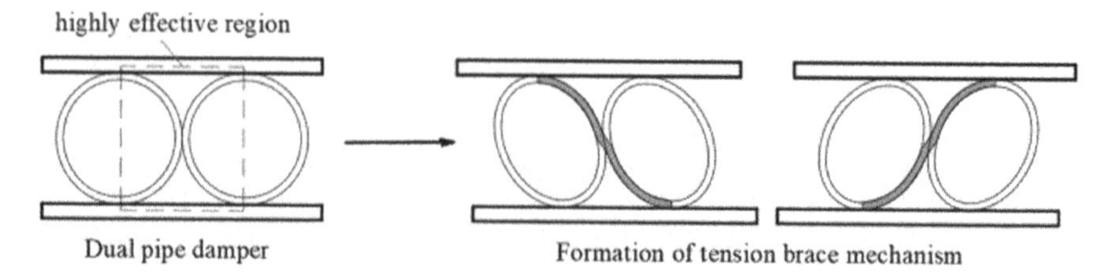

Fig. 1 X-shaped pipe damper from the most effective parts of dual pipe damper and the tension brace mode [9]

against seismic actions [9]. The x-shaped pipe damper (XPD) is a recently proposed metallic damper, energy dissipation takes place through the flexural bending of the pipe plates tensile stretching at composite pipe halves, the x-shaped pipe damper provides lateral resistance and energy dissipation. Cyclic test /quasi static cyclic test performed with 100 kN actuator with two layer x-shaped pipe dampers was considered for the study using only one directional cyclic [9]. Loading protocol was according to FEMA461 [9]. Stable Hysteretic loops were obtained and higher strength, initial stiffness and better deformation ability was displayed by circumferential weld than the fillet welded ones [9].

This study primarily focuses on determining the influence of diameter, length, thickness and number of x-shaped pipe damper when placed in an inverted v shaped/chevron braced single storey frame using pushover analysis to find the better model applicable to the building frame. Material properties were not changed and only the influence of changes in diameter, length, thickness and number of XPD was studied. The effect of x-shaped pipe damper only for inverter v-shaped/ chevron braced single storey frame was examined. Effects of damper system in a multi-storey/building arrangement are not explored. Study was conducted on two dimensional frame using static non-linear pushover analysis. Prior study show pushover analysis as effective [11].

3 Numerical Analysis

Previous experimental studies have focussed only on x-shaped pipe damper (XPD) [9]. In this study to stimulate the experimental research, numerical models were created using finite element software. The model dimensions and all other detailed parameters were selected from XPD journal [9] for simulation using the finite element software ANSYS 19.0. For the current work, two layered x-shaped pipe damper geometry is considered as per the journal [9]. The specimen named S133T5-C was selected for the numerical study with diameter 133 mm, 5 mm thickness and circumferential weld of 20 mm. The material properties of pipe includes elastic modulus of 202.03 GPa, yield stress is 341 MPa and ultimate stress of 491 MPa [9]. Cyclic loading is applied to the created model after meshing with proper standard loading protocol and boundary conditions adopted from the journal [9]. After experimental study, ultimate force, ultimate displacement, yield displacement and ductility obtained from journal are 70.1 kN, 25.8412 mm, 1.54 mm and after the validation study using finite element analysis in ANSYS, ultimate force, ultimate displacement, yield displacement and ductility obtained are 67.081 kN, 24.049 mm, 1.5032 mm. Comparing the results obtained from finite element analysis and experimental study, the percentage of error was minimal and may be utilised for further studies.

3.1 Modeling of XPD on Chevron Braced Single Storey Building Frame

The structural performance of x-shaped pipe damper with chevron bracing on a single steel frame using developed models was investigated using ANSYS 19.0 WORKBENCH. Engineering data sections in ANSYS software was used to assign the material properties of models, these engineering data sets already has preassigned values for each material. Table 1 [10] shows the material properties and properties of single storey steel frame with bracing are shown in Table 2 [10].

The columns are fixed supported to the base and to the base of columns braces are connected [10]. Plates are placed on top and bottom of pipe halves for proper transfer of loads and connections. Parametric study was conducted with two numbers of dampers on chevron bracing within a single storey frame. By comparing the performance of dampers, the more effective damper number and dimension was selected. The geometry of models used in the study is shown in Fig. 2. The minimum diameter, length, thickness and number of XPD model are selected according to the suggestions made by journal [9]. The minimum diameter, length and thickness of XPD provided are 133 mm, 40 mm and 5 mm respectively. Pushover analysis proved effective in the previous study, it was undertaken [11].

Figure 3 shows the support and loading conditions for XPD base model. Lateral load is applied to the top corner of the frame, fixed supports are provided at the bottom of frame as per journal [10], as the length of selected beam is 9.2 m lateral supports are provided at two vertex positions of beam and at the top face of XPD plate. Similar conditions were provided when number of dampers is increased in a frame.

3.2 Analysis of Model

Pushover analysis was carried out in each different combination of selected dimensions to identify the better suitable parameter for XPD. The parameters varied are diameter (399, 532 mm), length (120, 130, 140 mm), thickness (15, 25 and 35 mm) and number of XPD (two, four and six). In this study the diameter, length, thickness and number of XPD values were changed to determine more effective

Table 1 Material properties [10]

Material	Steel damper	Steel beam/column	Steel brace
Yield strength (MPa)	345	344	289
Modulus of elasticity (Pa)	2×10^{11}	2×10^{11}	2×10^{11}
Poisson's ratio	0.3	0.3	0.3
Density (kg/m^3)	7850	7850	7850

Table 2 Properties of single storey steel frame with chevron brace [10]

Storey height (m)	Storey width (m)	Beam size (inxlb/ft)	Column size (inxlb/ft)	Brace size (inxin)
4.27	9.2	W12 × 106	W8 × 40	HSS12.75 × 0.5

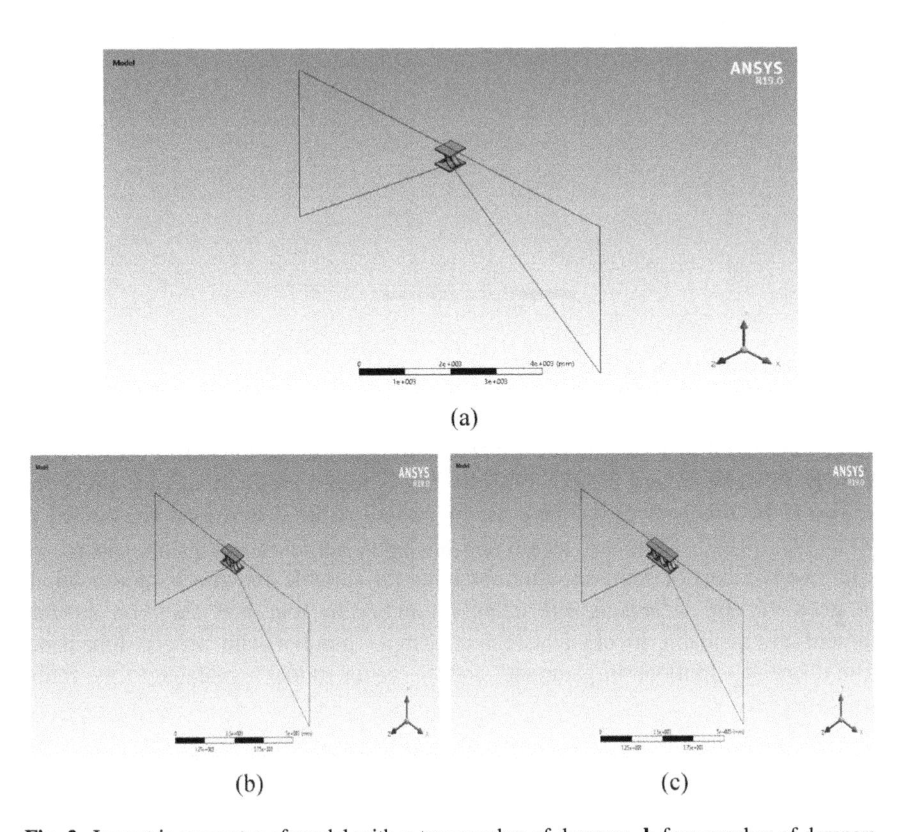

Fig. 2 Isometric geometry of model with **a** two number of dampers, **b** four number of dampers, **c** six numbers of damper

model. Thereby fixing the diameter value in the first case, length and thickness values are changed in other separate cases. Models represented using notations 2-XPD-133-40-5 indicates two number of XPD with 133 mm diameter, length 40 mm and 5 mm thickness.

In the first step, to study the effects of size on damper performance, the overall dimension of the base model is increased to two, three and four times with dimensions like 2-XPD-133-40-5, 2-XPD-266-80-10, 2-XPD-399-120-15, 2-XPD-532-160-20. In the second step, by fixing the dimension of diameter and thickness, length values are changed. The models created are 2-XPD-399-120-15, 2-XPD-399-130-15 and 2-XPD-399-140-15. In the third step, the diameter, length

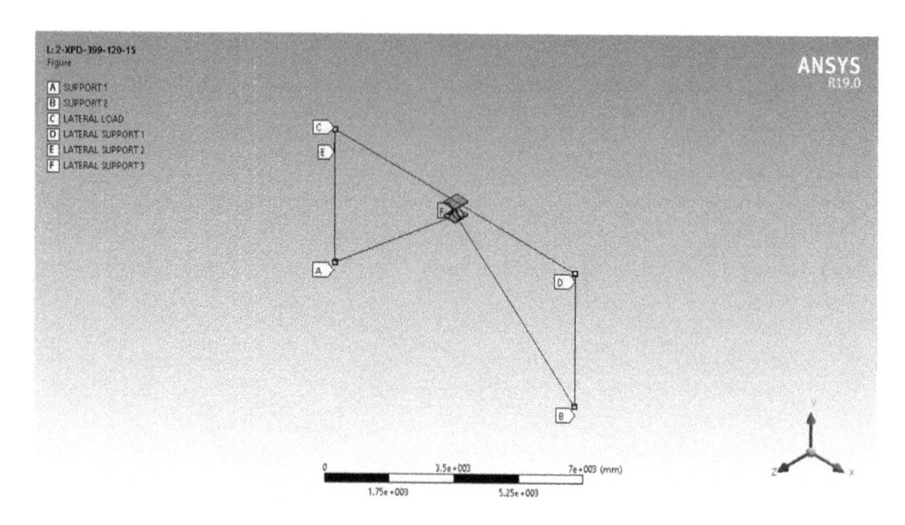

Fig. 3 Isometric view of support and loading conditions

values are fixed and the thickness values are changed. The models obtained are 2-XPD-399-140-25 and 2-XPD-399-140-35. As final step, in order to improve the stiffness, the numbers of dampers are improved as the dimensions are limited to XPD-399-140-35. Diameter, length and thickness variation studies are carried out with two numbers of dampers, after obtaining the suitable dimension, modifications in geometry by increasing the damper number to four and six with suitable dimensions available to obtain effective damper number with proper dimension. The effect of variations in parameter and the better model applicable to the frame was selected.

4 Results and Discussions

From the parametric study of models the ultimate load value is increased as the diameter, length, thickness and number of XPD dampers was increased. As other output parameters like ultimate deflection, yield load, yield deflection and ductility are equally important in the study their performance should also be evaluated, but they show difference in behaviour thus are not predictable. Analysis indicates that the ultimate strength of two numbers of XPD was not comparable to frame with braces, in spite of higher ductility parameters. Hence 4-XPD-399-140-35 was seemed suitable. Finally on analyzing the results obtained by increasing the damper number to four and six, XPD with six numbers of dampers comes with added cost because of extra material required with no significant improvement in ductility or other parameters selected for the study.

Table 3 Results obtained

Model	Ultimate load (kN)	Ultimate deflection (mm)	Yield deflection (mm)	Yield load (kN)	Ductility (μ)
2-XPD-133-40-5	283.23	132.39	4.12498	44.225	32.09
2-XPD-266-80-10	597.15	83.997	12.087	167.78	6.95
2-XPD-399-120-15	1143.1	193.07	13.882	256.81	13.91
2-XPD-532-160-20	1170.9	194.89	23.334	558.59	8.35
2-XPD-399-130-15	1144.9	189.74	20.351	325.07	9.32
2-XPD-399-140-15	1180.2	212.34	20.286	345.94	10.47
2-XPD-399-140-25	1760.9	177.76	18.906	968.79	9.40
2-XPD-399-140-35	2096.5	166.71	17.648	1487.7	9.45
4-XPD-399-140-35	5416.6	196.53	17.991	3593.9	10.92
4-XPD-399-140-25	2787.9	148.36	18.652	1943.8	7.95
4-XPD-399-140-30	3547.8	135.63	18.088	2583.2	7.49
6-XPD-399-140-35	7288.2	160.03	18.371	5158.2	8.71
2-XPD-399-120-25	1617.4	181.89	19.087	863.75	9.53
2-XPD-399-120-35	2036.2	144.23	17.637	1386.1	8.18
4-XPD-399-120-15	1214.2	166.69	6.698	293.87	24.89
6-XPD-399-120-15	1538.9	157.67	19.916	723.24	7.92
With braces	4084.6	28.447	6.0223	1059.6	4.72
Bare frame	212.62	116.01	40.703	111.13	2.85

Hence, the dimensions are limited to 399 mm diameter, 140 mm length and 35 mm thickness for the selected frame. On increasing the diameter, the x-shaped portion of damper was extruded greater than the plate size of XPD. Also when length and thickness values were increased greater than the limit, the results of output parameters were found more than the braced frame model (Table 3). The required force was reached at 399 mm diameter, 140 mm length and 35 mm thickness, therefore it does not need further improvement in dimensions. Also other possibilities with decrease in length and thickness from the limited dimensions with two and four numbers of dampers are also studied in Table 3 and suitable model is selected as 4-XPD-399-140-35.

5 Conclusions

The influence of diameter, length, thickness and number of x-shaped pipe damper when placed in an inverted v shaped/chevron braced single storey frame was determined. The increase in dimensional parameters of dampers shows an increase in the load carrying capacity of the single storey frame. When dampers were introduced in a braced frame the overall performance of selected frame improved when comparing the output parameters like ultimate load, ultimate deflection, yield load, yield deflection and ductility with the braced frame model developed. The model with four numbers of XPD-399-140-35 may be selected as the more effective model under pushover analysis in ANSYS 19.0 for a single storey frame. Thus this model may be implemented to building frame for further investigations on XPD dampers and their effectiveness.

References

1. Sabelli R, Mahin S, Chang C (2003) Seismic demands on steel braced frame buildings with buckling restrained braces. Eng Struct 25:655–666
2. Soong TT, Spencer BF (2002) Supplemental energy dissipation: state-of-the-art and state-of-the practice. Engineering Structures 24:243–259
3. Utomo J, Moestopo M, Surahmana A, Kusumastutia D (2015) Estimating the ultimate energy dissipation capacity of steel pipe dampers. Procedia Eng 125:1101–1107
4. Maleki S, Bagheri S (2010) Pipe damper, part i: experimental and analytical study. J Constr Steel Res 66:11088–11095
5. Mahjoubi S, Maleki S (2016) Seismic performance evaluation and design of steel structures equipped with dual-pipe dampers. J Constr Steel Res 122:25–39
6. Maleki S, Bagheri S (2010) Pipe damper, part ii: application to bridges. J Constr Steel Res 66:1096–1106
7. Maleki S, Mahjoubi S (2013) Dual-pipe damper. J Constr Steel Res 85:81–91
8. Maleki S, Mahjoubi S (2014) Infilled-pipe damper. J Constr Steel Res 98:45–58
9. Guo W, Wang X, Yu Y, Chen X, Li S, Fang W, Zeng C, Wang Y, Bud D (2020) Experimental study of a steel damper with X-shaped welded pipe halves. J Constr Steel Res 170:106087
10. Ashwin Kumar PC, Sahoo DR, Kumar A (2018) Seismic response of concentrically braced frames with staggered braces in split-x configurations. J Constr Steel Res 142(2018):17–30
11. Bincy V, Usha S (2020) Dual-pipe damper with bracing system for seismic retrofitting. Springer Nature, Switzerland AG

Progressive Collapse Analysis of a Multistoried Building with Flat Slab

G. Anandakrishnan and Jiji Antony

Abstract Progressive collapse causes the failure of a primary structural member to propagate to the adjoining members, thus causing the collapse of the entire structure. The need to mitigate the impact of progressive collapse is necessary as it leads to catastrophic economic loss and causalities. The flat slab buildings are more prone to progressive collapse due to the absence of beams which are used to transfer the load initially resisted by the removed column. The main cause of failure in a flat slab building is the punching shear failure occurring at the slab column connections. The aim of this study is to investigate the performance of a multistoried flat slab buildings subjected to progressive collapse. Also, the study investigates the effect of drop panel on the progressive collapse of the building and also the impact of progressive collapse of structure designed based on Indian codes for different locations of column removal. The linear static progressive collapse analysis of the flat slab building is done using ETABS.

Keywords Progressive collapse · Flat slab · Drop panel · ETABS

1 Introduction

The failure of a primary structural element spreads to the adjacent member causing the complete or partial failure of the structure is called progressive collapse. The major reason for progressive collapse is aircraft impact, design/construction error, fire, gas explosions, accidental overload, hazardous materials, vehicular collision, bomb explosions, etc. Flat slabs are reinforced concrete slab building in which the slabs are directly supported on the column without the use of beams. The advantages of providing flat slab structures include ease of construction, reduced storey height and ease of routing of services. The load transferring mechanism of flat slab building is different from the RCC framed building. In the conventional RCC framed building the beam members transfer the load from the slab to the column but

G. Anandakrishnan (✉) · J. Antony
Department of Civil Engineering, FISAT, Angamaly, Kerala, India

G. C. Marano et al. (eds.), *Proceedings of SECON'21*, Lecture Notes in Civil Engineering 171, https://doi.org/10.1007/978-3-030-80312-4_20

flat slab building relies on the slab column connection for the load transfer. During the column removal, the additional load gets redistributed to the adjacent slab column connection and causes punching shear failure at that junction. The failure of the adjacent slab column connection in punching shear causes the load to further redistribute and causes the punching shear failure to spread over a large portion of the structure, leading to the collapse of the entire structure [1]. The General Service Administration (GSA) [2] and Department of Defense (DoD) [3] guidelines are used for the design and analysis of structures subjected to progressive collapse.

Qian and Li [1] studied on the effect of parameters like drop panel, service load and slab reinforcement ratio on the progressive collapse of a flat slab structure. The addition of drop panel tends to reduce the effect of progressive collapse and the slab reinforcement ratio must be kept below 1.1% for good punching shear behavior. The study by Weng et al. [4] suggests to provide an integrity reinforcement of minimum 0.63% for good post punching behavior of flat slab building. It was also observed that the increase in the thickness of flat slab increases the load resisting capacity. The study by Senthil et al. [5] showed that increase in the slab thickness decreases the deflection in the slab and the increase in the length of slab increases the deflection in the slab. Russell et al. [6] suggests that the most critical column removal depends on the slab geometry and higher strength concrete has the capacity to reduce the effect of progressive collapse. Russell et al. [7] recommends a dynamic amplification factor between 1.39 and 1.62. Also, it was observed that the change in strain rates and material property had limited influence on the progressive collapse of flat slab building. Olmati et al. [8] study shows that the irregular geometry tends to increase the shear demand of flat slab building by 70%. The study by Peng et al. [9] showed that the static loading causes an average of 54% of the failure load to remain even after the occurrence of punching shear failure in the structure. Qian et al. [10] suggested the use of GFRP strips to be used as tie elements at the center of each panel to increase the integrity and ductility of the member. The provision of GFRP strips increased the punching shear capacity of the slab column connection and reduced the impact of progressive collapse on the structure. Qian et al. [11] suggested the use of CFRP strips across the flat slab and Mohamed et al. [12] suggested to provide integrity reinforcement continuously across the slab at the bottom region to mitigate the impact of progressive collapse.

The aim of the study is to compare the effect of progressive collapse of a typical flat slab building and flat slab building with drop panel by performing a linear static progressive collapse analysis.

2 Building Modelling

The structures considered for the study is a 11 storey RC flat slab building with six bays in longitudinal direction and four bays in the transverse direction. The span spacing along longitudinal direction is 6 m and in transverse direction is 4 m. The plan for the basic models of flat slab building and flat slab building with drop are

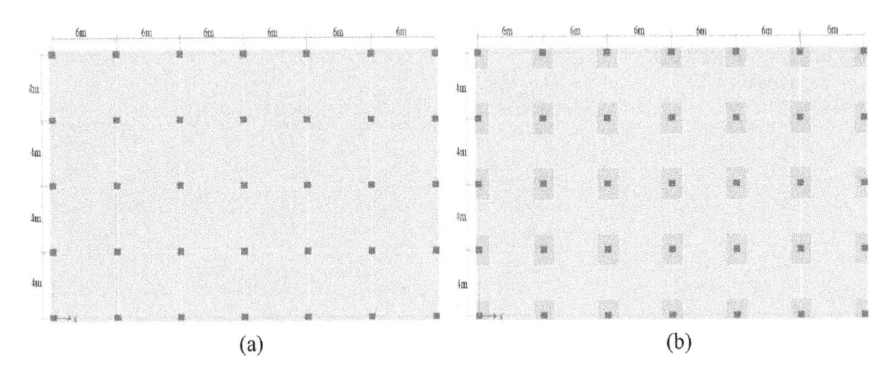

Fig. 1 Plan of the modelled building **a** Flat slab building. **b** Flat slab building with drop

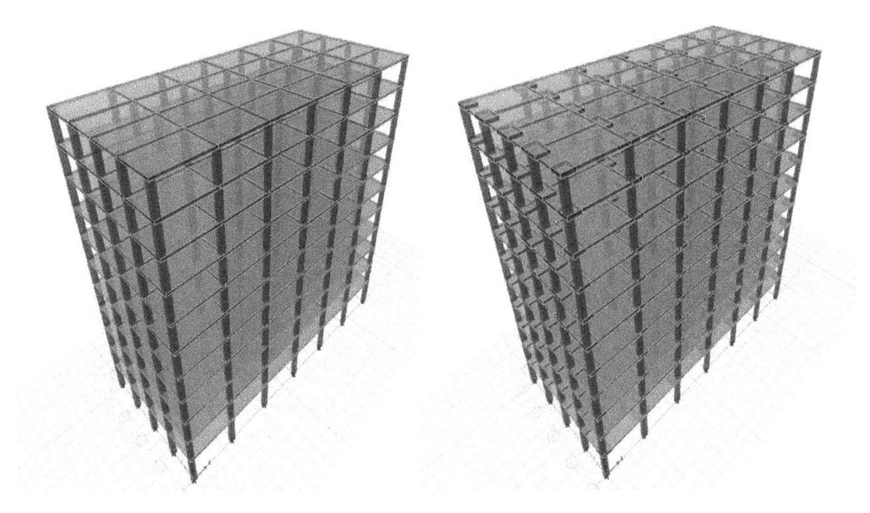

Fig. 2 3D rendered view of flat slab building and flat slab building with drop

shown in Fig. 1. The height of floor in ground floor is taken as 4 m and the height of remaining floor is taken as 3 m. The size of column is taken as 400 mm 600 mm and the flat slab of thickness 280 mm. The drop panel of size 2000 mm × 2000 mm and thickness 100 mm is provided. The concrete and steel of grade M30 and Fe415 are taken for the study. Figure 2 shows the 3D model of the building studied. The building was designed in accordance with IS 456:2000.

2.1 Load and Load Combinations

A live load of 4 kN/m^2 is applied as per IS: 875(Part 2): 1987 [13], a floor finish of 2 kN/m^2 is applied as per IS: 875 (Part 1): 1987 [14] and dead load is assigned by

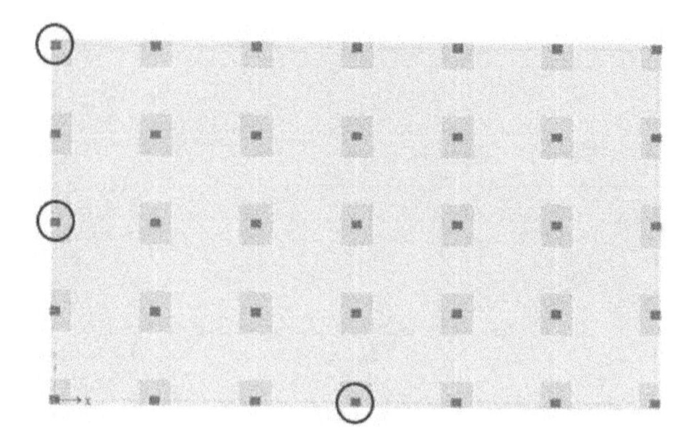

Fig. 3 Typical locations of column removal for both building models

the software itself. The GSA guidelines provided load combination for the linear static analysis of a building for progressive collapse

$$2(\mathrm{DL} + 0.25\,\mathrm{LL})$$

where DL is dead load, which includes self-weight of the building and loads due to floor finish; and LL is the live load.

3 Progressive Collapse Analysis

In this study, a linear static analysis was conducted to analyze the potential of progressive collapse as per the guidelines provided by GSA [2] in the ETABS software. The progressive collapse is achieved by the removal of a critical column and studying its effect on the structure. The criterion for the removal of columns is provided by GSA guidelines. In this study, 3 different locations of column removal at the ground floor are considered as per the GSA guidelines recommendation [2]:

(a) Corner column removal (CC)
(b) Middle column removal along the shorter side (MSC)
(c) Middle column removal along the longer side (MLC)

The Fig. 3 shows the locations of column removal considered in the study. The different column removal cases are analyzed on separate building models.

4 Results and Discussions

The linear static progressive collapse analysis on the both the building model for the three cases of column removal was done. The results of the study are in terms of Demand Capacity Ratio (DCR), Axial force in the column adjacent to the removed column and the vertical displacement at the location of column removal. The GSA guidelines suggests to have the Demand Capacity Ratio of a member to be less than 2 to resist progressive collapse. The Demand Capacity Ratio is defined as the ratio of the demand (Q_{UD}) coming to the member after the column removal to the capacity (Q_{CE}) of the member [2]. The DCR value for the flat slab building is taken in terms of the axial forces in the column members. The capacity of the member is calculated from Eq. 1 provided in IS 456:2000 [15] for the determining the capacity of column member for axial force.

$$Q_{CE} = 0.4 \times f_{ck} \times A_c + 0.67 \times f_y \times A_{sc} \quad (1)$$

where Q_{CE} = Ultimate capacity of the column section; f_{ck} = Compressive strength of concrete; f_y = Yield strength of steel; A_c = Area of concrete; A_{sc} = Area of longitudinal reinforcement bars.

The expected ultimate axial force (Q_{CE}) for the column is taken as 3663.6 kN.

4.1 *Effect of Corner Column Removal on Progressive Collapse*

The first case considered is the corner column removal at the ground floor of both the building models. Figure 4 shows the DCR value at the columns adjacent to the removed column due to corner column removal for flat slab building and flat slab building with drop. It is observed that the maximum DCR value observed is 1.31 and 1.38 for flat slab and flat slab with drop building. The DCR values at the adjacent columns is more for the building with drop panel compared to the flat slab building.

Figure 5 shows the axial force diagram obtained for corner column removal in flat slab building and flat slab building with drop. Figure 6 shows the axial force in the adjacent columns and vertical displacement at floors above the removed column for flat slab building and flat slab building with drop before and after column removal. The maximum axial force in the flat slab building is 3226.09 kN and the maximum axial force in the flat slab building with drop is 3433.71 kN in the adjacent column before the corner column removal. The maximum axial force in the flat slab building is 4794.4 kN and the maximum axial force in the flat slab building with drop is 5058.4 kN in the adjacent column for corner column removal. The maximum axial force in the adjacent column is observed for flat slab building with drop. The results show that the axial force in the adjacent column decreases with the increase in the floor level for both building models.

The vertical displacement before the column removal for flat slab building is 1.27 mm and for flat slab building with drop is 1.41 mm. The vertical displacement at the location of column removal for flat slab building is 19.83 mm and for flat slab building with drop is 16.94 mm. It can be observed that the incorporation of drop panel tends to decrease the vertical displacement at all the stories above the removed column.

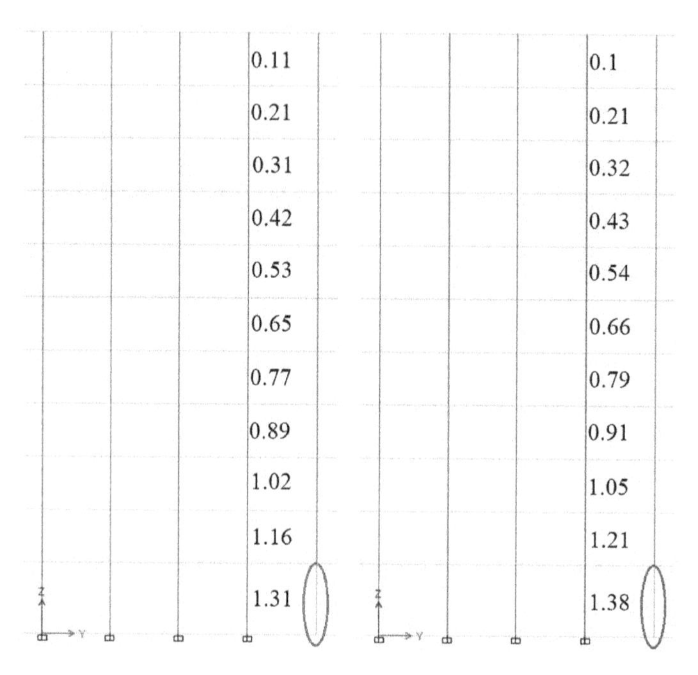

Fig. 4 DCR value of flat slab building and flat slab building with drop for corner column removal

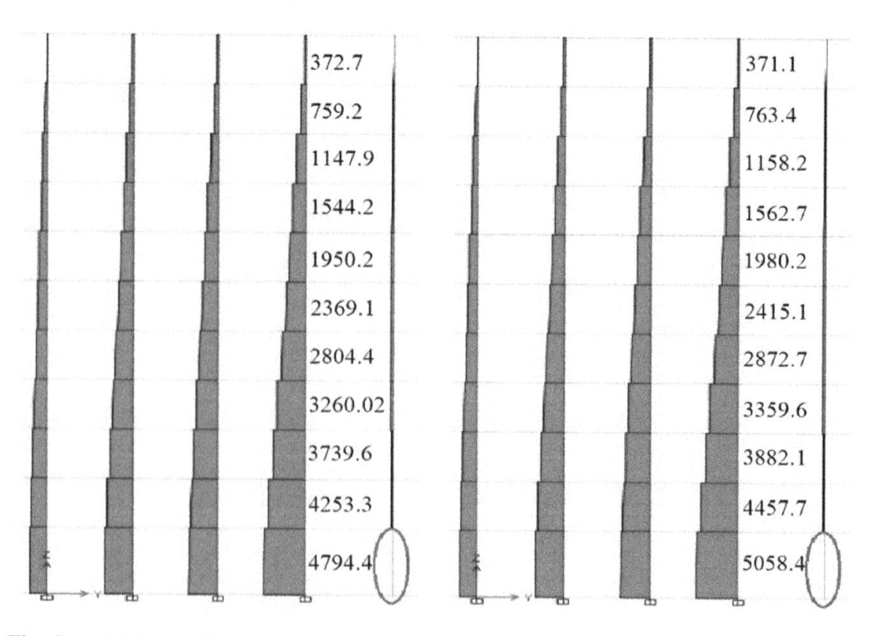

Fig. 5 Axial forces diagram for the corner column removal in flat slab building and flat slab building with drop

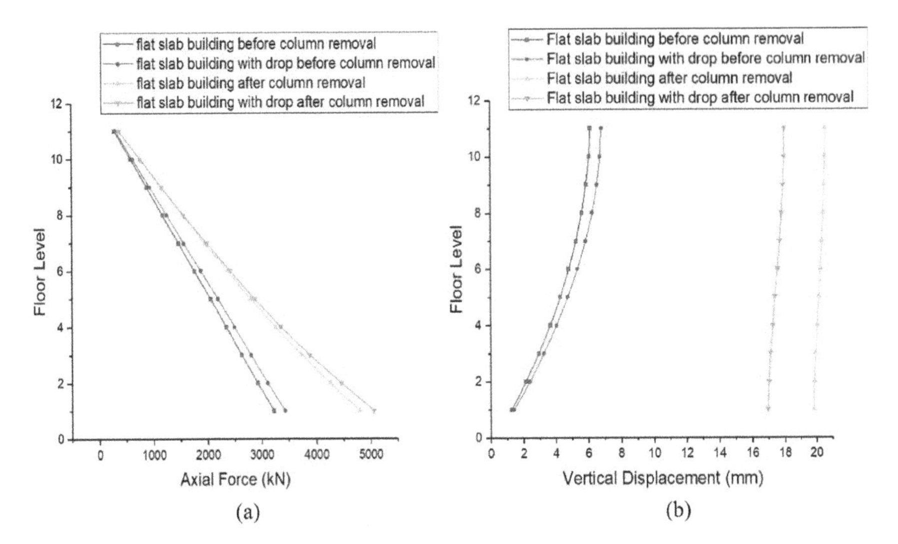

Fig. 6 **a** Axial forces in the adjacent column. **b** Vertical Displacements in the floors above the removed column for the corner column removal

4.2 Effect of Middle Short Column Removal on Progressive Collapse

The second case considered is the middle short column removal at the ground floor of both the building models. Figure 7 shows the DCR value at the columns adjacent to the removed column due to middle short column removal for flat slab building and flat slab building with drop. It is observed that the maximum DCR value observed is 1.25 and 1.31 for flat slab building and flat slab with drop building. The DCR values at the adjacent columns is more for the building with drop panel compared to the flat slab building.

Figure 8 shows the axial force diagram obtained for middle short column removal in flat slab building and flat slab building with drop. Figure 9 shows the axial force in the adjacent columns and vertical displacement at floors above the removed column for flat slab building and flat slab building with drop before and after column removal. The maximum axial force in the flat slab building is 3226.1 kN and the maximum axial force in the flat slab building with drop is 3433.7 kN in the adjacent column before the middle short column removal. The maximum axial force in the flat slab building is 4586.1 kN and the maximum axial force in the flat slab building with drop is 4828.1 kN in the adjacent column for middle short column removal. The maximum axial force in the adjacent column is observed for flat slab building with drop. The results show that the axial force in the adjacent column decreases with the increase in the floor level for both building models.

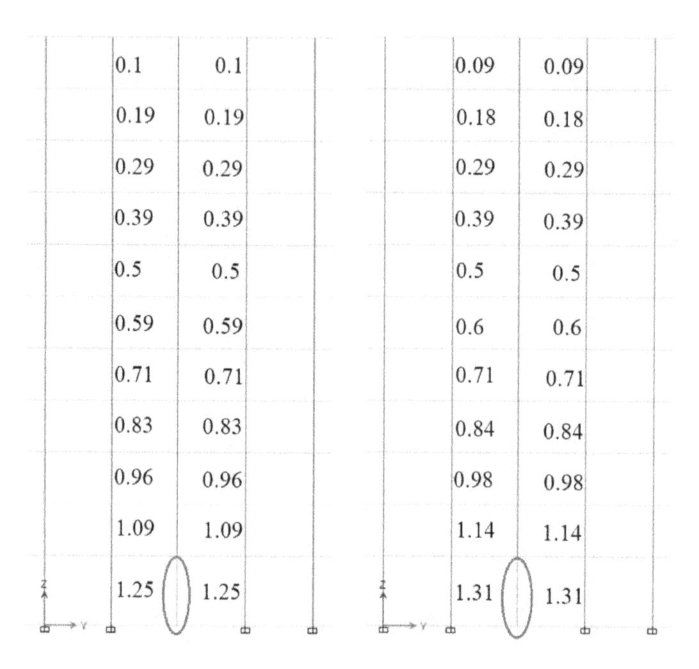

Fig. 7 DCR value of flat slab building and flat slab building with drop for middle short column removal

The vertical displacement before the column removal for flat slab building is 2.05 mm and for flat slab building with drop is 2.2 mm. The vertical displacement at the location of column removal for flat slab building is 15.59 mm and for flat slab building with drop is 13.55 mm. It can be observed that the incorporation of drop panel tends to decrease the vertical displacement at all the stories above the removed column.

4.3 Effect of Middle Long Column Removal on Progressive Collapse

The third case considered is the middle long column removal at the ground floor of both the building models. Figure 10 shows the DCR value at the columns adjacent to the removed column due to middle long column removal for flat slab building and flat slab building with drop. It is observed that the maximum DCR value observed is 1.26 and 1.32 for flat slab building and flat slab with drop building. The DCR values at the adjacent columns is more for the building with drop panel compared to the flat slab building.

Figure 11 shows the axial force diagram obtained for middle long column removal in flat slab building and flat slab building with drop. Figure 12 shows the

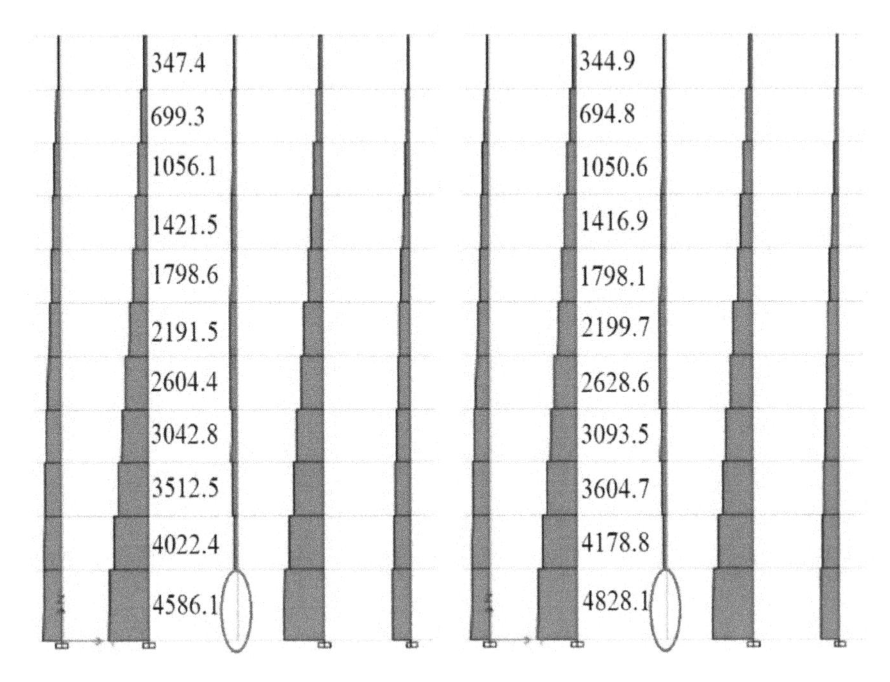

Fig. 8 Axial forces diagram for the middle short column removal in flat slab building and flat slab building with drop

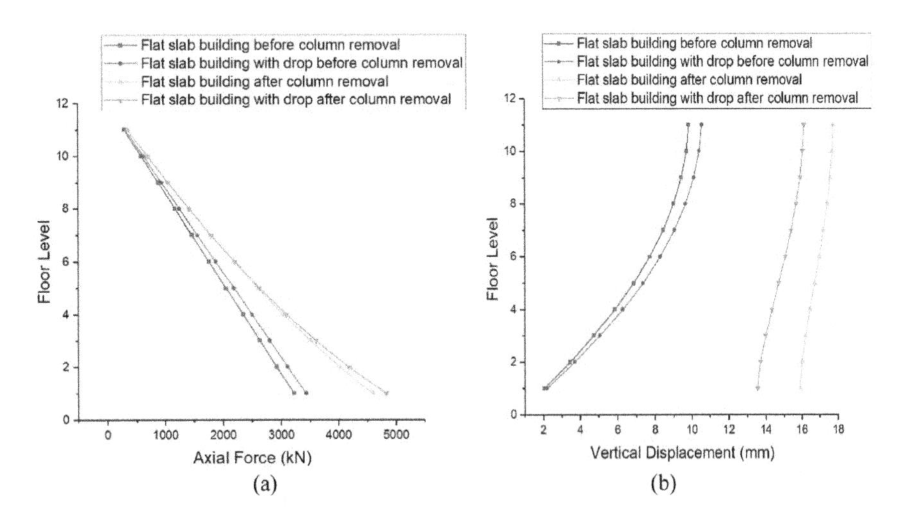

Fig. 9 **a** Axial forces in the adjacent column, **b** Vertical Displacements in the floors above the removed column for the middle short column removal

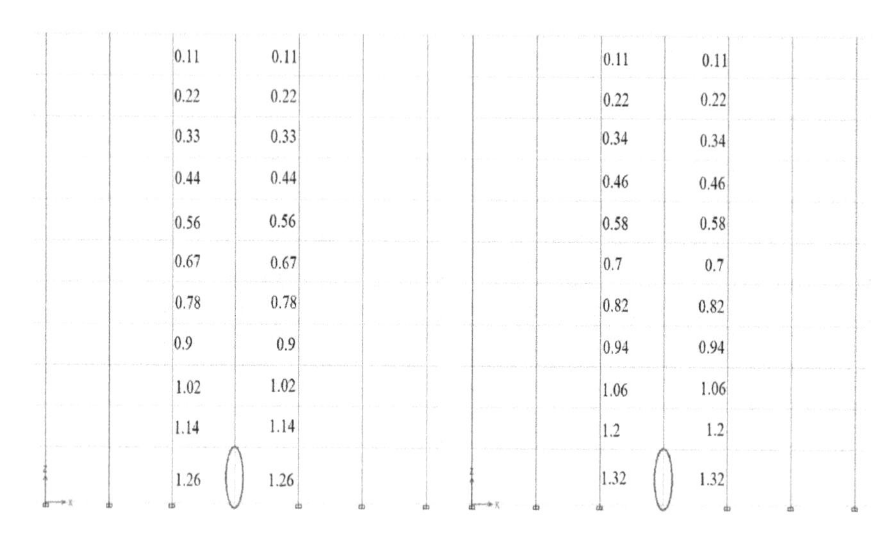

Fig. 10 DCR value of flat slab building and flat slab building with drop for middle long column removal

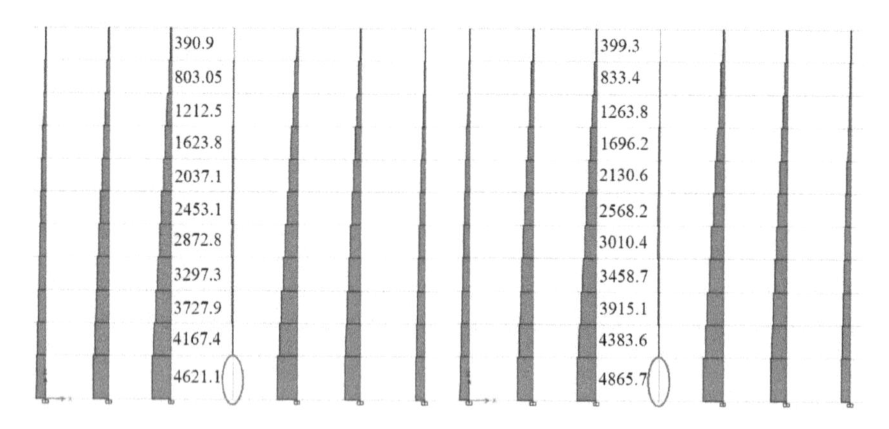

Fig. 11 Axial forces diagram for the middle long column removal in flat slab building and flat slab building with drop

axial force in the adjacent columns and vertical displacement at floors above the removed column for flat slab building and flat slab building with drop before and after column removal. The maximum axial force in the flat slab building is 3711.02 kN and the maximum axial force in the flat slab building with drop is 3943.33 kN in the adjacent column before the middle short column removal. The maximum axial force in the flat slab building is 4621.1 kN and the maximum axial force in the flat slab building with drop is 4865.7 kN in the adjacent column for middle long column removal. The maximum axial force in the adjacent column is

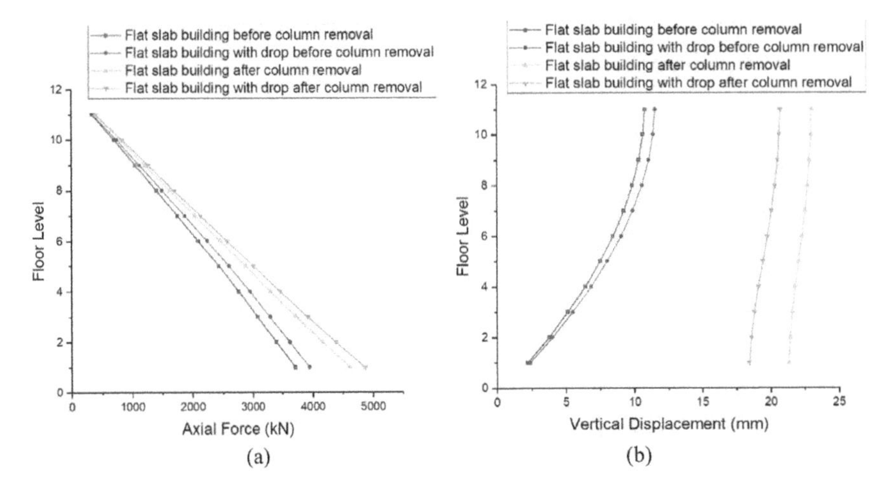

Fig. 12 **a** Axial forces in the adjacent column. **b** Vertical Displacements in the floors above the removed column for the middle long column removal

observed for flat slab building with drop. The results show that the axial force in the adjacent column decreases with the increase in the floor level for both building models.

The vertical displacement before the column removal for flat slab building is 2.23 mm and for flat slab building with drop is 2.38 mm. The vertical displacement at the location of column removal for flat slab building is 21.3 mm and for flat slab building with drop is 18.41 mm. It can be observed that the incorporation of drop panel tends to decrease the vertical displacement at all the stories above the removed column.

4.4 Comparison of Results

Table 1 shows the summary of the DCR values for the three cases of column removal. GSA guidelines [2] suggests that the members having DCR value greater than 2 for regular building tends to undergo progressive collapse. The results show that both flat slab building and flat slab building with drop is able to resist the progressive collapse as the DCR value of all the members is less than 2. The maximum value of DCR is obtained for the corner column removal case for both the building models. So, the corner column removal is the most critical location of column removal in the building. This is because the middle column has got more neighboring column for the load transferring compared to the corner column [16]. The results show that the DCR value obtained for middle short column removal is 4.83% less than the DCR for corner column removal and the DCR value obtained for the middle long column removal is 4.1% less than DCR value obtained for

Table 1 Summary of DCR value for the three cases of column removal

Location of column removal	Flat slab building	Flat slab building with drop
	DCR value	DCR value
Corner column removal	1.31	1.38
Middle short column removal	1.25	1.31
Middle long column removal	1.26	1.32

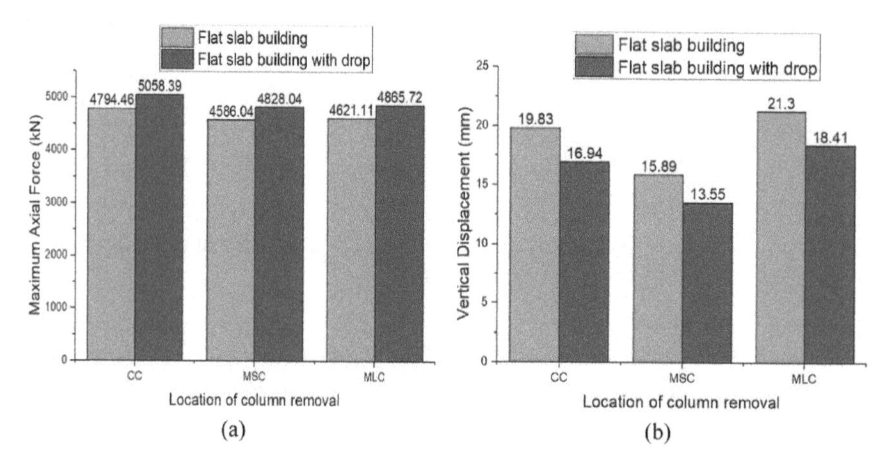

Fig. 13 Comparison of **a** Max. axial force in the adjacent column, **b** vertical displacement for all three column removal scenarios in flat slab building with and without drop

corner column removal. In all the cases, the flat slab building is less susceptible to progressive collapse than flat slab building with drop as the DCR values are less in the lower floor for the flat slab building.

Figure 13 shows the comparison of maximum axial force in the column adjacent to the removed column and the vertical displacement at the location of the removed column for all three column removal cases. The axial force in the adjacent column was increased by 48, 41 and 24% for column removal at the corner, middle short side and middle long side. But the increase in the axial force at the corner location away from the removed column is only 2.46, 0.22 and 2.25% for column removal at the corner, middle short side and middle long side. The maximum increase in the axial force was observed during the corner column removal and minimum increase in the axial force was observed in the middle long side column removal. The results also showed that the max force is transferred to the adjacent column as compared to the corner column away from the removed column in the flat slab building [1].

Figure 13b shows that the middle column along the longer side produces the maximum displacement and middle column along the shorter side produces the minimum displacement. The result also shows that the drop panel reduces the vertical displacement at the location of column removal for all cases of column removal. The provision of drop panel in flat slab building decreases the vertical displacement by 14.6, 14.7 and 13.6% for column removal at the corner, middle short side and middle long side. The results show that the provision of drop panel reduces the impact of progressive collapse at the locations of column removal by reducing the vertical displacement.

5 Conclusion

The linear static progressive collapse analysis on a 11-storey flat slab building was conducted. The following conclusion can be made from the study:

- The study on the RC flat slab building showed that the building was resistant to progressive collapse as all the members have DCR value less than 2.
- The corner column removal is the most critical column removal case in both the buildings as the maximum DCR value was obtained for corner column removal.
- The DCR value for the flat slab building with drop was higher than flat slab building making it more critical in progressive collapse analysis.
- The maximum axial force in the adjacent column was obtained in the case of the corner column removal for both the building models.
- The axial force in the adjacent column is increased by 48, 41 and 24% for column removal at the corner, middle short side and middle long side, but the increase in the axial force at the corner location away from the removed column is only 2.46, 0.22 and 2.25% for column removal at the corner, middle short side and middle long side.
- The max load is transferred to the adjacent column compared to the corner column away from the removed column in the building during the progressive collapse.
- The maximum vertical displacement was observed in the longer side middle column removal case as compared to corner and short side middle column removal.
- The addition of drop panel in the building tends to decrease the vertical displacement at the removed location in all the column removal case.
- The addition of drop panel in flat slab building decreased the vertical displacement by 14.6, 14.7 and 13.6% for column removal at the corner, middle short side and middle long side.

References

1. Qian K, Li B (2015) Dynamic disproportionate collapse in flat-slab structures. J Perform Constr Facil 29(5):B4014005
2. GSA (2013) Progressive collapse analysis and design guidelines for new federal office buildings and major modernization projects, The US General Services Administration

3. Unified Facilities Criteria (UFC) (2009) Design of buildings to resist progressive collapse. Department of Defense
4. Weng YH, Qian K, Fu F, Fang Q (2020) Numerical investigation on load redistribution capacity of flat slab substructures to resist progressive collapse. J Build Eng 2:101109
5. Senthil K, Jain S, Bawa S (2018) Study on influence of thickness and span length of flat slab for progressive collapse. ASMMCE—2018, S-101
6. Russell JM, Owen JS, Hajirasouliha I (2018) Nonlinear behaviour of reinforced concrete flat slabs after a column loss event. Adv Struct Eng 21(14):2169–2183
7. Russell JM, Owen JS, Hajirasouliha I (2019) Dynamic column loss analysis of reinforced concrete flat slabs. Eng Struct 198:109453
8. Olmati P, Sagaseta J, Cormie D, Jones AEK (2017) Simplified reliability analysis of punching in reinforced concrete flat slab buildings under accidental actions. Eng Struct 130:83–98
9. Peng Z, Orton S, Tian Y (2015) Experimental dynamic response of reinforced concrete flat plate sub-structure under collapse scenario. In: Structures congress 2015, pp 1090–1100
10. Qian K, Li B (2015) Strengthening of multibay reinforced concrete flat slabs to mitigate progressive collapse. J Struct Eng 141(6):04014154
11. Qian K, Li B (2013) Strengthening and retrofitting of RC flat slabs to mitigate progressive collapse by externally bonded CFRP laminates. J Compos Constr 17(4):554–565
12. Mohamed O, Khattab RA, Mishra A, Isam F (2019, February) Recommendations for reducing progressive collapse potential in flat slab structural systems. In: IOP conf. ser. mater. sci. eng, vol 471, p 052069
13. IS 875(Part 1)-1987 Indian Standard Code of Practice for Design Loads (Other than earthquake) for Building and Structures. Part 1: Dead Loads. Bureau of Indian Standards, New Delhi
14. IS 875(Part 2)-1987 Indian Standard Code of Practice for Design Loads (Other than earthquake) for Building and Structures. Part 2: Imposed Loads. Bureau of Indian Standards, New Delhi
15. IS 456: 2000 Indian Standard—"Plain and Reinforced Concrete—Code of practice. Bureau of Indian Standards, 2000, New Delhi
16. Garg S, Agrawal V, Nagar R (2020) Progressive collapse behavior of reinforced concrete flat slab buildings subject to column failures in different storeys. Mater Today: Proc

Numerical Study on Perforation Characteristics of Carbon-Fiber Reinforced Composite Laminates Subjected to Impact Loading

Gyanesh Patnaik, Anshul Kaushik, Abhishek Rajput, and Guru Prakash

Abstract Carbon fiber reinforced composite has gained a huge popularity as a protective material against impact loading due to its excellent properties such as light weight, high specific strength and high specific modulus. Determining perforation characteristics of fiber reinforced composite laminates is an integral part for the design of protective civil and military structures. In this work, the perforation characteristics (ballistic limit, residual velocity, perforation energy) of carbon fiber reinforced polymer (CFRP) laminates were determined with the help of experimental tests and numerical simulations. CFRP laminates with four different fiber orientations, each of two different thicknesses (1 and 2 mm) were considered in this study. These laminates were impacted by conical, hemispherical and ogive nose projectiles made of steel for a wide range of velocities. The high velocity impact experiments were carried out using pneumatic gun setup. A numerical model was developed in ABAQUS/EXPLICIT using Hashin damage criteria to understand the perforation behavior of CFRP sheets under high velocity impact. The accuracy of the numerical model was assessed by comparing its prediction with experimental results of cross-ply laminates. The perforation characteristics predicted with the help of numerical simulations were in good agreement with the experimental test results. Effective ply configuration was achieved in terms of energy absorption and damage resistance for better performance under impact loading. The influence of shape of the projectile, ply orientation, thickness at different impact velocities on energy absorbing capacity of CFRP laminates were discussed.

Keywords Ballistic limit · CFRP laminates · Hashin damage · Perforation characteristics · Energy absorption

G. Patnaik (✉) · A. Kaushik · A. Rajput · G. Prakash
Department of Civil Engineering, Indian Institute of Technology Indore, Indore, India
e-mail: phd1901204003@iiti.ac.in

G. C. Marano et al. (eds.), *Proceedings of SECON'21*, Lecture Notes in Civil Engineering 171, https://doi.org/10.1007/978-3-030-80312-4_21

1 Introduction

The utilization of carbon fiber reinforce polymer (CFRP) composite materials stretches over various industries such as infrastructures, aerospace, automobiles, defense sector, sports equipment due to its high tensile strength, high strength to weight ratio, resistant to fatigue, high energy absorbing capacity and many more. Properties like high modulus makes CFRP structurally suitable to replace traditional materials such as aluminum and titanium. Formation of structures from these composite fibers are subjected to various types of impact. Thus, it becomes relevant to understand the perforation process for effective design of structures. Hence, interpretation of the response of CFRP composite against impact becomes important. Researchers in the past has worked on determining the ballistic limit velocity, energy absorbing capacity and residual energy of the plates [1, 2]. Various factors are identified which affects the energy absorbing capacity and extent of damage in fiber reinforced polymer composites. Factors such as impactor velocity, its shape and size, stacking sequence, thickness of laminates and more [3–8]. Changing fiber orientation can be one of the effective factors in improving resistance against impact [9–11]. Many researchers investigated the effect of orientation of fibers on low velocity impact both in experimental and numerical studies. Hitchen and Kemp [10] investigated the placing of $\pm 45°$ plies on the outer surface to minimize delamination areas. Dorey [12] determined that the laminates having $\pm 45°$ as surface layers gives higher resistance to impact and improved strength as compared to layers of 0° orientation at surface. Results indicated that there was an increase in ability of composites in absorbing energy elastically. Velmurugan [13] studied the effect of thickness and fiber orientation of CFRP composites against high velocity impact by a conical steel bullet. Results indicated that [0/90] absorbs more energy and have higher ballistic limit than other layup sequences presented in the paper. Guoqi et al. [14] investigated Kevlar sheets for impact studies. They impacted the laminates by a conical projectile at and above ballistic limit and observed that the change in volume fraction does not affect the impact resistance of laminates. Shi et al. [15] used finite element (FE) method to simulate the performance of composite under low velocity impact. They applied Hashin damage model to estimate the resistance of cross ply laminates against impact. They predicted delamination area both experimentally and numerically at lower and higher levels of energy and both experimental and numerical results are in good concordance. Corran et al. [16] considered the effect of nose shape, mass of the projectile on steel and aluminum plates. Blunt and conical projectile were used on different thickness of plates and for range of velocities. They found that the ballistic velocity of the plate changes with varying shape and mass of the projectile. Radin and Goldsmith [17] developed analytical models to find out residual velocity of layered and monolithic targets of aluminum and polycarbonate by varying different parameters such as bullet shape, diameter, and impact velocities. Although few works are available in literature, yet it becomes essential to analyse the CFRP laminates under impact. In addition, due to complexity in modeling the constitutive behavior of composites, few studies are available that solve the problem numerically. Hence present study is focused on the effect of orientation of fibers, thickness, and projectile shape on the performance of CFRP laminates.

2 Damage Model for Fiber Reinforced Composites

Out of various failure model used for damage modelling of composites, Hashin damage model has been widely used due to its ease of use. It is also preferred as it can model four failure modes such as tensile and compressive failure of fibers and matrix separately. The model is quadratic in nature and the general form of equation as per Hashin [18] is:

Case 1: Tensile fiber failure: $\sigma_{11} > 0$From standard uniaxial tensile test, $\sigma_{11} = \sigma_{\mathrm{A}}^{+}$ and for combined loading of σ_{11}, σ_{22}, the approximate equation for tensile fiber failure mode was given by.

$$\left(\frac{\sigma_{11}}{\sigma_{A}^{+}}\right) + \frac{1}{\tau_{A}^{2}}\left(\sigma_{12}^{2} + \sigma_{13}^{2}\right) = 1 \tag{1}$$

Tensile matrix failure: $\sigma_{22} + \sigma_{33} > 0$

$$\frac{1}{\sigma_{\mathrm{T}}^{+2}}(\sigma_{22} + \sigma_{33})^{2} + \frac{1}{\tau_{\mathrm{T}}^{2}}\left(\sigma_{23}^{2} - \sigma_{22}\sigma_{33}\right) + \frac{1}{\tau_{A}^{2}}\left(\sigma_{12}^{2} + \sigma_{13}^{2}\right) = 1 \tag{2}$$

Case 2: Compressive fiber failure: $\sigma_{11} < 0$

$$\sigma_{11} = -\sigma_{\mathrm{A}}^{-} \tag{3}$$

Compressive matrix failure: $\sigma_{22} + \sigma_{33} < 0$

$$\frac{1}{\sigma_{T}^{-}}\left[\left(\frac{\sigma_{T}^{-}}{2\tau_{T}}\right)^{2} - 1\right](\sigma_{22} + \sigma_{33}) + \frac{1}{4\tau_{\mathrm{T}}^{2}}(\sigma_{22} + \sigma_{33})^{2} + \frac{1}{\tau_{\mathrm{T}}^{2}}\left(\sigma_{23}^{2} - \sigma_{22}\sigma_{33}\right) + \frac{1}{\tau_{A}^{2}}\left(\sigma_{12}^{2} + \sigma_{13}^{2}\right) = 1 \tag{4}$$

σ_{A}^{+} and σ_{A}^{-} are tensile failure stress and compressive failure stress in fiber direction respectively. σ_{T}^{+} and σ_{T}^{-} are tensile failure stress and compressive failure stress transverse to fiber direction respectively. $\sigma_{11}, \sigma_{22}, \sigma_{23}, \sigma_{12}, \sigma_{13}$ are stress components where first and second subscript represents the face and direction respectively in which stress acts. τ_{A}andτ_{T} are axial failure shear and transverse failure shear respectively.

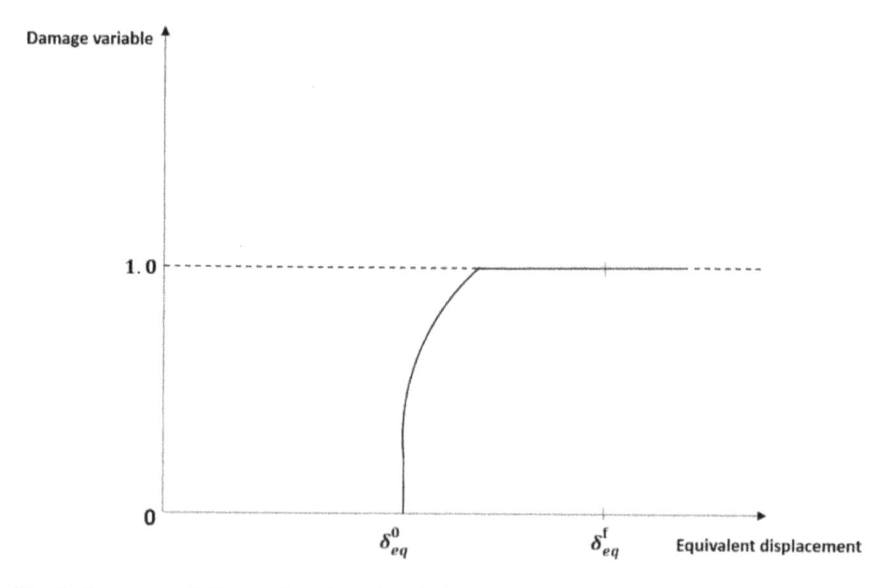

Fig. 1 Damage variable as a function of equivalent displacement

2.1 Damage Initiation, Evolution and Element Removal

The beginning of stiffness degradation marks the initiation of damage and the evolution of damage for each of the failure modes takes place after the damage initiation. The material behaves as linear elastic before damage and as the damage took place and progresses the curve slope gets negative. Damage evolution depends on various factors such as strength and stiffness of the material. After damage evolution, material degradation takes place. As the material starts degrading, the elements of the material needs to be removed as it leads to mesh convergence difficulties. So, by default, in ABAQUS [19], those elements got deleted as the damage variable at different points of the material reaches upper limit (1.0) and they no longer contributes to the stiffness of the material as shown in Fig. 1.

3 Experimental Investigation

High velocity impact tests has been performed on carbon fiber laminates of orientation $[0/90]_2$ and $[0/90]_4$ with a conical shaped projectile. Single stage pneumatic gun (Fig. 2) has been used to perform high velocity impact test. Carbon fiber composite laminates of span 200 mm × 200 mm were prepared with epoxy resin and hardener. In this setup, CFRP laminate of 1 mm and 2 mm thick was impacted with 150 m/s and 113 m/s respectively and the angle of impact was normal.

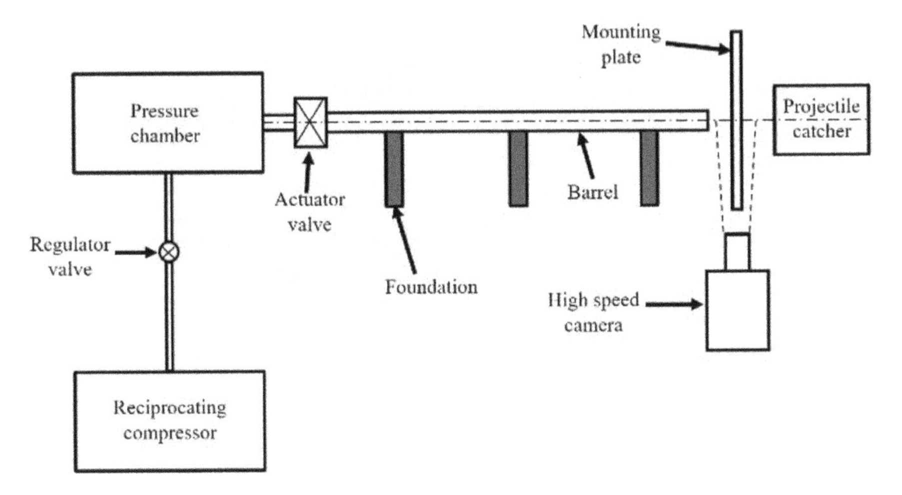

Fig. 2 Single stage pneumatic gun setup

The target was clamped at edges. All the processes were recorded with a high speed camera which helps in understanding the impact process.

The damage pattern of 1 mm and 2 mm thick plates were shown in Figs. 3 and 4 respectively. The damage occurred at the rear face was more than the front face. Out of various energy absorbing mechanism, delamination and plug formation was observed in CFRP targets. The experimental results of the CFRP target at different velocity was presented in Table 1.

4 Numerical Modelling and Its Validation

A numerical model was developed and executed using ABAQUS/Explicit. Parts were modelled using different modules which gives rise to an input file and then job was submitted to run the analysis. In ABAQUS/CAE 'Part Module', plate and impactor were modelled as 3D deformable solid and as three dimensional analytical rigid shell respectively. CFRP plates of span 200 mm × 200 mm and thickness (1 mm and 2 mm) was modelled. The projectile used in this analysis was of conical shape of steel with properties as given in Table 2.

The material properties of carbon composites used in this study were taken from literature [20] and is tabulated below (Table 3). Properties of material, fiber orientation, ply count and each layer thickness were assigned in 'property' module while in case of analytical rigid body, property has been assigned at the centroid of the projectile.

Each laminate consists of four layers and eight layers, each layer of thickness 0.25 mm, in case of 1 mm and 2 mm respectively. The orientation of fiber in 1 mm laminate is as shown in Fig. 5.

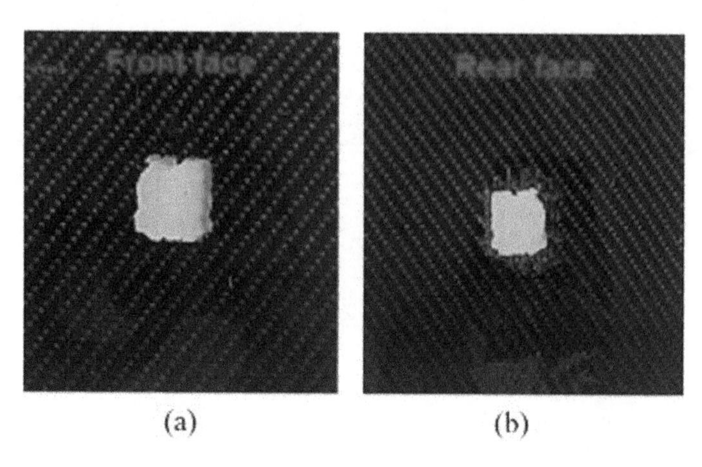

(a) (b)

Fig. 3 Damage pattern in front and rear face of laminates with stacking sequence $[0/90]_2$

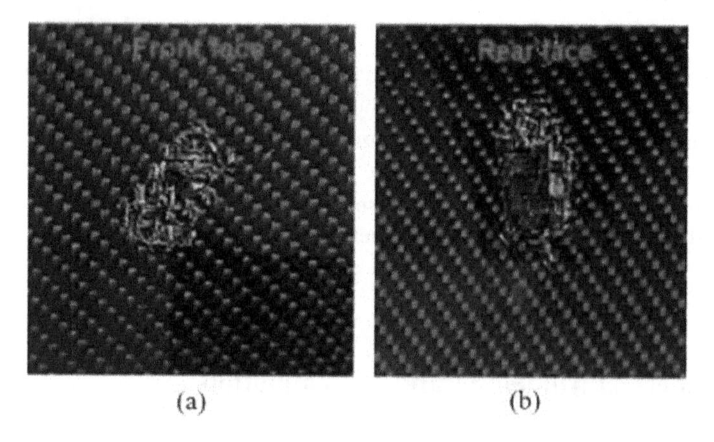

(a) (b)

Fig. 4 Damage pattern in front and rear face of laminates with stacking sequence $[0/90]_4$

Table 1 Experimental results of $[0/90]_2$ and $[0/90]_4$

Thickness (mm)	Impact velocity (m/s)	Residual velocity (m/s)	Ballistic limit (m/s)
1	150	125	51
2	113	88	83

Table 2 Properties of steel projectile

Young's modulus (GPa)	Mass (gm)	Poisson's ratio	Density (kg/m^3)
210	15	0.3	7850

Table 3 Properties of CFRP plate

Property	Value
T300/5208 composites	
Elastic modulus (GPa)	
E_{11}	136
E_{22}	9.8
E_{33}	9.8
Poisson's ratio	
μ_{12}	0.28
μ_{31}	0.28
μ_{23}	0.15
Shear moduli (GPa)	
G_{12}	4.7
G_{13}	4.7
G_{23}	4.261
Tensile strength (MPa)	
X_t	1550
X_c (Mpa)	1090
Y_t	59
Compressive strength (MPa)	
X_c	1090
Y_c (Mpa)	207
Shear strength (MPa)	
S_{12}	128
S_{23}	75
In plane fracture toughness (kJ/m^2)	
G_{1c}^T	91.6
G_{1c}^C	79.9
G_{2c}^T	0.22
G_{2c}^C	1.1
Density, ρ (kg/m^3)	1540

In 'assembly' module, both parts (plate and bullet) were assembled together and a gap of 1 mm between plate and impactor was maintained to apply contact algorithm. The field output and history output variables required were selected in 'step' module. Depending upon the velocities, the time taken for simulation was varied between 0.0002 and 0.002 s. In order to constraint the degree of freedom of plate in all directions, ENCASTRE option has been used and predefined velocity was given at reference point of bullet. The interaction between plate and bullet was defined using surface-to-surface contact interaction type as shown in Fig. 6. Kinematic contact method was used to obtain a stable master (projectile)—slave (target plate) contact pair. Tangential behavior and normal behavior has been assigned in contact property between bullet and plate. In contact property, friction

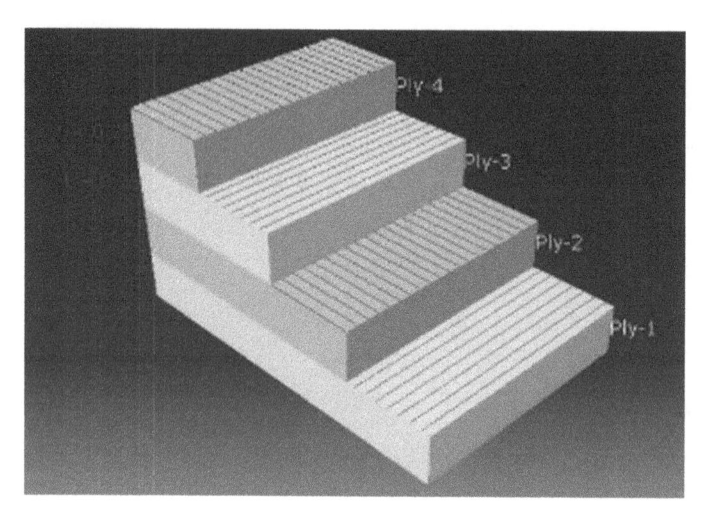

Fig. 5 Ply stacking sequence of fiber orientation $[0/90]_2$

Fig. 6 Surface-to-surface contact interaction type

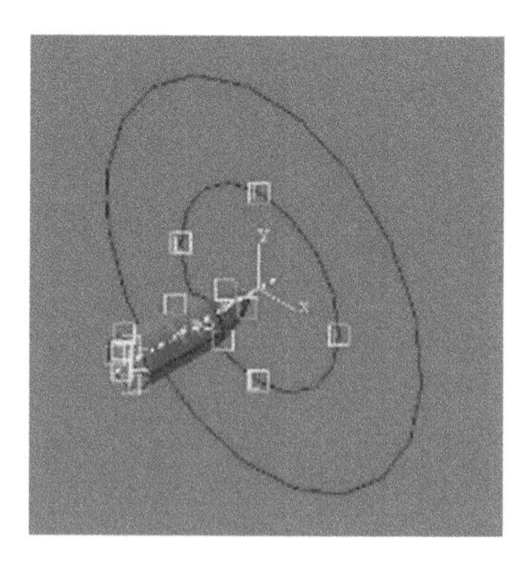

coefficient of 0.1 was employed between plate and bullet. Primary impact zone, secondary impact zone and outer impact zone were marked on the CFRP plate and fine meshing was applied in the primary zone. The mesh density gets coarser as we move towards outer impact zone.

SC8R (8-node quadrilateral continuum shell, reduced integration) elements of size 1 mm and 6 mm were used in primary and outer impact zone respectively. In the secondary impact zone, SC6R (6-node triangular continuum shell, reduced integration) elements were used of size 4 mm. Size of the mesh was determined through mesh convergence study. Figure 7 shows the mesh convergence study by varying element size of the target plate at a particular velocity (V_i = 100 m/s) and the mesh size was varied till the result converges. Job was submitted after meshing and after completion of job, Abaqus provides an output file (*.ODB) which carry

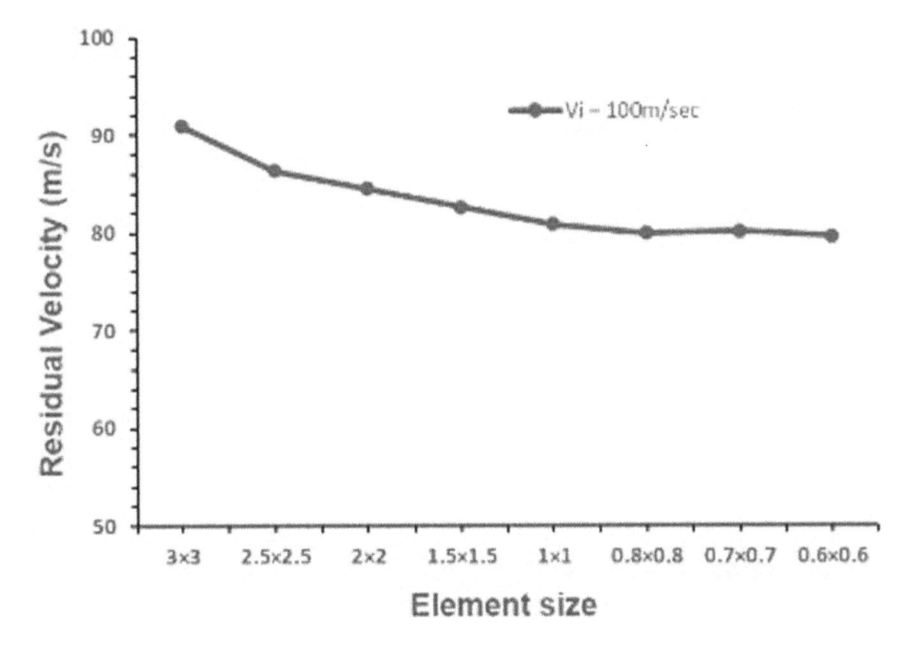

Fig. 7 Mesh convergence study of CFRP laminate of 1 mm thick at 100 m/s

results from the analysis and ‘Visualization’ module allows display of analysis results in the form of graphical representation.

4.1 *Model Validation*

FE model was validated with the experimental results obtained from high velocity impact experiments. The experimental values of ballistic limit and residual velocity were compared with numerically obtained results. Figures 8 and 9 shows the comparison of ballistic limit and residual velocity at different impact velocity respectively. From both the tables, it can be observed that the results obtained from both numerical as well as experimental were in good concordance with each other. The error percentage in ballistic limit and residual velocity calculation comes out to be 7.35% and 8.13% respectively in case of 1 mm plates. This verifies the reliability of the numerical model and hence this numerical model can be used in obtaining perforation characteristics by varying different parameters.

After the model was validated, perforation characteristics of different fiber orientations was determined through the validated numerical model. Laminates with different orientations considered in this study are shown in Fig. 10.

Figure 11 shows Mises stress on front and rear face of CFRP targets at different step time when impacted with an ogive nose projectile.

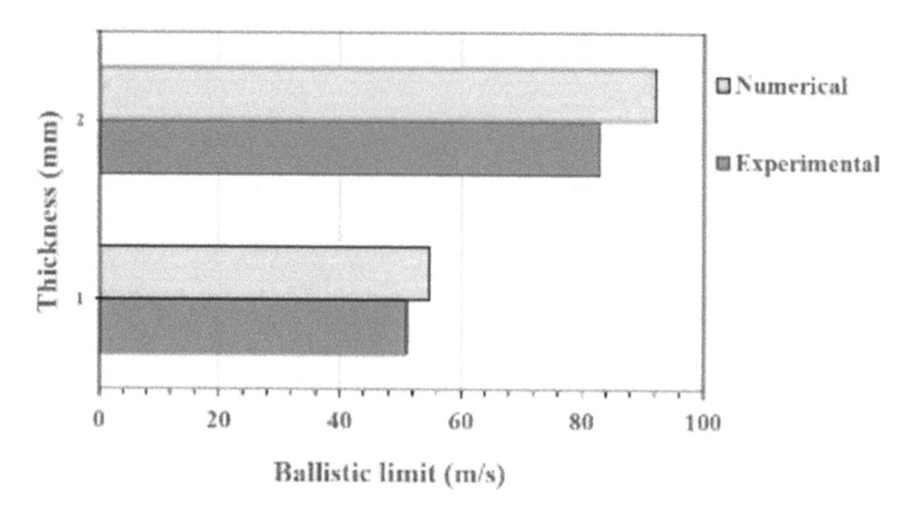

Fig. 8 Comparison of ballistic limit

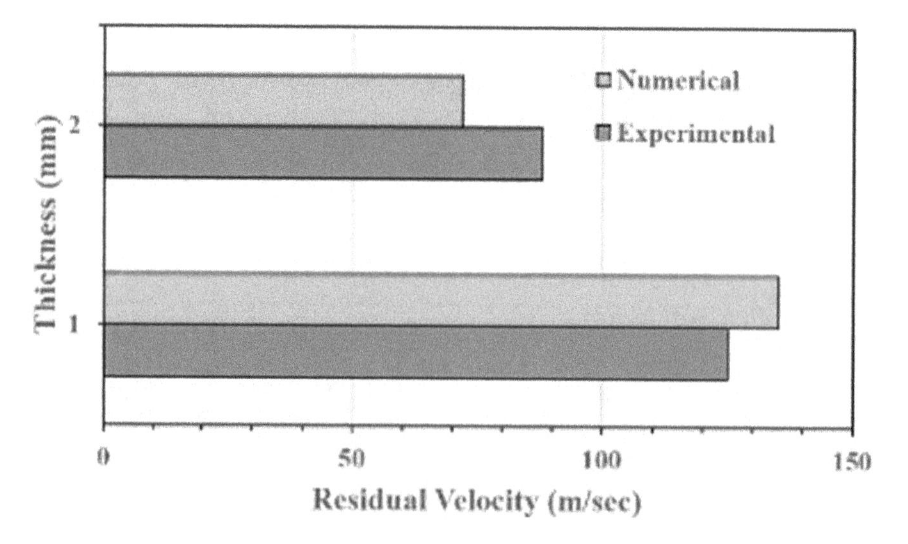

Fig. 9 Comparison of residual velocity

5 Results and Discussion

In this section, simulation results were presented and discussed. The aim of the numerical investigation was to find out the ballistic limit velocity, residual velocity and energy absorption of CFRP targets of different fiber orientations such as $[0/90]_2$, [0/90/30/−60], [0/90/45/−45], [30/−60/60/−30]. Symmetric stacking sequence of these fiber orientations were considered for 2 mm thick target plates. These four

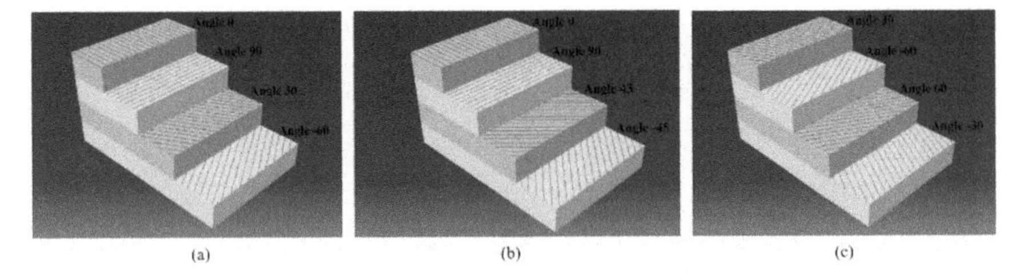

Fig. 10 Stacking sequence of 1 mm laminates with different fiber orientations

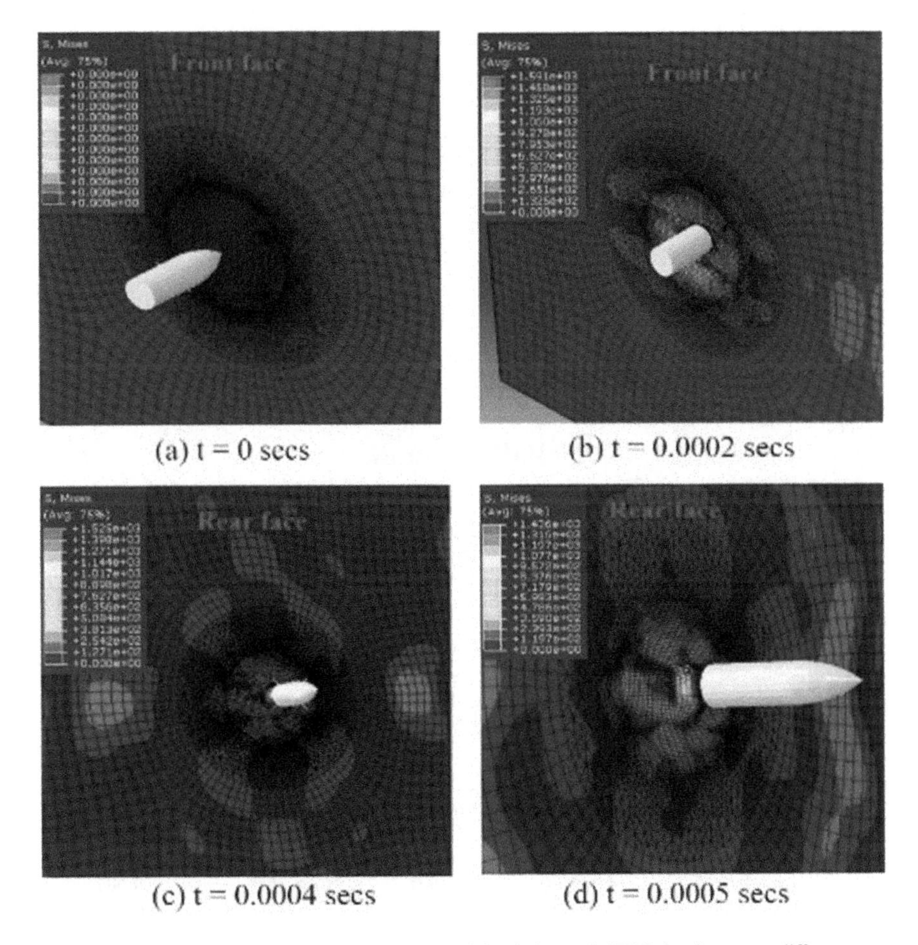

Fig. 11 Mises stress representation on front and back face of CFRP laminates at different step time

different fiber orientations were impacted with three different types of projectile for a wide range of velocities.

5.1 Ballistic Limit

Ballistic limit is an important parameter in perforation study as it helps in preventing perforation. Ballistic limit can be defined as the maximum velocity at which the projectile got stuck or the minimum velocity at which the projectile gets perforated through target plate. In this study, ballistic limit has been obtained as average of maximum stuck velocity and minimum perforated velocity for accurate results.

Figure 12 shows the velocity time curve at different impact velocities. From the above figure, it can be observed that at 91.5 m/s, the projectile get stuck onto the target and at 92.5 m/s, the projectile penetrates the plate completely. So, average of these two velocities gives the ballistic limit for [0/90]2 s, impacted by an ogive shaped projectile.

It can be observed from Table 4 that the fiber orientation along with thickness has its influence on the perforation characteristics. It can also be noticed that with increase in sharpness of projectile, the ballistic limit decreases.

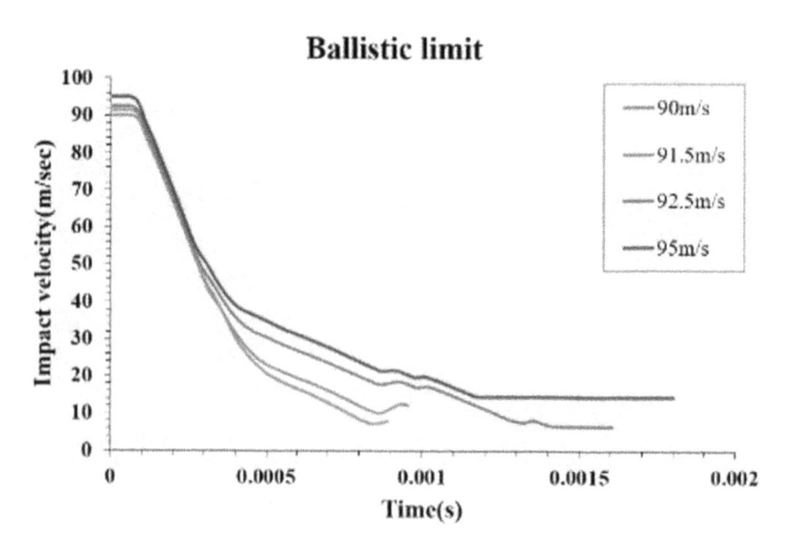

Fig. 12 Calculation of ballistic limit

Table 4 Ballistic limit of different laminates subjected to different projectile

Ballistic limit (m/s)	Hemispherical		Conical		Ogive	
	1 mm	2 mm	1 mm	2 mm	1 mm	2 mm
$[0/90]_2$	59.5	98	54.75	96.25	54	92
[0/90/30/-60]	58	97.25	53.25	95.5	52.75	91.5
[0/90/45/-45]	60.25	99.5	56	97.5	55	94.75
[30/-60/60/-30]	57	96	52.5	94.25	51	90

Table 5 Residual velocity of 1 mm thick plate at 100 m/s

Residual velocity (m/s)	Hemispherical	Conical	Ogive
$[0/90]_2$	75.12	77.2	78.6
[0/90/30/-60]	76.9	78.8	79.2
[0/90/45/-45]	73.8	76.3	77.9
[30/-60/60/-30]	77.1	79.5	80.2

5.2 *Residual Velocity*

It can be defined as the velocity of the projectile with which it exits after complete penetration. CFRP laminates of 1 and 2 mm thick with different fiber orientations were impacted with different projectiles at 100 and 150 m/s and the residual velocity are recorded and tabulated in Tables 5 and 6 respectively. It can be observed from both the tables that after ballistic limit, fiber orientation has no significant effect on residual velocity and with varying nose shape, the residual velocity increases with increase in sharpness.

5.3 *Energy Absorption of Target Laminates*

Energy absorption by the plates is an important parameter in the performance of the material. Energy absorption can be visualized in many forms such as fiber breakage, matrix cracking, plug formation, delamination, etc. Energy absorption characteristics can be directly related to the resistance against impact. More energy absorbed represents high damage resistance against impact. Tables 7 and 8 presents the energy absorption by the targets of 1 mm and 2 mm thick impacted by different projectile at 100 m/s and 150 m/s respectively.

Table 6 Residual velocity of 2 mm thick plate at 150 m/s

Residual velocity (m/s)	Hemispherical	Conical	Ogive
$[0/90]_{2\ s}$	110.5	113	115.4
$[0/90/30/\text{-}60]_s$	111.2	112.3	116.8
$[0/90/45/\text{-}45]_s$	109.4	110.8	113.6
$[30/\text{-}60/60/\text{-}30]_s$	112.8	114.6	117

Table 7 Energy absorption by 1 mm thick plate at 100 m/s

Energy absorbed (J)	Hemispherical	Conical	Ogive
$[0/90]_2$	32.7	30.3	28.6
[0/90/30/−60]	30.6	28.4	27.9
[0/90/45/−45]	34.1	31.3	29.4
[30/-60/60/−30]	30.4	27.6	26.8

Table 8 Energy absorption by 2 mm thick plate at 150 m/s

Energy absorbed (J)	Hemispherical	Conical	Ogive
$[0/90]_{2\ s}$	77.1	72.9	68.9
$[0/90/30/-60]_s$	76	74.1	66.4
$[0/90/45/-45]_s$	78.9	76.6	71.9
$[30/\text{-}60/60/\text{-}30]_s$	73.3	70.2	66

6 Conclusion

In this work, ballistic limit, residual velocity, energy absorbed by the targets were determined with the help of a FE model. The model was developed using Abaqus. The reliability of the model was verified through comparison of results obtained from numerical simulation with experimental results. Both were found in good concordance with each other. After validation of the model, the perforation characteristics were determined by varying different parameters. From the results, it can be concluded that at a particular velocity, with increase in sharpness of the projectile, the ballistic limit decreases and the residual velocity increases. Stacking sequence has very little influence on the perforation characteristics of the laminate. The orientation of fibers in different directions has significant effect on ballistic limit and with increase in thickness the ballistic limit increases, which proves to be helpful in preventing perforation. Among four different fiber orientations, [0/90/45/−45] performs better against impact as it consists of unidirectional plies in different directions which helps in resisting multidirectional loads. [30/−60/60/−30] proves to be least effective against impact. The results obtained from this study can be used for design purpose.

References

1. Balaganesan G, Velmurugan R, Srinivasan M, Gupta NK, Kanny KJIJIE (2014) Energy absorption and ballistic limit of nanocomposite laminates subjected to impact loading. Int J Impact Eng 74:57–66
2. Pandya KS, Dharmane L, Pothnis JR, Ravikumar G, Naik NK (2012) Stress wave attenuation in composites during ballistic impact. Polym Testing 31(2):261–266
3. Dutta PK, Farrell D, Taylor S, Tadayon A, Hui D (1996) Ballistic perforation of graphite/epoxy composite, Special report 96–29. US Army Corps of Engineers
4. Cantwell WJ, Morton J (1989) The influence of varying projectile mass on the impact response of CFRP. Compos Struct 13(2):101–114
5. Cantwell WJ, Morton J (1990) Impact perforation of carbon fibre reinforced plastic. Compos Sci Technol 38(2):119–141
6. Pernas-Sánchez J, Artero-Guerrero JA, Viñuela JZ, Varas D, López-Puente J (2014) Numerical analysis of high velocity impacts on unidirectional laminates. Compos Struct 107:629–634
7. Guoqi Z, Goldsmith W, Dharan CKH (1992) Penetration of laminated Kevlar by projectiles—II. Analytical model. Int J Solids Struct 29(4):421–436
8. Mines RAW, Worrall CM, Gibson AG (1998) Low velocity perforation behaviour of polymer composite sandwich panels. Int J Impact Eng 21(10):855–879
9. Fuoss E, Straznicky PV, Poon C (1998) Effects of stacking sequence on the impact resistance in composite laminates—part 1: parametric study. Compos Struct 41(1):67–77
10. Hitchen SA, Kemp RMJ (1995) The effect of stacking sequence on impact damage in a carbon fibre/epoxy composite. Composites 26(3):207–214
11. Lopes CS, Seresta O, Coquet Y, Gürdal Z, Camanho PP, Thuis B (2009) Low-velocity impact damage on dispersed stacking sequence laminates. Part I: Experiments. Compos Sci Technol 69(7–8):926–936
12. Dorey G (1975) Failure mode of composite materials with organic matrices and their consequences in design. In: AGARD conference proceedings
13. Velmurugan R, Naresh K, Shankar K (2018) Influence of fibre orientation and thickness on the response of CFRP composites subjected to high velocity impact loading. Adv Mater Process Technol 4(1):120–131
14. Guoqi Z, Goldsmith W, Dharan CH (1992) Penetration of laminated Kevlar by projectiles—I. Experimental investigation. Int J Solids Struct 29(4):399–420
15. Shi Y, Swait T, Soutis C (2012) Modelling damage evolution in composite laminates subjected to low velocity impact. Compos Struct 94(9):2902–2913
16. Corran RSJ, Shadbolt PJ, Ruiz C (1983) Impact loading of plates—an experimental investigation. Int J Impact Eng 1(1):3–22
17. Radin J, Goldsmith W (1988) Normal projectile penetration and perforation of layered targets. Int J Impact Eng 7(2):229–259
18. Hashin Z (1980) Failure criteria for unidirectional fiber composites
19. Abaqus V (2014) 6.14-1. Abaqus/standard user's manual and Abaqus CAE manual. Providence, RI, USA: Dassault Systemes Simulia Corp
20. Barbero EJ (2013) Finite element analysis of composite materials using AbaqusTM. CRC press

Fatigue Life Prediction of Reinforced Concrete Using Artificial Neural Network

B. S. Vishnu, Keerthy M. Simon, and Bharati Raj

Abstract Fatigue is a phenomenon of gradual, permanent internal changes in a material due to repeated or cyclic loading. The fatigue failure of structural elements may decreases the life of infrastructures, therefore the fatigue life of those structures should be considered. Highway and airfield pavements, bridge decks, offshore supporting structure, machinery foundation etc. are subjected to high cycle repeated loading. The randomness in parameters due to the heterogeneous nature of concrete due to fatigue loading leads to complexities in analysing fatigue failure of reinforced concrete. Probabilistic approach is more dependable for the prediction of fatigue life of reinforced concrete than deterministic approach as it can include variations and uncertainties. In recent years, artificial neural network emerged as a new promising computational tool which adopts a probabilistic approach for modelling complex relationships. The purpose of this study is to extract the data from fatigue tests conducted on reinforced concrete beam to create an artificial neural network predictive model. The developed model can able to predict the critical crack length of reinforced concrete members at which failure occurs by considering the fracture mechanics properties and material properties accountable for the softening behaviour of concrete as input. The developed ANN model and analytical model is capable of predicting the fatigue life of reinforced concrete with reasonable accuracy and in a faster approach.

Keywords Artificial neural network (ANN) · Fatigue life · Probabilistic approach · Reinforced concrete

1 Introduction

Concrete is the most commonly used materials used for the construction purpose. Although it is generally used it has some weakness also like low tensile strength, poor ductility, crack under tension, brittle failure, which leads to cracking. Cracking

B. S. Vishnu (✉) · K. M. Simon · B. Raj
Department of Civil Engineering, NSS College of Engineering, Palakkad, India

G. C. Marano et al. (eds.), *Proceedings of SECON'21*, Lecture Notes in Civil Engineering 171, https://doi.org/10.1007/978-3-030-80312-4_22

of concrete will affect the durability of structures and crack propagation will be more when structures subjected to fatigue loading.

Fatigue is the phenomenon of gradual, perpetual internal changes happening in a material due to repeated or cyclic loading. This leads to crack formation from the existing flaws within the material and results in fatigue fracture of materials. These may reduce the life of infrastructure and therefore the fatigue life of structures should be considered. Highway and airfield pavements, bridge decks, offshore supporting structure, machinery foundation etc. are subjected to high cycle repeated loading.. The randomness in parameters due to the heterogeneous nature of concrete due to fatigue loading leads to complexities in analysing fatigue failure of reinforced concrete. Fatigue life prediction of reinforced concrete is done by deterministic approach and probabilistic approach. Deterministic approach includes Fatigue life models, Fracture mechanics model and Fatigue damage model. Fatigue life method is not a proper method to find fatigue life concrete because it suffers from distinct property. For brittle and ductile material, fatigue life model is more applicable [1]. Fracture mechanics model is also extended from metals and comparing to fatigue life models, fracture mechanics gives a good prediction of fatigue failure of concrete. If varying parameters are their fracture mechanics model become more complex and is not dependable [2]. By the simultaneous presence of microcracks and macrocracks fatigue damage in concrete can be identified [3]. This model is applicable only for viscoelastic concrete.

Due to the disadvantages of the above-mentioned fatigue models, the importance of probabilistic approach came into importance for the fatigue life prediction of reinforced concrete. Probabilistic approach uses distributed values rather than fixed values. These distributed values have variations and uncertainty. Commonly used probabilistic approaches are Weibull distribution, Bayes rule and Artificial Neural Network (ANN). ANN is an effective computational tool developed by inspiring from biological nervous system. ANN can be used for fatigue life study by analyzing the given input data and output data [4].

2 Artificial Neural Network

ANN is developed as a part of artificial intelligence and having several interconnected artificial neurons. Neuron is a nonlinear unit which receives input signals yielding an output. ANN has three layers of neurons, Input layer which receives information, Hidden layer which perceives and evaluate the information from input layer, and Output layer which provides the final output. Based on problem to be solved ANN architecture is divided into Single layer Neural Network and Multilayer Neural Network.

2.1 Modelling of Neural Network

In this work, fatigue life of reinforced concrete in terms of relative crack depth is validated by collecting data from literature. These data were collected by considering the fracture mechanics properties and the material that accounts for the softening behaviour of concrete as input.

Neural network toolbox of MATLAB 2018a has been used to develop and train the model. The experimental and analytical data's reported by Sonalisa and Kishen [5] were used to develop, train and validate the model. The developed neural network architecture is based on multilayer perceptron network (MLPN) or multilayer feed-forward back propagation network. Levenberg-Marqaurdt back propagation algorithm was used for training algorithm as it gives an accurate result with lesser time and is more dependable for complex modelling problems.

The neural network is developed using the experimental data sets of reinforced concrete beam under four-point bending test of a fatigue load ΔP 24.75 kN. The geometrical properties and material parameters of the specimen are given in Table 1.

Considering the material and fracture mechanics properties necessary for the crack propagation to occur, the input parameters are chosen and is given in Table 2.

Relative crack depth at which failure occurs is chosen as output data. For modelling, the data sets were divided randomly and used as 70% for training, 15% for validation and 15% for testing. The network configuration adopted for developing the model was created according to specification given in Table 3.

Mean squared error (MSE) and regression value (R) were selected as performance criteria in order to identify the optimum number of hidden neurons which gives the best performing ANN model. The model with least MSE value and R value close to 1 was selected as the optimum neural network architecture and in this work the network with 7 neurons in hidden layer gives better results. Thus, the best architecture obtained gives MSE value as 0.0000003 and R value as 0.999 and have 6 input neuron, 7 hidden neuron and 1 output neuron (N 6-7-1) as shown in Figs. 1 and 2 shows neural architecture model.

Table 1 Geometry and material property of specimen

Sl no	Input parameter	Input range
1	Span S (mm)	1200
2	Depth D (mm)	250
3	Thickness B (mm)	150
4	Notch size a (mm)	30
5	Fracture toughness G_f (N/mm)	0.3
6	Elastic modulus of concrete E (MPa)	16,500
7	Tensile stress σ_t (MPa)	2.86

Table 2.2 Details of input parameter

No	Input parameter	Description
1	Number of cycles to failure	Number of cycles required for crack development
2	Structural size	Depth of the beam
3	Area of steel	Area of steel reinforcement in beam
4	Tensile stress	When major principal stress exceeds this value crack occur
5	Modulus of elasticity of concrete	Ratio of applied stress to corresponding strain
6	Energy release rate	Energy required by crack to propagate

Table 2.3 Neural network configuration for validation

Parameter	Specification
No: of neurons in input layer	6
No: of neurons in hidden layer	2–10
No: of neurons in output layer	1
Training function	Levenberg–Marquardt (trainlm)
Activation function	Tan-sigmoid
Performance function	Mean squared error (MSE) Regression value

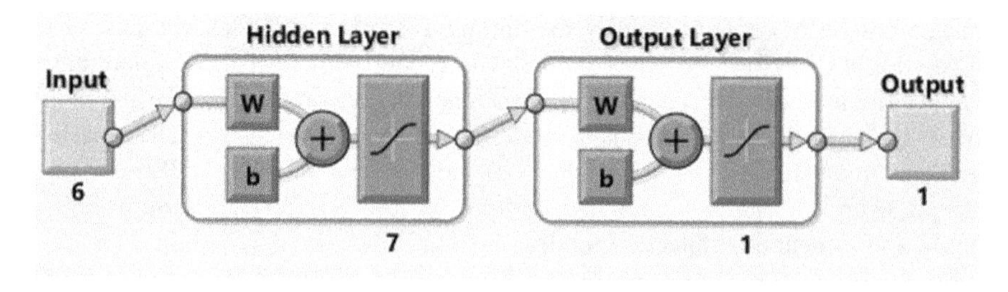

Fig. 1 N 6-7-1 neural network architecture in MATLAB R2018a

2.2 *Evaluation of Network Model*

The performance of the developed neural network is evaluated based on Regression value (R). The model with R value closer to 1 represents the best model. Also, it is important to decide the number of neurons in the hidden layer to give the optimum network architecture. An optimal neural network can be obtained based on R value.

The reliability of the ANN model in predicting the relative crack depth at failure for data sets of reinforced concrete beam with R value of 0.999 is indicated in Fig. 3. The predicted output is mostly fit, indicating the accuracy of prediction in

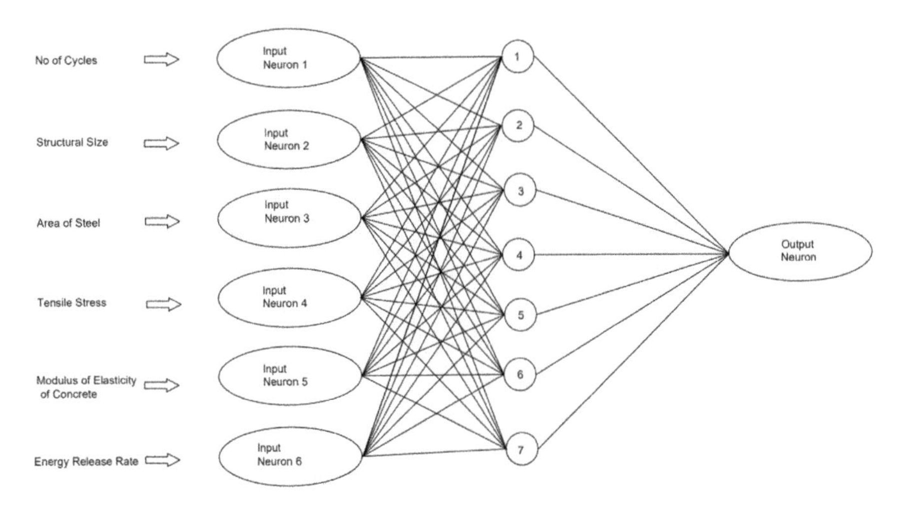

Fig. 2 Neural Network Architecture Model

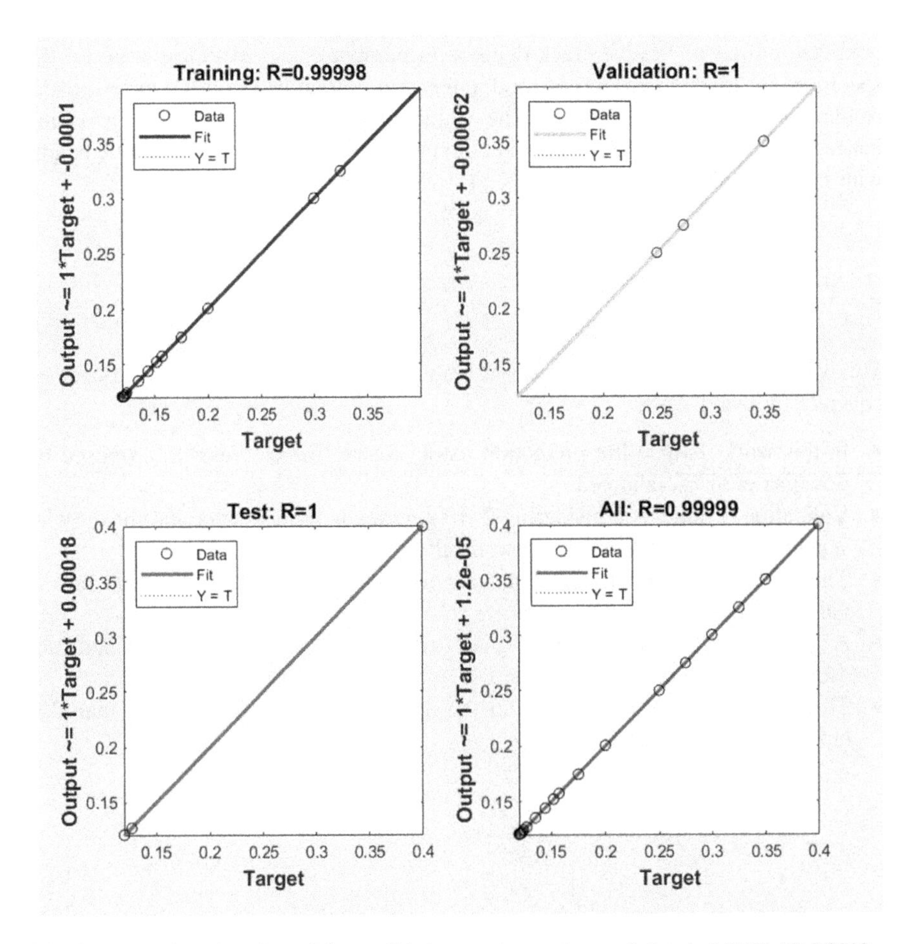

Fig. 3 Regression plots for training, validation, testing, and overall data in MATLAB R2018a

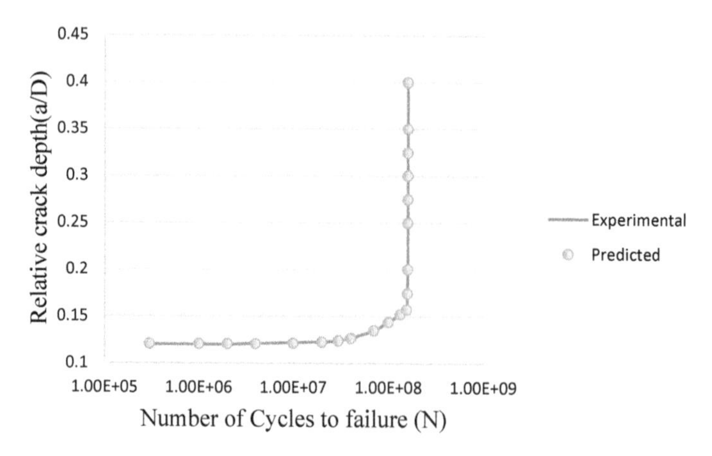

Fig. 4 Comparison of relative crack depth with number of cycles using the predicted output with experimental results

ANN tool. The graph is plotted comparing the experimentally obtained result and predicted output of relative crack depth with number of cycles in log scale for the specimen are given in Fig. 4. A good agreement is seen between the experimental results and the predicted output. The validation results are within 1% error and hence the developed model can predict the fatigue life of reinforced concrete beams with considerable accuracy.

3 Conclusions

This work deals with the validation of fatigue life prediction model for reinforced concrete by using ANN.

- In this work, fatigue life prediction model for reinforced concrete proposed by Sonalisa et al. is validated
- Validation is done by considering 2–10 neurons in hidden layer and model with 7 neurons gives the most accurate result
- The best architecture model obtained gives MSE value as 0.0000003 and R value as 0.999
- A graph is plotted for comparing the experimentally obtained result and predicted output of relative crack length with number of cycles to failure
- This developed model can predict the relative crack depth with less than 1% error in a faster approach.

References

1. Miner MA (1945) Cumulative damage in fatigue. J Appl Mech 12(3):159–164
2. Slowik V, Plizzari GA, Saouma VE (1996) Fracture of concrete under variable amplitude fatigue loading. ACI Mater J 93(3):272–283
3. Pervaiz Fathima KM, Chandra Kishen JM (2015) A thermodynamic correlation between damage and fracture as applied to concrete fatigue. Eng Fract Mech 145:1–20
4. Hajela P, Berke L (1991) Neurobiological computational modes in structural analysis and design. Comput Struct 41:657–667
5. Ray S, Chandra Kishen JM (2014) Analysis of fatigue crack growth in reinforced concrete beams. Mater Struct 1:47(1–2):183–198

Effects of Loading Pattern on Critical Temperature of Cold-Formed Compound Flexural Members

Ravikant Singh and Avik Samanta

Abstract Within the past few years, the utilization of Cold-formed steel (*CFS*) as structural element has grown significantly. High strength to weight ratio, efficiency in construction, and a wide range of applications make it a perfect alternative to its hot-rolled counterparts. Apart from its many advantages, *CFS* members have some disadvantages too. High section factor (parameter to area ratio) makes these members more vulnerable to failure at elevated temperatures. So far, very limited work has been done on *CFS* flexural members under fire; thus, an attempt has been made in this work to bridge this gap. This paper reports ongoing numerical parametric study results on a lipped channel beam, connected back to back to form a compound beam (*I-section*). In this work, the finite element (*FE*) model was developed and validated against experimental and numerical results available in literature using commercially available program (*ABAQUS*). Utilizing the developed *FE* model, further investigation was conducted. The effect of different loading pattern on the critical temperature of the compound section is studied. To simulate fire on the member standard *ISO* fire curve was used, and non-linear thermal profile was defined in the model using nodal thermal data. As findings of this ongoing research, it can be concluded that the critical temperature of a beam is strongly affected by the loading pattern on the member. The critical temperature of 350 °C defined in Eurocode 3, Part 1.2 was found to be over-conservative for Lateral torsional buckling at higher slenderness range.

Keywords Compound beam · Cold-formed steel · Fire · Flexural members · Thin-walled members

R. Singh (✉) · A. Samanta
Indian Institute of Technology Patna, Patna, Bihar 801106, India
e-mail: ravikant_1921ce01@iitp.ac.in

G. C. Marano et al. (eds.), *Proceedings of SECON'21*, Lecture Notes in Civil Engineering 171, https://doi.org/10.1007/978-3-030-80312-4_23

1 Introduction

Since the last decade with the advancement in constructional materials and practices, cold-formed steel (*CFS*) has emerged as a leading material for a cost-effective, efficient alternative to its hot-rolled counterpart. *CFS* members are adopted widely in commercial, personal, and public spaces. Apart from serving as a secondary member, these members are widely used as primary load-bearing structural elements in low-rise to mid-rise buildings [1]. Due to limited cross-sectional thickness, these members are very prone to stability failure such as local, distortional, or lateral torsional buckling. Thin cross-section also increases section factor of the *CFS* member leading to poor performance of these members in case of fire [2]. A number of research work have been conducted in the past related to *CFS* members at ambient temperature [3–9]. However, relatively marginal work has been done on *CFS* members at elevated temperatures [10–12]. Close inspection of these studies indicates that most of these work involved similar loading pattern on the members, and mostly 4-point loading setup was used in these members. It was also observed that for study of member exposed to fire, uniformly elevated thermal profile was used [10, 11], whereas in actual fire scenario, such idealized conditions will not always exist. It is worth highlighting that none of the research found in the referred literatures compared the effects of loading pattern on critical temperature of the members; in the present research work authors attempts to bridge that gap.

Finite element (*FE*) based software package have been adopted to successfully in structural simulations in past research, and appropriately developed *FEM* models can be used to replicate actual experiments close to reality. This study involves numerical *FE* simulations using commercially available program *ABAQUS/CAE 6.14* [13]. Developed *FE* model first validated from available experimental [14] and numerical work [15] and then further adopted in this study. Two lipped channel members attached in the back-to-back assembly were used in this study. Three different loading patterns, 3-Point loading (*3PL*), 4-Point loading (*4PL*), and uniformly distributed loading (*UDL*) pattern, were considered. In order to implement the fire action on member *ISO-834* [16] standard fire-time curve was used.

2 Material Properties

In this study non-linear stress–strain model was adopted based on Ramberg–Osgood [17] relation as presented in Eq. (1). The value of strain (ε_T) corresponding to the stress (σ_T) at T°*C* temperature can be obtained using the following equation.

$$\varepsilon_T = \frac{\sigma_T}{E_T} + \beta\left(\frac{f_{yT}}{E_T}\right)\left(\frac{\sigma_T}{f_{yT}}\right)^{\eta_T} \quad (1)$$

In the above equation, the term f_{yT} represents yield strength of the material, whereas E_T denoted Modulus of elasticity of the material at T°*C*. β in the above equation was taken as 1.5 [18]. The equation for Ramberg–Osgood parameter (η_T) was taken from Ranawaka and Mahendran [19]. High strength steel of grade G450 was used, yield strength, and the ultimate strength of the material was taken as 514.5 MPa and 542.5 MPa, respectively, from Kankanamge and Mahendran [18]. Again the modulus of elasticity was taken as 206.328 GPa from the literature [18].

Yield strength and Modulus of elasticity of steel deteriorate with increasing temperature; thus, appropriate reduction factors were introduced in order to account for the change in material properties with temperature. Reduction factors for both yield strength and modulus of elasticity were taken from Eurocode 3, Part 1.2 for class 4 cross-sections [20].

3 Load and Boundary Conditions

The ultimate capacity of the member was calculated using Eurocode 3, Part 1.3 [21] and 50% of the obtained capacity of the member was applied as load on the members. Three different loading patterns were selected in this study, namely 4-Point loading (*4PL*), 3-Point loading (*3PL*) and uniformly distributed load (*UDL*). In case of *4PL* the load was applied at *L/3* location from supports, where *L* represents effective span of the member. For *3PL*, the load was applied at the center of the beam. In point loading scheme, the load was transferred to the member through plates attached to it in order to avoid load concentration. Simply supported boundary condition was used with one end being roller whereas another end was hinged. For thermal load, *ISO-834* [16] standard fire-time curve was used to simulate the worst fire scenario.

4 Numerical Model

This study involves two-lipped channel beams of span 2.5 m connected in back-to-back assembly. In order to simulate connected beams using screw, the members were connected using "TIE" constrain at specific locations along the length of beam, whereas across cross-section screws were applied at *h/5* (*h* = depth of cross section in mm) distance from both top and bottom of the cross-section. Dimensions of member cross-section and location of the screw is presented in Fig. 1.

Two numerical *FE* models were developed in this research work. First *2D* heat transfer model was developed in order to acquire possible temperature distribution pattern on the member due to fire action. Total time of 1800 s was considered for heating phase. Diffusive heat transfer quadrilateral element with 4 nodes, and 11 degree of freedom, commonly referred as *DC2D4* was used for heat transfer

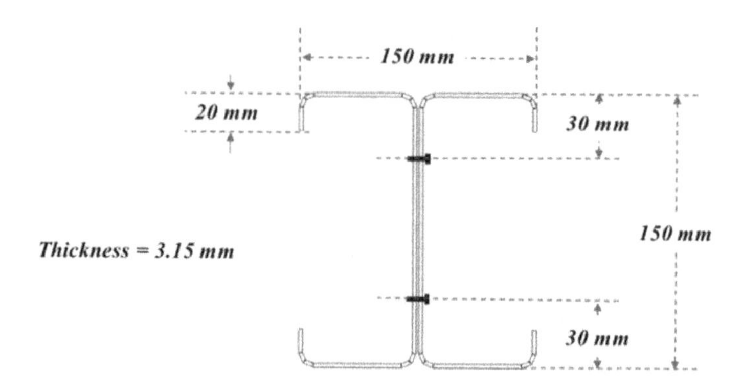

Fig. 1 Member dimension and Screw location on considered member

simulation. To study member structural behavior at elevated temperature *3D* non-linear thermomechanical model was developed. In thermo-mechanical simulation shell, 4 nodal element with reduced integration approach (*S4R*) was adopted. In order to handle large hourglass deformation arising due to large strain, hourglass control was turned "ON" in *ABAQUS*. Effects due to large deformations were incorporated using "NLGEOM, ON" through the command line. Mesh size of 10 mm × 10 mm, found to provide the best results in reasonable computational resources. Obtained nodal thermal data from heat transfer analysis was used as nodal input in the structural model.

5 Validation Study

Developed FE model was validated against the experimental [14] and numerical work [15] from the available literature. 2D heat transfer analysis was validated first. Thermal profile of experimental work by Laím et al. [14] and numerical work of and Laím and Rodrigues [15] was compared with thermal profile obtained from developed model. As represented from Fig. 2a–c, developed model was capable to predict the temperature of the member cross-section accurately. Solid line in presented figures represent developed model temperature, which at all point of time stays within experimental and numerical work, indicating a strong validation of developed model.

3D thermo-mechanical analysis was also validated with results from available literature [14, 15]. Vertical displacement of the center of span at the web-flange junction was recorded and plotted with average temperature of the cross-section (Fig. 3). Critical temperature recorded with the developed model was 667 °C. A good agreement between experimental and available numerical result can be seen with developed FE model. A difference of 3.47% with experimental temperature was observed, supplementing validity of the developed model.

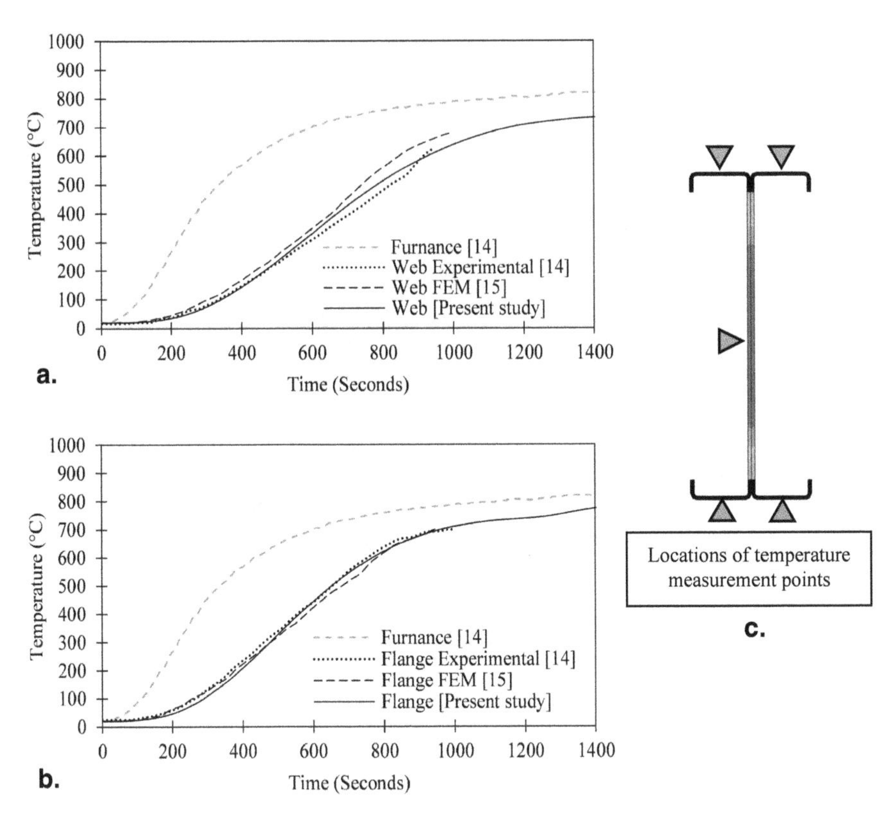

Fig. 2 Temperature profile on 2D validation model **a.** Web temperature, **b.** flange temperature, and **c.** location of temperature measurement points

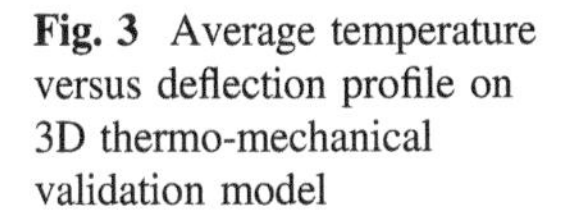

Fig. 3 Average temperature versus deflection profile on 3D thermo-mechanical validation model

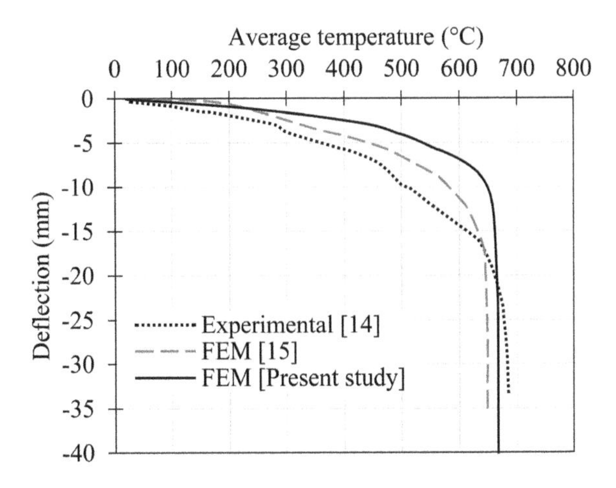

6 Results

6.1 2D Heat Transfer Simulation

Results of heat transfer analysis at equal intervals have been presented in Fig. 4. Thermal profile at 360 s interval is presented till the total considered time of 1800 s (30 min). In time intervals of 360 and 720 s, as shown in Fig. 4b, c, the maximum difference in thermal profile was observed with difference in temperature magnitude of 78 and 88 °C. This maximum variation in thermal profile was observed due to the fact that between these time the rate of heating was highest, due to which extended parts of member such as flanges and lips got heated rapidly as compared to narrowed compact portion such as web. After these time interval as heating rate became constant (720–1440 s), and as whole section came close to equilibrium temperature, thermal variation across cross-section also diminished from 88 to 56 °C and 22 °C, respectively for 1080 and 1440 s of simulation interval. Figure 4f represents thermal profile of members at the end of the simulation.

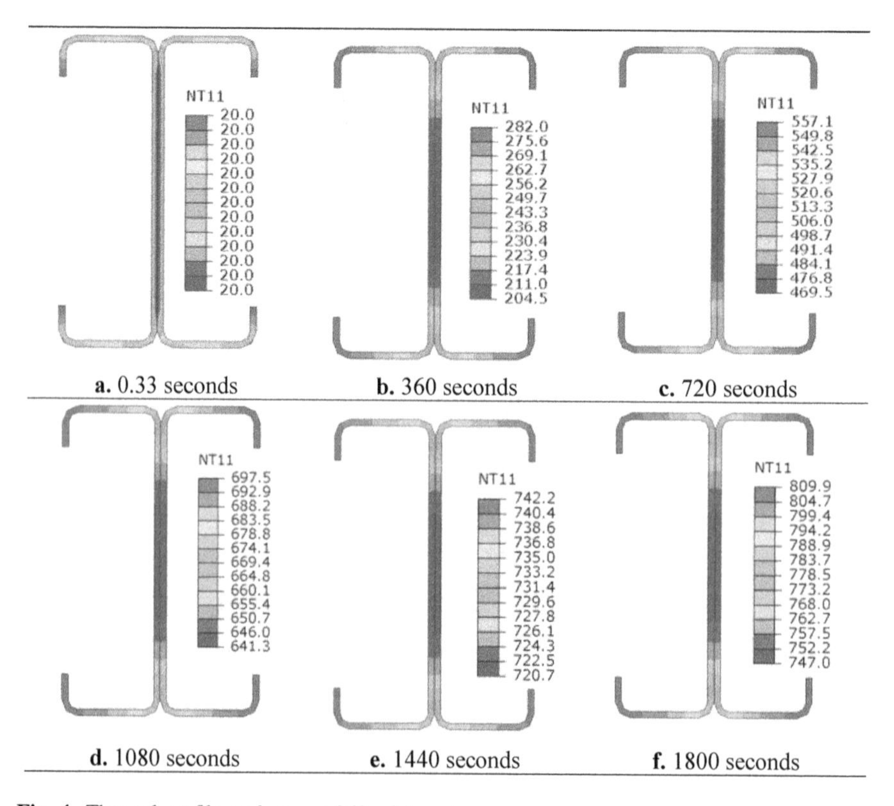

Fig. 4 Thermal profile on beam at 360 s interval

6.2 3D Thermo-mechanical Simulation

Nodal thermal data obtained from heat transfer analysis was used as input nodal temperature for the thermo-mechanical part of the analysis. The result of thermo-mechanical simulation is presented in Fig. 5. Average temperature (°C) is plotted on horizontal axis of curve whereas on vertical axis displacement (mm) in negative vertical direction at top mid of the cross section at web-flange junction is plotted. As highlighted in Fig. 5, in initial part of the curve some deflection is already present, which is due to the initial applied load on the member before start of fire simulation. Which in this study was 50% of the ultimate load bearing capacity of the member, calculated in accordance with Eurocode 3, Part 1.3 [21].

At the same applied load level of 50% of capacity, a clear difference in the member behavior was visible. Member under uniformly distributed load was found to behave best under elevated temperature, for which critical temperature was 651.7 °C. Having Critical temperature of 552.9 °C lowest critical temperature was recorded in case of 3-Point loading setup. Whereas critical temperature for 4-Point loading was found to be 593.3 °C. It is worth highlighting that in case of *UDL* not much deflection was visible till 500 °C, after which more deflection started appearing. Whereas in case of point loading this was not the case, as in these member deflection started just after the fire exposure.

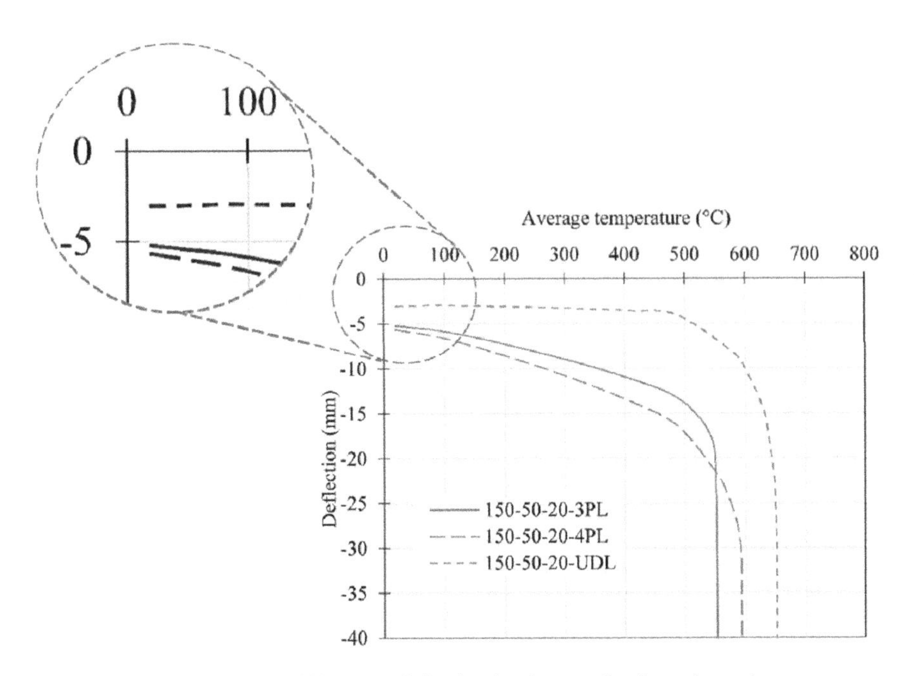

Fig. 5 Average temperature (°C) versus deflection (mm) curve for flexural member

7 Summary and Conclusion

Critical temperature limit of 350 °C available in Eurocode 3, Part 1.2 [20] was found to be over-conservative, as members were capable of withstanding much higher temperature than limiting temperature. Eurocode prediction was 202.9, 243.3 and 301.7 °C lower than the temperature observed in this research work, indicating the extremely conservative design of the Eurocode 3, Part 1.2 [20] for cold-formed steel.

With this preliminary observation of ongoing research, it can be concluded that while designing a member under fire it is crucial to consider appropriate loading pattern on the member, as considering *UDL* over *3PL* can produce unsafe critical temperature prediction of magnitude up to 100 °C. It can also be concluded that Eurocode limiting temperature of 350 °C is highly over-conservative for *CFS* members.

References

1. Moen CD, Schafer BW (2011) Direct strength method for design of cold-formed steel columns with holes. J Struct Eng ASCE 137:559–570
2. Javed MF, Hafizah N, Memon SA, Jameel M, Aslam M (2017) Recent research on cold-formed steel beams and columns subjected to elevated temperature: a review. Constr Build Mater 144:686–701
3. Martins AD, Camotim D, Gonçalves R, Dinis PB (2018) On the mechanics of local-distortional interaction in thin-walled lipped channel beam. Thin-Walled Struct 125:187–202
4. Szymczak C, Kujawa M (2016) On local buckling of cold-formed channel members. Thin-Walled Struct 106:93–101
5. Bui HC (2009) Buckling analysis of thin-walled sections under general loading conditions. Thin-Walled Struct 47:730–739
6. Put BBM, Pi Y, Trahair NS (1999) Lateral buckling tests on cold-formed channel beams. J Struct Eng ASCE 125:532–539
7. Pham CH, Hancock GJ (2013) Experimental investigation and direct strength design of high-strength, complex C sections in pure bending. J Struct Eng ASCE 139:1842–1852
8. Laím L, Rodrigues JPC, Da Silva LS (2013) Experimental and numerical analysis on the structural behaviour of cold-formed steel beams. Thin-Walled Struct 72:1–13
9. Anbarasu M (2016) Local-distortional buckling interaction on cold-formed steel lipped channel beams. Thin-Walled Struct 98:351–359
10. Dolamune Kankanamge N, Mahendran M (2012) Behaviour and design of cold-formed steel beams subject to lateral-torsional buckling at elevated temperatures. Thin-Walled Struct 61:213–228
11. Landesmann A, Camotim D (2016) Distortional failure and DSM design of cold-formed steel lipped channel beams under elevated temperatures. Thin-Walled Struct 98:75–93
12. Laím L, Rodrigues JPC (2018) Fire design methodologies for cold-formed steel beams made with open and closed cross-sections. Eng Struct 171:759–778
13. ABAQUS/Standard Version 6.14 (2014) ABAQUS/CAE User's manual. Dassault Systèmes: Simulia Corp., Providence: RI, USA

14. Laím L, Rodrigues JPC, Da Silva LS (2014) Experimental analysis on cold-formed steel beams subjected to fire. Thin-Walled Struct 74:104–117
15. Laím L, Rodrigues JPC (2016) On the applicability and accuracy of fire design methods for open cold-formed steel beams. J Build Eng 8:260–268
16. ISO 834-1 (1999) Fire resistance tests—elements of building construction, part 1: general requirements. International Organization for Standardization Geneva, Switzerland
17. Ramberg WR, Osgood W (1943) Description of stress–strain curves by three parameters, NACA technical note, 902
18. Kankanamge ND, Mahendran M (2011) Mechanical properties of cold-formed steels at elevated temperatures. Thin-Walled Struct 49:26–44
19. Ranawaka T, Mahendran M (2010) Numerical modelling of light gauge cold-formed steel compression members subjected to distortional buckling at elevated temperatures. Thin-Walled Struct 48:334–344
20. EN1991-1-2 (2002) Actions on structures exposed to fire. European committee for standardization Brussels, Belgium
21. EN1993-1.3 (2004) Design of steel structures. general rules. European Committee for Standardization Brussels, Belgium

Bearing Capacity of Square Footing on Hybrid Geosynthetic Reinforced Granular Fill Over Soft Soil

Dona Abraham, R. Rakendu, and Dipty Isaac

Abstract In soft soils, ground improvement techniques are the widely adopted methods for improving the load bearing capacities. Hybrid geosynthetic is utilized to improve the load carrying capacity of the soft soil in this study. The conclusions of this project are suitable to all problems where the load carrying capacity of the soil is a major issue (e.g., foundations, embankments, pavements). In this report, the load carrying capacity of square footings on geosynthetic reinforced granular fill placed over the soft soil is studied. The footing consisted of a rigid square steel plate with a dimension 120 × 120 mm (B). On both homogeneous and layered foundation systems, a series of laboratory model tests were conducted. The layered systems were made up of dense sand of different layer thicknesses (H = 0.5–1.1B) overlying the soft soil subgrades. The tests used planar geogrid, nonwoven coir geotextile and hybrid geosynthetic reinforcements. The effectiveness of hybrid geosynthetic in improving the bearing capacity of soft soil was evaluated by conducting laboratory plate-load test. The findings revealed that footing settlement, layer thickness (H), and reinforcement type all had a significant impact on foundation efficiency. In this analysis, a maximum of about 4.67-times increase in the ultimate bearing capacity was noticed. The findings showed that using hybrid geosynthetic reinforcement at the sand-soft soil interface can enhance the foundation beds performance.

Keywords Geogrid · Coir geotextile · Hybrid geosynthetic · Plate load model test

---

D. Abraham (✉) · R. Rakendu
Department of Civil Engineering, Saintgits College of Engineering, Kottayam, Kerala, India

D. Isaac
Rajiv Gandhi Institute of Technology, Kottayam, Kerala, India

G. C. Marano et al. (eds.), *Proceedings of SECON'21*, Lecture Notes in Civil Engineering 171, https://doi.org/10.1007/978-3-030-80312-4_24

1 Introduction

For low to moderate load applications on soft soils, soil improvement techniques are strongly recommended. Because of its low load-bearing capacity and excessive settlement, building a structure directly on soft soil is very difficult. As a result, soft soil must be treated to increase its carrying capacity. Several researchers have investigated the reinforcement effects in foundation applications, taking into account various factors such as the type of reinforcements, strength of reinforcements, geometry, placement depths, and the number of layers etc. [1–5]. Yang et al. [5] investigated the behavior of clay reinforced with geotextile as well as the impact of sandwiching geotextile in a thinner sand layer to increase the reinforced clay shear strength. As for the technique of sandwiching, the test results concluded that the sand layers that encapsulate the reinforcement can enhance the soil-geotextile interaction, resulting in improved reinforced clay shear behavior.

Because of its good engineering properties and longevity, Lal et al. [4] identified coir geotextile made of coir fibres, as a better substitute to geosynthetics for reinforcement applications. Numerous studies have shown the nature of geogrid reinforced sand-clay foundation systems with varying clay subgrade strengths [1, 2]. According to the findings, geogrid reinforcement at the sand-clay interface will significantly increase the efficiency of foundation beds in terms of subgrade strength and layer thickness. The use of a hybrid geosynthetic permeable layer can significantly reduce groundwater table level and pore water pressures by draining water into the hybrid geosynthetics plane [3]. In this study, hybrid geosynthetic with the combination of non-woven coir geotextile and geogrid was utilized to improve the load carrying capacity of the soft soil. The findings of laboratory plate load tests on hybrid geosynthetic reinforced granular fill over soft soil are discussed in this study.

2 Methodology

2.1 Materials

Soil

The foundation bed/subgrade was built with locally available clayey silt soil and river sand was utilized as the filling material for the hybrid geosynthetic. The soil sample was collected from Mannanam, Kottayam district. The sample was black coloured silty soil.

Hybrid geosynthetic

Biaxial geogrid and non-woven coir geotextile are combined together to form a hybrid geosynthetic as shown in Fig. 1. Uniform stitches using nylon thread is used for fabrication of hybrid geosynthetic. Drainage and reinforcement criteria can be met with a combination of geogrid with appropriate opening dimensions and a suitable non-woven geotextile.

2.2 Testing Program

A total of five different series of laboratory plate load tests were conducted. Unreinforced configurations such as homogeneous clayey silt and layered were used in two series of tests, while various reinforcements such as geotextile, geogrid, and hybrid geosynthetic were used in three series of tests. The reinforcements were placed at the sand-soil sample interface. The unreinforced or reinforced sand layer thickness (H) overlying the soft soil bed was varied in the experiments (0.5, 0.7, 0.9, 1.1B). Plate load model tests were conducted on a square footing of dimension 120 × 120 mm resting on 0.6 × 0.6 × 0.3 m soil bed. Figure 2a, b shows the schematic diagram and photographic view of test setup for model test respectively. The soil beds were created in a steel tank with 0.6 × 0.6 × 0.6 m dimensions. Dimensions of tank are five times greater than the foundation width to maintain that the rupture of footing generated within the tank and to depreciate scaling issues as per IS: 1888 (1982). The soil bed was prepared by compacting locally available clayey silt soil. To obtain same properties, soft soil was compacted into three layers (each layer with a thickness of 10 mm). The granular fill was dry sand, which was put at the desired depth on the clayey silt surface in the test tank.

The footing was positioned in the center of the soil bed. The footing was loaded by pushing it towards the soil. During loading, dial gauge measurements and loading on the proving ring were noted at equal intervals of settlements of footing. The mean value of the two dial readings was considered as final displacement. The Loading was kept going until the settlement approached a value of about 25% of width of footing.

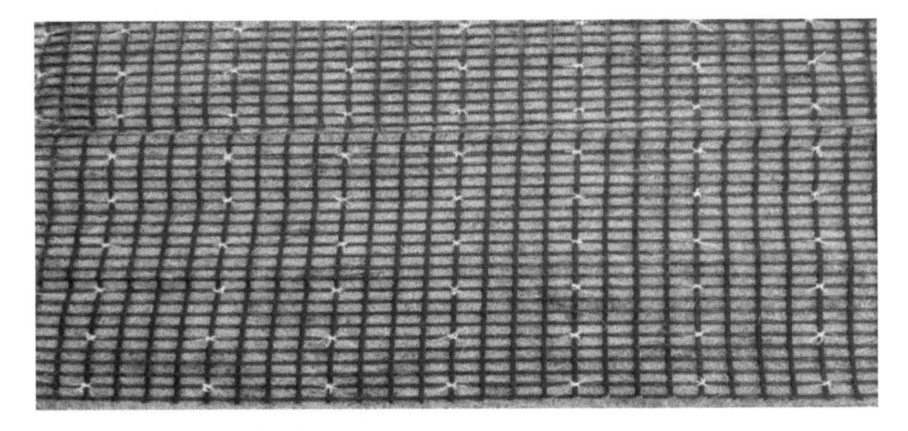

Fig. 1 Hybrid geosynthetic using coir geotextile and polyester geogrid

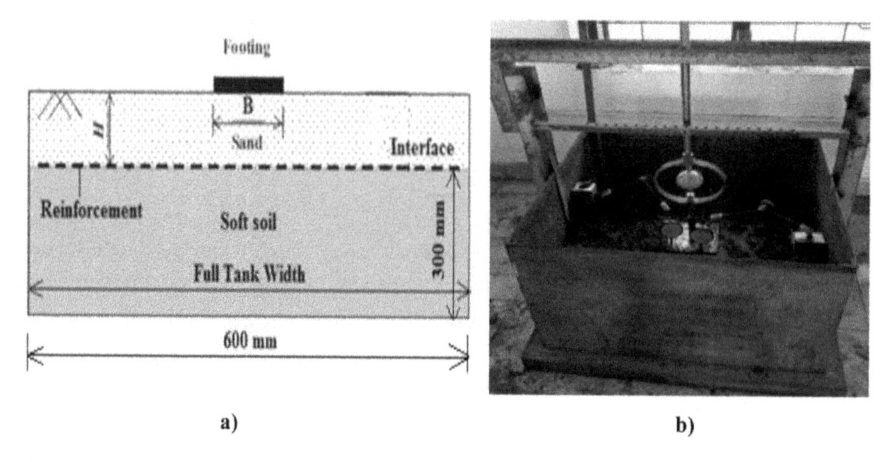

Fig. 2 **a** Schematic diagram of test setup for model test. **b** Photographic view of test setup for model test

2.3 Initial Properties of Soft Soil

Figure 3a, b shows the particle size distribution curve of soft soil and sand respectively. Table 1 shows the basic properties of soft soil and sand.

Natural clayey silt was used to prepare the soft soil bed in this analysis. As per Indian Standard specification, the soft soil can be classified as clayey silt with high plasticity (MH). The soil had a silt content of 50% and a clay content of 16%. The plasticity index of soil is 101.6%. This experiment was conducted using locally available river sand. The classification according to the Indian Standard specification is SP (poorly graded sand). All experiments were performed at a relative density of 72%.

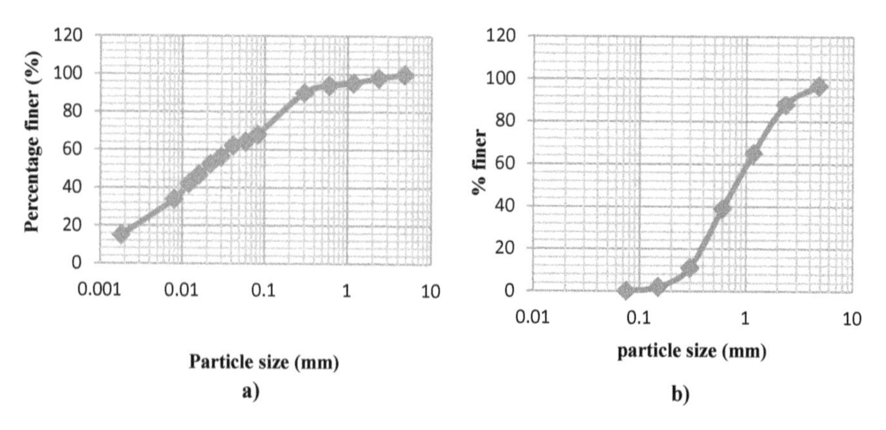

Fig. 3 **a** Particle size distribution of soft soil. **b** Particle size distribution of sand

Table 1 Properties of soft soil and sand

Materials	Properties	Values
Soft soil	Specific gravity	1.93
	Water content	134%
	Liquid limit	159.6%
	Plastic limit	58%
	Shrinkage limit	16%
	Field density	1.58 g/cc
	Undrained shear strength, cu	6 kN/m^2
Sand	Specific gravity	2.67
	Uniformity coefficient, Cu	3.3
	Coefficient of curvature, Cc	1
	Classification	SP
	Angle of internal friction, Φ	36°
	Relative density, Id	72%

2.4 Properties of Hybrid Geosynthetics

Testing of the developed hybrid geosynthetic was carried out in ASTM lab and testing lab of Central Coir Research Institute, Kalavoor. The geogrid was collected from Strata Geosynthetics, Daheli. The non-woven coir geotextiles were procured from Charangattu Coir Mfg. Co. (P) Ltd, Cherthala. The Properties of geogrid, coir geotextile and hybrid geosynthetic are listed in Table 2.

3 Results and Discussions

Test results were evaluated based on the load settlement behavior of foundation systems having different layer thicknesses (H) and reinforcements.

3.1 Homogeneous Foundation (Clayey Silt)

On homogeneous foundation beds of silty soil, plate load model tests were conducted. The test on the unreinforced clayey silt bed was conducted to compare the load bearing capacity of the unreinforced bed to that of the reinforced bed. The load settlement curves of the homogeneous clayey silt bed for square footing (120 × 120 mm) are shown in Fig. 4. Settlements varied non-linearly with load intensity, as can be seen. The curve shows a peak value at 28 kN/m^2. An ultimate bearing pressure of about 28 kN/m^2 can be noticed from the curve.

Table 2 Properties of geogrid, coir geotextile and hybrid geosynthetic

Materials	Parameters	Quantity
Geogrid	Type	Biaxial (BX 80)
	Material	Polyester
	Tensile strength	80 kN/m
	Elongation	< 15%
	Aperture MD	20 mm in both directions
	Roll width	2.5 m
	Design life	120 years
Coir geotextile	Fibre	Coir
	Fibre Content	100%
	Width	2.4 m
	Length	50 m
	Area	120 m^2
	Weight per roll	72 kg
	Type	Coir Stitched Blanket type (non-woven)
	Weight	600 g/m^2
	Bottom netting type	Light
	Top netting type	Light
	Stitching thread	HDPE Monofilament
	Slope recommendation	>1:1
	Tensile strength ISO-10319	2.8 kg-force per 20 cm
	D4595 Elongation	34 × 20%
	D4491 Flow velocities, short term	4.26 m/s
	Maximum permissible shear stress	1.100 g/m^2
Hybrid geosynthetic	Thickness at 2 kPa	13.175 mm
	Mass per unit area	1240 gsm
	Tensile strength	20.25 kN/m
	Elongation at break	9.59%
	Soil geosynthetic interface	220
	Transmissivity	3.0×10^{-6} m^2/s
	Cross-plane permeability	1.43×10^{-4} m/s

3.2 Layered-Foundation Systems (Soft Soil + Sand)

In this case, the tests varied the unreinforced sand layer thickness (H) overlying the soft soil as 0.5, 0.7, 0.9, and 1.1B [1, 2]. In order to find out the optimum thickness of the sand fill, plate load model tests were performed. The findings were almost similar to each other regardless of thickness of layers. The load-settlement responses of the layered foundation (sand- clayey silt bed) are presented in Fig. 5.

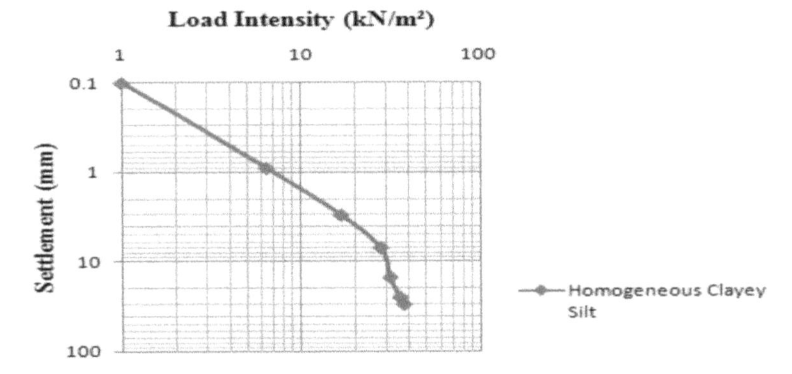

Fig. 4 Load settlement plot of unreinforced soil bed

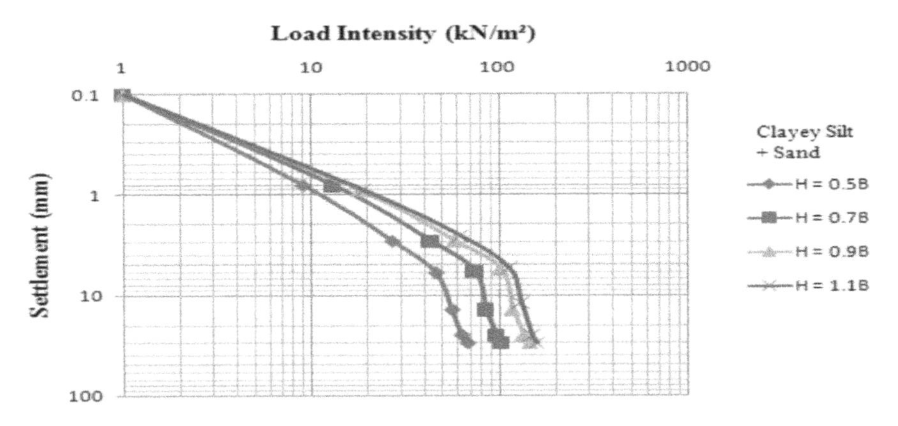

Fig. 5 Load-settlement responses of the unreinforced sand-clayey silt bed

An improvement in ultimate bearing capacity from 28 to 46 kN/m^2 can be noticed from the curve, for clayey silt bed with H = 0.5B thick. In addition, higher bearing capacity values were observed with increase in layer thickness (H). With an increase in H from 0.5 to 1.1B, the ultimate bearing capacities were increased from 46 to 112 kN/m^2. However, a reduction in the improvement rate was noticed beyond H = 0.9B. In Fig. 5, an improvement in bearing capacity of about 102–112 kN/m^2 was observed when H was increased from 0.9 to 1.1B; while, the improvement was in the range of 46 to 102 kN/m^2 for H = 0.5–0.9B. Based on the findings, the optimum thickness for the layered structure is H = 0.9B, which allows the clayey silt bed and the overlying sand layer to contribute the most. For H = 0.5–0.9B, the failure surface was extended to the silty soil bed, resulting in increased resistance. While, for H > 0.9B, the surface of failure forms within the layer of sand and sand behavior dominates layered responses [2].

3.3 Layered Foundation System (Soft Soil + Sand + Non-Woven Coir Geotextile)

The performances of layered configurations having a non-woven coir geotextile at the sand-clayey silt interface were studied in this test. The load-footing settlement curves of clayey silt + sand + coir geotextile system are presented in Fig. 6.

The ultimate bearing capacity of the foundation system with clayey silt, sand and coir geotextile is enhanced from 51 to 77, 107, and 115 kN/m^2 when the layer thickness (H) is varied from 0.5 to 0.7, 0.9, and 1.1B, respectively as shown in Fig. 6. The bearing capacities were increased as the layer thickness was increased. Whereas, for H > 0.9B improvement rate were decreased. The ultimate bearing capacity of clayey silt + sand + geotextile system is 107 kN/m^2 at H = 0.9B.

3.4 Layered Foundation System (Soft Soil + Sand + Geogrid)

The behavior of foundation systems with clayey silt, sand and geogrid were tested. The load-footing settlement responses of these layered foundation system reinforced with geogrid are presented in Fig. 7.

In Fig. 7, the ultimate bearing capacity of the geogrid reinforced foundation is enhanced from 55 to 82, 114, and 119 kN/m^2 for the changes in the layer thickness (H) from 0.5 to 0.7, 0.9, and 1.1B, respectively. For H > 0.9B, improvement rate of bearing pressures were reduced. At H = 0.9B, the ultimate bearing capacity for gogrid-reinforced system with clayey silt bed is 114 kN/m^2, whereas the homogeneous silty bed has a value of 28 kN/m^2 and the unreinforced layered bed has a value of 102 kN/m^2.

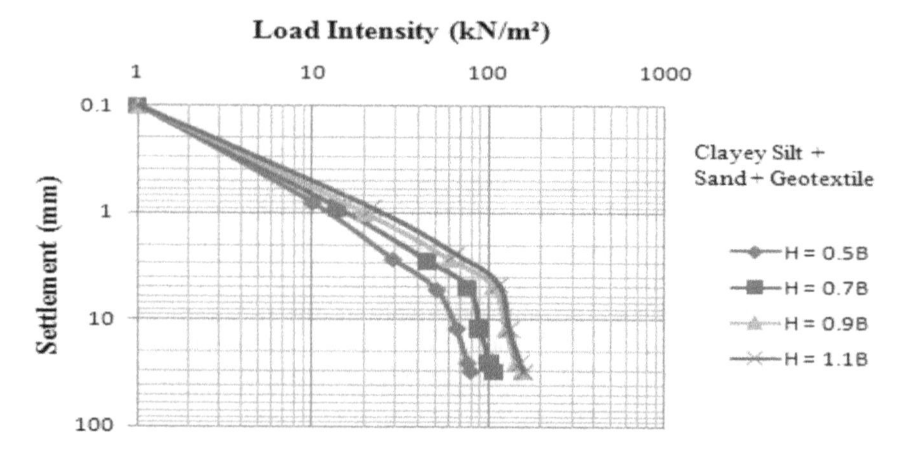

Fig. 6 Load-settlement responses of foundation systems with clayey silt, sand and geotextile

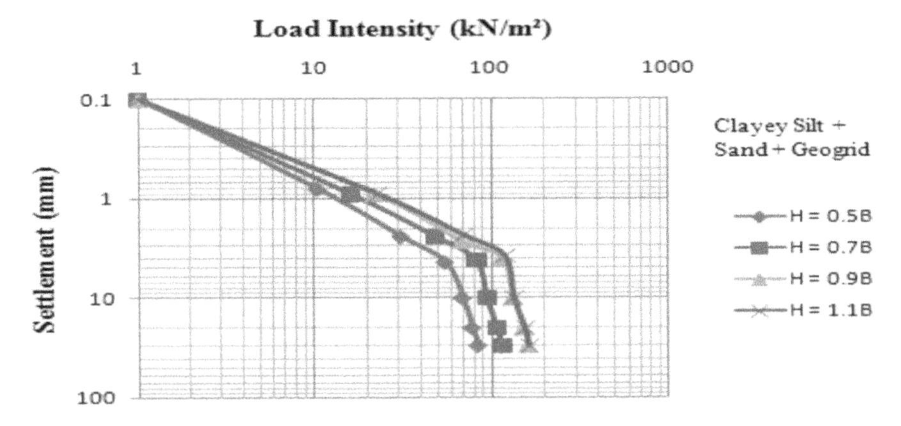

Fig. 7 Load-settlement responses of foundation systems with clayey silt, sand and geogrid

3.5 Layered Foundation System (Soft Soil + Sand + Hybrid Geosynthetic)

The responses of clayey silt + sand + hybrid geosynthetic foundation systems were studied in this test. The load-footing settlement responses of these layered foundation system reinforced with hybid are presented in Fig. 8. The ultimate bearing capacity of the hybrid geosynthetic reinforced foundation is enhanced from 60 to 90, 131, and 135 kN/m^2 for the layer thickness (H) variation from 0.5 to 0.7, 0.9, and 1.1B, respectively. In Fig. 8, an increase in bearing pressure of about 131–135 kN/m^2 can be seen when H is increased from 0.9 to 1.1B; while, the increase was in the range of 60–131 kN/m^2 for H = 0.5–0.9B.

3.6 Effect of Layer Thickness (H)

The load-settlement findings revealed that sand layer thicknesses (H) ranging from 0.5 to 1.1B had a major impact on foundation performance. The ultimate bearing capacities for the unreinforced layered foundations are found to be varied as 46–112 kN/m^2 for the variation of layer thickness from 0.5 to 1.1B. For an increase in H from 0.5 to 0.9B, the ultimate bearing capacities for the layered foundations reinforced with geotextile, geogrid, and hybrid geosynthetic are found to be varied as 51–107, 55–114 and 60–131 kN/m^2 respectively; whereas, the increase was in the range of 107–115, 114–119, 131–135 kN/m^2 for H = 0.9–1.1B. Bearing capacities were increased with increase in H/B values. This is due to the sand layer being punched into the clayey silt bed that enhances the strength of the soft soil underneath. The rate of improvement were reduced for H > 0.9B (i.e., thicker layers). This is due to inadequate strain developed at planar reinforcement to generate

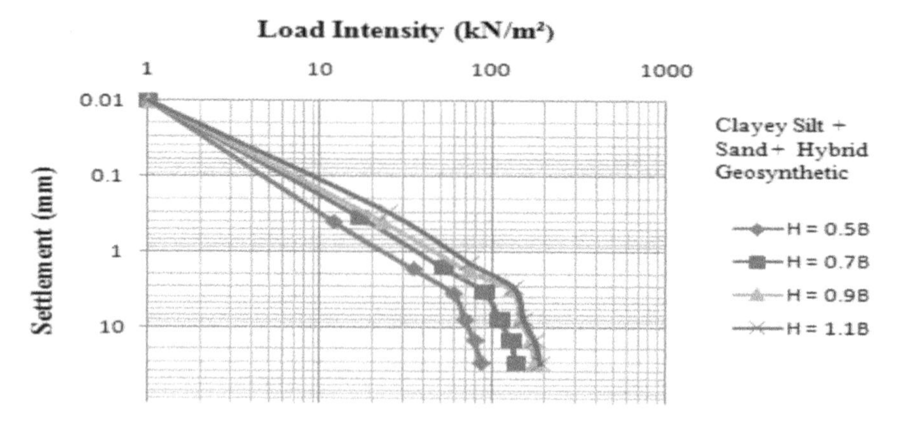

Fig. 8 Load-settlement responses of foundation systems with clayey silt, sand and hybrid geosynthetic

membrane resistance, and that may be the cause of squeezing out of sand and a decrease in load intensity as layer thickness increases [2]. When H > 0.9B, localized settlements were occurred, and the interface - reinforcement does not produce adequate extra benefits. From the observations, the optimum thickness of the sand layer for all the layered configurations can be considered as H = 0.9B.

3.7 *Effect of Different Reinforcement*

Figure 9 indicated the comparative behavior of load-settlement responses with various systems at H = 0.9B. The bearing capacities of reinforced foundations are significantly higher than those of homogeneous and unreinforced layered systems. At H = 0.9B, the hybrid geosynthetic-reinforced system (131 kN/m^2) has a 1.3 times higher ultimate bearing capacity than the unreinforced layered system (102 kN/m^2). The ultimate bearing capacity of the soil beds also increased with increasing superiority of the type of reinforcements such as, clayey silt < unreinforced < geotextile < geogrid < hybrid geosynthetic. The ultimate bearing capacity of the clayey silt bed reinforced with hybrid geosynthetic is significantly greater than that reinforced with geotextile and geogrid alone. The settlement of the fill was decreased when hybrid geosynthetic was placed at the interface.

Limitations—Long term effects are not taken into consideration in the present study.

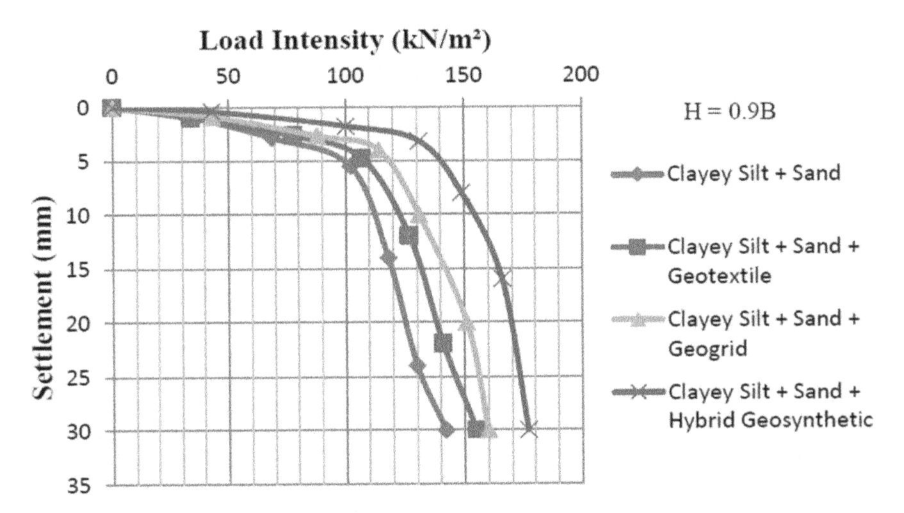

Fig. 9 Comparative load settlement responses of different foundation systems at H = 0.9B

4 Conclusions

- Load settlement curves have showed that the foundation performance was greatly affected by layer thickness (H), settlement of footing and type of reinforcement.
- For unreinforced sand layer + clayey silt bed, the maximum increase in ultimate bearing capacity was found to be in the range of 3.64 times, as compared to the respective homogeneous clayey silt bed.
- In the case of clayey silt bed + sand + hybrid geosynthetic foundation, the ultimate bearing capacity is 4.67 times greater than that of the homogeneous soft soil bed.
- The ultimate bearing capacity of the hybrid geosynthetic reinforced layered foundation is 1.3 times higher than that of the unreinforced layered foundation and 1.14–1.26 times more than that of the layered foundation reinforced with geotextile and geogrid alone. It is always beneficial to use the hybrid geosynthetic rather than using geotextile and geogrid alone.
- From the findings, it can be concluded that load carrying capacities of the unreinforced and reinforced layered foundations increase as layer thickness increases upto H = 0.9B. The improvements in the performances are due to the membrane and interlocking effect of the planar reinforcement at sand-silty soil interface. The improvements were decreased with further increases in the sand layer thickness. This is due to the inadequate strain developed at planar reinforcement with increased layer thickness. For all the configurations, the optimum thickness of sand layer for maximum performance improvement can be concluded as 0.9B.

- The ultimate bearing capacity increased as the form of reinforcement improved, such as clayey silt < unreinforced < geotextile < geogrid < hybrid geosynthetic.

References

1. Biswas A, Krishna AM, Dash SK (2016) Behavior of geosynthetic reinforced soil foundation systems supported on stiff clay subgrade. Int J Geomech ASCE 04016007:1–15
2. Biswas A, Ansari MA, Dash SK, Krishna AM (2015) Behavior of geogrid reinforced foundation systems supported on clay subgrades of different strengths. Int J Geosyn Ground Eng 1(20):1–10
3. Bhattacherjee D, Viswanadham BVS (2019) Centrifuge model studies on performance of hybrid geosynthetic–reinforced slopes with poorly draining soil subjected to rainfall. J Geotech Geoenviron Eng 145(12):04019108:1–13
4. Lal D, Sankar N, Chandrakaran S (2017) Effect of reinforcement form on the behaviour of coir geotextile reinforced sand beds. Soils Found 57:227–236
5. Yang K-H, Yalew WM, Nguyen MD (2016) Behavior of geotextile-reinforced clay with a coarse material sandwich technique under unconsolidated - undrained triaxial compression. Int J Geomech ASCE 16(3):04015083:1–15

Compressive Strength Prediction with Boundary-Defined Datasets

P. Hrishikesh and G. Unni Kartha

Abstract Concrete is one of the leading sources of carbon emission to the atmosphere. Cement being the major part of that offset. Every concrete mix that doesn't meet the strength requirements will be demolished creating further damage to the environment along with loss of time & money. Early & accurate prediction of the Concrete Compressive Strength (CCS) improves quality control & scheduling of the operations. Soft-computing techniques like Machine Learning (ML) have long been used to create a reliable model to predict CCS by finding linear or nonlinear relations among the parameters. A dataset containing the compressive strength measurement of sufficient number of samples along with details about the proportion of each ingredient corresponding to each sample is necessary to create a ML model that will have high prediction capability. Dataset containing observations from laboratory based studies from national sources are collected to analyze the performance of different ML models. The dataset is further developed by adding boundary conditions to the dataset. The performance of each model with and without the boundary values datasets are compared.

Keywords Machine learning · Concrete compressive strength · Dataset · Parameters · Boundary conditions

1 Introduction

1.1 General

Cement is the most consumed man-made material in the world. While the invention of concrete has boosted infrastructure development, it has also created a negative impact in the environment on a life-threatening level. The environmental damage is mainly attributed to the greenhouse gas emissions during manufacturing and consumption of ingredients of concrete. In 2019, concrete accounted for about

P. Hrishikesh (✉) · G. Unni Kartha
FISAT, Angamaly, Ernakulam, Kerala, India

G. C. Marano et al. (eds.), *Proceedings of SECON'21*, Lecture Notes in Civil Engineering 171, https://doi.org/10.1007/978-3-030-80312-4_25

8 percent of the carbon dioxide emitted into the atmosphere (www.ecori.org). Majority of the infrastructure is built on either conventional reinforced concrete or high performance concrete. When the recently cast concrete doesn't provide the required compressive strength, the member has to be demolished and cast with freshly mixed concrete, making the previously used cement & other ingredients to waste along with the additional carbon offset the materials had on the environment. From 2003 to 2014, there was a 65% increase in construction and demolition waste (CDW) just in the United States (according to the Environmental and Protection Agency) and in 2014, 535 million tons of CDW were generated. It is worth noting that roughly 70% of CDW can be classified as concrete. Numerous researches are conducted all around the world to reduce the carbon offset in the construction industry. One of the many approaches is to reduce concrete waste [1].

Accurately predicting the compressive strength of concrete would improve the quality control and thereby reduce wastage of concrete. Also, a robust, predictive model that could estimate compressive strength as a function of the mixture proportions would be useful in enabling high throughput mixture design, and reducing the empirical, labour intensive nature of trial batching approaches that are the basis of industrial practice today [2]. To train the ML models, informations are provided in the form of spreadsheets hereby referred in the paper as "datasets". In the case of predicting compressive strength of concrete, the above mentioned dataset contains observations of CCS at different ages and information about properties and quantities of each ingredient in the corresponding mix.

1.2 Previous Works

As mentioned above, the scope of the topic inspired numerous authors employed ML models to create a strength model. Initially, the works to the compressive strength analysis were based on laboratory level tests and experiments. Among those who worked on understanding the influence of ingredients on concrete's mechanical properties, Yeh's works provided one of the first reliable databases for future research. Regarding his works, Yeh in [3] used artificial neural networks (ANN) to predict the compressive strength of high-performance concrete (HPC). Using observations obtained from the trial batches of HPC, produced in laboratories. The results from the study proposed that a strength model based on ANN is more accurate than a regression analysis. A major contribution of his work is a well consolidated dataset, available at UCI dataset repository accessible to everyone [4].

Numerous authors employed his datasets to analyse the results of various ML models, sometimes exclusively. Ali Behnood et al. used the dataset to analyze the M5P model algorithm to predict CCS. Chou et al. [4] exclusively used it to optimize the prediction accuracy of various models by comparing data mining methods. Their study showed MART (Multiple Additive Regression Trees) as the most

Table 1 Performances of various models in previous works

Author	Model	Dataset	RMSE	R^2
Yeh et al.	ANN	Data from 17 different sources	–	0.922
Young et al. [2]	Linear regression	Yeh et al.	8.8	0.66
		VIP suffolk	5	0.49
	Neural network	Yeh et al.	6.3	0.82
		VIP suffolk	4.8	0.54
	Random forest	Yeh et al.	5.7	0.86
		VIP suffolk	4.4	0.6
	Boosted tree	Yeh et al.	5.8	0.85
		VIP suffolk	4.5	0.59
	SVM	Yeh et al.	6.4	0.83
		VIP suffolk	4.5	0.59
Tuan et al. (2020)	XGB	Yeh et al.	4.27	0.93
Avinash et al. (2020)	Linear regressor	Dataset from Kaggle (Yeh et al.)	10.29	0.61
	Ridge regressor		10.29	0.61
	Lasso regressor		10.91	0.57
	Decision tree		7.5	0.79
	Random forest		5.3	0.9
	AdaBoost regressor		7.75	0.77
	Gradient boosting		5.2	0.91

reliable model & it also points out to consider MAPE (Mean Absolute Percentage Error) & training time along with RMSE & R^2. Table 1 shows the results from major studies and it is clear that irrespective of the size of datasets or type of individual models used in the ensemble, they provide more accurate results.

Until recently researchers were limited in terms of dataset availability. They usually employ datasets obtained from laboratories, which limited their size of observations to around 1000 CCS measurements or less [2]. Benjamin et al. provided one of the early analyses of a large dataset. The author in [2] collected measurements from a concrete manufacturer along with the dataset from Yeh et al. The models had lower levels of accuracy when predicting compressive strength of job-site concrete samples, but they were still able to predict compressive strength with the average relative error within the limit of 10%. RF models had the best performance measures [2].

In this paper the possibility of increasing the accuracy & reliability of prediction models by defining & including the boundaries to the datasets are reviewed.

2 Methodology

Data collection. Datasets containing information about the CCS of a mixture & ingredients associated with it are collected directly from various concrete manufacturers and are consolidated into a single database. The dataset is further developed by adding observations at boundary conditions.

Setting up ML Models. A hosted Jupyter notebook service called colaboratory is used to execute various models and data extraction procedures using python code.

Employing individual & ensembled ML models. Various standalone & ensembled models including but not limited to, various ANNs, SVM, decision trees, random forest, etc. are either built or imported from various libraries and trained using a subset of the dataset (training dataset) collected earlier.

Evaluating the influence of boundary conditions. The level of influence of boundary conditions in the data on prediction accuracy of each ML model can be measured using performance evaluation metrics.

2.1 Models Used

The models used for the review are of 2 types, stand-alone models and ensemble ML models. stand-alone models include regression models, ANNs, decision trees, support vector machines, etc. Ensemble model is a combination of 2 or more stand-alone models in predefined logic. The basic concept of ensembled models is to train several stand-alone models using the training data, and then integrate them to form an ensemble model. By doing so, the accuracy and robustness of ensembled models will be improved [5]. There are 4 major groups of algorithms for ensemble learning—bagging, boosting, voting and stacking [5].

Linear Regression

Linear regression is a model that provides the output or target as a linear combination of the input variables. In addition to critically analysing the data, linear regression also serves as a good baseline from which other techniques can be evaluated [2].

ANN

An artificial neural network is a computational model that attempts to simulate the structure and/or functional aspects of biological neural networks. Various ANN applications can be categorized as classification or pattern recognition or prediction and modelling. ANNs are widely used in many fields, including technical diagnosis and non-destructive testing, power systems, robotics, etc.

The advantages ANN have compared to other models is that there is an unrestricted number of inputs and outputs and the clearly defined number of hidden layers and hidden neurons. The major drawback of ANNs is the relatively more time needed to determine the number of layers and hidden neurons, which requires repetitive trial and error-tuning processes. The training algorithms commonly used in ANNs are Levenberg–Marquardt, gradient descent, gradient descent with momentum, gradient descent momentum and adaptive learning rate, and gradient descent with adaptive learning rate. The backpropagation algorithm is the most used algorithm to vary the connection weights and bias values for training [4].

Decision Trees

In decision trees, the data is split continuously according to a predetermined parameter. The tree mainly consists of two elements, decision nodes and leaves. The data is split at decision nodes and the leaves provide final outcomes [6].

Random forest

Random forest is an ensemble model where multiple regression trees are integrated in order to increase the stability of individual trees. Random forest employs two strategies to improve the instability. It initially uses bootstrap aggregation (sampling with replacement) to create multiple datasets that were sampled from the original dataset. Each of these datasets are given to an individual tree within the ensemble. Second, randomness is implemented during tree-learning in order to decrease the correlation between each tree within the ensemble [7, 8].

Gradient Boosted trees

The main concept of boosting is to build a model using a stand-alone model. Then another model is trained on the residuals from the previous model. This process of modelling continued for a fixed level of iteration. Like random forest, boosted trees are an ensemble method for dealing with the instability and poor predictive performance of simple regression trees. Therefore, a boosted tree is a regression model where a regression tree is used in each iteration. Boosted trees is observed to perform reliably on highly nonlinear datasets and is found to be more effective on problems where each input parameters have different level of influence [7, 9].

2.2 *Dataset*

The dataset consists of 1030 lab based observations, featuring 278 unique concrete mixture designs and their age-dependent compressive strengths. Yeh collected the data from 17 different sources and consolidated it into a single dataset [3]. It is accessed through the UCI repository [10].

3 Boundary Conditions

The boundary conditions applied mainly consist of giving values to a single input parameter & therefore the CCS's value will always be 0 MPa. The values given to each parameter is based on the range of each parameter in the given dataset. The 2 boundary conditions considered are:

Table 2 Description of Yeh's dataset

	Count	Mean	Std	Min	25%	50%	75%	Max
Cement (kg/m^3)	1030	281.17	104.51	102.00	192.38	272.90	350.00	540.00
BFS (kg/m^3)	1030	73.90	86.28	0.00	0.00	22.00	142.95	359.40
Fly Ash (kg/m^3)	1030	54.19	64.00	0.00	0.00	0.00	118.30	200.10
Water (kg/m^3)	1030	181.57	21.35	121.80	164.90	185.00	192.00	247.00
Superplasticizer (kg/m^3)	1030	6.20	5.97	0.00	0.00	6.40	10.20	32.20
Coarse Aggregate (kg/m^3)	1030	972.92	77.75	801.00	932.00	968.00	1029.40	1145.00
Fine Aggregate (kg/m^3)	1030	773.58	80.18	594.00	730.95	779.50	824.00	992.60
Age (days)	1030	45.66	63.17	1.00	7.00	28.00	56.00	365.00
CCS (MPa)	1030	35.82	16.71	2.33	23.71	34.45	46.14	82.60

BFS = Blast Furnace Slag

I. when every input parameter is zero, corresponding compressive strength is zero.
II. when the age of concrete is zero, corresponding compressive strength is zero.

Minimum, mean & maximum of each parameter from Table 2 is provided as various scenarios in between the 3 above mentioned boundary conditions. A secondary dataset created by appending observations in Table 2 to Yeh's dataset. The secondary dataset can be termed as a "boundary-defined dataset" (B + dataset). This can be further explained in Table 3.

Table 4 gives the description of the boundary defined dataset.

4 Performance Analysis and Discussion

After running the program with 80% of both datasets (training dataset) individually, the performance of the same model on testing datasets is compared using performance evaluation metrics. After employing the B + dataset for training, the linear regression and neural network models show improvement in R^2 scores whereas the remaining models R^2 scores are slightly decreased. The performance of the RF models on each testing datasets is shown in graphs of predicted v/s true outputs. (Fig. 1).

The performance of each model with Yeh's & B + datasets are recorded as root mean square error (RMSE) & R^2 in Table 5. According to this table, the decrease in R^2 after employing B + dataset is within the range of 0.5–2.5%, but incase of DNN & Linear regression the increase in R^2 score is upto 16%.

Table 3 Boundary-defined dataset

Cement (kg/m^3)	BFS (kg/m^3)	Fl A (kg/m^3)	Water (kg/m^3)	SP (kg/m^3)	CA (kg/m^3)	FA (kg/m^3)	Age (days)	CCS (MPa)
0.00	0.00	0.00	0.00	0.00	0.00	0.00	0.00	0.00
Minimum	0.00	0.00	0.00	0.00	0.00	0.00	0.00	0.00
Mean	0.00	0.00	0.00	0.00	0.00	0.00	0.00	0.00
Maximum	0.00	0.00	0.00	0.00	0.00	0.00	0.00	0.00
0.00	Minimum	0.00	0.00	0.00	0.00	0.00	0.00	0.00
0.00	Mean	0.00	0.00	0.00	0.00	0.00	0.00	0.00
0.00	Maximum	0.00	0.00	0.00	0.00	0.00	0.00	0.00
0.00	0.00						Minimum	0.00
0.00	0.00	–	–	–	–	–	Mean	0.00
0.00	0.00						Maximum	0.00
540.00	359.4	200.1	247	32.20	1145	992.60	0.00	0.00

Fl A = Fly Ash, SP = Superplasticizer, CA = Coarse Aggregate, FA = Fine Aggregate

Table 4 Description of B + dataset

	Count	Mean	Std	Min	25%	50%	75%	Max
Cement (kg/m^3)	1056	275.63	111.48	0	190.30	266.00	350.00	540.00
BFS (kg/m^3)	1056	72.83	86.79	0	0.00	20.00	142.50	359.40
Fly Ash (kg/m^3)	1056	53.28	64.02	0	0.00	0.00	118.23	200.10
Water (kg/m^3)	1056	177.85	33.61	0	164.60	184.20	192.00	247.00
Superplasticizer (kg/m^3)	1056	6.12	6.08	0	0.00	6.10	10.10	32.20
Coarse Aggregate (kg/m^3)	1056	952.81	159.11	0	931.20	967.10	1028.40	1145.00
Fine Aggregate (kg/m^3)	1056	757.71	136.45	0	719.93	778.40	822.20	992.60
Age (days)	1056	44.93	63.53	0	7.00	28.00	56.00	365.00
CCS (MPa)	1056	34.94	17.41	0	22.70	33.73	45.45	82.60

When tested manually using a "null set", i.e., all input parameters are equal to zero & hence CCS is 0 MPa, as shown in Figs. 2 and 3, the model becomes more reliable to entries of a null set after training with B + datasets. The results from each ML model clearly indicates that the B + datasets provided better training for the models compared to primary datasets.

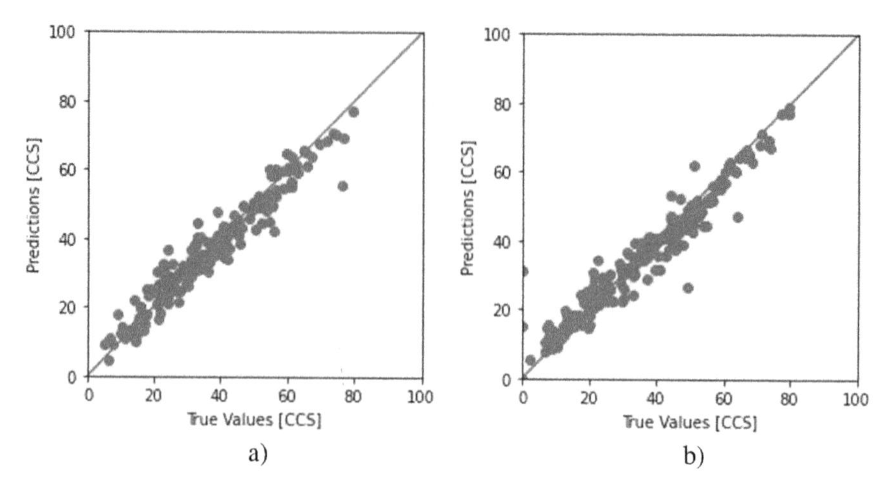

Fig. 1 Plot showing the variation of predicted & true outputs in the validation of RF model trained using **a)** Yeh's dataset and **b)** B + dataset

Table 5 Overview of the models performance

ML model	Metrics	Yeh	B + Yeh	Error (%)
Linear regression	RMSE	11.7165	12.0444	2.80
	R^2	0.4673	0.5462	16.87
DNN	RMSE	6.5895	6.3590	−3.50
	R^2	0.8315	0.8735	5.05
Decision trees	RMSE	5.8804	6.7015	15.42
	R^2	0.8658	0.8595	−0.55
Random forest	RMSE	4.2032	4.8515	13.96
	R^2	0.9314	0.9264	−0.73
Gradient boosting	RMSE	4.2637	5.3687	25.92
	R^2	0.9295	0.9098	−2.11

```
print(A)
prediction = forest.predict(A).flatten()
print('CCS =', prediction, 'MPa')

[[0. 0. 0. 0. 0. 0. 0. 0.]]
CCS = [11.258] MPa
```

Fig. 2 Manual testing of ML model (Yeh's dataset)

```
print(A)
prediction = forest.predict(A).flatten()
print('CCS =', prediction, 'MPa')

[[0. 0. 0. 0. 0. 0. 0. 0.]]
CCS = [0.] MPa
```

Fig. 3 Manual testing of ML model (B + dataset)

5 Conclusion

In supervised learning, the machine learning algorithms are trained to provide an output according to the provided input parameters. In case of regression models, the training is achieved by using a dataset that contains a wide array of features and outputs. Usually the dataset is created by extracting real world observations.

The ML models are used to identify the underlying patterns and relationships between the features and the resulting output. In this paper, when trained with Yeh's dataset, even the highest performing regression model, random forest (according to R^2 score) makes errors when dealt with a null set. The same model predicted precisely for the same set of inputs after training with a boundary defined dataset. The RF model was able to identify the significance in not only the proportion of each ingredient, but also in the quantity of them. This was achieved by introducing a set of new observations which does not occur naturally in the real world. In this new approach the author has guided the models to learn the behaviour of features at boundary conditions, thereby terming the approach "guided supervised learning".

References

1. Han T, Siddique A, Khayat K, Huang J, Kumar A (2020) An ensemble machine learning approach for prediction and optimization of modulus of elasticity of recycled aggregate concrete. Constr Build Mater
2. Young BA, Hall A, Pilon L, Gupta P, Sant G (2018) Can the compressive strength of concrete be estimated from knowledge of the mixture proportions? New insights from statistical analysis and machine learning methods. Cem Concr Res (Elsevier)
3. Yeh L-C (1998) Modeling of strength of high-performance concrete using artificial neural networks. Cem Concr Res (28):1797–1808 (Copyright © 1998 Elsevier Science Ltd)
4. Chou J-S, Chiu C-K, Farfoura M, Al-Taharwa I (2011) Optimizing the prediction accuracy of concrete compressive strength based on a comparison of data-mining techniques. J Comput Civil Eng
5. Feng D-C, Liu Z-T, Wang X-D, Chen Y, Chang J-Q, Wei D-F, Jian Z-M (2020) Machine learning-based compressive strength prediction for concrete: an adaptive boosting approach. Constr Build Mater 230

6. Chou J-S, Tsai C-F, Pham A-D, Lu Y-H (2014) Machine learning in concrete strength simulations: multi-nation data analytics. Constr Build Mater 73
7. DeRousseau MA, Laftchiev E, Kasprzyk JR, Rajagopalan B, Srubar WV (2019) A comparison of machine learning methods for predicting the compressive strength of field-placed concrete. Constr Build Mater 228 (Elsevier Ltd)
8. Cook R, Lapeyre J, Ma H, Kumar A (2019) Prediction of compressive strength of concrete: critical comparison of performance of a hybrid machine learning model with standalone models. J Mater Civil Eng (ASCE)
9. Shariati M, Mafipour MS, Ghahremani B, Azarhomayun F, Trung NT, Ahmadi M, Shariati A (2020) A novel hybrid extreme learning machine–grey wolf optimizer (ELM GWO) model to predict compressive strength of concrete with partial replacements for cement. Springer Nature, Springer-Verlag London Ltd.
10. UCI Machine Learning Repository Homepage. https://archive.ics.uci.edu

Residual Life Assessment of Reinforced Concrete Considering Tension Softening Behaviour

S. Ajimi, M. Simon Keerthy, and J. Bharati Raj

Abstract Under fatigue loading, concrete exhibits softening behaviour due to the formation of an inelastic zone ahead of the crack tip. The formation of this inelastic zone is due to the heterogeneities present in the concrete. This inelastic zone is dominated by various toughening mechanisms. Current design practices for reinforced concrete is by assuming a zero tensile strength for concrete which is over conservative. Actually concrete can withstand a reasonable tensile stress and strain. Therefore study of RC member should consider the tension softening response of concrete. Under fatigue loading, strength and stiffness decrease progressively according to the maximum amplitude and the number of cycles. Fracture plays an important role in failure of normally and lightly reinforced beam. Since FPZ mechanisms and fibre bridging action resist crack propagation, we need to consider those while predicting remaining life of RC member. Fatigue failure occurs when applied load is much less than the moment capacity. Such structures susceptible to fatigue load need to be monitored and residual life is to be predicted. This paper aims to predict the residual life of reinforced concrete members are assessed in terms of moment carrying capacity by considering various tension softening laws.

Keywords Residual life · Fracture process zone · Tension softening laws · Reinforced concrete

1 Introduction

Many concrete structures like bridges, pavements, highways, airports, flyovers and other infrastructural engineering structures undergo repeated loading. Therefore, a structural fatigue failure may occur because of this cyclic loading and significant changes on the characteristics of materials such as stiffness, toughness and durability etc. may occur which will in turn affect residual life of concrete member. Fatigue is a phenomenon that is taken place in a material in a gradual, permanent,

S. Ajimi (✉) · M. Simon Keerthy · J. Bharati Raj
Department of Civil Engineering, NSS College of Engineering, Palakkad, India

G. C. Marano et al. (eds.), *Proceedings of SECON'21*, Lecture Notes in Civil Engineering 171, https://doi.org/10.1007/978-3-030-80312-4_26

micro-structural way due to the application of repeated loading. Under cyclic loading, the stresses near the crack tip are high enough to lead to failure even if, the nominal stresses are well below the yield limit of the material, Even though concrete is a heterogeneous material, it is treated as homogeneous material from design perspectives. But under fatigue loading, concrete may exhibit a softening response due to the existence of heterogeneities when it is tested under displacement control. Quasi- brittle materials like concrete will show a strain-softening behaviour. This softening behaviour is due to the formation Fracture Process Zone (FPZ) ahead of the crack tip.

Fracture Process Zone (FPZ) is dominated by various toughening mechanisms. Out of these toughening mechanisms, micro cracking and aggregate bridging are the major mechanisms that are responsible for the softening behaviour [1]. Fatigue failure will occur even when the applied load is much lower than the yield value (ie, the tensile strength for concrete). So it is necessary to assess remaining life of existing structure to prevent catastrophic failure. As the crack within the concrete member after initiation, propagate continuously under fatigue loading, its strength and stiffness decreases progressively. Hence residual strength assessment of members are also an important issue that has to be looked upon. Due to the presence of FPZ in front of crack tip, post peak softening response acutely influence crack propagation and therefore it is appropriate to consider FPZ mechanisms while assessing residual strength of concrete. There are various tension softening models to mathematically represent the post peak softening mechanisms of concrete [2]. FPZ in ductile and brittle materials are shown in Fig. 1.

2 Residual Life Assessment

The residual life of concrete structures under fatigue loading can be found out using different principles. Some of them are Linear Elastic Fracture Mechanics Principle (LEFM) and Non Linear Fracture Mechanics Principle (NLFM).

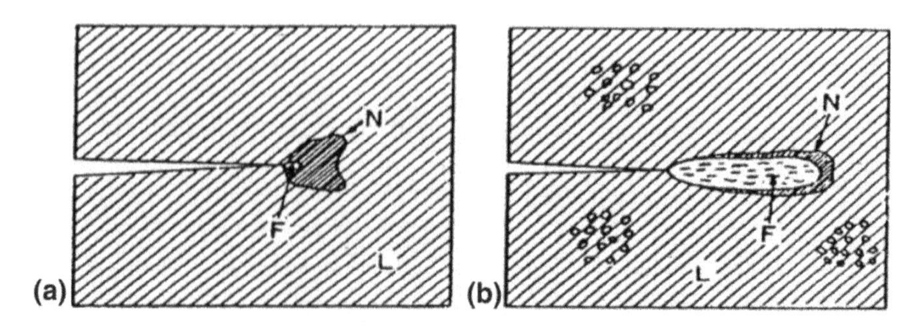

Fig. 1 FPZ in ductile and brittle materials. **a** Ductile–brittle (metals). **b** Quasi-brittle (concrete) [2]

2.1 Residual Life Assessment Using LEFM

Linear Elastic Fracture Mechanics (LEFM) is a method to study the crack growth rate and the mechanism of crack growth in different components of a structure. This method is originally developed by Griffith and is modified by Irwin. LEFM is applicable to almost every material but it should satisfy certain criteria such as material should be elastic except at some infinitesimally small area near the crack tip. LEFM is based on the stress intensity factor K, which can be determined by using stress analysis. When the stress intensity at the crack tip (K_C) reaches a critical value K_{IC}, the crack starts to propagate. In this principle, it is assumed that when a major fatigue crack is formed, the bridging behaviour within that region will govern the rate of fracture crack propagation.

According to classical theory, when a structural component is subjected to high amplitude cyclic loading, the bottom of the members will experience an in-plane loading. But the current design practices for reinforced concrete assumes a zero tensile strength for concrete which is actually overly conservative. In fact, concrete can bear significant tensile stress and strain, in case of plain concrete member, the residual life of a pre-notched beam can be computed in terms of moment carrying capacity [2]. LEFM based fatigue law is used to study the fatigue behaviour of reinforced concrete beams after suitable modifications to incorporate the various load frequencies that applied on beam and the effect of reinforcement present. For the reinforcement present in the beam, a closing force is used at that level. The residual life of the reinforced members subjected to fatigue loading is a function of increasing crack length for assuring safety [3].

2.2 Mathematical Modelling of FPZ

There are various toughening mechanisms like crack shielding, crack deflection, aggregate bridging and micro cracking are experienced in fracture process zone. Out of these, micro cracking and aggregate bridging are the major mechanisms that are responsible for softening behaviour. This FPZ mechanisms can be can be mathematically modelled.

Once the crack start to propagate, aggregate in concrete will provide a bridging resistance along the cracked surface. The bridging force development can mathematically be represented using tension softening laws. The tension softening models are obtained by defining the stress as a function of crack opening displacement w. The different tension softening laws represented as linear, bilinear, trilinear, exponential and power curves [2]. Bridging resistance offered at the crack tip is found out by connecting mesoscale properties with the macroscale such as fracture toughness and elastic modulus at interface. This model is called cyclic aggregate bridging model [1]. Apart from the tension softening models, the FPZ can be mathematically modelled using Cohesive Crack Model. It presents a new cohesive stress—crack opening relation by introducing the effect of micro–macro crack interactions. The study reveals a transient behaviour of macro crack opening as the micro crack is formed in the surrounding [4].

2.3 Residual Life Assessment Using NLFM

It is not appropriate to use LEFM principles while predicting the fracture behaviour of concrete since a large process zone is existing there. So, nonlinear fracture mechanics theory is used to predict the residual life accurately. Here, the existence of an inelastic zone called fracture process zone is considered. And also secondary crack formation is considered.

A fictitious crack model is used to study the tensile cracking behaviour of plain concrete member under bending which assumes the existence of long and infinitesimally narrow fracture process zone at the crack tip. The predicted remaining life using power and bilinear law shows better result [2]. The residual life of plain concrete member can also be found by cyclic aggregate bridging model in terms of moment carrying capacity by considering the aggregate bridging action. The stress–strain distribution along the crack profile is assumed to be linear [1]. An inverse method is used to find the moment carrying capacity or load carrying capacity corresponding to each crack size or tensile strain or crack opening at the tip of crack. This examined the effect of tension-softening on the moment carrying capacity of reinforced concrete beam. A linear crack opening profile is assumed to find the ultimate moment carrying capacity using the crack tip opening displacement w_c criteria [5].

3 Residual Life Assessment of Reinforced Concrete Beam

The residual life of concrete beam is calculated in terms of moment carrying capacity. The process zone is modelled by using bilinear softening law. The experimental data represented by Sain and Kishen [5] is used to develop the model. The geometrical and fracture properties of the RC beam used for validation is given in Table 1. They used linear tension softening law to find the moment carrying capacity while bilinear tension softening law is used in this paper. The residual life of the RC beam is determined as a function of increasing crack length in the stable region. The crack opening profile is assumed as linear. The moment carrying capacity is obtained by using equations of equilibrium ie, total tensile force is equal to total compressive force. The stress- strain distribution of reinforced concrete beam is shown in Fig. 2. For the concrete in compression the constitutive relation is given in Eq. 1,

Table 1 Details of RC beam

Depth (mm)	150
Width (mm)	100
Length (mm)	1200
Steel area (mm^2)	113.09
Yield stress (MPa)	544
E (MPa)	35.6e3
f_{ck} (MPa)	45
f_t (MPa)	3.75
G_f (N/mm)	0.0725
w_c	0.037

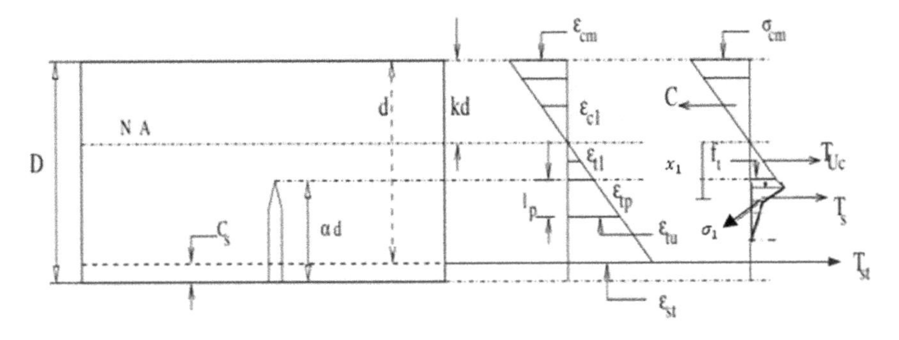

Fig. 2 Stress–strain distribution of the RC beam

$$\sigma = \begin{cases} E \in_C & \in_c \leq \in_{cp} \\ f_{ck} & \in_{cp} \leq \in_c \leq \in_{cu} \end{cases} \tag{1}$$

where, f_{ck} is crushing strength of concrete, E elastic modulus of concrete, $\in_{cu}$ plastic strain and $\in_{cp}$ is ultimate strain.

For the uncracked concrete in tension which is following a linear relation given in Eq. 2,

$$\sigma = E \in_t \tag{2}$$

For the reinforcement in tension it follows a bilinear relation as Eq. 3,

$$\sigma_{st} = \begin{cases} E_S \in_{st} < f_y \\ f_y > f_y \end{cases} \tag{3}$$

For the reinforcement in tension, the strain is given by Eq. 4,

$$\epsilon_{st} = \frac{\epsilon_{tu}\,(1-k)}{1-\alpha-k} \tag{4}$$

E_S is the elastic modulus of steel and f_y is yield stress. For computing moment carrying capacity, equations of equilibrium is used in an incremental procedure. The uncracked region is divided into number of segments x each having a depth of dx = (d − kd − αd − lp)/x. Using trial and error procedure, neutral axis depth factor k is determined. Using this k, the length of softening zone lp is calculated as Eq. 5,

$$l_p = \left(1 - \frac{\epsilon_{tp}}{\epsilon_{tu}}\right)(1 - k - \propto)d \tag{5}$$

Depth of neutral axis is calculated as total compressive force equals the total tensile force. Once k is determined, moment of resistance can be computed as Eq. 6

$$M_R = M_{soft} + M_{UT} + M_{st} \tag{6}$$

Moment of resistance of softening zone given in Eq. 7,

$$\mathrm{M_{soft}} = T_s x \; lever\; arm = T_s\left[\left(\frac{2kd}{3}\right) + y_1 + x_g\right] \tag{7}$$

where x_g is the centroidal distance of softening zone given in Eq. 8,

$$x_g = \frac{(2\sigma_1 + f_t) + (l_p - y)^2}{3 * \left[(\sigma_1 + f_t) * y + (l_p - y)\sigma_1\right]} \tag{8}$$

$$y = x_1 - y_1$$

Resistance provided by the softening zone given in Eq. 9,

$$T_s = \frac{1}{2}B(f_t - \sigma_1)(x_1 - y_1) + \frac{1}{2}B\sigma_1(l_p - x_1 - y_1) \tag{9}$$

Moment of resistance of uncracked tension concrete given in Eq. 10,

$$M_{UT} = T_{Ut}\left[\left(1 - \propto - \frac{k}{3}\right)D - \frac{y1}{3} - l_p\right] \tag{10}$$

T_{Ut} Is tensile resistance provided by uncracked tension concrete and y1 is length of corresponding uncracked section.

Moment of resistance due to reinforcements given in Eq. 11

$$M_{st} = T_{st}\left[\left(1 - \frac{k}{3}\right)d\right] \tag{11}$$

Resistance provided by the reinforcement given in Eq. 12,

$$T_{st} = \sigma_{st}A_{st} \tag{12}$$

The same steps are done for various crack lengths till the equilibrium is satisfied. The computed moment is normalised. Normalised moment carrying capacity is the plotted against and relative crack depth.

3.1 Result

The plot of normalised moment capacity versus relative crack depth using linear and bilinear methods are shown in Fig. 3. From the graph, it is seen that, the moment carrying capacity is increasing as the crack length increases upto yield point. Later, moment capacity decreases with the crack propagation. Both laws following the similar trend.

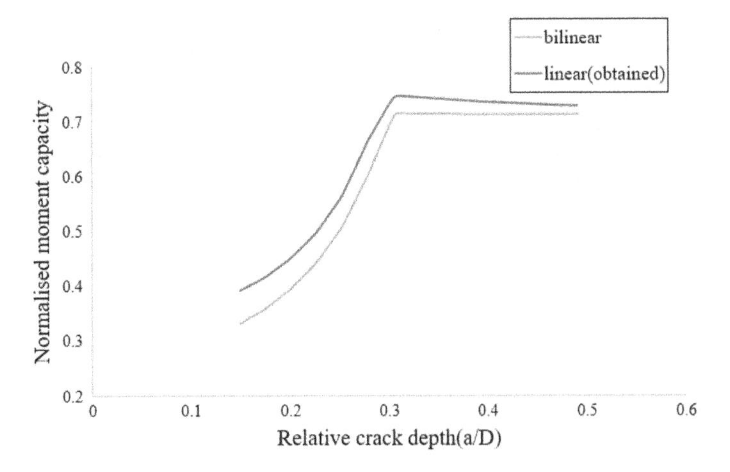

Fig. 3 Normalised moment capacity v/s relative crack depth using present method

4 Conclusions

This paper include determination of residual life of reinforced concrete structures in terms of moment carrying capacity. It also validates the proposed method using MATLAB software.

- The life of concrete structures can be determined using LEFM principles. But LEFM over predicts the life of concrete like quasi brittle material. LEFM approach ignores the fracture process zone mechanisms. So Remaining life of concrete structures can be predicted accurately by NLFM principles by considering FPZ mechanism.
- The post-peak softening behaviour has been mathematically modelled by using distinct tension softening laws. Out of the various tension softening models, bilinear law predicts fatigue behaviour accurately for plain concrete members.
- For reinforced concrete, the FPZ is modelled using linear law. It is seen that the moment carrying capacity is increasing as the crack length increases upto yield point. Later, moment capacity decreases with the crack propagation.

References

1. Simon KM, Chandra Kishen JM (2016) Influence of aggregate bridging on the fatigue behavior of concrete. Int J Fatigue 90:200–209
2. Rama Chandra Murthy A, Palani GS, Iyer NR, Gopinath S, Raghu Prasad BK (2012) Residual strength evaluation of concrete structural components under fatigue loading. Sadhana 37(1): 133–147
3. Sain T, Chandra kishen JM (2007) Residual strength assessment of concrete considering tension softening behaviour. Int J Fatigue 29:2138–2148
4. Alam SY, Loukili A (2020) Effect of micro-macro crack interaction on softening behaviour of concrete fracture. Int J Solids Struct 34–45
5. Sain T, Chandra kishen JM (2008) Fracture stability and residual strength assessment of reinforced concrete beams. Mater Struct 41:1451–1463
6. Ray S, Chandra Kishen JM (2014) Analysis of fatigue crack growth in reinforced concrete beams. Mater Struct 47:183–198

Seismic Performance Evaluation of Seesaw Braced Energy Dissipation System

Anjanet Anto and Asha Joseph

Abstract A large amount of energy is dissipated to the structure during a major earthquake causing structural damage to the building. Based on the understanding about seismic activities, the need for seismic energy dissipation has become an important factor in structural engineering. A new vibration system based on seesaw mechanism characteristics and pretension effect is the seesaw energy dissipation system. The system comprises braces, seesaw members, and dampers. The frame moves right and left under a lateral load, and the tensile force generated in the braces is delivered to the seesaw member and the dampers dissipate the energy. As the tensile force occurs in bracing members, the issue of brace buckling problem can be avoided, such that it enables the use of steel rods for bracing members. Long steel rods can be used as bracing between the seesaw members and the moment frame connections, by introducing pretension in rods. This study focuses on the analysis of the seismic performance of the seesaw braced steel structure using different bracing configurations in different seismic zones as per IS code using the SAP 2000. This type of vibration control device provides an alternate solution for typical low rise steel frames with substantial damping and seismic response reduction.

Keywords Seesaw bracings · Steel structure · Seismic performance · Base reaction · Storey displacement · Storey drift

1 Introduction

From the past years moderate to severe earthquakes have occurred at intervals of 5–10 years. Many seismic codes are based on the concept of 'accepting heavy losses to the structure during an earthquake, such that the building is prevented from collapse. Earthquakes are likely to have the following effects: a large number of

A. Anto (✉) · A. Joseph
Department of Civil Engineering, Federal Institute of Science and Technology, Angamaly, Kochi 680577, India

G. C. Marano et al. (eds.), *Proceedings of SECON'21*, Lecture Notes in Civil Engineering 171, https://doi.org/10.1007/978-3-030-80312-4_27

people will be displaced from their homes or places of work for an extended period of time; severely damaged structures will be difficult to demolish; debris recovery will be a massive burden; and the amount of repair work will be time consuming and challenging. As a result, it is desirable to design and construct the buildings in such that they either don't sustain significant damage, or can be restored easily. It can be accomplished by techniques such as seismic isolation and control [1].

After a major earthquake, some of the input seismic energy may be transferred into energy dissipation devices placed within a structure, allowing damaged components to be easily replaced or retrofitted [2]. To dissipate earthquake induced energy in elements, energy dissipation or damping systems are increasingly being used in new and retrofit building. These elements are either hysteretic (yielding steel devices) or viscoelastic or viscous [3]. Despite the fact that these components behave in a variety of ways, they are all used with the goal of limiting or preventing structural frame damage. Yielding steel devices, friction devices, viscoelastic devices and most recently, fluid viscous devices have become the most widely used devices for seismic protection of structures [4].

Based on several researchers, study on steel structures with tension only components, such as cables along with other devices or elements such as dampers, tendons, etc. is proposed with the aim of significantly reducing earthquake induced displacement and forces. The aim of the study is to compare the storey displacement, drift and base reaction of a seven storey seesaw braced and unbraced steel structures in different seismic zones of India.

2 Description of 3D Steel Structure

A 7 storey steel structure with a rectangular plan configuration of 9 × 26.2 m and storey height of 3.15 m having special structures such as shear wall and stairs are considered for the study. Figure 1 shows the beam column layout of the building. At every floor, the diaphragm action is assumed. As per IS 875 the dead and live load on the slabs are 4.0 kN/m^2 and 2.0 kN/m^2, respectively [5, 6]. The building is designed initially as typical concentrically braced frames according to IS 800 with fixed base [7]. The design seismic load is calculated using IS 1893 2000 that corresponds, soil type 3 in the four seismic zones of India.

2.1 Seesaw Braced Energy Dissipation System

Jae Do Kang and Tagawa et al. suggested a new vibration control device based on seesaw mechanism and using long rods. The proposed vibration control system comprises pin supported seesaw plates, braces and a couple of dampers [8]. The brace member comprises two tension rods, turnbuckles and a cross turnbuckle. As shown in Fig. 2. The two tension rod braces intersect at the edges of the seesaw

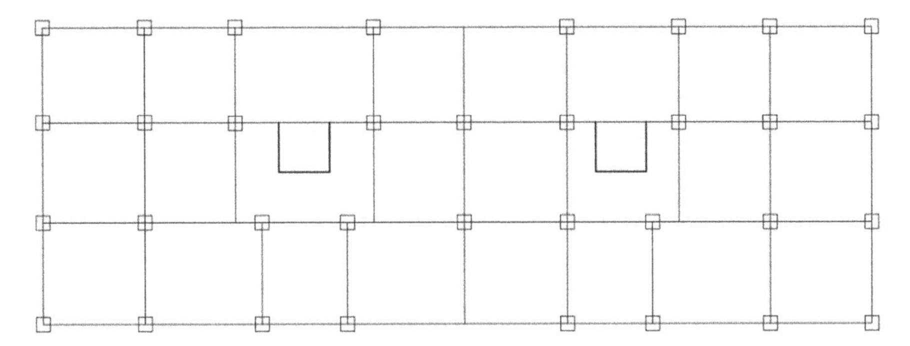

Fig. 1 Beam column layout of the building

member. The pretension is introduced in the bracing by turning up the turnbuckles, so that only tensile force occurs in the bracings during frame deformation. As a result the issue of brace buckling is minimal. Because of these advantages, the steel rods or cables can be used as bracing members between the seesaw member and the moment frame connections over some stories. To avoid the contact with the bracing members, a cross turnbuckle is provided at the intersection part [9].

The bracing members should have sufficient strength to maintain their elasticity during an earthquake. The damper undergoes shear deformations and dissipates the seismic energy as the frame deforms under the lateral load. Since the rods are attached to the ring joint, only the tensile force is generated in the rods. There are two mechanical hinges on the ring joint. As the lateral load direction reverses, the tensile axial force will immediately get generated in the opposite rod. During an earthquake, the frame top swings right and left and the bracing members deliver the

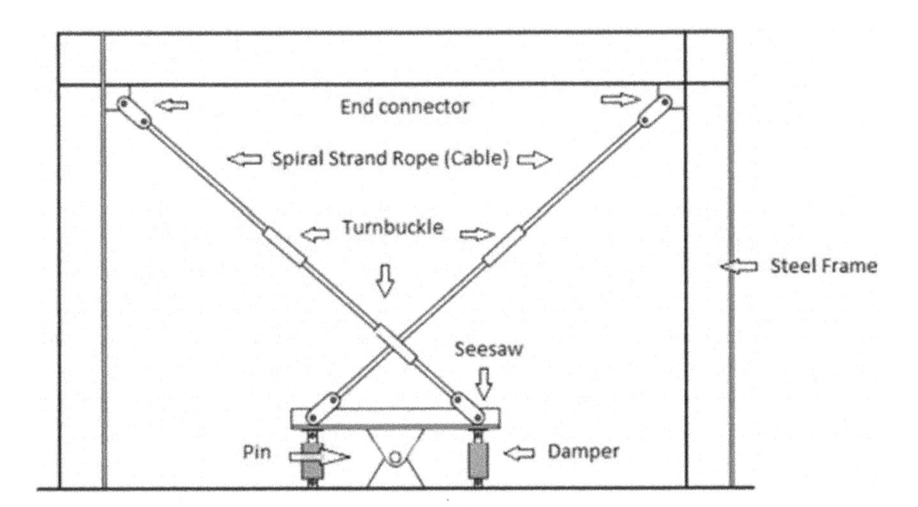

Fig. 2 A typical frame with seesaw system [10]

earthquake energy to the seesaw member. The damper deformations are forced by the seesaw member. This part works similar to a seesaw. Therefore, the term 'seesaw braced energy dissipation' is used for this system. Thus as a result of lateral load, the frame deforms and the dampers deforms in tension and compression [2, 11–14].

2.2 *Seesaw Braced Steel Structures*

In the seven storey steel structure, a typical seesaw system is installed. As shown in Fig. 3. The study considers two types of bracing configurations. The seesaw system is mounted in the mid span of the perimeter frame and the spiral strand ropes are connected at both ends of the beam, on the same floor level. In configuration I, the pretensioned spiral strand ropes are attached to the ends of the top perimeter beam, and to the top beam of the middle bay in configuration II as in Fig. 4. To avoid issues related to eccentricity, the seesaw system is centrally placed. A pretension of about 10% of the spiral strand ropes' tensile breaking strength is applied. In this system, the nonlinear viscous dampers with a damping coefficient of 250 kNs/m are vertically placed.

The vertical steel plate of the seesaw is 870 mm in height and horizontal plate is 1600 mm in length. Fe 345 steel is used for the beam, column and the steel plate. All connections are moment resisting. The slab and stair concrete grades are M25, and the shear wall is M30. The seesaw-braced steel structures are designed and analysed using SAP 2000 software [15]. Table 1 summarizes the sections for beams and columns, as well as the diameter and the minimum tensile strength of the spiral strands used in the analysis. The green inclined lines in Fig. 4, are spiral strand ropes, modeled as cable elements and dampers are modeled as link elements.

2.3 *Seismic Analysis of Seesaw Energy Dissipation System (SEDS)*

A response spectrum is the plot of earthquake related responses such as velocity, deflection, acceleration of a building for a given earthquake motion and a percentage of critical damping. In the frequency domain of the response spectrum, multiple modes of vibrations are used. The response from the design spectrum is recorded for each mode based on the modal mass and modal frequency and then they are combined to obtain the total response of the structure.

For rigid structure, the damping ratio was set to 5%, with a response reduction factor of 5 (for moment resisting frames) and an importance factor of 1 (for residential buildings). As per IS 1893: 2016, the zone factor for the respective seismic zones are considered [16]. The response spectrum for the soft soil was used in the analysis as in Fig. 5.

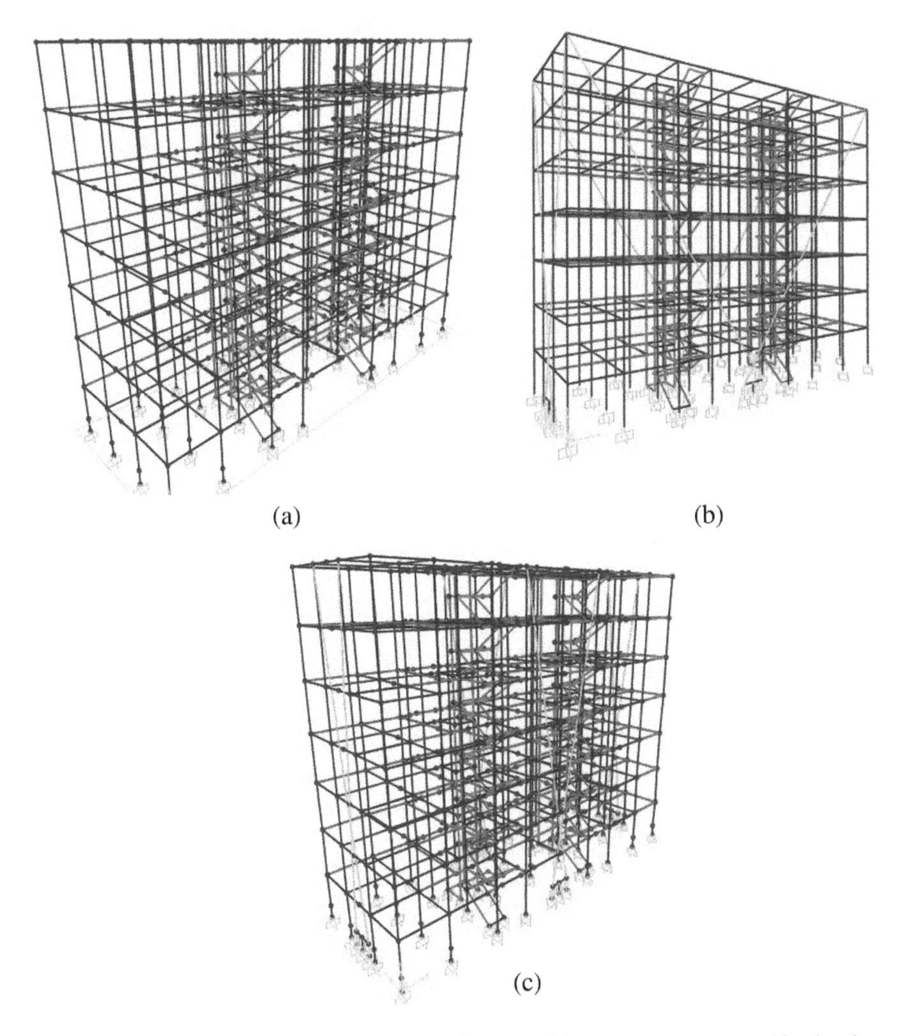

Fig. 3 Various configuration of 3D steel frames with seesaw system **a** No bracing. **b** Configuration I. **c** Configuration II

3 Response to Dynamic Load

Modal analysis is used to evaluate the natural mode shapes and time period of the steel structure during free vibration. Tables 2, 3 and 4 shows the modal properties of the first few dominant modes of vibration, i.e. the time period and the mass participation factors for the cases of unbraced and two types of bracing configuration respectively. From the tables, it can be observed that the bracing configuration I has a shorter time period. In bracing configuration I, the seesaw system with pretensioned spiral strands attached to the ends of the top perimeter beam makes the

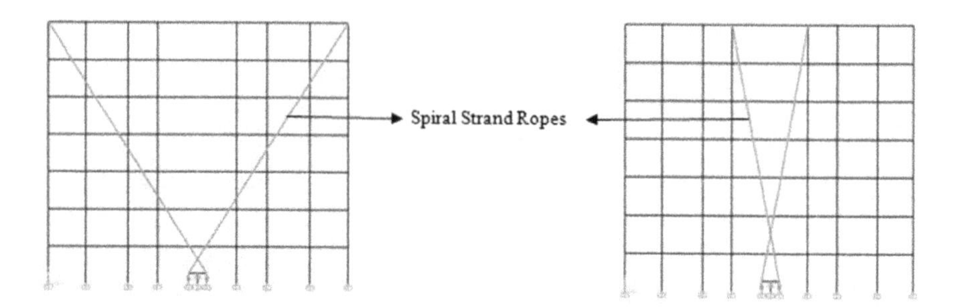

Fig. 4 Various configurations of the steel structure with seesaw system **a** Configuration I. **b** Configuration II

Table 1 Section properties of structural elements

Structure	Section
Beam	ISMB 350
Column	ISHB 400
Cable diameter	115 mm—Fig. 3b
	50 mm—Fig. 3c

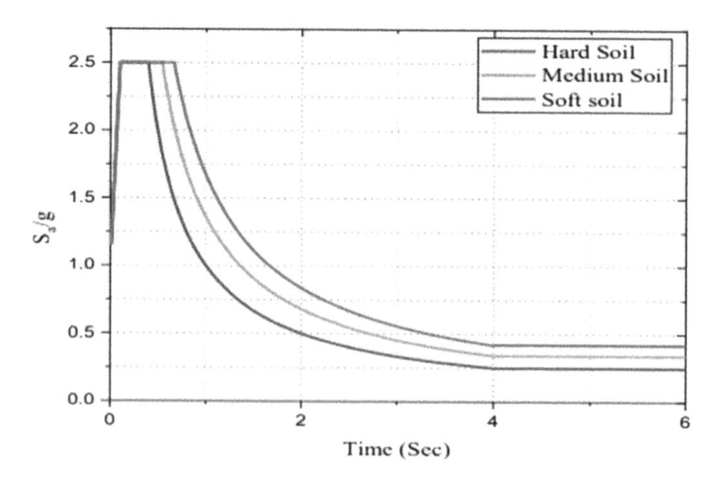

Fig. 5 Design response spectrum for different soil (5% damping) [16]

structure more stable under earthquake ground motions, which results in a shorter time period. Figure 6 shows the mode shape of steel structure with bracing configuration I for the first three dominant modes of vibration.

Table 2 Modal properties of unbraced steel structure

Mode No	Period (s)	Mass participation X	Mass participation Y	Mass participation Z
1	1.792	0.006	0.036	0.789
2	1.504	0	0.782	0.032
3	0.992	0.816	0	0

Table 3 Modal properties of steel structure with bracing configuration I

Mode No	Period (s)	Mass participation X	Mass participation Y	Mass participation Z
1	1.456	0	0.760	0.076
2	1.287	0.002	0.066	0.799
3	0.8	0.852	0	0.005

Table 4 Modal properties of steel structure with bracing configuration II

Mode No	Period (s)	Mass participation X	Mass participation Y	Mass participation Z
1	1.736	0.004	0.058	0.778
2	1.506	0	0.764	0.053
3	0.985	0.822	0	0

3.1 Base Reaction

Seismic ground motion causes lateral force at the base of the structure. The measure of the overall expected lateral force at the base of a structure is the base reaction. Figure 7 and Table 5 depicts the base reaction of the structures under consideration in various seismic zones. The maximum base reaction for unbraced steel structure is 1506.62 kN. For the seesaw braced steel structure with configuration I and II, the base reaction is 1365.61 and 1501.49 kN. The steel structure with no bracing and the steel structure with bracing configuration II have very similar base reactions. Configuration I however, indicate a substantial reduction of 9.36% in the base reaction when compared to the other two cases.

3.2 Storey Displacement

Tables 6, 7 and 8 shows the displacement of stories in different seismic zones. For the steel structure with no bracing and bracing configuration 2, the displacement is increasing from lower to upper storey. The maximum storey displacement of

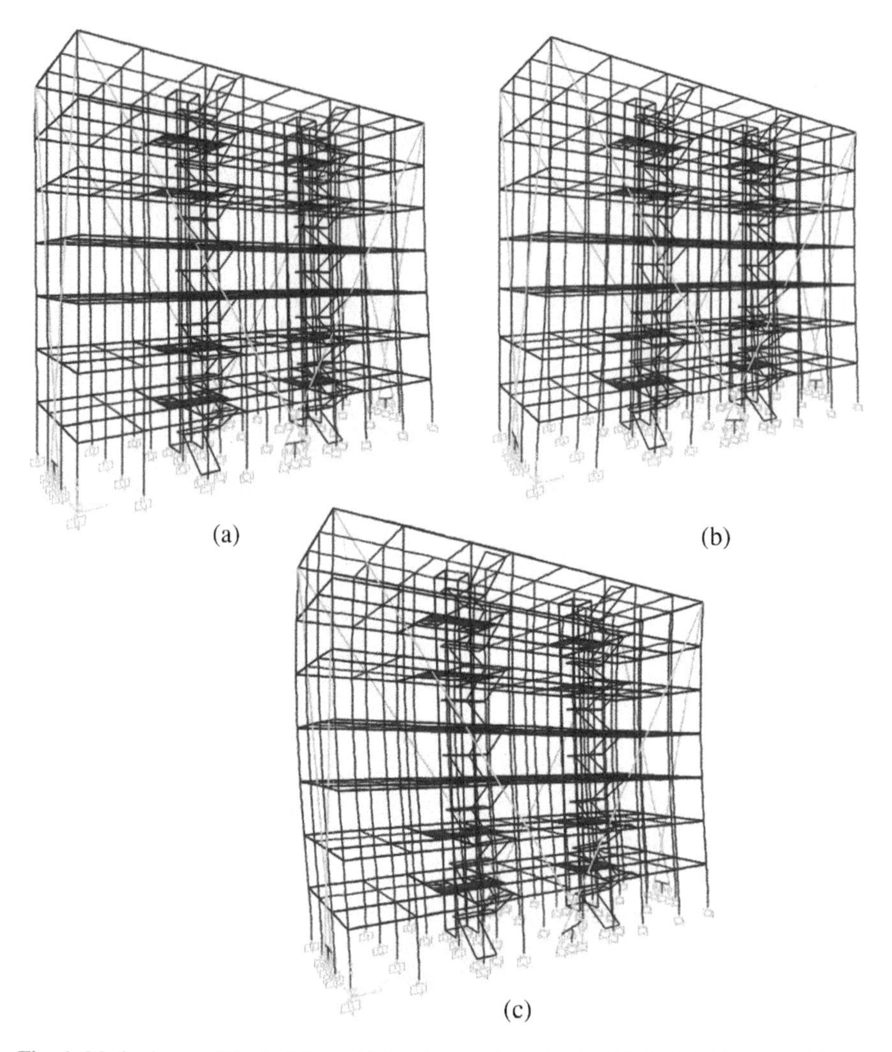

Fig. 6 Mode shapes of the building with bracing configuration I **a** Mode 1. **b** Mode 2. **c** Mode 3

seesaw braced steel structure with bracing configuration I is at the storey level six. The maximum storey displacement for unbraced steel structure is 17.99 mm in the seismic zone V, while the maximum storey displacement obtained for seesaw braced steel structures with bracing configurations I and II is 13.74 mm and 17.44 mm respectively. The storey displacement was reduced by 21.25% at the sixth storey level for bracing configuration I, and by 3.05% at the seventh storey level for bracing configuration II.

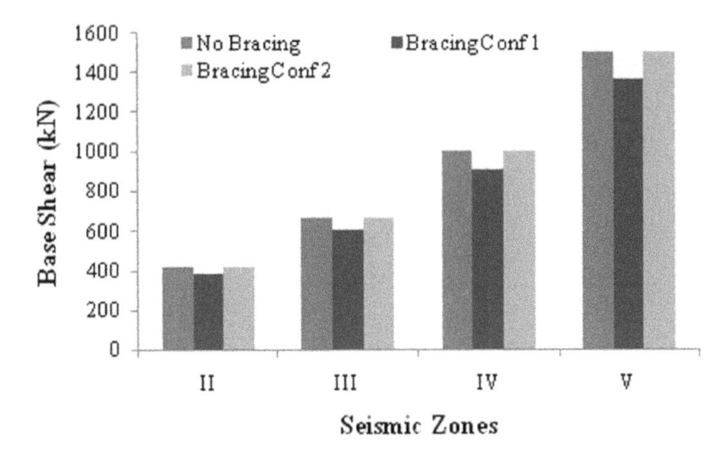

Fig. 7 Base reactions in different seismic zones

Table 5 Base Reaction in different seismic zones (kN)

Seismic zones	No bracing	Bracing configuration I	Bracing configuration II
II	418.50	379.34	417.1
III	669.61	606.94	667.33
IV	1004.41	910.41	1000.99
V	1506.62	1365.61	1501.49

Table 6 Storey Displacement of steel structure with no bracing (mm)

Storey	Zone II	Zone III	Zone IV	Zone V
7	5.02	8.03	12.05	17.99
6	4.85	7.75	11.62	17.44
5	4.39	7.03	10.55	15.83
4	3.70	5.92	8.88	13.32
3	2.80	4.47	6.71	10.07
2	1.76	2.81	4.21	6.32
1	0.69	1.11	1.66	2.49

Table 7 Storey displacement of steel structure with bracing configuration I (mm)

Storey	Zone II	Zone III	Zone IV	Zone V
7	3.60	5.76	8.64	12.97
6	3.81	6.10	9.15	13.74
5	3.73	5.98	8.97	13.46
4	3.32	5.31	7.97	11.95
3	2.60	4.17	6.25	9.38
2	1.67	2.68	4.02	6.03
1	0.66	1.07	1.60	2.40

Table 8 Storey displacement of steel structure with bracing configuration II (mm)

Storey	Zone II	Zone III	Zone IV	Zone V
7	4.98	7.98	11.96	17.44
6	4.85	7.77	11.66	17.09
5	4.45	7.12	10.69	16.03
4	3.77	6.03	9.05	13.58
3	2.86	4.58	6.88	10.32
2	1.80	2.88	4.33	6.49
1	0.70	1.13	1.69	2.54

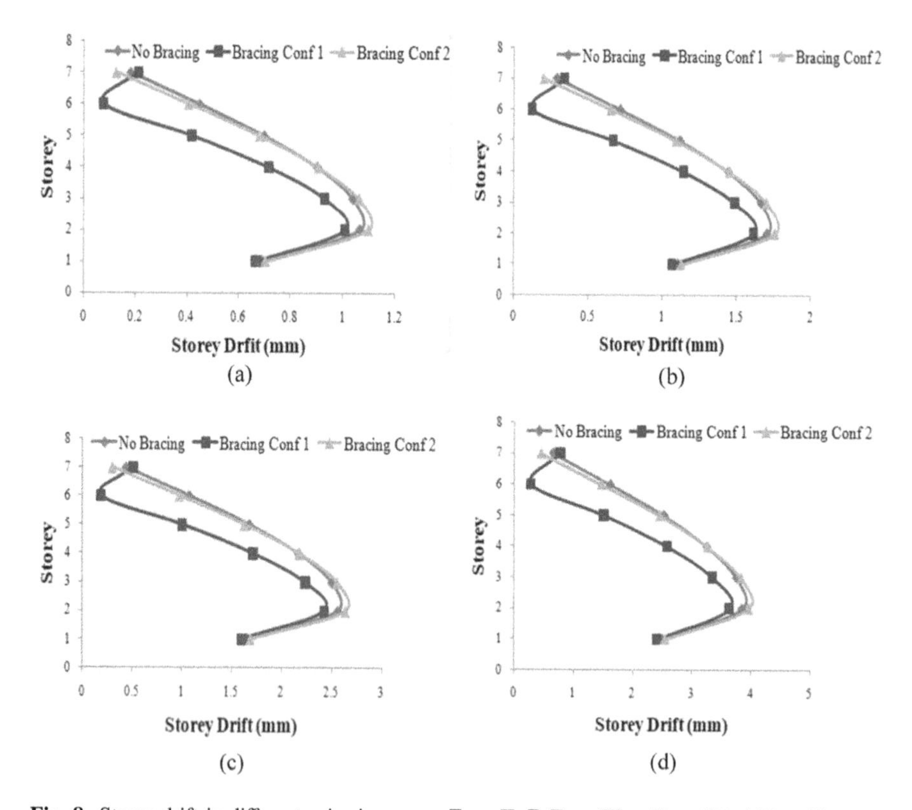

Fig. 8 Storey drift in different seismic zone **a** Zone II. **B** Zone III. **c** Zone IV. **d** Zone V

3.3 *Storey Drift*

On comparing the drift response results in Fig. 8 obtained from the response spectrum seismic analyses, taking into account the effects of seismic zones, it is clearly obtained that the seesaw equipped steel structure with spiral strand ropes attached at the perimeter beam has a substantial reduction in the storey drift. About 3–80% reduction in storey drift is observed by using the first type of bracing configuration in all the seismic zones and 2–30% reduction in storey drift is observed by using the bracing configuration II in all the seismic zones.

The maximum storey drift for unbraced steel structure is 3.84 mm in the seismic zone V, while the maximum storey drift obtained for seesaw braced steel structures with bracing configurations I and II, is 3.628 mm and 3.955 mm respectively. As per IS 1893 2002, clause 7.11.1, the storey drift in any storey due to the minimum specified design lateral force, with partial load factor of 1, shall not exceed 0.004 times the storey height such that maximum permissible storey drift in this case is 12.6 mm. From the seismic analysis performed on the steel structure with and without seesaw energy bracing configurations, it is found that the storey drift of steel structure is within the safe limit. In all the seismic zones, the steel structure with seesaw system with pretension cable attached to the centre bay and without seesaw system have similar drift concentrations.

4 Conclusion

The study presents the seismic analysis of a seesaw braced seven storey steel structure with fluid viscous dampers in different seismic zones of India. In seesaw braced steel structure, the pretensioned spiral stands are anchored to the top perimeter beam in configuration I and the spiral strands are attached to the top beam of the middle bay in the configuration II. The study compares base reaction, storey displacement and drift for the models under consideration.

Base reaction reduction is an important factor in the multistorey buildings under seismic loading. It is the estimation of total horizontal load acting on a structure. From the response spectrum analysis conducted on the steel structure, the following are noted:

- For the bracing configuration I, the base reaction got reduced by 9.36% while a reduction of 0.34% was obtained for bracing configuration 2.
- Both the unbraced steel structure and steel structure with configuration II shows similar results.
- By providing the pretensioned spiral strands at the top perimeter beam and seesaw system at the middle bay, the base reaction can be reduced and this configuration can be considered for practical purposes.

The total displacement of any storey level with respect to the ground is the storey displacement. From the present study it is observed that:

- The bracing configuration I have the maximum displacement smaller than the unbraced steel structure and steel structure with bracing configuration II.
- A reduction of 21.25% in maximum storey displacement is achieved by steel structure with configuration I.

- The maximum storey displacement is observed in top storey in unbraced and braced configuration II steel structure, while maximum storey displacement of bracing configuration I is at the storey level six.

Storey Drift is the displacement of two consecutive floors. It has a very important role in earthquake engineering. The maximum permissible storey drift is 0.004 h, where h is storey height. The following observations are found from the study:

- The comparative seismic drift response results demonstrate that, in all seismic zones, using the first type of bracing configuration reduces storey drift by about 3–80%, and using the second type of bracing configuration reduces storey drift by about 2–30%.
- The storey drift obtained in all the cases are within the permissible limit.

Therefore the seesaw system with spiral strand attached to the ends of the top perimeter beam can be considered as an alternative for ordinary steel braces such that it can avoid the buckling problem and allows a controlled seismic behavior.

References

1. Hosseini M, Fekri M, Yekrangnia M (2016) Seismic performance of an innovative structural system having seesaw motion and columns equipped with friction dampers at base level. Struct Design Tall Spec Build 25(16):842–865
2. Tagawa H et al (2016) Cyclic behavior of seesaw energy dissipation system with steel slit dampers. J Constr Steel Res 117:24–34
3. Kang J-D, Tagawa H (2013) Seismic response of steel structures with seesaw systems using viscoelastic dampers. Earthquake Eng Struct Dynam 42(5):779–794
4. Symans MD et al (2008) Energy dissipation systems for seismic applications: current practice and recent developments. J Struct Eng 134(1):3–21
5. IS 875-2 (1987) Code of practice for design loads (other than earthquake) for buildings and structures, part 2: imposed loads. Bureau of Indian Standards
6. IS 875-3 (1987) Code of practice for design loads (other than earthquake) for buildings and structures, part 3: wind loads. Bureau of Indian Standards
7. IS 800 (2007) General construction in steel-code of practice. Bureau of Indian Standards
8. Kang J-D, Tagawa H (2013) Seismic performance of steel structures with seesaw energy dissipation system using fluid viscous dampers. Eng Struct 56:431–442
9. Kang J-D, Tagawa H (2014) Experimental evaluation of dynamic characteristics of seesaw energy dissipation system for vibration control of structures. Earthquake Eng Struct Dynam 43(12):1889–1895
10. Kang J-D, Tagawa H (2016) Comparison between experimental and analytical results for seesaw energy dissipation systems using fluid viscous dampers. Earthq Eng Eng Vib 15 (1):79–90
11. Papagiannopoulos GA (2018) On the seismic behaviour of tension-only concentrically braced steel structures. Soil Dyn Earthquake Eng 115:27–35
12. Katsimpini PS, Papagiannopoulos GA, Sfakianakis MG (2018) On the seismic response and damping capacity of low-rise plane steel frames with seesaw system. Soil Dyn Earthq Eng 107:407–416

13. Katsimpini PS et al (2020) Seismic drift response of seesaw-braced and buckling-restrained braced steel structures: a comparison study. Soil Dyn Earthquake Eng 129:105925
14. Katsimpini PS et al (2020) Seismic response of low-rise 3-D steel structures equipped with the seesaw system. Soil Dyn Earthquake Eng 128:105877
15. SAP 2000 (2016) Static and dynamic finite element analysis of structures: version 20.0. Comput Struct [Berkeley, California]
16. IS 1893 (2002) Criteria for earthquake resistant design of structures. Bureau of Indian Standards

Performance of Steel Building Equipped with Crescent Shaped Bracing System under Seismic Excitation

T. P. Neethu and M. Preethi

Abstract The majority of death and building damages in natural disasters are caused by earthquake. The seismic performance of structure depends on its structural properties and ground motions. An effective lateral load resisting system should be provided in all structures to ensure safety. Recently, a new hysteretic device were developed by researchers known as Crescent Shaped Bracing (CSB) within the frame work of stiffness, strength and ductility design. CSB has a special "ad-hoc" shape which allows to choose lateral stiffness independently from the yield strength of the device. This made CSB as an alternative for the conventional energy dissipating devices. In this study, dynamic response of steel building with crescent shaped bracing system with various angles and configuration were investigated. Nonlinear time history analysis was carried out in ETABS to evaluate the performance based on maximum storey displacement and maximum storey drift.

Keywords Earthquake · Crescent shaped bracing · Steel structure · Nonlinear time history analysis · ETABS

1 Introduction

The amount of damage caused by earthquakes could increase in the future. This is not because of the increase in number and magnitude of the earthquake but because of the increase in land value, linked to a growing number of densely populated metropolitan areas and an increase in the value of goods and services in these areas. Over the third quarter of major cities in the world (cities with a population of over 10 million) exist in high-risk areas. The most effective way to prevent earthquake disaster is to use an earthquake resistant design in earthquake regions. In order to obtain structural adequacy against earthquake, steel bracings can be provided. Which will provide the required lateral strength and stiffness to the structure.

T. P. Neethu (✉) · M. Preethi
Federal Institute of Science And Technology, Angamaly, Ernakulam, India

G. C. Marano et al. (eds.), *Proceedings of SECON'21*, Lecture Notes in Civil Engineering 171, https://doi.org/10.1007/978-3-030-80312-4_28

The braces are of various types such as X bracings, V bracing, eccentric bracing, diagonal bracing. The ultimate aim of bracing is to resist tensile forces or to resist both axial compressive forces and tensile forces. The experimental studies shows that, tension compression braces provide better performance under cyclic loading such as earthquake as compared to tension only braces having negligible compressive strength.

The multiple seismic objectives in terms of ductility, stiffness and strength can be fulfilled by incorporating Crescent shaped bracing system in a structure [1]. A case study was conducted in Bologna (Italy), in which first storey seismic isolation was provided to a 5 storey structure by using crescent shaped bracing system which showed a better seismic performance [1]. CSBs has a special geometric configuration known as "ad-hoc" defined in order to provide the predefined multiple seismic objectives for a structure. Commercial steel profiles were used to made CSBs and are connects two points of a frame (i.e. two opposite corners of consecutive stories of a frame when used as diagonal braces). CSBs are characterized by an initial ductile and final hardening behaviour. The initial ductility capacity of CSB lies within 3–3.5. The hardening behaviour prevents the large displacements that may cause the collapse of structure due to P-D effect [2]. The use of CSB as diagonal braces in a structure will be an alternative to the conventional hysteretic devices like Scorpion YBS devices and buckling restrained braces [2]. Experimental studies shows that the better performance is obtained by CSBs made form a unique element such as from laser cut manufacturing technique, if it is not available welding can be used. But the welding point shouldn't be at the knee sections and ends of CSB in order to prevent sudden premature fragile failure [3].

Figure 1 shows a frame with CSB placed diagonally. In which the limb length is kept equal. And θ represents the angle between CSB and diagonal of the frame. In this study angle is varied from 5° to 30° to find the seismic performance. Also, the performance of various CSB configurations in steel building were also studied. Seismic performance of single crescent shaped bracing system and double crescent shaped bracing system in RCC building shows that the CSBs are capable to resist earthquake when the structures having different seismic demands and are of different occupancy level [4, 5]. Some of the analytical studies shows that the seismic performance of CSBs were more efficient than K and V type bracings [6].

2 Methodology

A 3D steel building were designed by using ETABS 2018 version with and without having crescent shaped bracings. The best suitable angle for crescent shaped bracing were determined. Also the effectiveness various configuration including single crescent shaped bracing system and double crescent shaped bracing system against earthquake were determined. The seismic behaviour of the structure were performed by a non-linear dynamic analysis known as time history analysis. The result parameters, maximum storey displacement and maximum storey drift are

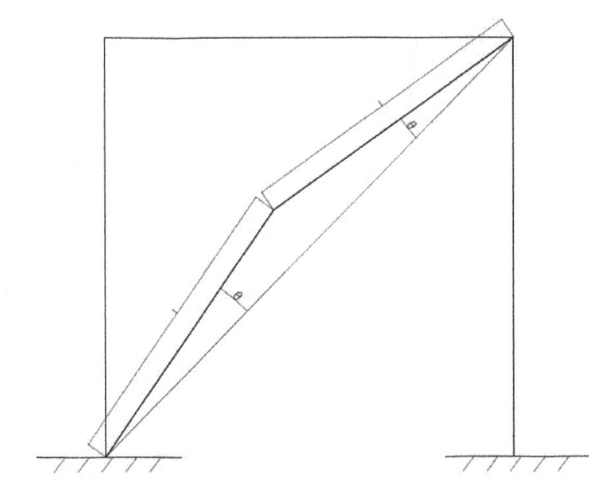

Fig. 1 Frame with diagonal CSB

evaluated to arrive conclusions. El Centro California was used as the function data in order to perform time history analysis.

3 Modelling of Structure

A G + 15 storey steel structure having a plan dimension 18 m × 9 m were modelled. The number of bays were 6 and 3 in X and Y directions respectively. The height of all the stories were kept as 3 m. The beams, columns and bracing were chosen as ISMB 450, ISWB 600 and ISMB 175. Structural steel of Fe345 and concrete of grade M25 were used for the structural members. Steel deck thickness is chosen as 100 mm. Models without CSB and same building layout having different CSB angles were modelled. CSB angles 5°, 10°, 15°, 20°, 25° and 30° were modelled to determine the best suitable angle. The equal limb length were considered for bracings. Figure 2 shows the 3D model of steel structure with and without CSB modelled for various angle study.

To find the seismic behaviour of CSB, the structure is modelled by using the best suitable angle. Two configurations from single crescent shaped bracing system (SCSB) and double crescent shaped bracing (DCSB) system were modelled. Figure 3 shows the elevation of various CSB configurations created for analysis.

3.1 Loading

Live load on the structure is applied as per IS: 875 (part II) 1987. A live load of 3 kN/m^2 and 1.5 kN/m^2 is applied on the floors and top storey respectively. The dead load is software assigned. Floor finish is chosen as 1 kN/m^2. Wall load is calculated as

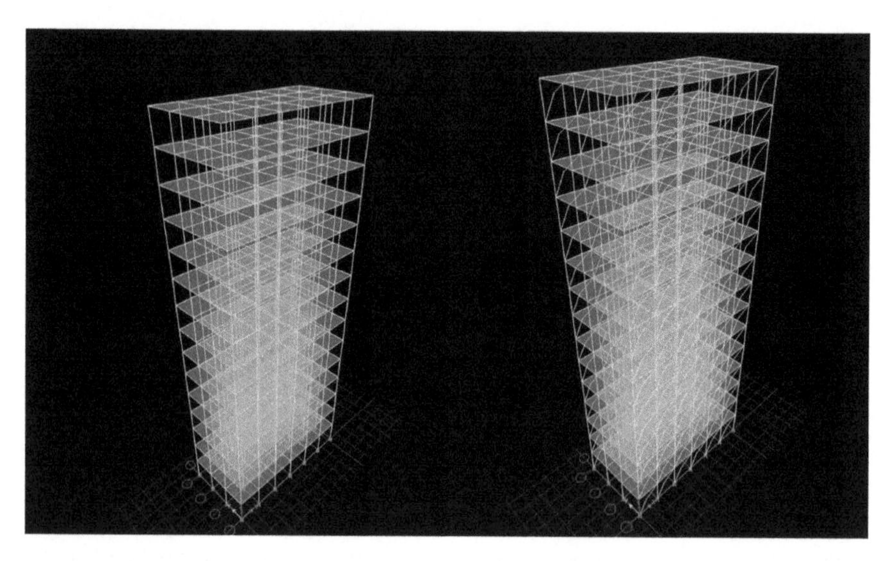

Fig. 2 3D model of bare frame and frame with CSB

13.8 kN/m and 6.9 kN/m on floors and top storey respectively. Seismic loads are applied in X and Y directions as per IS 1893 (Part I) 2016. For the design earthquake load computation zone factor of 0.16, hard soil, importance factor of 1 and response reduction factor of 5 were chosen. The support conditions are assumed as fixed.

4 Analysis and Discussion

Maximum storey displacement and maximum storey drift were determined by using time history analysis at different angle for CSB. Figures 4 and 5 shows the maximum storey displacement and maximum storey drift obtained along longer direction of bare frame and CSB with various angles.

Maximum storey displacement obtained for the bare frame is 4.052 mm. Except the angle 30° the maximum storey displacement obtained is less than bare frame.

Maximum storey drift of the frame is obtained as 0.000135. The maximum storey drift value found to be equal to bare frame model when the angle is 30°.

Maximum storey displacement and maximum storey drift is found to be the least for CSB angle 15°. Figure 6 shows the Maximum sorey displacement and maximum storey drift plot obtained after analysis in ETABS. For angle 15° maximum storey displacement and maximum storey drift obtained after analysis are 3.03 mm and 0.000081 respectively.

The best suitable angle for CSB was found as 15°. By using this angle four various configurations SCSB-1, SCSB-2, DCSB-1 and DCSB-2 were modelled. Figures 7 and 8 shows the maximum storey displacement and maximum storey drift

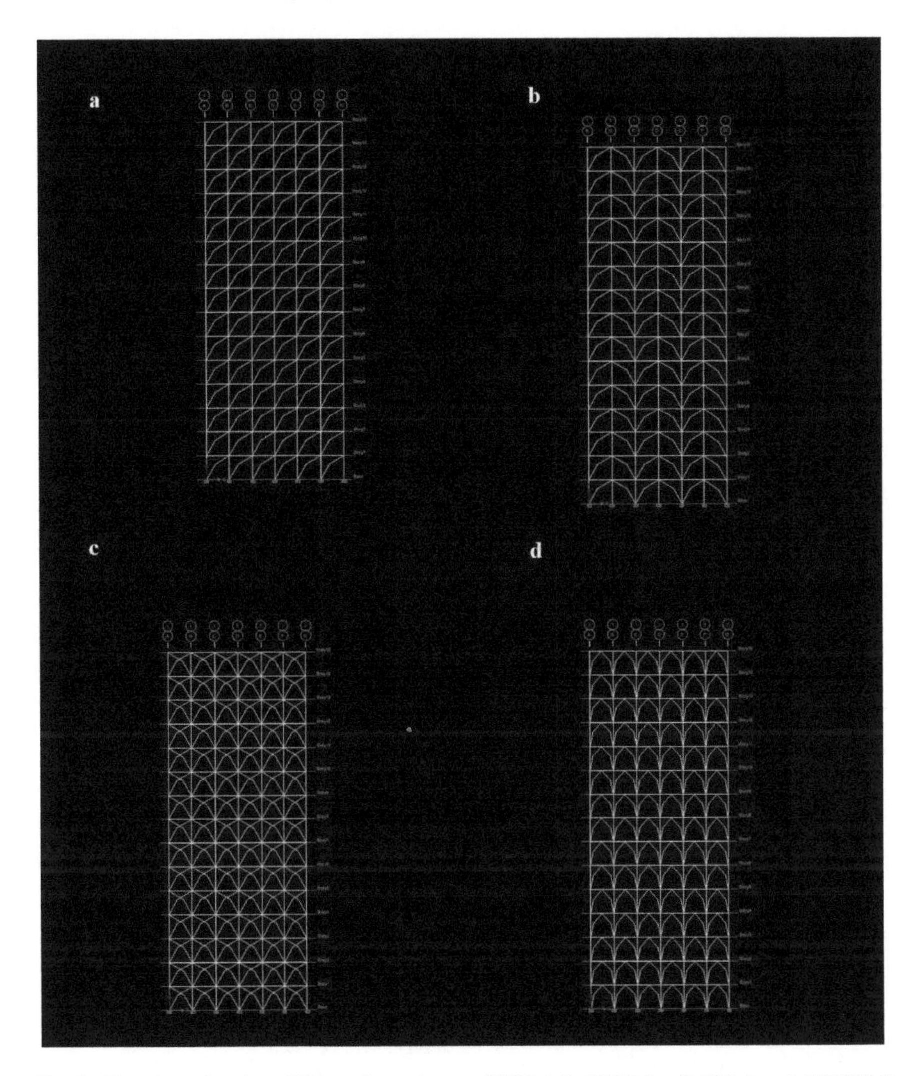

Fig. 3 Elevation of various CSB configurations: **a** SCSB-1, **b** SCSB-2, **c** DCSB-1 and **d** DCSB-2

obtained along the longer direction for the bare frame and for various CSB configurations.

Maximum storey displacement and drift is decreased for all 4 configurations. Figure 9 shows the maximum storey displacement and maximum storey drift plots obtained after analysis for DCSB-2. The maximum displacement and drift were found to be the least for model DCSB-2 and the values are 2.53 mm and 0.000067 respectively.

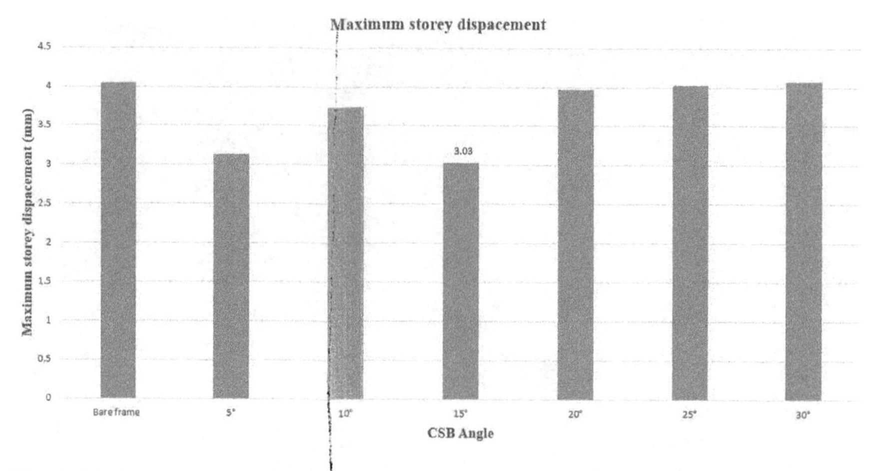

Fig. 4 Maximum storey displacement for models with various CSB angles

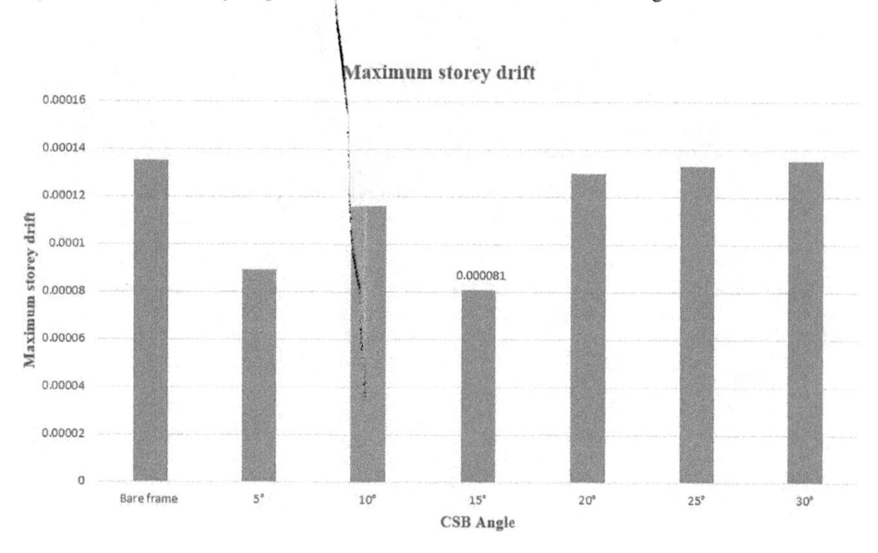

Fig. 5 Maximum storey drift for models with various CSB angles

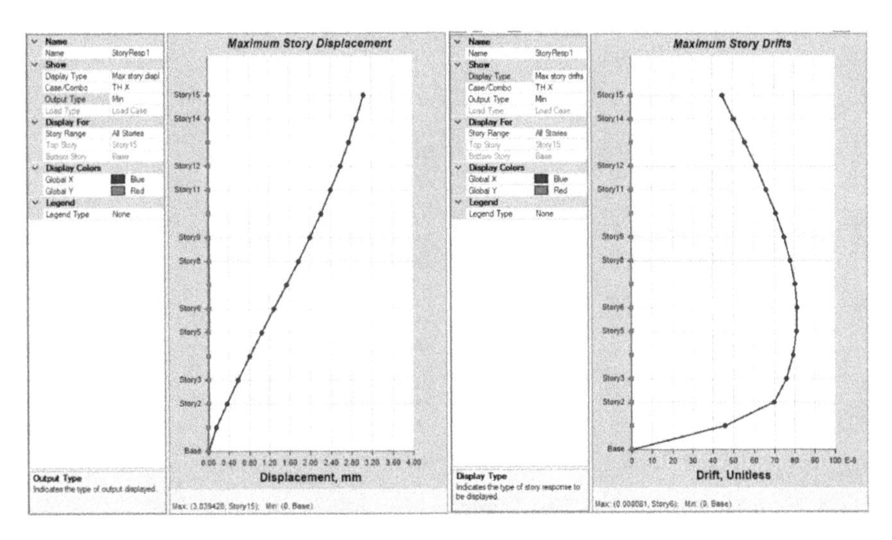

Fig. 6 Maximum storey displacement and maximum storey drift plot for angle 15°

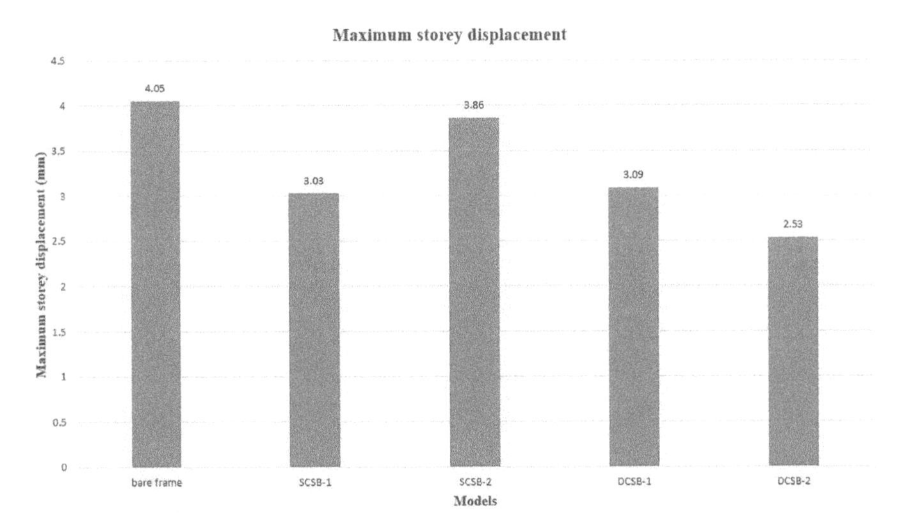

Fig. 7 Maximum storey displacement for models with various CSB configurations

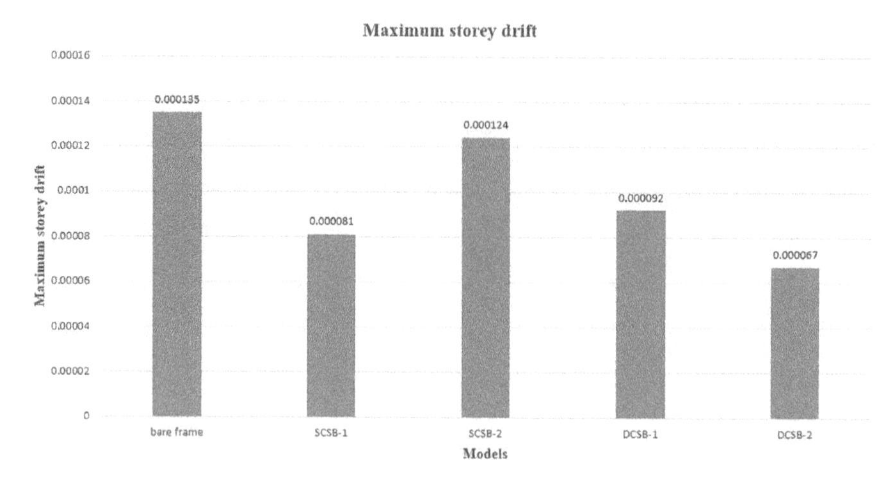

Fig. 8 Maximum storey drift for models with various CSB configurations

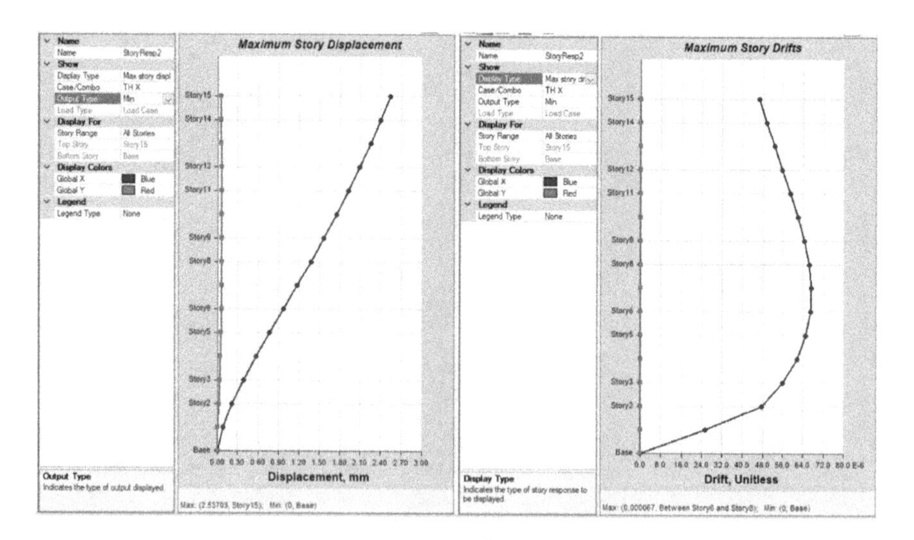

Fig. 9 Maximum storey displacement and maximum storey drift for configuration DCSB-2

5 Conclusion

Based on the study of seismic performance of G + 15 storey steel building equipped with CSB, following conclusions can be drawn:

- The earthquake resistance of the structure can be improved by adding CSB system.

- The best suitable angle for CSB system with equal limb length is found to be 15°. 25% and 40% reduction in maximum storey displacement and maximum storey drift were obtained.
- When angle is increased beyond 15°, maximum storey displacement and drift values slightly increases and for 30° the obtained maximum storey displacement and drift is greater than bare frame model. So angle above 30° is not suitable for CSB system.
- For single crescent shaped bracing system, SCSB-1 shows better performance than SCSB-2 configuration. The obtained maximum storey displacement and drift for SCSB-1 is 22% and 35% less than that of SCSB-2 model.
- For double crescent shaped bracing system, DCSB-2 shows better performance than DCSB-1 configuration. The obtained maximum storey displacement and drift for DCSB-2 is 18% and 27% less than that of DCSB-1 model. Overall decrease in maximum storey displacement and drift for DCSB-2 is 38% and 50% when compared to bare frame model.

Results shows significant similarity with the work done by Pallavi Partani [5]. In which 12–14% reduction in maximum storey displacement and 20% reduction in maximum storey drift were obtained when a ground soft storey concrete building is strengthened by CSB.

References

1. Trombetti T, Silvestri S, Gasparini G, Ricci I (2009) Stiffness-strength-ductility-design approaches for crescent shaped braces. Open Constr Build Technol J 3:127–140
2. Palermo M, Silvestri S, Gasparini G, Trombetti T (2015) Crescent shaped braces for the seismic design of building structures. Mater Struct 48(5):1485–1502 (Springer)
3. Palermo M, Pieraccini L, Dib A, Silvestri S, Trombetti T (2017) Experimental tests on Crescent Shaped Braces hysteretic device. Eng Struct 144:185–200 (Elsevier)
4. Kammouh O, Silvestri S, Palermo M, Cimellaro GP (2018) Performance-based seismic design of multistory frame structures equipped with crescent-shaped brace. Struct Control Health Monit 25(2):e2079
5. Pallavi P, Roshni J (2019) Seismic analysis of multi-storied rcc building with crescent shaped bracing in ground soft storey. Int J Eng Appl Sci Technol 4(3)
6. Hari Narayanan MA, Gopika M (2020) Seismic analysis of a high rise building provided with crescent shaped braces. Int Res J Eng Technol (IRJET) 07(03)

Numerical Study on Web Crippling of Aluminium Lipped Channel Sections

Winny Balan and R. Miji Cherian

Abstract Aluminium alloys are nowadays used abundantly in the building construction due to its advantages such as lightness, resistance to corrosion, ease of manufacture and more strength to weight ratio. Aluminium sections are used in roof structures, curtain wall and other structural applications. Thin-walled structural members are extremely vulnerable to buckling instabilities, including web crippling, which is one of the failure that occur in Aluminium sections under transverse concentrated loads or reactions. Aluminium members are highly prone to this kind of failure because of their low elastic modulus. This paper presents the details of a numerical investigation conducted on the web crippling of fastened Aluminium lipped channel (ALC) sections. Furthermore, the effect of connection contacts and the effect of various frictional coefficients on the web crippling behaviour are also included in this investigation. From the study it is observed that modelling of connection contact plays a significant role in the numerical study of web crippling behaviour. At the same time it is realised that the change in frictional coefficients does not considerably influence the force displacement relationship.

Keywords Aluminium lipped channel sections · Restrained flanges · Frictional coefficients · Web crippling

1 Introduction

Over the last two decades, the use of Aluminium sections in structural industries has been enlarged due to the advanced mechanical properties and growing demand for thin walled structural member. Aluminium can be used as a structural material because of its greater strength to weight ratio compared to steel. In construction industries, Aluminium members are used as load carrying elements such as purlins, joists, rafters and columns. Aluminium alloys are divided into 2 types, Wrought

W. Balan (✉) · R. Miji Cherian
Department of Civil Engineering, Government Engineering College Trichur, Thrissur, Kerala 680009, India

G. C. Marano et al. (eds.), *Proceedings of SECON'21*, Lecture Notes in Civil Engineering 171, https://doi.org/10.1007/978-3-030-80312-4_29

alloys and Cast alloys. Aluminium member have large variety of advantages like high strength to unit weight ratio, excellent corrosion resistance, light in weight and high workability, easy to handle, transport and erect, good appearance, require only less maintenance, connections may be riveting, bolting or welding and also have some disadvantages like high initial cost, non availability, lower modulus of elasticity which results in stiffness reduction, more liable to buckling, large factor of safety compared to that of steel, high coefficient of thermal expansion. Lipped and unlipped channels are used for the construction of thin structural member. Aluminium members are subjected to different modes of failure such as shear failure, bending failure and crippling failure on web. When subjected to concentrated forces and reactions under different loading conditions, thin-walled Aluminium members undergo web crippling failures. The loading conditions are in 4 types, based on the position of load or reaction through one flange or both flanges. These are End one flange (EOF), End two flange (ETF), Interior one flange (IOF) and Interior two flange conditions (ITF) [1].

Gunalan and Mahendran [2] conducted an experimental study on the web crippling performance and strength of unlipped steel channels under EOF and IOF loading. A detailed comparison of current and previous test results [3] with founded by equations in AS/NZS 4600, AISI S909 was done by the researchers. It concluded that equations are un-conservative for those sections under EOF and IOF loading. Hence new design equations were proposed to accurately predict the crippling load. Alsanat et al. [4] carried out an experiment to test the ALC sections subjected to web crippling under the ETF and ITF loading. The results then compared with the nominal web crippling capacity predicted using the equations provided by the Australian, American and European Standards. The comparison shows that equations are unsafe and unreliable to calculate the capacity for ALC sections under both loading. Alsanat et al. [5] conducted a numerical investigation of ALC sections subjected to web crippling under two loading. The non linear FE analysis using ABAQUS was carried out. Using the validated numerical model, the parametric study was performed. Zhou et al. [6] investigated the web crippling of Aluminium plane and lipped sections with bolted flange. The experimental results and the capacity predicted numerical analysis were compared with the design strengths. American, Australian/New Zealand and European specifications design rules were used for calculating the design strength. The researchers found that, the codes are generally un-conservative for IOF and EOF and conservative for ETF and ITF. Unreliable for both one and two flanges conditions. Alsanat et al. [1] conducted experimental study on fastened ALC Sections Subjected to web crippling. On comparing the experimental results with the guidelines taken from AS/NZS 1664.1, AS/NZS 4600 and Eurocode 3 are mostly un-conservative and unreliable. Finally the equations were modified to calculate the ultimate crippling load of sections by the researchers.

The current study aims to develop a non linear finite element model for predicting web crippling performance of ALC section under two flange conditions. Using the experimental results reported in Alsanat et al. [1], the model was validated. The influence of different type of contact conditions and frictional coefficients of the contact surface during loading, on crippling load of ALC sections are also to be investigated.

2 Methodology

2.1 Experimental Setup

Experimental study conducted on the Aluminium Channel section is selected for the numerical investigation by Alsanat et al. [1]. The sections were rolled formed using the Aluminium grade 5052H36 [5]. Figure 1a shows the test set up of fastened sections under the ITF conditions. Figure 1b shows the section sketch of a ALC section [1]. The flanges and the bearing plates were connected using M12 bolt in two flanges. Table 1 gives the properties of materials and the specimen details of the section.

2.2 Numerical Study

ANSYS software was used to simulate the lipped channel sections subjected to crippling on web. The dimensions and the properties of material of section were taken from the experiment given in Sect. 2.1 for modelling the section.

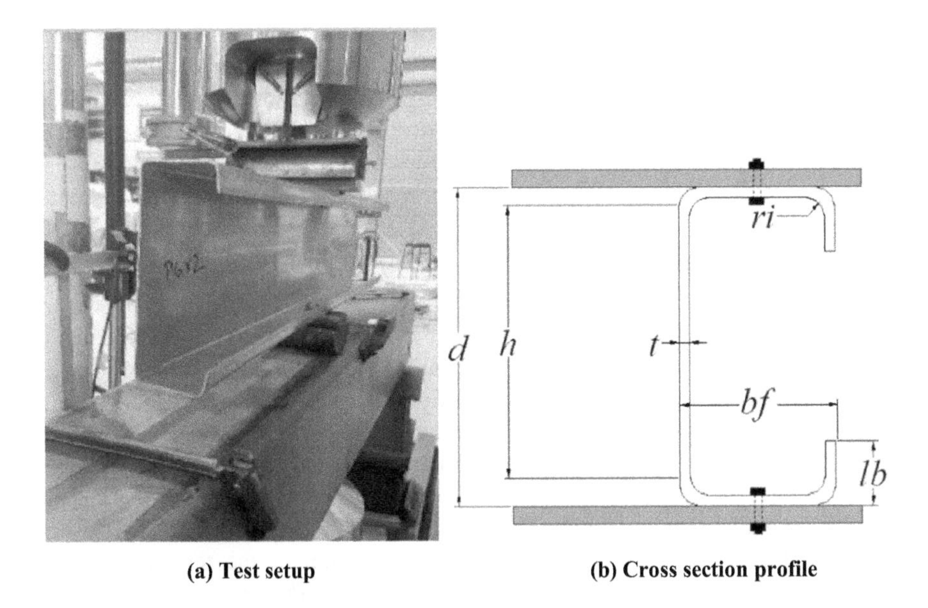

(a) Test setup **(b) Cross section profile**

Fig. 1 Experimental set-up and section profile [1]

Table 1 Material properties and specimen details [1]

Aluminium alloy	5052H36	Depth (d)	156.7 mm
Modulus of elasticity (E)	63.55 GPa	Thickness (t)	2.93 mm
Poisson's ratio (μ)	0.33	Radius (r_i)	4.9 mm
Yield strength (f_y)	206	Flange (b_f)	62.5 mm
Ultimate stress (f_u)	248	Lip (l_b)	22.8 mm
Ultimate strain (ε_u)	5.55%	Length (L)	774 mm

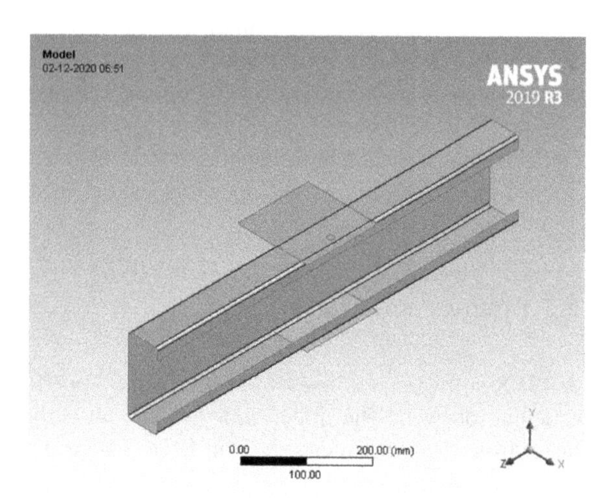

Fig. 2 Model of the section

All the sections were modelled using SHELL181 finite elements. Bearing plates were modelled as shell elements. A mesh size of the beam was 10 × 10 mm. CONTA174 and TARGE170 elements were used for modelling the interface between the flanges and the bearing plates. Figures 2 and 3 show the model and the meshing of the section. Figure 4 given the boundary conditions and loading on the section.

2.3 *Grid Independence Study*

Grid independence study was performed to remove/decrease the influence of the number of grid size on the computational results. The grid size was reduced from 50 to 10 mm and the corresponding forces were calculated. The numerical results found to approach the experimental results as the mesh size becomes finer. Details of grid independence study conducted for numerical study is shown in Fig. 5.

From Fig. 5 it was found that the numerical results approach to the experimental results for a grid size of 10 mm. Hence, for further study, 10 mm mesh size was adopted. Figure 6a shows the crippling failure region of the test specimen in experimental set-up and Fig. 6b shows the crippling failure regions in finite element model.

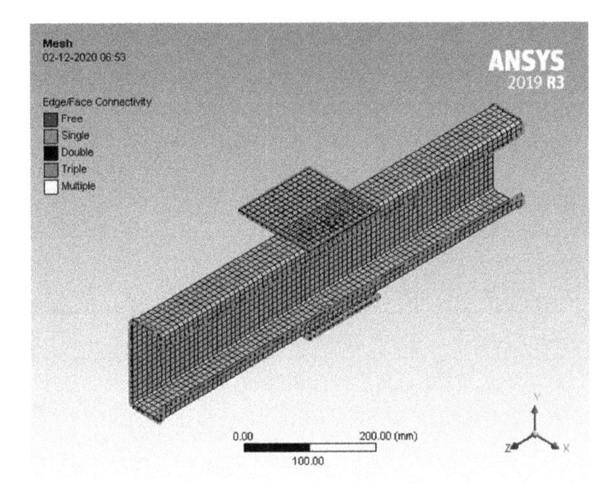

Fig. 3 Meshing of the section

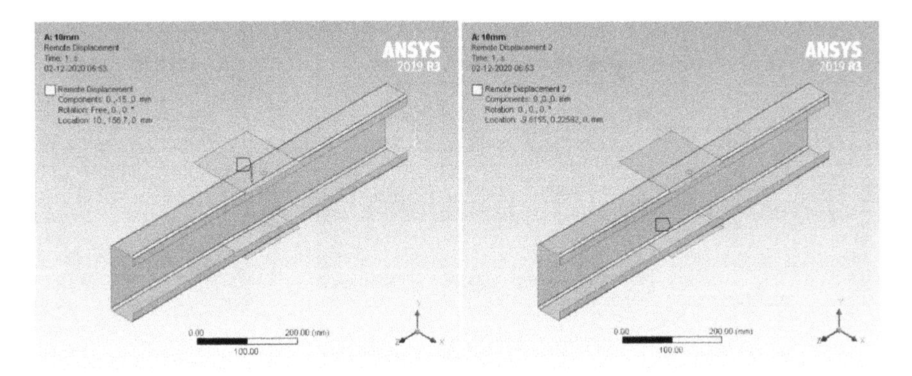

(a) Top flange and bearing plate **(b) Bottom flange and bearing plate**

Fig. 4 Loading and boundary condition

3 Effect of Different Contacts Options Available in ANSYS Software in Web Crippling Capacity

Selection of the proper contact type from the available options of the software highly influences the accuracy of the result. It depends on the physical setup during the experiment. Five types of contacts are available in ANSYS software and they are bonded contact, No separation contact, Frictionless contact, Rough contact and Frictional contact.

In bonded contact, there is no sliding or separation between faces/edges is allowed. The region is assumed as glued. This type of contact length/area will not

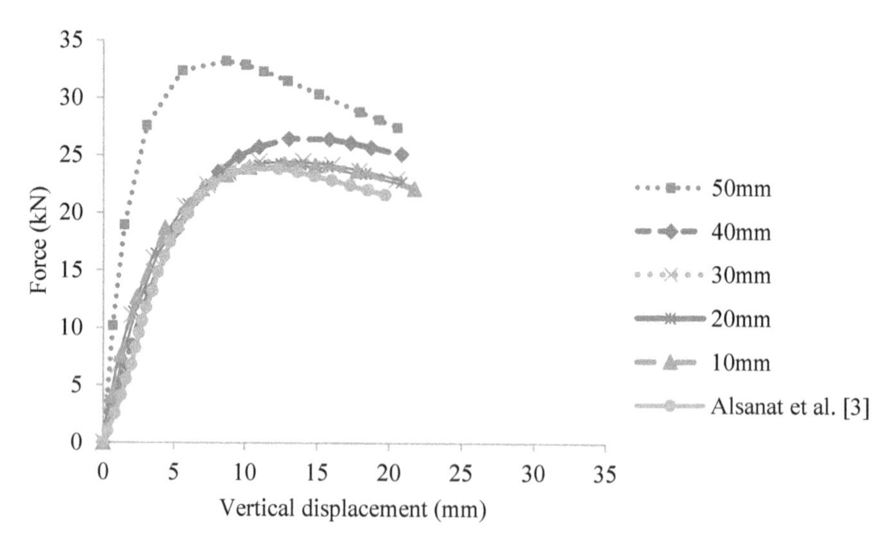

Fig. 5 Force versus vertical displacement graph of grid independence study

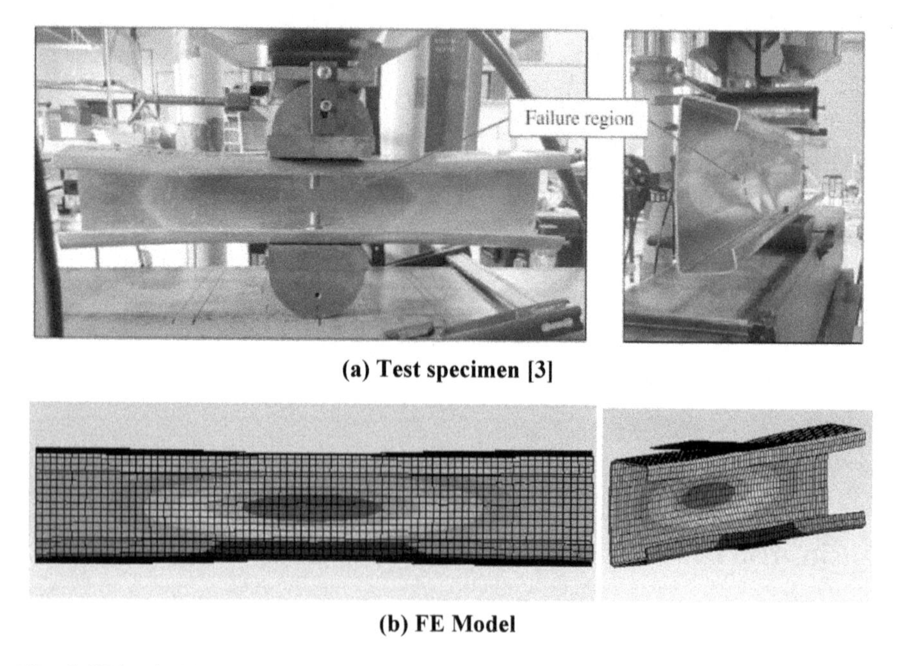

(a) Test specimen [3]

(b) FE Model

Fig. 6 Web crippling failure under ITF loading

vary during the load application. No Separation contact is similar to the bonded case. It not allowed the separation of the geometries. In Frictionless contact, when separation is occurs, the normal pressure become zero. This make gaps between

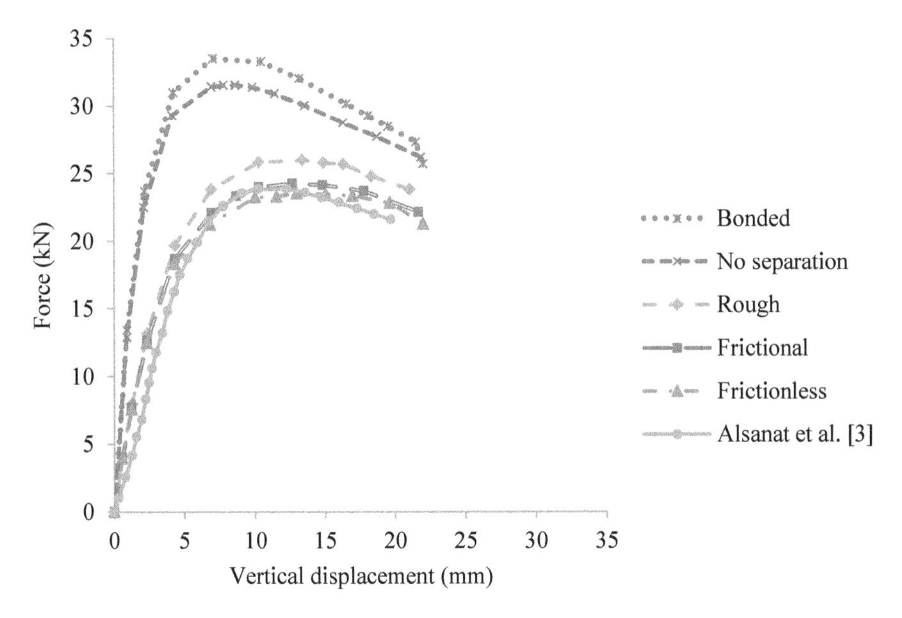

Fig. 7 Effect of contacts in web crippling capacity with different contacts

bodies depending on the loading. It most suitable for solution is nonlinear because the area of contact may vary as the load is applied. Rough contact is similar to the frictionless contact. Here no sliding is allowed. In Frictional contact, before they start sliding relative to each other, the contacting geometries can take shear stresses up to a certain magnitude.

For studying the influence of the variety of contact surface on the crippling load, the contact type was changed one by one keeping all other parameters as constant. The results of the effect of contacts in the web crippling capacity are shown in Fig. 7. From Fig. 7 it is understandable that the type of contact area has high influence on the prediction of web crippling load. For the experimental study under consideration frictional contact is found to be most suitable for modelling the section.

4 Effect of Frictional Contact by Changing the Frictional Coefficients in Web Crippling Capacity

The roughness of contact surface is another factor which is likely to influence the crippling load. The study was conducted to determine the influence of different frictional coefficients in the web crippling of ALC section under interior two flange conditions, keeping all other parameters as constant. For this study, the frictional contact is used to connect the flanges and the bearing plates. The frictional coefficient values selected are 0.4, 0.3, 0.2, 0.1 and 0.

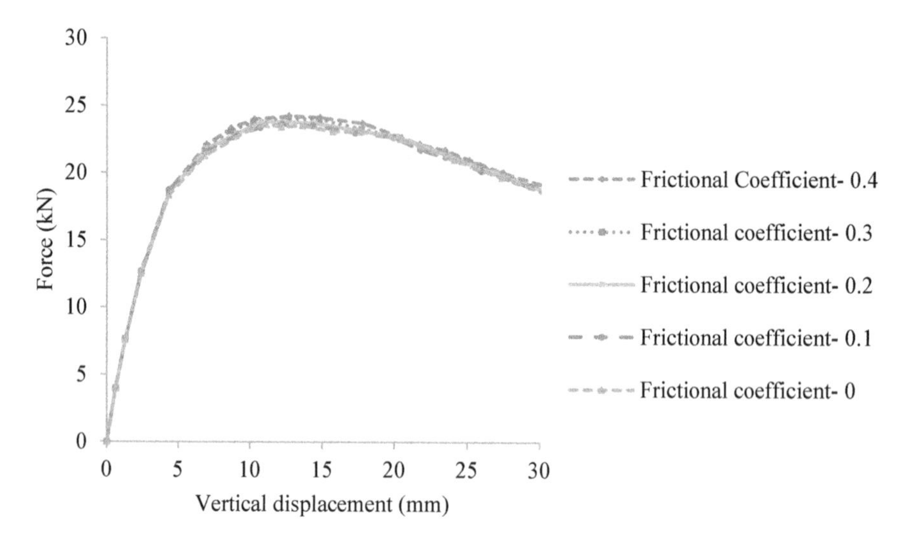

Fig. 8 Effect of frictional contacts in the web crippling capacity with different frictional coefficients

From this study it was found that the frictional coefficient has negligible influence in predicting the crippling capacity. Details of displacement and corresponding force reactions are shown in Fig. 8. From Fig. 8 it can be clearly identify that the variation of the frictional coefficient from 0.4 to 0 does not changed the crippling load significantly. This is due to the reason that as the loading is vertical, there is no relative horizontal movement between the contact surfaces in this case and hence, negligible influence for the coefficient of friction.

5 Conclusions

A numerical investigation was conducted on ALC sections subjected to web crippling under loading conditions. The model was developed in ANSYS software and validated with the experimental result in Alsanat et al. [1]. Furthermore the effect of different contacts and frictional coefficients on crippling capacity of Aluminium sections was studied and discussed in detail. From this study, it was found that, from the available contact surface models, frictional contact is most suitable for modelling the section of the experimental study under consideration. The developed model predicted the failure behaviour and the crippling load accurately. At the same time from the current study it was found that the frictional coefficient has negligible influence on predicting the web crippling load for the study under consideration, as there is no considerable relative movement between the contact surfaces.

References

1. Alsanat H, Gunalan S, Poologanathan K, Guan H, Baniotopoulos C (2020) Fastened aluminum-lipped channel sections subjected to web crippling under two-flange loading conditions: experimental study. J Struct Eng 146(4):04020023
2. Gunalan S, Mahendran M (2019) Experimental study of unlipped channel beams subject to web crippling under one flange load cases. Adv Steel Constr 15(2):165–172
3. Young B, Hancock GJ (2001) Design of cold-formed channels subjected to web crippling. J Struct Eng 127:1137–1144
4. Alsanat H, Gunalan S, Guan H, Keerthan P, Bull J (2019) Experimental study of aluminium lipped channel sections subjected to web crippling under two flange load cases. Thin Walled Struct 141:460–476
5. Alsanat H, Gunalan S, Keerthan P, Guan H, Tsavdaridis KD (2019b) Web crippling behaviour and design of aluminium lipped channel sections under two flange loading conditions. Thin Walled Struct 144:106265
6. Zhou F, Young B (2020) Web crippling of aluminium alloy channel sections with flanges restrained. Thin Walled Struct 148:106576

Parametric Study of High Rise Building with Skybridge

Christy Baby and Jiji Antony

Abstract The limited land availability in populated cities give rise to build, high rise buildings in relatively near as linked building system which consist of several structures connected by skypools, skygardens, skybridge. Skybridge is elevated walkway consisting of an enclosed bridge between two or more buildings. Skybridge provides additional escape routes, horizontal evacuation at height, convenience for communication under emergencies and fire. The parameters such as storey drift, displacement, and storey shear that influence response of high rise building with skybridge, under the seismic load were studied. And also extends to compare its influence by varying the skybridge connection type (fixed, hinge and roller) and effect of skybridge at various floor level. The results shows skybridge located at three quarter height and fixed connection have better performance. Analysis is done using ETABS Software.

Keywords Linked building system · Skybridge · Storey drift · Storey displacement · Storey shear

1 Introduction

Now a days in populated cities, high rise buildings have been built adjacent because of insufficient available land and are designed as linked building. Several examples are there such as Island tower sky club, The Gate of the Orient in Suzho, Saudi Arabia. More than just pedestrian link bridges, skybridges relieve ground—level congestion and evacuation at height when fire or other emergencies occur.

Many researches have studied the dynamic response behavior under lateral loading conditions. Imad Shakir et al. [1] studied the influence of skybridge on seismic responses and suggested that structural responses reduced with skybridge compared to building without skybridge. This is because the additional link stiffness due to the skybridge which helps to resist the seismic loads. Sayed [2] studied

C. Baby (✉) · J. Antony
Department of Civil Engineering, FISAT, Angamaly, Kerala, India

G. C. Marano et al. (eds.), *Proceedings of SECON'21*, Lecture Notes in Civil Engineering 171, https://doi.org/10.1007/978-3-030-80312-4_30

the influence of location of skybridge on the seismic response and suggested that top location is the most unfavorable location. The displacement was less in top location but contrary base shear force is more. Sun et al. [3] showed that the connecting parameters of skybridge interactively affect response of two towers. John and McCafferty [4] provides survey of various typologies of sky bridges, introduces key considerations for their design. Song et al. [5] studied the correlation between wind forces on linked building by changing gap distances. Along cross wind force, gap ratio 0.5 have more negative correlation, that decrease wind responses. Kim et al. [6] studied the correlation between wind forces applying on the linked building systems by changing gap ratio between the building and suggested gap ratio 0.5. Mccall et al. [7] studied about connection of skybridge and suggest roller connected allow buildings to sway independently, Fixed constrain them to deflect as cantilever unit and hinge connection constrain to sway unison.

The aim of this study is to find out the influence of skybridge connection type (roller, hinge and fixed) and location of skybridge at various floor level on seismic performance high rise building with skybridge.

2 Modelling and Analysis

2.1 Building Configuration

Linked building system having two analogous towers of G+35 storey of plan dimension 30 m × 30 m is considered. Structural elements dimension and plan view are shown in Table 1 and Fig. 2. Storey height is taken as 3 m. M30 concrete and Fe415 steel is taken. Slab thickness is 120 mm. Two building were connected by a 15 m long skybridge with 18 m width. The skybridge is modelled with ISWB600 steel section for beam. In order to find out the influence of location of skybridge, 5 cases were considered as shown in Table 2 and elevation of buildings is shown in Fig. 1. The influence of roller, hinge and fixed connection is also considered.

Table 1 Structural elements dimension (mm)

Element	Storey no.				
	1–4	5–10	11–15	16–26	27–36
Column (b × w)	1400 × 1400	1000 × 1000	800 × 800	600 × 600	500 × 500
Beam (b × d)	300 × 700	300 × 700	300 × 700	300 × 700	200 × 500
Shear wall (t)	550	450	350	250	200

Table 2 Location of skybridge

Case no.	Link elevation
1	Without skybridge
2	Floor 9 (quarter height)
3	Floor 18 (middle height)
4	Floor 27 (three quarter height)
5	Floor 36 (top)

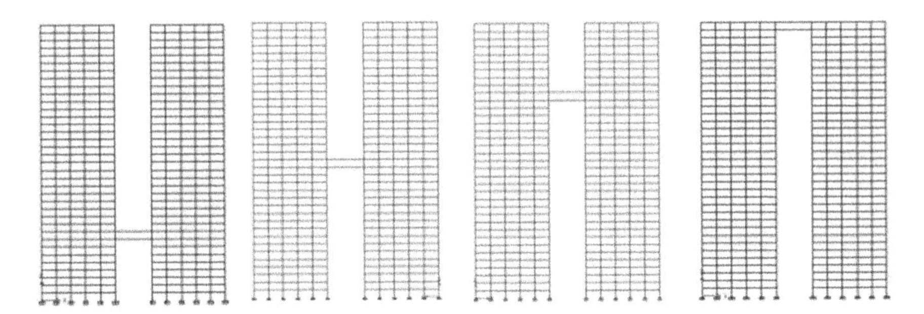

Fig. 1 Elevation of linked building system with skybridge at different location

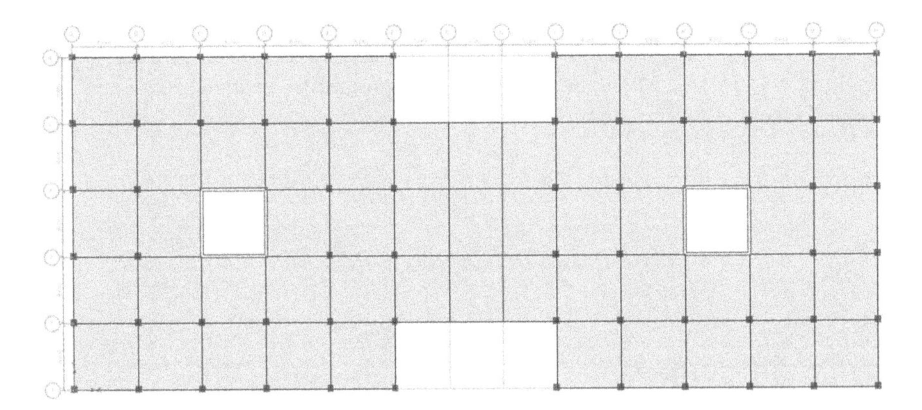

Fig. 2 Plan view of linked building system

2.2 *Seismic Analysis*

Seismic analysis of linked building is done using non-linear dynamic analysis using time history analysis. Time history function data entered for analysis was El Centro California. It had moment magnitude of 6.9 and maximum perceived intensity of X (Extreme) on Mercalli intensity scale. The ground motion records are taken from PEER ground motion records. Live load of 4 kN/m^2 is applied as per IS: 875(Part II) 1987 [8].

3 Results and Discussions

Time history analysis is performed to find out seismic performance based on parameters such as storey displacement, storey drift, and storey shear.

3.1 Effect of Skybridge Location

The maximum storey displacement of buildings linked by skybridge at various location are shown in Fig. 3. Maximum displacement is when skybridge located at three quarter height is 345.918 mm. The maximum storey displacement is reduced to 3% when skybridge is located at top. The displacement is reduced in all locations of skybridge compared to building without skybridge and reduces as the skybridge is at higher locations. This is because the extra link stiffness due to the skybridge which helps to resist seismic loads.

The storey drift of buildings linked by skybridge at various locations are shown in Fig. 4. The storey drift of all models are within the permissible limit as per IS1893 (Part 1): 2016 [9]. Maximum storey drift is when skybridge located at top is 0.005527.

The maximum storey shear of building for different cases of skybridge location is shown in Fig. 5. Maximum storey shear for skybridge located at three quarter height is 63,615 kN. Maximum storey shear is found to be increased when skybridge is located at top is 64,630 kN.

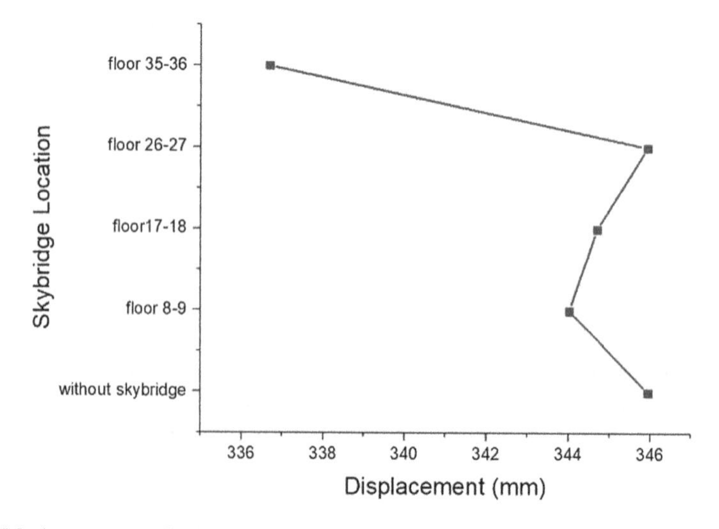

Fig. 3 Maximum storey displacement of buildings linked by skybridge at various

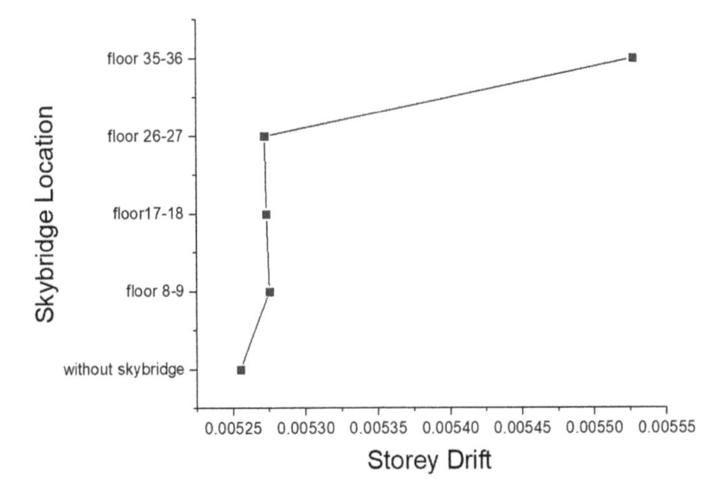

Fig. 4 Maximum storey drift of buildings linked by skybridge at various location

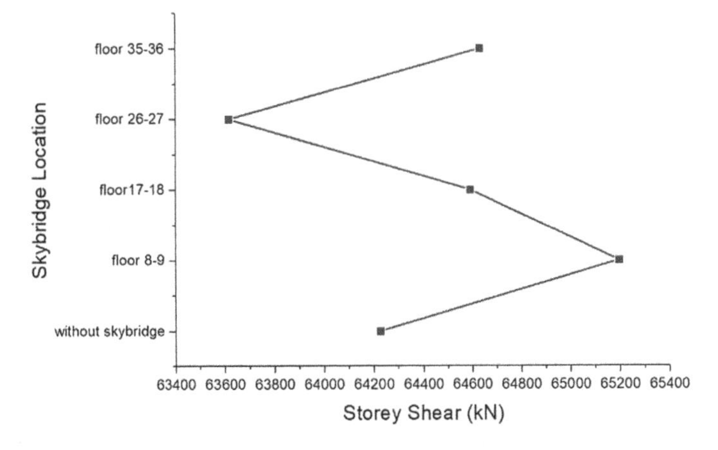

Fig. 5 Maximum storey shear of buildings linked by skybridge at various location

3.2 Effect of Skybridge Connection

The seismic analysis is done with buildings linked by skybridge at three quarter height of building with different connection. The maximum storey displacement of buildings linked by skybridge with different connection is shown in Fig. 6. Among three connections, roller connection has minimum displacement of 331.42 mm and reduced to 4%. Roller connection allow buildings to sway independently under lateral loading.

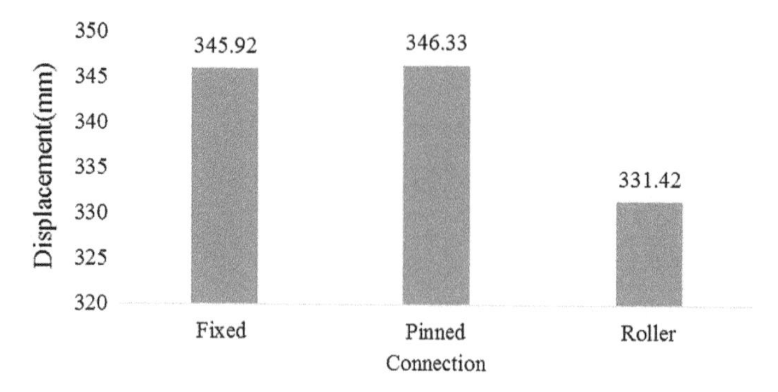

Fig. 6 Maximum storey displacement of buildings linked by skybridge with different connection

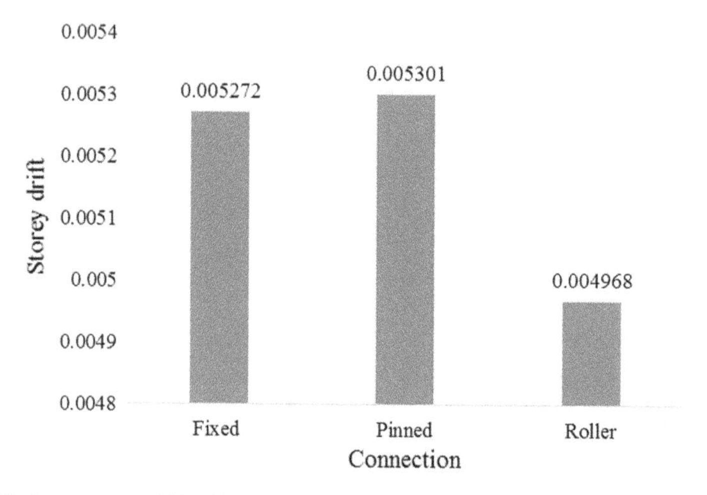

Fig. 7 Maximum storey drift of buildings linked by skybridge with different connection

Maximum storey drift of buildings linked by skybridge with different connection are shown in Fig. 7. The storey drift is reduced for roller connection compared with two connections.

Maximum storey shear of buildings linked by skybridge with different connection are shown in Fig. 8. The storey shear is minimum for fixed connection and has maximum storey shear of 64,466 kN for roller connection. Fixed connection are flexural stiff and prevents buildings to deflect as cantilever unit.

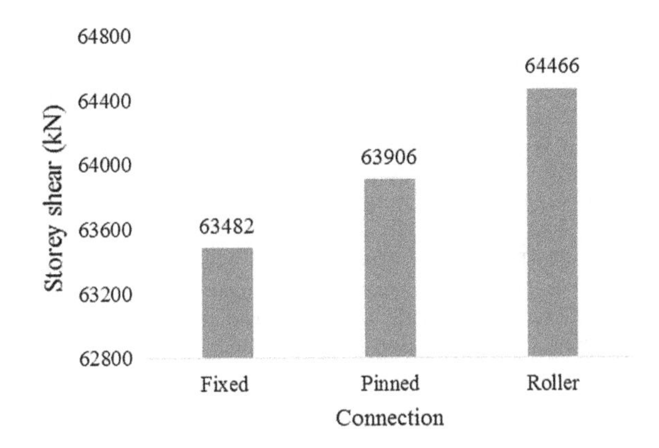

Fig. 8 Maximum storey shear of buildings linked by skybridge with different connection

4 Conclusion

This study presents the seismic performance of G+35 high rise building with skybridge by varying skybridge location and connection and following conclusion can be made from the study:

- The maximum storey displacement of building with skybridge at top is lesser than all other location and reduced to 3%. Contrary to displacement storey shear is higher than all other locations.
- The displacement is reduced in all locations of skybridge compared to building without skybridge and reduces as the skybridge is higher locations. This is because the extra link stiffness due to the skybridge which helps to resist the seismic loads.
- The maximum storey shear is lower than all other locations when skybridge is located at three quarter height of building and reduced to 2.5%.
- Considering different location skybridge at three quarter height will be effective and top is unfavorable location.
- Among three connection maximum displacement is reduced to 4% for roller connection.
- The fixed connection have lesser storey shear response. This makes the fixed connection better choice over others.

References

1. Imad Shakir A, Mahmod M, Hanoon AN, Jaafar MS, Mussa MH (2018) Sesimic response analysis of linked twin tall buildings with structural coupling. Int J Civil Eng 9:208–219
2. Mahmoud S (2019) Horizontally connected high rise buildings under earthquake loadings. Ain Shams Eng J 10:227–241
3. Sun H-s, Mohan L, Hong-ping Z (2014) Connecting parameters optimization on unsymmetrical twin—tower structure linked by skybridge. J Central South Univ 21(6):2460–2468

4. John H, McCafferty P (2014) Navigating the complexities of rooftop sky bridge construction. Struct Congr
5. Song J, Tse KT, Tamura Y, Kareem A (2016) Aerodynamics of closely spaced buildings with application to linked buildings. J Wind Eng Ind Aerodyn 149:1–16
6. Kim B, Tse KT, Chen Z, Park HS (2020) Multi-objective optimization of a structural link for a linked tall building system. J Build Eng 101382
7. Mccall et al (2013) Structural analysis and optimization of skyscrapers connected with skybridges and artria
8. IS 875(Part 2)-1987, Indian standard code of practice for design loads (other than earthquake) for building and structures. Part 2: imposed loads. Bureau of Indian Standards, New Delhi
9. IS 1893(Part 1)-2016, Indian standard criteria for earthquake resistant design of structures. Bureau of Indian Standards, New Delhi

Comparative Study of Monopod and Tripod Suction Caisson Foundation for an Offshore Wind Turbine

P. H. Jasna, Beena Mary John, and Rajesh P. Nair

Abstract Suction bucket foundation is one of the reliable types of foundation for an offshore wind turbine due to its cost-effectiveness and ease of installation. A three-dimensional numerical simulation analysis was conducted in this paper using ABAQUS 6.16 to determine the load bearing action of a monopod and tripod suction caisson foundation in sand under drained monotonic loading conditions. A comparison between the lateral load bearing capacity of monopod and tripod suction caisson under the same loading condition was made. The effects of related parameters i.e., the diameter (D) and horizontal loading conditions were discussed. The findings show that the loading conditions and bucket geometries influence the ultimate horizontal load and overturning moment capacities. An increase in the capacities is found with the increment in both the parameters. Load–moment interaction diagrams helps in the selection of suitable geometry for preliminary design for the foundation of an offshore wind turbine (OWT).

Keywords OWT · Suction bucket · Ultimate load and overturning moment capacities · Load–moment interaction diagrams

1 Introduction

Offshore wind energy has proved to be a powerful renewable energy source as compared to onshore wind energy. The major challenge of offshore wind energy industry is to lower the cost. In offshore wind energy projects, the cost of turbine only account for around 28% of the capital cost while foundations and substructures account for around 25% [1]. The development of foundation and installation technologies helps to reduce the cost of offshore wind turbine. Suction bucket foundations could be a dependable and cost-effective option for the next generation of offshore wind turbine substructures.

P. H. Jasna (✉) · B. M. John · R. P. Nair
Department of Ship Technology, Cochin University of Science and Technology, Cochin, India

G. C. Marano et al. (eds.), *Proceedings of SECON'21*, Lecture Notes in Civil Engineering 171, https://doi.org/10.1007/978-3-030-80312-4_31

Suction caissons are massive steel cylinders with a closed top and an open bottom. The bucket skirt is the cylindrical portion, and the bucket lid is the upper plate that closes the bucket. A tripod caisson foundation is made up of three individual buckets arranged in a triangular shape. The foundation's bearing capacity is increased with this form of foundation.

A series of FE tests were performed on suction caissons mounted in clay and sand to examine the effects of various factors on the load bearing behavior and deformation characteristics of suction buckets. Bearing capacity of a caisson foundation is an important component in the modelling of an OWT and it depends on the bucket dimensions and the loading [2–4].

A very few numerical analyses are conducted on tripod suction foundation. Load and moment carrying capacities of tripod caisson foundations are found to be higher when compared to the monopod foundation.

The load bearing behaviour of monopod and tripod suction bucket foundations for various bucket geometries and loading conditions is investigated in this paper using numerical analysis.

2 Numerical Simulation

The simulation modelling was done with the ABAQUS 6.16 finite element software. To reduce the computational effort and considering the symmetrical nature of the problem only just half of the entire system was modelled. Preliminary tests were carried out to assess mesh fineness and model dimensions in order to achieve sufficient accuracy of the results and avoid the effect of boundary conditions. Soil and bucket were modeled with 8 node volume linear brick elements with reduced integration and hourglass control (C3D8R). To simulate soil behaviour, the Mohr–Coulomb failure model is employed [2].

A location at Rameswaram, Tamil Nadu was selected based on environmental data obtained [5]. Table 1 lists the soil characteristics used in the model. A steel plate element is used to model the bucket. The skirt thickness was 30 mm in the modelling, which is the typical bucket thickness in use and the thickness of the caisson lid was taken as 100 mm [3]. Bucket properties include modulus of elasticity (E) of 210 GPa and Poisson's ratio of (υ) 0.3. The steel's unit weight (γ) was also set to 77 kN/m^3.

The displacements are restrained horizontally out of the plane direction along the vertical boundaries of the soil domain. Displacements are restrained at the soil boundaries in all directions at the bottom.

The FEM analysis was carried out stepwise [2, 3]. In the initial step, geostatic stresses consisting only of soil were determined by applying gravity loading. Subsequently the caisson and soil contact are initiated. To provide contact between the bucket and the soil, the "master–slave" modelling principle was enabled. In the interaction properties, the contact friction angle, which was calculated as two-thirds of the internal friction angle, was used to describe contact surfaces. In the next step,

Table 1 Soil characteristics [5]

Soil description	Depth (m)	Submerged unit weight (kN/m^3)	Internal friction (Φ)	Dilation angle (Ψ)	Poisson ratio (υ)	Cohesion (kN/m^2)
Grey fine sand	3	7.5	33	3	0.25	1
Grey silty fine sand	10	8.5	39	9	0.25	1
Crushed pieces of rock	12	8.5	44	14	0.25	1
Fine silty sand	18	8.5	41	11	0.25	1
Silty fine sand	30	8.5	42	12	0.25	1

a typical vertical load of 10 MN for a large wind turbine tower which represents the own weight of the superstructure of the wind turbine was applied to the reference point and was kept constant. The center of the bucket lid was considered as the reference point for the monopod foundation as shown in Fig. 1. For a tripod foundation the reference point was considered to be the point of intersection from the top center of each bucket directly below the center of the wind turbine tower as shown in Fig. 2. Then, lateral monotonic load (H) simulating the wind and wave load is applied through a constant rate of lateral displacement at an eccentricity h above the center of bucket lid which results in an overturning moment at the bucket in addition to the lateral load. The lateral bearing capacities and overturning moment capacities were found out from the lateral load displacement curves by

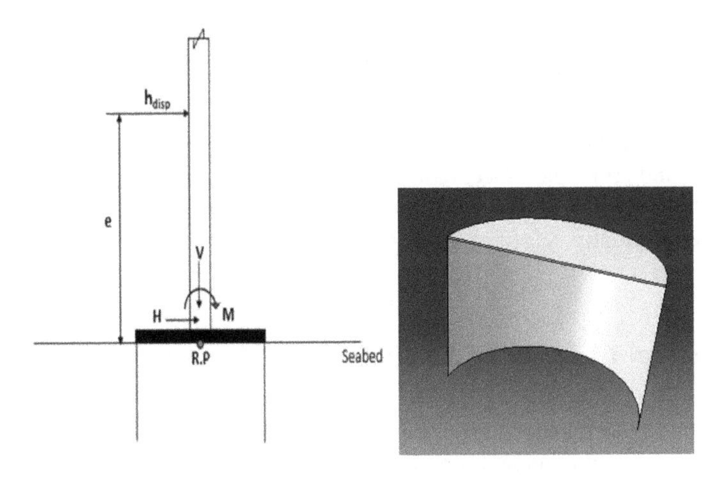

Fig. 1 Sketch of suction bucket foundation

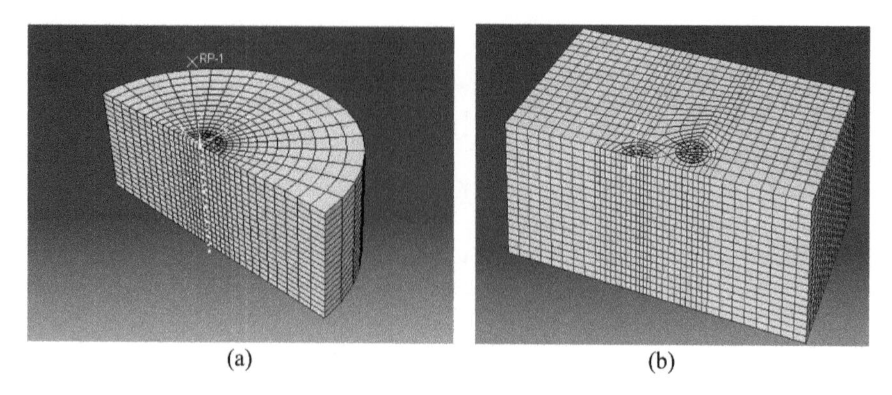

(a) (b)

Fig. 2 Finite element model **a** monopod suction caisson foundation, **b** tripod suction caisson foundation

using tangential intersection method. In the analysis load and moment capacities was obtained for different load eccentricities.

3 Comparison with Previous Study

To validate the numerical modelling, the results of the finite element analysis were compared with the published results on the load carrying capacity of the suction caisson foundation in sand [2]. The comparison of the achieved results along with that published result [2] for a length/diameter ratio L/D = 0.75, diameter D = 12 m and under different eccentricities (h = 0 m, 20 m, 100 m) is shown in Fig. 3. There is good agreement in the results of load vs displacement obtained from both the results.

4 Results and Discussion

Numerical simulation was performed on different bucket geometries with diameters between 8 and 20 m and aspect ratio L/D of 0.75. For tripod foundation the spacing ratio S/D = 1 was also considered as the distance from the center of each bucket to the reference point.

4.1 Lateral Capacity of Suction Bucket Foundation

The results of lateral load–displacement and lateral overturning moment-rotation curves for both monopod and tripod caisson foundations for a diameter D = 12 m, L/D = 0.75, vertical load V = 10 MN as shown in Figs. 4, 5, 6 and 7.

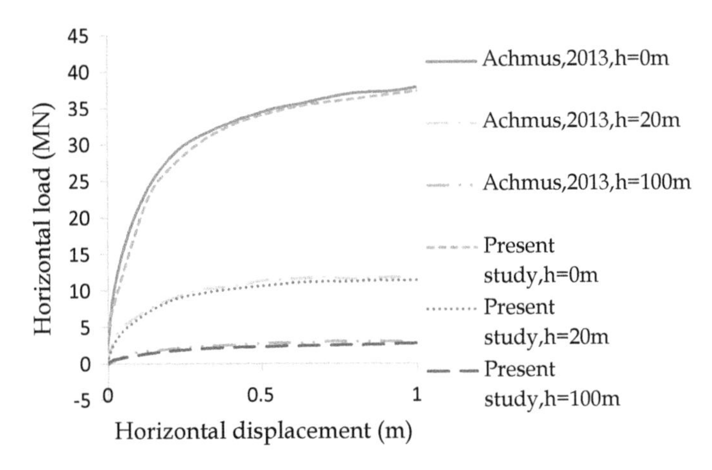

Fig. 3 Verification with [1] when L/D = 0.75, vertical load V = 10 MN with different load eccentricity

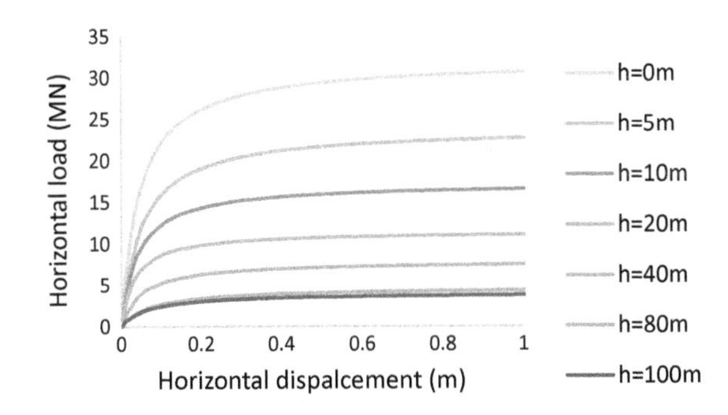

Fig. 4 Horizontal load–displacement interaction diagram for monopod (D = 12 m, L/D = 0.75)

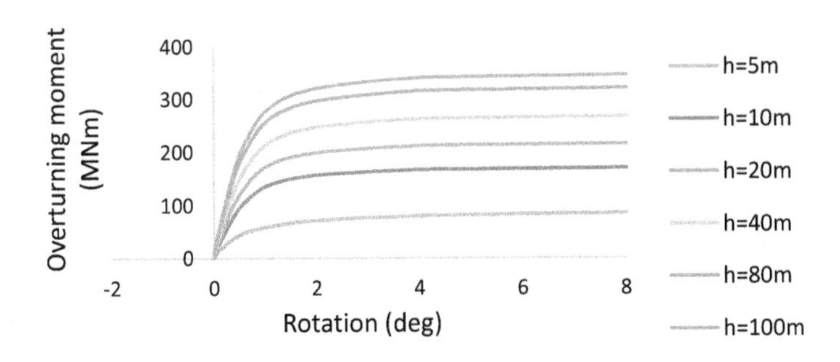

Fig. 5 Moment–rotation interaction diagram for monopod (D = 12 m, L/D = 0.75)

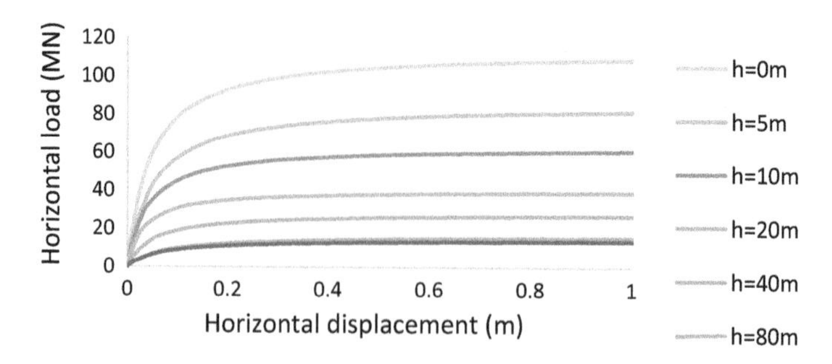

Fig. 6 Horizontal load–displacement interaction diagram for tripod (D = 12 m, L/D = 0.75)

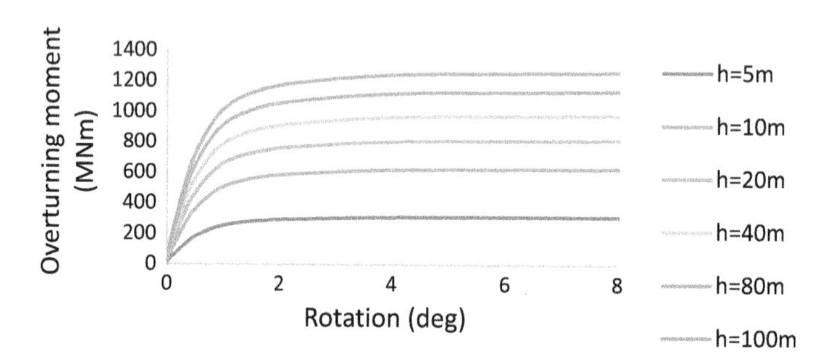

Fig. 7 Moment–rotation interaction diagram for tripod (D = 12 m, L/D = 0.75)

As depicted from the results the horizontal eccentricity significantly affects the lateral load capacity and lateral overturning moment capacity of the monopod and tripod foundation. The ultimate lateral load capacity monopod foundation is observed to decrease from 31 MN to 3.6 MN and 110 MN to 13 MN for tripod foundation with the increase of eccentricity from h = 0 m i.e., pure lateral load to h = 100 m. The ultimate overturning moment capacity increases from 86 to 342 MNm for monopod foundation and from 310 to 1248 MNm for a tripod foundation when the eccentricity is increased from h = 5 m to h = 100 m.

With increase in the load eccentricity (h) the horizontal load capacity of the bucket foundation decreases meanwhile the moment increases. The maximum lateral load capacity was obtained when there is no horizontal eccentricity (h = 0 m) i.e., pure lateral load.

Ultimate horizontal load and horizontal overturning moment capacity of a tripod caisson foundation is almost 3.6 times higher ultimate load and lateral overturning moment capacities of monopod caisson foundation for similar bucket geometries and loading conditions.

4.2 Lateral Load Overturning Moment Interaction Diagram

The ultimate horizontal load (H_u) and horizontal overturning moment capacities (M_u) were found for different bucket geometries with diameters between 8 and 20 m for an aspect ratio L/D of 0.75 for different horizontal load eccentricities (h = 0 m, 5 m, 10 m, 20 m, 40 m,80 m, 100 m) for both monopod and tripod suction bucket foundation. Load-moment interaction diagram is shown in Figs. 8 and 9.

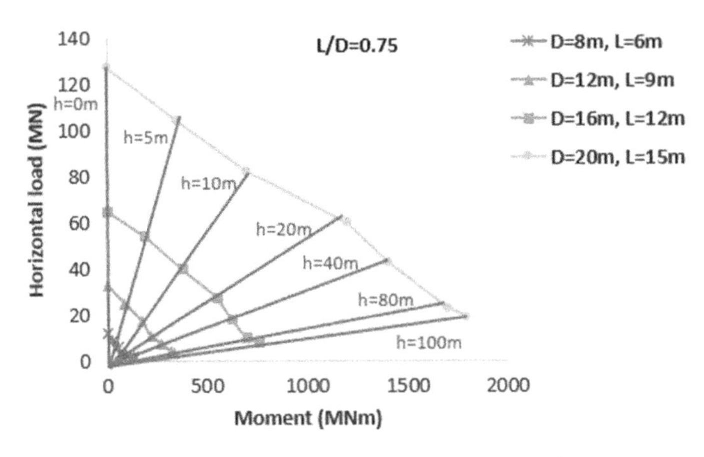

Fig. 8 Ultimate capacity interaction diagram L/D = 0.75 for monopod

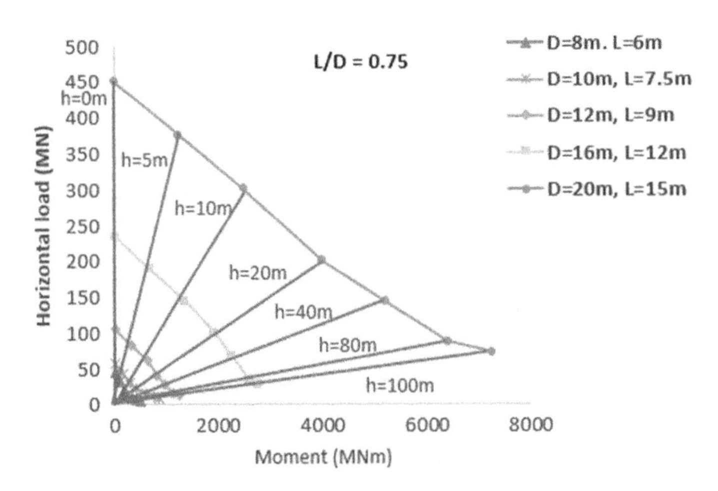

Fig. 9 Ultimate capacity interaction diagram L/D = 0.75 for tripod

Bucket geometries also affect the ultimate lateral capacity and overturning moment capacity. An increase in the caisson dimensions i.e., the diameter and skirt length, increases the ultimate lateral load and lateral overturning moment.

5 Conclusions

Numerical simulation of monopod and tripod caisson foundations have been simulated by a three-dimensional numerical analysis using ABAQUS software. Following conclusions are derived from the analyses:

- The caisson geometry and load eccentricity significantly affect the ultimate lateral load and moment capacities. The bucket foundation's ultimate lateral load capacity decreases as load eccentricity increases, while the overturning moment capacity increases.
- The ultimate lateral capacity increased as the skirt length and diameter of the caisson increased.
- The ultimate lateral load and overturning capacity of a tripod foundation is almost 3.6 times monopod foundation for same bucket geometries.
- The lateral load-overturning moment interaction diagram allows you to choose the caisson geometry needed for the preliminary design for a specific load combination of superstructure load and lateral loads caused by waves, wind and currents.

References

1. Musial W, Ram B (2010) Large-scale offshore wind power in the United States: assessment of opportunities and barriers. Nat Renew Energy Lab. https://doi.org/10.2172/990101
2. Achmus M, Akdag CT, Thieken K (2013) Load bearing behaviour of suction bucket foundations in sand. Appl Ocean Res 43:157–165
3. Bagheri P, Son SW, Kim JM (2013) Investigation of the load-bearing capacity of suction caissons used for offshore wind turbines. Appl Ocean Res 67:148–161
4. Bagheri P, Yoon JC, Park D, Kim JM (2019) Numerical analysis of suction bucket foundations used for offshore wind turbines. In: Vietnam symposium on advances in offshore engineering, pp 336–341. Springer, Singapore
5. Jose NM, Mathai A (2018) A study on natural frequency of offshore wind turbine in a layered soil. Int Res J Eng Technol (IRJET) 5
6. Tran NT, Hung LC, Kim S (2017) R: Evaluation of horizontal and moment bearing capacities of tripod bucket foundations in sand. Ocean Eng 140:209–221
7. Bagheri P, Kim JM (2019) Evaluation of cyclic and monotonic loading behavior of bucket foundations used for offshore wind turbines. Appl Ocean Res 91:101865
8. Deb TK, Singh B (2016) Behaviour of bucket foundations in sandy bed subjected to eccentric lateral loading. In: Geotechnical conference, pp 313–321. Springer, Singapore

9. Houlsby GT, Byrne BW (2000) Suction caisson foundations for offshore wind turbines and anemometer masts. J Wind Eng 24:249–255
10. Zhan YG, Liu FC (2010) Numerical analysis of bearing capacity of suction bucket foundation for offshore wind turbines. Electron J Geotech Eng 15(10):76–81

Modelling Tour Chaining Patterns of Urban Commercial Vehicles

S. M. Vinaya and M. V. L. R. Anjaneyulu

Abstract Freight transportation is an important part of urban transportation system and contributes to pollution, traffic congestion, infrastructure damage, etc. This paper presents a tour chain-based approach to modelling urban commercial vehicle daily activity patterns of Kozhikode district. A daily tour chain refers to a sequence of tours made by a vehicle in a day. Latent class analysis is used to cluster individuals based on their unobserved heterogeneity. Tour patterns are modelled with Multinomial logit model for cluster-based data and whole data. In which cluster-based data is outperformed. Results of study indicated that host of factors including operator, vehicle, commodity and tour characteristics affect choice of tour chain strategy with good prediction accuracy and correlation value.

Keywords Tour-chain based modelling · Daily activities of commercial vehicles · Latent class · MNL

1 Introduction

Freight transportation plays a vital role in modern urbanized civilization. Massive flow of goods needs to be efficiently transported to support the urban economy and lifestyle [1]. The formulation of urban logistics and policies requires good understanding of urban goods and commercial vehicle movements.

Commercial vehicle movement in an urban area has a number of discriminating features [2]. Tour chaining is an important element in urban settings. A tour is made when a vehicle makes one or more intermediate stops before returning to a base depot. Commercial vehicle tour often consists of number of stops for performing various actions likes goods pickup, delivery, service and maintenance [3]. Tour strategies can be classified into single direct, single peddling, multiple direct, multiple peddling and mixed one [4]. Direct and peddling strategies differ by number of intermediate stops served per load. Direct shipping serves only one

S. M. Vinaya (✉) · M. V. L. R. Anjaneyulu
Department of Civil Engineering, NIT Calicut, Calicut, India

G. C. Marano et al. (eds.), *Proceedings of SECON'21*, Lecture Notes in Civil Engineering 171, https://doi.org/10.1007/978-3-030-80312-4_32

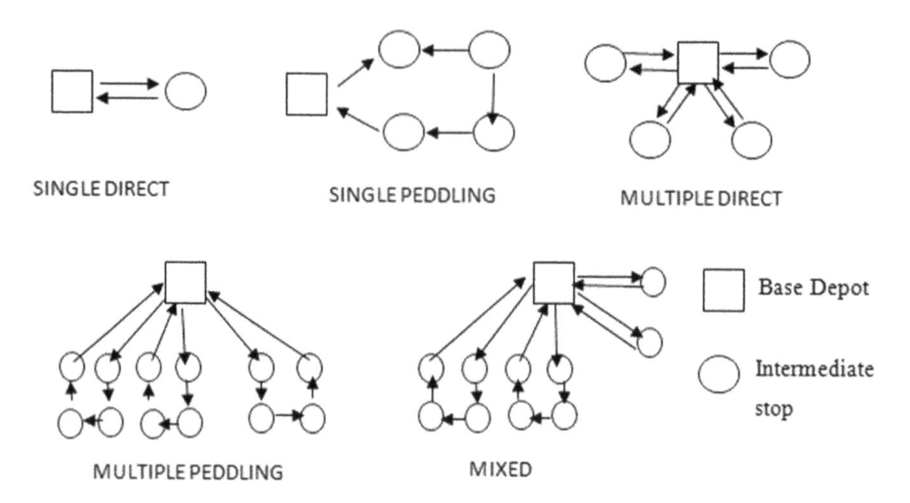

Fig. 1 Commercial vehicles daily tour chaining strategies

customer or intermediate stop per vehicle load. Peddling shipping serves more than one customer or intermediate stops per vehicle load. If vehicle is making more than one journey in a day it can be termed as multiple direct or peddling based on the number of intermediate stops. A mixed is the one which combines both direct and peddling tour strategies. Different tour chaining strategies of commercial vehicles are shown in Fig. 1.

Individual's behaviour will be heterogenic in nature. Hence latent class analysis is used to cluster individuals of similar behaviour. Latent Class Model (LCM) is for the analysis of discrete choice data depicts a semi parametric version of the Multinomial Logit Model (MNL). The LCM is based on the concept that individuals belong to different segments or classes with identical preferences within classes [5]. It estimates the class membership probability together with the class preference parameters. Multinomial logit model is used for modeling the clustered and non-clustered data and their results are caparisoned. The data is collected by means of revealed preference survey from the operators carrying commodities from markets of Kozhikode district. Objective of the study is to model tour chaining patterns of urban commercial vehicles in terms of influencing variables with and without latent class analysis.

2 Data Description

The questionnaire survey is conducted during 2018–2019 in Kozhikode. The carriers of the commodity who operate on a daily basis are surveyed using the questionnaire prepared. Revealed preference survey by face to face interview method is adopted to collect the details. Valiyangadi Fish market, fruits and

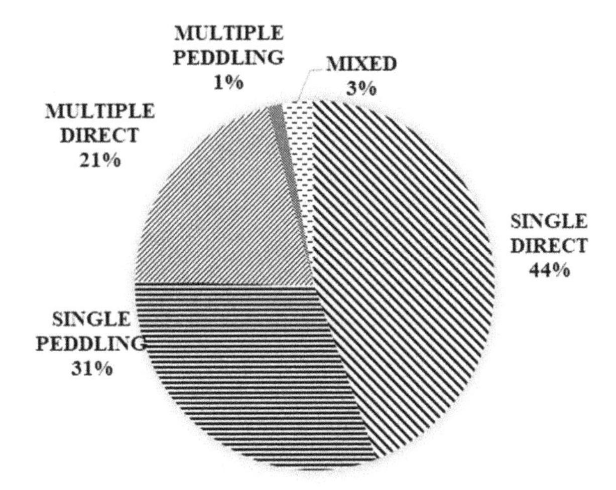

Fig. 2 Composition of tour pattern

vegetables market at Palayam and Valiyangadi grocery market are the places of data collection. A total of 605 drivers are interviewed, which includes 89 drivers from Fish market, 214 drivers from Palayam and 302 drivers from Valiyangadi. From a total of 605 samples single direct, single peddling, multiple direct, multiple peddling and mixed tour categories are of 269, 187, 124, 8 and 17 respectively in number. Composition of tour pattern is shown in Fig. 2. Description of some of the collected data is shown in Table 1.

3 Latent Class Analysis

Latent Class Analysis (LCA) is a method of analysing the relationship among collected data when some variables are unobserved. Latent class analysis is done in R software using poLCA package. poLCA uses expectation–maximization and Newton–Raphson algorithms to find maximum likelihood estimates of the parameters of the latent class. This package will retain a model which gives lowest BIC (Bayesian Information Criterion) on among the iterations for each number of clusters. The operator's characteristics are considered for the latent class analysis. Various trials of latent class analysis are done with variables which are having significant correlation values. When the number of categories for a variable is decreasing, significance of correlation values is found to be increasing and give better results for latent class analysis. Minimum BIC with positive degrees of freedom is taken as the criteria for selecting optimum number of clusters. Least BIC value is obtained with the variables ownership, taxi driver and loading points with 2 clusters of population shares of 300 and 305. Estimated and predicted population shares are found to be in almost same range. The individual belonging to each cluster of latent class model was analysed from the posterior probability.

Table 1 Description of data

Variable	Description
TOTAL DIS	Total distance travelled in a day
DW LOAD	Dwell time at the loading point (min)
DW UNLOAD	Dwell time at the unloading point (min)
LOAD	Total weight of commodity carried by the vehicle (ton)
SPEED	Speed of the vehicle with load (km/hr)
LOAD DIS	Total distance travelled with load (km)
DIS BTW	Average distance between the stops (km)

In the cluster 1, all the operators are working as drivers at the loading points. Out of which 11.1% of operators are using hired vehicle and 88.8% are using own vehicles. In the cluster 2, none of the operators are working as driver at the loading points. Out of which 13.6% of the operators are using hired vehicles and remaining are using own vehicles.

4 Multinomial Logit Model

Multinomial logit models are used for developing tour chaining patterns of urban commercial vehicles for the clustered data and whole data. The model results are shown in Table 2.

Multicollinearity in the logistic regression solution is detected by examining standard error for coefficients except for intercept value. A standard error larger than 2 indicates numerical problems. Variables with numerical problems should not be interpreted. None of the variables in this analysis had a standard error larger than 2.

4.1 Interpretation of Results of Cluster 1

While looking in to 'average dwell time at the unloading points' it can be seen that 0.284 is the coefficient and 1.329 is the odds ratio. It can be said that if the dwell time at unloading point is increased by one unit, then the odds of operators who choose tour type 1 (single direct) with reference to tour type 5 (mixed one) increases by 32.9% (i.e., (1.329-1) * 100) i.e., operators tend to use single direct tour type. Likewise, coefficients of each choice can be interpreted. Also, the positive sign for the variable indicates that dwell time at unloading point is more for single direct. But it is negative in case of multiple peddling which indicates that dwell time at unloading point decreases as operator choses multiple peddling tour strategy.

Table 2 Parameter estimates of two clusters

Cluster 1					Cluster 2			
Tour type		B	Std. error	Exp (B)	Variables	B	Std. error	Exp (B)
Single direct	Intercept	−9.504	2.368		Intercept	−2.572	3.34	
	DW UNLOAD	0.284	0.077	1.329	DIS BTW	0.134	0.218	1.143
	LOAD DIS	−0.066	0.019	0.937	DW UNLOAD	0.008	0.036	1.008
	LOAD	−1.282	0.33	0.277	SPEED	0.236	0.201	1.266
	SPEED	0.256	0.08	1.291				
	DW LOAD	0.031	0.01	1.031				
Single peddling	Intercept	−4.331	2.201		Intercept	−2.636	3.349	
	DW UNLOAD	0.154	0.077	1.167	DIS BTW	0.035	0.219	1.035
	LOAD DIS	−0.031	0.015	0.969	DW UNLOAD	−0.026	0.037	0.974
	LOAD	−0.52	0.29	0.594	SPEED	0.302	0.202	1.352
	SPEED	0.104	0.077	1.11				
	DW LOAD	0.028	0.01	1.028				
Multiple direct	Intercept	−3.578	2.098		Intercept	−4.295	3.781	
	DW UNLOAD	0.166	0.075	1.181	DIS BTW	0.124	0.223	1.132
	LOA DIS	−0.016	0.012	0.984	DW UNLOAD	−0.009	0.045	0.991
	LOAD	−0.03	0.222	0.97	SPEED	0.207	0.214	1.23
	SPEED	0.148	0.073	1.16				
	DW LOAD	0	0.01	1				
Multiple peddling	Intercept	−2.946	4.033		Intercept	−0.771	4.137	
	DW UNLOAD	−0.342	0.214	0.711	DIS BTW	−0.183	0.305	0.833
	LOAD DIS	−0.011	0.03	0.989	DW UNLOAD	−0.139	0.114	0.87
	LOAD	0.651	0.453	1.918	SPEED	0.246	0.22	1.279
	SPEED	0.173	0.168	1.189				
	DW LOAD	0.005	0.018	1.005				

Sign for the variable 'distance travelled with load' is negative in all cases. If the distance travelled with load is increased by one unit, the odds of people choosing 1 with reference to 5 will decrease by 6.3%. This may be due to that, the distance travelled with load in single direct tour type is less compared to mixed tour type. If the distance travelled with load is increased by one unit, then odds of people choosing tour type 2, 3 and with reference to 5 decreases by 3.1%, 1.6% and 1.1% respectively.

Sign for the variable 'total load' is negative for the tour types 1, 2 and 3 but it is positive in case of tour type 4. If the total load carried by the operator in a day is increased by one unit the odds of choosing tour type 1, 2 and 3 with reference to tour type 5 decreases by 72.3, 40.6 and 3%. But if the load carried by the operator in day increased by one unit, the odds of choosing tour type 4 with reference to tour type 5 will increases by 91.8%. The total load carried in the single direct, single peddling and multiple direct will be less compared to mixed one but it will be more in case of multiple peddling.

4.2 Interpretation of Results of Cluster 2

While looking at the variable 'distance between stops' has positive sign in case of choices 1, 2 and 3 but it is negative for choice 4. The average distance between the stops is more for single direct, single peddling and multiple direct but less for multiple peddling when compared to mixed tour chain strategy. For a unit change in average distance between stops, the odds for choosing choice 1, 2 and 3 with reference to the choice 5 increases by 14.3, 3.5 and 13.2% but decreases for choice 4 with reference to the choice 5 by 16.7%.

Variable 'dwell time at the unloading point' has positive sign for choice 1 but it is negative for all other choices. For a unit change in the variable 'dwell time at unloading point' the odds of the operator for choice 1 with reference to the choice 5 increases by 0.8% and for the choices 2, 3 and 4 with reference to the choice 5 decreases by 2.6%, 0.9% and 13% respectively. The operators in cluster who are following single direct tour type has more dwell time at the unloading point when compared to mixed tour chain strategy.

The variable 'speed' has positive sign for all choices. Single direct, single peddling, multiple direct and multiple peddling follows more speed than mixed tour chain strategy. For a unit change in speed the odds of the operator for choosing choice 1, 2, 3 and 4 with reference to the choice 5 increases by 26.6%, 35.2%, 23% and 27.9% respectively.

4.3 Comparison of Results Cluster Based Data and Whole Data

Average dwell time at the loading point, average dwell time at the unloading point, speed, load and total distance travelled with load are the variables obtained for cluster 1. For the cluster 2, the variables used are average dwell time at the

Table 3 Comparison of model results

Sl no	Goodness of fit statistics	Multinomial logistic regression models			
		Cluster 1	Without lc	Cluster 2	Without lc
1	Variables	DWLOAD, DWUNLOAD, LOAD, SPEED, LOAD DIS		DWUNLOAD, SPEED, DIS BTW	
2	Pseudo R square (nagelkerke)	0.725	0.538	0.702	0.267
3	Chi-square (deviance)	312.628	697.811	298.371	854.887
4	BIC	440.263	841.726	383.144	950.830
5	AIC	360.628	745.811	330.371	886.887
6	Model validation results (prediction accuracy) (%)	71.6	66.4	68.5	53

unloading point, speed and average distance between the stops. Same variables are used for the multinomial logistic regression modelling of whole data. By comparing goodness of fit statistics such as Pseudo R square, chi-square, BIC and AIC, it is observed that latent class gives best results. Prediction accuracy is also better for latent cluster-based models. Comparison of multinomial logistic regression results of cluster-based data and whole data is shown in Table 3.

5 Conclusions

This paper has put forward a tour chain concept and presented a study of tour-chain-based modelling approach for understanding the complex process of urban commercial vehicle movements. Tour chaining represents the inter connectiveness of daily delivery tours and daily activities carried out by a single commercial vehicle [6]. The findings of the study confirm that intuition. Furthermore, the tour-chain model provides much more detailed analysis of delivery strategy preference than the tour-based model. The tour-chain model may provide useful policy suggestions.

The conclusions derived from the study are included in the sections. Single direct tour pattern is the most observed tour pattern and which shared 44% of total data collected. Multiple peddling and mixed tour chaining strategies are observed to be less. In case of fish transport, single direct and single peddling was equal in proportion and mixed one is observed to be least with only 1%. Most of the operators are using mini truck for fish transport. The preference for single direct tour pattern is high in case of vegetables and fruits transport from Palayam market and that accounts for 46% of total. Single peddling and multiple direct are equally preferred with a share of 24%. For the transport of grocery items single direct tour pattern is most preferred and this is followed by single peddling. For the transport of grocery items goods auto rickshaw was most preferred with a share of 46% and is followed by truck.

Optimum number of clusters is obtained with variables such as ownership, taxi driver and loading points are the variables with a cluster membership of 305 and 300. Cluster 1 consists of operators working as drivers at the loading points. Dwell time at the loading and unloading point, speed, total distance travelled while carrying the load and total load carried are variables used in the modelling of tour chaining pattern. Results obtained for cluster based on latent class analysis are much better than that for the whole data.

However, there are limitations for the study. First, Model assumes that there is no price variation across different tour chaining patterns. Since the logistic cost related with each tour chain is not obtained in the survey. Second, other variables like employment size and income are not collected in the survey and hence could not be used for modelling.

References

1. Lin JJ, Zhou W (2012) How urban commercial vehicle trips are chained: case studies in Texas and Idaho. Transportation research board annual meeting
2. Holguín-Veras J, Patil GR (2005) Observed trip chain behavior of commercial vehicles. Transportation Research Board 1906:74–80
3. Khan M, Machemehl M (2017) Analysing tour chaining pattern of urban commercial vehicles. Transportation research part A: policy and practice, vol 102(C). Elsevier, pp 84–97
4. Ruan M, Lin JJ, Kawamura K (2012) Modeling urban commercial vehicle daily tour chaining. Transportation Research Part E: Logistics and Transportation Review 48(6):1169–1184
5. Atasoy B, Glerum A, Bierlaire M (2011) Mode choice with attitudinal latent class: a Swiss case-study. Second international choice modelling conference
6. Figliozzi MA (2007) Analysis of the efficiency of urban commercial vehicle tours: data collection, methodology, and policy implications. Transportation Research Part B Methodological 41(9):1014–1032

Study of Plan Irregularity on Different Seismic Zones

G. Nair Anjana and B. R. Beena

Abstract Population growth as well as shortage of land increased the demand of high rise building and aesthetic perspective of construction activities resulted in plan irregular buildings. Structural irregularity have several effects on the response of structures. Structures with configuration irregularity possess a serious threat from ground motion. This study focused on the analysis of different plan irregular buildings in different seismic zones using response spectrum method. The attempt was to know the effects of plan irregularity in high rise buildings under different seismic zones using ETABS 2018 software and comparing the responses of plan irregular buildings and selecting a seismic-resistant plan. The results indicate that the value of displacement and drift have effects on plan irregularity.

Keywords Seismic zones · Response spectrum method · ETABS 2018

1 Introduction

Natural disasters creates a huge threat to the world. Most devastating and unpredictable one among natural disaster is the Earthquake. Earthquake causes loss of life and damage to the properties. We cannot resist or prevent the earthquake but we can design a building to be earthquake resistant [1].

A structure should possess ductility, stiffness, regular configuration and lateral strength. Irregular building show higher damage compared to that of regular building. Nowadays irregular buildings are getting emerged due to their aesthetic purposes and also due to the increase in population demands. Due to their non-uniform distribution of masses, damages are more resulted in irregular buildings. Seismic performance of a buildings are mainly connected to configuration irregularity effects. Plan irregularity is mainly responsible for the failure of structures during earthquake. The choice of location of irregularity is helpful for improving the aesthetic aspects of a building [2].

G. N. Anjana (✉) · B. R. Beena
Federal Institute of Science and Technology, Angamaly, Eranakulam, India

G. C. Marano et al. (eds.), *Proceedings of SECON'21*, Lecture Notes in Civil Engineering 171, https://doi.org/10.1007/978-3-030-80312-4_33

Diaphragm discontinuity and re-entrant corners are two type of plan irregularities. The diaphragm discontinuity may occur with abrupt variations in stiffness. The re-entrant corners, are created by reducing certain percentage of area from a regular model and shapes may be in form of L, H, T, C and I [3].

Buildings having lack of resistance to seismic forces are subjected to seismic analysis. Dynamic effects are usually considered during seismic analysis hence it may become complex. However for simple regular structures equivalent linear static analysis sufficient one. The analysis are mainly carried out on irregular as well as regular buildings. Dynamic analysis are performed as per IS 1893-2016 (part 1). Response spectrum or time history analysis are the dynamic analysis carried out on buildings [4].

Many researchers give out ideas related to buildings with irregular geometry having change in shape and plan configuration. STAAD PRO software was used for studying the effects of irregularity. In my study, attempt has been made to study the behaviour of plan irregular buildings in different seismic zone and to identify a seismic- resistant irregular plan building using ETABS 2018 software [5].

2 Methodology Adopted

Different plan irregular buildings having a dimension 36 × 36 mm have been modelled in software ETABS 2018. Response spectrum analysis were carried out on different plan irregular buildings like L shaped building, T shaped building and C shaped building under different seismic zones. Different parameters are studied to understand the seismic behaviour of plan irregular buildings. Type of soil, seismic zone factor, standard response spectra were applied to building as per IS 1893 (part I) 2016. After the analysis MS excel is used to process the outputs form the software. All the buildings chosen for my study were multi -storied reinforced buildings. The analysis is carried along only in one horizontal direction i.e. x direction.

3 Modelling of Structures

Three buildings of G+9 storey with different shapes having a plan dimension 36 m × 36 m were modelled using ETABS 2018 software. Height of the storey was 3 m for all plan irregular buildings. The building configurations were listed in Table 1. Live load provided was 3 kN/m^2. As per IS 875 (Part II) 1987 and dead load was software assigned. Seismic design factors were assumed as: Importance factor-1, Response reduction factor-5, Soil type-II and Zone factor varies as per different seismic zones. The building configuration were listed on Table 1. The plan view of L shape, C shape and T shape building were shown in Fig. 1.

Table 1 Building configuration of three models

Number of storey	10
Size of beam	300 × 600 mm
Size of column	500 × 500 mm
Slab thickness	150 mm
Concrete mix for column	M30
Concrete mix for beam/slab	M25
Typical storey height	3 m
Grade of steel	Fe415
Support	Fixed

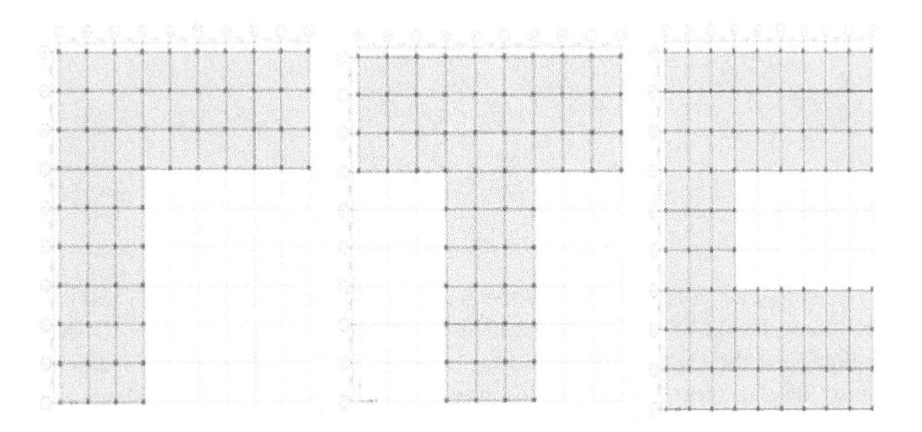

Fig. 1 Plan view of L shape, T shape and C shape model

4 Analysis

Response spectrum analysis was carried out on three models under different seismic zones using ETABS 2018 software. Maximum storey displacements, storey drifts and base shear were determined for three models under different seismic zones such as Zone II, Zone III, Zone IV, and Zone V. In this study, response spectrum function was set as per IS 1893: 2016.

5 Results

5.1 Maximum Storey Displacement

Storey displacement is the lateral displacement of the storey relative to the base. The variation of displacement for T shape, C shape and L shape buildings are shown in Fig. 2 for Zone V.

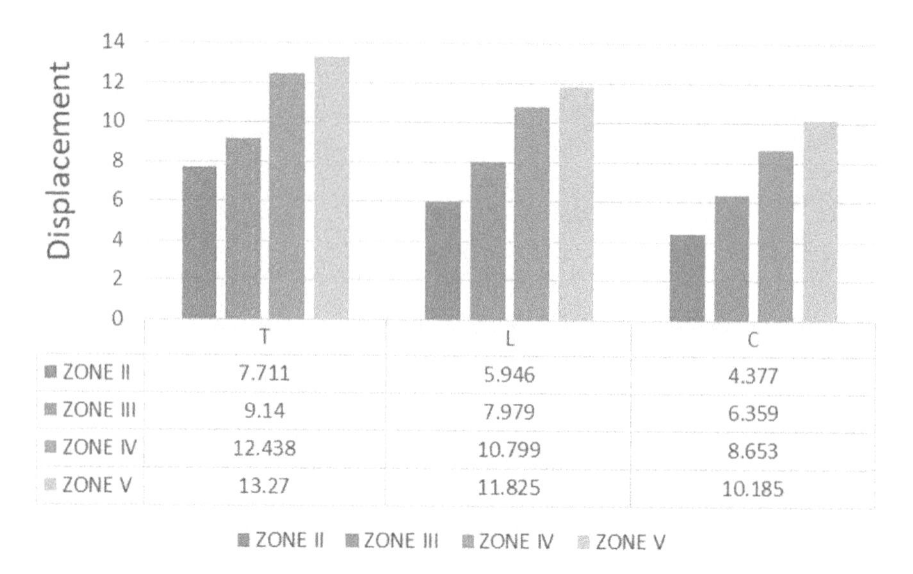

Fig. 2 Displacement in all three models

In all models, it is observed that the displacements increased in similar way with the storey height of the building and it is maximum at the top storey. C shape building has the least storey displacement and T shape building has the maximum storey displacement. Similar trends were followed by the three buildings with varying displacement values for other seismic zones also. C shape building has a storey displacement of 4.377 mm at zone II and T shape building has a maximum displacement of 7.711 mm at Zone II.

5.2 *Storey Drift*

Storey drift is defined as the relative sway of frame to the adjacent floor. Figure 3 shows the storey drift in different seismic zones for the three buildings.

In all models, it is observed that T shape building has highest storey drifts values whereas the C shaped building has the least values of storey drift. Maximum storey drift is 0.0062 for T shape building and the minimum storey drift is 0.000212 for C shape building. The value of storey drift decreases from higher zone to lower zones for all the three models. The effect of irregularity on storey drift is more depicted at Zone V. So the irregular configuration have effect on change in drift values.

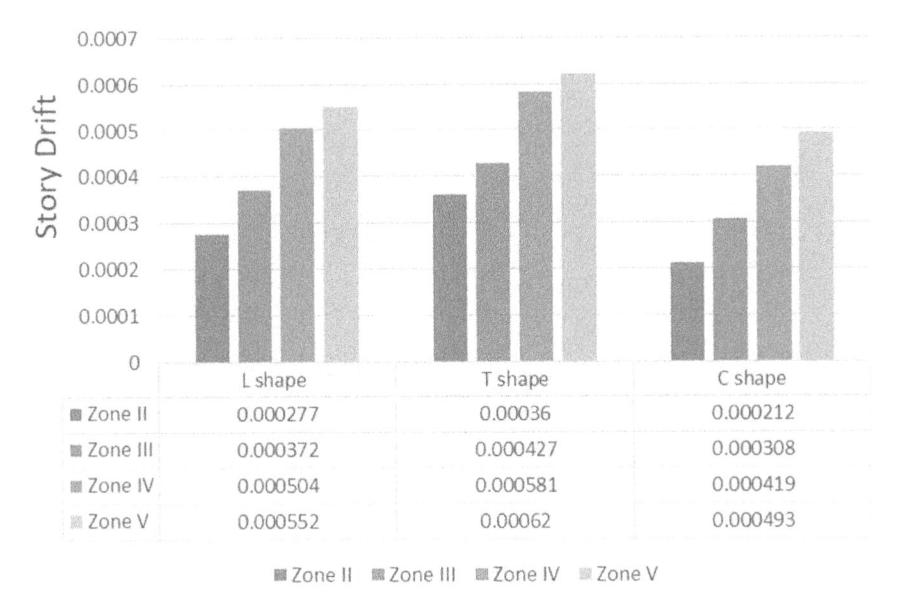

	L shape	T shape	C shape
Zone II	0.000277	0.00036	0.000212
Zone III	0.000372	0.000427	0.000308
Zone IV	0.000504	0.000581	0.000419
Zone V	0.000552	0.00062	0.000493

Fig. 3 Storey drift in L shape, T shape and C shape models

5.3 *Base Shear*

Base shear is the force which is acts in the base of any structure to overturn the building when earthquake load acts in a structure. The base shear values for the three buildings are shown in Fig. 4.

In all models it is observed that the base shear value is highest for Zone V and lowest for Zone II for all the three buildings. L shape building has the least base shear value and C shape building has the maximum base shear value which is observed at Zone V. Change in shape of building leads to the variation in value of base shear. But for Zone II, the base shear value for C shape building is 1408.69 kN which is lesser than base shear value of T shape building i.e. 1428.88 kN.

6 Conclusion

Conducting seismic performance of G+9 story plan irregular building under different seismic zones, the findings are summarised below.

C shape building has the least storey displacement and storey drift compared to that of L shape and T shape building for all the seismic zones. Maximum storey displacement for T shaped building is 76.17% more than that of C shaped building at Zone II. There is 30.66% decrease in storey drift found in C shape building when compared with L shape building and 69.811% decrease in storey drift with that of T

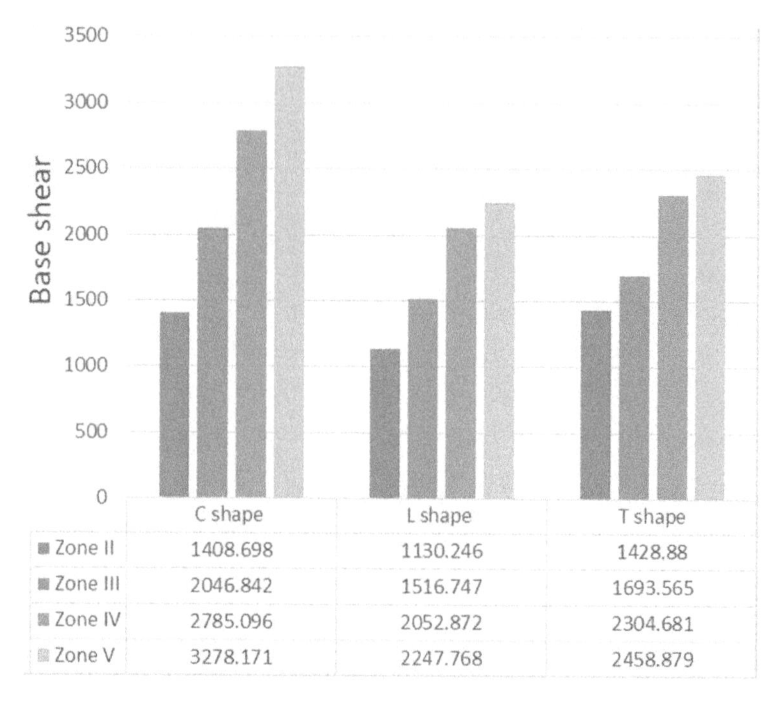

Fig. 4 Base shear for C shape, L shape and T shape buildings

shape building. Base shear value reduces by 1.4% for C shape building compared to that of T shape building. C shape building can be chosen as seismic resistant irregular plan when it is compared to other two models as it shows least value for displacement and drift for different seismic zones. The irregularity in buildings have effect on variation in value of response of structures.

References

1. Abrahamb NM et al (2019) Analysis of irregular structures under earthquake loads. Proc Struct Integrity 14:806–819
2. Yousuf M, Shimpale PM (2013) Dynamic analysis of reinforced concrete building with plan irregularities. Int J Emerg Technol Adv Eng 3(9)
3. Mohod MV (2015) Effect of shape and plan configuration on Seismic response of structure. Int J Sci Technol Res 4
4. Khanal B et al (2020) Seismic elastic performance of L-shaped building frames through plan irregularities. Structures 27:22–36
5. Wood SL (1992) Seismic response of RC frames with irregular profiles. J Struct Eng 2(118): 545–566

Performance Analysis of CFS Encased Recycled Plastic Laminate Beam

Issra Mohmed Ayoob and P. E. Kavitha

Abstract Plastic lumber is a sustainable alternative to lightweight structural wood material which is produced from virgin or recycled plastics. With the use of recycled plastic as the construction material, the amount of plastic waste going into landfills and incineration can be reduced. The raw material used in this study being recycled plastic is available in ample quantities. This paper is intended to propose a sustainable composite beam made with recycled plastic lumber blocks strengthened with cold-formed steel. Cold form steel is widely utilised in the manufacture of lightweight structural elements. Both the material can be used to produce lightweight building elements at a much lower manufacturing cost compared to conventional materials. Plastic lumber beams strengthened internally as well as externally using CFS encasement are statically analysed and optimised. Study of strength characteristics of plastic lumber beams by varying orientations, B/D ratio and thickness of laminates is also conducted. Ultimate load, ultimate displacement, ductility and failure patterns are also computed and analysed using Ansys Workbench 19.0 software. Optimisation of result is obtained by comparing the performance of compsite recycled plastic lumber beams with CFS beam and plastic lumber beam with similar dimensions. This paper proposes an innovative sustainable lightweight composite beam over a short span.

Keywords Plastic lumber · Lightweight · Flexural strength · Composite beam · Encasement

I. M. Ayoob (✉) · P. E. Kavitha
Department of Civil Engineering, Federal Institute of Science and Technology, Angamaly, Kerala, India

G. C. Marano et al. (eds.), *Proceedings of SECON'21*, Lecture Notes in Civil Engineering 171, https://doi.org/10.1007/978-3-030-80312-4_34

1 Introduction

1.1 General Background

The invention of plastic can be said as one of the milestones in the history of development. Non-corrodible, lightweight, flexible, cost-efficient and many other properties made it so popular that plastic has solved many crises faced in almost all fields such as electronics, transportation, food packaging, construction, etc. [1]. Versatile plastics have supported innovations that have made life easier, safe and compactable. Plastic can be classified based on its decomposition: biodegradable and non-biodegradable plastics. Biodegradable plastic is produced from microorganisms, petrochemicals and renewable materials, or with all three above mentioned raw material combinations. It gets degraded by the action of living organisms. Meanwhile, non-biodegradable plastics cannot be converted into safer components by the action of microorganisms. They prevail in nature for a very long time causing pollution of soil, water and air. From the dawn of development, one of the main problems human beings face is managing the plastic waste they produce [2]. It is the responsibility of an engineer to come up with a solution. Many measures were proposed and implemented in this field. Some of the disposal methods are incineration, landfill, 5 R's which includes reduce, reuse, recycle, recover and residual management [2]. The landfill method has been the oldest and commonly used method of waste disposal. Piles of garbage are dumped on landfill sites without even segregating. This has many ill effects such as leaching of harmful chemicals into groundwater and provides a breeding ground for rodents and pathogenic microorganisms [1]. It also reduces the aesthetic value of that area. So engineers have to find a way out of this problem. Recycling plastic waste can reduce a large amount of garbage going into landfill and incineration. Recycled plastic products are available in the market [3]. It is also used in the construction industry. One such product is Plastic Lumber blocks [4].

1.2 Plastic Lumber

Plastic lumber can be defined as wood-like material which is produced from virgin or recycled plastics. Plastic lumber is mostly made from high-density polyethene (HDPE), polyvinyl chloride (PVC), polylactic acid (PLA), acrylonitrile butadiene styrene (ABS), polystyrene (PS) and polypropylene (PP) [2]. Recycled plastic lumbers available in the market are high-density polyethene recycled plastic lumber, commingled recycled plastic lumber, wood-filled recycled plastic lumber, fibre-reinforced recycled plastic lumber [1]. Miscellaneous combinations such as glass-reinforced plastic lumber, rubber-plastic lumber, mixed plastic, multiple laminations of oriented HDPE and cross-linking of PE molecules by thermoset processes are also available [5]. Plastic lumber is water and termite-proof, which will impart longer life compared to wooden lumber. Bacterial and fungus growth is forbidden over the surface, hence reducing the cost of maintenance [4]. It is a clean, nontoxic, nonporous, and recyclable material. It is resistant to shock and chemical reaction [6]. Irregular and joint-less shapes are possible when using plastic lumber. It is flexible, splinter-free and higher shear strength compared to wood, therefore, it does not crack when curved and shaped [7]. Plastic lumber is graffiti resistant and requires no sanding, painting, staining or sealing. It is also popular for its wood-like properties like hammering, sawing, chiselling, grooving, carving, laminating, screwing and nailing. Plastic lumber is more susceptible to warping and discolouration. It is likely to sag under a heavy load application. Temperature can be a limiting factor for its use. It can melt under a tabletop camping grill [4]. It has increased rates of thermal expansion and contraction, hence reducing the soundness of the material. The slip resistance property of plastic lumber is very low and can cause danger if it is used outdoors especially during monsoon. So a slip-resistant cover has to be provided or the surface finish has to be altered for its application as decking material. Compared to wooden lumber, plastic lumber has lower stiffness and axial strength hindering its application as a structural material [7]. In the construction industry, plastic lumber is mostly used in non-structural applications. It is commonly used in outdoor elements like tables, benches, trash bins, fencing, decks, construction curbs, removable speed bumps, parking lot, stop bollards and railroad ties [3]. Plastic lumber is also used as door and window frames, siding, cladding of wall, cribbing [2, 5]. To use plastic lumber in the structural field, the axial strength has to be improved [8]. Increasing the cross-sectional area can strengthen the element but this remedy is not feasible. Another option is to encase the plastic lumber structural element with a material that can impart the required strength. This paper is intended to come up with a lightweight structural beam element made up of plastic lumber blocks encased with cold-formed steel [9].

2 Analytical Program

2.1 Material Properties

Materials used for the creation of the structural light-weight composite beam elements were mainly plastic lumber laminates strengthened using cold-formed steel. Coach screws and adhesives were used to combine all the components of the beam element. Properties of plastic lumber, cold-formed steel and coach screw assigned for analysis is tabulated in Table 1. Polypropylene with flexural modulus 2.1 GPa and tensile strength of 25 MPa is the plastic lumber used in this work [10]. M10 screws with grade 8.8 are used for connection purposes [11]. The tensile strength of the screws is 640 N/mm^2. A friction coefficient of 0.6 was assigned to depict the adhesive connection between the laminates as well as laminate and cold-formed steel.

Table 1 Material properties of plastic lumber, cold-formed steel and coach screw [10–12]

	Plastic lumber	Cold-formed steel	Coach screw
Density (kg/m^3)	860	7850	7850
Young's modulus (MPa)	2100	2×10^5	2×10^5
Shear modulus (MPa)	875	76,923	76,923
Poisson's ratio	0.2	0.3	0.3
Yield strength (MPa)	25	541	640

2.2 Modelling and Analysis

Based on the studies proposed in this paper, various models have to be created for undertaking the static analysis. Table 2 illustrates the configuration of the models created. Each model has a span of 1400 mm [13] which serves the purpose of a light-weight structural beam element for a shorter span. The thickness of cold-formed steel used in all the studies is 1.53 mm.

Encasement Study. The first set of models created in Ansys DesignModeler based on the encasement study as PLB will be encased with cold-formed steel. Internal encasement (IE), as well as external encasement (EE), are considered in this study. Laminates used will be of equal thickness to 20 mm. Non-linear static analysis (flexural bending using a four-point loading test) was conducted. In the internal encasement study, two types of arrangements are taken into consideration. In the first configuration, two cold-formed steel sections are placed back to back in the middle portion of the beam. The gap is then filled with plastic lumber blocks. At first, all the laminates are stuck with adhesive then it is screwed internally with self-driven screws. Another type of configuration coming under internal encasement is the hatched section arrangement. In this case, the CFS section is bent into required dimensions and made into a hatch section. Plastic laminates are used to fill the rest of the area inside the beam.

External encasement study also has two types of arrangements under consideration. In the first arrangement, two C-sections are taken to externally encase the PLB for the strengthening purpose and the middle portion is filled with recycled plastic laminates. The second type of arrangement consists of a single cold-formed steel section around the beam. As in the above-mentioned arrangement, the middle section of the beam is filled with recycled plastic lumber laminates.

Parametric Study. For the parametric study, the most effective design is considered out of the four models analysed in the above study. Three parameters are considered under this case. They are variations in the orientation of plastic laminates, aspect ratio and thickness of laminates within the plastic lumber beams.

Laminates Alignment. Horizontal, vertical, and a combination of vertical and horizontal arrangements of laminates externally encased with a single C-section are the different orientations considered [14]. Out of the various alignments mentioned above, the best arrangement is determined.

Table 2 Description of the models

Models	Illustration	Size of the model			Thickness of laminates (mm)	C/S area B × H (mm^2)
		L (mm)	B (mm)	H (mm)		
I-section		1400	120	120	–	14,400
Plastic lumber beam		1400	120	120	20	14,400
EE-C-F2F		1400	120	120	20	14,400
EE-C		1400	120	120	20	14,400
IE-C-B2B		1400	120	120	20	14,400
IE-C-H		1400	120	120	20	14,400
EE-C-VH		1400	120	120	20	14,400
EE-C-H		1400	120	120	20	14,400
IE-C-10-40-10		1400	120	120	10, 40	14,400
IE-C-25-10-25		1400	120	120	10, 25	14,400
IE-C-10-20-30		1400	120	120	10, 20, 30	14,400

(continued)

Table 2 (continued)

Models	Illustration	Size of the model			Thickness of laminates (mm)	C/S area B × H (mm^2)
		L (mm)	B (mm)	H (mm)		
IE-C-1.44		1400	100	144	17	14,400
IE-C-2.25		1400	80	180	13	14,400

Note Specimen naming was based on certain naming conventions which composed of short forms such as IE which represents internal encasement, EE represents external encasement, C represents cold-formed steel, F2F represents face-to-face, B2B represents back-to-back. H and V represents horizontal and vertical respectively, VH represents combination of vertical and horizontal. 10-40-10, 25-10-25 and 10-20-30 represents various thickness of laminates provided. 1.44 and 2.25 represents the aspect ratio assigned

The Thickness of Laminates Used. A parametric study is also done by the varying thickness of laminates. Outer and inner laminates with different thickness are taken under consideration. The first model coming under this parametric study has laminates of thickness 10 and 40 mm arranged alternatively. In the second model, the thickness of laminates used were 25 and 10 mm whereas, in the third model 10, 20 and 30 mm are the thickness adopted for the laminates. The above three models were compared with plastic lumber beam internally encased with cold-formed steel I-section and the best performing model was determined.

Aspect Ratio. The third parameter is the aspect ratio. Under this parameter, the behaviour of the composite is studied when the depth of the beam is varying. 1, 1.44 and 2.25 are the three cases of aspect ratio that have been taken into consideration.

3 Results

3.1 Encasement Study

An encasement study is conducted to determine the variation in the properties of the composite beam elements when the arrangement of cold-formed steel is altered. Stress Distribution in Fig. 1. shows the stress formed on plastic lumber externally encased with two C-section, externally encased with a single C-section, plastic lumber internally encased with two C-section and internally encased with hatched section is 695.95 N/mm^2, 854.88 N/mm^2, 599.15 N/mm^2 and 554.11 N/mm^2 respectively. All four models are bent with a maximum deflection in the mid-span.

The load–deflection graph comparing all the four models is depicted in Fig. 2. In this graph, it is visible that plastic lumber internally encased with two C-sections has a better performance compared to other models carrying a load of 139.17 kN with a deflection of 69.717 mm. The first model depicting plastic lumber externally encased with two C-sections has the second-best performance, with a load-carrying capacity of 125.753 kN with a deflection of 84.49 mm. The least performing model among the four models is plastic laminates encased with a hatch section with a deflection of 98.184 mm for a load of 101.954 kN.

3.2 Parametric Study

For the parametric study, the most effective design is considered out of the four models analysed in the above study. Three parameters considered under this case are variations in the orientation of plastic laminates, aspect ratio and thickness of laminates within the plastic lumber beams. All the results obtained under the study are tabulated in Table 3.

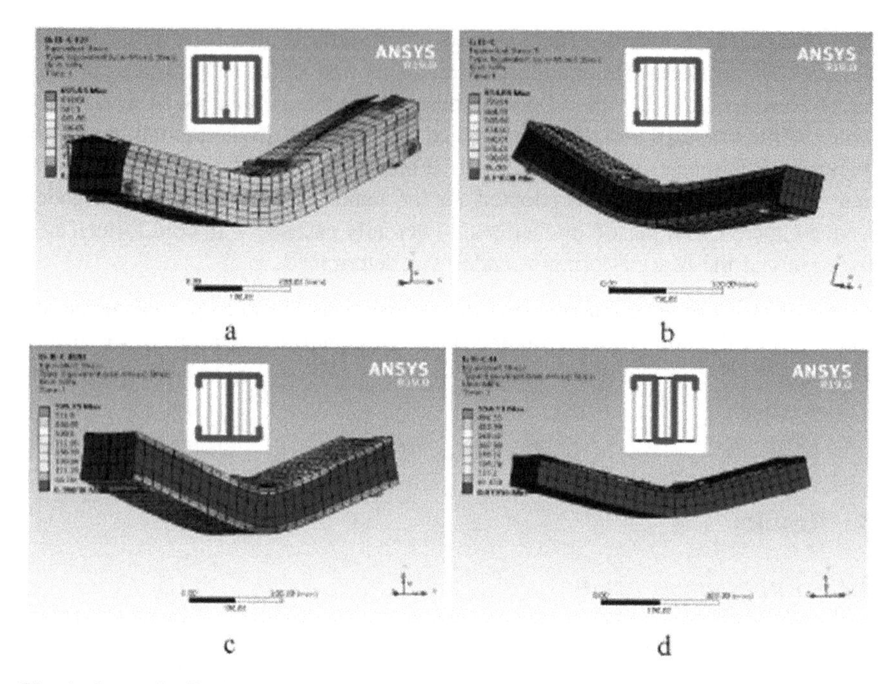

Fig. 1 Stress distribution: **a** plastic lumber externally encased with two C-section, **b** externally encased with single C-section, **c** plastic lumber internally encased with two C-sections, **d** encased with hatched section

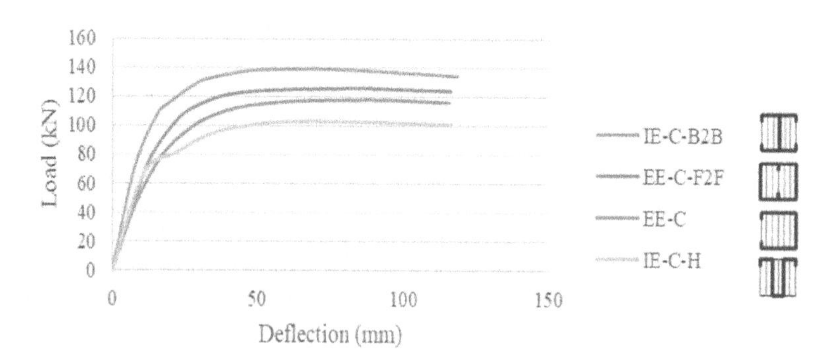

Fig. 2 Load-distribution graph for various encasement studies

Variation in laminate orientation. For different orientations of laminates provided in the composite beam, it was recognised that the most appropriate base model was externally encased plastic lumber with a single C-section from the encasement study as it has a design suitable for providing different laminates orientations and it is one of the least performing models in encasement study. So this model is chosen to examine whether there will be any change in the performance of the alignment of

Table 3 Results from the analysis of models under the proposed studies

Model	P_u (kN)	Δ_u (mm)	k_u (kN/mm)	P_y (kN)	Δ_y (mm)	k_y (kN/mm)	σ_u (N/mm^2)
EE-C-F2F	125.7	84.49	1.49	48.44	7.80	6.21	695.95
EE-C	117.8	89.96	1.31	45.54	8.13	5.60	854.88
IE-C-B2B	139.1	69.72	2.00	72.45	7.94	9.13	599.15
IE-C-H	101.9	98.18	1.04	66.89	10.94	6.12	554.11
EE-C-V	117.8	89.96	1.31	45.54	8.13	5.60	854.88
EE-C-VH	111.1	141.2	0.79	63.06	18.40	3.43	825.23
EE-C-H	100.1	111.2	0.90	62.19	23.84	2.61	720.8
IE-C-B2B	139.1	69.72	2.00	72.45	7.94	9.13	599.15
IE-C-10-40-10	139.0	62.24	2.23	73.18	7.95	9.21	577.69
IE-C-25-10-25	139.1	74.48	1.87	77.92	8.49	9.18	576.83
IE-C-10-20-30	139.3	73.72	1.89	109.76	15.57	7.05	560.03
IE-C-B2B	139.1	69.72	2.00	72.45	7.94	9.13	599.15
IE-C-1.44	160.8	51.05	3.15	132.08	16.74	7.89	606.96
IE-C-2.25	184.7	27.77	6.65	147.45	11.83	12.46	540.89

Note P_u: ultimate load, Δ_u: ultimate deflection, k_u: ultimate stiffness, P_y: yield load, Δ_y: yield deflection, k_y: yield stiffness, σ_u: ultimate stress

the laminates provided is varied. This model had its laminates of thickness 20 mm placed vertically. Figure 3 shows stress distribution of all the three models: single C-section externally encasing vertically oriented laminates, vertical and horizontal laminates and horizontal laminates with ultimate stresses of 854.88 N/mm^2, 825.23 N/mm^2 and 720.8 N/mm^2 respectively. All three models encounter bending failure. But in the case of the horizontal orientation of laminates, a torsional failure is also noticed as the laminates slip out of the encasement.

A graph was plotted with the deflections of all the models on the x-axis and the ultimate load on the y-axis is depicted in Fig. 4. Graph of the vertical laminates model was more curved than that of the horizontal as well as the combination of vertical and horizontal laminates model which is more of a linear shape. Vertically oriented laminates model can be identified as the best performing model under this study as it has a load-carrying capacity of 117.839 kN with a deflection of 89.96 mm. Whereas, the vertical-horizontal combination and horizontal orientation have load-carrying capacities of 111.172 kN and 100.133 kN with a maximum deflection of 141.2 mm and 111.2 mm respectively.

Variation in thickness of laminates. In this study, a composite beam of plastic laminates of equal thickness of 20 mm internally encased with two C-sections to form an I-section is taken as the base model. Other models are created by varying the thickness of laminates which were initially of the same thickness. The second model under consideration was I-section internally encasing laminates of thickness 10 mm, 40 mm, 10 mm placed alternatively on both sides of the web of I-section. The third model consisted of plastic laminates of thickness 25 mm, 10 mm, 25 mm

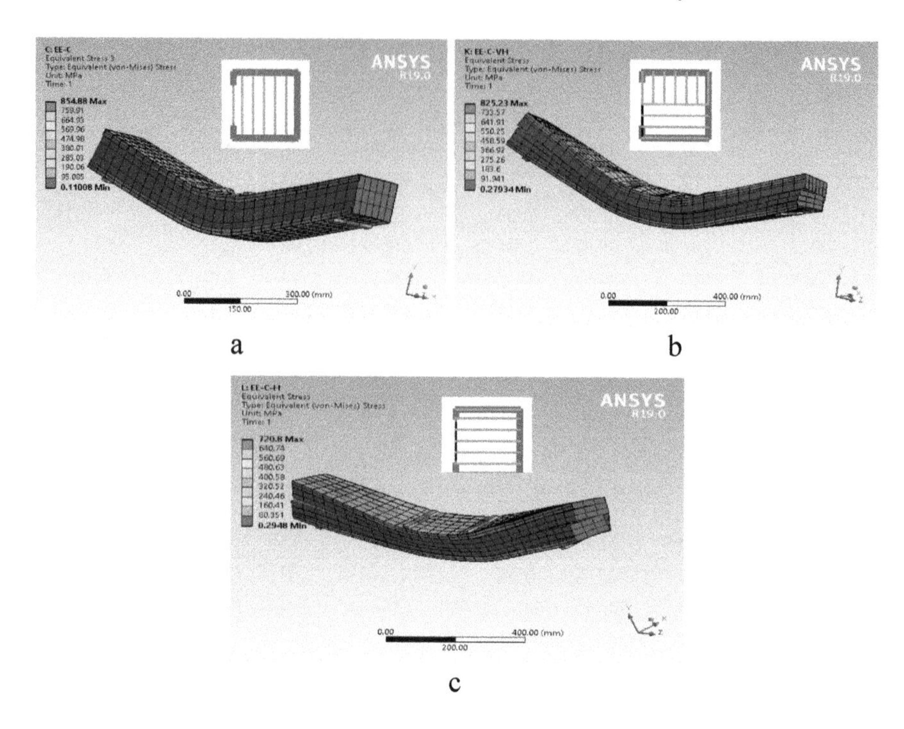

Fig. 3 Stress distribution: **a** single C-section externally encasing vertically oriented laminates, **b** vertical and horizontal laminates, **c** horizontal laminates

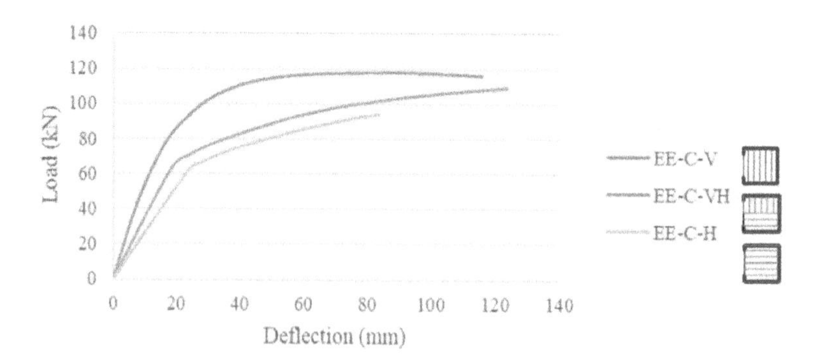

Fig. 4 Load-distribution graph for various alignment studies

placed adjacent to each other on both sides of the web section. The fourth case understudy was the laminate thickness of 30 mm, 20 mm and 10 mm alternatively on both sides of the web of I-section. Figure 5 shows the equivalent stress variation in the models. All the above-mentioned cases have a bending failure with a maximum deflection in the mid-span. The equal thickness laminated model has the ultimate stress of 599.15 N/mm^2. IE-C-10-40-10, IE-C-25-10-25 and

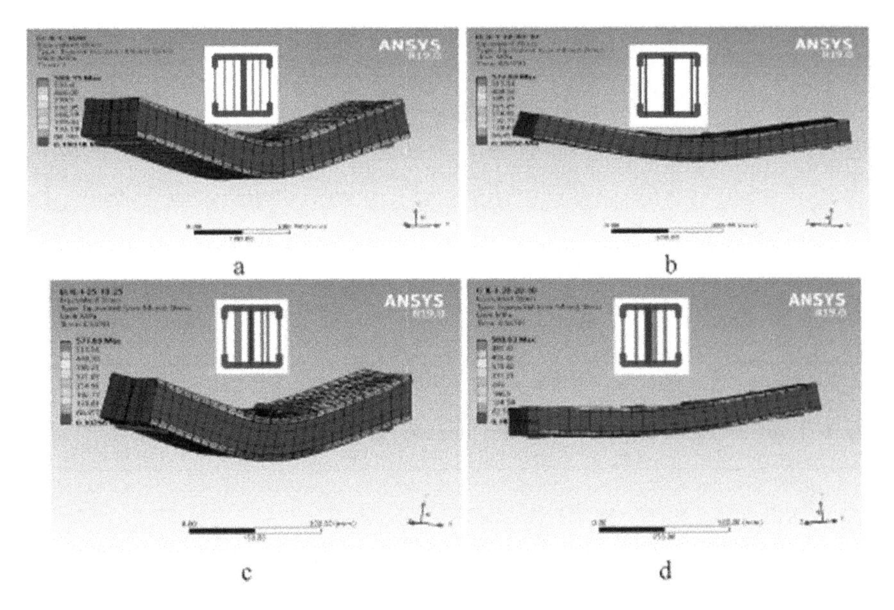

Fig. 5 Stress distribution: **a** I-section internally encasing laminates of thickness 20 mm; **b** laminates of thickness 10 mm, 40 mm, 10 mm; **c** laminate thickness of 25 mm,10 mm, 25 mm; **d** laminate thickness of 10 mm, 20 mm, 30 mm

IE-C-10-20-30 models have ultimate stresses of 577.69 N/mm^2, 576.83 N/mm^2, and 560.03 N/mm^2 respectively. It is therefore noted that as long as the width of the cross-section is constant, variation in the thickness of laminates has very less influence on the strength of the beam elements.

From the load–deflection graph shown in Fig. 6, it is seen that the thickness of laminates does not influence much on the characteristic strength of the beams. IE-C-B2B, IE-C-10-40-10, IE-C-25-10-25 and IE-C-10-20-30 models have ultimate load-capacity of 139.17 kN, 139.041 kN, 139.157 kN and 139.325 kN corresponding to deflections 69.717 mm, 62.238 mm, 74.483 mm and 73.72 mm.

Variation in the aspect ratios. The third criteria under parametric study are aspect ratio. The aspect ratio can be defined in this study as the proportion of breadth and depth of the beam element. The base model for this study is the model IE-C-B2B which has an aspect ratio of 1. The breadth and depth of the beam are 120 mm. Other aspect ratios considered for the study are 1.44 and 2.25. Figure 7 shows the stress distribution of models IE-C-B2B, IE-C-1.44 and IE-C-2.25 which has ultimate stress of 599.15 N/mm^2, 606.96 N/mm^2 and 540.89 N/mm^2 respectively. In all the cases, there is bending in the beams. But as the width of the beam reduces, there is crippling in the flanges of the I-section. It is also visible that as the depth of the beam increases, deflection in the beams is also reduced as the moment of inertia is increasing.

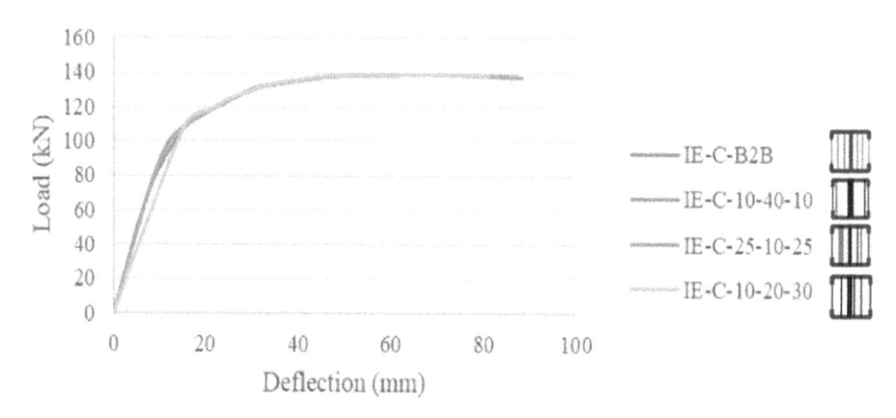

Fig. 6 Load–deflection graph for laminates thickness variation study

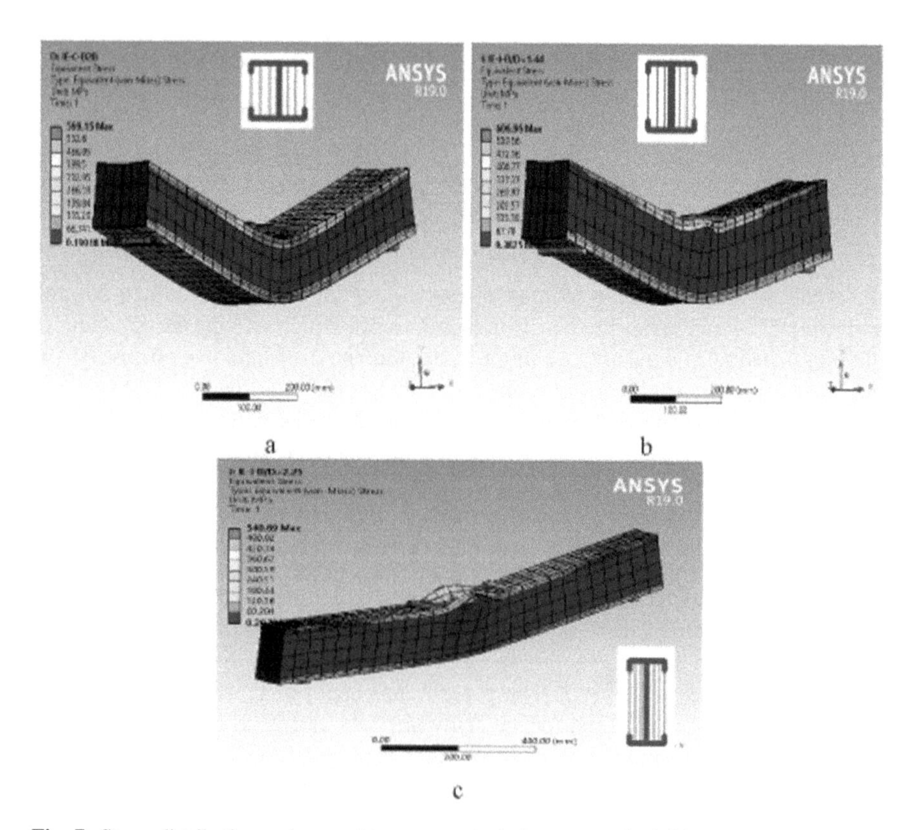

Fig. 7 Stress distribution: **a** beam with aspect ratio 1, **b** aspect ratio 1.44, **c** aspect ratio 2.25

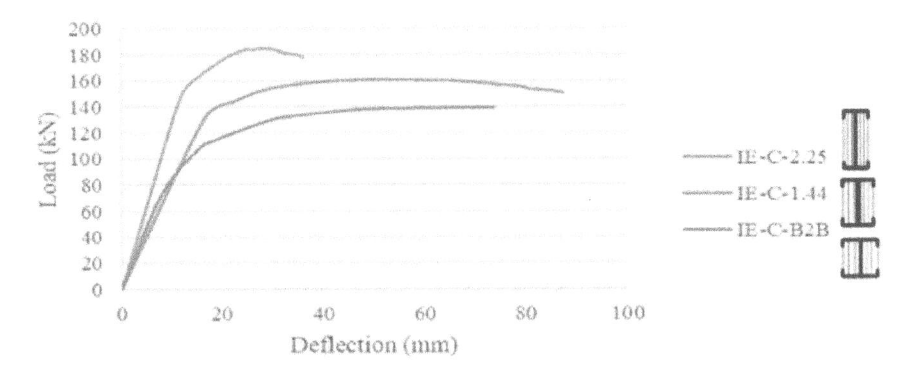

Fig. 8 Load–deflection graph based on variation in the aspect ratio

The load–deflection graph for the models of different aspect ratios is depicted in Fig. 8. IE-C-B2B, IE-C-1.44 and IE-C-2.25 have a load-carrying capacity of 139.17 kN, 160.84 kN and 184.793 kN corresponding to deflections 69.717 mm, 51.049 mm and 27.771 mm respectively. It is clear from the graph that as the aspect ratio of the beam increases, the load-carrying capacity of the beam increases whereas the deflection decreases. This is because of the increase in depth which in turn increases the moment of inertia of the beam. So, it is inferred that as the aspect ratio of the beam increases, the strength also improves.

All the results obtained from the encasement study and parametric studies are tabulated under Table 3. It consists of ultimate load (P_u), yield load (P_y), ultimate deflection (Δ_u), yield deflection (Δ_y), ultimate stiffness (k_u), yield stiffness (k_y) and ultimate stress (s_u) of all the models analysed under each study proposed in this paper.

4 Comparison of Structural Behaviour of Models

The ultimate load carried by all the models taken into consideration in the encasement and parametric study is compared with the ultimate load-carrying capacity of the I-section beam as well as the plastic lumber beam. Figure 9 depicts a graph for the comparison of percentage increase in ultimate load in different models compared to I-section and plastic lumber beam.

When the case of the I-section is considered, model IE-C-2.25 has the highest percentage of increase in the ultimate load of 175.07% and comparing it with the plastic lumber beam, it has an increment of 295%. This is the model which performed best in the aspect ratio study. Compared to the plastic lumber beam, I-section has 30.36% more strength. In the whole of the study, the hatch section composite model IE-C-H has the least increment of 49.05% compared to the I-section and 114.04% concerning plastic lumber beam. It is proven from the graph that all the composite beam models have better performance than both the I-section beam as well as plastic lumber beam.

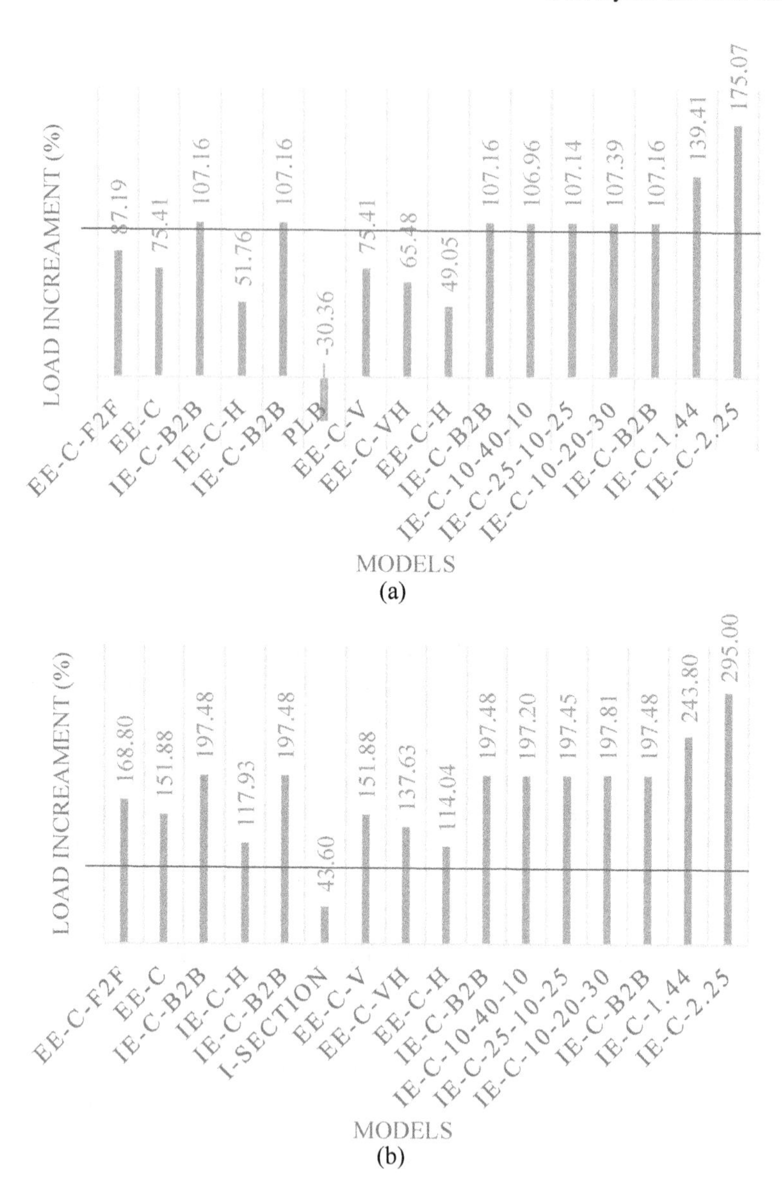

Fig. 9 Comparison of percentage increase in ultimate load in different models compared to **a** I-section, **b** plastic lumber beam

5 Conclusions

The flexural behaviour of light-weight composite plastic lumber beams encased with cold-formed steel under four-point bending was investigated analytically in this paper. One of the goals was to develop a composite beam with improved structural performance. Another goal was to develop a sustainable structural element that is made out of renewable, amply available, cost-efficient material. The use of plastic lumber as a structural element reduces the extensive use of wood for construction purposes. It was also intended to use recycled plastic to reduce plastic debris going into nature causing all forms of pollution. Based on the analytical results obtained from the encasement study and parametric studies, the conclusions can be summarised as follows:

1. In the encasement study, it is observed that plastic lumber internally encased with two C-sections (IE-C-B2B) has a better performance compared to the plastic lumber model (PLB) with a percentage load increment of 197.48%.
2. In the encasement study, it is observed that plastic lumber internally encased with two C-sections (IE-C-B2B) has a better performance compared to the I section model with a percentage load increment of 107.16%.
3. In the orientation study, the vertically oriented laminates model (EE-C-V) can be identified as the best performing model under the laminates alignment study, with 175.41% load-carrying capacity than the I-section.
4. In the orientation study, the vertically oriented laminates model (EE-C-V) can be identified as the best performing model under the laminates alignment study, with 151.88% load-carrying capacity than the horizontally oriented plastic lumber beam (PLB).
5. In the laminate thickness variation study, it is inferred from the results that as long as the width of the beam is kept constant, variation in the thickness of laminates used has least influence on the strength of the beam elements.
6. In the variation of aspect ratio study, it can be concluded that composite beam with aspect ratio 2.25 (IE-C-2.25) has the highest percentage of increase in the ultimate load of 175.07% when compared to I-section and comparing it with the plastic lumber beam (PLB), it has an increment of 295%.
7. In the variation of aspect ratio study, it is inferred that as the aspect ratio (depth to breadth ratio) of the beam increases, the load-carrying capacity of the beam increases and the deflection decreases. This is due to the increase in depth which in turn magnifies the moment of inertia of the beam.

References

1. Keskisaari A, Kärki T (2017) Raw material potential of recyclable materials for fiber composites: a review study. J Mater Cycles Waste Manage 19(3):1136–1143
2. Goli V, Sai SN, Mohammad A, Singh DN (2020) Application of municipal plastic waste as a manmade neo-construction material: issues and wayforward. Res Conserv Recycl 161: 105008
3. Carroll DR, Stone RB, Sirignano AM, Saindon RM, Gose SC, Friedman MA (2003) Structural properties of recycled hdpeplastic lumber decking planks. Int J Polym Mater 52(8): 709–724
4. Krishnaswamy P, Francini R (2000) Long-term durability of recycled plastic lumber in structural applications
5. Mengeloglu F, Kurt R, Gardner DJ (2007) Mechanical properties of extruded high density polyethylene and polypropylene wood flour decking boards 477–487
6. Townsend T, Tolaymat T, Solo-Gabriele H, Dubey B, Stook K, Wadanambi L (2004) Leaching of CCA-treated wood: implications for waste disposa. J Hazard Mater 114(1–3): 75–91
7. Dias BZ, de Alvarez CE (2017) Mechanical properties: wood lumber versus plastic lumber and thermoplastic composites. Ambiente Construído 17.2:201–219
8. Krishnaswamy P, Lampo R (2001) Recycled-plastic lumber standards: from waste plastics to markets for plastic-lumber bridges. Standard News 2
9. Adil DM et al (2018) Structural performance of cold-formed steel composite beams. Steel Compos Struct 27.5:545–554
10. dos Santos FA, Canto LB, da Silva ALN, Visconte LLY, Pacheco EBV (2018) Processing and properties of plastic lumber. In: Thermosoftening plastics. IntechOpen
11. ISO 898-1:2013, https://www.iso.org/standard/60610.html
12. Sundararajah L, Mahendran M, Keerthan P (2017) Web crippling studies of SupaCee sections under two flange load cases. Eng Struct 153:582–597
13. Qi Y et al (2017) Bending performance of GFRP-wood sandwich beams with lattice-web reinforcement in flatwise and sidewise directions. Constr Build Mater 156:532–545
14. Nadir Y, Nagarajan P (2014) The behavior of horizontally glued laminated beams using rubber wood. Constr Build Mater 55:398–405

Structural Optimisation of Hyperbolic Paraboloid Shell Foundation

Amrutha Joseph and P. E. Kavitha

Abstract Foundations are the structural elements used to transfer super structural loads to ground without failure. It is important that the foundation should be strong enough to resist all kinds of failures. Shell foundations are proved to be performing much better than conventional flat footings especially when transmitting heavy superstructural loads to poor bearing soil. Shells have larger stiffness and strength compared with plane surface structural elements which ensure minimum material and maximum structural performance. So, it is more economical where the materials are scarce and overpriced. Different types of shells are used in foundations such as conical shell, triangular shell, pyramidal shell, hyperbolic paraboloid shell, cylindrical shell inverted dome, elliptic paraboloid, inverted spherical shell etc. In this research the structural performance of hyperbolic paraboloid (hypar) shell foundation is investigated using finite element software. The thickness of the shell is optimised based on the stress distribution behaviour of hypar foundation.

Keywords Hypar foundation · Shell thickness · Bearing capacity · Settlement

1 Introduction

Foundation is the least visible aspect of an engineering structure, but still the most significant since it is responsible for supporting the structure's layers. The foundation is also recognized as the most basic prerequisite for transferring all load components from the superstructure to the substructure. After a great deal of scientific study and innovation, various types of foundation were established and put into effect. The idea of using a shell type foundation design is not new, as it has been used in the building industry since the inverted brick arch foundation was first used. In developing parts of the world, in addition to traditional foundations, an alternative foundation has become popular. Shell foundations outperform traditional flat footings, especially when heavy superstructural loads must be transferred

A. Joseph (✉) · P. E. Kavitha
Department of Civil Engineering, FISAT, Angamaly, Kerala, India

G. C. Marano et al. (eds.), *Proceedings of SECON'21*, Lecture Notes in Civil Engineering 171, https://doi.org/10.1007/978-3-030-80312-4_35

to poor bearing soil [1]. Shells are the structures which show maximum structural properties as compared with the materials they used. They derive their strength from 'form' rather than 'mass'. These shell foundations have shorter spans and a lower rise to thickness ratio than shell roofs. It's important to remember that the loads that the shells would have to bear as base structures will be much greater than in roofs. Shells when used as elements in the foundation are decisively more economical where labour is cheap but cost of materials are high [2]. In developing countries around the world, where construction materials are scarce and expensive but labour is relatively cheap and plentiful, the resulting economy is important [3]. To improve strength and work efficiency in soil, shell footings as foundations rely heavily on their geometrical shape and streamlined continuity. As a result, shells are thin slab structures whose performance as a supporting element is highly dependent on their shape and the quality of the material used in their construction. Shell foundations are commonly constructed in arched, circular, square, triangular, spherical, conical, cylindrical, hyperbolic, parabolic, pyramidal and strip shapes to achieve optimum structural efficiency [4]. Because of their single point of discontinuity, hypar shells are well suited to support single column loads. The most adaptable feature of its geometry is its straight-line property, which gives it all the benefits of both a shell and a plain surface [5]. Here the hyperbolic paraboloid shell foundation is studied based on stress distribution by varying the shell thickness.

2 Literature Review

Syed Jalaludeen Shah et.al studied the effect of varying ratios of rise to lateral dimension of shell, soil state and interface roughness of hypar shell footing and were also compared with ordinary rectangular footing by conducting non-linear static analysis. He considered both the clayey and sandy soils to determine the influence of the rise of the shell in terms of settlement and stress [6]. Ahmed et al. has designed and compared hypar footing with conventional footing based on the economic aspect. The hypar shell footing was found more economical than that of conventional footing, and it saves up to 43.78% of concrete and 4.76% of steel, respectively. It gives minimum materials consumption over the conventional footing [1] Aziz et al. has studied the influence of varying parameters like ridge beam cross sectional dimension, edge beam dimension and shell thickness on the bending moment, shear force and axial force of the structure [7]. Krishnan et al. has evaluated the change in geotechnical behaviour of Hypar shell footing with varying edge beam configuration and compared with an equivalent flat slab footing. The addition of edge beam at the bottom of the shell footing was found to increases the load carrying ability of the footing [8]. According to Mohammed et al. study, he considered embedment ratio R = 0, 0.5, 1. based on the load settlement study he inferred that completely embedded shell footing is shown to have a better load carrying capacity compared with the footing with no embedment [9].

By reviewing the various previous researches, it has been found that the concept of shell foundation is a better alternative as compared to the normal foundation especially in case of foundation structures in low bearing capacity soil. compared to the conventional footing it shows better performance in settlement characteristics and also cost effective in the aspect of material saving [10]. Even though there are various types of shell foundations there, conical and hyperbolic paraboloid shell foundations are mostly constructed shell foundations. Research in structural behaviour is considerably less compared to geotechnical behaviour. Hence a detailed analysis on this topic found to be necessary.

3 Geometric Model and Material Property

3.1 Geometry of Foundation

For the analysis the base model of the hyperbolic paraboloid shell footing was modelled with an external dimension of the footing as 2.4 m × 2.4 m with a ratio of rise to the lateral dimension as 0.6 m. The dimension of column provided was 500 mm × 500 mm. The shell is completely embedded in the soil. The dimension details of the hypar shell footing are taken as per the details mentioned in [8]. The allowable minimum thickness of the shell as per IS 9456-1980 is 150 mm and allowable maximum thickness is as per thin plate criteria (The ratio of thickness to the least dimension should not exceed 1/10) is 240 mm. So, the present study is conducted by varying the shell thickness within the above range i.e., 160, 170, 180, 190, 200 and 220 mm. Figure 1 shows the geometric model of the hypar foundation modelled in the ANSYS Workbench.

3.2 Geometry of the Soil Model

Table 1 list the properties of reinforced concrete and soil properties. The soil of dimension 1730 cm × 1780 cm × 1450 cm was modeled and the property of soil mentioned in the table was assigned.

3.3 Material Property

The material properties of the reinforced concrete and sand used for the analysis are given in Table 1. The footing material used is reinforced concrete with Modulus of Elasticity of concrete, Ec = 2.7×10^7 and Poisson's ratio = 0.3. The medium dense sand used has a modulus of elasticity 3×10^4 kN/m^2 and poisson's

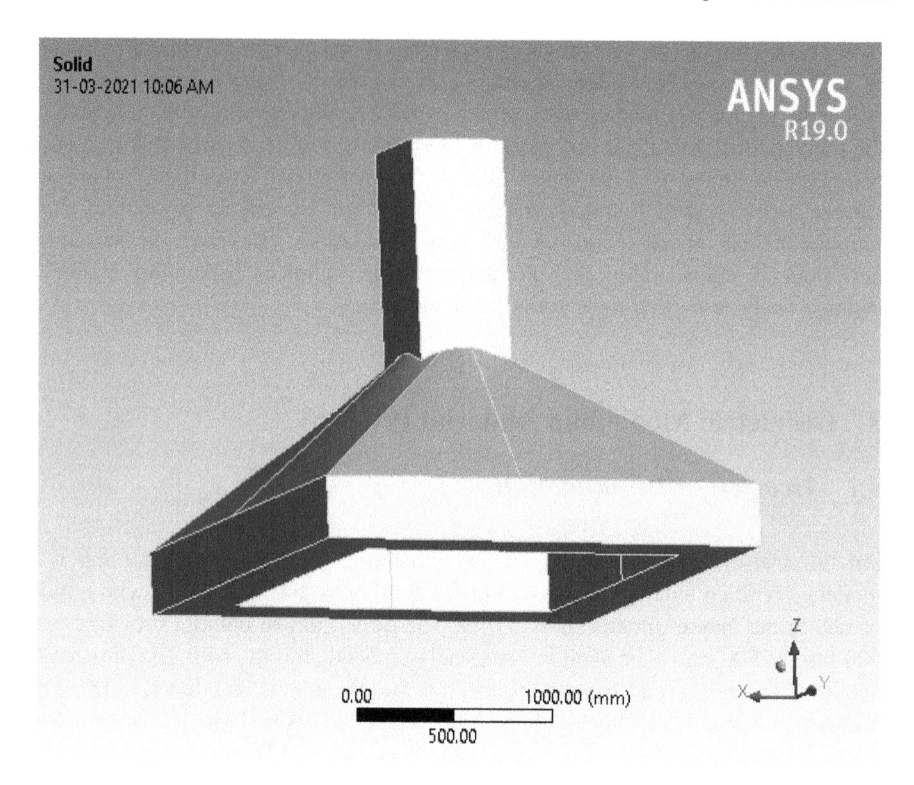

Fig. 1 Modelled hypar foundation

Table 1 Material properties

Sl. No	Properties	Reinforced concrete	Medium dense sand [8]
1	Modulus of elasticity (kN/m^2)	2.7×10^7	3×10^4
2	Poisson's ratio	0.18	0.3
3	Angle of internal friction (ϕ)	–	33
4	Cohesion (kN/m^2)	–	5
5	Density (kN/m^3)	2300	1700

ratio = 0.3. Hypar shell footing modelled has a dimension of 2.4 m × 2.4 m. The dimension of soil considered is 1.75 m × 1.75 m and the depth of the soil block considered is 1.45 m. The orientation of the model is according to the coordinate system shown in Fig. 2. The height of the soil model is directed along the Z axis and XY plane is chosen as the plane for footing. Different properties of the reinforced concrete and medium dense sand were assigned to various parts of the model in the static analysis.

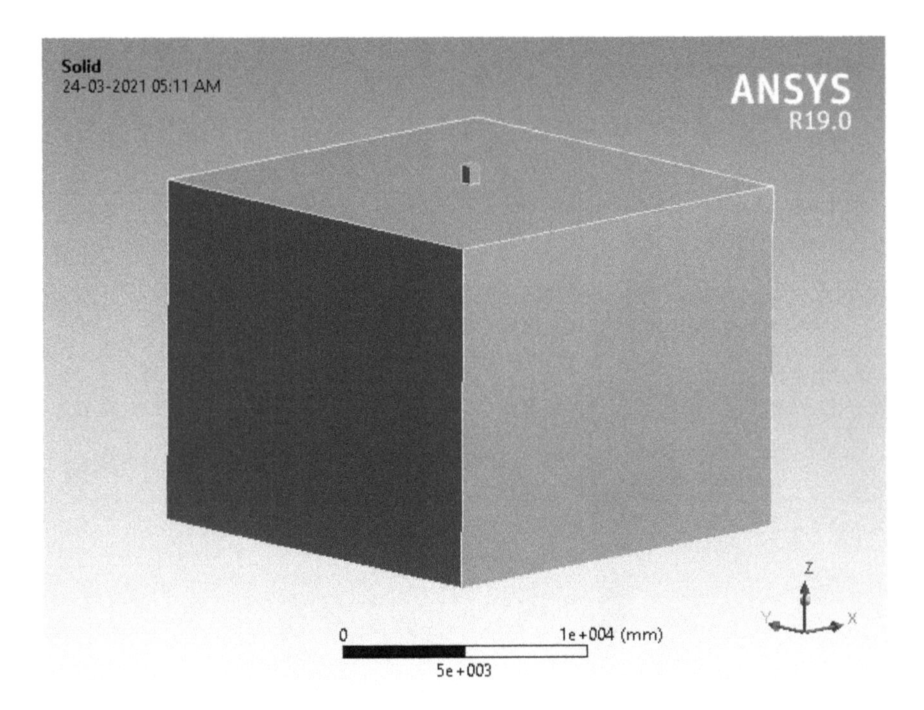

Fig. 2 Modelled hypar foundation fully embedded in the soil block

4 Numerical Analysis

The analysis was done by varying the thickness of shell such as 160, 180, 200 and 220 mm in order to analyses the effect of this on the stress distribution pattern. The four different models with varying thickness were created with the material properties as mentioned in Table 1. The shell is fully embedded in the soil. The footing and the soil are connected by bonded connection.

An incremental load is applied on the column to determine the stress distribution on the structure. The side boundaries of the soil model are limited to allow for soil movements at the model edges while preventing lateral movement, as shown in Fig. 3. The load carried by the structure and the settlement corresponding to the maximum elastic stress value of 10 MPa of the M30 concrete (IS 456-2000 Table 21) is tabulated in Table 2 and the optimum thickness of the shell is suggested. Also the maximum load value with respect to the allowable settlement of our soil (50 mm as per IS 1904: 1986 Table 1) is also tabulated and compared.

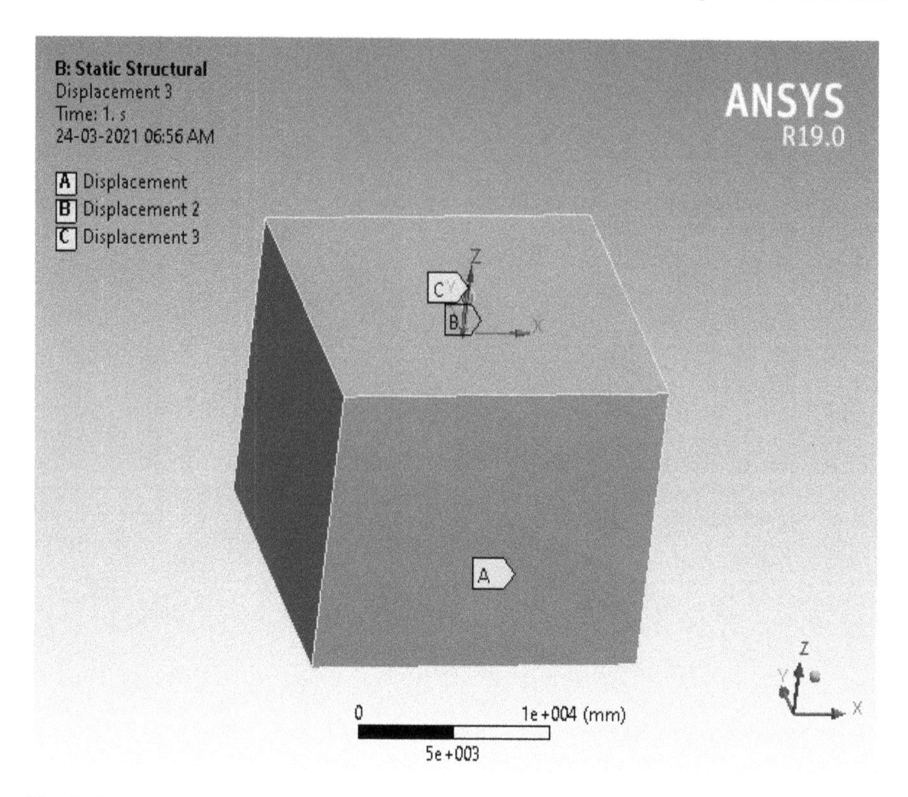

Fig. 3 Boundary condition provided

Table 2 Consolidated load settlement value

Sl. No	Thickness of shell (mm)	Load corresponding to max. allowable stress in concrete (kN)	Displacement corresponding to max. allowable stress in concrete (mm)	Load corresponding to allowable settlement (kN)
1	160	36.57	29.00	50.10
2	180	44.52	24.25	76.83
3	200	59.26	25.50	109.4
4	220	56.81	31.00	80

5 Result and Discussion

The analysis was done by varying the thickness of the shell as 160, 180, 200 and 220 mm and the stress distribution on the shell is found as shown in Fig. 4. The allowable stress on concrete is used to calculate the ultimate load carrying capacity. The load corresponding to the equivalent elastic stress value of concrete exceeds 10 MPa is considered as the ultimate load carrying capacity.

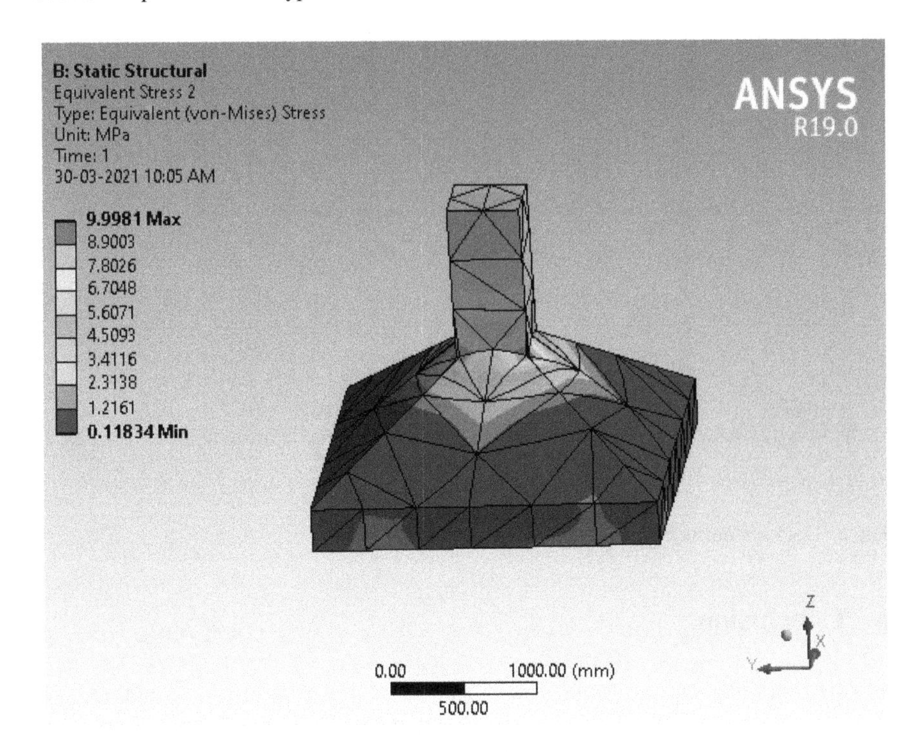

Fig. 4 Elastic stress contours for 200 mm shell thickness

The hyperbolic paraboloid shell foundation with varying thickness160, 180, and 200 mm showed an increase in the load carrying capacity as shown in Table 2. But as the thickness increased to 220 mm there was a sudden decrease in load carrying capacity. Which indicates that the behaviour of structure changes from shell to beam when thickness is increased beyond optimum thickness. The load settlement curve in Fig. 5 depicts the pattern of variance in load carrying capacity. It is clear that the load carrying capacity of the shell foundation with thickness 180, 200, and 220 mm showed an increment of 21.73, 62.04 and 55.34% from that of 160 mm thick shell foundation. The same pattern of increase in load carrying capacity is observed corresponding to maximum allowable settlement as per the codal provisions also. This shows that as the thickness increases, the ultimate load carrying capacity also increases which is due to the increase in rigidity of the foundation. But beyond the thickness of 200 mm, the foundation showed a reduction in load carrying capacity at the thickness of 220 mm.

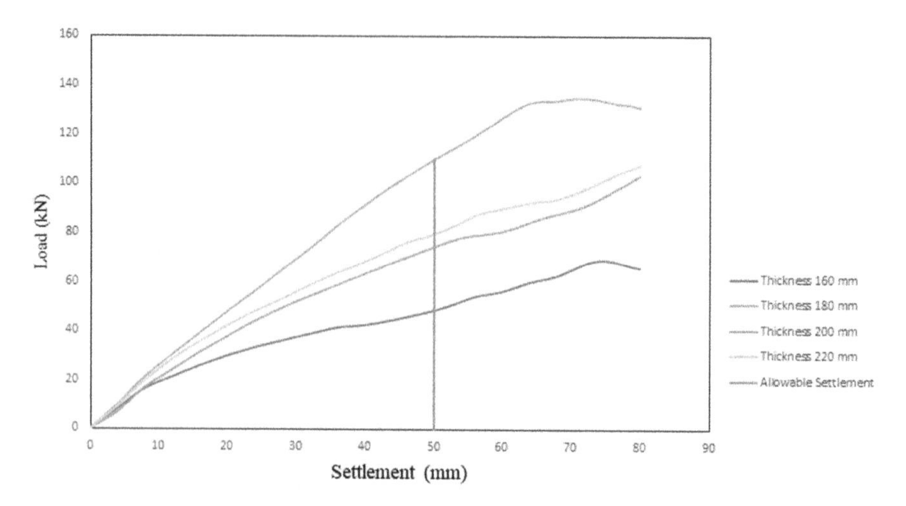

Fig. 5 Load settlement curve

6 Conclusion

The present study discusses the static analysis on the hyperbolic paraboloid shell foundation by varying the thickness of the shell from 160, 180, 200 and 220 mm. The load carrying capacity and settlement of the hyperbolic paraboloid shell foundation corresponding to the permissible stress value of the concrete showed that, as the thickness of the shell increases the load value also increases upto a shell thickness of 200 mm and on increasing the shell thickness further the load carrying capacity decreases. It may be due to the reduction in membrane stress as the shell structure and the loss of its shell behavior. Hence it can be concluded that an increase in thickness of shell improves the load carrying capacity upto a certain optimum thickness.

References

1. Ahmed SM, Kewate S (2015) Analysis and design of shell foundation: IS: 9456–1980 provision. Int J Sci Eng Res 6(12)
2. Salunkhe M, Yadav M, Pal S, Patil S, Mhatre V (2016) Conical and hyperbolic paraboloid shell foundation. Int J Recent Innov Trends Comput Commun 4(4). ISSN: 2321-8169
3. Kurian NP, Devaki VMJ (2005) Analytical studies on the geotechnical performance of shell foundation. Can Geotech J 42:562–573
4. Stavridis LT, Armenakas AE (1988) Analysis of shallow shells with rectangular projection: applications. J Eng Mech ASCE 114(6)
5. Shoukath S, Rajesh AK (2017) Seismic performance of hyperbolic paraboloid and inverted spherical shell foundation. Int Res J Adv Eng Sci 2(2):148–151

6. Jalaludeen Shah S, Sageer T, Sidharthan A (2016) Behavior of hyperbolic paraboloid shell footings under point loading. Int J Sci Eng Res 7(10)
7. Aziz RJ, Al-Azzawi AA, Al-Ani AA (2011) Finite element elastic analysis of hypar shells on winkler foundation. J Serb Soc Comput Mech 5(1):1–18
8. Krishnan AV, Sivapriya SV, Nagarajan V (2017) Finite element analysis of HYPAR shell footings with variation in edge beam dimensions and embedment ratio. Int J Earth Sci Eng 10 (02):150–154. ISSN: 0974-5904
9. Huat BBK, Mohammed TA (2006) Finite element study using FE code (PLAXIS) on the geotechnical behavior of shell footings. J Comput Sci 2(1):104–108. ISSN: 1549-3636
10. Faisal Mohammed M, Singh BK, Pandey VK (2018) Shell foundations: a complete review. Int J Technol Res Eng 5(7)

Evaluation of Modulus of Elasticity of Plastic Aggregate Concrete

Aleesha Anna Saju and K. K. Smitha

Abstract Modulus of elasticity of a material is the ratio of applied stress to the corresponding strain within elastic limit. It indicates the resistance of a material to deformation. The elastic modulus of concrete can be calculated using empirical formulae provided by different codes. These formulae are based on the relation between elastic modulus of concrete and characteristic compressive strength. The strength of concrete depends on the proportions and modulus of elasticity of the aggregate. The codes are providing formulae for the determination of elastic modulus of conventional concrete. Replacement of materials in concrete is gaining attention all over the world nowadays. The applicability of these empirical formulae for such concretes are not clear. So, evaluating the elastic modulus of concrete with replacement components become essential. In this experimental study, the modulus of elasticity of plastic aggregate concrete is evaluated. Plastic aggregate concrete refers to the concrete mixed with PET bottle waste as replacement for coarse aggregates on various proportions. 5, 10, 15% of coarse aggregate are replaced with plastic aggregates and elastic modulus is studied.

Keywords Modulus of elasticity · Concrete · Plastic aggregates · Plastic aggregate concrete

1 Introduction

The modulus of elasticity is the measurement of a substance's resistance to being deformed elastically when a stress is applied to it. The empirical equations for elastic modulus of concrete are derived with relation between elastic modulus and compressive strength of concrete. The applicability of these formulae for different concretes developed by replacing or addition of various materials are not clear. So, it is important to evaluate the elastic modulus of such types of concrete. With

A. A. Saju (✉) · K. K. Smitha
Toc H Institute of Science and Technology, Arakkunnam, Kochi, India
e-mail: smithakk@tistcochin.edu.in

G. C. Marano et al. (eds.), *Proceedings of SECON'21*, Lecture Notes in Civil Engineering 171, https://doi.org/10.1007/978-3-030-80312-4_36

increase in the population, the use of plastic has increased which created problems in its disposal. Including plastic in concrete can be one way of eliminating the disposal problem of plastic wastes and the excessive extraction of natural aggregates. Even though the suitability of plastic wastes as aggregates in the concrete are studied, its structural behaviour is not seriously noticed. Polyethylene Terephthalate (PET) bottle waste is used as partial replacement for coarse aggregate in this study. The modulus of elasticity of concrete containing plastic aggregates is evaluated and an equation is proposed for modulus of elasticity of plastic aggregate concrete in terms of compressive strength.

2 Literature Survey

There are several studies regarding the suitability of plastic as aggregates in concrete. Plastic can be incorporated in concrete as fine aggregates, coarse aggregates and fibres. Studies show that plastic aggregates showed similar strength development properties like conventional concrete [1]. Coarse aggregate replacement with plastic waste in 10 and 15% showed increased flexural and tensile strength of concrete [1–5]. The plastic incorporation made concrete more ductile and were able to take additional load even after they crack [1, 4]. Comparable compressive strength was observed for the plastic incorporated concrete [1, 4, 6]. Increased split tensile strength was observed when plastic used as coarse aggregate and up to 20% replacement showed good results [3, 6, 7]. Slump value and compressive strength showed increase with increase in percentage replacement with plastic [4, 8, 9]. Incorporation of plastic made concrete more lightweight and it helps to reduce the earthquake forces [4, 9].

Fine aggregate replaced with PET waste in 5, 7.5, 15% by volume showed 31.6% reduction in drying shrinkage and 13% decrease in density. 3.6% reduction in compressive strength was observed for 7.5% replacement with fine plastic waste. 3.5% reduction in water absorption was observed in 7.5% flaky plastic aggregate mix [7]. Inclusion of rubber latex along with plastic waste showed decrease in workability by 3.1%. 17.8% decrease in slump value and 9% increase in impact energy was observed in 2% plastic mixed concrete. The acid and sulphate ingress reduced up to 20% and 28% for plastic mixed concrete [2]. 10% of sand was replaced with plastic aggregate having different sizes. 23.5% reduction in density of concrete was observed for 2–4 mm sized PET aggregate concrete. 1.2% increase in compressive strength was observed in PET mixed concrete [5].

A study about the structural behaviour of concrete beams and columns reinforced with waste plastic incorporated rebars indicate that use of plastic powder in the rebars which leads to increased tensile and compressive strengths of rebars by 17.2% and 19.3% respectively, hence proving to be a strength additive material. 50% increase in bond strength of this plastic incorporated rebars with concrete was observed in the pull-out test [10].

Many studies are carried out about the evaluation of elastic modulus of different types of concrete. Elastic modulus of concrete is usually expressed in terms of compressive strength. Mechanical properties of concrete mainly depend on the properties and proportions of binder and aggregates used [11–13]. Modulus of elasticity is proportional to square root of compressive strength and square of unit weight of concrete [14]. From the study using silica fume and three different aggregates, new equations were proposed for modulus of elasticity in terms of compressive strength [12].

$$E_C = 10.25 f_c^{0.316} \text{ (limestone)}$$

$$E_C = 8 f_c^{0.352} \text{ (Andesite)}$$

$$E_C = 10.75 f_c^{0.312} \text{ (Quartzite)}$$

The growth of modulus of elasticity depends on compressive strength increase though this assumption is not satisfying for all concretes. Elastic modulus increased with age of concrete [15]. Statistical analysis study on relationship between modulus of elasticity and compressive strength of recycled aggregate concrete showed that up to 30% replacement of recycled aggregates has minimal effects on elastic modulus. Utilization of recycled aggregates as coarse aggregate and fine aggregates showed 15 and 40% reduction in elastic modulus. Elastic modulus is influenced by cement paste, aggregate and age of concrete and new equations were proposed for elastic modulus of recycled aggregate concrete [16].

$$E_C = 12.96\left(\frac{\text{fc}}{10}\right)^{0.3} \text{ (50\% replacement)}$$

$$E_C = 18.26\left(\frac{\text{fc}}{10}\right)^{0.3} \text{ (100\%replacement)}$$

Investigation of performance of empirical and composite models for predicting the modulus of elasticity of high strength concretes incorporating ground pumice and silica fume used compressive strength and unit weight of concrete to predict elastic modulus by common empirical models. Model predictions were then compared to experimental elastic modulus. It was found that prediction equations depend on temperature, aggregate type, w/c ratio. Compressive strength has a significant role in predicting elastic modulus of high strength concrete [17]. In another study, modulus of elasticity of ultra-high-performance concrete was evaluated. Data were collected from literatures and also tested to evaluate its accuracy. When compressive strength increased elastic modulus also increased but not at the same rate as conventional concrete. Compressive strength varied from 124 to 162 MPa and elastic modulus from 37 to 46 GPa. New equation was proposed to predict elastic modulus with reasonable prediction accuracy [18].

$$E_C = 8010 f_c^{0.36}$$

Evaluation and comparison of analytical models to estimate elastic modulus in normal strength concretes were done in another study. In majority of existing models, elastic modulus is being underestimated. Modulus of elasticity and compressive strength are parameters which are important in design of structures [19, 20]. The proposed model was based on the equation $E_C = 9.19 f_c^{0.354}$, gives good estimation of elastic modulus of concrete for compressive strength from 20 to 50 MPa [19].

3 Methodology

The methodology adopted is as follows:

1. Selection of topic for the study: The research problem was identified.
2. Literature Review: The information required for the selected topic was collected from literature survey.
3. Collection of Materials: The materials such as cement, coarse aggregate, fine aggregate, PET bottle wastes, superplasticizers were collected.
4. Tests on Materials: The material testing was carried out for determining its properties.
5. Mix Design: Mix was designed for M25 grade of concrete.
6. Casting of Cylinder Specimens: Cylinder specimens (150 × 300 mm) were casted for control mix M25 grades of concrete and for plastic aggregate concrete.
7. Testing of Specimen: Compressive strength for 28 day and stress–strain readings using compressometer were done on cylinder specimens.
8. Data Analysis: Results from the tests were analysed and a relation between elastic modulus and compressive strength of concrete was developed.
9. Interpretation of Results: Conclusion and discussions were derived from the analysis of test results.

4 Materials and Properties

The materials used in the experimental study are cement (PPC), fine aggregate (M sand), coarse aggregate (20 mm), PET bottle flakes as replacement material for coarse aggregate and superplasticizer. In the present experimental work, Portland Pozzolana cement was used confirming to IS 1489:1991 (Part 1). Cement is generally used as the main binder material.

Fine aggregates usually consist of river sand. Due to the environmental impact of river excavations, M-sand was used as the fine aggregate in the present study. M-sand used was confirming to Zone I of IS 383:1970 (Reaffirmed 2016). The coarse aggregates are generally crushed stones with size ranging from 20 to 4.75 mm. They occupy around 70% of the total volume of concrete. The coarse aggregates used were with a nominal size of 20 mm downgraded crushed aggregates. The properties of coarse aggregates were confirming to IS 383:1970 (Reaffirmed 2016).

Plastic used in this research is PET (polyethylene terephthalate) bottle flakes. The bottles were collected from the defective bottle waste in manufacturing companies. The bottle neck and bottom were cut using the plastic bottle neck cutting machine and it is crushed into the required size using plastic bottle crushing machine. The size ranges from 14 to 18 mm.

To increase the workability of concrete superplasticizer is used. The superplasticizer used in the present study is Conflo LN. It is a high molecular weight polymer based on lignosulphonate and is chloride free. It conforms to BS 5075 Part 1:1982 and IS 2645: 1978. A dosage range of 0.15% to 0.5% by weight of cement is recommended. Water is an important ingredient of concrete, as it participates actively in the chemical reaction with cement. The strength of cement concrete is obtained from the bonding action of the hydrated cement gel. Table 1 gives the various tests and properties of the materials.

5 Mix Design

The mix was designed as per IS 10262:2019. The coarse aggregate is replaced with plastic flakes by volume. Table 2 give the proportions of the components in the plastic aggregate concrete in each mix.

Table 1 Materials and properties

Material	Experiment	Result
Cement	Specific gravity	3.2
	Initial setting time	40 min
	Consistency	33%
Fine aggregate	Specific gravity	2.65
	Water absorption	0.72%
Coarse aggregate	Specific gravity	2.63
	Water absorption	1.23%
Plastic Flakes	Specific gravity	1.29
	Water absorption	0

Table 2 Mix proportions

Mix	Nomenclature	Proportion
Concrete mix of grade M25	CM25	1: 2.36: 3.57: 0.005
Concrete mix of grade M25 with coarse aggregate replaced by 5% plastic flakes	CM25P5	1: 2.36: 3.39: 0.087: 0.005
Concrete mix of grade M25 with coarse aggregate replaced by 10% plastic flakes	CM25P10	1: 2.36: 3.218: 0.174: 0.005
Concrete mix of grade M25 with coarse aggregate replaced by 15% plastic flakes	CM25P15	1: 2.36: 3.039: 0.26: 0.005

A sample of the mix design is as follows;

Mix Design of CM25P5

Stipulations for Proportioning

(a) Grade designation: M25
(b) Cement used: PPC
(c) Maximum nominal size of aggregate: 20 mm
(d) Minimum cement content: 300 kg/m^3
(e) Maximum water- cement ratio: 0.50
(f) Exposure condition: moderate
(g) Workability: 100 mm
(h) Concrete placing method: Non pumping
(i) Degree of supervision: Good
(j) Aggregate type: Crushed angular
(k) Maximum cement content: 450 kg/m^3
(l) Chemical admixture type: Superplasticizer

Test Data for Materials

1. Cement used: PPC
2. Specific gravity of cement: 3.2
3. Specific gravity of

   - Coarse aggregate: 2.65
   - Fine aggregate: 2.63

4. Water absorption

   - Coarse aggregate: 0.72%
   - Fine aggregate: 1.23%

Target strength for mix proportion

$$\begin{aligned} f'_{ck} &= f_{ck} + 1.65\,s \\ &= 30 + 1.65 \times 4 \\ &= 31.6\,N/mm^2 \end{aligned}$$

Selection of Water Cement Ratio

Taking moderate exposure condition (Table 5, IS 456:2000)
Maximum water cement ratio = 0.5
Adopt water cement ratio = 0.48 (IS 10262:2019, fig 1)

Selection of Water Content

(Table 4, IS 10262:2019)

Maximum water content for 20 mm aggregate = 186 l
Estimated water content for 100 mm slump $= 186 + ((6/100) \times 186)$
$= 197\,l$
Water content can be reduced by 20 to 30% as superplasticizer is used
Water content = $197 \times 0.80 = 157.6l$

Calculation of Cement Content

w/c ratio = 0.48
Cementitious content = 157.6/0.48
= 329 kg/m^3 > 300 kg/m^3

Proportion of coarse aggregate and fine aggregate content

(Table 5, IS 10262:2019)

The ratio of volume of coarse aggregate and volume of total aggregate for 20mm size of coarse aggregate and fine aggregate (zone I,IS 383:1970) for w/c ratio = 0.5 is 0.60.
Proportion of volume of coarse aggregate is increased at a rate of −/+ 0.01 for every +/− 0.05 change in w/c ratio.

Volume of total coarse aggregate = $0.62 + [0.01 \times 0.02]$
= 0.6002
Volume of plastic flakes = $\left(\frac{5}{100}\right) \times 0.6002$
= 0.03001
Volume of normal coarse aggregate = $0.6002 - 0.03001$
= 0.5702
Volume of fine aggregate = $1 - 0.6002 = 0.3998$
Mix calculations

Mix calculations per unit volume of concrete shall be as follows:

(a) Volume of concrete $= 1\ m^3$
(b) Volume of cement $= \frac{mass\ of\ cement}{specific\ gravity\ of\ cement} \times (1/1000)$
$= (329/3.2) \times (1/1000)$
$= 0.102\ m^3$
(c) Volume of water $= (157.6/1) \times (1 \times 1000)$
$= 0.157\ m^3$
(d) Volume of superplasticizer $= (1.645/1.136) \times (1/1000)$
$= 0.0014\ m^3$
(e) Volume of all in aggregate $= [a - (b + c + d)]$
$= [1 - (0.102 + 0.157 + 0.0014)]$
$= 0.7396\ m^3$
(f) Mass of coarse aggregate = d × volume of coarse aggregate × specific gravity of coarse aggregate × 1000
$= 0.7396 \times 0.5702 \times 2.65 \times 1000$
= 1117.55kg
(g) Mass of plastic flakes $= 0.7396 \times 0.03001 \times 1.29 \times 1000$
= 28.63 kg
(h) Mass of fine aggregate $= d \times volume\ of\ fine\ aggregate \times specific\ gravity\ of\ fine\ aggregate \times 1000$
$= 0.7396 \times 0.3998 \times 2.63 \times 1000$
= 777.67kg

Mix Proportion

Cement $= 329\ kg/m^3$
Water $= 157.6\ l$
Fine aggregate $= 777.67\ kg/m^3$
Coarse aggregate $= 1117.55\ kg/m^3$
Plastic flakes $= 28.63\ kg/m^3$
Superplasticizer $= 1.645\ kg/m^3$
Water cement ratio = 0.48
Mix proportion = 1 : 2.36 : 3.39 : 0.087 : 0.005

6 Casting and Testing

Cylinder specimens of 300 mm height and 150 mm diameter were casted for M25 grade of concrete and tested for 28th day compressive strength and the modulus of elasticity was evaluated. The deformations were obtained directly from longitudinal compressometer attached to the specimen. Figure 1 show the casted specimens and the testing of cylinder specimen.

Fig. 1 **a** Casted cylinder specimen, **b** testing of specimen, **c** tested specimen

The compressive strength of concrete was determined by crushing of the test specimen by means of compression testing machine according to IS 516:1959 (Reaffirmed 2013). Cylinders of 150 mm diameter and 300 mm height were used for the testing. Failure load divided by area of specimen gives the compressive strength of concrete.

The longitudinal compressometer apparatus is used for determination of the strain and deformation characteristics of cement concrete cylindrical specimens of 150 mm dia × 300 mm length. The test is carried out as per ASTM C469. Place the concrete cylinder upright. Assemble the apparatus in the cylinder specimen. Unscrew the seven contact screws (3 at the top and 3 at the bottom ring). Place the compressometer over the concrete specimen locating the specimen at the centre of the ring. The length of the cylinders should be vertical to provide the correct height. Hand-tighten the 3 contact screws in the lower lock ring and 3 at the top ring. Remove the two spacer rods. The axial strain dial indicator is zeroed with the stem close to the fully extended position. Zeroing can be done by rotating the dial face. When the loading starts, the dial starts to show the deformation readings and the values can be obtained from the dial.

Compressive strength = *Load applied/Cross sectional area.*
Actual deformation = Obtained deformation/2

Strain and elastic modulus are calculated from the values of deformation and stress obtained.

The tested specimen showed good crack resistance properties. The plastic flakes bridged the crack propagation and prevented the splitting of specimen into two pieces even after the failure load.

7 Results and Discussions

A graph is plotted between cylinder compressive strength and modulus of elasticity of concrete. Figure 2 shows the graph plotted between compressive strengths and elastic modulus of M25 grade concrete. Figures 3, 4 and 5 indicates the graph plotted between cylinder compressive strength and elastic modulus of plastic aggregate concrete with 5, 10, 15% plastic respectively.

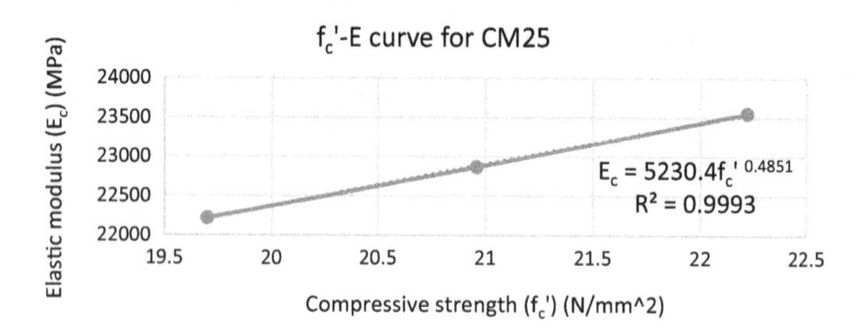

Fig. 2 Relation between cylinder compressive strength and elastic modulus of CM25

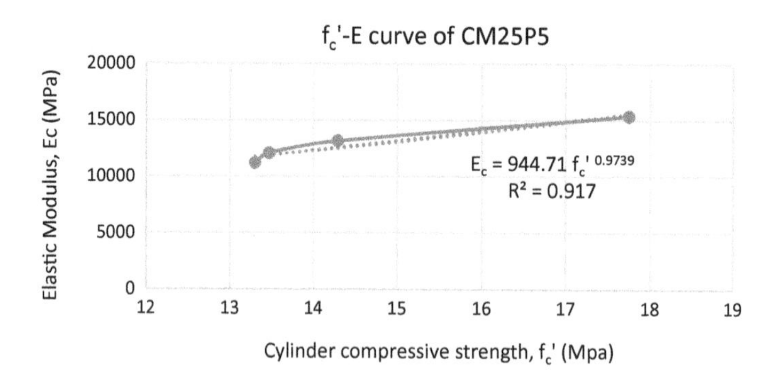

Fig. 3 Relation between cylinder compressive strength and elastic modulus of CM25P5

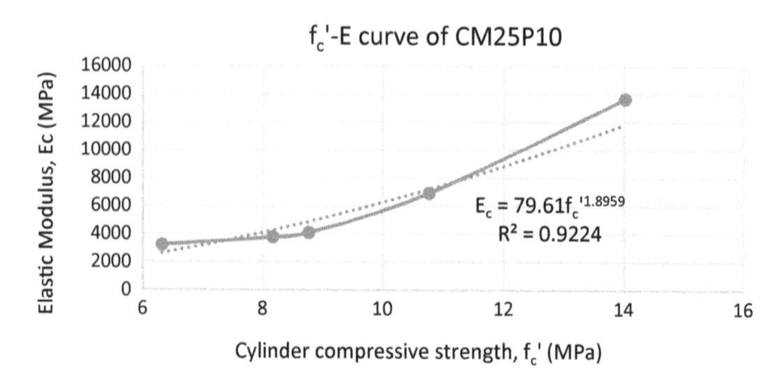

Fig. 4 Relation between cylinder compressive strength and elastic modulus of CM25P10

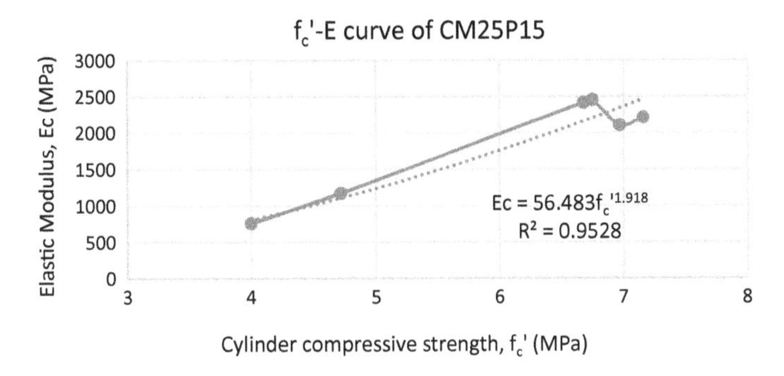

Fig. 5 Relation between cylinder compressive strength and elastic modulus of CM25P15

The concrete without plastic aggregates showed similar relation between compressive strength and modulus of elasticity of concrete as per IS 456: 2000. In this experiment the parameter used are compressive strength of cylinder and its elastic modulus.

The following graphs of plastic aggregate concrete shows different pattern and the equation when compared with control mix.

The plastic aggregate concrete showed irregular pattern in the relation between compressive strength and modulus of elasticity when compared to conventional concrete. As the percentage of plastic aggregates increased, the compressive strength and modulus of elasticity values decreased. Reduced bonding between the plastic aggregate and cement paste can be the reason for the reduction in the compressive strength while the elastic modulus value depends on the aggregate used.

8 Conclusion

The modulus of elasticity has a vital role in the structural behaviour of the materials in construction industry. It displays the deformation characteristics of a material. Various replacements are studied in the concrete, but those are not used widely in the construction field. That may be due to the uncertainty in the behaviour of those new materials in the structures. It is important to evaluate and study the structural behaviour of these materials along with the strength properties. One of the basic characteristics of the materials is its young's modulus or the elastic modulus. Plastic as a waste material, it has got durability and chemical resistance and therefore it can be utilized in the concrete as a replacement material. 70–80% increase in the elastic modulus value is observed from the study. This indicates lesser deflection, cracking and an improved serviceability characteristic of plastic incorporated concrete. 10% replacement with plastic showed higher elastic modulus values. 40–60% reduction in the compressive strength values is observed when plastic is used. Incorporation of plastic in concrete make it eco-friendly and the plastic aggregate concrete showed good crack bridging behaviour. To improve the compressive strength, the surface of plastic aggregate can be made rougher using sand or other methods.

References

1. Saikia N, de Brito J (2013) Waste polyethylene terephthalate as an aggregate in concrete. Mater Res 16(2):341–350
2. Ankur BC, Arora NK (2018) Workability, strength and durability of concrete containing recycled plastic fibres and styrene-butadiene rubber latex. Constr Build Mater 180:382–395
3. Tamang LWT, Wangmo T (2017) Karma Tshering Darjay. Karma Sangay Phuntsho, Use of Plastics in Concrete as Coarse Aggregate, IJEAR 7(2):1–5
4. Rafiq Ahmad Pirzada (2018) Tapeshwar Kalra, Fayaz Ahmad Laherwal, Experimental Study on Use of Waste Plastic as Coarse Aggregate in Concrete with Admixture Superplasticizer Polycarboxylate Ether. International Research Journal of Engineering and Technology (IRJET) 5(3):1–6
5. Thorneycroft J, Orr J, Savoikar P, Ball RJ (2018) Performance of structural concrete with recycled plastic waste as a partial replacement for sand. Constr Build Mater 161:63–69
6. Subramani T, Pugal VK (2015) Experimental study on plastic waste as a coarse aggregate for structural concrete. Int J Appl Innov Eng Manag (IJAIEM) 4(5):1–9
7. Silva RV, de Brito J (2013) Nabajyoti Saikia, Influence of curing conditions on the durability related performance of concrete made with selected plastic waste aggregates. Cement Concr Compos 35:23–31
8. Mathew P, Varghese S, Paul T, Varghese E (2013) Recycled plastics as coarse aggregate for structural concrete, International Journal of Innovative Research in Science. Eng Technol 2 (3):1–4
9. Yun Wang Choi (2009) Dae Joong Moon, Yong Jic Kim, Mohamed Lachemi, Characteristics of mortar and concrete containing fine aggregate manufactured from recycled waste polyethylene terephthalate bottles. Constr Build Mater 23:2829–2835
10. Fayaz J (2019) Adarsha C Y, Raghavendra T, Udayashankar B C, Natarajan K, Structural behaviour of concrete beams and columns reinforced with waste plastic incorporated GFRP (WPGFRP) rebars. J Build Eng 23:172–184
11. Jacek G, Wojciech P, Budzynski W (2017) Stress-strain relationship and modulus of elasticity of rocks and of ordinary and high-performance concretes. Constr Build Mater 153:728–739
12. Mostofinejad D, Nozhati M (2005) Prediction of the modulus of elasticity of high strength concrete. Iran J Sci Tech 29(B3):311–321
13. Nemati KM, Noguchi T (1995) Relationship between compressive strength and modulus of elasticity of high strength concrete. J Struct Constr Eng 474(1):1–10
14. Takafumi N, Fuminori T, Nemati KM, Chiaia BM, Fantilli AP (2009) A practical equation for elastic modulus of concrete. ACI Struct J 106(5):690–696
15. Krizova K, Hela R (2016) Evaluation of modulus of elasticity of different types of concrete compared with Eurocode 2. Key Eng Mater 677:181–185
16. de Jorge B (2016) Silva Vasco Rui, Ravindra Kumar Dhir, Establishing a relationship between modulus of elasticity and compressive strength of recycled aggregate concrete. J Clean Prod 112:2171–2186
17. Turhan B (2016) Investigation of performance of some empirical and composite models for predicting the modulus of elasticity of high strength concretes incorporating ground pumice and silica fume. Constr Build Mater 127:850–860
18. Ali A (2017) Dang N Canh, Gray S Prinz, Micah Hale W, Evaluation of modulus of elasticity of ultra-high-performance concrete. Constr Build Mater 153:918–928

19. Behnam V, Shami N (2018) Empirical models and design codes in prediction of modulus of elasticity of concrete. Struct Civil Eng 13:38–48
20. Behnam V (2018) Modulus of elasticity of concrete in design codes and empirical models: analytical study. Pract Period Struct Des Constr 23(4):04018022–04018031

Study on Strengthening of Deteriorated Corrugated Steel Beams Using Post Tensioning

Helna Varghese and P. E. Kavitha

Abstract Steel beams have been a part of the construction industry since the last decades of the nineteenth century and they have gained importance due to their improved functionality compared to conventional concrete structural elements. Among various types of steel beams corrugated steel beams are special light-weight beams with corrugations at the web portions. Weight reduction aspects of corrugated steel beams aids in considerable savings in cost of structure. Corrugated beams suffer deteriorations mainly due to unexpected excessive loading and exposure to extreme surroundings. Various researches in this field suggest that the strengthening of deteriorated corrugated steel beams can be done by post tensioning techniques in the locations of deterioration. The post tensioning technique is recommended in the areas where damages are mainly due to corrosion. The deteriorations are provided at different locations of the beam. Post tensioning is achieved through reinforcement bars of suitable diameter. The result includes the ultimate load carrying capacity, total deformation, and strain and stress distribution of strengthened beams. The present study focuses on the effect of post tensioning as an innovative strengthening technique for retrofitting the deteriorated corrugated steel beams. Strengthening of corrugated steel beams is expected to increase the ultimate load carrying capacity as well as improve the functional utility and life span of the structure.

Keywords Corrugated steel beams · Post tensioning · Reinforcement bars · Load carrying capacity

1 Introduction

Steel beams are beams which are used to carry heavy loads which are made up of structural steel. Steel beams come in different sizes and types; hence they have different applications in the construction of structures and buildings. The geometry,

---

H. Varghese (✉) · P. E. Kavitha
Civil Department, Federal Institute of Science and Technology, Angamaly, Kerala, India

G. C. Marano et al. (eds.), *Proceedings of SECON'21*, Lecture Notes in Civil Engineering 171, https://doi.org/10.1007/978-3-030-80312-4_37

size and shape of beams are determined based on the specifications of the structure. Corrugated steel beams are a special form of steel beams usually available as I shaped beams which have corrugations at the web region. Corrugated steel beams are gaining their importance in construction practices due to their advantages like weight reduction, savings in cost, reduced roof dead loads, prevention of buckling failure of the web due to corrugation and huge architectural capabilities. Corrugations can be provided as different corrugated web profiles like trapezoidal, rectangular, triangular, sinusoidal and square corrugations. Use of corrugated webs helps to achieve adequate out of plane stiffness and bending resistance without using stiffeners. The corrugated web beams have few weaknesses due to their geometric characteristics. The local buckling strength of the flange can be less than that of plane web beams. Also as a result of web eccentricity, additional in-plane transversal moment occurs in flanges. This moment reduces the flexural strength of corrugated web beams. Corrugated beams are susceptible to deteriorations when exposed to surroundings and subjected to loading. When exposed to surroundings there will be corrosion in steel beams and when subjected to loading there will be crack formations. Corrosion of steel beams is due to improper design, usage of uncoated steel, exposure to different surroundings, improper maintenance and so on. In order to regain the strength of beams suitable rehabilitating measures should be taken to maintain the strength and serviceability characteristics of the corrugated steel beam. Proper strengthening techniques help in reducing the chances of failure to certain extent. The strengthening method is chosen based on the effect of deteriorations in the load carrying capacity of beams.

The strengthening of deteriorated corrugated steel beams can be done by post tensioning technique using reinforcement bars of suitable diameter. Post tensioning is a technique which is found to be effective in strengthening steel beams [1]. Post tensioning is the method of stressing the elements before the application of load so that it compensates for the tensile stresses formed when the element is subjected to loading. Post tensioning requires high tensile steel and tensioning equipment. It requires skilled laborers and proper supervision. Post tensioning is an effective method to strengthen the deteriorated corrugated steel beam, especially when the damages are in the form of corrosion. The effect of post tensioning varies as the diameter of bars changes. The tension force on the bar is increased by increasing the diameter of the bar. The tension force of bars can be maintained with the help of providing a suitable rigid support between the bar and the corrugated steel beam. Analysis of strengthening of deteriorated corrugated beams can be done using experimental and analytical methods. The analytical method is followed by model development and finite element analysis using software's like ABACUS and ANSYS. ANSYS has a wide range of applications and can be used in different fields of engineering. The beams can be modeled and analyzed using ANSYS. The ultimate load, total deformation, stress distribution, failure pattern and various such parameters can be investigated with the help of ANSYS 19.0.

2 Literature Review

Corrugated beams with different profiles were studied by Raiza et al. and it was observed that such beams are having 18.41% increase in load carrying capacity and more bending strength compared to the equivalent area plane web beams [2]. Steel beam with trapezoidal corrugated web beam is studied for finding out the load carrying capacity in comparison with plane web beams. The trapezoidal corrugated web beam had more load carrying capability than plane web beam. Trapezoidal corrugated beams had more resistance to torsional buckling and chances of failure at shear zone were minimum [3]. Shear buckling and stress distribution of trapezoidal corrugated beams were investigated by Moussa et al. [4]. The beams were tested to find their behavior under shear loading. The load carrying capacity of beams was affected by shear buckling and the tested beams showed post buckling residual strength which was around 50% of ultimate load of tested beams. A study on accordion effect of prestressed steel beams with corrugated web is done by theoretical and experimental researches which showed that prestressing efficiency was observed more in corrugated beams than beams with plane web [5].

Parametric study of the external strengthening of composite beams using post tensioned tendons was conducted by El-Zohairy and Salim [6]. From the study it was concluded that post tensioning by means of high-strength tendons can be used to effectively increase the ultimate capacity of steel–concrete composite beams to about 21%. Behavior of prestressed beams was also investigated by Belletti and Gasperi [7], Park et al. [8] and Siwowski et al. [9]. The studies showed that prestressed structures showed more advantages compared to non prestressed structures. The conclusions proved that an application of high performance steel tendons reduces the size of steel members and the total amount of steel required, and helps to increase the load carrying capacity of structures. Strengthening of steel beams using post tensioning was studied by Taoum et al. [1]. The prestressing was carried out with the help of reinforcement bars. The post tensioning can be external post tensioning or internal post tensioning. In external post tensioning the bars were welded to the bottom flange and in internal post tensioning the bars were welded to the web portion of beam. A reverse bending moment was formed at the mid-span of the beam when the bars were pulled using the manual screw jack which increases its load carrying capacity. The post tensioning method induces stress in both tension and compression flange of the beam and this stress will counteract the stresses formed during loading. The tension force was varied by changing the diameter of bars. The external post tensioning method was more effective in increasing the load-carrying capacity of the steel I-beams than the internal post tensioning method. The post tensioning aided in increasing the load carrying capacity of beams up to 60% more than that of control beam. The behavior of prestressed composite beams was studied by Saadatmanesh et al. [10]. The observations showed that the prestressed members have advantages including ability to enlarge the elastic range of behavior, to increase the ultimate capacity, to reduce the structural steel weight and to improve the fatigue and fracture strength.

3 Methodology

3.1 General

Modeling and analysis of the specified corrugated and prestressed beams were done using the finite element software ANSYS 19.0. Specimens were modeled according to the specifications and requirements based on the validation study [1]. The properties of the specimens were provided in engineering data of the software as mentioned in Table 1. Model was created using geometry based on the journal of validation. In validation, a steel beam strengthened with external post tensioning technique was modeled using software. The prestressing was done with the help of 12 mm reinforcement bars. Post tensioning bars are welded at both ends of the bottom flange. A prestressing force of 56,200 N is equally given by two reinforcement bars at bottom flange and a displacement control loading was provided. The ultimate load from finite element analysis was compared with that of experimental ultimate load from the journal. The percentage difference of ultimate load obtained from finite element analysis with respect to experimental result was 4.5%. Detailed parametric study was carried out on modeling and analysis of corrugated beam, deteriorated beams and strengthened beams with help of ANSYS 19.0.

3.2 Corrugated Beam

Corrugated steel I beams are modeled with corrugations in the web region. Trapezoidal profile is adopted for web corrugations. Overall depth of the beam is 150 mm. The flange width is 75 mm. The web thickness is 2.66 mm [11]. Loading

Table 1 Properties of corrugated beam, deteriorated beams and strengthened beams

Properties	Values
Overall length of beam	1400 mm
Overall depth of beam	150 mm
Thickness of flange	7 mm
Width of flange	75 mm
Thickness of web	2.66 mm
Elastic modulus of steel beam	2 × 10^5 MPa
Yield strength of steel beam	411 MPa
Yield strength of reinforcement bar	500 MPa
Poisson's ratio	0.3
Element	Solid 186
Loading	Displacement control loading
Boundary condition	Simply supported at both ends

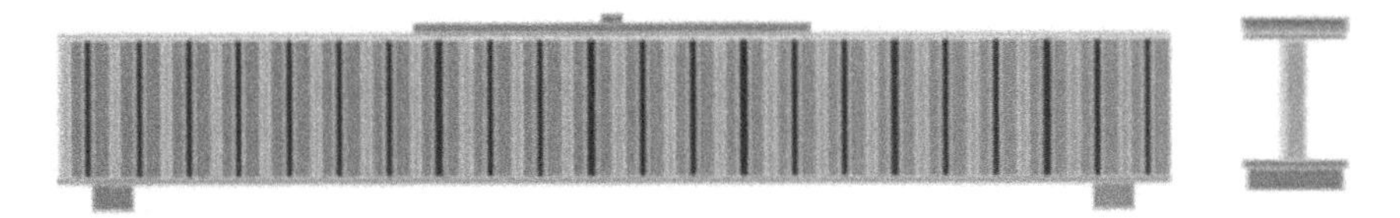

Fig. 1 Corrugated steel beam

is given as axial loading. Displacement control loading with displacement restricted to 25 mm is adopted. The boundary condition is provided as simply supported at both ends with the help of two support plates provided at bottom flange. A steel plate is provided in the top to reduce the effect of buckling at centre when subjected to loading. The mesh size is adopted as 10 mm for the corrugated web portion. Modeling is followed by analysis to find out the load carrying capacity of the beam. The ultimate load is obtained as 107.73 kN. The ultimate load of the normal beam is taken as reference to find out the effectiveness of post tensioning technique. Figure 1 shows model of corrugated steel beam (control beam).

3.3 Deteriorated Beams

Deteriorations mainly in the form of corrosion affect the strength of beams. Damages gradually reduce the load carrying capacity of beams and cause failure of beams. The loss in strength depends on the location and extent of damage caused by corrosion. Different models were developed by varying the location and intensity of damage. The deteriorations are visible in flange as well as web. The effect varies as the location changes. The beam which was mostly affected by deterioration is taken for strengthening. Totally six deteriorated beams were modeled by varying the percentage of damage in thickness throughout the length of web as well as flange. The beams are analyzed and load carrying capacity of deteriorated beams were compared with the load carrying capacity of normal beam to find out the percentage reduction in load carrying capacity of deteriorated beams with respect to normal beam. The specimens with damage in the flange region experience more reduction in load carrying capacity than beam with damage in the web portion. Among all the models the specimen with 75% loss in the flange thickness has maximum reduction in strength and this specimen is taken for strengthening using post tensioning. Table 2 shows the comparison on percentage reduction in load of the deteriorated beams with corrugated beam. Model with 0% corrosion in flange thickness and 0% corrosion in web thickness is named as FC-0-WCT-0 where FC denotes Flange Corrosion and WCT denotes Web Corrosion in Thickness.

The load deflection curve for control beam and deteriorated beams is shown in Fig. 2. At allowable deflection also the reduction in load carrying capacity is same pattern as discussed above.

Table 2 Percentage reduction in load of deteriorated beams

Model ID	Section of model	C/S area of beam (mm^2)	Percentage reduction in C/S area	Ultimate load (kN)	Percentage reduction in load	Load corresponding to allowable deflection
FC-0-WCT-0 (control beam)		1412	Nil	107.73	Nil	97
FC-0-WCT-25		1321	6.44	105.79	1.8	90
FC-0-WCT-50		1231	12.81	90.73	15.78	75
FC-0-WCT-75		1140	19.26	45.21	58.03	42
FC-25-WCT-0		1281	9.27	82.54	23.39	50
FC-50-WCT-0		1149	18.62	58.02	46.15	40
FC-75-WCT-0		1018	27.90	34.38	68.09	25

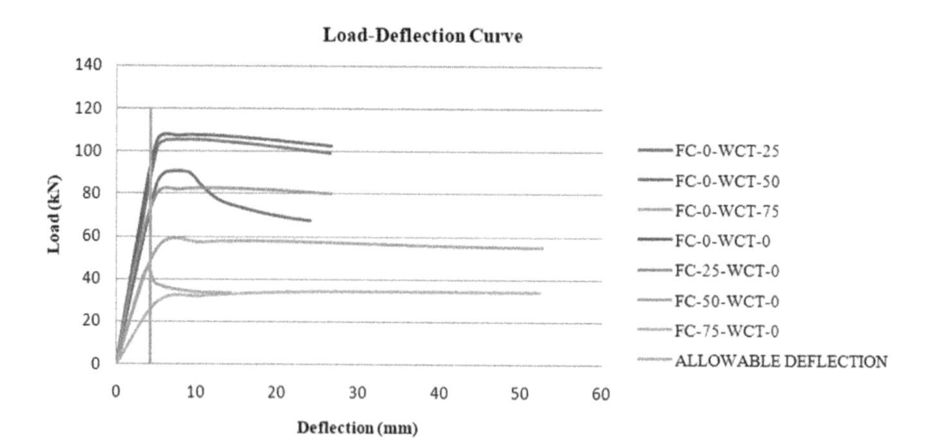

Fig. 2 Load deflection curve for control beam and deteriorated beams

3.4 Strengthening of Deteriorated Corrugated Steel Beam Using Post Tensioning

Considering the above analysis, strengthening of deteriorated corrugated steel I beam was done by post tensioning using a reinforcement bar of suitable diameter. Beam which was worst affected by deterioration was taken for strengthening method. The beam which had 75% loss in thickness at bottom flange throughout its entire length has more than 60% reduction in load carrying capacity compared to other damaged beams and is selected for rehabilitation study. The model selected has an ultimate load of 34.38 kN and was taken for strengthening using post tensioning as shown in Fig. 3

The reinforcement bar of diameter 12 mm is used for post tensioning technique. The bar has yield strength of 500 MPa. The bottom flange is provided with two bars of 12 mm diameter in which prestressing was applied. The bars were welded to the support plates at both ends of the bottom flange. The post tensioned bars were kept in position with the help of a rigid support provided between the bottom flange and bars at centre. The beam is simply supported on both ends. Boundary conditions were provided at support plates provided at both ends. Displacement control loading was provided on the top plate. The loading was provided as two steps. The external loading is provided in the second step and prestressing was provided in first step. The tension force was equally distributed on two reinforcement bars, with equal force on two sides of the single bar. The total tension force is equally distributed on two bars. 50% of calculated prestressing force that is 28100 N was provided to find out the optimum level of prestressing required to regain the strength of the damaged beam to the strength of the normal beam. The stress distribution and strain of strengthened beam is shown in Figs. 4 and 5.

4 Results and Discussions

The load carrying capacity of the beams is determined by the ultimate load of the beam. Failure location is indicated by the strain diagram. The first analysis was carried out on corrugated steel beams without deterioration to find out the strength of the normal beam. The ultimate load of a normal beam is 107.73 kN. The beams provided with 25, 50 and 75% loss in web thickness showed reduction in ultimate

Fig. 3 Corrugated steel beam strengthened with post tensioned bar

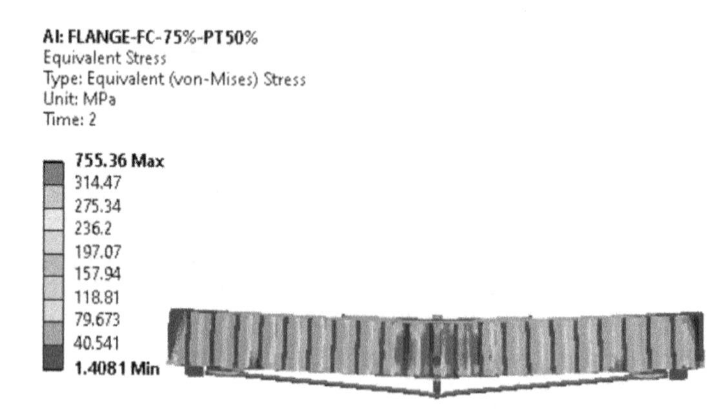

Fig. 4 Stress distribution of the strengthened beam

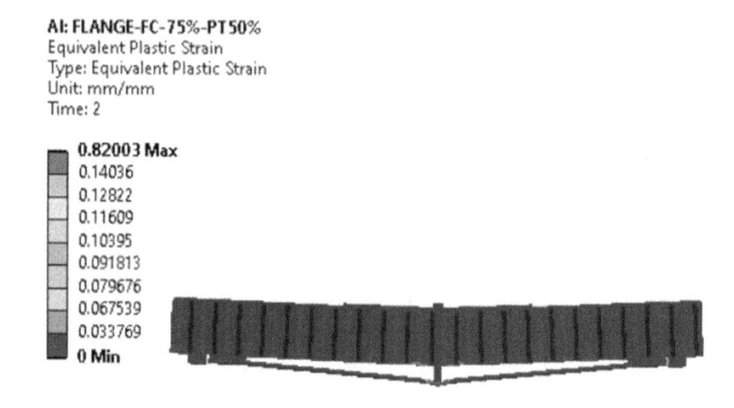

Fig. 5 Strain diagram showing failure location

load in the range of 1.8–58%. The beams provided with 25, 50 and 75% loss in flange thickness showed loss in ultimate load in the range of 23–68%. This indicates that deterioration in flange is more critical than that of the web.

Strengthening technique was carried out on the beam which has 75% deterioration in thickness of bottom flange throughout the length of the beam to regain the strength of the damaged beam to that of the normal beam. The ultimate load of the deteriorated beam was obtained as 34.38 kN. The failure of the beam was due to yielding of flange. The strengthening of the beam was carried out by post tensioning and prestressing force is provided on bars of 12 mm diameter. From the analysis 50% of prestressing was sufficient to regain the strength of the damaged beam to the original beam. The ultimate load is obtained as 112.36 kN (greater than that of control beam). The failure of the beam is due to the yielding of the post tensioned bar at the centre. Figure 6 shows the graph representing ultimate load of control beam, deteriorated beams and strengthened beam.

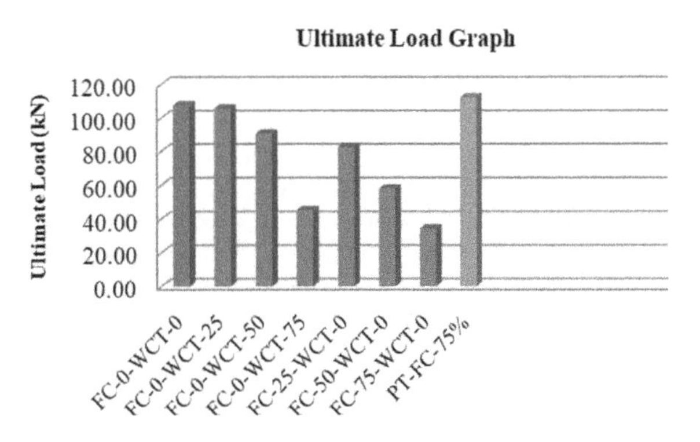

Fig. 6 Ultimate load of control beam, deteriorated beams and strengthened beam

5 Conclusion

The corrugated steel beams are widely used in the construction industry due to its increased stiffness compared to conventional steel beams. Exposure to extreme environmental conditions, the beams are vulnerable to corrosion. In order to prevent the failure or loss of strength of beams due to the effect of corrosion suitable strengthening methods are to be implemented. The choice of method depends on location and depth of corrosion, economical aspects and the feasibility of application. Strengthening of corrugated beams using post tensioning technique is an emerging method of rehabilitation of beams damaged due to corrosion. The modeling and analysis can be done efficiently and accurately with the help of ANSYS 19.0. The control beam was modeled and analyzed and the ultimate load was obtained as 107.73 kN. By providing deteriorations in flange as well as web, the loss in ultimate load of deteriorated beam was compared with that of control beam and then the rehabilitation technique was adopted and following conclusions were made:

- The beams with 25, 50 and 75% loss in thickness of the web have 1.8–58% reduction in load carrying capacity.
- The beams with 25, 50 and 75% loss in thickness of the flange have 23–68% reduction in load carrying capacity.
- The strength of beams was greatly affected by deteriorations at flange when compared with the web.
- The damaged beam with the lowest ultimate load of 34.38 kN was taken for strengthening.
- The ultimate load of the strengthened beam was obtained as 112.36 kN. The strengthening technique by post tensioning aided in regaining the strength of the damaged beam to the strength of the normal beam.
- The result indicates that strengthening technique by post tensioning proved to be efficient in regaining the strength of beams which has more than 50% of loss in its load carrying capacity due to corrosion.

References

1. Taoum A, Jiao H, Holloway D (2015) Upgrading steel I beams using local post tensioning. J Constr Steel Res 113:127–134. https://doi.org/10.1016/j.jcsr.2015.06.012
2. Raiza Ashrawi MA, Sunitha Rani CM, Smitha KK (2016) Load carrying capacity of corrugated web beam. Int Res J Eng Technol (IRJET) 03(09)
3. Karote A, Ghude A, Fulpagare V, Jadhav K, Patil V (2016) Experimental investigation of steel beam with trapezoidal corrugated web beam. Int Res J Eng Technol (IRJET) 03(09)
4. Moussa Leblouba M, Junaid T, Barakat S, Altoubat S, Maalej M (2017) Shear buckling and stress distribution in trapezoidal web corrugated steel beams. Thin Walled Struct 113:13–26
5. Oh J-Y, Lee DH, Kim KS (2012) Accordion effect of prestressed steel beams with corrugated web. Thin-Walled Struct 57:49–61
6. El-Zohairy A, Salim H (2017) Parametric study of the external strengthening of composite beams using post tensioned tendons. Struct Congr
7. Belletti B, Gasperi A, Behavior of prestressed steel beams. https://doi.org/10.1016/ASCE.ST.1943-541X.0000208
8. Park S, Kim T, Kim K, Hong S-N (2010) Flexural behavior of steel I-beam prestressed with externally unbonded tendons. J Constr Steel Res 66:125–132
9. Siwowski T, Piątek B, Siwowska P, Wiater A (2020) Development and implementation of CFRP post-tensioning system for bridge strengthening. Eng Struct 207:110266
10. Saadatmanesh H, Albrecht P, Ayyub BM (1989) Analytical study of prestressed composite beams. J Struct Eng 115(9)
11. Elgaaly M, Seshadri A, Hamilton RW (1997) Bending strength of steel beams with corrugated webs. J Struct Eng 123(6)

Parametric Study of Mechanically Stabilized Earth Wall Using GEO 5

V. S. Krishnapriya, Swetha Maria Charles, Sharon Shogu, T. C. Nithul, and K. Panjami

Abstract A mechanically stabilized earth (MSE) wall is an engineered system consisting of alternating layers of soil reinforcement and compacted backfill material fixed to the wall facing and supported on a foundation. The present study is focused on parametric analysis of MSE walls using an analytical method, GEO5 MSE to determine the factor of safety of the wall. The factor of safety of a 10 m high MSE wall was computed with respect to varying environments of backfill soil, the length of reinforcement to height of the wall (L/H), reinforcement characteristic strength and spacing of reinforcement. This study mainly focuses on finding out how variations in the factors affect the behaviour of the wall and its factor of safety. It was observed that improvement in the angle of friction leads to an increase the factor of safety. It was also noted that, increase in L/H ratio from 0.4 to 1.0 showed an increase in factor of safety by 15.92% and increase in reinforcement spacing from 0.2 m to 1 m showed a decrease in factor of safety by 74.52%. Factor of safety was also found to increase with an increase in characteristic strength of embedded reinforcement. From the parametric analysis, MSE walls were developed to replace the failed retaining walls at various locations in India.

Keywords MSE retaining wall · Factor of safety · Reinforcement · GEO5

1 Introduction

Conventional retaining structures were made from concrete and were designed as gravity or cantilever walls which are essentially rigid structures and cannot accommodate significant differential settlements unless founded on deep foundations. With increasing height of soil to be retained and poor subsoil conditions, the cost of reinforced concrete retaining walls increases rapidly and structures try to

V. S. Krishnapriya (✉) · S. M. Charles · S. Shogu · T. C. Nithul · K. Panjami
Federal Institute of Science and Technology, Angamaly, Ernakulam, Kerala, India
e-mail: panjamik@fisat.ac.in

G. C. Marano et al. (eds.), *Proceedings of SECON'21*, Lecture Notes in Civil Engineering 171, https://doi.org/10.1007/978-3-030-80312-4_38

overturn due to the earth pressure. In order to deal with this problem Mechanically Stabilized Earth Technique is employed [1].

French engineer Henri Vidal invented the modern form of MSE, termed Terre Armee (reinforced earth) using reinforcements made of steel strips in the 1960s. Since the 1980s the development of reinforced soil has been considerable using a range of construction forms and reinforcements including metallic and polymeric anchors, strips and grids.

The first contemporary forms of MSE walls were constructed in Europe in the late 1960s.

A mechanically stabilized earth (MSE) wall is a composite system consisting of soil reinforcement in the form of steel strips or bars, welded wire mats or geotextile sheets, backfill material which acts as a retaining structure, a facing element in the form of precast concrete panels, dry cast modular blocks and panels, and wrapped sheets of geosynthetics which acts as a supporting system and a foundation. In the construction of MSE wall, reinforcements are placed in layers in the backfill soil, and this reinforced mass resists the earth pressure caused by the retained soil using the relative motion between reinforcement and soil. Tensile capacity is built in the soil through friction and confinement by interaction between steel or synthetic reinforcing elements and soil. This combination of soil and reinforcement combine their best characteristics and behave as a gravity mass and resist lateral earth pressure [1].

Mechanically stabilized earth (MSE) retaining walls are the foremost suitable design alternatives to the traditional retaining walls because of their simple, rapid and cost-effective construction, reduced right-of-way acquisition and can bear large settlements compared to conventional concrete retaining walls [1]. They have a wide range of applications which are bridge abutments, overpasses and underpasses for conveyors, roadways, railways, vertical walls, ramps and steepened slopes for haul roads, highways and railways. The ability of MSE walls to withstand extreme settlement and seismic forces has increased its usage. Various methods are available for the analysis and design of an MSE wall. Analytical methods of analysis include Limit Equilibrium analysis like Bishop, Spencer etc. and finite element analysis are done with the help of computer programs like PLAXIS, GEO5 etc.

The present study is concentrated on parametric sensitivity analysis of MSE walls employing an analytical method, which uses the GEO5 MSE to work out the factor of safety of the wall. The MSE walls have been analyzed for horizontal and vertical movements with respect to length of the reinforcement, type of soil, height of wall and type of reinforcement. External stability analysis have also been executed for the overturning, sliding and bearing capacity. This study mainly focuses on the factors affecting the wall and its effect on the factor of safety of the wall.

2 Design Concept of MSE Wall

Ultimate limit state and serviceability limit states are the two limit states considered in the design of Mechanically Stabilised Earth walls. The stability of the MSE structures is checked for two conditions: External stability and Internal stability. Internal and external stability of MSE walls depend upon interaction between the backfill soil and reinforcement, reinforcement characteristic strength and vertical spacing. When determining internal stability, the strength of the reinforcement and its interrelation to the facing, as well as its pullout resistive length are checked. When determining external stability, the reinforced soil mass subjected to loading exerted by the retained soil is taken as a coherent mass, and implicitly treated as a rigid body [1, 2]. The fundamental mechanisms of mechanically stabilized earth and reinforced concrete are alike.

2.1 *Methodology*

The basic structure of the MSE wall is modelled using GEO5 MSE with the required specifications, which includes the height of blocks and the number of blocks required. Specifications of various soils are entered and assigned to the respective interfaces. Reinforcements are selected and are provided as per the specified length and spacing between them. Surcharge of required magnitude is provided at the top of the fill. Analysis is performed and stability is checked to obtain the factor of safety (see Fig. 1).

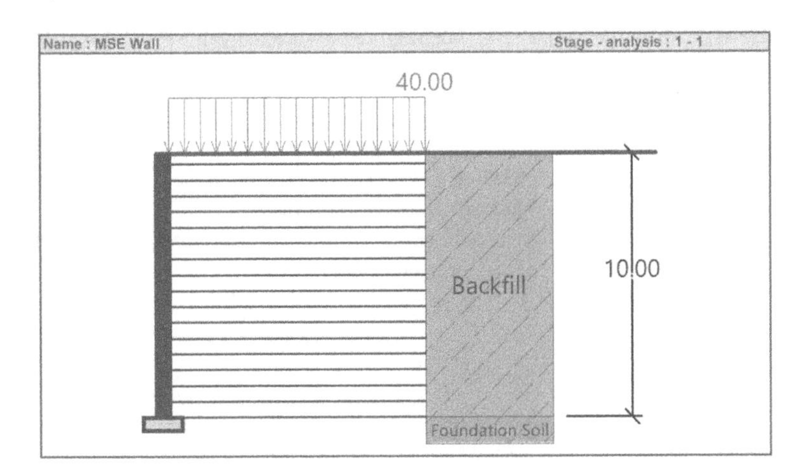

Fig. 1 MSE wall used in modelling

3 Modelling of MSE Wall

An MSE Wall was modelled to check the influence of various parameters on the factor of safety of the MSE Wall. Tables 1 and 2 lists the geometrical and soil properties that were used for modelling the MSE wall using Mohr Coulomb method. The 10 m high MSE wall was provided with a 0.5 m thick facing of standard concrete. Granular soil was proposed in the design of backfill material as high soil content can lead to drainage problems and thus failure. Geogrid of 10 m length was provided as reinforcement (see Fig. 1).

4 Results and Discussion

Four important parameters were chosen to study its effect on the factor of safety of an MSE wall and they are reinforcement characteristic strength, reinforcement spacing, type of backfill soil and L/H ratio. Influences of these parameters are as mentioned below.

4.1 Effect of Reinforcement Characteristic Strength on Factor of Safety

Geogrids are used in the construction of retaining walls and soil slopes. It reinforces the soil by containing the lateral earth pressures. They are manufactured from high quality HDPE which is non-biodegradable and resistant to soil chemicals [2]. Characteristic strength of the reinforcement used varied from 13.24 to 43.59 kN/m

Table 1 Properties of MSE wall

Properties	Values
Height of wall (m)	10
Thickness of facing panel (m)	0.1
Reinforcement characteristic strength (kN/m)	43.59
Length of reinforcement (m)	10

Table 2 Properties of soil at site

Properties	Weathered rock	Reinforced fill
Unit weight (kN/m^3)	20	19
Cohesion (MPa)	20	–
Friction angle (°)	40	35
Elastic modulus (MPa)	100	80
Poisson's ratio	0.2	0.3

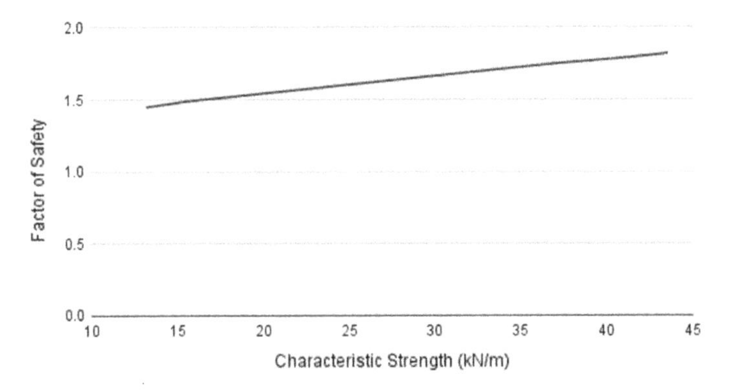

Fig. 2 Factor of safety versus characteristic strength

Table 3 Factor of safety with respect to reinforcement characteristic strength

Characteristic strength (kN/m)	Factor of safety
13.24	1.45
15.67	1.5
16.4	1.75
37.16	1.79
41.19	1.49
43.59	1.82

and the highest factor of safety of 1.82 was seen for geogrid having 43.59 kN/m as the characteristic strength which makes it a good choice of reinforcement (see Fig. 2). Reinforcements of characteristic strengths 13.24, 16.40, 37.16, 41.19, 15.67 and 43.59 kN/m were used to analyse the wall to obtain the corresponding factor of safety (Table 3).

4.2 Effect of Reinforcement Spacing on Factor of Safety

The reinforcement spacing was identified as a major factor controlling the behavior of MSE Walls. Two types of spacing were considered in studying the effects of spacing: small (less than or equal to 0.4 m) and large (larger than 0.4 m). Increasing reinforcement spacing decreased the factor of safety (see Fig. 3). A reinforcement spacing of 0.1–1 m was provided along the length of the wall and it was seen that as spacing increased, the factor of safety decreased making 0.5 m of reinforcement spacing an affordable option in this case considering the economic point of view [2] (Table 4).

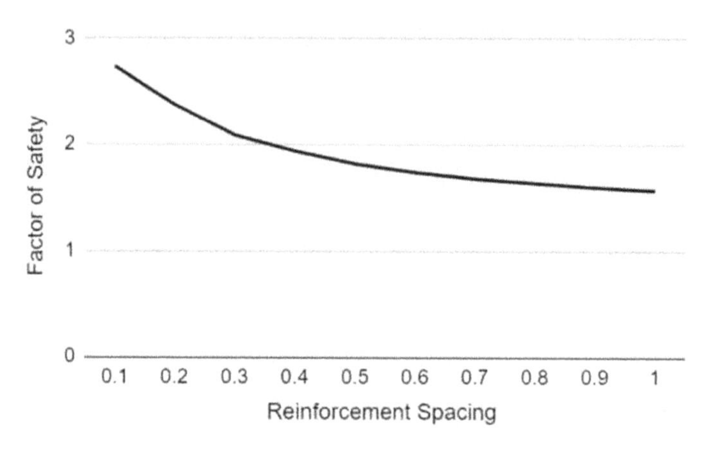

Fig. 3 Factor of safety versus reinforcement spacing

Table 4 Factor of safety with respect to reinforcement spacing

Reinforcement spacing (m)	Factor of safety
0.2	2.38
0.4	1.94
0.6	1.74
0.8	1.64
1	1.57

4.3 Effect of Backfill Soil on Factor of Safety

Stability of an MSE wall is dependent on the shear strength properties of the backfill soil [3] and shear strength is described as a function of Cohesion(c) and Angle of internal friction (φ).Therefore, these two physical parameters play an important role in the stability of an MSE wall. Cohesion is a force that holds like particles together and angle of internal friction is the capability of soil to withstand shear stress. Change in environmental factors can affect cohesive property of soil and they are poorly drained, thus it isn't desirable.

Characteristics of the backfill are as represented in Table 5. It was found that as the friction angle increases, the factor of safety increases whereas cohesion beyond a certain limit could lead to failure of the wall. Example: Soil 3 and Soil 4 though having an angle of internal friction of 34° but shows a variation in the factor of safety. This variation in factor of safety is due to the presence of a cohesive force of 1 MPa in soil 4 whereas cohesive force in soil 3 is zero (see Fig. 4) which implies that presence of cohesive force can affect the factor of safety of the Mechanically Stabilized Earth wall.

From the analysis, it can be inferred that soils with better internal friction contribute to the stability of MSE walls, on the other hand cohesive soils can reduce the factor of Safety [4].

Table 5 Factor of safety with respect to backfill parameter

Material	Unit weight (kN/m^3)	Cohesion (MPa)	Friction angle (°)	Elastic modulus (MPa)	Poisson's ratio	F.O.S
Soil 1	18	–	25	60	0.33	1.82
Soil 2	17	–	30	60	0.33	2.01
Soil 3	15	–	34	60	0.33	2.21
Soil 4	18.88	1	34	12.5	0.32	2.06
Soil 5	18	–	30	60	0.32	1.97
Soil 6	19	–	28	60	0.32	1.88
Soil 7	20.3	–	25	60	0.32	1.75
Soil 8	18.88	1	30	10	0.3	1.97
Soil 9	20	10	30	60	0.3	2.01

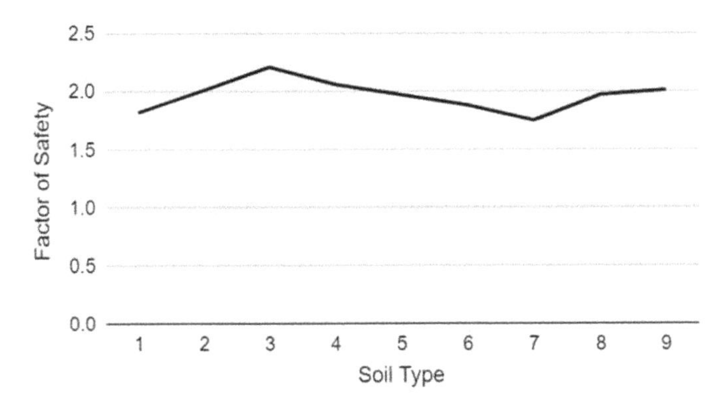

Fig. 4 Factor of safety versus soil type

4.4 *Effect of Ratio of Length of Reinforcement to the Height of the Wall on Factor of Safety*

Many agencies restrict the length of reinforcement as 0.7 times the height of the wall [5, 6]. However, these restrictions are not followed everywhere. Therefore an analysis was carried out to study the variation in factor of safety with respect to the length of reinforcement to height of the wall (L/H). The L/H ratio was considered and values from 0.4 to 1 were taken by referring to the literature review and the corresponding factor of safety values was obtained. An L/H of 1 gave the highest value of factor of safety of 1.82 (see Fig. 5). It can be inferred that with an increase in the L/H ratio, the factor of safety increases thereby improving the performance of the MSE wall (Table 6).

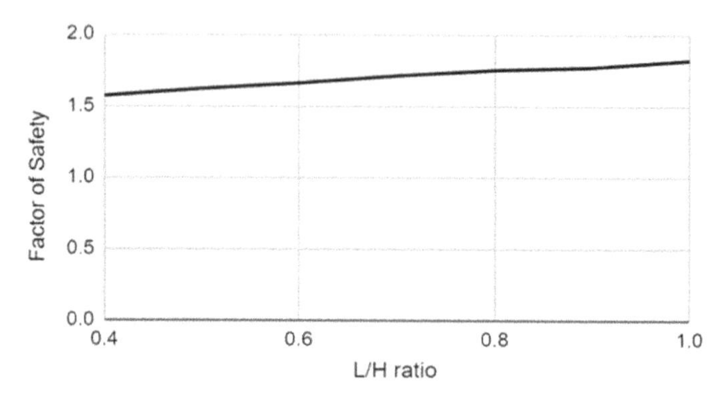

Fig. 5 Factor of safety versus L/H

Table 6 Factor of safety with respect to L/H

L/H	Factor of safety
0.4	1.57
0.5	1.62
0.6	1.66
0.7	1.71
0.8	1.75
0.9	1.77
1	1.82

5 Rehabilitation of Failed Retaining Walls

Retaining walls fail due to improper design, lack of proper reinforcement, insufficient drainage, expansion of soil, overloading etc. Analysis of the retaining walls as mentioned below using GEO5 gave a factor of safety below the prescribed limit whereas an MSE wall gave a satisfactory result which implies that construction of an MSE Wall could have been a better alternative instead of a retaining wall.

CASE 1: Retaining wall at Dwarakanagar, Visakhapatnam
The retaining wall of a multi storeyed building located at Dwarakanagar, Visakhapatnam failed on 3rd November, 2012. Severe cracks were formed around the corner. The wall was about 6.1 m high. The subsoil profile at the site consisted of 4.5 m deep yellowish brown clayey soil with a cohesion value (C) of 5 kN/m^2 and an internal friction (ϕ) of 35°. Backfill soil belonged to the IS classification SC. The retaining wall produced a factor of safety of 1.33 [7]. Since the obtained factor of safety was less than the permissible limit of 1.5, it was planned to replace the retaining wall with an MSE wall using a reinforcement having characteristic strength of 43.59 kN/m with a spacing of 0.5 m and L/H ratio of 0.5. A factor of Safety of 1.92 was obtained after analysis.

CASE 2: Retaining wall at Bypass road in India
Geotechnical investigations on retaining walls located at a bypass road in India showed clear signs of distress immediately after completion of the work and hence it couldn't be opened for the public. Cracks were visible on the pavement and this was due to excessive vertical settlement and lateral displacement. The wall considered was approximately 11.8 m high. Analysis of the retaining wall showed that the factor of safety of the wall was 1.05, making it unsafe for use [8]. As a rehabilitation measure, replacement of the retaining wall with an MSE wall of 11.8 m high with a reinforcement length of 11.8 m and spacing of 0.5 m gave a factor of safety of 1.54.

6 Conclusions

This paper summarises the analysis of a mechanically stabilized earth wall considering four of its parameters. The four parameters used for this study were the reinforcement spacing, reinforcement characteristic strength, L/H ratio and the backfill soil parameters. The various parameters were used in the GEO5 MSE software to find out the factor of safety accordingly and results were obtained.

The spacing of reinforcements differed from 0.2 m to 1 m which showed a decrease in factor of Safety by 74.52%. The characteristic strength of reinforcement were varied from 13.24 to 43.59 kN/m and a factor of safety above 1.5 was observed. Variation in L/H ratio of reinforcements from 0.4 to 1 showed an increase in factor of safety by 15.92%. Type of soil on the basis of its cohesiveness and angle of internal friction had a great influence on the factor of safety. MSE walls with appropriate dimensions were suggested at the end of the paper to replace two failed retaining walls of India.

References

1. Golakiya HD, Lad MD (2015) Design and behaviour of mechanically stabilized earth wall. J Emerg Technol Innov Res 2(11):145–153
2. Hulagabali A, Dodagoudar GR (2019) Evaluation of safety factors due to reinforcement stiffness and evaluation of safety factors due to reinforcement stiffness and its interaction with structures because of their cost-effectiveness and ability to withstand much larger no. November, 2019
3. Hossain MS, Kibria G, Khan MS, Hossain J, Taufiq T (2012) Effects of backfill soil on excessive movement of MSE wall. J Perform Constr Facil 26(6):793–802
4. Koerner RM, Soong TY (2001) Geosynthetic reinforced segmental retaining walls. GeotextGeomembr 19(6):359–386
5. Kibria G, Hossain MS, Khan MS (2014) Influence of soil reinforcement on horizontal displacement of MSE Wall. Int J Geomech 14(1):130–141
6. Hulagabali AM (2018) Behaviour of MSE wall with different soil properties and reinforcement length. Int J Res Appl Sci Eng Technol 6(1):1491–1494

7. Abhishek SV, Tarachand V (2021) Case study of failure of retaining wall at Dwarakanagar, Visakhapatnam. Conference 2021, Proc. Indian Geotech.
8. Sivakumar Babu GL, Raja P, Raghuveer Rao P (2015) Forensic analysis of failure of retaining wall. 15th Asian Reg Conf Soil Mech Geotech Eng ARC 2015 New Innov Sustain

Numerical Investigation of Steel Concrete Hybrid Beam

Anjali M. Nair and S. Sreerath

Abstract Hybrid beam is a progress to the construction industry. When a steel section such as an I section is attached to a concrete component such that there is a transfer of forces and moments between them, then a steel concrete hybrid member is formed. Motivation behind employing the hybrid system was to make up the mutual deficiencies of RC and Steel systems. An analytical model of steel concrete hybrid beam will be developed using ANSYS software. In this study, reinforced hollow concrete beams, which weigh about 15% less than steel–concrete composite frames, preserving the merits of solid concrete beams, including reduced floor depth and steel I beams are used to form a hybrid beam subjected to eccentric loading which includes the performance evaluation of lipped I section in hybrid beam and study of rib portions in I beam.

Keywords Hybrid beam · Ribbed beam · Lipped I section

1 Introduction

Hybrid construction blends better steel and concrete qualities with lower costs, faster construction, fire protection etc. [1]. Because of their advantages over traditional concrete and steel buildings, steel concrete hybrid buildings are nowadays very common. In comparison to conventional beams, steel–concrete hybrid precast beams minimize construction time while improving efficiency and quality [2]. Encased steel section have high resistance to fire and corrosion. In any type of structure, the basic purpose of all kinds of structural system is to transfer the loads effectively. A beam is a structure that mainly resists loads that are laterally applied to the axis of the beam. Its mode of deflection is primarily by bending. It is a one dimensional horizontal flexural member which resists load by bending and which provides support to slab and vertical walls [3]. Buildings are often constructed for

A. M. Nair (✉) · S. Sreerath
Department of Civil Engineering, FISAT, Angamaly, Kerala 683577, India
e-mail: sreeraths@fisat.ac.in

G. C. Marano et al. (eds.), *Proceedings of SECON'21*, Lecture Notes in Civil Engineering 171, https://doi.org/10.1007/978-3-030-80312-4_39

architectural purposes without using a column for a very wide period. In such situations they may cause failure, such as flexural failure, if ordinary beams are given. The use of deep beams in the construction of some very long span halls, for example, is very effective and robust. Some properties associated with the construction of deep beams includes the non-linearity of strain distribution, plane section does not remain plane after deformation [4], shear deformation cannot be neglected as in case of ordinary beams, stress distribution is not linear even in the elastic stage. Due to the heavy emphasis on economic aspects of the newly designed buildings, newer solutions for structural construction are sought [5]. The latest technology "hybrid beam" is given more importance and hence in an important position in improving the overall performance of the building. The use of hybrid beams has many benefits including cost reduction, safety, sustainability, less deformations, better shear strength, high stiffness, faster construction. The fact that each material is used to the fullest advantage makes composite Steel-Concrete construction very efficient and economical [6].

2 Finite Element Model of Hybrid Beam

This work is focused on finite element modeling and analysis of steel concrete hybrid beam in ANSYS v16.

2.1 *Geometry and Connection Details*

An I beam and the hollow reinforced concrete beam together makes the steel concrete hybrid beam. Beam length is 4400 mm. Figure 1 shows the section details of I beam and reinforced hollow concrete beam. Hollow portions have reduced weight, which affects especially the cost of transport, handling and erection for precast cross sections. Other than hollow portion (2.8 m length) the entire beam is concreted [7]. Width of 450 mm is provided for the composite beam. Upper and lower rebars is 2-HD25. Stirrups are provided at a spacing of 40 with 10 mm diameter. Upper and lower headed studs are provided at 400 mm spacing where H = 50 mm. Bonded connections are provided.

2.2 *Material Property*

Linear analysis may not provide the true behaviour of structure subjected to ultimate loads. In such cases, it is necessary to model the material non linearity to study the structure subjected to large deformations. The material properties of concrete and steel from ANSYS are shown in Table 1.

Fig. 1 Section details

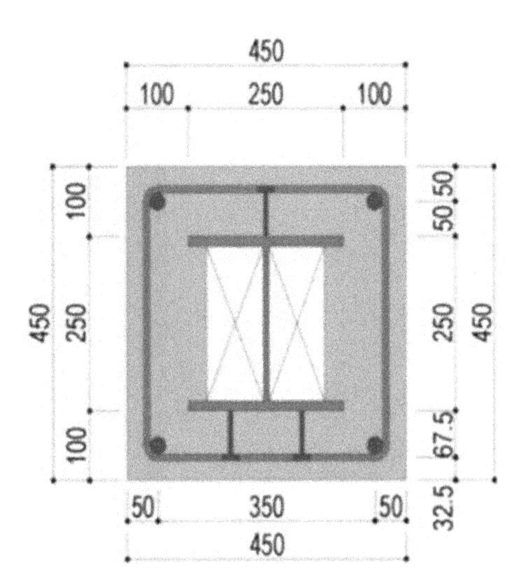

Table 1 Material properties

Properties	Concrete	I steel	Rebar steel
Density (kg/m^3)	2300	7850	7850
Poisson's ratio	0.18	0.3	0.3
Young's modulus (Pa)	30,000	200,000	200,000
Ultimate compressive strength (Mpa)	41	0	0
Tensile yield strength (MPa)	0	250	250
Compressive yield strength (Mpa)	0	250	250
Ultimate yield strength (MPa)	5	460	460

2.3 Model Configuration

The beam and loading plates are modelled using higher order three-dimensional, twenty-node solid element that exhibits quadratic displacement behavior (SOLID186) available in ANSYS. A hollow precast beam composited with a wide-flange steel portion is shown in cross-section in Fig. 1. At the mid-length of the hybrid beam, the 2.8-m-long hollow portion was filled with Styrofoam [2]. Contacts between two elements are provided in order to act as a single unit. The contact between the reinforcement bars and concrete is also made bonded contact. The reinforcing elements, including longitudinal reinforcement, transverse reinforcement, are modelled using three-dimensional two-node beam elements (BEAM188) with six DOFs in each node. The modelled hybrid beam in ANSYS software.

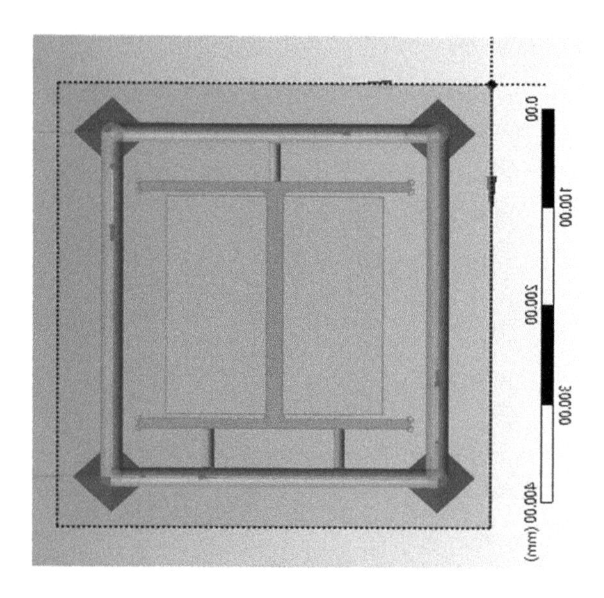

Fig. 2 Finite element model

2.4 *Loading and Boundary Conditions*

The finite element modelling and analysis of present study is carried out using ANSYS Workbench 16.0. Boundary conditions of the finite element model is shown in Fig. 2. Fixed base is considered in the study. Displacement control incremental loading is provided on the two plates. The plates are placed at a distance of 1200 mm as shown in Fig. 2. For determining the structural behavior, non-linear static is to be performed. Loading was applied by displacement control method until failure.

3 Parametric Study

3.1 *Effect of Rib Portion in I Beam*

Ribs are steel components used to keep structural members in position. It prevents buckling of the beam's web and flange. Ribs take less stress as compared to web and flange. Steel rib plates are commonly used to increase the strength and stiffness of beam-to-column joints [8]. Thick plates for ribs are not used. The thicker the plate, the thicker the weld and more twisted the beam. Figure 3 shows I beam with 4 ribs of each of thickness 9 mm, width 118 mm, height 238 mm and mass of ribs as 8.138 kg.

The plot (Fig. 4) shows displacement and its corresponding force of beam with and without ribs. It is clear that I beam with rib (red curve) carries more load comparing to the beam without ribs (Fig. 5).

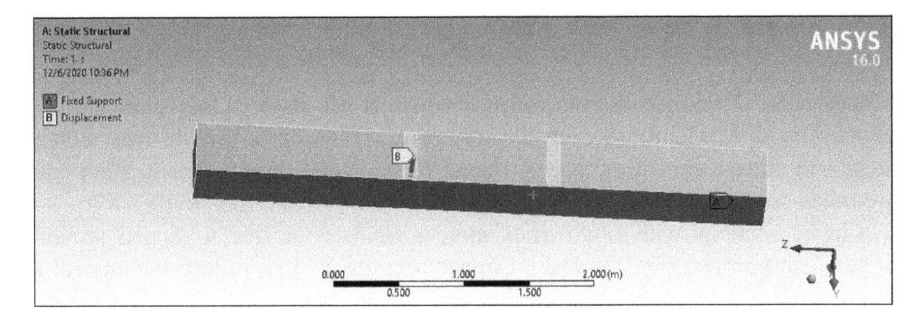

Fig. 3 Test set up

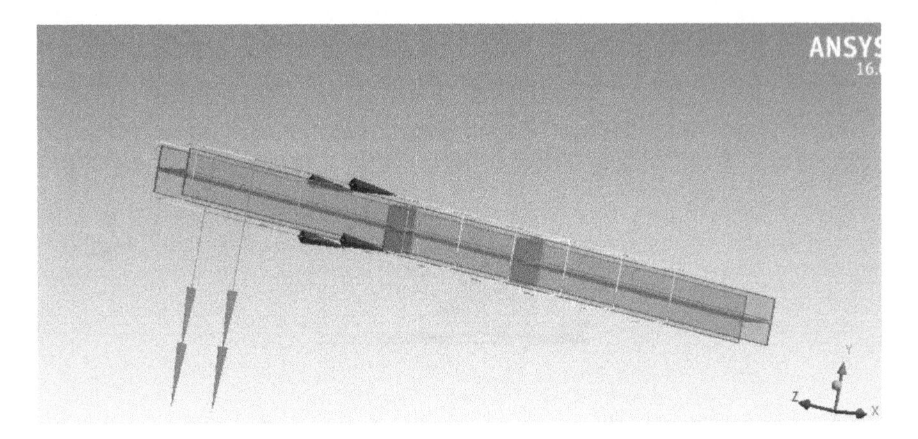

Fig. 4 I section with rib

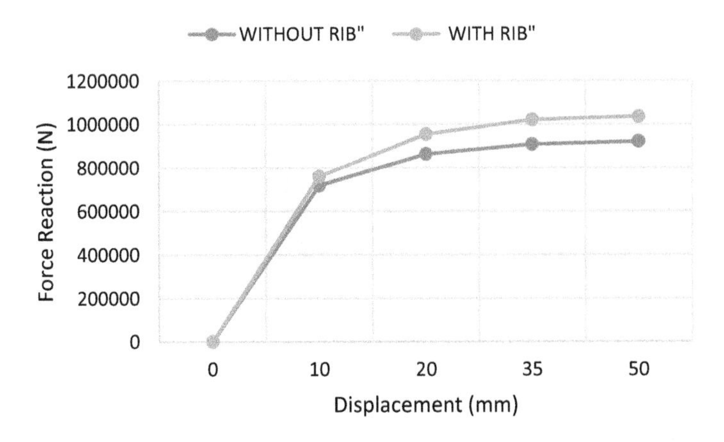

Fig. 5 Plot of force reaction versus displacement for I section with and without rib

3.2 Lipped Section with Rib

Lipped is an alternative of cold rolled channel. It has reduced thickness and cost effectiveness [9]. Mainly used in steel structures, pollution control devices, storage areas and other architectural needs. Figure 6 shows the lipped section. Both ends are made cervical in order to form a lipped section (AISI S100-16w/S1-18).

Lipped sections with ribs carries more load as compared to lipped sections without rib portions (Fig. 7). Use of lipped sections makes it more economical as these sections have reduced thickness and carries more load compared to conventional steel sections as shown in Fig. 7 [9].

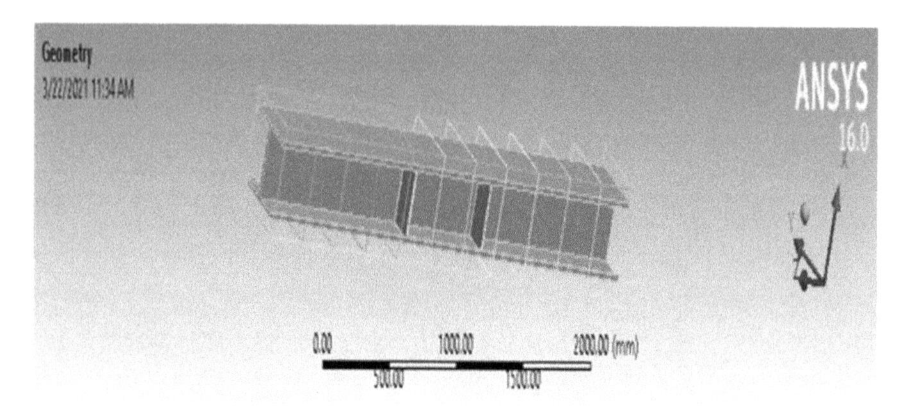

Fig. 6 Lipped section with ribs

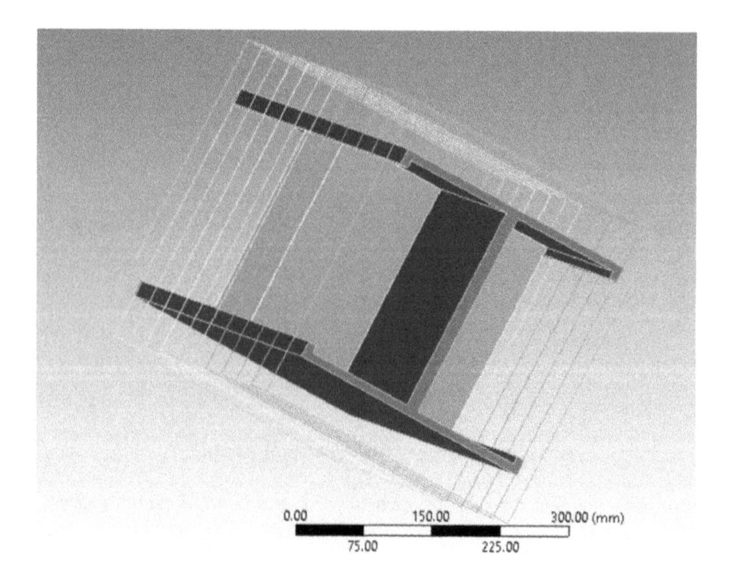

Fig. 7 Lipped section with curvical ends

Table 2 Summary of analysis results

Parts name	I section rib	Lipped I-section rib	I section without rib
	Mass (kg)	Mass (kg)	Mass (kg)
Line body (rebar steel)	69.54	69.54	69.54
Solid body (concrete)	1749.24	1740.84	1751.66
Solid Body (I-steel)	326.5	358.37	326.5
Ribs 4 Nos	8.138	8.138	0
Total mass	2153.4	2176.88	2147.7

3.3 Comparison of I Section with Rib, Lipped I Section with Rib and I Section Without Rib

Comparison of mass of I section with and without rib and lipped section with rib are shown in Table 2. After analysis it was observed that lipped section with ribs has more mass comparing with the other two specimens.

4 Results and Discussions

Non- linear static analysis is performed in terms of load and displacement. Lipped I section with ribs (Fig. 8) is obtained from ANSYS 16.0. 24 edges with 104 faces and 224 vertices are provided (Fig. 9).

Ribs are provided at 9 mm thickness. Providing 4 nos. of ribs gave a total mass of 2153.4. I steel is about 326.5 kg.

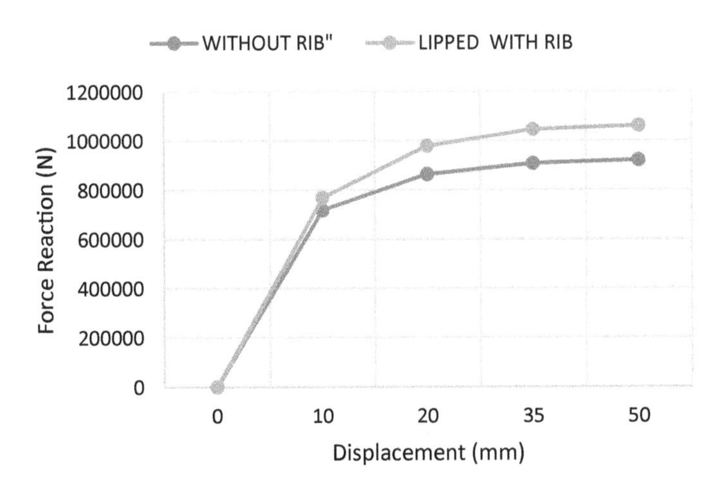

Fig. 8 Plot of force reaction versus displacement for lipped section with ribs

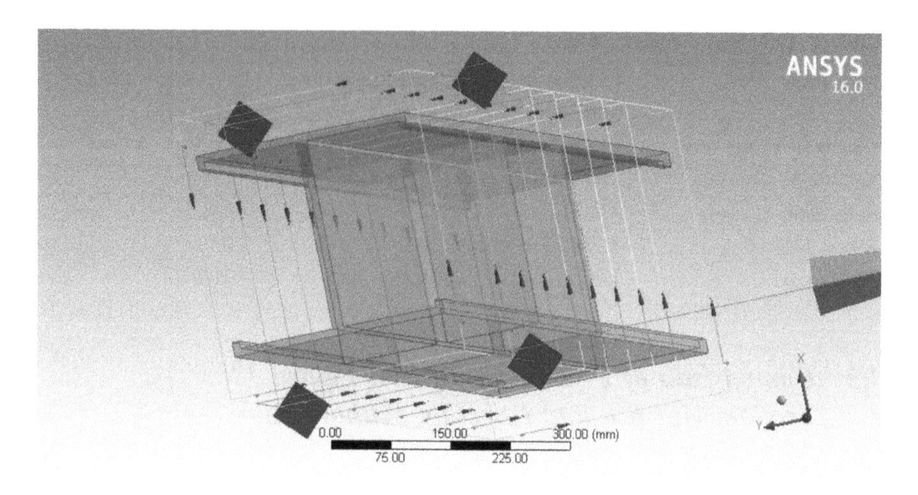

Fig. 9 Lipped I section with ribs (ANSYS)

Figure 5 shows the load versus displacement plot for I beam with and without ribs. From the analysis it is clear that the ultimate load carrying capacity is more in case of I beam with ribs compared with that without ribs. The load carrying capacity of I beam with ribs was increased to 15.67% comparing it with no rib condition.

Figure 8 shows the load displacement curve of lipped section with and without ribs. The load carrying capacity of lipped section with ribs was greater than of about 13.45% comparing with no rib condition. Figure 9 shows the lipped section with ribs(I section replaced with a lipped section) from ANSYS along with the reinforcement details.

5 Conclusions

The following conclusions can be made from the study,

- The presence of ribs plays an important role in the structural performance on the hybrid beam.
- The positioning of ribs also affects the overall performance of the structure. In this study providing 2 sets of ribs was desirable.
- Load carrying capacity of I beam with ribs was increased to 15.67% comparing with no rib condition.
- By providing lipped section, the thickness of the structure can be reduced.
- Lipped section with ribs have more load carrying capacity. Load carrying capacity of lipped section with ribs was about 13.45% comparing it with lipped section without ribs.

- The provision of lipped section along with ribbed section significantly improves the structural performance of the beam.
- Load carrying capacity of lipped section with ribs was increased to 15.68% more than that of I beam with ribs.
- The addition of ribs will not increase the dead weight of the structure significantly. Hence there won't be any need for the provision of additional load carrying members.
- Hence steel concrete hybrid beams are having sufficient load carrying capacity and as they are aesthetically pleasant.

References

1. Morino S (1998) Recent developments in hybrid structures in Japan—research, design and construction. Eng Struct 20(4–6):336–346
2. Kima S-K, Kimb J-M, Hong W-K (2020) Material nonlinear finite element analysis of hybrid hollow concrete beams encasing steel sections. Structures 25:500–519
3. Bhattarai BP, Bhattarai N (2017) Experimental study on flexural behavior of reinforced solid and hollow concrete beams. Int J Eng Res Adv Technol (IJERAT) 3(11)
4. Mohammadhassani M et al (2011) Failure modes and serviceability of high strength self compacting concrete deep beams. Eng Fail Anal 18.8:2272–2281
5. Kozioł P (2015) Modern design of steel-concrete composite structures. Zeszyty Naukowe Politechniki Częstochowskiej. Budownictwo 21(171):118–127
6. Tedia A, Maru S (2014) Cost, analysis and design of steel-concrete composite structure Rcc structure. IOSR J Mech Civil Eng 11.1:54–59
7. Hussain Alshimmeri AJ, Ghadhban Al-Maliki HN (2014) Structural behavior of reinforced concrete Hollow beams under partial uniformly distributed load. J Eng 20(7)
8. Tartaglia R, D'Aniello M, Landolfo R (2018) The influence of rib stiffeners on the response of extended end-plate joints. J Constr Steel Res 148:669–690
9. Keerthan P, Mahendran M (2013) Experimental studies of the shear behaviour and strength of lipped channel beams with web openings. Thin-Walled Struct 73:131–144

Structural Performance of Splice Connection for Tubular GFRP Section Members

Mittu Kuriakose and S. Sreerath

Abstract The use of fiber reinforced polymer (FRP) composites in the repairing and strengthening of structures has been the focus of extensive research in recent years. In the construction industry, glass fiber reinforced polymer (GFRP) has emerged as a new building material. The various advantages of GFRP include high strength, light weight, low maintenance, durability, resists salt water, chemicals and the environment. GFRP has been the focus of limited study due to its difficulty in connection. Two bonded sleeve joints (BSJ) and a bolted flange joint (BFJ) combined together to form a splice connection was proposed for tubular GFRP section members. The steel tube is inserted into the tubular GFRP member by bonded sleeve joint and the endplates are attached through a bolted flange joint. The bonded sleeve joint reduces stress concentration in the FRP members and allows easy on-site installation, whereas the bolted flange joint reveals ductile failure. An analytical model of splice connection will be developed using ANSYS software. The study focuses on axial and bending performance of splice connection under varying aspect ratio.

Keywords GFRP · Splice connection · Bonded sleeve joint · Bolted flange joint

1 Introduction

Steel, concrete, timber and masonry are four conventional building materials that have dominated the construction industry in the past. Masonry and timber were the two materials used a few hundred years ago. But, timber, steel and concrete are the most commonly used building materials today. Corrosion, deterioration and other ageing issues are some of the demerits of these conventional building materials [1]. As a result, there is a growing demand for alternatives to traditional building materials these days. Fiber reinforced polymer (FRP) composite materials have emerged as a modern building material in the last 30 years due to their superior

M. Kuriakose (✉) · S. Sreerath
Department of Civil Engineering, FISAT, Angamaly, Kerala, India
e-mail: sreeraths@fisat.ac.in

G. C. Marano et al. (eds.), *Proceedings of SECON'21*, Lecture Notes in Civil Engineering 171, https://doi.org/10.1007/978-3-030-80312-4_40

mechanical, chemical and physical properties [2]. Due to their high strength and light weight, FRP composites were first produced for military and aerospace applications in 1940. FRP composites have high strength-to-weight ratios, excellent durability in a variety of environments and are extremely simple and fast to install [3, 4]. The ability to control the mechanical properties of FRP composites by choosing the right fibers and orienting them in the right direction is a significant benefit of using FRP materials. The FRPs are dimensionally stable and corrosion resistant. Some of the applications of this material include the repair of existing structures and the building of new ones [5].

Modular buildings are prefabricated structures made up of repeating units known as modules. It is a construction method in which the sections are prefabricated and are transported to the construction site for installation. Wood, Fiber Reinforced Polymer, concrete, steel and other materials are used to build these structures. Fiber Reinforced Composite Polymer is a composite material made up of polymer matrix and fiber reinforcement. Glass (GFRP), carbon (CFRP), aramid and basalt are commonly used fibers. Epoxy, vinyl ester or polyester thermosetting plastics make up the polymer. Fiber Reinforced Polymer Composites have shown increased usage in civil engineering constructions over the last two decades [6]. These lightweight, corrosion-resistant materials have gained worldwide recognition for their use in the restoration and strengthening of existing structures. Because of their light weight, high strength, corrosion resistance and low maintenance requirements glass fiber reinforced polymer has been used as load-bearing structural members in civil engineering construction [7]. Pultrusion methods are used to produce I-shaped and tubular sections having more scope for frame and construction applications. However, a proper joint between GFRP beams and columns are needed for such implementations. GFRP composites provide adequate strength and stiffness at economical cost. FRP composites have great potential as load-bearing members in new construction due to various advancements in pultrusion manufacturing technique. Bridge decks, beams, columns and floors are some of the examples. The brittle and anisotropic behavior of FRP materials must be considered when designing connections for these members [8]. Hollow section profiles (I or channel profiles) are more resistant to torsional and global buckling than closed section profiles (circular or rectangular tubular profiles). These advantages of tubular GFRP members counterbalanced the difficulty of connecting them due to material anisotropy and the closed section form. FRP materials are currently being used to build hybrid FRP bridges for both vehicular and pedestrian use. In the FRP bridge industry, connecting FRP members is a huge challenge. Due to the need to produce in sections and assemble on site, FRP bridges cannot be manufactured off-site. Mechanical joints, bonding and other methods of connecting these elements are in demand. A steel bolted flange joint connects two tubular steel GFRP bonded sleeve joints to form the splice connection [9, 10].

In this study, structural performance of the splice connection for tubular GFRP section members is evaluated. To understand the bending and axial performance of the splice connection, a finite element model was created using ANSYS 16.0. The splice connections were analyzed under varying aspect ratio.

2 Finite Element Modelling

2.1 General

ANSYS workbench 16.0 software is used for modelling splice connection for tubular GFRP section members. SOLID186 element was used to create finite element model to investigate the bending performance and axial performance of the splice connection. In the present model 20 node SOLID 186 is adopted.

2.2 Geometry

To investigate the structural performance of splice connections for tubular GFRP section members, a finite element model was created using the SOLID186 element of ANSYS 16.0. The splice connection for bending performance are indexed as B-X-Y where B refers to bending load, X denotes GFRP tube length and Y denotes GFRP tube width. The length and width of GFRP tube is varied by 10%. Two sets of splice connection each with three types are considered: B-790-79, B-790-88, B-790-102 and B-790-102, B-918-102, B-1020-102. The splice connections for axial performance are indexed as A-X-Y where A refers to axial loading, X denotes GFRP tube length and Y denotes GFRP tube width. Two sets of splice connection each with three types are considered: A-790-79, A-790-88, A-790-102 and A-790-102, A-918-102, A-1020-102. The thickness of the GFRP tube is taken as 9.5 mm. Figures 1 and 2 shows the loading and support condition of beam and column respectively.

Fillet welds with a leg length of approximately 6 mm were used to connect a steel square tube of grade 355 (80 × 80 × 6 mm) to a 6 mm thick steel flange plate of grade 250. The steel tube was then coaxially connected and bonded to a square glass fiber reinforced polymer (GFRP) tube. The steel flange plates are then bolted together using M12 grade 8.8 bolts that have been pretensioned to 65 kN each. Table 1 lists the material properties.

3 Results and Discussion

Detailed three-dimensional finite element modelling is conducted for evaluating the bending performance and axial performance of splice connection. In case of bending, the boundary condition to simulated four point bending. The four point bending is simulated by applying constrain in y direction at one end and applying constrain in x and y direction at the other end. Therefore, the beam acts as a simply supported beam and displacement loading is applied. Load is applied by displacement control method until failure. Then, non-linear static analysis is carried

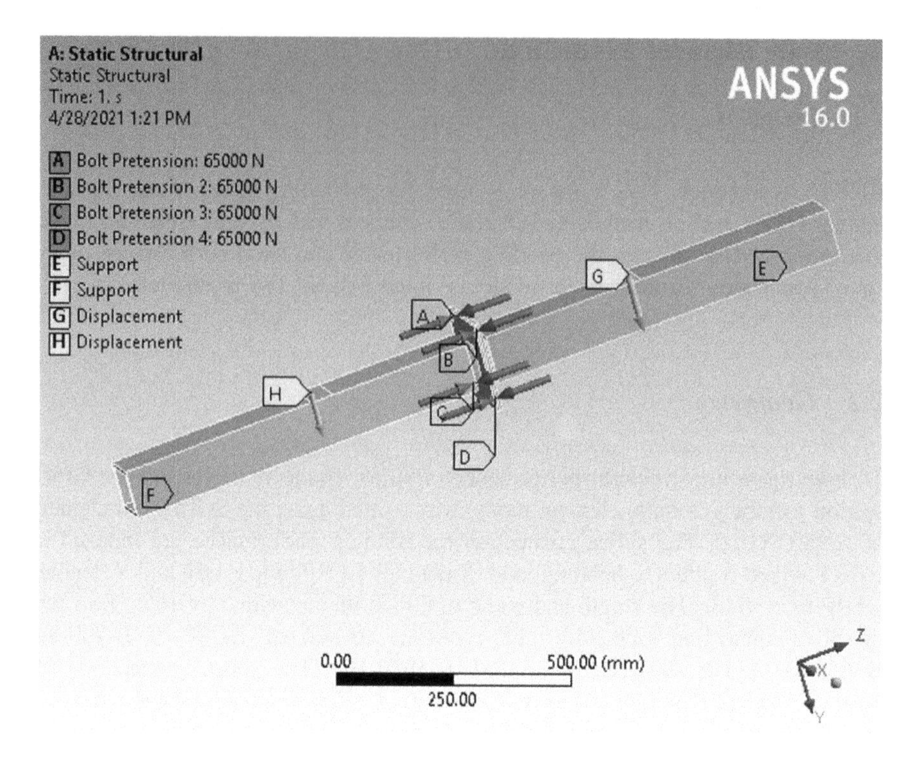

Fig. 1 Loading and support condition of beam

out to determine the structural behavior. Two sets of splice connection each with three types are considered for bending. First set include B-790-79, B-790-88, B-790-102 by varying the width of GFRP tube and second set include B-790-102, B-918-102, B-1020-102 by varying the length of GFRP tube.

Figure 3 shows the bending performance of splice connection by varying the width. From the moment-displacement curve we can see that B-790-102 provides higher moment carrying capacity when compared to B-790-102 and B-790-88. Figure 4 shows the bending performance of splice connection by varying the length. From the moment-displacement curve we can see that B-1020-102 provides higher moment carrying capacity than B-790-102 and B-918-102. Table 2 shows equivalent stress results obtained after bending performance.

For axial performance, the boundary condition is to simulated axial loading. Both ends of the column is pinned and axial load is applied. Two sets of splice connection each with three types are considered for axial performance. First set include A-790-79, A-790-88, A-790-102 by varying the width of GFRP tube and second set include A-790-102, A-918-102, A-1020-102 by varying the length of GFRP tube.

Figure 5 shows the axial performance of splice connection by varying the width. From the moment-displacement curve we can see that A-790-102 provides higher load carrying capacity when compared to A-790-88 and A-790-79. Figure 6 shows

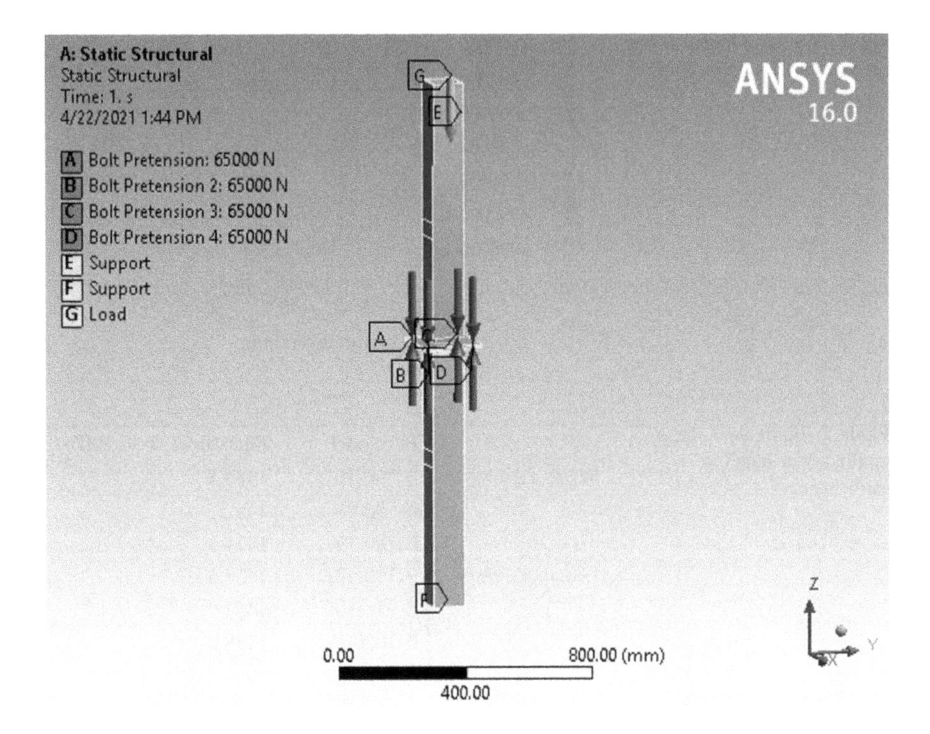

Fig. 2 Loading and support condition of column

Table 1 Material properties

	Yield strength (MPa)	Young's modulus (GPa)	Poisson's ratio
Steel square tube	420.1	209.5	0.28
Steel flange plate	311.8	201.2	0.28
GFRP square tube	330.6	25.2	0.3
Bolt	1043	235	0.28
Epoxy	32.2	4.25	0.28

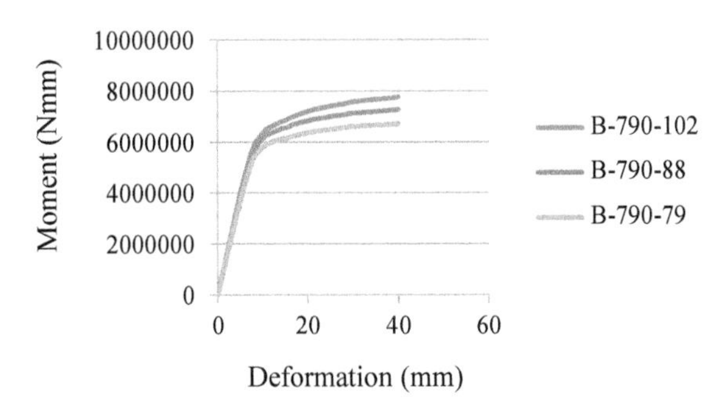

Fig. 3 Moment-displacement curve for splice connection of varying width

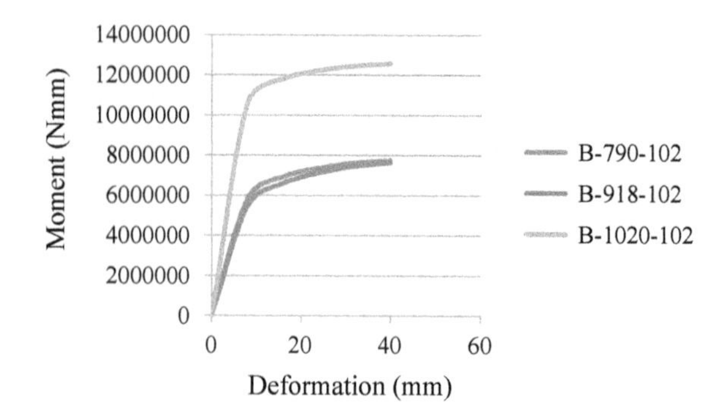

Fig. 4 Moment-displacement curve for splice connection of varying length

Table 2 Equivalent stress results under bending performance

	FE model	Equivalent stress (MPa)
Width varying	B-790-102	1487.5
	B-790-88	1452.7
	B-790-79	1471.3
Length varying	B-790-102	1487.5
	B-918-102	1487.9
	B-1020-102	1488.0

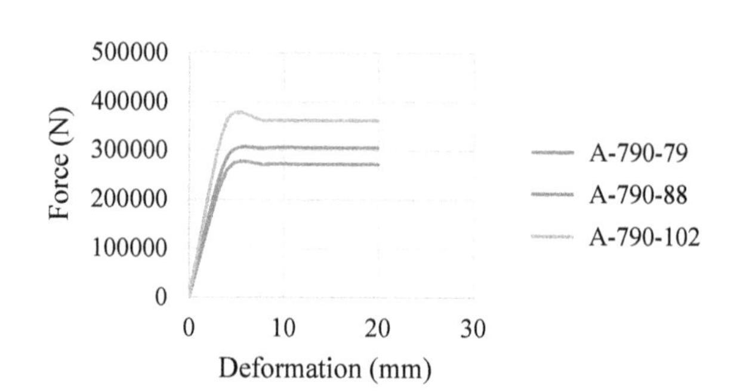

Fig. 5 Force-displacement curve for splice connection of varying width

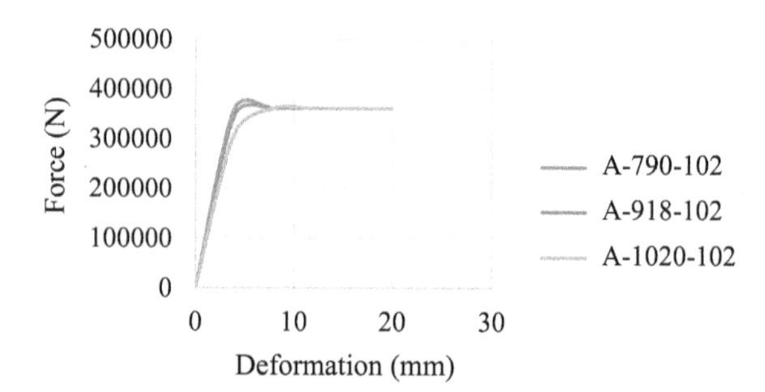

Fig. 6 Force-displacement curve for splice connection of varying length

Table 3 Equivalent stress results under axial performance

	FE model	Equivalent stress (MPa)
Width varying	A-790-102	1492.9
	A-790-88	1459.4
	A-790-79	1470.9
Length varying	A-790-102	1492.9
	A-918-102	1492.7
	A-1020-102	1492.4

the axial performance of splice connection by varying the length. From the force-displacement curve we can see that A-790-102 provides higher load carrying capacity than A-1020-102 and A-918-102. Table 3 shows equivalent stress results obtained after axial performance.

4 Conclusions

The bending performance and axial performance of splice connection for tubular GFRP section members was investigated in this paper. The following conclusions are drawn based on the results of finite element modelling:

1. B-790-102 showed better performance than other two under bending with higher moment carrying capacity and stress
2. When the width was increased by 10%, there was increase in moment carrying capacity and stress
3. B-1020-102 with higher moment carrying capacity and stress showed better performance than other two under bending
4. When the length was increased by 10%, there was increase in moment carrying capacity and stress
5. A-790-102 showed better performance than other two under axial loading with higher load carrying capacity and stress
6. When the width was increased by 10%, there was increase in load carrying capacity and stress
7. A-790-102 with higher load carrying capacity and stress showed better performance than other two under axial loading
8. When the length was increased by 10%, showed increase in load carrying capacity and stress

References

1. Razaqpur AG et al (2019) GFRP hollow column to built-up beam adhesive connection: Mechanical behaviour under quasi-static, cyclic and fatigue loading. Compos Struct 224:111069
2. Xie L et al (2019) Pultruded GFRP square hollow columns with bolted sleeve joints under eccentric compression. Compos B Eng 162:274–282
3. Qiu C et al (2020) Cyclic performance of splice connections for hollow section fibre reinforced polymer members. Compos Struct 112222
4. Fang H et al (2019) Connections and structural applications of fibre reinforced polymer composites for civil infrastructure in aggressive environments. Compos Part B Eng 164: 129–143
5. Zhang ZJ, Bai Y, Xiao X (2018) Bonded sleeve connections for joining tubular glass fiber–reinforced polymer beams and columns: experimental and numerical studies. J Compos Constr 22.4:04018019
6. Luo FJ et al (2019) Development of latticed structures with bolted steel sleeve and plate connection and hollow section GFRP members. Thin Walled Struct 137:106–116
7. Ascione F et al (2017) Strength and stiffness of adhesively bonded GFRP beam-column moment resisting connections. Compos Struct 160:1248–1257
8. Zhang Z et al (2018) Cyclic performance of bonded sleeve beam-column connections for FRP tubular sections. Compos B Eng 142:171–182
9. Qiu C et al (2019) Bending performance of splice connections for assembly of tubular section FRP members: Experimental and numerical study. J Compos Constr 23(5):04019040
10. Qiu C et al (2018) Axial performance of steel splice connection for tubular FRP column members. Compos Struct 189:498–509

Identification of Groundwater Recharge Potential Zones and Groundwater Quality Mapping Using Remote Sensing and GIS

Aiswarya R. Nair, Ajay Suresh, Aman Saju, Ganga Sankar, and Rinu J. Achison

Abstract Groundwater is a very important natural capital for the consistent and economic provision of potable water supply for rural and urban environs. The present scenario of climate change has posed a challenge to the fundamental well-being of all countries, and the situation is even worse in countries or regions which are already suffering from water scarcity. Due to the decreasing trend in rainfall in certain regions and increasing water demands, the surface water sources are depleting and therefore, there is a need to locate alternate sources of water. The development of a reasonable model for groundwater potential can be a better solution. This project aims to generate a groundwater potential zonation map using the integrated use of remote sensing and geographic information system (GIS) for the selected study area. Different thematic maps of information, such as geomorphology, geology, soil type, land slope and land use/land cover, drainage density, lineament density, hydrogeomorphology were integrated with weighted overlay in GIS to generate groundwater potential zonation maps of the area. Based on the groundwater potential index values, the study area is classified as 'high', 'moderate', or 'low'. Then the pertinent sites for artificial recharge structures have to be identified by the overlay analysis of the drainage layer and geomorphology of the area through GIS technologies. These structures will be capable of fulfilling the demand for groundwater for future generations.

Keywords Groundwater potential zones · ArcGIS 10.3

1 Introduction

Groundwater is one of the most important natural resources of the earth which is a very essential source for drinking, irrigation, and industrialization purposes. In the absence of good quality sub-surface water and the growing need for water for

A. R. Nair · A. Suresh (✉) · A. Saju · G. Sankar · R. J. Achison
Federal Institute of Science and Technology (FISAT), Hormis Nagar, Mookkannoor, Angamaly, Ernakulam 683577, India

G. C. Marano et al. (eds.), *Proceedings of SECON'21*, Lecture Notes in Civil Engineering 171, https://doi.org/10.1007/978-3-030-80312-4_41

domestic, agricultural, and industrial uses, groundwater is an attractive source for good quality and replenishment. In the present scenario of urbanisation, industrialisation and population increase, the rate of consumption of groundwater is increasing continuously. So it is important to identify groundwater potential zones and also monitor and conserve these sources. Efficient management of groundwater in these areas is also very important. The occurrence and movement of groundwater in an area is influenced by different parameters such as geology, soil, geomorphology, lineament density, slope, drainage density, land use/land cover, and the interrelationship between these parameters [1]. Artificial recharge systems are engineered systems which impound surface water to infiltrate through soil layers and percolate into the aquifers to augment groundwater resources. In the unsaturated areas, where these cannot be implemented, shafts and trenches can be directly injected through wells to the aquifers. Groundwater potential zones are the zones that mark the availability of groundwater. Identification of groundwater potential zones can benefit in proper development and utilization of both surface water and groundwater resources by eliminating water scarcity. This can enhance agricultural practices and their income. Uneven distribution of groundwater and misuse of water resources in a year often increases the chances of flooding during monsoon. Also significant decrease in water during non-monsoon periods. Appropriate and efficient techniques should be used to find potential sites for drilling of wells and for providing artificial recharge structures. This is to ensure the sustainability of well yields and to overcome the crisis of water table depletion. For efficient and successful studies of groundwater, GIS provides spatial data integration and natural resource management tools [2].

Groundwater quality is one of the major environmental aspects which is to be studied and supervised depending upon its spatial distribution [2]. Due to the lack of effective management, a reduction in quantity and also quality of groundwater can be a problem. The aim of this study is to evaluate groundwater quality in the study area as a case study and the application of spatial distribution by using Geographic Information System (GIS) and geostatistical algorithms. The groundwater quality parameters such as pH, total dissolved solids, total hardness, alkalinity, chloride, nitrate, sulfate, calcium, magnesium, and fluoride, were collected for this. Using the geostatistical (Kriging) method, maps of each parameter were created [3]. This was used to create the groundwater quality index (GWQI) map using the index method. In general, this integrated method is a sufficient assessment tool for environmental spatially distributed parameters.

2 Study Area

Muvattupuzha is located in the east of Kochi in Ernakulam district. Muvattupuzha is the originating point of Muvattupuzhayar. This place is famous for the meeting point of three rivers together namely, Thodupuzhayar, Muvattupuzhayar, and Kaliyar, hence the name Muvattupuzha. The town shares the border with Kottayam district on the south and Idukki on the east. Muvattupuzha Block has an area of 206.07 km^2.

3 Objectives and Scope

The objectives of this project are:

- To generate a groundwater potential zonation map of the selected study area (Muvattupuzha block).
- Groundwater potential map validation.
- To propose a suitable artificial recharge method for the areas identified with poor groundwater potential.

The scope of this project includes:

- Increasing the sustainability of wells.
- Arresting the declining groundwater level trends.
- Reducing the reliance on government water supply.

4 Methodology [1, 3–6]

See Fig. 1.

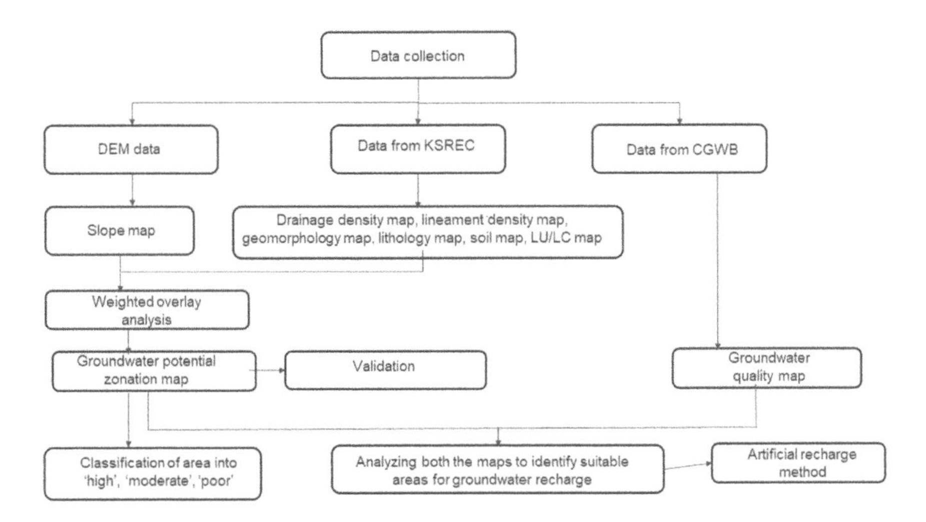

Fig. 1 Methodology flow chart

4.1 Results and Discussions

Slope Map

The slope is the rate of change of elevation and it is considered as the governing factor of superficial water flow, since slope determines the gravity effect on it. The slope is directly proportional to run-off. Groundwater recharge will be less on steep slopes. It is slow in gently undulating plains. Therefore, enough time is available for infiltration. Slope maps were derived from Aster DEM data, which were downloaded from the Earth data, NASA, with 30 m resolution. The study area was masked out from DEM and the slope was derived using the spatial analyst tool in ArcGIS 10.3. From the map generated, most of the region lies under 0–0.42°. So it's clear that the altitude is much lesser and thus it contains vast flatlands and paddy fields spread around the region. Only a small amount of the region comes under the slopes greater than 15° so that the areas surrounding the Muvattupuzha block are less hilly and highly fertile (Fig. 2).

Land use/land cover map

Land use refers to man's activities in the land, numerous uses which are carried out on land, etc. Land cover denotes the natural vegetation, water bodies, rocks, etc. that resulted due to land transformation. Thematic map of land use/land cover was collected from Kerala State Remote Sensing and Environmental Centre (KSREC),

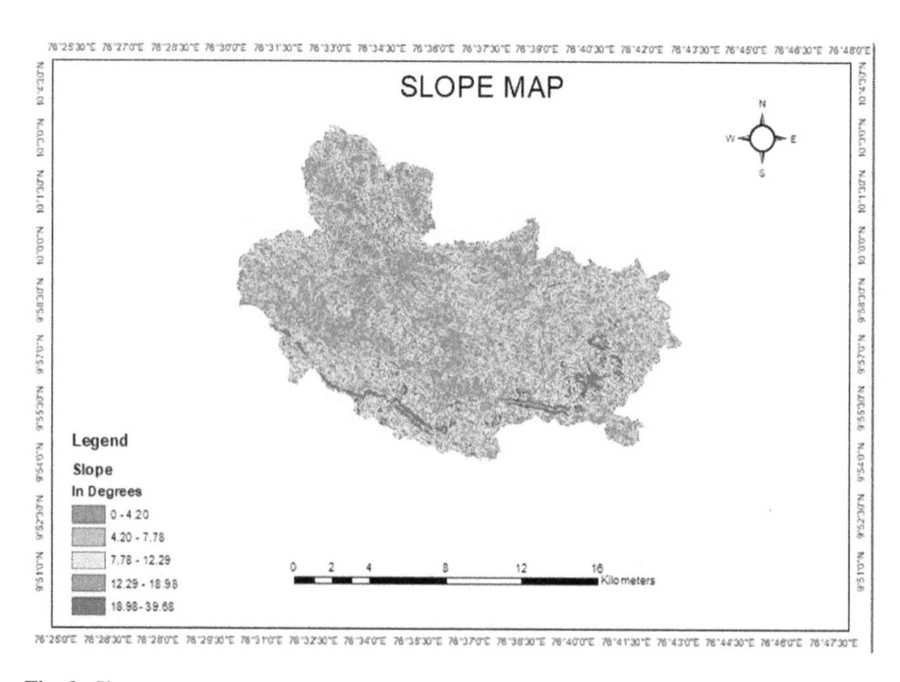

Fig. 2 Slope map

Trivandrum. From the map, it is evident that almost all of the region is used for agriculture plantation due to the extremely fertile soil encompassing the region. And only less than 20% of the region is used for built-up spares. less than 10% occupied by deciduous forest. Less than 5% used for industrial purposes (Fig. 3).

Geomorphology map

Geomorphological maps portray important geomorphic units, landforms, and underlying geology and aid understanding of the processes, materials/lithology, structures, and geologic controls relating to groundwater occurrence, likewise on groundwater potentials. It is thought about as surface indicators for the identification of subsurface water conditions. Thematic map of geomorphology was obtained from KSREC, Trivandrum. Geomorphology of the region is such that, More than 60% is a plateau region and the rest is the piedmont zone and pediplain region. Less than 5% are residual hills (Fig. 4).

Drainage density map

Drainage density is an inverse function of permeability. The line density tool of the spatial analyst is used to get a density map from the drainage map. Drainage density map was obtained by using the Line Density tool of ArcGIS 10.3. The Line Density tool calculates the density of linear features in the neighborhood of each output raster cell (Fig. 5).

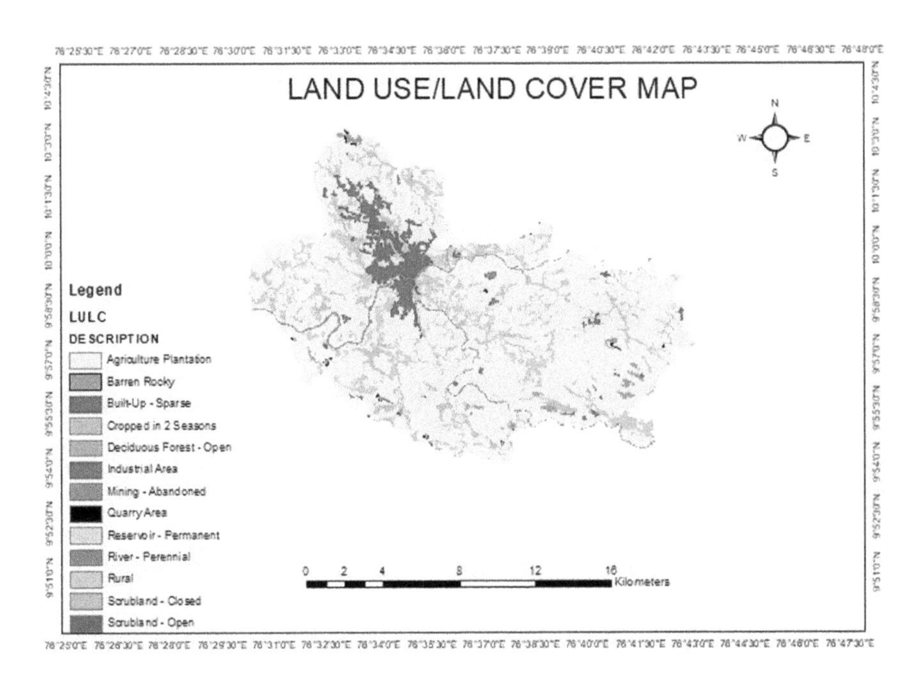

Fig. 3 Land use/land cover map

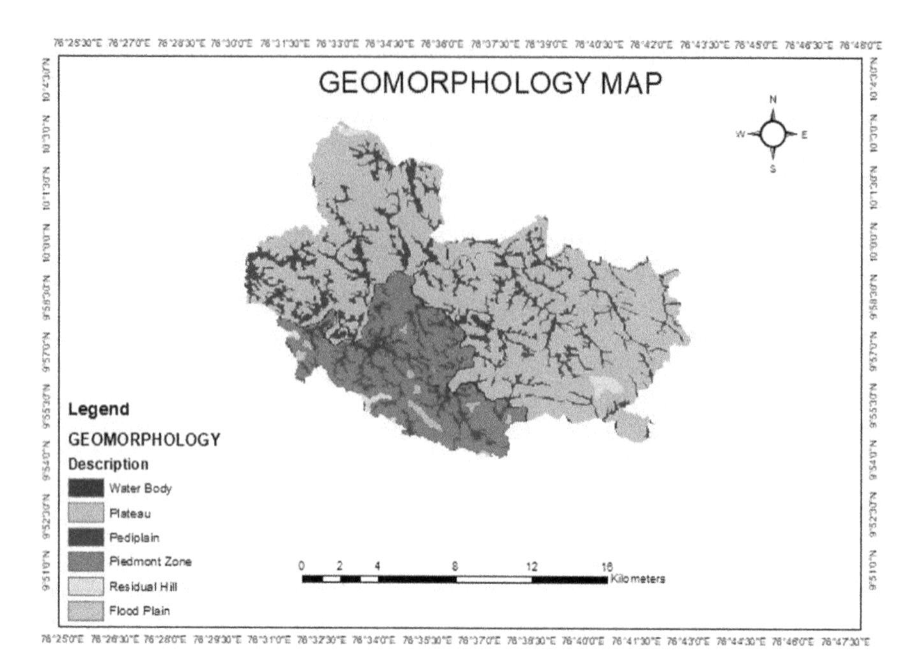

Fig. 4 Geomorphology map

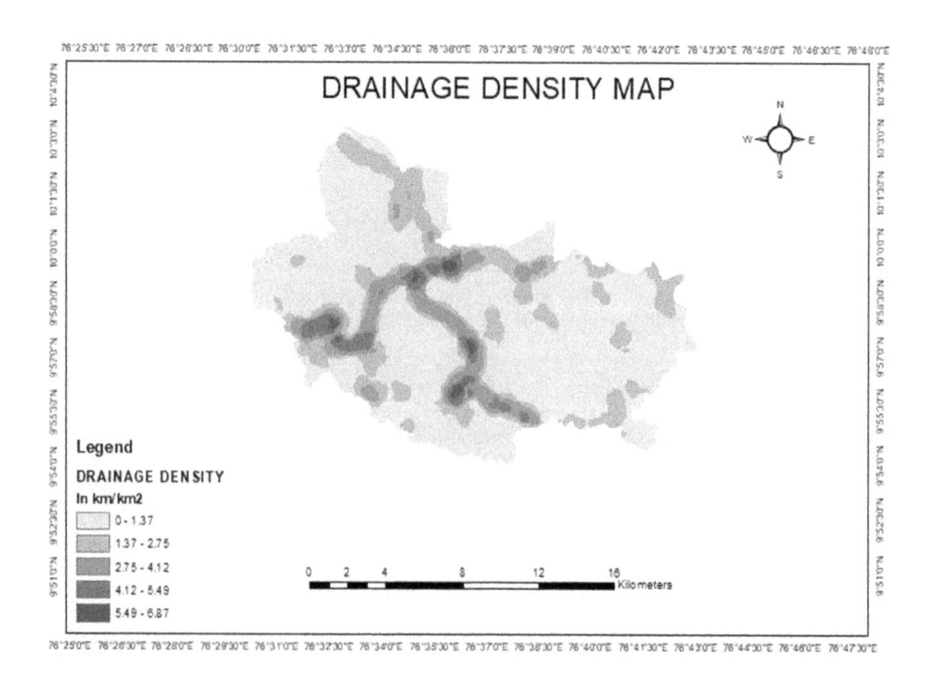

Fig. 5 Drainage density map

Soil map

Soil characteristics have a substantial role in the infiltration of water. The rate of infiltration largely depends on the grain size of the soil. Thematic map of soil was obtained from KSREC, Trivandrum. Most of the region has a gravel soil formation. Less than 10% is occupied by clayey soil near the North-Eastern region. Most of the region has a fertile soil formation (Fig. 6).

Lineament density map

Lineaments represent zones of faulting and fracturing resulting in increased permeability. They are hydro-geologically vital since they act as pathways for groundwater movement. The lineament density of an area can directly reveal the groundwater potential since the presence of lineaments typically denotes a permeable zone. Areas with high lineament density are good for groundwater potential zones. Lineament map was collected from KSREC. Lineament density was then calculated by spatial density analysis in ArcGIS (Fig. 7).

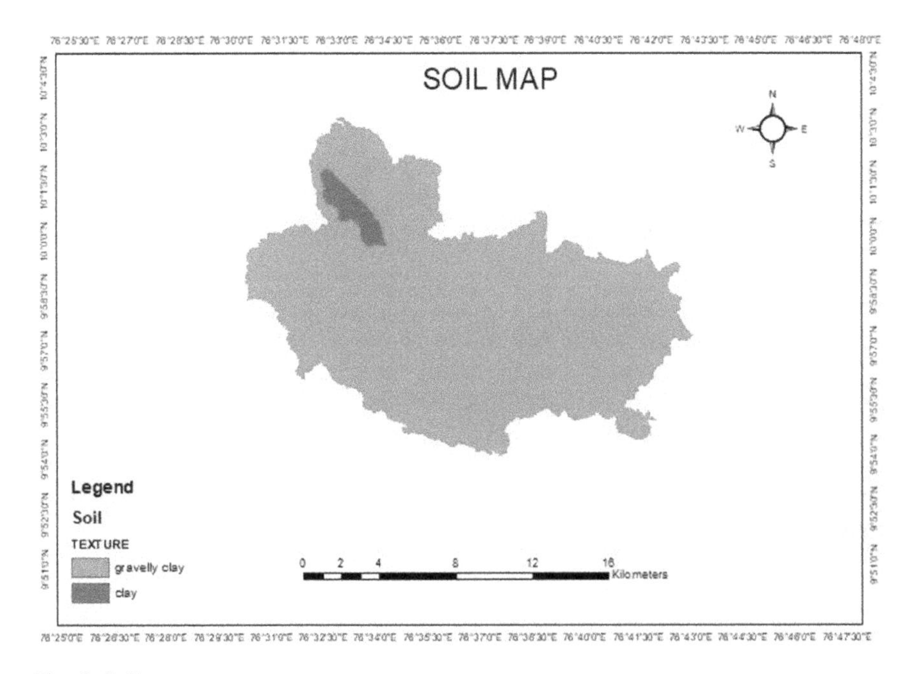

Fig. 6 Soil map

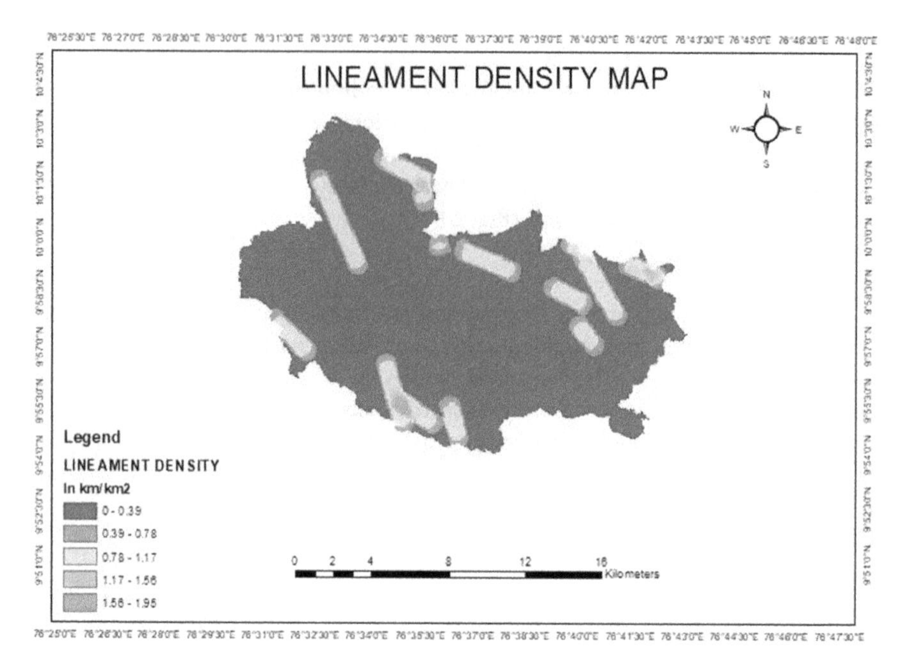

Fig. 7 Lineament density map

5 Derivation of Normalised Weights of Thematic Layers by AHP Method

Analytic Hierarchy Process (AHP) is adopted here for the derivation of normalized weights of thematic layers. The AHP method is preferred for decision-making of a problem consisting of several parameters that can affect the result. The stronger the influence of one factor over other factors, the higher its relative importance. Which gives a larger weight (Tables 1, 2 and 3) [5, 2, 7].

Table 1 Pairwise comparison matrix For AHP [5]

	GM	DD	LD	Slope	GG	LU/LC	Soil
GM	1	2	3	5	7	8	9
DD	1/2	1	2	3	4	8	7
LD	1/3	1	1	3	7	7	7
Slope	1/5	1/3	1/3	1	5	5	5
GG	1/7	1/4	1/7	1/5	1	7	4
LU/LC	1/8	1/8	1/7	1/5	1/7	1	3
Soil	1/9	1/7	1/7	1/5	1/4	1/3	1
Column total	2.41	4.85	6.76	12.60	24.39	36.33	36.00

Table 2 Determining the normalised weight for thematic layers [5]

	GM	DD	LD	Slope	GG	LU/LC	Soil	Normalized weights
GM	0.41	0.41	0.44	0.40	0.29	0.22	0.25	0.35
DD	0.21	0.21	0.30	0.24	0.16	0.22	0.19	0.22
LD	0.14	0.21	0.15	0.24	0.29	0.19	0.19	0.20
Slope	0.08	0.07	0.05	0.08	0.20	0.14	0.14	0.11
GG	0.06	0.05	0.02	0.02	0.04	0.19	0.11	0.07
LU/LC	0.05	0.03	0.02	0.02	0.01	0.03	0.08	0.03
Soil	0.05	0.03	0.02	0.02	0.01	0.01	0.03	0.02

Table 3 Well data

Parameters	Influence factor	Rank
Slope	11	
0–4.20		5
4.20–7.78		4
7.78–12.29		3
12.29–18.98		2
18.98–36.98		1
Land use/landcover	3	
Agricultural plantation		4
Barren Rocky		1
Built-up sparse		1
Cropped in two seasons		4
Deciduous forest		3
Industrial area		1
Mining abandoned		1
Quarry area		1
Reservoir permanent		5
River perennial		5
Rural		3
Scrubland closed		1
Scrubland open		1
Geomorphology	35	
Waterbody		5
Plateau		1
Pediplain		3
Piedmont zone		4
Residual hill		1

(continued)

Table 3 (continued)

Parameters	Influence factor	Rank
Flood plain		5
Drainage density	22	
0–1.37		5
1.37–2.75		4
2.75–4.12		3
4.12–5.49		2
5.49–6.87		1
Soil	2	
Clay		1
Gravelly soil		4
Lineament Density	20	
0–0.39		1
0.39–0.78		2
0.78–1.172		3
1.172–1.56		4
1.56–1.95		5
Geology	7	
Charnockite		3
Migmatite complex		5
Basic rocks		1
Khondalite rocks		5

6 Groundwater Quality Data [3, 6]

See Tables 4 and 5.

7 Spatial Distribution of Chemical Parameters in the Pre-monsoon Stage

See Figs. 8, 9, 10, 11, 12, 13, 14, 15, 16, 17, 18, 19 and 20.

8 Spatial Distribution of Chemical Parameters in the Post-monsoon Stage

See Figs. 21, 22, 23, 24, 25, 26, 27, 28, 29, 30, 31 and 32.

Table 4 Well data

Well no.	Well type	Tahsil/Taluk	Block/Mandal	Village	Hamlet	Latitude	Longitude
E81	Dug Well	Muvattupuzha	Muvattupuzha	Koothattukulam	Koothattukulam, Ward 2	09°51′10″	76°35′15″
E82	Dug Well	Kunnathunad (perumbavur)	Muvattupuzha	Ramamangalam	Ramamangalam	09°56′24″	76°28′53″
E83	Dug Well	Muvattupuzha	Muvattupuzha	Piravom	Piravom	09°52′15″	76°29′11″
E95	Dug Well	Kunnathunad (perumbavur)	Muvattupuzha	Perumbavoor	Perumbavoor	10°06′53″	76°28′28″
BW112	Bore Well	Muvattupuzha	Muvattupuzha	Velloorkunnam (vazhapilly)	Vazhapilly	09°59′45″	76°33′50″
BW113	Bore Well	Muvattupuzha	Muvattupuzha	Muvattupuzha	Thirkkalathur	10°01′58″	76°32′56″
E80	Dug Well	Muvattupuzha	Muvattupuzha	Muvattupuzha	Muvattupuzha	09°58′41″	76°34′52″
GWE-22	Dug Well	Muvattupuzha	Muvattupuzha	Eranellur	Kalambore	09°59′51″	76°37′28″

Table 5 Groundwater quality data

Well no.	Sampling date	Latitude	Longitude	pH	Ec	Total hardness	Alkalinity	Total dissolved oxygen	cl	SO_4	NO_3	Ca	Mg	F	Fe
E81	Mar-18-19	09°51′10″	76°35′15″	7.9	260	75	44	156	28	5.3	7.6	16	8.5	0.05	0.37
E82	Mar-18-19	09°56′24″	76°28′53″	7.5	165	45	16	99	26	21.4	1.8	10	4.9	0	0.06
E95	Mar-16-19	10°06′53″	76°28′28″	8.2	100	50	44	60	10	2.9	0	10	6.1	0.25	1.46
BW112	Mar-14-19	09°59′45″	76°33′50″	8.3	174	60	74	104	12	0	0	10	8.5	0.3	4.93
BW113	Mar-14-19	10°01′58″	76°32′56″	8	220	75	56	132	19	9.6	4.2	16	8.5	0	0.29
E80	Mar-18-19	09°58′41″	76°34′52″	7.6	250	85	40	150	36	26	3.45	16	11	0.05	0.76
GWE-22	Mar-18-19	09°59′51″	76°37′28″	6.9	53	20	10	32	10	0	0	4	2.4	0	0.27

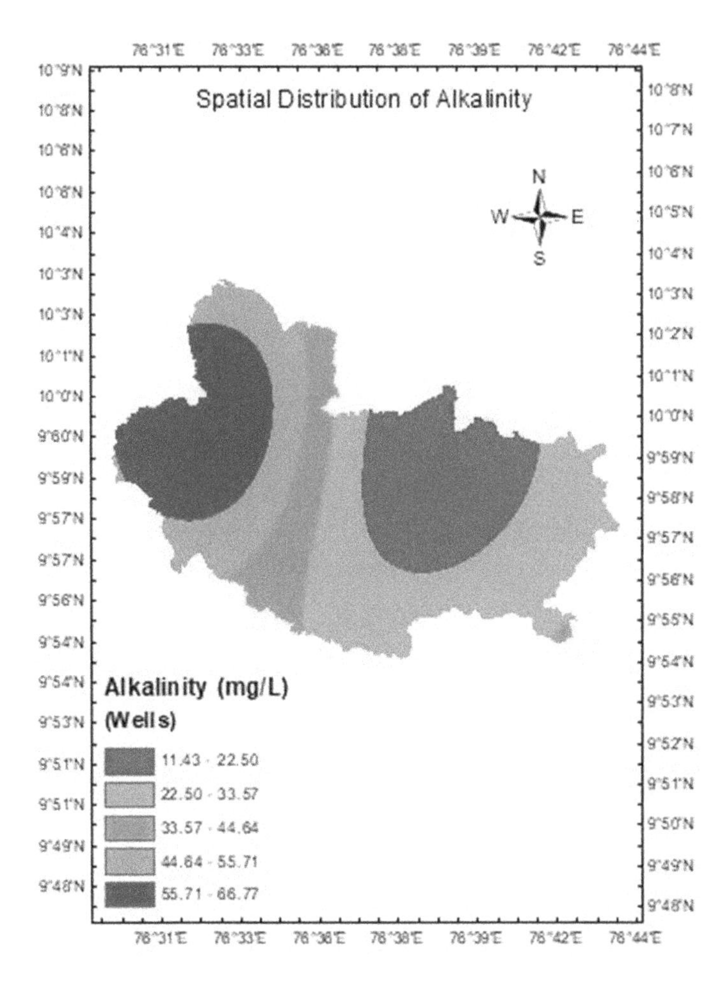

Fig. 8 Spatial distribution of alkalinity

9 Groundwater Quality Index

Water Quality Index is one of the most effective tools that can compare water quality parameters with the respective regulatory standards (WHO) and provide a single number that can express overall water quality at a certain location based on several water quality parameters. WQI helps in summarising the water quality into simple terms like good, bad, excellent etc. which are easily understandable by the public. By combining multiple parameters into a single index, a more comprehensive picture of the pollution state is provided. When mapping the index, the areas of high and low water quality can be easily specified. The water quality index is calculated in three steps. First, a weight (wi) was assigned to each of the ten

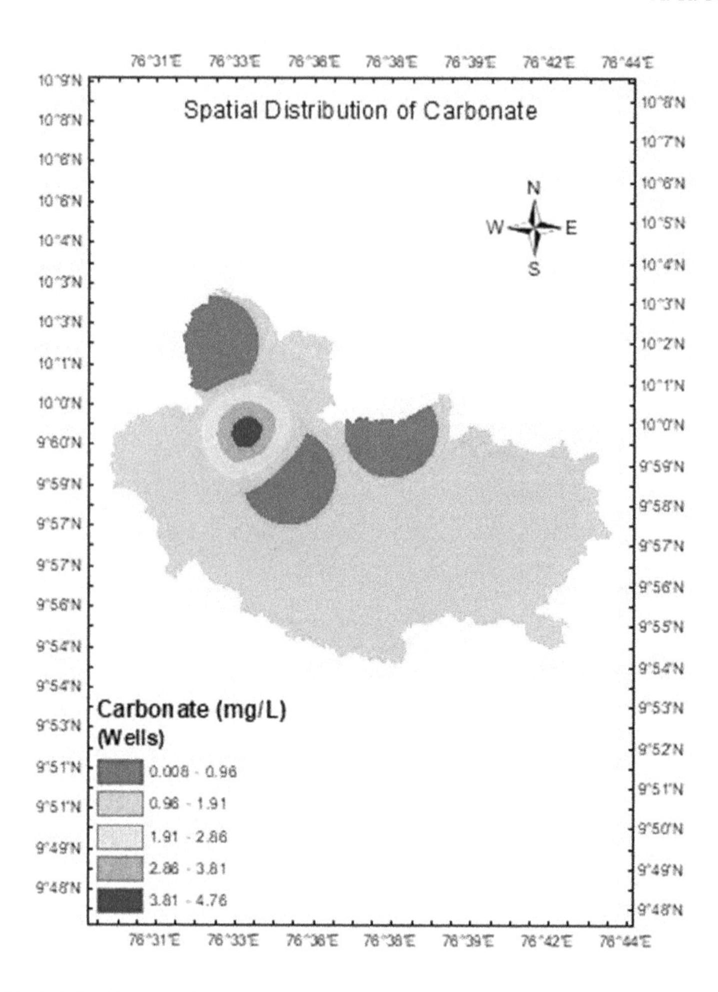

Fig. 9 Spatial distribution of carbonate

parameters according to its relative importance in the overall quality of water. The weight 5, which is the maximum weight assigned to nitrate, since it is the most affecting component on public human health. A weight of 2 was assigned to magnesium due to its low effect on human health. In the second step, the relative weight (Wi) is computed by:

$$W_i = w_i + \Sigma_{i=1}^{n} w_i$$

W_i Relative weight;

w_i Weight of each parameter;

n No. of parameters;

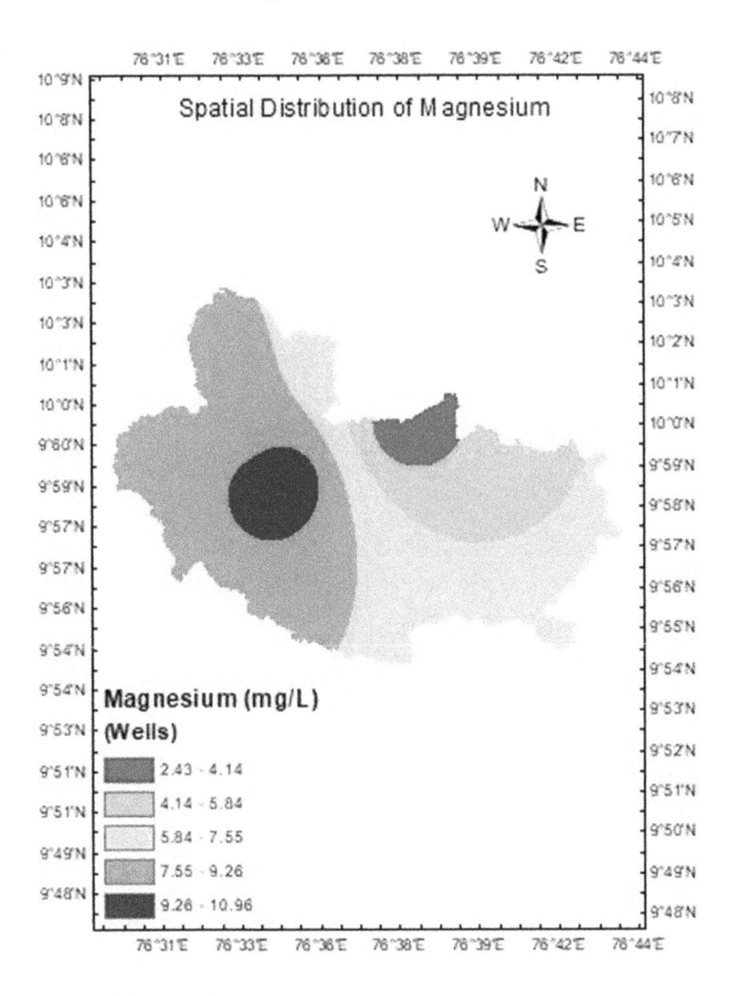

Fig. 10 Spatial distribution of magnesium

In the 3rd step, a quality rating scale (q_i), for each parameter was assigned by dividing its concentration in each water sample by its respective standard (WHO standard). And to express it in percentage, the result was multiplied by 100.

$$q_i = \frac{C_i}{S_i} \times 100$$

q_i Quality rating scale;
C_i Concentration of each pollutant in water sample in mg/L;
S_i WHO standard concentration;

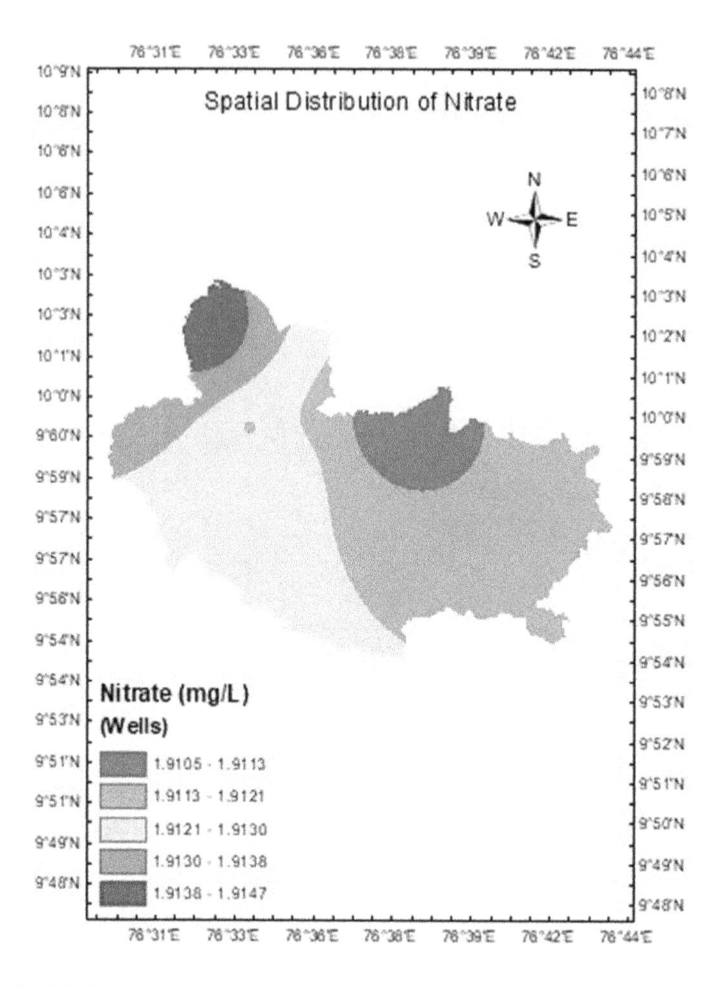

Fig. 11 Spatial distribution of nitrate

For computing the WQI, the S_i was determined for each parameter. The sub-index of *i*th quality parameter can be determined by:

$$SI_i = W_i \times q_i$$

$$WQI = \sum_{i=1}^{n} SI_i$$

The computed WQI values are classified into five types as shown in Table 6.

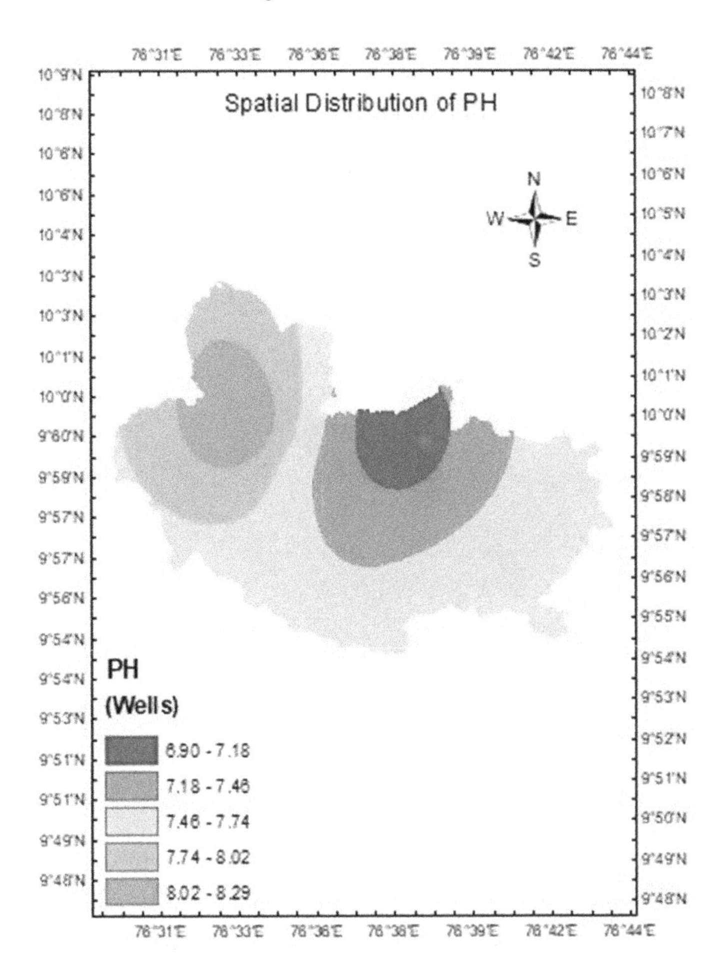

Fig. 12 Spatial distribution of pH

10 Results and Conclusion

10.1 Groundwater Potential Zonation Map

From the different maps studied above we can successfully conclude that GIS and remote sensing can provide a suitable platform for proper analysis of large volumes of data and decision making for the studies of groundwater. This approach has been successfully used and illustrated for the estimation of groundwater potentiality of the Muvattupuzha block, Kerala, India. The use of the weighted overlay method was found to be very convenient in the mapping of groundwater potential zones of the study area (Fig. 33).

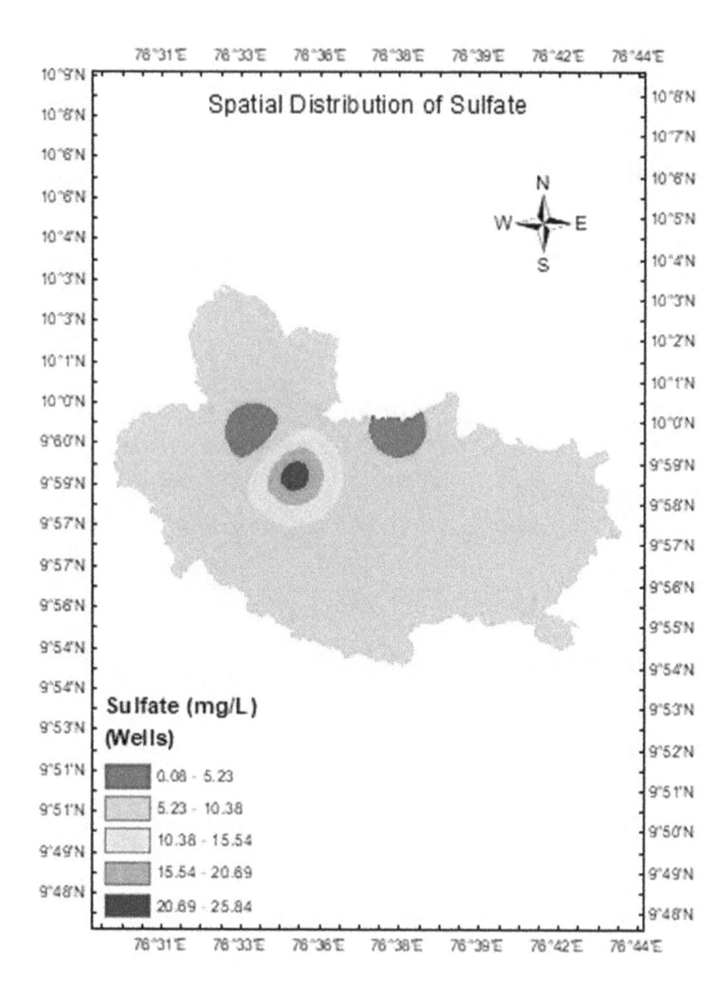

Fig. 13 Spatial distribution of sulphate

10.2 Groundwater Quality Index

Water quality classification based upon WQI value obtained from water quality data (Tables 7 and 8).

From the table it is clear that water drawn from the wells E95, BW112 and the surrounding areas, that are Perumbavoor and Vazhapilly respectively, are unsuitable for drinking purposes. The water quality Index is greater than 300, which shows that the water cannot be used for drinking purposes. The water quality index of the well GW-22 is in between 200–300, hence the water is of poor quality (Figs. 34 and 35).

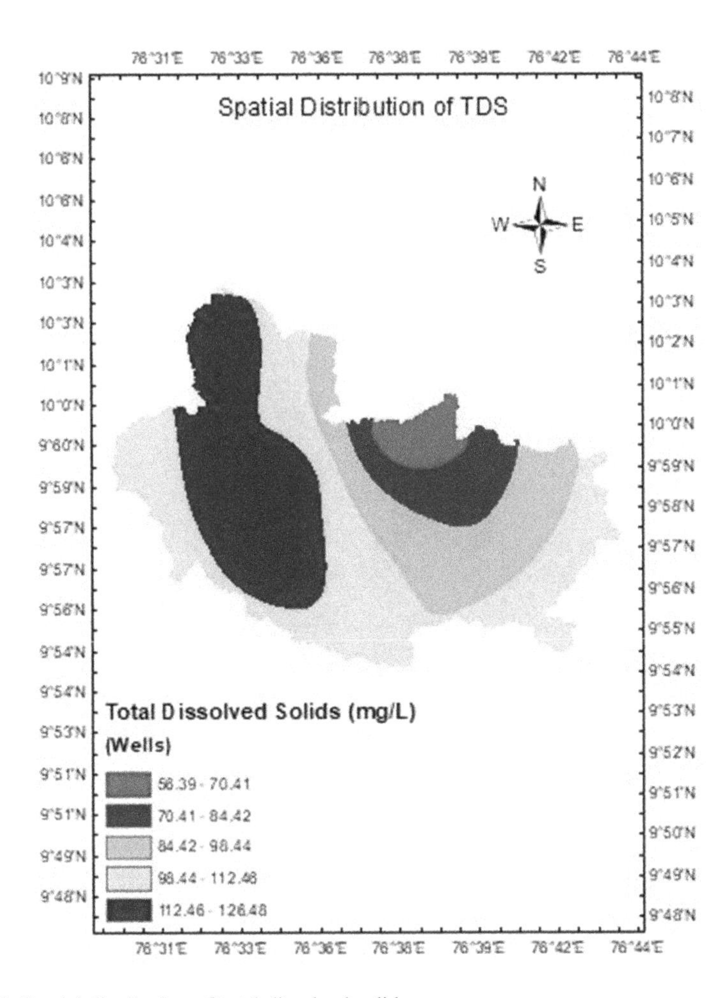

Fig. 14 Spatial distribution of total dissolved solids

11 Artificial Recharge Methods

Site characteristics of the study region:

- The average annual rainfall of of the region varies from minimum of 2779 mm to maximum of 4526 mm.
- The pH of groundwater lies between 6.5 and 8.5.
- The region consist of irrigable land.
- Drainage density varies from 0 to 6.87 km/km^2.

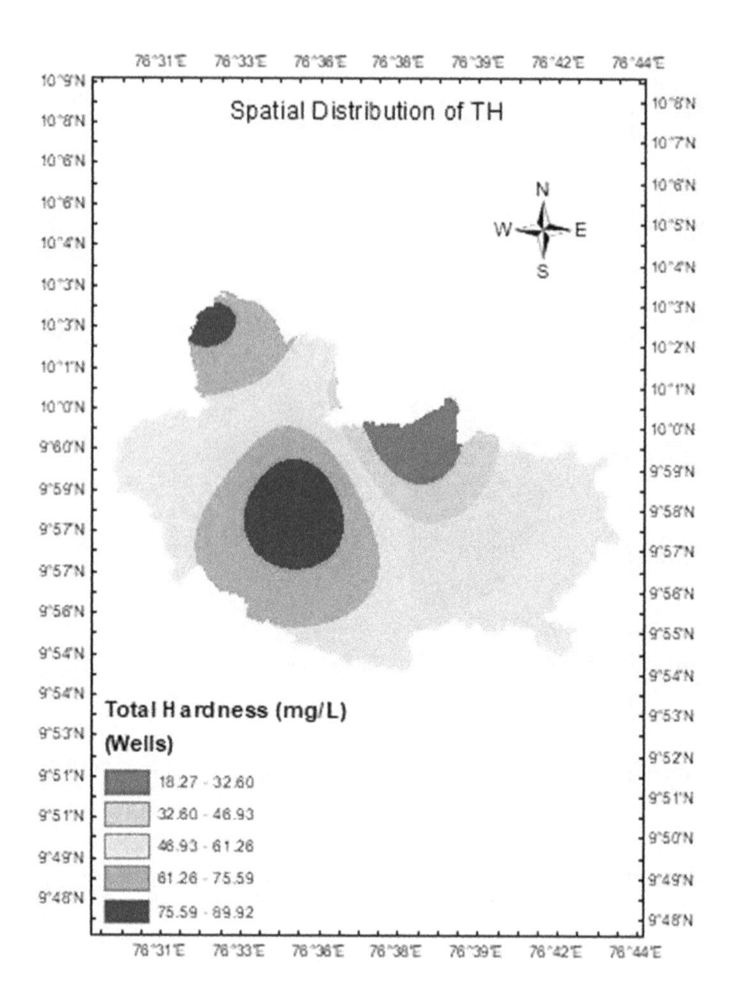

Fig. 15 Spatial distribution of total hardness

The artificial recharge methods which are proposed for the study area based on the site suitability are check dams, contour bunding, rooftop rainwater harvesting and gully plugging. The above methods were selected considering the suitability of the area for recharge in terms of climate, topography, soil, land use characteristics, hydrogeologic set-up, available technical skills, manpower, and management capabilities, and the capacity and willingness of the user community to bear the costs.

Check dams can be constructed across small streams having gentle slopes that are feasible both in hard rock as well as alluvial formation. The site selected for check dam should have sufficient thickness of permeable bed or weathered formation to facilitate recharge of stored water within a short span of time. The water

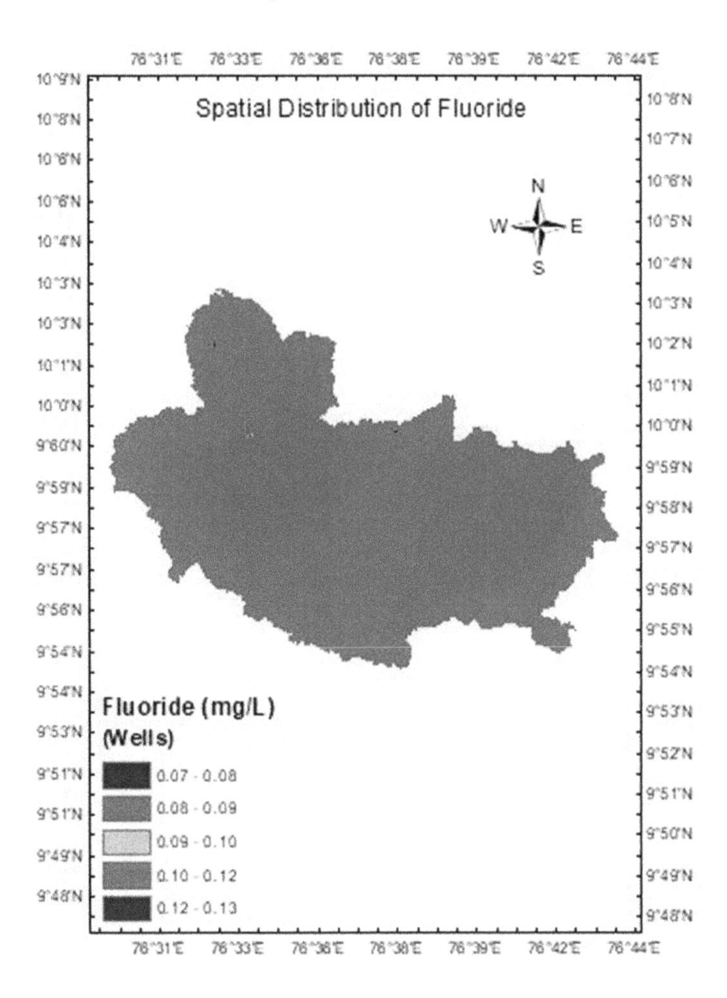

Fig. 16 Spatial distribution of fluoride

stored in these structures is mostly confined to stream courses and the height is normally less than 2 m.

Contour bunds involve a watershed management practice so as to build up soil moisture storages. This technique is generally adopted in areas receiving low rainfall.

Gully plugs are the smallest run-off conservation structures built across small gullies and streams rushing down the hill slopes carrying drainage of tiny catchments during the rainy season. Usually, the barrier is constructed by using local stones, earth and weathered rock, brushwood, and other such local materials.

In Urban areas, the roof top rainwater can be conserved and used for recharge of ground water. This approach requires connecting the outlet pipe from the rooftop to divert the water to either existing wells/tubewells/borewell or specially designed

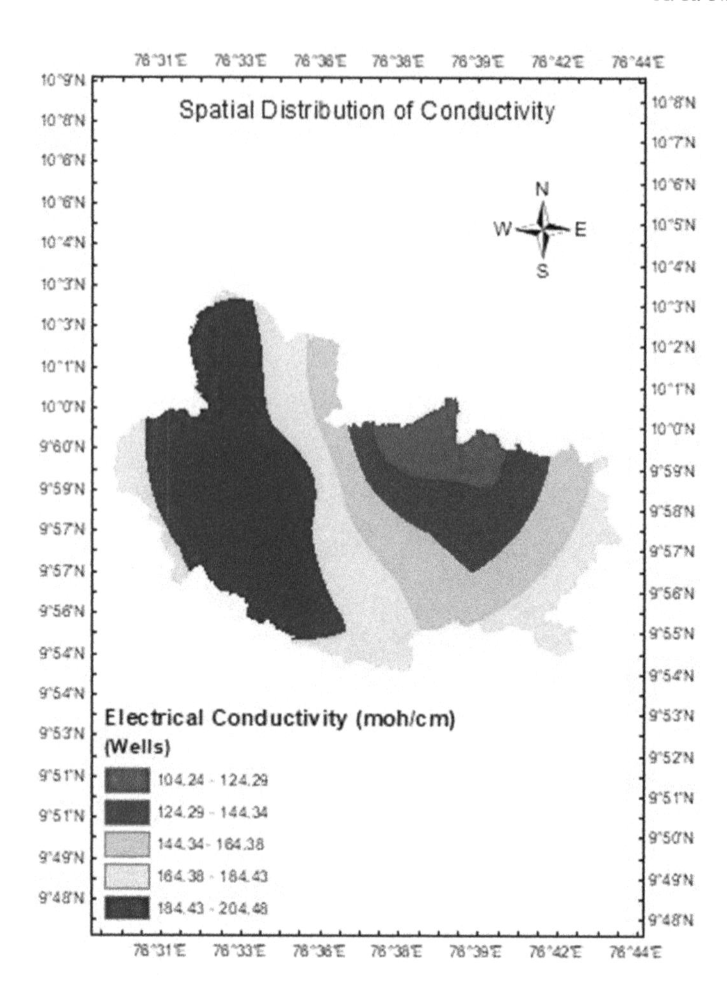

Fig. 17 Spatial distribution of conductivity

wells. The urban housing complexes or institutional buildings have a large roof area and can be utilising for harvesting roof top rainwater to recharge aquifers in urban areas.

These are the suggestions for artificial recharge methods suitable for the study area on the basis of its characteristics.

The GIS and remote sensing techniques have helped in:

- Minimising use of cost, time and labour.
- Integration of eight thematic maps such as drainage density, rainfall, LULC, slope, geology, geomorphology, lineament density, and soil which gives information to local authorities and planners about the areas suitable for groundwater exploration.

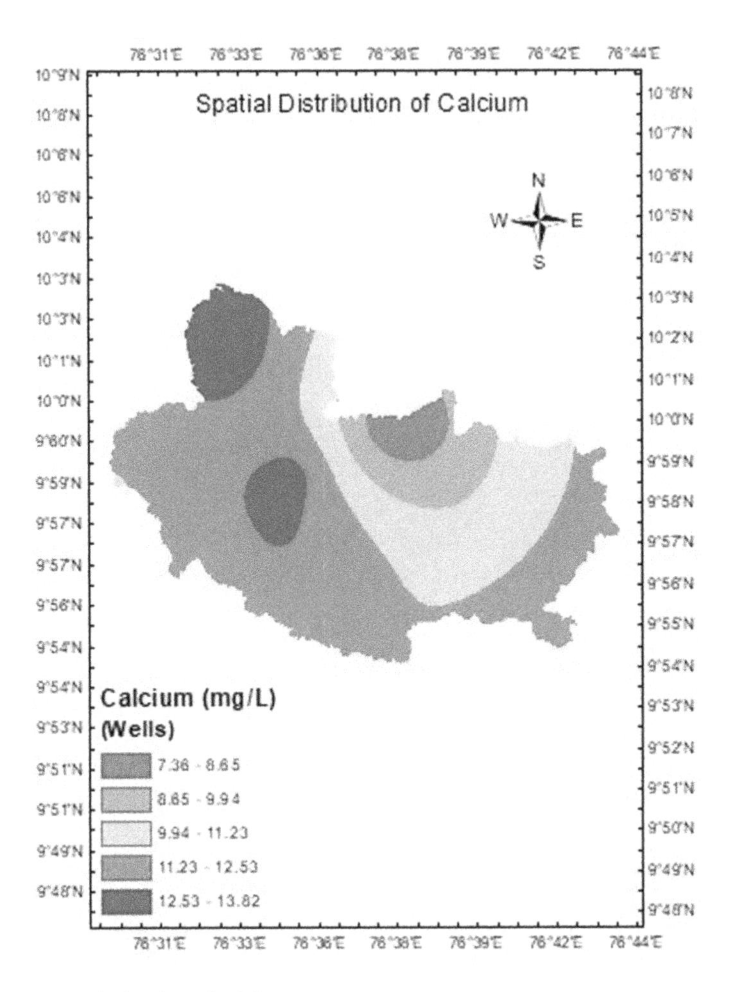

Fig. 18 Spatial distribution of calcium

- Generation of groundwater potential map of the area.
- To extract useful information from satellite images.

Conclusion:

- Three groundwater prospect zones were identified and classified as low, moderate, and high potential zones.
- Low potential zones cover 0.28 km^2 from the total area, 118.93 km^2 is of the moderate potential zone and high potential zones are of 94.45 km^2.
- Moderate groundwater potential zones were found to be more in this region.
- High potential zones were found to be along the lineaments.
- The places with a flat topography were identified to have moderate potential zones.

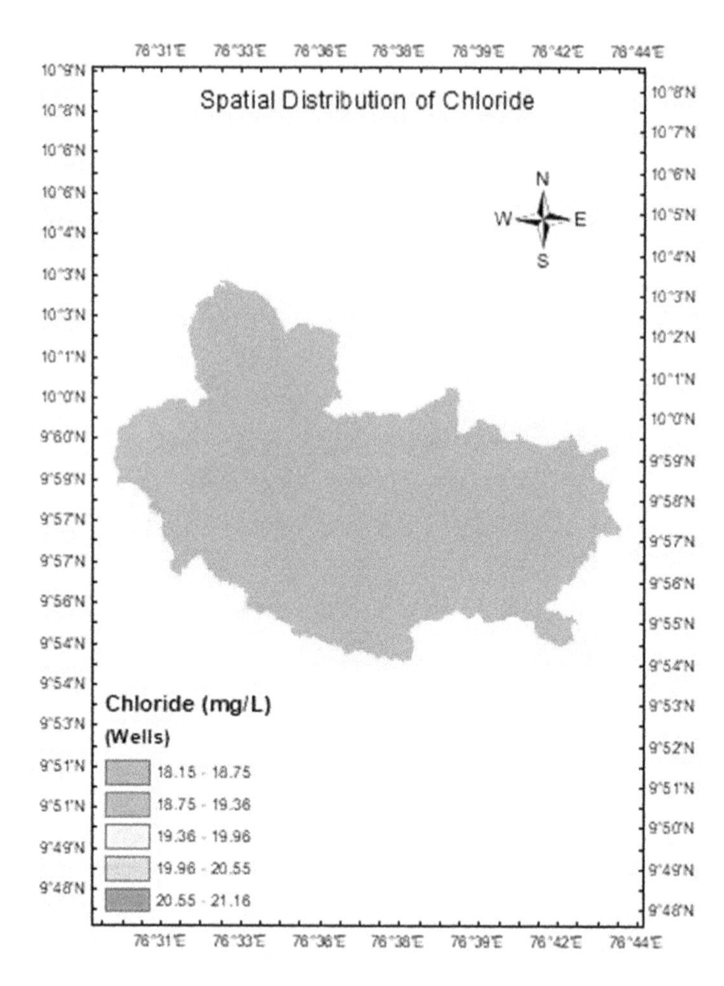

Fig. 19 Spatial distribution of chloride

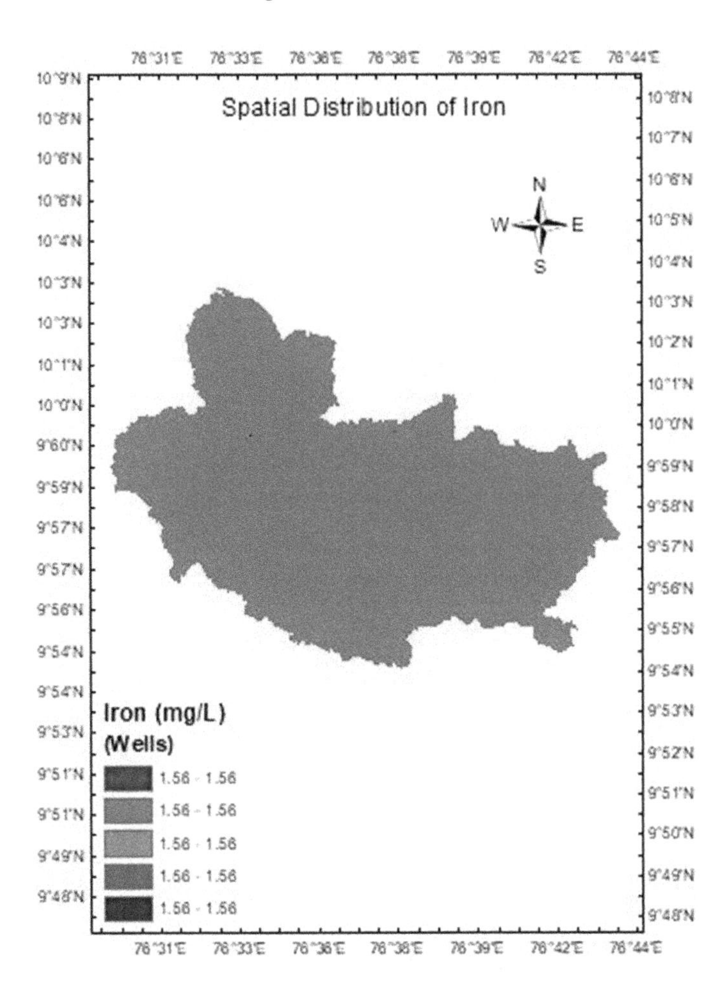

Fig. 20 Spatial distribution of iron

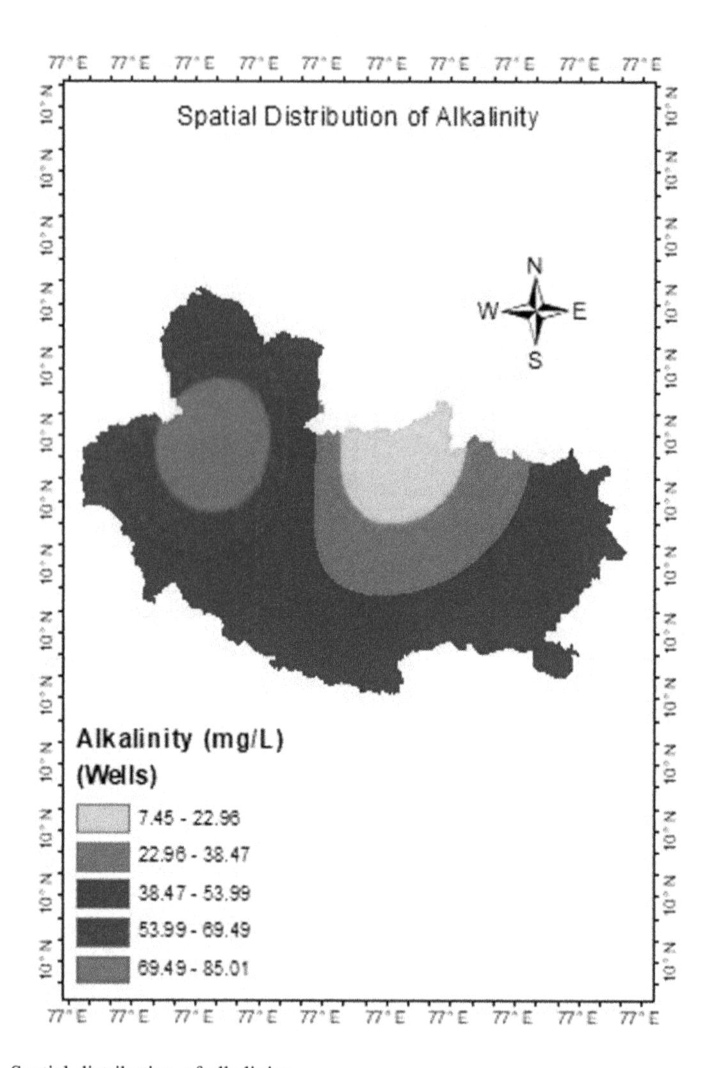

Fig. 21 Spatial distribution of alkalinity

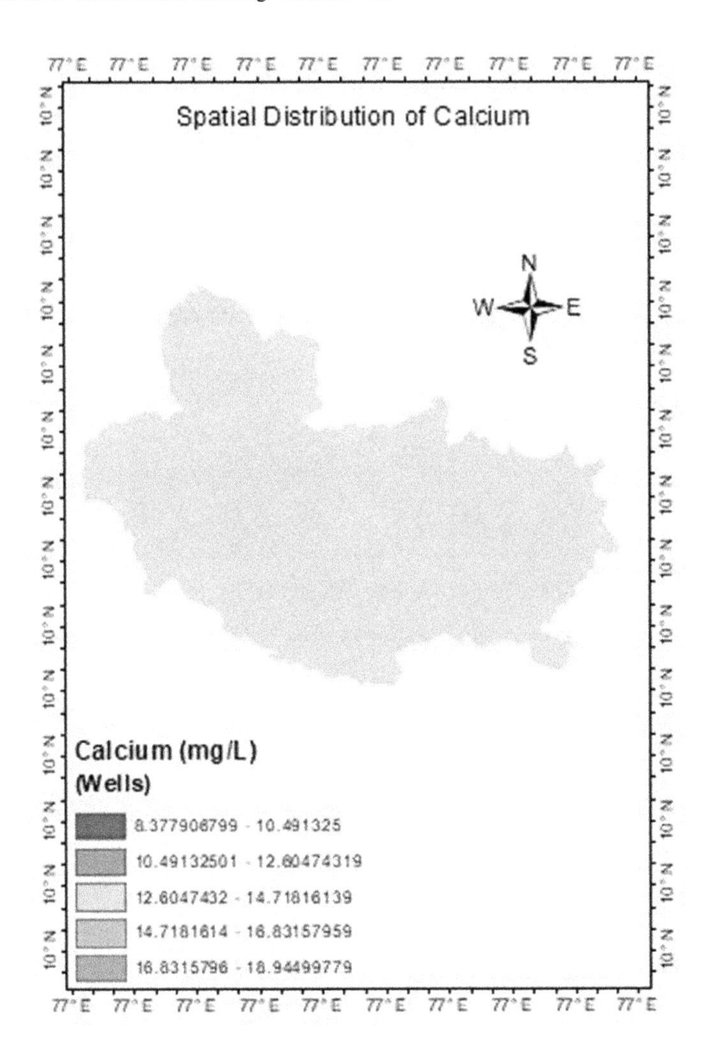

Fig. 22 Spatial distribution of calcium

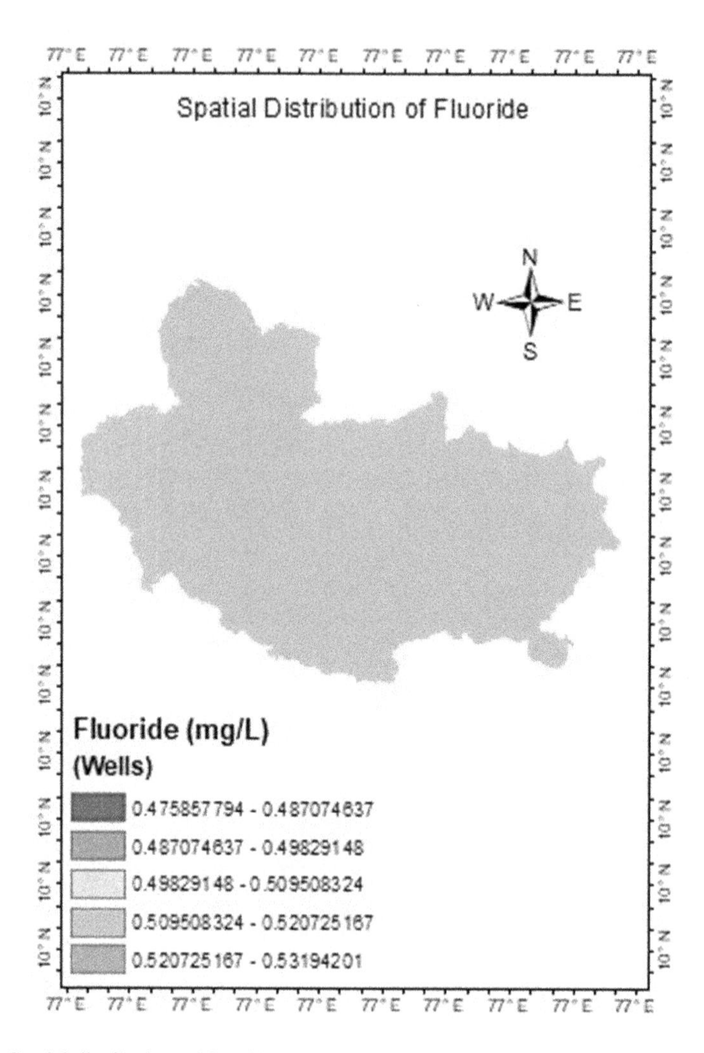

Fig. 23 Spatial distribution of fluoride

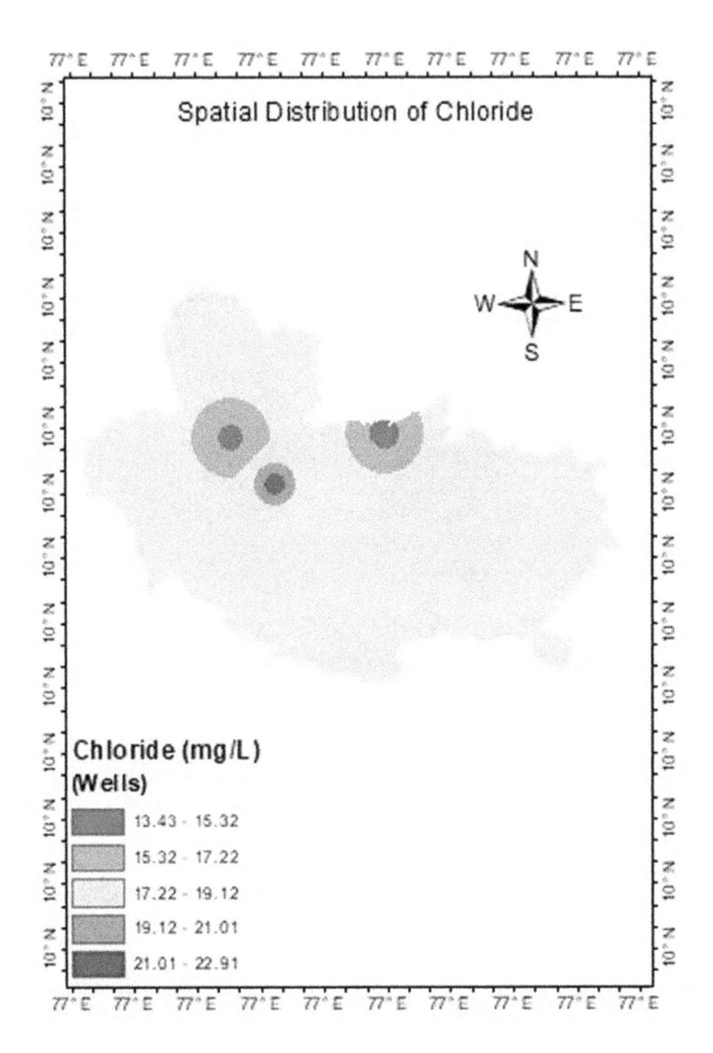

Fig. 24 Spatial distribution of chloride

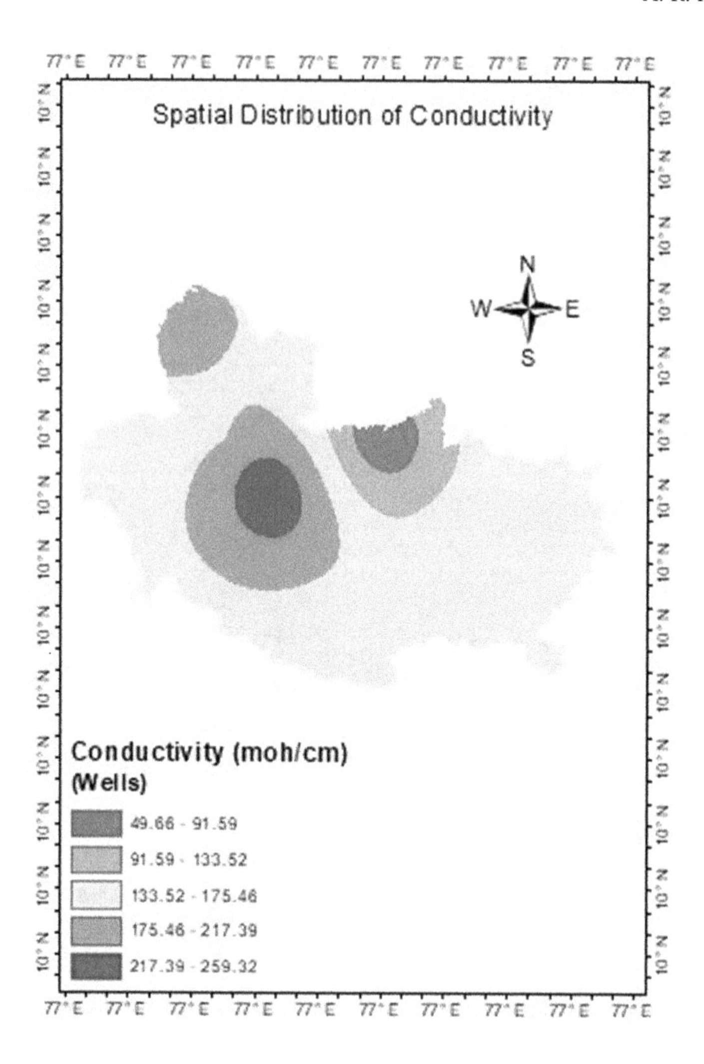

Fig. 25 Spatial distribution of conductivity

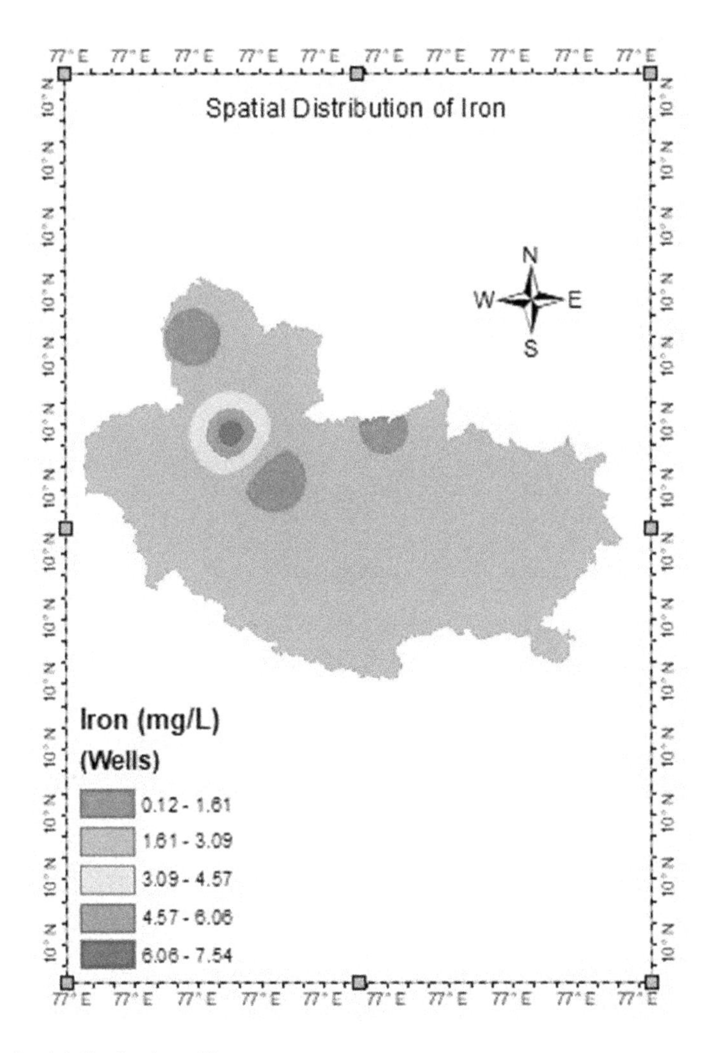

Fig. 26 Spatial distribution of iron

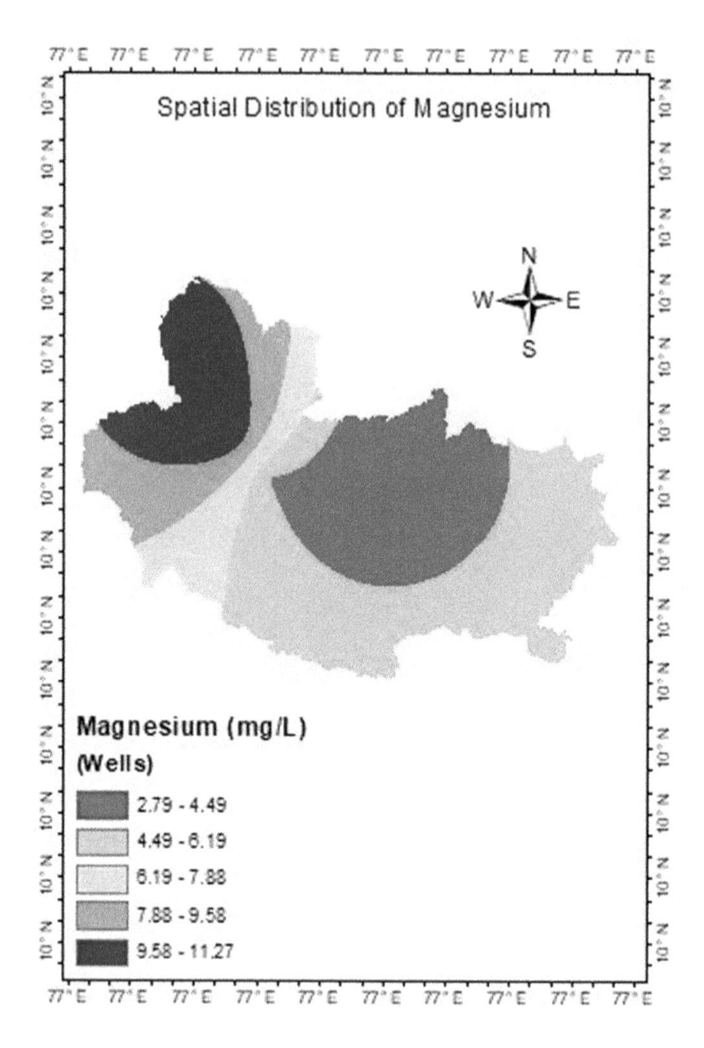

Fig. 27 Spatial distribution of magnesium

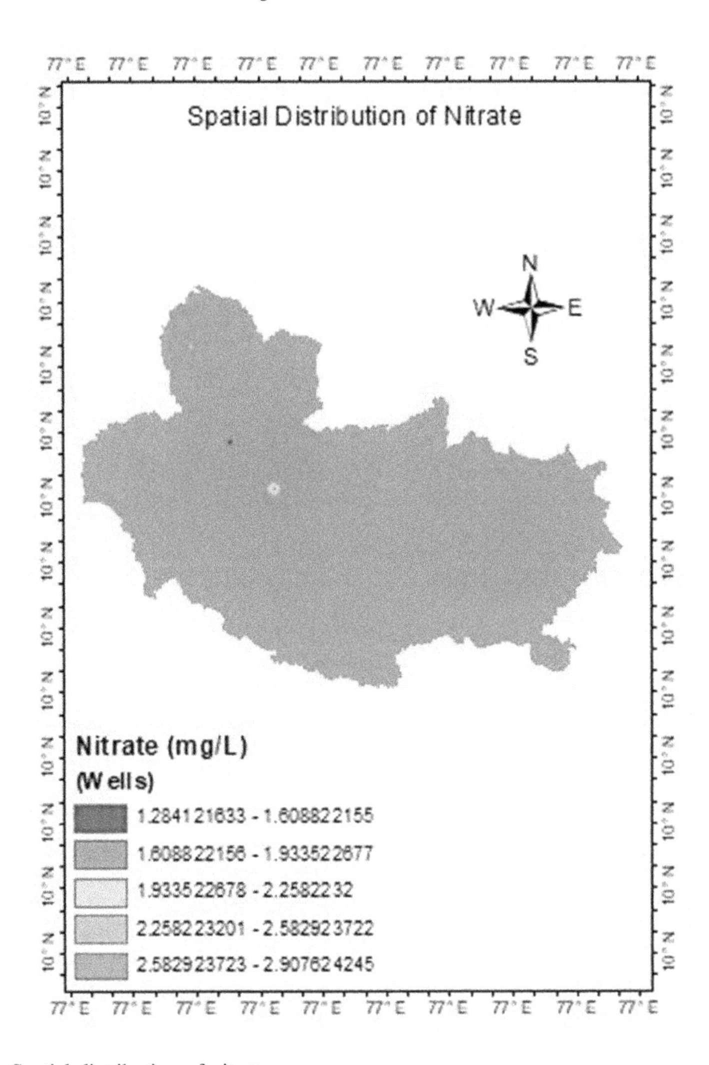

Fig. 28 Spatial distribution of nitrate

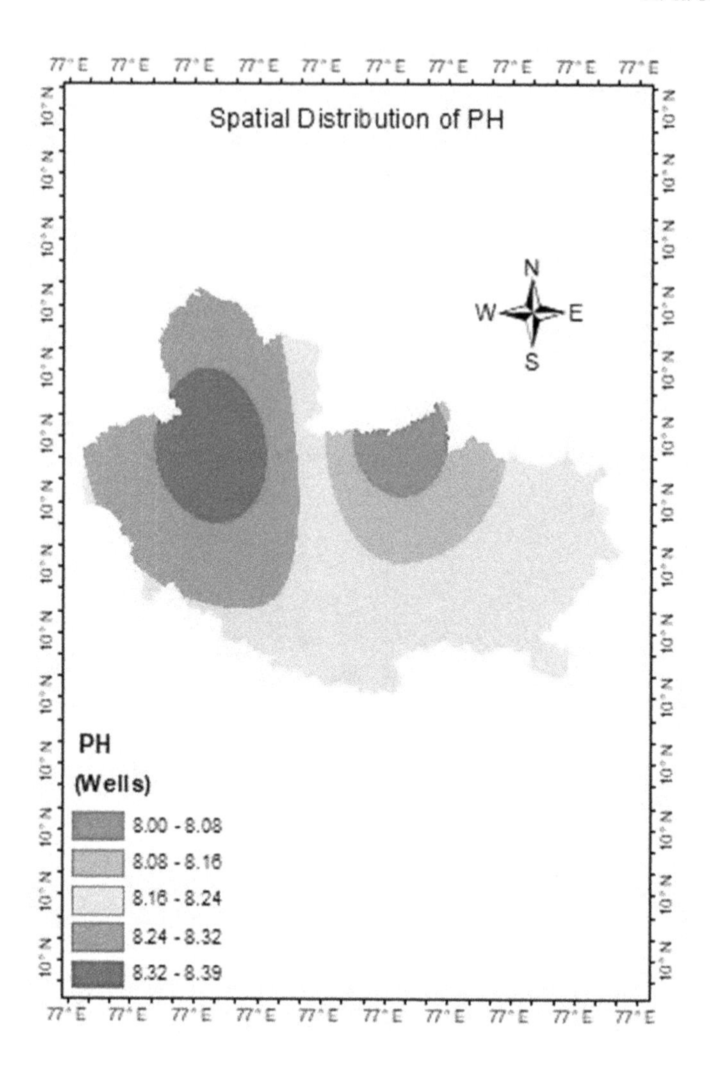

Fig. 29 Spatial distribution of pH

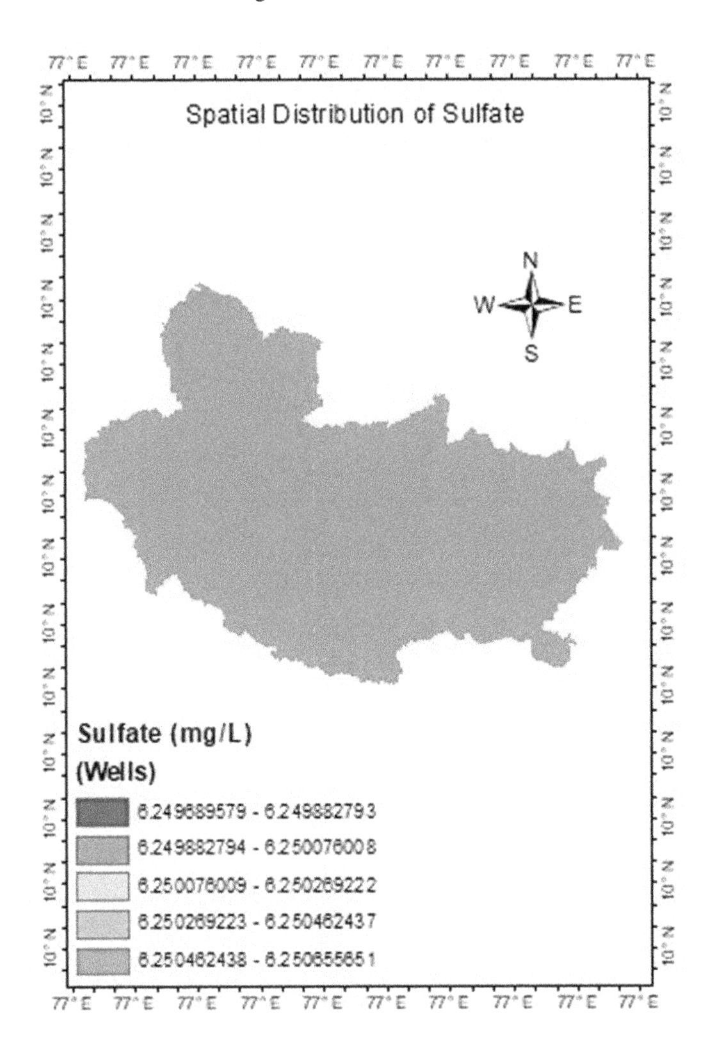

Fig. 30 Spatial distribution of sulphate

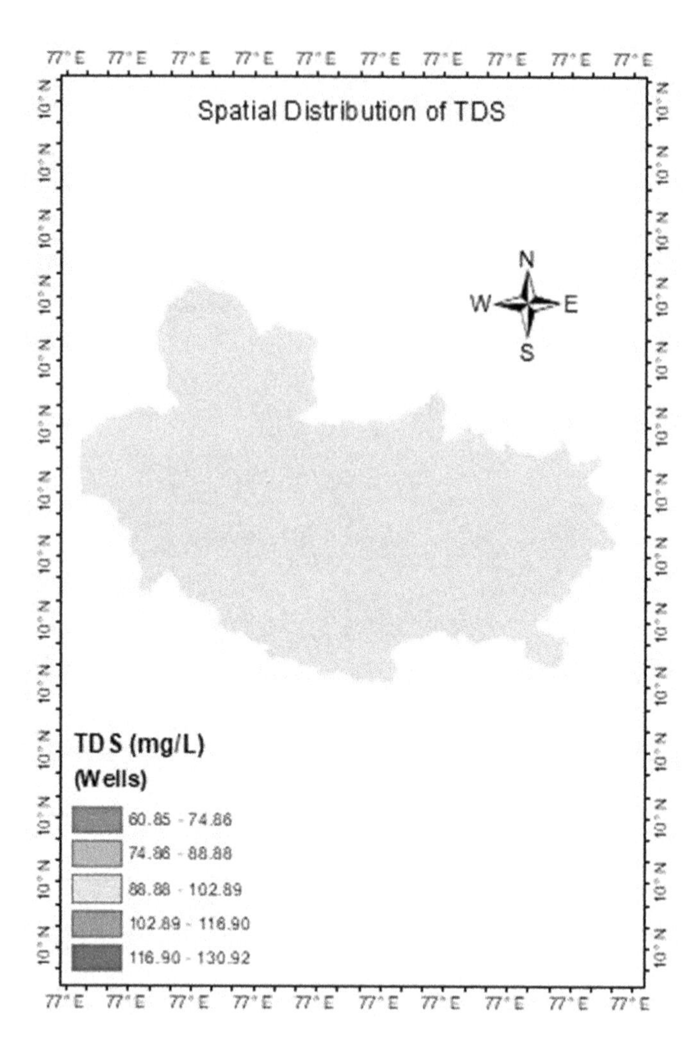

Fig. 31 Spatial distribution of total dissolved solids

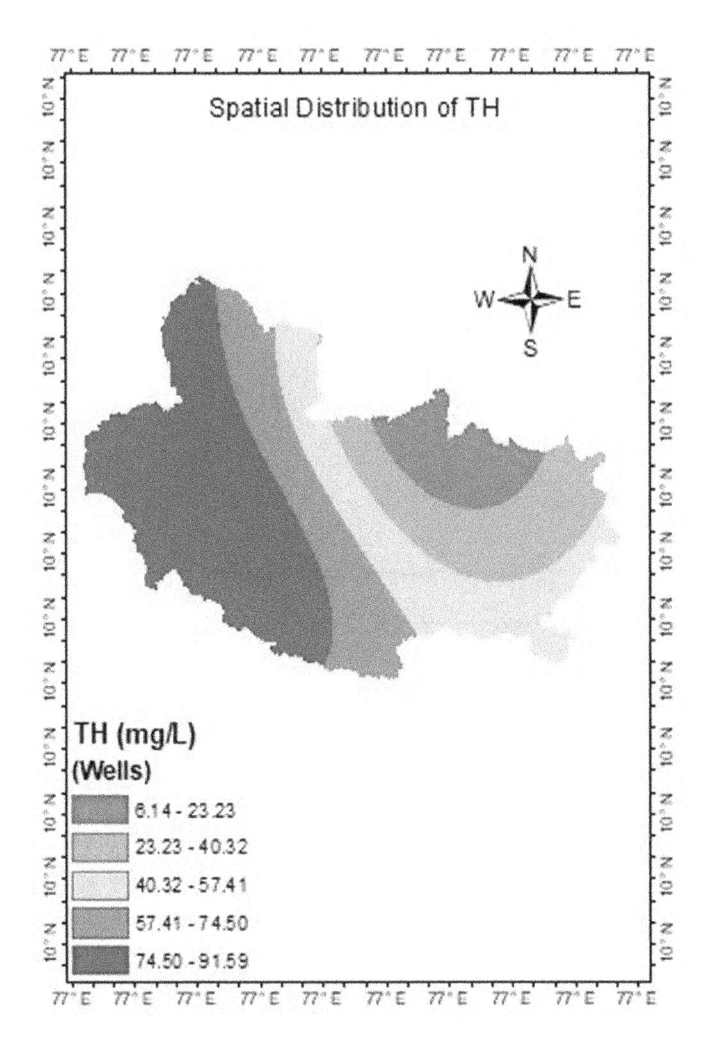

Fig. 32 Spatial distribution of total hardness

Table 6 Classification ranks

WQI value	Water quality
<50	Excellent
50–100	Good water
100–200	Poor water
200–300	Very poor water
>300	Water unsuitable for drinking

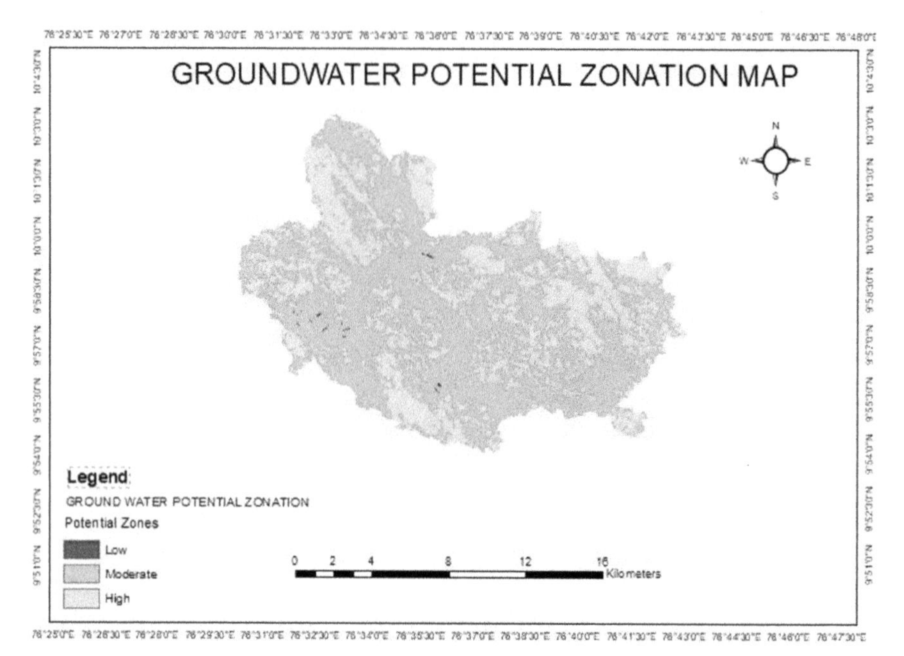

Fig. 33 Groundwater potential zonation map

Table 7 Water quality index of pre-monsoon

Well no.	WQI
E81	93.64
E82	15.81
E95	364.95
BW112	1215.09
BW113	73.1
E80	188.55
GW-22	65.97

Table 8 Water quality index of post-monsoon

Well no.	WQI
E81	54.24
E82	25.71
E95	44.52
BW112	1879.96
BW113	143.97
E80	29.29
GW-22	203.39

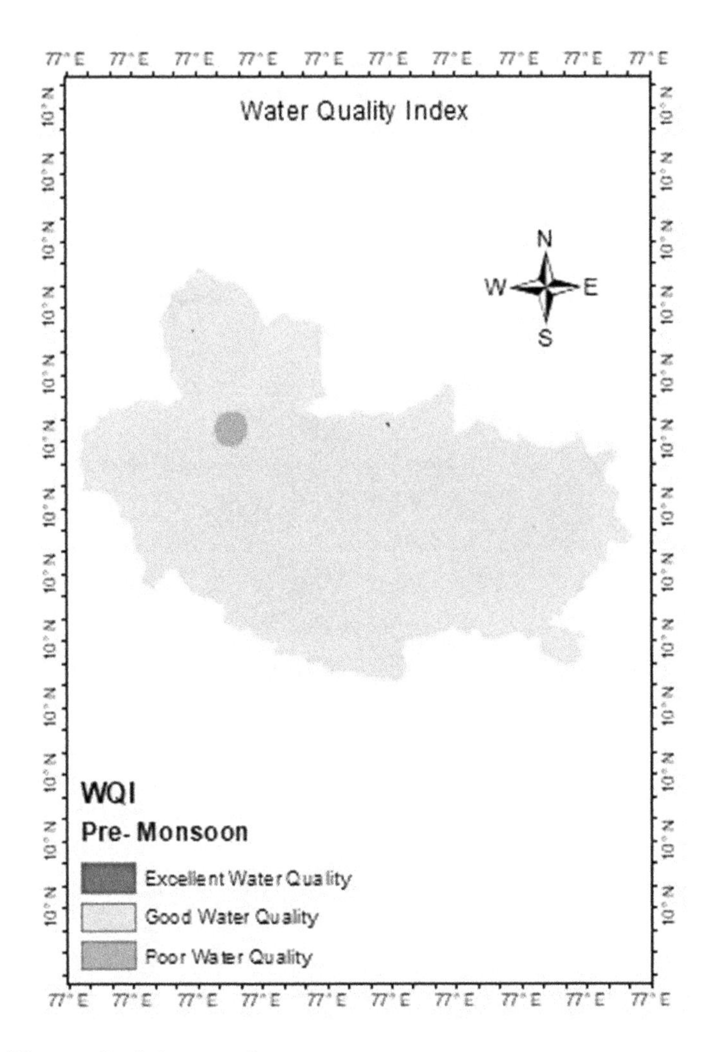

Fig. 34 Water quality index map of pre-monsoon

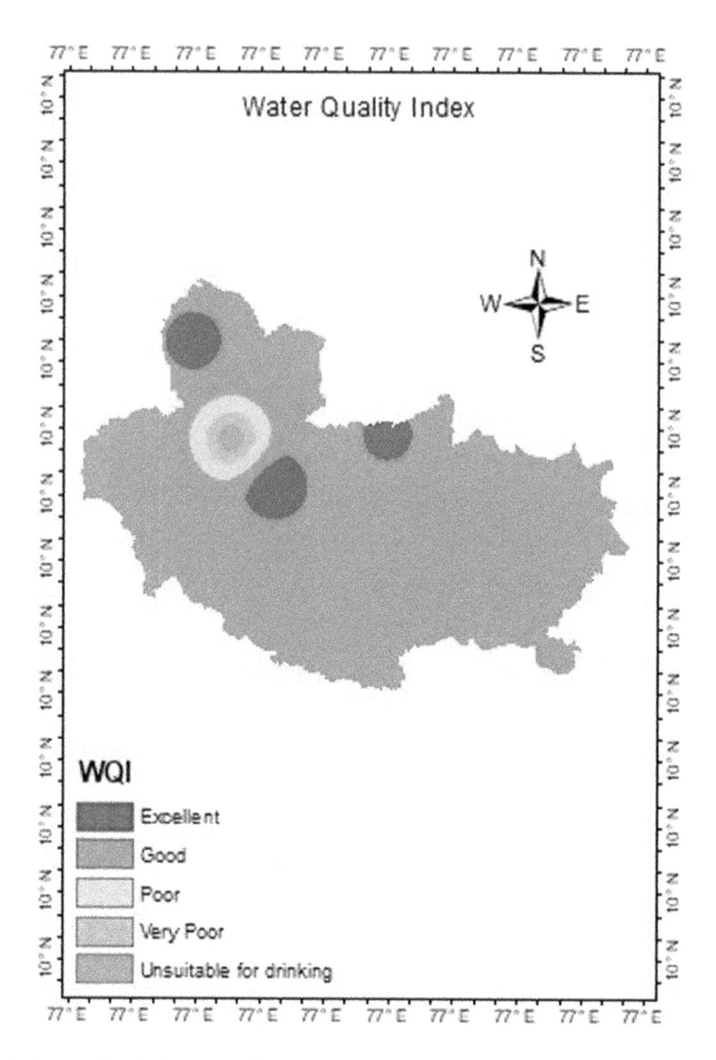

Fig. 35 Water quality index map of post-monsoon

References

1. Andualem TG, Demeke GG (2019) Groundwater potential assessment using GIS and remote sensing: a case study of Guna tana landscape, upper blue Nile Basin, Ethiopia. J Hydrol Reg Stud 24:100610
2. Njeban HS et al (2018) Comparison and evaluation of GIS-based spatial interpolation methods for estimation groundwater level in AL-Salman District|Southwest Iraq. J Geogr Inf Syst 10.04:362
3. Delbari M, Amiri M, Motlagh MB (2016) Assessing groundwater quality for irrigation using indicator kriging method. Appl Water Sci 6.4: 371–381
4. Dashtpagerdi MM, Vagharfard H, Honar-bakhsh A (2013) Application of cross-validation technique for zoning of groundwater levels in Shahrekord plain
5. Diaw et al (2018) Recharge potentials mapping using remote sensing and GIS techniques: case of shallow aquifers in the Senegal River Delta. J Environ Toxicol Stud 2.2
6. Nas B, Berktay A (2010) Groundwater quality mapping in urban groundwater using GIS . Environ Monit Assess 160(1):215–227
7. Senthilkumar M, Gnanasundar D, Arumugam R (2019) Identifying groundwater recharge zones using remote sensing & GIS techniques in the Amaravathi aquifer system, Tamil Nadu, South India . Sustain Environ 29(1):1–9

Progressive Collapse Analysis of Masonry Infill RCC Structures

S. Vishal Karthik, V. Vasugi, and G. Senthil Kumaran

Abstract The Masonry infills in RCC structures have become quite common in construction however their effect of initial stiffness is not considered generally. The adversity or the potential of progressive collapse can be studied only by considering the effect of infills in the structure. During the collapse of a structure, the properties of infills also affect the potential or gravity of collapse. In this study a G + 8 storey structure with cellular light weight concrete masonry block were modelled as diagonal strut by equivalent diagonal strut approach in order to consider the effect of masonry infill in the structure. By GSA guidelines columns susceptible to more damage were removed. The critical columns were assessed by demand capacity ratio by P-M curve. By P-M interaction curve the structure with corner column removal and short side column scenario exhibited more critical columns.The critical column in the structure were removed and assessed by displacement coefficient method of Push over analysis in FEM software ETABS. The total energy capacity of short side column removal structure was found to be 21% greater than corner column removal structure and the effect of damage of corner column removal structure were much higher than the short end column removal struct*ure*.

Keywords Demand capacity ratio · P-M curve · GSA guidelines · Total energy capacity · Cellular concrete block

1 Introduction

Each structure might be subjected to small local damage as a result of some unforeseen actions during its lifetime. The initial local damage might be very minimal, can be constrained and brought to a standstill in some cases or it can also

---

S. Vishal Karthik (✉) · V. Vasugi
School of Civil Engineering, VIT Chennai, Tamil Nadu, Chennai 600127, India

G. Senthil Kumaran
Department of Civil Engineering and Construction, The Copperbelt University, Kitwe, Zambia

G. C. Marano et al. (eds.), *Proceedings of SECON'21*, Lecture Notes in Civil Engineering 171, https://doi.org/10.1007/978-3-030-80312-4_42

result in increase to a higher level of damage. The latter scenario is commonly known as the progressive collapse. Progressive collapse refers to the spread of initial failure into a chain of failure of elements within the structure. The potential of failure is generally higher when a structure loses its one or many of its major vertical load carrying element. This failure may arise due to an unexpected loading like blast explosion attack by terrorist, vehicle accident, fire accident or even design error in structures. Even though the masonry infill do not directly take part in load resistance, their physical interaction with the horizontal and vertical load bearing elements results enhancing the rigidity and overall stiffness of the structure. Studies have shown that the presence of infill in the structures has a substantial influence on the seismic response of the reinforced concrete frames and also on the energy dissipation mechanism of the structures. Neglecting the effect of infill leads to inaccurate assessment of the structural rigidity, stiffness of the structure which in turn reflects in inaccurate assessment in understanding the structure on event of a local damage causing a collapse. This research seeks to achieve the following three major objectives: (i) To model the masonry infill into a double diagonal strut by equivalent strut approach, (ii) To determine the demand capacity ratio by P-M interaction curve, and (iii) To determine total energy capacity, Robustness coefficient of structure for different critical column removal scenarios by Pushover analysis.

2 Literature Review

Barrossa and Cavoco [1] studied the impact of non-load bearing masonry infills on the robustness of a RC framed structure during the occurrence of a landslide. It was concluded that the collapse of the damaged structure relies majorly on the strength of the masonry infill wall and it is fundamental part in restricting the damage progression of the RC structure.

Aniendhita and Data Iranata [2] presented a comparative study of different diagonal methods of masonry infill. The basic theory of this study was that the masonry panel itself has a predominant effect on the strength and stiffness of any RC structure and its presence influences the energy dissipation capacity of the structure. Konstantinos and Asimina [3] examined the abnormalities in the structure caused by arbitrary placement of masonry infills on seismic effect in RCC structures. Ning Ning and John Ma [4] presented an experimental study on the impact of masonry infill in the seismic response of reinforced concrete structures at low frequency cyclic loading. Sankhla and Bhat [5] presented a comparative study on the influence of masonry infill walls on reinforced concrete structures Apart from studying the effect of infill it also emphasized on the influence of soft storey on the framed structures due to lateral loading. Liberatore and Noto [6] studied the in-plane response of masonry infills by deterministic and probabilistic analysis concluded from the study that the single strut approach model cannot be used to simulate the effect of masonry infill and the frame.

3 Modeling

Cellular lightweight concrete were used as masonry infill material. Owing to their light weight, low density durability, strength, structural stability, fire resistance and insulation cellular light weight blocks are being used as masonry infills in official buildings. Cellular light weight blocks contains silica, Portland cement, pozzolanic or mortar pastes consisting of all the above ingredients with a homogeneous voids in the cell structure that are usually attained stable preformed foam. The voids in the blocks are completely harmless that also enables in reducing the self weight of the concrete blocks. Both load bearing and non load bearing blocks can be obtained in accordance with IS-2185 Part 4 Code for concrete masonry units. They exhibit better sound insulation, thermal property than the autoclaved aerated blocks Cellular light weight blocks of size 600 mm × 200 mm × 200 mm were used in this study. The physical properties of cellular light weight blocks are shown in Table 1.

The structure used in this study is a G + 8 storey reinforced concrete commercial building with special moment resisting frame. The plan area of the building is 28 m × 12 m with a typical storey height of 3.6 m and a total height of 32.4 m. The grade of concrete and steel used are M30 and Fe415 respectively. The structural elements beams, columns and slabs are designed in accordance with IS456:2000 [7]. The thickness of slab is 150 mm, the size of all the beams and columns in the structure are kept same. In order to distribute larger moments to the columns, rectangular beams are used in the structure and also to avoid local eccentricity the beams rest centrally on all the columns. The cross section of beams and columns are 250 mm × 350 mm and 450 mm × 300 mm respectively. The floor diaphragms are assumed to be rigid. The plan area of the building is shown in Fig. 1.

Masonry infill develop a strong interaction with the frames and to simulate the effect of cellular light concrete blocks, the masonry infill are transformed into braced fames in the structure. The masonry struts are modeled as double diagonal struts. The geometric properties of the masonry double diagonals struts are parameters of the length of contact between the beam and wall 'α_L' and between the column and the wall 'α_h'. To model the equivalent strut the width of the strut is to be calculated. Width of the equivalent strut is mainly related to the length of contact between the beam and the wall 'λ_l' and the length of the contact between the column and the wall 'λ_h'.

Table 1 Physical properties of cellular concrete blocks

Properties	Values
Density	1000 kg/m^3
Size	600 mm × 200 mm × 200 mm
Compressive strength	10 N/mm^2
Water absorption percentage	10

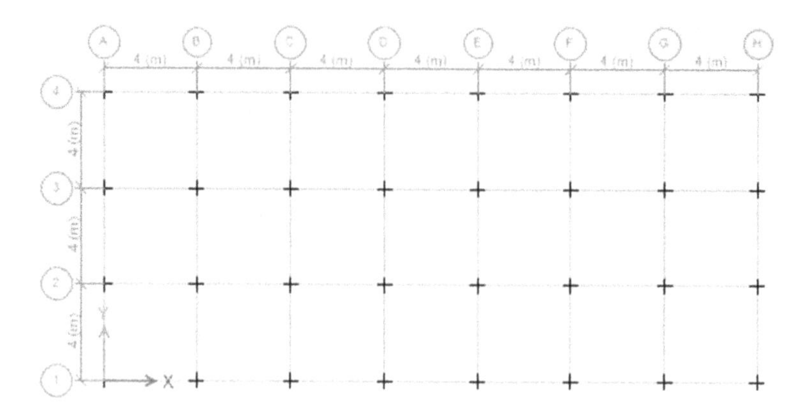

Fig. 1 Plan area of the building

- $w = 1/2\ \sqrt{\alpha_h^2 + \alpha_L^2}$
- $\alpha_h = \pi/2\lambda_h$
- $\alpha_L = \pi/\lambda_l$
- $\alpha_h = 0.598$ m
- $\alpha_L = 1.323$ m
- Width of the strut = 0.725 m

The length and cross section area of the diagonal strut are dependent on the height, thickness of the infill wall and width of the strut.

- $L_d = \sqrt{h^2 + L^2}$
- $L_d = 4.96$ m
- $A_d = t \times w$
- $A_d = 0.145\ m^2$

The structure is modelled and analyzed in finite element software Etabs 2017 (extended three dimensional analysis of building systems). Etabs is effective and accurate for modelling and non linear analysis of frames. The structure is first modelled as a bare frame and then the double equivalent struts braces are placed in between the beams and columns in order to define the infill panel. The infill frame model and elevation is shown in Figs. 2 and 3 respectively.

The self weights of the all structural components in the structure are automatically added in the ETABS software 2017 by giving the self weight multiplier factor as one. The live load in the structures is assigned referring to IS875 Part 2 [8] code of practice for design loads. The building is considered to be in zone III. The seismic loading specifications are adopted from IS 1893:2016 [9] and are given in Table 2.

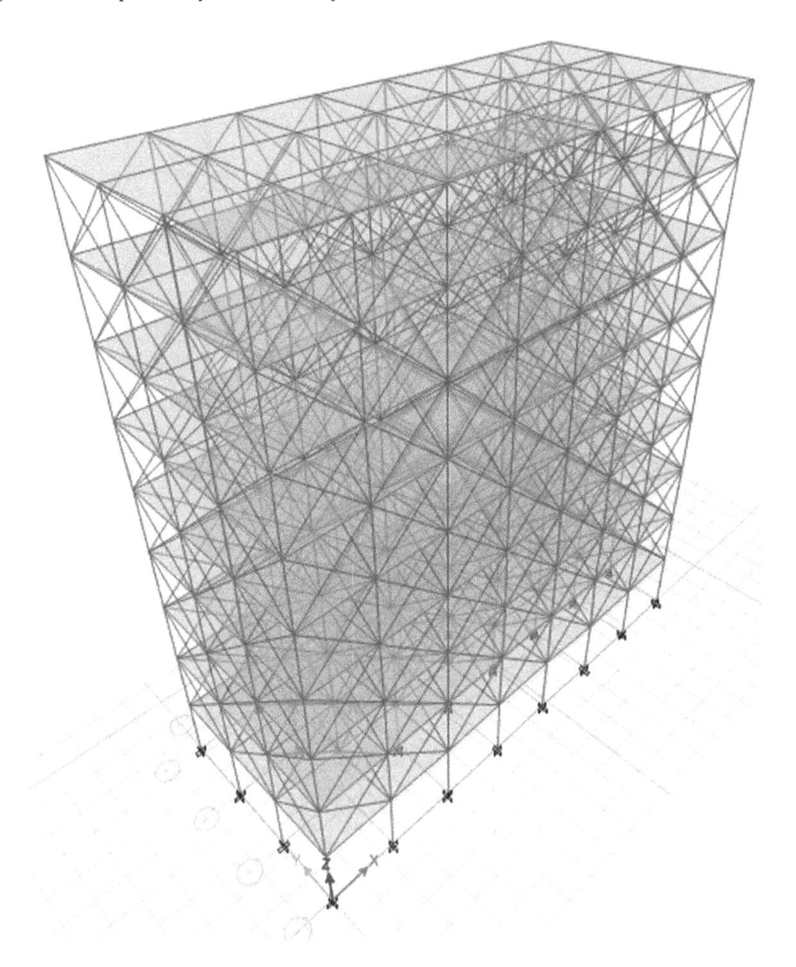

Fig. 2 Infill frame modelled in Etabs

4 Column Removal Scenario

The detailed guidelines recommended in the US General Service administration (GSA) [10] for the loading conditions and column removal scenarios are followed in this study. For progressive collapse failure analysis, GSA guidelines mandates four major column loss scenarios in industrial structure shown in Fig. 4. The GSA guidelines recommend removal of four major columns in the first storey of the structure. To determine the extent of collapse all four column loss scenarios in GSA design guidelines were considered in this study and the columns were removed from the modelled infill structure. The demand capacity ratio is assessed for the vertical elements in the in filled frame for all column removal scenarios by P-M interaction curve to study the extent of damage after a local failure.

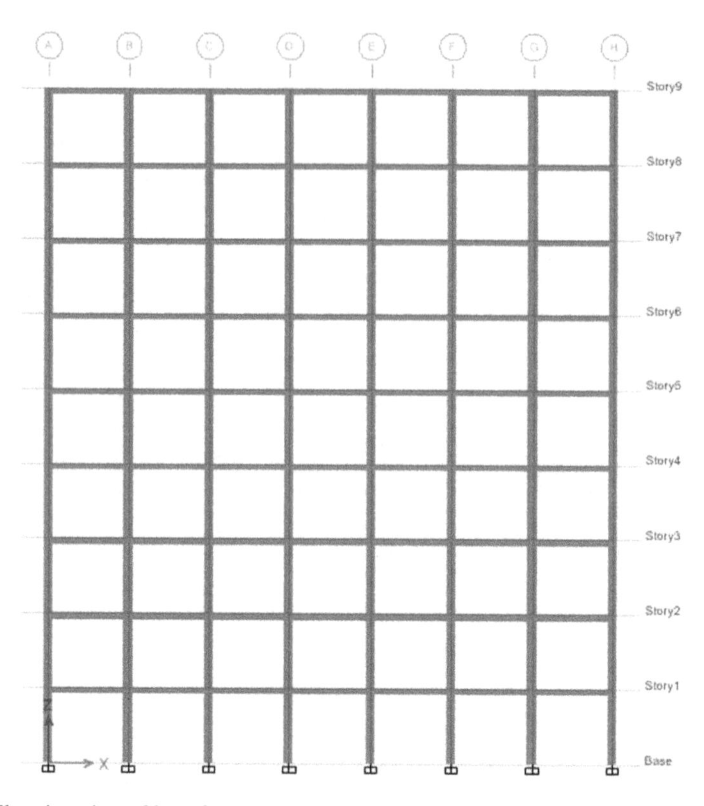

Fig. 3 Elevation view of bare frame

Table 2 Seismic parameters

Parameter	Consideration
Zone	III
Seismic zone factor	0.16
Type of soil	Type II medium soil
Seismic importance factor I	1.5
Damping coefficient	0.05
Response reduction factor	5

- CASE 1: An exterior column close to the centre of long side of the structure
- CASE 2: An exterior column close to the centre of short side of the structure
- CASE 3: A column at any corner of the structure
- CASE 4: An interior column for stilt ground storey structures

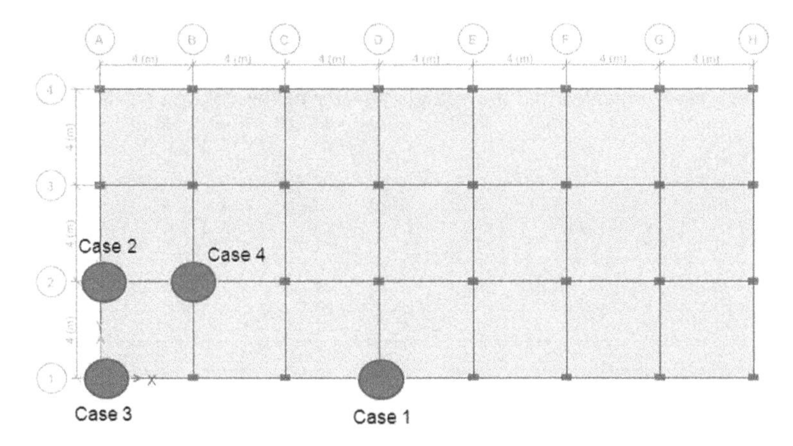

Fig. 4 GSA column loss scenario

5 Results and Discussions

5.1 Demand Capacity Ratio by P-M Interaction Curve for Infill Frame

The infill frame structure is first analyzed for all load combinations referring to IS456:2000 [7] and general service administration guidelines for progressive collapse. Demand capacity ratio is the ratio of the member force of element that is actually required in the structure to the actual strength of the element in the structure (member force/member strength). Demand capacity ratio of all the columns are first assessed before any column is removed from the infill structure by means of P-M interaction curve (moment and axial force). P-M interaction curve for column representing the relationship between the capacity of axial force and capacity of ultimate bending moment in the structure is designed referring to IS 456:2000. The axial force and moment along both axis for columns for the load combination 1.2 (DL + LL + ELy) obtained from Etabs software is shown in Table 3. From the plan, the columns are numbered conveniently from left to right in the ground storey from C1 to C32. Column C2 showed the highest axial force and moment in the structure, the typical moment-axial force curve for the column C2 along both the axis is shown is shown in Figs. 5 and 6.

Table 3 Demand axial forces and bending moment of infill structure

Story	Column number	Axial force (kN)	Moment (kN m)
Story 1	C2	2434	55.4
Story 1	C30	2421	53
Story 1	C13	2411	52
Story 1	C17	2399	50

The red curve indicates the capacity of the column element and blue dot represents the actual demand of axial force and moments in the structure. The blue dot inside the curve indicates that the actual capacity of the column C2 is greater than the demand.

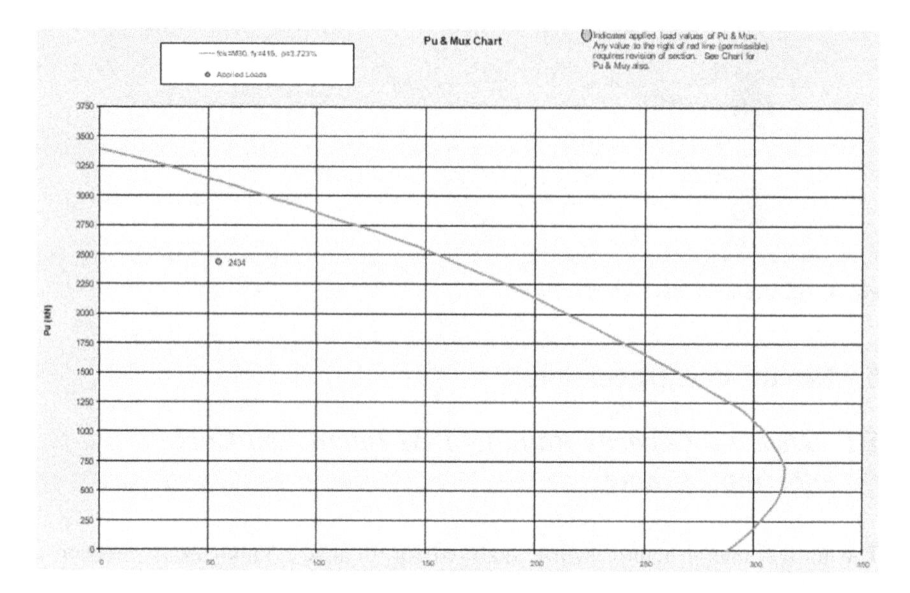

Fig. 5 Pu and Mux curve for column C2

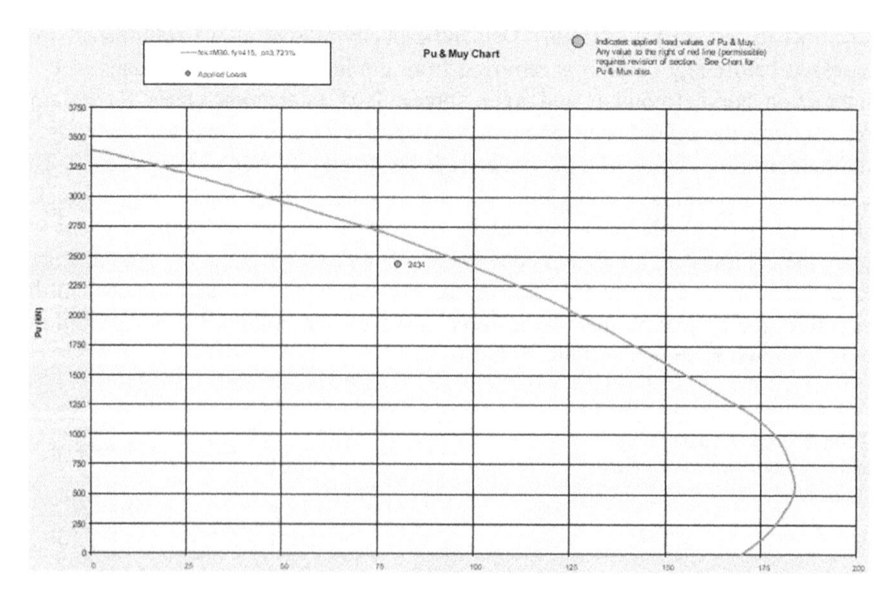

Fig. 6 Pu and Muy curve for column C2

All the four column loss scenarios as per GSA guidelines are followed and columns are removed from the infill structure and analyzed for axial forces and moments. For Each and every vertical element in the structure demand capacity ratio is studied by P-M interaction curve in order examine the effect of all the column removal scenarios. The removal of corner column (case 3) and short end column (case 2) exhibited more critical failure columns in the structure.

5.2 Demand Capacity Ratio by P-M Interaction Curve for Corner Column Removal Scenario

The axial forces and moments of the critical column after removal of corner columns is shown in Table 4. The P-M curve of the critical column C4 after removal of corner is shown in Figs. 7 and 8.

After removal of corner column in the structure the column C4 and C8 failed in both the axis (Mux and Muy) and the column C12 failed along the Muy axis.

Table 4 Demand axial forces and bending moment of critical columns after corner column removal

Column no.	Axial force (kN)	Moment (kN m)	Load combinations
C4	3016	90	1.2(DL + LL-Ely)
C8	3038	85	1.2(DL + LL-Ely)
C12	2891	84	1.2(DL + LL-Ely)

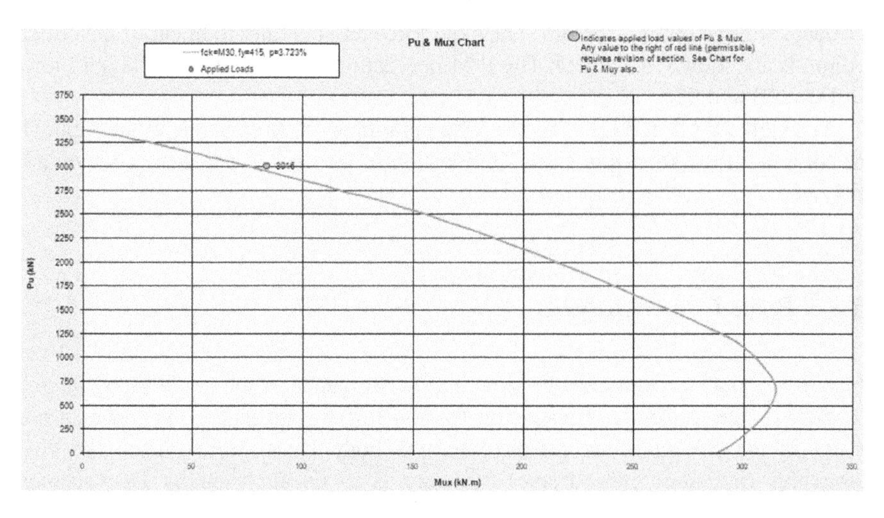

Fig. 7 Pu and Mux curve for critical column C4

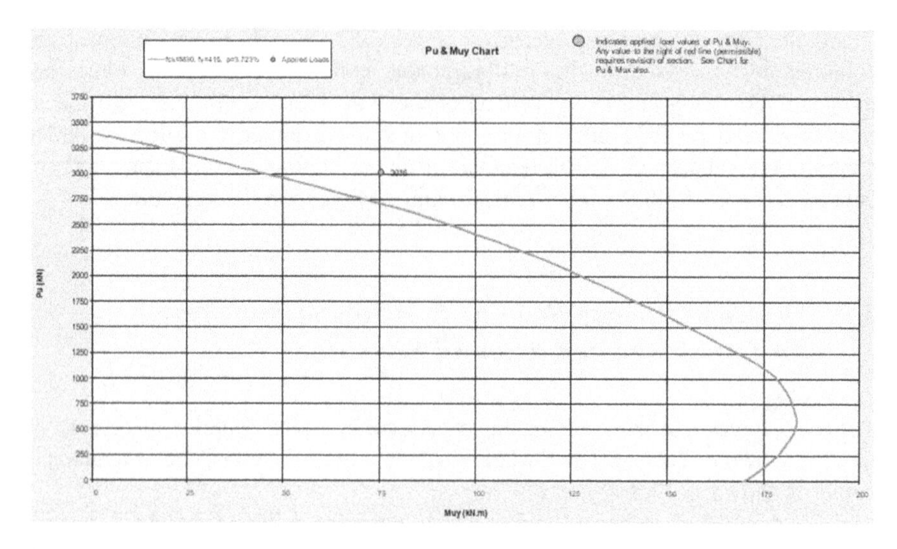

Fig. 8 Pu and Muy curve for critical column C4

All the critical columns were observed in the ground storey and the location of critical columns is shown in Fig. 9.

5.3 *Demand Capacity Ratio by P-M Interaction Curve for Short End Column Removal Scenario*

After the removal of short end column from the infill structure, there were three critical columns in the structure. The axial force and bending moment of the critical columns are shown in Table 5. The P-M interaction curve of column C3 are shown in Figs. 10 and 11.

The Column C3 and C11 failed on the both the axis (Mux and Muy) on removal of column on the short end side. The location of the critical columns are shown in Fig. 12.

5.4 *Push Over Analysis*

Push over analysis estimates the structural performance of the structure by seismic demand, force–deformation capacity, by non linear static curve. The critical columns in the structures are removed for the both the scenarios. Then the three structures (including infill frame) Structure is to be analyzed by Displacement

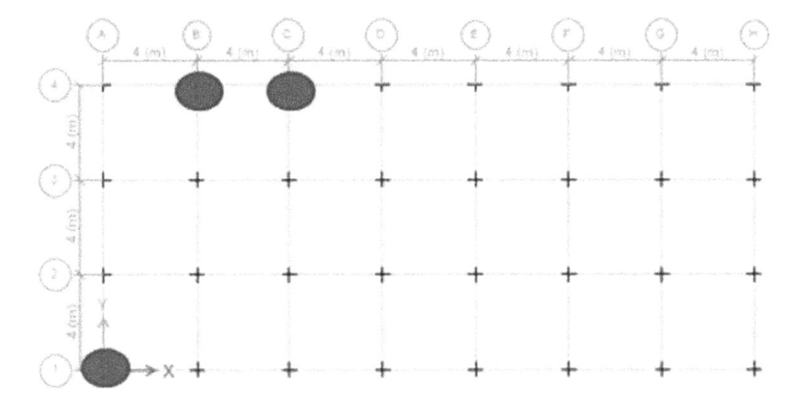

Fig. 9 Location of critical columns (after corner column removal)

Table 5 Demand axial forces and bending moment of critical columns after short end column removal

Column No	Axial force (kN)	Moment (kN m)	Load combinations
C11	3018	88	1.2(DL + LL-Ely)
C3	3004	87	1.2(DL + LL-Ely)
C8	271	83	1.2(DL + LL-Ely)

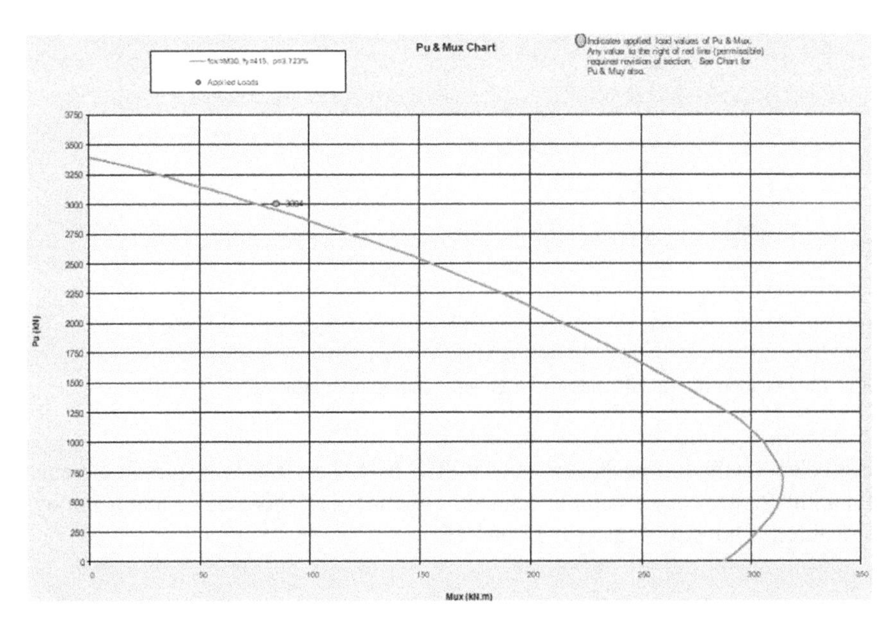

Fig. 10 Pu and Mux curve for C3

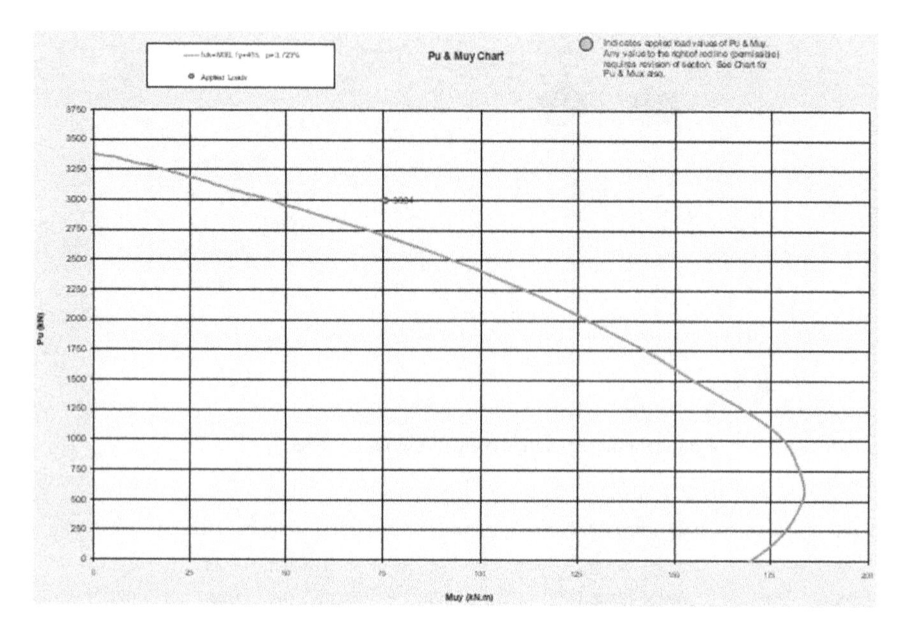

Fig. 11 Pu and Muy curve for C3

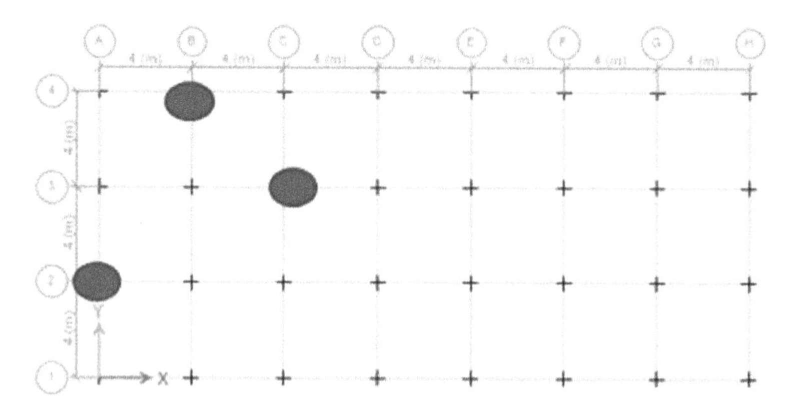

Fig. 12 Location of critical columns (after short end column removal)

coefficient method of pushover analysis. The base shear versus displacement curve for infill frame, corner column removal structure and short end column removal structure is shown in Figs. 13, 14 and 15.

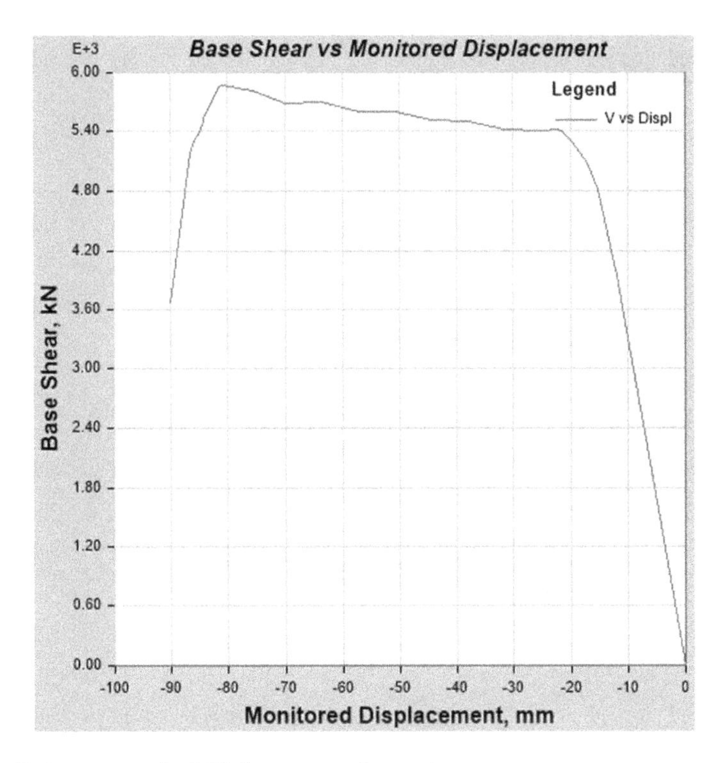

Fig. 13 Pushover curve for infill frame removal-scenario

5.5 *Global Ductility*

Ductility ratio is "**μ**" the ratio of ultimate displacement of the structure to the yield displacement. Ductility ratio is calculated for all the scenarios and shown in Table 6.

$$\text{Ductility ``}\boldsymbol{\mu}\text{"} = \Delta_{max}/\Delta_y$$

where

Δ_{max} Ultimate displacement.
Δ_y Yield displacement.

The Global ductility of Short end column removal structure is greater than the corner column removal structure but not more than the infill structure.

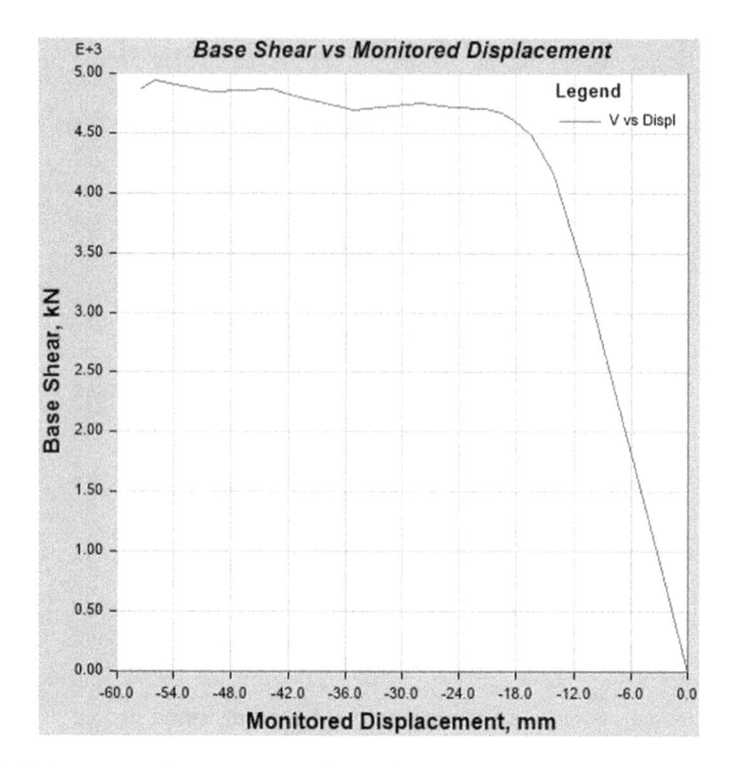

Fig. 14 Pushover curve for corner-removal scenario

5.6 Total Energy Capacity

Total energy capacity is calculated for all the scenarios by calculating the area under the load displacement curve from the pushover analysis. The total energy capacity values are shown in Table 7. The total energy capacity of Short end column removal structure is greater than the corner column.

5.7 Robustness Coefficient

Robustness is defined as the ability of a structure to resist a local damage in the structure. Robustness coefficient 'R' is ratio of base shear of structure that is damaged to the base shear of intact infill structure. The Robustness coefficient of short end column removal scenario structure and corner column removal scenario structure is shown in Table 8.

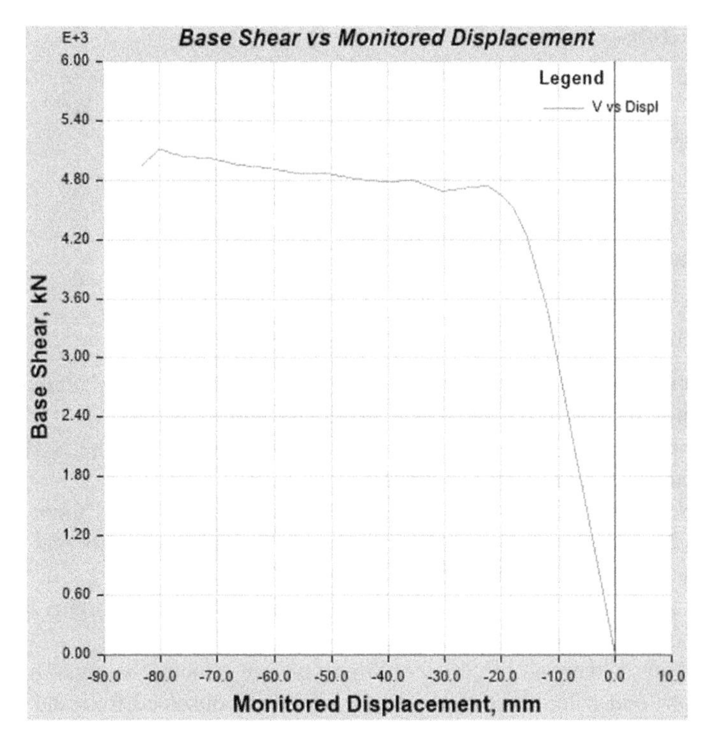

Fig. 15 Pushover curve for short end column removal scenario

Table 6 Global ductility

S.No	Type of structure	Global ductility
1	Infill structure	4.1
2	Corner column removal structure	3.2
3	Short end column removal structure	3.7

Table 7 Total energy capacity

S.No	Type of structure	Total energy capacity (N mm)
1	Infill structure	398,426
2	Corner column removal structure	217,640
3	Short end column removal structure	340,350

Table 8 Robustness coefficient

Type of structure	V_d base shear of damaged structure (kN)	Robustness coefficient $R = V_d/Vi$
Corner column removal structure	4940	0.89
Short side column removal structure	5108	0.92

6 Conclusions

The followings were concluded from the analysis:

- The structure with corner column removal and short side column scenario exhibited more critical columns by P-M interaction curve.
- The total energy capacity of short side column removal structure is 21% greater than corner column removal structure.
- The global ductility of short side column removal structure is found to be 3.7 and 3.2 for corner column removal structure, which represents that short side column structure is more ductile.
- The Robustness coefficient for short side removal structure is higher (0.92) than the corner column removal structure (0.89).
- The effect of damage of corner column removal structure is much higher than the short end column removal structure which is observed from the push over analysis.

References

1. Barrosa, Cavaco E (2019) Effect of non-structural masonry brick infill walls on the robustness of a RC framed building severely damaged due to a landslide. Elsevier Eng Struct 180 (2019):274–283. www.elsevier.com/locate/engstruct
2. Aniendhita, Data Iranata (2017) Comparative study on diagonal equivalent methods of masonry infill panel. AIP conference Proceedings 1855, 030011. https://doi.org/10.1063/1.4985481
3. Kostinakis K, Asimina (2019) Effect of in-plan regularities caused by masonry infills on the seismic behaviour of R/C buildings. Soil Dyn Earthq Eng
4. Ning N, Ma J (2019) Influence of masonry infills on seismic response of RC frames under low frequency cyclic load. Elsevier Eng Struct 183:70–82. www.elsevier.com/locate/engstruct
5. Sankhla SS, Bhati D (2016) A comparative study on the effect of infill walls on RCC frame structures. J Mech Civil Eng 13:01–08
6. Liberatore L, Noto F (2018) A in plane response of masonry infill walls. Eng Struct 167:533–548
7. IS 456 (2000): Plain and reinforced concrete—code of practice [CED 2: cement and concrete]
8. IS875 (Part 2) Code of practice for imposed loads
9. IS 1893:2016 Criteria for earthquake resistant design of structures
10. General Service Administration for Alternate Path Analysis and Design Guidelines for Progressive Collapse Resistance Revision 1, January 28, 2016

Dynamic and Seismic Evaluation of Curved Cable Stay Bridges

Gautham Sudhakar and Reshma Prasad

Abstract Cable-stayed bridges are the highly flexible bridges and it is popular due to its economy for longer spans and good aesthetics. Cable stayed bridges are mainly designed based on the flexural strength of the deck, the compressive strength of the pylons, and the tensile strength of the cable stays. By the combination of these parameters, these bridges show excellent structural performance. Though these bridges show good structural efficiency, they exhibit complex non-linear characteristics. Implementing longer spans, horizontal curvature and increasing column skewness are some of the recent trends in modern bridge engineering; making the bridge an irregular structure with irregular behaviour under horizontal and vertical loads. The aim of this research is to study the structural performance of cable stay bridges with curvature of $\theta = 11°$ and $22°$ under seismic excitations with the aid of modal and time history analyses and also to compare the performance of curved bridge alignments with that of the straight bridge alignment by comparing several parameters. The modelling is done in AutoCAD and exported to SAP 2000 software to analyze. From this research work, it is noticed that single curve bridge with $\theta = 11°$ exhibit similar performance to that of straight bridges. So, providing curvature within that range is highly feasible when required. It was also noted that curved bridges with high values of curvature exhibit dissimilar patterns from that of straight bridges in some cases.

Keywords Cable stay bridges · Curved · Skewed · Time history analysis · Modal analysis

G. Sudhakar · R. Prasad (✉)
Department of Civil Engineering, FISAT, Angamaly, Kerala, India
e-mail: reshmaprasad@fisat.ac.in

G. C. Marano et al. (eds.), *Proceedings of SECON'21*, Lecture Notes in Civil Engineering 171, https://doi.org/10.1007/978-3-030-80312-4_43

1 Introduction

The cable stay bridges are one of the most frequently used bridge types in the world. Cable-stayed bridges have been a competitive bridge type for long bridges, especially for spans in a range of 500 and 1000 m. Cable-stayed bridges are flexible bridges and are popular because of its economic feasibility for longer spans and good aesthetics [1]. The bridges are designed taking into consideration several parameters such as compressive strength of the pylons, tensile strength of the cable stays and flexural strength of the deck [2]. By effective action of each main member, cable-stayed bridges show amazing structural performance [3]. Stay cables has the role of intermediate vertical supports for the superstructure rather than conventional piers or columns [4]. So, cable-stayed bridges are a good option for long-span bridges.

Cable-stayed bridges with many distinctive styles are increasing worldwide. These bridges are now being built in different styles for distinct purposes [5]. So, the accurate assessment of these bridges using dynamic methods has become an increasing concern due to their role in infrastructure development [6]. Implying longer spans, curvature and increased column skewness are some of the modern trends in bridge engineering [7].

Challenges in the various analysis techniques used for a curved cable stayed bridge might be due to its complex nature as a continuous and large structure, which makes it highly difficult in representing it as a lumped-mass system. Also, due to this complexity, many modes are closely spaced in case of time period/frequency. And due to the long span, spatial variability of earthquake motions may also be highly significant.

Wilson et al. detailed about 3D modelling of the cable-stayed bridges and the bridge used for this study is the Quincy Bayview Bridge in Illinois, USA [8]. Serdar et al. studied the static non-linear analysis of curved RC bridges [9]. Significant mode shapes are obtained through modal analysis [10]. Biondini et al. studied the uncertainty effects on the structural performance of curved cable-stayed bridges in its lifetime taking into consideration the bridges like Certosa Bridge and Malpensa Bridge (curved bridge) in Italy [11, 12]. ACI codes and guidelines for design of cable stay bridges (ASCE) [13] were used as a base for modelling purposes [14]. From these literature studies, it is found that there are not many studies regarding the curved long span cable stay bridges.

So, the aim of the study is to compare the performance of straight cable stay bridges and curved cable stay bridges by means of modal analysis and time history analysis.

2 Bridge Modelling

The cable stay bridge considered for the study was the Quincy Bay View Bridge, Illinois, USA. The bridge includes two H-shaped concrete towers, double semi fan type cable arrangement, and a concrete-steel girder bridge deck. The middle span is

274 m and two equal side spans of 134 m each, which sum up to a total length of 542 m. The height of the pylon is 71 m from the waterline. There are a total of 56 cables, 28 cables support the middle span and 14 cables supporting each side span. The width of the deck from centre-to-centre distance between the cables is 12 m. Referring to Wilson et al. [8], the sections used for pylons, cables and deck are tabulated in Table 1 along with the material properties (Fig. 1).

The deck was made up of concrete with steel I section girders below it. The schematic diagram of pylon and cable arrangement of the bridge is given in Figs. 2 and 3. To understand the effectiveness of curvature on the bridge, curvature is introduced to the bridge model with $\theta = 11°$ and 22° as in Table 2. Thus, four bridge prototypes were modelled in the AutoCAD software as shown in Figs. 4, 5 and 6.

Concrete is assigned for deck, pylons and strut sections and steel is assigned for girders and cable sections. The bridge is aligned in X direction and the stay cables are connected to the deck with rigid links at thirty nodes provided at a c/c distance of 19 m along the deck. The base of both the pylons were kept fixed and the deck is kept to be simply supported.

These models are exported to SAP 2000 software and materials, sections, links and restraints were assigned for the analysis purposes as in Fig. 7.

Table 1 Section and material properties

Pylon						
Pylon sections	Cross sectional area (m^2)	Moment of inertia about z-z axis (m^4)	Moment of inertia about y-y axis (m^4)	Moment of inertia about x-x axis (m^4)	Young's modulus (MPa)	Mass density (kg/m^3)
Tower part 1	17.88	32.75	1422.42	27.64	30,787	2400
Tower part 2	8.77	30.62	1239.4	19.76	30,787	2400
Tower part 3	7.06	28.05	531.67	15.39	30,787	2400
Lower strut	6.36	13.37	13.83	7.37	30,787	2400
Upper strut	7.24	14	14.98	8.23	30,787	2400

Cables			
Cable no	Cross sectional area (m^2)	Young's modulus (MPa)	Cable weight (N/m)
1	0.0089	205,000	888.7
2	0.0067	205,000	769.1
3	0.0053	205,000	707.8
4	0.0034	205,000	519.5

Deck		
Modulus of elasticity	Steel	2.068×10^5 N/mm^2
	Concrete	3.07×10^4 N/mm^2
Unit weight	Steel	7849.05 kg/m^3
	Concrete	2400 kg/m^3
Poisson's ratio	Steel	0.30
	Concrete	0.25

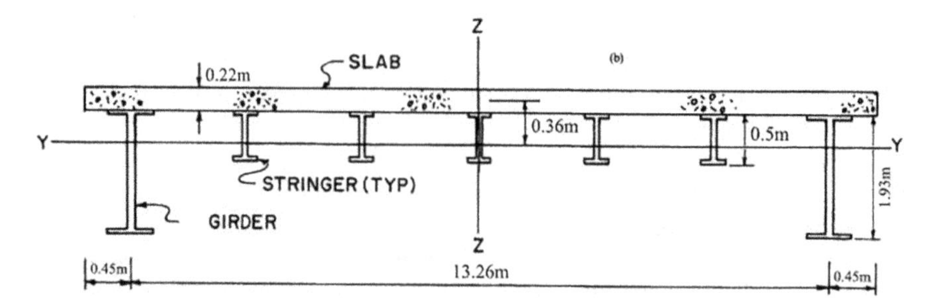

Fig. 1 Deck section [8]

Fig. 2 Schematic diagram of pylon

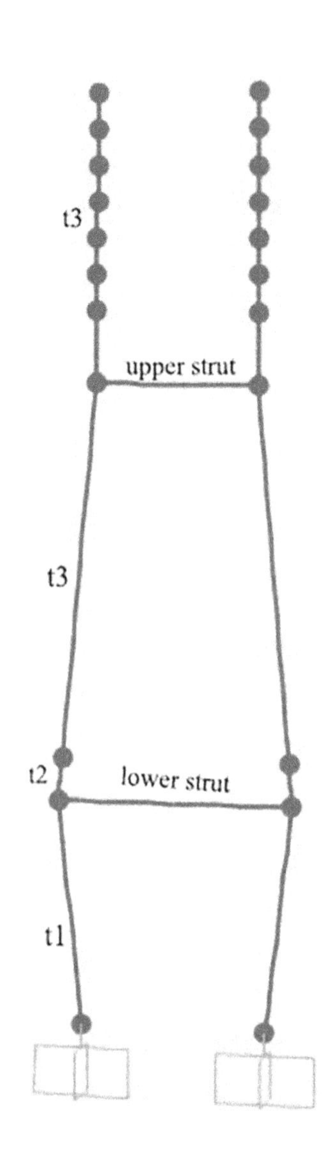

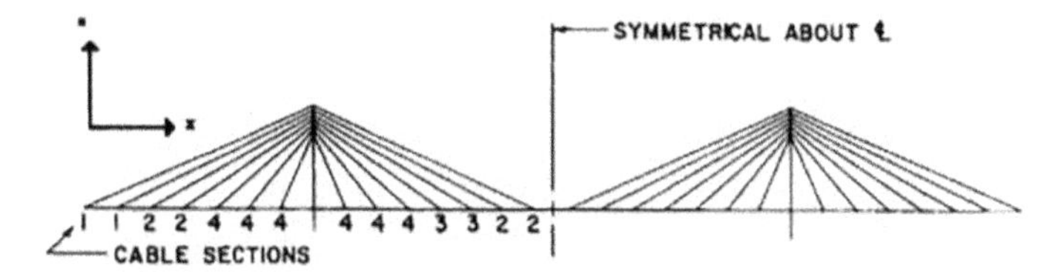

Fig. 3 Cable arrangement [14]

Table 2 Bridge prototypes

Model name	Description	Angle	Radius	Arc length (m)
SB	Straight bridge	0°	∞	542
SCB1	Single curve bridge 1	11°	1373.5 m	545.6
SCB2	Single curve bridge 2	22°	693 m	556
DCB	Double curve bridge	22°	353 m	556

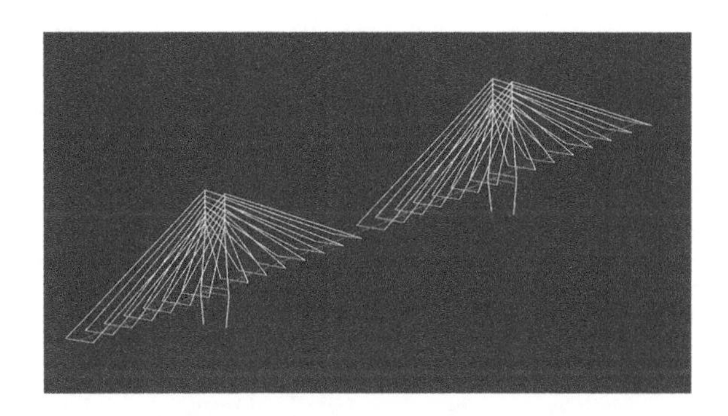

Fig. 4 Straight bridge prototype

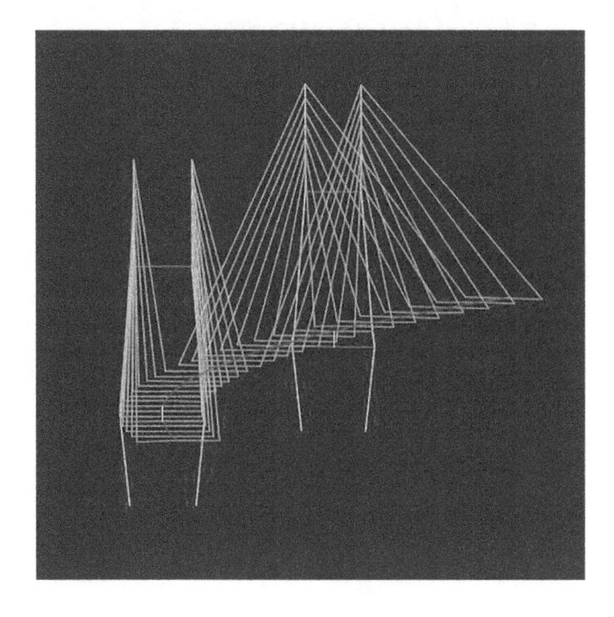

Fig. 5 Single curve bridge 1 prototype

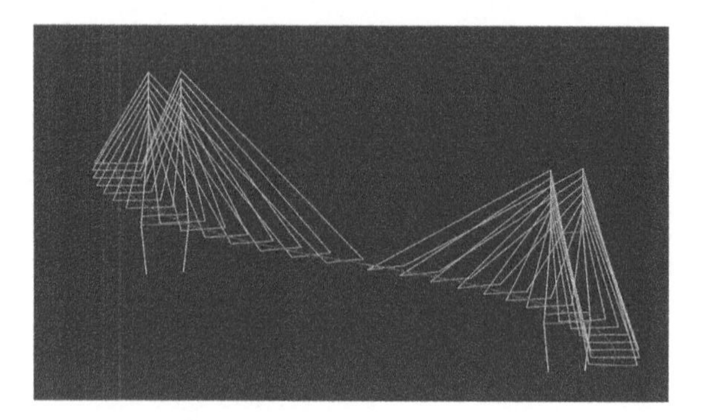

Fig. 6 Double curve bridge prototype

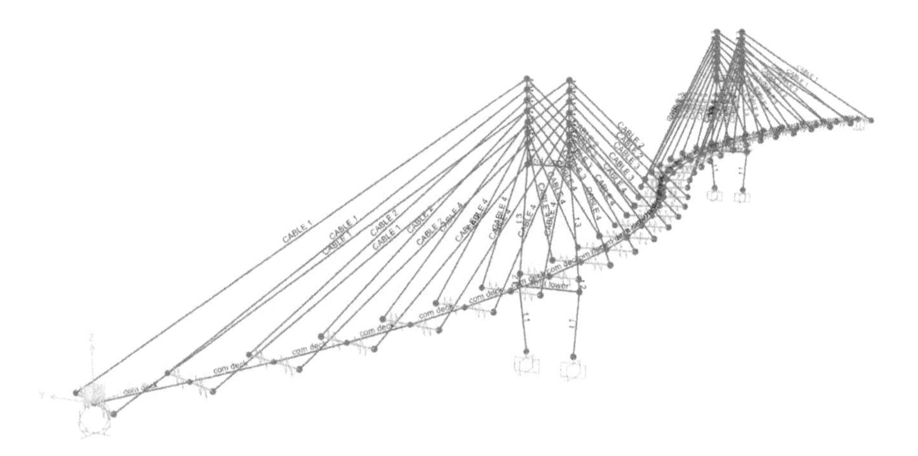

Fig. 7 Sections assigned in SAP2000

3 Analysis

3.1 Modal Analysis

The dynamics of any structure is analyzed by a simple equation of motion (Eq. 1), assuming the damping coefficient as zero. The basic equation gives mode shapes based on the degree of freedom [15]. Modal analysis is very important in analyzing the boundary conditions and accuracy in the modelling by observing the frequencies obtained [16].

$$M\ddot{U} + KU = F(t) = 0 \tag{1}$$

Modal analysis is done for different cases to calculate the time periods and frequencies. The mode shapes for SB, SCB1, SCB2, DCB are analyzed. Modal analysis imparts a vital role in analyzing the possible behavior of any structure.

3.2 *Time History Analysis*

To understand the dynamic behavior of the cable-stay bridge, the earthquake ground motions of Imperial Valley station of El Centro earthquake have been considered. Mainly the X component (TH-X) and Y components (TH-Y) are taken into consideration.

The results of the time history analysis have been presented below to discuss the effect of earthquake on several different cases. The results have shown different behavior than what is expected logically. The study also limited itself to the discussion based only on the maximum displacements of pylons and deck sections.

4 Results and Discussions

4.1 *Mode Shapes (Time Period/Frequency)*

Mode Shapes for first and second modes are shown in Figs. 8, 9, 10 and 11. The results are also tabulated in Table 3.

From the analysis, the mode shapes for all the four prototypes tends to show a familiar pattern. An increase in the time period was observed when curvature is introduced. Also, time period increased when the double curvature is introduced instead of single curvature. On increasing the curvature, the length of the deck spine and the mass of the same increases. Hence it exhibits a decrease in stiffness values and thus an increase in the time period.

4.2 *Time History Analysis*

Pylon Displacement: To analyze the pylon displacement, displacement values at the bottom and top of the pylon and also at the strut portions were considered. Displacement at both earthquake components TH-X and TH-Y were analyzed. For both the analysis, displacement corresponding to the same direction was considered.

From the graph (Fig. 12), it is noticed that at the bottom part of pylon there is no displacement and usually as the height of the pylon increases the displacement also increases.

Fig. 8 Mode shapes of SB

Fig. 9 Mode shapes of SCB 1

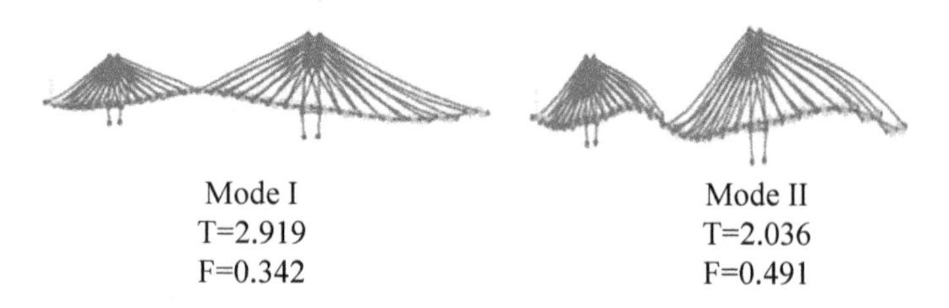

Fig. 10 Mode shapes of SCB2

Fig. 11 Mode shapes of DCB

It is also observed that as curvature of deck increases, the displacement of pylon also increases. All prototypes have similar displacement patterns and in that, the DCB exhibit maximum displacement at various locations as it has maximum

Table 3 Time period/frequency

Modal analysis results						
	Mode I		Mode II		Mode III	
	T (s)	F (Hz)	T (s)	F (Hz)	T (s)	F (Hz)
Straight bridge	2.735	0.365	1.917	0.521	1.28	0.78
Single curve bridge1	2.798	0.357	1.951	0.512	1.3	0.768
Single curve bridge2	2.919	0.342	2.036	0.491	1.352	0.739
Double curve bridge	2.984	0.335	2.085	0.479	1.363	0.733

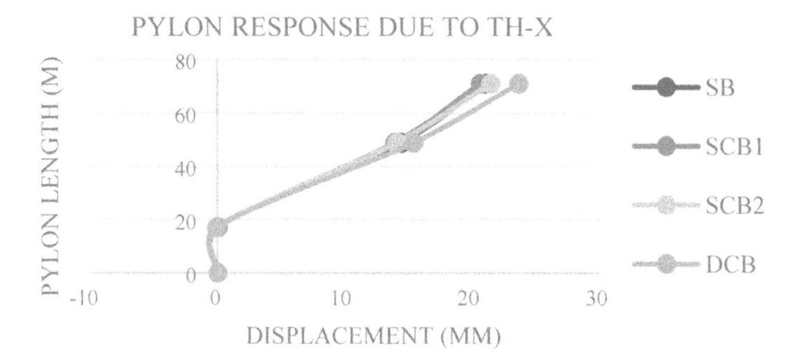

Fig. 12 Graph representation of Pylon displacement due to TH-X

variation in the position of pylon due to the curvature. In the pylon below the deck portion, the deflection is in the -X direction while above the deck portion there is +X displacement. DCB has a displacement variation of 10% compared to SCB2.

In the graph showing pylon response due to TH-Y (Fig. 13), the displacement is very small compared to the displacement of pylon in -X direction (Fig. 12). Considering the geometry of the bridge, the axis of the pylon is perpendicular to the X direction, so the stiffness will be less and the displacement will be more in X direction; and the vice versa will occur in case of Y direction.

Deck Displacement: To analyze the deck displacement, displacement values at all the thirty nodes along the deck spine element were considered. The deck is aligned along the X-direction. Displacement at both earthquake components TH-X and TH-Y were analyzed. For the analysis, when considering the response due to TH-X, values of displacement along the Z-direction (vertical direction) were considered. When considering the response due to TH-Y, values of displacement along the Y-direction were considered.

Due to the TH-X component, considering the displacement in Z direction (Fig. 14), all the bridges show similar displacement patterns thus all the four prototypes exhibit similar performance and there aren't any considerable changes in its behavior to the seismic excitations. There is slight variation in the case of curved bridges with highest curvature (SCB 2 and DCB) in the deck portion between the

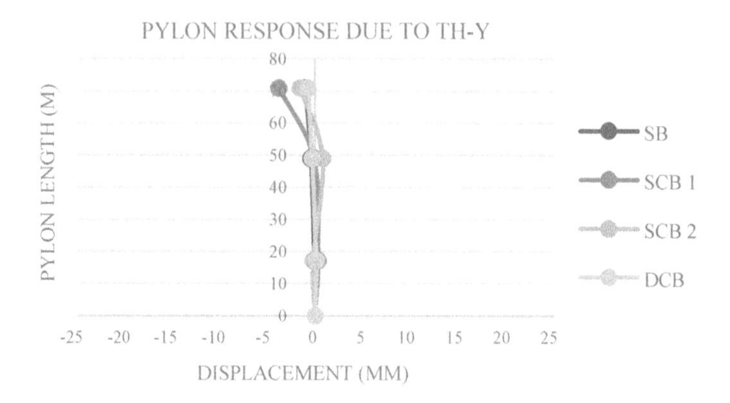

Fig. 13 Graph representation of Pylon response due to TH-Y

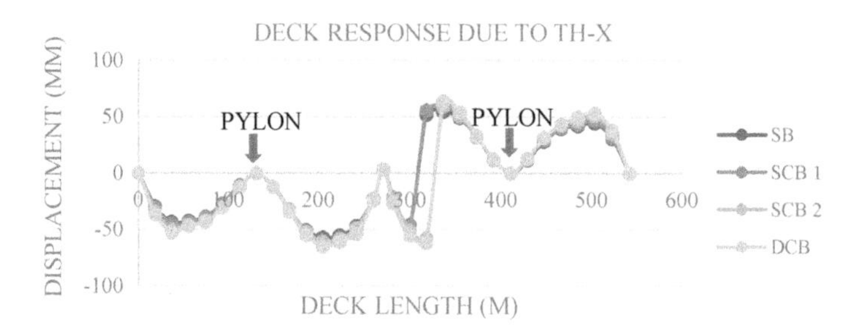

Fig. 14 Deck displacement due to TH-X

two pylons due to changes in position of deck and pylons due to high amount of curvature. Also, at the end portion of the deck and at the position of the two pylons there is zero displacement. Comparatively high values of displacement were observed at the mid span which is mainly due to high force concentration at the joints of mid span between pylons. Also, the displacements in all prototypes are almost similar within a range of ± 7 cm with small variations in the values irrespective of curvature of the bridges.

While comparing the lateral displacement of the bridge deck due to the TH-Y component (Fig. 15), from the graph, it is inferred that there is symmetric variation about the midpoint of the bridge. Between the two pylons, in all cases, there is a high value of displacement compared to the displacement between the pylons and the end points. As curvature increases, the displacement tends to increase. Between the two pylons at the mid portion, the SB prototype exhibits maximum displacement while the SCB1 exhibits minimum displacement. As the SCB1 shows minimum deck displacement while the straight bridge exhibits high displacement between the two pylons.

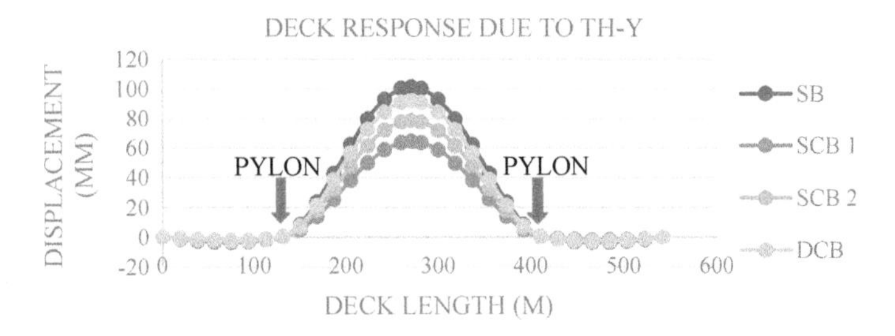

Fig. 15 Deck displacement due to TH-Y

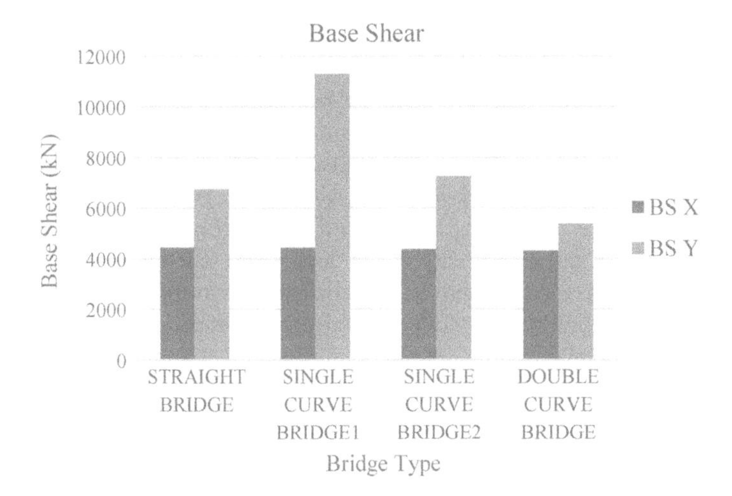

Fig. 16 Response to base shear

These variations are mainly attributed to torsional effect due to the increasing curvature and eccentricity of the deck due to the curvature induced.

The values of the base shear in the corresponding time history analysis obtained from SAP 2000 is shown in Fig. 16. Base shear in X direction is very less compared to Y direction. The SCB1 exhibits the highest value of base shear. Double curve bridge exhibit least base shear. As the curvature increases, stiffness decreases, the base shear also decreases.

5 Conclusions

In this study, linear elastic finite element models of cable-stayed bridge are formulated with different curvature conditions like straight, single curved and double curved bridge conditions with included angle ranging from 0°, 11° and 22°. Time

History Analysis and Modal Analysis were conducted on the four-bridge prototype to analyze its performance when inducing curvature.

- In the Modal analysis, as the curvature increases, due to the variation in stiffness, there is a proportionate increase in the time period. The time period is high in case of double curve bridge and low in case of straight bridge.
- In the time history analysis, Due to the TH-X component, the pylon displacement and the deck upliftment shows similar behavior with slight variations in the values. In the case of pylon, maximum displacement was obtained by DCB and in the case of deck displacement maximum value was obtained by SCB2 and DCB.
- Due to the TH-Y component, the lateral displacement in the pylon and the deck were analyzed. Because of the eccentricity in the position and due to the changes in the alignment of the pylon, in case of pylon, similar graph patterns were observed in all cases with very small displacement values. When considering the lateral displacement of the deck, maximum displacement is observed between the two pylons. Here the maximum displacement was observed in SB and least displacement was observed in SCB1.Also, there is very less displacement observed between the pylons and the end points. It is mainly attributed to the torsional effects and eccentricity of the deck due to curvature provided.
- In the case of base shear, there is a significant decrease in base shear in curved bridges which might be due to the decrease in stiffness.

Though there are variations in the displacement patterns in some cases, it is mainly due to the eccentricity in the position of the pylons and the change in the alignment. Also, there will be torsion due to the provided curvature. For a layout to have better configuration in resisting earthquakes, the structure should have lower base shear and displacements. Curved bridges with high values of curvature exhibit dissimilar patterns from that of straight bridges in some cases. So, for more specific conclusions, more parametric research and studies are to be carried out in this area. As there are variations in the displacement patterns and base shear in most cases, the curvature effects should be considered at the design procedure of such bridges.

References

1. Zadeh OS (2012) Comparison between three types of cable stayed bridges using structural optimization, The University of Western Ontario
2. Valdebenito GE, Aparicio AC (2006) Seismic behaviour of cable-stayed bridges: a state-of-the-art review. In: 4th international conference on earthquake engineering, Taipei, Taiwan
3. George HW (1998) DISCUSSION: cable-stayed bridges—parametric study. J Bridg Eng 3(3):148–150
4. Yang Y, Wang X, Wu Z (2020) Long-span cable-stayed bridge with hybrid arrangement of FRP cables. Compos Struct 237:111966

5. Ren WX, Obata M (1999) Elastic-plastic seismic behavior of long span cable-stayed bridges. J Bridg Eng 4(3):194–203
6. CSI (2009) SAP2000 Ultimate 15.0.0: Static and dynamic finite element analysis of structures, computers and structures. Berkeley, California
7. Polepally G, Pasupuleti VDK, Dongre A (2020) Comparison of different types of pylon shapes on seismic behaviour of cable-stayed bridges. In: Emerging trends in civil engineering. Springer, Singapore, pp 69–80
8. Wilson JC, Gravelle W (1991) Modelling of a cable-stayed bridge for dynamic analysis. Earthq Eng Struct Dynam 20(8):707–721
9. Serdar N, Folić R (2016) Comparative analysis of modal responses for reinforced concrete (RC) straight and curved bridges. Procedia Eng 156:403–410
10. Domaneschi M, Martinelli L (2010) Seismic mitigation of the ASCE cable-stayed bridge. In: 34th IABSE symposium. IABSE-AIPC-IVBH, pp 1–8
11. Ali HM, Abdel Ghaffar AM (1994) Seismic energy dissipation for cable stayed bridges using passive devices. Earthq Eng Struct Dynam 23:877–893
12. Biondini F, Frangopol DM, Malerba PG (2008) Uncertainty effects on lifetime structural performance of cable-stayed bridges. Probab Eng Mech 23(4):509–522
13. Committee on Cable-Stayed Bridges (1992) Guidelines for the design of cable-stayed bridges. American Society of Civil Engineers
14. Singh PK (1998) Discussion of "cable-stayed bridges—parametric study" by PK Singh. J Bridg Eng 3(3):149–150
15. Soneji BB, Jangid RS (2007) Passive hybrid systems for earthquake protection of cable-stayed bridge. Eng Struct 29(1):57–70
16. Dutta AK et al (2008) Design of an active controller for Quincy Bayview Bridge, Illinois, U.S.A., Against seismic excitation Part I: model updating. Structural control and health monitoring, 1057–1077

Seismic Performance of Modular Steel Braced Building Equipped with BRB Braces

Anat Antony and N. Neeraja

Abstract Modular steel construction (MSB) is a novel construction technique that helps in reducing the time spent on the construction site. The demand for modular steel buildings has increased because of their improved quality, less material waste and fast on-site installation. Bracing systems are used to resist the lateral loads of MSBs especially in low rise buildings. A regular 4-storey MSB was compared with torsionally irregular MSB using ETABS software and its performance was evaluated for different bracing system such as external braces, internal braces and BRB. Time history analysis was utilized to judge the benefits in terms of the response parameters lateral displacement, storey drift and base shear. Torsional Irregularity came into context when the load eccentricity was 15%. The results showed a good improvement in the seismic resistance of MSB with incorporation of additional braces. External braces and external BRB braces shows better performance among the considered bracing systems.

Keywords Modular steel building · BRB braces · Time history analysis · Irregular structures

1 Introduction

The concept of prefabricated modular structures has arisen as an effective solution to achieve both speedy construction as well improved and sustainable quality of the final product. A modular structure is made up of a number of prefabricated units known as "modules." Modular building is constructed with prefabricated modules at a factory, then transported it to the site were the foundation is laid and finally are installed at site. At the plant, each unit is often completely fitted with plumbing,

A. Antony (✉)
Federal Institute of Science and Technology, Angamaly, Kerala, India

N. Neeraja
Department of Civil Engineering, Federal Institute of Science and Technology, Angamaly, Kerala, India

G. C. Marano et al. (eds.), *Proceedings of SECON'21*, Lecture Notes in Civil Engineering 171, https://doi.org/10.1007/978-3-030-80312-4_44

flooring, and lighting. Modular construction is popular and practised for buildings with repetitive layouts such as low-rise schools, homes, hotels, dormitories, hospitals etc. Adding braces or supplying shear walls to the entire modular steel building (MSB) provides lateral stability. For lateral stability, one to six storey MSBs typically depend on bracing elements [1, 2]. On-site, the individual module units are attached to form the entire structure. Their usage has a variety of benefits, including increased assembly speed, increased quality control at a factory, increased work safety as the time required for high-altitude work decreases, testing and rapid adoption of new technology at the plant, and a reduction in noise and construction waste at a construction site etc. Modular buildings, on the other hand, vary from conventional structures in terms of conduct, detailing specifications, and construction method [3]. Because of its novel structural nature, the construction industry is hesitant and inexperienced with adopting this technology, which is preventing more people from using it [4].

Buckling restrained braces (BRB) are a new form of concentric braced frames that address the buckling problem of braces [5]. A central core plate is encased in a mortar-filled tube in traditional configuration of BRB braces, which prevents core plate from buckling in compression. The core plate's compressional behaviour is governed by yielding rather than buckling, which is analogous to tensional loading and results in a stable hysteretic curve with increased ductility. The heavy weight and healing issue of the mortar are disadvantages of traditional BRB configurations. A new form of BRB, known as all-steel BRBs, is being developed to overcome these inefficiencies. The idea of the new configuration remains unchanged; however, it doesn't require any filling material; that is, the core plate will be encased in a steel tube without any mortar or unbounding material, making all-steel BRBs lighter, simpler, and faster to fabricate. As a result, this style is more cost-effective and practical than traditional BRBs [6, 7].

In this study the torsional irregularity of the modular building is investigated and it's strengthening by providing additional braces and BRB is also carried out. Torsional response in building is caused by two factors: (a) eccentricity between the centre of mass and centre of stiffness of vertical lateral force resisting elements and (b) accidental torsion due to the rotational component of ground motion about a vertical axis, eccentricity between assumed and actual stiffness and mass, uncertainty in live load distribution, and uncertainty in dead loads due to variations in labour and materials. The seismic performance of irregular MSB is examined in terms of storey drift, base shear, and maximum lateral displacement in order to obtain a detailed overview of structural response for torsional irregularities.

2 Design, Modeling and Analysis of MSB

The lateral loads in an MSB get transferred by the corners that are tied that help the MSB to act as a single unit. Whereas horizontal forces are transferred by these corner ties via tension and compression developed as well as through the connections provided between them. This happens because of the diaphragm action of floor and ceiling transfers loads to these corners. Relative displacements and rotations may occur between the modules. This may be both horizontally and vertically between the bolts and connecting plates [8].

A 4-storey MSB was selected for the study. It was designed by Fathieh and Mercan [9]. Figure 1 depicts a typical plan and an elevation of the MSB. ETABS software was used for the modelling and analysis of the building. There are 12 modules dimensioned 3.5 m × 4 m with a height of 3.5 m at each level. When the modules are installed next to one another on-site, there could be more than one column at each axis because each unit has its own columns installed off-site (Fig. 2a). The horizontal centre to centre distance between the columns is 0.35 m. In some situations, a gap between the floor and ceiling beams is required for corner supported modules to enable bolting or welding and to allow mechanical and electrical facilities to run throughout the building [9]. Therefore, in this model a 0.15 m clear space between the floor and ceiling beam is provided. To catch the independent rotations that may occur due to partial field welding at ends of columns, a short column section is placed between these beams. The section properties of this short column are the same as the lower column to which it is connected. At the top end of the short column, a joint is provided to simulate rotation at vertical connections which is assumed to be pinned connection [10]. Steel plates or shop-welded clip angles are used for the horizontal connections of individual units [9]. Superimposed dead loads of 0.75, 0.32, and 0.7 kN/m^2 are introduced to account for additional loads on floor, roof, and ceiling respectively [10]. Design live loads of 3 kN/m^2 are assumed according to IS 875 (Part 2: 1987). Seismic loads are assumed as per IS 1893 (Part 1: 2016). Table 1 lists the frame sections for the columns, beams and braces for the model. All of the columns and braces are made of square hollow structural sections (HSS), which are usually used in modular

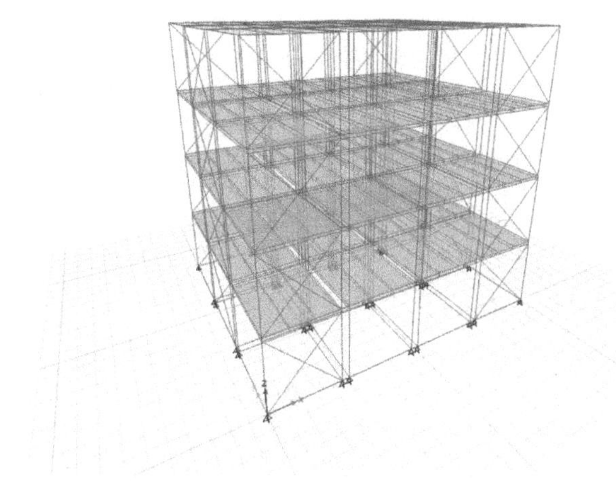

Fig. 1 3D view of 4-storey MSB

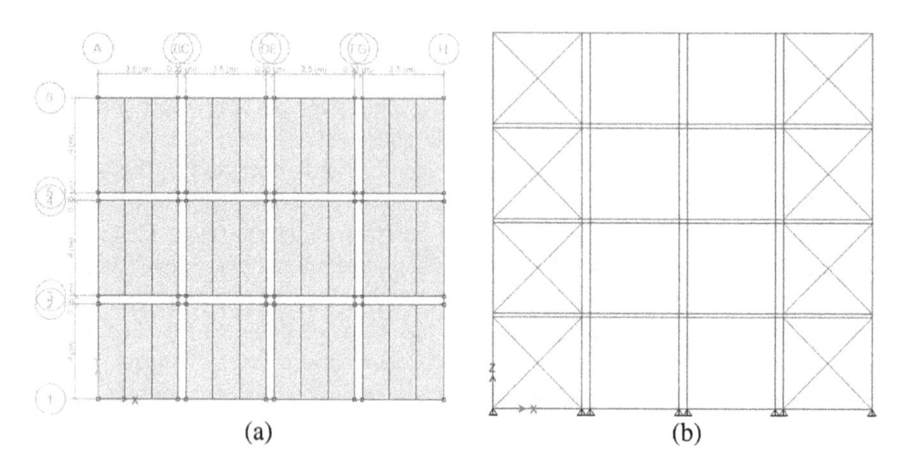

Fig. 2 Four storey modular steel building a floor plan, b elevation

Table 1 Section properties of MSB

Storey	Columns	Braces	Beams
4	HSS 76 × 76 × 5	HSS 51 × 51 × 5	W 100 × 19
3	HSS 102 × 102 × 6	HSS 51 × 51 × 5	W 100 × 19
2	HSS 102 × 102 × 6	HSS 76 × 76 × 5	W 100 × 19
1	HSS 127 × 127 × 5	HSS 76 × 76 × 5	W 100 × 19

systems, and the ceiling and floor beams are made of wide flange sections. The specification for all-steel BRB was selected from the study of Hosseinzadeh and Mohebi [7] corresponding to P_e/P_y ratio equal to 1.52.

The time history data used was that of ground motion recorded at Hollister Diff. Array during the 1989 Loma Prieta earthquake [9]. The structure was analysed for different eccentricities—0, 5, 10, 15 and 20%. Additional X bracings were provided externally and internally as different cases to study their contribution towards the performance of the structure. To analyse the performance of strengthening by BRB braces, the additional braces were replaced by all-steel BRB braces. The existing braces in the building were kept as such.

3 Results and Discussions

From time history analysis performed for MSB with different eccentricities and bracing systems following results are obtained. In this study the parameters like base shear, displacement, storey drift and time period were analysed.

3.1 Torsional Irregularity

To make the structure torsionally irregular the load on floor diaphragm were made to eccentrically act at different percentage—5, 10, 15 and 20%.

Displacement

Maximum displacement was observed in models with eccentricities is greater than that of base model. Maximum displacement value was obtained at top storey. Figure 3a, b shows the maximum displacement values obtained in the x and y direction. 2, 4, 6 markings in the y direction of graph indicate the short column segment between the floor and ceiling beams. As per IS 1893 (Part 1):2016, torsional irregularity is to be considered when max horizontal displacement of one end

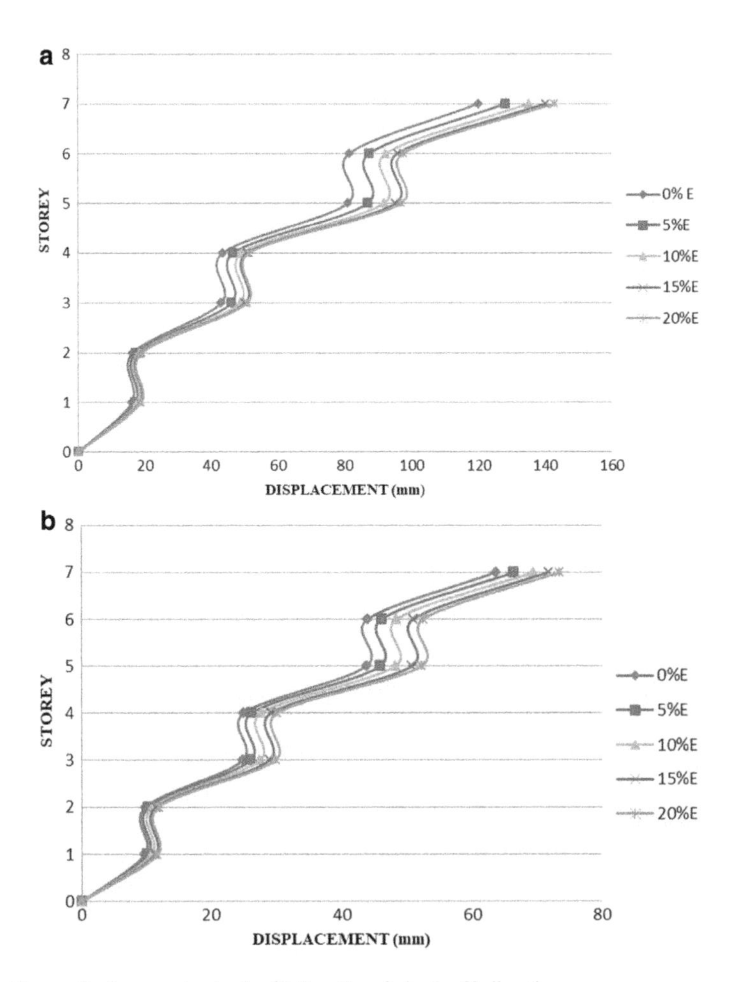

Fig. 3 Storey displacement **a** in the X direction, **b** in the Y direction

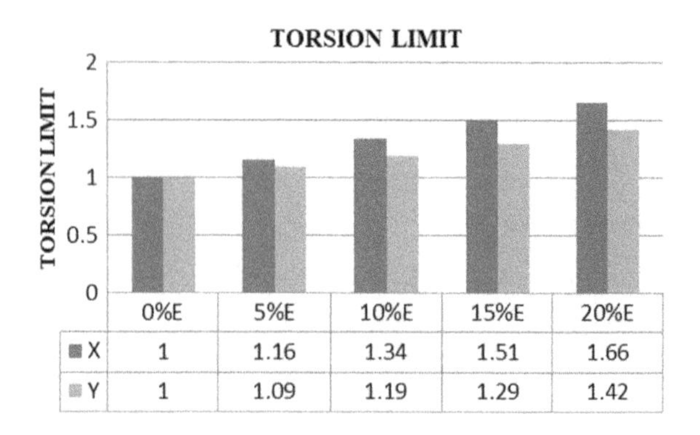

	0%E	5%E	10%E	15%E	20%E
X	1	1.16	1.34	1.51	1.66
Y	1	1.09	1.19	1.29	1.42

Fig. 4 Torsion limit for different eccentricities

of any floor in the direction of lateral force >1.5 times min horizontal displacement at far end of same floor. The model having 15% eccentricity exceeds the above limit. Figure 4 shows the torsion limit obtained for different percentage of eccentricities. Therefore this model was chosen for strengthening.

Storey Drift

Storey drift increases as the no. of storey increases. It along the Y direction is less than that in the X direction. Figure 5 displays the maximum storey drif values obtained for the models in both X and Y direction.

Base Shear

Base shear values shows a decrement in the X direction and slight increment in the Y direction as the eccentricity increases. Figure 6 shows the obtained base shear values in the X and Y direction for models.

3.2 Strengthening by Bracings

The model having 15% eccentricity is chosen for strengthening because this model exceeds the torsion limit. It was analysed with external braces, external BRB

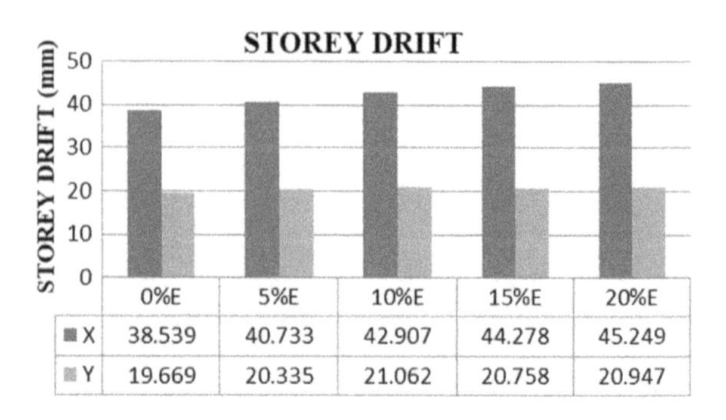

	0%E	5%E	10%E	15%E	20%E
X	38.539	40.733	42.907	44.278	45.249
Y	19.669	20.335	21.062	20.758	20.947

Fig. 5 Maximum drift values in the X and Y direction

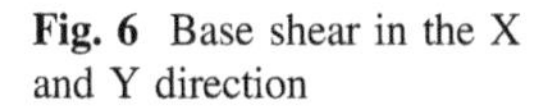

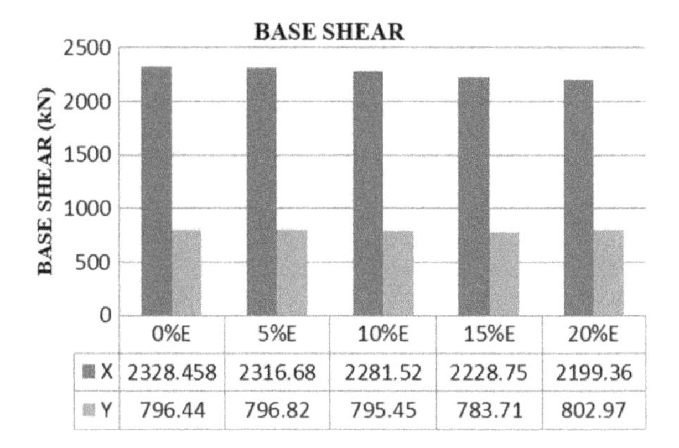

	0%E	5%E	10%E	15%E	20%E
X	2328.458	2316.68	2281.52	2228.75	2199.36
Y	796.44	796.82	795.45	783.71	802.97

Fig. 6 Base shear in the X and Y direction

braces, internal braces and internal BRB braces. The parameters are compared with above bracing systems and the selected eccentric model.

Displacement

The displacement values are lowered when provided with additional bracing. It is minimum for external brace in the X direction and internal braces in the Y direction. While checking the torsion limit, external braces and external BRB braces are within the limit. Figure 7a, b shows the maximum displacement values obtained in the X and Y direction. The torsion limit for the models with different bracing system is shown in Fig. 8.

Storey Drift

Storey drift is reduced for models having additional bracings when compared to the base model. It is minimum for model with external bracing. Figure 9 shows the maximum storey drift values obtained in X and Y direction.

Base Shear

The base shear values are increased in both the X and Y direction for models with additional braces. Figure 10 shows the obtained base shear values in the X and Y direction for models.

4 Conclusions

Modular construction has a lot of potential for increasing productivity and quality in the construction industry. Modular buildings are expected to have a major impact on the construction industry in the future. The successful launch of numerous projects in various regions has demonstrated its viability. Many building codes do not cover their design and detailing, and modular buildings continue to be designed according to conventional theories.

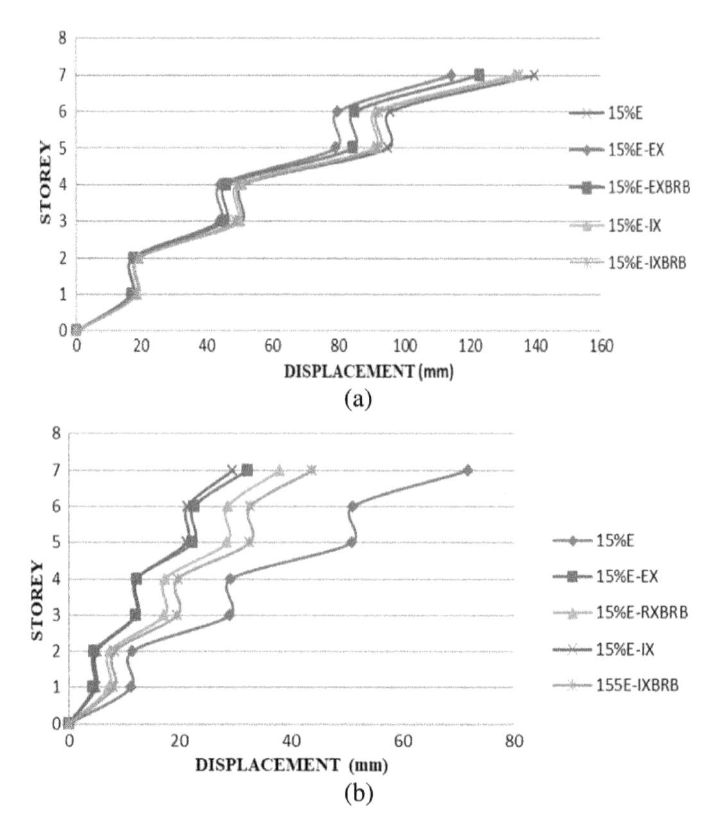

Fig. 7 Storey displacement **a** in the X direction, **b** in the Y direction

Fig. 8 Torsion limit for different bracings

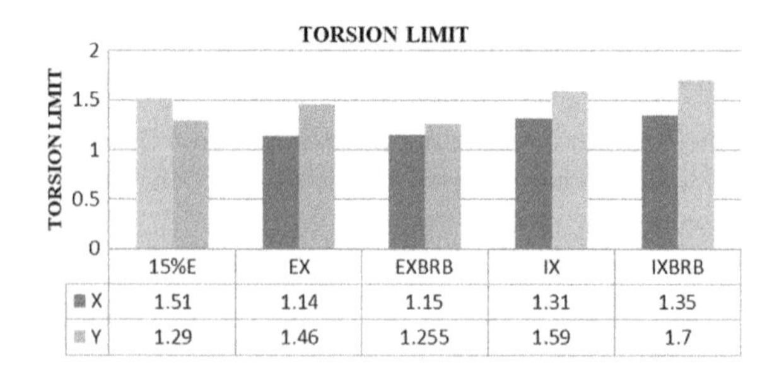

	15%E	EX	EXBRB	IX	IXBRB
X	1.51	1.14	1.15	1.31	1.35
Y	1.29	1.46	1.255	1.59	1.7

The study attempted to analyse a torsionally irregular modular building and stabilizing it with different bracing system. The model with 15% and above eccentricity exceeds the torsion limit, therefore the model corresponding to 15% eccentricity was chosen for strengthening. From the above parametric study following conclusions are inferred:

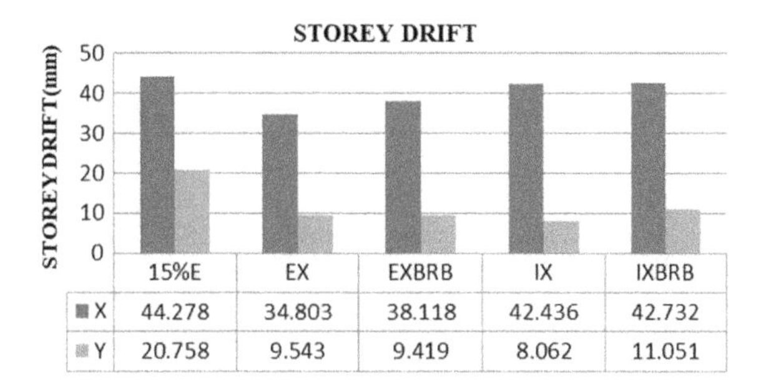

Fig. 9 Maximum drift values in the X and Y direction

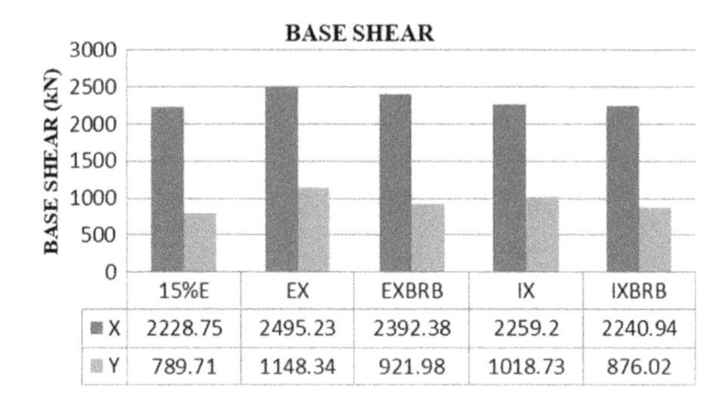

Fig. 10 Base shear in the X and Y direction

- The lateral displacement was reduced for all the models when compared to base model. But the models with internal braces and internal BRB braces fail to meet the torsion limit.
- External braces and external BRB braces shows a similar seismic response. BRB braces pose several advantages than the conventional braces in the context to their seismic performance.
- The maximum displacement was reduced by 18.18% and 55.17% for external bracing in the X and Y direction respectively and that for BRB bracing was 12.17% and 46.99% in the X and Y direction respectively.
- The results showed a good improvement in the seismic resistance of frames with incorporation of additional braces. The results indicate a reduction in the drift values as compared to the base model.
- The base shear value shows an increment in models with external bracing and external BRB bracing as compared to the base model.

References

1. Sultana P, Youssef MA (2018) Seismic performance of modular steel frames equipped with shape memory alloy braces. Bull Earthq Eng 16(11)
2. Annan CD, Youssef MA, El Naggar MH (2009) Seismic vulnerability assessment of modular steel buildings. J Earthq Eng 13(8)

3. Lacey AW, Chen W, Hao H, Bi K (2018) Structural response of modular buildings–an overview. J Build Eng 16
4. Deng E-F, Zong L, Ding Y, Zhang Z, Zhang J-F, Shi F-W, Cai L-M, Gao S-C (2020) Seismic performance of mid-to-high rise modular steel construction-a critical review. Thin Walled Struct 155
5. Fahnestock LA, Sause R, Ricles (2007) Seismic response and performance of buckling-restrained braced frames. J Struct Eng 133(9)
6. Ebadi Jamkhaneh M, Ebrahimi AH, Amiri MS (2018) Seismic performance of steel-braced frames with an all-steel buckling restrained brace. Pract Periodical Struct Design Constr 23(3)
7. Hosseinzadeh S, Mohebi B (2016) Seismic evaluation of all-steel buckling restrained braces using finite element analysis. J Constr Steel Res 119
8. Chua YS, Richard Liew JY, Pang SD (2020) Modelling of connections and lateral behavior of high-rise modular steel buildings. J Constr Steel Res 166
9. Fathieh A, Mercan O (2016) Seismic evaluation of modular steel buildings. Eng Struct 122
10. Annan CD, Youssef MA, El Naggar MH (2008) Seismic overstrength in braced frames of modular steel buildings. J Earthq Eng 13(1)

Improving Highway Alignment Using Openroads Software

Shubham Sharma and Amardeep Boora

Abstract Highway alignment is the layout of the centerline of the road on the ground level. An ideal alignment should be safe, economical, short, and comfortable. This paper presents the improved design standards of the existing road of 4.5 km, which is the part of National Highway Authority of India (NHAI) project using computational design software (OpenRoads). Drivers' speed behavior is investigated on the curves using adhesive tape. Stretch of 500 m on the curve is taken in which 400 m is the transition before and after the curve and 100 m is the curve length and the same was done for the other cures. It reviled that the operating speed is constant along the curves with rapid acceleration and deceleration on the tangents/transitions. Various other parameters i.e. traffic volume, horizontal radius, the cross slope is also calculated manually. This paper shows various alignment models of the road, helping in determining a constant design speed by modifying the alignment to straighter sections and minimizing the speed variations along the stretch. The distance also got reduced after the elimination of curves from the actual alignment. This paper also gives the cost estimation of different alignments that are designed and is compared with the actual structure.

Keywords Highway alignment · Drivers' speed behavior · Operating speed

1 Introduction

The position of the centerline of the highway in-ground is called highway alignment. It consists of horizontal and vertical alignment. Earlier times the design used to take a lot of time as there was not much software that supported the designing of the infrastructures. The software was buggy and very complex to use. The design data used to be on sheets and those designs were created by hand and hence the designs prepared can't be analyzed. Presently many design and analytical software

S. Sharma (✉) · A. Boora
Department of Civil Engineering, Jaypee University of Information Technology, Himachal Pradesh, Solan 173234, India

G. C. Marano et al. (eds.), *Proceedings of SECON'21*, Lecture Notes in Civil Engineering 171, https://doi.org/10.1007/978-3-030-80312-4_45

are present that covered the gap of doing manual work. The data that is required to design a highway like a survey file or the terrain file, is created easily which gives a replica of the site, to the designer making it more simple and easy to design the highway while keeping all the negative and positive obligatory points in mind.

This software also allows the user to analyze the design which helps to achieve three goals: increase the efficiency of the construction, improve the infrastructure's safety, and extend their service life [1]. Softwares automatically detect any flaws in the design and warn about any feature that exceeds the permissible limits given in the respective codes. The most dominating crash factor is the sight distance which should be provided adequately especially on curves. Introducing spiral curves with circular curves provides the available sight distance (ASD) greater than stopping sight distance (SSD) [2]. Poor sight distance also affects the fuel efficiency due to the rapid acceleration and deceleration. Kang et al. [3] created a modal that automatically calculates ASD and SSD and also evaluates site distance deficiencies of the proposed alternative alignments and the most suitable alignment is chosen. Ashraf et al. [4] conducted a study to improve highway alignment. Cures were difficult to reduce or eliminate hence road widening up to 1.43 m and increasing the Superelevations to 7% was suggested at the curves to increase the ASD, hence making the alignment more comfortable to drive.

1.1 Project Location

The stretch selected is 4.4 km is the bypass to the Ner-chowk city constructed on plain and rolling terrain [5]. It is an important road as it is a four-lane divided road which reduces the traveling distance by 2.5 km (approx.). The maximum allowable speed that can be achieved on this bypass also increased by 60 km/h reducing the travel time.

1.2 Software Description

OpenRoads Designer is a comprehensive and fully functioned detailed design application for surveying, drainage, subsurface utilities, and roadway design that supersedes all capabilities previously delivered through InRoads, GEOPAK, MX Road, and PowerCivil. It supports all aspects of a detailed roadway design including survey, geotechnical, drainage, subsurface utilities, terrain, road, roadway furniture, and more. The incorporation of reality meshes, imagery, point cloud, and other geocoordinated data sources can provide the continuous context of real-world conditions throughout the design and construction lifecycle.

2 Objectives of the Study

- Road inventory to know the existing condition of the road
- Designing alternative horizontal alignment
- Introducing super elevations on curves
- Cut and Fill Calculations based on earthwork
- Traffic volume survey.

3 Field Investigations

3.1 Road Inventory

The length of the road is measured along with the land used. The terrain was determined and elevation points were noted along with the chainage length. The width of the carriageway and shoulder was measured and the present pavement condition is checked. The measurements are given in Table 1.

3.2 Traffic Volume Survey

It is the number of vehicles crossing a section of road per unit time at any particular period (usually peak hours are considered) (Table 2).

Total number of vehicles in both directions at peak hour = 369 + 425 = 794

50% of above value = 397

Average daily traffic = 3970 vehicles/day.

Table 1 Road inventory

Geometric features	Existing road
Length	4587 m
Width	22.0 m
Number of lanes	4
Traffic moment	2-way
Divided/undivided	Divided (4.5 m divider width)
Shoulder type	Earthen shoulder 2 m wide
Pavement drainage condition	Moderate
Road side drainage	Poor
Land use	Commercial and residential

Table 2 Traffic volume stats

Time	Car	Mini bus	Bus	2-axle truck	JCB	Tractor	Total
Morning peak hour							
8:00–10:00	280	15	13	40	4	17	369
Evening peak hour							
5:00–7:00	320	22	15	55	2	11	425

4 Design Procedure

3 alignments are designed in total. Existing alignment (E) and modified alignments A1 and A2.

4.1 Collecting the Topographical Data

1. Plotting polygon points on the map to extract map data and creating the paths in the polygon as in Fig. 2. The red path is the bypass section and green path is the road passing from the Ner-chowk city (for which the bypass is sanctioned). The polygon as kml file format (Fig. 1).
2. Importing the corridor discription according to the actual road shown in Fig. 3. Creating terrain file from global mapper as in Fig. 3.

Fig. 1 Capturing the site location from google earth

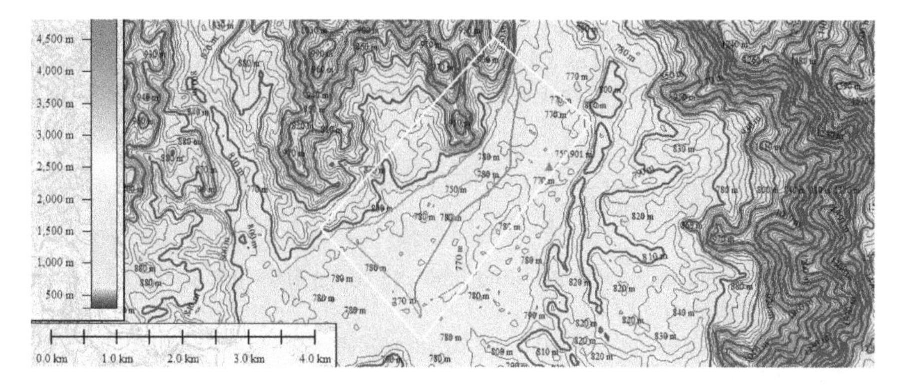

Fig. 2 Creating terrain modal in global mapper

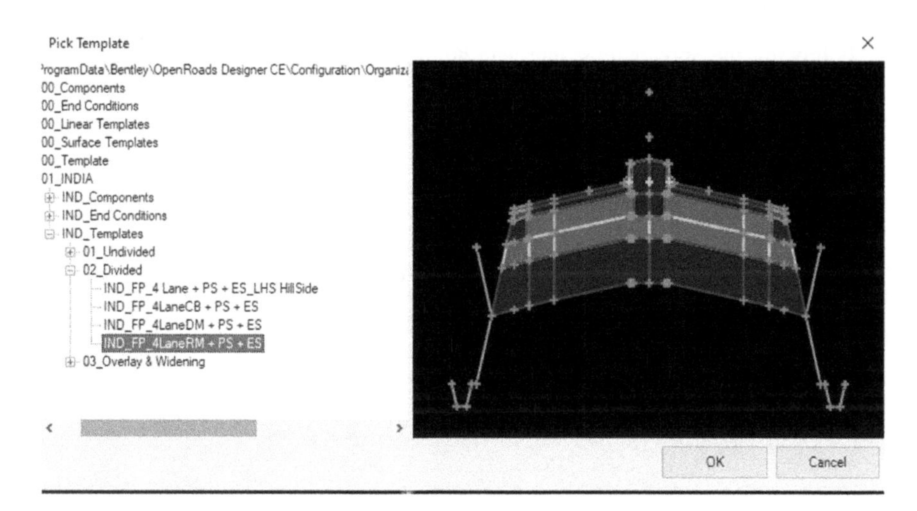

Fig. 3 Corridor description

4.2 Geometric Design Using OpenRoads Software

The existing alignment is designed following the IRC parameters.

4.2.1 Horizontal Alignment

1. The terrain file is attached matching the coordinates of the existing alignment.
2. An identical existing horizontal alignment is prepared using the line and arc tool defining the feature as the centerline.
3. The alignment is 4.2 km and the design speed is locked at 100 km/h and Superelevation is provided to the entire section according to IRC SP 99 [6] and IRC 38 [7].

4. Vertical alignment is prepared before creating the corridor.
5. The corridor is designed as a 4 lane divided road with earthen shoulders same as the actual existing alignment as shown in Fig. 4 according to IRC SP 84 [8].

4.2.2 Vertical Alignment

1. This alignment should be prepared after the center lines are drawn of the horizontal alignment.
2. Profile model tool is selected from the geometry ribbon and the terrain is displayed as the vertical section.
3. The alignment is prepared by the line and curve tools incorporating suitable sag and k values according to IRC SP 23 [9].

Following the above steps, alternate alignments are also designed for the same 100 km/h design speed parameters.

4.3 Output from Software

4.3.1 Horizontal Alignment According to IRC 38 [7]

$$\text{Radius of curve} = V^2/127(e+f) \tag{1}$$

Minimum Radius = 150 m

Maximum Radius = 1000 m (Figs. 5 and 6).

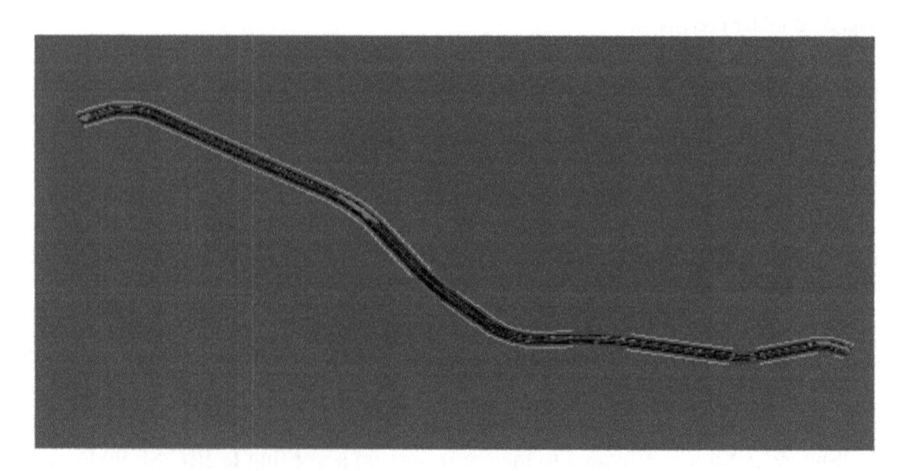

Fig. 4 Horizontal alignment of the existing road E

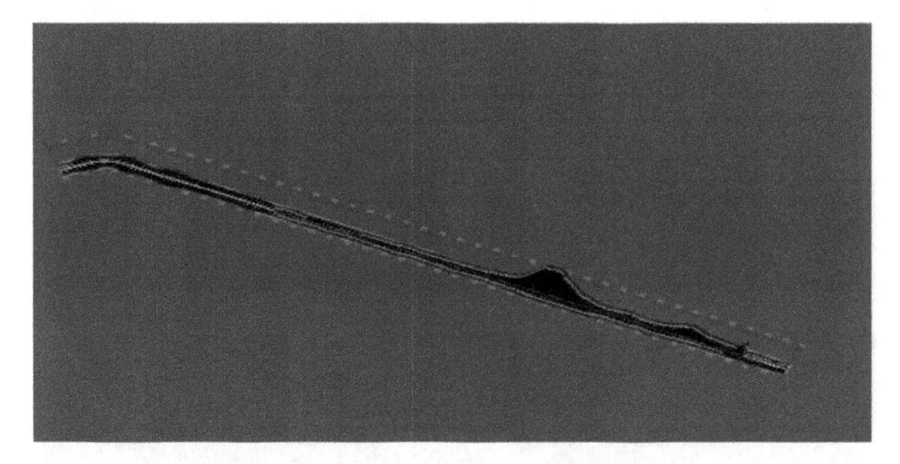

Fig. 5 Modified horizontal alignment A1 (having straight sections)

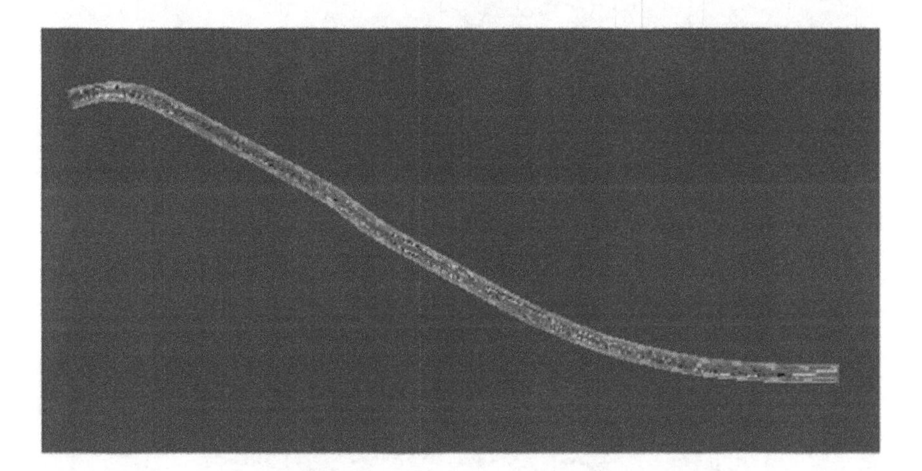

Fig. 6 Modified horizontal alignment A2 (increased curved radius)

4.3.2 Vertical Alignment According to IRC SP 23 [9]

The green dotted line is the actual ground profile and the red line is the vertical profile according to which the highway is constructed. Figures 7, 8 and 9 shows the different vertical profile of all the three horizontal alignments respectively.

4.3.3 Superelevations According to IRC 38 [7]

Maximum E = 0.07 provided $f < 0.15$ (here f = 0.147) [7]

Radius beyond which no super elevation is required = 800 m (Figs. 10, 11 and 12).

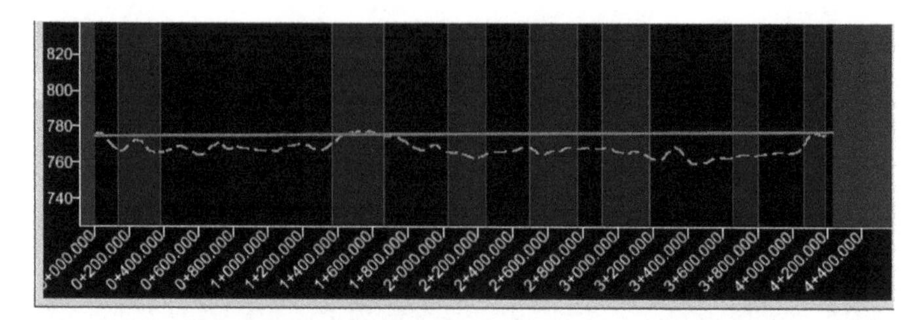

Fig. 7 Vertical alignment of existing road E

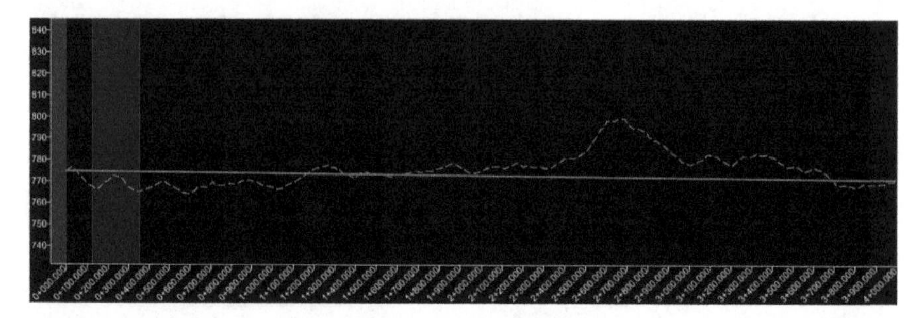

Fig. 8 Vertical alignment of the modified road A1 (having straight sections)

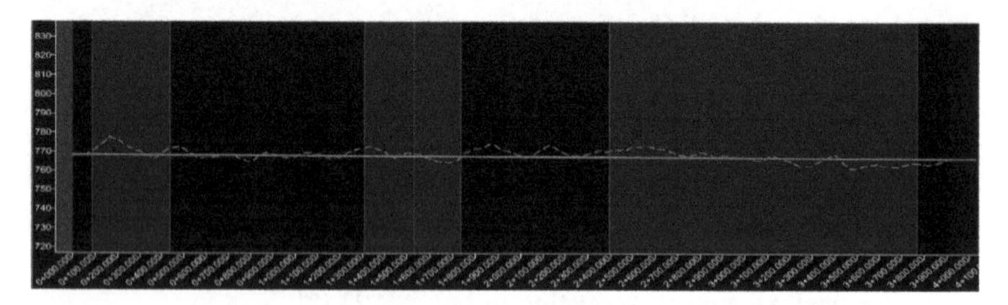

Fig. 9 Vertical alignment of the modified road A2 (with increased curved radius)

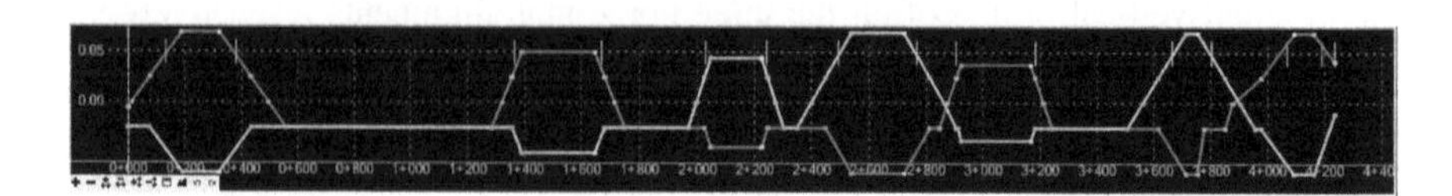

Fig. 10 Superelevation sections of the existing alignment E

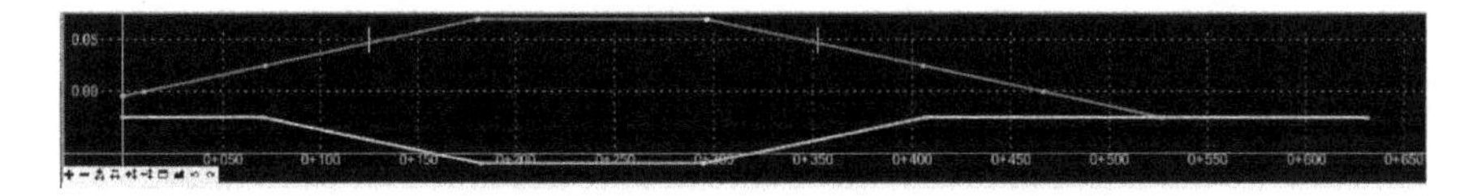

Fig. 11 Superelevation section of the modified alignment A1 (having straight sections)

Fig. 12 Superelevation sections of the modified alignment A2 (with increased curved radius)

4.3.4 Adding the Unit Price of the material used during the construction, in the Software

See Figs. 13, 14 and 15.

5 Result and Discussion

The most suitable 3 alignment designs are taken. These alignments are E1 (the existing alignment), A1 (the modified alignment with straighter sections), and A2 (the modified alignment with increased curve radius.)

Material	Surface Area	Volume	Units	Unit Cost	Total Cost/Material
Cut Volume	0.0000	221517.1481	CuM	115.00	25474472.03
Fill Volume	0.0000	781113.1183	CuM	190.00	148411492.48
Mesh\1_Asphalt/Flexible (IRC:37-2018)\Bituminuous Concrete (BC)	0.0000	4192.0876	CuM	920.00	3856720.59
Mesh\1_Asphalt/Flexible (IRC:37-2018)\Dense Bituminous Macadam (DBM)	0.0000	6707.3402	CuM	760.00	5097578.55
Mesh\2_Aggregate & Soil (IRC:37-2018)\Granular Sub Base (GSB)	0.0000	30413.5959	CuM	900.00	27372236.31
Mesh\2_Aggregate & Soil (IRC:37-2018)\Subgrade	0.0000	63971.2575	CuM	215.00	13753820.36
Mesh\2_Aggregate & Soil (IRC:37-2018)\Wet Mix Macadam (WMM)	0.0000	22436.5142	CuM	975.00	21875601.35
Mesh\3_Concrete/Rigid (IRC :58-2015)\TC_Conc Misc	0.0000	34.4185	CuM	400.00	13767.40
Mesh\3_Concrete/Rigid (IRC :58-2015)\TC_Curb	0.0000	755.7788	CuM	370.00	279638.16
Mesh\Asphalt\TC_Asph Conc Wearing Cse	1089.9428	0.0000	SqM	32.00	34878.17
Mesh\Grading\TC_Cutslope	33583.6812	0.0000	SqM	195.00	6548817.83
Mesh\Grading\TC_Ditch	16565.8255	0.0000	SqM	40.00	662633.02
Mesh\Grading\TC_Earth Shoulder	0.0000	1776.6067	CuM	215.00	381970.44
Mesh\Grading\TC_Fillslope	72128.0341	0.0000	SqM	90.00	6491523.07
Mesh\Grading\TC_Topsoil	0.0000	12351.3406	CuM	190.00	2346754.71

ort Total Estimated Cost: 262601904.47

Fig. 13 Total quantities of the existing alignment E with detailed unit cost

Material	Surface Area	Volume	Units	Unit Cost	Total Cost/Material
Cut Volume	0.0000	38729.0401	CuM	115.00	4453839.61
Fill Volume	0.0000	967706.8736	CuM	190.00	183864305.98
Mesh\1_Asphalt/Flexible (IRC:37-2018)\Bituminuous Concrete (BC)	0.0000	4020.8042	CuM	920.00	3699139.85
Mesh\1_Asphalt/Flexible (IRC:37-2018)\Dense Bituminous Macadam (DBM)	0.0000	6433.2867	CuM	760.00	4889297.90
Mesh\2_Aggregate & Soil (IRC:37-2018)\Granular Sub Base (GSB)	0.0000	29170.9344	CuM	900.00	26253840.96
Mesh\2_Aggregate & Soil (IRC:37-2018)\Subgrade	0.0000	61357.4719	CuM	215.00	13191856.47
Mesh\2_Aggregate & Soil (IRC:37-2018)\Wet Mix Macadam (WMM)	0.0000	21519.7863	CuM	975.00	20981791.66
Mesh\3_Concrete/Rigid (IRC :58-2015)\TC_Conc Misc	0.0000	33.0122	CuM	400.00	13204.87
Mesh\3_Concrete/Rigid (IRC :58-2015)\TC_Curb	0.0000	662.2812	CuM	370.00	245044.06
Mesh\Asphalt\TC_Asph Conc Wearing Cse	1045.4091	0.0000	SqM	32.00	33453.09
Mesh\Grading\TC_Cutslope	7673.5277	0.0000	SqM	195.00	1496337.90
Mesh\Grading\TC_Ditch	20236.9824	0.0000	SqM	40.00	809479.30
Mesh\Grading\TC_Earth Shoulder	0.0000	1704.0168	CuM	215.00	366363.62
Mesh\Grading\TC_Fillslope	85831.2215	0.0000	SqM	90.00	7724809.93
Mesh\Grading\TC_Topsoil	0.0000	11846.6802	CuM	190.00	2250869.25

ort Total Estimated Cost: 270273634.45

Fig. 14 Total quantities and cost of modified alignment A1 with detailed unit cost

Material	Surface Area	Volume	Units	Unit Cost	Total Cost/Material
Cut Volume	0.0000	425596.8855	CuM	115.00	48943641.83
Fill Volume	0.0000	54665.5076	CuM	190.00	10386446.44
Mesh\1_Asphalt/Flexible (IRC:37-2018)\Bituminuous Concrete (BC)	0.0000	4034.4217	CuM	920.00	3711667.95
Mesh\1_Asphalt/Flexible (IRC:37-2018)\Dense Bituminous Macadam (DBM)	0.0000	6455.0747	CuM	760.00	4905856.77
Mesh\2_Aggregate & Soil (IRC:37-2018)\Granular Sub Base (GSB)	0.0000	29269.7293	CuM	900.00	26342756.41
Mesh\2_Aggregate & Soil (IRC:37-2018)\Subgrade	0.0000	61565.2750	CuM	215.00	13236534.11
Mesh\2_Aggregate & Soil (IRC:37-2018)\Wet Mix Macadam (WMM)	0.0000	21592.6687	CuM	975.00	21052851.94
Mesh\3_Concrete/Rigid (IRC :58-2015)\TC_Conc Misc	0.0000	33.1240	CuM	400.00	13249.59
Mesh\3_Concrete/Rigid (IRC :58-2015)\TC_Curb	0.0000	776.3598	CuM	370.00	287253.13
Mesh\Asphalt\TC_Asph Conc Wearing Cse	1048.9496	0.0000	SqM	32.00	33566.39
Mesh\Grading\TC_Cutslope	48622.0806	0.0000	SqM	195.00	9481305.72
Mesh\Grading\TC_Ditch	4132.9059	0.0000	SqM	40.00	165316.23
Mesh\Grading\TC_Earth Shoulder	0.0000	1709.7879	CuM	215.00	367604.40
Mesh\Grading\TC_Fillslope	5778.6784	0.0000	SqM	90.00	520081.06
Mesh\Grading\TC_Topsoil	0.0000	11886.8021	CuM	190.00	2258492.40

ort Total Estimated Cost: 141706624.37

Fig. 15 Total quantities and cost of the modified alignment A2 with detailed unit cost

Table 3 Various outputs from the software

Alignment	Total length KM	Safe design speed (km/h)	Total cut volume m^3	Total fill volume m^3	The total estimated cost of alignment (CRs.)
E1	4.23	80–100	221,517.14	181,113.17	26.26
A1	3.97	100	38,729.04	967,706.87	27.02
A2	4.02	100	425,596.88	54,665.50	14.17

1. Table 3 clearly illustrates that the total length of M1 and M2 decreased by 240 m and 200 m respectively.
2. The fluctuating design speed of the existing alignment i.e. 80–100 km/h got stabled to 100 km/h with the elimination of the curves hence reducing the travel time.
3. The total cut and fill volume also got reduced. In existing alignment, the fill volume is greater than the cut volume that indicates the fill material had to be transported which results in the increased overall cost of the structure. In both modified alignments the cut volume is greater than the fill volume hence no filling material is to be transported and the excess cut volume could be used as the pavement aggregate that would also result in cost minimization of the project.
4. The estimated cost of the modified alignment i.e. M1 is greater because the cutting area is from mountainous terrain and to stabilize or prevent landslides, retaining walls are provided that increased the overall cost on the other side the alignment M2 has less estimated cost as no superstructure is to be provided.
5. The actual cost of the road project is 26.64 Cr. and as per software calculations it cost comes out to be 26.27 Cr of alignment E which is close enough to the actual cost, hence the software precision is good.
6. The sight distance errors of E are eliminated by introducing straighter sections or by increasing curve radius as done in M1 and M2 alignments respectively.

From these results, it is depicted that there are errors in the existing alignment which can be minimized just by increasing the curve radius. If the curve radius could not be increased, then there should be road widening provided on the curves to make those curves safe. The road widening could also result in the increased overall cost of the alignment. The modified alignment M2 which has increased curve radius results in decreasing the cost of the existing alignment, and has a good sight distance, hence the modified alignment M2 could be the most favorable for the area. It is also concluded that switching over to Openroads software, opens up more features in the designing of roads with good precision. Moreover, the analysis time of the software is also less and many features can be analyzed with varying values within the permissible limits of respective country codes.

References

1. Maier F (2017) Utilizing 3D digital design data in highway construction—case studies. no.
2. You QC, Easa SM (2016) Innovative roadside design curve of lateral clearance: roadway simple horizontal curves. J Transp Eng 142(11):1–11. https://doi.org/10.1061/(ASCE)TE.1943-5436.0000889
3. Kang MW, Shariat S, Jha MK (2013) New highway geometric design methods for minimizing vehicular fuel consumption and improving safety. Transp Res Part C Emerg Technol 31:99–111. https://doi.org/10.1016/j.trc.2013.03.002
4. Ashraf A, Singh N, Shrivastava Y, Vishwas JS (2018) Design of sub-arterial urban road using mxroad software. Int Res J Eng Technol 5(5):2125–2134

5. IRC 52 (2001) Recommendations about the alignment survey and geometric design of hill roads. Bureau of Indian Standards, New Delhi
6. IRC SP 99 (2013) Manual of specification and standards for expressways. Bureau of Indian Standards, New Delhi
7. IRC:38 (1988) Guidelines for design of horizontal curves for highways and design tables. Bureau of Indian Standards, New Delhi
8. IRC SP 84 (2014) Manual of specifications and standards for four laning of highways. Bureau of Indian Standards, New Delhi
9. IRC SP 23 (1993) Vertical curves for highway. Bureau of Indian Standards, New Delhi

Drying Shrinkage of Alkali-Activated Paste, Mortar and Concrete—A Comparison Study for Varying Activator Modulus

Kruthi Kiran Ramagiri, Sai Yoshitha Kudumala, Ravali Chintha, and Arkamitra Kar

Abstract Alkali-activated binder (AAB) is gaining popularity as a viable substitute for Portland cement (PC). A reaction between aluminosilicate-rich precursors, generally industrial residue including fly ash and/or slag, and alkaline activators like sodium/potassium silicate and/or hydroxide produces AAB. Previous research on AAB shows their superior mechanical performance compared to PC. However, AAB with a greater proportion of slag tends to exhibit higher shrinkage than PC. Owing to the complex chemistry of AAB, its final performance as a binder depends on several factors like the type and combination of precursors used, chemistry of activators, and curing conditions. The present study, therefore, is an experimental investigation to evaluate the drying shrinkage of AAB paste, mortar and concrete with varying precursor and activator combinations. In this study, the drying shrinkage strains of AAB prepared using two different slag: fly ash ratios of 30:70 and 40:60 are determined for paste, mortar, and concrete specimens. The silicate modulus (silica to sodium oxide ratio or Ms) of the activating solution is varied as 1.0 and 1.4, and the corresponding effect on shrinkage is evaluated. The shrinkage strains were calculated from length changes measured using a length comparator and height gauge. The results from the present study show that optimum proportioning precursors reduce the drying shrinkage of AAB. Explanations were presented for the possible mechanisms which improve the dimensional stability of AAB.

Keywords Alkali-activated binder · Drying shrinkage · Ms · Ambient curing · Durability

K. K. Ramagiri (✉) · S. Y. Kudumala · R. Chintha · A. Kar
Department of Civil Engineering, BITS-Pilani Hyderabad Campus, Hyderabad, India
e-mail: p20170008@hyderabad.bits-pilani.ac.in

S. Y. Kudumala
e-mail: h20191430105@hyderabad.bits-pilani.ac.in

R. Chintha
e-mail: h20191430099@hyderabad.bits-pilani.ac.in

A. Kar
e-mail: arkamitra.kar@hyderabad.bits-pilani.ac.in

G. C. Marano et al. (eds.), *Proceedings of SECON'21*, Lecture Notes in Civil Engineering 171, https://doi.org/10.1007/978-3-030-80312-4_46

1 Introduction

Alkali-activated binders are emerging as a possible substitute for Portland cement (PC) owing to their low carbon footprint and associated energy consumption. AAB exhibits enhanced mechanical performance and durability in comparison to PC [1]. The microstructure, mechanical performance, and durability of AAB depend predominantly on the composition of the precursors used. To promote the practical use of AAB on a large-scale, research on its long-term durability is necessary. One of the durability parameters of paramount importance is the shrinkage of AAB.

Previous research on drying shrinkage of AAB reported that using slag as a precursor aggravates shrinkage [2–6]. One possible way to increase the drying shrinkage resistance of AAB concrete is to use a combination of slag and fly ash as the precursor. Previous research investigated the effects of blending fly ash with varying proportions of slag (obtained from a blast furnace in granulated form, followed by grinding) on the strength development and drying shrinkage up to 180 days for concrete with AAB [7]. The results concluded that shrinkage is inversely proportional to slag content for AAB concrete cured under ambient conditions. However, only two slag:fly ash ratios of 10:90 and 20:80 are implemented, and the results are reported in accordance with the existing Australian standards only [7].

Shrinkage of concrete leads to cracking under restrained conditions, which affect the durability of concrete. AAB exhibit superior mechanical properties but poor shrinkage behaviour compared to PC [8]. Owing to its complex chemistry and tortuous pore network, the prediction of the shrinkage performance of AAB is difficult. Several factors, viz. oxide composition and specific surface of the precursors, chemistry and concentration of the activator, and curing conditions, influence the shrinkage behaviour of AAB [8].

Drying shrinkage in AAB is due to loss of unbound water upon exposure to an atmosphere with low relative humidity. Despite more significant moisture loss in PC, it has always been observed that AAB has higher drying shrinkage than PC [9–13]. It is postulated that a higher proportion of mesopores (1.25–25 nm) increases the capillary tension and silica gel containing high unbound water content, causing these observations. These findings conclude that their mass loss and shrinkage of AAB are not directly correlated. Decreasing the activator modulus (Ms) reduces the drying shrinkage but increases the autogenous shrinkage. Ms is calculated as the ratio of SiO_2/Na_2O in the alkaline activating solution. An increase in autogenous shrinkage is attributed to increased polymerisation and reaction product formed owing to higher pH. Reduction in the drying shrinkage of AAB with lower Ms is related to increased strength, providing enhanced ability to resist more stresses [14]. It is evident from the previous studies that both the precursor and the activator composition significantly affect the final shrinkage of AAB. However, limited research is available on the shrinkage behaviour of ambient-cured blended AAB. Therefore, the present study aims to address this issue by evaluating the shrinkage behaviour of AAB paste, mortar, and concrete with varying precursor proportion and activator modulus. The materials used and the experimental procedures followed are presented in the following section.

2 Experimental Methodology

2.1 *Raw Materials and Design Mixes*

The raw ingredients used as precursors used in AAB preparation are low calcium fly ash from NTPC Ramagundam and slag from JSW Cement Ltd. Natural river sand and 10 mm crushed granite conforming to IS 383 (2016) are used as fine and coarse aggregate, respectively. The activator is a liquid blend containing sodium silicate solution (29.5% SiO_2 and 14.7% Na_2O) and sodium hydroxide pellets (commercial grade lye). AAB with slag:fly ash varied as 30:70 (S3F7) and 40:60 (S4F6) is used in the present study. The water-to-solids ratio is maintained at 0.3 for both mixes. The relevant information is provided in Table 1.

2.2 *Methods*

The drying shrinkage of AAB concrete is evaluated as per the specification of ASTM C157/157 M [15]. Shrinkage strains of AAB paste and mortar are evaluated using prismatic specimens of dimensions 285 × 25 × 25 mm, whereas prismatic specimens of 285 × 75 × 75 mm are cast for AAB concrete. The specimens are allowed to cure in the moulds until they harden. The length measurement

Table 1 Proportions of ingredients to prepare AAB mixes (kg/m^3)

Mix		Slag	FA	Ms = 1.0		Ms = 1.4		C.A	F.A	Water
				NaOH	Na_2SiO_3	NaOH	Na_2SiO_3			
S3F7	Paste	382	891	79	366	–	–	–	–	272
S4F6		513	891	80	369	–	–	–	–	274
S3F7		392	770	–	–	35	423	–	–	253
S4F6		527	915	–	–	35	426	–	–	255
S3F7	Mortar	165	790	34	158	–	–	–	1509	117
S4F6		220	384	34	158	–	–	–	1513	117
S3F7		166	330	–	–	15	179	–	1525	107
S4F6		223	388	–	–	15	180	–	1530	108
S3F7	Concrete	120	334	25	115	–	–	1209	651	85
S4F6		160	280	25	115	–	–	1209	651	85
S3F7		120	240	–	–	11	129	1209	651	77
S4F6		160	280	–	–	11	129	1209	651	77

commenced immediately after the specimens are demoulded. Following the initial length measurement, subsequent readings are taken at a regular interval of 24 h for 90 days. The length change is reported with respect to the original length of the specimen. Constant ambient conditions (23 °C and 50% RH) are maintained for the entire period of specimen storage till testing.

The readings are taken using both length comparator conforming to the specifications of ASTM C490 and height gauge [16]. The height gauge is to determine the influence of the testing procedure on the shrinkage strain measurement of concrete. The shrinkage strains are calculated by dividing the change in length by the initial length [17].

The length change measurements for concrete samples are obtained only through a height gauge owing to a constant arrangement of the length comparator throughout the testing period. However, length changes in mortar specimens are recorded using both length comparator and height gauge.

3 Results

The shrinkage values at 90 days for paste, mortar and concrete specimens of S3F7 and S4F6 mixes are presented in Table 2 for two different Ms. The drying shrinkage variations for these specimens with age are presented in Figs. 1, 2, and 3, respectively.

The common observation for paste, mortar and concrete specimens is the increase in drying shrinkage with slag content. The addition of slag entails additional Ca^{2+} ions in the system, leading to the production of additional compounds and an increased degree of polymerisation [6, 18]. The increase in shrinkage can be correlated to the production of calcium silicate hydrate (C–A–S–H) matrix due to the reaction between slag and the activator. This co-exists with the polymeric sodium aluminosilicate hydrate (N–A–S–H) network produced by the reaction between fly ash with the activator [18]. It is observed that varying the Ms affects the drying shrinkage of paste specimens primarily.

In AAB paste specimens, a higher Ms increases the concentration of $[SiO_4]^{-4}$ in the systems, leading to an enhanced rate of reaction [19]. The slight increase of shrinkage strains in S3F7 mortar and concrete is related to the pore size distribution

Table 2 Drying shrinkage strains (µε) at 90 days

Mix	Ms = 1.0			Ms = 1.4		
	Paste	Mortar	Concrete	Paste	Mortar	Concrete
S3F7	14,149	2349	734	16,667 (17.8%)	2411 (2.7%)	787 (7.1%)
S4F6	17,893	2784	874	21,000 (17.4%)	2576 (−7.5%)	874 (0%)

(%) indicate the percentage variation with respect to specimens with Ms of 1.0

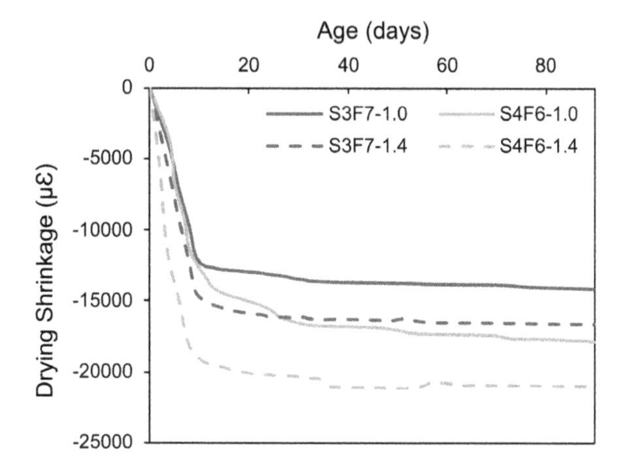

Fig. 1 AAB paste shrinkage variation with age

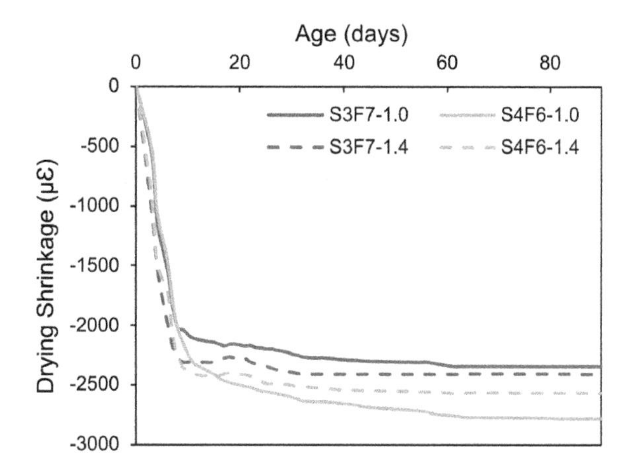

Fig. 2 AAB mortar shrinkage variation with age

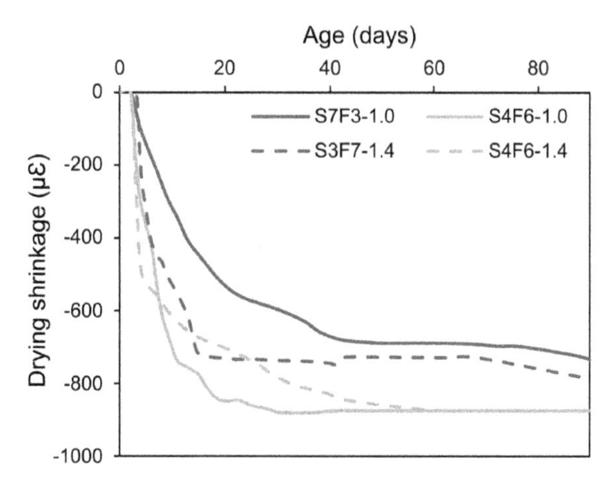

Fig. 3 AAB concrete shrinkage variation with age

[20]. Increasing Ms decreases porosity resulting in fine pores and higher capillary stresses hence augmenting shrinkage strains [21]. As the slag proportion increases, the reaction rate also increases, and therefore, the difference in the drying shrinkage strains observed in S4F6 specimens is negligible.

It is well-established from the previous research that concrete exhibits lower shrinkage than mortar, and mortar exhibits lower shrinkage than paste. This behaviour is attributed to the inclusion of aggregates that restrain the shrinkage of the matrix, increasing the dimensional stability of concrete [22].

4 Summary

The conclusions from this study are summarised as follows:

- Approximately one-fourth increase in shrinkage strains is observed in AAB paste with an additional one-tenth of slag, whereas in AAB mortar, this variation is observed to be one-fifth and one-fourteenth for specimens with Ms 1.0 and 1.4, respectively.
- Shrinkage strains are observed to increase by one-fifth and one-tenth with 10% additional slag content in AAB concrete specimens with Ms of 1.0 and 1.4, respectively. S3F7 with Ms of 1.0 results in lower shrinkage.
- No significant changes are observed with an increase in Ms from 1.0 to 1.4. Slightly reduced shrinkage at low Ms is attributed to higher strength and consequently enhanced ability of AAB to resists stresses.

References

1. Turner LK, Collins FG (2013) Carbon dioxide equivalent (CO_2-e) emissions: a comparison between geopolymer and OPC cement concrete. Constr Build Mater 43:125–130
2. Bakharev T, Sanjayan JG, Cheng YB (2000) Effect of admixtures on properties of alkali-activated slag concrete. Cem Concr Res 30(9):1367–1374
3. Palacios M, Puertas F (2007) Effect of shrinkage-reducing admixtures on the properties of alkali-activated slag mortars and pastes. Cem Concr Res 37(5):691–702
4. Atiş CD, Bilim C, Çelik Ö, Karahan O (2009) Influence of activator on the strength and drying shrinkage of alkali-activated slag mortar. Constr Build Mater 23(1):548–555
5. Collins F, Sanjayan JG (2000) Cracking tendency of alkali-activated slag concrete subjected to restrained shrinkage. Cem Concr Res 30(5):791–798
6. Collins F, Sanjayan JG (2000) Effect of pore size distribution on drying shrinking of alkali-activated slag concrete. Cem Concr Res 30(9):1401–1406
7. Deb PS, Nath P, Sarker PK (2014) The effects of ground granulated blast-furnace slag blending with fly ash and activator content on the workability and strength properties of geopolymer concrete cured at ambient temperature. Mater Des 62:32–39
8. Kumarappa DB, Peethamparan S, Ngami M (2018) Autogenous shrinkage of alkali activated slag mortars: Basic mechanisms and mitigation methods. Cem Concr Res 109:1–9

9. Collins FG, Sanjayan JG (1999) Workability and mechanical properties of alkali activated slag concrete. Cem Concr Res 29(3):455–458
10. Krizan D, Zivanovic B (2002) Effects of dosage and modulus of water glass on early hydration of alkali–slag cements. Cem Concr Res 32(8):1181–1188
11. Lee NP (2007) Creep and shrinkage of inorganic polymer concrete. BRANZ study report SR 175, BRANZ Ltd, Judgeford, New Zealand
12. Chi MC, Chang JJ, Huang R (2012) Strength and drying shrinkage of alkali-activated slag paste and mortar. Adv Civil Eng 579732
13. Ye H, Radlińska A (2016) Shrinkage mechanisms of alkali-activated slag. Cem Concr Res 88:126–135
14. Thomas JJ, Allen AJ, Jennings HM (2012) Density and water content of nanoscale solid C–S–H formed in alkali-activated slag (AAS) paste and implications for chemical shrinkage. Cem Concr Res 42(2):377–383
15. ASTM: ASTM C157/C157M-11 (2011) Standard test method for length change of hardened hydraulic-cement mortar and concrete. ASTM International, West Conshohocken, PA
16. ASTM: ASTM C490/C490M-11 (2011) Standard practice for use of apparatus for the determination of length change of hardened cement paste, mortar, and concrete. ASTM International, West Conshohocken, PA
17. Alex A, Ramagiri KK, Kar A (2019) Correlation between microstructural characteristics and specimen-level shrinkage of alkali-activated binder concrete. In: UKIERI Concrete Congress 2019, India
18. Ramagiri KK, Kar A (2019) Effect of precursor combination and elevated temperatures on the microstructure of alkali-activated binder. Indian Concr J 93(10):34–43
19. Chithiraputhiran SR (2012) Kinetics of alkaline activation of slag and fly ash-slag systems. Arizona State University. Ms thesis
20. Lee NK, Jang JG, Lee HK (2014) Shrinkage characteristics of alkali-activated fly ash/slag paste and mortar at early ages. Cem Concr Res 53:239–248
21. Awoyera P, Adesina A (2019) A critical review on application of alkali activated slag as a sustainable composite binder. Case Stud Constr Mater 11, e00268
22. Idiart A, Bisschop J, Caballero A, Lura P (2012) A numerical and experimental study of aggregate-induced shrinkage cracking in cementitious composites. Cem Concr Res 42(2): 272–281

Strengthening of Deep Beam Using External Bonding of Carbon Fiber Reinforced Polymer

S. Hemanth, Neelam Rani, and M. Abdul Akbar

Abstract In deep beams, the primary cause of failure is shear and the corresponding strain distribution is non-linear. The load is transferred through arch action in deep beams and as a result of the transfer of load, widening of the strut happens and due to the strain developed in the strut, cracks are formed. This paper reports the results of finite element analysis carried out on deep beams with external bonding of carbon fibre reinforced polymer (CFRP) in ABAQUS and the optimum configuration that leads to increase in the shear capacity of deep beams using CFRP. After validating the model with experimental results, simulation was done by modification and grouping of various parameters such as thickness, number of plies and increasing the modulus of elasticity of CFRP. CFRP thickness of 0.222, 0.333, 0.444 mm per ply was used. Different number of plies used are 3, 5, 7 and modulus of elasticity was increased in proportions of 10, 20 and 30%. The results of the study indicate that, compared to increase in thickness of CFRP plies and increase in the modulus of elasticity of CFRP, most effective method of CFRP strengthening is obtained when the number of plies of CFRP is increased.

Keywords Deep beams · Carbon fiber reinforce polymer · Shear strengthening · External bonding

1 Introduction

Retrofitting and strengthening of concrete structures has gained importance due to the economical concern as well as the environmental conservation, i.e. by retrofitting of old structures, life span and strength of structures can be increased by

S. Hemanth (✉) · N. Rani · M. A. Akbar
Department of Civil Engineering, Dr. B R Ambedkar National Institute of Technology, Jalandhar, Punjab, India
e-mail: ranin@nitj.ac.in

M. A. Akbar
e-mail: akbarma@nitj.ac.in

G. C. Marano et al. (eds.), *Proceedings of SECON'21*, Lecture Notes in Civil Engineering 171, https://doi.org/10.1007/978-3-030-80312-4_47

avoiding construction of new buildings and thereby decrease the carbon foot print and depletion of natural resources. CFRP can be used in strengthening of new structures and repair of old structures, the application of CFRP in retrofitting provides promising results [1]. Shear strengthening of structures using external bonding of CFRP is receiving increased popularity due to advantages like easiness in handling on account of its light weight, corrosion resistance etc. [2]. According to IS-456(2000), a beam is said to be a deep beam if the effective span is less than twice the depth for simply supported beam and effective span less than 2.5 times the depth for continuous beam [3]. In reinforced concrete (RC) beams, the failure is caused mainly due to flexure and shear [4]. However, in deep beams the failure is caused due to shear. Deep beam consists of two regions namely, B-region (Bernoulli) and D-region (Discontinuity). In order to analyse complex state of stress in the D-region, strut and tie model can be used, so that the complex state of stress will be reduced into simple uniaxial stress paths just like that of a truss [5]. When the deep beam is loaded within a distance of twice shear span to depth ratio, proper strut formation will take place. Due to loading, compressive force will be created in the strut, which will result in the formation of transverse strain and it will further result in split cracking [6]. The external bonding of CFRP helps to reduce the growth of the diagonal cracks and increase the load carrying capacity of the deep beam [7].

As a structural strengthening material, CFRP has been in use as a method of flexural strengthening since decades [8]. Now it is being used in shear strengthening also. The high strengthening capacity of external bonded CFRP laminates attracted the attention of researchers all over the world, resulting in increased research works since 1990s. The durability of concrete beams that are externally bonded with FRP was studied and it was found that the failure was mainly due to the debonding of FRP from the concrete surface [8]. This experiment shows the significance of proper selection of the resin used in bonding the FRP to the concrete surface [8]. Shear strengthening of RC beams using epoxy bonded FRP composites were studied both analytically and experimentally [9]. The study concluded that the effectiveness in FRP increases with the increase in the axial rigidity of it up to a certain limit after which the increment is not linear [9]. The improvement in shear capacity of existing T-beams using CFRP was studied and determined the importance of proper anchorage of CFRP to the concrete surface. The experiment also indicates that u-wrap CFRP will provide better anchorage than the side face CFRP [10].

Shear strengthening of deep beams using FRP was studied and it was observed that shear strength increased up to 40% [7]. The paper also discusses the possibility of using external bonding of CFRP where addition of further shear reinforcement is not possible and concluded that the method is a possible way of increasing the shear strength [7]. A study on prediction of behaviour of deep beams with openings was carried out using finite element analysis [11]. This work concludes that openings should not be provided in areas of compression strut and the decrement in the strength depends upon the size of the opening provided [11]. The deformation of strut in deep beams strengthened with CFRP was studied and it was observed that the application of CFRP reduces the strut widening and thereby increases the ultimate load carrying capacity of the beam [5]. A comparative study was carried out for deep beam using CFRP and glass fibre reinforced polymer (GFRP) and it was observed that although both showed an improvement in strength, the rate of increment was higher for CFRP [12].

In the last decades, a lot of analytical as well as experimental research works were carried out in this topic but the number of numerical analysis using finite element software's are less. This research focuses on simulation of deep beams with CFRP using finite element analysis software, ABAQUS 2017. As it was determined that CFRP is better than other fibres [12], the study emphasis to improve CFRP in shear strengthening.

2 Methodology

The geometric and material properties of the deep beam used for simulation were taken from an experimental work carried out to study the deformation of strut in deep beams strengthened using CFRP [5]. The model developed in ABAQUS was validated using strains obtained at the mid-shear span of the model with the corresponding strains obtained from experiments for loads within elastic limit and for shear-span to depth (a/d) ratios of 0.75, 1, 1.25, 1.5. The validation is shown in Table 1. The validation is limited to the linear range of loading and hence the study reported through this work is also limited to loads which does not go beyond the linear range.

When the deep beam is loaded, it will transfer the load to the support through the strut. As a result of the load transfer, compression will occur in the strut and it will result in the shortening along the length and widening of the strut in the transverse direction. The main reinforcement present in the deep beam will act as a tie, so that load is transferred through tied arch action. Due to the widening of the strut, diagonal cracks will be formed at mid span and it will propagate towards the support and loading edge. If CFRP is used in the shear spans of the deep beam, widening of the strut and the diagonal crack formation can be limited.

CFRP was applied on the side face of the deep beam along the shear span, in the form of sheet. The parameters that are varied in the study are number of plies,

Table 1 Model validation data

Shear span to depth ratio	Strain		% Error
	Own model	Experiment [5]	
0.75	0.0012418	0.00104554	18.7
1	0.0010047	0.000867	15.8
1.25	0.0010407	0.00110221	5.5
1.5	0.0016632	0.00196924	15.5

thickness of CFRP layers and modulus of elasticity along its different directions. The study was limited to a/d ratio 1. The orientation of the fibre was in the horizontal direction. Twenty-seven models were made with variations in the above parameters. Different CFRP thickness used are 0.222, 0.333, 0.444 mm/ply. The number of plies used are 3, 5 and 7 and modulus of elasticity of CFRP was also increased in proportion of 10, 20 and 30% along all directions from the values specified in the base model (model used for validation). The load applied on all models were 250 kN (within elastic range as observed from [5]). Finite element analysis was carried out on the models to observe the variations in stress, strain and deflection values to identify the parameter and the combination of parameters that increase the shear strength of the beam strengthened using CFRP.

2.1 *Properties of Materials*

The density, compressive strength (28 days) and tensile strength of concrete are 2420 kg/m^3, 37.02 MPa and 3.31 MPa respectively [5]. Properties of CFRP used for modelling is shown in Table 2, where E1 is the modulus of elasticity along the fiber direction and E2 is the modulus of elasticity along the transverse directions of the fiber. G12, G13 and G23 are modulus of rigidities of CFRP along different planes (following the standard notation of nomenclature). μ12 is the Poisson's ratio across the X–Y plane.

2.2 *Assembly*

The assembly of the model adopted [1] and the location of the strain measurement, stress measurement and displacement measurement is shown in Fig. 1. The deep beam is simply supported and it is loaded with 4-point loading. The components in the deep beam were assembled as shown in Fig. 1. The overall length, overhang (along both ends), width and depth of the beam are 1840 mm, 200 mm, 140 mm and 350 mm respectively (as adopted from [5]). Three layers of 16 mm diameter bars (with 3 numbers in each layer) is provided as main reinforcement and 6 mm diameter bars as horizontal and vertical stirrups along the two sides of the beam with a spacing of 100 mm. Anchor plates of 120 mm height and 10 mm thickness

Table 2 Properties of CFRP

Modulus of elasticity (MPa)		Poisson's ratio	Modulus of rigidity (MPa)		
E1 [5]	E2 [13]	μ12 [13]	G12	G13	G23
230,000	9500	0.17	98,290.6	98,290.6	3544.78

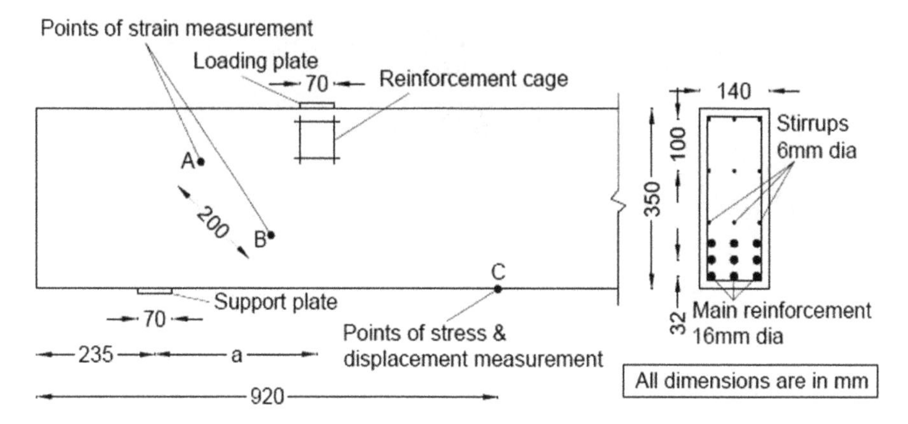

Fig. 1 Deep beam and its cross-section

were provided at the ends of longitudinal reinforcement for better anchorage. Additional 2 numbers of steel cages (consisting of 6 mm diameter bars) were provided under the loading plates to avoid local bearing stresses. The size of loading plate and support plate is 70 mm × 140 mm × 10 mm.

2.3 Modelling

For creating the model in ABAQUS, different parts of deep beam were formed by assigning their geometric details (Fig. 1) and mechanical properties. Parts were assembled together to form deep beam with CFRP (Fig. 2). Interaction between steel and concrete is defined as embedded region. CFRP and concrete interaction was modelled as a tie. Load of 250 kN is applied on the loading plate as a concentrated force. All the parts of deep beam were meshed together with a mesh size of 20 mm. Strains were measured at points (A and B) which are 200 mm apart in the mid shear span by creating path between nodes (as shown in Fig. 1). Stress (S33) and vertical displacement (U2) was measured at mid-point (C) along bottom face of deep beam (Fig. 1).

3 Results and Discussion

The models were analyzed in ABAQUS (Fig. 3) after assembling for a refined mesh size of 20 mm. As expected, all the three parameters that are used in simulation increased the strength of the deep beams (Fig. 4). Three factors are considered for the comparative study. Three factors are strain (Ɛ) at the mid shear span, vertical displacement (U2) at the mid span of the tension face and bending stress (S33) at the mid span of the tension face.

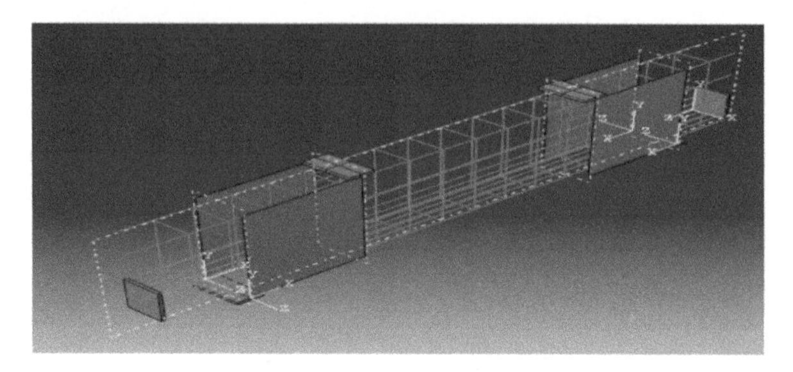

Fig. 2 Assembly of internal reinforcement and CFRP in model

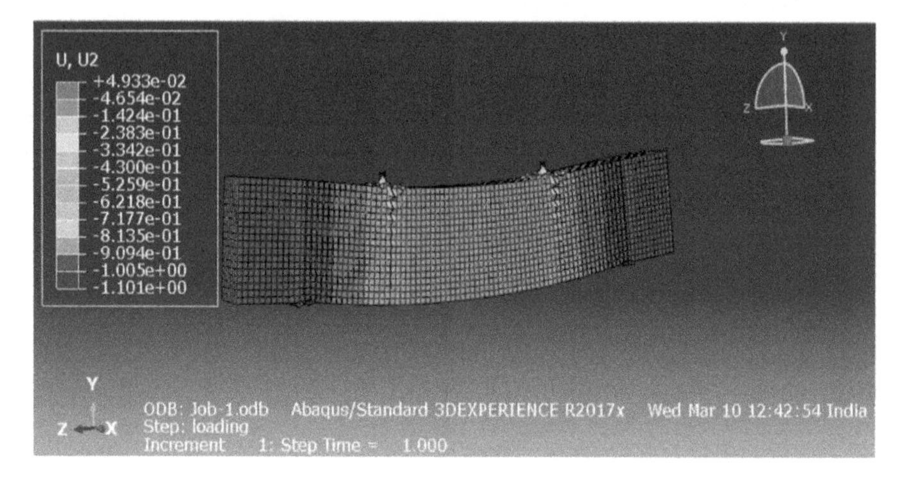

Fig. 3 Contour diagram obtained for vertical displacement in deep beams

All 27 models were named similarlly with abbreviations representing CFRP thickness, number of plies and increase in modulus of elasticity (ie. for model t2n3e10 it represents a CFRP thickness of 0.222 mm, number of plies 3 and increase in modulus of elasticity by 10% compared to the base model properties). The maximum decrease in the strain was observed in the model t4n7e30 and it was about 10.42% with respect to the base model used for validation. Compared to model t2n7e30, 4% more decrement in strain was observed in model t4n7e30 and compared to the model t3n7e30, 1.94% more dcrement in the strain was observed in model t4n7e30 (Fig. 4).

From Fig. 4, it is observed that, upon comparison of models t4n7e30 and t4n5e30, the percentage decrease in strain of latter was less by 2.16%. However, the decrease in strain reduced to 4.83% upon comparison for the case with 3 number of plies ie. t4n3e30 with the corresponding model with 7 plies (t4n7e30). A stark

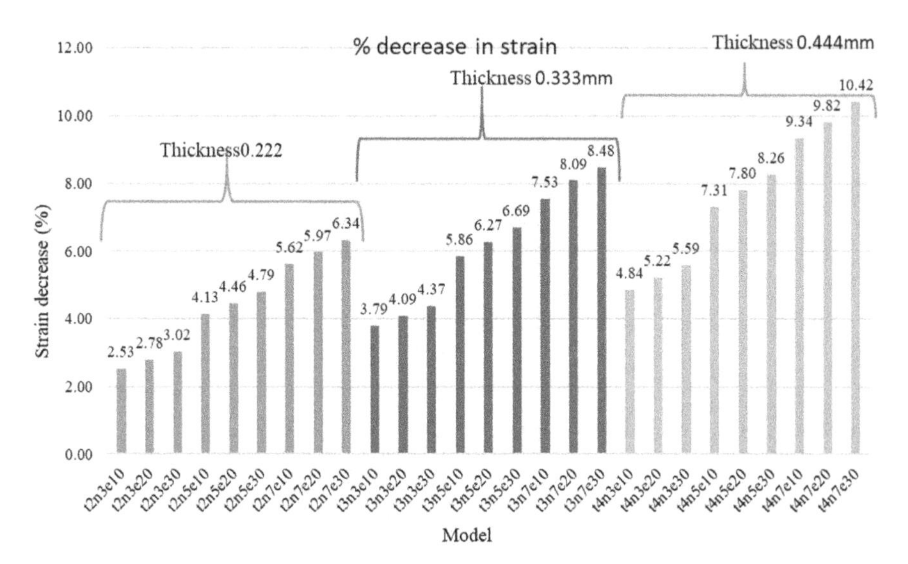

Fig. 4 Percentage decrease in strain for cfrp thickness

contrast in results were observed when the modulus of elasticity was varied from 30% (t4n7e30) compared to 20 and 10% increment (t4n7e20 and t4n7e10). Least reduction in strain was observed when the number of plies are decreased (compared to decrease in other parameters). Therefore, it can be concluded that increasing the number of plies is the most effective approach in reducing strain compared to increasing the modulus of elasticity and increasing the thickness of CFRP.

In order to identify the best parameter among the three, two different combinations were considered. One combination consists of keeping two parameters minimum and the third parameter maximum and the second combination consist of keeping the two parameters minimum and the third parameter as medium. Displacement (U2) and stress (S33) at the mid tension face of the deep beam are the factors considered here. In the first combination, the models considered are t2n3e30, t2n7e10 and t4n3e10. In the second combination, the models considered are t2n3e20, t2n5e10 and t3n3e10.

From Fig. 5, it was observed that stress (S33) is maximum in model t2n3e30 and minimum in model t2n7e10. In model t2n3e30, thickness and number of plies were kept to minimum and increment in modulus of elasticity was maximum i.e. thickness of 0.222 mm and 3 number of plies, increment in modulus of elasticity was 30% which shows higher value of stresses but the model t2n7e10 with minimum thickness and minimum increment in modulus of elasticity and maximum number of plies shows minimum value of stresses.

In combination 2 (Fig. 5), the model t2n5e10 shows lesser stress where CFRP thickness and increment in modulus of elasticity is less and number of plies are more. This indicates that models with maximum number of plies will create lesser stress, when compared with the models having more CFRP thickness and more

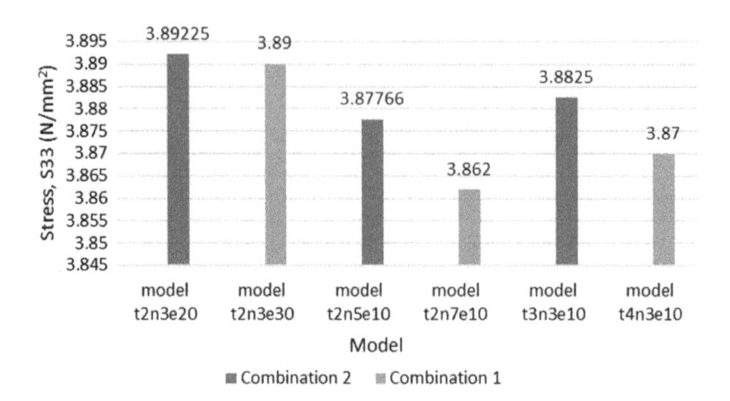

Fig. 5 Stress S33 in combination 1 and 2

increment in modulus of elasticity. When comparing model t2n3e30 with minimum CFRP thickness and maximum modulus of elasticity shows higher value of stresses and model t4n3e10 with maximum thickness and minimum modulus of elasticity shows lesser stresses. Thus, it can be concluded that for reducing the stresses, increasing the thickness of CFRP is better when compared to increasing the value of modulus of elasticity.

The vertical displacement obtained at the midspan of the tension face for combinations 1 and 2 is shown in Fig. 6. Among the three models of combination 1, model t2n7e10 shows minimum displacement and model t2n3e30 shows maximum displacement. In model t2n7e10, the number of plies is more and CFRP thickness as well as the increment in the modulus of elasticity is less. Similarly, for combination 2, the model with maximum number of plies shows minimum vertical displacement. In case of decrease in the vertical displacement also, the models having maximum number of CFRP plies shows better performance. When comparing the parameters, thickness of CFRP and increment in the modulus of elasticity of CFRP, the model t4n3e10 shows less displacement compared to that of t2n3e30. So, in reducing the displacement, increasing the thickness of CFRP is a better parameter when compared with the increasing the modulus of elasticity of the CFRP.

From the finite element analysis carried out using ABAQUS, out of three parameters considered (thickness of CFRP, number of plies of CFRP and increment in the modulus of elasticity of the CFRP) the best parameter was increasing the number of plies of CFRP, as it reduces the strain at the mid shear span, stress (S33) and vertical displacement at the midspan of the tension face, when compared with the other two parameters. Increase in the thickness of CFRP shows better performance when compared to increase in modulus of elasticity. From the results obtained from simulation a best combination of these parameter was arrived at by applying maximum number of plies of CFRP, medium increment in thickness of CFRP and minimum increment in the modulus of elasticity (model t3n7e10). While comparing the 27 models made with changing the three parameters, model t3n7e10

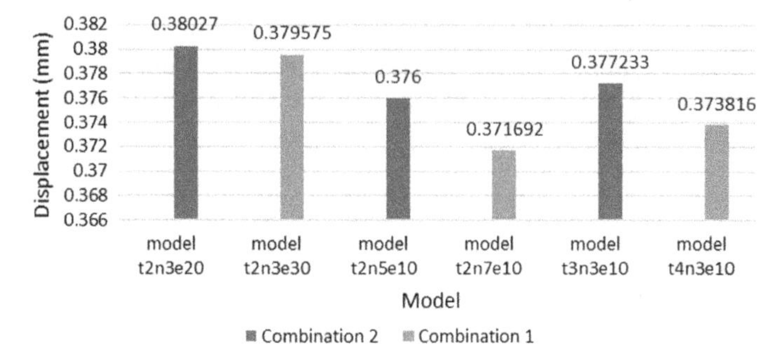

Fig. 6 Displacement in combination 1 and 2

shows better performance than 18 models including all the models made with 0.222 mm thick CFRP and shows less performance than the remaining 8 models with high value parameters. Hence, the above approach of combination of parameters can be used for optimum and economical shear strengthening of deep beams using CFRP.

4 Conclusions

The need for shear strengthening and retrofitting of deep beams are increasing. As CFRP was found to be effective in shear strengthening, the following conclusions were arrived at based on finite element analysis of 27 deep beam models in ABAQUS carried out in this study. The results of this study are based on value of thickness of 0.111 mm (base model), 0.222, 0.333, 0.444 mm, range of increase in modulus of elasticity in increments of 10, 20, 30% and number of plies as 3, 5 and 7.

- As expected, increasing the thickness of CFRP, increasing the number of plies of CFRP and increasing the modulus of elasticity will be effective in rehabilitation of shear span of deep beam.
- Increasing the number of plies of CFRP shows better performance when compared to the other two parameters (modulus of elasticity and thickness of CFRP), as the models with 3 number of plies shows least reduction in strain (4.86% less than model with 7 number of plies).
- However, the main difference in increasing the number of plies and increasing the thickness of CFRP arises when debonding failure occurs between the plies in non-linear range. As the study was limited to linear range, debonding failure was not considered in this study.
- When comparing the thickness of CFRP and modulus of elasticity of CFRP, increasing the thickness gives better result, as the variation in the modulus of elasticity of CFRP results in considerably less effect on strain reduction compared to that of the variation in the thickness of CFRP.

- In the combination of the three parameters, the optimum result was obtained with maximum number of plies, medium increment in thickness and smallest increment in the modulus of elasticity of the CFRP where strain, stress and displacement values are less when compared to other combinations.

References

1. Javed MA, Irfan M, Khalid S, Chen Y, Ahmed S (2016) An experimental study on the shear strengthening of reinforced concrete deep beams with carbon fiber reinforced polymers. 1–9. https://doi.org/10.1007/s12205-016-0739-3
2. Murad Y (2018) An experimental study on flexural strengthening of RC beams using CFRP sheets. Int J Eng Technol 7:2075–2080. https://doi.org/10.14419/ijet.v7i4.16546
3. IS 456 (2000) Concrete, plain and reinforced. Bur Indian Stand Delhi 1–114
4. Siddiqui NA (2009) Experimental investigation of RC beams strengthened with externally bonded FRP composites. Lat Am J Solids Struct 6:343–362
5. Panjehpour M, Chai HK, Voo YL (2014) Strut deformation in CFRP-strengthened reinforced concrete deep beams. Sci World J 2014:1–9. https://doi.org/10.1155/2014/265879
6. American Concrete Institute (ACI 318–99): ACI 318–99 (1999) Building code requirements for structural concrete
7. Islam MR, Mansur MA, Maalej M (2005) Shear strengthening of RC deep beams using externally bonded FRP systems. Cem Concr Compos 27:413–420. https://doi.org/10.1016/j.cemconcomp.2004.04.002
8. Toutanji HA, Gómez W (1997) Durability characteristics of concrete beams externally bonded with FRP composite sheets. Cem Concr Compos 19:351–358. https://doi.org/10.1016/S0958-9465(97)00028-0
9. Triantafillou TC (1998) Shear strengthening of reinforced concrete beams using epoxy-bonded FRP composites. ACI Struct J 95:107–115. https://doi.org/10.14359/531
10. Khalifa A, Nanni A (2000) Improving shear capacity of existing RC T-section beams using CFRP composites. Cem Concr Compos 22:165–174. https://doi.org/10.1016/S0958-9465(99)00051-7
11. Mohamed AR, Shoukry MS, Saeed JM (2014) Prediction of the behavior of reinforced concrete deep beams with web openings using the finite element method. Alexandria Eng J 53:329–339. https://doi.org/10.1016/j.aej.2014.03.001
12. Zaher AH, Elsonbaty MM (2020) Strengthening and repairing of RC deep beams using Cfrp and Gfrp 11:64–85
13. Burtscher SL (2008) Wedge anchorage for CFRP strips. 12:446–453. https://doi.org/10.1061/(ASCE)1090-0268(2008)12

Development of Virtual Lab to Solve Problems in Structural Dynamics

Anupriya Sakya, Arpit Asim Verma, K. Haripriya, and S. Dhanush

Abstract The rapid development of internet technology has pushed mankind to shift towards virtual platforms, and this increasing popularity has paved its path into the conventional teaching-learning process. Hence, it has now become essential to extend the traditional hands-on laboratories to the online platform. With this idea of incorporating e-learning, a virtual structural dynamics laboratory has been developed. Virtual laboratory (VL) acts as an interface on which users can perform simulated experiments and obtain authentic results without limitation of trials. At present, many technologies are being used to develop virtual labs. In the current study, MATLAB was used to develop this tool. For ease of understanding, this tool has been divided into different modules. Data such as material and cross-sectional properties are obtained from the user which is analysed to produce the responses of the structure. This VL intends to perform all the experiments mentioned in the manual of NPEEE, MHRD like Free Vibration of Single degree of freedom (S.D.O.F) System, Vibration of Multi degree of freedom (M.D.O.F) System, Torsional Response of Building, Continuous Systems, etc. With the help of this platform, a learner with basic knowledge of structural dynamics can visualize and interpret the dynamics of a structure.

Keywords Virtual lab · Structural dynamics · MATLAB · Degrees of freedom · Simulations

---

A. Sakya (✉) · A. A. Verma · K. Haripriya · S. Dhanush
RV College of Engineering, Mysore Rd, RV Vidyaniketan Post, Bengaluru, Karnataka 560059, India
e-mail: anupriyasakya.cv17@rvce.edu.in

A. A. Verma
e-mail: arpitasimverma.cv17@rvce.edu.in

K. Haripriya
e-mail: kharipriya.cv17@rvce.edu.in

S. Dhanush
e-mail: dhanushs@rvce.edu.in

G. C. Marano et al. (eds.), *Proceedings of SECON'21*, Lecture Notes in Civil Engineering 171, https://doi.org/10.1007/978-3-030-80312-4_48

1 Introduction

Earthquake is a natural tremor or shaking of the earth's crust that occurs at or below the surface. In the past, the Indian subcontinent has seen some of the world's most strong earthquakes. According to the seismic zonation map in IS: 1893: 2002, more than 60% of India is vulnerable to earthquakes. India has experienced eight moderate earthquakes in the last decade, all of which have resulted in significant casualties and innumerable property loss. Since the likelihood of an earthquake recurrence in India can be predicted in the future, it is critical to raise awareness about the after-effects of earthquakes and to understand the response of structures subjected to dynamic loads. There is a need to emphasize on reducing the seismic vulnerability of structures by ascertaining its safety.

The study of the behaviour of a system subjected to dynamic loading, such as earthquakes, wind, and waves, is known as structural dynamics. Structural dynamics deals about the analysis of the time dependent loads when subjected to external acceleration. The subject is commonly not seen and taught as a core subject of the civil engineering curriculum [1]. The unpopularity of this course is because of its mathematical nature as well as illustrating structural dynamics using chalk board isn't an effective way [2]. This difficulty in teaching and learning structural dynamics has been a barrier stopping undergraduate and graduates from opting to learn this course. Only few institutions have an independent structural dynamics laboratory. An efficient way to draw attention to learn this course would be to frame a set of experiments which could enable students learn the basics of the subject and perceive the significance of dynamic loads on structure. Technological advances have significantly aided structural analysis and structural dynamics for Earthquake engineering applications.

2 Virtual Structural Dynamics Laboratory

Setting up a real time structural dynamics laboratory as well as repeating the experiments becomes expensive. Thus, there is a need for a tool to augment teaching learning process. The best way to enable this is by setting up a virtual structural dynamic laboratory (VSDL) which includes all the real time experiments. In Virtual labs there is no fear of instrument breakage and it is also feasible to repeat the experiment innumerable times. Muniram Budhu in his findings observed that the interactions coded on a virtual platform can prove effective in recalling prior knowledge, to stimulate learning and to test learning outcomes [3]. In order to address this problem, a virtual laboratory was developed with a series of simple experiments to carry out a single or multi-storey linear and nonlinear structural analysis. The user can enter data such as material and cross-sectional properties, and the programme will calculate and display accurate results using graphical simulations.

3 Literature Review

As the novel Coronavirus pandemic hit the globe, it brought with it some grave unprecedented challenges that impacted all sectors. One of the major industries to be hit was the education system. School closures due to the pandemic affected approximately 1.077 billion students in 186 countries as of 30 September 2020 [4]. With the advancement of computer and other technologies it is possible to set up a virtual lab which is like real world physical lab and perform complex and time-consuming equations in a fraction of second [5]. Many studies have demonstrated the benefits of virtual and interactive exploration of observable phenomena compared with physical experiments [6]. Many dynamics laboratories have been coded using different languages. One of them being Visual Basic Application (VBA). Ramancharla et al. in his studies built a virtual laboratory for better understanding of the behaviour of lateral load for the learners using VBA [7]. B.F. Spencer et al. used JAVA to develop a VSDL that could run online by installing a Java run time environment and testing it on a multi-storey shear building. A VSDL was developed by J. R. Red-Horse for using it as a validation exercise in Scandia National Laboratory [8]. SISMILAB an online laboratory programmed with the assistance of MATLAB was created to understand the concepts of earthquake engineering [9].

4 Methodology

4.1 App Design

All the applications were designed using the App Designer feature of MATLAB. These modules were deployed as Standalone Desktop applications and as Web App using MATLAB Compiler. The Standalone application will run on any device without the need to instal MATLAB software, while the Web Apps are hosted on a MATLAB Web App Server. End users can use a browser to access and run the web apps by entering a specific URL to the server's home page or the app itself. ETABS was used to validate the results of the simulations performed in MATLAB.

4.2 Data Acquisition

The virtual laboratory is programmed using MATLAB. Various applications will be developed to perform simulated experiments. This will allow users to experiment with control system properties such as the number of storeys, mass, stiffness, damping, and nonlinear properties, as well as control objectives, to better understand the control design process and improve earthquake response [10]. When users click on the button "Calculate," VL performs a virtual experiment with the linear and non-linear models of the multi-story building, calculating the various spectral estimates such as natural frequencies, mode shapes, time-related responses, etc.

4.3 Data Processing

The data obtained by the user was processed with the help of MATLAB. Due to its limited reliance on an operating platform, the MATLAB programming language provides considerable benefits. It can perform extensive data analysis, visualization and develop an application with a graphical user interface.

5 VSDL Modules

The tool developed is divided into the following modules.

1. Multi Degree of Freedom
2. Duhamel's integral
3. Continuous System

5.1 Experiment 1: Multi Degree of Freedom

The user may enter control system properties such as the number of storeys, mass, stiffness, damping, and nonlinear properties in this module. The application designed analyses the data and calculates the spectral estimates such as natural frequencies and mode shapes with respect to time.

5.1.1 Modal Analysis

The app is coded for 3 storeys. The user must first choose the number of storeys for the building on which the analysis will be performed and enter the mass and stiffness for the corresponding floors. The application then evaluates the natural frequencies by calculating the square root of the Eigen values obtained by performing arithmetic calculations on mass and stiffness matrix.

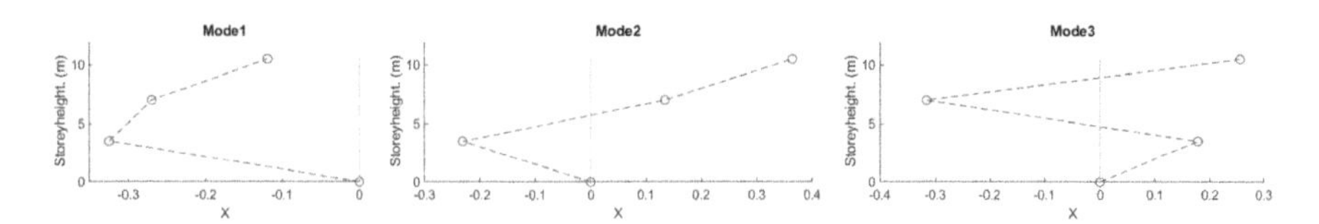

Fig. 1 Demonstration of mode shapes

5.1.2 Interpretation of Mode Shape

User can vary the values of input parameters and check the changes in the values of mode shape and the effect on total response of the structure as shown in Fig. 1.

5.1.3 Mathematical Model

Figures 2 and 3 shows the mathematical model of the MDOF system.

The equation for MDOF system is given as:

$$m\ddot{x} + c\dot{x} + kx = 0 \tag{1}$$

where,

m = mass.
c = damping coefficient.
k = stiffness.
$\ddot{x}$ = acceleration.
$\dot{x}$ = velocity.
x = displacement.

For undamped system c = 0

$$m\ddot{x} + kx = 0 \tag{2}$$

The above equation for 3DOF can be represented in the form of

$$\begin{bmatrix} m_1 & 0 & 0 \\ 0 & m_2 & 0 \\ 0 & 0 & m_3 \end{bmatrix} \begin{bmatrix} \ddot{x}_1 \\ \ddot{x}_2 \\ \ddot{x}_3 \end{bmatrix} + \begin{bmatrix} k_1 + k_2 & -k_2 & 0 \\ -k_2 & k_2 + k_3 & -k_3 \\ 0 & -k_3 & k_3 \end{bmatrix} \begin{bmatrix} x_1 \\ x_2 \\ x_3 \end{bmatrix} = 0 \tag{3}$$

$$[k - \lambda m] = 0$$

$$\omega_1 = \sqrt{\lambda_1},\ \omega_2 = \sqrt{\lambda_2},\ \omega_3 = \sqrt{\lambda_3}.$$

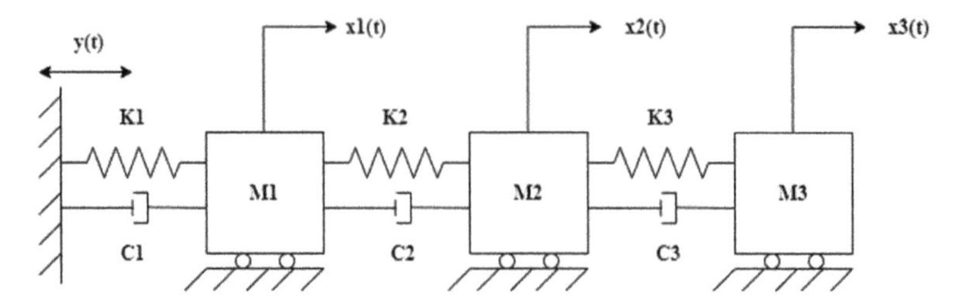

Fig. 2 Multi mass-damper-spring model representation of a three-storey building

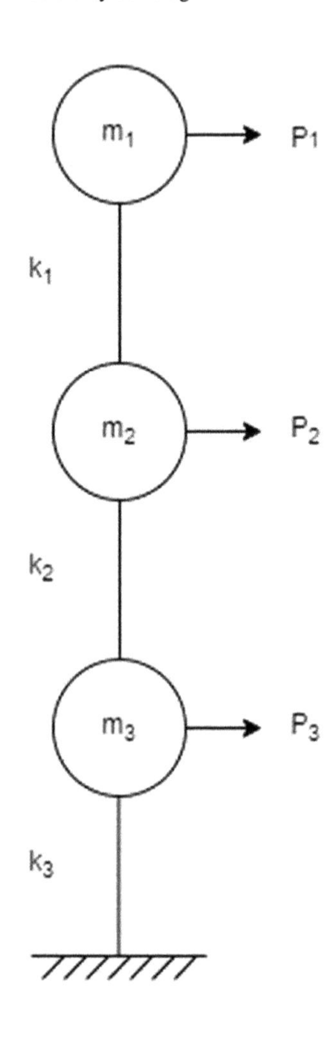

Fig. 3 Lumped mass idealization of three-storey building

Where,

λ = Eigen Value

ω = Angular Frequency

Square root of λ gives us angular frequency [11].

5.2 *Experiment 2: Duhamel's Integral—Damped System*

Structures are often exposed to non-harmonic loads. One popular method of analysing such linear systems and structures subjected to external vibration is by performing Duhamel's integration [12].

In this experiment, the user has the option of entering structure properties such as mass, stiffness, and damping coefficient. As shown in Fig. 4, the application performs calculations and determines the structure's natural frequency, damped natural frequency, and natural time. With the initial conditions set to zero, the application iterates for up to 0.1 s.

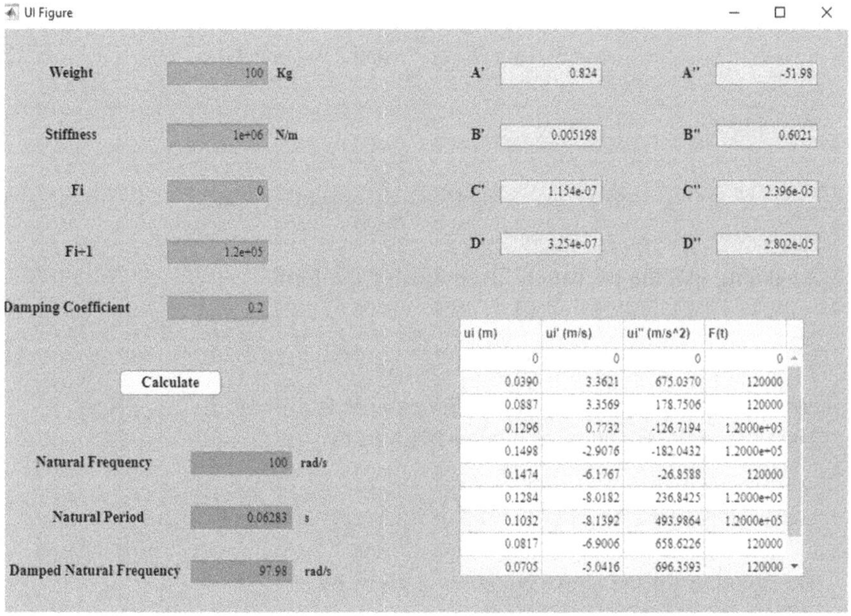

UI Figure

Weight 100 Kg

Stiffness 1e+06 N/m

Fi 0

Fi+1 1.2e+05

Damping Coefficient 0.2

Calculate

Natural Frequency 100 rad/s

Natural Period 0.06283 s

Damped Natural Frequency 97.98 rad/s

A' 0.824 A'' -51.98

B' 0.005198 B'' 0.6021

C' 1.154e-07 C'' 2.396e-05

D' 3.254e-07 D'' 2.802e-05

ui (m)	ui' (m/s)	ui'' (m/s^2)	F(t)
0	0	0	0
0.0390	3.3621	675.0370	120000
0.0887	3.3569	178.7506	120000
0.1296	0.7732	-126.7194	1.2000e+05
0.1498	-2.9076	-182.0432	1.2000e+05
0.1474	-6.1767	-26.8588	120000
0.1284	-8.0182	236.8425	1.2000e+05
0.1032	-8.1392	493.9864	1.2000e+05
0.0817	-6.9006	658.6226	120000
0.0705	-5.0416	696.3593	120000

Fig. 4 Duhamel's integral-damped system in MATLAB

5.2.1 Mathematical Model

Duhamel's integral describes the response of a damped system which is obtained by substituting initial velocity.

$dv = F(\tau)\frac{d\tau}{m}$ produced by the impulse $F(\tau)d\tau$ in the equation of damped free vibration

$$u(t) = e^{-\varepsilon\omega t}(u_0 \cos\omega_D t + \frac{v_0 + u_0\varepsilon\omega}{\omega_D}\sin\omega_D t) \quad (4)$$

where,

ε = damping ratio.
ω = natural frequency.
ω_D = damped frequency.
u_0 = initial displacement.
v_0 = initial velocity.
t = time.

Substituting

$$u_0 = 0 \;\; v_0 = F(\tau)\frac{d\tau}{m} \text{ and t for } (t - \tau) \quad (5)$$

We obtain

$$du(t) = e^{-\varepsilon\omega(t-\tau)}F(\tau)\frac{d\tau}{m\omega_D}\sin\omega_D(t-\tau) \quad (6)$$

Assuming that the excitation function is of the form

$$F(t) = \left(1 - \frac{t-t_i}{\Delta t}\right)F_i + \left(\frac{t-t_i}{\Delta t}\right)F_{i+1} \quad t_i \le t \le t_{i+1} \quad (7)$$

where $t_i = i.\Delta t$ for equal intervals of duration Δt and i = 1, 2, 3…. , N

The differential equation of motion is given by

$$m\ddot{u} + c\dot{u} + ku = \left(1 - \frac{t-t_i}{\Delta t}\right)F_i + \left(\frac{t-t_i}{\Delta t}\right)F_{i+1} \quad t_i \le t \le t_{i+1} \quad (8)$$

The solution for the above equation is given by

$$u = e^{-\varepsilon\omega(t-t_i)}[C_i\cos\omega_D(t-t_i) + D_i\sin\omega_D(t-t_i)] + B_i + A_i(t-t_i) \quad (9)$$

velocity $\dot{u}$ is given by

$$\dot{u}_{i+1} = \mathrm{A}u_i + \mathrm{B}\dot{u}_i + C''F_i + D''F_{i+1} \tag{10}$$

acceleration is given by

$$\ddot{u}_{i+1} = \frac{1}{m}(F_{i+1} - c\dot{u}_{i+1} - ku_{i+1}) \tag{11}$$

where,

$$\mathrm{A}_i = \frac{F_{i+1} - F_i}{k\Delta t} \quad B_i = \frac{F_i - c\mathrm{A}_i}{k} \quad C_i = u_i - B_i \quad D_i = \frac{\dot{u}_i - \mathrm{A}_i - \varepsilon\omega C_i}{\omega_D} \tag{12}$$

Substituting all the coefficients with, $c = \frac{2\varepsilon k}{\omega}$, the displacement, velocity, and acceleration at time step $t_{i+1} = t_i + \Delta t$ is given by

$$\begin{aligned} u_{i+1} &= \mathrm{A}'u_i + \mathrm{B}'\dot{u}_i + C'F_i + D'F_{i+1} \\ \dot{u}_{i+1} &= \mathrm{A}u_i + \mathrm{B}\dot{u}_i + C''F_i + D''F_{i+1} \\ \ddot{u}_{i+1} &= -\omega^2 u_{i+1} - 2\varepsilon\ddot{u}_{i+1} + \frac{F_{i+1}}{m} \end{aligned} \tag{13}$$

where,

$$\begin{aligned}
\mathrm{A}' &= e^{-\varepsilon\omega\Delta t}\left(\frac{\varepsilon\omega}{\omega_D}\sin\omega_D\Delta t + \cos\omega_D\Delta t\right) \\
B' &= e^{-\varepsilon\omega\Delta t}\left(\frac{1}{\omega_D}\sin\omega_D\Delta t\right) \\
C' &= \frac{1}{k}\left\{e^{-\varepsilon\omega\Delta t}\left[\left(\frac{1-2\varepsilon^2}{\omega_D\Delta t} - \frac{\varepsilon\omega}{\omega_D}\right)\sin\omega_D\Delta t - \left(1 + \frac{2\varepsilon}{\omega\Delta t}\right)\cos\omega_D\Delta t\right] + \frac{2\varepsilon}{\omega\Delta t}\right\} \\
D' &= \frac{1}{k}\left\{e^{-\varepsilon\omega\Delta t}\left[\left(\frac{2\varepsilon - 1}{\omega_D\Delta t}\right)\sin\omega_D\Delta t + \left(\frac{2\varepsilon}{\omega\Delta t}\right)\cos\omega_D\Delta t\right] + \left(1 - \frac{2\varepsilon}{\omega\Delta t}\right)\right\} \\
\mathrm{A}'' &= -e^{-\varepsilon\omega\Delta t}\left(\frac{\omega^2}{\omega_D}\sin\omega_D\Delta t\right) \\
B'' &= e^{-\varepsilon\omega\Delta t}(\cos\omega_D\Delta t) - \frac{\varepsilon\omega}{\omega_D}\sin\omega_D\Delta t \\
C'' &= \frac{1}{k}\left\{-e^{-\varepsilon\omega\Delta t}[\left(\frac{\omega^2}{\omega_D} + \frac{\omega\varepsilon}{\omega_D\Delta t}\right)\sin\omega_D\Delta t + \frac{1}{\Delta t}\cos\omega_D\Delta t] - \frac{1}{\Delta t}\right\} \\
D'' &= \frac{1}{k\Delta t}\left\{-e^{-\varepsilon\omega\Delta t}\left(\frac{\omega\varepsilon}{\omega_D}\sin\omega_D\Delta t + \cos\omega_D\Delta t\right) + 1\right\}
\end{aligned} \tag{14}$$

5.3 Experiment 3: Continuous Systems

In this experiment the structural dynamics problem for one dimensional system with distributed mass is formulated. This application is divided into 3 sub modules based on the support conditions of the beam.

1. Simply supported beam
2. Cantilever beam
3. Fixed end beam

Each application allows user to enter the values of flexural rigidity, moment of inertia, mass, and length of the beam. The application performs calculations based on the support conditions chosen by the user and gives three modal shapes for three angular frequencies obtained as shown in the Figs. 5, 6 and 7.

For case of free vibration, the equation of undamped motion is given by:

$$m(x)\frac{\partial^2 u}{\partial t^2} + \frac{\partial^2}{\partial x^2}\left[EI(x)\frac{\partial^2 u}{\partial x^2}\right] = 0 \tag{15}$$

The solution of the above equation for the case of uniform beams is of the form

$$EI\emptyset^{IV}(x) - \omega^2 m\emptyset(x) = 0 \;\; or \;\; \emptyset^{IV}(x) - \beta^4\emptyset(x) = 0 \tag{16}$$

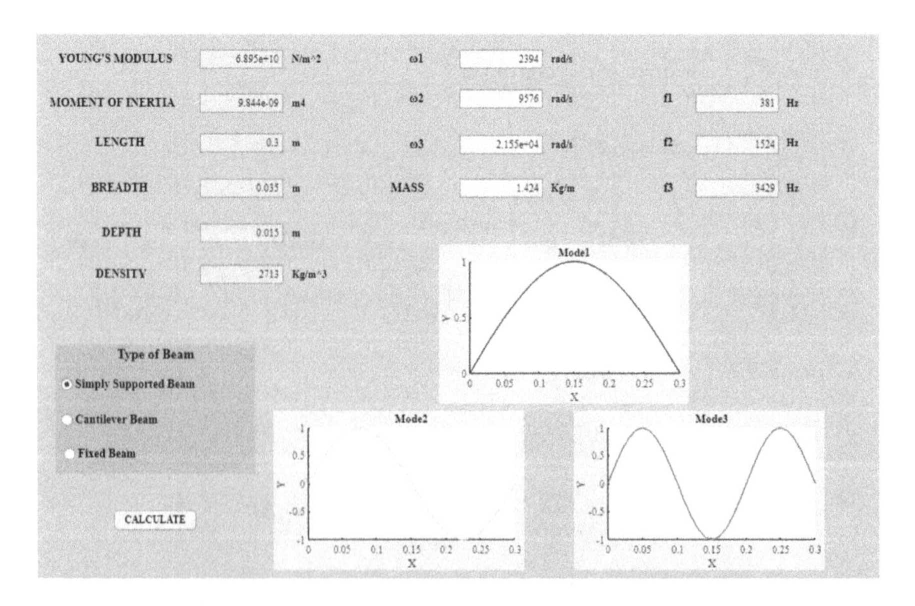

Fig. 5 Continuous systems for simply supported beam in MATLAB

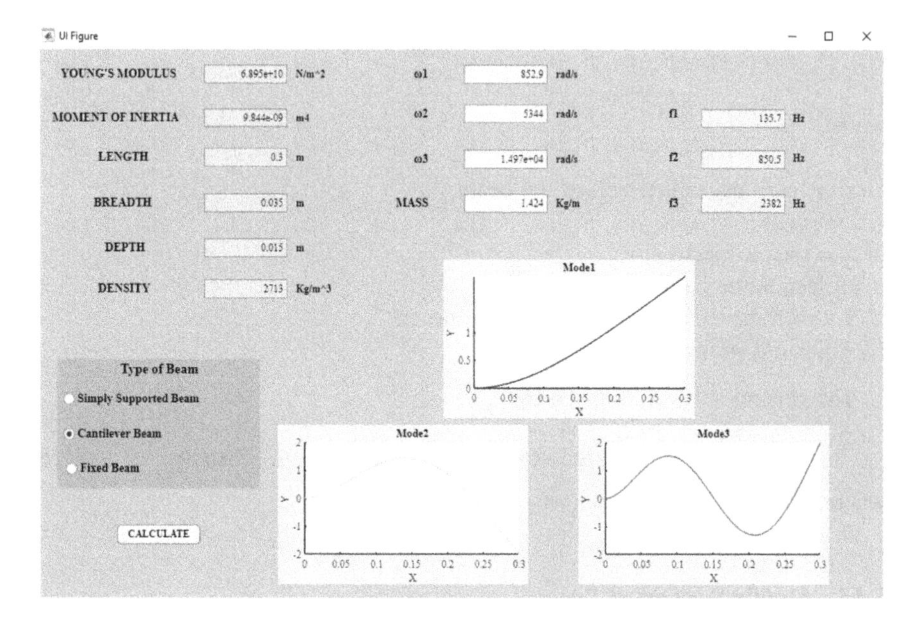

Fig. 6 Continuous systems for cantilever beam in MATLAB

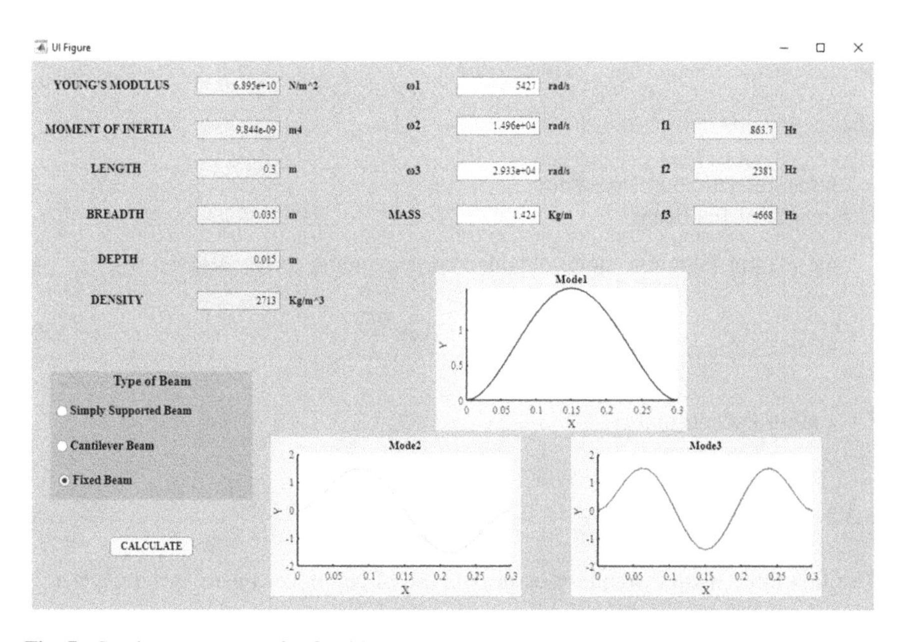

Fig. 7 Continuous systems for fixed beam in MATLAB

$$\beta^4 = \frac{\omega^2 m}{EI} \tag{17}$$

where,

$m(x)$ = mass per unit length of the beam.
x = position.
$\emptyset(x)$ = spatial function.
β = Eigen value parameter.
u (x, t) = transverse displacement of the beam.
EI = flexural strength of the beam,

The general solution is given by:

$$\emptyset(x) = C_1 \sin \beta x + C_2 \cos \beta x + C_3 \sinh \beta x + C_4 \cosh \beta x \tag{18}$$

where, C_1, C_2, C_3, C_4 = unknown constants [13].

5.3.1 Simply Supported Beam

In the case of simply supported beams, the equation for finding the natural frequency is given by:

$$\omega_n = \frac{n^2 \pi^2}{L^2} \sqrt{\frac{EI}{m}} \quad n = 1, 2, 3, \ldots \tag{19}$$

where,

ω_n = natural vibration frequency.
L = length of the beam.

The natural vibration mode obtained corresponding to ω_n is given by

$$\emptyset_n(x) = C_1 \sin \frac{n\pi x}{L} \tag{20}$$

C_1 = arbitrary constant.

5.3.2 Cantilever Beam

The first three natural frequencies in the case of cantilever beam are given by:

$$\omega_n = (\beta_n L)^2 \sqrt{\frac{EI}{mL^4}} \tag{21}$$

And the natural vibration mode is given by:

$$\emptyset_n(x) = C_1\left[\cosh \beta_n x - \cos \beta_n x - \frac{\cosh \beta_n L + \cos \beta_n L}{\sinh \beta_n L + \sin \beta_n L}(\sinh \beta_n x - \sin \beta_n x)\right] \quad (22)$$

5.3.3 Fixed Beams

The equation for finding the natural frequency is given by

$$\omega_n = (\beta_n L)^2 \sqrt{\frac{EI}{mL^4}} \quad (23)$$

The normal modes corresponding to the given frequency is given by:

$$\emptyset_n(x) = \cosh \beta_n x - \cos \beta_n x - \sigma_n(\sinh \beta_n x - \sin \beta_n x) \quad (24)$$

where,

$$\sigma_n = \frac{\cos \beta_n L - \cosh \beta_n L}{\sinh \beta_n L - \sinh \beta_n L} \quad (25)$$

6 Validation of Multiple Degree of Freedom Analysis in Matlab, Etabs and Experimental Procedure

Determine the lump mass, storey stiffness, natural frequency, and mode shape for the three-storey system shown in Fig. 8 with following dimensions:

a. Plan (B*W) = 0.15 m * 0.3 m.

   Where,
   B = Slab length.
   W = Slab width.

b. Storey height

   Ground floor = 0.4 m.
   First floor = 0.4 m.
   Second floor = 0.4 m.

c. Slab thickness = 0.0127 m.
d. Column = 0.003 m * 0.02511 m.
f. Material—Aluminium.

   E = 68.948GPa.
   Density = 2712.629 kg/m^3.

6.1 Experimental Analysis

See Figs. 8 and 9.

6.2 Analysis Using MATLAB

See Figs. 10, 11 and 12.

6.3 Analysis Using ETABS

The three-storey system is considered to be a 3 degree of freedom system and analyzed. Fast Fourier Transform was performed on the data obtained from experimental analysis. As a result, frequency plots were obtained as shown in

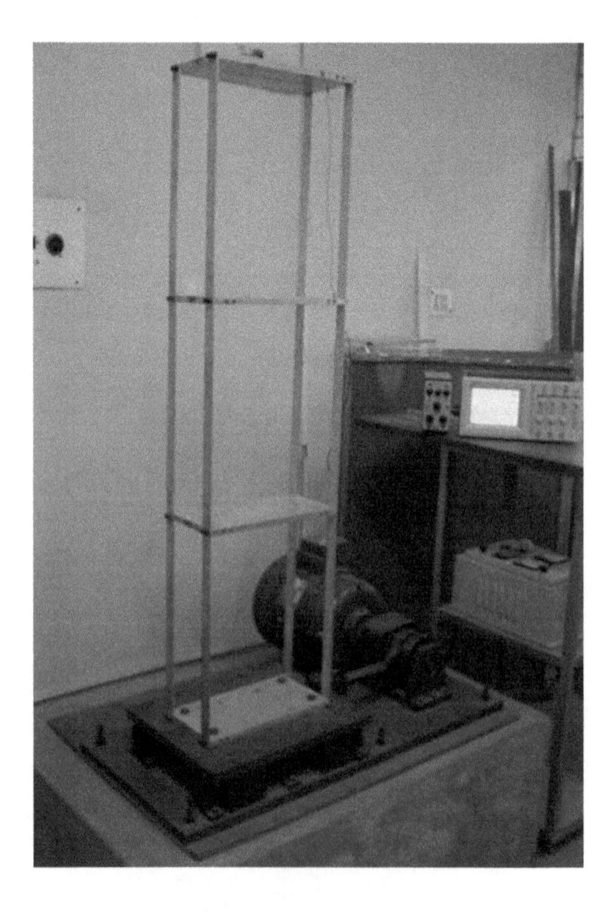

Fig. 8 Experimental setup for three-storey building frame

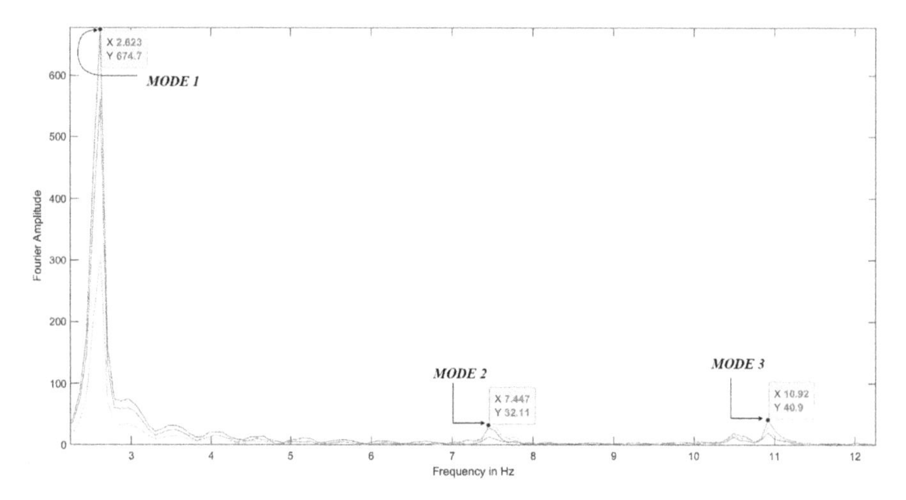

Fig. 9 Frequency plots indicating various modal frequencies

Fig. 10 Lumped mass analysis

Fig. 9. The dimensions and material properties of the three-storey system were taken as the input values in the app developed as shown in Fig. 10. The output was obtained in the form of natural frequency and mode shapes. The findings from MATLAB is illustrated in Fig. 11. Figures 12 and 13 depicts the mode shapes of vibration obtained for the system (Fig. 13).

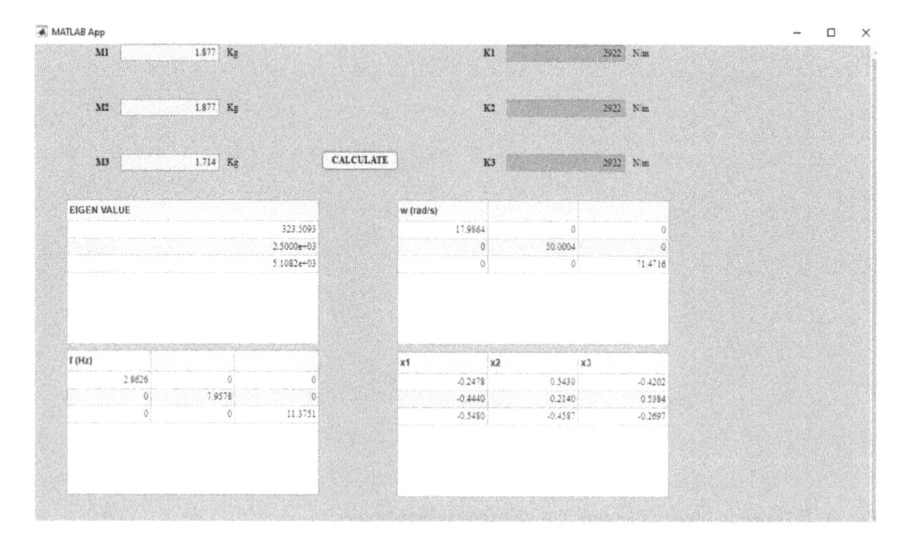

Fig. 11 Eigen value, natural frequency

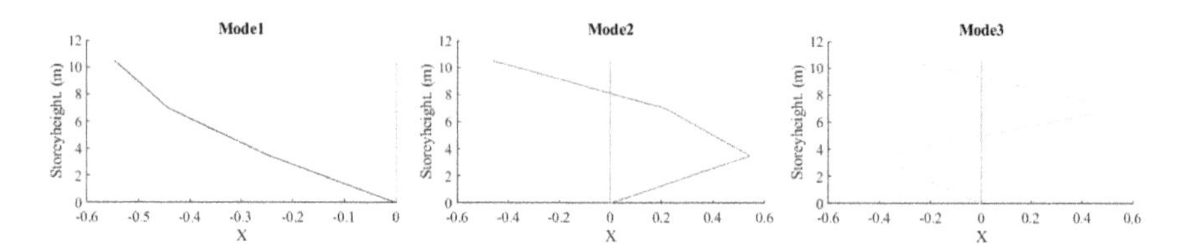

Fig. 12 Mode shapes of vibration

Table 2 shows the comparison table which lists the results obtained from the experimental analysis, app developed and ETABS. Since ETABS is an established software, the findings from MATLAB as shown in the Fig. 11 and experimental analysis as shown in the Fig. 9 are compared to the results from ETABS as shown in the Table 1. The findings obtained from MATLAB and experimental analysis differed by 0.28% and 6.5%, respectively. The slight deviation in the results obtained from the app developed was because of the following reasons:

1. Mass of accelerometer equal to 225 g attached to the frame at each storey was not included in calculations.
2. Model that was used did not provide 100% fixity in the joints.

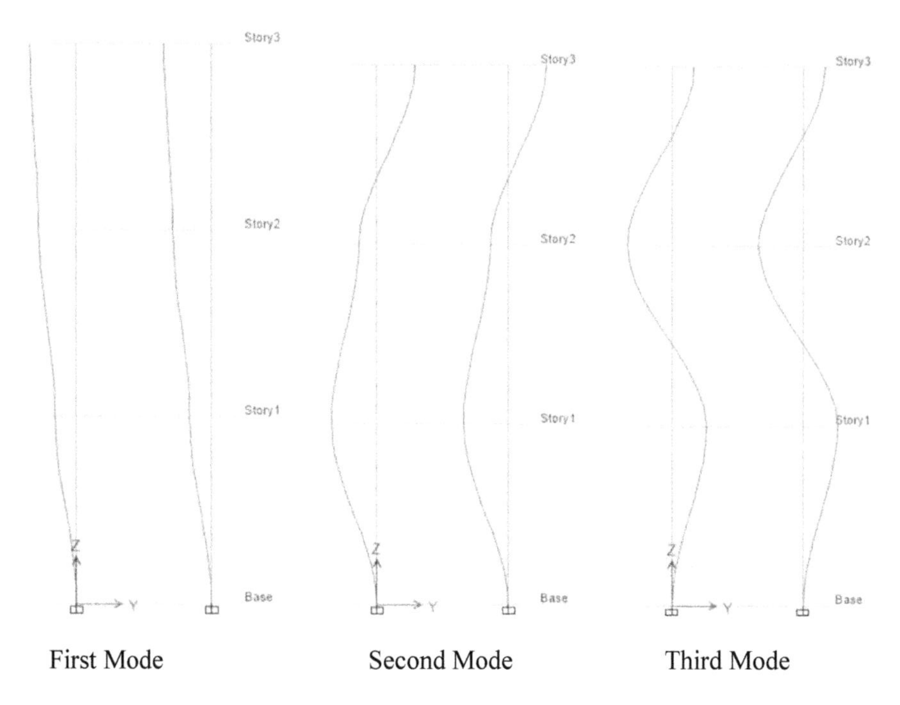

Fig. 13 Mode shape of vibration in ETABS

Table 1 Modal frequencies table

Case	Mode	Period sec	Frequency cyc/sec
Modal	1	0.349	2.868
Modal	2	0.125	7.98
Modal	3	0.088	11.42

Table 2 Comparison table

	Experimental			MATLAB			ETABS		
	MODE1	MODE2	MODE3	MODE1	MODE2	MODE3	MODE1	MODE2	MODE3
Frequency (Hz)	2.623	7.447	10.92	2.8626	7.9578	11.3751	2.868	7.98	11.42
Percentage Error % (w.r.t ETABS)	8.54	6.67	4.37	0.19	0.28	0.39	–	–	–

7 Conclusion

In an attempt to stimulate learning through online platform and to increase the awareness regarding the effects of earthquake, a software tool was developed to act as a virtual laboratory to solve problems in structural dynamics. The tool currently comprises of 3 Modules which enables the user to understand the fundamentals of structural dynamics. The tool developed has Standalone Desktop Applications that has zero dependency on the software of the host computer and can run independently on any platform.

8 Scope for Future Work

Currently, efforts are being undertaken to incorporate graphical simulations into the virtual lab. Graphical simulations can help the users to observe the actual behaviour of the structure under different loads virtually. Following the deployment of the apps on the college server, a user-based survey will be conducted. This would aid in the virtual lab's improvement.

References

1. Thambiratnam DP (2013) Structural dynamics education in the new millennium. In: Proceedings of the international engineering and technology education conference, pp 1–12. Central Queensland University and the University of Technical Education
2. Gao Y, Yang G, Spencer BF, Lee GC (2005) Java-powered virtual laboratories for earthquake engineering education. Comput Appl Eng Educ 13(3):200–212. https://doi.org/10.1002/cae.20050
3. Asplund L (2001) Session. ACM SIGAda Ada Lett XXI(1):11–13. https://doi.org/10.1145/374369.374373
4. Impact of the COVID-19 pandemic on education—Wikipedia. https://en.wikipedia.org/wiki/Impact_of_the_COVID-19_pandemic_on_education. Accessed on 9 Jan 2021
5. Gupta MK, Singh AJ. Development of virtual structural dynamics lab (VSDL): modal analysis tool
6. Physical and virtual laboratories in science and engineering education—PubMed. https://pubmed.ncbi.nlm.nih.gov/23599479/. Accessed on 10 Jan 2021
7. Ramancharla P (2005) Virtual structural dynamics laboratory, pp 1–7
8. Red-Horse JR, Paez TL (2008) Sandia national laboratories validation workshop: structural dynamics application. Comput Methods Appl Mech Eng 197(29–32):2578–2584. https://doi.org/10.1016/j.cma.2007.09.031
9. Guerrero-Mosquera LF, Gómez D, Thomson P (2018) Development of a virtual earthquake engineering lab and its impact on education. DYNA 85(204):9–17. https://doi.org/10.15446/dyna.v85n204.66957
10. Vibrations, dynamics and structural systems 2nd edition—Madhujit Mukhopadhyay—google books. https://books.google.co.in/books?id=-InwPwAACAAJ&printsec=copyright#v=onepage&q&f=false. Accessed on 10 Jan 2021

11. Report F (2006) Development of experimental setups for earthquake engineering education. Eng Educ
12. Chopra AK (2007) Dynamics of structures. Pearson Education India
13. Paz M (2012) Structural dynamics: theory and computation. Springer Science & Business Media

Lateral Load Behavior of Unreinforced Masonry Spandrels

Jacob Alex Kollerathu and Arun Menon

Abstract Spandrels, are usually classified as secondary elements and even though their behaviour has not received adequate focus unlike piers, they significantly affect the seismic capacity of the structure. Masonry spandrels are often damaged and the first structural components that crack within Unreinforced Masonry structures. Despite this, existing analytical methods typically consider a limit case in which the strength of spandrels is either neglected, considered to be infinitely rigid and strong or treated as rotated piers. It is clearly evident that such an assumption is not plausible. Hence, reliable predictive strength models are required. This thesis attempts to re-examine the flexural behaviour of spandrels and proposes an analytical model. The model is based on the interlocking phenomena of the joints at the end-sections of the spandrel and the contiguous masonry. The proposed analytical model is incorporated within a simplified approach to account for the influence of spandrel response on global capacity estimate of URM buildings.

Keywords Lateral load behavior · Spandrel · Storey-shear mechanism

1 Introduction

In an earthquake, masonry spandrels are in general the first structural components that crack within Unreinforced Masonry (URM) buildings. Recognizing that the masonry piers are the most important elements both for vertical load-carrying capacity and for seismic resistance, there has been considerable research in the past, which has yielded significant understanding of the force-deformation characteristics of masonry piers. Only few models for spandrel strength have been developed, and few codes for e.g. [1] and [2] provide basis for lateral load strength estimate of the

J. A. Kollerathu (✉)
CHRIST (Deemed To Be University), Bengaluru, Karnataka 560060, India
e-mail: jacob.alex@christuniversity.in

A. Menon
Indian Institute of Technology Madras, Chennai, Tamil Nadu 600036, India

G. C. Marano et al. (eds.), *Proceedings of SECON'21*, Lecture Notes in Civil Engineering 171, https://doi.org/10.1007/978-3-030-80312-4_49

spandrel. Existing simplified analysis methods such as Storey-Shear Mechanism (SSM) idealises the spandrel to behave as a rigid coupler undergirding no damage. Results from experimental tests such as [3–5] indicate that it is not generally correct to idealize the lateral load behaviour of the spandrels as rigid couplers and that cracking in the spandrels leads to significant energy dissipation. To account for the effects of a spandrel at the global level, there has to be consistent models to describe the failure criteria of the spandrels (shear/flexure) with force (or moment)-deformation (or rotation) models. In this paper, available models to describe the failure mechanisms of the spandrel are reviewed, and the predicted strength values from each of these models are compared. An analytical model is developed to study the lateral load behaviour of URM spandrels. The proposed model is validated against experimental results and existing models.

2 State-Of-The-Art of Research

2.1 Experimental and Numerical Research

Gattesco et al. [6] studied the behaviour of masonry spandrels with wooden lintels (with and without axial load demand). The compressive strength of mortar, Young's modulus of masonry, tensile strength of masonry and cohesive strength of the joints were estimated as 2.17 MPa, 4626 MPa, 0.44 MPa and 0.19 MPa respectively. A diagonal tensile mode of failure was observed. It is observed that an increase of the axial load demand in the spandrel shifts the shear mechanism at failure from flexural rocking to diagonal tension. Using a similar test setup, Graziotti et al. [7] conducted full-scale displacement controlled cyclic quasi-static tests on stone masonry spandrels both with and without horizontal tensile resistant elements. It was observed that the spandrel without any tensile resistant features had a flexural mode failure, while in the presence of a tie rod the spandrel element failed in shear. The tie rod also had beneficial effects on the shear capacity of the spandrel. Beyer [8] studied the lateral load behaviour of spandrel elements supported by a timber lintel and a masonry arch. Results from the tests indicate that the timber lintel element had negligible influence on the shear and flexural capacity of the spandrel. Similar inferences are observed in the numerical models described in Sect. 2. Results from [8] observed that while for spandrels with very axial compression loads in the absence of tensile resistant elements, the governing mode of failure follows a flexural behaviour, under significant axial loads the spandrel has shear behaviour. Experimental investigation [6] were conducted to understand the cyclic response of two masonry spandrels, one with a timber lintel and the other with masonry arch. The spandrels were loaded to failure and experimental observations indicated that spandrels with the arch demonstrated an inferior response in terms of shear capacity compared to the specimen with the timber lintel. These specimens were later strengthened with horizontal steel bars, and this led to an increase of 30% in the shear capacity. Additionally, the authors reviewed existing codal provisions describing the shear and flexural behaviour of masonry spandrels. Results indicate that existing provisions were able to predict the flexural and shear capacities obtained experimentally for unreinforced and reinforced masonry spandrels well. Additionally, results indicated that timber lintels had negligible influence on the peak strengths of URM spandrels. Therefore, it is observed that cut lintel do not influence the lateral load resistance of spandrels. A numerical model was developed by Rinaldin et al. [9] developed a numerical model was developed to represent the cyclic behaviour of the masonry spandrel. Results from NLFEM analyses were compared with the results from [6]. Results obtained indicate that the shear capacity estimates from NLFEM analysis consistently gives a lower bound of the capacities obtained experimentally, to the tune of 5–15%. Parisi et al. [10] through monotonic and cyclic in-plane loading experimentally investigated the lateral load behaviour of URM walls with different spandrels types. Results indicate that the specimen with the wooden lintel resisted higher base shear than the URM wall with the flat arch. This is because the wooden lintel did not slip off from the piers causing it to rock between the piers. The introduction of the RC bond beam on top increased the peak resistance of the URM wall by nearly 50%.

3 A Model for the Flexural Behaviour of Spandrels

The following section describes an analytical model proposed to describe the flexural behaviour of unreinforced masonry spandrels. The proposed model (described in Eqs. 1–3) accounts for the behaviour of the bed joints with a Mohr-Coulomb criterion as prescribed in (1). Earlier models also account for the equivalent tensile strength of head joints based on the cohesive strength of mortar. However, none of the models account for the tensile strength of the mortar while addressing this phenomenon. In the analytical model proposed in this thesis, the strength of the head joints will be the minimum of the tensile strengths of the brick, mortar and the brick-mortar interface and the cohesive strength of the joints (Figs. 1 and 2).

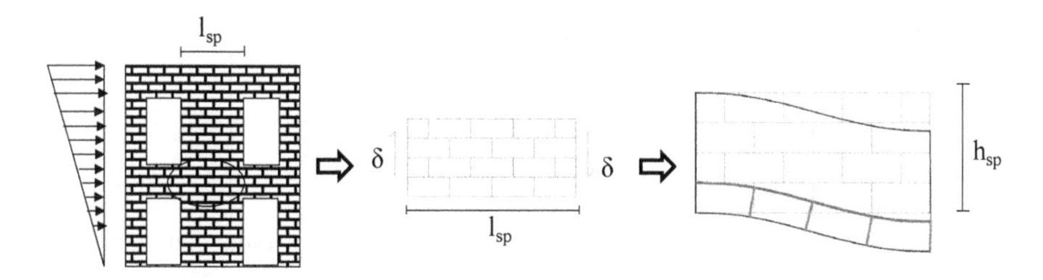

Fig. 1 Deformations in a spandrel in a building under lateral loading

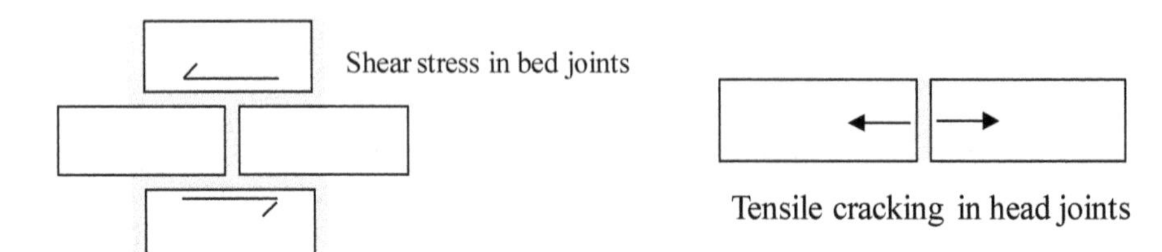

Fig. 2 States of stress in bed and head joints

The strength of the bed joints, V_{bj} and head joints, V_{hj} are determined as:

$$V_{bj} = \sigma_{bj} t_b \frac{l_b}{2} \tag{1a}$$

σ_{bj}: stress in the bed joints which is limited to

$$c + \mu v_{sp} \sigma \tag{1b}$$

$$V_{hj} = f_t h_b \frac{t_b}{2} N_b \tag{1c}$$

The capacity of the spandrel is a sum of the resistances provided by the bead joints and head joints and is given as per

$$V_u = \left(\sigma_{bj} t_b \frac{l_b}{2} + f_t h_b \frac{t_b}{2} N_b \right) N_b \tag{2}$$

where:

c: cohesive strength of the joints;
μ: coefficient of friction;
v_{sp}:coefficient representing the axial stress on spandrel from the vertical stress on the pier;
f_t: Tensile stress in head joints;
l_b, t_b, h_b: length, thickness and thickness of the brick;
N_b: Number of bed joints

$$\frac{0.5 h_{sp}}{(h_b + t_j)} \tag{3}$$

h_{sp}: height of spandrel, t_j: thickness of joint.

It is observed from experimental results that it is the head joints that crack first. Results from (7) indicate that at an element rotation of 0.1%, there is cracking leading to a loss in stiffness. From the proposed model, it is observed that at a rotation of 0.09% there is a loss of strength and stiffness when the contact of the head joints is lost (see Fig. 3). It is proposed that once the head joints crack, they do not participate in resisting deformations, and there is a loss in strength (close to 6%) in the spandrel. Thereafter, the capacity of the spandrel is:

$$V_u = \left(\sigma_{bj} t_b \frac{l_b}{2}\right) \frac{N_b}{2} \tag{4}$$

The bed joints resist the deformations until the cohesive strength of the bed joints are lost.

The peak capacity of the spandrel is hence obtained as

$$V_u = \sigma_{p,bj} t_b \frac{l_b}{2} N_b \tag{5a}$$

where, $\sigma_{p,bj}$: stress in the bed joints which is given as

$$\sigma_{p,bj} = c + \mu v_{sp} \sigma \tag{5b}$$

Once the cohesive strength of the bed joints is exceeded, they resist the lateral deformations with a residual strength provided by the un-cracked length of the bed joints. The spandrel continues to resist the deformations with its residual strength until the bricks start to crack in tension. This observation is consistent with existing experimental research. The ultimate displacement of the spandrel is defined as the displacement at which the maximum stress in the spandrel corresponds to the tensile strength of the brick. The spandrel displacement, δ_u corresponding to the

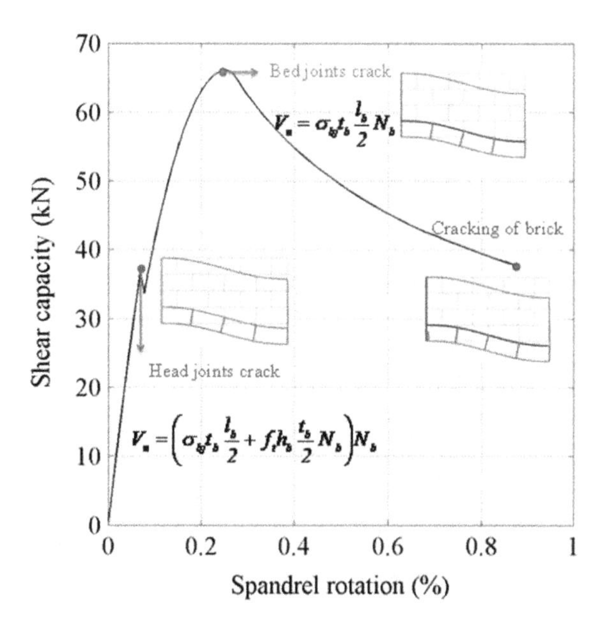

Fig. 3 Example of a shear-displacement graph for a spandrel

tensile cracking of the bricks is obtained as 9.10 mm. The rotation of the spandrel is given as

$$\theta = \frac{\delta}{l_{sp}} \tag{6}$$

This is obtained as 0.0076 which is close to the currently empirically assumed ultimate displacement of 0.008. Figure 6 represents the non-linear force-displacement relation of spandrel element developed using the proposed analytical model.

4 Validation of Proposed Model

The model developed to describe the flexural behaviour of spandrels was validated with NLFEM analysis (TNO-Diana 9.6®), with experimental results (7) and (9) and existing closed-form solutions. Unreinforced masonry spandrels were modelled using the simplified micro-modelling approach. The non-linear behaviour of the unit-mortar interface has been modelled as per the combined-shearing-crushing model.

The setup for estimating spandrel strength in NLFEM is same as that in the experiment with the pier on the left being supported at the top and bottom, while the right pier is free. The piers are subject to an axial stress of 0.5 and 0.17 MPa, to be consistent with the experiments. The rotation on the top of the two piers was not constrained. The axial compression on the spandrels is zero to be consistent with the experimental specimen. The vertical displacement is applied along the bottom the right pier. As results from neither experimental setup reported damage in the timber lintel beam, it is modelled as an elastic isotropic one-dimensional beam element with a bearing length of 200 mm on either side. The right pier is subject to a monotonic incremental uplift, whereas in the experiments the spandrel is subject to a cyclic displacement.

The peak flexural capacity from the analytical model is compared with strength values from existing solutions. From Fig. 4a–b the following observations are made:

- The initial elastic stiffness and the peak shear capacities obtained from the model compare well with NLFEM analysis and experimental results.
- The cracking of the head joints at an element rotation of 0.09% has not been reported in the experiments. However in the experimental study by (7) and (9), at an element rotation of 0.1%, there is the first observation of flexural cracking in the head joints leading to a drop of stiffness. From the proposed non-linear force-deformation, one observes that at a rotation of 0.09 there is a loss of strength and stiffness when the contact of the head joints is lost.
- The analytical model underestimates the peak capacity by 5% and 7%, respectively, when compared to results from experiments and NLFEM analysis.

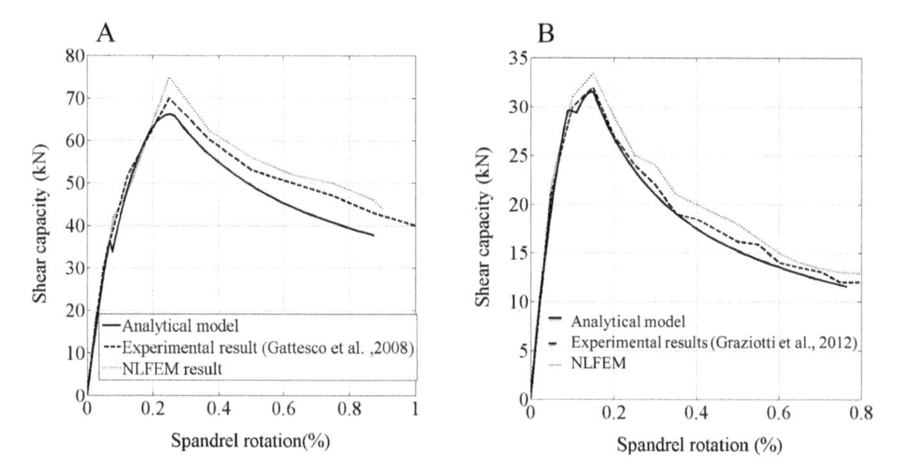

Fig. 4 Validation of analytical model with **a** experimental Gattesco et al. [6] and NLFEM analysis **b** experimental Graziotti et al. [7] and NLFEM analysis

- In the post-peak phase the model underestimates the flexural capacity by 10% and 12%, respectively, when compared to experimental and NLFEM results. This deviation in the analytical model maybe due to fact that in the experiments the loading was cyclic while in the analytical model it is uni-directional.
- The ultimate spandrel rotation from the model 0.9% is very close to the experimental results 1.0% (7) and NLFEM results 0.9%. The rotation of the spandrel is obtained as per Eq. 6.
- The ultimate spandrel rotation from the model 0.76% is very close to the experimental results 0.8% (8) and NLFEM results 0.8% (Figs. 5 and 6).

5 Results and Conclusions

Realistic seismic assessments of URM buildings need to account for the effect of the masonry spandrels on their global behaviour. In this paper, a simple analytical model that describes the flexural mechanisms of brick masonry spandrels is proposed. The model is based on the assumption that the interlocking provided by the joints contribute to the flexural strength of the spandrel. A feature of the proposed model is the ineffectiveness of the head joints to shear resistance post cracking. This is supported from experimental observations. It is also observed from existing models (e.g. FEMA 306) that considering the strength of head joints overestimates the strength of the spandrel obtained from experimental results by nearly 11%. The conclusion from results obtained from research carried out can be listed as follows:

1. Inclusion of a lateral strength model for spandrels even in simplified analysis such as SSM provides a closer estimate of initial elastic stiffness, peak shear

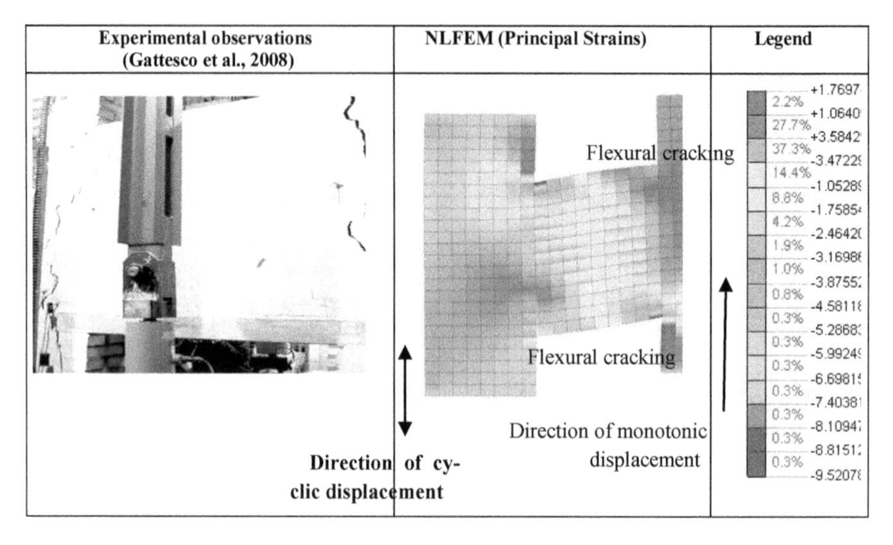

Fig. 5 Comparison of deformed shapes at ultimate

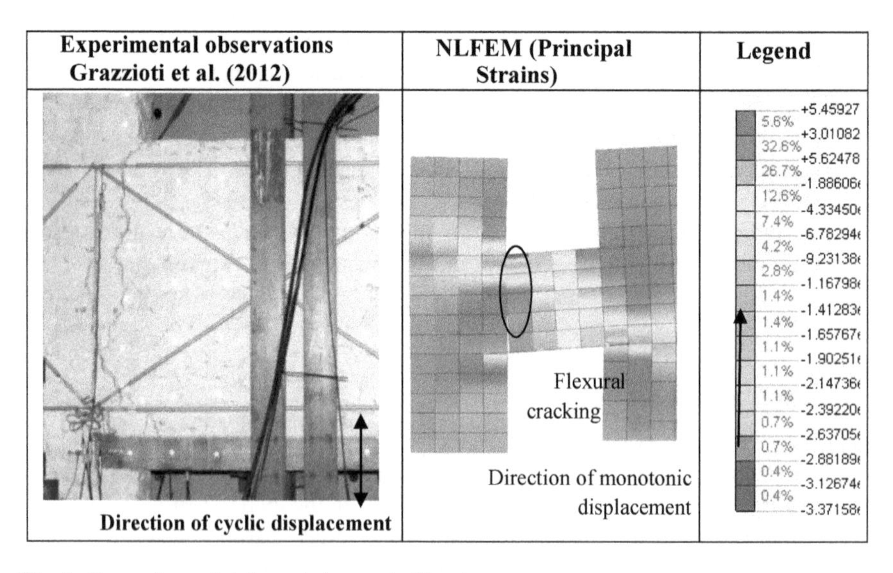

Fig. 6 Comparison of deformed shapes at ultimate

capacity and the ultimate displacement in comparison to more rigorous methods, such as NLFEM analysis.

2. The strength model used in the existing EFM approach to represent lateral load behaviour of the spandrel overestimates the shear capacity by nearly 13% of experimentally determined spandrel capacity. The proposed flexural model provides a closer estimate of the lateral load capacity of the spandrel.

References

1. FEMA 306 (1998) Evaluation of earthquake damaged concrete and masonry wall buildings
2. OPCM 3431 (2005) Further modifications and integrations on OPCM in 3274/07
3. Bendetti D, Carydis P, Limongelli M (2001) Evaluation of the seismic response of masonry buildings based on energy functions. Earthq Eng Struct Dyn 30(7):1061–1081
4. Magenes G, Kingsley GR, Calvi GM (1995) Seismic testing of a full-scale, two-story masonry building: test procedure and measured experimental response, experimental and numerical investigation on a brick masonry building prototype. Pavia
5. Yi T, Moon FL, Leon RT, Kahn LF (2006) Lateral load tests on a two-story unreinforced masonry building. J Struct Eng 132(5):643–652
6. Gattesco N, Macorini L, Dudine A (2016) Experimental response of Brick-Masonry spandrels under in-plane cyclic loading. ASCE J Struct Eng 2:1–14
7. Graziotti F, Magenes G, Penna A, Fontana D (2012) Experimental cyclic behaviour of stone masonry spandrels. In: 15th world conference on earthquake engineering
8. Beyer K (2012) Peak and residual strengths of brick masonry spandrels. Eng Struct 41:533–547
9. Rinaldin G, Amadio C, Gattesco N (2017) Review of experimental cyclic tests on unreinforced and strengthened masonry spandrels and numerical modelling of their cyclic behavior. Eng Struct 132:609–623
10. Parisi F, Augenti N, Prota A (2014) Implications of the spandrel type on the lateral behavior of unreinforced masonry walls. Earthq Eng Struct Dyn 43:1867–1887

Severity Prediction of Construction Site Accidents Using Simple and Ensemble Decision Trees

M. Rijo George, Madhusudana Rao Nalluri, and K. B. Anand

Abstract Workplace safety is always a concern of utmost importance in any organization. Studies have shown that the fatality rate is highest in the construction industry among all other industries. The construction project managers need to understand the risk status of each of their projects and thus implement preventive measures. The introduction of digital tools into construction sites not only reduces the health and safety hazards among workers but also paves the way to the economic growth of the industry. The present study implements techniques of data analytics and Machine Learning (ML) into the construction safety sector. For the analysis, a dataset with 4847 incident reports during 2015–2017 from Occupational Safety and Health Administration (OSHA) database is used. Initially, the major attributes contributing to the incident are identified. Based on these identified factors, they were classified as Before Accident and After Accident attributes, and ML algorithms are used for the prediction of the construction fatality. Performance evaluation of these ML algorithms shows us that Random Forest (RF) has better prediction results for Before Accident attributes, whereas Decision Tree (DT) performed well for After Accident attributes. From a broader perspective, this study will help the safety management team to understand the severity of safety risks faced in each of their construction projects and also facilitate the implementation of proper preventive safety mechanisms.

Keywords Machine learning · Construction safety · Workplace injuries · Prediction · Safety management · Accident prevention

M. Rijo George · K. B. Anand (✉)
Department of Civil Engineering, Amrita School of Engineering, Coimbatore, Amrita Vishwa Vidyapeetham, Coimbatore, India
e-mail: kb_anand@amrita.edu

M. R. Nalluri
Department of Computer Science and Engineering, Amrita School of Engineering, Coimbatore, Amrita Vishwa Vidyapeetham, Coimbatore, India

G. C. Marano et al. (eds.), *Proceedings of SECON'21*, Lecture Notes in Civil Engineering 171, https://doi.org/10.1007/978-3-030-80312-4_50

1 Introduction

Construction is an ever-growing industry and this sector has the highest rate of fatal work injuries among all other industries which can result in both human as well as economic sufferings. When compared to other countries a huge lag could be observed in the Indian construction sector concerning safety standards and safety inspection as well [1]. The construction companies must also try to understand the level of safety in each of its projects, thus implementing better or preventive actions wherever required proactively [2]. Many safety practices are implemented at construction sites, but traditional methods are showing their limitations and innovations in construction safety is to be instigated [3]. Understanding the working environment helps to increase productivity and also reduces both direct and indirect costs incurred in worksites [4]. Construction safety datasets are released by various organizations and are also available at some government releases, for example, OSHA. Analysis of big data, complex data, and real-time data from online sources is a challenge for conventional statistical methods, thus utilization of this data is less efficient [5]. Computational techniques help increase the efficiency of the projects as well as make the whole process and information transparent [6]. Thus, ML algorithms are used to analyze these large datasets, which understand the data and automatically improves by learning from it. The predictive mechanism of ML algorithms is widely utilized in the field of medical science [7, 8]. Even though such applications are limited in the field of construction, Safety researchers in the construction industry by application of ML methods into accident reports, have detected hidden patterns among the incidents and was able to efficiently predict the accident outcomes like the severity of injury and construction accidents [9], cause of delay [10], understanding the constructability levels [11], severity and occurrence of accidents [12], unsafe behaviour of scaffolders [13], etc. When predicting the severity of an injury or its related attributes different methods are practiced in each study. It could be using ML algorithms like DT [9], RF [9, 14], Logistic Regression (LR) [9, 15], K-nearest neighbour (KNN) [9], and Support Vector Machines (SVM) [9] or Stochastic Gradient Tree Boosting (SGTB) [14] or AdaBoost [16] or XGBoost [17]. Studies have shown that the efficiency of the management is largely affected when they are not aware of the depth of severity of risk involved in each of their projects [18]. This inefficiency leads to poor safety practices leading to accidents or fatalities. The same dataset used in this study was utilized by Passmore et al. [19]. Even though the dataset is not a representative sample of the whole U.S workers in the construction industry, the study helps to comprehend the severity of each factor influencing construction injuries. Rather than focusing more on non-fatal injuries, the study focused on the estimation of the relative frequency of fatal injuries. The identification of the most probable risk factors helps to lessen the degree of injuries as well as bring about a serious decline in the fatality rate as well. Further studies on this dataset using ML techniques would help to understand more about the safety outcomes.

The main objective of this paper is to analyze the datasets collected and use ML algorithms for the prediction of fatal and non-fatal cases in construction sites. This study employs predictive algorithms in the given attributes and their performance is evaluated. This enables the safety management team to understand the severity of safety risks faced in each of their construction projects. The comprehension of these factors could help to adopt preventive mechanisms and also in developing safety training, thus efficiently reducing accidents.

2 Methodology

2.1 Dataset

A brief methodology of the sequence of execution of work is detailed in Fig. 1. The dataset used in this study is extracted from the OSHA database and it contains abstracts of the accidents that occurred in the construction industry alone between 2015 and 2017. It includes details of 4847 accident reports, both fatal and non-fatal. The datasets contain unstructured text abstracts along with structured data, which were extracted from these reports using Natural Language Processing (NLP) [14], a computational language analysis technique. Since the structured dataset is available, there is no need to use NLP techniques to extract the data. The structured dataset used in this study contains 11 major OSHA attributes, which includes 153 elements, helping to understand the incident in detail and also used for prediction effectively.

2.2 Data Understanding

Data Understanding helps us to comprehend more on the attributes or features involved, the nature of these attributes and also aids to visualize these datasets to make more sense into the data. The identification of noise, outliers, and missing values during this stage aids in the ease of data pre-processing [20]. By plotting the bar chart of each attribute, the understanding of the distribution frequency of each element as well as the identification of outliers involved in each attribute was obtained. It was observed that some of the elements of the data were not available and many were left blank. The large percentage of such unavailable data can affect the efficiency of the model and thus they could be eliminated (Table 1).

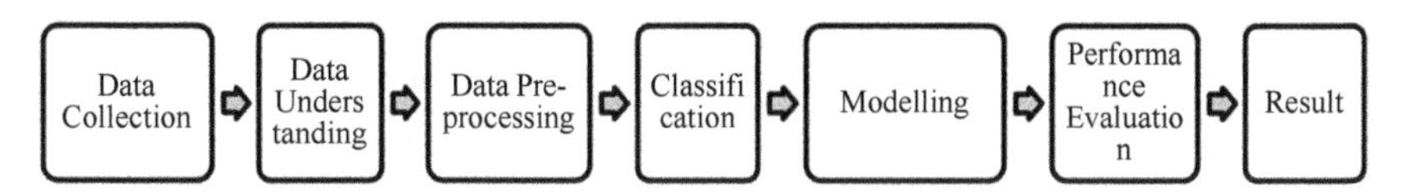

Fig.1 Methodology of work

Table 1 Outliers in the dataset

S. No	Attribute	No. of blanks	% (in terms of total data)	Remarks
1	Construction end use	3820	78.82	Omitted from further analysis
2	Building stories	4148	85.58	
3	Project cost	4261	87.92	
4	Project type	3767	77.72	
5	Degree of injury	0	0	Considered for data pre-processing
6	Nature of injury	2	0.05	
7	Part of body	2	0.05	
8	Event type	2	0.05	
9	Environmental factor	7	0.15	
10	Human factor	7	0.15	
11	Task assigned	0	0	

2.3 *Data Pre-Processing*

Since real-world databases are huge in size and originate from multiple and heterogeneous sources, they are extremely vulnerable to missing, noisy and inconsistent data. Transformation of raw data into pre-processed good quality data is vital for obtaining the best results when applied to the ML algorithms. Some major steps followed in data preprocessing are:

Data Reduction: The attributes involved in a particular data may be huge and the size of the dataset can slowdown the analytical process [20]. Therefore, not all the features involved in the dataset are put in for prediction. Thus, the 4 attributes containing a large percentage of blanks were omitted from the analysis, and 7 attributes were considered for data pre-processing (Table 1).

Data Cleaning: Necessary replacements were performed and removal of missing values and inconsistencies [20] were executed. All the observations containing missing values were omitted for a clean dataset.

Data Transformation: Transformation of the raw data, a form of the data reduction process, so that all the processes involved in the analytical stages could be more proficient, helping to comprehend all the patterns identified [20]. Transformation of categorical data to factor or numerical attribute for modeling.

Feature Selection: Boruta feature selection method could be applied to identify significant input variables [9]. Boruta package of R was performed with 20 iterations in 25.8683 secs and no attributes were deemed unimportant. Similarly, several packages of R are employed during the study (Table 2).

Thus, a new clean data set of 4838 observations and 7 attributes are obtained for modeling.

Table 2 Major R packages used

Task name	Packages used	List of functions
Data pre-processing	dplyr	complete.case(), na.omit ()
	Boruta	Boruta()
Classification	party	ctree()
	rpart.plot	rpart.plot()
Modelling	caret	confusionMatrix ()
	rpart	rpart()
	randomForest	randomForest()

2.4 *Classification*

Classification, a supervised learning approach, helps to recognize the major attributes and group them into categories for predictive analysis, based on the training dataset. The classification of the input variables could be risk status [9] or injury type, energy type and body part [14] or injury severity [17] or even general classification as fatal or non-fatal [16] as well. These outcomes help the construction managers and safety engineers to understand the risk status of each of their sites and implement preventive measures. DT is a powerful and frequently used classification tool that produces a tree-like model with the attributes and generates their outcomes [21]. Party package and rpart package of R was used for DT classification. A DT model was trained using the 7 selected variables. As the attributes in the higher nodes of a DT are more influential in classifying the dataset, they could be used as significant input variables for modeling. It was seen that *Part of Body, Nature of Injury, Event Type, and Environmental Factor* are the attributes majorly contributing to the outcome in one package and *Part of Body and Nature of Injury* was deemed important in another package (Fig. 2). Thus, we could eliminate the attributes like *Human Factor and Task Assigned* for prediction and use the other four attributes for further analysis, i.e., *Part of Body, Nature of Injury, Event Type, and Environmental Factor.*

2.5 *ML Modeling and Performance Evaluation*

The original datasets are segregated for training and testing and the predicted attributes of the trained model are compared with the testing set [14]. The ML algorithms use the training datasets during the training phase for predictions. The test datasets are not available during the training phase. Afterward, when the testing phase is executed the test dataset is utilized by the algorithm for predictions. Confusion matrices (Table 3) are generated for each model based on the testing dataset, which helps to understand the performance of the respective ML algorithms in predicting these safety outcomes.

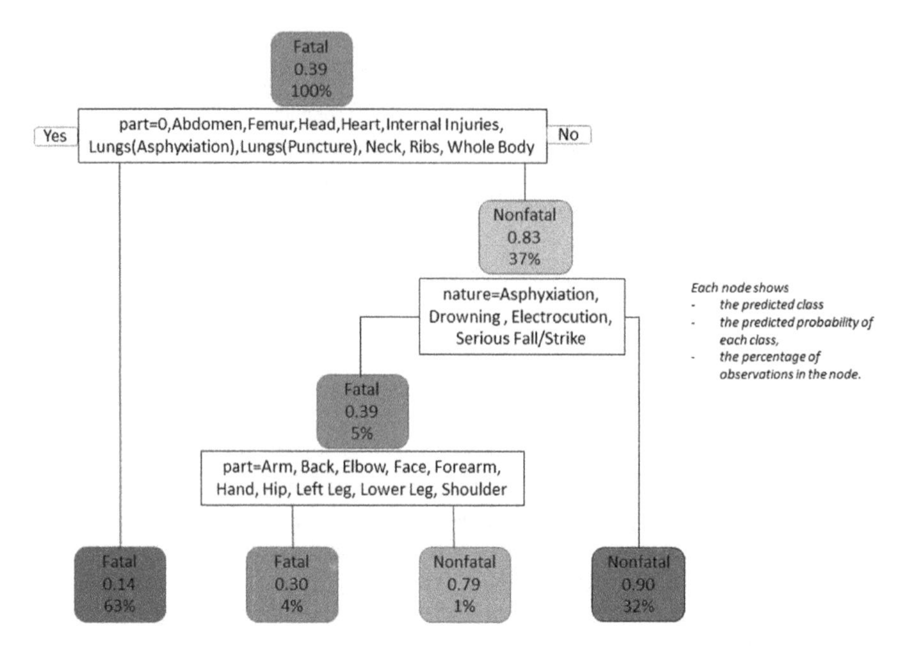

Fig. 2 DT classification with rpart package

Table 3 Confusion matrix

Predicted case	Actual case	
	Fatal	Non-fatal
Fatal	True Positive (TP)	False Positive (FP)
Non-fatal	False Negative (FN)	True Negative (TN)

Recall, precision and accuracy [9], or F-1 score [17, 16], etc., are some commonly used performance measures. In this study, the following performance measures were evaluated from the confusion matrix generated: Accuracy, Recall (also known as Sensitivity or True Positive Rate (TPR)), Specificity (also known as True Negative Rate (TNR), Type 1 error (False Positive Rate (FPR)), Type 2 error (False Negative Rate (FNR)), Precision, F1 score, Kappa value (Table 4). Accuracy shows the measure of the correct predictions made by the model. But false predictions can have catastrophic impacts on construction sites. Thus, Type 1 error and Type 2 error also need to be considered. Whenever the false-negative rate has a serious impact on the predicted results, recall and precision is used. F1 score combines the values of recall and precision to measure the performance of the model. Kappa score helps us to understand the inter-rater reliability, i.e., the degree of agreement between the predicted and the actual cases. It measures the homogeneity of values of the predicted and actual model.

Table 4 Performance measures

Accuracy	(TP + TN)/(TP + TN + FP + FN)
Recall, sensitivity, TPR	TP/(TP + FN)
Specificity, TNR	TN/(FP + TN)
FPR, Type 1 error	FP/(TP + FN)
FNR, Type 2 error	FN/(FP + TN)
Precision	TP/(TP + FP)
F-1 Score	2 × (Precision × Recall)/(Precision + Recall)
Kappa*	(Po–Pe)/(1–Pe)

*Po—observed proportion of agreement, Pe—expected agreement by chance

3 Results and Discussions

By classification, it was possible to identify the major attributes in the dataset contributing to understanding the *Degree of Injury*. Based on the features involved in it, they are further classified as following (Table 5) Prediction of the *Degree of Injury*, fatal or non-fatal, is executed based on these attributes. Since the dataset available to us is categorical in nature, two ML algorithms; DT and RF are used for predictive analysis. A train-test split of 80–20% was used in the predictive models.

A comparative study of DT and RF models was conducted on Before Accident attributes (Fig. 3a, b). The performance of both models was evaluated using the eight performance measures discussed. When the Before Accident attributes, i.e., *Event Type* and *Environmental Factor*, were evaluated, we were able to understand that the RF model showed better results when compared to the DT model.

Later a similar study was conducted on the After Accident attributes, i.e., *Part of Body* and *Nature of Injury*. By evaluating the performance measures, it was observed that the DT model showed better results when compared to the RF model. It was also seen that there was a considerable reduction in the FNR value as compared to the Before Accident analysis. The *Environmental Factor* involves features like Materials Handling, Sound Level, Temperature, Weather, etc., which are the factors detailing the working environment condition which has led to the accident. Also, caught in or between, Fall, Struck-by, etc., are the features involved in the *Event Type*. By understanding the Before Accident attributes, it will be possible to identify the attributes contributing towards an accident and thus can be

Table 5 Risk factors affecting fatality

Occurrences	Risk factors
Before Accident: Factors leading to accidents	Environmental factor
	Event type
After Accident: Factors which are consequences of accidents	Nature of injury
	Part of body

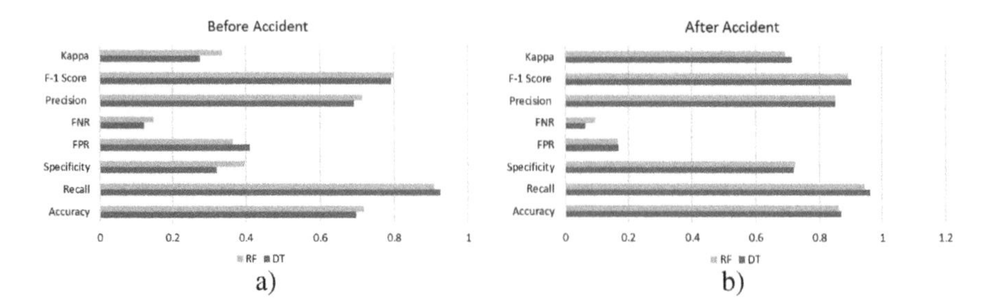

Fig. 3 **a** Comparison of RF and DT models in before accident attributes, **b** comparison of RF and DT models in after accident attributes

prevented by taking appropriate measures. The situations leading to accidents/ incidents could be avoided by proper safety measures at sites. For this analysis, the application of RF models was seen as effective. Even though the accuracy of the result is less compared to the After Accident attributes, the understanding helps safety managers to predict the severity index of their respective construction projects and thus to prevent fatal accidents at projects and also increases the safety standards of construction sites. While the '*Nature of Injury*' details the physical characteristics of an injury like Amputation, Crushing, Dislocation, etc., '*Part of Body*' indicates the different parts of the body affected like Abdomen, Arm, Back, etc. Even though these After Accident data could be used only after an accident has occurred, it significantly helps for safety analysis. The previous safety records of a construction company could be utilized for this analysis and the results aids to identify the construction projects of high risk and initiate preventive mechanisms. The application of DT models was seen effective than the RF models in this case. Based on the situation or the attributes involved the performance of the ML algorithms may vary, thus a construction management team handling digital datasets must understand the precise tool required for their purpose.

4 Conclusions

When multiple projects and stakeholders are involved in construction, the information generated is numerous and brings in large documentations as well. A manager can analyse a small dataset of a small construction site using simple analysis tools. But when it comes to bigger projects large and vibrant datasets are available and the analysis of these datasets requires special tools. When conventional tools are used it affects the quality of knowledge extracted and is also a time-consuming process. ML techniques help to extract the hidden knowledge inside the large databases and transform it into useable and valuable data. This study was based on the dataset collected from the OSHA database. They were run through several processes (like data understanding, data pre-processing, classification), and the dataset was reduced to 7 out of 29 attributes. They were further classified as Before Accident and After Accident attributes and ML algorithms were used to predict the *Degree of Injury*, as fatal and non-fatal. The DT and RF models were used for prediction and it was seen that the latter performed better in the Before Accident case and the former in the After Accident case. Thus, the application of data analytics and ML techniques into the construction industry was seen as effective. The results produced would help the management understand that a proper analysis of the data would aid to identify projects of high risk and analysis of Before Accident attributes will help them to prevent occurrences of such accidents as well. Also, the importance of data collection and safety records is also evident in the study which is mostly procrastinated in the Indian construction sector.

References

1. Govind AN, Vivek S (2019) Analysis of construction safety management in India and UAE. Int J Innovative Technol Exploring Eng 8(6):848–852
2. Houchen CAO, Yang Miang GOH (2019) Analyzing construction safety through time series methods. Front Eng Manag 6(2):262–274
3. Esmaeili B, Hallowell MR (2011) Diffusion of safety innovations in the construction industry. J Constr Eng Manag 138(8):955–963
4. Renganath K, Suresh M (2016) Analysing the drivers for safety practices using interpretive structural modelling: a case of Indian manufacturing firms. IEEE Int Conf Comput Intell Comput Res (ICCIC)
5. Bilal M, Oyedele LO, Qadir J, Munir K, Ajayi SO, Akinade OO, Owolabi HA, Alaka HA, Pasha M (2016) Big data in the construction industry: a review of present status, opportunities, and future trends. Adv Eng Inform 30(3):500–521
6. Ganesh J, Suresh M (2016) Safety practice level assessment using multigrade fuzzy approach: a case of Indian manufacturing company. IEEE international conference on computational intelligence and computing research (ICCIC), pp 1–5
7. Gupta D, Khare S, Aggarwal A (2016) A method to predict diagnostic codes for chronic diseases using machine learning techniques. In: International conference on computing, communication and automation (ICCCA), pp 281–287
8. Ani R, Augustine A, Akhil NC, Deepa OS (2016) Random forest ensemble classifier to predict the coronary heart disease using risk factors. In: Proceedings of the international conference on soft computing systems. Advances in intelligent systems and computing, vol 397, pp 701–710
9. Clive Poh QX, Chalani UU, Yang MG (2018) Safety leading Indicators for construction sites: a machine learning approach. Autom Constr (93)375–386
10. Soibelman L, Kim H (2002) Data preparation process for construction knowledge generation through knowledge discovery in databases. J Comput Civ Eng 16(1):39–48
11. Skibniewski M, Arciszewski T, Lueprasert K (1997) Constructability analysis: machine learning approach. J Comput Civ Eng 11(1):8–16
12. Goh YM, Chua D (2013) Neural network analysis of construction safety management systems: a case study in Singapore. J Constr Manag Econ 31(5):460–470
13. Goh YM, Binte Sa'Adon NF (2015) Cognitive factors influencing safety behavior at height: a multimethod exploratory study. J Constr Eng Manag 141(6)

14. Antoine Tixier JP, Matthew Hallowell R, Rajagopalan B, Bowman D (2016) Application of machine learning to construction injury prediction. Autom Const 69:102–114
15. Esmaeili B, Hallowell MR, Rajagopalan B (2015) Attribute-based safety risk assessment. II: predicting safety outcomes using generalized linear models. J Constr Eng Manag 141(8): 04015022
16. Gholizadeh P, Esmaeili B, Memarian B (2018) Evaluating the performance of machine learning algorithms on construction accidents: an application of ROC curves. Constr Res Congr 8–18
17. Baker H, Matthew Hallowell R, Antoine Tixier JP (2020) AI-based prediction of independent construction safety outcomes from universal attributes. Autom Constr 118:103–146
18. Goh YM, Soon WT (2014) Safety management lessons from major accident inquiries. Pearson, Singapore, (ISBN: 9789814598699)
19. Passmore D, Chae C, Borkovskaya V, Baker R, Yim JH (2019) Severity of U.S. construction worker injuries, 2015–2017. E3S Web Conferences 97:06038
20. Witten IH, Frank E, Hall MA, Pal CJ (2016) Data mining: practical machine learning tools and techniques, 4th ed. Morgan Kaufmann, Boston, (ISBN: 978-0-12-374856-0)
21. Sankar A, Divya Bharathi P, Midhun M, Vijay K, Senthil Kumar T (2016) A conjectural study on machine learning algorithms. In: Suresh L, Panigrahi B (eds) Proceedings of the international conference on soft computing systems. Advances in intelligent systems and computing, vol 397, pp 105–116

Study on Effect of GGBS, Molarity of NaOH Solution and Curing Regime on Compressive Strength and Microstructure of Fly Ash-GGBS Based Geopolymer Concrete

Kunal Pradhan and Bulu Pradhan

Abstract In this study, the effects of ground granulated blast furnace slag (GGBS) and molarity of NaOH solution on compressive strength (CS) and microstructure of fly ash-GGBS based geopolymer concrete (GPC) have been investigated under different curing regimes. Class F fly ash and GGBS were used as aluminosilicate source materials where fly ash was replaced with 20, and 40% GGBS. A combination of varying concentrations of NaOH solution (8, 10, and 12 M) and Na_2SiO_3 solution was used as the alkaline solution. Cube specimens of 150 mm size were cast for determining CS of GPC at the ages of 7 and 28 days. The curing regimes adopted for fly ash-GGBS based GPC specimens were ambient curing and water curing. The microstructure of GPC was studied by X-ray diffraction (XRD) and Field emission scanning electron microscopy (FESEM) analyses. From the results, it is observed that the workability of GPC mixes reduced when GGBS content and molarity of NaOH solution was increased. The CS of GPC mixes increased as the amount of fly ash replaced by GGBS and the molarity of NaOH solution increased, under both curing regime. GPC cube specimens subjected to water curing achieved higher CS as compared to ambient curing at both the ages. Results of XRD and FESEM analyses confirmed the presence of N–A–S–H and N–(C)–A–S–H gel, which are responsible for the development of CS of geopolymer concrete.

Keywords Geopolymer concrete · Fly ash · GGBS · Ambient curing · Water curing · Compressive strength · Microstructure

K. Pradhan (✉) · B. Pradhan
Department of Civil Engineering, Indian Institute of Technology Guwahati, Guwahati 781039, India
e-mail: kunal176104015@iitg.ac.in

B. Pradhan
e-mail: bulu@iitg.ac.in

G. C. Marano et al. (eds.), *Proceedings of SECON'21*, Lecture Notes in Civil Engineering 171, https://doi.org/10.1007/978-3-030-80312-4_51

1 Introduction

Concrete is the most widely used building material on earth. Portland cement is the main binder in the traditional concrete. The manufacture of Portland cement aids considerably to the greenhouse effect due to emission of high volume of carbon dioxide with 0.81 kg of CO_2 released per kg of Portland cement produced [1, 2]. The manufacturing process of Portland cement itself requires significant amount of energy [3]. In order to preserve the natural resources as well as to lower the Portland cement consumption, the use of waste products from energy and minerals industry as supplementary cementitious materials has been the focus of many research works [4, 5]. GPC can be developed as a promising alternative to conventional concrete, which uses industrial waste such as fly ash, GGBS etc. in place of Portland cement. An alkaline solution, usually a blend of NaOH solution and Na_2SiO_3 solution, is used to trigger alumina and silica rich source material (i.e. fly ash and GGBS etc.) to produce geopolymer binder. The production of GPC consumes huge quantity of industrial waste and improves the sustainability [6].

The geopolymer reaction mechanism is summarized into three steps i.e. dissolution of silica and alumina rich source material, gel formation and polycondensation [7]. The drawback of using fly ash based GPC in in-situ construction is its requirement for heat curing for early age strength development. Therefore, its use has been limited to precast construction industry [8]. Various additives have been used in fly ash based GPC to develop suitable early age strength under ambient curing condition [9]. Addition of a calcium based raw material by partial replacement of fly ash has shown a positive effect on GPC with improvement in mechanical properties and decrease in setting time [10]. The influence of GGBS on fresh and mechanical properties of GPC has been investigated by many researchers. Lee and Lee [11] reported a decrease in setting time of geopolymer paste with increase in GGBS content (15, 20, 25 and 30%) and increase in concentration (4, 6 and 8 M) of NaOH solution. Nath and Sarker [12] incorporated 10, 20 and 30% of GGBS by replacing fly ash and reported reduced setting time and workability, and increased compressive strength (CS) of geopolymer concrete under ambient temperature at higher GGBS content. Deb et al. [13] reported a rise in 28 days CS of GPC, prepared with fly ash and GGBS, by 8% when Na_2SiO_3/ NaOH solution ratio decreased from 2.5 to 1.5. The 28 days CS of GPC increased from 43 to 54 MPa when fly ash replacement by GGBS was increased from 10 to 20% at Na_2SiO_3/NaOH solution ratio of 1.5. Mehta et al. [14] reported an increase in CS of fly ash based alkali activated concrete with incorporation of GGBS (0–20%). A maximum CS of 65 N/mm^2 was reported with 20% GGBS content at 90 days. A denser matrix was reported with inclusion of 20% GGBS as compared to control alkali activated concrete with 0% GGBS. Ghafoor et al. [15] prepared GPC with different NaOH solution concentration (8–16 M), Na_2SiO_3/NaOH ratio (1.5–2.5), L/ B ratio (0.4–0.6) and reported an increase in workability when L/B ratio was increased and concentration of NaOH solution was decreased. The CS and flexural strength of GPC increased when the concentration of NaOH solution was increased and decreased when Na_2SiO_3/NaOH ratio and L/B ratio were increased.

It is observed from the reported literature that ambient curing has been used in the development of fly ash-GGBS based GPC in most of the research works. However, there is limited research work in the reported literature on the impact of water curing on the properties of fly ash-GGBS based GPC. Therefore, in this research work, the effects of different types of curing regime as well as that of GGBS replacement and NaOH solution molarity on the properties of GPC made with fly ash and GGBS have been evaluated.

2 Experimental Details

2.1 *Materials*

Class F fly ash and GGBS were used as silica and alumina rich source materials in the production of GPC. For alkaline solution, a blend of NaOH solution and Na_2SiO_3 solution was used. The NaOH solution was prepared by dissolving the required quantity of NaOH pellets in water as per the required molarity. The NaOH solution was made 48 h before the preparation of GPC mixes and was mixed with Na_2SiO_3 solution 24 h before the preparation of GPC. River sand was used as fine aggregate. For coarse aggregate, a combination of 10 mm MSA (maximum size of aggregate) and 20 mm MSA crushed aggregates was used.

2.2 *Mix Proportion, Casting and Curing*

The unit weight of fresh GPC mix for all the mixes was fixed at 2310 kg/m^3. The total binder content and L/B (liquid-to-binder) ratio used were 410 kg/m^3 and 0.5 respectively. GGBS was incorporated into fly ash at 20% and 40% in the GPC mix. The NaOH solution of different molarity such as 8, 10, and 12 M was used in the preparation of GPC cube specimens. The ratio of Na_2SiO_3/NaOH solution, by mass, was fixed at 1.5. The coarse aggregate and fine aggregate proportion of GPC mix was fixed at 64% and 36% respectively by weight of total aggregate.

A laboratory drum mixer was used to prepare of fresh GPC mixes. The mixing process consists of dry mixing of coarse and fine aggregate in the mixer followed by addition of binder (fly ash + GGBS) for further mixing. After dry mixing, alkaline solution was added with the dry mixture followed by further mixing to obtain the fresh GPC mix. The freshly prepared geopolymer concrete mix was cast into 150 mm cube moulds in 3 layers by subjecting each layer to adequate compaction. After 24 h of preparation, the GPC cubes were demoulded. The GPC cubes were divided into two groups. One group was subjected to ambient laboratory curing while the other group was subjected to normal water curing till the age of testing.

2.3 *Test Procedures*

Slump test was performed to determine the workability of fresh GPC mix. The CS test of GPC cube specimens was performed at 7 and 28 days. At the end of the test, the core of the broken cube was further crushed in a pulverizer. The crushed material was sieved through 75 μm sieve followed by storing in air tight containers for use in microstructure study. To observe the phase composition of GPC, XRD analysis was carried out using Rigaku model SmartLab 9 KW X-ray diffractometer. The GPC powder sample was scanned in a range from 5° 2θ to 60° 2θ at a rate of 20° 2θ per minute with 0.03° 2θ step size. The FESEM analysis was conducted to observe the morphology of GPC. An aluminum stub with a double stick carbon tape was used to mount the GPC powder sample. Before analyzing, the powder sample was gold coated by sputtering method.

3 Results and Discussion

3.1 *Slump Values of Fresh GPC*

From the obtained slump values of GPC mixes, as shown in Fig. 1, it is noted that the slump value of GPC mix decreased as the amount of fly ash replaced by GGBS increased. The reduction in slump value of the mix could be attributed to the hastened reaction of calcium present in GGBS as well as due to the effect of angular shape of GGBS particles over the effect of spherical shape of fly ash particles [16]. The slump value of the GPC mix decreased when there is an increase in the concentration of NaOH solution from 8 to 12 M. This could be due to the fact that at lower molarity of NaOH solution, there is reduction in solids content of the alkaline solution resulting in an increase in the consistency of the GPC mix.

3.2 *Compressive Strength (CS)*

Figures 2 and 3 show the effect of GGBS replacement, concentration of NaOH solution, and curing regime on 7 and 28-day CS of GPC mixes respectively. An improvement in CS of GPC mix was observed when there is an increase in GGBS content irrespective of concentration of NaOH solution, curing regime, and age. This could be ascribed to the effect of formation of more amount of N–A–S–H and C–A–S–H gels as well as formation of C–S–H gel [16], due to greater extent of geopolymerization process in GPC at higher GGBS content. Further, the additional heat liberated during the reaction of calcium with water enhanced the polymerization process, which increased the formation of N-A–S–H gel [17, 18] to a comparatively greater extent in the GPC mix made with more GGBS content. When

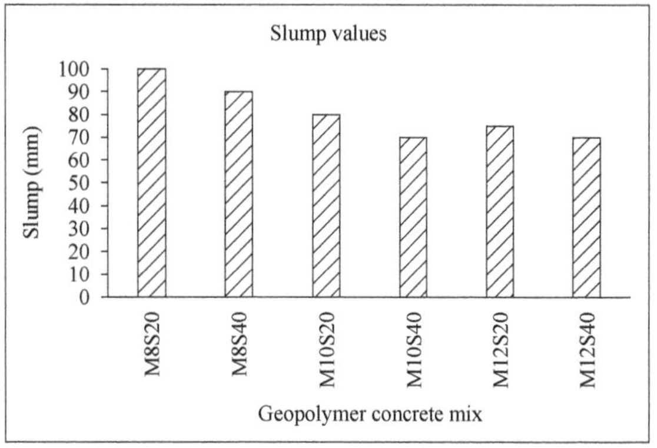

Fig. 1 Slump values of geopolymer concrete mixes

M8S20: GPC mix with 8 M NaOH solution and 20% GGBS

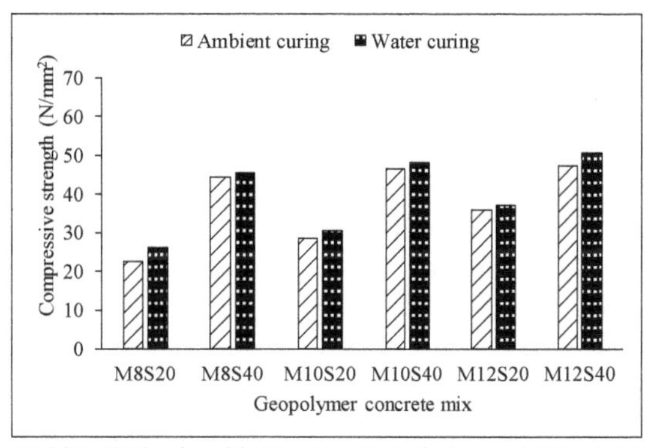

Fig. 2 7-day CS of GPC mixes under water curing and ambient curing

M10S20: GPC mix with 10 M NaOH solution and 20% GGBS

the concentration of NaOH solution was increased from 8 to 12 M, the CS of geopolymer concrete mixes increased both at 7 and 28 days for both GGBS replacement levels, and both curing type as noted from Figs. 2 and 3. This is due to the effect of enhanced dissolution of aluminosilicates in the alkaline solution with higher molarity of NaOH, which resulted in greater extent of geopolymerization reaction [19]. It was observed from Figs. 2 and 3 that the GPC cube specimens subjected to water curing achieved higher CS than those subjected to ambient curing at both 7 and 28 days irrespective of GGBS replacement level and concentration of NaOH solution. This may be due to the development of more amount of geopolymer gel and C–S–H gel in the GPC mixes subjected to water curing. A maximum 28-day CS of 69.33 N/mm^2 was achieved for the GPC mix made with 12 M NaOH solution and 40% GGBS content under water curing regime.

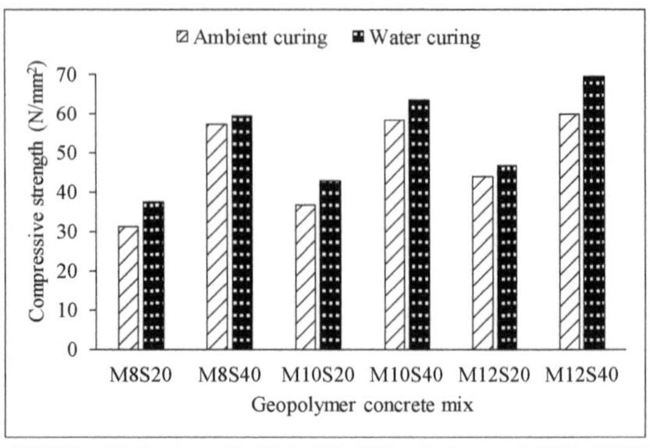

Fig. 3 28-day CS of GPC mixes under water curing and ambient curing

M12S40: GPC mix with 12 M NaOH solution and 40% GGBS

3.3 XRD Analysis

Figure 4 shows the XRD patterns of GPC prepared with 12 M NaOH solution and varying GGBS contents of 20 and 40% under ambient curing and water curing, at the age of 7 days.

From the XRD patterns, different phases in the GPC mixes were quartz (SiO_2) at 20.78° 2θ, 26.65° 2θ, 36.71° 2θ, 39.45° 2θ, 42.45° 2θ, 45.83° 2θ, 50.2° 2θ and 54.85° 2θ; mullite ($Al_{4.59}Si_{1.41}O_{9.7}$) at 16.43° 2θ, 33.22° 2θ and 40.56° 2θ; muscovite ((K, Na) Al_2(Si, Al)$_4O_{10}(OH)_2$) at 8.8° 2θ; anorthoclase (($Na_{0.75}$ $K_{0.25}$) ($AlSi_3O_8$)) at 27.6° 2θ and 35.3° 2θ; albite ((Na, Ca) Al (Si Al)$_3O_8$) at 28.02° 2θ; and C–S–H at 29.4° 2θ.

From the XRD patterns shown in Fig. 4, the formation of N–A–S–H gel in GPC mixes was indicated by the peaks corresponding to anorthoclase and that of N–(C)–A–S–H gel was indicated by the peak corresponding to albite. In addition, the XRD patterns showed the formation of C–S–H in the GPC mixes. Mostly higher peaks of albite, anorthoclase and C–S–H were observed in the GPC mixes with increase in GGBS content from 20 to 40% irrespective of curing regime as noted from Fig. 4. This indicates greater extent of geopolymerization in the geopolymer concrete mixes prepared with higher GGBS content. However, no systematic variation in peak intensity of albite, anorthoclase and C–S–H was observed from the XRD pattern (Fig. 4) with type of curing regime.

3.4 FESEM Analysis

The FESEM images of geopolymer concrete mixes made with 20% GGBS content and NaOH solution molarity of 8 M and 10 M, at 28 days are shown in Fig. 5. The FESEM images showed the presence of geopolymer gels, partially reacted fly ash particles, and voids in the GPC mixes. Denser microstructure was observed in

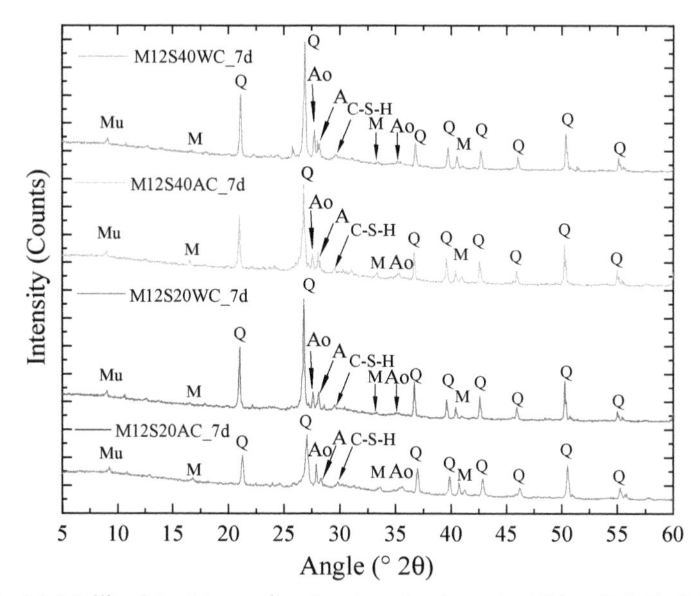

Q: Quartz; M: Mullite; Mu: Muscovite; Ao: Anorthoclase; A: Albite; C-S-H: Calcium silicate hydrate

Fig. 4 XRD patterns of GPC mix made with 20 and 40% GGBS contents at 12 M NaOH solution and subjected to ambient curing and water curing for 7 days

the GPC mix prepared with 10 M NaOH solution in contrast to that prepared with 8 M NaOH solution as noted from Fig. 5, which may be ascribed to the effect of geopolymerization process to a higher extent in the GPC mix made with higher NaOH solution molarity than that made with lower NaOH solution molarity.

4 Conclusions

The conclusions obtained from this study are presented below.

- The workability of the GPC mix reduced as the GGBS content and concentration of NaOH solution were increased.
- The CS of GPC, at 7 and 28 days, increased as fly ash replacement by GGBS increased. A maximum gain in CS of 59.11% was observed at the age of 28 days when the GGBS content in GPC mix was increased from 20 to 40% with 10 M NaOH solution under ambient curing condition.
- The CS of GPC, at 7 and 28 days, increased with the increase molarity of NaOH solution. A maximum gain in CS of 26.28% was observed at the age of 7 days, when the concentration of NaOH solution increased from 8 to 12 M with 20% GGBS content under ambient curing condition.

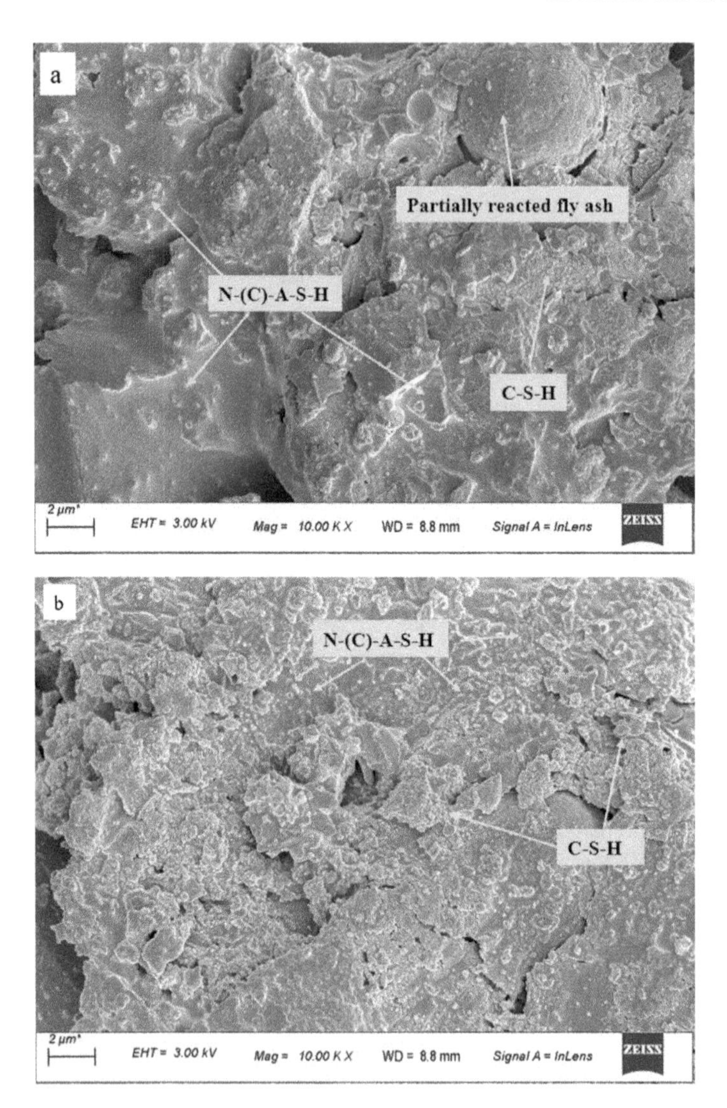

Fig. 5 FESEM images of GPC made with 20% GGBS content under ambient curing condition: **a** 8 M NaOH solution, and **b** 10 M NaOH solution, at the age of 28 days

- In comparison to GPC cured under ambient condition, water cured GPC showed higher 7 and 28-day CS.
- The microstructure study through XRD and FESEM analyses indicated the formation of geopolymer gels such as N–A–S–H gel and N–(C)–A–S–H gel along with the formation of C–S–H gel in the GPC mixes.

References

1. Zhuang HJ, Zhang HY, Xu H (2017) Resistance of geopolymer mortar to acid and chloride attacks. Procedia Eng 210:126–131
2. Huntzinger DN, Eatmon TD (2009) A life-cycle assessment of Portland cement manufacturing: comparing the traditional process with alternative technologies. J Clean Prod 17:668–675
3. Mermerdas K, Manguri S, Nassani DE, Oleiwi SM (2017) Effect of aggregate properties on the mechanical and absorption characteristics of geopolymer mortar. Eng Sci Technol Int 20:1642–1652
4. Saha S, Rajasekaran C (2017) Enhancement of the properties of fly ash based geopolymer paste by incorporating ground granulated blast furnace slag. Constr Build Mater 146:615–620
5. Mclellan BC, Williams RP, Lay J, Riessen AV, Corder GD (2011) Costs and carbon emissions for geopolymer pastes in comparison to ordinary portland cement. J Clean Prod 19:1080–1090
6. Zhang P, Gao Z, Wang J, Guo J, Hu S, Ling Y (2020) Properties of fresh and hardened fly ash/slag based geopolymer concrete: a review. J Clean Prod 270:122389
7. Soutsos M, Boyle AP, Vinai R, Hadjierakleous A, Barnett SJ (2016) Factors influencing the compressive strength of fly ash based geopolymers. Constr Build Mater 110:355–368
8. Singh B, Rahman MR, Paswan R, Bhattacharyya SK (2016) Effect of activator concentration on the strength, ITZ and drying shrinkage of fly ash/slag geopolymer concrete. Constr Build Mater 118:171–179
9. Rashad AM (2014) A comprehensive overview about the influence of different admixtures and additives on the properties of alkali-activated fly ash. Mater Des 53:1005–1025
10. Diaz EI, Allouche EN, Eklund S (2010) Factors affecting the suitability of fly ash as source material for geopolymers. Fuel 89:992–996
11. Lee NK, Lee HK (2013) Setting and mechanical properties of alkali-activated fly ash/slag concrete manufactured at room temperature. Constr Build Mater 47:1201–1209
12. Nath P, Sarker PK (2014) Effect of GGBFS on setting, workability and early strength properties of fly ash geopolymer concrete cured in ambient condition. Constr Build Mater 66:163–171
13. Deb PS, Nath P, Sarker PK (2015) Drying shrinkage of slag blended fly ash geopolymer concrete cured at room temperature. Procedia Eng 125:594–600
14. Mehta A, Siddique R, Ozbakkaloglu T, Shaikh FUA, Belarbi R (2020) Fly ash and ground granulated blast furnace slag-based alkali-activated concrete: mechanical, transport and microstructural properties. Constr Build Mater 257:119548
15. Ghafoor MT, Khan QS, Qazi AU, Sheikh MN, Hadi MNS (2021) Influence of alkaline activators on the mechanical properties of fly ash based geopolymer concrete cured at ambient temperature. Constr Build Mater 273:121752
16. Xie J, Wang J, Rao R, Wang C, Fang C (2019) Effects of combined usage of GGBS and fly ash on workability and mechanical properties of alkali activated geopolymer concrete with recycled aggregate. Compos Part B 164:179–190
17. Chindaprasirt P, Phoo-Ngernkham T, Hanjitsuwan S, Horpibulsuk S, Poowancum A (2018) Effect of calcium-rich compounds on setting time and strength development of alkali-activated fly ash cured at ambient temperature. Case Stud Constr Mater 9:e00198
18. Mehta A, Siddique R (2017) Properties of low-calcium fly ash based geopolymer concrete incorporating OPC as partial replacement of fly ash. Constr Build Mater 150:792–807
19. Huseien GF, Ismail M, Khalid NHA, Hussin MW, Mirza J (2018) Compressive strength and microstructure of assorted wastes incorporated geopolymer mortars: effect of solution molarity. Alexandria Eng J 57:3375–3386

Analysis of Interface Bond Strength for Thin-Whitetopping Pavements Using Finite Element Method

Annapurna M. Nambiar and K. Jayakesh

Abstract Thin-Whitetopping is a relatively new method of pavement rehabilitation, where a fresh layer of Portland Cement Concrete is placed over a moderately damaged asphalt layer. These layers are bonded together using different interface methods and it plays a vital role on the pavement performance. The service lifetime of thin whitetopping pavements depends on interface bond strength, type of material, size and thickness of slab. Loads due to moving vehicles can cause distress, which results in debonding increasing stress in the concrete layer and causes cracks. In the present study, a 3 D Finite Element Model subjected to dynamic loading is performed using ANSYS to analyze the influence of interface bond strength for Thin-Whitetopping pavements. The model is analyzed for different PCC layer thicknesses. It is observed that as the thickness of overlay increases, maximum shear stress and deformation at the bond interface of PCC and bituminous layer decreases. Due to excessive stress at the interface, early failure was observed.

Keywords Thin-Whitetopping · Debonding · Finite element · Dynamic loading · Interface bond strength

1 Introduction

1.1 General

The method of application of overlay on the existing pavements with moderate damage is a commonly adopted method since a couple of years [5]. When a portland cement concrete layer is placed over a bituminous pavement, it is known as "Whitetopping". It is used as a rehabilitation method to act as a structural strengthening alternative [3, 8]. Based on the interface type it is classified as

A. M. Nambiar (✉) · K. Jayakesh
Department of Civil Engineering, Amrita School of Engineering, Amrita Vishwa Vidyapeetham, Coimbatore, India

G. C. Marano et al. (eds.), *Proceedings of SECON'21*, Lecture Notes in Civil Engineering 171, https://doi.org/10.1007/978-3-030-80312-4_52

conventional, thin and ultra-thin. Construction using this technique, on asphalt pavements having comparatively lesser damage has been found to be cost-effective and is becoming popular [5]. If the thickness of PCC overlay is greater than 100 mm and less than 200 mm it is referred as "Thin Whitetopping". This type ensures durability, better service lifetime, lesser life cycle cost, improved safety and structural capacity [15]. Design service life involves many variables, such as interlayer bonding, slab size and thickness, toughness and present condition of the prevailing material. Volume change is also identified as one of the major reasons leading to delamination at the interface between old and new concrete or surface cracking. Distress such as rutting and cracking may occur at the joints due to the moving traffic and it will result in debonding [5, 6].

Interface bonding between the PCC and asphalt layer is the main criteria affecting the performance of whitetopping. Different interface treatment techniques are used so that proper bonding is ensured. Hence these layers act as a composite structure distributing the traffic and environmental loading in a different manner compared to normal flexible pavements. According to the monolithic behavior, the load is distributed by moving the neutral axis from the center of the slab to the base. The TWT will be able to endure the bending stresses, and shear stress will be higher at the neutral axis. Shear stresses can increase, as the vehicles drive continuously over time, leading to accumulation of strain at the joints [13]. From literature study, it is found that the interface bonding can be classified as bonded, unbonded and partially bonded. As a fully bonded condition is an ideal case, we cannot ensure there is full bonding. Therefore, partial bonding is often figured out. Bonding generally depends on surface preparation, type of tack coat, application rates, temperature, pavement material, movement of vehicles etc. [9] (Fig. 1).

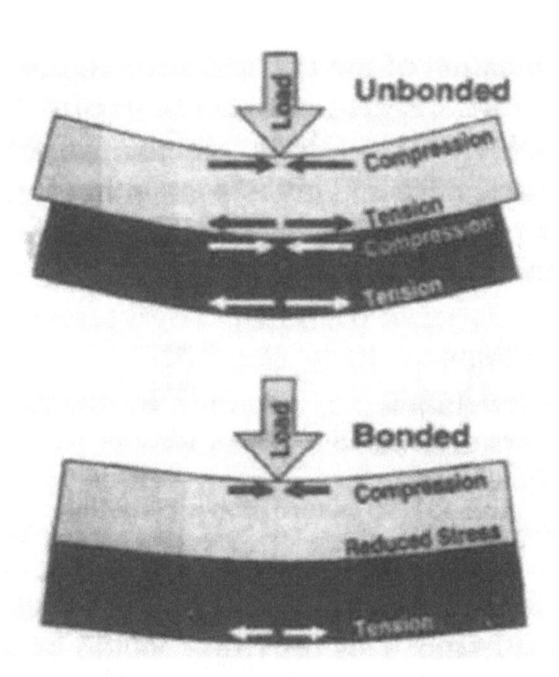

Fig. 1 Stress diagram for bonded and unbonded whitetoppings [13]

Several studies have been done in relation to bonded concrete overlays and whitetopping, both experimentally and analytically. Performance of overlays depend on the surface preparation, materials, and curing. Delaminations were determined using tests such as rebar sounding, falling weight deflectometer and rolling dynamic deflectometer [10]. Deflections at the critical load positions are found out using Benkleman beam deflection test [4, 14, 18]. To evaluate the bonding strength, stiffness, and fatigue resistance at the interface, experimental studies were performed on asphalt and composite cylindrical samples subjected to different loads. This study gave an insight to the fact that the major causes leading to the failure of BCOA section is water. Adverse behavior of water in cases happens before debonding, when the interface and asphalt weaken to the point that the two layers' monolithic action is lost. Even though the bond strength was sufficient, excessive bond stress formed at the bonded overlays' interface. As a result, numerical analysis was used to investigate into the other probable causes of early distress [17]. Analysis of the bond characteristics, thickness of the layers, effect of materials on the performance of whitetopping has been done with the help of finite element softwares such as ANSYS, ABAQUS, AASHTOWARE etc. [11]. Research has been done on bonded and unbonded overlays with varying thicknesses of asphalt and cement concrete layer. The effect on critical stress by the change in flexural rigidity, overlay thickness and impact of load transfer on debonding was studied [16, 20]. Debonding is found to increase the load induced stress in concrete by 40–55% [5]. Non-linear temperature analysis with varying temperatures was done and the difference of temperature between mid-depth and the pavement surface was found to be double of that between the mid-depth and the bottom [19]. Studies were done on the role of friction at the interface on stresses, different CBR values and induced deflections of Thin Whitetopping pavements [12]. Performance of BCOA pavements on accelerated loading and interface bond strength was also evaluated [18]. It was found that with the amount of load repetitions, bond strength decreased. So it is considered as the major parameter to predict failure due to debonding and fatigue analysis for BCOA pavement design.

Moving loads over the pavement causes rutting, reflective cracking, debonding, volume changes etc. Taking a broad view of the study, it can be observed that the bond strength and thickness of whitetopping are the most crucial factors influencing its performance. To understand the feasibility and behavior of stress, further studies are to be done. Finite element method allows understanding the interconnection of the variables and helps in saving the time used for trial experiments. In this paper, a 3D Finite Element Model subjected to dynamic loading is performed using ANSYS 2020 R to analyse the influence of interface bond strength for Thin-Whitetopping pavements. The model is analysed for different PCC layer thicknesses. It is observed that as the thickness of overlay increases, stress at the bond interface of PCC and asphalt layer decreases. Due to excessive stress at the interface, premature failure was observed.

2 Methodology

An extensive study of the available literature was done to understand the behaviour of Thin whitetopping when it is subjected to vehicular loads. Studies have been conducted to assess the impact of bond stress, tack coat and influence of interface bond on pavement life and it was found that distress in pavements can be due to many reasons such as axle load, temperatures, thickness of overlay, thickness of existing layer, joint spacing etc. From this it was noticed that studies have not been done for different interface bonding conditions and different cement concrete layer thickness. The datas ie; material and geometric properties were collected from journals. The modelling and analysis were done using ANSYS 2020 R. The subgrade, granular, asphalt and concrete layers were modelled using box/solid elements and it was meshed for analysis [2]. The bonding between the asphalt and concrete layers were modelled using contact elements. The wheel loads were applied at the centre of the pavement section as uniformly distributed pressure. Analysis was done and the obtained results were interpreted.

3 Fem Modelling Features

In the present study, finite element models were created and analysed using ANSYS 2020 R2. Pavement sections consisting of subgrade, subbase, asphalt layer and concrete layer were modelled with varying parameters. They were analysed to understand the behaviour of the interface bonding when subjected to moving vehicle load. These models were analysed for four different thicknesses and two different bonding conditions. Fully bonded and 50% bonded conditions were considered for analysis. In the case of partially bonded, the unbonded portion was assumed to be at the centre.Size of each layer was taken as 1000 mm × 1000 mm. Thickness of the subgrade layer taken was 500 mm, sub-base 450 mm and asphalt layer 100 mm. Table 1 shows the details of material and geometric properties [7] (Figs. 2 and 3).

Table 1 Material and geometric properties

Layers	Properties			
Description	Modulus of elasticity	Poisson's ratio	Density	Thickness (mm)
Cement concrete	30,000 MPa	0.15	2400 kg/m^3	100,125,150,175,200
Asphalt	1000 MPa	0.35	2002 kg/m^3	100
Subbase	200 MPa	0.4	2066 kg/m^3	450
Subgrade	60 MPa	0.4	1600 kg/m^3	500

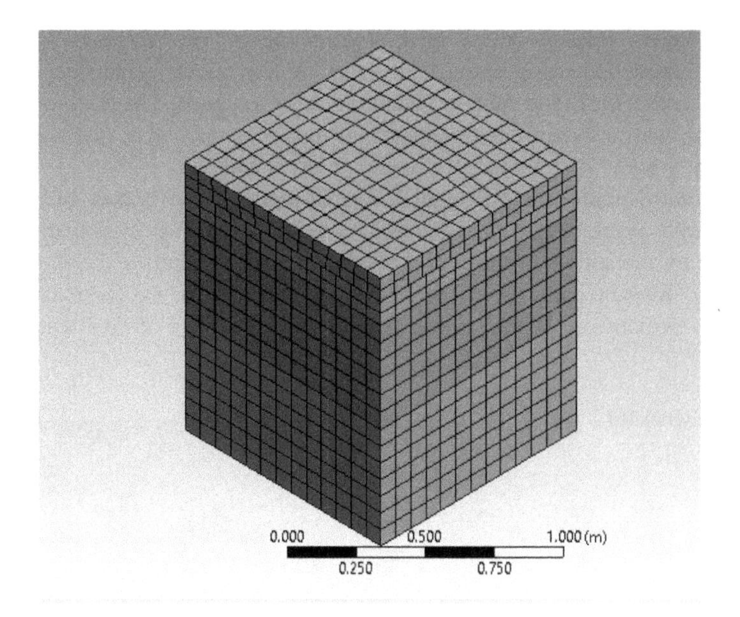

Fig. 2 3D model considered for analysis

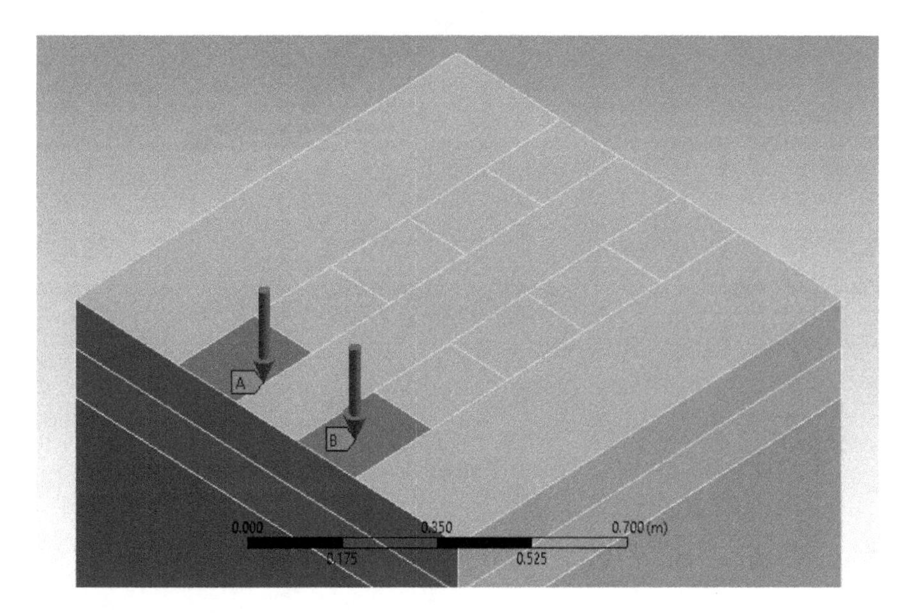

Fig. 3 Loading position considered

These layers were modelled using solid elements. The interface between the layers are modelled using contact elements. It was given as surface to surface "bonded" contact [1]. The support conditions were assigned. The bottom portion of the section was assigned fixed support and displacements were given at the four sides in x, y and z directions respectively (Figs. 4 and 5).

The bottom portion ie; foundation was assigned fixed support and the other portions were given displacements in the respective directions. Automatic meshing was done by the software. Mesh size obtained was 71.42 mm × 71.42 mm. After modelling, transient analysis was performed to determine the effects of moving load on the performance of Whitetopping. Single axle consisting of dual wheels which

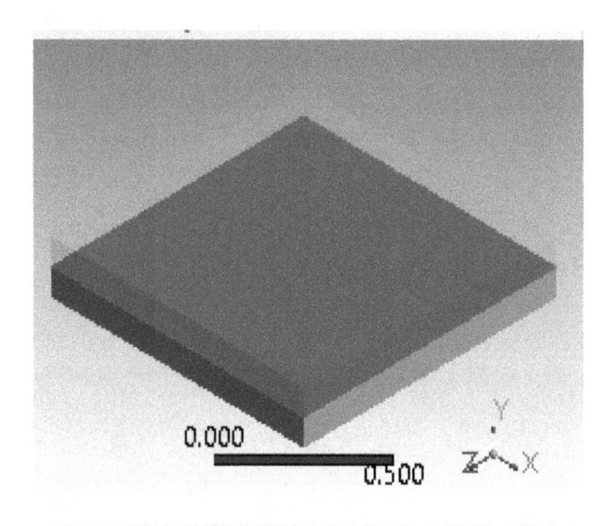

Fig. 4 Fully bonded

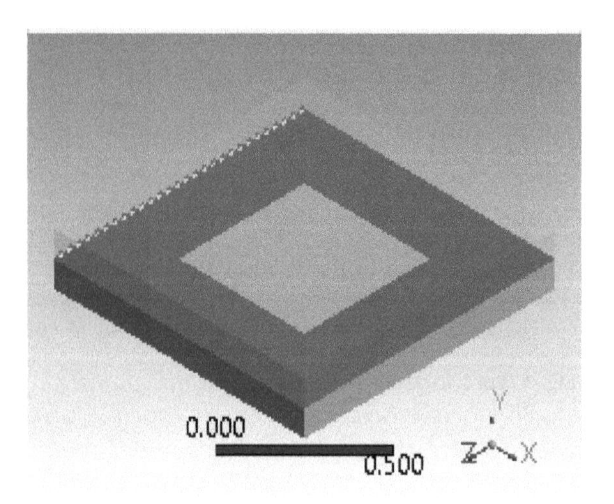

Fig. 5 50% bonded

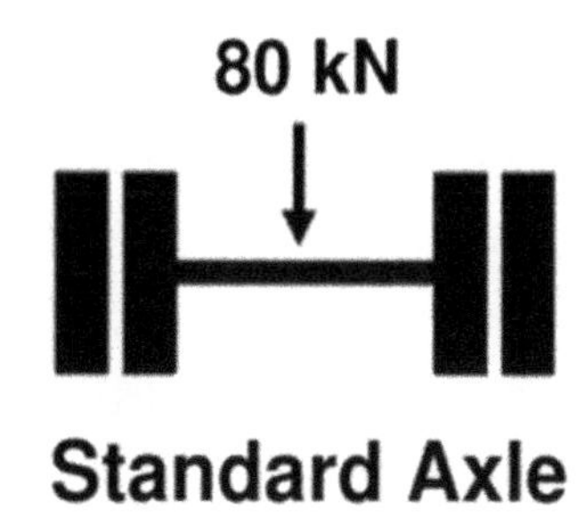

Fig. 6 Standard axle load

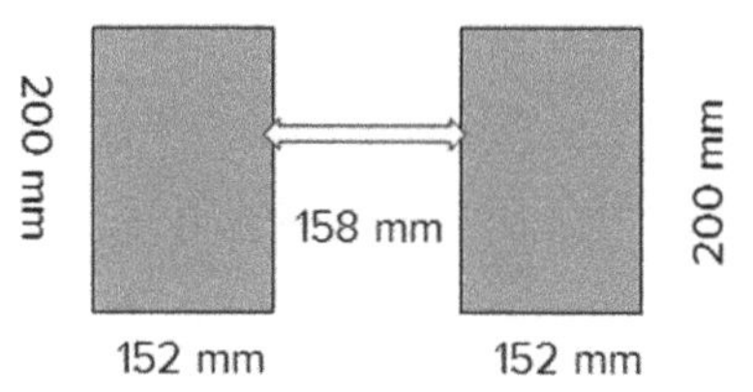

Fig. 7 Tyre imprints

carries a load of 80 KN ie; 8 tonnes is considered as a standard axle. The magnitude of load acting on the pavement section was considered equal to that of 80 KN, where 40 KN was applied to each of the two wheels. Load is applied as evenly distributed pressure of 0.65 MPa over the entire loading area. It was applied over two rectangular areas of dimension 152 mm × 200 mm. Figure 7 shows the tyre imprints. The centre to centre distance between two wheels is 310 mm as per IRC: 58–2015. The load was applied at the centre (Fig. 6).

4 Results and Discussion

A total of ten models were analyzed to determine the stress and deformation developed at the interface of Thin-whitetopping. These models were analyzed with 5 different thickness parameters and two interface bonding conditions. The changes in stresses developed under these parameters were analyzed and then evaluated.

4.1 Effect of Thickness and Interface Bonding on Pavement Deformation

Total deformation versus time graph for fully bonded and partially bonded thin whitetoppings when wheel load is applied at the centre is shown in Figs. 8 and 9. The deformation induced is greater at the edges than the interior portion. Deformation induced due to the moving loads does not show any significant changes in fully bonded and 50% bonded condition.

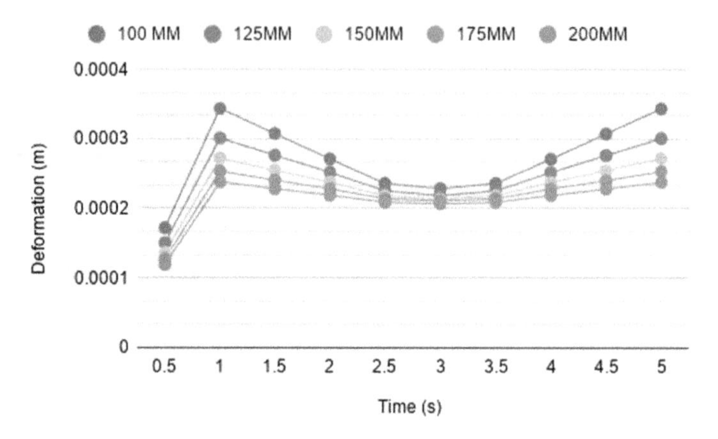

Fig. 8 Total deformation versus time chart for 100% bonded

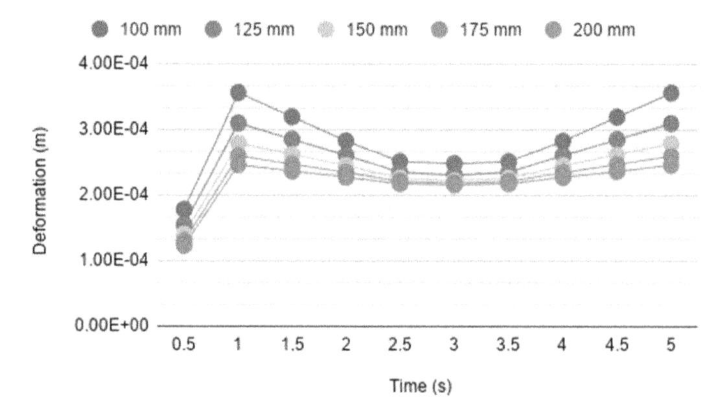

Fig. 9 Total deformation versus time chart for 50% bonded

4.2 Impact of Thickness and Interface Bonding on Loading Stress

Figures 10 and 11 shows variation of stress when different thicknesses of cement concrete layer are used under fully bonded and 50% bonded cases. When the vehicles move over the pavement, load is distributed, shifting the neutral axis from the middle of the slab to the bottom as the fully bonded TWT shows monolithic action. Shear stress will be maximum at the neutral axis. In case of partially bonded, neutral axis shifts downwards and the reduced stress may occur somewhere in between bonded and unbonded cases. It is found that the stress developed decreases as the thickness of the PCC layer increases. The stress developed in case of 200 mm thick slab is insignificant when it is partially bonded.

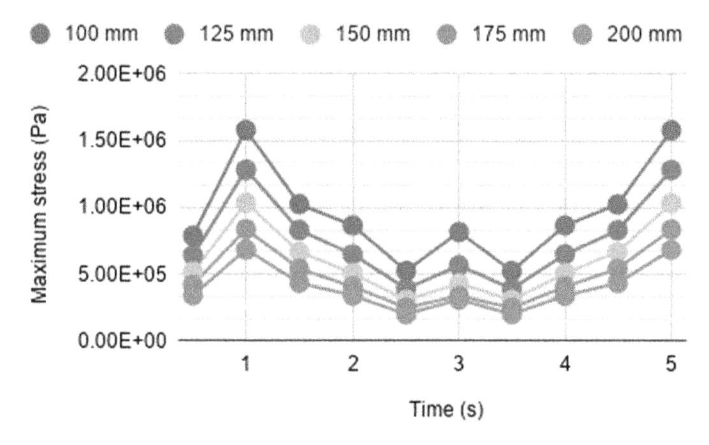

Fig. 10 Maximum stress versus time chart for fully bonded

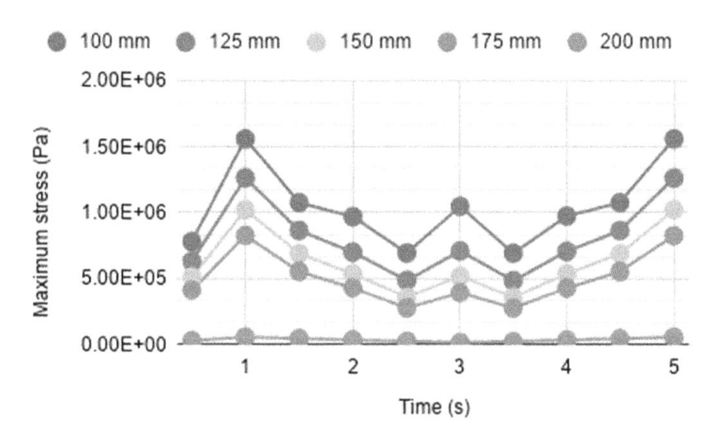

Fig. 11 Maximum stress versus time chart for 50% bonded

4.3 Impact of Thickness and Interface Bonding on Frictional Stress

The variation of frictional stress at the contact surface for varying thickness under two different bonding conditions are shown in Figs. 12 and 13. Frictional stress at the interface is greater for fully bonded than 50% bonded. It is found that stress induced is greater for 100 mm thick slab for both the cases.

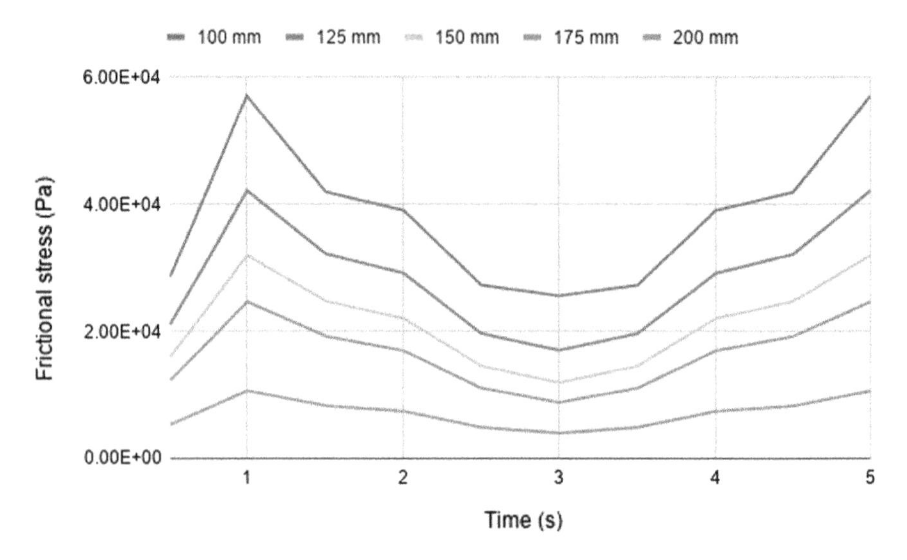

Fig. 12 Frictional stress versus time chart for fully bonded

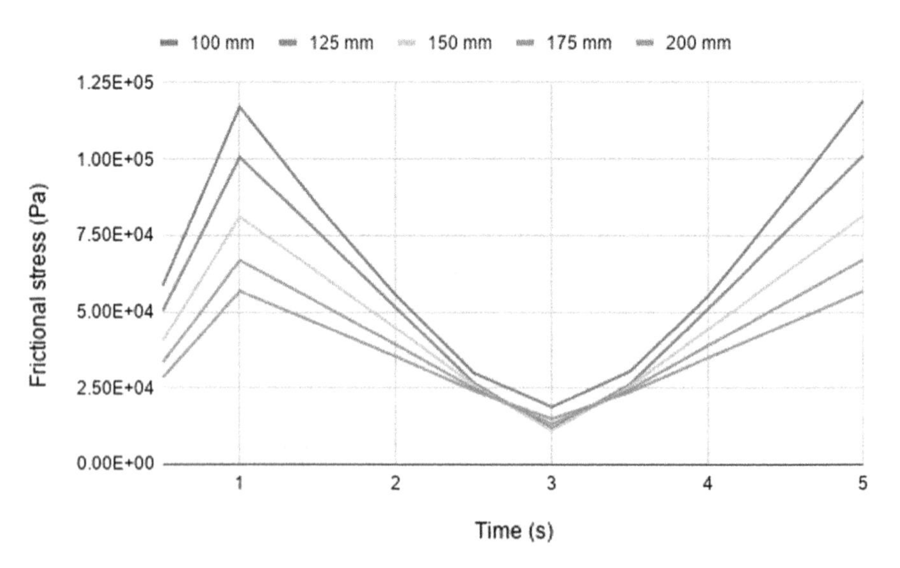

Fig. 13 Frictional stress versus time chart for 50% bonded

5 Conclusion

When Thin whitetopping is subjected to moving loads of vehicles, stresses are developed at the interface between asphalt and PCC layer. The objective of the study was to analyze the behavior of interface bonding under varying thicknesses.

- The deformation induced for 100, 125, 150, 175 and 200 mm for fully bonded condition has no significant variation compared to partially bonded condition. Deformation is more for 100 mm i.e.; it decreases as the thickness increases.
- Shear stress is found to be maximum at the neutral axis. For partially bonded condition, neutral axis shifts downwards and the reduced stress may occur somewhere in between bonded and unbonded cases. It is found that the stress developed decreases as the thickness of the PCC layer increases and it is more at the edges than at the centre. Frictional stress also showed a similar behavior. It could be seen that for partially bonded condition, the edge stresses developed for 200 mm thick overlay was more than fully bonded whereas the stress at the interior were found to have a very little variation.

References

1. ANSYS contact technology guide—ANSYS release 9.0
2. ANSYS mechanical APDL element reference
3. Arthi S, Narasimha VL, Pandurangan K (2015) Finite element analysis of thin whitetopping pavements. Int J Innovative Res Scie Eng Technol
4. Ashish PA, Shishir SG, Shrikant SK, Utkarsha BI, Yogesh YG, Yogita BD (2016) Pavement rehabilitation using thin white topping. Int J Innovative Res Sci Eng Technol 5(5)
5. Barman M, Vandenbossche JM, Li Z (2017) Influence of interface bond on the performance of bonded concrete overlays on asphalt pavements. J Transp Eng Part B: Pavements 143 (3):04017008
6. Bhattacharya BB, Gotlif A, Darter MI, Khazanovich L (2019) Impact of joint spacing on bonded concrete overlay of existing asphalt pavement in the AASHTOWare pavement me design software. J Transp Eng, Part B: Pavements 145(3):04019018
7. Bulusu VJ, Kusam SR, Muppireddy AR (2019) Influence of interface friction on stresses and deflections of thin white-topping pavements. In: International conference on transportation and development 2019: smarter and safer mobility and cities, pp 395–401. American Society of Civil Engineers, Reston, VA
8. Burnham T (2009) Whitetopping: concrete overlays of asphalt pavements. Minnesota Department of Transportation, Office of Materials and Road Research
9. Covey D, Coleri E, Mahmoud A (2017) Tack coat rheological properties and the effects on interlayer shear strength. J Mater Civ Eng 29(11):04017221
10. Delatte NJ Jr, Fowler DW, McCullough BF, Gräter SF (1998) Investigating performance of bonded concrete overlays. J Perform Constr Facil 12(2):62–70
11. Dormohammadi A, Tandon V, Rodarte AA (2019) Evaluating influence of thickness and interface bonding on overlay service life using 3D FEM. Int J Geomech 19(4):04019012
12. Gupta A, Kumar A (2014) Comparative structural analysis of flexible pavements using finite element method. Int J Pavement Eng Asphalt Technol 15(1):11–19
13. IRC:SP:76-2008—Tentative guidelines for conventional, thin and ultra thin whitetopping
14. Jundhare DR, Khare KC, Jain RK (2012) Structural analysis of ultrathin whitetopping under wheel loading using the computer code 'ANSYS.' Int J Struct Eng 3(3):208–228
15. Jundhare DR, Khare DK, Jain DR. Performance evaluation of ultrathin whitetopping in India by BBD test. Int J Eng Re Technol (IJERT), ISSN, 2278–0181
16. Jundhare DR, Khare KC, Jain RK (2010) Study of edge stresses and deflections in whitetopping overlay on winkler foundation. In Indian geotechnical conference, pp 877–880

17. Kim YK, Lee SW (2013) Performance evaluation of bonded concrete overlay. Constr Build Mater 49:464–470
18. Mahdi M, Wu Z, Rupnow TD (2020) Evaluation of bonded concrete overlay of asphalt under accelerated loading. transportation research record, 0361198120961385
19. Teja SS, Reddy KS, Reddy MA, Pandey BB. Analysis of bonded concrete pavements using 3D FEM
20. Zhu M, Weng X, Zhang J (2019) Three-dimensional FEM loading stress analysis on a new airport concrete pavement ultra-thin whitetopping overlay structure. Struct Concr 20(2): 628–637

Strengthening and Compressive Behavior of Composite Multilimbed Light Weight CFS Built-Up Column

Irene Joy and Nincy Jose

Abstract In recent times, cold formed steel (CFS) is extensively used in residential and commercial buildings due to its high strength to self-weight ratio and easiness of construction. The manufacturing of CFS sections are done at the room temperature by rolling or pressing. Built-up composite box sections are becoming popular for column members in CFS construction. The structural elements can be of columns, studs, beams, floor decking, built up sections and other any parts of the structure. In this paper, multilimbed built-up box sections are made with identical lipped channels, connected at their flanges with self-drilling screws. In this arrangement, buckling of the individual channel is prevented by the screws. To make the columns more stronger, these multilimbed hollow sections are filled with ultra light weight concrete. This study is to investigate the axial performance of composite multilimbed light weight CFS built-up column in ANSYS software. Initially three limbed I section is modeled as hollow and analyzed under axial load. Then partially filled and fully filled sections are modeled and analyzed. Finally the behaviour of each section is compared with a conventional steel column (ISMB150).

Keywords Axial capacity · Multilimbed composite build-up sections · Cold formed steel · Ultra light weight concrete · Screws

1 Introduction

Cold formed steel (CFS) is extensively used in residential and commercial buildings due to its high strength to self-weight ratio and easiness of construction. Cold-formed steel (CFS) sections are manufactured at room temperature by rolling or pressing thin steel sheets into CFS sections. Both the structural as well as the nonstructural parts of a building are made using the thin steel sheets. The building materials can be of columns, studs, beams, floor decking, built up sections and other

I. Joy (✉) · N. Jose
Civil Department, Federal Institute of Science and Technology, Angamaly, Kerala, India

G. C. Marano et al. (eds.), *Proceedings of SECON'21*, Lecture Notes in Civil Engineering 171, https://doi.org/10.1007/978-3-030-80312-4_53

any parts of the structure. CFS construction materials differ in many aspects than hot rolled steel sections. The strength of elements is analyzed by the buckling property of the element. Two or more cold-formed single sections are attached together inorder to get greater cross-sectional properties. The parts of standard shapes (for example—C, Z) is connected by a bolt, screw, or weld to make new shapes. The closed box sections allow greater distances between supports and carries greater load than single sections. Due to the excellent structural behavior, built-up sections are used in the steel buildings. Cold formed steel sections are widely used as trusses member or space frames. In recent construction, the usage of CFS columns in frame structure has been used in residential construction for building up to double story.

The use of concrete filled steel columns reduces production time with the aid of using getting rid of the formwork and the want for tying of reinforcing bars. Also the use of CFS members can improve the seismic performance of the structure by providing high strength, ductility and energy absorption capacity. Concrete filling inside the column delays the global buckling and also resist the axial loads. Concrete contributes its inherent mass, stiffness, and economy. Steel is useful for its high construction speed, strength, long-span capabilities, and light weight. Therefore, it is highly advantageous to use concrete filled steel columns to the axial loading. As the number of limb increases, the ductility of the section increases and viceversa. This research was proposed to evaluate the performance of cold formed steel sections with and without filled with light weight concrete using ANSYS software.

2 Literature Review

Roy et al. [1] investigated the axial capacity of cold-formed steel built-up box sections. Here single C section and builtup box sections of both short and slender columns are considered. From the experimental results, it was clear that built-up box sections showed more load carrying capacity than single channels.

Li et al. [2] investigated flexural strength of cold-formed thin-walled steel beams with built-up box section. From the study, the equivalent box section can be used to calculate the flexural strength of built-up box beams bending about weak axis.

Younga et al. [3] concluded based on the experimental results that design recommendations were proposed for concrete-filled high strength stainless steel tube columns.

Lau et al. [4] conducted research on effect of screw spacing on behavior of axially loaded back-to-back cold formed steel built-up channel sections.

Reyes et al. [5] conducted study on evaluation of the slenderness ratio in built-up cold-formed box sections. Lim et al. [6] studied the nonlinear behavior of axially loaded back-to-back built-up cold-formed steel un-lipped channel sections. The axial strength of built-up sections are 31% lesser on average than the built-up lipped channel sections. Muftah et al. [7] conducted research on the ultimate load of builtup cold formed steel column. Tested 3 specimens of CFS sections, single C, BTB (back to back), and BU (box-up) sections. From the results it was clear that ultimate load of back to back section is the highest when compared with other sections.

Almamooria et al. [8] conducted study on effect of section shape on the behaviour of steel columns filled with light weight aggregate concrete. From the experiments it was concluded that steel tubes filled with lightweight concrete are more liable to local buckling than sections filled with normal weight concrete.

Aravindhan et al. [9] done experimental investigation on CFS long and short composite columns infilled with M-sand concrete. From the test results, it was concluded that columns filled with conventional concrete showed lesser strength when compared with columns filled with M-sand concrete.

Ting et al. [10] presented the behaviour of back-to-back gapped builtup CFS channel sections under compression.

Yang et al. [11] conducted experiments on rectangular concrete-filled steel tubes loaded axially on a partially stressed cross-sectional area. The test results showed that, while being loaded on partial cross-sectional area, RCFST stub column had a favourable bearing capacity and ductility compared with the corresponding plain concrete specimens.

Abhilash et al. [12] conducted study on axial behaviour of semi-light weight aggregate concrete-filled steel tube columns. From the experiments, the increase of thickness of steel section will increase both the ultimate axial load and axial deformation corresponding to the ultimate load This improves the ductile behaviour of the columns. Elangovan et al. [13] studied the structural behaviour of concrete filled steel tubular columns. Various steel materials such as stainless steel, mild steel and cold formed steel tubular sections were analysed using varying parameters and results were compared.

Valipour et al. [14] studied the nonlinear static and cyclic analysis of concrete-filled steel columns.

3 Methodology

3.1 General

Analysis is done using the software called ANSYS. Specimens are modeled according to the desired specifications and requirements. Specimens are modeled with solid 186 elements. The properties of the specimens are provided in engineering data of the software. Model is created using geometry. The column is pinned at top and bottom. All specimens have yield stress of 559 MPa and poisson's ratio is taken as 0.3. Columns are 1500 mm in length. Modeling and analysis of steel columns, partially filled columns and fully filled columns are carried out with help of ANSYS 19.0. These CFS multilimbed composite sections are considered as primary members hence its load carrying capacity is compared with conventional I section (ISMB150).

3.2 Hot Rolled Steel Section

Figure 1 is the hot rolled steel I section is analysed in ANSYS software. Steel grade Fe250 is taken for analyzing. Overall heigtht and depth of the column is 1500 and 150 mm. Top and bottom flange width is 80 mm and thickness is 7.6 mm. Web thickness is 4.8 mm. The column is pinned at top and bottom. Finite element meshing can be provided as coarse, medium and fine. Column is analysed to find out the ultimate load. The ultimate load is obtained as 458.43 kN. The ultimate load of a hot rolled steel column taken as reference to find out the performance of multilimbed hollow steel, partially filled and fully filled I sections.

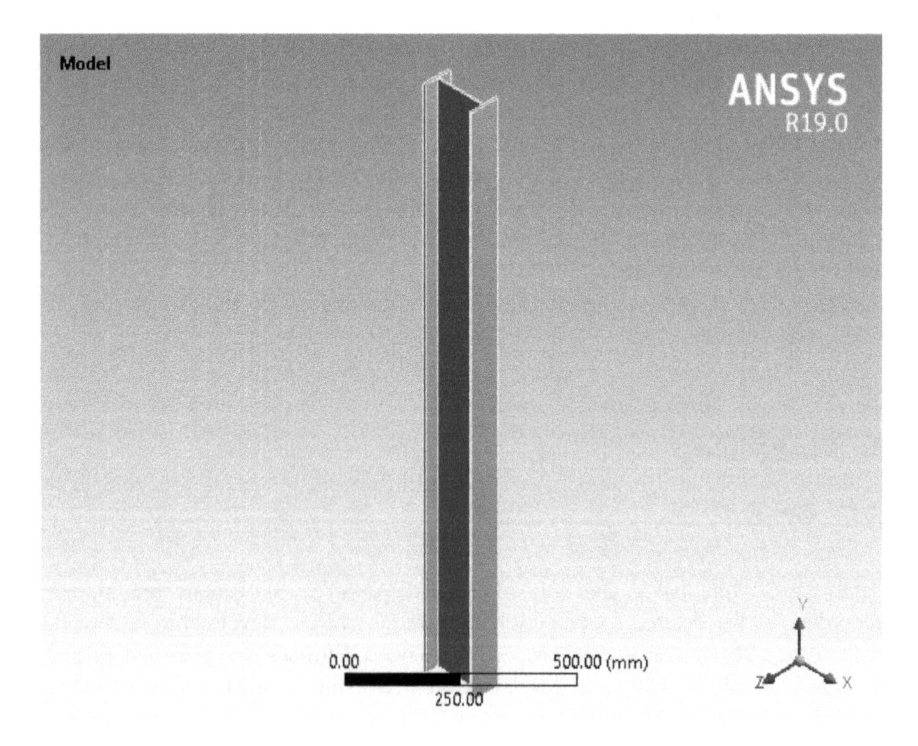

Fig. 1 Model of ISMB150 from the software

3.3 Multilimbed Steel Column

Multilimbed steel I sections (3 L-I-STEEL) are modeled in ANSYS software shown in Fig. 2. Dimensions of the column is shown in the Table 1. Material properties of the column is shown in Table 2. Loading is given as axial loading. The column is pinned at top and bottom. Boundary conditions are provided at support plates provided at both ends. Displacement control loading is provided on the top plate. Finite element meshing can be provided as coarse, medium and fine. The adaptive meshing method is adopted. The shape of mesh is hexahedron and the sizing of mesh is 20 mm. Analysis is carried out to find the ultimate load of the column.

3.4 Partially and Fully Filled Sections

Inorder to increase the load carrying capacity hollow sections are filled with ultra light weight concrete (ULWC). Partially filling is done as 6 types. All the sections are shown in the Table 3.

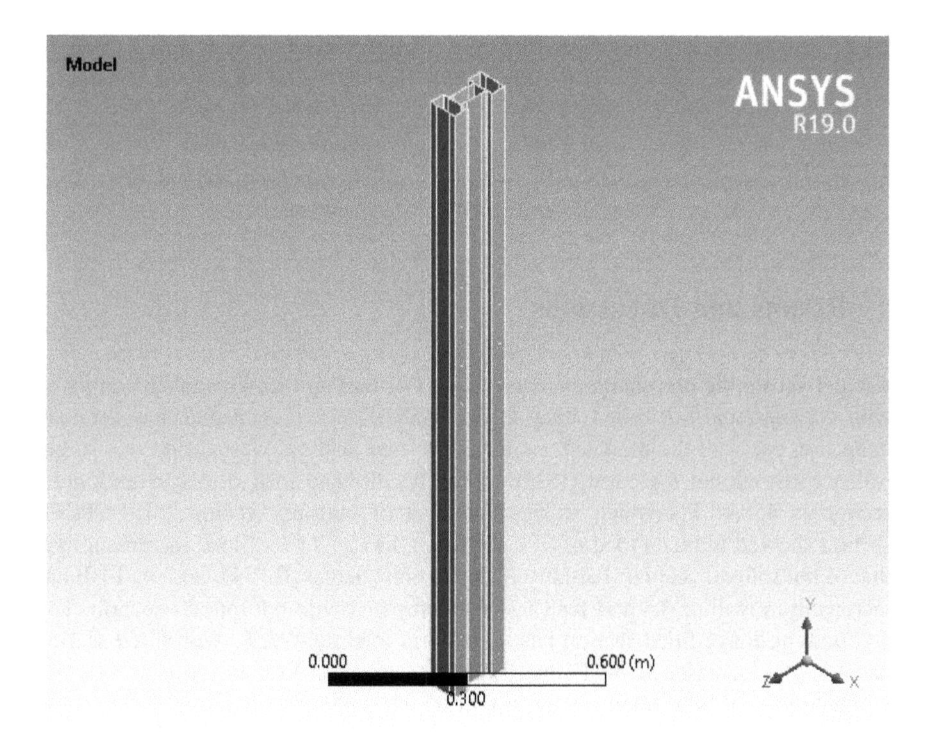

Fig. 2 Multilimbed hollow steel section

Table 1 Dimensions of the column

Property	Value
Height	1500 mm
Depth	158.7 mm
Flange width	76.3 mm
Flange thickness	41.2 mm
Web thickness	41.2 mm

Table 2 Material properties of column [1]

Property	Value
Nominal thickness	1.01 mm
Yield stress	559 MPa
Gauge width	12.5 mm
Ultimate stress	678 MPa
Young's modulus	207 MPa
Poisson's ratio	0.3

Table 3 Specimen type and labeling

Specimen label	Specimen type
3L-I-Steel	Hollow section
3L-I-TF&BF	Top and bottom flange filled
3L-I-WF	Web filled
3L-I-TF50% BF50%	50% filled top and bottom flange
3L-I-TF	Top flange filled
3L-I-BF	Bottom flange filled
3L-I-BF & WEB	Bottom flange and web filled
3L-I-TF,BF,WEB	Fully filled

4 Results and Discussions

Table 4 shows the percentage increase in load of builtup members when compared with conventional hot rolled steel section (ISMB150). Figs. 3 and 4 is the load deflection curve of the modeled sections. The first analysis was carried out on hot rolled conventional I section (ISMB150). The ultimate load of a conventional I section is 458.43 kN which is less than that of built-up sections. 3L-I-STEEL section showed ultimate load of 617 kN which has 34.71% of load increment than that of hot rolled I section. Partially filled section such as 3L-I-TF and 3L-I-BF has increment in load of 4.4 and 5.88% when compared with hot rolled I section. Top and bottom flange filled section has increase in load by 49.2%. Web filled section has 40.72% increase. 3L-I-TF50% BF50% section has load increase of 41.11%. Ultimate load of fully filled section is 971.07 kN and increase in load is 111.83%.

Table 4 Percentage increase in load

Model	Weight (kg)	U-DEF (mm)	PU-LOAD (kN)	% of increase in load
ISMB150	21.363	8.1253	458.43	1.00
3L-I-STEEL	12.785	12.92	617.55	34.71
3L-I-TF&BF	22.67	11.79	683.97	49.71
3L-I-WF	17.728	9.52	645.12	40.20
3L-I-TF50% BF50%	17.728	12.84	646.90	41.11
3L-I-TF	17.728	15.07	478.61	4.40
3L-I-BF	17.728	14.807	485.37	5.88
3L-I-BF & WEB	22.67	14.951	553.36	20.71
3L-I-TF,BF & WEB	27.613	11.555	971.07	111.83

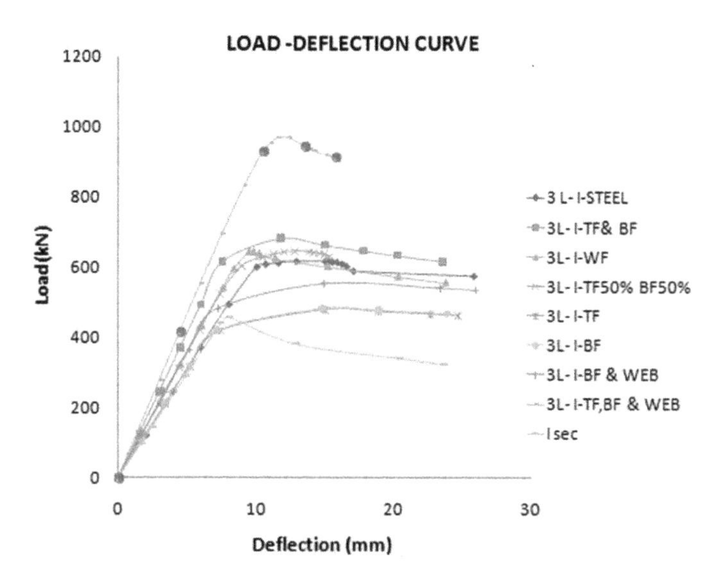

Fig. 3 Load-deflection curve of all the models

Load deflection curve for one limb filled sections are shown in the Fig. 4 web only filled section showed more ultimate load than other two sections (top flange filled and bottom flange filled).

From the percentage of increase in load chart Fig. 5, except 3L-I-TF, 3L-I-BF and 3L-I-BF&WEB all other models showed more than 30% of increase in load than hot rolled steel section. Multilimbed column with fully filled limbs showed 111.3% of increase in load.

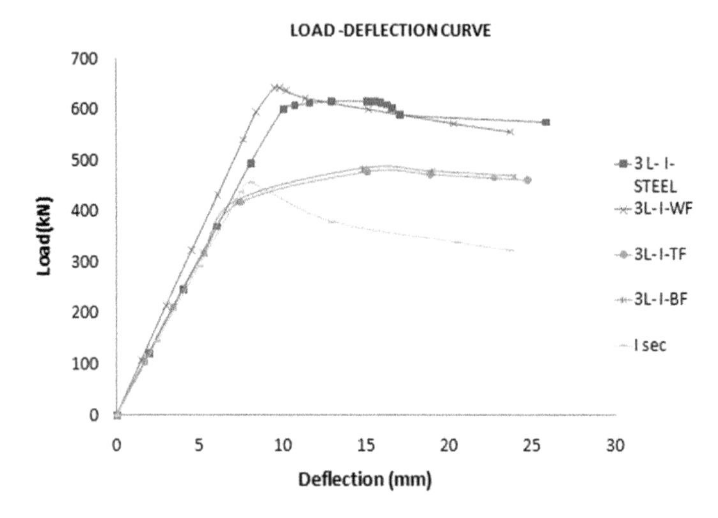

Fig. 4 Load deflection curve of sections with only one limb is filled

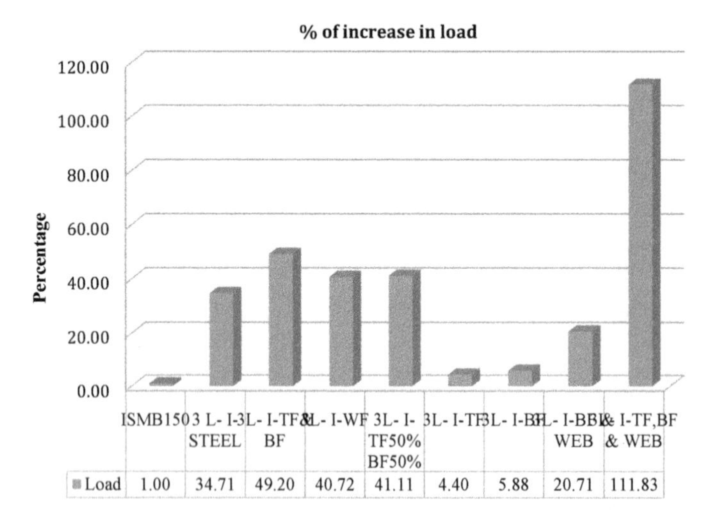

Fig. 5 Percentage of increase in ultimate load

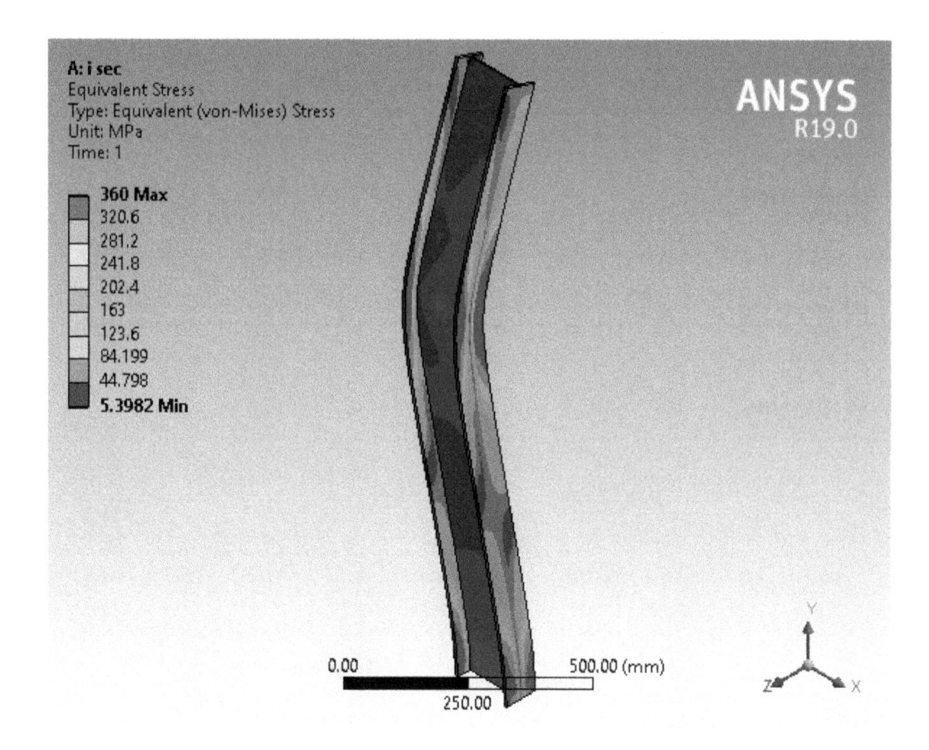

Fig. 6 Stress distribution of ISMB150

4.1 Stress Distribution and Type of Failure

From the stress distribution, maximum stress for hot rolled section (ISMB150) is 360 MPa which is shown in the Fig. 6. Maximum stress is at the buckled area and at the lower part of the flanges. Stress distribution for 3L-I-STEEL is shown in the Fig. 7. Maximum stress is 641.44 MPa and minimum stress is 175.59 MPa. This builtup hollow CFS section has almost double the maximum stress as that of ISMB150. Maximum stress for 3L-I-TF&BF is 677.99 MPa and minimum stress is 1.0548 MPa which is shown in the Fig. 8. Maximum stress for 3L-I-WF and 3L-I-TF50% BF50% is 677.85 and 677.86 MPa which is almost same value. Maximum stress for 3L-I-TF, 3L-I-BF, 3L-I-BF&WEB is 677.73, 677.68 and 677.68 MPa. Maximum stress on 3L-I-TF,BF&WEB as shown in Fig. 9 is 677.99 MPa which is greater than hollow CFS steel section. Type of failure of each column is shown in the Table 5.

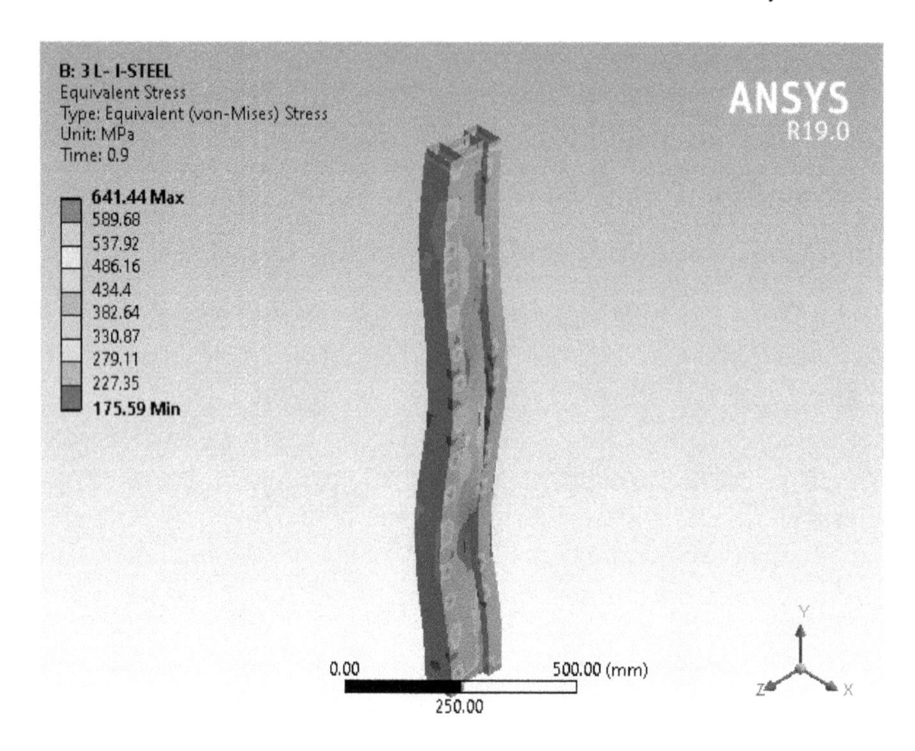

Fig. 7 Stress distribution of 3L-I-STEEL

5 Conclusion

In recent times, Cold Formed Steel (CFS) is widely used in residential and commercial buildings because of its greater strength to self-weight ratio and ease of construction. Builtup composite box sections are becoming popular for column members in CFS construction. To make the columns more stronger, these multilimbed hollow sections are filled with ultra light weight concrete. The modeling and analysis can be done efficiently and accurately with the help of ANSYS 19.0. ISMB150 is modeled and analyzed and the ultimate load was obtained as 458.43 kN. Multilimbed sections are modeled and analysed and compared with the conventional hot rolled section and following conclusions are made:

- Multilimbed CFS hollow steel column showed load increase of 34.71% than that of hot rolled section.
- Both flange filled section has 49.2% of increase in load carrying capacity.

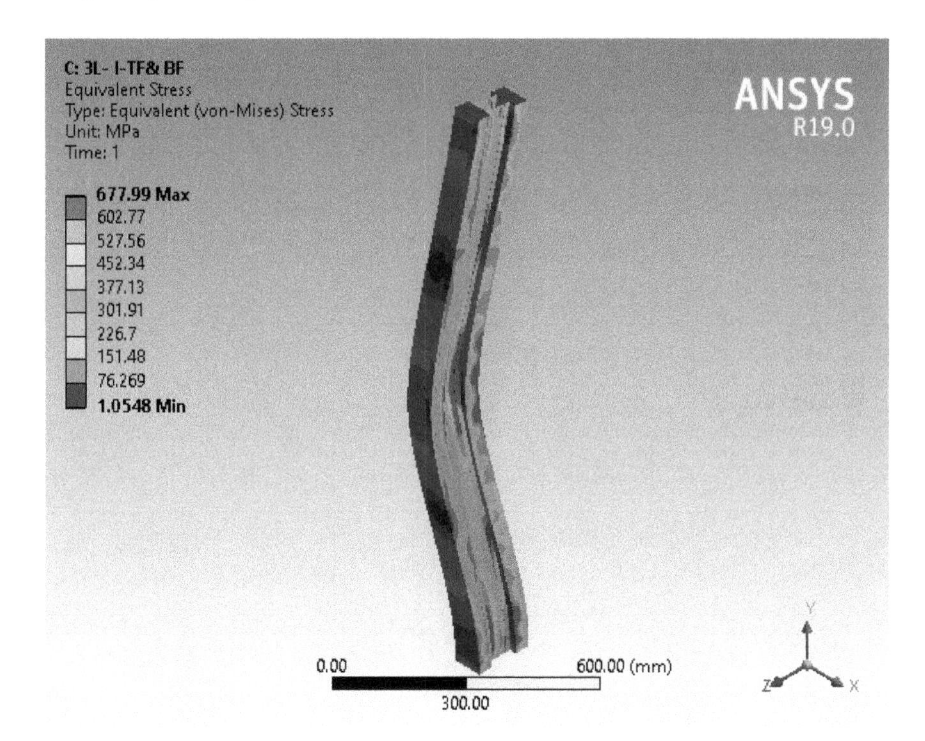

Fig. 8 Stress distribution of 3L-I-TF&BF

- Fully filled section has 111.83% increase in load than ISMB150. Since it showed more than 100% (971.07 kN) of increase in load, it can be compared with the ultimate load of ISMB300 (ultimate load of 916.86 kN).
- Symmetrically filled sections showed more load capacity than assymetrically filled sections.
- In case of single filled sections, web only filled showed more load capacity. All others (BF only and TF only) are assymetrically filled hence showed less load capacity when compared to hollow CFS section.
- CFS hollow section is lighter by 40.16% than ISMB150 so that it can be used as primary member to reduce the dead load of the structure.
- The result indicates that use of ultra light weight concrete filling in the limbs proved to be efficient in increment of ultimate load of the columns.

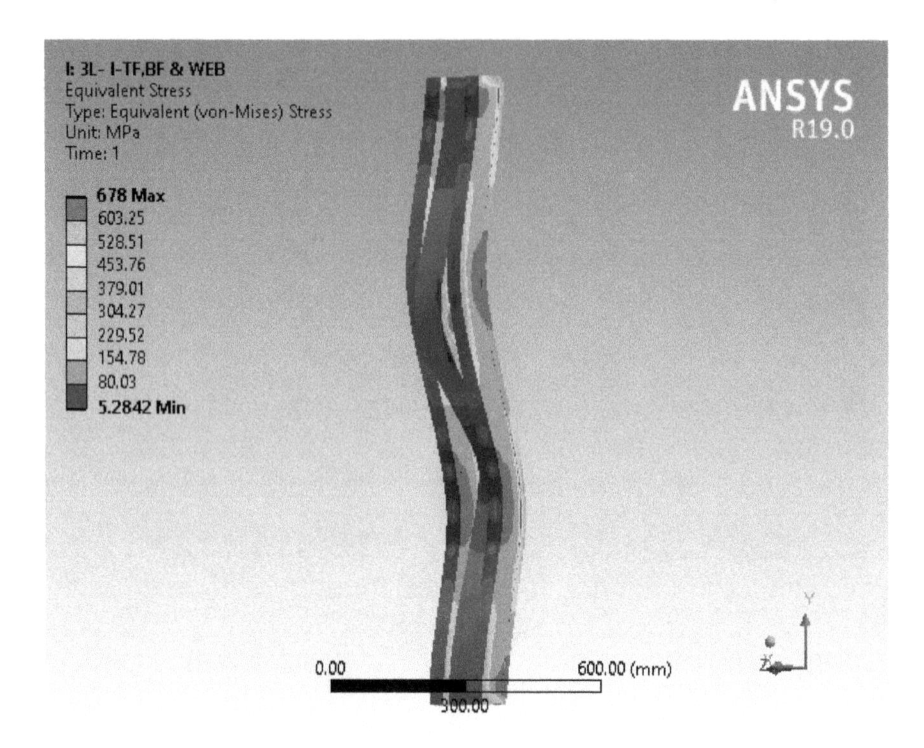

Fig. 9 Stress distribution of 3L-I-TF,BF&WEB

Table 5 Type of failure of columns

Model	Type of failure
ISMB150	Lateral global buckling
3L-I-Steel	Inward outward lateral buckling of limbs
3L-I-TF&BF	Global buckling
3L-I-WF	Local buckling of limbs
3L-I-TF50% BF50%	Inward outward buckling
3L-I-TF	Lateral local buckling
3L-I-BF	Lateral local buckling
3L-I-BF & WEB	Global buckling of limbs
3L-I-TF,BF,WEB	Inward outward lateral buckling of limbs

References

1. Roy K, Ting TCH, Lau HH, Lim JBP (2019) Experimental and numerical investigations on the axial capacity of cold-formed steel built-up box sections. J Constr Steel Res 160:411–427
2. Li Y-L, Li Y-Q, Shen Z-Y (2016) Investigation on flexural strength of cold-formed thin-walled steel beams with built-up box section. Thin-Walled Struct 107:66–79

3. Younga B, Ellobody E (2006) Experimental investigation of concrete-filled cold-formed high strength stainless steel tube columns. J Constr Steel Res 62:484–492
4. Lau HH, Lim JBP, Roy K, Ting TCH (2017) Effect of screw spacing on behavior of axially loaded back-to-back cold formed steel built-up channel sections. Adv Struct Eng
5. Reyes W, Guzman A (2011) Evaluation of the slenderness ratio in built-up cold-formed box sections. J Constr Steel Res 67:929–935
6. Roy K, Lau HH, Lim JBP, Ting TCH (2018) Nonlinear behavior of axially loaded back-to-back built-up cold-formed steel un-lipped channel sections. Steel Comp Struct
7. Muftah F, Sani MSHM, Mohammad S, Tahir M (2014) Ultimate load of built-up cold formed steel column 9(11)
8. Almamooria AHN, Naserb FH, Dhahirc MK (2020) Effect of section shape on the behaviour of thin walled steel columns filled with light weight aggregate concrete: Experimental investigation. Case Stud Constr Mater 13
9. Aravindhan A, Satheesh VS, Veeraragavan S (2017) Experimental investigation of cold formed steel long and short composite columns infilled with m-sand. Int J Sci Technol Eng 3
10. Ting TCH, Roy K, Lau HH, Lim JBP (2018) Experimental investigation into the behaviour of back-to-back gapped builtup cold-formed steel channel sections under compression
11. Yang Y-F, Han L-H (2009) Experiments on rectangular concrete-filled steel tubes loaded axially on a partially stressed cross-sectional area. J Constr Steel Res 65:1617–1630
12. Abhilash M, Jhanjhari S, Parthiban P, Karthikeyan J (2019) Axial behaviour of semi-lightweight aggregate concrete-filled steel tube columns—a DOE approach. J Constr Steel Res 162
13. J. Lavanya J, Elangovan R (2017) The structural behaviour of concrete filled steel tubular columns. Int Res J Eng Technol
14. Valipour HR, Foster SJ (2010) Nonlinear static and cyclic analysis of concrete-filled steel columns. J Constr Steel Res 66:793–802

Drought Intensity Mapping of Kannur District, Kerala, India

Nivya Sekharan, Mariya Kurian, and S. Jawahar Saud

Abstract An event of prolonged shortage in supply of water is called drought. It is considered to be a weather event with severe economic implications after hurricanes. Altered weather patterns, climate change, fluctuating ocean and land temperatures, reduced soil moisture, deforestation, soil degradation and other manmade interventions are the predominant reasons for drought events. This study examines the intensity of drought in Kannur district of Kerala, India. With an increase in population density and improper urbanization activities in place, human interventions are considered to be one of the primary causes of drought events in the district. Uneven rainfall patterns over the past 10 years have helped intensify the scenario for the worst. It eventually leads to decrease in available water which results in disappearance of wetlands, groundwater depletion and impact on water quality. Drought causes significant yield reductions both for rain fed and irrigated crops. Agriculture in Kannur district of Kerala, India is the worst affected by drought events. One of the other primary causes of droughts in Kannur district of Kerala is improper use of available water. Irrigation water is either over supplied or under supplied in most cultivable areas. Our study will focus on identifying such areas by creating a drought intensity map.

Keywords Drought · Drought in Kannur · Primary causes · Yield reduction · Agriculture in Kannur · Drought intensity map

1 Introduction

Drought is the phenomenon of having prolonged water shortages in an area [1]. The second-most costly weather events after hurricanes are droughts [2]. Agriculture is the most vulnerable sector which weakens due to the influence of droughts. Considerably decreased precipitation, depletion of surface water and decrease in groundwater supplies are the major reasons for the reduction in crop production.

N. Sekharan · M. Kurian (✉) · S. Jawahar Saud
Federal Institute of Science and Technology, Angamaly, Kerala, India

G. C. Marano et al. (eds.), *Proceedings of SECON'21*, Lecture Notes in Civil Engineering 171, https://doi.org/10.1007/978-3-030-80312-4_54

Increase in atmospheric temperature which is followed by drought will also adversely affect agriculture. This in turn leads to decrease in crop production and increase in production costs. Studies show that climate changes and human activities are the main causes of frequent and severe droughts [3].

Kannur district of Kerala, India is a multi-hazard prone district not only for drought, but for other natural calamities such as floods, cyclones, earthquake, coastal erosion, etc. The district faces many kind of natural or manmade disaster every year. Kannur is very much vulnerable to disasters due to its unique geo-climatic and social condition [4]. Agriculture in Kannur is the worst affected by the drought events. However, there will be certain areas which will be more water stressed than other areas. Crops grown over there have more chances to get dry and get lost. There will be variation in the water needed. Thus our study focuses on analyzing the drought intensity pattern in the agricultural fields of Kannur and identifying areas which are more susceptible to extreme condition of drought by creating a drought intensity map of Kannur and then comparing it with LULC (Landuse Land Cover). The use of GIS and multi-criteria decision tools in the fields of monitoring and forecasting of natural disasters is increasing over the recent years. According to [5–13], it has been understood that for mapping various natural hazards, the relevant data sets required needed to be figured out and all these data sets needed to be incorporated in the GIS software. With this background, the present objective of generation of drought intensity map of Kannur, Kerala, India is done by using the software ARCGIS 10.3. The drought zone mapping is carried out by using a range of parameters such as drainage density, rainfall, digital elevation model (DEM), and land use land cover (LULC) and by using the technique of weighted overlay analysis [14–17] It has been classified to extreme, moderate, slight and no drought areas.

We have also validated the generated with drought prone map of 2016 obtained from KSDMA (Kerala State Disaster Management Authority) to indicate the changes that had occurred over the years. The main reason behind such changes may be due to human induced modifications such as deforestation, urbanisation which results in uneven rainfall, less drainage and elevation variations. This shows that drought conditions can increase over the years. Thus in order to mitigate drought and reduce crops from getting dry there is a need for water saving, effective water distribution and utilization to mitigate drought and reduce crops from getting dry.

2 Study Area

Kannur district of Kerala, India has got its name from the place of its headquarters which was located at Kannur town. The word Kannur was derived from the old name ‘Cannanore’. The geographical division of the district can be described as highlands, midlands and lowlands. Highlands mainly comprises of mountainous areas, the midlands lie between mountains and lowlands comprises of undulating hills and valleys. Lowland is narrow when compared to the other regions and this is the region that comprises of seashore, deltas and rivers. The total area of the district is 2966 km^2. The latitudes of Kannur are 11° 40’ to 12° 48’ north and 74° 52’ to 76° 56’ east.

Kannur district is surrounded by Kasargod district on the north, Kozhikode and Wayanad districts on the south, Lakshadweep Sea on the west and the Western Ghats on the east. The general climate of the district is humid. Kannur district suffers from summer during the time period from March to May. After that the South-West monsoon is followed by this till September. The next 2 months will be post-monsoon. Then the following season will be North-East monsoon up to February. The maximum daily temperature of Kannur during summer is 35°. Average annual precipitation of the district is 3438 mm. In Kannur, more than 80% of rainfall is obtained from South-West monsoon. The district obtains 65% of annual precipitation from the heavy rainfall of July. Kannur is very rich in vegetation. The natural vegetation in Kannur consists of various types of forests. But the vegetation is not uniform in spite of generally favorable climatic conditions. In certain restricted regions, plant formations assume different characters with their own micro climate or special soil properties. The main crops of Kannur are paddy, coconut, arecanut, tapioca, rubber, pepper, cashew etc. Paddy is the one that occupies the largest area among these crops. The average yield of paddy is 1.74 tons per hectare. After paddy, the next important crop is coconut. It is extensively grown throughout the district. Cashewnut is a major cash crop grown in Kannur. Kannur plays a great role in the production and cultivation of cashew. Pepper is the major one of Kannur among the spices. It can be generally seen as an intercrop with coconut, arecanut and also other trees. The most important industrial cash crop of Kannur is rubber. In Kannur, drought occurs mainly in the month of March–May due to which loss of crop occurs. As a result of which 42.34 Ha of cultivated area is affected. Kannur district has been faced drought recurrently for the past decades. But comparing to other natural disasters, prediction of drought is quiet easier. Hence it could be managed [4] (Fig. 1).

3 Data Sources and Methodology

In order to create the drought intensity map of Kannur, ARCGIS integrated with technique of weight overlay analysis using the weighted overlay tool [14–17] are used by considering the factors which influence the drought vulnerability, viz., Drainage density, Rainfall, DEM and land use land cover. The details of different data sources, data preparation in GIS environment and application of weight overlay analysis are shown below (Fig. 2).

3.1 Generation of Thematic Layers

BASE MAP PREPARATION: Shapefile of Kerala has been downloaded from the NASA Indian village boundary [18]. From the shapefile of Kerala, shapefile of Kannur district of Kerala, India has been clipped out in order to obtain the base map.

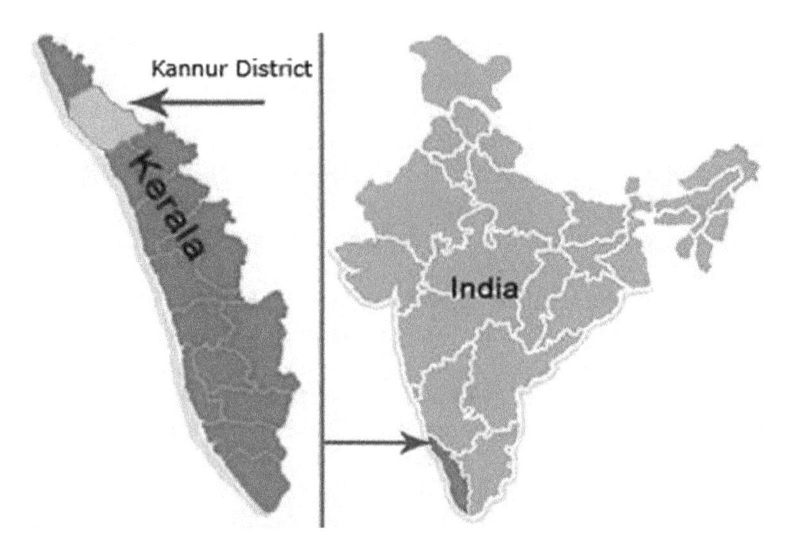

Fig. 1 Study area. *Source* http://kannurtourism.com

DIGITAL ELEVATION MODEL (DEM): From the Earth data search [19], Digital elevation model (DEM) has been downloaded. Various processes such as mosaic and conversion of DEM to Universal Transverse Mercator (UTM) Coordinate System, along with the application of fill tool has been carried out to obtain the final DEM. Using extraction tool, the Final DEM corresponding to the base map has been extracted out. Green colour in the map depicts that the area is of lower elevation and red colour indicates the area of higher elevation. This dem gives an idea of geographic regions of Kannur district such as highland, midlands and lowlands.

DRAINAGE DENSITY: From the final DEM, drainage line has been generated using certain tools such as hydrology tool, line density tool from which drainage density has been developed. The red portion in the map depicts areas having very less drainage density whereas the dark green portion in the map depicts high drainage density areas.

FINAL ANNUAL RAINFALL MAP: From CRU (Climatic Research Unit) DATA [20], CRU monthly climate data set (2011–2019) has been downloaded and by using certain tools such as data management tools the annual rainfall map of 2019 has been generated. By interpolating to the study area and by cutting the map, the Final annual rainfall map of Kannur district of Kerala, India in the year 2019 has been generated. As the colour increases the intensity of rainfall also increases that is light blue colour depicts the area receiving a much lower rainfall corresponding to other areas. The dark blue colour depicts the area receiving a higher value of rainfall.

LAND USE LAND COVER: The land use land cover of 2015–2016 has been obtained from the Kerala state disaster management authority (KSDMA), which depicts the physical substances that cover the Earth's surface at a given location

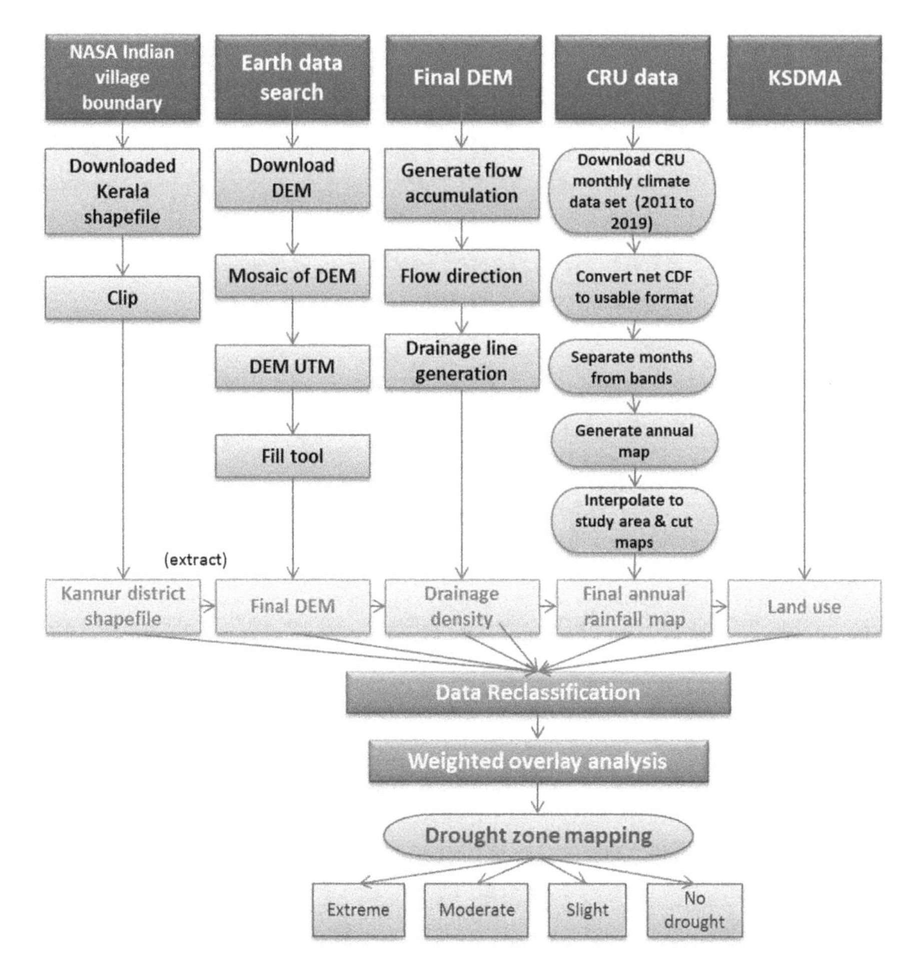

Fig. 2 Methodology

such as agricultural area, waterbody, wastelands, forest areas, urban areas etc. (Figs. 3, 4, 5, 6).

3.2 *Data Reclassification*

It is mandatory to reclassify data for overlay analysis. Reclassification makes same value to range of pixels. As example 1–50 (Actual value of range of pixels) = 1 (New single value for range) 0.51–110 = 2111 to 250 = 3 like that [21, 22]. Without this we cannot perform the overlay analysis. We classified our data such as Rainfall, drainage density and DEM into many classes depending upon the no of drought zones we require.

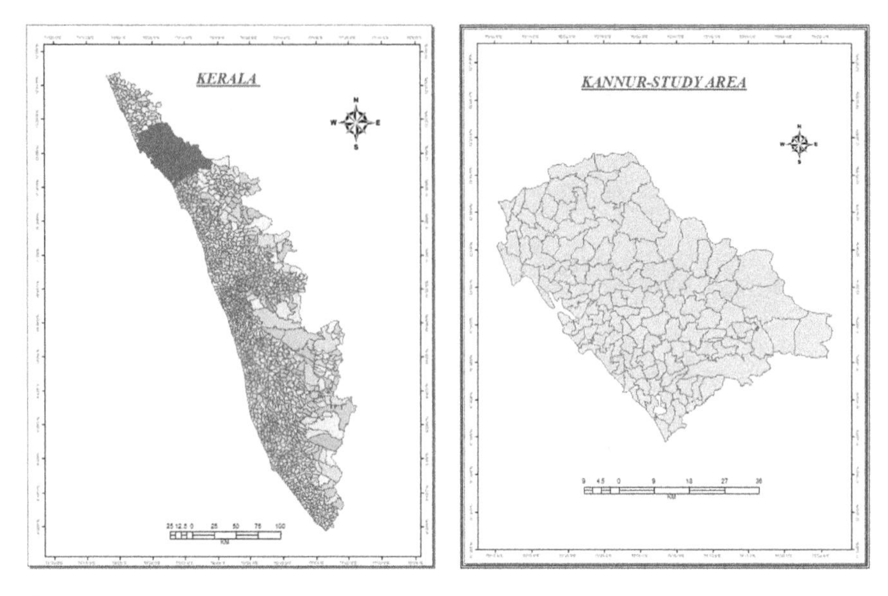

Fig. 3 Shapefile of Kannur clipped from shapefile of Kerala. *Source* https://sedac.ciesin.columbia.edu/data/set/india-india-village-level-geospatial-socio-econ-1991-2001)

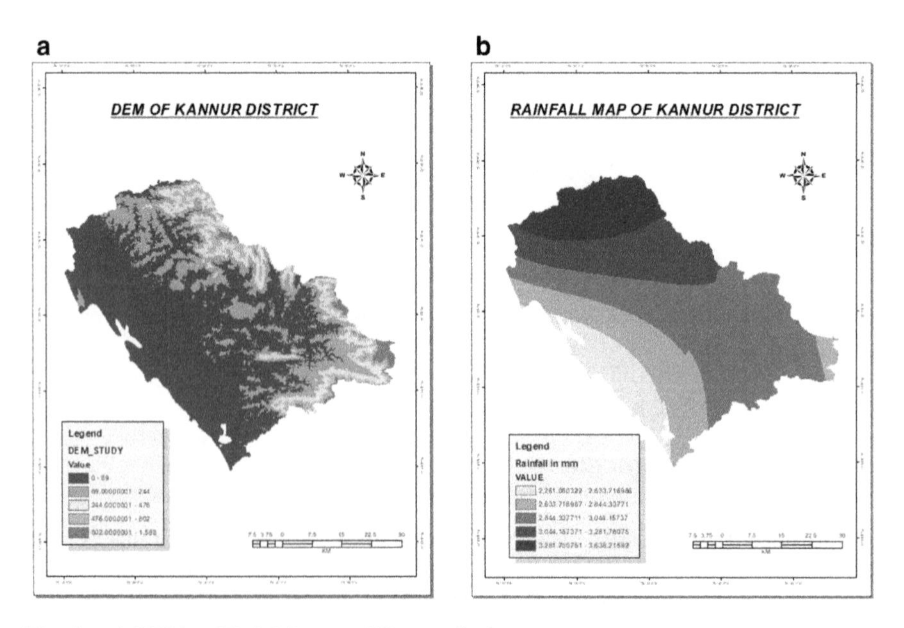

Fig. 4 **a**, **b** DEM and Rainfall map of Kannur district

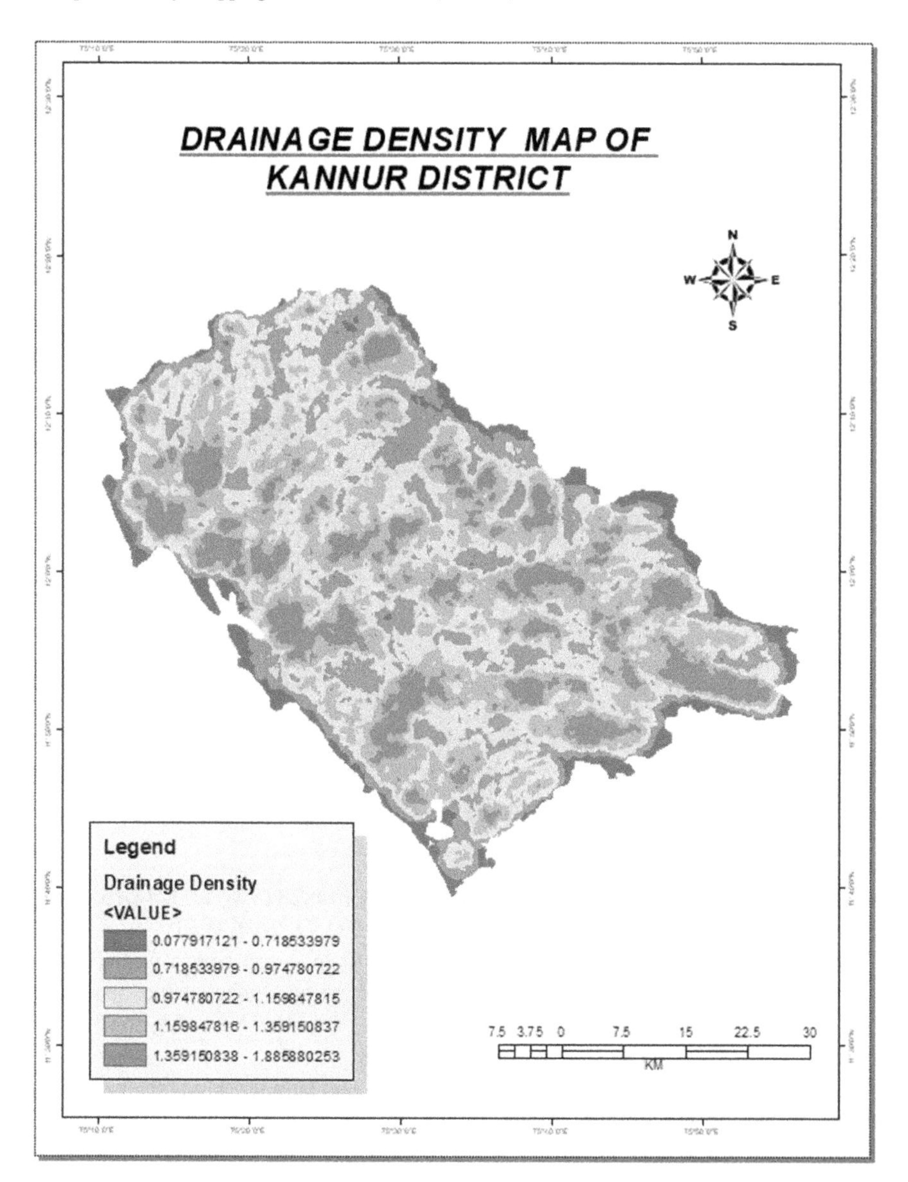

Fig. 5 Drainage density map of Kannur

DIGITAL ELEVATION MODEL (DEM): This dem is divided into five classes where each class represents certain values. First class takes values from 0 to 89, second class from 89 to 244, third class from 244 to 476, fourth class from 476 to 800, fifth class from 802 to 1568.

RAINFALL: The rainfall intensity a map has been divided into 5 classes each class taking certain values. The first class takes values from 2261.08 to 2633,

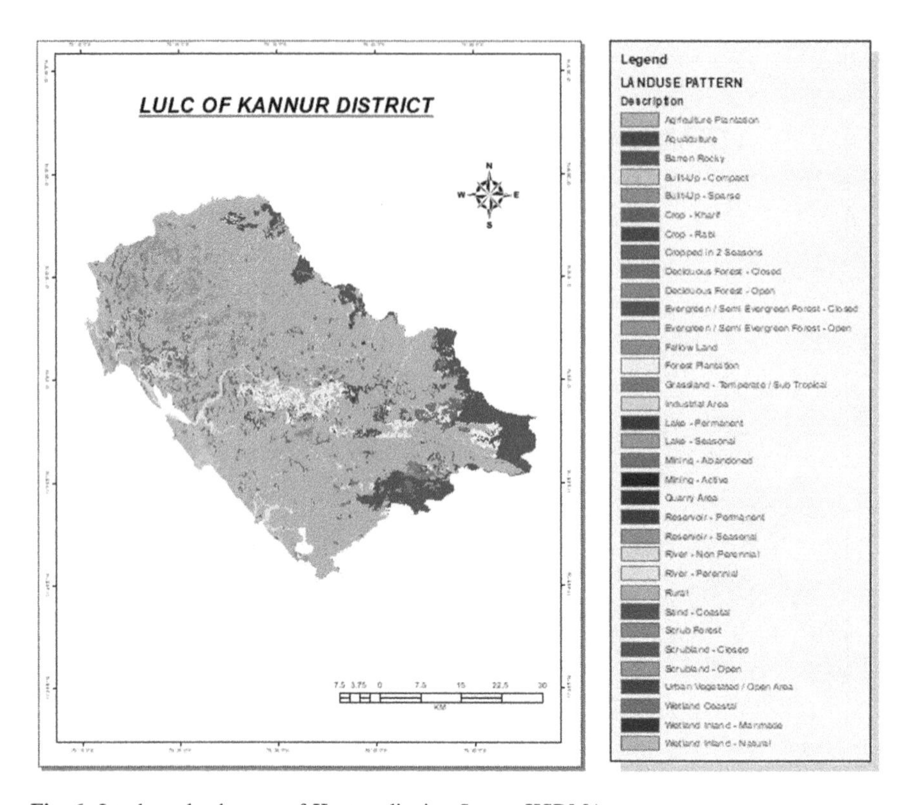

Fig. 6 Land use land cover of Kannur district. *Source* KSDMA

second class from 2263.71 to 2844.33, third class from2844.33 to 3044.15, fourth class from 3044.15 to 3281.78, fifth class from 3281.78 to 3638.21.

DRAINAGE DENSITY: The drainage density map has been divided into 5 classes where each class represent certain values. First class takes values from 0.07 to 0.71, second class from 0.71 to 0.97, third class from 0.97 to 1.15, fourth class from 1.159 to 1.359, fifth class from 1.35 to 1.88.

LAND USE LAND COVER OF KANNUR: The land use land cover obtained from KSDMA containing 34 groups such as agricultural plantations, aquaculture, barren rocky lands etc. has been grouped into five classes namely waterbody, wasteland, forests, agriculture and urban area (Table 1).

3.3 Weighted Overlay Analysis

The Weighted Overlay tool is one of the most used approaches for overlay analysis to solve multi-criteria problems. The Weighted Overlay tool helps to implement several of the steps in general overlay analysis process within a single tool. This tool has two important terms "influence" and scale value. Influence is overall

Table 1 Reclassification of DEM, rainfall, drainage density and Land use land cover

DEM		Rainfall		Drainage density		Land use land cover
Old values	New	Old values	New	Old values	New	Class
0–89	1	2261.80322–2633.71698	1	0.077917–0.718534	1	Waterbody
89–244	2	2633.716986–2844.3371	2	0.718534–0.974781	2	Wasteland
244–476	3	2844.3371–3044.15737	3	0.974781–1.159848	3	Forest
476–800	4	3044.15737–3281.78075	4	1.159848–1.35915	4	Agriculture
802–1568	5	3281.7807–3638.215825	5	1.35915–1.88588	5	Urban area

importance of layer and its sum should be equal to 100. The influence value assigns different value to each layer based on the importance. The layer which is most important among the layers is assigned a higher value and which is less important is assigned a lower value. Scale is importance of attribute of layers. It is a method of rating value based on importance [14–17]. Here 1 is less important and 5 is more important. Based on the influence and scale value we define theme of our analysis. Here Land use is considered as the most important layer in assessing drought vulnerability. Thus, it is given the highest influence value of 40% and in the Land use itself drought mostly affect urban area and agricultural area. Thus, all agricultural area and urban area from the land use land cover of Kannur is assigned a ranking of 4 and 5 respectively. All the wasteland areas and forest areas are assigned a ranking of 2 and 3 respectively. It affects the waterbody less thus it is given the least scale value of 1. The next important layer is Drainage density. Thus it is assigned an influence value of 30%. Less drainage has more possibility of drought thus scale values are given in an reverse order that is 5,4,3,2,1. The next important layer is Rainfall and thus it is assigned an influence value of 20% and in this lower rainfall contributes more to drought. So its given a reverse value that is 5,4,3,2,1. The least important layer is the Dem which is assigned a influence value of 10%. Drought will happen in lower elevation more and higher elevation less. Thus it is given a reverse value of 5,4,3,2,1 (Table 2).

4 Results and Discussion

Drought intensity map of Kannur has been generated after overlaying Digital Elevation Model (DEM), Rainfall, Drainage density and land use land cover using weight overlay tool which revealed the area under variable degree of drought. The resultant map which was reclassified using natural breaks divided the whole study area into 4 zones namely severe drought areas, moderate areas, slight drought and no drought areas. After generating drought intensity map we compare it with the land use land cover map obtained from KSDMA and analysed 130 villages of Kannur district (Fig. 7).

Table 2 Weighted overlay analysis

	Layer	Influence	Classification	Scale
1	Drainage density	30%	1	5
			2	4
			3	3
			4	2
			5	1
2	Rainfall	20%	1	5
			2	4
			3	3
			4	2
			5	1
3	DEM	10%	1	5
			2	4
			3	3
			4	2
			5	1
4	Land use land cover	40%	Waterbody	1
			Wasteland	2
			Forest	3
			Agriculture	4
			Urban area	5

After analysing the agricultural pattern and drought intensity of 130 villages, we found that in a village itself having agricultural plantations there are areas susceptible to moderate and slight drought. For eg. in Karivellur village having agricultural plantation there are areas susceptible to slight and moderate droughts. Thus in a agricultural village itself having same plantation there will be variation in water needed. If this is not taken into account some areas will be more water stressed than other areas. Thus there is a need for efficient water supply system that distributes water effectively which means that when drought season occurs water from the areas susceptible to slight droughts can be diverted or utilized to supply water to those areas susceptible to moderate or extreme conditions. Thus analysing this has made us understood about the fact that drought could be mitigated if we utilise and distribute water accordingly without the need of any extra tanks. It has made us understand that in order to mitigate the problem of drought one needs to utilise and distribute water accordingly. Extreme drought conditions conditions could be found in some agricultural areas of some villages such as Pallikunnu, Azhikode South, Kannapuram, Cherkkunu, Azhiyoor, Thiruvangadu, Thalassery, Muzhappilangad, Matool, Ramanthali, Chokli, Edakkad, Kannur and Dharmadam. However when we validated with the drought prone area map of Kannur of 2016 obtained from KSDMA, there were many other areas having extreme drought. This rapid decrease in Kannur for areas susceptible to drought is due to the 2018 floods of Kerala.

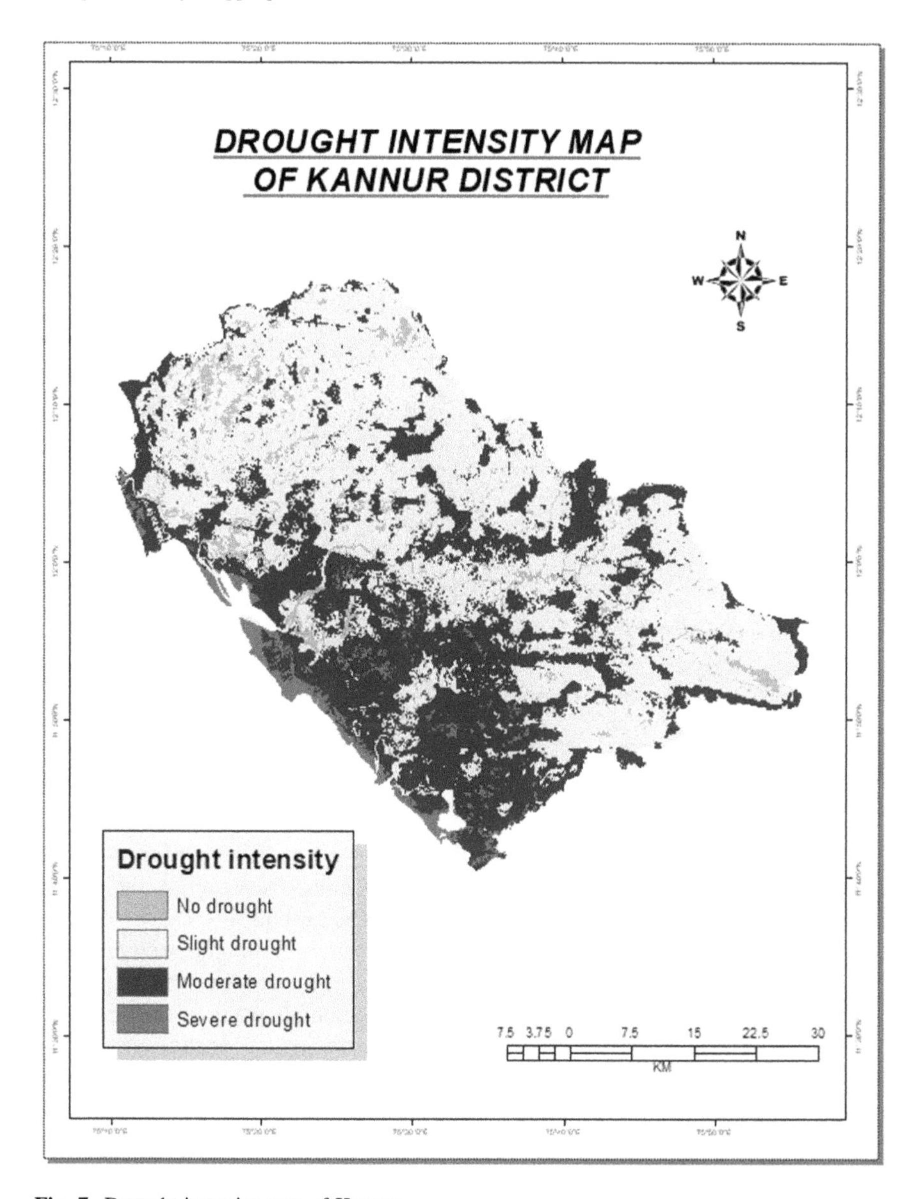

Fig. 7 Drought intensity map of Kannur

Kannur was one of the districts in Kerala which was adversely affected by the flood. So once the things get back to normal, these areas in Kannur will suffer by severe drought issues. This depicted that with the coming years one would face extreme conditions of drought issue as a result no water would be there to supply thus urging the need for saving water (Fig. 8).

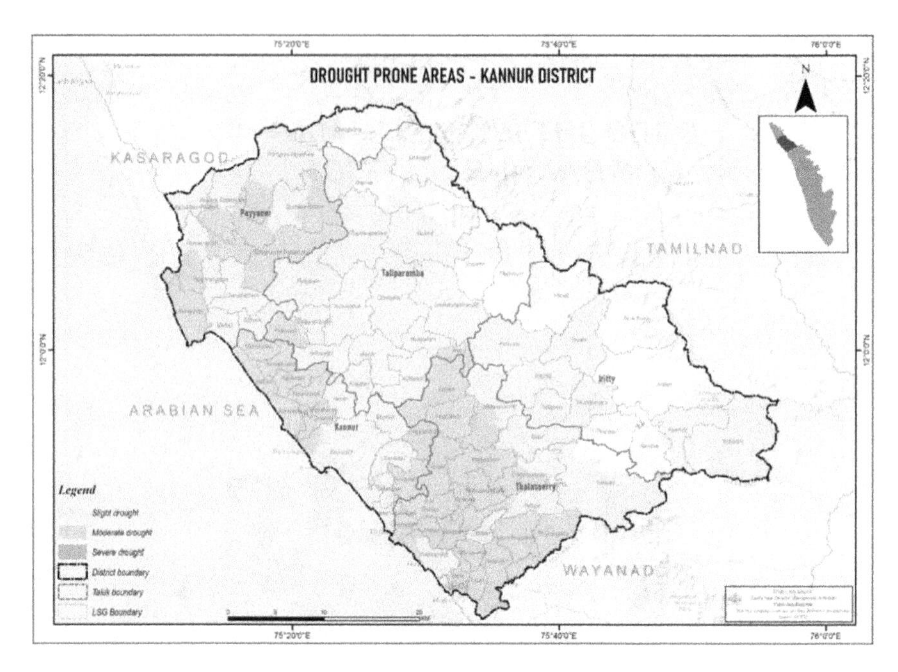

Fig. 8 Drought prone area map of Kannur district 2016. *Source* KSDMA

5 Conclusion

Drought which are the second-most costly weather events after hurricanes are an event of prolonged shortages in the water supply. Kannur district of Kerala, India is a multi-hazard prone district where floods, cyclones, earthquakes, coastal erosion, etc. can occur. With an increase in population density and improper urbanization activities in place, human interventions are considered to be one of the primary causes of drought events in the district. Uneven rainfall patterns over the past 10 years have helped intensify the scenario for the worst. Agriculture in Kannur district of Kerala, India is the worst affected by drought events. So, to identify the agricultural areas which are more prone to drought, a drought intensity map of Kannur has been generated using the technique of weighted overlay analysis considering the factors that are influencing drought vulnerability, namely Drainage density, Rainfall, DEM and land use land cover with the help of ArcGIS 10.3. The base map of Kannur was prepared from the shapefile of Kerala downloaded from NASA Indian village boundary. Digital Elevation Model (DEM) has been prepared from the data obtained from Earth data search and by using various tools, the drainage density and annual rainfall map of 2019 has been generated. The land use landcover has been obtained from KSDMA. After reclassifying all the data and applying the weighted overlay tool the drought intensity map of Kannur has been generated. The resultant map which was reclassified using natural breaks divided the whole study area into 4 zones namely severe drought areas, moderate areas, slight drought and no drought areas. So to analyze the agricultural areas which are more prone to drought, we have analysed the agricultural land obtained from land use land cover and drought intensity map of the 130 villages in Kannur. After analyzing the agricultural pattern and drought intensity of 130 villages, we found that in a village itself having agricultural plantation there are areas susceptible to moderate and slight drought. Thus in an agricultural village itself having same plantation there will be variation in water needed. Thus there is a need for efficient water supply system that distributes water effectively which means that when drought season occur water from the areas susceptible to slight droughts can be diverted or utilized to supply water to those areas susceptible to moderate areas. Some areas in some villages such as Pallikunnu, Azhikode South, Kannapuram, Cherkkunu, Azhiyoor, Thiruvangadu, Thalassery, Muzhappilangad, Matool, Ramanthali, Chokli, Edakkad, Kannur and Dharmadam showed extreme drought conditions. After analyzing this condition we have also validated with the drought prone area map of Kannur of 2016 and found that there were many other areas having extreme drought. This difference in our map is due to the 2018 floods of Kerala. Once the things get back to normal, these areas in Kannur will suffer by severe drought issues. Thus in order to mitigate drought there is a need for an efficient water management system that ensures better distribution, utilisation and saving of water. Providing better distribution utilisation and saving of water eliminates the need for extra tanks to be established. It tells the fact that one can mitigate the issue of drought by just utilizing the resource effectively without spending money on establishing big tanks etc.

Acknowledgements The authors would like to thank the Hazard & Risk analyst of State Emergency Operation Centre (SEOC), Kerala Disaster Management Authority (KSDMA), for providing the required data for the study as mentioned in the paper.

References

1. It's a scorcher -and Ireland is officially 'in drought' Irish Independent, 18 July 2013
2. https://www.nationalgeographic.org/
3. Kuwayama Y, Thompson A, Bernknopf R, Zaitchik B, Vail P (2018) Estimating the impact of drought on agriculture using the US drought monitor. Am J Agr Econ 101(1):193–210
4. District Disaster Management Plan Kannur June 2015
5. Chopra P (2006) Drought risk assessment using remote sensing and GIS: a case study of Gujarat. Master thesis submitted to international institute for geo-information science and earth observation, ITC, Enschede, The Netherlands
6. Han P, Wang PX, Zhang SY, Zhu DH (2010) Drought forecasting based on the remote sensing data using ARIMA models. Math Comp Model 51:1398–1403
7. Pogarčić I, Frančić M, Davidović V (2008) Application of AHP method in traffic planning. In: 16th International symposium on electronics in traffic planning
8. Prathumchai K, Honda K, Nualchawee K (2001) Drought risk evaluation using remote sensing and GIS: a case study in Lop Buri Province. In: 22nd Asian conference on remote sensing, Singapore

9. Chen YR, Yeh CH, Yu B (2011) Integrated application of the analytic hierarchy process and the geographic information system for food risk assessment and food plain management in Taiwan. Nat Hazards 59(3):1261–1276
10. Pandey S, Pandey AC, Nathawat MS, Kumar M, Mahanti NC (2012) Drought hazard assessment using geoinformatics over parts of Chotanagpur plateau region Jharkhand. India Nat Hazards 63(2):279–303
11. Stefanidis S, Stathis D (2013) Assessment of food hazard based on natural and anthropogenic factors using analytic hierarchy process (AHP). Nat Hazards 68(2):569–585
12. Palchaudhuri M, Biswas S (2016) Application of AHP with GIS in drought risk assessment for Puruliya district India. Nat Hazards 84(3):1905–2192
13. Zagade N, Kadam AK, Umrikar BN, Maggirwar BC (2018) Remote sensing-based assessment of agricultural droughts in sub-watersheds of upper Bhima basin India. J Remote Sens Land 2(2):105–111
14. https://desktop.arcgis.com/en/arcmap/10.3/tools/spatial-analyst-toolbox/how-weighted-overlay-works.html
15. https://pro.arcgis.com/en/pro-app/latest/help/analysis/raster-functions/weighted-overlay-function.htm
16. Basharat M, Shah HR, Hameed N (2016) Landslide susceptibility mapping using GIS and weighted overlay method: a case study from NW Himalayas Pakistan. Ar J Geosci 9(4):1–19
17. Shit PK, Bhunia GS, Maiti R (2016) Potential landslide susceptibility mapping using weighted overlay model (WOM). Model Earth Syst Environ 2(1):21
18. https://sedac.ciesin.columbia.edu/data/set/india-india-village-level-geospatial-socio-econ-1991-2001
19. http://search.earthdata.nasa.gov
20. https://sites.uea.ac.uk/cru/data
21. http://desktop.arcgis.com/en/arcmap/latest/extensions/geostatistical-analyst/data-classification.htm
22. https://desktop.arcgis.com/en/arcmap/10.3/tools/spatial-analyst-toolbox/an-overview-of-the-reclass-to-ols.htm

Behavior of Harp and Perimetral Bracing System in Pre-engineered Building Subjected to Lateral Loads

Merin Thomas and Gayathri Krishnakumar

Abstract Pre-engineered buildings (PEB) are pre- fabricated in the factory and the components are assembled at site thereby reducing the time of completion. In PEB the frame geometry matches the shape of the bending moment (internal stress) diagram so the total weight of the structure can be reduced due to optimization in material usage. Horizontal (seismic/wind) load is the unreliable load that is coming on the structure, which results in production of storey drift, overturning moment, storey displacement etc., causing failure of the structure. In PEB there is series of two-dimensional frames repeated at every certain distance. These frames are stabilized longitudinally by providing of bracings. The frames in the transverse direction bears loads in its plane with no help from the longitudinal bracings, so the columns act as bending brackets, requiring large-size columns and large foundations. Against this arrangement, special on-roof bracing arrangements like Perimetral and Harp Bracing arrangement can be used as an alternative. The present work deals with the suitability of type of bracings arrangement by performing Time history analysis on 3D industrial structures with different configurations of bracings arrangement using ETABS. Also the suitability of different type of bracings arrangement for varying span is studied.

Keywords Pre-engineered building · Bracing · Harp bracing · Perimetral bracing · Lateral loads

1 Introduction

Pre Engineered Building (PEB) type of structures are extensively used as industrial structures which consists of channel steel purlins spanned between the rigid frames and also with the wall cladding at sides. The steel moment resisting frames in PEB are susceptible to undergo lateral displacement under lateral loads. Horizontal load like seismic and wind loads are the unreliable load that is coming on the structure.

M. Thomas (✉) · G. Krishnakumar
Department of Civil Engineering, FISAT, Angamaly, Kerala, India

G. C. Marano et al. (eds.), *Proceedings of SECON'21*, Lecture Notes in Civil Engineering 171, https://doi.org/10.1007/978-3-030-80312-4_55

Such structure should be designed to resist both gravity and lateral loads. Gravity loads includes dead load, live load, dust load etc. Whereas lateral load includes seismic load, wind load and blast load. Due to this lateral loads, high stresses are produced which then leads to sway or vibration. Horizontal or lateral loading results in production of storey drift, overturning moment, storey displacement etc., which are responsible for failure of the structure. To inhibit these responses bracings can be used in structures. Structures with bracings, show better performance in reducing structural parameters (stress ratio) and systematic parameters (time period, base shear and lateral displacement).

Shaik [1], explained the concept of PEB. In PEB the frame geometry matches the shape of bending moment diagram. Optimization in material usage reduces the total weight of the structure. On performing an analytical study on PEB and conventional steel frame PEB found to be more economical than conventional building. Staad Pro Jinsha and Mathew [2], studied the effect of change in bay spacing in Pre-Engineered Buildings and it was observed that the weight of the PEB depends on the bay spacing and the increase in the bay spacing up to certain spacing weight reduces and further increase in bay spacing makes weight heavier. PEB structures with 8 m spacing are found to be more economical. Saleem [3], performed a design-based optimisation for pre-engineered steel frames made of built-up sections and studied the effect of unbraced length ratios on flange and web slenderness ratios. The use of appropriate values of flange and web slenderness according to the member unbraced length can reduce the section weight and improves its efficiency significantly. Dhandel [4], studied different types of bracings namely, Inverted V Bracing (IVB), Vertical V bracing (VVB) and Cross Diagonal Bracing (CDB). Providing bracing configurations in frames reduce the natural time period of vibration. Pablo [5], performed a comparative study between rigid frames with traditional bracings, and harp bracings. Spans from 10 to 35 m, as well as bay lengths from 20 to 50 m were considered. Economical comparison were made. Harp and perimetral bracing reduces the deflection at eave level and thus the column and foundation size can be reduced. Vasugi [6] studied effect of wind load on industrial building with different bracing patterns. For PEBs without any bracing, the displacement of the structure for wind load in the longitudinal direction of the building, this is because the roof and the ridge cannot alone provide rigidity to PEB. For wind load in the transversal direction of the building, considerable displacement can be prevented by tapered frames itself.

The present work deals with the suitability of type of bracings arrangement by performing Time history analysis from the El-Centro Data on 3D industrial structures with different configurations of bracings arrangement using ETABS. Also the suitability of different type of bracings arrangement for varying span is studied. The results based on the parameters maximum storey displacement, storey drift, and time period are evaluated.

2 Building Modelling

The structures considered for the study is Pre-engineered industrial steel buildings with harp, perimetral and conventional bracings. The effect of harp, perimetral and conventional bracing under seismic loading is studied for 20, 30 and 40 m span. Six numbers of bays are provided at 6 m spacing in longitudinal direction, keeping the length of building as 48 m. The height of the buildings from the base to eave level is 10 m. Roof slope provided is 1:10. The grade steel is taken as Fe345. The columns and beams are designed as non prismatic sections (Figs. 1, 2 and 3).

2.1 Load and Load Combinations

The configuration of building and loading conditions are developed according to IS 800:2000 [9], IS 875:1987 (Part 1, 2) [7, 8] and IS 1893 (Part 1):2002 [10]. A dead load of 0.10 kN/m^2 is applied as per IS: 875 (Part 1): 1987 [7], and a live load of 0.75 kN/m^2 is applied as per IS: 875 (Part 2): 1987 [8]. Self weight is assigned by the software itself. The design earthquake load considered is as per IS 1893 (Part 1): 2002 [10], Zone–III, R = 5 (Steel structure), I = 1, Z = 0.16. The support conditions are assumed as pinned.

The building is designed for following load combinations as per IS 800:2000 [9]

1.5DL
1.5(DL + LL)
1.5(DL + EQX)
1.5(DL + EQY)
1.5(DL-EQX)
1.5(DL-EQY)

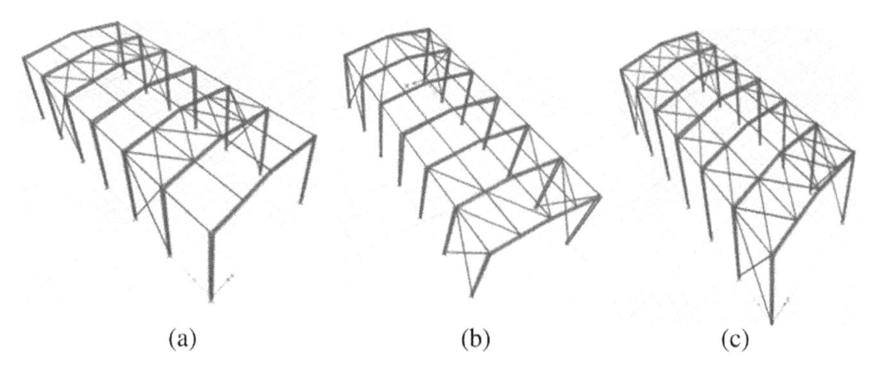

Fig. 1 3D view of 20 m span PEB (**a**) conventional bracing, (**b**) harp bracing, (**c**) perimetral bracing

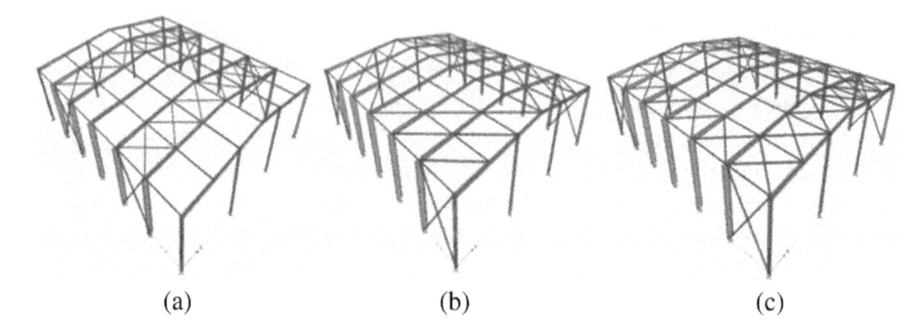

(a) (b) (c)

Fig. 2 3D view of 30 m span PEB (**a**) conventional bracing, (**b**) harp bracing, (**c**) perimetral bracing

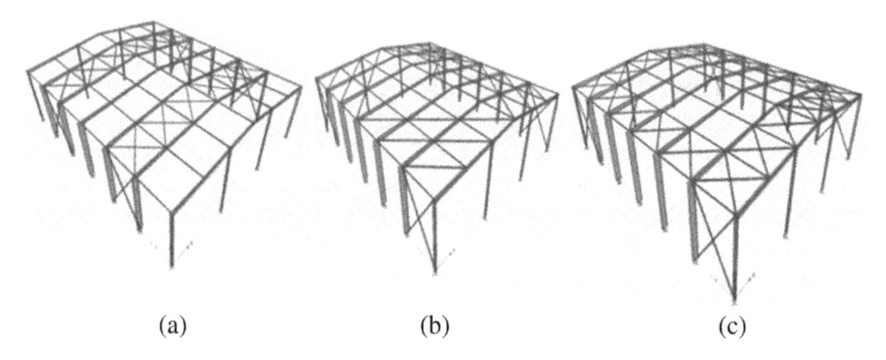

(a) (b) (c)

Fig. 3 3D view of 40 m span PEB (**a**) conventional bracing, (**b**) harp bracing, (**c**) perimetral bracing

1.2(DL + LL + EQX)
1.2(DL + LL + EQY)
1.2(DL + LL-EQX)
1.2(DL + LL-EQY)
0.9DL + 1.5EQX
0.9DL-1.5EQX
0.9DL + 1.5EQY
0.9DL-1.5EQY.

2.2 Time History Analysis

Linear modal time history analysis is performed from the El-Centro Data to study the suitability of type of bracings arrangement on 3D industrial structures for 20, 30 and 40 m spans.

3 Results and Discussions

3.1 Displacement

The models of 20, 30 and 40 m span are analyzed with conventional bracing, harp bracing and perimetral bracing to avoid high displacement. Displacement obtained for different bracing for varying spans is shown in Tables 1 & 2.

For 20 m span, the displacement of PEB with conventional, and perimetral bracing almost same and the least displacement is obtained for PEB with conventional bracing. For 30 and 40 m span, the displacement of PEB with harp and perimetral bracing almost same and is very much lower than conventional bracing. The least displacement in 30 and 40 m is obtained for PEB with harp bracing.

3.2 Storey Drift

Drift obtained for different bracing for different spans is shown in Table 3 & 4 shows the variation of displacement for different bracings.

For 20 m span, the drift in X and Y direction less for PEB with conventional bracing. The drift of PEB with harp bracing is less compared to perimetral and conventional bracing for 30 and 40 m span.

3.3 Time Period

Drift obtained for different bracing for different spans is shown in Table 5 shows the variation of displacement for different bracings.

Table 1 Displacement in X direction

Span	Conventional bracing	Harp bracing	Perimetral bracing
20 m	22.85	23.82	24.35
30 m	42.17	22.36	22.74
40 m	20.6	10.79	10.99

Table 2 Displacement in Y direction

Span	Conventional bracing	Harp bracing	Perimetral bracing
20 m	0.88	1.25	0.873
30 m	5.23	1.99	1.86
40 m	7.56	1.68	2.38

Table 3 Drift in X direction

Span	Conventional bracing	Harp bracing	Perimetral bracing
20 m	0.002285	0.002382	0.002436
30 m	0.00421	0.002169	0.002206
40 m	0.001996	0.00104	0.00106

Table 4 Drift in Y direction

Span	Conventional bracing	Harp bracing	Perimetral bracing
20 m	0.000059	0.000083	0.00007
30 m	0.000093	0.000104	0.000087
40 m	0.000096	0.00009	0.00105

Table 5 Time period

Span	Conventional bracing	Harp bracing	Perimetral bracing
20 m	1.331	1.125	1.227
30 m	2.148	1.163	1.179
40 m	0.913	0.644	0.652

The time period of PEB with harp bracing is less compared to perimetral and conventional bracing for 20, 30 and 40 m span.

4 Conclusion

The Linear Modal Time History analysis on a PEB with harp, perimetral and conventional bracing arrangement for varying span is studied. The following conclusion can be made from the study:

- For larger spans PEB with harp and perimetral bracing shows less displacement, drift and time period compared to conventional bracing when subjected to seismic loading. On comparing harp and perimetral bracing, harp bracing shows lower displacement compared to perimetral bracing. Therefore, for larger span harp and perimetral bracings are suitable
- For smaller spans conventional bracing shows better performance than perimetral and harp bracing.

References

1. Shaik K, Ratnamala Reddy BSS, Jagarapu DCK (2020) An analytical study on pre engineered buildings using staad pro. Mater Today Proc 33(1):296–302
2. Jinsha MS, Mathew LA (2013) Analysis of pre-engineered buildings. Int J Sci Res 5(7): 1049–1051
3. Saleem MU, Khurram N, Qureshi HJ, Kazmi ZA, Siddiqui ZA (2018) Optimisation of flange and web slenderness for pre-engineered built-up steel sections. Proc Instit Civil Eng Struct Build 172(8):599–608
4. Dhande SN (2015) Industrial building design on seismic issues. IJIRSET 4(5):2840–2856
5. Martínez JM, Montero PG, Peguero CN, de Prado FD (2005) Behavior of harp bracings in rigid frame metallic bays (ASCE) 0733–9445, 131(9):1397
6. Vasugi K, Somasekharan S (2017) Wind load analysis for industrial building with different bracing patterns and its comparison with pre engineered building. Int J Civil Eng Technol 8(4):1074–1085
7. IS 875(Part 1)-1987. Indian standard code of practice for design loads (other than earthquake) for building and structures. part 1: dead loads. Bureau of Indian Standards, New Delhi
8. IS 875(Part 2)-1987. Indian standard code of practice for design loads (other than earthquake) for building and structures. part 2: imposed loads. Bureau of Indian Standards, New Delhi
9. IS 800-2007. General construction in steel—code of practice. Bureau of Indian Standards, New Delhi
10. S 1893 (Part I)-2002. Criteria for earthquake resistant design of structures. Part 1 General Provisions and Buildings. Bureau of Indian Standards, New Delhi

Challenges and Benefits of BIM Adoption in a Metrorail Project

Sulakshya Gaur and Abhay Tawalare

Abstract Improper integration of different practices in the projects has always been a critical challenge in project delivery. Building Information Modelling (BIM) emerges as one of the important tools that help to facilitate this process. However, its implementation is not devoid of challenges and bottlenecks. With a case study on one of the Indian megaprojects employing 5-D BIM, this work enlisted eight major challenges encountered in its implementation. Further, these were grouped into three categories i.e., personnel issues, process issues, and technological limitations. Along with this, these challenges were mapped with the practices/steps adopted by the organization to overcome these obstacles. The detailed analyses also revealed that mutual consent between the four most important stakeholders i.e., client, consultant, designer, and contractor is of utmost importance for the implementation of BIM. The findings also revealed that BIM as a process can integrate various processes in the construction project and thus can pave a way for better management of stakeholders too.

Keywords BIM · Megaproject · Metro rail project · Project management · Stakeholders

1 Introduction

The disintegrated nature of the construction industry presents a humongous challenge for the efficient delivery of the projects. To overcome this pertinent crisis, several improvement tools have been developed to enhance the overall performance [1]. One of such tools that became popular among construction professionals was Building Information Modelling (BIM). BIM acts as an important tool in the modern-day construction process, that allows users to create multidimensional models for efficient project management during the life cycle of the project [2].

S. Gaur · A. Tawalare (✉)
Visvesvaraya National Institute of Technology, Nagpur, Maharashtra, India
e-mail: abhaytawalare@civ.vnit.ac.in

G. C. Marano et al. (eds.), *Proceedings of SECON'21*, Lecture Notes in Civil Engineering 171, https://doi.org/10.1007/978-3-030-80312-4_56

BIM, by its sheer ability to promote effective collaboration between groups and teams, enhances the flow of information thus promoting better communication [3]. This not only reduces the overall project time and in turn the cost but also helps a great deal in the reduction of the overall life cycle cost of the project.

However, BIM suffers from numerous challenges in its adoption and several previous studies have focused on identifying these factors. The previous studies have generally focused on the projects in various developed countries such as the United States and the United Kingdom. These research works have pointed out lack of human, technological and financial resources along with poor technical and managerial skills as major challenges to BIM adoption [1]. A comprehensive analysis of these factors remains absent concerning the type of project and also specifically about the implementation of 5-D BIM.

Taking these into account, this work uses an exploratory study to determine/ assess the challenges the project team faced in BIM adoption in Mass Rapid Transport System (MRTS) i.e., a metro project in India. This becomes more important as the project considered in this study is the first to use 5-D BIM and SAP-ERP (Enterprise Resource Planning) in India and across Asia [4].

2 Literature Review

Collaboration between teams is described as the major link to achieve common project goals and hence certain practices need to be adopted to promote it efficiently. A shift from the traditional practice that heavily depends on 'non-intelligent/non-interoperable data' along with 2-D drawings needs to be undertaken [5]. Hence, to facilitate this process use of BIM as a tool becomes important. BIM greatly helps in the distribution/exchange of information between the participants and different teams, such as data transfer to teams during the O&M phase of the project [6]. The ability of BIM to integrate cost data provides an opportunity to designers and contractors to look into various alternatives and appraise them accurately during the initial stages itself [3].

Although the advantages of using BIM in the construction process are very clear, reluctance in its adoption is one of the major barriers to its adoption along with many other challenges it faces [7]. The role of top management (support from the top leadership) [8] along with the availability of proper financial resources is essential for the adoption of any type of innovation in the construction industry [1]. A coalescence of such previous studies reveals that the unavailability of resources and lack of skills among the project teams are the critical barriers to the implementation of BIM [1]. Moreover, these studies have a major focus on the projects in developed countries such as the USA, the UK, Canada, France Australia, Japan, and New Zealand [9] and but it is still a new technology in developing markets and suffers a slow adoption rate [10].

Hence, this study by focusing on the first metro project to employ 5-D BIM and SAP-ERP system in India tries to find out the primary challenges in the adoption of the whole system along with the benefits it has provided. 5-D BIM model is the one that integrates 3-D model with time (4th dimension) and cost (5th dimension). Cost-related information is provided using SAP-ERP. The challenges enumerated in this study along with the possible solutions may pave way for the implementation of BIM by highlighting substantial BIM advantages and hence its ability in managing some key stakeholders.

3 Research Methodology

The study uses a semi-structured interview approach to understand the ideas and challenges faced in the adoption of BIM in a Mass Rapid Transport System (MRTS) i.e., a metro project. An ABC (fictitious name) metro project in one of the major cities of central India was undertaken for the study. Since this project is the first one to employ 5-D BIM along with the SAP-ERP system, it was found suitable for this study. Data was collected with the help of semi-structured interviews and discussions with the project participants. The use of semi-structured interviews helped in keeping the discussion open with the respondents. In some instances, a detailed insight was sought from them to understand some of the challenges and obstructions in better ways. Moreover, using this method they provided detailed information about their experience in using BIM as a tool and a process.

A thorough literature study was done before the preparation of questions for the interviews. This helped in obtaining some initial ideas about the basic bottlenecks in BIM adoption and also helped in keeping the conversations with the respondents fluent. The interviews were conducted for a period of 2 months and a single interview lasted for a duration of 45 min–1.5 h. Eight project personnel were interviewed during this duration and all of them belonged to the senior managerial positions and had hands-on experience in using BIM as well as managing the overall project. These participants belonged to different sections i.e. planning, design, procurement, and IT. Moreover, these respondents were limited to general manager, executive director, assistant general managers. All the respondents had extensive experience in their relevant fields and were deemed suitable for this study.

The interview data were analyzed using NVivo 10. All the challenges towards BIM implementation were coded as child nodes during the initial analysis. Once the initial coding process was completed these child nodes were grouped under the constructs/parent nodes. These parent nodes were then named based on respondents' input and previous literature.

4 Results and Discussions

The analysis of the interviews presented major challenges the project team faced in the BIM adoption. The analysis of the interview data presented eight major challenges they witnessed in the adoption/implementation of BIM in their project. These challenges were grouped under three constructs i.e., personnel issues (1st construct), process-based issues (2nd construct), and technological limitations (3rd construct). The inhibition of personnel towards technological advancement i.e., unwillingness to accept change was the highest coded factor in the 1st construct followed by lack of skilled personnel for using BIM. In the 2nd construct, both, difficulties in adopting new technology and huge upfront cost in BIM adoption were equally important challenges. The third construct dealing with the technological limitations had the poor IT (HW, SW, and network) infrastructure as the most coded factor.

Apart from this, the respondents mentioned few strategies that they had adopted or that need to be brought in place to facilitate the BIM adoption process. In this, the most important were training and education of employees and a need for strong government mandate and BIM standards. These two factors were coded in all the conducted interviews. The next strategies were having strict contractual provisions and laws followed by publicizing the advantages of BIM and having indigenous software and better IT infrastructure. The content analysis of the interviews done through NVivo is presented in Table 1 below.

The interview respondents along with discussing the challenges also detailed the benefits BIM has to offer. The advantages are witnessed in better collaboration of different teams and groups for reaching an optimum design and solutions. This along with the ability of BIM integrated with ERP platform provides for quick approval of bills and reduction in manual cataloging of office documents. By providing a virtual workspace, BIM brings different teams together during the planning and design stage itself. This leads to better creation of designs and reduces any future reworks. Because of numerous advantages, it becomes important to better address the inefficient adoption of BIM. To present the challenges along with the critical strategies that can be adopted to overcome those challenges a conceptual model is prepared and is presented in Fig. 1 below. This is followed by a discussion on the challenges faced in the BIM adoption in a metro rail project.

Challenges to BIM adoption:

4.1 Personnel Issues

The lack of a sufficient BIM skilled workforce in India calls for an extensive training regime of construction personnel with the new technology to provide them with hands-on experience of software usability. Not only the training of existing employees but to develop a better future workforce, BIM needs to be incorporated

Table 1 Results of the content analysis of the conducted interviews

Constructs	BIM Challenges	Sources	References	% Articles coded
Personnel Issues	Unwillingness towards technological advancements	6	10	75%
	Lack of skilled personnel with BIM usage	5	13	62.5%
	Draining and outsourcing of workforce	2	2	25%
Process-based issues	Adaptability with the new platform	5	7	62.5%
	Huge upfront cost in the project	5	8	62.5%
	Lack of mutual consent between important stakeholders	2	5	25%
Technological limitations	Poor laws pertaining to data protection and privacy	4	6	50%
	Poor hardware, software, and networking infrastructure	5	9	62.5%
Strategies (that can be) undertaken to facilitate BIM adoption				
Strategies		**Sources**	**References**	**% Articles coded**
Training and education of employees		8	13	100%
Publicizing the advantages of BIM		4	7	50%
Strong government mandate and BIM standards		8	15	100%
Strict contractual provisions and laws		6	11	75%
Development of indigenous software and better technical infrastructure		3	5	37.5%

in the curriculum too. One of the other issues that act as an add-on to this lack of a skilled workforce is the draining and outsourcing of the workforce. India being a diversified country, possess a skilled workforce too who are well accustomed to BIM and its uses, but their shifting to work for international companies and projects makes it difficult to find suitable designers and consultants for indigenous projects. This area was explained by one of the experts as:

> It is not that there is a lack of talent pool in India. A lot of young professionals who are coming out are very competent and extremely skillful, but the lack of job opportunities and less pay means they are easily lost to our foreign counterparts. This is where governments' role comes to the forefront, they must strive to provide better job and working conditions to young professionals to keep them attracted locally so that it will provide a great impetus to the Indian construction industry at large.

Apart from the lack of knowledge about a particular platform or the software the mere adamant nature of senior employees or management poses a strong obstacle for application or bringing in any new technology in the system. The unaccepting

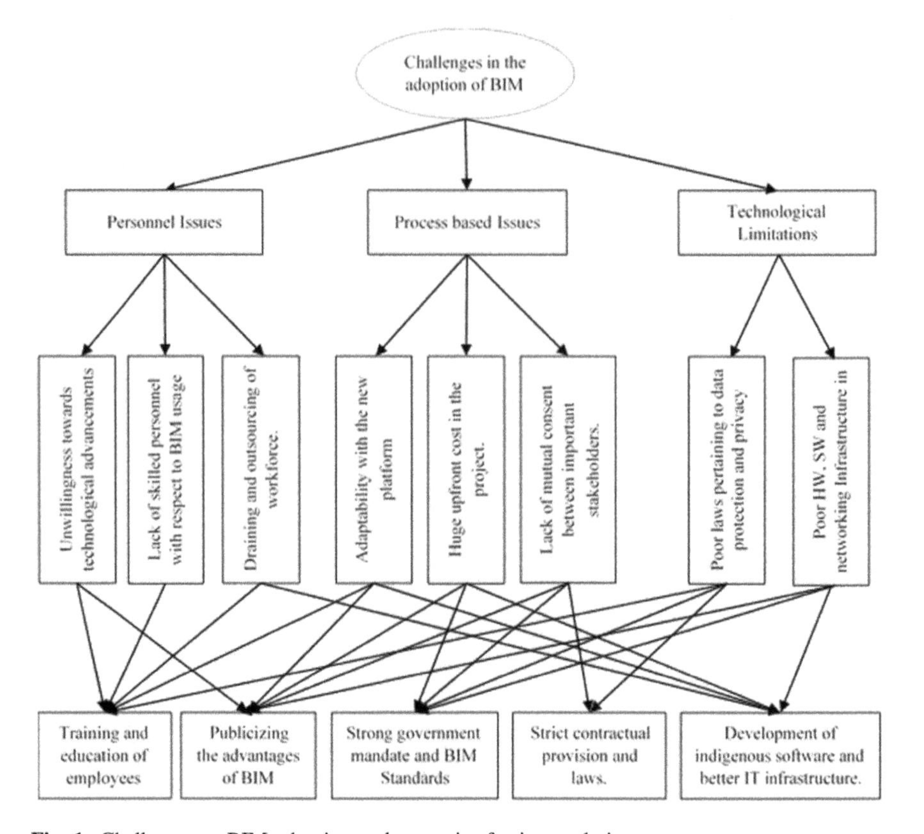

Fig. 1 Challenges to BIM adoption and strategies for its resolution

nature of the top management or the senior employees in the system makes the overall process of changing a system or inculcating a new methodology in a set process very daunting. Although this can be the result of both, either their inability to adopt a new process or it can be a result of their rigorous attitude to not change.

These major roadblocks to the adoption of BIM can be resolved to some extent by having better training and education of the employees. The advantages BIM is going to offer in their day-to-day task can also be used as a motivating measure for facilitating its adoption.

4.2 Process-Based Issues

The second set of challenges towards BIM adoption was more related to the BIM process itself, although the BIM as a platform is quite user-friendly and is very advantageous, due to the integration of different tasks, a lot of users face difficulty in adapting to it. Secondly, the regular training of the existing workforce along with developing new technical infrastructure that comes as a mandatory prerequisite for its adoption makes it a costly affair. This means, that a lot of small and medium-sized firms tend to avoid it. Apart from all these, the mutual consent between the four most important stakeholders in construction projects i.e., client, contract, designer, and consultant are extremely important for the BIM adoption. This was explained in more detail by the respondent as:

> This group (i.e., client, general contractor, DDC, and consultant) acts as the driving wheel to the project. Any conflict between these groups leads to delay in and in any extreme case leads to the project being shelved. Hence, for any new technology to be incorporated it becomes extremely essential for their mutual consent. A lot of time, the client (who is accustomed to the benefits of BIM) approves for its use but then the contractor opposes it on the part of huge investments that need to be done to accommodate its usage in the project.

This issue can be resolved by better educating the associated parties on the long-term benefits of using BIM in the project so that they are more than ready to use it and harness its complete benefits. Also, in such cases, the role of government comes out to be extremely important. The government should mandate the BIM usage in certain projects so that it becomes mandatory on the part of stakeholders involved to follow the set directions. Accordingly, the contract should be designed to take into account these mandatory mandates and standards so that there is no scope of conflicts at a later stage.

4.3 Technological Limitations

The advancements in the technologies have resulted in automatic protection of privacy owing to the increase in technicalities and reduction in human intervention. However, these still sometimes threaten privacy infringement [11]. Collaboration between different groups and participants along with a huge flow of data serves as a backbone of BIM. However, in this process of collaboration and data sharing, it becomes extremely difficult to decide the ownership of data [12]. Therefore, it becomes extremely important to manage these two important legal risks i.e., Intellectual property and having a clear picture of data ownership [13] associated with BIM. Apart from the data protection problem, the investment in developing a huge technical infrastructure for BIM also poses a serious challenge to its adoption. The collaboration of different groups on a single platform results in the churning of huge data that needs to be stored efficiently and directly results in huge upfront costs. This can be dealt with by having clear coordination between the client and contractors where they can come in agreement and share the fees and expenditure. An excerpt from the interview with the expert revealed certain strategies they adopted towards data protection is:

> We had a strict policy about data transfer and data ownership. To further protect it, access to the particular group was only provided who were associated with that task. For example, the design team and the DDC were only provided access to the designs and once the designs were approved by the team, then the contractor was provided access to that particular drawing and it was marked as Good for Construction (GFC).

As for the increased cost in the establishment of infrastructure, the stakeholders need to be able to witness the long-term benefits that can be obtained by the BIM usage and hence willing to make the initial investment upfront.

5 Conclusion

This article through a case study on a metro rail project presents the challenges encountered in the implementation of BIM in the project. Through a generated conceptual model, it provides certain strategies to overcome those barriers and efficiently aid in its implementation. The findings of the interviews with the experts revealed results related to lack of standard manual and documents for BIM implementation lack of attitude of managers, managers insufficient knowledge of BIM, increased cost as a result of BIM adoption as the important challenges faced in its adoption. The major findings suggested overcoming those challenges are the training and education of the existing workforce and also the new professionals. Also, a stakeholder-based ecosystem presented the advantage BIM has to offer in bringing different teams and groups together for better coordination, and collaboration of the workforce. This advantage of BIM will certainly aid the stakeholder management process and will lead to better execution of the project.

References

1. Ozorhon B, Karahan U (2017) Critical success factors of building information modeling implementation. J Manag Eng 33(3):04016054
2. Tomek A, Matějka P (2014) The impact of BIM on risk management as an argument for its implementation in a construction company. Procedia Eng 85:501–509
3. Okakpu A et al (2020) Risk factors that influence adoption of building information modelling (BIM) for refurbishment of complex building projects: stakeholders perceptions. Int J Constr Manag 1–13
4. Pakhale PD, Pal A (2020) Digital project management in infrastructure project: a case study of Nagpur Metro Rail Project. Asian J Civil Eng 21(4):639–647
5. Vidalakis C, Abanda FH, Oti AH (2020) BIM adoption and implementation: focusing on SMEs. Constr Innov 20(1):128–147
6. Papadonikolaki E, Vrijhoef R, Wamelink H (2016) The interdependences of BIM and supply chain partnering: empirical explorations. Architectural Eng Design Manag 12(6):476–494
7. Alreshidi E, Mourshed M, Rezgui Y (2018) Requirements for cloud-based BIM governance solutions to facilitate team collaboration in construction projects. Requirements Eng 23(1): 1–31

8. Ozorhon B, Cinar E (2015) Critical success factors of enterprise resource planning implementation in construction: case of Turkey. J Manag Eng 31(6):04015014
9. Sinoh SS, Othman F, Ibrahim Z (2020) Critical success factors for BIM implementation: a Malaysian case study. Eng Constr Archit Manag 27(9):2737–2765
10. Ahuja R et al (2016) Adoption of BIM by architectural firms in India: technology-organization-environment perspective. Architectural Eng Design Manag 12(4):311–330
11. Chatterjee S (2019) Is data privacy a fundamental right in India? Int J Law Manag
12. Alwash A, Love PED, Olatunji O (2017) Impact and remedy of legal uncertainties in building information modeling. J Legal Aff Dispute Resolut Eng Constr
13. Arshad MF et al (2019) Contractual risks of building information modeling: toward a standardized legal framework for design-bid-build projects. J Constr Eng Manag 145(4): 04019010

Thermal Insulation by Fiber Added High Volume Flyash Aerated Concrete Wall Cladding

Amritha Raj, Maheshwar Kotwal, Dhanya Sathyan, and K. M. Mini

Abstract The goal of this research is to develop a sustainable environment-friendly thermal insulation aerated concrete cladding using rice straw, silica fume, fly ash, and cement. Silica fume and rice straw were added to improve the strength and functional properties of aerated concrete. The strength and thermal properties of aerated concrete with various foam volumes are investigated in this study. A numerical analysis of thermal insulation capacity is also carried out using the COMSOL software. For all specimens, the w/s ratio is fixed to 0.55. The result shows that the mix with 20% foam volume is best in terms of desirable compressive strength and thermal conductivity.

Keywords COMSOL multiphysics · Building simulation · Rice straw · Aerated concrete · Thermal insulation

1 Introduction

Before 1950, most of the houses and buildings were made of mud and stones and coating of walls was also done with mud and straw. But these construction materials take more area (walls width) and lack strength. With the advancement in construction technology, concrete is mainly utilised as a construction material because of its good construction performance. Aerated concrete is an innovative, economical solution for thermal insulation in building [1]. The main source of energy in modern times is from thermal power plants. When the energy (electricity) is generated from coal combustion, some waste products are formed, known as,

A. Raj · M. Kotwal · D. Sathyan · K. M. Mini (✉)
Department of Civil Engineering, Amrita School of Engineering, Amrita Vishwa Vidyapeetham, Coimbatore 641112, India
e-mail: k_mini@cb.amrita.edu

G. C. Marano et al. (eds.), *Proceedings of SECON'21*, Lecture Notes in Civil Engineering 171, https://doi.org/10.1007/978-3-030-80312-4_57

Coal combustion products (CCP). Coal combustion products include bottom-ash, gypsum, fly-ash, etc. The disposal of coal combustion products (CCP) has become a very big threat to the environment these days [2]. Researchers and Scientists are trying enormously to find different ways towards the utilisation of fly ash. One of the ways is utilizing this waste product in the construction industry effectively. The advantages of using fly ash in construction purpose include energy-saving, cost-saving, low CO_2 content, etc. If some proportion of cement is replaced with fly-ash, it will help to reduce the greenhouse gas emission. Also, the concentration of poisonous components like lead and cadmium is lower in fly ash [3]. The replacement of sand with fly ash in cement outcomes an increase in strength for a given density [4]. According to the research by Jones and McCarthy [5], the utilization of fine fly ash as a partial substitution to the cement and use of coarse fly ash as a substitution to fine aggregate reduces the drying shrinkage and heat of hydration to a great extent. Majorty of the structures in India are not using thermal insulation materials in buildings results in higher energy consumption to achieve the desired temperature inside a room.[6]. Materials having low thermal conductivity have a high resistance to heat stream and thus their use in construction can decrease the amount of transfer of heat into buildings. Aerated concrete is known for its better properties of thermal insulation. Aerated concrete is light in weight and its thermal conductivity lies in the reach between 0.10 to 0.70 W/m K [7, 8]. The use of aerated concrete on the global scale was about 5.6% in Australia, North America, and Africa and in Europe it is about 33.3%. In Asia, it is because of the lack of awareness and confidence of people in this material [9]. Also the proper availability of required technology to manufacture is not available. [10]. But now in recent times, builders, civil engineers, contractors, all over the world are giving more importance to the use of foam-concrete because now people are more concerned about the environment and energy conservation. A well-insulated building reduces energy consumption by keeping rooms warm in the winter and cool during the summer. So, materials having lower thermal conductivities are preferred to reduce heat loss. Since the aerated concrete is made by blending foam into the cement-based slurry, the density and the porosity are the variables that affect the thermal conductivity of the aerated concrete [11]. Despite having many benefits of foam-concrete, the main drawbacks of aerated concrete are shrinkage and lower strength. To overcome this drawback of aerated concrete, rice straw can be effectively used for improving the thermal insulation and shrinkage resistance properties of foam-concrete [12]. Studies were done on the thermal conductivities of the wall the panels produced using the combination of rice husk and gypsum revealed that the conductivity of the panels decreases and almost diminishes with the increase of the amount of rice husk [13, 14]. So our research work aims to develop a sustainable environmental-friendly thermal insulated aerated concrete wall cladding using cement, silica fume, rice straw and fly ash.

2 Experimental Details

2.1 Materials and Mix Proportions

Fresh clean potable water, Class F fly ash of specific gravity 2.1, Silica fume, protein-based foaming agent (Ethoxylate of vegetable protein), and Ordinary Portland cement (OPC) 53 grade satisfied the codal provision of IS 12269:2013 [15] are used to make aerated concrete cladding. Compositions of mixes used are shown in Table 1.

2.2 Details of Study

Aerated concrete specimens are produced by mixing the mortar (mix of fly ash, cement, silica fume, and water) with the measured volume of foam. A foam generator is used to produce foam. To make a stable foam, foaming agent and water in a weight-to-weight proportion of 1: 30 is combined and 450 kPa pressure is applied. The 28-day compressive strength of aerated concrete cubes of size 50 × 50 × 50 mm is determined using a universal compression test machine according to Appendix A of IS 2250:1981 [16]. Thermal conductivity tests are carried out using hot guarded equipment in accordance with ASTM C 177–97 [17].

3 Results and Discussion

3.1 Compressive Strength

Figure 1 shows the compressive strength of aerated concrete cubes with and without fibre after 28 days of curing. It's been observed that as the percentage of

Table 1 Mixture compositions used

Mixes	20FR0	30FR0	40FR0	50FR0	20FR1	30FR1	40FR1	50FR1
Density	925	800	675	425	925	800	675	425
Water (kg/m^3)	636.9	558	478.5	398.2	636.9	558	478.5	398.2
Cement (kg/m^3)	492	431	370	308	492	431	370	308
Rice straw(kg/m^3)	0	0	0	0	5.79	5.79	5.79	5.79
Silica fume (kg/m^3)	87	76	65	54	87	76	65	54
Fly ash (kg/m^3)	579	507	435	362	579	507	435	362
Foam (% of total volume)	20	30	40	50	20	30	40	50

*20FR1—20%foam, F—fly ash, R1—With rice straw

foam volume increases, strength decreases. There is 55 to 88.1% reduction in strength is observed when the percentage of foam content increased from 20 to 50. At the point when the foam volume is 20%, 30%, 40%, and 50% corresponding 28 days strength acquired are 5.44, 4.48, 3.472, and 2.4 MPa, respectively (with-out fiber). Compressive strength is discovered to be reduced with the fiber addition. The percentage reduction in compressive strength due to straw addition found to be increased with % foam volume. The percentage reduction in strength because of the addition of straw is, 18.7%, 58.5% and 72.5% respectively for 30%,40%, and 50% foam volume. The strength reduction is due to the results of increased porosity. Mix with 20% foam volume got desirable compressive strength.

3.2 Thermal Conductivity

Thermal conductivity is observed to be diminished with foam volume (Fig. 2). Higher-density aerated concrete shows higher conductivity. Also observed that the addition of rice straw reduced the conductivity further. The cellulose content, and hollow structure of rice straw results reduction in conductivity. Thermal conductivity is found to be reduced from 0.099 to 0.073 W/mK when foam volume increased from 20 to 50%. For fiber added mix the corresponding reduction is from 0.09 to 0.069 W/mK.

3.3 Rate of Water Absorption

Figures 3 and 4 shows the variation of water absorption with time. It is observed that aerated concrete with lower density absorbs more water than higher density. Therefore, it can be concluded that aerated concrete with lower density is less durable than aerated concrete with higher density. Also, it is observed that addition fiber increases the porosity there by the water absorption. The mixes without fiber

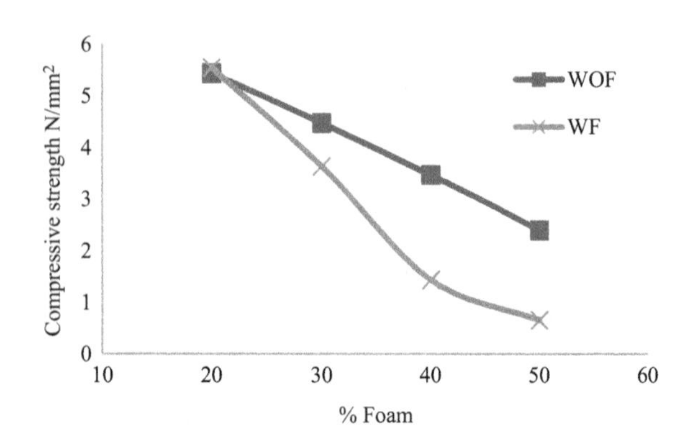

Fig. 1 Variation of compressive strength with percentage foam volume

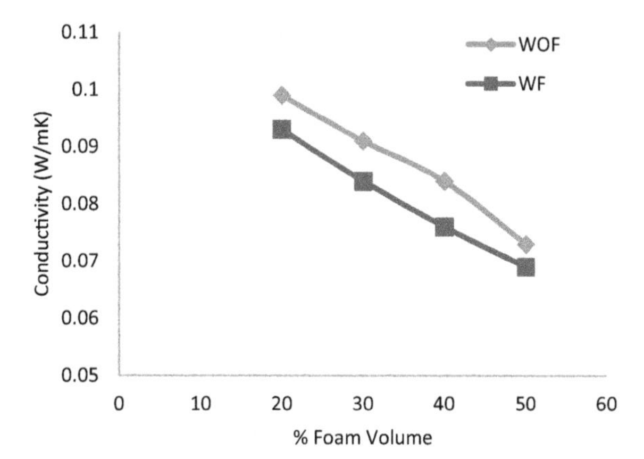

Fig. 2 Variation of thermal conductivity with percentage foam volume

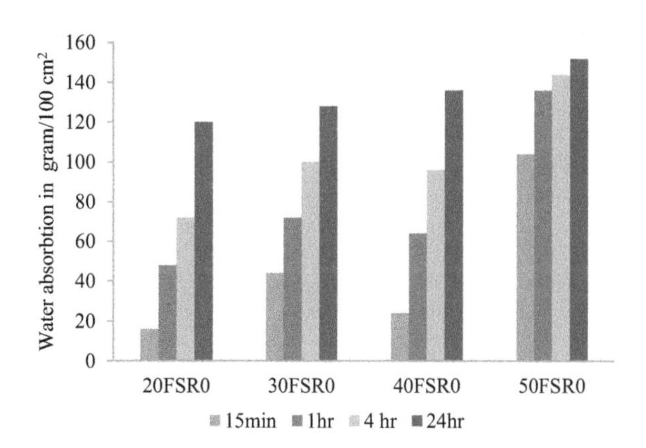

Fig. 3 Variation of water absorption with time

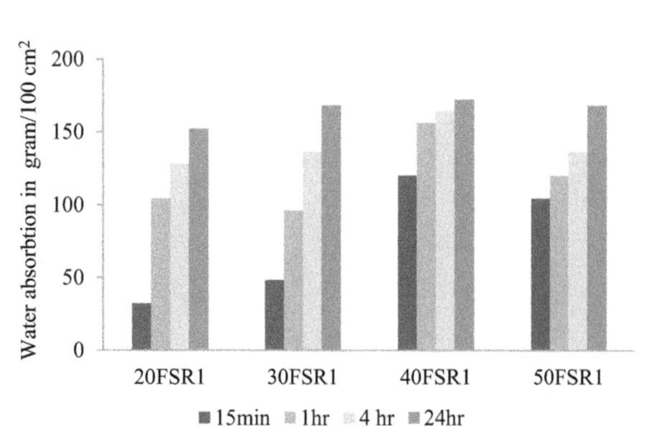

Fig. 4 Variation of water absorption with time

have a 24 h water absorption value in the range of 120–152 gm/100 cm^2 and with fiber the range increased to 152–168 gm/100 cm^2. In Fig. 4 it is observed that when foam volume increases, there is a slight increase in water absorption only. This increase in water absorption regardless of foam volume reveals that only some of the air voids are fully or partially filled.

3.4 Numerical Analysis

To find the effectiveness of the insulation panel, a study was done using COMSOL Multiphysics software. A 3D single-room building is modelled. Figure 5 shows the dimensions of the model developed. A 20 mm insulation panel, 230 mm thick brick masonry, Concrete roofing, wooden windows and doors are also provided. Parameters used to model building are the convective heat transfer coefficient inside the room is 8.29 W/(m^{2}K) and for Outside 34 W/(m^{2}K). The outcome shows that the mix 20FR1 is optimum in terms of thermal conductivity and desirable compressive strength. Thermal conductivity of 0.30 W/(mK), 0.806 W/(mK), 2.10 W/(mK) 0.093 W/

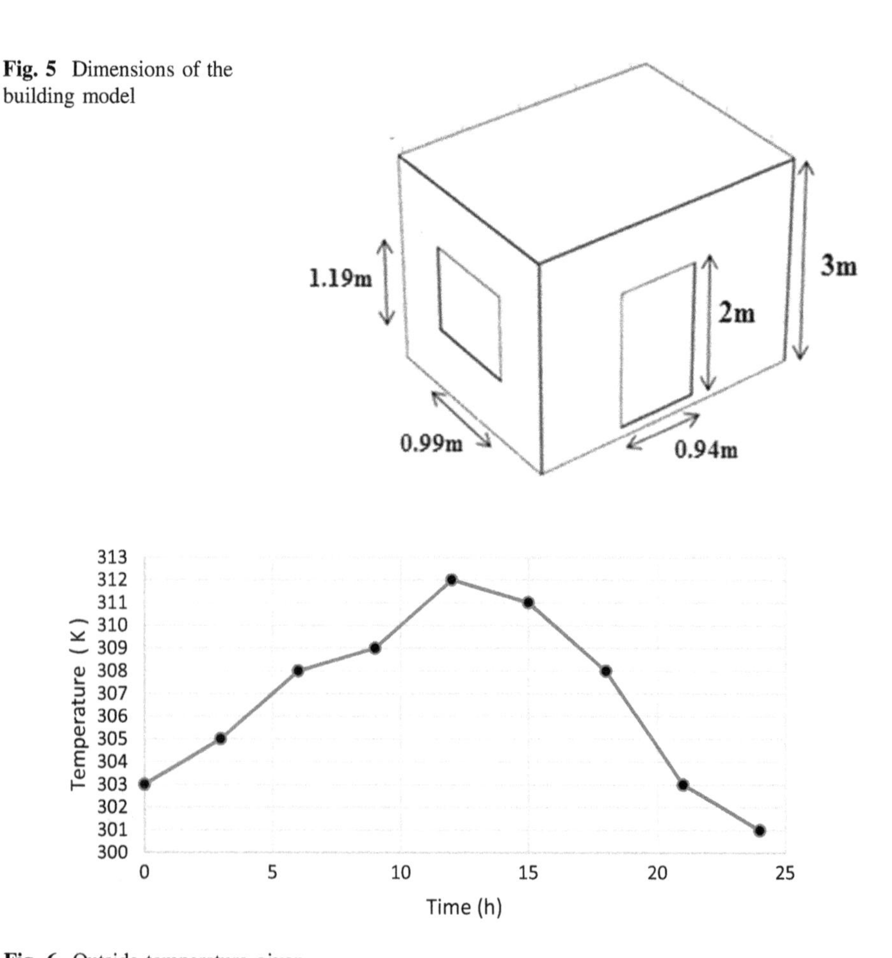

Fig. 5 Dimensions of the building model

Fig. 6 Outside temperature given

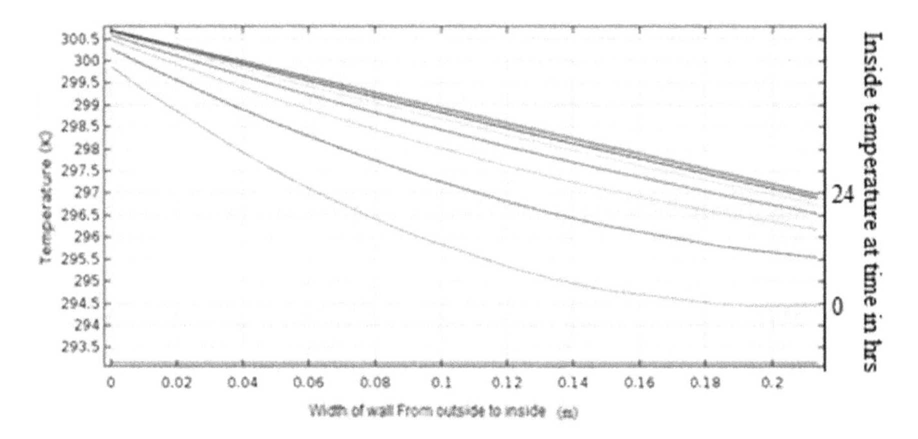

Fig. 7 Temperature variation across the wall

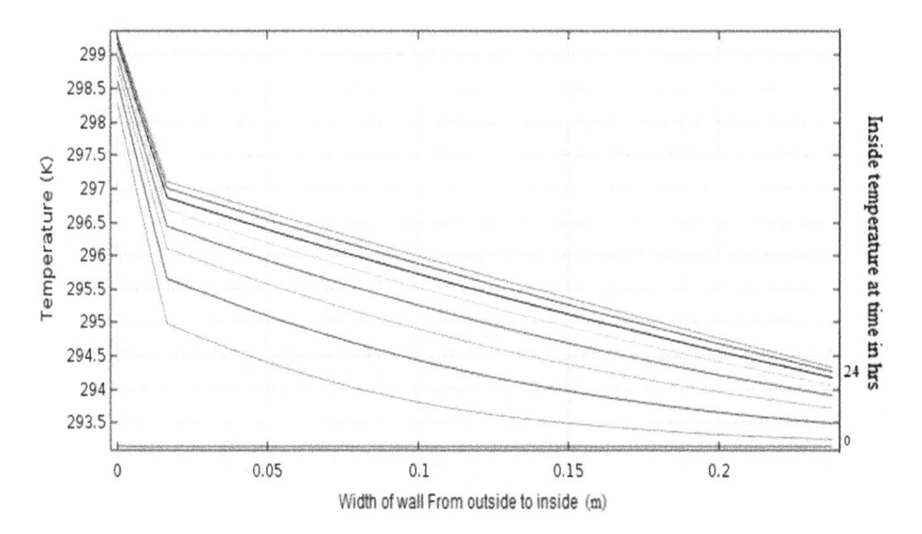

Fig. 8 Temperature variation across the wall with insulation

(mK), and Thermal capacity of 2390 J/(kgK), 800 J/(kgK), 1050 J/(kgK),757 J/(kgK) are given for Wood, Brick wall, concrete and Insulation panel respectively [18]. Heat transfer by conduction convection and radiation are considered.

Figure 6 shows the outside climatic temperature with time. (randomly taken daytime temperature in Chennai India). Figures 7 and 8 shows the corresponding temperature variation across the wall with and without insulation panel. In Fig. 8 sudden linear drop of temperature up to 0.02 m due to the presence of insulation. In Figs. 7 and 8 yellow colour line shows the temperature across the building at the highest outside temperature of 312 K. there was a 3 K difference in temperature was observed for a building with and without an insulation panel. For a building without an insulation panel, temperature variation across the wall (brick wall) corresponding to the input temperature profile varies from 294.5 to 297 K. And for the insulated wall, it will range from 293.5 to 294.5 K.

4 Conclusions

According to the research, the panel developed can be beneficially used as external wall cladding in the building by providing proper water resisting coat.

- Compressive strength was found to be reduced with foam volume. A mix with 20% foam volume was found to give better heat resistance and desirable compressive strength.
- The rate of water absorption and thermal resistance was found to increase with foam volume and fibre content. Aerated concrete with lower density was found to be less durable than higher density.
- It was discovered that the addition of rice straw to foam concrete enhanced its thermal properties.
- In terms of thermal and strength properties, the 20FR1 mix can be effectively used for wall cladding.
- From the numerical simulation, it was discovered that by using aerated concrete wall cladding, we can maintain a comfortable temperature inside the structure.

Acknowledgements The authors are thankful for the monetary help given by the Department of Science and Technology (TMD / CERI / BEE/2016/032).

References

1. Weigler H, Karl S (1980) Structural lightweight aggregate concrete with reduced density-lightweight aggregate foamed concrete. Int J Cem Compos Lightweight Concr 2 (2):101–104
2. Abdullah MH, Rashid AS, Anuar UH, Marto A, Abuelgasim R (2019) Bottom ash utilization: a review on engineering applications and environmental aspects. IOP Conf Ser Mater Sci Eng 527(1):012006
3. Kim HK, Lee HK (2015) Coal bottom ash in the field of civil engineering: a review of advanced applications and environmental considerations. KSCE J Civ Eng 19(6):1802–1818
4. Nambiar EK, Ramamurthy K (2006) Influence of filler type on the properties of foam concrete. Cem Concr Compos 28:475–480
5. Jones MR, McCarthy A (2005) Behaviour and assessment of foamed concrete for construction applications. Thomas Telford Publishing, Dundee, Scotland, UK, pp 61–88
6. Chunekar A, Varshney S, Dixit S (2016) Residential electricity consumption in India: What do we know? Prayas (Energy Group), Kothrud, Pune
7. BCA (1994) Foamed concrete: composition and properties. Report Ref. 46.042, Slough, BCA
8. Jones MR, McCarthy A (2005) Preliminary views on the potential of foamed concrete as a structural material. Mag Concr Res 57(1):21–31

9. Raj A, Sathyan D, Mini KM (2019) Physical and functional characteristics of foam concrete: a review. Constr Building Mater 221:787–799
10. Zhao X, Lim SK, Tan CS, Li B, Ling TC, Huang R, Wang Q (2015) Properties of foamed mortar prepared with granulated blast-furnace slag. Materials (Basel) 8:462–473. https://doi.org/10.3390/ma8020462
11. MdAzree OM (2011) Effective thermal conductivity of foamcrete of different densities. Concr Res Lett 2(1):181–219
12. Bhagyasree R, Raj A, Sathyan D, Mini KM (2020) Mechanical and durability properties of hybrid fiber reinforced foam concrete. Constr Building Mater 245:118373
13. Musa MN, Abdul Aziz MF (2016) Thermal conductivity for mixture of rice husk fiber and gypsum. Appl Mech Mater 819:69–73
14. Raj A, Sathyan D, Mini KM (2021) Performance evaluation of natural fiber reinforced high volume fly ash foam concrete cladding. Adv Concrete Constr 11(2):151
15. IS 12269 (2013) Ordinary Portland cement 53 grade—specification. Bureau of Indian Standards, New Delhi, India
16. I.S:2250–1981 (n.d.) Code of practice for preparation and use of masonry mortars. Bureau of Indian Standards, New Delhi, India
17. Test method for steady-state heat flux measurements and thermal transmission properties by means of the guarded-hot-plate apparatus .Title Am Soc Test Mater ASTM C 177 (1997)
18. Raj A, Sathyan D, Balaji K, Mini KM. Heat transfer simulation across a building insulated with foam concrete wall cladding. Materials Today: Proceedings, S2214-7853(21)00331-X

Inhibitors to Interface Management Implementation: Perceptions of Construction Stakeholders

Vijayeta Malla and Venkata Santosh Kumar Delhi

Abstract Interfaces are omnipresent facets of a project management and their presence in construction domain is copious. Owing to inter-disciplinary interactions at the cross points, it is of paramount importance to apprehend complications which emerge at those interfaces. However, at interface point, possibility of conflicts are obvious owing to interactions, transfer of responsibilities, and exchange of information or resources. At this cross over, in case of misalignment of requirements from either sides of interfaces, then there is a scope for yielding of interface problems which ultimately result in disputes. Nevertheless, optimum stakeholder understanding through unified process such as Interface Management possibly eradicates such interface issues. The current study explores inhibitors to Interface Management adoption and through analysis of variance (ANOVA), exploration of viewpoints of contractors, owners, consultants and academicians through survey of 99 respondents was performed. Additionally, the study identifies top 10 critical inhibitors from initial thirty inhibitors identified from content analysis of the literature. Subsequently, the mitigation measures have been mapped from 4 experts using Nvivo. Also, the responses gathered instituted impact and frequency measures with regards to capturing of inhibitors towards IM adoption. The findings indicate that out of 60 possible cases (both frequency and impact), 53 cases show agreement over all inhibitors to IM adoption from all four categories of stakeholders. One key differing opinion among stakeholders is "Novice at dealing with regional and government specific legal aspects". Thus, this study provides a roadmap for strategic orientation of organization which intends to institutionalize Interface Management.

V. Malla (✉) · V. S. K. Delhi
IIT Bombay, Mumbai, India
e-mail: vmalla@nicmar.ac.in

V. S. K. Delhi
e-mail: venkatad@iitb.ac.in

V. Malla
National Institute of Construction Management and Research (NICMAR), Hyderabad, India

G. C. Marano et al. (eds.), *Proceedings of SECON'21*, Lecture Notes in Civil Engineering 171, https://doi.org/10.1007/978-3-030-80312-4_58

Keywords Interface management · Barriers · Interfaces · Construction · Project management · Stakeholders · Built environment · ANOVA · Complexity · Systems engineering · Nvivo · Mitigation measures · Coordination · Collaboration · BIM

1 Introduction

Globally construction sector's growth is poised to reach $8 trillion USD by 2030 where the prime drivers of such growth is significantly contributed by nations such as United States of America, China and India [1]. Along with envisaging of such copious construction activity, there is a tremendous emphasize on digital transformation of construction landscape with inclusion of digital infusion initiatives (DII) for gaining competitive market advantage. Apart from this, current uncertain milieu of austere adherence to schedule, cost and scope of managing projects becomes even more complex. A typical infrastructure project outsources partial scope to multiple contracting players through multiple contract agreements to minimize the size and complexities. In such scenarios, Systems engineering (SE) assists as a promised avenue in managing complex projects of trans-disciplinary and colossal scope through systematized process. SE inherently decomposes complex systems into sub-components/systems for enhanced management of work through delegating to individual discipline engineers [2].

Despite, intact schedules and continued efforts, unforeseen conflicts arise on the confluence on components of systems integration [3]. Thus, early identification of interfaces between components of the systems is essential for prevention of integration issues. Interface Management is defined as envelopment and managing the confluence of entities at battery limits [4]. The entities might comprise of physical, phases, systems, organizations, individual, machinery and so on. As a result, practitioners of complex construction projects advocate the institution of Interface Management to ameliorate coordination and integration among the stakeholders with minimal clashes [5]. In absence of such systematized process such as Interface Management, it ultimately yields skewed triple constraints or defects prone of facility at post project close stage. Hence, Interface Management steers projects efficiently over the project life cycle through managing interfaces resulting in enhanced project performance attributes.

This study investigates the perceptions of four chief stakeholders in adoption of Interface Management (IM) from the construction industry purview. Extant literature captured minimally on holistic perspectives of interface issues from varied construction stakeholders. In order to address this gap, the study captures perceptions of inhibiting factors to IM on utilizing severity and occurrence measures. Therefore, this study attempts to answer two research questions as follows:

1. What are the perceptions of construction players on inhibiting factors to IM implementation?

2. What are the critical inhibiting factors to IM implementation in Built Environment sector?

2 Research Methodology

The research design administered comprises of three phases. In the first phase, a thorough content analysis of extant scholarly articles from impactful journals and reputed conferences was referred and resulted in preliminarily selection of thirty inhibiting factors. The search strategy deployed keywords such as 'interfaces', 'interface issues', interface problems', 'interface management', 'interface points', 'interface agreement' and 'interface control documents' in construction industry while sourcing research articles from varied databases. In the second phase, with aid of a structured questionnaire survey the responses were solicited from contractors, owners, consultants and academicians on the measures like impact and occurrence/frequency of inhibiting factors on a Likert scale of 1–5, where 1 indicates least impact or occurrence and 5 value indicates high impact or occurrence. The survey was circulated to 325 professionals enveloping diverse construction stakeholders through online mode using google forms. In the third phase, analysis of responses was performed to find out insightful findings on the variance of identified factors. The statistical based analysis was performed using SPSS software version 26.0. Further, suggestive actions for identified inhibitors were derived from qualitative study.

3 Results and Discussions

Overall 99 respondents completed the survey (the rate of response is 30.4%) covering 25 projects.

The salient characteristics of the respondents were depicted Table 1 and the structured survey responses were analyzed using SPSS version 26. Cronbach's alpha is a metric which ascertains the internal consistency and reliability of deployed scale [15]. Current study returned values of 0.964 and 0.965 for impact and frequency measures respectively. Moreover, alpha value above 0.7 is indicative of acceptability of the instrument and respective outcomes approve that the structured survey is reliable. In order to analyze and observe any substantial differing opinions of the four varied groups-Contractors, Owners, Consultants and Academicians, a one way ANOVA test was conducted. The outcomes of the test are outlined in Table 2, and out of an aggregate 60 likely cases for impact and frequency measures, 53 cases (88%) were observed to possess statistically no significant variation where the $p > 0.05$. This can be inferred that the four groups of stakeholders substantially agree on the list of inhibitors to Interface Management (IM).

Table 1 Characteristics of respondents

Type of construction stakeholder	Frequencies	% of aggregate
Academicians/Researchers	13	13%
General contractors	53	54%
Project management/Design consultants	20	20%
Owner/Client	13	13%
	99	
Experience		
0–5 years	11	11%
6–10 years	24	24%
11–15 years	29	29%
16–20 years	11	11%
>20 years	24	24%
	99	
Educational credentials		
Diploma	14	14%
Graduate	41	41%
Post graduate	38	38%
Doctorate	6	6%
	99	
Occupational level		
Top management	18	18%
Senior manager	11	11%
Mid-level manager	48	48%
Academicians	13	13%
Engineers	9	9%
	99	

The top critical inhibitors (w.r.t Impact measure) to interface management implementation in construction projects are depicted in Table 2. Those ten critical inhibitors are Interface issues at engineering and construction process phases, Poor soft skills among the project stakeholders, Unaware of project site conditions by the project team, Wrong selection of construction materials by designer, Change orders, Improper work packaging, Change resistant towards embracing of technological processes by veteran construction professionals, Novice at dealing with region and government specific legal aspects, Novice at utilizing modern methods of construction and Overloading of extra cost on contractors. The outcomes indicate a combination of planning phases and Use of Technology related inhibitors were prominent.

Table 2 ANOVA results of preliminary inhibitors to Interface Management and Ranking of inhibitors

	Inhibiting factors to interface management implementation	ANOVA(Sigma)		(RII)-(Impact)	Rank
		Impact	Frequency		
1	Improper drawings, specifications and guidelines [6, 7]	0.602	0.354	0.527	29
2	Delays attributable to client owing to non-approval drawings and non-assessment of performance [6]	0.661	0.698	0.622	27
3	Delays in finalizing construction schedule and engineering works [6]	0.215	0.290	0.633	25
4	Ambiguities in contract document such as unclear details and specific skills for labour deployment [7, 8]	0.598	0.792	0.524	30
5	Unspecified Liquidated Clauses in contract [6]	0.344	0.434	0.637	24
6	Change orders [7]	0.319	0.587	0.722	4
7	Delays in design services completion [7]	0.141	0.869	0.686	14
8	Improper schedule of payments [6]	0.202	0.900	0.665	20
9	Absence of cost indices for material, workforce and machinery to be used by owners and designers for cost estimation [6]	0.120	0.795	0.684	16
10	Inaccurate project costs and maintenance costs by designers and main contractors [6]	0.786	0.153		28
11	Overloading of extra cost on contractors [6]	0.030*	0.718	0.698	10
12	Client's low allocation of budget for design and execution services [7]	0.753	0.226	0.688	14
13	Exorbitant design costs [7]	0.424	0.756	0.684	17
14	Novice at utilizing modern methods of construction [6]	0.294	0.816	0.710	7
15	Change resistant towards embracing of technological processes by elder construction professionals [6]	0.443	0.375	0.716	6
16	Novice at dealing with region and government specific legal aspects [8–10]	0.006*	0.016*	0.706	8
17	Novice at encountering audits, compliance to building codes and trade union protocols [8, 10]	0.087	0.069	0.694	12
18	Improper work packaging [11]	0.912	0.158	0.724	3
19	Wrong selection of construction materials by designer [7]	0.530	0.328	0.700	9
20	Ignorance of designer of non-availability of construction materials [7]	0.036*	0.061	0.676	19
21	Institution of fast track delivery project method [12]	0.867	0.285	0.665	21
22	Augmented conflicts of multiple contractors on thrusting their views [10]	0.524	0.659	0.657	22
23	Interface issues at engineering and construction process [9, 11–14]	0.445	0.513	0.755	1

(continued)

Table 2 (continued)

	Inhibiting factors to interface management implementation	ANOVA(Sigma)		(RII)-(Impact)	Rank
		Impact	Frequency		
24	Deficient information flow between client's facility management department and finance department [7]	0.113	0.023*	0.696	11
25	Poor soft skills among the project stakeholders [7–11, 14]	0.824	0.620	0.751	2
26	Lack of knowledge of environmental conditions by owners for communicating to designers while designing phase [7]	0.680	0.186	0.678	18
27	Unaware of project site conditions by the project team [12]	0.634	0.029*	0.720	5
28	Poor writing communication skills between main contractor and supervisors [6]	0.671	0.226	0.624	26
29	Poor design communication tools between designers and clients [7]	0.489	0.010*	0.690	13
30	Unawareness in envisaging regional weather and environmental calamities by contractor [8, 9, 11]	0.584	0.092	0.655	23

* indicates significance value $p < 0.05$

3.1 Mitigation Measures for Critical Inhibitors to Interface Management (IM)

Mitigation measures for the top ten critical inhibitors were investigated from a qualitative study through expert interview consisting of four members whose average experience is 23 years for an average duration of approximately 35 minutes. The rationale for choosing experts were those construction professionals whose minimum work experience is ten years and substantially worked for interface oriented projects. On the basis of set criteria, four experts were chosen for interviewing.

The coding of the interviews was performed using Nvivo version 1.3 by adapting coding protocol [16] and a priori codes were grounded from PESTEL (Political, Economic, Social, Technological and Legal) framework.

The authors used a-priori coding (PESTEL framework) which is a deductive coding. This PESTEL framework was then validated with three subject matter experts (who possessed average work experience of 12 years) on its usage in our existing study. Then the coding strategy was performed using individual-based sorting strategy which involved fives steps. Based on the anchor codes of the research question the five step coding protocol used were (a) compilation of the codes (open-coding), (b) arranging them alphabetically, (c) consolidating the codes, (d) sorting the codes and (e) labeling the cluster codes. For the first cycle of coding, 'attribute', 'in-vivo' and 'descriptive' coding were used. In the second cycle, 'focussed' and 'axial' coding was used [17–19]. The Figs. 1, 2, 3 and 4 depicts visual

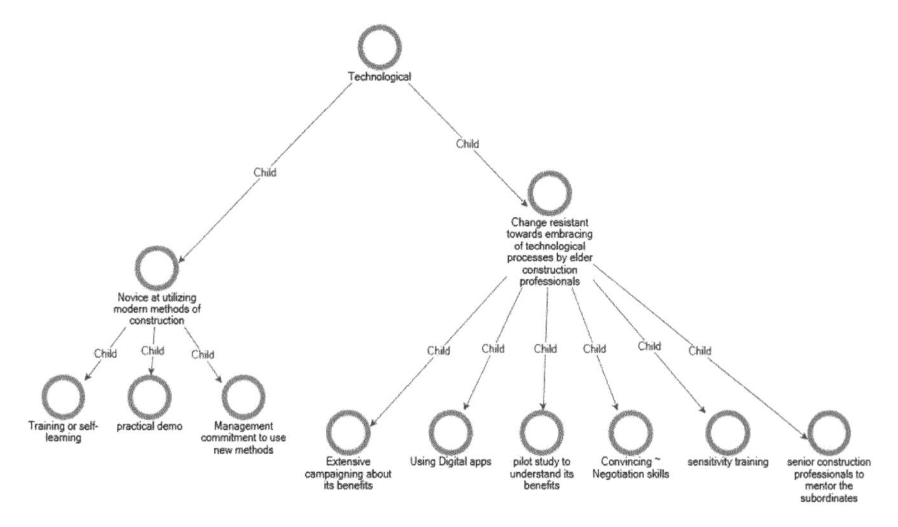

Fig. 1 Mitigation measures for 'Technological' dimension of PESTEL framework using Nvivo

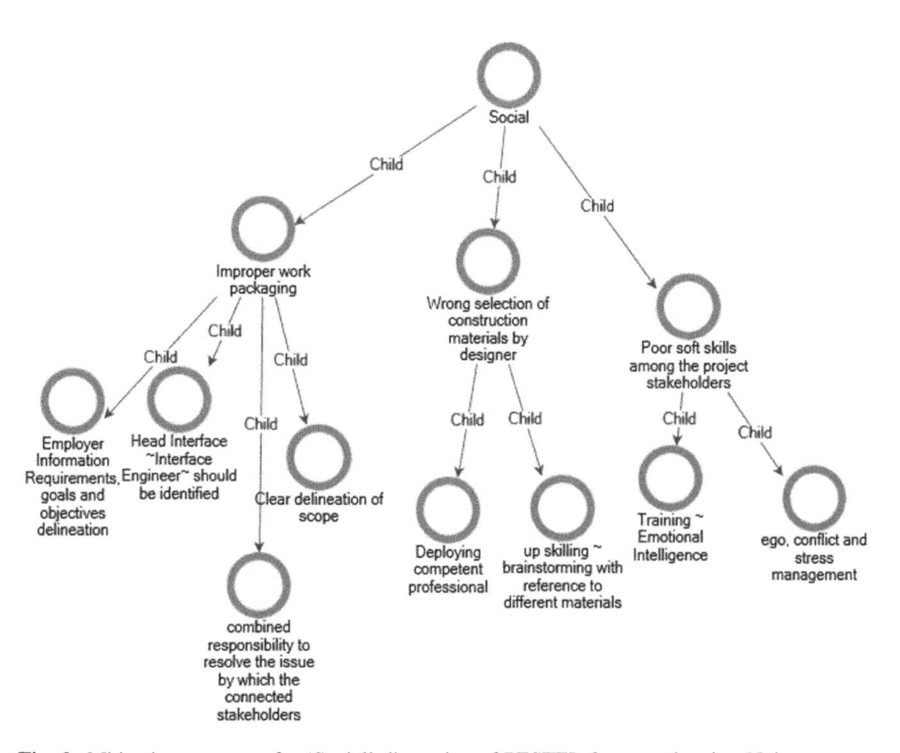

Fig. 2 Mitigation measures for 'Social' dimension of PESTEL framework using Nvivo

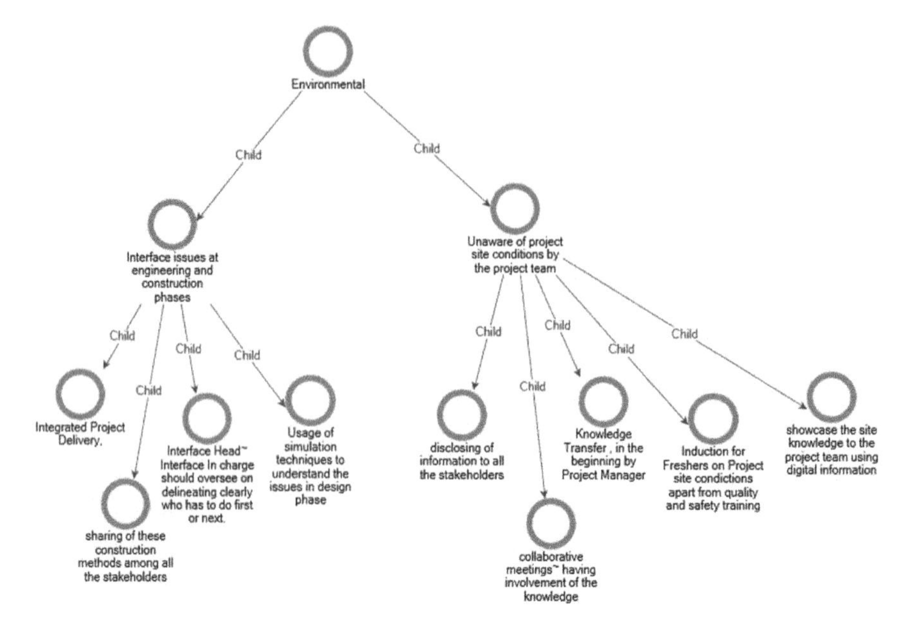

Fig. 3 Mitigation measures for 'Environmental' dimension of PESTEL framework using Nvivo

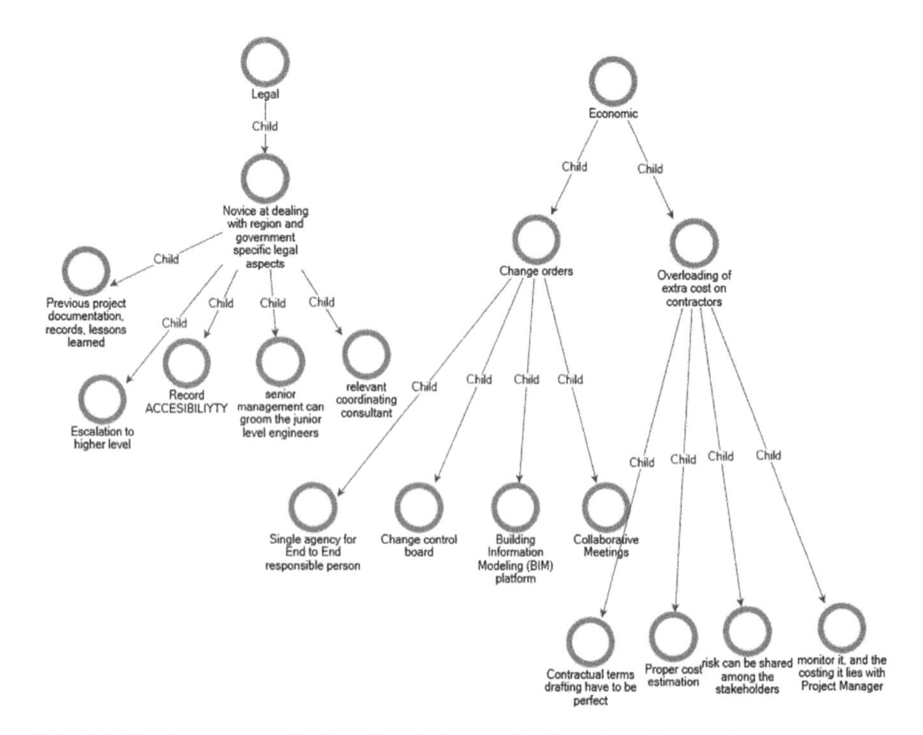

Fig. 4 Mitigation measures for 'Legal' and 'Economic' dimension of PESTEL framework using Nvivo

representation of coding using project maps where mitigation measures suggested by experts for each inhibitor were mapped under each dimension of PESTEL framework. Few key recommended measures highlighted by experts are institution of information management tool such as Building Information Modeling (BIM), Lean tenets, Digitalized Interface Management System, simulation tools and techniques, dedicated Interface Management roles, central repository of lessons learned archives and Knowledge transfer to all stakeholders.

4 Conclusions

Understanding critical interface prone problems and their perceptions from stakeholders is essential for prioritization. This study analyzed varied perceptions of clients, consultants, contractors and academicians through ANOVA and explored suggestive measures from subject matter experts using Nvivo. The outcomes especially recommended measures from subject matter experts provides a roadmap to Interface Managers/Heads and planning managers as a strategic orientation to IM implementation.

References

1. Economics O (2015) A global forecast for the construction industry to 2030
2. Zummo KJ (2010) A methodology for the integration of design teams for the development of complex Systems. PhD Thesis, Southern Methodist University, Dallas, USA
3. Kossiakoff A, Sweet WN, Seymour SJ, Biemer SM (2011) Systems engineering principles and practice, 2nd edn. John Wiley & Sons, New Jersey, S.M
4. Godinot M (2003) The work breakdown structure matrix: a tool to improve interface management. Master of Engineering Thesis, National University of Singapore
5. Shokri S, Safa M, Haas CT, Haas RC, Maloney K, MacGillivray S (2012) Interface management model for mega capital projects. In: Construction research congress, construction challenges in a flat world, pp 447–456
6. Al-Hammad A (1995) Interface problems between owners and maintenance contractors in Saudi Arabia. J Perform Constr Facil 9(3):194–205
7. Al-Hammad A, Al-Hammad I (1996) Interface problems between building owners and designers. J Perform Constructed Facil 10(3):123–126
8. Huang R, Huang C, Lin H, Ku W (2008) Factor analysis of interface problems-a case study of MRT. J Mar Sci Technol 16(1):52–63
9. Al-Hammad A (2000) Common interface problems among various construction parties. J Perform Constr Facil 14(2):71–74
10. Ku H, Lin J, Huang C, Shiu R (2010) Development interface knowledge management system for the mass rapid transit system construction. In: Proceedings of the 8th international conference on supply chain management and information systems, pp 1–6
11. Chen Q, Reichard G, Beliveau Y (2007) Interface management-a facilitator of lean construction and agile project management. Int Group Lean Constr 1(1):57–66
12. Al-Hammad A, Assaf S (1992) Design-Construction interface problems in Saudi Arabia. Building Res Inf 20(1):61–63

13. Mortaheb MM, Rahimi M (2010) Interface management in mega oil refinery projects. In: 6th International project management conference, Tehran, pp 1–19
14. Al-Hammad A (1993) Factors affecting the relationship between contractors and their subcontractors in Saudi Arabia. Building Res Inf J 21(5):269–273
15. Santos JA (1999) Cronbach's alpha: a tool for assessing the reliability of scales. J Ext 37(2): 1–5
16. Braun V, Clarke V (2020) One size fits all? What counts as quality practice in (reflexive) thematic analysis? Qualitative research in psychology, pp 1–25
17. Saldaña J (2021) The coding manual for qualitative researchers, Sage
18. Creswell JW (2012) Qualitative inquiry and research design: choosing among five approaches, 3rd edn. Sage, Thousand Oaks, CA
19. LeCompte MD, Schensul JJ (1999) Analyzing and interpreting ethnographic data. AltaMira Press, Walnut Creek, CA

Effect of Alkaline Liquid and GGBS Replacement Level on Corrosion Behaviour of Reinforcing Steel in Chloride Contaminated Geopolymer Concrete Made with Fly Ash and GGBS

Suresh Chandra Sadangi and Bulu Pradhan

Abstract In this paper, the effect of molarity of sodium hydroxide (NaOH) solution on corrosion behaviour of reinforcing steel in chloride contaminated geopolymer concrete (GPC) prepared with fly ash and GGBS was studied for different GGBS replacement levels. In this work, GPC was prepared with various molarity of NaOH solution such as 8, and 14 M with GGBS replacement levels of 0, 15, and 45%. Sodium chloride concentrations of 0 and 3% of geopolymer solids (by mass) were added to the alkaline liquid during the preparation time of GPC mixtures. Prismatic reinforced GPC specimens were made for half-cell potential measurement and corrosion current density (I_{corr}) by linear polarization resistance (LPR) measurement. The obtained results showed that the 28 days compressive strength of GPC increased with raise in molarity (i.e. from 8 to 14 M) of NaOH solution. Further, higher compressive strength was achieved in all the GPC mixes made with 45% GGBS replacement level as compared to other replacement levels. In addition, decrease in compressive strength was observed in GPC mixes added with 3% sodium chloride (NaCl). The higher probability of occurrence of corrosion and higher I_{corr} of steel reinforcement were observed in 3% NaCl admixed geopolymer concrete. The increase in I_{corr} of steel reinforcement was observed with raise in molarity of NaOH solution. Further, the I_{corr} of steel reinforcement in GPC decreased with increase in GGBS replacement level i.e. from 15 to 45%.

Keywords Geopolymer concrete · Fly ash · GGBS (Ground granulated blast furnace slag) · Admixed chloride · Corrosion current density (I_{corr}) · Half-cell potential

S. C. Sadangi (✉) · B. Pradhan
Department of Civil Engineering, Indian Institute of Technology Guwahati, Guwahati 781039, India
e-mail: sures176104023@iitg.ac.in

G. C. Marano et al. (eds.), *Proceedings of SECON'21*, Lecture Notes in Civil Engineering 171, https://doi.org/10.1007/978-3-030-80312-4_59

1 Introduction

Concrete is most widely utilized in the construction of infrastructure projects. The primary binder utilized in the manufacture of concrete is the Portland cement. However, the manufacture of cement has led to environmental challenges due to the emissions of CO_2. Cement manufacture contributes about 5 and 7% of CO_2 emissions worldwide [1]. In addition, the total industrial energy of about 10–15% are consumed worldwide by the cement industry [2]. Thus, the manufacture of Portland cement has been considered as one of the greatest energy intensive processes as well as it results in significant environmental pollution. Alternative binding materials have been studied to determine a low carbon green binder to substitute the Portland cement. Geopolymer has been developed as one of the most promising alternative binding materials [3]. Geopolymer concrete (GPC) has superior cementitious characteristics with reduced greenhouse gas emissions as well as lower consumption of energy and non-renewable natural resources. The geopolymer technology uses a large amount of industrial trash materials like ground granulated blast furnace slag (GGBS), fly ash, silica fume etc. as source materials to avoid the negative impact on environment and ecology [4]. The selection of source material for preparation of GPC depends on its cost, availability, and application [5]. GGBS and fly ash are commonly used in the production of GPC. Based on the source material used, the GPC can be cured at elevated temperature or under ambient condition [6].

Reinforced concrete structures durability is of high significance as it determines the performance during the service life. Chloride produced corrosion in reinforced concrete structures is one of the significant durability problems, which results in spending of billions of dollars globally for repair and maintenance of reinforced concrete structures [7]. Due to the limited knowledge on the durability properties of GPC, currently there is comparatively less application of GPC in the field. The long-term durability properties as well as corrosion behaviour of steel reinforcement in GPC are yet to be investigated fully [8]. Just a few studies on corrosion performance of reinforcing steel in geopolymer concrete admixed with chloride have been published.

The objective of the present research effort is to study the influence of molarity of sodium hydroxide (NaOH) solution and GGBS replacement on workability, compressive strength and corrosion behaviour of reinforcing steel in chloride admixed GPC produced from fly ash and GGBS.

2 Experimental Details

2.1 *Materials and Mix Proportions*

In this experimental study, Class-F fly ash and GGBS were considered as the aluminosilicate source materials for preparation of GPC. Both GGBS and fly ash were sieved through 150 μm sieve before use in the production of GPC. Locally available river sand with highest grain size of 4.75 mm was considered as fine aggregate. Crushed aggregates of 10 mm MSA (maximum size of aggregate) and 20 mm MSA were considered as coarse aggregate for preparation of GPC mixes. To activate source materials, the combination of sodium hydroxide (NaOH) solution and commercially available sodium silicate (Na_2SiO_3) solution (Na_2O: 8%, and SiO_2: 26.5% by weight) was considered as alkaline liquid and it was prepared 24 h before preparation of GPC mixes. The NaOH solution of required molarity was made first by dissolving commercially available sodium hydroxide pellets in laboratory tap water 48 h before preparation of GPC mixes. In this study, two different molarity of sodium hydroxide solution such as 8 and 14 M, and three different GGBS replacement levels i.e. 0, 15 and 45% of total binder by mass were used in the preparation of GPC. To investigate the impact of admixed chloride on corrosion performance of steel reinforcement in GPC, sodium chloride (NaCl) concentration of 3% of geopolymer solids by mass was added with the alkaline liquid during production of geopolymer concrete mixes. The mix proportion of GPC mixes are presented in Table 1. For steel reinforcement, Tempcore TMT steel bars with a length of 340 mm and a diameter of 12 mm were used.

2.2 GPC Sample Preparation and Curing

Cube samples of dimension 150 mm were made for obtaining compressive strength of GPC mixes. Prismatic reinforced GPC samples of dimension 72 mm × 72 mm × 300 mm were made for half-cell potential and LPR (linear polarization resistance) measurements. The reinforcing steel rod was scrubbed by wire brush to extract surface scale before centrally embedded in prismatic sample. To avoid crevice corrosion, the reinforced steel specimen was covered in insulating tape and then coated with epoxy at the points where the steel specimen separated from the surrounding concrete. The prismatic sample had a 30 mm cover to the reinforcing steel rod at the bottom and all sides. The exposure length of steel rod in prismatic sample was 180 mm. For preparation of samples, the freshly prepared GPC was placed in respective moulds of cube and prismatic samples in different layers, every layer being compacted on a vibrating table. The outline diagram of reinforced GPC sample is presented in Fig. 1.

Table 1 Details of mix quantity for GPC mixes

GPC mix	Fly ash (kg/m^3)	GGBS (kg/m^3)	GGBS (%)	Molarity of NaOH solution	NaOH solution (kg/m^3)	Na_2SiO_3 solution (kg/m^3)	FA** (sand) (kg/m^3)	CA*** 20 mm MSA (kg/m^3)	CA 10 mm MSA (kg/m^3)
FS0*	400	–	0	8 and 14 M	80	120	567.6	576.2	576.2
FS15	340	60	15						
FS45	220	180	45						

* FS0 for GPC mix with 100% fly ash and 0% GGBS, ** FA for fine aggregate, *** CA for coarse aggregate

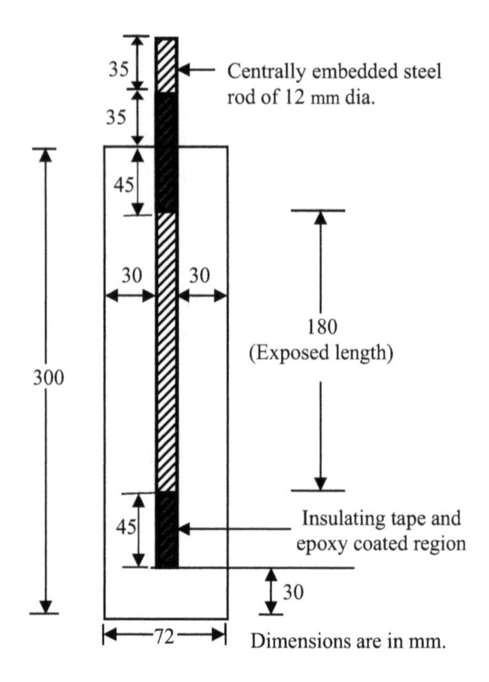

Fig. 1 Outline diagram of reinforced GPC sample

The GPC samples prepared in fly ash and GGBS (FS15 and FS45) were removed from the moulds after twenty-four hours of casting, whereas the GPC samples prepared by only fly ash (FS0) were held in forty-eight hours in ambient laboratory conditions after casting followed by oven-dry curing for another forty-eight hours at a temperature of 80 °C before demoulding. All the demoulded GPC samples were held in ambient laboratory conditions up to the testing age. In present research work, three replicate prismatic samples and three replicate cube samples were made from each GPC mix.

2.3 *Different Tests Conducted on GPC*

Test for Workability

For evaluating the fresh property of GPC, slump test was performed on the fresh GPC mix immediately after preparation.

Test for Compressive Strength

Compressive strength tests were performed on cube samples of different geopolymer concrete mixes in a compression testing machine (CTM) of capacity 2000 kN at 28 days of casting. For each GPC mix, there replicate cube samples were tested, and the average value represented the compressive strength of that mix.

Test for Corrosion Parameters

The corrosion performance of prismatic reinforced GPC samples was monitored in a corrosion monitoring instrument (ACM (make), Gill AC-1542). The electrochemical techniques such as half-cell potential and LPR measurements were conducted on the prismatic samples at 60 days of casting. In the present investigation, saturated calomel electrode (i.e. SCE) was used as a reference electrode, and two stainless steel plates were used as auxiliary electrode. During testing, the prismatic reinforced GPC sample was partially immersed in test solution with similar concentration of NaCl as that admixed in preparation of GPC. The embedded reinforcing steel in the prismatic specimen (working electrode), auxiliary electrode, and reference electrode, were linked to the corrosion monitoring instrument as presented on Fig. 2. The half-cell potential of the reinforcing steel in the prismatic sample was determined using SCE as reference and the possibility of occurrence of reinforced steel corrosion in GPC was assessed according to ASTM: C876 [9]. During the measurement of I_{corr} by LPR technique, the steel bar reinforced in prismatic sample was polarized to ± 20 mV from the corrosion potential at 6 mV per minute (i.e. scan rate). The I_{corr} of steel reinforcement was calculated from Stern-Geary equation [10].

$$I_{corr} = \frac{B}{Rp} \tag{1}$$

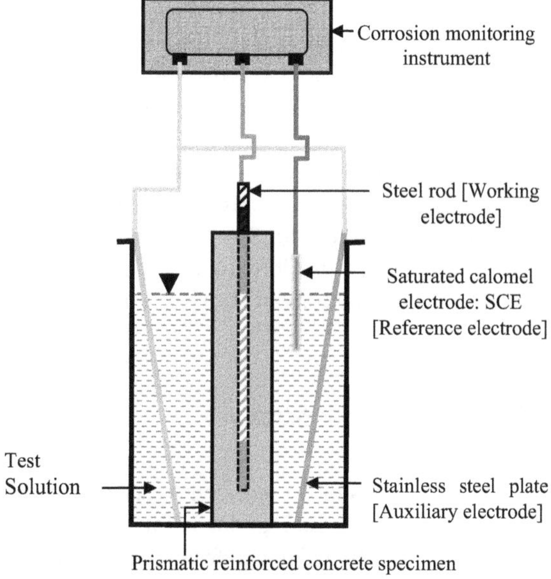

Fig. 2 Experimental set-up diagram for half-cell potential, and LPR measurements

where, R_p: polarization resistance of reinforcing steel, and B: Stern-Geary constant, Here B = 26 mV for reinforcing steel bar in active condition [10].

3 Results and Discussion

The following are the outcomes of slump tests (workability), compressive strength tests, and corrosion tests performed on various GPC mixes.

3.1 Workability

The observed slump values of GPC mixes prepared from 8 and 14 M NaOH solution are reported in Fig. 3a, b. The slump value of GPC mixes decreased as the molarity of NaOH solution increased, as seen in these figures. The slump value of GPC mixes decreased up to 38.8% by increase in molarity of sodium hydroxide solution from 8 to 14 M. The decrease in slump value at 14 M molarity of sodium hydroxide solution may be attributed to the increase in viscosity of the GPC mix due to available of higher quantity of solids in NaOH solution. Further, slump value of the GPC mixes was influenced by the GGBS content (%). Higher slump value was observed in only fly ash (FS0) GPC mixes (i.e. 0% GGBS) and the slump value decreased with raise in percentage of GGBS in the GPC mixes. This can be due to the effect of angular shape of GGBS particles which decreased the particle mobility effect of spherical shape of fly ash particles in the geopolymer concrete mixes. The slump value decreased up to 26.6% at 45% GGBS in GPC mixes. The lowest slump value of 55 mm was observed at 45% GGBS in GPC control mix (i.e. at 0% NaCl) prepared with 14 molarity NaOH solution. Higher slump value was observed in 3% NaCl admixed geopolymer concrete mixes for all replacement levels of GGBS and both 8 and 14 M NaOH solution (Fig. 3a, b) thereby indicating increased consistency of the GPC mix with available of chloride ions. Maximum slump value of 130 mm was observed in only fly ash (FS0) GPC mix made with 8 M NaOH solution and admixed with 3% NaCl. The increase in consistency of NaCl added GPC mixes can be ascribed to the effect of increased particle mobility due to available of chloride ions in the GPC mixes.

3.2 Compressive Strength

The 28 days results of average compressive strength (three replicate cubes) of GPC mixes are presented in Fig. 4a, b for NaOH solution molarity of 8 and 14 M respectively. It can be noted from Fig. 4a, b that the compressive strength of geopolymer concrete mixes increased with increase in molarity of sodium

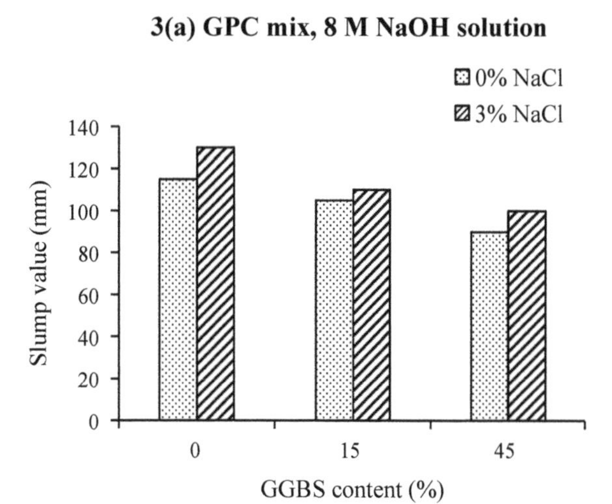

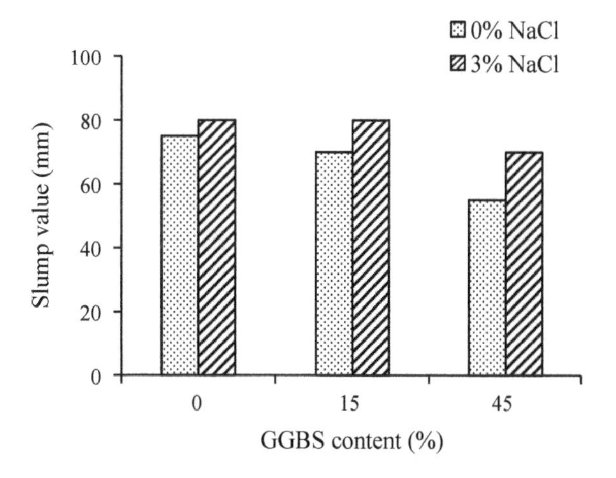

Fig. 3 Slump value of GPC mix made with **a** 8 M NaOH solution and **b** 14 M NaOH solution

hydroxide solution from 8 to 14 M. The increase in compressive strength at higher molarity of sodium hydroxide solution may be due to the effect of more leaching of Si, Al and Ca species from the source materials (GGBS and fly ash) thereby leading to greater degree of polycondensation process and resulting in development of higher compressive strength of geopolymer concrete [11]. The increase in compressive strength was more in FS0 (only fly ash) GPC mixes with respect to that made with GGBS and fly ash (both replacements) when sodium hydroxide solution molarity increased from 8 to 14 molarity. Further, higher compressive strength was observed in all the geopolymer concrete mixes made with 45% GGBS replacement level as compared to other replacement levels. The incorporation of slag as an

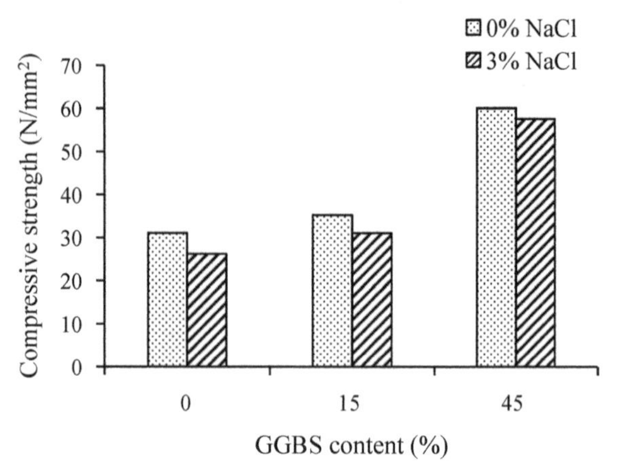

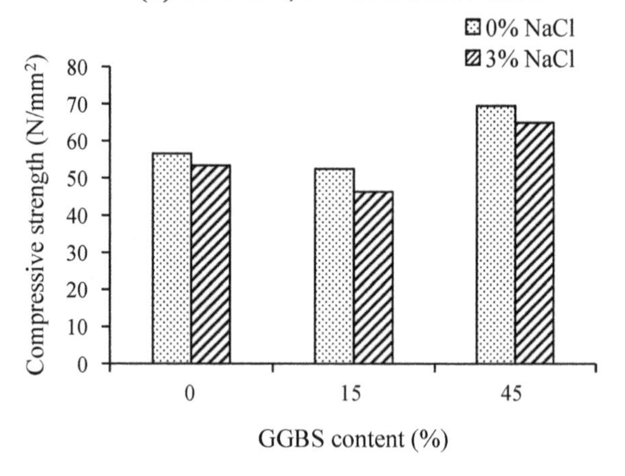

Fig. 4 Compressive strength of GPC made with **a** 8 M NaOH solution and **b** 14 M NaOH solution

additional source of alumina, silica, and calcium improved the degree of geopolymerization process, subsequently enhanced compressive strength development of geopolymer concrete [12] and the effect was more significant at higher GGBS replacement (45%). The 28 days compressive strength of GPC mix prepared in fly ash and 15% GGBS was lesser than FS0 (100% fly ash) GPC at 14 M NaOH solution as shown on Fig. 4b. This can be attributed to the effect of alteration in the extent of geopolymerization reaction under ambient condition in GPC mix prepared with fly ash and lower amount of slag at higher molarity of NaOH solution. Further from Fig. 4a, b, it is concluded that the NaCl added GPC mixes exhibited lesser compressive strength than that made without NaCl for all GGBS replacements and both molarity of sodium hydroxide solution. The decrease in compressive strength of GPC mixes admixed with NaCl can be due to the effect of precipitation and crystallization of sodium chloride in the aluminosilicate gel [13].

3.3 *Half-Cell Potential*

The half cell potential values of steel reinforcement embedded in different GPC mixes, at 60 days of casting are presented in Fig. 5a, b. This is observed from Fig. 5a that the values of half-cell potential in steel reinforcement became less negative with increase in GGBS percentage in GPC mixes made with 8 M NaOH solution whereas the reverse variation was noticed in the GPC mixes prepared with 14 M NaOH solution for control GPC mix as well as that admixed with NaCl. This can be due to the effect of alteration in oxygen and moisture content near steel reinforcement in GPC mixes prepared from different molarity of NaOH solution. Further, in NaCl added GPC mixes, the value of half-cell potential was more negative than −270 mV (w.r.t SCE) for entire GGBS (%) contents and both molarity (i.e. 8 and 14 M) of sodium hydroxide solution (Fig. 5a, b), thereby indicating greater possibility of occurrence of steel corrosion in accordance with ASTM-C876 [9] in chloride added GPC mixes. This is due to the fact the existent of chloride ions nearby steel reinforcement affected its passivity. In control GPC mixes (without NaCl), the values of half-cell potential were less negative than −270 mV (w.r.t SCE) at 8 M NaOH solution whereas at 14 M NaOH solution, the values of half-cell potential were higher negative than −270 mV (w.r.t SCE) in GPC mixes prepared from both fly ash and GGBS except that prepared from only fly ash. It is worth noting that the values of half-cell potential may be higher negative than −270 mV (w.r.t SCE) without causing major corrosion, this can be due to the polarization phenomena caused by oxygen deficiency [8, 10].

3.4 *Corrosion Current Density*

The obtained values of corrosion current density (I_{corr}) of steel bar from LPR measurement at 60 days of casting are presented in Fig. 6a, b. It observed from these figures, that the I_{corr} of steel reinforcement was lower in control geopolymer concrete specimens (without NaCl) as compared to the GPC specimens prepared in 3% sodium chloride for all GGBS replacement levels and both molarity of NaOH solution. The higher I_{corr} of steel reinforcement in NaCl added GPC can be attributed to the effect of improvement in conductivity of geopolymer concrete in available of chloride ions. Moreover, the corrosion current density of steel reinforcement in NaCl added GPC mixes reduced with raise in GGBS replacement level (i.e. from 0 to 45%). This may be due to the effect of formation of compacted microstructure at higher GGBS replacement that resulted in higher resistivity of GPC. Further, there was no uniform variation in I_{corr} values with raise in GGBS

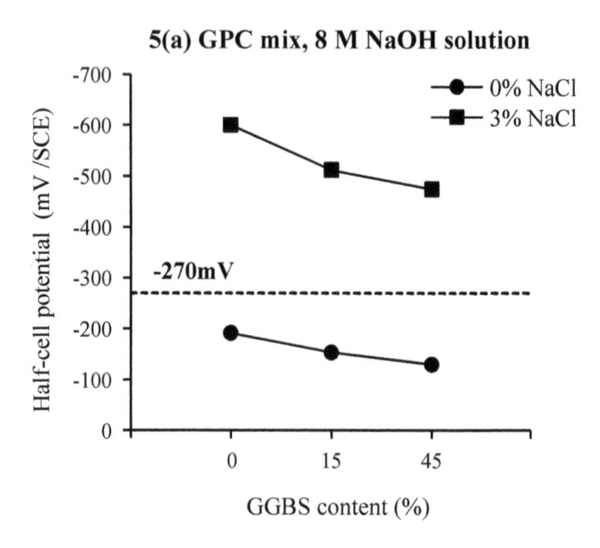

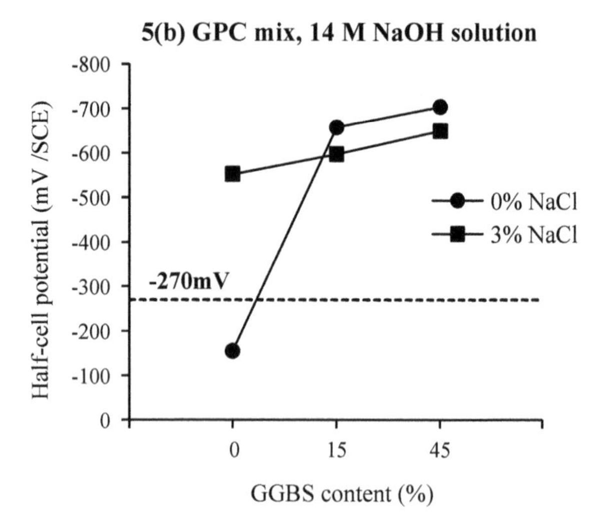

Fig. 5 Half-cell potential of steel reinforcement in GPC made with **a** 8 M NaOH solution and **b** 14 M NaOH solution

replacement level (i.e. from 0 to 45%) in case of control GPC mixes. From Fig. 6a, b, an increase in I_{corr} of steel reinforcement was noticed with raise in molarity sodium hydroxide solution (i.e. from 8 to 14 M), which may be due to alteration in the electrolytic pore solution of GPC at greater molarity of sodium hydroxide solution which affected the passivity of steel reinforcement.

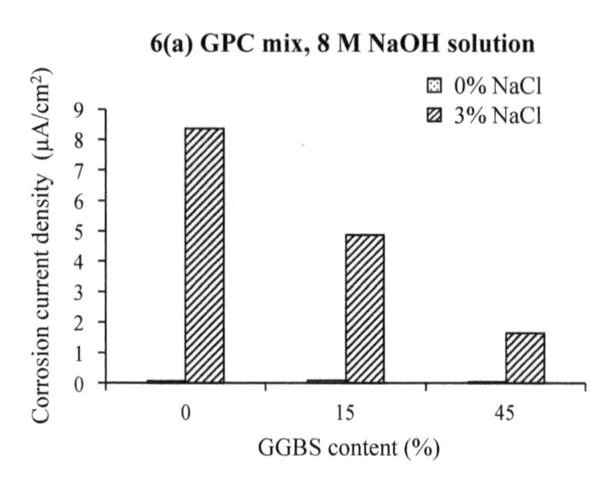

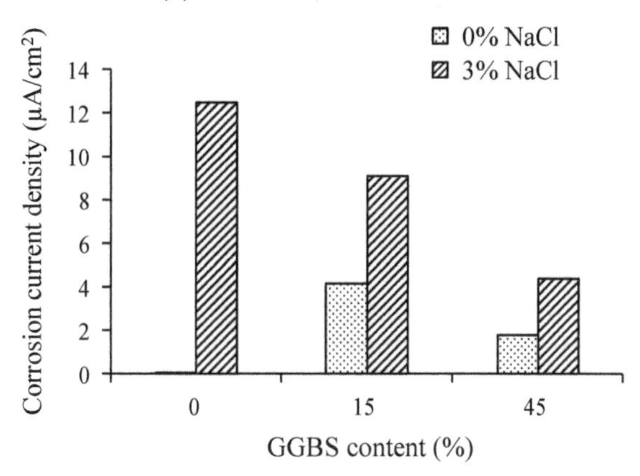

Fig. 6 Corrosion current density (I_{corr}) of steel reinforcement in GPC made with **a** 8 M NaOH solution and **b** 14 M NaOH solution

4 Conclusions

The following findings were observed from the present investigation.

- Higher workability was revealed in NaCl admixed GPC mixes as compared with control GPC mixes.
- An increase in molarity of sodium hydroxide solution resulted in higher compressive strength of GPC. GPC mixes made with 45% GGBS replacement exhibited higher compressive strength as compared with other replacement levels.
- Addition of NaCl in the mixes reduced the compressive strength of GPC.
- Higher probability of occurrence of steel reinforcement corrosion was observed in NaCl admixed geopolymer concrete. The half-cell potential values more negative than −270 mV (w.r.t SCE) in case of control GPC mixes made with

both fly ash and GGBS at 14 M NaOH solution may be due to the effect of polarization phenomena induced as a result of oxygen deficiency.

- The presence of NaCl in GPC significantly increased the corrosion current density (I_{corr}) of steel reinforcement.
- The I_{corr} of steel reinforcement in GPC reduced with increase in GGBS replacement and increased with increase in molarity of NaOH solution.

References

1. Wardhono A, Gunasekara C, Law DW, Setunge S (2017) Comparison of long term performance between alkali activated slag and fly ash geopolymer concretes. Constr Build Mater 143:272–279
2. Madlool NA, Saidur R, Hossain MS, Rahim NA (2011) A critical review on energy use and savings in the cement industries. Renew Sustain Energy Rev 15(4):2042–2060
3. Shi C, Jiménez AF, Palomo A (2011) New cements for the 21st century: the pursuit of an alternative to Portland cement. Cem Concr Res 41(7):750–763
4. Valencia-Saavedra W, Mejía de Gutiérrez R, Gordillo M (2018) Geopolymeric concretes based on fly ash with high unburned content. Constr Build Mater 165:697–706
5. Hadi MNS, Farhan NA, Sheikh MN (2017) Design of geopolymer concrete with GGBFS at ambient curing condition using Taguchi method. Constr Build Mater 140:424–431
6. Olivia M, Nikraz H (2012) Properties of fly ash geopolymer concrete designed by Taguchi method. Mater Des 36(1):191–198
7. Gunasekara C, Law D, Bhuiyan S, Setunge S, Ward L (2019) Chloride induced corrosion in different fly ash based geopolymer concretes. Constr Build Mater 200:502–513
8. Tennakoon C, Shayan A, Sanjayan JG, Xu A (2017) Chloride ingress and steel corrosion in geopolymer concrete based on long term tests. Mater Des 116:287–299
9. ASTM C876-15 (2016) ASTM C 876 standard test method for corrosion potentials of uncoated reinforcing steel in concrete. Annu B ASTM Stand 1–8
10. Pradhan B (2014) Corrosion behavior of steel reinforcement in concrete exposed to composite chloride-sulfate environment. Constr Build Mater 72:398–410
11. Fang G, Ho WK, Tu W, Zhang M (2018) Workability and mechanical properties of alkali-activated fly ash-slag concrete cured at ambient temperature. Constr Build Mater 172:476–487
12. Zaheer M, Khan N, Shaikh A, Hao Y, Hao H (2016) Synthesis of high strength ambient cured geopolymer composite by using low calcium fly ash. Constr Build Mater 125:809–820
13. Giasuddin HM, Sanjayan JG, Ranjith PG (2013) Strength of geopolymer cured in saline water in ambient conditions. Fuel 107:34–39

Introducing a Simplified Approach for 3D Simulation of Hydration of Cementitious Binder Systems Using MATLAB (MATHYD3D)

Swetha Prabhakar, Hridyesh Raj Tewani, Arkamitra Kar, Saptarshi Sengupta, and Nipun Tiwari

Abstract This study proposes MATHYD3D as a simplified approach for 3D simulation of the hydration of Portland cementitious systems containing supplementary cementitious materials (SCMs) using MATLAB coding. The binder composition, proportion of SCM, water-cementitious materials ratio, and desired age are taken as input from the user. In the first step, the corresponding volume fractions of the hydration products are calculated using stoichiometric equations. The MATLAB coding then creates a representative volume element (RVE). Based on the calculated volume fractions of the hydration products, the RVE is filled up with several smaller elements, each assigned a particular color corresponding to the component it is representing. The proportion of each color within the RVE is directly proportional to the volume fraction of the corresponding hydration product. Thus, at "zero" age, the RVE is a completely grey cube corresponding to unreacted binder and no other products. As hydration progresses over time, the code generates the elements with different colors proportional to the hydration products. Spherical, cubic, and Delaunay triangle-based tetrahedrons are assigned to the individual elements. Computational speed and accuracy show that tetrahedrons are optimal. Volume fraction results at 1-day, 7-day, 28-day, and 90-day age are found to deviate around 20% w.r.t. NIST's VCCTL-CEMHYD3D.

S. Prabhakar (✉) · A. Kar · S. Sengupta · N. Tiwari
Department of Civil Engineering, BITS-Pilani, Hyderabad Campus, Hyderabad, Telangana, India
e-mail: f20170987@hyderabad.bits-pilani.ac.in

A. Kar
e-mail: arkamitra.kar@hyderabad.bits-pilani.ac.in

S. Sengupta
e-mail: f20170917@hyderabad.bits-pilani.ac.in

N. Tiwari
e-mail: f20180119@hyderabad.bits-pilani.ac.in

H. R. Tewani
Department of Civil and Environmental Engineering, University of Wisconsin-Madison, Madison, WI, USA
e-mail: htewani@wisc.edu

G. C. Marano et al. (eds.), *Proceedings of SECON'21*, Lecture Notes in Civil Engineering 171, https://doi.org/10.1007/978-3-030-80312-4_60

Keywords Cement hydration · SCM · 3D simulation · MATLAB

1 Introduction

The substitution of Portland cement (PC) in construction with industrial waste and by-products as supplementary cementitious materials (SCMs) can reduce the CO_2 emissions resulting from PC manufacture and prevent the disposal of SCM such as fly ash, slag, and silica fume as solid waste. Addition of SCM to PC enhances the strength characteristics of hardened concrete made with these binders. But these additions increase the complexity of the chemistry of hardened cementitious paste. The focus of this study is thus to develop a readily available simplified visual representation of the hydration reactions in MATLAB (MATHYD3D). This will enable better understanding of the hydration mechanism of cementitious systems containing SCM and encourage their widespread practical application [1–3]. Calcium silicate hydrate (C–S–H(I)) matrix is predominant component of the hydrated cementitious systems. SCM inclusions result in pozzolanic reaction between the reactive silica present in them and calcium hydroxide (CH) produced by PC hydration to further complicate this composition by forming secondary or pozzolanic C–S–H (C–S–H(II)) having different Ca/Si ratios and specific gravities compared to C–S–H(I) [4]. Additionally, the degree of hydration of PC and the degrees of reaction of the SCM are starkly different from each other, leading to differential formations of the two types of C–S–H [5]. Thus, this study focuses on developing a visual tool to distinguish between C–S–H (I) and C–S–H (II) in a simplified manner, for different combinations of PC and SCM binders. Simulations of the hydration mechanism through numerical modelling approach using various computational models have been proposed in the past couple of decades despite several challenges [1, 6].

The popular approaches adopted are namely the discretization approach and the continuous (or vector) approach. In the discretization approach, the three-dimensional cement hydration and microstructure development model, CEMHYD3D is popularly used [7]. It produces approximate visualisations of the hydration products obtained through a digitised collection of voxels. A probabilistic simulation model, HydratiCA, considers hydration kinetics and cellular automation on a regular computational lattice to visualise hydration reactions [8]. The Virtual Cement and Concrete Testing Laboratory (VCCTL) demonstrated that particle shapes of the hydration products govern the early-age characteristics of the fresh and hardened paste as their specific surfaces influence the formation of nucleation sites for hydration reactions [9]. The discretization models generate relatively coarse output images with greater computation times for the simulations [1]. The continuous approach is used in tools like HYMOSTRUC3D [10] and Integrated Particle Kinetics Model (IPKM) [11], where the hydrating binder particles are visualised as expanding spheres. DuCOM includes the effects of varying oxide compositions but fails to distinguish between C–S–H (I) and C–S–H (II) [12].

HYMOSTRUC3D-E simulates the hydration and microstructural formation of PC systems using fly ash and slag as SCM [13]. CEMHYD3D v3.0 provides the input for most of the aforementioned studies [14]. It generates multiple simulation models for PC hydration under different conditions (temperature, water cement ratios, and age in days) and for varying PC + SCM mix proportions using C programming. It is beneficial for users with a fundamental knowledge of the C language. The present study proposes a relatively simplified simulation regime named MATHYD3D using a user-friendly interface created in MATLAB while maintaining the fundamental principles of hydration. The specific objectives of this study are: (i) to develop a discrete elements-based user-friendly 3-D simulation of cement hydration using MATLAB coding; (ii) to experimentally validate the results through comparison with VCCTL and CEMHYD3D; and (iii) to incorporate the results of taking into account the effects of SCM on the hydration of binary and ternary cementitious mix systems.

2 Materials and Methodology

The cement and SCM composition used for performing stoichiometric volume fraction calculations to formulate the fundamental equations for developing MATHYD3D is obtained from [5]. The cement used for validation of the proposed simulation model is the CCRL cement 152 in the NIST database [15]. The relevant Bogue's compositions with standard ceramic and cement chemistry notations are provided in Table 1. For this study, MATLAB is used to write the codes to calculate the stoichiometric volume fractions of cement hydration products and to generate the corresponding 3D simulation of cement hydration. The codes are compiled and executed in MATLAB.

Analogous to CEMHYD3D, a user-interactive code MATHYD3D is developed in this study to represent the hydration of cementitious binder systems with age. It requires the user to enter the Bogue's composition for their cement, water-cementitious materials ratio (w/cm), desired age, Ca/Si for C–S–H (I) (n), and for C–S–H (II) (n'). MATHYD3D then computes the volume fractions of the hydration products and the output is visualised as a 3D image. The steps are summarised in Fig. 1. The 3D image is composed of a representative volume

Table 1 Bogue's composition of cements used in this study

Clinker Phase (Bogue's compounds)	Volume Fraction (Cement used for MATHYD3D)	Volume Fraction (CCRL cement 152)
C_3S	0.5055	0.7344
C_2S	0.2524	0.0938
C_3A	0.12	0.1311
C_4AF	0.08	0.0407

element (RVE) representing the hydrated paste at any age. Thus, raw cementitious binder before the beginning of hydration is visualised as a grey cube of size 1 × 1 × 1 MATLAB units. The hydration products include C–S–H (I), CH, AF_t, AF_m, C_3AH_6, FH_3, C–S–H (II), and unhydrated binder residue (conventional ceramic and cement chemistry symbols used). The corresponding volume fraction expressions are obtained from [5]. Discretization approach is then used to simulate the growth of these compounds inside the RVE. The values of 'α' are obtained from [15]. The respective colours assigned to these components are presented in Fig. 1. Based on the stoichiometric volume fractions of the hydration products, the RVE is filled up with several discrete constituent elements, each assigned a particular colour corresponding to the component it represents. A random pattern for generation and orientation of the individual elements is selected to resemble heterogenous chemistry of the hydrated cementitious paste. The proportion of elements corresponding to each colour within the RVE is directly proportional to the volume fraction of the corresponding hydration product. Thus, at zero-age, the RVE is a single grey cube filled with zero smaller elements within it, representing unhydrated PC. As hydration progresses over time, MATHYD3D generates several elements of different colours corresponding to the different hydration products. Separate iterations are performed assuming spherical, cubic, and tetrahedral shapes of the individual elements. The optimal one is selected based on mathematical complexity, computation times, and output accuracy.

The first shape considered for the individual elements is spherical, the most effective shape in terms of reducing surface area and surface energy [12]. Distances between two adjacent elements are maintained to avoid overlapping of adjacent spherical elements. These distances are computed and their values stored assigned respective variables. Additionally, it is ideal to assign non-uniform diameters to each sphere to simulate real-life scenarios. The equivalent diameters of these spheres are directly proportional to the cube root of their corresponding volume fractions, thereby increasing mathematical complexity and computation times substantially. Hence, for simplicity, cubic shape is considered for the elements inside the RVE. Using MATLAB, a coordinate matrix consisting of all the coordinates of the cubic elements is used to make a connectivity matrix where each row represents the connectivity between coordinates to form a single cubic element. Each cubic mesh is assumed to represent one particle corresponding to a particular hydration product and assigned the corresponding colour. Thus, the number of individual cubic elements corresponding to each colour within the RVE is directly proportional to the volume fractions of these products. These individual cubic elements within the RVE are generated randomly to resemble the heterogeneous microstructure of the hydrated cementitious paste. A sample output for PC hydration using cubic elements in MATHYD3D is shown in Fig. 2. Finer mesh sizes increase the accuracy but also the computation times. So, Delaunay triangle-based tetrahedral shape is considered for the individual elements. No relevant studies are reported on the application of this shape. Uniformly shaped cubic elements do not consider the different particle sizes of the different hydration compounds. The adjacent tetrahedrons generated by the Delaunay tetrahedrons can

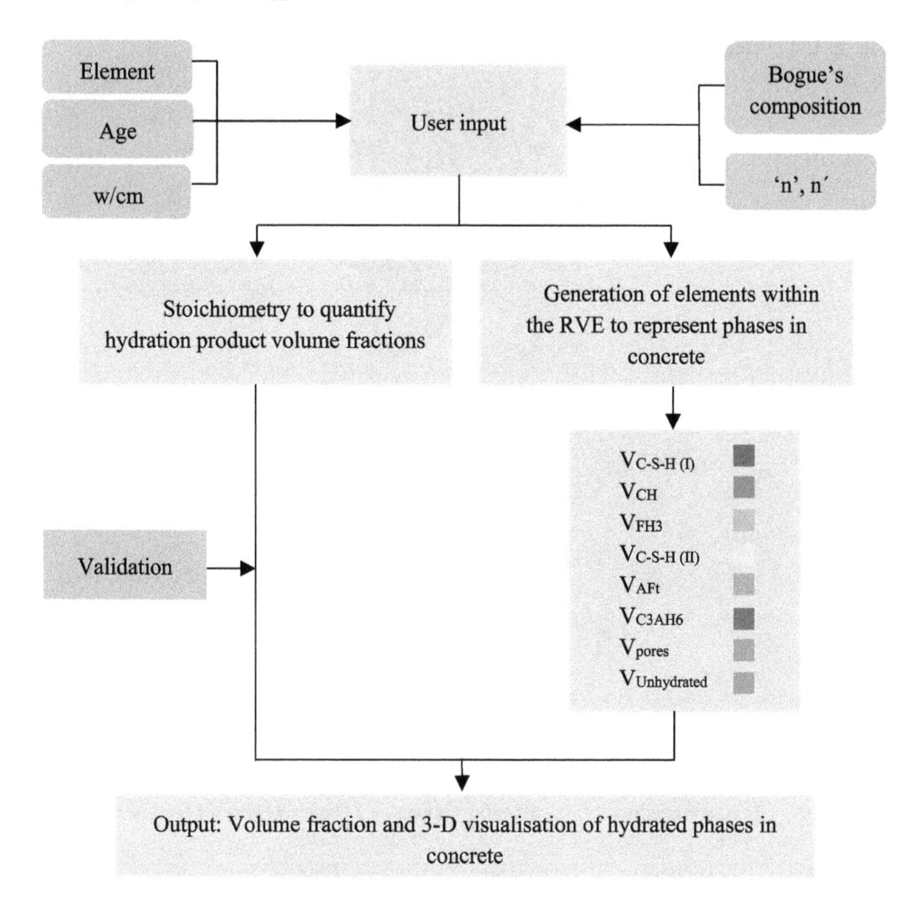

Fig. 1 Schematic operation of MATHYD3D

be adjusted to adopt different sizes based on the compound they are representing. Moreover, they do not suffer from any lack of contact at the tangential surfaces, unlike the spherical elements. This non-uniform generation of elements in three-dimensional space can be considered a more practical simulation of real-life hydration, compared to the other two shapes. The output for tetrahedral individual elements is presented in Fig. 3, for the same PC mix proportions and ages used for the cubic elements. Table 2 presents qualitative observations on the relative characteristics of each shape and the basis for selection of an optimal shape.

The efficiency of the stoichiometric equations used for MATHYD3D output is determined through comparison with the outputs obtained from VCCTL 9.5.1 [4]. For comparison, the reference cement is CCRL cement 152 (Table 1) from NISTIR7232 [14]. A uniform w/cm of 0.35 is maintained for the mixes. The degree of hydration data corresponding to CCRL cement 152 obtained from [15] is then used to generate the hydration product volume fractions for four different ages, viz.

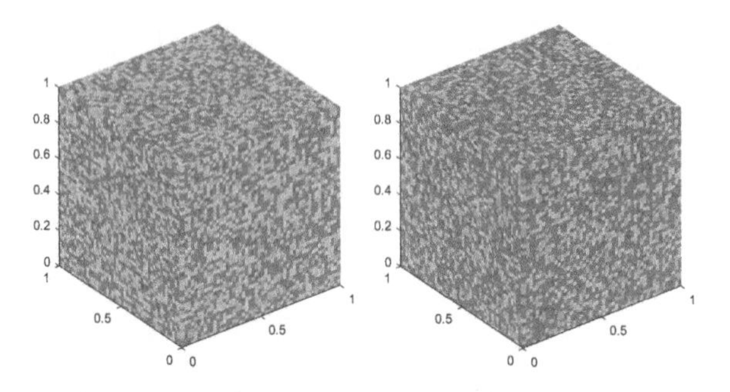

Fig. 2 Cubic elements

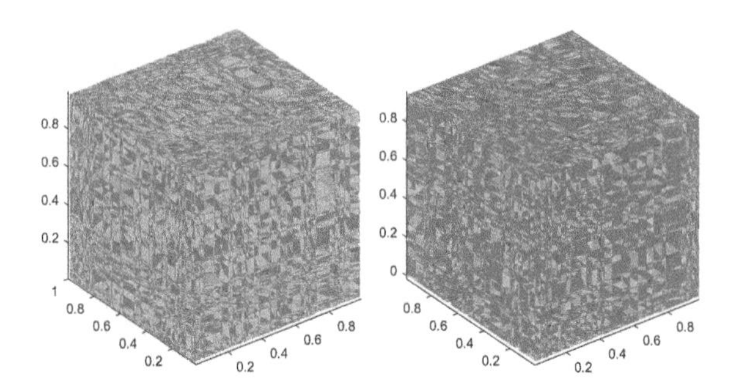

Fig. 3 Tetrahedral elements

Table 2 Comparing the different shapes of elements

Shape of elements	Mathematical complexity	Execution time and computing power	Resemblance to real-life heterogeneity of hydrated cement paste
Spherical (varying diameter)	Highest	Highest	Highest
Cubic	Least	Moderate to High	Least
Tetrahedral	Moderate	Lowest	Moderate to high

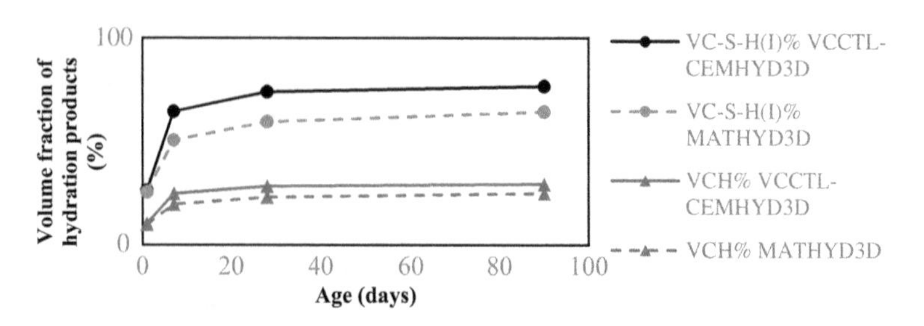

Fig. 4 Validation of MATHYD3D using hydrated volume fractions of CCRL cement 152

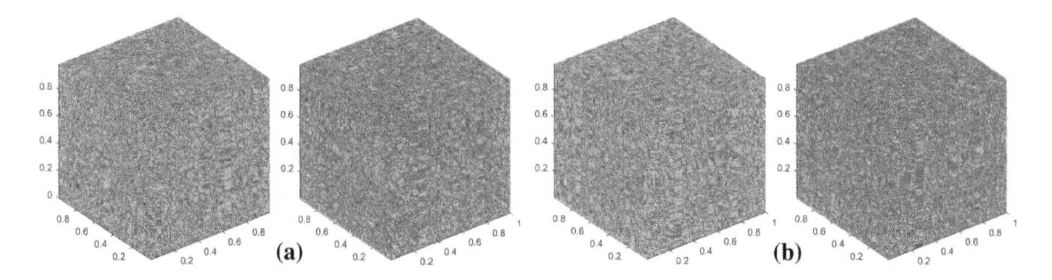

Fig. 5 MATHYD3D output for **a** 80% PC + 20% fly ash, w/cm 0.4 and **b** 80% PC + 10% fly ash + 10% silica fume, w/cm 0.4

1, 7, 28, and 90 days. The results are shown in Fig. 4. Then, the visual output from MATHYD3D is generated for blended cement binder systems (Fig. 5). The results are shown only for 7- and 90-days ages for the sake of brevity.

3 Results and Discussions

Using cubic elements, the gradual increase in the proportion of red coloured cubes representing C–S–H (I) is clearly visible at 7 and 90 days of hydration (Fig. 2). Figure 3 presents the 3D visualisation of hydration at 7- and 90-days age using tetrahedral elements. Similar trend is observed as in case of cubic elements. In case of cubic elements, adjacent elements are placed in discrete manner. However, in case of tetrahedral elements, the distribution of the individual elements is more randomized and closer to the real-life heterogeneity of the hydrated cementitious paste. The colour scheme is same as in Fig. 1.

The relative characteristics of the different shapes are presented in Table 2.

Subsequently, the computation times for cubic shapes is found to increase rapidly with increasing number of elements within the RVE. Tetrahedral shape is thus suggested as the optimal shape for MATHYD3D. Figure 4 shows the relative quantitative outputs from CEMHYD3D and MATHYD3D for the volume fractions of C–S–H (I) and CH for CCRL cement 152 at ages of 1, 7, 28, and 90 days.

The predicted values from CEMHYD3D and MATHYD3D show approximately 20% variation for ages from 1 through 90 days. Figures 5a, b present 3D visualisation of the progress of hydration for blended cement systems using MATHYD3D. Majority of the RVE is occupied by C–S–H (I), followed by C–S–H (II) in matured pastes. CH is observed to diminish due to conversion to C–S–H (II) through pozzolanic reactions.

4 Conclusions

MATHYD3D can predict the hydrated product volume fractions with approximately 20% deviation from VCCTL-CEMHYD3D. Tetrahedral RVE elements exhibit optimal computational time and resemblance to heterogeneity of hydrated cement paste. MATHYD3D in its present form provides a relatively simple and swift 3D simulation of cement hydration with satisfactory level of accuracy. It can be further refined through inclusion of more types of cement from different sources in the database to account for the uncertainty in composition.

References

1. Zhu Z, Xu W, Chen H, Tan Z (2020) Evolution of microstructures of cement paste via continuous-based hydration model of non-spherical cement particles. Compos Part B: Eng 185, 107795 (2020).
2. Hu Q, Aboustait M, Kim T, Ley MT, Hanan JC, Bullard J, Winarski R, Rose V (2016) Direct three-dimensional observation of the microstructure and chemistry of C_3S hydration. Cem Concr Res 88:157–169
3. Lothenbach B, Kulik DA, Matschei T, Balonis M, Baquerizo L, Dilnesa B, Miron GD, Myers RJ (2019) Cemdata18: a chemical thermodynamic database for hydrated Portland cements and alkali-activated materials. Cem Concr Res 115:472–506
4. Taylor HF (1997) Cement chemistry, 2nd edn. Thomas Telford, London
5. Kar A, Ray I, Unnikrishnan A, Davalos JF (2012) Microanalysis and optimization-based estimation of C–S–H contents of cementitious systems containing fly ash and silica fume. Cement Concr Compos 34(3):419–429
6. Holmes N, Kelliher D, Tyrer M (2020) Simulating cement hydration using HYDCEM. Constr Building Mater 239:117811
7. Bentz DP (1997) Three-dimensional computer simulation of Portland cement hydration and microstructure development. J Am Ceram Soc 80(1):3–21
8. Bullard JW (2007) A three-dimensional microstructural model of reactions and transport in aqueous mineral systems. Modell Simul Mater Sci Eng 15(7):711
9. Liu C, Huang R, Zhang Y, Liu Z, Zhang M (2018) Modelling of irregular-shaped cement particles and microstructural development of Portland cement. Constr Build Mater 168:362–378
10. Van Breugel K (1995) Numerical simulation of hydration and microstructural development in hardening cement-based materials (I) theory. Cem Concr Res 25(2):319–331
11. Navi P, Pignat C (1999) Three-dimensional characterization of the pore structure of a simulated cement paste. Cem Concr Res 29(4):507–514
12. Maekawa K, Ishida T, Kishi T (2008) Multi-scale modeling of structural concrete. Taylor & Francis, London and New York
13. Gao P, Ye G, Wei J, Yu Q (2019) Extension of the Hymostruc3D model for simulation of hydration and microstructure development of blended cements. Heron 64(1/2):125
14. Bentz DP (2005) CEMHYD3D: a three-dimensional cement hydration and microstructure development modelling package. Version 3.0 NISTIR 7232. US Department of Commerce, National Institute of Standards and Technology
15. Bentz DP (2006) Capillary porosity depercolation/repercolation in hydrating cement pastes via low-temperature calorimetry measurements and CEMHYD3D modeling. J Am Ceram Soc 89(8):2606–2611

Random Forest-Based Algorithms for Prediction of Compressive Strength of Ambient-Cured AAB Concrete—A Comparison Study

Kruthi Kiran Ramagiri, Sriman Pankaj Boindala, Md Zaid, and Arkamitra Kar

Abstract The compressive strength of alkali-activated binder (AAB) concrete is governed by several factors like type, chemical composition and proportion of raw materials, and curing regime. Empirical prediction of such a relationship entails the utilisation of robust and intelligent algorithms. Random Forest (RF) is an advanced ensemble algorithm with advantages of effective generalisation ability, strong resistance to overfitting, and importance analysis. Several approaches to improve the prediction accuracy of RF are available in the literature. The present study aims to compare the accuracy of RF and its different configurations to predict the compressive strength of ambient-cured AAB concrete. Five different RF packages, RF, regularised RF, cforest, ranger, and random forest-SRC, are used in the present study. Three attribute evaluators from the feature selection module: classifier attribute evaluator, correlation attribute evaluator, and reliefF are investigated to find the relative importance of the input parameters. A total of 364 data sets with seven input parameters (fly ash, slag, sodium silicate (SS), sodium hydroxide (SH), fine aggregate, coarse aggregate, the molarity of SH) are used to evaluate the compressive strength of AAB concrete. The results show that the ranger algorithm with reliefF feature selection developed using the selected database exhibits the most accurate prediction based on MAE and RMSE values. It is also observed that the molarity of SH is the most critical factor affecting the compressive strength of ambient-cured AAB concrete.

K. K. Ramagiri (✉) · A. Kar
Department of Civil Engineering, BITS-Pilani Hyderabad Campus, Hyderabad, Telangana, India
e-mail: p20170008@hyderabad.bits-pilani.ac.in

A. Kar
e-mail: arkamitra.kar@hyderabad.bits-pilani.ac.in

S. P. Boindala
Civil and Environmental Engineering, Technion – Israel Institute of Technology, Haifa, Israel

M. Zaid
Terracon Consultants Inc., Kansas, USA

G. C. Marano et al. (eds.), *Proceedings of SECON'21*, Lecture Notes in Civil Engineering 171, https://doi.org/10.1007/978-3-030-80312-4_61

Keywords Alkali-activated concrete · Strength prediction · Random forest · Machine learning · Feature selection

1 Introduction

Alkaline activation of aluminosilicate-rich precursor(s) results in the formation of an alkali-activated binder (AAB) [1]. As a result, a wide variety of materials have the potential to be used as activators or precursors. Sodium silicate (SS) and/or sodium/potassium hydroxide are typical AAB activators [2]. Fly ash is the most extensively used precursor in AAB due to its wide availability. However, usage of fly ash entails thermal curing for desired strength gain. An efficient way to minimise thermal-curing of AAB is to incorporate ground granulated blast furnace slag (slag). When fly ash and slag are activated with SS and/or sodium hydroxide (SH), the final performance of AAB concrete is governed by several factors, including fly ash/slag ratio [3], fine aggregate/total aggregate ratio [4], the molarity of SH, SS/SH [5], and the curing regime. The available references addressing the design and testing of AAB are RILEM TC-224 and IS 17452: 2020 [6, 7]. The lack of codified mix design provisions, accompanied by the complexities associated with raw materials, and the time-consuming method of deciding an acceptable mix proportion impedes the functional application of AAB concrete. This highlights the importance of an accurate prediction of compressive strength.

There is extensive literature reported on prediction models of PC concrete using regression analysis. However, to enhance the accuracy of strength prediction, machine learning (ML) models like artificial neural network adaptive neural network-based fuzzy inference systems [8], neuro-fuzzy polynomials [9], and support vector machine was used [10]. There are a few published studies on AAB compressive strength prediction [11, 12]; however, they adopted the thermal curing regime. Random forest (RF) model improved strength prediction accuracy of high-performance concrete given feature selection, and parameter optimisation are performed [13]. RF is one of the most optimised algorithms, with benefits such as a limited number of model parameters, significance analysis, powerful predictive capacity, and efficient tolerance to overfitting. Therefore, the present study aims to compare the accuracy of RF and its different configurations to predict the compressive strength of ambient-cured AAB concrete. Five different RF packages, `randomForest`, regularised RF (`RRF`), `cforest`, `ranger`, and random forest-SRC (RF-SRC) (package: `rfsrc`) are used in the present study. Three attribute evaluators (AE) from the feature selection module: classifier attribute evaluator, correlation attribute evaluator, and ReliefF are investigated to find the relative importance of the input parameters.

2 Methodology

2.1 Dataset Collection and Description

A dataset containing 364 mix proportions was prepared from the existing literature to predict the compressive strength [14]. This study specifically focuses on the prediction of compressive strength of ambient-cured AAB concrete with fly ash or slag or a combination of both as precursors along with usage of both SS and SH as activators.

As illustrated in Table 1, the input and output parameters show a wide range of data. Furthermore, Fig. 1 shows a weak correlation among the parameters, thereby corroborating their consideration as mutually exclusive.

2.2 Random Forest Packages

Random forest (RF) is a nonlinear, robust approach for enhancing predictive precision by fitting an ensemble of trees to optimise model estimates. The classic RF uses Breiman's Classification and Regression Tree (CART) algorithm to grow trees [15]. Splits in RF are recognised to be skewed against variables with several potential splits when building a tree. RRF memorises previous split features and penalises new feature if their gain is similar to the previous node's features, thus alleviating redundant features [16]. Unlike RF, Conditional inference forest (Cforest) can be used for larger random sub-sample, and it employs conditional inference for growing trees, i.e., permutation tests are for splits [17]. Random forest generator (ranger) is used for rapid execution of RF or recursive partitioning and is preferable for high-dimensional data [18]. Random Forest-SRC creates a single Breiman's random forest approach for survival, regression, and classification problems [19].

Table 1 Statistical assessment of parameters in the dataset

Attribute	Fly ash	Slag	SS	SH	FA	CA	M	Compressive strength
Units	kg/m^3	kg/m^3	kg/m^3	kg/m^3	kg/m^3	kg/m^3	M	MPa
Minimum	0	0	18	3.5	459	331	1	8
Maximum	523	450	373	160	1360	1298	20	86.1
Average	180.9	218	110	52.29	723	1042	8.2	42.4
Median	120	255	107	53	721	1111	10	41
SD	168.1	163	52.6	32.7	126	217	4.63	17.3

FA Fine aggregate; *CA* Coarse aggregate; *M* Molarity of SH; *SD* Standard deviation

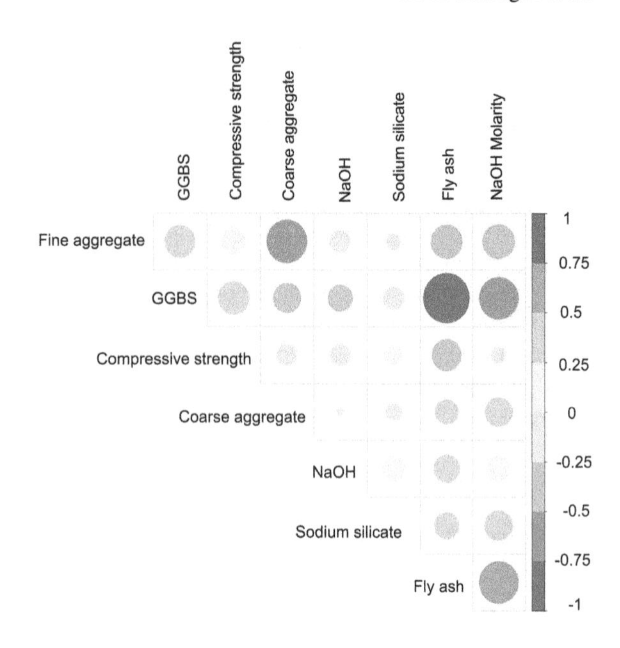

Fig. 1 Correlation matrix of the dataset

2.3 *Feature Selection*

Three feature selectors or attribute evaluators viz. classifier attribute evaluator, correlation attribute evaluator, and reliefF are used in the present study to assess the relative importance of independent variable (attribute) with respect to the dependent variable (class). Classifier attribute evaluator assesses the worth of attribute using a user-specified classifier. The correlation attribute evaluator estimates the significance of the attribute by measuring Pearson's correlation between it and the class. When reliefF is used, the merit of the attribute is evaluated by sampling an instance several times and comparing the given attribute score with the closest instance of a similar or different class.

2.4 *Methodology*

The RPlugin (v.1.3.30) available on Weka (v. 3.9.4) software is used for the implementation of the models [20]. RPlugin enables the use of R modules by Weka. It includes an R console, a 'Knowledge Flow' portion for running R scripts, and a wrapper classifier for the MLR (machine learning in R) package. For the selected five RF models, default input parameters available on Weka are selected. The models are evaluated through three performance metrics, adjusted R^2, mean absolute error (MAE), and root-mean-square error (RMSE). Three AEs are investigated in this study and based on the ranking of independent attributes a sensitivity analysis is performed. In this study, RF is selected as the classifier for the classifier attribute evaluator to maintain uniformity in selecting models. For the three attribute evaluators, the Ranker search method is employed.

3 Results

Table 2 shows the results of models developed using different RF packages for the dataset obtained. The performance of the ML models is illustrated in Fig. 2. Without any feature selection, the RF model developed using the selected dataset exhibited better prediction accuracy than others. Furthermore, RRF, ranger, and RF-SRC models demonstrated comparable prediction performance to RF. The Cforest model built using the chosen database achieved the lowest adjusted R^2 and highest MAE and RMSE values. The variation between predicted and actual values is apparent when cforest is used. When dealing with a combination of categorical and numerical independent variables, Cforest excels, but when dealing with pure independent variables, it tends to overestimate most likely outcomes and underestimate the likelihood of less likely outcomes [21]. Since all the independent variables are of numerical category, this could be the reason for the poor performance of Cforest.

Feature selection is a significant pre-processing phase of the Knowledge Data Discovery process. The primary use of feature selection is in data reduction by identifying redundant data or data exhibiting a weak correlation with the dependent variable. The attributes are ranked based on their average merit.

According to the analysis using classifier AE and reliefF AE on the assembled dataset of 364 mix proportions, the molarity of SH has the highest merit and is rated as highly affecting compressive strength (Table 3). The molarity of SH in AAB concrete is directly related to its water content. Hence, analogous to PC concrete, the compressive strength of AAB concrete is also significantly influenced by its water content. The other significant attributes governing the compressive strength are fly ash, slag, and fine aggregate contents. However, based on the selected database SS and SH are the least ranked by classifier AE and reliefF AE. Correlation AE estimated that slag has a high correlation with compressive strength in the selected database, while fly ash has the lowest correlation. According to correlation AE, the other governing attributes are fine aggregate, SS contents, and

Table 2 Performance of different RF models

Package	Adjusted R^2	MAE	RMSE
RF	0.791	6.014	8.291
RRF	0.788	6.037	8.338
Cforest	0.680	7.847	10.209
ranger	0.789	6.032	8.321
RF-SRC	0.774	6.388	8.634

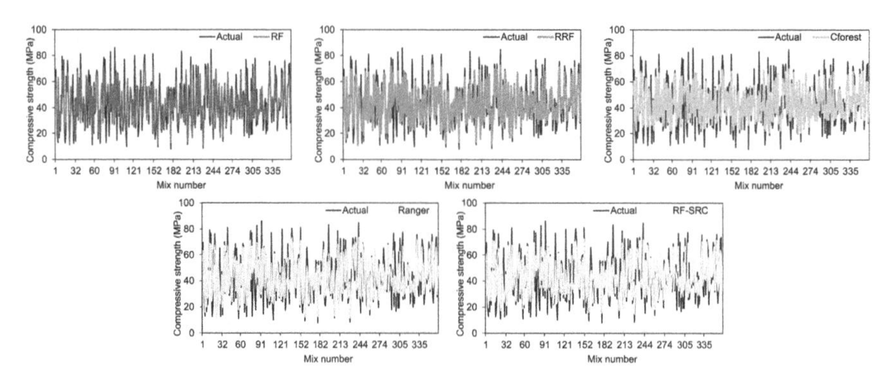

Fig. 2 Prediction performance of different RF models

Table 3 Feature selection results

Attribute	Classifier		Correlation		ReliefF	
	rank	Merit	rank	Merit	rank	Merit
Fly ash	2.7	2.369	7	−0.371	3.6	0.01
Slag	1.9	2.603	**1**	**0.365**	3.7	0.01
SS	7	−0.629	3	0.056	2.8	0.011
SH	5.5	0.478	6	−0.077	7	0.002
Fine aggregate	4	0.971	2	0.211	4	0.01
Coarse aggregate	5.5	0.472	4.7	−0.035	5.9	0.008
Molarity of SH	**1.4**	**2.703**	4.3	−0.026	**1**	**0.022**

molarity of SH. The difference in assessing the merit of attributes by the selected methods is due to the underlying mechanism by which they function. Table 4 presents the model performance in predicting the compressive strength of AAB concrete after excluding the attribute compromising prediction accuracy. Models with best and abysmal performance are highlighted in Table 4. When feature selection is used to exclude a variable, RRF with reliefF AE and Cforest with correlation AE performed the best and poorest of all models developed using the chosen database. The performance of these two models is graphically represented in Fig. 3.

Owing to the use of RF as a classifier in Classifier AE, it is inclined to biased variable selection, preferring variables with many possible splits or categories [22]. Correlation AE is a multivariate filter method of feature selection, and one of its limitations is the tendency to ignore the association between features in predicting the output. For example, if two attributes have a similar influence on the class, using correlation AE might obscure an attribute's actual effect and inaccurately interpret its contribution to the class. Usually, embedded methods have the capability of capturing feature interaction [23]. ReliefF is an individual evaluation filter method. In addition to the advantages of filter algorithms (faster computations and selected

Table 4 Model performance after feature selection

Attribute	FS	Excluded variable	Adjusted R^2	MAE	RMSE
RF	Cl	SS	0.773	6.118	8.527
	Co	Fly ash	0.789	6.040	8.312
	Rf	SH	0.792	5.914	8.241
RRF	Cl	SS	0.771	6.163	8.566
	Co	Fly ash	0.782	6.123	8.422
	Rf	**SH**	**0.793**	**5.901**	**8.220**
CForest	Cl	SS	0.669	7.947	10.318
	Co	**Fly ash**	**0.659**	**8.076**	**10.411**
	Rf	SH	0.668	7.989	10.365
ranger	Cl	SS	0.772	6.138	8.549
	Co	Fly ash	0.784	6.071	8.387
	Rf	SH	0.793	5.867	8.214
RF-SRC	Cl	SS	0.750	6.844	9.095
	Co	Fly ash	0.761	6.839	8.984
	Rf	SH	0.770	6.591	8.825

FS Feature Selection; *Cl* Classifier AE; *Co* Correlation AE; *Rf* reliefF AE

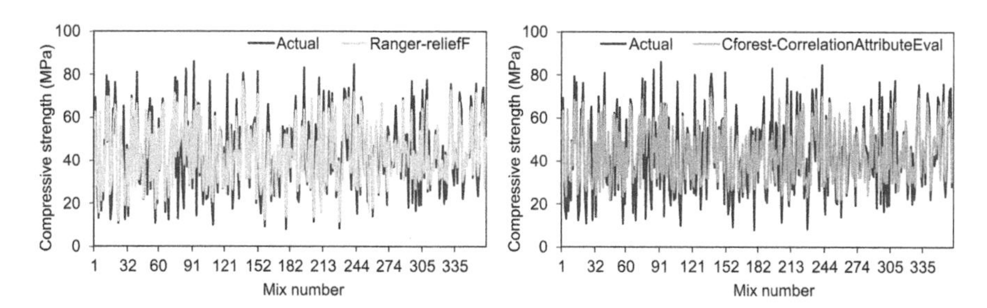

Fig. 3 Performance of the best and poor prediction models

attributes are independent of induction algorithm), reliefF is the sole known filter algorithm that can recognise the feature dependencies [24].

4 Summary

In this study, five different ML models are developed using RF packages to predict the compressive strength of fly ash and/or slag-based AAB concrete activated with a combination of sodium silicate and sodium hydroxide and cured under ambient conditions. 364 datasets are collected. The significance of attributes is evaluated using three attribute evaluators, and a sensitivity analysis is performed based on their ranking. The findings from the present study are summarised as follows:

1. The random forest model has an adjusted R^2 of 0.79 and RMSE of 8.29, and Cforest has an adjusted R^2 of 0.68 and RMSE of 10.21.
2. Without opting for feature selection and considering all the attributes, for the selected database of 364 mix proportions, the developed RF performs the best compared to other ML models. Cforest built using the chosen database performed poorly in predicting the compressive strength of AAB concrete.
3. For the five ML models used in the present study, the adjusted R^2 value increased, and RMSE decreased when ReliefF feature selection was used.
4. According to the sensitivity analysis performed using input from the compiled dataset, the sodium hydroxide content and molarity are the least and most governing factors affecting the compressive strength of AAB concrete. This finding indicates that water content is the most critical factor governing the compressive strength of AAB concrete, similar to PC concrete.
5. Ranger with reliefF feature selection outperformed other models with an adjusted R^2 value of 0.793 and RMSE of 8.22.
6. For the selected database of 364 mix proportions, all the variants of Cforest resulted in lower adjusted R^2 and higher RMSE values. Compared to the ranger with reliefF, the RMSE of Cforest is 26.65% higher, and the adjusted R^2 value is 17% lower. Hence, based on the results from the present study, Cforest is not recommended for predicting the compressive strength of AAB concrete.

Owing to the interdependencies of attributes, the reliefF attribute evaluator is recommended to identify the least contributing attribute as it will reduce the prediction accuracy. Ranger, RRF, and RF with reliefF attribute evaluator can efficiently predict the compressive strength of AAB concrete.

References

1. Provis JL (2018) Alkali-activated materials. Cem Concr Res 114:40–48
2. Al Bakri AMM, Kamarudin H, Bnhussain M, Rafiza AR, Zarina Y (2012) Effect of Na_2SiO_3/NaOH ratios and NaOH molarities on compressive strength of fly-ash-based geopolymer. ACI Mater J 109(5):503–508
3. Ramagiri KK, Chauhan DR, Gupta S, Kar A, Adak D, Mukherjee A (2021) High-temperature performance of ambient-cured alkali-activated binder concrete. Innovative Infrastruct Solutions 6(2):1–11
4. Yang KH, Hwang HZ, Lee S (2010) Effects of water-binder ratio and fine aggregate-total aggregate ratio on the properties of hwangtoh-based alkali-activated concrete. J Mater Civ Eng 22:887–896
5. Puertas F, Martínez-Ramírez S, Alonso S, Vázquez T (2000) Alkali-activated fly ash/slag cements. Strength behaviour and hydration products. Cem Concrete Res 30:1625–1632
6. Provis JL, Van Deventer JS (eds) (2013) Alkali activated materials: state-of-the-art report, RILEM TC 224-AAM, Springer Netherlands
7. Bureau of Indian Standards (2020) IS 17452: 2020-Use of alkali activated concrete for precast products—Guidelines, New Delhi, India

8. Khademi F, Akbari M, Jamal SM, Nikoo M (2017) Multiple linear regression, artificial neural network, and fuzzy logic prediction of 28 days compressive strength of concrete. Front Struct Civ Eng 11:90–99
9. Zarandi MF, Türksen IB, Sobhani J, Ramezanianpour AA (2008) Fuzzy polynomial neural networks for approximation of the compressive strength of concrete. Appl Soft Comput 8 (1):488–498
10. Abd AM, Abd SM (2017) Modelling the strength of light-weight foamed concrete using support vector machine (SVM). Case Stud Constr Mater 6:8–15
11. Nazari A, Sanjayan JG (2015) Modeling of compressive strength of geopolymers by a hybrid ANFIS-ICA approach. J Mater Civ Eng 27:1–8
12. Nazari A, Sanjayan JG (2015) Modelling of compressive strength of geopolymer paste, mortar and concrete by optimised support vector machine. Ceram Int 41:12164–12177
13. Pengcheng L, Xianguo W, Hongyu C, Tiemei Z (2020) Prediction of compressive strength of high-performance concrete by random forest algorithm. In: IOP conference series: earth and environmental science, 552, 012020, IOP Publishing
14. Ramagiri KK, Boindala SP, Zaid Md (2021) Compressive strength of alkali-activated binder concrete cured at ambient temperature. Mendeley Data V1
15. Breiman L, Friedman J, Stone CJ, Olshen RA (1984) Classification and regression trees. CRC press
16. Deng H, Runger G (2012) Feature selection via regularised trees. In: The 2012 international joint conference on neural networks, pp 1–8
17. Wright MN, Wager S, Probst P (2013) Ranger: a fast implementation of random forests. R package version 0.5. 0
18. Ishwaran H, Kogalur UB (2014) Random forests for survival, regression and classification (RF-SRC). R package version 1.6
19. Hothorn T, Hornik K, Zeileis A (2006) Unbiased recursive partitioning: a conditional inference framework. J Comput Graph Stat 15:651–674
20. Witten IH, Frank E, Trigg LE, Hall MA, Holmes G, Cunningham SJ (1999) Practical machine learning tools and techniques with Java implementations, Weka
21. Sage A (2018) Random forest robustness, variable importance, and tree aggregation. PhD Thesis, Iowa State University, Iowa
22. Strobl C, Boulesteix AL, Zeileis A, Hothorn T (2007) Bias in random forest variable importance measures: Illustrations, sources and a solution. BMC Bioinf 8(1):1
23. Bolón-Canedo V, Sánchez-Maroño N, Alonso-Betanzos A (2013) A review of feature selection methods on synthetic data. Knowl Inf Syst 34(3):483–519
24. Urbanowicz RJ, Meeker M, La Cava W, Olson RS, Moore JH (2018) Relief-based feature selection: introduction and review. J Biomed Inform 85:189–203

Dynamic Analysis of Tall Building with Cross, Diagonal, V Type and Inverted-V Bracing

Ch. Chaitanya Manikanta, S. Elavenil, and V. Vasugi

Abstract The enhancement of reinforced concrete structures against horizontal loading is done by adopting steel bracing system. Steel bracing incorporates a potential advantage over other plans like higher quality and stiffness, economical, less space, includes much less weight compared to the existing structure. In this study, analytical investigation is carried out using ETABS software on a R.C irregular building with distinctive types of bracing (Diagonal, V type, inverted V type, X type) the response spectrum method is carried out for the analysis of structures. The buildings are located in the zone III region. The main parameters considered in this paper are lateral displacement, story drift, axial force and base shear. It is concluded that the maximum story displacement of the building is reduced by 45% by using X and Inverted V type bracing compared to the normal structure. The displacement decreases from the top story to the base, and the story shear is reduced by 8% by using X and Inverted V braced buildings compared to other types.

Keywords Dynamic analysis · Bracings · Story shear · Story drift · Displacement · Stiffness

1 Introduction

India's population is increasing day by day, mainly in cities and towns, but in present conditions, we don't have more arrival sources in cities and towns. So we need to develop the construction of high-story buildings and tall structures in major cities [1].

Earthquakes are natural disasters so we cannot stop them. When an earthquake happens, harm will cause higher in tall building structures compared with multi-story buildings [2, 3]. So we need to provide extra safety for the tall buildings

Ch.Chaitanya Manikanta (✉) · S. Elavenil · V. Vasugi
School of Civil Engineering, VIT Chennai, Chennai, India
e-mail: chaitanya.manikanta2019@vitstudent.ac.in

G. C. Marano et al. (eds.), *Proceedings of SECON'21*, Lecture Notes in Civil Engineering 171, https://doi.org/10.1007/978-3-030-80312-4_62

to carry the external loads on buildings. By utilizing the bracing system in tall buildings, extra safety can be achieved for building against the external loads [4, 5], the performance of bracing is to provide strength and stability of the building and resist lateral loads [6], either from steel members or from a concrete 'core'. For the bracing system, beams and columns are arranged as they were to support the vertical load since the bracing system got to carry all horizontal loads [7]. Bracings are different sorts like Cross bracing, diagonal bracing, knee bracing, K-type bracing, V-type bracing, inverted-V bracing etc. Selection of bracing system is important for the stiffness of structure [3, 8]. The strength and stiffness of a braced building mainly depends on the bracing type and portion of the bracing adopted in the buildings [9].

In this present work, analytical investigation is performed on the dynamic analysis of 50 storied R.C irregular building structures with bare frames and four different bracing systems (Cross, Diagonal, V, and Inverted-V) [10, 11].and the results are compared for all five different building structures. By performing the Response Spectrum Method [12], conclusion is arrived for the most suitable bracing, from the maximum story drift, shear, displacement, and building stiffness [13].

2 Building Data

In this research work, modeling, and analysis are carried out for the fifty-storied reinforced concrete vertical irregular building structure with Cross, Diagonal, V and Inverted V typed bracings [14, 15]. ETABS Software is used in this analytical investigation [16–18]. It is a more user-friendly software for designing the braced buildings with different types of bracings, and to provide bracings at any portion of the building structure and can assign bracings quickly by utilizing quick assign tools. It has more shortcuts compared with other software available in the market.

The area of the building is reduced for each ten floors of height. The building information and modeling parameters data are shown in Tables 1 and 2.

In this present work for knowing the dynamic response of building, assumptions are made that 50 storied vertical irregular building is located in Mumbai [19]. The seismic data for the selected location is given in Table 3.

Table 1 Building dimensions for 50 storied R.C vertical irregular building structure

Floor	Area
From base to 10th floor	66 mts × 66 mts
From 11th floor to 20th floor	54 mts × 54 mts
From 21st floor to 30th floor	42 mts × 42 mts
From 31st floor to 40th floor	30 mts × 30 mts
From 41st floor to 50th floor	18 mts × 18 mts

Table 2 Preliminary data for 50 storied R.C vertical irregular building structure

S. no	Variable	Data
1	No of storey	50
2	Building location	Mumbai
3	No of bay in X- direction	12
4	No of bays in Y-direction	12
5	Length along x direction	6 mts
6	Length along y direction	6 mts
7	Height of each floor	3 m
8	Size of beam	300 × 500 mm
9	Size of column	600 × 600 mm
10	Thickness of slab	200 mm
11	Grade of concrete for slab	M40
12	Grade of steel	Fe 500
13	Type of bracings	Cross, Diagonal, V and Inverted V
14	Grade of Steel for bracings	Fe 345
15	Load Cases	D.L, L.L,EQ X,EQ Y,RSP X and RSP Y

Table 3 Seismic data for 50 storied R.C building structure

S. no	Model description	
1	Building located in zone	3
2	Zone factor	0.16
3	Soil type	2 (medium soil)
4	Response reduction factor (R)	4 (building with ordinary braced frame)
5	Impact factor	1.2
6	Damping ratio	5%

3 Modeling and Analysis

A total of five different models are analyzed with and without bracings [20]. The height of the building in the Z-direction is 153 m with each floor height 3 m. The length of the building in X-direction and Y-direction is 66 m.

The geometrical configuration of the building structure in the Plan view and 3D view of 50 storied RC building are shown in Figs. 1 and 2.

The four different bracing systems (Cross, Diagonal, V, and Inverted-V) are provided for 50 storied vertical irregular buildings. The bracing system is a vital parameter in braced buildings. Since in tall building structures wind loads and earthquake loads will cause more effect at building corners compared to the middle portion, hence the provision of good strength at building corners become necessary to bare the external loads on building structure. In this research work, steel bracing system is provided at the corners of the building structure as per IS 800-2007 [21, 22]. Figures 3, 4, 5 and 6 shows the bracing system in the elevation view up to the 10th floor.

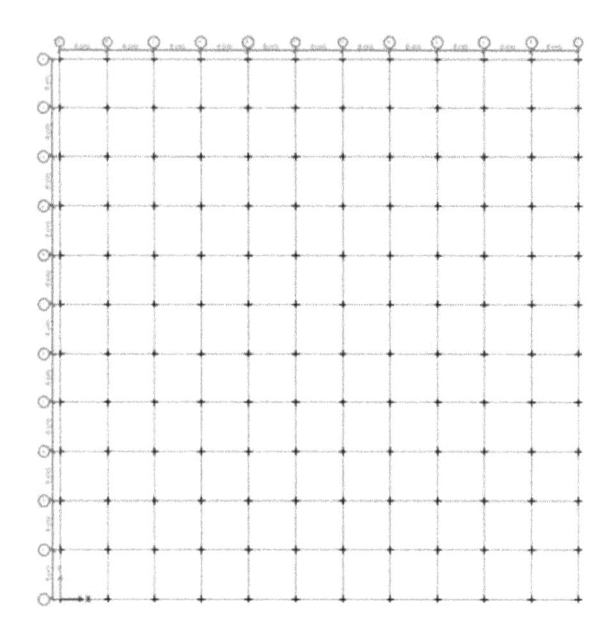

Fig. 1 2D plan view of 50 storied R.C building

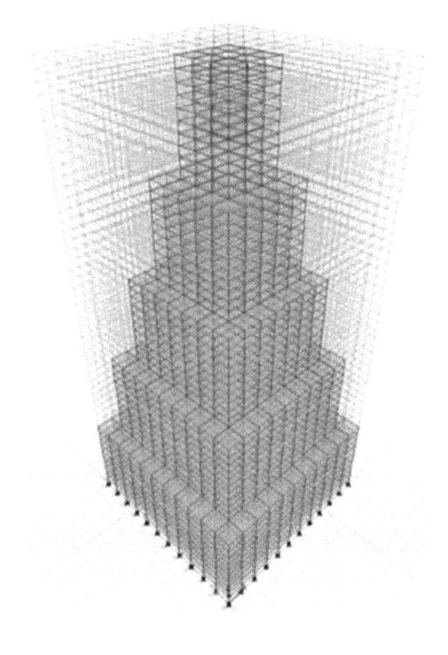

Fig. 2 3D View of 50 storied R.C building With Bare Frame

4 Results and Discussion

The results are evaluated for the different types of bracings such as maximum displacement, shear, story drift, and story stiffness. Figures 7, 8 shows the storey displacement in the X and Y direction. The maximum storey displacement of the

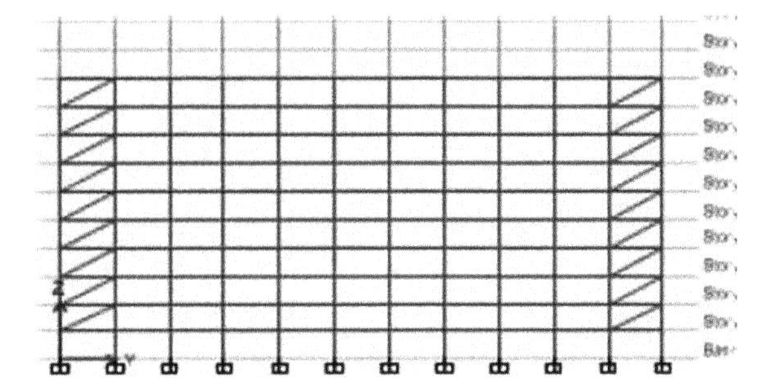

Fig. 3 Elevation—A in Cross type braced building

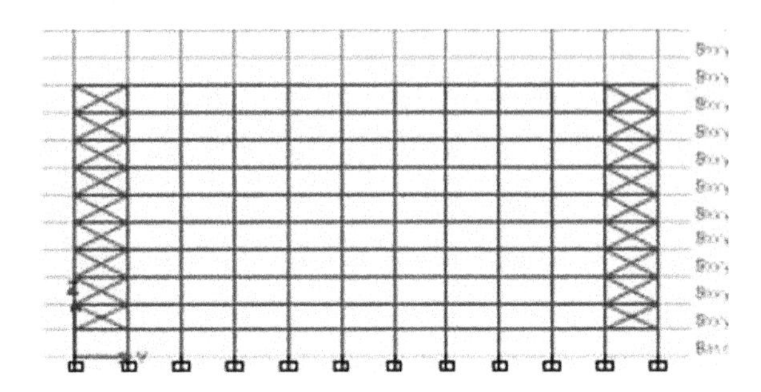

Fig. 4 Elevation—A in Diagonal type braced building

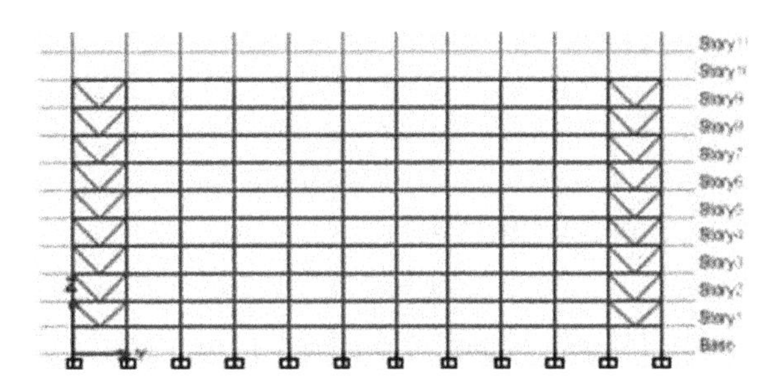

Fig. 5 Elevation—A in V-type braced building

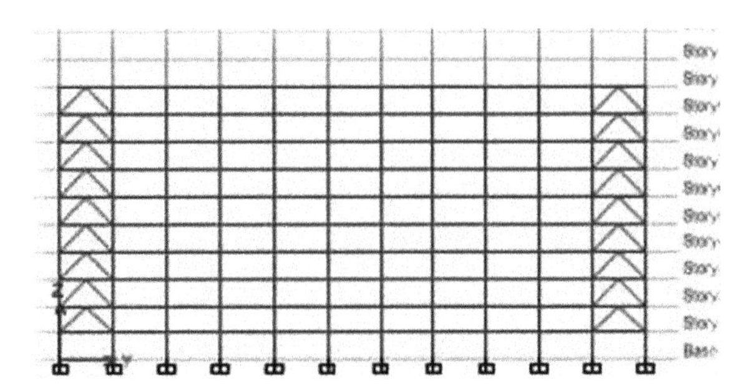

Fig. 6 Elevation—A in Inverted-V type braced building

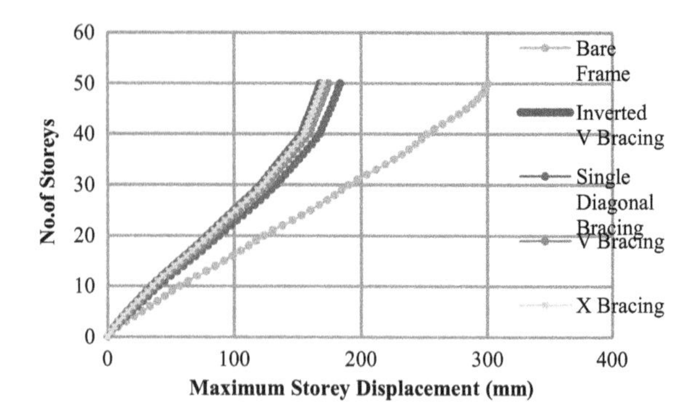

Fig. 7 Storey displacement in X-direction

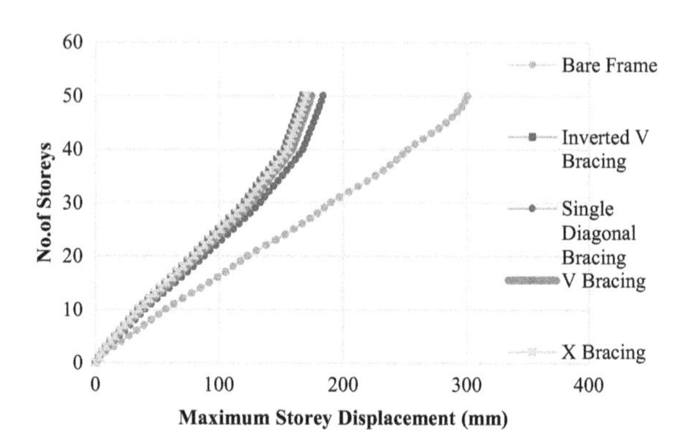

Fig. 8 Storey displacement in Y-direction

building is reduced by 45% by using Cross and Inverted V type bracing compared to the normal structure.

4.1 Storey Displacement

According to Indian standard codes IS: 1893-2016 and IS 456-2000, the maximum permissible story displacement is H/500, where H is the total height of the building structure. In this work our building height H = 153 m, so 153/500 = 306 mm. we got less than 306 mm displacement in all 5 different building structures. Hence safe.

From Table 4, the lateral displacement is decreased to the largest extent in Cross and Inverted V typed bracing buildings, whereas the displacement is more in the bare frame building.

4.2 Storey Drift

Figures 9 and 10 shows the storey drift in the X and Y direction. Storey-drift is higher at intermediate storey levels and lower at the top storey.

Table 4 Displacement of storey in X and Y direction for different types of bracings

Frame	RSP-X (mm)	RSP-Y (mm)
Bare frame	300.30	300.30
Cross bracing	169.93	170.01
Diagonal bracing	183.56	183.56
V bracing	174.11	174.31
Inverted V bracing	167.61	167.77

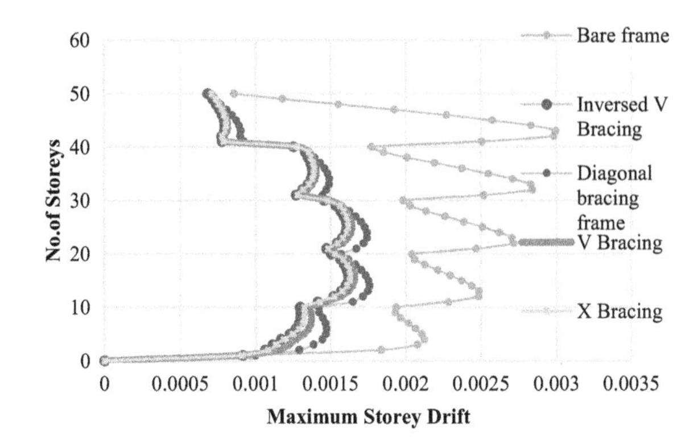

Fig. 9 Storey drift in X-direction

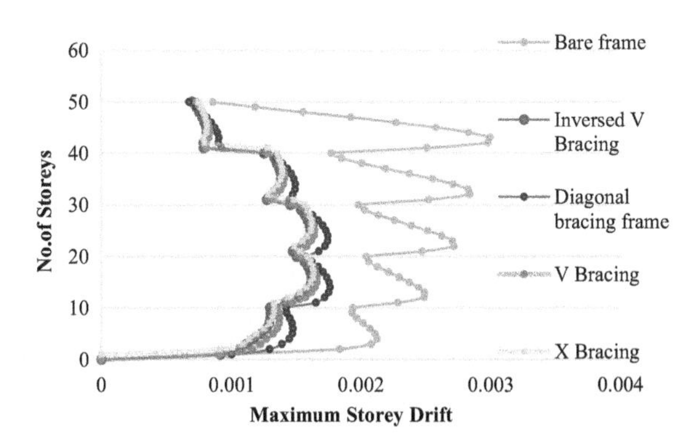

Fig. 10 Storey drift in Y-direction

From IS: 1893-2016 Clause 7.11.1 the maximum permissible story drift is $0.004 \times h$, where h is the height of the story. Here the building storey height h = 3 m so $0.004 \times 3 = 0.012$. We got less than 0.012 storey drift in all 5 different building structures. Hence safe.

From above graphs it is observed that the story drift is Minimum in Inverted V and Cross-typed braced buildings compared to other, and the maximum story drift occurs in diagonal braced and Bare Frame buildings. The maximum drift occurs in middle stories.

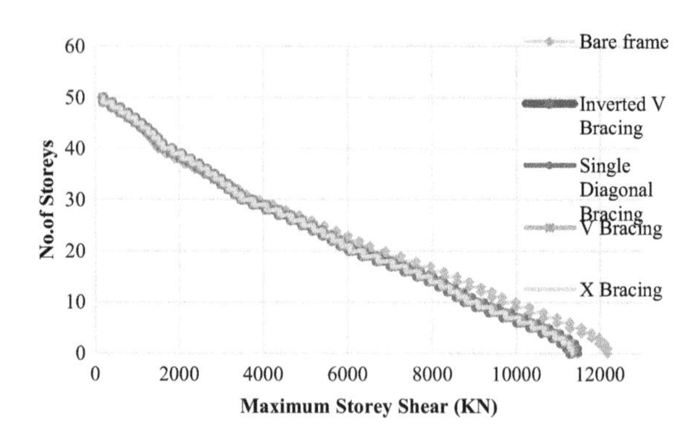

Fig. 11 Storey shear in X-direction

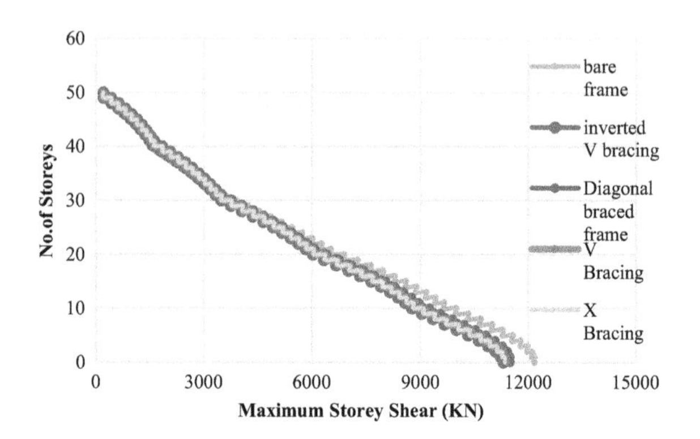

Fig. 12 Storey shear in Y-direction

4.3 Storey Shear

Figures 11 and 12 shows the storey shear in the X and Y direction. Storey shear is reduced by 8% by using X and Inverted V braced buildings compared to others. The maximum storey shear occurs for bare frame and diagonal braced buildings.

From Table 5 it is evident that story shear is minimum in Cross and Inverted-V braced buildings compared to others, and the highest story shear happens in bare frame and diagonal braced buildings. Moreover, the value of shear decreases from bottom to top story, the maximum story shear happens at the ground floors.

4.4 Storey Stiffness

Figures 13 and 14 shows the storey stiffness in the X and Y direction. Storey stiffness is increased by 13% by using X and Inverted-V type bracings.

From Table 6 it is observed that Storey stiffness shows higher in Cross and Inverted-V type braced buildings compared to other bracings.

Table 5 Storey shear values from above graphs

Frame	RSP-X (KN)	RSP-Y (KN)
Bare frame	12,174.96	12,174.96
Cross bracing	11,335.84	11,336.88
Diagonal bracing	11,484.65	11,484.65
V bracing	11,375.34	11,378.17
Inverted V bracing	11,288.81	11,291.04

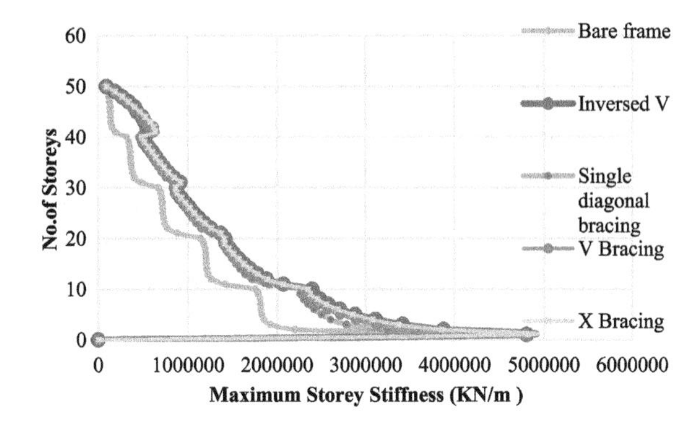

Fig. 13 Storey stiffness in X-direction

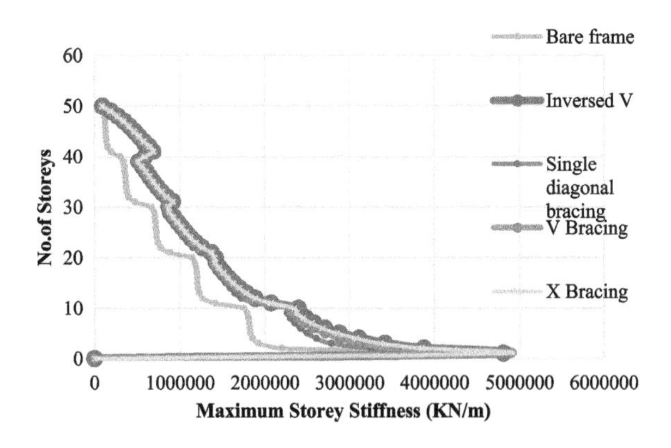

Fig. 14 Storey stiffness in Y-direction

Table 6 Storey stiffness values from above graphs

Frame	RSP-X (KN/m)	RSP-Y (KN/m)
Bare frame	4,225,372.6	4,225,373.1
Cross bracing	4,816,585.1	4,816,446.4
Diagonal bracing	4,397,033.4	4,397,033.4
V type bracing	4,779,652.2	4,779,144.1
Inverted-V bracing	4,806,242.7	4,805,948.6

5 Conclusion

The dynamic analysis of tall structure with vertical irregular building having Cross, Diagonal, V and Inverted-V type bracings were investigated and the following conclusions are arrived based on the dynamic analysis of five different building structures.

- The maximum storey displacement is decreased by 45% by using Cross and Inverted-V type bracing compared to the normal structure. Displacement value decreases from the top storey to base.
- Storey shear is reduced by 8% by using X and Inverted V braced buildings compared to others. The maximum storey shear occurs for bare frame and diagonal braced buildings.
- Storey drift is higher at intermediate storey levels and lower at the top storey.
- Storey stiffness is increased by 13% after using X and Inverted-V type bracings. It is observed that storey stiffness is higher in Cross and Inverted V type bracing.,

From the analytical investigation, it is concluded that the buildings with X type and Inverted-V type bracing are found to be most effective.

References

1. Xiaohu M, Zaigen Mu, Guoyue L (2014) Study on seismic performance of a super high-rise building based on the rare earthquake. In: Fifth international conference on intelligent systems design and engineering applications 978-1-4799-4261-9/14
2. Eber Alberto G, Arturo T, Luis Eduardo P (2012) Case studies on the seismic behavior of reinforced concrete chevron braced framed buildings. Eng Struct 45:78–103
3. Kadid A, Yahiaoui D (2011) Seismic Assessment of Braced RC Frames. Procedia Eng 14:2899–2905
4. Hendramawat A, Kristiawan SA, Basuki A (2013) Evaluation of the use of steel bracing to improve seismic performance of reinforced concrete building. Procedia Eng 54:447–456
5. Sabouri-Ghomi S, Ebadi P (2008) Concept improvement of behavior of X-bracing systems by using easy-going steel. In: The 14th world conference on earthquake engineering, Beijing, China
6. Mahmoud Maheri R, Akbari R (2003) Seismic behavior factor, R, for steel X-braced and knee-braced RC buildings. Eng Struct 25:1505–1513
7. Priyanka T, Shilpa VB (2016) Effect of bracing systems on storey height by using response spectrum analysis. Int Res J Eng Technol 3(7)
8. Preethi B, Mallika A (2015) Comparative study of different bracing patterns for different plan irregularities of the buildings. J Civil Eng Environ Technol 2(14):36–41
9. Siva Naveen D, Nimmy Mariam A, AnithaKumari S (2019) Analysis of irregular structures under earthquake loads. Procedia Struct Integrity 14:806–819
10. AkhilaLekshmi NH, Aswathy Kumar S (2015) Dynamic analysis of an irregular RC building with different bracing systems. Int J Sci Res, Index Copernicus Value (2013): 6.14, Impact Factor (2015): 6.391

11. Made Sukrava (2017) Staged analysis of RC frame retrofitted with steel braces in low and medium-rise buildings. Procedia Eng 171:1002–1009
12. Dhruvil Y, Piyush Jain, Dr. Patel VR (2017) Response spectrum analysis of tall building having different structural system. Int J Adv Eng Res Dev 4(5)
13. Brunesi E, Nascimbene R, Casagrande L (2016) Seismic analysis of high-rise mega-braced frame-core buildings. Eng Struct 115:1–17
14. IS 800:1984: Indian Standard Code of Practice for General Construction, in Steel (first revision)
15. IS 456: 2000 (2000) Indian Standard Code of Practice for plain and reinforced Concrete, Bureau of Indian Standards, New Delhi
16. Reddy APK, Kumar MP (2017) Analysis of G+30 high-rise buildings by using Etabs for various frame sections in zone IV and zone V. Int J Innovative Res Sci Eng Technol 6(7)
17. Harshitha MK, Vasudev MV (2018) Analysis of R.C framed structure with structural steel braces using Etabs. Int Res J Eng Technol 5(1)
18. Khaleel MT, Kumar D (2016) Seismic analysis of steel frames with different bracings using ETSBS software. Int Res J Eng Technol 3(8)
19. IS 1893 (Part-1) (2016) Indian standard criteria for earthquake resistant design of structures. Bureau of Indian Standards, New Delhi
20. Grishma M, Hardik J (2017) Comparative seismic performance and evaluation of X-bracing and off-diagonal bracing systems in RC buildings. Nat Conf Recent Res Dev Core Disciplines Eng
21. Anitha M, Divya K (2015) Study on seismic behavior of knee braced steel frames. Int Res J Eng Technol 2:40–45
22. Massumi A, Tasnimi AA (2008) Strengthening of low ductile reinforced concrete frames using steel X bracings with different details. In: World conference on earthquake engineering, Beijing, China

Flexural Torsional Buckling Behaviour of I-Section Beams with Longitudinally Profiled Flanges

S. Sukanya and B. Rajeevan

Abstract Longitudinally profiled flanged I-beams are having flanges with varied thickness along the length of the beam. Since the flanges offer most of the moment of resistance in an I section, the provision of varied thickness of flange along the span is found to be more efficient. Various studies have been conducted on beams with lateral bracing systems that prevents flexural torsional buckling. The bracing system may not exist in all cases and hence the beam may be subjected to flexural torsional buckling. Members under compression are subjected to combined bending and twisting, accompanied with local buckling of members. In the present study, a finite element model was developed to examine the performance of steel I-beams with longitudinally profiled flanges under flexural torsional buckling. The combined effect of bending and torsion develops bending about the minor-axis due to torsional rotation. The key factors affecting the moment carrying capacity under flexural torsional buckling were found to be the sectional geometry, namely, flange width-to-thickness ratio, ratio of depth of beam to thickness of web and span of the beam. The paper highlights the influence of these parameters in more detail. It was observed that by varying the depth-to- thickness ratio from 42 to 125 that the moment capacity increased. The flange width to thickness ratio was varied from 8 to 15, the span was differed from 1 to 6 m, which decreased moment capacity.

Keywords Buckling · Flexure · Torsion · Profile · Beam · Steel · Flange

S. Sukanya (✉) · B. Rajeevan
Department of Civil Engineering, Government College of Engineering Kannur, Kannur, India
e-mail: 19m525@gcek.ac.in

B. Rajeevan
e-mail: rajeevan@gcek.ac.in

G. C. Marano et al. (eds.), *Proceedings of SECON'21*, Lecture Notes in Civil Engineering 171, https://doi.org/10.1007/978-3-030-80312-4_63

1 Introduction

The construction industry aims at developing structures with maximum efficiency and minimum material. The field of structural engineering has witnessed tremendous developments in this regard over the years. One such development is the use of beams with tapered webs [1–3]. The beams with tapered webs distribute stresses evenly throughout the beam. Buckling is more significant in profiled beams with flatter and wider sections of profile when subjected to axial loading. The provision of stepped beams leads to uniform stress distribution in a beam [4]. Steps are provided at regions of higher moment, to resist moment with minimum amount of steel. Buckling moments were found to be more dependent on flange thickness than on flange width.

Sudden buckling may occur in a member leading to its lateral displacement, in addition to twisting of the member about its axis. This is referred to as flexural torsional buckling [5]. The design of a steel member is based on the elastic critical buckling moments when unrestrained, thus highlighting the importance of analysis with regard to flexural-torsional buckling. Numerous studies on lateral buckling of steel beams and the factors that affect buckling were done [6–10]. The studies found that the existing equations for estimating buckling moment needs modification so that the elastic critical moment from the experiments are closer to that obtained from the existing equations. There are various factors that affect the response of an element when subjected to flexural-torsional buckling. The influence of these factors, namely the geometry, support condition [11], length of the support, type of stiffeners and bracings [12–14], end restraints [15, 16], geometrical imperfection [17], residual stresses [9] were studied and the equations were modified considering these factors.

Innovations in the type of material used lead to the addition of other metals [18]. The usage of steel along with other metals or ceramic improves the strength of the element, which in turn increases the buckling load. The type of material and the amount of material affect flexural-torsional buckling. A model with non-uniform cross-section had much more buckling capacity than a section with uniform cross section [19]. The longitudinally profiled beam has non-uniform cross section, as the flange thickness varies throughout the span [20]. The moment in a beam being nonuniform throughout, the varying thickness leads to economy by restricting the amount of steel required, when compared to a beam of uniform cross-section throughout. The longitudinally profiled flanged beam could be used instead of the conventional I-section beams with uniform flange thickness for the construction of roads and bridges. The flexural strength of beams with longitudinally profiled flanges [21] were analyzed, providing lateral bracing in the beam section to prevent flexural torsional buckling of beam. The beams deform when stressed and this behavior of longitudinally profiled beams under serviceability limit state was also evaluated [22]. Past studies were mainly focused on developing equations for the moment and shear of longitudinally profiled beams. Studies on flexural torsional buckling and the effect of geometrical parameters in flexural torsional buckling for these beams are limited in number. In this paper, a longitudinally profiled beam was modeled using ANSYS, a finite element modelling and analysis (FEM) software. The paper presents flexural torsional buckling behavior of longitudinally profiled flanged beam and the geometrical factors that influence its behavior.

2 Validation

2.1 Experimental Model

The model was validated using the data from [20]. The total span of the beam was 6000 mm. The flange thickness at both ends were 15.5 mm and varied linearly up to the midspan where a thickness of 32 mm was provided. The beam was simply supported and subjected to three-point loading. The cross-section of the longitudinally profiled beam is shown in Fig. 1 and the cross-sectional dimensions of the modeled beam are given in Table 1.

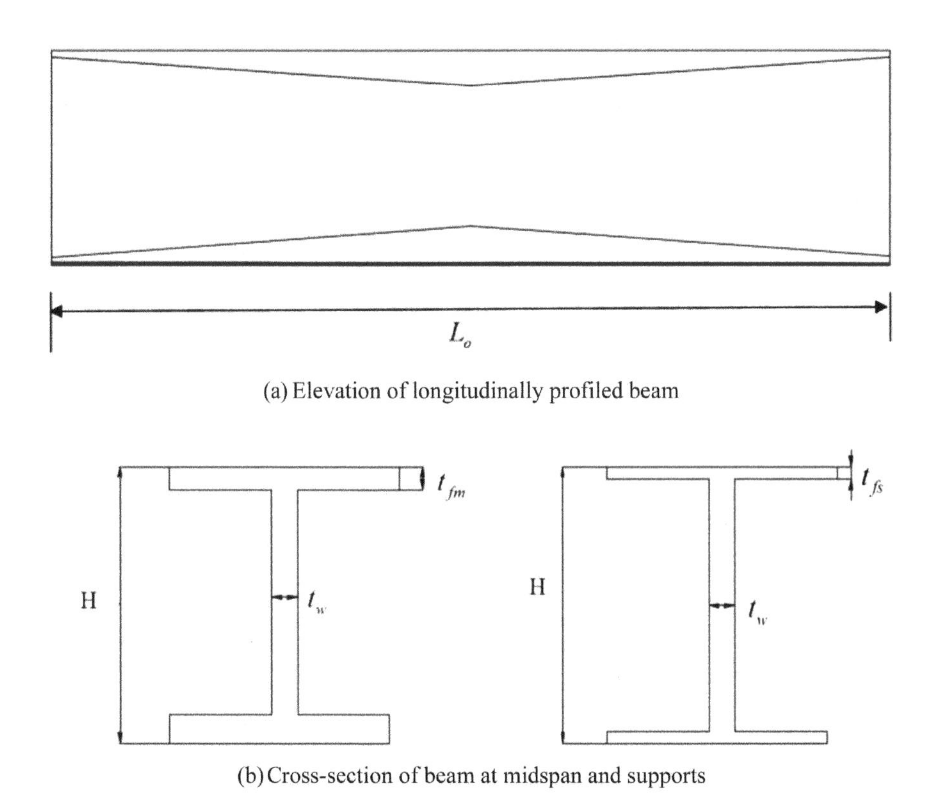

Fig. 1 Model of beam with dimensions [20]

Table 1 Cross-sectional dimensions of beam

Parameter	Dimension (mm)
Overall Length (*Lo*)	6000.5
Shear span (*Ls*)	5600.5
Overall Depth (*H*)	500.5
Flange Width (*wf*)	202.5
Flange thickness at end of span (*tfs*)	15.5
Flange thickness at midspan (*tfm*)	32
Web thickness (*tw*)	12

2.2 Finite Element Model and Analysis

Multilinear isotropic hardening was used to incorporate the material property in the linear as well as nonlinear range with available tensile-coupon test results [20]. All the sections of the beam were modelled using 8-noded 3D element SOLID185. It is generally adopted for steel elements to capture the flexural behavior accurately. Figure 2 depicts the SOLID185 element with its degree of freedom.

Two supports were provided at 200 mm from both the ends, hence resulting in a beam with overhangs on both sides. The beam was subjected to 3-point loading, as shown in Fig. 3. The same condition was stimulated by applying a displacement of 29.55 mm at the point of loading [20]. The mesh size was altered to determine the size of mesh which yield converging results for moment versus beam-end rotation curve. For a mesh size of 30–35 mm, the results were similar and converging. The results of experiment and that of the finite element analysis is graphically represented in Fig. 4.

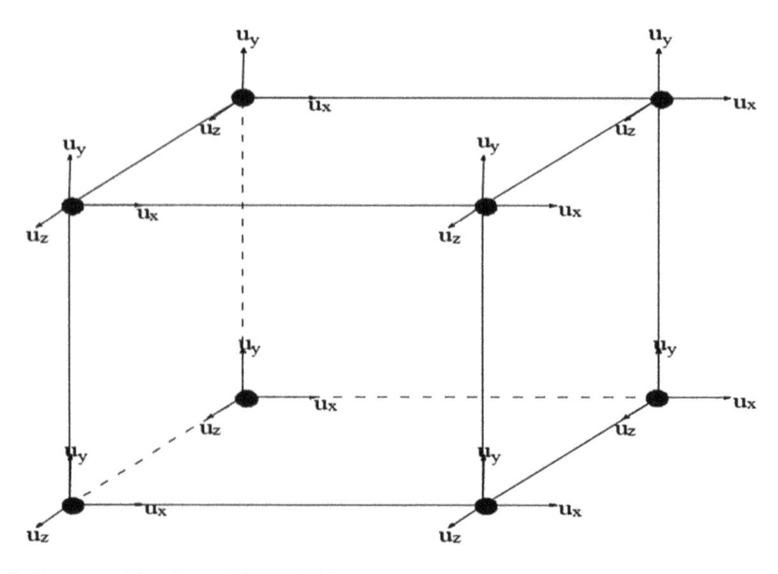

Fig. 2 Degrees of freedom of SOLID185

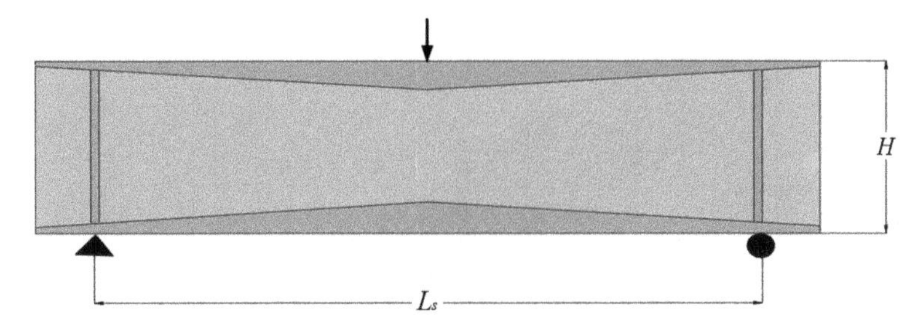

Fig. 3 Support conditions and midspan loading

Both the experimental and FEA results is in close agreement up to the point of yielding. The stress-strain characteristics and the modulus of elasticity vary with the thickness of the flange along the span of the beam. This cannot be adequately captured in ANSYS. The longitudinally flanged beam, having varied flange thickness, would mean that the material property varies throughout the span. The deviation in the moment-rotation curve of the finite element analysis and the experimental curve is attributed to this fact.

3 Flexural Torsional Buckling

The normalized moment-beam end rotation curves were initially determined by preventing flexural torsional buckling. The moment capacity of the beam of the beam is reduced due to flexural torsional buckling. The various factors that govern the flexural torsional buckling of these beams were also studied in detail. Non-linear

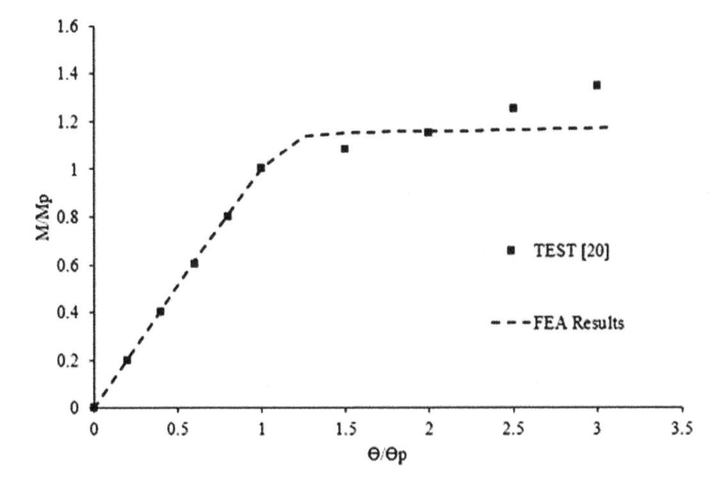

Fig. 4 Moment versus beam end rotation curves of longitudinally profiled flanged beam

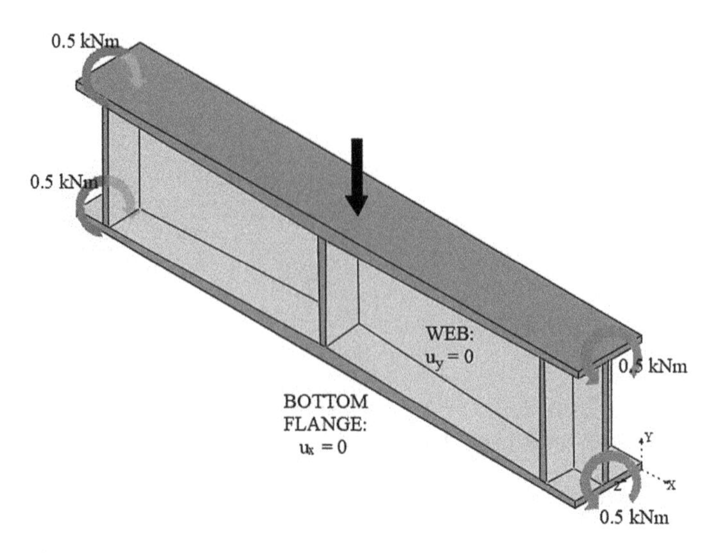

Fig. 5 Loading and support conditions

structural steel was used to model the material. The dimensions of the beam are shown in Table 1 and is schematically represented in Fig. 1. For validation, simply supported ends were provided and the displacement at the top flange was restrained, hence preventing flexural torsional buckling. Further, to investigate the response of beam to flexural-torsional buckling, the following support conditions were provided: horizontal displacement of the flanges at the bottom and vertical displacement of web were prevented. Moment equal in magnitude (equal to unity), but opposite in direction was applied at the two ends of the beam along with displacement at midspan. The loading and support conditions of the beam are schematically represented in Fig. 5 and that modeled in ANSYS is shown in Fig. 6. The deformed shape of the longitudinally profiled beam with the magnitude of deformation is shown in Fig. 7. The mode of failure by both flexure and torsion is also clearly depicted by the figure.

The load-strain relationship of the web, the top and bottom flanges subjected to flexural torsional buckling obtained by analysis in ANSYS is shown in Fig. 8. The load-strain was linear until the ultimate load was reached for the top flange, web, and the bottom flange. The negative and the positive value of strain indicates compression and tension, respectively. Upon reaching the ultimate load, the strain in the top flange increased rapidly with load. This is due to the lateral displacement occurred due to flexural torsional buckling. Such a load-strain relationship is not obtained for the web and bottom flange since it is the top flange that is primarily subjected to lateral displacement, not the bottom flange.

Fig. 6 Loading and support conditions modeled in ANSYS

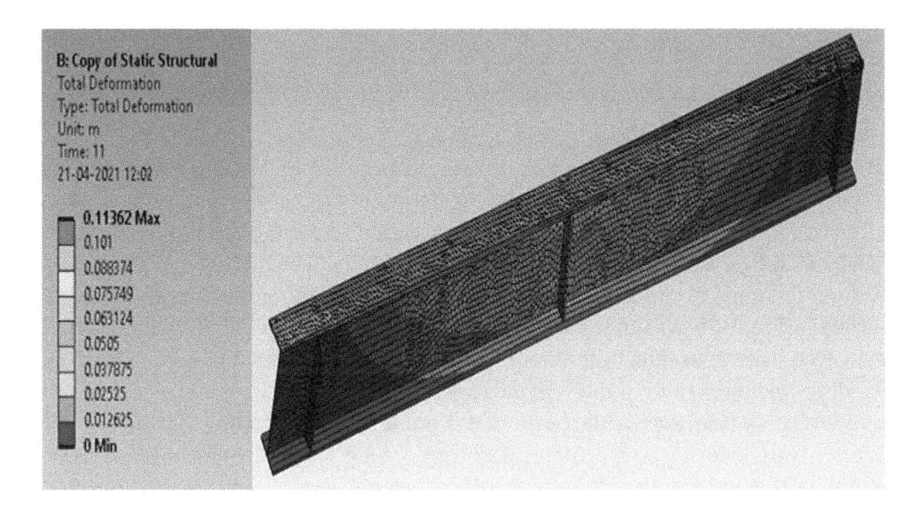

Fig. 7 Deformed shape of longitudinally profiled flanged beam

4 Parametric Study

The effect of depth-to-thickness ratio of the web (hw/tw), width-to-thickness ratio of the flange (bf/tf) and the total span (Lo) on flexural-torsional buckling of longitudinally profiled beams were studied. By altering the thickness of the web, the effect of depth-to-thickness ratio was investigated from a range of 42–125. The width to thickness ratio of flange was varied from 8 to 15, where the thickness refers to thickness of the flange at the midspan. The spans were varied from 1 to 6 m and their effect has also been studied.

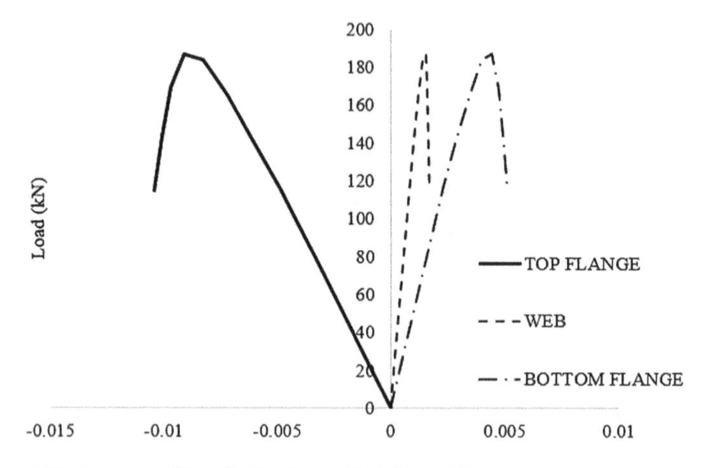

Fig. 8 Load-Strain curve of longitudinally profiled flanged beam

4.1 *Effect of Depth-To-Thickness Ratio of Web (hw/tw)*

The web depth to thickness ratio were varied from 42 to 125. The web depth to thickness ratio of 42 was adopted from the design of critical section of beam for the ultimate moment. The depth of the web was varied as per the codal provisions and hence the ratios were so obtained. The influence of depth-to-thickness of the web on flexural torsional buckling at constant width and thickness of the flange and span is shown in Fig. 9. It was observed that the plastic moment has slightly increased as the web depth-to-thickness ratio was increased from 42 to 125. The depth-to-thickness ratio of the web decreases as the web thickness increases. As the web thickness is increased, the beam is less prone to local buckling than beams with thinner webs and hence the moment corresponding to yield point increases. The normalized value of moment to the plastic moment is hence reduced. It is observed that the ratio of moment to plastic moment increased by 17% as the web depth-to-thickness ratio was decreased from 42 to 125.

4.2 *Effect of Width-To-Thickness Ratio of Flange (bf/tf)*

The effect of width-to-thickness ratio of flange was evaluated by differing the width of the flange and is shown by Fig. 10. The flange thickness refers to the thickness of the flange at midspan wherein the moment and hence the thickness is greater than at the ends. By varying the ratios from 8 to 15 and keeping the web thickness and depth a constant, the plastic moment was reduced with increase in width-to-thickness ratio of the flange. The ratio of 8 was adopted from the design for the critical flange width of beam. The width of the flange was adequately

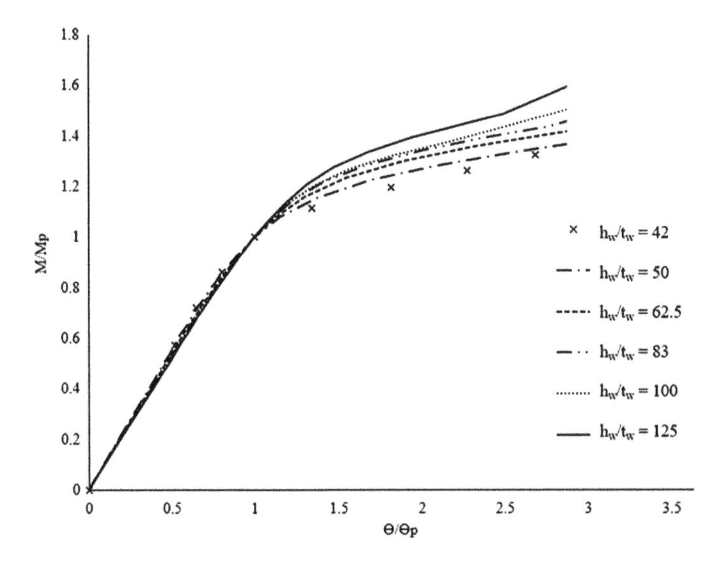

Fig. 9 Effect of depth-to-thickness ratio of web

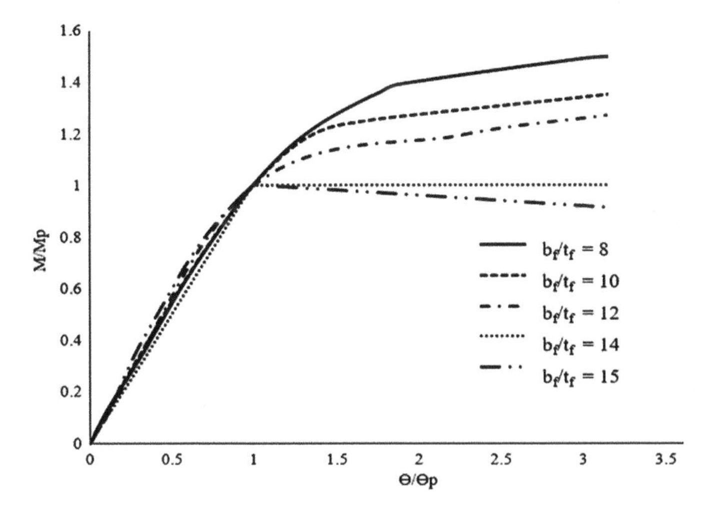

Fig. 10 Effect of width-to-thickness ratio of flange

increased until the flange width was equal to the depth of the web. The moment capacity of the section was reduced after the yield point. As the width-to-thickness ratio of the flange is increased, local buckling is initiated early [23]. The moment capacity of the beam reduced by 66.6% as the width-to-thickness ratio was increased from 8 to 15. Buckling occurring early indicates that the capacity of the beam would be reduced, hence width-to-thickness ratio and moment capacity thus have an inverse relationship.

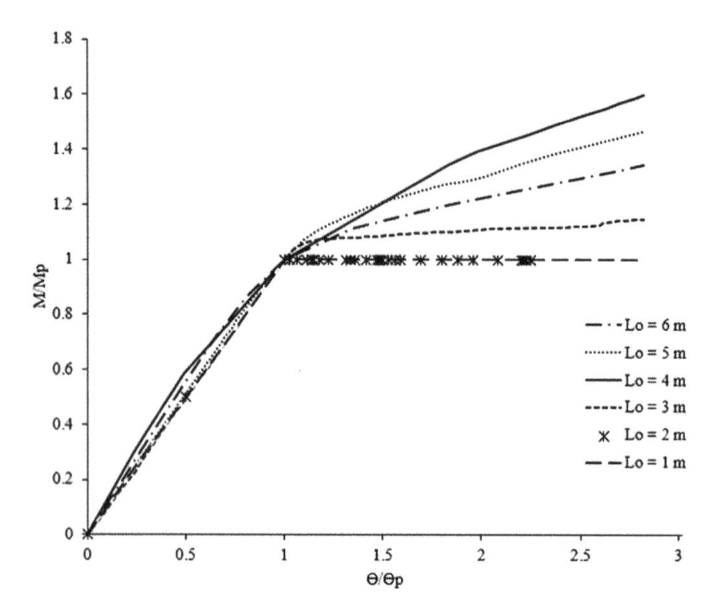

Fig. 11 Effect of span

4.3 Effect of Span

The span of the beams was increased from 1 to 6 m whereas the other cross-sectional dimensions were kept constant. The effect on flexural torsional buckling was analyzed is shown in Fig. 11. As the span of the beam was increased from 1 to 3 m, the moment capacity of the beam rapidly increased by 70%. Beyond a span of 4 m, the capacity of the section was reduced with increase in span by 17.6%. The extent of buckling of beam is highly influenced by the unbraced length of the beams. For beams with larger spans, the flexural torsional buckling becomes more significant with increasing span. Thus, the moment capacity is reduced beyond a span of 4 m.

5 Conclusions

The behavior of longitudinally profiled beams subjected to flexural torsional buckling were studied. A numerical model was developed to validate the data from [20]. A detailed study on flexural-torsional buckling of longitudinally profiled steel beam was done. Parametric studies were also conducted to determine the influence of various factors on flexural torsional buckling of longitudinally profiled beams. The following conclusions were devised:

i. The geometry of the longitudinally profiled flanged beam has considerable effect on the moment capacity.
ii. The top flange is predominantly subjected to buckling than the bottom flange and the web. Thus, the displacement at the top flange increases rapidly with load after yielding.
iii. Increasing the depth to thickness ratio of web increases the moment capacity subsequent to yielding by 17% as the web depth-to-thickness ratio was increased from 42 to 125.
iv. Increasing the span and width-to-thickness ratio of flange decreases the moment capacity of the section after yielding due to flexural torsional buckling mode. The moment capacity of the beam reduced by 66.6% as the width-to-thickness ratio was increased from 8 to 15. Also, as the span of the beam was increased from 1 to 3 m, the moment capacity of the beam increased by 70%. As the span was increased beyond 4 m, the capacity of the section was reduced with increase in span by 17.6%.

References

1. Yuan WB, Kim B, Chen CY (2013) Lateral-torsional buckling of steel web tapered tee-section cantilevers. J Constr Steel Res 87:31–37
2. Tankova T, Martins JP, da Silva LS, Marques L, Craveiro HD, Santiago A (2018) Experimental lateral-torsional buckling behaviour of web tapered I-section steel beams. Eng Struct 168:355–370
3. Zhang L, Tong GS (2008) Lateral buckling of web-tapered I-beams: a new theory. J Constr Steel Res 64(12):1379–1393
4. Park JS, Kang YJ (2004) Flexural torsional buckling of stepped beams subjected to pure bending. KSCE J Civil Eng 8(1):75–82
5. Papangelis JP, Trahair NS, Hancock GJ (1998) Elastic flexural-torsional buckling of structures by computer. Comput Struct 68(1–3):125–137
6. Pi YL, Trahair NS (1994) Inelastic bending and torsion of steel I-beams. J Struct Eng 120(12): 3397–3417
7. Taras A, Greiner R (2008) Torsional and flexural torsional buckling—a study on laterally restrained I-sections. J Constr Steel Res 64(7–8):725–731
8. Bradford MA, Liu X (2016) Flexural torsional buckling of high-strength steel beams. J Constr Steel Res 124:122–131
9. Kang L, Meng L, Lin Y (2020) Experimental and numerical investigation of lateral torsional buckling behavior and capacity of welded Q460 beams. J Constr Steel Res 172:106166
10. Gonçalves R (2019) An assessment of the lateral-torsional buckling and post-buckling behaviour of steel I-section beams using a geometrically exact beam finite element. Thin-Walled Struct 143:106222
11. Vacev T, Nešović I, Milić M, Zorić A, Paunović S. Influence of support conditions on single-span profiled sheet metal strength through nonlinear FEM analysis. Int J Steel Struct 1–17
12. Azhari M, Bradford MA (1993) Local buckling of I-section beams with longitudinal web stiffeners. Thin-walled Struct 15(1):1–13
13. Chen S, Wang X (2012) Finite element analysis of distortional lateral buckling of continuous composite beams with transverse web stiffeners. Adv Struct Eng 15(9):1607–1616

14. Nguyen CT, Joo HS, Moon J, Lee HE (2012) Flexural torsional buckling strength of I- girders with discrete torsional braces under various loading conditions. Eng Struct 36:337–350
15. Valentino J, Pi YL, Trahair NS. Inelastic buckling of steel beams with central torsional restraints. J Struct Eng 123(9):1180–1186
16. Rajkannu JS, Jayachandran SA (2020) Flexural torsional buckling strength of thin-walled channel sections with warping restraint. J Const Steel Res 169:106041
17. Agüero A, Pallarés L, Pallarés J (2015) Equivalent geometric imperfection definition in steel structures sensitive to flexural and/or torsional buckling due to compression. Eng Struct 96:160–177
18. Nguyen TT, Thang PT, Lee J (2017) Flexural torsional stability of thin-walled functionally graded open-section beams. Thin-Walled Struct 110:88–96
19. Ruta GC, Varano V, Pignataro M, Rizzi NL (2008) A beam model for the flexural torsional buckling of thin-walled members with some applications. Thin-Walled Struct 46(7–9): 816–822
20. Liu X, Wang Y, Ban H, Liu M, Veljkovic M, Bijlaard FS (2020) Flexural strength and rotation capacity of welded I-section steel beams with longitudinally profiled flanges. J Constr Steel Res 173:106255
21. Wang Y, Liu X, Ban H, Liu M, Shi Y, Wang Y (2018) Deformation behavior at SLS of welded I-section steel beams with longitudinally profiled flanges. J Constr Steel Res 146: 122–134
22. Wang Y, Xu D, Bu Y, Liu X, Ban H, Liu M (2020) Elastic buckling analysis of rectangular longitudinally profiled (LP) steel plates under uniform compression. Thin-Walled Struct 157:107003
23. Iguchi T, Tabuchi M, Tanaka T, Kihara S (2020) Effects of beam flange width-to-thickness ratio on beam flange fracture caused from scallop root. In: Advances in steel structures (ICASS'02), pp 197–204. Elsevier

Risk Uncertainty Quantification for Various Occupancy Classes Using Stochastic Ground Motion

Leanda J. Payyappilly and Surendra Nadh Somala

Abstract Seismic events are amongst the natural hazards that are known to produce excessive economic losses and human casualties throughout history. Seismic risk can be observed as a function of hazard, exposure and vulnerability. While hazard and exposure are dependent on the site of interest and are unchangeable, studies on vulnerability can help in reducing the seismic risk by adopting retrofitting methods. One of the features that affect the damage and loss characteristics of a building is its occupancy class (commercial, residential, educational, hospitals etc.). The occupancy class of a building mainly affects the number and severity of the casualties as well as the contents damage produced during a seismic event. This work aims to compare the damage and loss properties, i.e., the number of casualties, the economic loss and the damage in properties along with the associated uncertainty of buildings having various occupancy classes by considering a single building model type and subjecting them to a set of stochastic ground motions. All building occupancy classes are according to HAZUS-MH. Normalized losses and injuries are computed using Performance Based Engineering (PBE) workflow developed by Natural Hazards Engineering Research Infrastructure (NHERI).

Keywords NHERI · SimCenter · Uncertainty quantification · Seismic risk

1 Introduction

A better practice in seismic risk assessment is to quantify uncertainty [1] along with the estimation of economic losses and injuries caused due to a seismic event. However, uncertainty quantification in risk assessment has received limited attention in the literature. Occupancy class of a building (commercial, residential, educational, hospitals etc.) is a feature that affects damage and loss characteristics of the building. Studies mainly focus on a particular occupancy class [2] when mul-

L. J. Payyappilly (✉) · S. N. Somala
Indian Institute of Technology Hyderabad, Hyderabad, India
e-mail: ce20mtech12003@iith.ac.in

G. C. Marano et al. (eds.), *Proceedings of SECON'21*, Lecture Notes in Civil Engineering 171, https://doi.org/10.1007/978-3-030-80312-4_64

titude of earthquakes are considered while those considering several occupancy classes are limited in terms of earthquake variability accounted for. This work aims to compare the damage and loss properties, i.e., the number of casualties, the economic loss and the damage in properties along with the associated uncertainty of buildings having various occupancy classes by considering a single building model type and subjecting them to a set of stochastic ground motions. The variability in ground motion is studied using stochastic earthquake simulations [3] for time histories of intensities that are of engineering interest, considering variability in the shallow shear wave velocity as uniform distribution and variability in close to site rupture distance as normal distribution.

Building models belonging to various occupancy class categories have been explored. Residential, commercial, industrial buildings have been considered. Also, categories like agriculture, educational institutes, non-profit buildings like churches are dealt with in this study. Rigorous nonlinear time history analysis performed for the various occupancy classes listed in the HAZUS is done using a large number of stochastically simulated earthquakes and the variability in economic losses and injuries is quantified in a normalized sense.

2 NHERI SimCenter Workflow

The Natural Hazards Engineering Research Infrastructure (NHERI) workflow for Performance Based Engineering [4] integrates the sophisticated time history analysis capabilities of OpenSees (Open System for Earthquake Engineering Simulation) together with uncertainty quantification engine Dakota, developed by the Sandia National Laboratories, USA. This seamless coupling allows us to understand how the variability in structural parameters (storey height, stiffness, hardening ratio etc.) influence the response (inter-storey drift) and risk (losses and injuries). Both HAZUS based [5] and FEMA P-58 based [6] damage and loss computations are available in NHERI tools. However, due to limitation of availability of input parameters for FEMA P-58, we use HAZUS based approach in this study.

2.1 Occupancy Classes

All main categories of occupancy classes i.e., residential, commercial and industrial are considered along with the sub categories. In addition to these other occupancy classes listed in HAZUS, such as agriculture, educational buildings, non-profit buildings are considered in this study. Emergency response services like police stations and fire stations are also considered. The symbolic notation of occupancy classes used in this study is listed in Table 1.

Table 1 Occupancy classes (adopted from HAZUS)

Label	Occupancy class	Example descriptions
Residential		
RES1	Single family dwelling	House
RES3	Multi family dwelling	Apartment/condominium
RES4	Temporary lodging	Hotel/motel
RES5	Institutional dormitory	Group housing (military, college), jails
RES6	Nursing home	
Commercial		
COM1	Retail trade	Store
COM3	Personal and repair services	Service station/shop
COM4	Professional/technical services	Offices
COM5	Banks	
COM6	Hospital	
COM7	Medical office/clinic	
COM8	Entertainment and recreation	Restaurants/bars
COM10	Parking	Garages
Industrial		
IND1	Heavy	Factory
IND2	Light	Factory
IND3	Food/drugs/chemicals	Factory
IND4	Metals/minerals processing	Factory
IND5	High technology	Factory
IND6	Construction	Office
AGR1	Agriculture	
REL1	Church/non-profit	
GOV2	Emergency response	Police/fire station/EOC
EDU2	Colleges/universities	Does not include group housing

Table 2 Building details

Storey height	3.65 m
Number of storeys	3
Plan area	83.6 m^2
Floor weight	667 KN
Storey stiffness	1.7×10^4 KN/m
Yield strength	6.89×10^9 KN/m^2

2.2 Building Details

All the building models used in this study are concrete moment frames. All building models have been limited to symmetric low-rise, designed for moderate code, throughout this work. The details of the building model used in the study are listed in Table 2.

2.3 Uncertainty Propagation

A Latin Hypercube Sampling approach [7] is adopted with a random seed simulating a large number of samples for each random variable used in this work. A multivariate log-normal distribution is fit to the engineering demand parameter to 2000 realizations of structural response. In particular, we explore the uncertainty on the most unconstrained parameter V_{s30}, the top 30 m shear-wave velocity [8] and close-to-site rupture distance and how it influences the damage and loss estimates.

2.4 Finite Element Method

An explicit Newmark's time integration scheme is used for transient analysis solved with unsymmetric multifrontal solver interfacing through scipy, a python library for scientific computation. Rayleigh damping is assumed. OpenSees, the engine behind structural response computation, gives inter-storey drift as well as peak floor accelerations. As the focus of this study is restricted to structural components, we only use the peak inter-storey drift due to base excitation.

3 Stochastic Ground Motions

Stochastic ground motion models are utilised to obtain simulated earthquake motions for analysing the seismic demand of the model building types. Simulation models are formulated to realistically incorporate the features of real earthquake ground motions such as frequency content and time-varying intensity. These characteristics can greatly influence the structural response [9]. Thus, using stochastic models synthetic ground motions can be created for a target seismic event. Two stochastic motion models available in the PBE workflow are Vlachoes et al. [10] and Dabaghi and Kiureghian [11]. The Vlachoes et al. model which is used in this study makes use of regression relations to generate seismic ground acceleration time histories for the site of interest when an earthquake scenario description is provided. The user can enter values for a number of earthquake scenarios describing parameters that are used to generate a seismic event. The various parameters include:

(1) moment magnitude of the earthquake M_w,
(2) closest-to-site rupture distance R_{rup}, and
(3) average shear-wave velocity in top 30 m V_{s30} at the site of interest.

The details of the ground motion used in this study are given in Table 3.

Table 3 Earthquake scenario describing parameters

Parameter	Symbol	Value
Shear-wave velocity in top 30 m	V_{s30}	Uniform distribution
		Min: 100 m/s
		Max: 300 m/s
Moment magnitude	M_w	7
Closest-to-site rupture distance	R_{rup}	Normal distribution
		Mean: 50 km
		Standard deviation: 5 km

4 Results and Discussion

Earthquake loss assessment studies have been conducted using PBE software considering the Hazus Earthquake Model. All structures are subjected to stochastic ground motions. Vlachoes et al. [10] model is used and uncertainty is introduced by varying V_{s30} and R_{rup}.

Figures 1, 2 and 3 show the variation of Peak Interstory Drift (PID), Economic loss and Injury percent, respectively, for the different occupancy class with moderate code design. All models are considered as three storey concrete moment frame (C1L) structures.

4.1 Peak Inter-Storey Drift

The Peak Inter-storey Drift (PID) parameter is used as the Engineering Demand Parameter (EDP). The EDP ranges from as low as 0.0108 to 0.0196. Buildings that provide professional/technical service like offices (COM4) are seen to have the highest median PID as well as variability. Retail trade buildings like stores (COM1) are having the lowest median PID.

4.2 Economic Losses

The economic losses considered in this study are normalized with respect to the repair costs. The lowest median economic losses are observed for structures used for parking purposes like garages (COM10). The highest median economic losses are observed for professional/technical service like offices (COM4). Incidentally, this is also the building type whose PID is considerably high compared to that of other occupancy classes.

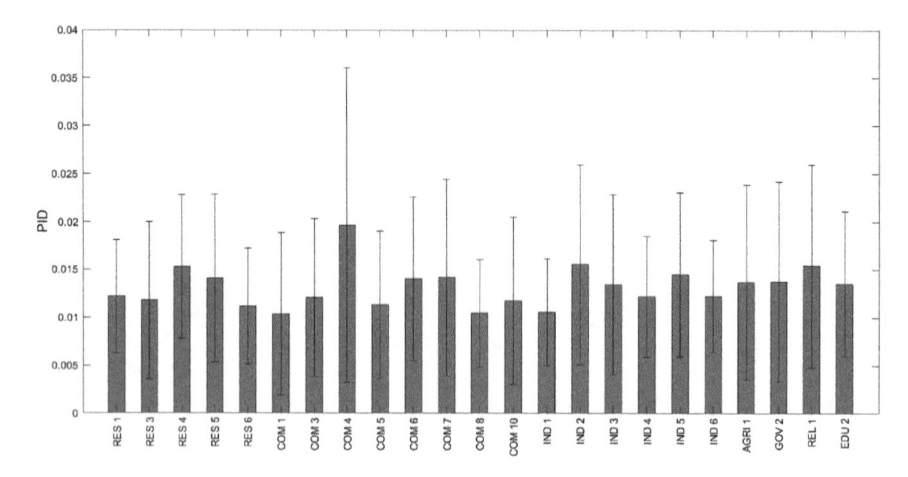

Fig. 1 Variation of peak inter storey drift for the various occupancy classes along with the associated uncertainty

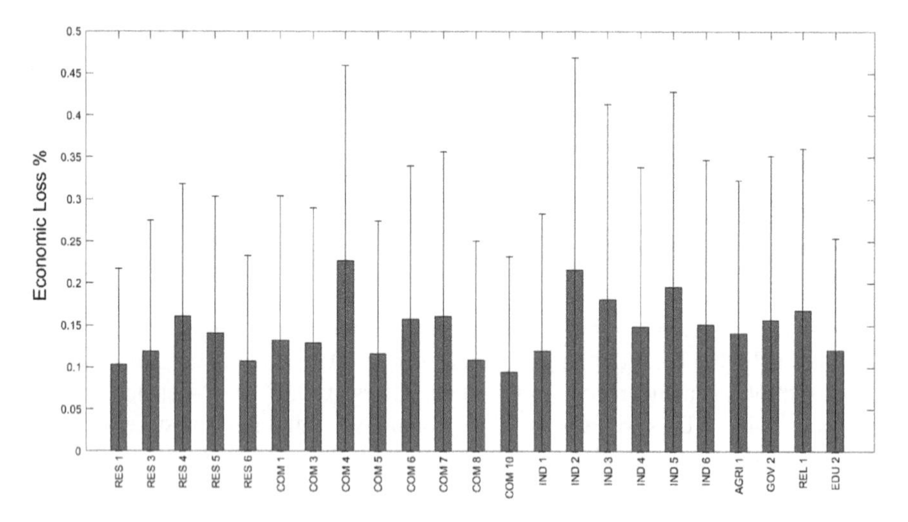

Fig. 2 Variation of economic loss (percentage of replacement cost) for the various occupancy classes along with the associated uncertainty

4.3 Injuries

The percentage of injuries for severity 1 (light injury) are also found to be quite high for professional/technical service like offices (COM4). Interestingly, entertainment and recreational occupancy class which includes restaurants (COM8) are found to have the lowest percentage of injuries.

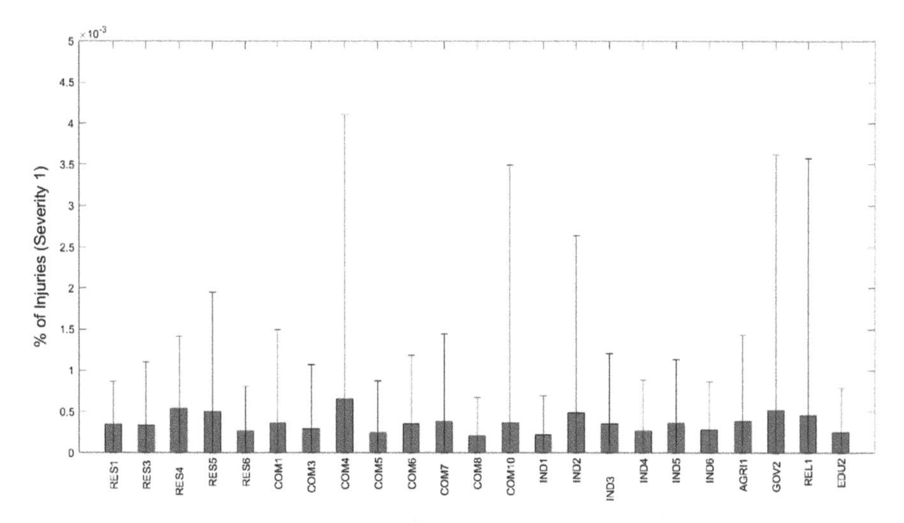

Fig. 3 Variation of percentage of injuries (severity 1) for the various occupancy classes along with the associated uncertainty

5 Conclusion

Our detailed study on rigorous nonlinear dynamic analysis of building having various occupancy classes, considering the uncertainty in building parameters and earthquake loading to understand their influence on seismic risk leads us to the following inferences:

- Among the various building models COM4 (Professional/technical services) has the highest economic loss percent followed by IND2 (Light Factory). COM10 (Parking) has the lowest economic loss percent followed by RES1 (Single family dwelling).
- Injury percent also follows a similar trend with COM4 having the highest percent followed by IND2. COM8 (Entertainment and recreation) has the lowest percent.
- PID also follows similar trend COM4 having the highest drift followed by IND2.
- Compared to economic loss the injury percent is very less. Compared to economic loss and injury percent peak inter-storey drift has lesser amount of uncertainty.

Acknowledgements Funding from the Ministry of Earth Sciences, India through grant MoES/P.O. (Seismo)/1(304)/2016 is greatly acknowledged.

References

1. Ellingwood BR, Kinali K (2009) Quantifying and communicating uncertainty in seismic risk assessment. Struct Saf 31:179–187
2. Dolce M, Prota A, Borzi B, da Porto F, Lagomarsino S, Magenes G, Moroni C, Penna A, Polese M, Speranza E, Verderame GM, Zuccaro G (2020) Seismic risk assessment of residential buildings in Italy. Bull Earthquake Eng 1–34
3. Pavel F (2021) Seismic risk assessment of on-ground circular reinforced concrete and prestressed concrete water tanks using stochastic ground motion simulations. Bull Earthquake Eng 19:161–178
4. Zsarnoczay A, McKenna F, Wang C, Elhaddad W, Gardner M (2019) NHERI-SimCenter/PBE: Release v2.0.0 (Version v2.0.0). Zenodo
5. Committee E, Whitman R, Borcherdt R, Brookshire D, Eisner R, Holmes W, Olson R, O'Rourke M, Lagorio H, Reitherman R, Solutions S (2013) HAZUS-MH 2.1 earthquake model technical manual
6. Agency F, Hamburger R, Bachman R, Heintz J, Hooper J, Whittaker A (2012) Seismic performance assessment of buildings, vol 1, Methodology
7. Scheingraber C, Käser M (2018) Uncertainty quantification for seismic risk assessment using Latin hypercube sampling and Quasi Monte Carlo simulation. In: 16th European conference on earthquake engineering
8. Passeri F, Foti S, Cox BR, Rodriguez-Marek A (2018) Influence of epistemic uncertainty in shear wave velocity on seismic ground response analyses. Earthq Spectra 35:929–954
9. Li Y, Conte JP, Barbato M (2016) Influence of time-varying frequency content in earthquake ground motions on seismic response of linear elastic systems. Earthquake Eng Struct Dynam 45:1271–1291
10. Vlachos C, Papakonstantinou KG, Deodatis G (2018) Predictive model for site specific simulation of ground motions based on earthquake scenarios. Earthquake Eng Struct Dynam 47:195–218
11. Dabaghi M, Kiureghian AD (2018) Simulation of orthogonal horizontal components of near-fault ground motion for specified earthquake source and site characteristics. Earthquake Eng Struct Dynam 47:1369–1393

Mapping of Soil Moisture Index Using Optical and Thermal Remote Sensing

Jeenu John, R. Jaganathan, and D. S. Dharshan Shylesh

Abstract Soil moisture has a significant influence in the occurrence of drought since it influences the energy exchange between the atmosphere and the land surface. This paper discuss about the significance of soil moisture and the estimation of soil moisture index (SMI) using remote sensing. Optical and the thermal spectra is processed in SAGA GIS for the computation of Land Surface Temperature (LST) using thermal band of Landsat 8 and the Normalized difference vegetation (NDVI) using the Red and Near Infrared Band of Landsat 8. A linear regression analysis is performed between LST and the NDVI is to estimate the soil moisture index. The derived soil moisture index is used to map the severity of drought from extreme drought to no drought condition and the results were validated with the NDVI results. The Results shows that the relationship between LST and the NDVI can be utilized to map drought risk areas with the help of Landsat-8 images at a higher resolution which is beneficial for hydrological studies.

Keywords Soil moisture · NDVI · LST · Drought · Landsat-8

1 Introduction

Soil Moisture controls the exchange of energy between the land and the atmosphere significantly affects the occurrence of the flood and the drought [1]. If the concentration of soil moisture is high in the soil then the chance of conversion of rainfall to the runoff is high. If the concentration of the soil moisture is less then chance of occurrence of drought and affects the health of the vegetation [2, 3]. Once Plants reaches the permanent wilting point then the soil is dry and the vegetation no longer cannot extract water from the soil [4].

There are different techniques available for the estimation of the soil moisture such as the time domain reflectometry, gravimetric techniques which are found to be the most accurate method for the estimation of the soil moisture [5]. If the terrain

J. John (✉) · R. Jaganathan · D. S. Dharshan Shylesh
Department of Geography, University of Madras, Guindy Campus, Chennai, India

G. C. Marano et al. (eds.), *Proceedings of SECON'21*, Lecture Notes in Civil Engineering 171, https://doi.org/10.1007/978-3-030-80312-4_65

is heterogeneous then the point measurements cannot be extrapolated to the whole region of the study [6, 7]. With the advent of remote sensing techniques the requirement of manpower for the estimation of soil moisture can reduce or completely. In remote sensing techniques thermal, optical and microwave spectra is used for the estimation of the soil moisture [7, 8]. The thermal remote sensing computes the Land surface temperature of a region is used for the estimation of the soil moisture requires ancillary data when they are used alone. The generation of the vegetation indices for the estimation of the soil moisture varies according to the ecosystem and the vegetation type [5, 9]. The time taken by the canopy to show changes with the change in the ecosystem is slow. Hence the combined use of optical and thermal bands for estimating soil moisture is found to accurate and does not require any ancillary parameters and they does not change with the change in the ambient conditions. Microwave remote sensing product are available at coarse resolution are not available for the regional studies the retrieval of the soil moisture from the backscattering value is found to be affected vegetation scattering, local incidence angle the surface roughness of the area [10, 11].

In this study the applicability of optical and thermal spectra of the Landsat 8 is used for the estimation of the soil moisture in the study area. The combined use of optical and thermal remote sensing neglects the requirement of ancillary parameters for the estimation of the soil moisture. Using Optical remote sensing NDVI is derived since NDVI denotes the vegetation health of a region [12, 13]. The NDVI and the Land Surface temperature is correlated to map the drought of a region with the estimation of the soil moisture Index. The decrease of soil moisture induces water stress in the vegetation which will affect the health of the vegetation is mapped to determine the severity of the drought in the Kovalam Basin [14, 15].

2 Study Area

Kovalam sub-basin is located in the southern side of the Chennai basin, covering an area of 701.63 km^2. The outlet for this sub-basin is the Muttukadu backwaters joining the Bay of Bengal. This sub-basin has no major river flowing into it except that the Buckingham canal is running through the basin from north to south, and carries the water from the other parts. Kovalam basin can be demarcated into two parts North and South, with junctions on either side of Muttukadu backwaters (See Fig. 1).

3 Data

Landsat 8 Operational Land Imager (OLI) and Thermal Infrared Sensor (TIRS) satellite data from 2014 to 2019 is used for the Kovalam basin (Path 152 and Row 51). Landsat 8 images unaffected by cloud and cloud shadows were used for the study.

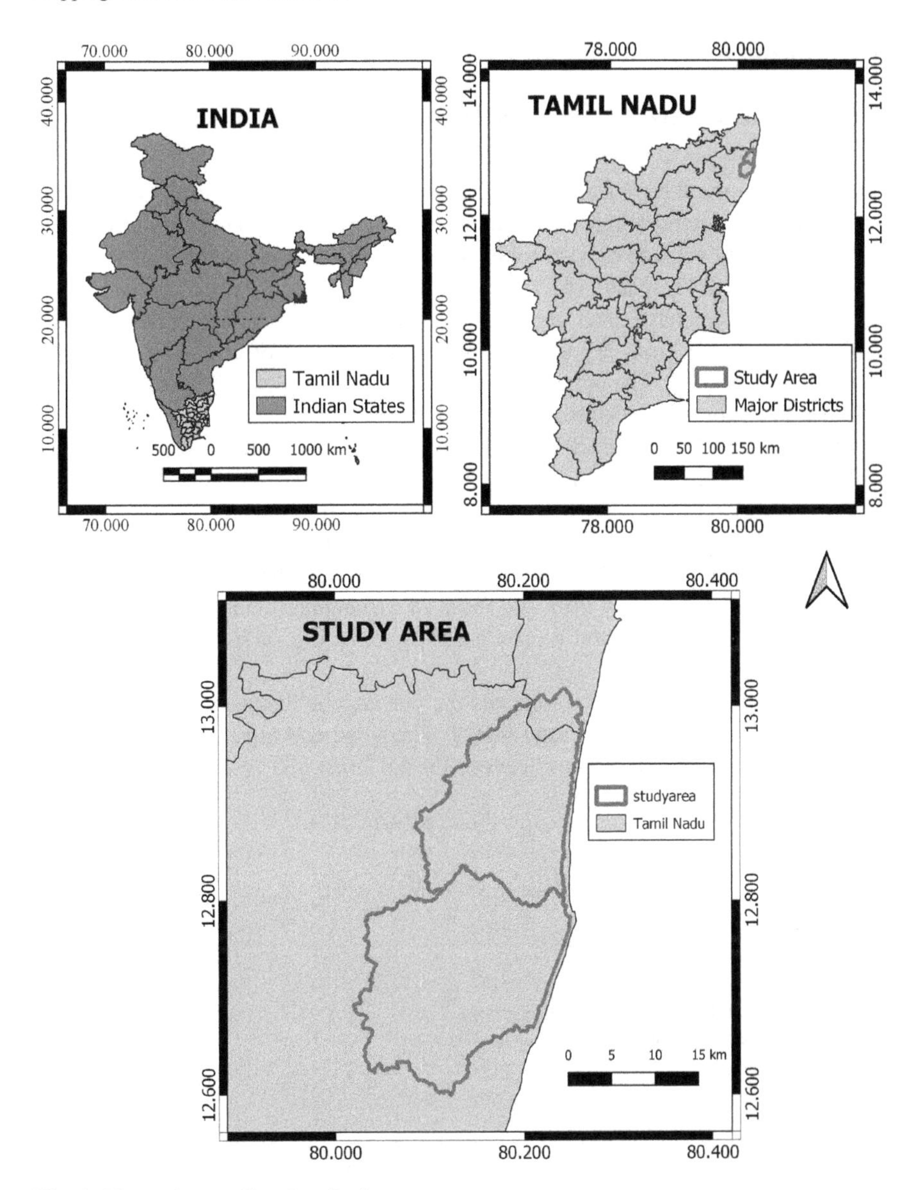

Fig. 1 The study area-Kovalam Basin

4 Methodology

4.1 Soil Moisture Estimation Using TOTRAM Model

The model proposed by Price estimated soil moisture on the basis of the water stress undergone by vegetation due to variation in LST. Spectral distribution of pixels is either in triangular shape or trapezoidal shape depends on the number of vegetated pixels in NDVI-LST space [16].

If the study area have pixel varying from bare soil to vegetated area then surface soil moisture varies from dry to wet. Spectral distribution of LST versus NDVI is shown below (Fig. 2). The upper edge represents the dry edge; the pixel varies from dry bare soil region to vegetated region with high moisture content, and lower edge represents the wet edge; pixel varies from wet bare soil to fully vegetated region with high moisture content (Fig. 2). LST max (dry edge) is the maximum LST corresponds to zero evapotranspiration, ideally, it is the point where.

Plants reach permanent wilting point. LST_{min} (Wet edge) is the minimum LST, where enough moisture content are available in the soil for evapotranspiration. In some cases the corners may meet and result in a triangular model due to land cover variation. NDVI-LST spectral plot is used for retrieving surface soil moisture by deriving dry edge and wet edge.

Hence estimation of the dry edge and the wet edge is the most important step as it varies with variation of LST and NDVI. The slope and regression are estimated through simple linear regression between the LSTmax and LSTmin (Eqs. 1–2).

$$LST_{max} = a_1 + NDVI * b_1 \tag{1}$$

$$LST_{max} = a_2 + NDVI * b_2 \tag{2}$$

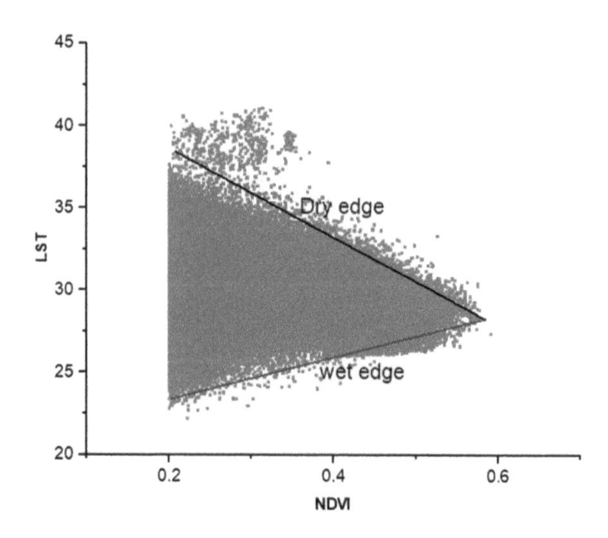

Fig. 2 The spectral distribution of pixels in the LST-NDVI space

a_1 and b_1 are the intercept and slope of the dry edge and a_2 and b_2 are the and intercept and slope of the wet edge respectively. NDVI can retrieve information of health of vegetation due to the water stress is obtained by using Eq. (3)

$$NDVI = \frac{NIR - R}{NIR + R} \tag{3}$$

Soil moisture index using the LST and NDVI is estimated using Eq. (4)

$$SMI = \frac{LST_{max} - LST}{LST_{max} - LST_{min}} \tag{4}$$

4.2 *Computation of LST*

The thermal infrared band 10 is sued to estimate the LST according to the L8 data user handbook. Atmospheric correction of the landsat8 data is achieved by converting Digital Number to sensor radiance using the following equation

$$L_\lambda = M_L \times Q_{cal} + A_L \tag{5}$$

where L_λ = spectral radiance at sensor's aperture in (W m^{-2} sr^{-1} μm^{-1}), M_L = band-specific multiplicative rescaling factor from Landsat metadata, A_L = band-specific additive rescaling factor from Landsat metadata and Q_{cal} = quantized and calibrated standard product pixel values.

In order to estimate LST, the top of atmospheric radiance is converted to sensor brightness temperature. LST of a region is the function of thermal emissivity, relies on fraction of vegetation is retrieved using Eq. (8) [17]

$$P_V = \left(\frac{NDVI - NDVI_{min}}{NDVI_{max} - NDVI_{min}}\right)^2 \tag{6}$$

NDVI = Digital Number (DN) values from NDVI Image, $NDVI_{min}$ = Minimum DN values from NDVI Image, $NDVI_{max}$ = Maximum DN values from NDVI Image. The Normalized Difference Vegetation Index (NDVI) is computed using Eq. (3), emissivity of the study area is computed using Eq. (7)

$$\varepsilon = 0.004 \times Pv + 0.986 \tag{7}$$

Finally, the LST was computed using Equation

$$LST = \frac{T_b}{1 + \left(\frac{0.00115 T_b \times ln\varepsilon}{1.4388}\right)} \quad (8)$$

h—Planck's constant, c—Velocity of light, s—Boltzmann constant.

5 Results and Discussion

Soil moisture is mapped in the years 2014, 2015, 2016, 2017, 2018 and 2019 on Kovalam river basin using the spectral relationship of LST versus NDVI (Fig. 3) NDVI values falling below 0.2 is excluded from clustering since they resulted in oversaturated pixels.

LST and NDVI were plotted against each other using Landsat 8 image and obtained a negative relationship. The relationship distinctly depicts the influence of LST on the health of vegetation which is correlated with the soil moisture of the region. NDVI is divided into different clusters and the maximum and minimum LST corresponding to each cluster is used to obtain the dry edge and wet edge from the spectral distribution of LST and NDVI (Table 1). Negative slope of dry edge shows that LST is decreasing with increase in vegetation and positive slope of wet edge increases with increase in vegetation (Table 1). Soil moisture is divided into four classes Very Low (0–0.25), low (0.25–0.5), Moderate (0.5–0.75) and high soil moisture (0.75–1) classes based on the spatial variability (Fig. 3). Using NDVI-LST spectral plot model it was found that aerial distribution of soil moisture in the year 2014–2018 in high class and moderate class is significantly increased due the reduction of land surface temperature in the year 2018 (Min 16.847 °C and Max 35.882 °C) (Fig. 4).

Aerial distribution of soil moisture in low soil moisture class in year 2014–2015 have declined by 32.4% in 2015 (507.3777 to 342.5976 km^2) and again increased by 3.34% by in the year 2016 (342.5976 to 354.0672 km^2) due to the reduction of LST (Fig. 4). During the year 2014–2017 LST is increased by 2.6932 °C (43–46.5932 °C) and settlement is increased by 23.75% (130.1985–161.1207 km^2) and a corresponding increase in soil moisture in very low class is observed. Increase of concrete structures reduces the infiltration of water to soil and increases runoff.

From 2014 to 2018 a significant increase in moderate soil moisture and a reduction in low and very low class is observed, since the analysis of soil moisture is performed in the month of January which comes after the North east monsoon (Fig. 3). In the year 2019 area covered by the settlement is increased by 38.772% and LST is increased by 2.621 °C (46.521–43.9).

Land cover and LST of a region significantly influences the slope of the dry edge and the wet edge. Slope of dry edge and wet edge is found to low in 2018 compared to the rest of the years (Table 1). Since January comes after the north east monsoon,

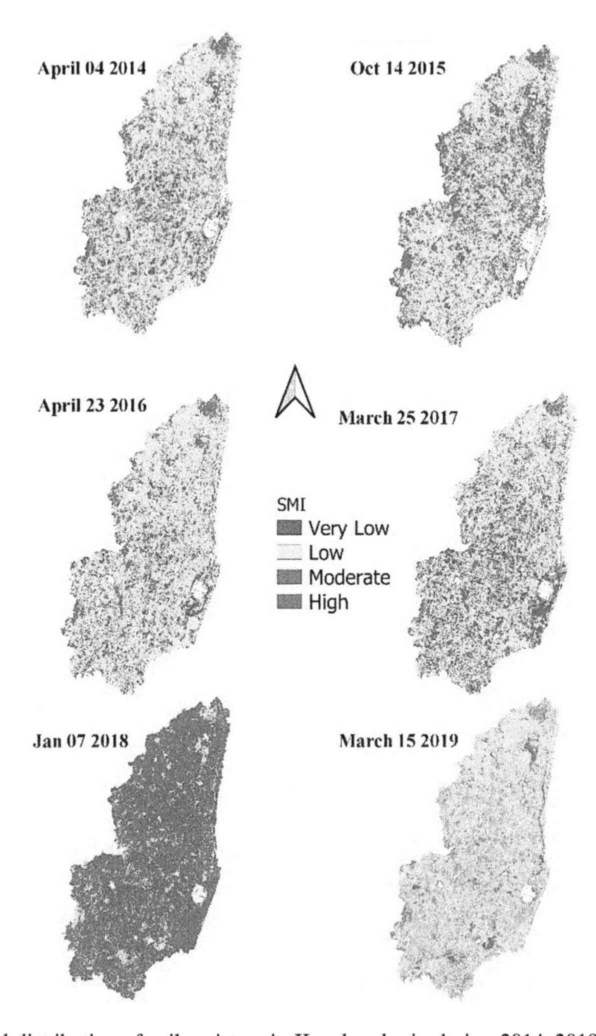

Fig. 3 Spatial distribution of soil moisture in Kovalam basin during 2014–2019

Table 1 Dry edge and wet edge of the model

Date of image acquired	Dry edge	Wet edge
2014	y = −17.618x + 40.664	y = 13.72x + 19.671
2015	y = −19.084x + 41.183	y = 13.341x + 19.41
2016	y = −19.683x + 40.857	y = 5.4676x + 25,634
2017	y = −15.747x + 46.007	y = 0.9244x + 28.775
2018	y = −0.0949x + 32.976	y = 11.864x + 18.3
2019	y = −14.196x + 42.674	y = 5.6352x + 23.96

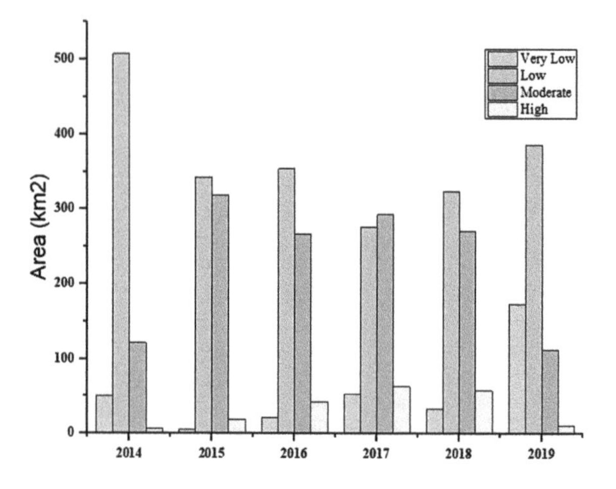

Fig. 4 Aerial distribution of soil moisture class from 2014 to 2019 in Kovalam watershed using LST-NDVI space

NDVI shows large variation and therefore slope decreases in wet season [18]. Slope of dry edge and wet edge is higher in summer season (Table 1) is due to abrupt variation in the vegetation with the increase in land surface temperature.

6 Conclusion

The estimation of soil moisture is inevitable as it plays an important role in hydrological models, ecological applications, and the field of agriculture. In this study soil moisture is retrieved using LST-NDVI in regional scale is and the variation of soil moisture during the year is studied. Soil moisture is found to high in the months after north east monsoon and found to low in the months where LST is high. However, the study could able to demonstrate the potentiality of Landsat imageries to determine soil moisture at a higher resolution instead of using global coarser resolution products.

References

1. Ghahremanloo M, Mobasheri MR, Amani M (2019) Soil moisture estimation using land surface temperature and soil temperature at 5 cm depth. Int J Remote Sens 40(1):104–117. https://doi.org/10.1080/01431161.2018.1501167
2. Emami H, Mojaradi B, Safari A (2017) The effect of soil salinity on the use of the universal triangle method to estimate saline soil moisture from landsat data: Application to the smapex-2 and smapex-3 campaigns. Int J Remote Sens 38(23):6623–6652. https://doi.org/10.1080/01431161.2017.1363431
3. Chen J, Wang C, Jiang H, Mao L, Yu Z (2011) Estimating soil moisture using temperature-vegetation dryness index (TVDI) in the Huang-huai-hai (HHH) plain. Int J Remote Sens 32(4):1165–1177. https://doi.org/10.1080/01431160903527421

4. Yue J, Tian J, Tian Q, Xu K, Xu N (2018) Development of soil moisture indices from differences in water absorption between shortwave-infrared bands. ISPRS J Photogramm Remote Sens 154:216–230 (September 2018). https://doi.org/10.1016/j.isprsjprs.2019.06.012
5. Zhang F, Zhang LW, Shi JJ, Huang JF (2014) Soil moisture monitoring based on land surface temperature-vegetation index space derived from MODIS data. Pedosphere 24(4):450–460. https://doi.org/10.1016/S1002-0160(14)60031-X
6. Carlson TN, Gillies RR, Perry EM (1994) A method to make use of thermal infrared temperature and NDVI measurements to infer surface soil water content and fractional vegetation cover. Remote Sens Rev 9(1–2):161–173. https://doi.org/10.1080/02757259409532220
7. Zhang D , Zhou G (2016) Estimation of soil moisture from optical and thermal remote sensing: a review. Sensors (Switzerland) 16(8). https://doi.org/10.3390/s16081308
8. Zhang N, Hong Y, Qin Q, Liu L (2013) VSDI: A visible and shortwave infrared drought index for monitoring soil and vegetation moisture based on optical remote sensing. Int J Remote Sens 34(13):4585–4609. https://doi.org/10.1080/01431161.2013.779046
9. Sharma PK, Kumar D, Srivastava HS, Patel P (2018) Assessment of different methods for soil moisture estimation: a review. J Remote Sens GIS 9(1):57–73
10. Koley S, Jeganathan C (2020) Estimation and evaluation of high spatial resolution surface soil moisture using multi-sensor multi-resolution approach. Geoderma 378:114618. https://doi.org/10.1016/j.geoderma.2020.114618
11. Sadeghi M, Jones SB, Philpot WD (2015) A linear physically-based model for remote sensing of soil moisture using short wave infrared bands. Remote Sens Environ 164:66–76. https://doi.org/10.1016/j.rse.2015.04.007
12. Zhang D, Tang R, Tang BH, Wu H, Li ZL (2015) A simple method for soil moisture determination from LST-VI feature space using nonlinear interpolation based on thermal infrared remotely sensed data. IEEE J Sel Top Appl Earth Obs Remote Sens 8(2):638–648. https://doi.org/10.1109/JSTARS.2014.2371135
13. Sruthi S, Aslam MAM (2015) Agricultural drought analysis using the NDVI and land surface temperature data; a case study of Raichur District. Aquat Procedia 4:1258–1264. https://doi.org/10.1016/j.aqpro.2015.02.164
14. Barzegar Z, Momeni M, Jozdani S (2016) Expanding the applicability of remotely sensed soil evaporation transfer coefficient to the estimation of soil moisture over bare soil and partially vegetated areas. Remote Sens Lett 7(4):368–377. https://doi.org/10.1080/2150704X.2016.1142679
15. Zhang D et al (2014) Surface soil water content estimation from thermal remote sensing based on the temporal variation of land surface temperature. Remote Sens 6(4):3170–3187. https://doi.org/10.3390/rs6043170
16. Carlson TN, Gillies RR, Schmugge TJ (1995) An interpretation of methodologies for indirect measurement of soil water content. Agric For Meteorol 77(3–4):191–205. https://doi.org/10.1016/0168-1923(95)02261-U
17. Sobrino JA, Jiménez-Muñoz JC, Paolini L (2004) Land surface temperature retrieval from LANDSAT TM 5. Remote Sens Environ. https://doi.org/10.1016/j.rse.2004.02.003
18. Sadeghi M, Babaeian E, Tuller M, Jones SB (2017) The optical trapezoid model: a novel approach to remote sensing of soil moisture applied to Sentinel-2 and Landsat-8 observations. Remote Sens Environ 198:52–68. https://doi.org/10.1016/j.rse.2017.05.041

Wind Effect on Closely Spaced Rectangular Tall Buildings of U Shape Geometrical Pattern

Shanku Mandal, Sujit Kumar Dalui, and Soumya Bhattacharjya

Abstract The modern cities of the developing world are facing challenges of extreme population and land availability. These challenges can easily overcome by utilizing the available areas properly with the development of closely spaced tall buildings. But most often, this solution attracts wind force that leads to severe wind interference. This study focuses on typical cases, where the three rectangular shapes with 75 m height, 45 m length and 15 m width each are closely spaced and formed a U shape geometrical pattern. Eight different cases have been considered in which one shape has been fixed at the same place and the other two shapes are moved both sideways and frontwards by 7.5 m, 15 m, 22.5 m and 30 m. Further, the two isolate U shape cases in which the three shapes are attached and a single isolated rectangular shape are also adopted. The wind flow at 0°, 45° and 90° has been generated by Computational Fluid Dynamics (CFD). The validation and mess sensitive study is attached to satisfy the requirements of a CFD method. The interference factor (IF) is calculated to demonstrate the impact of the interference. The wind pressure and force discrepancy have been observed due to the interference and flow angle. In most cases, the rise in mean pressure coefficients has been observed when the interference buildings are present at the sidewards. The pressure contour plots reveal that the location of the buildings and the flow angle has a significant impact on the face pressure distribution.

Keywords Closely spaced · Tall building · Wind interference · Wind effect · Computational fluid dynamics · Interference factor · Mess sensitive study · Flow angle

S. Mandal (✉) · S. K. Dalui · S. Bhattacharjya
Department of Civil Engineering, Indian Institute of Engineering Science and Technology, Shibpur, Howrah, India
e-mail: soumya@civil.iiests.ac.in

G. C. Marano et al. (eds.), *Proceedings of SECON'21*, Lecture Notes in Civil Engineering 171, https://doi.org/10.1007/978-3-030-80312-4_66

1 Introduction

In modern cities, open land availability is a preliminary issue. So, those land must be used completely is always the answer in modern construction practices. Thus most often, closely placed tall buildings are constructed. The nature of the wind flow is quite complex in closely spaced structures than a stand-alone structure since wind interference is the critical factor. Although the assessment of interference effect is the important factor for wind-resistant design, a very limited section is available in various guidelines, which are not sufficient. A few manuscripts had highlighted some typical cases of wind interference. The numerical study of Sohankar [1] on twin square shapes, illustrated the effect of gap spacing and Reynolds numbers in wind force and pressure variations. Yu et al. [2] measured the effect of interference on pressure distribution in the various arrangement of the two buildings having different height and breadth ratios. Kar and Dalui [3] investigated the response of an octagonal shaped building in the presence of three square shapes. Zu and Lam [4] arranged two tall buildings at an arbitrary position and identified the across wind responses. Ma et al. [5] studied the differences in moment and force coefficients of twin rectangular shapes due to wind vibration. Korobkov et al. [6] investigate wind pressure and thermal effect between two interfered square shapes. Quan et al. [7] studied the interference effect on the surrounded building due to the presence of a tall building. Wu et al. [8] presented the wind induced effects of two circular shapes using Large eddy simulation. Behera et al. [9] concluded that the increment in plan ratio of the building extends the interference zone. Du et al. [10] presented the wind coefficients of two square shapes on diagonal and horizontal arrangement placed in close proximity.

The residential and academic buildings layout plan in modern cities are mostly following the U shape geometrical pattern where three rectangular buildings are placed in close proximity or a single isolated U shape where three rectangular buildings are attached to each other. The benefits of this geometrical pattern are that the open space in between two limbs can be utilized as a parking area, commercial requirements or a playground. Sometimes in the case where the buildings are not attached to each other, a street can be placed in between the open space of the buildings. Despite numerous studies, none of the research has considered this typical scenario. The uniqueness of this research is that in this study this typical case has been considered and the wind effect has been presented. Different cases have been adopted and the wind effect has been measured in the 0°, 45° and 90° wind angle using CFD. The variation in wind force, flow characteristics, mean pressure coefficients and pressure contour has been illustrated.

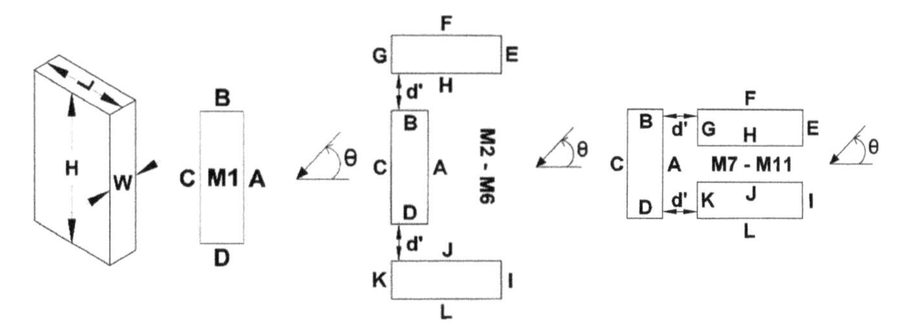

Fig. 1 The details of building models

2 Details of Building Models

The wind responses of an isolated rectangular building (M1) and the different interference cases (M2–M11) has been calculated at 0°, 45° and 90° wind angle. For the simulation, the dimensions of the buildings are reduced by following the 1:300 scale. The building models are shown in Fig. 1.

The height (H), length (L) and width (W) of the rectangular buildings are considered 250 mm, 150 mm and 50 mm respectively. 'd'' is the distance between an isolated rectangular building and the interference rectangular buildings. 'd'' is varying from 0 to 100 mm at an interval of 25 mm. At the M1 model, only the rectangular building is considered where no interference building is present. In the case of M2 model, the three similar rectangular building is considered in which the buildings are attached in such a way that it forms a U shape geometrical pattern. At M3, M4, M5 and M6 model cases the two rectangular buildings are shifted sidewards by 25 mm, 50 mm, 75 mm and 100 mm respectively. In the case of M7 model the U shape pattern is formed where two rectangular buildings are attached at the front side of the isolated rectangular building. The frontward shift of two rectangular buildings by 25 mm, 50 mm, 75 mm and 100 mm has been considered for M8, M9, M10 and M11 models respectively. The M2–M11 model cases are formed a closely spaced U shape geometrical patterns.

3 Solution Methodology

The numerical simulation utilized the Ansys CFX module [3], where the wind flow has been produced through computational fluid dynamics (CFD). The solution methodology involves a few steps which started with the modelling of the building, then selection of suitable domain and meshing. After that, the boundary conditions have been implemented for the building and the domain and the turbulence model and the equations have been selected. Finally, the wind responses have been calculated after the successful completion of the numerical analysis.

3.1 Computational Domain Setup

The extreme boundary of the domain is placed at a distance of 15H, 5H and 5H from the model's back face, two side faces and the front faces respectively [11]. The sidewalls and roof of the boundary are governed by the free slip boundary conditions, whereas no slip is adopted for the model faces (Fig. 2). The simulation has been executed on k-ε turbulence model. 10 m/s wind velocity is produced at the domain inlet. 1 m height is considered for the Atmospheric Boundary Layer (ABL), which is generated by the power-law equation (Eq. 1).

$$\frac{V}{V_0} = \left(\frac{h}{h_0}\right)^{\alpha} \tag{1}$$

In the equation, 'h' and 'h_0' signify the reference height and the building depth above the ground respectively. 'V' and 'V_0' indicates the wind velocity at the 'h' and 'h_0' respectively. In this study, the exponent 'α' is taken as 0.133.

3.2 Generation of Mesh

The entire domain has meshed with tetrahedral elements but the fine layers of square elements are provided at the nearby locations of the building models (see Fig. 3) to measure the accurate wind responses of the building. This meshing technique is capable of generating reliable results without involving extensive computational resources.

3.3 Mesh Sensitivity Study

The suitable meshing technique has been determined by the mesh sensitivity study on the M2 type model at a normal wind angle. Seven mesh elements size has been adopted for this study. The total element sizes are 1,457,621 (MS1), 4,865,743 (MS2), 8,498,327 (MS3), 13,874,521 (MS4), 19,845,372 (MS5), 24,657,894

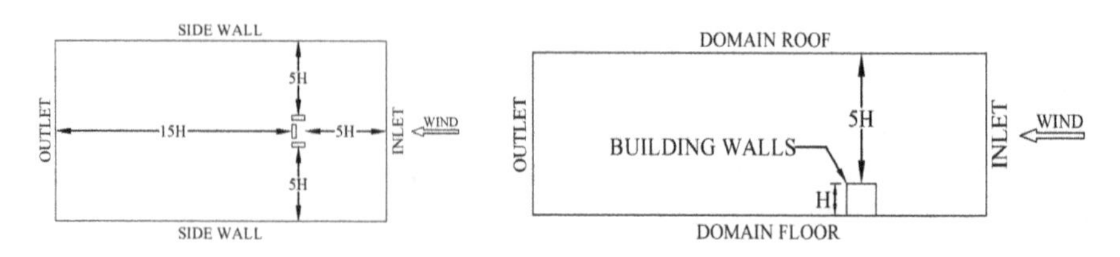

Fig. 2 Plan and elevation view of the domain

Fig. 3 The typical mesh pattern for the M5 model

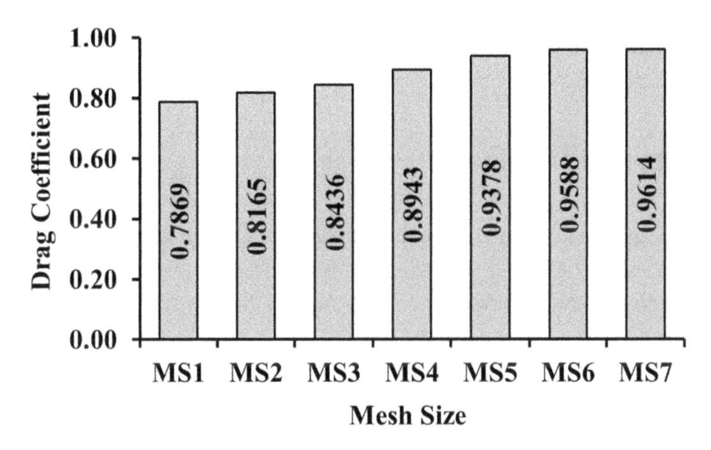

Fig. 4 The mesh sensitivity study of the M2 model

(MS6) and 31,287,459 (MS7), which represents course to fine meshing. From the drag coefficient comparison (see Fig. 4), the MS6 type mesh has been selected for this study since the computational error in the MS6 type is approx. 0.27% as compare to MS7 mesh type.

3.4 Validation

The validity of the measured responses of the current study has been investigated by comparing the pressure coefficient graphs from the literature [12] (see Fig. 5). Exactly the same model dimensions and the other analysis techniques have been

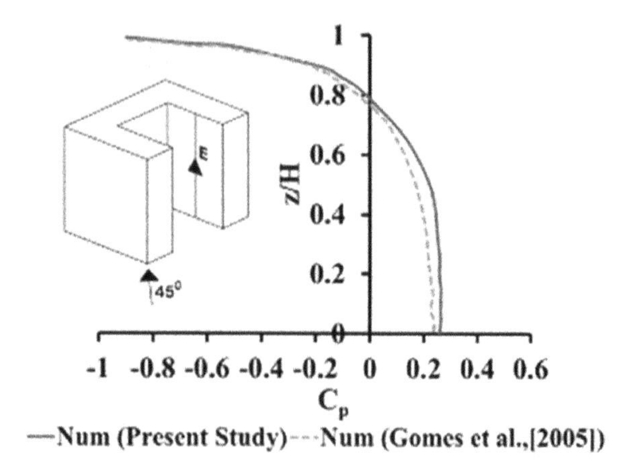

Fig. 5 The comparison of the vertical pressure profile

considered. The comparison of pressure plot shows almost similar trends and that validates the current study.

4 Results and Discussion

The wind responses of the isolated and closely spaced model cases have been demonstrated in 0°, 45° and 90° wind angle. The comparison of the responses on different model cases indicates the influence of the closely spaced interference buildings.

4.1 Variation in Wind Flow Streamline

Some typical wind flow patterns have been demonstrated in Fig. 6. The free flow of the wind largely affected by the presence of closely spaced buildings. The formation of vortices at the leeward and side faces of the interference building causes remarkable variation in wind responses compared to the isolated rectangular case. The prolonged vortices have been formed at the backside of the rectangular shape in the presence of interference buildings. It is also observed that the streamlines dependent on the location of the interfering buildings and the angle of the wind.

4.2 Variation in Force Coefficients

The force coefficients have been calculated at different wind angle for the isolated rectangular building (M1). When interferences cases have been considered the

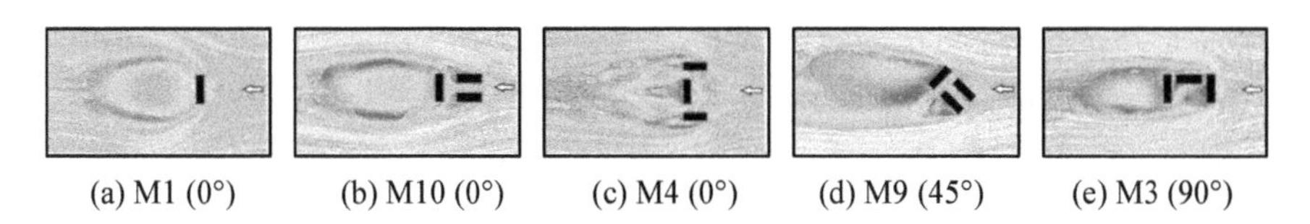

Fig. 6 The flow patterns around the typical building models

Table 1 The comparison of force coefficients at the various wind angle

	Wind angle (0°)		Wind angle (45°)		Wind angle (90°)	
Model	C_{fx}	C_{fy}	C_{fx}	C_{fy}	C_{fx}	C_{fy}
M1	1.036	0.011	0.786	−0.878	−0.052	−0.871
M2	0.959	0.011	0.817	−1.187	−0.092	−0.932
M3	1.125	0.005	0.741	−0.144	−0.002	0.074
M4	1.124	0.001	0.940	−0.186	−0.020	0.141
M5	1.024	0.007	1.062	−0.164	−0.047	0.165
M6	1.003	0.003	1.146	−0.036	−0.040	0.185
M7	0.932	0.004	0.689	−1.564	−0.030	−1.286
M8	0.393	0.048	0.209	−0.686	0.277	−1.169
M9	0.512	0.005	0.344	−0.568	0.306	−1.210
M10	0.519	0.021	0.469	−0.563	0.309	−1.221
M11	0.507	0.032	0.509	−0.589	0.342	−1.219

changes in the drag (C_{fx}) and lift (C_{fy}) coefficient of that building is tabulated in Table 1. At 0° the reduction in drag coefficient is measured when the three rectangular shapes are attached with each other to form a U shape (M2 and M7). The sidewards shifting of the two buildings (M3 and M4) attracts more drag force. However, when the distance of interference building is increased (M5 and M6) the darg force reduces. When the interference buildings are located at the front side of the rectangular building (M7–M11) the drag coefficient decreases. The lift coefficient at 0° is almost insignificant in all the model cases. In the case of 45° wind angle, when the interference buildings are located at sidewards the C_{fx} increases and C_{fy} decreases but both the coefficients increases in the attached U shape (M2). In the M7 case the C_{fy} is maximum but the frontward shifting of the interference buildings attracts less C_{fx} and C_{fy}. Almost negligible lift and drag force are noted in the M3, M4, M5 and M6 models but at M8, M9, M10 and M11 model cases the rectangular building experiences more drag and lift force as compared to the isolated rectangular model (M1) case. When the interference buildings are attached to the rectangular building (M7), it is observed that the lift force is critical at 45° and 90° angle. The maximum C_{fx} is noted at the rectangular building when the interference building is located sidewards at a distance of 7.5 m. In most cases, the interference caused a huge increment in drag and lift responses of the rectangular building.

Table 2 The Interference factor of the different faces of the isolated rectangular building

	Wind angle (0°)				Wind angle (45°)			
Model	A	B	C	D	A	B	C	D
M1	1.000	1.000	1.000	1.000	1.000	1.000	1.000	1.000
M2	1.338	–	0.739	–	2.001	–	0.337	–
M3	1.058	1.326	1.160	1.332	0.827	−2.537	0.984	0.999
M4	0.946	1.457	1.369	1.454	1.231	−2.475	1.070	1.053
M5	0.949	1.238	1.055	1.245	1.392	−2.422	1.229	1.058
M6	0.966	1.117	1.569	0.687	1.486	−2.381	1.299	1.067
M7	0.923	1.027	0.950	1.027	1.110	2.348	0.884	0.873
M8	0.212	0.607	0.651	0.709	0.256	0.775	0.754	0.802
M9	0.376	0.716	0.675	0.704	0.024	0.657	0.835	0.647
M10	0.403	0.725	0.647	0.693	0.226	0.516	0.943	0.687
M11	0.398	0.708	0.625	0.659	0.378	0.437	0.920	0.736

4.3 Variation in Interference Factor of Pressure Coefficients

The mean pressure coefficients have been calculated on the different faces of the rectangular building in both isolated and interference conditions. The mean local pressure at the faces greatly influenced by the interference building since the presence of the building changes the wind flow pattern. The wind angle shifts also contribute to the pressure variation. The Interference Factor (IF) is the ratio of mean pressure of the particular face on interference condition to that of the mean pressure of this face at isolate condition. The interference factor and the mean pressure coefficient of isolated rectangular shape have been tabulated in Tables 2 and 3. On interfering conditions, the mean pressure coefficient (C_p) on the faces of the rectangular model can be calculated by the following formula.

$$C_{p,\text{Interfering}} = \text{Interference Factor (I.F)} * C_{p,\text{Isolate}} \tag{2}$$

Almost all the faces have the positive IF but in some of the cases (M3, M4, M5 and M6) at face B, the negative IF is noted in 45° and 90° angle. The negative IF indicates the generation of high turbulence on that face because of interference building.

4.4 Variation in Pressure Contour

The pressure variation at different locations of the building faces has been illustrated through contour plots. The pressure contour on all the faces of the isolated

Table 3 The Interference factor of the different faces of the isolated rectangular building at a 90° angle and the mean pressure coefficient on the different faces of the isolated rectangular building

	Interference factor (90°)				Mean pressure coefficient			
Model	A	B	C	D	Face	0°	45°	90°
M1	1.000	1.000	1.000	1.000	Face A	0.641	0.349	−0.440
M2	1.209	–	1.058	–				
M3	1.140	−1.086	1.128	1.944	Face B	−0.447	0.209	0.487
M4	1.213	−1.118	1.159	1.795				
M5	1.222	−1.051	1.118	1.590	Face C	−0.319	−0.404	−0.440
M6	1.167	−1.082	1.092	1.573				
M7	1.734	0.978	1.819	1.903	Face D	−0.447	−0.629	−0.236
M8	0.968	0.926	1.623	2.491				
M9	0.954	0.917	1.633	2.658				
M10	0.932	0.929	1.613	2.661				
M11	0.883	0.943	1.625	2.620				

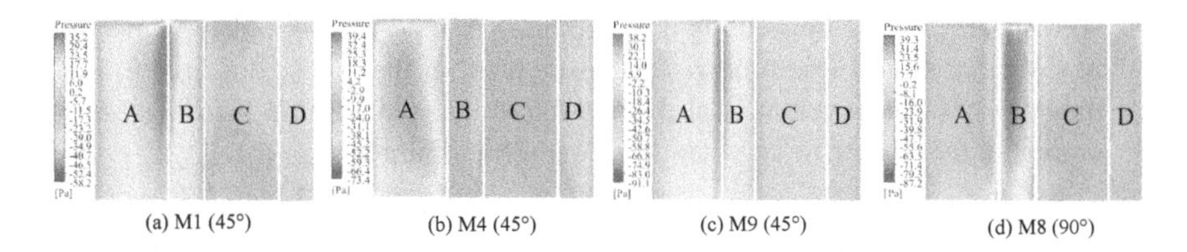

(a) M1 (45°) (b) M4 (45°) (c) M9 (45°) (d) M8 (90°)

Fig. 7 The comparison of pressure contour plot on the faces of the rectangular building

rectangular shape (M1) at 45° has been presented. To visualize the pressure variation due to interference buildings the M4 and M9 model has been adopted with the same wind angle. The pressure contour profile of the M8 model at a 90° angle has been attached to show the pressure variation when the wind angle shifts. The comparison of pressure contour has been given in Fig. 7. The comparison clearly shows that the pressure on the faces of rectangular building altered in the presence of interference buildings. The location of the interference building and the wind angle also contributes to pressure differences on the building faces. In isolated condition the corners of face A and B exerted positive pressure and face C and D has negative pressure but when interference building is present in sidewards (M4), most of the location of face A experiencing the positive pressure and all other faces have negative pressure. When interfering buildings are present at frontwards the corners of face A and B have positive pressure but the locations of these pressure are very limited compare to isolated condition. When the angle shifts, the positive pressure is observed in most of the locations at face B of the M8 model and the other faces are experiencing the negative pressure.

5 Conclusions

The influence of closely spaced interference buildings on wind responses in the existing structure has been demonstrated at 0°, 45° and 90° wind angle using CFD. The comparison of pressure contour, lift, drag and interference factor of pressure has been illustrated. The findings point out some important factors.

- The nonhomogeneous flow pattern yields since the location of interference buildings and wind angle changes. Thus, a significant variation in responses has been recorded. It indicates that the location, wind angle and distance between buildings should be considered for understanding the interference effect.
- The structural elements of the buildings should be designed by taking the effect of the presence of interference buildings because if the distance between the buildings is less, it causes a significant variation in drag and lift coefficient.
- The pressure on the building faces shows the critical variable in the presence of interference buildings. Hence the strong clad elements are required for the safety of the structure.
- Among all the building configurations, the M8 type is best suitable since this model type attracts less force and pressure most of the times compared to other model types. However, it is also observed that the sidewards building configurations draw lesser force when the flow angle charges.

As the wind effect on a rectangular building in the presence of closely spaced tall buildings in U shape geometrical pattern is not stated in earlier studies, this study provides a piece of information and a reasonable idea about the interference effect of the wind.

References

1. Sohankar A (2012) A numerical investigation of the flow over a pair of identical square cylinders in a tandem arrangement. Int J Numer Methods Fluid. 70:1244–1257. https://doi.org/10.1002/fld.2739
2. Yu XF, Xie ZN, Zhu JB, Gu M (2015) Interference effects on wind pressure distribution between two high-rise buildings. J Wind Eng Ind Aerodyn 142:188–197. https://doi.org/10.1016/j.jweia.2015.04.008
3. Kar R, Dalui SK (2016) Wind interference effect on an octagonal plan shaped tall building due to square plan shaped tall buildings. Int J Adv Struct Eng 8:73–86. https://doi.org/10.1007/s40091-016-0115-z
4. Zu GB, Lam KM (2018) Across-wind excitation mechanism for interference of twin tall buildings in staggered arrangement. J Wind Eng Ind Aerodyn 177:167–185. https://doi.org/10.1016/j.jweia.2018.04.019
5. Ma K, Hu C, Zhou Z (2019) Investigation on vortex-induced vibration of twin rectangular 5:1 cylinders through wind tunnel tests and POD analysis. J W Eng Ind Aero 187:97–107. https://doi.org/10.1016/j.jweia.2019.02.014

6. Korobkov SV, Terekhov VI, Koshin AA, Gnyrya AI, Mikhailov DA (2019) Dynamic and thermal interference effects on two neighbouring building models. J Phys Conf Ser 1382:012017. https://doi.org/10.1088/1742-6596/1382/1/012017
7. Quan Y, Chen J, Gu M (2020) Aerodynamic interference effects of a proposed taller high-rise building on wind pressures on existing tall buildings. Struct Des Tall Spec Build 29:1–17. https://doi.org/10.1002/tal.1703
8. Wu G, Du X, Wang Y (2020) LES of flow around two staggered circular cylinders at a high subcritical Reynolds number of 1.4×10^5. J Wind Eng Ind Aerod 196:104044. https://doi.org/10.1016/j.jweia.2019.104044
9. Behera S, Ghosh D, Mittal AK, Tamura Y, Kim W (2020) The effect of plan ratios on wind interference of two tall buildings. Struct Des Tall Spec Build 29. https://doi.org/-10.1002/tal.1680
10. Du X, Chen R, Dong H, Ma W, Xu H, Zhao Y (2021) Aerodynamic characteristics of two closely spaced square cylinders in different arrangements. J Wind Eng Ind Aerodyn 208:104462. https://doi.org/10.1016/j.jweia.2020.104462
11. Revuz J, Hargreaves DM, Owen JS (2012) On the domain size for the steady-state CFD modelling of a tall building. Wind Struct An Int J 15:313–329. https://doi.org/-10.12989/was.2012.15.4.313
12. Gomes MG, Moret Rodrigues A, Mendes P (2005) Experimental and numerical study of wind pressures on irregular-plan shapes. J Wind Eng Ind Aerodyn 93:741–756. https://doi.org/10.1016/j.jweia.2005.08.008

Soil Erosion Risk Analysis in Landslide Zones Using Remote Sensing, GIS and RUSLE

Mohd Talha Anees, Ahmad Farid Bin Abu Bakar, and Mohammad Zaid

Abstract The present study was conducted in the Gua Musang of Kelantan state, Peninsular Malaysia. The surface of the terrain is exposed due to anthropogenic and deforestation activities. Soil erosion from exposed surfaces increased the risk and vulnerability of landslides in the study area. Therefore, for soil erosion risk assessment in landslide zones, the objectives are to analyze pixel-based soil erosion risk using Revised universal soil loss equation (RUSLE), remote sensing, and Geographic Information System (GIS). Also, identification of soil erosion conditions in landslide zones. Results showed that very high soil loss in the landslide zone. 42% of the area comes under landslide zones, where 43% of landslide zone area comes under very high soil loss. Whereas, overall very high soil loss in the area is 39%. The results also indicated that settlement areas are under landslide zones, which could be at risk. Overall, soil erosion risk analysis in landslide zones using remote sensing, GIS, and RUSLE are useful and important.

Keywords GIS · Landslide · RUSLE · Remote sensing · Soil erosion

1 Introduction

Soil erosion is a serious concern among several researchers because it is the second biggest environmental problem, after population growth, in the world [1]. It is led by agricultural activities, industrialization, deforestation, and construction activities that cause land degradation and landslide due to the action of high rainfall-runoff. Soil erosion also influences water quality and fertility, and sedimentation rate in a river [2]. High sedimentation rate further reduce river depth and hence causes flash

M. T. Anees (✉) · A. F. B. A. Bakar
Department of Geology, Faculty of Science, University of Malaya, 50603 Kuala Lumpur, Malaysia

M. Zaid
Department of Civil Engineering, Aligarh Muslim University, Aligarh 202002, India

G. C. Marano et al. (eds.), *Proceedings of SECON'21*, Lecture Notes in Civil Engineering 171, https://doi.org/10.1007/978-3-030-80312-4_67

flood during high-intensity rainfall. Furthermore, due to high-intensity rainfall, a landslide can be triggered in regions of slopes greater than 30 degrees.

Remote sensing and geographic information system (GIS) are important techniques in analyzing Spatio-temporal land-use changes. These are used in analyzing land degradation due to soil erosion [3]. These can also use in monitoring major changes of post landslide or geo-hazard. Digital Elevation Models (DEMs) are generally used in pixel-based soil erosion analysis [2].

The Revised Universal Soil Loss Equation (RUSLE) is an important and widely used model for the long-term prediction of average annual soil loss [2, 4]. The focus of the present study is to use RUSLE in the study area to identify soil loss conditions in the landslide zone. Therefore, the main objectives of this study are (i) to analyze pixel-based soil erosion using remote sensing and RUSLE, and (ii) to correlate soil erosion with the landslide zones in the study area.

2 Material and Methods

2.1 Study Area

The study was conducted in Gua Musang of Kelantan state, Peninsular Malaysia. The size of the area is 130 km^2 that lie within coordinates of 5.52–5.47 N and 102.19–102.24 E (Fig. 1).

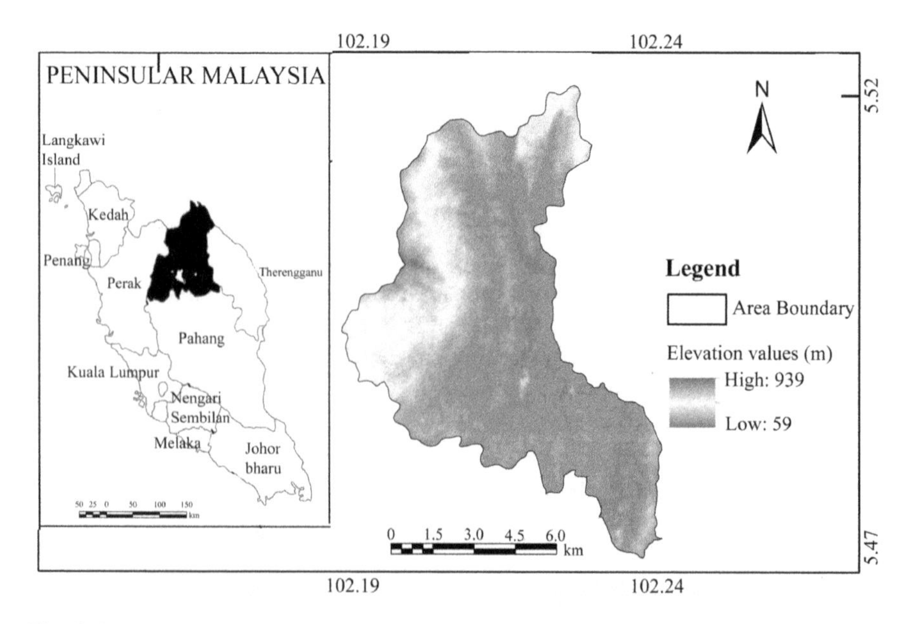

Fig. 1 Location of the study area with elevation values

The highest elevation and slope points are at 939 m and 75.8 degrees respectively. Northern and Eastern side of the study has a hilly region that is the upstream side. Whereas, the downstream side is towards the west and south-western side of the area. The climate of the area is humid and tropical with an average annual rainfall is 3017 mm. The average temperature is 25 °C while the daily average annual wind speed is 1.50 m/s [2].

2.2 Data

ASTER DEM was downloaded from the USGS website. Google Earth image was used for land use land cover map preparation. Rivers and tributaries were extracted from the topographic map with a scale of 1:200,000. Rainfall erosivity factor and soil erodibility factor were obtained from [2]. Landslide zones were extracted from [5].

2.3 Methodology

The methodology used by Anees et al. [2] to estimate soil loss and sediment yield was adopted and applied in the study area. Soil loss was estimated from a widely used RUSLE (Revised Universal Soil Loss Equation) given by Renard et al. [4]. Sediment yield (SY) was estimated with the help of sediment delivery ratio (SDR) and soil loss (SL). All equations used in this study are given in Table 1.

R factor was calculated using 30-min rainfall intensity data. Three soil types were found in the study area. K factor of these soil types are shown in Fig. 2. Slope length and steepness factor were extracted using ASTER GDEM (30 m resolution) in the GIS environment. Values of C and P factors were obtained from Anees et al. [2] based on the land used land cover map.

Landslide zones in the study area, extracted from Hashim et al. [5], are shown in Fig. 2.

Correlation between soil erosion and landslide zones. The criteria for correlation between soil erosion and landslide zones was based on the percentage of different categories of soil erosion. Different categories are the classification of soil erosion given in the guideline for soil erosion and sediment control in Malaysia issued by the Department of Irrigation and Drainage, Malaysia. Generally, correlation values vary between −1 (perfect negative correlation) to 1 (perfect positive correlation). Soil erosion cannot be negative, therefore, the range in this study is 0 (very low correlation) to 1 (very high correlation). For each soil erosion category, soil erosion is under 50% in the study area. Therefore, in terms of soil erosion percentages, the criteria for correlation are divided as very low (10%), low (20%), average (30%), high (40%), and very high (50%).

Table 1 Equations used for soil erosion estimation

S. No.	Equations	Reference
1.	$SL_{(xy)} = R_{(xy)} \times K_{(xy)} \times L_{(xy)} \times S_{(xy)} \times C_{(xy)} \times P_{(xy)}$ where SL is estimated soil loss, R is rainfall erosivity factor, K is soil erodibility factor, L is slope length factor, S is steepness factor, C is cover management factor, and P is support practice factor	[4]
2.	$R = \frac{1}{y}\sum_{j=1}^{y}\sum_{i=1}^{d_j}(EI_{30})_i$ where, y is the number of years of the record, d_j is the total number of days for a given year j, EI_{30} is the rainfall erosivity index of a single day I, and I_{30} is the maximum rainfall intensity in a 30 min period during the event (mm h^{-1})	[6]
3.	$L_{(xy)} = (m_{(xy)} + 1) * \left(\frac{X_{(xy)}}{22.1}\right)^{m_{(xy)}}$ where, $X_{(xy)}$ is flow accumulation of a grid cell and $m_{(xy)}$ a variable slope-length exponent	[7]
4.	$S_{(xy)} = -1.5 + \frac{17}{\left(1 + e^{\left(2.3 - 6.1 * \sin\theta_{(xy)}\right)}\right)}$	[8]
5.	$m_{(xy)} = \beta_{(xy)} / \left(1 + \beta_{(xy)}\right)$ where $\beta_{(xy)}$ is a grid cell ratio of rill to interrill erosion	[9]
6.	$\beta_{(xy)} = (\sin\theta_{(xy)}/0.0896) / \left(3 * (\sin\theta_{(xy)})^{0.8} + 0.56\right)$ where $\theta_{(xy)}$ is slope angle in degrees of a grid cell	[10]
7.	$SDR = 0.51 \times A^{-0.11}$ where A is the area of the watershed	[11]

3 Results and Discussions

3.1 RUSLE Factors

Rainfall erosivity factor ranged from 16,490 to 21,330 MJ mm ha^{-1} h^{-1} $year^{-1}$ with an average of 18,560 MJ mm ha^{-1} h^{-1} $year^{-1}$. High R factor values were observed in the western part of the study area and decreased towards the east direction. It should be noted here that western and eastern parts are high and low elevation areas. The K factor varies from 0.0365 to 0.0514 Mg h MJ^{-1} mm^{-1}. These are associated with clay loam and silt clay loam on the soil series Kangar and Cherang Hangus-Lubok Kiat respectively (Fig. 3).

Seven land use and land cover classes were identified by visual classification. These are dense forest, forest, mixed vegetation, cultivated land, scrub, settlement area, and waterbody (Fig. 4). Cover management factor and support practice factor values and maps are shown in Table 2 and Fig. 5.

Elevation values vary from 59 to 939 m with an average of 243 m. The slope of the study area varies from 0 to 75.8 degrees with an average of 14 degrees. Most of the high slope values are observed in the east associated with hilly terrain. The steepness factor ranged from 0.04 to 15.1, while the slope length factor ranged from 0 to 737 (Fig. 6).

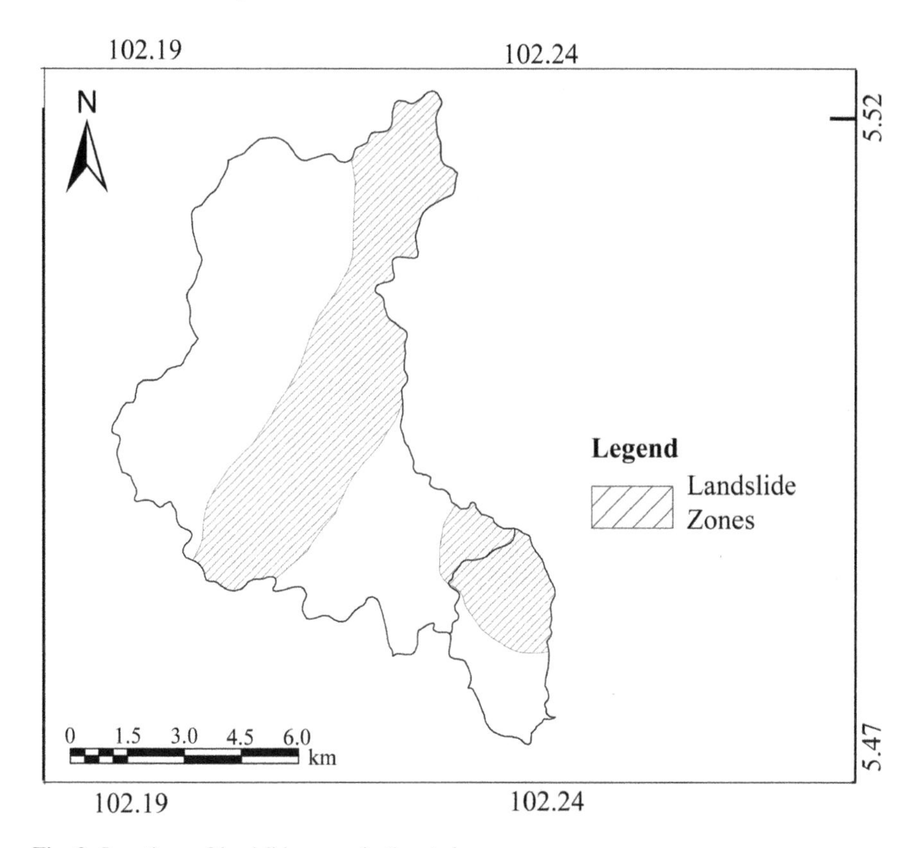

Fig. 2 Locations of landslide zones in the study area

Higher values are associated with the downstream side situated at the east of the study area. 81% of the study area has slope length and steepness values less than 10. Very high values are only 2.7% which are associated with low elevation areas.

3.2 Average Annual Soil Loss and Sediment Yield

Average annual soil loss ranged from 0 to 1,429,431.0 t ha−1 year−1 with an average of 3185.0 t ha−1 year−1. 38% of the area has zero soil loss, while 40% area has very high soil loss. Rest 22.5% area has low to high soil loss. Both zero and very high soil loss are randomly distributed in the study area. Most of the very high soil loss is associated with low elevation areas. Low soil loss is associated with dense forest.

A sediment delivery ratio is required to calcu-late sediment yield. For the study area, the sed-iment delivery ratio is 0.298498. Sediment yield varies from 0 to 426,682.0 t ha−1 year−1. 38% of the area has zero sediment yield, 25% has low

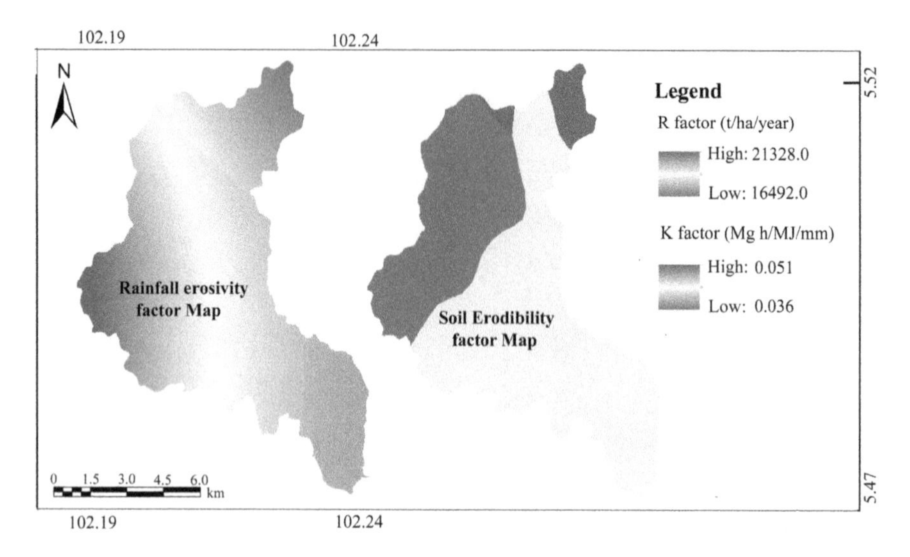

Fig. 3 Rainfall erosivity and soil erodibility map of the study area

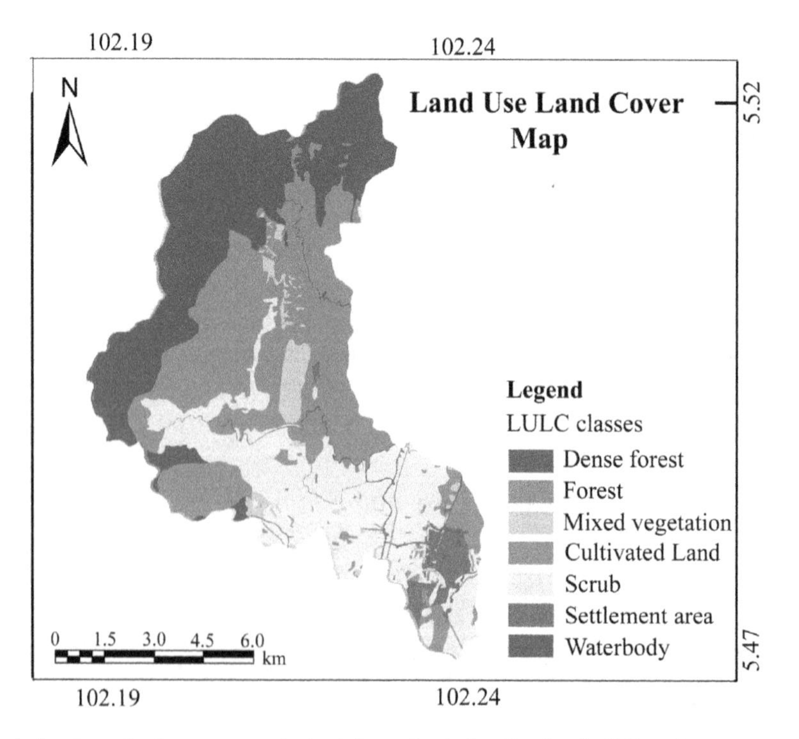

Fig. 4 Land use land cover map obtained from visual classification in GIS environment

Table 2 Land use land cover (LULC) classes with cover management (C) and support practice (P) factor values

LULC categories	C factor values	P factor values
Dense forest	0.03	0.8
Forest	0.03	0.8
Mixed vegetation	0.25	0.4
Cultivated land	0.85	0.7
Scrub	0.35	0.8
Settlement area	0.15	1

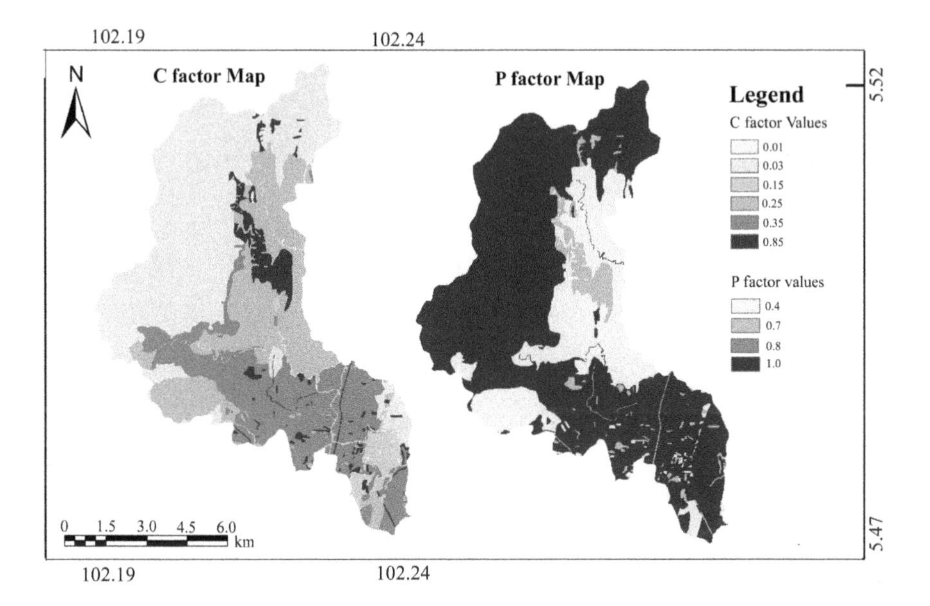

Fig. 5 Cover management and support practice factor maps

sediment yield, 30% has high sediment yield, and 8.5% has medium sediment yield. As compared to soil loss, high sediment yields are concentrated in the southern part of the study area (Fig. 7).

3.3 Correlation Between Soil Erosion and Landslide Zones

42% of the study area is under landslide zones. Soil loss in the landslide zone varies from 0 to 1,429,431.0 t ha^{-1} $year^{-1}$, which is the same as discussed above. It indicates that high soil loss occurred in landslide zones. Zero soil loss covered 39% of the area while 43% covered very high soil loss in landslide zones. Settlement areas are either in a landslide zone or in nearby areas, which could be at risk. The correlation according to different soil categories is given in Table 3.

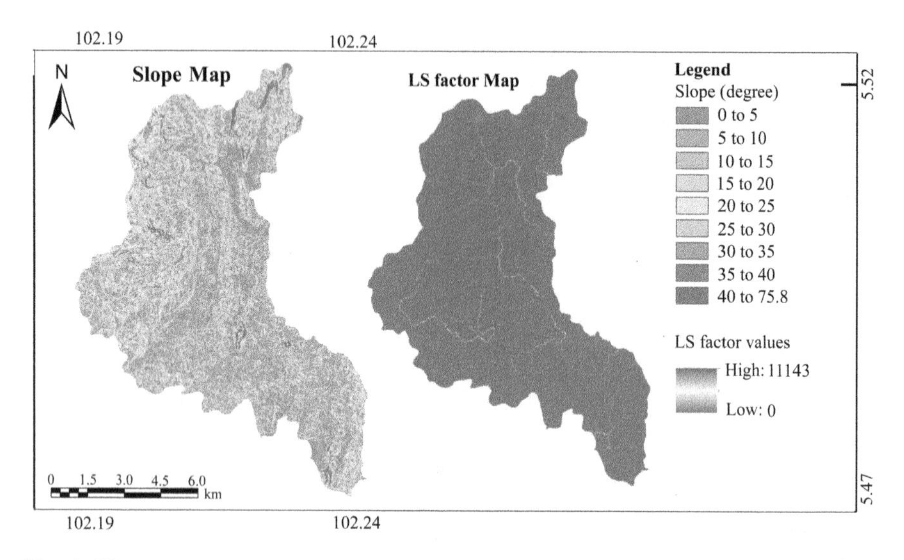

Fig. 6 Slope map and slope length and steepness factor maps

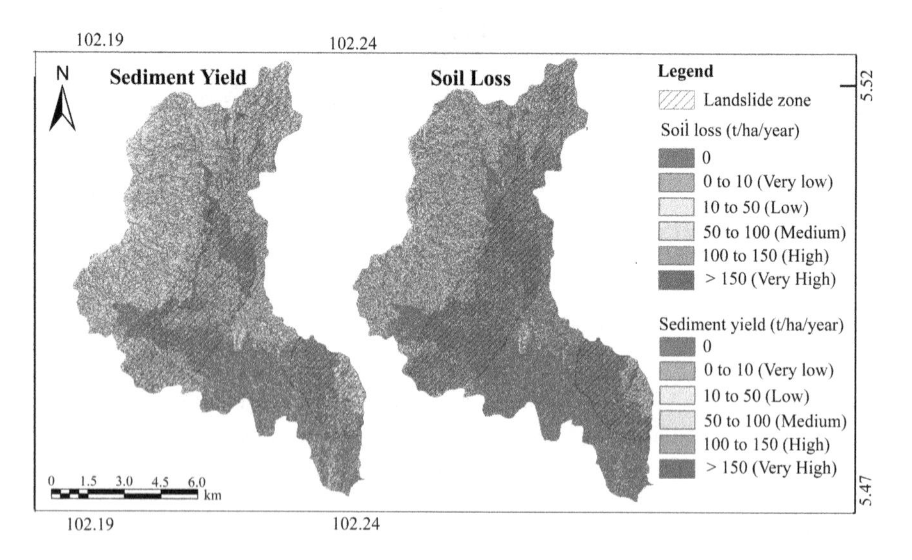

Fig. 7 Soil loss and sediment yield in study area and landslide zones. The classification of very low to very high is according to soil erosion guideline of Malaysia [12]

Table 3 Correlation between soil erosion and landslide zones based on soil erosion percentages of different categories

Zones	Soil erosion categories (t ha^{-1} $year^{-1}$)	Soil erosion area (percent)	Correlation
Landslide zone	0	39.05	High
	0–10	1.10	Very low
	10–50	4.78	Very low
	50–100	6.29	Very low
	100–150	5.65	Very low
	>150	43.14	Very high
Non-landslide zone	0	36.91	High
	0–10	1.93	Very low
	10–50	7.42	Very low
	50–100	11.00	Low
	100–150	5.51	Very low
	>150	37.22	High

4 Conclusion

The present study was conducted in a hilly area to identify the impact of soil erosion on landslide vulnerability. Rainfall erosivity is very high due to high precipitation. The highest elevation and slope in the area are 939 m 75.8 degrees, which high and vulnerable to the landslide. Following high elevation and slope values, soil loss and sediment yield in the area are also high. 40% very high soil loss is calculated. However, only 25% very high sediment yield is calculated. A very high correlation between soil erosion category (>150 t ha^{-1} $year^{-1}$) and landslide zone in the study area is found, while a high correlation for zero t ha^{-1} $year^{-1}$. It showed high landslide vulnerability in the study area. Furthermore, more than half of the settlement area comes under the landslide zone, which could be vulnerable to people of the study area.

Acknowledgements We gratefully acknowledge the government agencies, who provide data to conduct this study. We are also acknowledge the University of Malaya for providing financial support through Grant GPF017B-2018 to carry out this work.

References

1. Pradhan B, Chaudhari A, Adinarayana J, Buchroithner MF (2012) Soil erosion assessment and its correlation with landslide events using remote sensing data and GIS: a case study at Penang Island, , Malaysia. Environ Monit Assess 184(2):715–727
2. Anees MT, Abdullah K, Nawawi MNM, Norulaini NAN, Syakir MI, Omar AKM (2018) Soil erosion analysis by RUSLE and sediment yield models using remote sensing and GIS in Kelantan state, Peninsular Malaysia. Soil Res 56(4):356–372

3. El Jazouli A, Barakat A, Khellouk R, Rais J, El Baghdadi M (2019) Remote sensing and GIS techniques for prediction of land use land cover change effects on soil erosion in the high basin of the Oum Er Rbia River (Morocco). Remote Sens Appl Soc Environ 13:361–374
4. Renard KG, Foster GR, Weesies G, Mccool D, Yoder D (1997) Predicting soil erosion by water: a guide to conservation planning with the revised universal soil loss equation (RUSLE). US Government Printing Office, Washington, DC
5. Hashim M, Misbari S, Pour AB (2018) Landslide mapping and assessment by integrating Landsat-8, PALSAR-2 and GIS techniques: a case study from Kelantan State, Peninsular Malaysia. J Indian Soc Remote Sens 46(2):233–248
6. Renard KG, Freimund JR (1994) Using monthly precipitation data to estimate the R-factor in the revised USLE. J Hydrol 157:287–306
7. Mitasova H, Hofierka J, Zlocha M, Iverson LR (1996) Modelling topographic potential for erosion and deposition using GIS. Int J Geog Inform Syst 10(5):629–641
8. Nearing M (1997) A single, continuous function for slope steepness influence on soil loss. Soil Sci Soc Am J 61:917–919
9. Foster GR, Meyer LD, Onstad CA (1977) A runoff erosivity factor and variable slope length exponents for soil loss estimates. Transactions of the ASAE. Am Soc Agric Eng 20:0683–0687
10. McCool DK, Foster GR, Mutchler C, Meyer L (1989) Revised slope length factor for the universal soil loss equation. Transactions of the ASAE. Am Soc Agric Eng 32:1571–1576
11. Ouyang D, Bartholic J (1997) Predicting sediment delivery ratio in Saginaw Bay watershed. In: Proceedings of the 22nd national association of environmental professionals conference, 19–23 May, Orlando, Florida. National Association of Environmental Professionals, USA, pp 659–671
12. Department of irrigation and drainage: guideline for erosion and sediment control in Malaysia. Department of Irrigation and Drainage (DID), Kuala Lumpur (2010)

GIS Based Circuity Analysis of Transport Network Structure of Calicut

M. G. Sreelekha, K. Krishnamurthy, and M. V. L. R. Anjaneyulu

Abstract Road transport network has a key role in the overall development of any city. Analysing the road network development can help to assess the transportation performance and efficiency. Most studies that attempt to characterise the transport network focussed on easily measured indices such as percent connections or network density. Circuity has not been addressed in most of the network studies. When travelling between two places, most individuals choose routes that help to minimise the time and cost of travel. Here comes the intricate role of network circuity in the transportation studies. This study analysed the variation in network circuity among the zones of Calicut city, Kerala using GIS. The analysis shows that network circuity has lower values in the central zones and higher values in the outer suburban zones of the city. This illustrates that people are forced to travel distance much greater than the straight line distance, particularly for travel within the suburban zones of the city. Further analysis is to identify the factors that cause a network to be circuitous. The study observed that network density together with nodal connectivity can have significant effect in lowering the network circuity. Other than the coverage of road length per unit area, road stretches are to be optimally connected for reducing the circuitness of the route. This forms a useful guideline for designing a better performing transport network, particularly in the developing countries.

Keywords Transport network · Circuity · Structure · Urban area · Travel

M. G. Sreelekha (✉)
GEC Trichur, Trichur, India

K. Krishnamurthy · M. V. L. R.Anjaneyulu
NIT Calicut, Calicut, India
e-mail: kk@nitc.ac.in

M. V. L. R.Anjaneyulu
e-mail: mvlr@nitc.ac.in

G. C. Marano et al. (eds.), *Proceedings of SECON'21*, Lecture Notes in Civil Engineering 171, https://doi.org/10.1007/978-3-030-80312-4_68

1 Introduction

The existing road network of urban areas of most developing nations are inefficient, as they are not properly designed so as to reduce the travel. As the road network exhibit an irregular pattern and the road segments exhibit variation in number and length, there should be an essential criterion for designing the network so as to reduce the route length. Researchers have estimated many network parameters as an initiative to assess the transportation efficiency [1, 2]. Easily quantifiable measures such as percent connections and network density are widely used to analyse the transport network [3]. To sustain the growth of urban centres and to improve the quality of life, an efficient network is essential [4]. Again, the network is to be efficient in terms of the transportation performance.

Many transportation studies state the preference of least-cost path for almost all trips, exclusive of the mode or purpose of the trip [1, 5]. This, in turn, emphasise the preference of direct routes by the individuals, for reaching the destinations. Many researchers have included travel distance as measured by the euclidean distance for measuring the travel, due to the effort required in obtaining the real network distance [6]. Actually, network distance will be greater than the euclidean distance, due to the presence of obstacles such as terrain, water-bodies or mountains. But this network distance along the real travel path will be precise and specific. Travel path becomes circuitous when the actual travel distance differs from the euclidean distance [7]. Levinson and El-Geneidy [1] reported circuity as the ratio between the actual network distance and the direct straight line distance for any travel. An index of 1.0 is the best possible rating of network circuity, when the network distance approximates the euclidean distance,which indicates that people can travel directly to reach the destination [1]. Gonçalves et al. [8] have reported values of circuity approximating 1.3 for urban travel. Ballou et al. [9] observed the circuity for inter-city travel between 1.12 and 2.10. A similar concept as that of circuity is explained by the detour index and the inverse concept is described by directness [10, 11].

Past researches on network circuity and its effects are detailed here. Barrington-Leigh and Millard-Ball [12] and Giacomin and Levinson [13] pointed out that the addition of new road stretches in suburban zones cause an increase of network efficiency. Huang and Levinson [14] argue that circuity of the transit network is higher than that of the real road network. Travel decisions such as choosing the destination, route, mode, and even the activity to be performed is based on the parameter of circuity of the network as per Parthasarathi et al. [15]. Parthasarathi [7] discussed the substantial role of network circuity in transportation system performance, measured on the basis of the congestion and system usage across the metropolitan cities of the United States. These studies demonstrate the strong relationship between circuity and travel, indicating the importance of network design in transportation system performance. Ballou et al. [9] argue that road network circuity is caused due to network-related factors such as road density and connectivity, but this thought was not supported by correlation analysis. Giacomin and Levinson (2015) is of the opinion that the route circuitness may be due to the presence of obstacles, but the critical factors that cause circuity need further attention. To address this, the present study attempts to explore the road network circuity of various traffic analysis zones of Calicut Corporation located in Kerala, India. The other objective is to identify the network parameters that cause the circuity of the road network. This can be useful as a reference for effective decision making while planning an efficient transport network.

2 Materials and Methods

2.1 Study Area

Calicut is a medium-sized city in Kerala and is one of the most dynamic cities with many business and commercial establishments. A building boom has witnessed in the city due to the development of malls and multi-storeyed apartments in the recent years. Presently, there are seventy-five electoral wards under Calicut Corporation. These electoral wards are considered as the Traffic Analysis Zones (TAZs) for the present study. The study area is characterised by an area of 118 square kilometres and a population of 6.2 lakhs (as per the 2011 census). The Central Business District (CBD) of Calicut is Palayam. The location map of Calicut Corporation, including the zonal delineation, is given in Fig. 1. The road network map of Calicut Corporation is shown in Fig. 2. The network length is about 521 km and the network density is 4.42 km per square kilometre. Functionally the road stretches can be classified into roads enabling intercity trips and roads enabling intra-city trips. National Highways function as the roads enabling intercity trips. Intra-city roads are the city roads connecting the residential environment to the CBD and other job centres.

2.2 Methodology and Data Description

The analysis of network circuity involves data collection, database development in GIS and finally estimating the circuity of the network of each zone using the geospatial techniques of GIS. For this, the data related to the seventy-five zones of the municipal corporation of Calicut are collected from relevant sources. The CAD format of the zone boundary and the road network details are collected from Calicut Corporation Office. Using GIS software, these files are converted to shapefiles. The attributes like name and type of road are collected by conducting a road inventory survey and then entered into the attribute table. All the maps are brought into the same coordinate system and later manipulated for further applications.

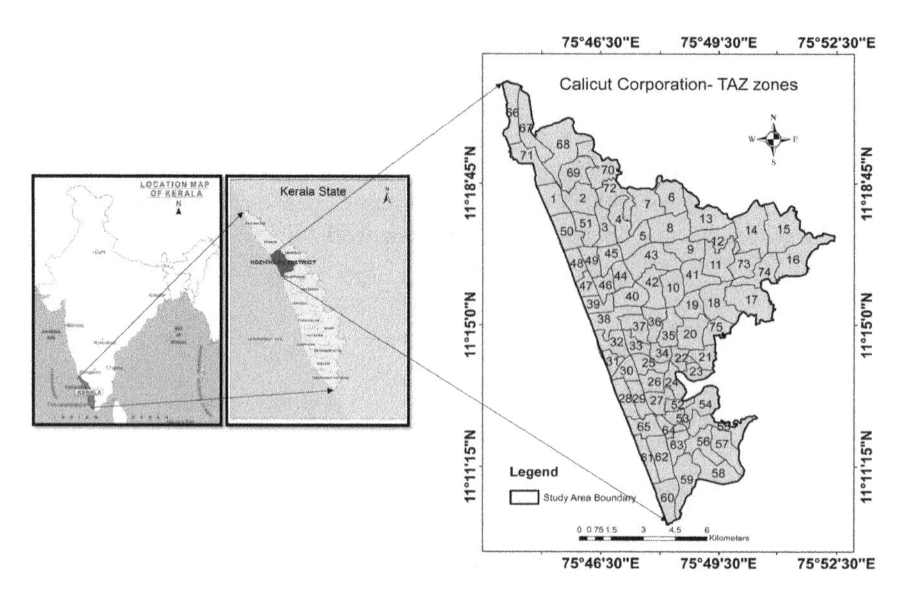

Fig. 1 Location map of Calicut corporation

The parameter of network circuity is calculated following a step by step procedure. The road network layer is clipped with the zonal boundary layer, so as to extract the road network in each zone. Each network is converted to a network dataset using GIS, by which the network gets split into nodes and edges. The network distance between each pair of nodes is obtained as a table. In this attempt, the nodal points get joined via the next adjacent link. With route distance as the impedance function, Dijkstra's [16] algorithm helps to find the shortest path. The implicit assumption is that travel takes place along the shortest path between the origin–destination pair. Accordingly, the shortest network distance between two nodal points (intersections) is identified as a proxy for the actual network distance. From the network distance between all such nodal pairs, cumulative network distance is calculated. Euclidean distance is calculated by making use of the point distance tool in GIS, taking the nodal intersection shape-file as the input feature as well as the near feature. Then the cumulative euclidean distance is calculated. Ratio of cumulative network distance and cumulative euclidean distance gives the network circuity of the zone. Following this technique, the values of road network circuity of all the zones are attained.

As a next step, the variation in network circuity of different zones is studied. Circuity estimates can be a better approximation of the distance travelled along the network. Further analysis included the identification of the potential factors capable of explaining network circuity. This is attained by performing correlation analysis and regression analysis taking circuity as the dependant variable and the probable network parameters as the explanatory variables.

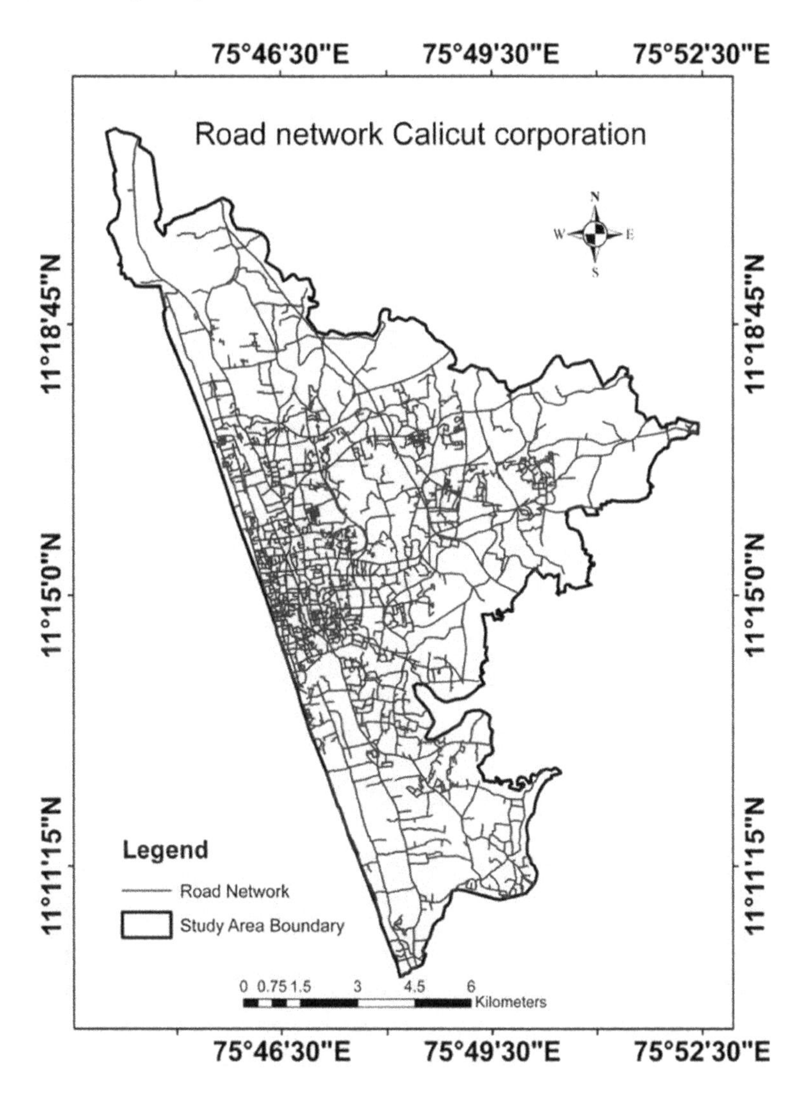

Fig. 2 Digitised road network of Calicut corporation

3 Result and Discussions

3.1 Analysis of Network Circuity

The variation in network circuity across the zones of Calicut is examined here. An attempt is also made to identify the network-related factors which make a network circuitous. Network circuity of all the zones was estimated as explained earlier. Network circuity varied from a minimum value of 1.018 (corresponding to CBD zone) to a maximum value of 2.063 (corresponding to suburban zone). This indicates that the central zones tend to have the least circuitous routes. Recently developed suburban zones with more dead ends and comparatively longer road stretches tend to have the most circuitous routes. Thematic map representation of circuity is given in Fig. 3.

A closer look at the figure indicates that the road network of the outer suburban zones shows higher network circuity than the segments falling in the central/inner zones. As the distance from CBD increases, the circuity of the network gets increased. Such variation can be explained by the intense network connections of

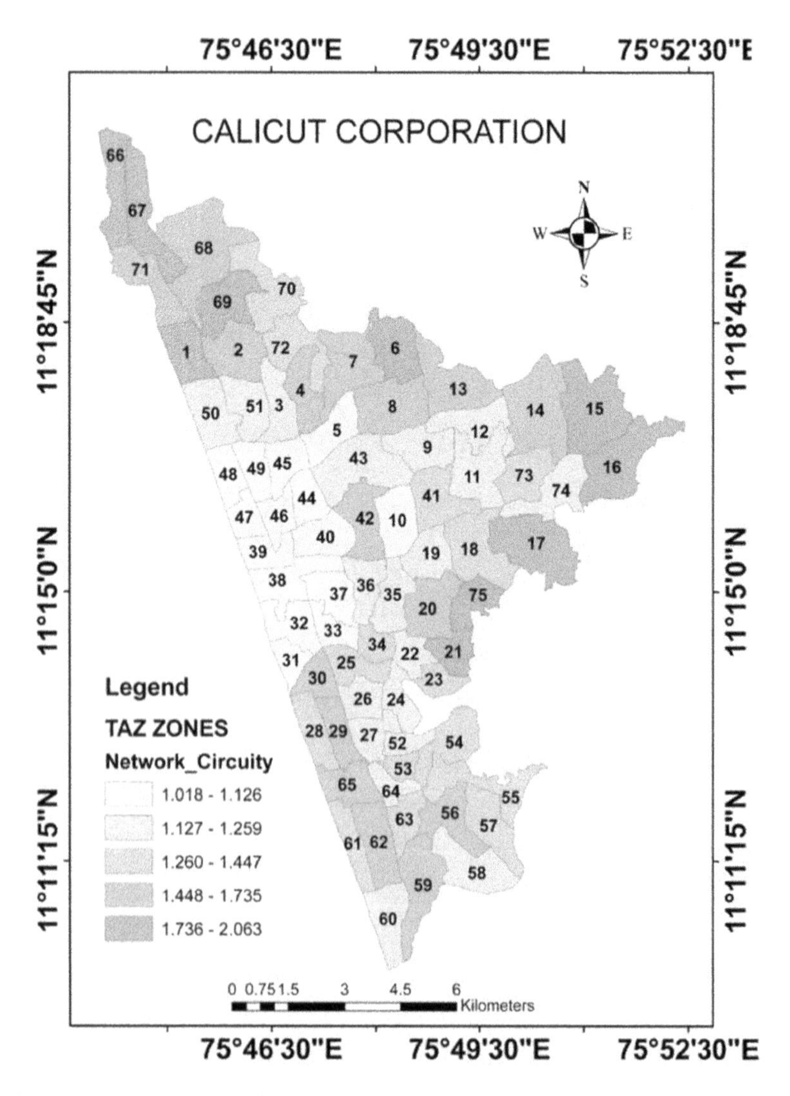

Fig. 3 Variation of road network circuity

the central city road network. The zones away from CBD have lesser connections and hence the road links will be long and mostly curved. This prompts the travellers to follow a circuitous route, for movement in zones away from CBD.

3.2 *Network Circuity and Its Influencing Factors*

In order to obtain the relationship between network circuity and its explanatory variables, regression analysis is performed. The purpose of such analysis is to predict network circuity, provided the local-level network parameters are known. The local-level network parameters as connectivity indices and coverage indices are proposed as the influencing factors.

Connectivity measures indicate the intensity of the connections among the road stretches. The indices used for estimating the network connectivity are number of links, number of nodes, average link length, ringness, percent cul-de-sac and beta index. Beta index is preferred over other graph theory indices considering its application in most network studies. Beta Index [17] compares the number of links to the number of nodes. It has a value of 'zero' when no edges, 'one' when included one circuit, and exceeds 'one' for a complicated network with numerous connections. Average link length [18] indicates the approximate length of a link. This index is used as a measure of speed in the network with the assumption that the longer the segment, the better the speed attained in the road stretch. Ringness [19] indicates the cumulative length of road stretches lying on ring-type network expressed as the ratio of the total network length. The percent of cul-de-sac refers to the nodes of degree one, which indicates the nodes having the connection to one and only one link [19]. Coverage measures indicate network growth and development and is assessed by network density and intersection density. Network density [3] is measured as the length of the road network per unit area. The greater the value of network density, the more the system will be developed. Intersection density [15, 20] refers to the count of intersections per unit area.

The dataset includes the network characteristics as the number of nodes, number of links, average link length, beta index, percent cul-de-sac, ringness, network density and intersection density confirming the seventy-five zones of the study area. The network dataset retrieved from GIS gives the number of nodes and the number of links. The cumulative length of the links within a network averaged over the number of links gives the average link length. The connectivity index of beta is obtained as the number of links to the number of nodes. The cumulative length of the ring type stretches gives the ringness of the network. The count of dead ends with the network is to deduce the percent cul-de-sac. The total length of roads and the total number of nodes within each zonal network, when divided by the corresponding land area of each zone, gives the network density and intersection density of each zone. The proposed statistical model is given in Eq. (1).

$$\text{Circuity} = \text{f (number of nodes, number of links, average link length, beta, ringness, percent cul-de-sac, network density, intersection density)}. \quad (1)$$

To obtain the correlation between each of the dependent and influencing variables, Pearson's correlation is performed using the statistical SPSS package. Correlation analysis table is not shown here because of space limitation. The correlation analysis indicated that the network characteristics such as number of links, number of nodes, beta index, ringness and network density have negative association with network circuity. Average link length and percent cul-de-sac have positive influence on network circuity.

The next step is to perform regression analysis to identify the parameters that have an effect on network circuity. First of all, the variable of highest correlation is entered into the model; then the next highest, but that having the least correlation with the one entered, and so on. Following this procedure, linear regression model is developed. Table 1 shows the parameter estimates of the model developed. R^2 value of 0.756 indicates that the model has excellent predictive ability. The t-statistic values higher than the critical value (1.96) at significance level of $\alpha = 0.05$, indicate that the coefficients of these variables are significantly different from zero. Number of links, average link length and beta index did not enter into the model as they are not significant variables. All the other variables have entered into the model. The final model of network circuity can be interpreted as in Eq. (2).

$$\text{Network circuity} = -0.013 * \text{Nodes} - 0.097 * \text{Ringness} + 0.059 * \text{Percent cul-de-sac} - 0.077 * \text{Network density} \quad (2)$$

Network density, which is having the highest correlation to network circuity, has entered as the most significant variable. Though beta index has the next highest correlation to network circuity, the variable of beta is having interdependency to

Table 1 Predicting network circuity

Dependent variable: network circuity (y)						
Explanatory variables (x)	Coefficient	Std. error	t-value	Sign	R^2-value	F-value
Nodes	− 0.013*	0.004	− 3.782	0.000	0.756	82.13
Average link length	0.395	0.317	1.275	0.217		
Beta index	− 0.471	0.079	1.247	0.308		
Ringness	− 0.097*	− 0.032	3.720	0.007		
Percent cul-de-sac	0.059*	0.028	2.563	0.003		
Network density	− 0.077*	0.009	− 8.305	0.000		

*Significant at 0.05 level

network density. Hence, the beta index has arrived as an insignificant variable. The same is the case for average link length and number of links. Other variables as the number of nodes, ringness and percent cul-de-sac have emerged as significant variables having influence on network circuity. Coefficients of all the variables, except that of percent cul-de-sac are negative, which implies that network circuity increases with a decrease in values of those variables.

That is, more the road length per unit area, the denser will be the network, allowing individuals to travel a direct straight distance. Nodal connections indicate that the road stretches are to be well-connected. As per the analysis, network density and connectivity alone may not guarantee the reduction of network circuity. Ringness and percent cul-de-sac are supplementary factors deciding the circuity. Ringness, the index demonstrating the presence of network rings, suggests the presence of closed paths within the network and this has a negative association with network circuity. The presence of cul-de-sacs (or dead ends) indicates the unconnected nodal intersections. On a network having many unconnected dead-ends, people will be forced to travel more distance to reach the destination, which indeed restricts the direct motion.

Further, the model has been validated by calculating the predicted values of circuity. Comparison of predictions of network circuity with the measured values of network circuity is included in Fig. 4. The map of predicted network circuity indicates a similar pattern as the map of measured network circuity. This ascertains the validation of the model developed.

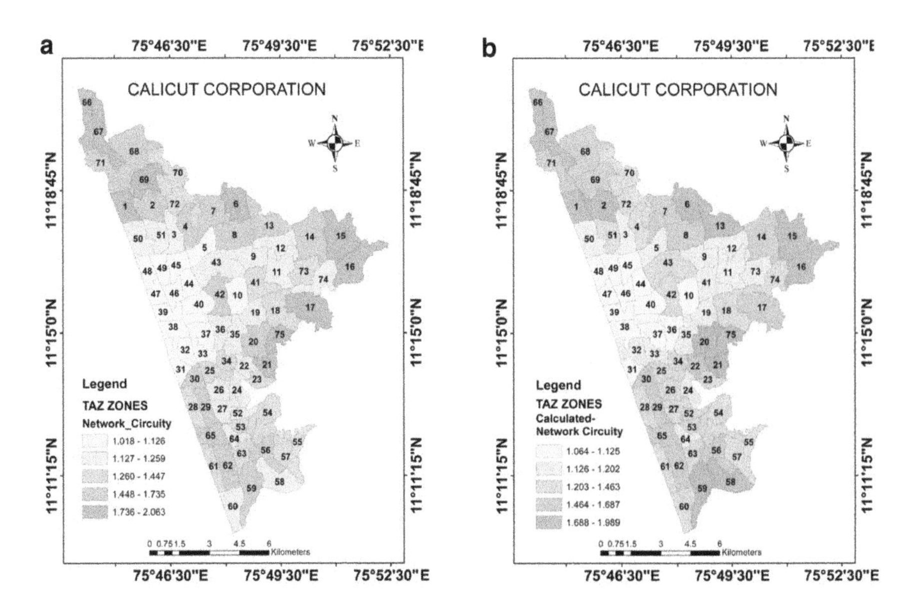

Fig. 4 **a** Measured network circuity **b** predicted network circuity

3.3 Discussion

As part of this study, the objective was to quantify the network circuity of the zones of Calicut Corporation. Circuity is the minimum for the zones near CBD, while the maximum for the suburban zones. This means that the road network of central zones of the city allows comparatively direct travel. The individuals are forced to cover greater network distance, for travel within zones farther from CBD. The denser the road network, the more connected the network will be, which minimises the need to travel a greater distance. For travel within suburban zones, individuals will have to cover the longer distance for work or non-work purpose, irrespective of the travel mode.

The study further investigates the micro-level network parameters for predicting the variation of network circuity. The present study has been restricted to identify the network-related factors alone, which can cause the circuitness of the network. The predicted model of network circuity illustrates that the parameters as ringness, number of nodes and network density are positively significant, while percent cul-de-sac influence negatively. This ascertains the effect of connectivity together with network density and ringness, on the circuity measure. For creating an efficient road network, the roads should be designed with minimum network circuitness. The result of the analysis explains that a denser network with optimal connections helps in reducing the circuity of travel routes. If the road network is ring type with smaller links and better connectivity, people can travel nearly directly to the destinations, resulting in lower circuity. Hence the road stretches are to be dense and optimally connected so as to make the travel route less circuitous. This can help to understand the design criteria of the transport network, so as to optimise the travel distance (or travel cost) within the urban locations. This forms the guideline support while designing or improving the existing road network. As the circuity factor gives an idea of the actual distance travelled in comparison with the direct distance between the origin–destination pair, such analysis may be generalised in other cities of similar character so as to help in measuring the performance of road network.

4 Circuity Evaluation Criteria for Network Design-Application

This section illustrates how the model prepared in the study can be applied to restrict the circuity, while designing the road network. Referring to the network circuity model as in Eq. (2), a road network having the following characteristics may be considered.

Number of links = 200, Average link length = 200 m.
Average land use area = 2.5 km^2, Number of nodes = 60.
Ringness = 0.6, Percent cul-de-sac = 2.1.

Then network density = 0.2 * 200/2.5 = 16 km/km^2.
Hence network circuity = −0.013 * 60 − 0.097 * 0.6 + 0.059 * 2.1 − 0.077 * 16 = 1.946.

As a further application, performance of the model has been examined based on sensitivity analysis. On changing the value of one of the model parameters, simultaneously keeping the other parameters constant, how the dependent variable changes have been studied here. 1% increase in the number of nodes decreases the network circuity by 0.78%. 1% increase of ringness decrease the network circuity by 0.05% and percent cul-de-sac increase the network circuity by 2.22%. Likewise, network circuity gets reduced by 0.11%, due to network density. By such detailed analysis, the network parameters to be changed so as to limit the network circuity and thereby improve the travel performance can be studied. This can help the planners to design the road network so as to assist the travellers to follow the most direct and high-performance routes.

5 Conclusions

Of the various measures for characterising the transport network, network circuity is an efficient measure capable of addressing the route travel distance. This study concentrated on estimating the network circuity of the seventy-five zones of Calicut, Kerala. Further, the micro-level network parameters that cause the network circuitness were identified.

As per the study, the average network distance approximates the direct straight line distance for travel within central zones. On the contrary, the suburban zones have more circuitous routes, indicating that recently developed areas are less connected than the central zones. Analysis in the further section indicates that circuitness of road network of local travel is defined not only by the presence of obstacles but also by some network-based topological parameters. Increased road network density together with better connectivity and ringness can help to reduce the path circuitness and thus increase the network efficiency. Designing a network considering these micro-level parameters can certainly help to improve the route directness and thus attain reduced travel. This study can help the planning authorities to design an efficient transport network that can favour the shortest route-based travel. Application services include the trips undergone by the school going students, local trips for shopping, delivery of online booking etc.

References

1. Levinson DM, El-Geneidy A (2009) The minimum circuity frontier and the journey to work. Reg Sci Urban Econ 39:732–738
2. Sreelekha MG, Krishnamurthy K, Anajaneyulu MVLR (2016) Assessment of topological pattern of urban road transport system of Calicut city. Transp Res Procedia 17:253–262

3. Bento A, Cropper M, Mobarak A, Vinha K (2003) The Impact of urban spatial structure on travel demand in the United States
4. Sreelekha MG, Krishnamurthy K, Anjaneyulu MVLR (2020) Urban road network and its topology: case study of Calicut, India. Eur Transp\Trasporti Europei, issue 78, Paper n 6. ISSN 1825–3997
5. Boisjoly G, El-Geneidy A (2016) Are we connected? Assessing bicycle network performance through directness and connectivity measures, a Montreal, Canada case study, Transportation Research Board. In: 95th annual meeting transportation research board
6. Hsu C-I, Wang H-M (2016) Strategies for green transportation while preserving mobility and accessibility: a case study of Taipei city. J Urban Plann Dev 142(1)
7. Parthasarathi P (2014) Network structure and metropolitan mobility. J Transp Land Use 7(2): 153–168
8. Gonçalves NS, Gonçalves DM, Assis CF, Marcelino TS (2014) Analysis of the difference between the euclidean distance and the actual road distance in Brazil. Transp Res Procedia 3
9. Ballou RH, Rahardja H, Sakai N (2002) Selected country circuity factors for road travel distance estimation. Transp Res Part A 36(9):843–848
10. Peponis J, Bafna S, Zhang Z (2008) The connectivity of streets: reach and directional distance. Environ Plann B 35:881–901
11. Ravulaparthy SK, Goulias KG (2014) Characterizing the composition of economic activities in central locations: graph-theoretic approach to urban network analysis. Transp Res Rec J Transp Res Board 2430:95–104
12. Barrington-Leigh C, Millard-Ball A (2015) A century of sprawl in the United States. Proc Natl Acad Sci 112(27):8244–8249
13. Giacomin DJ, Levinson DM (2015) Road network circuity in metropolitan areas. Environ Plann B: Plann Design 42(6):1040–1053
14. Huang J, Levinson DM (2015) Circuity in urban transit networks. J Transp Geogr 48:145–153
15. Parthasarathi P, Levinson D, Hochmair H (2013) Network structure and travel time perception. PLoS ONE 8(10)
16. Dijkstra EW (1959) A note on two problems in connexion with graphs. Numerische Mathematik 1:269–271
17. Kansky K (1963) Structure of transportation networks: relationships between network geometry and regional characteristics. Ph.D. thesis, University of Chicago, Research Paper No. 84
18. Jenelius E (2009) Network structure and travel patterns: explaining the geographical disparities of road network vulnerability. J Transp Geogr 17:234–244
19. Xie F, Levinson D (2007) Measuring the structure of road networks. Geog Anal 39:336–356
20. Cervero R, Kockelman K (1997) Travel demand and the 3Ds: density, diversity, and design. Transp Res D 2(3):199–219

Study on Labour Evacuation Time Using BIM and Agent Based Simulation During Construction of High-Rise Building

Sirajudeen Rahima Shabeen and S. Muhammad Ashraf Ali

Abstract Construction sites are labour intensive locations. During emergency, safe evacuation of labour from the construction site is necessary. The labour evacuation time depends on many factors namely number of workers working at a time, location of the worker and physical and psychological characteristics of the labour. The objective of this research is to calculate the evacuation time for labour during the construction of a multi-storey building using Agent Based Simulation (ABS) approach. The analysis is done using BIM and agent-based evacuation simulation in a high-rise residential building construction project consisting of G +10 floors. The project schedule is input into the model to know the exact number of labour at any given time point. Various time-points were selected from the schedule and the labour evacuation time was calculated. Further, the influence of change in schedule on the evacuation time was analyzed. Based on the simulation, it is observed that the labour evacuation time is influenced by number of labour in higher floors and the labour density.

Keywords Evacuation time · Agent based simulation · Construction site · High rise buildings · Labour

1 Introduction

Rapid urbanization has led to the spur in construction activities especially construction of multi-storied buildings. Construction is a highly labour-intensive activity involving many different types of labour performing various activities. Emergency situations like fire, explosion, collapse etc. during construction require safe and rapid evacuation of labour from the work location to the safe assembly

S. Rahima Shabeen (✉) · S. Muhammad Ashraf Ali
College of Engineering Guindy, Anna University, Chennai 600025, India
e-mail: rahima@annauniv.edu

G. C. Marano et al. (eds.), *Proceedings of SECON'21*, Lecture Notes in Civil Engineering 171, https://doi.org/10.1007/978-3-030-80312-4_69

point. However, the number of labour and their location in the building at different stages of construction varies greatly according to the work schedule. Hence, there is a need to dynamically evaluate the evacuation time at different stages of construction.

To this end, this research paper aims at evaluating the evacuation time at different stages of work schedule using simulation software and also finding the influence of work schedule on the evacuation time.

The calculation of evacuation time of labour in a construction site in this research paper essentially consists of the following important stages. (1) Development of three-dimension (3D) model of the building using Building Information Modelling (BIM) (2) Creation of construction schedule using Microsoft (MS) Project (3) Creation of a four-dimension (4D) model by linking time schedule and 3D building model developed in BIM (4) Development of evacuation simulation model using Agent based approach using Oasys MassMotion software.

Many studies have been done on evacuation time using simulations. Xiaoping et al. [1] studied the various methodologies for crowd evacuation and highlighted the need to include psychological and physiological elements affecting individual and collective behaviors in the development of the simulation model. Marzouk and Daour [2, 3] presented the profile of total evacuation time of construction labour that changed based on change of project activities over the duration of the project. Hua et al. [4] proposed a methodology to involve risk analysis in evacuation simulation model. Aleksandrov et al. [5] proposed a mathematical model for finding an optimal evacuation strategy for tall buildings employing Mixed-integer linear programming (MILP). Marzouk and Mohamed [6] evaluated the qualitative and quantitate aspects of evacuation simulation using Agent-Based simulation and Design Support Tools. Row [7] incorporated evidence based characteristics of labour in evacuation simulation model.

2 Details of the Structure

A typical reinforced concrete framed residential building of G+10 storey is chosen for the study. The floor height of each storey is 3 m and the floor area is 298 m^2.

Each storey comprises of four housing units and one staircase. The floor plan of each storey is similar with four housing units in each. Building is of height 33 m. The typical floor plan of one of the storey and elevation of the building is shown in Figs. 1 and 2 respectively.

The floor plan is drawn using the tool Revit indicating the main elements which will be constructed namely the walls, columns, beam and staircase. The 3D model of the building developed in Building Information Model (BIM) using Revit is shown in Fig. 3. The building has open storey at the ground floor with only columns and no

Fig. 1 Plan of the building

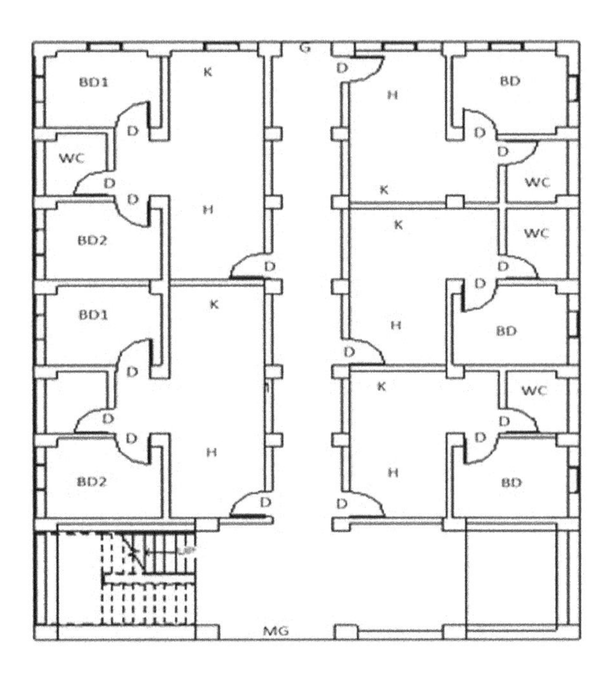

walls. This concept of using ground floor for parking is followed in most of the residential apartment complexes. There is a provision for lift but it is not considered as an exit point during construction. Also, lift is generally not preferred exit route in the case of emergencies. The only exit point in each floor is the staircase located at the corner of the building.

3 Construction Schedule

The construction activities for the construction of the G+10 storey RCC building are identified. The initial works are the clearing of the site and foundation works. The construction activities of each floor are similar. The first step is the construction of columns, beams and slabs which involves placement of formwork, bar bending and placement of reinforcement followed by concreting. The next is the construction of masonry walls. Followed by electrical and plumbing work, plastering and floor and wall finishing works. The duration of each activity is determined. The construction sequence is planned. A schedule is drawn based on the predecessor/ successor activities. The activities are scheduled using MS project. Figure 4 shows one part of the project schedule. Similarly, the project schedule was drawn for the whole of the project from start to finish (Appendix 1). The number of skilled and unskilled labours for each activity was found (Table 1). The total duration of the project is found to be 410 days. The number of labours involved in construction each day during the project schedule is shown in Fig. 5.

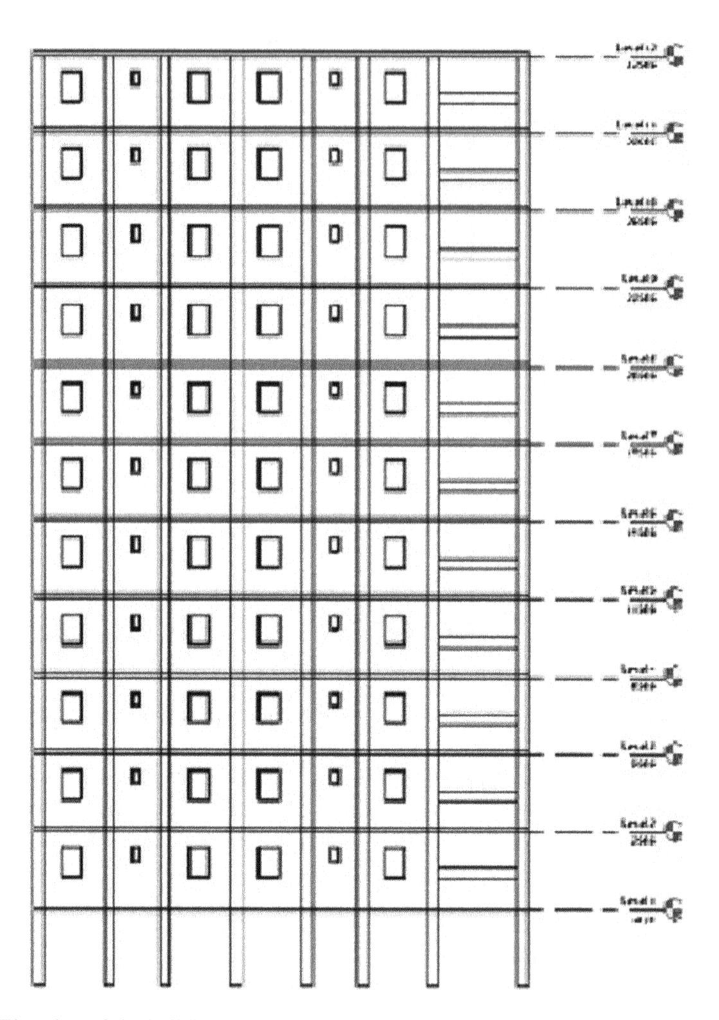

Fig. 2 Elevation of the building

4 Emergency Evacuation Model

Emergency evacuation can be simulated based on macroscopic or microscopic approach. Macroscopic approach models consider the crowd as a whole whereas microscopic approach considers each individual as a separate entity. Agent based simulation is a one of the microscopic evacuation models. The evacuation study in this research is done based on Agent simulation model using the crowd simulation software, Oasys MassMotion. The software, Oasys MassMotion is used to create the virtual building environment and to represent individual labour. The developed 4D building model is imported from BIM to MassMotion software. The evacuation simulation is done at twenty distinct time points selected from the schedule.

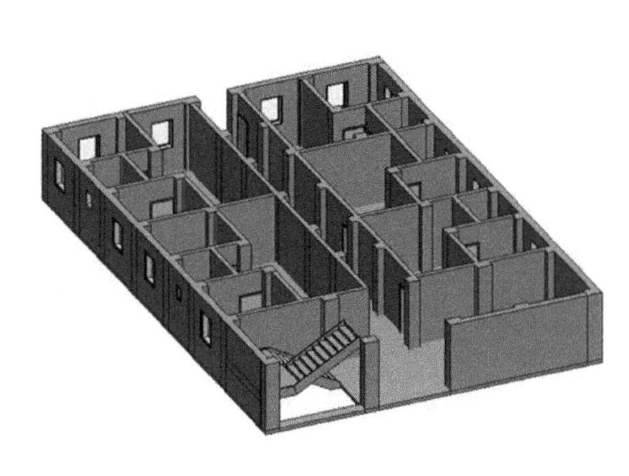

Fig. 3 3D model of the building

START OF PROJECT

Task ID	Task Name	Duration	Start	Finish	Predecessors
1	START OF PROJECT	1 day	Tue 05-03-19	Tue 05-03-19	
2	SURVEYING AND MARKIN	3 days	Wed 06-03-19	Fri 08-03-19	1
3	EXCAVATION	5 days	Sat 09-03-19	Fri 15-03-19	2
4	CEMENT CONCRETE FOUN	15 days	Fri 15-03-19	Tue 02-04-19	3
5	EARTH FILLING	4 days	Wed 03-04-19	Sat 06-04-19	4
6	− GROUND FLOOR	93 days	Mon 08-04-19	Tue 30-07-19	
7	PCC FLOORING	3 days	Mon 08-04-19	Thu 11-04-19	5
8	COLUMNS	8 days	Thu 11-04-19	Sat 20-04-19	7
9	SLAB FLOOR 1	14 days	Sat 20-04-19	Wed 08-05-19	8
10	SLAB CASTING	1 day	Wed 08-05-19	Thu 09-05-19	9
11	BRICK WORKS	7 days	Thu 06-06-19	Fri 14-06-19	21,10
12	ELECTRICAL WORKS	5 days	Fri 14-06-19	Thu 20-06-19	11
13	PLUMBING WORKS	3 days	Fri 14-06-19	Tue 18-06-19	11
14	PLASTERING WORK!	11 days	Thu 20-06-19	Thu 04-07-19	12,13
15	TILING WORKS	11 days	Thu 04-07-19	Wed 17-07-19	14
16	PAINTING WORKS	10 days	Wed 17-07-19	Tue 30-07-19	15
17	DOORS AND WINDO'	3 days	Wed 17-07-19	Sat 20-07-19	15
18	− FLOOR 1	90 days	Thu 09-05-19	Tue 27-08-19	
19	COLUMNS	8 days	Thu 09-05-19	Sat 18-05-19	10
20	SLAB FLOOR 2	14 days	Sat 18-05-19	Wed 05-06-19	19
21	SLAB CASTING	1 day	Wed 05-06-19	Thu 06-06-19	20
22	BRICK WORKS	7 days	Thu 04-07-19	Fri 12-07-19	32
23	ELECTRICAL WORKS	5 days	Fri 12-07-19	Thu 18-07-19	22
24	PLUMBING WORKS	3 days	Fri 12-07-19	Tue 16-07-19	22
25	PLASTERING WORK!	11 days	Thu 18-07-19	Thu 01-08-19	23,24
26	TILING WORKS	11 days	Thu 01-08-19	Wed 14-08-19	25
27	PAINTING WORKS	10 days	Wed 14-08-19	Tue 27-08-19	26,16
28	DOORS AND WINDO'	3 days	Wed 14-08-19	Sat 17-08-19	26,17

Fig. 4 Project schedule in MS project

The labour is the agent in the software. The agents are assigned physical characteristics according to the characteristics of a typical labour. The body radius of the individual labour agent is taken as 250 mm. The speed of agent is normally distributed in a range from 1.25 to 1.75 m/s with average 1.5 m/s and a standard deviation of 0.25 m/s. Various objects such as portals, floors, stairs and barriers are

Table 1 Duration and resources required

Activity	Duration	Number of crew	Number of labour in each crew	Total number of labour for each activity
Column rising	8	4	2 skilled labour, 3 labour	20
Slab work	14	4	2 skilled labour, 3 labour	20
Slab concreting	1	2	6 skilled labour, 14 labour	40
Brick work	7	4	2 skilled labour, 4 labour	24
Electrical works	5	2	1 skilled labour, 3 labour	8
Plumbing works	3	2	1 skilled labour, 2 labour	6
Plastering works	11	4	1 skilled labour, 3 labour	16
Tiling works	11	4	1 skilled labour, 2 labour	12
Painting works	10	4	2 skilled labour, 2 labour	16
Doors and windows installation	3	3	2 skilled labour, 4 labour	18

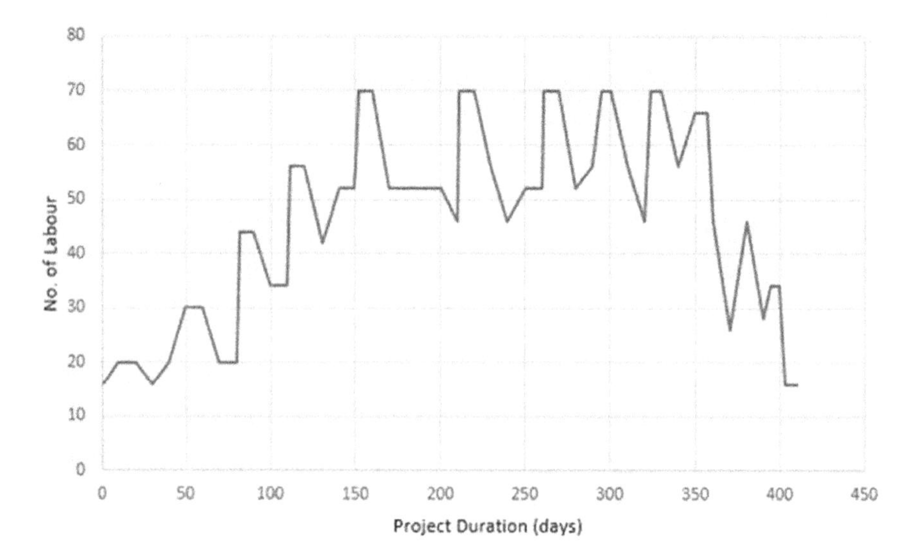

Fig. 5 Number of labour on each day of the project

created in the simulation model. The portals mark the areas where individual agents can enter and exit. Also, portals serve as intermediate destinations or final exits for agents. Further, it is assumed that the labours are familiar with the exits and barriers in the building construction site.

Agents created by an evacuation event will first wait a specified amount of time, then attempt to exit the simulation through the best of the destination portals. Only agents created by this event will evacuate the given zones or head to the destination

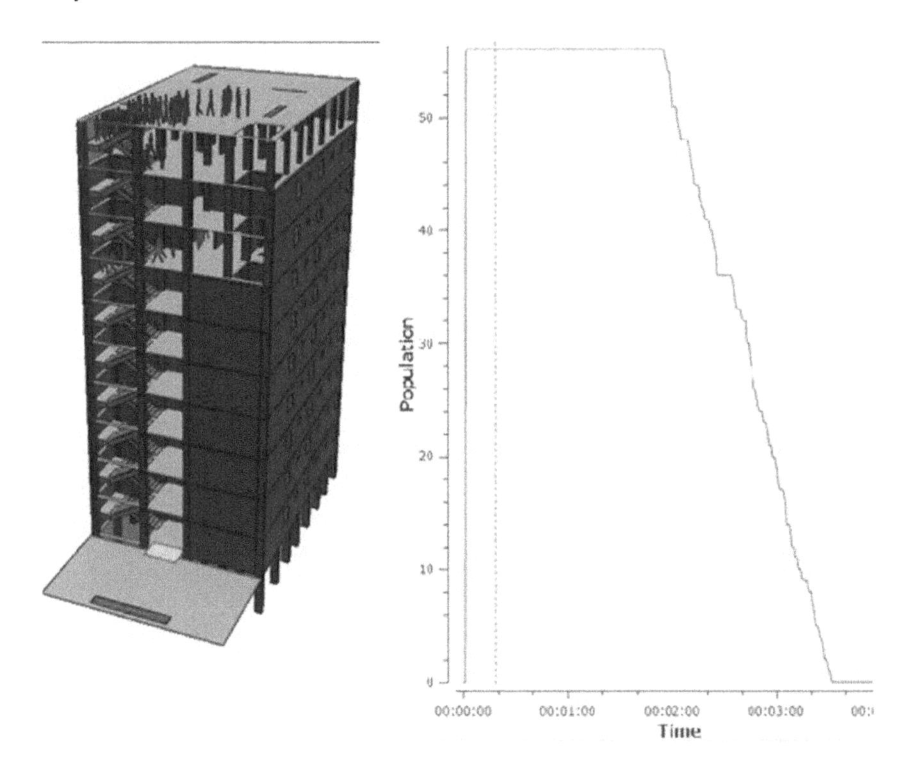

Fig. 6 Simulation at a selected time point

portals. The evacuation event is created by manual input of the starting time of evacuation, agent profile, number of agents to be generated in each portal, origin and exit. Agents are assigned different colours to differentiate different events. Graphical representation of the labour evacuation time at one of the selected time point and screen images of labour location using MassMotion modelling is shown in Fig. 6. Similar images and data are generated for all the twenty time points. All the time points are simulated and evacuation time of selected twenty time points are shown in Table 2.

5 Emergency Evacuation Model—Modified Scenario

The simulation results indicate that the labour evacuation time positively corelates with the expansion of the number of activities and labour density through the project execution. It can be observed that the duration of project execution is small and the number of labour in each time point is large in the current project schedule. The labour density at any given point of time has a significant effect on the labour evacuation time. Hence, a modified schedule is developed with reduced number of labour at any given point of time.

Table 2 Evacuation time for selected time points

Time point	Description of activities	Floor	Total labour	Number of labour in each time points	Evacuation time (s)
1	Brickwork	GF	24	44	63
	Columns	F2	20		
2	Electrical work	GF	8	34	80
	Plumbing work		6		
	Slab work	F2	20		
3	Tiling work	GF	12	46	96
	Electrical work	F1	8		
	Plumbing work		6		
	Slab work	F3	20		
4	Painting works	F1	16	70	108
	Doors and windows install		18		
	Plastering work	F2	16		
	Slab work	F4	20		
5	Tiling	F2	12	56	110
	Brickwork	F3	24		
	Columns	F5	20		
6	Plastering	F3	16	56	151
	Slab casting	F5	40		
7	Painting works	F3	16	70	144
	Doors and windows install		18		
	Plastering work	F4	16		
	Slab work	F6	20		
8	Tiling work	F4	12	46	157
	Electrical work	F5	8		
	Plumbing work		6		
	Slab work	F7	20		
9	Tiling	F5	12	56	160
	Brickwork	F6	24		
	Columns	F8	20		
10	Plastering	F7	16	56	211
	Slab casting	F9	40		
11	Tiling	F7	12	56	190
	Brickwork	F8	24		
	Columns	F10	20		
12	Tiling work	F7	12	46	188
	Electrical work	F8	8		
	Plumbing work		6		
	Roof slab	F10	20		

(continued)

Table 2 (continued)

Time point	Description of activities	Floor	Total labour	Number of labour in each time points	Evacuation time (s)
13	Painting works	F7	16	70	202
	Doors and windows install		18		
	Plastering work	F8	16		
	Roof slab	F10	20		
14	Plastering	F8	16	56	233
	Slab casting	F10	40		
15	Painting works	F8	16	66	216
	Doors and windows install		18		
	Electrical work	F9	8		
	Brickwork	F10	24		
16	Painting works	F8	16	74	223
	Doors and windows install		18		
	Plastering work	F9	16		
	Brickwork	F10	24		
17	Painting works	F8	16	46	196
	Doors and windows install		18		
	Plastering work	F9	16		
	Electrical work	F10	8		
	Plumbing work		6		
18	Painting works	F9	16	46	208
	Doors and windows install		18		
	Tiling work	F10	12		
19	Painting works	F10	16	34	211
	Doors and windows install		18		
20	Painting works	F10	16	16	185

Duration of activities can be extended up to the free float of the corresponding activities. Free float is the amount of time that an activity can be delayed without delaying the early start time of any successor activity. In the finishing stage column rising and slab work activities are delayed. Doors and windows installation and plumbing work activities are delayed throughout the project execution so as to have reduced labour density. Table 3 shows the modified resources and duration of the project.

The project is rescheduled for modified resources and durations of activities. However, the total project duration is still 410 days. Figure 7 represent the number

Table 3 Changed duration and resources required

Activity	Duration	Number of crew	Number of labour in each crew	Total number of labour for each activity
Column rising	10	3	2 skilled labour, 3 labour	15
Slab work	18	3	2 skilled labour, 3 labour	15
Slab concreting	1	2	6 skilled labour, 14 labour	40
Brick work	9	3	2 skilled labour, 4 labour	18
Electrical works	5	2	1 skilled labour, 3 labour	8
Plumbing works	6	1	1 skilled labour, 2 labour	3
Plastering works	14	3	1 skilled labour, 3 labour	12
Tiling works	14	3	1 skilled labour, 2 labour	9
Painting works	14	3	2 skilled labour, 2 labour	12
Doors and windows installation	9	2	1 skilled labour, 2 labour	6

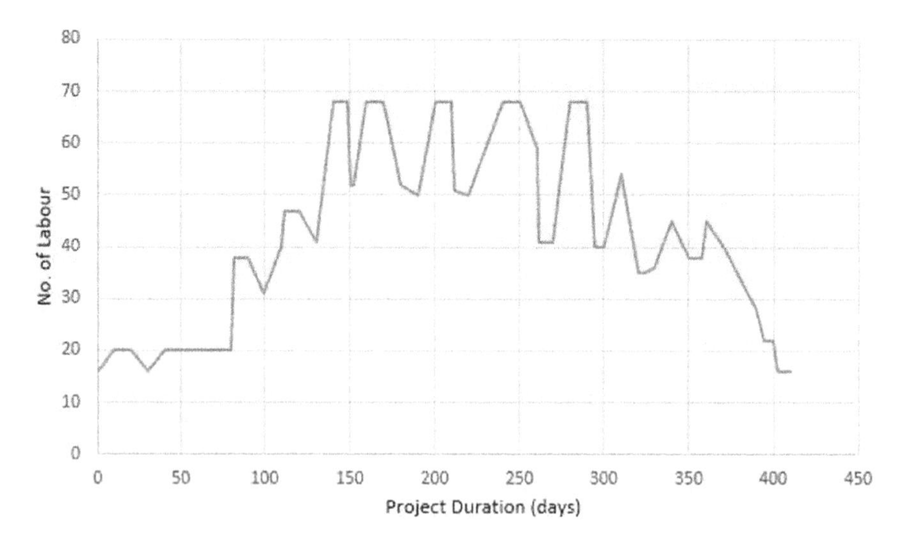

Fig. 7 Number of labour on each day of the project in the changed scenario

of labour for each day during execution of the project. Figure 8 represent the comparison of number of labour for each day in different scenario. Twenty time points were selected from the changed scenario to run the evacuation simulation model. The evacuation time of selected time points in changed scenario were comparing with scenario-1 and shown in Table 4.

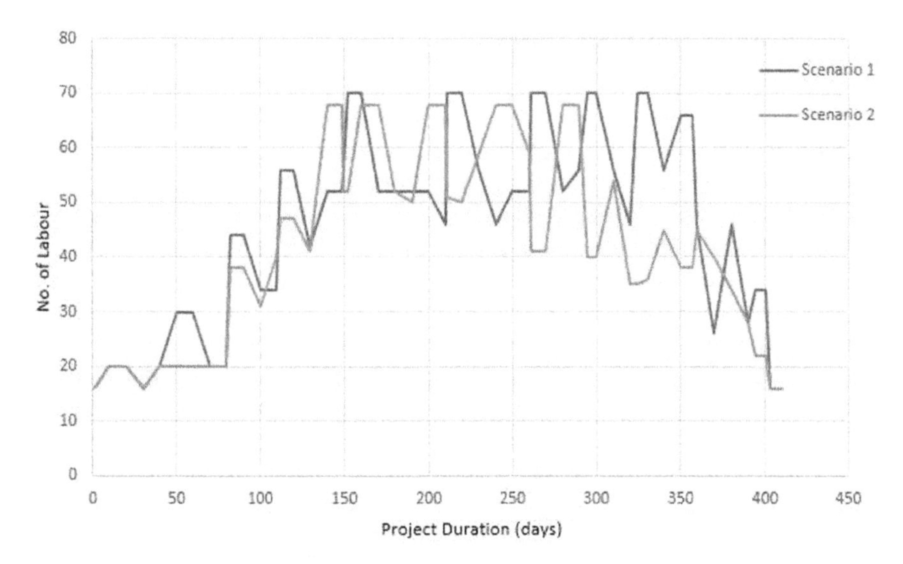

Fig. 8 Comparison of the number of labour on each day of the project in the two scenarios

Table 4 Comparison of evacuation time at different scenarios

Time points	Scenario-1		Scenario-2	
	Number of labour evacuated	Evacuation time (s)	Number of labour evacuated	Evacuation time (s)
1	44	63	38	63
2	34	80	31	67
3	46	96	41	82
4	70	108	52	101
5	56	110	52	115
6	56	151	68	129
7	70	144	52	129
8	46	157	52	147
9	56	160	59	162
10	56	211	63	183
11	56	190	54	185
12	46	188	35	184
13	70	202	36	185
14	56	233	45	184
15	66	216	38	191
16	74	223	43	196
17	46	196	45	194
18	46	208	34	196
19	34	211	22	194
20	16	185	16	185

6 Summary and Conclusions

The geometry of high-rise building model was created using Autodesk Revit Architecture. The construction activities were scheduled using MS Project. Twenty time points were selected from the project to perform the evacuation simulation. Autodesk Navisworks Manage software was used to integrate the files from Revit and MS project. All the selected time points were simulated by using MassMotion software and the evacuation time was calculated. Further, the duration of upper floors activities was extended up to the free float for minimizing the labour density of the corresponding activities. The resources and duration of the project were rescheduled. Again, an evacuation simulation model was run for the modified scenario at the selected twenty time points. Evacuation time of modified scenario (scenario-2) were compared with that of scenario-1.

The following observations and conclusions are made based on the study.

1. The number of labour at any given point of time in the construction schedule has a significant effect on the evacuation time. Labour evacuation time increases with the increase in the quantity of labour.
2. The simulation results indicate that the labour evacuation time increases when greater number of construction activities are scheduled at a given point of time and the labour density.
3. Also, in the construction of high-rise buildings, the labour density at the upper floors has a significant effect on overall evacuation time.

Further, the study demonstrates the possibility of planning the work schedule and labour density with the aim of ensuring faster labour evacuation during emergency. The research can be further extended to incorporate other features in the simulation model such as the presence of other staffs, localized obstructions due to material and machinery and different construction techniques.

Appendix 1

Project Schedule—Scenario 1

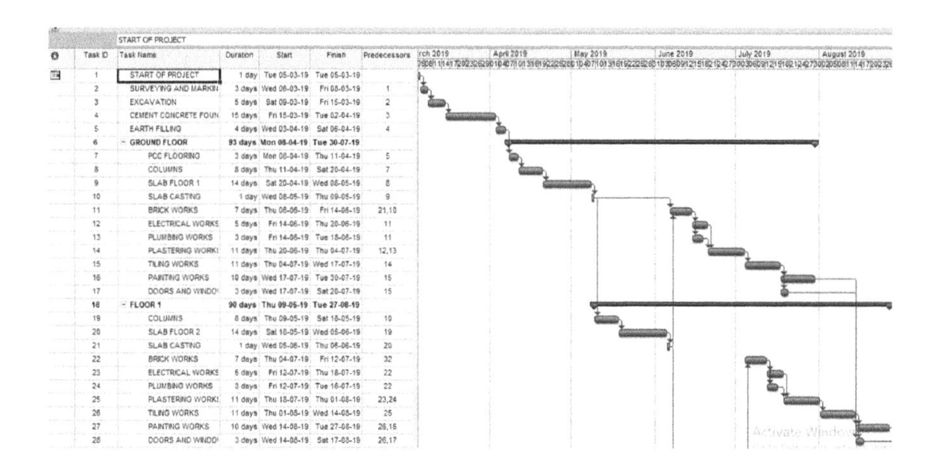

START OF PROJECT

Task ID	Task Name	Duration	Start	Finish	Predecessors
1	START OF PROJECT	1 day	Tue 05-03-19	Tue 05-03-19	
2	SURVEYING AND MARKIN	3 days	Wed 06-03-19	Fri 08-03-19	1
3	EXCAVATION	5 days	Sat 09-03-19	Fri 15-03-19	2
4	CEMENT CONCRETE FOUN	15 days	Fri 15-03-19	Tue 02-04-19	3
5	EARTH FILLING	4 days	Wed 03-04-19	Sat 06-04-19	4
6	**− GROUND FLOOR**	**93 days**	**Mon 08-04-19**	**Tue 30-07-19**	
7	PCC FLOORING	3 days	Mon 08-04-19	Thu 11-04-19	5
8	COLUMNS	8 days	Thu 11-04-19	Sat 20-04-19	7
9	SLAB FLOOR 1	14 days	Sat 20-04-19	Wed 08-05-19	8
10	SLAB CASTING	1 day	Wed 08-05-19	Thu 09-05-19	9
11	BRICK WORKS	7 days	Thu 06-06-19	Fri 14-06-19	21,10
12	ELECTRICAL WORKS	5 days	Fri 14-06-19	Thu 20-06-19	11
13	PLUMBING WORKS	3 days	Fri 14-06-19	Tue 18-06-19	11
14	PLASTERING WORKS	11 days	Thu 20-06-19	Thu 04-07-19	12,13
15	TILING WORKS	11 days	Thu 04-07-19	Wed 17-07-19	14
16	PAINTING WORKS	10 days	Wed 17-07-19	Tue 30-07-19	15
17	DOORS AND WINDO	3 days	Wed 17-07-19	Sat 20-07-19	15
18	**− FLOOR 1**	**90 days**	**Thu 09-05-19**	**Tue 27-08-19**	
19	COLUMNS	8 days	Thu 09-05-19	Sat 18-05-19	10
20	SLAB FLOOR 2	14 days	Sat 18-05-19	Wed 05-06-19	19
21	SLAB CASTING	1 day	Wed 05-06-19	Thu 06-06-19	20
22	BRICK WORKS	7 days	Thu 04-07-19	Fri 12-07-19	32
23	ELECTRICAL WORKS	6 days	Fri 12-07-19	Thu 18-07-19	22
24	PLUMBING WORKS	3 days	Fri 12-07-19	Tue 16-07-19	22
25	PLASTERING WORKS	11 days	Thu 18-07-19	Thu 01-08-19	23,24
26	TILING WORKS	11 days	Thu 01-08-19	Wed 14-08-19	25
27	PAINTING WORKS	10 days	Wed 14-08-19	Tue 27-08-19	26,16
28	DOORS AND WINDO	3 days	Wed 14-08-19	Sat 17-08-19	26,17

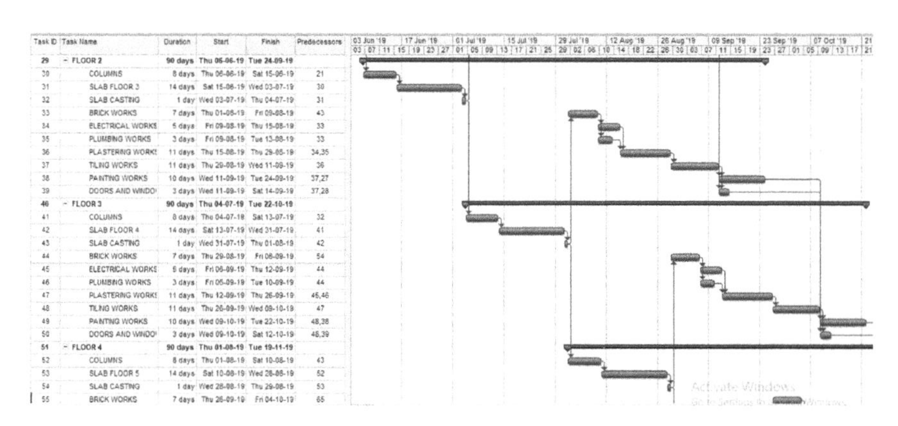

Task ID	Task Name	Duration	Start	Finish	Predecessors
29	**− FLOOR 2**	**90 days**	**Thu 06-06-19**	**Tue 24-09-19**	
30	COLUMNS	8 days	Thu 06-06-19	Sat 15-06-19	21
31	SLAB FLOOR 3	14 days	Sat 15-06-19	Wed 03-07-19	30
32	SLAB CASTING	1 day	Wed 03-07-19	Thu 04-07-19	31
33	BRICK WORKS	7 days	Thu 01-08-19	Fri 09-08-19	43
34	ELECTRICAL WORKS	5 days	Fri 09-08-19	Thu 15-08-19	33
35	PLUMBING WORKS	3 days	Fri 09-08-19	Tue 13-08-19	33
36	PLASTERING WORKS	11 days	Thu 15-08-19	Thu 29-08-19	34,35
37	TILING WORKS	11 days	Thu 29-08-19	Wed 11-09-19	36
38	PAINTING WORKS	10 days	Wed 11-09-19	Tue 24-09-19	37,27
39	DOORS AND WINDO	3 days	Wed 11-09-19	Sat 14-09-19	37,28
40	**− FLOOR 3**	**90 days**	**Thu 04-07-19**	**Tue 22-10-19**	
41	COLUMNS	8 days	Thu 04-07-19	Sat 13-07-19	32
42	SLAB FLOOR 4	14 days	Sat 13-07-19	Wed 31-07-19	41
43	SLAB CASTING	1 day	Wed 31-07-19	Thu 01-08-19	42
44	BRICK WORKS	7 days	Thu 29-08-19	Fri 06-09-19	54
45	ELECTRICAL WORKS	5 days	Fri 06-09-19	Thu 12-09-19	44
46	PLUMBING WORKS	3 days	Fri 06-09-19	Tue 10-09-19	44
47	PLASTERING WORKS	11 days	Thu 12-09-19	Thu 26-09-19	45,46
48	TILING WORKS	11 days	Thu 26-09-19	Wed 09-10-19	47
49	PAINTING WORKS	10 days	Wed 09-10-19	Tue 22-10-19	48,38
50	DOORS AND WINDO	3 days	Wed 09-10-19	Sat 12-10-19	48,39
51	**− FLOOR 4**	**90 days**	**Thu 01-08-19**	**Tue 19-11-19**	
52	COLUMNS	8 days	Thu 01-08-19	Sat 10-08-19	43
53	SLAB FLOOR 5	14 days	Sat 10-08-19	Wed 28-08-19	52
54	SLAB CASTING	1 day	Wed 28-08-19	Thu 29-08-19	53
55	BRICK WORKS	7 days	Thu 26-09-19	Fri 04-10-19	65

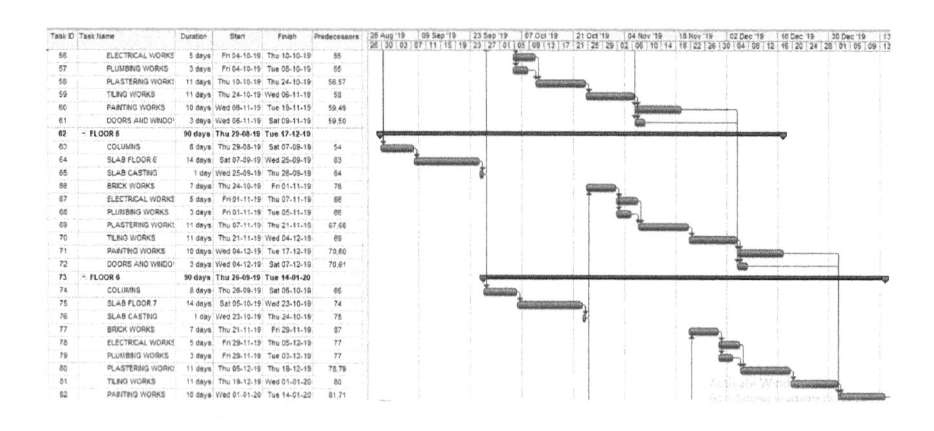

Task ID	Task Name	Duration	Start	Finish	Predecessors
56	ELECTRICAL WORKS	5 days	Fri 04-10-19	Thu 10-10-19	55
57	PLUMBING WORKS	3 days	Fri 04-10-19	Tue 08-10-19	55
58	PLASTERING WORKS	11 days	Thu 10-10-19	Thu 24-10-19	56,57
59	TILING WORKS	11 days	Thu 24-10-19	Wed 06-11-19	58
60	PAINTING WORKS	10 days	Wed 06-11-19	Tue 19-11-19	59,49
61	DOORS AND WINDO	3 days	Wed 06-11-19	Sat 09-11-19	59,50
62	**- FLOOR 5**	**90 days**	**Thu 29-08-19**	**Tue 17-12-19**	
63	COLUMNS	8 days	Thu 29-08-19	Sat 07-09-19	54
64	SLAB FLOOR 6	14 days	Sat 07-09-19	Wed 25-09-19	63
65	SLAB CASTING	1 day	Wed 25-09-19	Thu 26-09-19	64
66	BRICK WORKS	7 days	Thu 24-10-19	Fri 01-11-19	76
67	ELECTRICAL WORKS	5 days	Fri 01-11-19	Thu 07-11-19	66
68	PLUMBING WORKS	3 days	Fri 01-11-19	Tue 05-11-19	66
69	PLASTERING WORKS	11 days	Thu 07-11-19	Thu 21-11-19	67,68
70	TILING WORKS	11 days	Thu 21-11-19	Wed 04-12-19	69
71	PAINTING WORKS	10 days	Wed 04-12-19	Tue 17-12-19	70,60
72	DOORS AND WINDO	3 days	Wed 04-12-19	Sat 07-12-19	70,61
73	**- FLOOR 6**	**90 days**	**Thu 26-09-19**	**Tue 14-01-20**	
74	COLUMNS	8 days	Thu 26-09-19	Sat 05-10-19	65
75	SLAB FLOOR 7	14 days	Sat 05-10-19	Wed 23-10-19	74
76	SLAB CASTING	1 day	Wed 23-10-19	Thu 24-10-19	75
77	BRICK WORKS	7 days	Thu 21-11-19	Fri 29-11-19	87
78	ELECTRICAL WORKS	5 days	Fri 29-11-19	Thu 05-12-19	77
79	PLUMBING WORKS	3 days	Fri 29-11-19	Tue 03-12-19	77
80	PLASTERING WORKS	11 days	Thu 05-12-19	Thu 19-12-19	78,79
81	TILING WORKS	11 days	Thu 19-12-19	Wed 01-01-20	80
82	PAINTING WORKS	10 days	Wed 01-01-20	Tue 14-01-20	81,71

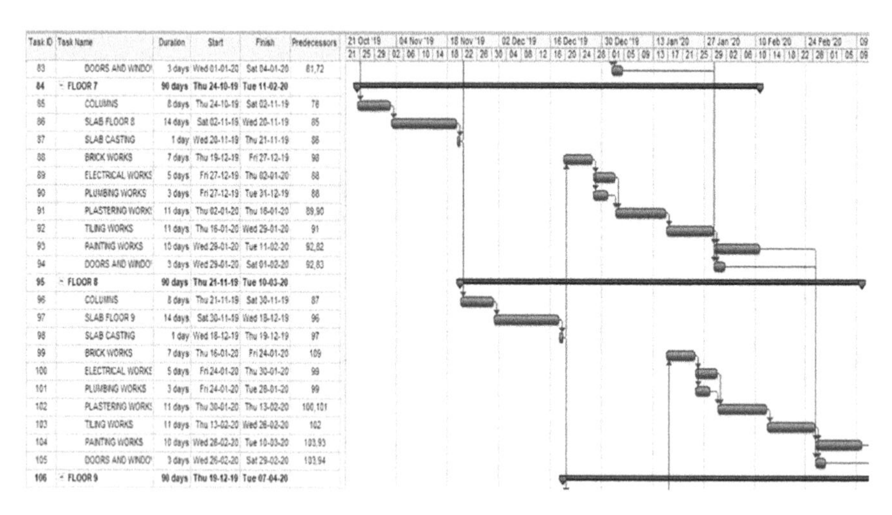

Task ID	Task Name	Duration	Start	Finish	Predecessors
83	DOORS AND WINDO	3 days	Wed 01-01-20	Sat 04-01-20	81,72
84	**- FLOOR 7**	**90 days**	**Thu 24-10-19**	**Tue 11-02-20**	
85	COLUMNS	8 days	Thu 24-10-19	Sat 02-11-19	76
86	SLAB FLOOR 8	14 days	Sat 02-11-19	Wed 20-11-19	85
87	SLAB CASTING	1 day	Wed 20-11-19	Thu 21-11-19	86
88	BRICK WORKS	7 days	Thu 19-12-19	Fri 27-12-19	98
89	ELECTRICAL WORKS	5 days	Fri 27-12-19	Thu 02-01-20	88
90	PLUMBING WORKS	3 days	Fri 27-12-19	Tue 31-12-19	88
91	PLASTERING WORKS	11 days	Thu 02-01-20	Thu 16-01-20	89,90
92	TILING WORKS	11 days	Thu 16-01-20	Wed 29-01-20	91
93	PAINTING WORKS	10 days	Wed 29-01-20	Tue 11-02-20	92,82
94	DOORS AND WINDO	3 days	Wed 29-01-20	Sat 01-02-20	92,83
95	**- FLOOR 8**	**90 days**	**Thu 21-11-19**	**Tue 10-03-20**	
96	COLUMNS	8 days	Thu 21-11-19	Sat 30-11-19	87
97	SLAB FLOOR 9	14 days	Sat 30-11-19	Wed 18-12-19	96
98	SLAB CASTING	1 day	Wed 18-12-19	Thu 19-12-19	97
99	BRICK WORKS	7 days	Thu 16-01-20	Fri 24-01-20	109
100	ELECTRICAL WORKS	5 days	Fri 24-01-20	Thu 30-01-20	99
101	PLUMBING WORKS	3 days	Fri 24-01-20	Tue 28-01-20	99
102	PLASTERING WORKS	11 days	Thu 30-01-20	Thu 13-02-20	100,101
103	TILING WORKS	11 days	Thu 13-02-20	Wed 26-02-20	102
104	PAINTING WORKS	10 days	Wed 26-02-20	Tue 10-03-20	103,93
105	DOORS AND WINDO	3 days	Wed 26-02-20	Sat 29-02-20	103,94
106	**- FLOOR 9**	**90 days**	**Thu 19-12-19**	**Tue 07-04-20**	

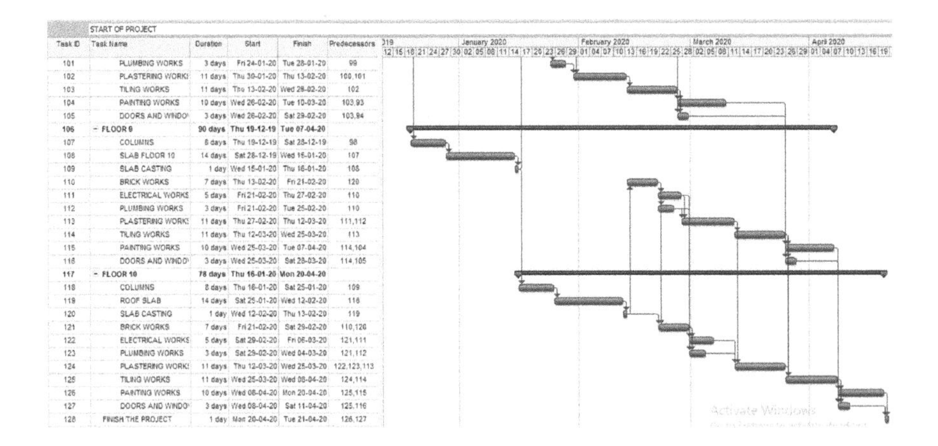

START OF PROJECT

Task ID	Task Name	Duration	Start	Finish	Predecessors
101	PLUMBING WORKS	3 days	Fri 24-01-20	Tue 28-01-20	99
102	PLASTERING WORKS	11 days	Thu 30-01-20	Thu 13-02-20	100,101
103	TILING WORKS	11 days	Thu 13-02-20	Wed 26-02-20	102
104	PAINTING WORKS	10 days	Wed 26-02-20	Tue 10-03-20	103,93
105	DOORS AND WINDO	3 days	Wed 26-02-20	Sat 29-02-20	103,94
106	**− FLOOR 9**	**90 days**	**Thu 19-12-19**	**Tue 07-04-20**	
107	COLUMNS	8 days	Thu 19-12-19	Sat 28-12-19	98
108	SLAB FLOOR 10	14 days	Sat 28-12-19	Wed 15-01-20	107
109	SLAB CASTING	1 day	Wed 15-01-20	Thu 16-01-20	108
110	BRICK WORKS	7 days	Thu 13-02-20	Fri 21-02-20	120
111	ELECTRICAL WORKS	5 days	Fri 21-02-20	Thu 27-02-20	110
112	PLUMBING WORKS	3 days	Fri 21-02-20	Tue 25-02-20	110
113	PLASTERING WORKS	11 days	Thu 27-02-20	Thu 12-03-20	111,112
114	TILING WORKS	11 days	Thu 12-03-20	Wed 25-03-20	113
115	PAINTING WORKS	10 days	Wed 25-03-20	Tue 07-04-20	114,104
116	DOORS AND WINDO	3 days	Wed 25-03-20	Sat 28-03-20	114,105
117	**− FLOOR 10**	**78 days**	**Thu 16-01-20**	**Mon 20-04-20**	
118	COLUMNS	8 days	Thu 16-01-20	Sat 25-01-20	109
119	ROOF SLAB	14 days	Sat 25-01-20	Wed 12-02-20	118
120	SLAB CASTING	1 day	Wed 12-02-20	Thu 13-02-20	119
121	BRICK WORKS	7 days	Fri 21-02-20	Sat 29-02-20	110,120
122	ELECTRICAL WORKS	5 days	Sat 29-02-20	Fri 06-03-20	121,111
123	PLUMBING WORKS	3 days	Sat 29-02-20	Wed 04-03-20	121,112
124	PLASTERING WORKS	11 days	Thu 12-03-20	Wed 25-03-20	122,123,113
125	TILING WORKS	11 days	Wed 25-03-20	Wed 08-04-20	124,114
126	PAINTING WORKS	10 days	Wed 08-04-20	Mon 20-04-20	125,115
127	DOORS AND WINDO	3 days	Wed 08-04-20	Sat 11-04-20	125,116
128	FINISH THE PROJECT	1 day	Mon 20-04-20	Tue 21-04-20	126,127

References

1. Xiaoping Z, Tingkuan Z, Mengting L (2009) Modeling crowd evacuation of a building based on seven methodological approaches. Build Environ 44:437–445. https://doi.org/10.1016/j.buildenv.2008.04.002
2. Marzouk M, Daour IA (2018) Planning labor evacuation for construction sites using BIM and agent-based simulation. Saf Sci 109:174–185. https://doi.org/10.1016/j.ssci.2018.04.023
3. Marzouk M, Daoor IA (2019) Simulation of labor evacuation : the case of housing construction projects simulation of labor evacuation: the case of housing construction projects. HBRC J 14:198–206. https://doi.org/10.1016/j.hbrcj.2016.07.001
4. Hua Y, He J, Gong J, Zhao J (2020) Hazardous area risk-based evacuation simulation and analysis of building construction sites. J constr Eng Manage 146:13. https://doi.org/10.1061/(ASCE)CO.1943-7862.0001798
5. Aleksandrov M, Cheng C, Rajabifard A, Kalantari M (2019) Modelling and finding optimal evacuation strategy for tall buildings. Saf Sci 115:247–255. https://doi.org/10.1016/j.ssci.2019.02.017
6. Marzouk M, Mohamed B (2019) Integrated agent-based simulation and multi-criteria decision making approach for buildings evacuation evaluation. Saf Sci 112:57–65. https://doi.org/10.1016/j.ssci.2018.10.010
7. Row P (2019) Construction site evacuation safety: evacuation strategies for tall construction sites. IOSH/Fire Saf Eng Group Univ Greenwich

Investigational Study of the Overtaking Behaviour on an Undivided State Highway

S. Harikrishnan, A. Akhil Krishnan, Jeevan Jewel, Siddharth Raj, and P. M. Lidiya

Abstract Overtaking is one of the major process in the driving operation. This process contributes to the majority of road accidents across the globe. The human behavior during overtaking influence the result of the whole overtaking manoeuvre. Overtaking studies are conducted across the world studying the driver's nature, overtaking parameters and other processes during an overtaking operation. In India these studies are conducted but is limited to the studies of National highways and major roads. This paper is an investigational study of the overtaking behavior on an undivided state highway. Overtaking involves lane-changings, speeding up and speeding down actions, speed of overtaking and overtaken vehicles, and also, speed and distance of vehicles in the opposite lane. In roads without any barrier or separation for the incoming and ongoing vehicles, the driver will have a tendency to use the opposite lane for the overtaking operation. The study is specifically on the undivided roads in mixed traffic conditions. Moving car method is used for data collection and overtaking parameters along with overtaking types are found out.

Keywords Overtaking manoeuvre · Overtaking parameters · Moving car method

1 Introduction

Indian highways including national and state highways mainly consist of two way mixed traffic and most of the vehicles have varying speed and are not following strict lane rules and discipline. Overtaking process is a major process in which the fast moving following vehicle will increase its speed to go past the vehicle before it. Different factors influence this overtaking operation such as the distance between the vehicles, vehicle speed, presence of vehicle in the opposing lane etc. In the case of two lane traffic conditions without any separation of the lanes, overtaking procedure is necessary as the vehicles will be moving with different speeds.

S. Harikrishnan (✉) · A. Akhil Krishnan · J. Jewel · S. Raj · P. M. Lidiya
Federal Institute of Science and Technology, Hormis Nagar, Mookkannoor, Angamaly, Ernakulam 683577, India

G. C. Marano et al. (eds.), *Proceedings of SECON'21*, Lecture Notes in Civil Engineering 171, https://doi.org/10.1007/978-3-030-80312-4_70

Overtaking studies are limited in India and the conducted studies are mainly of national highways and major roads. Asaithambi [1] conducted overtaking studies on undivided roads in mixed traffic. Flying and accelerative overtaking types were observed in the roads and analyzed the details of overtaking data on a two-lane two-way undivided road using moving car observer method and registration plate method. Analysis were done on the overtaking types and types of vehicles that performed overtaking. It was found that accelerative overtaking was used more often than other overtaking types and vehicles like car, two wheelers and LCVs conducted overtaking faster compared to other heavy vehicles.

Chandra [2] used floating car method to study the acceleration and overtaking characteristics of different types of vehicles. A relation was formed between acceleration and overtaking rates. Detailed study of flying overtaking was also done. Analysis of flying overtaking obtained. It was observed that the longitudinal safe distance maintained by a vehicle in front is a linear function of the speed. Vehicles maintained a minimum average lateral gap of 0.5 m. Llorca [3] used video cameras to observe overtaking on highways and the cameras were at fixed positions. A sample of 234 overtaking process were analyzed. They also developed multiple linear regression models and using this, explained distance and passing time duration. Overtaking analysis was done using video recordings and studied. Use of indicator in overtaking operation was studied. Study was also done and estimated the relation of speed and collision.

Indian roads consist mostly of undivided roads and highways and overtaking in these types of roads depends on various factors like drivers mindset, type of overtaking, overtaking parameters etc. Overtaking process become essential because of the speed variations of different types of vehicles in a traffic condition. In mixed traffic condition this variation exist largely. Also the driver have to consider the vehicle from the opposite lane. All these make overtaking a complex process which require detailed study. Also overtaking have significant impact on road traffic accidents.

This study was conducted because overtaking study in heterogeneous and mixed traffic conditions are sparse in the state of Kerala and this process of overtaking is a complex process which deserves a detailed study. There were only few studies based on the overtaking process and majority of the overtaking studies are conducted in national highways and other major roads. This study put emphasis on the overtaking process in undivided state highway with objectives like collecting time data of overtaking process of different types of vehicles and find out various overtaking parameters, identify the type of overtaking conducted by different type of vehicles etc.

2 Types of Overtaking

Asaithambi [1] and Chandra [2] explains the steps for a complete overtaking operation. It can be divided into three stages: the lane changing before overtaking, the acceleration stage and the lane changing stage after overtaking. The fast moving

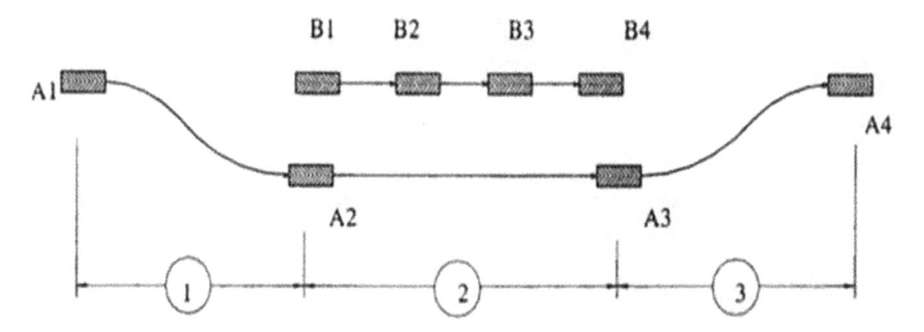

Fig. 1 Overtaking operation

vehicle, travelling from A1 to A2 is in the lane changing stage before overtaking, is in the acceleration stage from A2 to A3 and is in the lane changing stage from A3 to A4, after overtaking. The corresponding positions of the overtaken vehicle are shown as B1, B2, B3 and B4, respectively (Fig. 1).

Accelerative overtaking, means that the vehicle which is overtaking follows a vehicle which it needs to overtake and waits for a sufficient gap to perform an overtaking operation. In flying overtaking the vehicle will not change its speed during the overtaking operation. Sometimes, when there is no vehicle in the opposite lane, the overtaker will overtake more than one vehicle and it is called multiple overtaking.

3 Site Selection and Methodology

The methodology used in this study is from Asaithambi [1] in which moving car method is used for overtaking study of a national highway road stretch. The area selected for the study is Thodupuzha-Muvattupuzha Road connecting the main towns of Thodupuzha and Muvattupuzha. The study area of Thodupuzha-Muvattupuzha road is a part of State Highway 08. The selected stretch is the straight road between Achankavala and Madakkathanam. The reason for the selection of the stretch is that it is almost a straight stretch having considerable traffic flow even in off peak hours. The road stretch does not have any by roads or intersections in between. After a reconnaissance survey a straight road stretch of 1.7 km was selected for the study. Moving car method is used for data collection. The road stretch is a two lane undivided stretch with a lane width of 20 m (Fig. 2).

The study was conducted in the selected stretch of road from Achankavala to Madakkathanam. Three cameras were used for the collection of time intervals. One camera was positioned at the back portion, one at the rear side and one at the front side to record the complete overtaking procedure and to extract different time intervals. Moving car method was conducted in the selected stretch. In this method a test car was used which will travel with a constant speed. This allows other

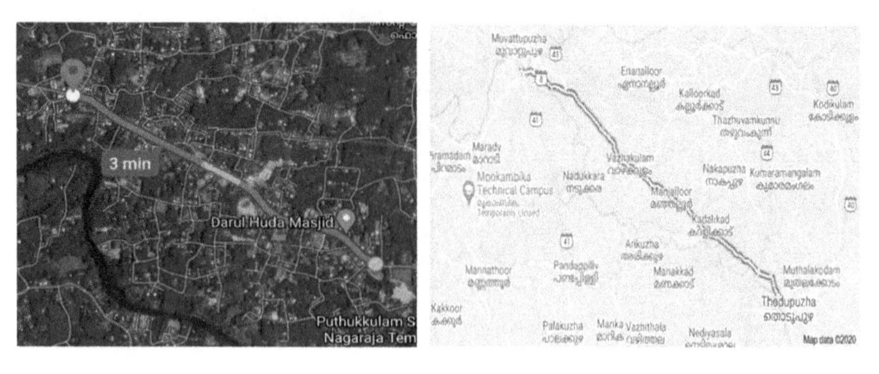

Fig. 2 Study stretch

vehicles to conduct the overtaking operation and can overtake the test car. Different categories of vehicles conducted the overtaking operation and overtaking parameters of these vehicles were evaluated. Video recordings of about 400 vehicles were collected which includes different categories of vehicles and the data of 88 vehicles were found out (Table 1).

The overtaking parameters explained in the above table is found out using moving car method and the calculations are done using the basic overtaking equations. Necessary assumptions are made such as the speed of the opposing vehicles. It is assumed as 45 km/h. Equations used in the calculation of overtaking parameters [1, 6]:

The minimum spacing of the vehicle depends on their speed and is given by empirical formula:

$$s = 0.7v_b + 6 \tag{1}$$

The distance covered b by the slow moving vehicle B travelling at a speed of v_b is calculated by the following:

Table 1 Events

Events in overtaking	Explanation
Event 1	When the overtaking vehicle moves to the opposing lane for initiating overtaking procedure
Event 2	When the front tyre of the overtaking vehicle is in line with the rear tyre of the test vehicle
Event 3	When the front tyre of the overtaking vehicle is in line with the front tyre of the test vehicle
Event 4	When the rear tyre of the overtaking vehicle is in line with the front tyre of the test vehicle
Event 5	When the overtaking vehicle either comes back to its original lane or continues to travel in the opposing lane after overtaking the test vehicle

$$b = v_b T \tag{2}$$

$$T = t_1 + t_2 + t_3 \tag{3}$$

$$d_2 = b + 2s \tag{4}$$

Acceleration,

$$a = 4s/T^2 \tag{5}$$

Speed of overtaking vehicle,

$$v_a = d_2/T \tag{6}$$

Distance travelled by opposing vehicle,

$$d_3 = v_c T \tag{7}$$

Distance travelled by overtaking vehicle during overtaking,

$$d_2 = v_b T + 2s \tag{8}$$

4 Data Collection and Calculation

The time data was collected from the videos recorded using moving car method. These time data were used to calculate different overtaking parameters using the equations discussed earlier. Data were collected in moderate traffic conditions. Some of the data collected and the parameters extracted are shown below in Tables 2, 3 and 4.

The data collected from the videos are time interval data between the different time intervals of the events of overtaking procedures and these time intervals can be used for finding the total overtaking time.

$$T = t_2 + t_3 + t_4 \tag{9}$$

where t_2, t_3 and t_4 are obtained from stopwatch or by video sinking. t_2 is obtained from the rear camera. t_3 is obtained by sinking the videos taken by the person seated at the front and rear. t_4 is obtained from the video taken by the person seated at the front.

The overtaking parameters that were evaluated includes.

- Speed of overtaking vehicle
- Acceleration
- Distance between overtaking vehicle and test vehicle before overtaking

Table 2 Overtaking parameters [1]

Parameters	Explanation
t_1	Time between event 1 and 2
t_2	Time between event 2 and 3
t_3	Time between event 3 and 4
t_4	Time between event 4 and 5
v_a	Overtaking vehicle speed
v_b	Test vehicle speed
v_c	Opposite lane vehicle speed
a	Acceleration
d_1	Distance between overtaking vehicle and overtaken vehicle before overtaking
d_2	Distance travelled during overtaking vehicle during overtaking
d_3	Distance between overtaking vehicle and overtaken vehicle after overtaking
T	Total overtaking time
b	Distance travelled by overtaken vehicle during overtaking time
s	Minimum spacing

Table 3 Sample of time data collected

Test vehicle speed (km/h)	Test vehicle speed (m/s)	Type of vehicle	t1 (s)	t2 (s)	t3 (s)	t4 (s)	T (s)
40	11.11	LMV	5.11	1.45	0.69	0.52	7.77
40	11.11	LMV	12.3	1.67	1.76	1.64	17.37
40	11.11	LCV	5.51	2.78	3.62	3.85	15.76
40	11.11	LMV	5.88	1.93	1.31	2.33	11.45
40	11.11	HMV	2.12	8.77	1.58	2.4	14.87
40	11.11	2 wheeler	2.69	2.57	0.88	1.01	7.15
40	11.11	LMV	0.63	1.63	1.17	1.08	4.51
40	11.11	LMV	2.7	2.41	0.69	0.58	6.38
40	11.11	HMV	1.24	4.96	2.46	3.91	12.57

- Distance travelled by the overtaking vehicle during overtaking
- Distance between overtaking vehicle and test vehicle after overtaking
- Distance travelled by test vehicle during overtaking time
- Minimum spacing between overtaking vehicle and test vehicle

All these overtaking parameters were evaluated using the empirical formulas used for estimating traffic characteristics.

Speed difference between test vehicle and the vehicle conducting overtaking operation were evaluated and the type of overtaking was identified from the video films captured by the persons sitting in the front and rear seat of the test vehicle. The type of overtaking is rechecked using the videos of the vehicles captured.

Table 4 Sample of overtaking parameters calculated

s (m)	b (m)	a (m/s)	d_2 (m)	v_a (km/h)	v_a (m/s)	v_c (km/h)	v_c (m/s)	d_3 (m)	d_1 (m)	D (m)
13.78	86.32	0.91	113.88	52.76	14.66	45.00	12.50	97.13	56.77	267.78
13.78	192.98	0.18	220.53	45.71	12.70	45.00	12.50	217.13	136.65	574.31
13.78	175.09	0.22	202.65	46.29	12.86	45.00	12.50	197.00	61.22	460.86
13.78	127.21	0.42	154.76	48.66	13.52	45.00	12.50	143.13	65.33	363.22
13.78	165.21	0.25	192.76	46.67	12.96	45.00	12.50	185.88	23.55	402.19
13.78	79.44	1.08	106.99	53.87	14.96	45.00	12.50	89.38	29.89	226.25
13.78	50.11	2.71	77.66	61.99	17.22	45.00	12.50	56.38	7.00	141.03
13.78	70.88	1.35	98.44	55.54	15.43	45.00	12.50	79.75	30.00	208.18
13.78	139.65	0.35	167.21	47.89	13.30	45.00	12.50	157.13	13.78	338.11
13.78	96.43	0.73	123.99	51.42	14.28	45.00	12.50	108.50	17.11	249.60

Table 5 Sample of speed difference and types of overtaking

Speed difference (m/s)	Speed difference (km/h)	Type of overtaking
3.55	12.77	Accelerative
1.59	5.71	Flying
1.75	6.29	Flying
2.41	8.66	Accelerative
1.85	6.67	Accelerative
3.85	13.87	Flying
6.11	21.99	Multiple
4.32	15.55	Accelerative
2.19	7.89	Flying
3.17	11.43	Accelerative
6.61	23.79	Flying
3.67	13.23	Flying
2.74	9.88	Accelerative

Videos were recorded of about 400 vehicles and the data (overtaking characteristics) were evaluated of 88 vehicles for further overtaking study. Of the 88 vehicles observed, 53 vehicles conducted accelerative overtaking, 28 vehicles conducted flying overtaking, and 7 vehicles conducted multiple overtaking (Table 5).

5 Result and Discussion

Data collected and calculated is analyzed to study types of overtaking occurring in the selected stretch and other overtaking parameters under mixed traffic conditions in a state highway. Accelerative, flying and multiple overtaking were observed. At Thodupuzha—Muvattupuzha road, accelerative overtaking by 61% of drivers, flying overtaking was done by 30.8% of drivers, and multiple overtaking by 8.2% of drivers was observed. Higher number of drivers preferred accelerative overtaking as they prefer a more safe way of overtaking procedure than conducting flying or multiple overtaking which is more rash and dangerous. This also shows the confidence levels of these drivers, Drivers conducting flying and multiple overtaking are more confident than the other 61% conducting accelerative overtaking. Higher number of accelerative overtaking and flying overtaking were conducted by LMV, LCV and two wheelers. As per Asaithambi [1] and Chandra [2] this may be due to higher operating characteristics. Buses and HMVs have larger size and aggressive behavior. Multiple overtaking is done highly by LMVs and two wheelers and this can be due to lane density which may be high (Fig. 3).

From the analysis accelerative overtaking performed by different category vehicles are: LMV (50.9%), LCV (20.7%), two wheelers (13.2%), buses (7.5%) and heavy vehicles (7.5%). Buses and heavy vehicles perform less number of accelerative type of overtaking compared to other categories. Flying overtaking analysis is as follows: LMV (53.5%), LCV (10.7%), two wheelers (17.8%), buses (7.14%), and heavy vehicles (10.71%). LMVs and two wheelers are performing more flying type of overtaking compared to other categories. As per Asaithambi [1] and Mahmud [4] these observations may be due to the difference in operating characteristics of different types of vehicles, size of vehicle and its functions. In the case of two-wheelers, both types of overtaking are almost equal which could be due to their higher mobility and limited size.

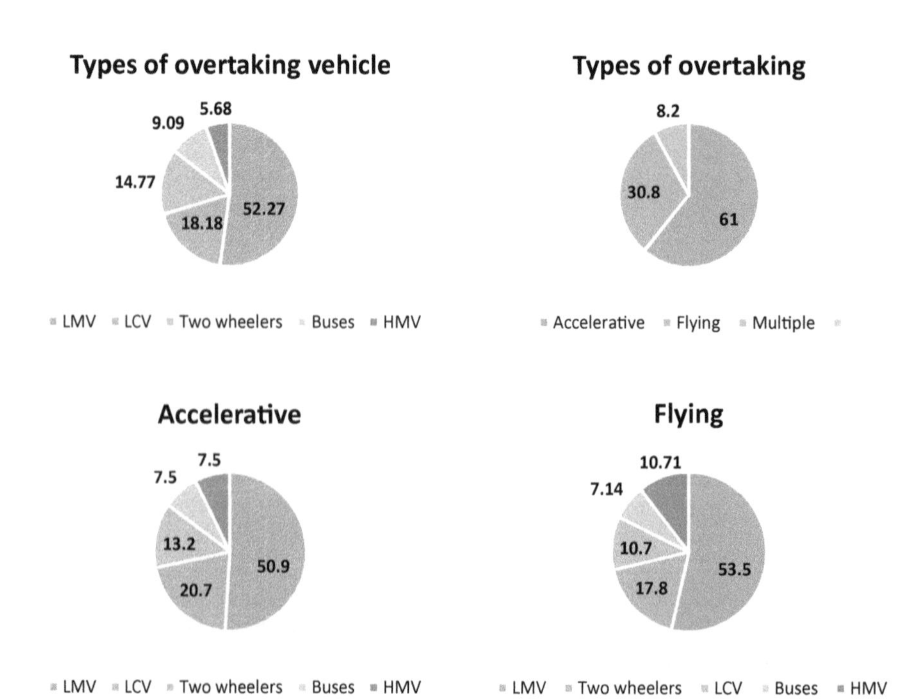

Fig. 3 Pie charts showing different observations

5.1 *Total Distance Covered During Overtaking*

Overtaking distances were found out and mean values were calculated. For accelerative overtaking the mean value is 245.93 m which is higher comparing to flying which has 178.35 m and multiple which has 220.73 m. Asaithambi [1] and Mocsári [5] explains this and it is believed to be due to the fact that, in accelerative type of overtaking the overtaking vehicle will follow the test vehicle until a gap is provided by the test vehicle to undergo the overtaking process. In the case of flying overtaking the drivers will not wait for the space in the opposite lane and will overtake the test vehicle without following it. In the cases of heavy vehicles and buses, their mean values are comparatively larger compared to other categories which can be contributed towards their operating characteristics and size as explained in Asaithambi [1]. Overtaking distances for two wheelers are less and can be explained using the same theory. Their size is small compared to other categories and also have higher manoeuvrability (Table 6).

5.2 *Total Time Required for Overtaking*

Accelerative overtaking have higher time followed by multiple overtaking and flying overtaking. As discussed earlier the overtaking vehicle will follow the test vehicle and this contributes to the higher value of time for accelerative overtaking. Total overtaking time for two wheelers and light motor vehicles are lesser. As per Asaithambi [1] and Mocsári [5], the reason may be due to the fact that these vehicles have higher operating characteristics and higher manoeuvrability (Table 7).

Table 6 Total overtaking time for different types of vehicles

Category	Mean	Min	Max
Two wheeler	213.3	109.79	229.11
LMV	209.35	74.63	531.13
LCV	226.86	105.38	390.75
Heavy vehicle	300.07	186.89	407.75
Bus	242.89	83.38	401.22

Table 7 Total overtaking time

Category	Mean	Min	Max
Two wheeler	6.45	2.57	13.57
LMV	5.81	1.53	18.92
LCV	6.72	2.54	12
Heavy vehicle	10.05	5.83	13.64
Bus	7.27	4.31	12.81

6 Conclusion

The study was carried out to understand the overtaking behavior and characteristics of vehicles on undivided roads in mixed traffic conditions in a state highway. Moving car method was used to collect overtaking data of different categories of vehicles on Thodupuzha-Muvattupuzha road. It was observed that 61% of the drivers conducted accelerative overtaking, 30.8% conducted flying overtaking and 8.2% conducted multiple overtaking.

Some of the conclusive points in moving car overtaking study:

- More than 50% of the drivers conducted accelerative overtaking and this may be due to the reason that most of the drivers are cautious and prefers a more safer approach.
- In the accelerative overtakings, 50.9% were LMVs, 20.7% were LCVs, 13.2% were two wheelers, 7.5% were buses and 7.5% were HMVs.
- In the flying overtakings, 53.5% were LMVs, 10.7% were LCVs, 17.8% were two wheelers, 7.14% were buses and 10.71% were HMVs.
- Higher number of accelerative and flying overtaking were conducted by LMVs, LCVs and two wheelers. This may be due to their static and operating characteristics.
- Mean overtaking time for flying overtaking is 4.90 s, multiple overtaking is 6.36 s and for accelerative overtaking it is 7.4 s. The difference in the time is comparable with the reaction time of the driver.
- In the cases of heavy vehicles and buses, their mean overtaking distance values are comparatively larger compared to other categories which may be due to their operating characteristics and size. Overtaking distances for two wheelers are less and can be due their smaller size compared to other categories and also because of higher manoeuvrability.

The findings from this study can be further used for constructing and evaluating graphical models of overtaking parameters like models of overtaking time and speed difference, speed and acceleration etc. These models can further be used for graphical analysis of the collected and evaluated data. It can also be further used for traffic simulation models. Time to collision study can also be conducted using the study data and findings which will give higher insights on traffic safety.

References

1. Asaithambi G, Shravani G (2017) Overtaking behaviour of vehicles on undivided roads in non-lane based mixed traffic conditions. J Traffic Transp Eng 4(3):252–261
2. Chandra S, Shukla S (2012) Overtaking behavior on divided highways under mixed traffic conditions. Procedia Soc Behav Sci 43:313–322
3. Llorca C, García A (2011) Evaluation of passing process on two-lane rural highways in Spain with new methodology based on video data. Transp Res Rec J Transp Res Board 42–51

4. Mahmud SM, Ferreira L, Hoque M, Hojati A (2018) Overtaking behaviour on rural highways under an heterogeneous traffic environment: evidence from a developing country
5. Mocsári T (2009) Analysis of the overtaking behaviour of motor vehicle drivers. Acta Technica Jaurinensis 97–106
6. Book of highway engineering by SKKhanna and CEG Justo

Wind Pressure Distribution on Multi-span Semi-circular Canopy Roofs

Neelam Rani, Ajay Pratap, and Ashok K. Ahuja

Abstract The present study attempts to examine the wind loads on multi-span semicircular canopy roofs. Experiment is performed inside open circuit atmospheric boundary layer wind. Experiments are undertaken for single, two, and three span semicircular canopy roofs. Wind pressures are evaluated on both upper and lower surface of the semicircular canopy roof surface model which is made up of Perspex sheet. These models are tested for seven different wind incidence angles from 0° to 180° at an interval of 30°. The wind velocity used for testing was 6.2 m/s, measured at the model height. In this study evaluated results are presented in contour plot form, graphs, and in the form of cross-sectional variation for the mean wind pressure coefficient. The roof is divided into three zones, and the resultant value for each zone is given in the form of a pressure coefficient. The study shows that maximum suction occurs at the apex on the upper roof surface for different wind incidence angle with its peak value estimated at an angle of 30°. Suction is developed at the lower canopy surface at all wind incidence angles with maximum suction observed at an angle of 0° wind incident.

Keywords Canopy roof · Pressure coefficient · Turbulence intensity · Velocity profile · Multi-span · Semi-circular

N. Rani · A. Pratap (✉)
Dr. B.R. Ambedkar National Institute of Technology, Jalandhar, India
e-mail: ajayp.ce.19@nitj.ac.in

N. Rani
e-mail: ranin@nitj.ac.in

A. K. Ahuja
Indian Institute of Technology, Jammu, India

G. C. Marano et al. (eds.), *Proceedings of SECON'21*, Lecture Notes in Civil Engineering 171, https://doi.org/10.1007/978-3-030-80312-4_71

1 Introduction

Canopy roofs find wide applicability at places such as railway stations, parking lots, restaurants, sports facilities, bus stand as well as solar panel installation [1, 2]. The maximum damaged occurred at the roof edges, corner and eves of low-rise buildings during wind-storm as it strikes on both upper and lower surfaces of roof [3–5]. While designing the canopy roof the wind load acting on it is considered to be the most important part of designing. However, current provisions in various national standards on wind loads deal with single and multi-span clad buildings, but recommendations about the wind loads on multi-span canopy roofs are not available in it. Even the information available for single-span canopy roofs is limited [6–8]. Therefore, the architects and structural engineers face difficulty while designing multi-span canopy roofs for wind loads.

Krishna [9] elaborated that only mean pressure coefficients are not adequate to ensure the safety of the building it is very important to know the variations and the peak values as well. Natalini et al. [2] carried out experiments on the models of vaulted canopy roof. The authors have used two types of models: (i) Deep model with area 60 × 15 cm and (ii) short model with area 30 × 15 cm; with varying eve height of 2, 4, 6 cm. Natalini et al. [4] undertook an experimental study on mean wind pressure coefficients for pitched canopy roofs. Paluch et al. [10] Studied the effect of the attached canopy on the wind pressure distribution on a rectangular plan building with an arch roof. The authors have used two types of models, deep (h = 40 mm) and short models (h = 20 mm) to find out the effect of model height.

Letchford et al. [11] studied the effect of porosity on wind pressure distribution on low rise canopy roof namely hip, gable, and mono-slope were chosen for the study. Letchford and Ginger [3] Presented the mean wind pressure distribution on the pitched canopy roof. Gumley [12] Investigated the distribution of wind loads on the pitched, trough, and mono-slope canopy roof with blockage (stacking). The author concluded that stacking produced a large effect on pressure coefficients on the canopy roof and, this is why the effect of stacking should be considered during the design of canopy roofs. Most roof failures originate in these regions [3].

Information available regarding the wind pressure coefficients on flat, mono-slope, pitched and trough canopy roofs in wind load code/standards [13–16] are very limited and up to some extent, and similar information on circular canopy roof is not available [1, 4, 17, 18]. So, it becomes very important to study the effect of wind load effect on the low-rise canopy structure such that it will be included in the wind standard/codes.

2 Experimental Details

2.1 Wind Tunnel

Experiment is undertaken inside open circuit atmospheric boundary layer wind tunnel in the Civil Engineering department at Indian Institute of Technology Roorkee, India. The wind tunnel has cross-sectional dimension of 2 m (width)

2 m (height) and has a length section of 15 m length approximately. At the upstream of wind tunnel Vortex generators and angular grids are installed to achieve atmospheric boundary layer airflow and cubical blocks of size 71, 50, and 37 mm are placed at upstream end of the test section to achieve mean wind velocity profile corresponding to the terrain category 2 with power law exponent (α) 0.3 as per [16]. The free stream mean wind flow velocity is taken to about 10 m/s in the experimentation. The wind velocity measured at the top of model having height 75 mm is 6.2 m/s used for calculating pressure coefficient values at the semi-circular canopy roof surfaces.

2.2 *Wind Flow Characteristic*

The velocity and turbulence intensity profile of the wind flowing inside the wind tunnel which is measured near the downstream end of the test section is shown in 1 and 2 respectively.

The instrument Testo-480 is used to measure wind velocity above the tunnel floor at different elevation. Values of pressure are measured using the Scanivalve ZOC33 pressure scanner by connecting plastic tubes coming from the pressure points on the instrumented model to the pressure scanner. The pressure scanner works at a frequency of 500 Hz, i.e., 500 readings are noted every second at each pressure point (Figs. 1 and 2).

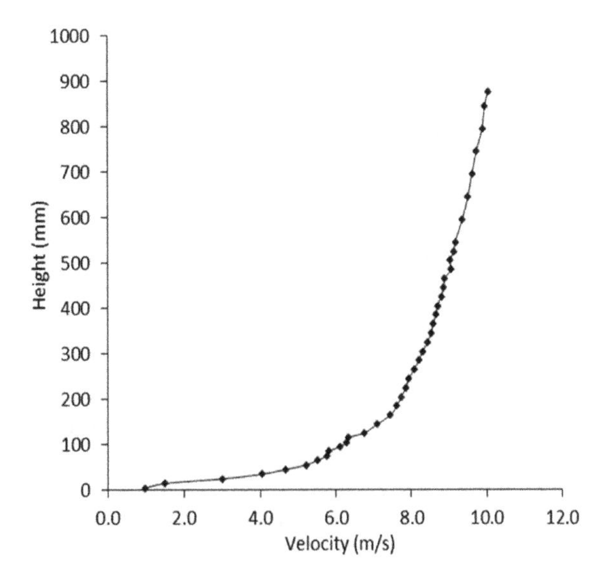

Fig. 1 Wind velocity profile

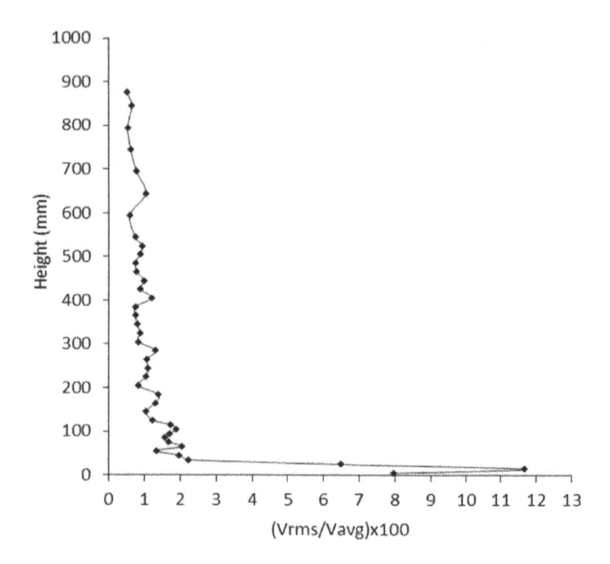

Fig. 2 Turbulence intensity

2.3 *Experimental Model Details*

Three models of semi-circular canopy roof are constructed for experimental study. First model is made up of perpex sheet which is used for instrument installation and consider as main model for measuring values. The remaining two model made up of plywood and it is treated as dummy model as shown in Fig. 3. A scaled model of 1:50 is used in the experiment with its dimensions are shown in Figs. 4 and 5 respectively. The plan dimensions of the models are 300 × 150 mm. The upper roof surface has seven pressure points on each section with a total of 49 pressure points (Fig. 5). The lower surface has six pressure points on each section with a total of 42 pressure points.

2.4 *Measurement Technique*

Single Span

Perspex sheet model of the semi-circular canopy roof with a rectangular plan is placed at the center of turntable inside the wind tunnel in such a way that points 1, 8, 15, 22, 29, 36, and 43 of the upper roof surface falls on the windward edge for 0° wind incidence angle (Fig. 6). For the single span semi-circular canopy roof model the pressure coefficient is evaluated for wind incidence angle ranging from 0° to 90° at an interval of 15°.

Fig. 3 Single, two and three span semi-circular canopy roof inside the wind tunnel

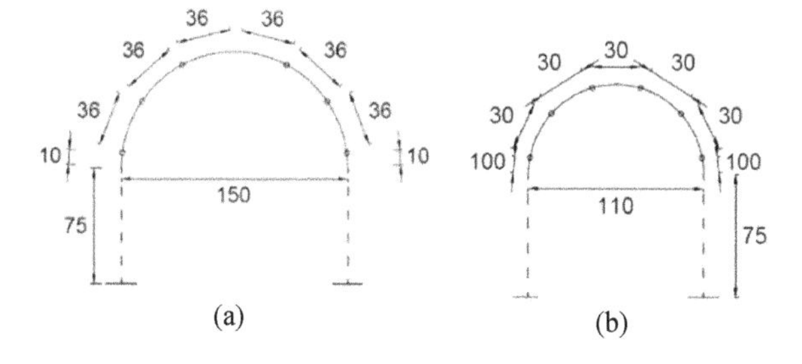

Fig. 4 Circumference dimensions of semi-circular canopy roof at **a** upper and **b** lower roof surfaces of the model (all dimensions are in mm)

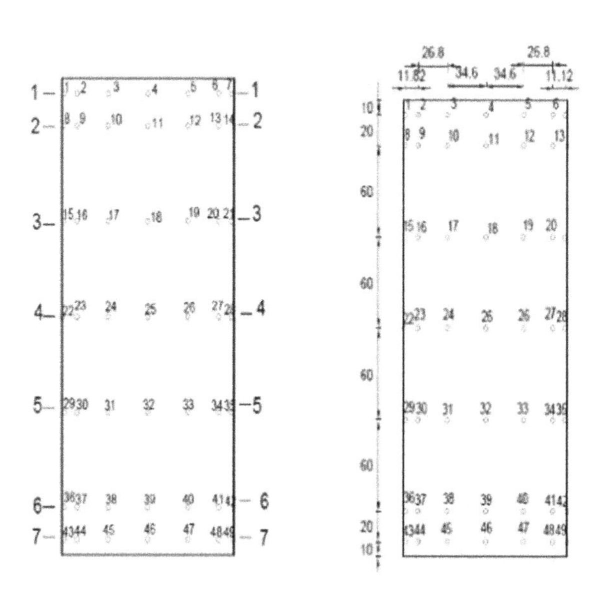

Fig. 5 Pressure points on upper roof surface of semi-circular canopy roof model (all dimensions are in mm)

Two Spans

Wind pressure on the two-span semi-circular canopy roof model is measured by placing a plywood model of identical dimensions on the turntable with the instrumented perspex sheet model. First, the model installed with instrumentation is placed on the upstream end, and then the pressure is measured for seven different

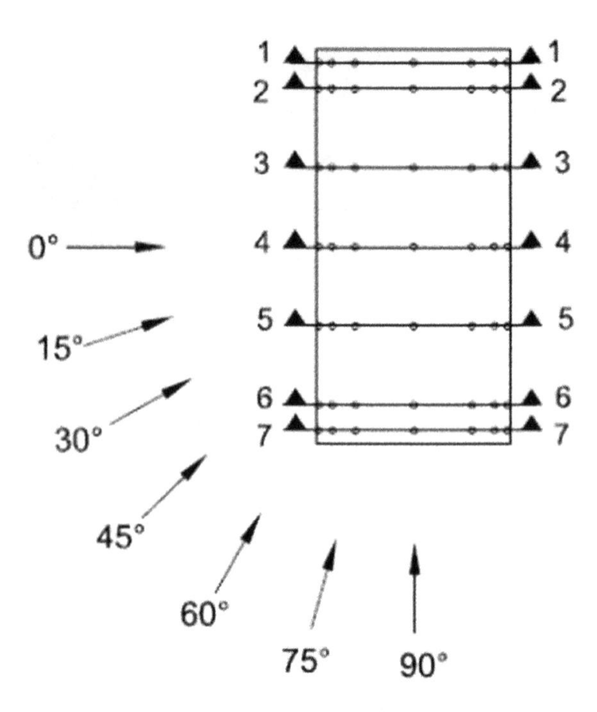

Fig. 6 Wind directions and sections on single span semi-circular canopy roof model

wind incidence angles. Then the positions of the models are interchanged and the instrumented model is placed on the leeward side (Error! Reference source not found.8).

Three Spans

To study the wind pressure distribution on a three-span semi-circular canopy roof, the instrumented model is kept at the center of the turntable and the other two non-instrumented models are placed on its downstream side. Wind strikes the model perpendicular to the longitudinal axis at 0° wind incidence angle. For the three-span semi-circular canopy roof model the wind pressure coefficient is obtained for wind incidence angle ranging from 0° to 90° at an interval of 15°.

3 Results

The measure wind pressure values are reported in the form of dimensionless pressure coefficients (C_P).

$$C_P = \frac{\bar{P}}{0.6V_{ref}^2} \tag{1}$$

where, V_{ref} = Wind velocity measured at the top of principal building model, and $\bar{P}$ = Mean pressure at a particular pressure tapping.

The pressure coefficients for each roof surface is distributed into three zones (A, B and C) according to [16]. For 0° wind incidence angle, Zone A is considered as windward side, Zone B is the middle portion (contain apex) and Zone C contains the leeward side. The recommended values of pressure coefficients for each zone are found out by taking the average of wind pressure coefficients for a particular zone for all wind incidence angles, the highest value obtained for the zone is reported.

The resultant pressure coefficient, C_{Pr}, is calculated by subtracting the pressure coefficient measured on the lower (C_{Pi}) and the upper surface (C_{Pe}) of semicircular canopy roof.

3.1 Single-Span Semi-circular Canopy Roof

Pressure developed at the upper roof surface of portion-A, i.e., the zone near windward eave (Figs. 7 and 8), decreases with an increase in wind incidence angle and changes to suction ($C_{Pr} = -0.12$) at wind incidence angle of 90°. Values of C_{Pr} in portions-A, B, and C at 0° wind incidence angle are obtained as 1.60, −0.68, and −0.17 respectively. Portion-B, i.e., zone near the apex of the roof, is subjected to maximum suction at 30° wind incidence angle with $C_{Pr} = -1.13$.

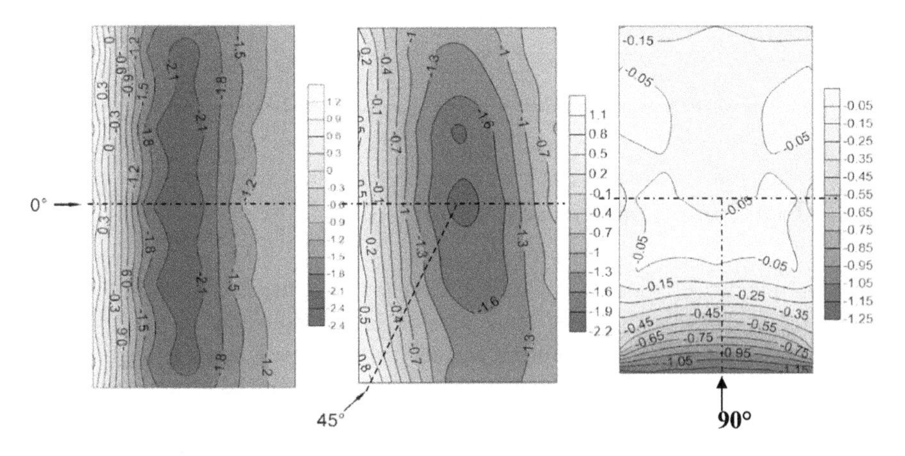

Fig. 7 Contours of mean wind pressure coefficients (***C*$_{P,mean}$**) on lower roof surface of single span semi-circular canopy roof

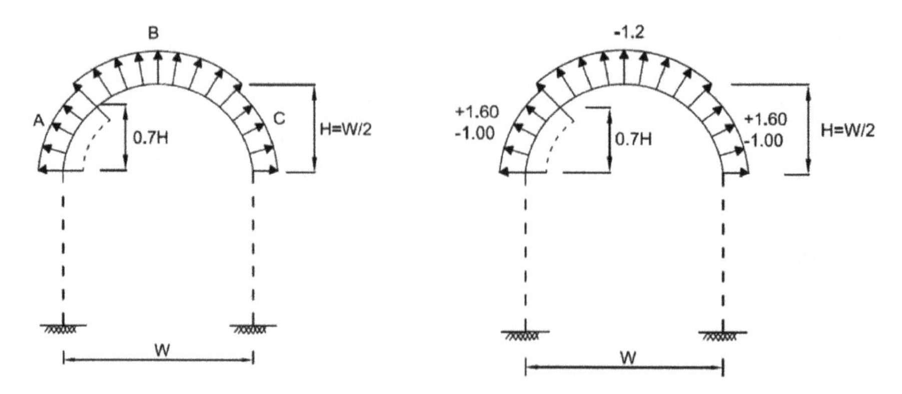

Fig. 8 Design wind pressure coefficient of single span semi-circular canopy roof

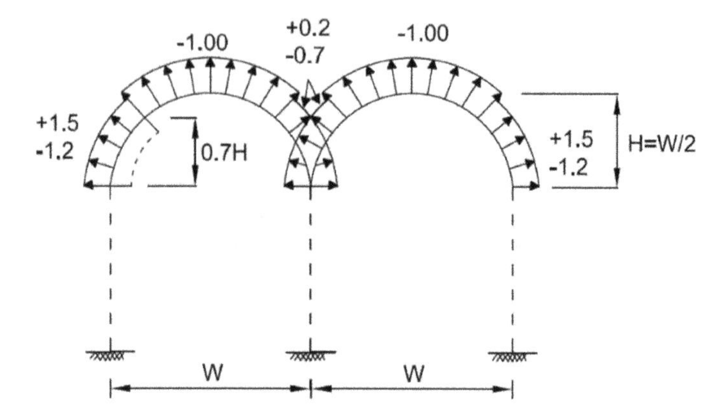

Fig. 9 Design wind pressure coefficient of single span semi-circular canopy roof

3.2 *Two-Span Semi-circular Canopy Roof*

At wind incidence angle of 0°, suction developed on the first span is higher as compared to the other span as separation of flow is dominant in the first span. However, in most cases, suction observed on the second span is higher when comparing to the first span. Positive pressure occurs only on the windward side and this pressure value decreases as wind incidence angle changes from 0° to 60° (Fig. 9). At 60° wind incidence angle, only section 4-4 has positive pressure, with C_{Pr} = 0.27 at the first span.

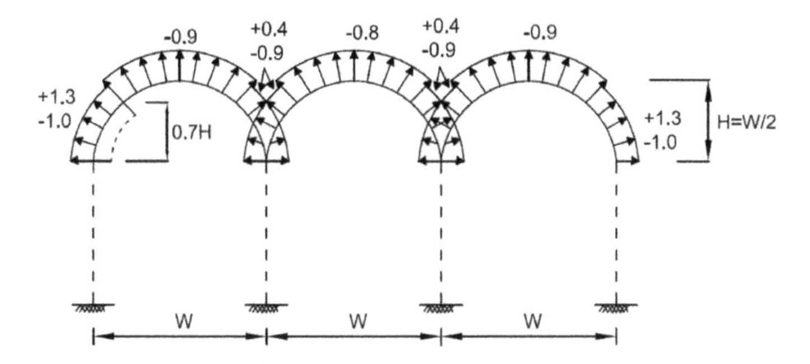

Fig. 10 Design wind pressure coefficient of three span semi-circular canopy roof

3.3 *Three-Span Semi-circular Canopy Roof*

It is noticed from the obtained experimental results of three-span semi-circular canopy roofs that portions A and C are subjected to both pressure and suction, while portion-B is subjected to suction only. Based on the experimentally obtained results in the current study, the recommended values of design wind pressure coefficients on a three-span semi-circular canopy roof are as shown in Fig. 10.

4 Conclusion

It is observed that pressure developed near the windward eave of the upper roof surface of a single-span semi-circular canopy roof for wind incidence angle between 0° and 60°, with maximum pressure occurring at 15° angle. Maximum suction on the upper roof surface always occurs at the apex with its maximum value obtained at wind incidence angle of 30°. The lower roof surface is always subjected to suction at each wind incidence angles.

Locally, magnitude-wise the highest value of $C_{P,min}$ noted for the upper roof surface is −2.55 for the single-span case for 30° wind incidence angle. Similarly, the highest value of $C_{P,min}$ noted for the lower roof surface magnitude-wise is −1.51 for the single-span case for 0° wind incidence angle. The highest value of $C_{P,max}$ noted for the upper roof surface is 2.70, which 2.10 times higher than the corresponding highest value for the single-span case. The highest value of $C_{P,max}$ noted for the lower roof surface is 3.22, which 2.75 times higher than the corresponding highest value for the single-span case. Pressure coefficient for three span is having lesser values as compared to single and two span. So, pressure coefficient decreases as number of span increases.

References

1. Abraham A, Arunachalam S, Knight GMS, Iyer NR (2012) Mean and fluctuating wind loads on a curved roof: effect of terrain conditions. Proc Int Conf Struct Civ Eng 2012:1–4
2. Natalini MB, Morel C, Natalini B (2013) Mean loads on vaulted canopy roofs. J Wind Eng Ind Aerodyn 119:102–113
3. Letchford CW, Ginger JD (1992) Wind loads on planar canopy roofs—part 1: mean pressure distributions. J Wind Eng Ind Aerodyn 45:25–45
4. Natalini B, Marighetti JO, Natalini MB (2002) Wind tunnel modelling of mean pressures on planar canopy roof. J Wind Eng Ind Aerodyn 90(4–5):427–439. https://doi.org/10.1016/S0167-6105(01)00205-7
5. Roy AK, Aziz A, Verma SK, Sharma SK (2018) Influence of surrounding buildings on canopy roof of low rise buildings in Abl By Cfd. no. March, 2018. https://doi.org/10.13140/RG.2.2.23274.62406
6. Roy AK, Ahuja AK, Gupta VK (2010) Variation of wind pressure on canopy-roofs. Int J Earth Sci Eng 03 (02spl.):19–30. https://doi.org/10.13140/2.1.3732.0480
7. Uematsu Y, Yamamura R (2019) Wind loads for designing the main wind-force resisting systems of cylindrical free-standing canopy roofs. Tech Trans 116(7):125–143. https://doi.org/10.4467/2353737XCT.19.076.10727
8. Singh J, Roy AK (2019) Effects of roof slope and wind direction on wind pressure distribution on the roof of a square plan pyramidal low-rise building using CFD simulation. Int J Adv Struct Eng 11(2):231–254. https://doi.org/10.1007/s40091-019-0227-3
9. Krishna P (1995) Wind loads on low rise buildings—a review. J Wind Eng Ind Aerodyn 54(55):383–396
10. Paluch MJ, Loredo-Souza AM, Blessmann J (2003) Wind loads on attached canopies and their effect on the pressure distribution over arch-roof industrial buildings. J Wind Eng Ind Aerodyn 91(8):975–994. https://doi.org/10.1016/S0167-6105(03)00047-3
11. Letchford CW, Row A, Vitale A, Wolbers J (2000) Mean wind loads on porous canopy roofs. J Wind Eng Ind Aerodyn 84(2):197–213. https://doi.org/10.1016/S0167-6105(99)00103-8
12. Gumley SJ (1984) A parametric study of extreme pressures for the static design of canopy structures. J Wind Eng Ind Aerodyn 16(1):43–56. https://doi.org/10.1016/0167-6105(84)90048-5
13. AS/NZS:1170.2 (2002) Australian/New Zealand standard: structural design actions Part-2: wind actions, vol 2
14. ASCE:7-98 (2013) American society of civil engineers (ASCE), vol 552
15. EN:1991-1-4-2005 (2011) Euro Code 1: actions on structures—wind actions, vol 1, no 2005
16. IS:875 (Part-3) (2015) Code of practice for design loads (other than earthquake loads), for building and structures—wind loads
17. Blackmore PA, Tsokri E (2006) Wind loads on curved roofs. J Wind Eng Ind Aerodyn 94(11):833–844. https://doi.org/10.1016/j.jweia.2006.06.006
18. Rani N (2018) Wind pressure distribution on multi-span canopy roofs, p 667

Analytical Study on Early—Age Crack Width in RCC Beam Structure

Binny Elza Ninan and Nivin Philip

Abstract Concrete can undergo early age cracking depending on the composition of the mix, environment exposure, rate of hydration and curing condition. Early age cracks generally progress after the placement of concrete within the first seven days. Early age cracking occurs within 60 days. Crack occurs when the tensile stress within the concrete reaches the tensile strength of the concrete. Early age cracks do not cause the failure of concrete structures unless the tolerance level is exceeded. Early age crack further leads to functionality, durability and aesthetical problems. Autogenous shrinkage is the main reason for early age cracking in high strength concrete. FE based simulation cracking performance of RCC beam specimens of size 700 mm × 150 mm × 150 mm with compressive strength of concrete as 70 MPa and characteristic strength of steel as 415 MPa were analyzed using ANSYS software in Explicit Dynamics. This study mainly focused on the reason for early age crack, types of crack formed and the effect of crack on RCC beam structure. Two types of beam, one without crack and the other with a crack width of 20 micron on the both faces were considered in this study. Flexural strength test were carried out. Parameters like equivalent stress, maximum principal stress, maximum principal elastic strain and equivalent plastic strain were compared with beam without crack and with crack.

Keywords FE simulation · Crack propagation · Solid 65-8 node solid element · Riedel-Hiermaier-Thoma (RHT) concrete

1 Introduction

1.1 General

Early age cracks occur when the strength of the coarse aggregate is more than the cementitious matrix. It is caused due to the combination of hydration heat and

B. E. Ninan (✉) · N. Philip
Saintgits College of Engineering, Pathamuttom, Kerala, India

G. C. Marano et al. (eds.), *Proceedings of SECON'21*, Lecture Notes in Civil Engineering 171, https://doi.org/10.1007/978-3-030-80312-4_72

shrinkage effect of concrete. Hydration heat effect of concrete takes place within 3–5 days after casting takes place [1]. At early age, concrete structures are subjected to deformation caused by shrinkage mechanisms [2]. Cracks affect the structure durability, aesthetics and the bearing capacity of the structure is reduced [3].

An eight node solid element (SOLID65) was used to model the concrete with three degree of freedom at each node, translation in the nodal x, y, z directions. The element is capable of plastic deformation, cracking in three orthogonal directions and crushing [4]. The inverse problem is solved iteratively for the crack location and crack size by the Newton–Raphson method in ANSYS [5].

Objective. The main objectives of the study include:

- To identify the reasons for early age crack
- To find width in early age crack
- To study the effect of early age crack in RC beam structure

Scope. Solid 65, 8 node solid element is used to model the concrete using ANSYS software. RHT Concrete model is used for analyzing concrete and Johnson Cook model is used to analyze steel in this study. Non-linear Dynamic analysis of RC structure (beam) is to be carried out. Crack width equal to 20 micron is considered in this study.

2 Modeling of RCC Beam

2.1 General

Finite element (FE) modeling of RCC beam was developed using the commercial software ANSYS Workbench. RCC beam was modeled using a three-dimensional solid part. A displacement condition was introduced such that the rotation and translations were restricted at the supports. In this study a RCC beam of 700 mm × 150 mm × 150 mm is analyzed as per IS 516:1959 [6]. The beam has 25 mm cover, 2 numbers of 8 mm diameter bars as main reinforcement for compression, 2 numbers of 8 mm diameter bars as hanger bars and stirrups of 6 mm diameter bars as per IS 456:2000 were analyzed [7]. The center to center spacing between the supports is 600 mm. Higher grade concrete has less cementitious matrix compared to the strength of the coarse aggregate and heat of hydration is high which leads to early age crack. M 70 concrete and Fe 415 steel are used. RHT concrete and Johnson Cook steel were used for analyzing. The material property of RHT concrete and Johnson Cook steel are shown in Tables 1 and 2 respectively.

RCC Beam Without Crack. Figure 1 shows the beam modeled using a three-dimensional solid part.

Figure 2 shows a fine mesh of maximum size 10 mm adopted in the model.

Table 1 Material property of RHT concrete

Density, kg/mm^3	2.4e−006
Specific heat, mJ kg^{-1} C^{-1}	6.54E+005
Solid density, kg/mm^3	2.75e−006
Compressive strength, MPa	70
Tensile strength	0.1
Shear strength	0.18
Intact failure surface constant, A	1.6
Intact failure surface exponent, n	0.61
Tension/compression meridian ratio, Q2.0	0.6805
Brittle ductile transition, BQ	1.05E−002
Hardening slope	2
Elastic strength/ft	0.7
Elastic strength/fc	0.53
Fracture strength constant, B	1.6
Fracture strength exponent, m	0.61
Compressive strain rate exponent, α	3.2e−002
Tensile strain rate exponent, δ	3.6e−002
Maximum fracture strength ratio, SFMAX	1.e+020
Damage constant, D1	4.e−002
Damage constant, D2	1
Minimum strain to failure	3.5e−003
Residual shear modulus fraction	0.13
Shear modulus, MPa	16,700

Table 2 Material property of Johnson Cook steel

Density, kg/mm^3	7.85e−006
Specific heat, mJ kg^{-1} C^{-1}	4.77e+005
Bulk modulus, MPa	1.59e+005
Initial yield stress, MPa	415
Hardening constant, MPa	510
Hardening exponent	0.26
Strain rate constant	1.4e−002
Thermal softening exponent	1.03
Melting temperature, C	1519.8
Reference strain rate (/s)	1
Shear modulus, MPa	81,800
Maximum tensile stress, MPa	480
Maximum shear stress, MPa	480
Maximum principal strain	3.e−002
Maximum shear strain	3.e−002

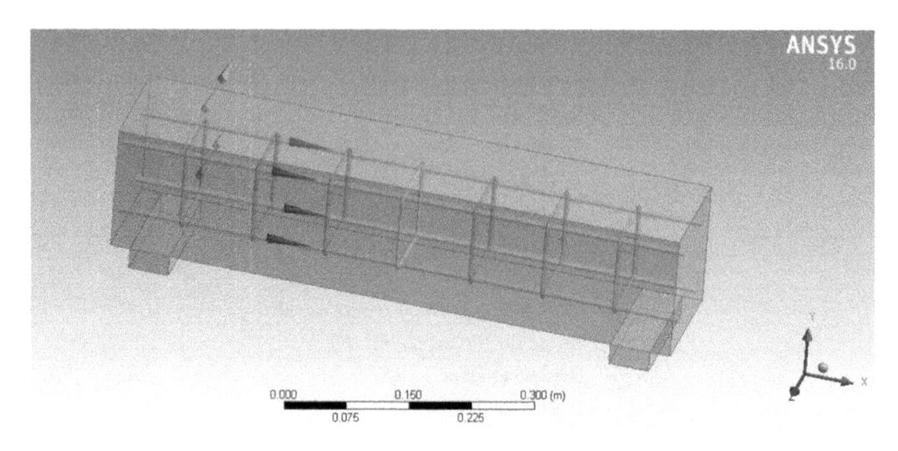

Fig. 1 Geometry of the beam without crack in ANSYS

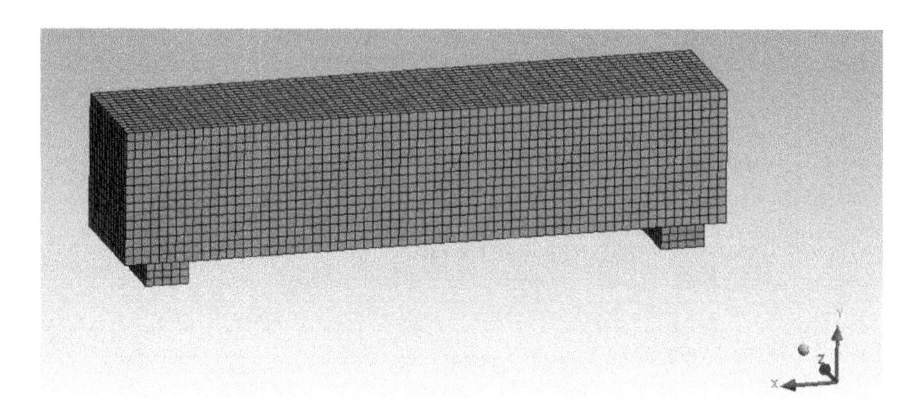

Fig. 2 Mesh diagram of beam without crack

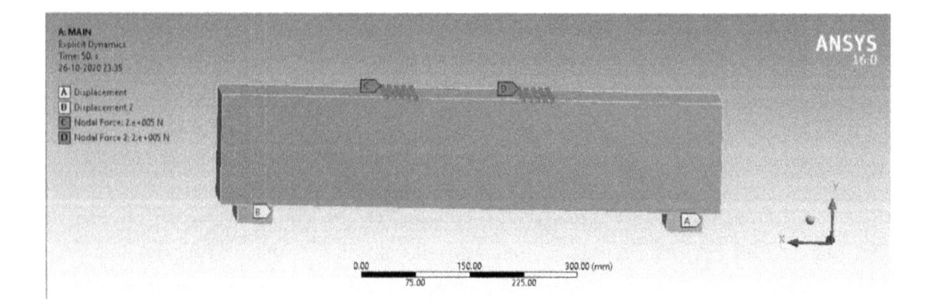

Fig. 3 Load and support condition of beam

The loads are applied on the nodes and displacement supports were applied with no translation and rotation Re (Fig. 3).

RCC Beam with Crack. The beam was modeled with 20 micron cracks on both the faces. The beam has crack length of 1.261955 m. Table 3 shows crack length used in the model.

$$\frac{Crack\ length}{Total\ length} = \frac{1.261955 \times 100}{0.7}$$
$$= 180.279\% \text{ (one side face)}$$

Figure 4 shows a beam modeled using a three-dimensional solid part with a crack of 20 micron.

Figures 5 and 6 show the mesh diagram and load and support condition of the beam respectively. The loads are applied on the nodes and displacement supports were applied with no translation and rotation.

Table 3 The crack length used in the model

SI No.	Crack length (m)
1	0.078375
2	0.057113
3	0.062883
4	0.044633
5	0.063422
6	0.073673
7	0.057557
8	0.082379
9	0.064948
10	0.086733
11	0.068432
12	0.085408
13	0.085589
14	0.069549
15	0.081362
16	0.062308
17	0.081246
18	0.056345
Total length	1.261955

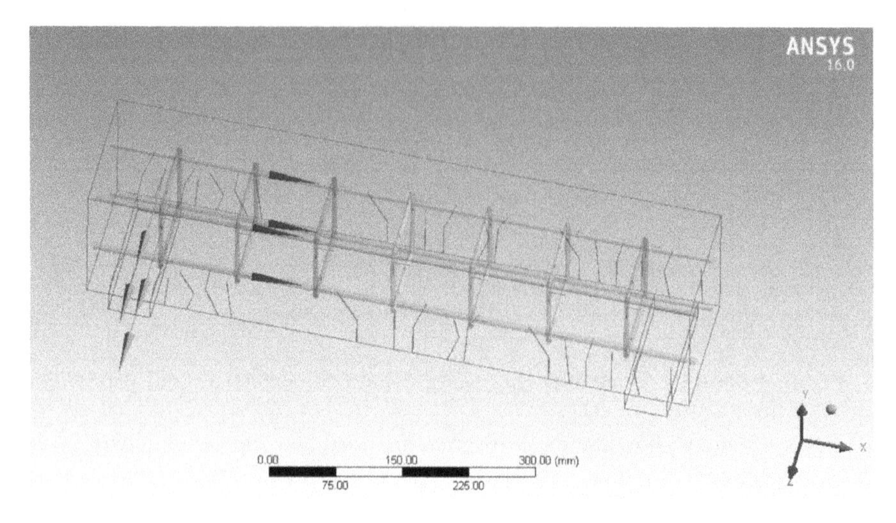

Fig. 4 Geometry of the beam with crack in ANSYS

Fig. 5 Mesh diagram of beam with crack

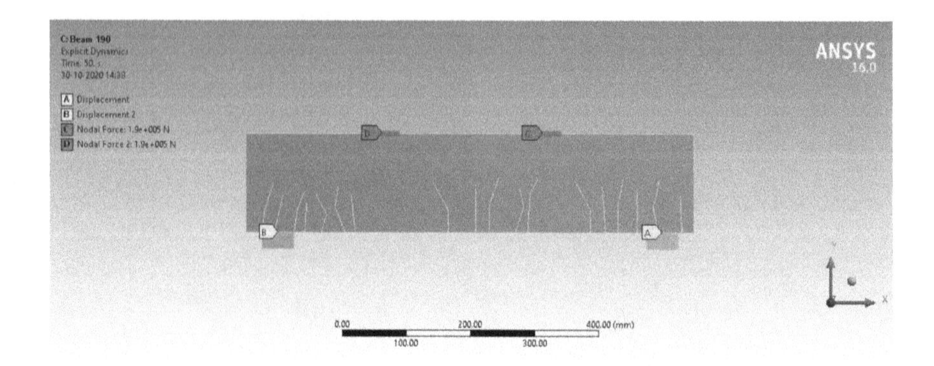

Fig. 6 Load and support condition of cracked beam

3 Result and Discussion

3.1 RCC Beam Without Crack

The beam was analyzed for different loads and the load resulting with the initial crack was considered. The load applied was 200 kN and the obtained results had end time of 50 s and maximum number of cycles 30,000. Figures 7 and 8 represents equivalent stress and maximum principal stress of the beam respectively.

Figures 9 and 10 show equivalent plastic strain and maximum principal elastic strain values of the beam respectively. Figure 11 shows the equivalent stress on the reinforcement of the beam.

RCC Beam with Crack. The beam was analyzed for different loads and the load resulting with the initial crack was considered. The load applied was 190 kN and the obtained results had end time of 50 s and maximum number of cycles 30,000. Figures 12 and 13 represent equivalent stress and maximum principal stress of the beam with crack respectively.

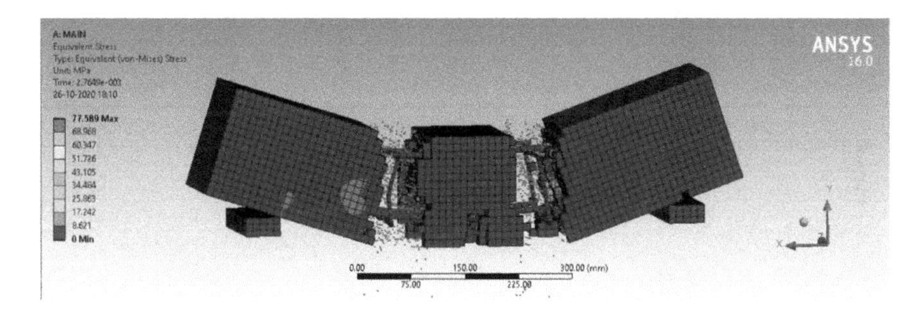

Fig. 7 Equivalent stress

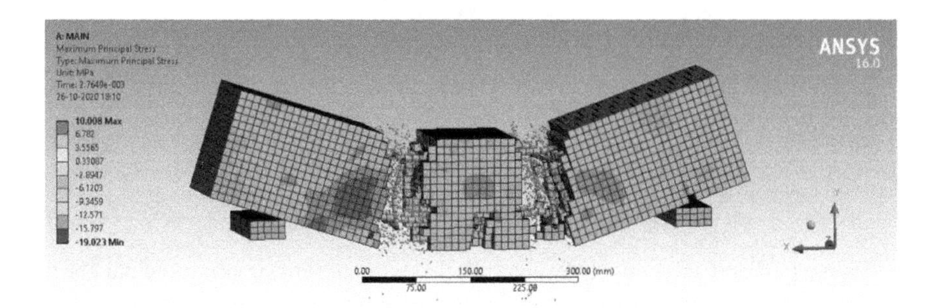

Fig. 8 Maximum principal stress

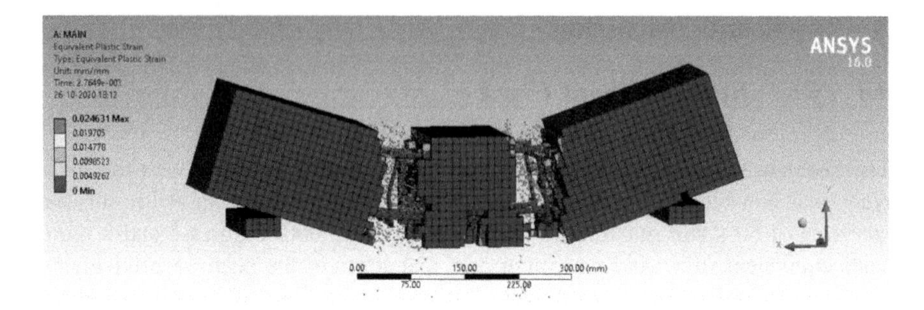

Fig. 9 Equivalent plastic strain

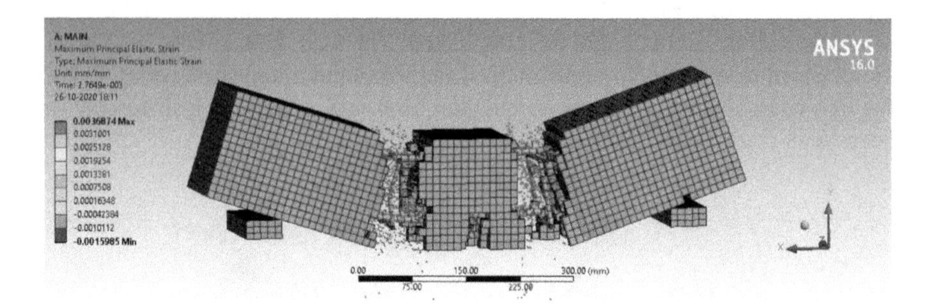

Fig. 10 Maximum principal elastic strain

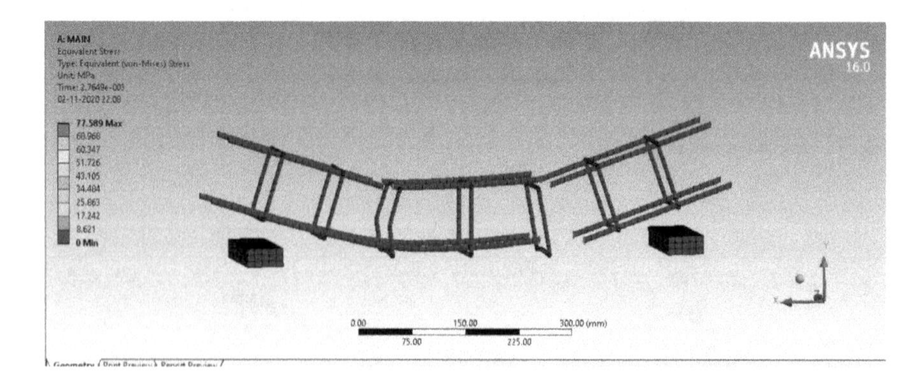

Fig. 11 Equivalent stress on the reinforcement

Figures 14 and 15 show equivalent plastic strain and maximum principal elastic strain values of the beam with crack respectively. Figure 16 shows the equivalent stress on the reinforcement of the beam with crack.

Discussion. Percentage variation for the beam without crack and the beam with crack for maximum principal stress from the results is as shown in Table 4.

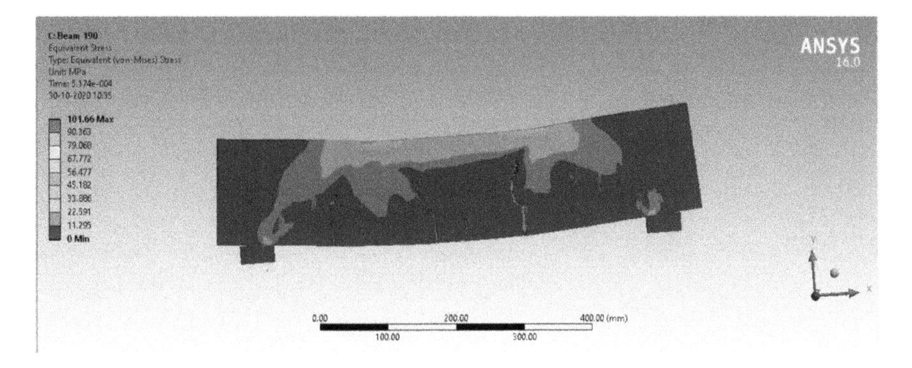

Fig. 12 Equivalent stress of the beam with crack

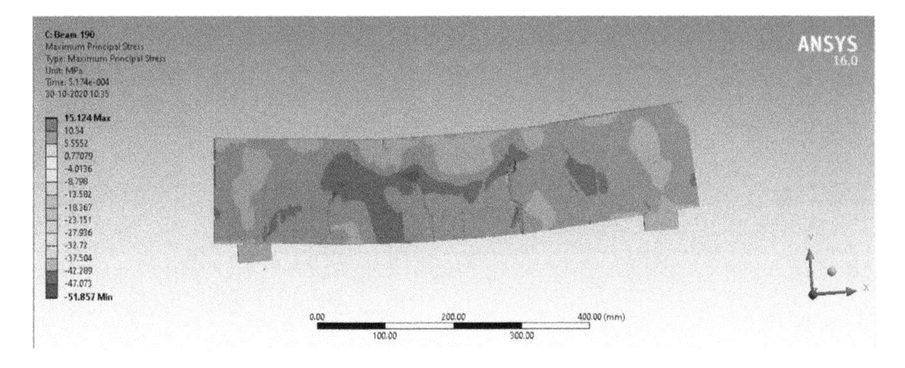

Fig. 13 Maximum principal stress of the beam with crack

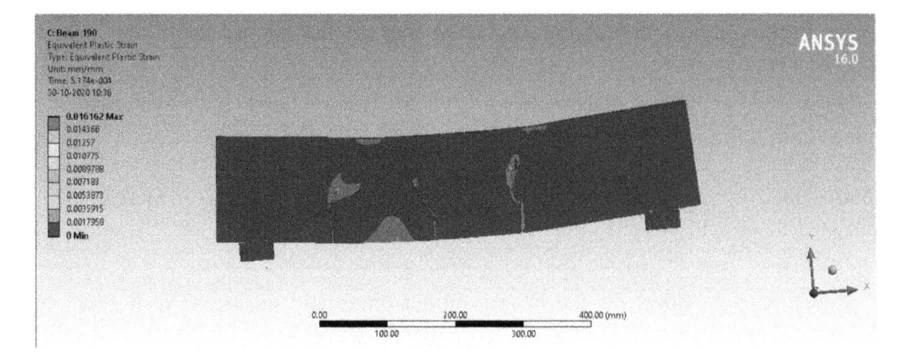

Fig. 14 Equivalent plastic strain of the beam with crackW

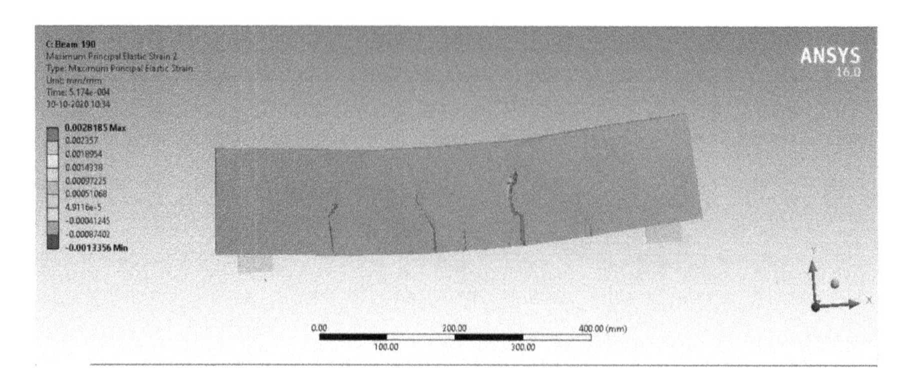

Fig. 15 Maximum principal elastic strain of the beam with crack

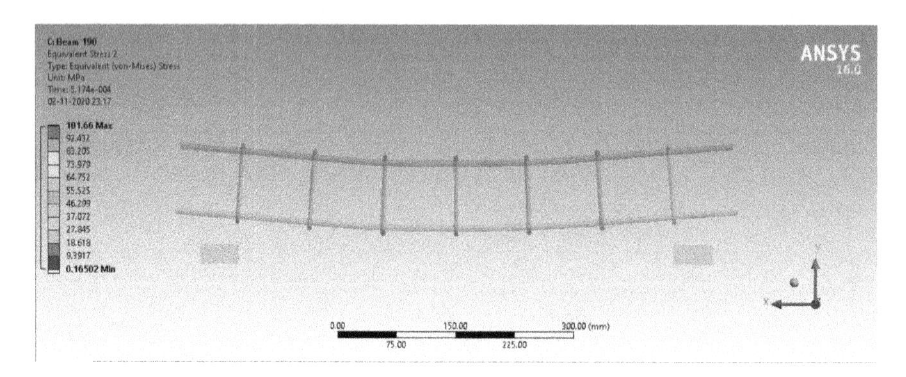

Fig. 16 Equivalent stress on the reinforcement of the beam with crack

Table 4 The percentage variation for the beam without crack and the beam with crack for maximum principal stress

	Stress of the beam without crack (MPa)	Stress of the beam with crack (MPa)	Variation
Initial value	0	0	0
Final value	10.008	15.124	51.12%

Percentage variation for the beam without crack and the beam with crack for equivalent stress from the results is as shown in Table 5.

Percentage variation for the beam without crack and the beam with crack for maximum principal elastic strain from the results is as shown in Table 6.

Percentage variation for the beam without crack and the beam with crack for equivalent plastic strain from the results is as shown in Table 7.

Table 5 The percentage variation for the beam without crack and the beam with crack for equivalent stress

	Stress of the beam without crack (MPa)	Stress of the beam with crack (MPa)	Variation
Initial value	0	0	0
Final value	77.589	101.66	31.02%

Table 6 The percentage variation for the beam without crack and the beam with crack for maximum principal elastic strain

	Strain of the beam without crack	Strain of the beam with crack	Variation
Initial value	0	0	0
Final value	0.0036	0.0028	22.22%

Table 7 The percentage variation for the beam without crack and the beam with crack for equivalent plastic strain

	Strain of the beam without crack	Strain of the beam with crack	Variation
Initial value	0	0	0
Final value	0.0246	0.0161	34.55%

4 Conclusions

Crack occurs when the tensile stress in the concrete reaches the tensile strength of the concrete. Temperature differences and stress development during hardening of concrete leads to crack in concrete structures. Early age crack further leads to functionality, durability and aesthetical problems. Analytical study of beam without crack and beam with crack were carried out. Loading capacity is reduced to 190 kN for the beam with a crack of 20 micron. Due to the effect of early age cracks, the bonding between the particles and matrix is not much effective compared to normal concrete thus the flexural capacity of concrete reached the early stage. The real criteria of failure for concrete are the limiting strain rather than the limiting stress.

The following are the conclusions made from the results:

1. The percentage variation for the beam without crack and the beam with crack conditions for maximum principal stress is about 51.12%.
2. Equivalent stress for the beam without crack and the beam with crack conditions has percentage variation about 31.02%.
3. The percentage variation for the beam without crack and the beam with crack conditions for maximum principal elastic strain is about 22.22%.
4. Equivalent plastic strain for the beam without crack and the beam with crack conditions has percentage variation about 34.55%.

References

1. Chang S, Yang M, Sun Y, Liu K (2019) Calculation method of early-age crack width in reinforced concrete bridge through nonlinear FEA model. KSCE J Civ Eng 23:3088–3096
2. Sule M, Breugel K (2001) Cracking behavior of reinforced concrete subjected to early age shrinkage. Mater Struct 34:284–292
3. Zhao W, Gong H, Wang Y, Xie Y (2014) Prediction on crack width based on ANSYS nonlinear analysis. ICMCE, Beijing, China
4. Subramani T, Manivannan R, Kavitha M (2014) Crack identification in reinforced concrete beams using Ansys software. Int J Eng Res Appl 6:133–141
5. Lee J (2009) Identification of a crack in a beam by the boundary element method. J Mech Sci Technol 24:801–804
6. IS 516:1959 method of test for strength of concrete. Bureau of Indian Standards, New Delhi
7. IS 456:2000 plain and reinforced concrete-code of practice. Bureau of Indian Standards, New Delhi

Motorized Two Wheeler Lateral Clearance Behavior in Heterogeneous Traffic: A Sensor Based Approach

Manisha Biswal, Kudururpaka Vamshikrishna, Sujeet Sahoo, and Prasant Kumar Bhuyan

Abstract In India, traffic flow behavior is highly heterogeneous and does not follow any lane discipline which encourages vehicles to occupy any available lateral position along the road width. In absence of lane discipline, vehicles not only interact longitudinally but also laterally. While in most of the developing countries, passenger cars and Motorized Two Wheelers (MTW) are major constituents of total traffic composition, the traffic operational analysis are focused mostly only on passenger car. Whereas, two wheeler possess unique characteristics thus behaving differently from passenger car. Therefore, understanding the behavior of MTW is necessary to study its impact on traffic stream. This research work focused on studying lateral clearance maintaining behaviors of motorized two wheelers in aspect of heterogeneous traffic context. Ultrasonic and Global Positioning System (GPS) sensors were used in this research for collecting detailed and accurate microscopic traffic data. The results showed that the subject vehicle tends to maintain smaller lateral clearance when road occupancy is high. Another objective of this study was to investigate the factors affecting lateral clearance. Lateral clearance varied with subject vehicle speed, interacting vehicle type. Lateral clearance was observed to exhibit positive gradient with speed of the subject vehicles. Furthermore, the lateral clearance maintained by vehicle followed General Extreme value and Weibull distributions.

Keywords Heterogeneous traffic · Motorized two wheeler · Ultrasonic sensor · GPS · Lateral clearance

1 Introduction

Traffic streams on Indian roads are different from traffic steam of other developed countries as it is composed of different kind of vehicles having diverse driving characteristics and maneuver abilities, which results into traffic streams with

M. Biswal (✉) · K. Vamshikrishna · S. Sahoo · P. K. Bhuyan
National Institute of Technology, Rourkela, India

G. C. Marano et al. (eds.), *Proceedings of SECON'21*, Lecture Notes in Civil Engineering 171, https://doi.org/10.1007/978-3-030-80312-4_73

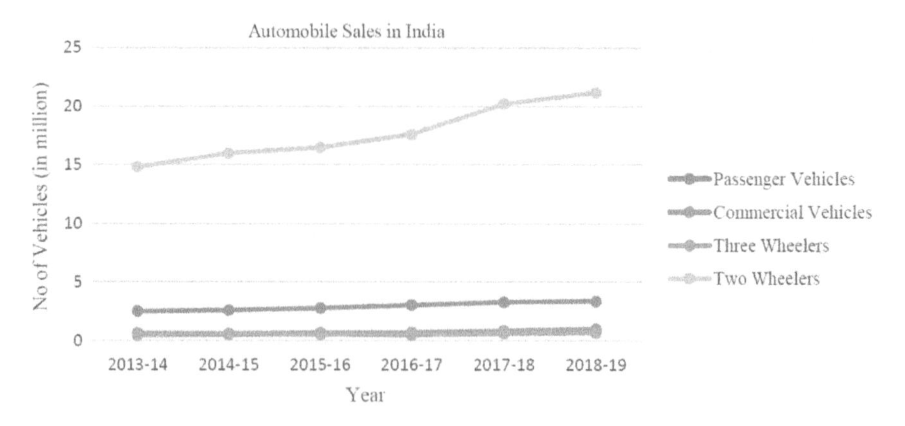

Fig. 1 Trend of automobile sales in India. *Source* http://www.siamindia.com

heterogeneous flow conditions. Indian traffic condition is heterogeneous in nature and lack of lane discipline, for which vehicles occupies any lateral position along the road width, hence vehicles not only interact longitudinally but it also interacts laterally with other vehicles by moving side by side. The survey conducted by Society of Indian Automobile Manufactures (SIAM) India, indicates Motorized Two-Wheelers (MTW) purchase increased from 11.8 to 21.18 Million from Financial year (FY) 2011 to FY 2019 is shown in Fig. 1. This statistic shows that MTWs contains major traffic composition in India. But the behavior of MTWs are very versatile in nature due to its small size and flexibility, that's why we cannot describe MTW behavior same as standard car behavior. Therefore, understanding MTWs behavior is necessary to know its impact of traffic operations.

There are lots of literature available on lane changing model to describe lateral movement behavior, however lane changing model are not sufficient to describe lateral movements in non-lane based traffic. In non-lane based traffic, lateral driving behavior is mostly explained by the lateral clearance maintained within the vehicles. Hence it is important to analyze the lateral clearance between the vehicle and its variation with the speed of vehicle. For current study, motorized two wheelers are selected as subject vehicles on which the sensors are mounted.

2 Background Literature

The literature review explained here is more in lateral clearance as this is the prime characteristic of the research work. There are some of the research carried out in advanced data collection methods that is by using sensor system.

2.1 Lateral Clearance

The Lateral Clearance (LC) is the lateral safety distance, which the vehicle maintained from interacting vehicles when it is travels through traffic stream. The maximum and minimum value of LC has been estimated for two lane road for different combinations of vehicles by taking data on LC from the physical barriers like kerb, median etc. into consideration [1]. The LC between interacting vehicles and subject vehicles was influenced by the speed of the subject vehicle being considered, speed of the interacting vehicle, and their respective vehicle types. The LC maintained by two interacting vehicles during flow was approximated as the sum of the two independent clearances distance, each contributed by the type and speed of the vehicles involved [2]. Some research was also conducted to analyze LC maintained between the vehicles by using area occupancy approach, and the research finds out that the LC is highly influenced by vehicle width and traffic composition [3]. It is also found that more frictional clearances are required with the increasing speed of passing vehicles. Also, it was found that increase in the center line separation reduces the time headway between the leader and the follower [4]. The previous research proposed a linear relationship between interacting vehicle speed and LC [5]. There are some simulation models developed for mixed traffic moving on two-lane roads. In these models they have used the data such as lateral placement, the arrival pattern, acceleration, and overtaking characteristics of different types of vehicles to model LC behavior. The research found that any vehicle was able to pass the lead vehicle when the LC was one and half times the width of the passing vehicle. Some simulation models developed a relationship between LC and speed profile which is linear by nature [6, 7]. LC for different type of two wheelers and analyzed the distribution of LCs in a shared traffic facility weak lane discipline. Computed mean variations and ninety percentile LC [8]. The validity of these models are limited to macroscopic analysis.

2.2 Use of Sensor in Previous Research

The research in traffic is moving towards high accuracy of data collection. LC in between vehicle should be more accurate and instantaneous which is quite challenging to obtained by traffic video recording techniques. In order to overcome this problem, modern sensors are utilized for more accurate analysis. Previously some research was conducted using optical sensor for collision avoidance of four wheelers [9]. Then ultra-sonic sensor was used to collect data. Ultrasonic sensor mounted on both side of vehicle to obtained intravehicular LC and relative speed of vehicles combined with GPS devices and multiple cameras. Some of the study concluded that, vehicles of similar class maintained less LC as compare to vehicles pair of different classes with similar speed [10]. The data obtained test-vehicle speed with a GPS data logger and with an accelerometer of 10 Hz frequency, and used ultrasonic sensor with 5 Hz frequency showed a minor variation in instantaneous LC. In this study, vehicle monitoring system consist of data logging shield to store the data obtained from microcontroller. Two ultrasonic sensors which are placed at 3.2 m above the ground and both are separated by a clearance of 1.0 m to prevent the detection of two vehicles simultaneously. This data is fed into a java based computer program which can determine number of vehicle's by analyzing number of raise and fall values obtained from ultrasonic sensor's and also vehicle type is found by finding maximum value between falling and rising values from height data series [11]. Previously some researcher also used vision based sensor for vehicle detection [12], through which types of vehicles can be determined.

In this research work, sound based sensors, that is ultra-sonic sensors are used to get more accurate data to study the LC behavior of MTW.

3 Methodology

The data required for this research was collected from the ring road of Rourkela city on peak hours of an average week day. The driver was explained of the experiment and a volunteer was present during the data collection. It was clearly explained to the driver that no personal data about him is being recorded to have an unbiased data collection. First few kilometer of data collection was ignored to remove biasing from the collected data. Three number of drivers with different vehicles were selected to conduct this study. Current data was collected form central urban area with under saturated traffic condition. The road was having fairly good pavement condition, median separated, raised kerbs and mostly consists of straight segments. The LC model parameters found in this study are limited for similar traffic and geometric conditions.

Ultrasonic sensors were mounted on subject vehicle with two number of sensors at each side and one near the front end of the vehicle and one at rear end of the vehicle such that they face outwards. Ultrasonic sensor module comprises of one transmitter and one receiver. Transmitter sends 40 kHz sound wave considering the effect of temperature and humidity (which was modulated) and the receiver waits for the echo to receive. Whenever any obstacles come ahead of the ultrasonic wave, it reflects back after striking the object. The total time taken for sending the signals and receiving them were noted down from which the distance between the two vehicles can be calculated. Upon receiving the signal, microcontroller will execute the program and calculates the total clearance (to and fro i.e., twice the actual clearance between the vehicles) travelled by the sound waves in centimeters. As the total setup is electronic and wired, there may be some problem in connection or improper functioning of the instrument. To know the type of problem occurring, live transmission of intravehicular clearance data is needed which can be achieved by employing a Bluetooth module. It can transmit data to an android application through Bluetooth.

The sensor emits ultrasonic waves, which are reflected back by the nearby vehicles and received via receiver fixed with the ultrasonic emitter. The motherboard calculated difference between the emitted wave and received wave which can give us accurate LC. The clearance between the vehicle is calculated by Eq. (1).

$$LC = V_{ult} * \Delta T/2 \quad (1)$$

where

LC = Lateral Clearance
V_{ult} = Ultrasonic Speed
ΔT = Time difference between emitted wave and received wave

The setup has a Bluetooth module that transmits the real time data to the mobile app which combines the data with time stamp and GPS co-ordinates. GPS co-ordinates are used to calculate the speed of subject vehicles. Three numbers of video camera are mounted on the subject vehicle, one facing front, other two are facing either side of vehicles. The video was used to find the classes of the interacting vehicles. The vehicle classes are tallied with data base with reference to time co-ordinate manually afterwards in laboratory. When the flow is congested, it was observed that, the vehicles are moving parallel to each other sharing the same lane but maintained a gap between them. This gap varies due to different traffic condition.

In this research, the data collected during peak hour to analyze the gap maintaining characteristics more accurately. When there are fewer number of vehicles interacting laterally, those data sets were rejected manually as they are not interacting with the vehicles. In statistics, sampling method from a population is partition into sub population which should mirror other sub-populations. The datasets sample are made from the total population using systematic sampling. In which the data sets are selected from total population at a constant interval. At first the starting member (position r) of the sample was randomly chosen from the total population. Then in the next step, a constant interval (i) was added to random number to get the next data point and the process was repeated to get the position r + i, r + 2i, r + 3i…r + (n − 1)i, which complete the sample (where n is the sample size).

Five types of distribution were chosen to test the LC namely, Weibull distribution, normal distribution, Log-Normal distribution, Generalized Extreme Value (GEV) distribution and Gamma distribution. Normal distribution being the simple distribution with uniform Kurtosis and centered skewed, Log-normal is left skew. GEV fits the distribution which have extreme event having low probability of occurrence but high impact on output. The main advantage of Weibull is that, it able to detect minute change LC with extremely small sample.

4 Result and Discussion

LC maintained between the vehicles have been analyzed for the vehicle classes of different combination of vehicle such as, Two Wheeler (TW) with TW, Three Wheeler (ThW) and Four Wheeler (FW). The subject vehicle (TW) speed was aggregated over bin size of 5 Kmph that is 0–5 Kmph, 5–10 Kmph, 10–15 Kmph, 15–20 Kmph, 20–25 Kmph, 25–30 Kmph, 30–35 Kmph and 35–40 Kmph. For all the three vehicle combination the mean lateral gap has been plotted against both subject and interacting vehicles with different vehicle combinations are shown in Fig. 2a–c.

From the figures, it can be seen that LC increases with increase in the speed of subject and interacting vehicle. The observations shows that the LC between TW-TW is minimum, since they both have similar acceleration or deceleration properties and high maneuverability properties.

The frequency plot was plotted for LC with varying vehicle speed. It can be observed that LC frequency distribution has a central peak and low values towards low and high LC. Hence, it was tested for normal, lognormal, Weibull, General Extreme Value (GEV) and gamma by Kolmogorov Smirnov test for a significance level of 5%. The results are shown in Table 1.

The p-value represents the acceptance or rejection of the null hypothesis that is, it checks whether the given date sets follows a specified distribution or not. Mean Value represent the whole data set in a single value.

From the observation, it was seen that, GEV, Normal and Weibull distribution follows similar trend and mean values of these two distributions are also similar.

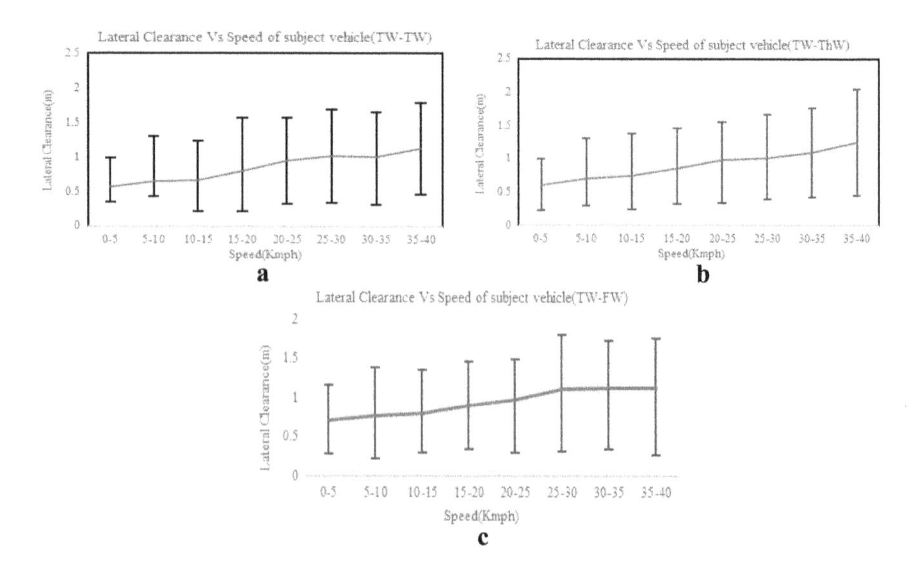

Fig. 2 **a** LC versus speed of subject vehicle (TW-TW); **b** LC versus speed of subject vehicle (TW-ThW); **c** LC versus speed of subject vehicle (TW-FW)

Table 1 KS test for different distribution far varying speed range

Mode	Speed range	KS test									
		Weibull		Normal		Log-normal		GEV		Gamma	
		p	mean	p	mean	p	mean	p	mean	p	mean
TW-TW	0–5	0.15	0.53	0.23	0.56	0.14	0.52	0.06	NAN**	0.18	0.56
	5–10	0.20	0.60	0.29	0.64	0.20	0.60	0.09	NAN**	0.23	0.64
	10–15	0.02	0.65	0.03	0.65	0.06	0.61	0.02	0.65	0.04	0.78
	15–20	0.02	0.78	0.02	0.79	0.08	0.74	0.01	0.79	0.05	0.79
	20–25	0.03	0.93	0.03	0.95	0.06	0.89	0.02	0.95	0.04	0.94
	25–30	0.01	1.01	0.01	1.01	0.05	0.96	0.01	1.01	0.03	1.01
	30–35	0.02	0.97	0.04	0.98	0.07	0.93	0.02	0.98	0.04	0.98
	35–40	0.01	1.11	0.01	1.12	0.04	1.08	0.01	1.12	0.03	1.12
TW-ThW	0–5	0.05	0.58	0.04	0.60	0.06	0.56	0.03	0.59	0.05	0.60
	5–10	0.02	0.67	0.02	0.70	0.06	0.66	0.02	0.70	0.04	0.70
	10–15	0.01	0.72	0.21	0.73	0.05	0.71	0.02	0.72	0.03	0.74
	15–20	0.01	0.83	0.02	0.85	0.05	0.82	0.02	0.85	0.03	0.85
	20–25	0.04	0.96	0.03	0.95	0.07	0.92	0.02	0.98	0.04	0.99
	25–30	0.02	0.98	0.02	1.00	0.06	0.95	0.01	1.01	0.03	1.00
	30–35	0.03	1.07	0.03	1.09	0.06	1.03	0.02	1.09	0.05	1.09
	35–40	0.07	1.23	0.04	1.25	0.1	1.18	0.03	1.25	0.06	1.25
TW-FW	0–5	0.03	0.69	0.03	0.71	0.06	0.67	0.04	0.70	0.05	0.70
	5–10	0.03	0.76	0.06	0.77	0.07	0.72	0.07	0.77	0.07	0.77
	10–15	0.07	0.79	0.06	0.80	0.08	0.76	0.07	0.81	0.07	0.80
	15–20	0.02	0.89	0.03	0.90	0.05	0.86	0.03	0.90	0.04	0.90
	20–25	0.10	0.96	0.06	0.98	0.09	0.93	0.07	0.98	0.08	0.98
	25–30	0.02	1.10	0.03	1.10	0.05	1.05	0.03	1.15	0.05	1.11
	30–35	0.03	1.11	0.03	1.13	0.06	1.05	0.02	1.13	0.04	1.13
	35–40	0.04	1.12	0.02	1.13	0.07	1.05	0.02	1.13	0.05	1.13

**GEV mean cannot be defined when shape factor >0

For example, mean values for Weibull, Normal and GEV are 0.6564, 0.6500 and 0.6546 respectively for speed range of 10–15 Kmph for TW-TW vehicle interaction. Hence, it can be said that they follow similar trend the graph for histogram with distribution curve fitted on them as shown in Fig. 3a–c.

The LC maintained by the interacting vehicles has been tried to modeled through applying different types of distribution. The results show that GEV, Normal distribution and Weibull distributions are able to model the gap more precisely.

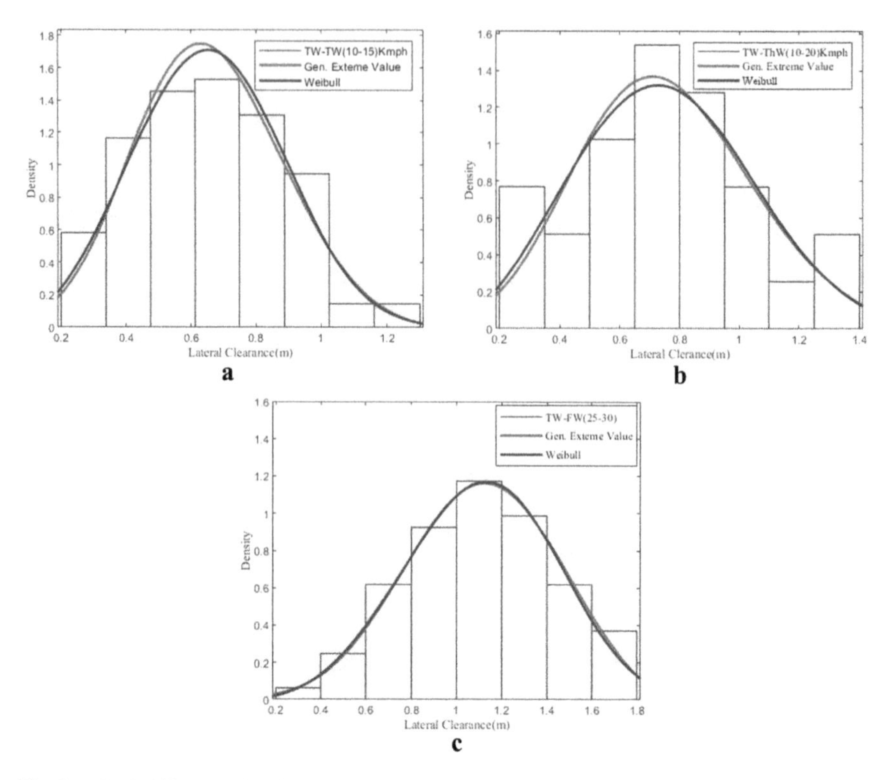

Fig. 3 **a** Probability density distribution of LC between TW-TW vehicle interaction; **b** probability density distribution of LC between TW-ThW vehicle interaction; **c** probability density distribution of LC between TW-FW vehicle interaction

5 Conclusion

The LC primarily affects the overtaking and filtering behavior. The vehicles tend to obtained the least LC when the road occupancy is high. In the present study, data were collected during peak hours and was manually filtered to obtained data only during high occupancy of road. The LC varies for different vehicle to vehicle combination and speed. The KS test shows that LC observed by MTW followed GEV, Normal and Weibull distribution for specific speed ranges. It was found that LC follows these distributions more accurately at higher speeds and shows more fluctuations to the distribution at lower speeds. Lower speed is not a desirable behavior of user, rather it is the result of forced condition. Hence, LC at lower speed ranges cannot be modeled into a distribution for complete data sets comprising of various forced conditions. Instead, it should be modeled individually with consideration of other factors responsible for lower speed ranges. Current study is based on assumption that driver maintain LC according to their safety perspective which can be directly linked with subject vehicle speed. Hence, LC data set should follow the pattern which gives a good fit to the distributions discussed above and is tested via KS hypothesis test. However, at lower speed, forced condition is prevalent causing change in driver's intuition behavior. This behavior causes chaotic and slapdash movement of vehicle. This unpredicted behavior of vehicles at lower speed leads to rejection of null hypothesis, "LC follows a specific distribution for which it is being tested". The LC model is a primary input parameter for any simulation tools for calibrating heterogeneous traffic or any mathematical model representing non-lane based traffic behavior.

References

1. Singh B (1999) Simulation and animation of heterogeneous traffic on urban roads. PhD thesis, IIT Kanpur
2. Arasan VT, Koshy RZ (2005) Methodology for modeling highly heterogeneous traffic flow. J Transp Eng ASCE 131(7):544–551
3. Pal D, Mallikarjuna C (2016) Analysis of the effect of variable lateral clearance maintaining behavior of vehicles on traffic flow modeling. Procedia Eng 142:198–204
4. Gunay B (2007) Car following theory with lateral discomfort. Transp Res Part B 41(7):722–735
5. Minh C, Sano K, Matsumoto S (2005) Characteristics of passing and paired riding Maneuvers of motorcycle. J Eastern Asia Soc Transp Stud 6(January):1496–1508
6. Maurya A, Chakroborty P (2007) CASIM: a realistic CA based traffic flow model. In: Proceedings of 11th world conference on transportation research, Berkeley, California, USA
7. Dey PP, Chandra S, Gangopadhyay S (2008) Simulation of mixed traffic flow on two-lane roads. J Transp Eng ASCE 134(9):361–369
8. Yanyong G, Tarek S, Mohamed ZH (2018) Analysis of the lateral distance between two wheelers and automobiles at shared traffic facility. TRB Annual Meeting
9. Wong C, Qidwai U (2004) Vehicle collision avoidance system [VCAS]. Sensors. In: Proceedings of IEEE, pp 316–319
10. Budhkar AK, Maurya AK (2017) Characteristics of lateral vehicular interactions in heterogeneous traffic with weak lane discipline. J Mod Transp 25:74–89
11. Stiawan R, Kusumadjati A, Aminah NS, Djamal M (2019) An ultrasonic sensor system for vehicle detection application. IOP Conf Ser J Phys Conf Ser 1204
12. Sun Z, Bebis R, Miller G (2004) On-road vehicle detection using optical sensors: a review. In: Proceedings 7th international IEEE conference on intelligent transportation systems, Washington, USA

Structural Application of Concrete Made of Recycled Aggregate Sourced from Construction and Demolition Waste

Vinay Kumar Singh, Ambuj Shukla, Rishi Kumar Sahani, Abhinav Raj Shekhar, and Ranjeet Singh

Abstract This paper presents the behavior of recycled aggregate concrete (RAC) made using field demolished concrete as aggregate. The experimental investigations were conducted on M25 grade concrete, 30 no. of cubes of size 150 mm × 150 mm × 150 mm and 15 no. of beam of size 700 mm × 150 mm 150 mm from which 20 cubes and 10 beams were prepared with 15 and 30% replacement of natural aggregate with recycled aggregate (RA) and rest with 0% replacement. The test parameters were specific gravity, water absorption, bulk density for aggregate and workability, compressive strength and flexure strength for RAC and natural concrete (NC). It was found that specific gravity of recycled concrete aggregate (RCA) was lower than natural aggregate (NA), water absorption of RCA was higher than NA and workability of fresh RAC was also low when compare to the NC. Test result shows the concrete needs increase in 3% of water for the use of RA. It was also investigated that for RAC with 15% replacement of NA with RCA, decrement in compressive strength is relatively lower than the RAC with 30% replacement and flexural strength of RAC was also got decrease as the percentage of RA in concrete increase. The result show that however, presence of RA decreases the strength of concrete but with limited replacement of concrete rubble could be transformed into useful RAC which not only make the construction economical, additionally it can be very useful in environment protection and also can be a major solution for the disposal problem of demolished concrete waste.

Keywords Recycled aggregate concrete (RAC) · Recycled concrete aggregate (RCA) · Natural concrete (NC) · Specific gravity · Water absorption · Workability · Compressive strength · Flexural strength

---

V. K. Singh · A. Shukla (✉) · R. K. Sahani · A. R. Shekhar · R. Singh
Department of Civil Engineering, Madan Mohan Malaviya University of Technology, Gorakhpur 273010, India

G. C. Marano et al. (eds.), *Proceedings of SECON'21*, Lecture Notes in Civil Engineering 171, https://doi.org/10.1007/978-3-030-80312-4_74

1 Introduction

From few decades, the use of concrete in countries like India, China, U.S.A. etc. has found to be increased tremendously due to their rapid growth of population and economy. As per the Chatham House Report in 2019, 4.4 billion tons of concrete produced worldwide annually and it will increase up to 5.5 billion tons per year by 2050 and some other reports also says the annual amount of construction and demolition waste in Indian is about 165–175 million tons. So, recycling of demolished concrete can be an effective way to solve the problem of disposal of concrete waste. Lots of research had already performed by many scholars [15], [17] and many more such as, Thomas et al. [1] and [21] had found that using RAC in concrete lowers the durability of concrete and loss of durability and mechanical properties was more for low water/cement ratios. Codal provision of different countries regarding the use of RAC was presented by Sonawane and Pimplikar [2] along with the basic concrete properties of concrete using RAC and its effect was also discussed. An over view report was presented by Xiao et al. [3] on RAC in China in which author had find out the mechanical property of concrete made using RAC and conclude that RAC is safe to be used in construction industries. Concrete made up to 25% replacement with RCA is found to be suitable for construction work is presented by Etxeberria et al. [4] and verified the experimental work with numerical model based on modified compression field theory and simplified models such as those proposed by Cladera and Mari, the Canadian standard CSA and the Eurocode-2.

2 Methodology

Reinforced and plain concrete were collected from different location (i.e., near Madan Mohan Malviya Technical University Gorakhpur, Uttar Pradesh, India). The collected materials were crushed in smaller size below 400 mm. Reinforcement bars and any unwanted impurities were separated from the crushed concrete using different techniques after that it was heated up to less than 110 °C [5] then washed and dried in sunlight at 24 h. Using sieve, Coarse and fine concrete aggregate was separated from crushed concrete. NA were replaced with obtained concrete aggregates by 0, 15 and 30% and specimens (cubes and beams) were casted. Different tests were performed in laboratory as per IS codes and previously available literatures like [16, 18–20] and [22]. The obtained compressive and flexural strength of 0% replacement were compared with the 15 and 30% replacement of NA with RCA.

3 Experimental Program

A total of 30 no. of cubes of size 150 mm × 150 mm × 150 mm and 15 no. of beams of size 700 mm × 150 mm × 150 mm were investigated. Compressive Strength of 15 cubes were determined after 7 days and remaining after 28 days whereas, flexure strength of all the 15 beams were determined after 28 days. Total 10 no. of cubes were prepared with 0, 15 and 30% of replacement of each. Similarly, 5 no. of beams were prepared with 0, 15 and 30% of replacement of each.

3.1 Materials

Ordinary Portland cement of 53 grade as per IS 12269-1987 [6] was used for concrete and beam casting. Locally available angular natural aggregate along with some amount of RCA was used as coarse aggregate (CA) and fine aggregates with angular particle shape having smooth texture was used. Table 1 shows the properties of ingredient used for casting of cubes and beams. Fineness modulus and zone of fine aggregate (FA) was 3.4 and 4 respectively.

3.2 Mix Design

Concrete mix proportion having aggregate partially replaced by RCA was designed on the basis of Indian Standard method IS:10262-2009 [7] guidelines. Mix design values for Cubes and beams for characteristic strength (f_{ck} = 25 MPa) are as follows:

Target Mean Strength: As per Indian standard guideline IS 10262:2009 [7] clause 3.2 target mean strength (f_m) can be defined as:

$$f_m = f_{ck} + 1.65\sigma$$

$$f_m = 25 + 1.65 \times 4$$

$$f_m = 31.6\text{MPa}$$

where, σ = Standard deviation in MPa and is derived using Table 1 of IS 10262:2009.

Water Cement Ratio: As per IS456:2000 [8] for moderate exposure to environment condition maximum free water cement ratio (w/c ratio) is 0.5 but for the studies purpose w/c ratio of 0.46 was adopted on the basis of experience and judgment.

Table 1 Material properties

Sr. No.	Material	Specific gravity	Surface moisture (before casting)
1	Cement	3.15	–
2	RCA	2	Nil
3	CA	2.74	Nil
4	FA	2.74	Nil

Table 2 Volume of cement and aggregate

Sr. No.	Material	Weight per 1 m^3 (kg)
1	Cement	418
2	Coarse aggregate	1222.04
3	Fine aggregate	630

Water Content for Fresh Concrete: As IS 10262:2009 [7] it is recommended that for nominal maximum size of 20 mm Coarse aggregate maximum water content is 186 L (slump range should for 25–50 mm) for slump more than 50 mm 3% of water shall be increased for every additional 25 mm slump. So, the required amount of water is 192 L.

Cement and Aggregate Content: Form IS 10262:2009 [7] for normal maximum size of coarse aggregate of 20 mm and Zone (iv) fine aggregate volume of coarse aggregate is taken 66% and volume of fine aggregate is 34% of total volume of aggregate. After performing calculation on the basis of Indian standard guidelines the volume of cement and aggregate is shown in Table 2 as follows:

Net Weight of RCA, NA, Cement and Fine Aggregate Used for Cubes and Beams: There were total 10 number of cubes for 15 and 30% replacement each and 5 No. of beams for 15 and 30% replacement each. So, The total volume of one cube = $0.150 \times 0.150 \times 0.150 = 0.003375\ m^3$.

The total volume of 10 no. cube = $0.150 \times 0.150 \times 0.150 \times 10 = 0.03375\ m^3$
The total volume of one beam = $0.700 \times 0.150 \times 0.150 = 0.01575\ m^3$
The total volume of 5 beams beam = $0.700 \times 0.150 \times 0.150 \times 5 = 0.07875\ m^3$

Using the weight per m^3 as provided in Table 2 and the above calculated volume for cubes and beams, the exact weight of 15 and 30% replacement of NA with RCA along with weight of other ingredient used for casting cubes and beams is given in Table 3.

Table 3 Weight of RCA, NA, cement and fine aggregate for preparation of cubes and beams

Items	Weight of RCA, NA, cement, required	Replacement of RCA with NA		
		0%	15%	30%
Cubes	Weight of RCA (kg)	0.00	4.52	9.03
	Weight of NA (kg)	41.24	35.06	28.87
	Weight of fine aggregate (kg)	21.26	21.26	21.26
	Weight of cement (kg)	14.11	14.11	14.11
	Water (L)	6.48	6.48	6.48
	Number of cubes	10.00	10.00	10.00
Beams	Weight of RCA (kg)	0.00	10.54	21.07
	Weight of NA (kg)	96.24	81.80	67.37
	Weight of fine aggregate (kg)	49.61	49.61	49.61
	Weight of cement (kg)	32.92	32.92	32.92
	Water (L)	15.12	15.12	15.12
	Number of cubes	5.00	5.00	5.00

4 Test Set-Up and Instrumentation

In order to measure the compressive strength of cube the loading setup was a Compression Testing Machine with a maximum loading capacity of 2000 kN. The compression test was carried out by following all the provisions as per recommended in IS 14858:2000 [9]. Surface of cube specimens were smooth and planar and were tested with care that the axis of specimen was aligned with the center of thrust of the spherically seated platen.

The flexural strength of the concrete was carried out using Flexural Testing Machine on the concrete beam specimens by three-point loading system in which beams were placed on roller with a c/c distance of 600 mm. The maximum loading of the machine was 100 kN. Specification for the apparatus for flexural test of concrete and the procedure to find out the flexural strength of concrete beam specimens is followed by the codal provision IS:9399 and IS:516 respectively [10, 11].

5 Result and Discussion

5.1 Specific Gravity and Water Absorption

The specific gravity and water absorption of RCA and NA was investigated by following all the procedure as recommended in IS: 2386 (Part III) codal provision [12]. It was found that the specific gravity of RCA is much lower than the specific gravity of NA, (see Table 1). It means that the density of NA is higher when compared to the density of RCA. Water absorption of NA was found to be less than RCA because old mortar adheres to the surface of NA make it more porous. It was found that water absorption of NA was 1% which is in permissible limit as per recommended in BS 8007:1987 [13] i.e., water absorption is less than 3% while water absorption for RCA was found to be 8%. Absorption of water of RCA was higher so it is advisable to maintain the surface of RCA saturated dry before start mixing of concrete.

5.2 Workability of Fresh Concrete

To measure the workability of fresh concrete slump test was performed, according to the codal provision IS1199-1959 [14]. It was investigated that with replacement of NA by RCA slump value of the concrete decrease as replacement increase from 15 to 30% because as the presence of RCA increase it absorb more water that's why workability decreases. RAC show slump of approx. 73 mm with 15% replacement which shows medium workability of concrete as per IS 456:2000 [8] on the water cement ratio of 0.5 with 3% increment of water.

5.3 Development of Compressive Strength

Concrete is good in compression so compressive strength of concrete is very important mechanical property and, in this section, obtained compressive strength from compression testing performed in laboratory on 30 No. of cubes of dimension (0.150 × 0.150 × 0.150)m is summarised in Tables 4 and 5.

The compressive strength of cubes was investigated after 7 and 28 days and it was found that, as the amount of RCA increase in the concrete, compressive strength of cubes get decrease. A comparison of obtained compressive strength is given in Fig. 1a and in Fig. 1b after 7 and 28 days of cube casting respectively at 0%, 15% and 30% replacement of NA with RCA. Figure 1a, shows 7 days compressive strength of RAC mix was 20.68 MPa when the 15% NA replace with RCA and 7 days compressive strength for the RAC mix was 19.46 MPa when the 30% NA replace with RCA. Figure 1b shows 28-days compressive strength for the RAC mix was 29.43 MPa when the 15% NA replaced with RA and 28 days compressive

Table 4 Compressive strength of cubes after 7 days at 0, 15 and 30% replacement of NA with RCA

	Test of cubes after 7 days								
	0% replacement of NA			15% replacement of NA			30% replacement of NA		
Sr. No.	Items	Load (kN)	Stress (MPa)	Items	Load (kN)	Stress (MPa)	Items	Load (kN)	Stress (MPa)
1	C0700a	538.4	23.92	C0715a	470	20.89	C0730a	430	19.11
2	C0700b	490.6	21.8	C0715b	460.6	20.47	C0730b	445.3	19.79
3	C0700c	565.2	25.12	C0715c	465.3	20.68	C0730c	438.3	19.48
4	C0700d	510.1	22.67	C0715d	462.6	20.56	C0730d	433.8	19.28
5	C0700e	552.15	24.54	C0715e	468	20.8	C0730e	441.9	19.64
	Avg. stress (MPa)		23.61	Avg. stress (MPa)		20.68	Avg. stress (MPa)		19.46

Table 5 Compressive strength of cubes after 28 days at 0, 15 and 30% replacement of NA with RCA

	Test of cubes after 28 days								
	0% replacement of NA			15% replacement of NA			30% replacement of NA		
Sr. No.	Items	Load (kN)	Stress (MPa)	Items	Load (kN)	Stress (MPa)	Items	Load (kN)	Stress (MPa)
1	C2800a	690.5	30.68	C2815a	677.8	30.13	C2830a	600	26.66
2	C2800b	710.6	31.58	C2815b	656.3	29.16	C2830b	659.1	29.29
3	C2800c	687.1	30.53	C2815c	652.5	29	C2830c	626.3	27.83
4	C2800d	688.95	30.62	C2815d	666.67	29.63	C2830d	605.47	26.91
5	C2800e	700.2	31.12	C2815e	657.67	29.23	C2830e	651.6	28.96
	Avg. stress (MPa)		30.906	Avg. stress (MPa)		29.43	Avg. stress (MPa)		27.93

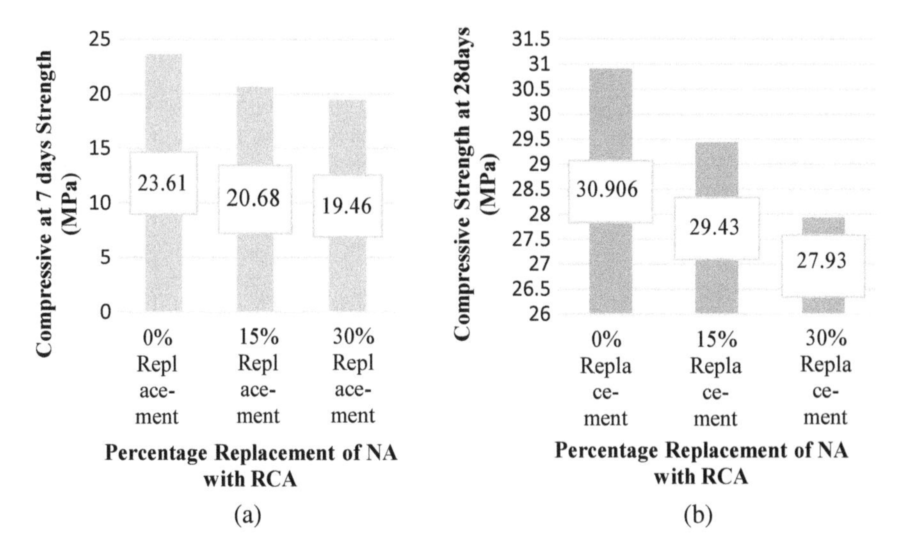

Fig. 1 **a** Comparison of compressive strength after 7 days, **b** comparison of compressive strength after 28 days

strength for the RAC mix was 27.93 MPa when the 30% NA replace with RCA. From comparison it is clear that the RAC gain strength slowly as compare to natural aggregate concrete and only 4.8 and 9.63% of decrement in compressive strength was observed for 15 and 30% replacement of NA with RCA respectively. The compressive strength at the age of 7-day of all the RAC mix cubes achieved at level of nearly 75% of 25 MPa.

5.4 *Development of Flexural Strength*

Flexural strength of concrete is an ability of concrete to withstand the load in bending. To obtained the Flexural strength author had used 3-point loading system in which 15 beams of dimension (0.700 × 0.150 × 0.150)m were tested and the obtained result was tabulated as shown in Table 6. Calculation of flexural strength was done by formula $\frac{Pl}{bd^2}$ for a > 200 mm and $\frac{3Pa}{bd^2}$ for 200 mm > a > 170 mm where, p is load (N), b and d are the cross-sectional dimension in mm, l is c/c distance between two rollers (i.e., in this case 600 mm) and a is the distance of starting of crack from the end of beam in mm.

The comparison of obtained flexure strength with and without replacement of NA with RCA is shown in Fig. 2a. It shows that at the age of 28 days flexural strength of beams were 3.34 MPa when 15% of NA was replaced with RCA and for 30% replacement flexural strength was 2.76 MPa. Drop of 22% and 35.5% in flexural strength was found for 15% and 30% replacement of NA with RCA respectively.

Table 6 Flexural strength of beams after 28 days at 0, 15 and 30% replacement of NA with RCA

Test after 28 days											
0% replacement of NA				15% replacement of NA				30% replacement of NA			
Items	Load (kN)	a (mm)	Stress (MPa)	Items	Load (kN)	a (mm)	Stress (MPa)	Items	Load (kN)	a (mm)	Stress (MPa)
B00a	26	267	4.62	B15a	19.6	211	3.48	B30a	18	177	2.83
B00b	22.4	193	3.84	B15b	21	224	3.73	B30b	14.3	243	2.53
B00c	24.7	233	4.39	B15c	17.2	185	2.82	B30c	17.2	192	2.93
B00d	25.76	228	4.58	B15d	21.13	197	3.7	B30d	15.41	238	2.74
B00e	22.33	259	3.97	B15e	17.67	189	2.97	B30e	15.58	183	2.77
Avg. stress (MPa) =			4.28	Avg. stress (MPa) =			3.34	Avg. stress (MPa) =			2.76

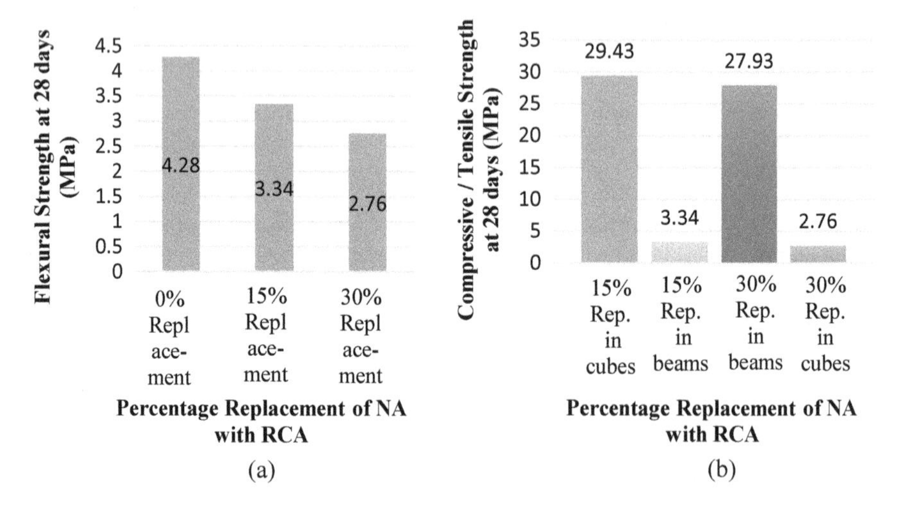

Fig. 2 **a** Comparison of flexural strength, **b** comparison of compressive and flexural strength

6 Conclusion

Inside the limit of investigation detailed previously, the following can be concluded:

(1) Specific Gravity of concrete having RA is lower than the concrete having only NA in it. So, it can be concluded that the density of concrete with RA is lower that the RAC. Water absorption of RAC was found to be 8% than the NA because old mortar adheres to the surface of NA make it more porous. It is advisable to maintain the surface of RCA saturated dry before start mixing of concrete.

(2) As the replacement of RA increases in the concrete workability decrease for a particular w/c ratio. In this report it was found that for 0.5 w/c ratio 3% increase in water required for replacement of 15 and 30% replacement of NA with RCA. So, more water required for workable fresh RAC.
(3) The compressive strength of RAC at the replacement of 15% was found to be decreases by 12% when tested at the age of 7 days while decreases by 4.8% when tested at the age of 28 days and when the replacement of NA was 30% the decrement was 17.6% when tested at the age of 7 days and 9.63% when the tested at the age of 28 days. Thus, it can be concluded that as the proportion of RCA increases in the mix, compressive strength of RAC decreases it is because RCA is rounded in shape when compared to NA and presence of deleterious materials and old mortar on the surface of RCA loosen its bonding with mix results in decrement of compressive strength is found for RAC.
(4) The flexural strength of RAC after 28 days decreases by 22% when the NA replaced with RA by 15% and decreases by 35.5% when NA replaced with RA by 30%. Decrement of flexural strength was found to be more than the compressive strength of concrete when compared with NC with 0% replacement because concrete is good in compression.

A comparison between the compressive and flexural strength at the age of 28 days concrete with replacement of 15 and 30% of NA with RCA is shown in Fig. 2b from where, it can be concluded that recycled aggregate is more suitable for compression member rather than any flexural member because concrete itself week in bending and RA decreases the flexural strength more whereas for compression member up to 30% replacement of NA with RCA is acceptable and it can be utilized in construction practice.

References

1. Thomas C, Setién J, Polanco J, Alaejos P, De Juan MS (2013) Durability of recycled aggregate concrete. Constr Build Mater 40:1054–1065
2. Sonawane TR, Pimplikar SS (2013) Use of recycled aggregate concrete. IOSR J Mech Civ Eng 52:59
3. Xiao J, Li W, Fan Y, Huang X (2012) An overview of study on recycled aggregate concrete in China (1996–2011). Constr Build Mater 31:364–383
4. Etxeberria M, Marí AR, Vázquez E (2007) Recycled aggregate concrete as structural material. Mater Struct 40(5):529–541
5. IS:2386 (Part II)-1963 Methods of test for aggregates for concrete (PART II) estimation of deleterious materials and organic impurities. (Eleventh Reprint September 2006) UDC 691.322: 543.86. Bureau of Indian Standards
6. IS:12269-1987, Indian standard specification for 53 grade ordinary Portland cement, UDC 666.942.32. Bureau of Indian Standards
7. IS 10262:2009, Indian standard concrete mix proportioning—guidelines, ICS 91.100.30. Bureau of Indian Standards
8. IS 456:2000, Indian standard plain and reinforced concrete—code of practice, ICS 91.100.30. Bureau of Indian Standards

9. IS 14858:2000 Compression testing machine used for testing of concrete and mortar—requirements. Bureau of Indian Standards
10. IS:9399-2004 Specification for apparatus for flexural testing of concrete. Bureau of Indian Standards
11. IS:516-2006 Methods of tests for strength of concrete. Bureau of Indian Standards
12. IS:2386 (Part III)-1963 Methods of test for aggregates for concrete-specific gravity, density, voids, absorption and bulking (Eighth Reprint March 1997) UDC 691.322: 531.75, Bureau of Indian Standards
13. BS 8007:1987. Code of practice for design of concrete structures for retaining aqueous liquids. British Standards
14. IS:1199-1959, Indian Standard Methods of sampling and analysis of concrete (Eleventh Reprint November 1991) UDC 666'IS. 456: 2000 97: 620'11. Bureau of Indian Standards
15. Padmini AK, Ramamurthy K, Mathews MS (2009) Influence of parent concrete on the properties of recycled aggregate concrete. Constr Build Mater 23(2):829–836
16. Xiao J, Li J, Zhang C (2005) On statistical characteristics of the compressive strength of recycled aggregate concrete. Struct Concr 6(4):149–153
17. Yong PC, Teo DCL (2009) Utilisation of recycled aggregate as coarse aggregate in concrete. J Civ Eng Sci Technol 1(1):1–6
18. Li Q, Zhang C (2017) A statistical study on the compressive strength of recycled aggregate concrete. Chem Eng Trans 59:421–426
19. Wang P, Zhang XG, Wang F, Deng MR, Qin WB, Tian Q (2016, June) Test study on cube compressive strength of recycled lightweight aggregate concrete. In: 2nd annual international conference on advanced material engineering (AME 2016). Atlantis Press, pp 630–634
20. Xiao J, Zhang K, Xie Q (2016) Reliability analysis for flexural capacity of recycled aggregate concrete beams. Struct Eng Int 26(2):121–129
21. Rahal K (2007) Mechanical properties of concrete with recycled coarse aggregate. Build Environ 42(1):407–415
22. Rakshvir M, Barai SV (2006) Studies on recycled aggregates-based concrete. Waste Manage Res 24(3):225–233

Effect of Side Ratio on V Plan Shaped High-Rise Building Under Wind Excitation

Satadru Bhattacharya and Sujit Kumar Dalui

Abstract The primary objective of the present study is to determine the effect of side ratio of V plan shaped tall building subjected to wind induced force. Different cases have been made by considering different side ratio of V plan shaped tall building. The side ratio is considered 1.5, 2, 3, 4 and 5 by adjustment of length and width of the building maintaining same plan area and building height. The wind incidence angle is 0°–90° with 30° regular interval for each case. Computational Fluid Dynamics (CFD) is adopted to perform the numerical simulation of 'V' plan shaped building which helps to create a similar wind environment as in urban terrain. Grid independence test is performed to improve the accuracy of result by adopting finer meshing of Computational Domain. Mean Pressure coefficient of each face, force coefficient in along and across wind direction, flow pattern around building, pressure variation on each face is obtained by numerical analysis in ANSYS CFX. Further, a comparison has been made with all the cases of V-plan shaped tall building model of same floor area to study the effectiveness side ratio and observations have been made on the suitability of side ratio variation based on the numerical result.

Keywords Side ratio · Computational fluid dynamics · Force coefficient · Wind incident angle

1 Introduction

Advanced design of extremely tall and irregular structure which is exposed to extreme wind environment is the challenging concern of structural designers and researchers due to height and nonuniformity. Researchers are serious about the improvement of structural as well as architectural aspects by incorporating different types of shapes, sizes, aspect ratio etc. and light in weight on the other side. Some

S. Bhattacharya (✉) · S. K. Dalui
Department of Civil Engineering, Indian Institute of Engineering Science Technology, Shibpur, Howrah, India

G. C. Marano et al. (eds.), *Proceedings of SECON'21*, Lecture Notes in Civil Engineering 171, https://doi.org/10.1007/978-3-030-80312-4_75

essential global and local modification factors like aspect ratio, wind angle, taper, setback, chamfer, corner modification to reduce the force and pressure co-efficient of these high-rise structures are considered by the Previous researchers. Shape modification of the building like corner modification, setback, tapering, opening etc. helps to reduce the wind load. Further, they incorporated wind tunnel test to find out the exact effect of wind load from different wind incident angle. Different types of geometrical shape like + , T, U, C, Y, H, E, L, O etc. are taken into consideration for study. Tapering of the sides of building and corner modification of the edges are very significant to reduce the force and pressure co-efficient among all the modification. Among those unconventional shapes of building, V-shape building is studied by a very few researchers. In addition to that, IS 875 (part-III): 2016 has not mentioned any special criteria for this type of irregular shapes of building. So, for present study, V-plan shaped building with different corner modification is considered to be studied rigorously.

Previous researchers have explored response due to various global and local modification of irregular plan shaped building model exposed to wind force. Jain and Mandal [1] have investigated the response of the dynamic Analysis of multi-storey V-shaped building with setback arrangement. Stathopoulos and Zhou [2] observed the numerical forecasting of turbulent wind pressure on the exposed surfaces of L shaped building due to various wind direction using the k-ε turbulence model and cartesian rectangular grid generation. Tamura and Miyagi [3] studied the consequence of corner modification (sharp, rounded, chamfered) on aerodynamic forces using experimental setup in wind tunnel. Gomes et al. [4] studied the result of the pressure distribution of irregular plan shapes on wind tunnel test. Experimental models used in this test are L shaped, U shaped and cube-shaped model with 1:100 ratio and compared with analytical result using a software PHOENICS considering various wind angle. Kwok and Bailey [5] investigated the effect of various aerodynamic devices such as slotted corner, small fins on the wind induced response of high-rise building. Kim et al. [6] have investigated the tapering effect on minimization of the rms across-wind displacement responses of a high-rise building using aeroelastic model tested in wind tunnel. Mara and Case [7] have studied the variation of overall building response with the corner modifications due to wind induced motion. They have considered corner cuts, chamfers, venting and fins as corner modification to reduce the wind force on building. Chakraborty et al. [8] have presented the result obtained from wind tunnel test and numerical investigation on a ‘ + ’ plan shaped building. The wind tunnel test was performed in an open circuit wind tunnel on rigid model with a 1:300 scale for wind incident angle from 0° to 45° and compared with numerical simulation with the help of ANSYS CFX. Li and Li [9] have developed the optimal design model of an L shaped building using Kuhn–tucker conditions and validated the model with wind tunnel test results. Elias et al. [10] have studied multi-mode wind response control of chimneys with SSI consideration using distributed multiple tuned mass dampers (d-MTMDs) and compared with single tuned mass damper (STMDs), arbitrarily installed distributed MTMDs (ad-MTMDs) to find out best effective mass damper and d-MTMDs gives best result among all the dampers. Kumar and Dalui [11] have studied the variation of pressure distribution due to variation of internal angle between limbs of a regular cross plan shaped building for various wind direction (0°–180°) using CFD with ANSYS CFX (k-ε turbulence model). Elshaer et al. [12] have studied the effect of corner modification using an optimization algorithm, large eddy simulation and ANN to reduce the along and across wind response of building considering wind directionality. Bhattacharya and Dalui [13] explored wind flow pattern and mean pressure coefficient of E plan shaped building using wind tunnel test and CFD simulations with the help of ANSYS software. Wind incident angles are varied from 0° to 180° with 30° regular interval. Elias and Matsagar [14] have investigated the effectiveness of multiple tuned mass dampers (MTMDs) to reduce the wind force of a tall building considering soil structure interaction (SSI) and compared the response with single tuned mass damper (STMDs) to find out suitable damper among both. A comparison has been made by Nikose and Sonparote [15] between the dynamic response (base shear and bending moment) of a building using ANN and IWC standards. They have developed design charts for the forecasting of dynamic wind response of high-rise buildings. Sanyal and Dalui [16] have investigated the effect of internal limb angle variation on Y plan shaped tall building. The have considered 30° interval of variation with respect to 0°–180° wind incidence angle with 30° regular interval and compare the data with wind tunnel test results.

2 Methodology of Work

2.1 Problem Formulation

In this study, a computational method is adopted to perform analysis of wind induced force acting on this type of irregular building. ANSYS CFX is used to create Model and perform fine meshing. Height of the V-shaped model is 180 m; both limbs of the model are 90 m long with 30 m width. In the present study, 1:300 scale is considered to model (Fig. 1) the building. Various cases have been made by considering different side ratio which is made by length and width variation of the 'V' plan shaped building model. The angle between the limbs is 90° which remain unchanged. Here side ratio (γ) which is the ratio of modified length and modified width of the limb. The percentage of side ratio ($\gamma = L/B$) is considered 1.5, 2, 3, 4 and 5. Wind incident angle is increased from 0° to 90° with 30° regular interval for each case. Modelling details for all the cases with side ratio is given in Table 1.

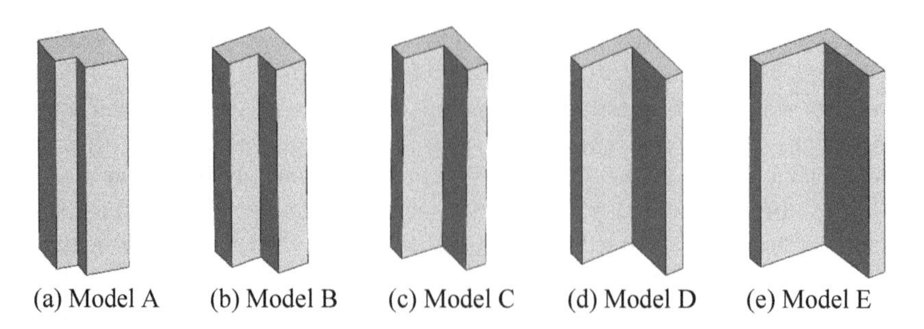

Fig. 1 Modelling details of V plan shaped building

Table 1 Modelling details of V plan shaped building

Model name	Length of limb (L in mm)	Width of limb (B in mm)	Side ratio γ = L/B	Limb angle (θ)	Wind angle (β)
Model A	237	158	1.5	90°	0°, 30°,60° and 90°
Model B	259	129	2	90°	
Model C	300	100	3	90°	
Model D	339	85	4	90°	
Model E	373	75	5	90°	

2.2 Solution Methodology

Computational Fluid Dynamics is the primary tool based on Navier–Stokes equation which is used in the simulation of the model. The k-ε turbulence model which is based on Reynolds Averaged Navier Stokes (RANS) equation is chosen among all the turbulence model to perform simulation in ANSYS CFX. Previously wind tunnel test is used to perform analysis of the aero-elastic model by most of the researchers but nowadays, the CFD method is being used widely by various researcher. Full-scale model is extremely time and resource-consuming so length scale is reduced to 1:300 which is adopted for all the cases of side ratio modified V plan shaped building model. The wind velocity profile is generated by power law equation and basic wind speed is reduced to 1:5 scale. So, inlet velocity is considered as 10 m/s.

Among all those RANS model, the standard k-ε model is widely used to solve computational fluid mechanics problems. Gradient diffusion hypothesis is adopted to relate the Reynolds stresses to the mean velocity gradients and the turbulent viscosity in k-ε turbulence models. The turbulent viscosity is product of turbulent length scale and turbulent velocity. The turbulent kinetic energy, k is defined as the variance of the fluctuation in velocity. ε is the turbulence eddy dissipation and has dimension per unit time.

According to the power law, the Wind velocity profile is generated by using the formula.

$$U_Z = U_0 * \left(\frac{Z}{Z_0}\right)^{\alpha} \quad (1)$$

where U_z is wind velocity at certain height Z, U_0 is wind velocity at boundary layer, Z_0 is boundary layer depth and value of α is taken as 0.133 for open terrain with well obstruction. Inlet velocity is considered as 10 m/s which is according to 1:5 scale (Basic Wind speed 50 m/s) as the building model is considered in 1:300 length scale. Figure 2 is mean wind velocity and turbulence intensity profile which shows very good agreement between the theoretical, analytical and wind tunnel test data.

Mean Pressure Coefficient (C_P)

The external mean pressure coefficient of each face is obtained with the formula

$$C_P = \frac{P_m}{0.5\rho U_H^2} \quad (2)$$

where P_m is the actual wind pressure, ρ is the air density (taken as 1.2 kg/m^3) and U_H is the mean wind velocity at building height H.

Force Coefficient (C_F) and Moment Coefficient (C_M).

The Global force coefficient (C_F) and Global Moment Coefficient (C_M) of the building model is obtained as follows [17]

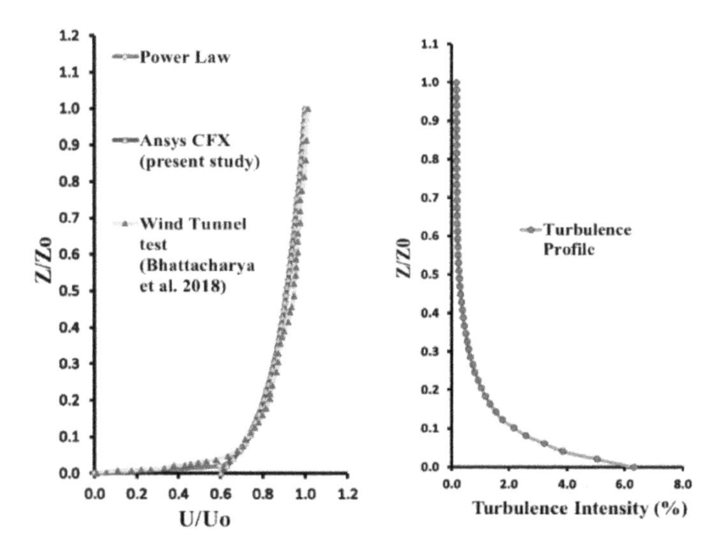

Fig. 2 Mean wind velocity profile and turbulence intensity profile

$$C_{FX} = \frac{F_X}{0.5\rho U_H^2 * L_X * H} \tag{3}$$

$$C_{FY} = \frac{F_Y}{0.5\rho U_H^2 * L_Y * H} \tag{4}$$

$$C_{FZ} = \frac{F_Z}{0.5\rho U_H^2 * L_X * L_Y} \tag{5}$$

$$C_F = \sqrt{C_{FX}^2 + C_{FY}^2} \tag{6}$$

where C_{FX}, C_{FY} and C_{FZ} are the aerodynamic force coefficients of the building along X, Y and Z direction respectively, C_F is the global force coefficient of the building, F_X, F_Y and F_Z are the total force along X, Y and Z direction respectively, ρ is the air density (taken as 1.2 kg/m^3), L_x and L_y are the projected wind facing length along X and Y direction respectively and U_H is the mean wind velocity at building height H.

3 Numerical Study

3.1 Computational Domain and Meshing

The meshing of the entire computational domain is required to maintain the accuracy and similarity with wind tunnel study. The computational Domain covers 5H in upstream, 5H from each side wall, 5H from top surface of the building model and 15H in downstream from the edges of the building model according to the recommendation of Franke et al. [18] where H is the height of the building. Walls of the domain are considered as free slip condition Surfaces of the building are considered as no-slip condition and Pressure of outlet is 0 pa (Fig. 3).

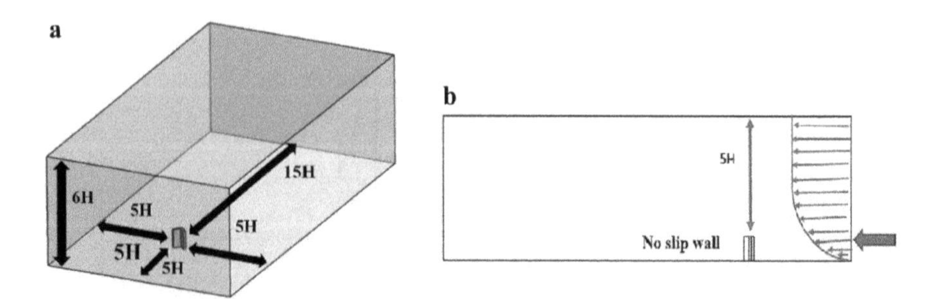

Fig. 3 Computational domain of V shaped building model **a** plan view **b** elevation

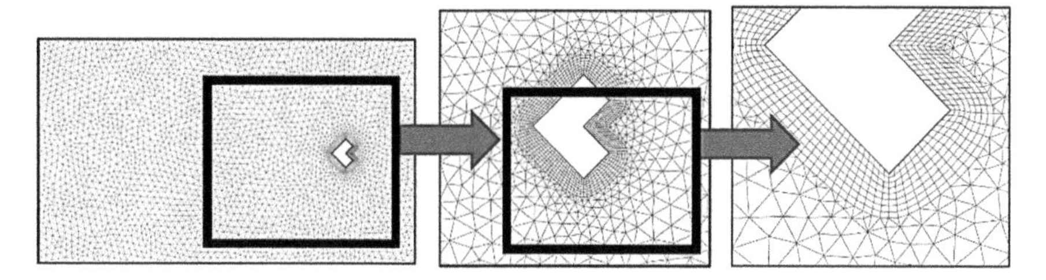

Fig. 4 Meshing arrangement of models (plan view and zoom in view)

The fineness of the grid arrangement of the domain is very important to attain accuracy in result. The element size is arranged such that the element size is smaller near the building region and gradually increases to the wall of the domain to generate the inflation. Inflation is given near building to achieve more accurate response from the simulation. Here tetrahedron [19] and triangular element is used to generate mesh (Fig. 4).

3.2 Grid Sensitivity Study

A grid sensitivity study is an essential and major study to achieve more accuracy in the response obtained from the entire work. This study is based on finer grid arrangement of the computational domain until the outcome such that means pressure coefficient on each face, force coefficient etc. converges with null error percentage. The result details of grid sensitivity study are given in Table 2.

It can be clearly observed from the study that, if meshing is finer then number of elements of meshing increases, then error percentage is found decreased so to reduce the error percentage in analysis this test should be performed. When the

Table 2 Grid convergence study details with error

Model no	Model name	Element number	Drag coefficient (C_{fx})	Error %
1	case 1	4599288	0.951	−1.929
2	case 2	7827509	0.906	2.894
3	case 3	10366750	0.959	−2.787
4	case 4	12546014	0.961	−3.001
5	case 5	15896489	0.958	−2.680
6	case 6	17150303	0.952	−2.036
7	case 7	20241094	0.941	−0.857
8	case 8	22117942	0.937	−0.429
9	case 9	27078552	0.933	0
10	case 10	33019015	0.933	–

response is converged with the previous value, then corresponding element size is considered as relevant which is used in further investigation of the side ratio modification.

4 Results and Discussion

4.1 Validation with Previous Work

Before proceeding to analysis of the present study, Validation of response obtained by previous researcher is performed with the experimental and numerical results obtained by Gomes et al. [4]. They explored the mean pressure response of L shaped building along vertical centreline with constant limb angle 90° in different wind incident angle from 0° to 180°. L shaped building acts as V shape when wind incident angle is 45° so in the present study, analysis of model maintaining same dimension as Gomes et al. [4] has been performed using software based on k-ε method to check the practicability of the software. Inlet velocity in considered 10 m/s. It is given in Fig. 5 which shows good agreement with numerical and experimental data.

4.2 Flow Pattern Variation Study

Wind Flow Pattern around the models A, B, C, D and E for 0° Wind Incidence Angle are shown in Fig. 6a, b. From the figure it is observed that the flow separates at the edge of the side faces of the model so velocity is extremely high at this region due to huge separation and side wash of flow. Due to this huge separation of wind

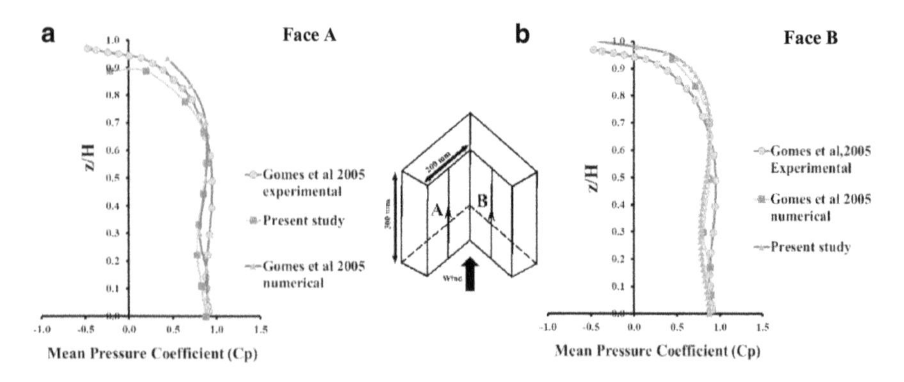

Fig. 5 Comparison of mean pressure coefficient (C_p) along vertical centreline **a** Face A **b** Face B (validation work)

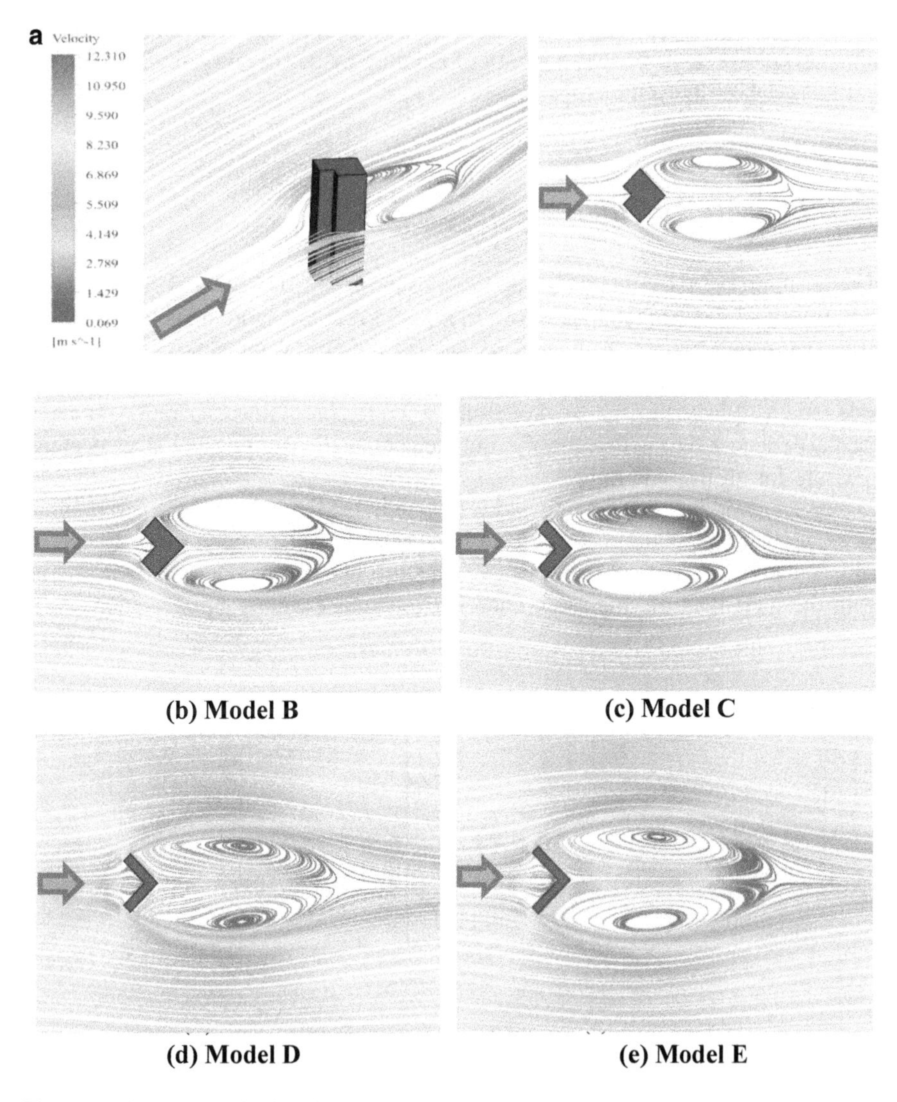

Fig. 6 **a** Flow pattern details of V plan shaped building (Model A), **b** flow pattern of V plan shaped building with various side ratio (plan view)

flow, suction pressure is generated at the wake region of model and two symmetrical and uniform vortices are generated. Size of vortices increases with the increment of side ratio of model.

At 0° wind incidence angle, vortices are almost symmetrical and even in the wake region of building model so pressure distribution is almost identical on the symmetrical faces.

At 30°, 60°, 90° wind incidence angle, vortices are unsymmetrical and nonuniform in size in the wake region of building model so pressure distribution is not identical on the symmetrical faces.

4.3 Pressure Variation Study

The mean pressure coefficient of the faces of model A, B, C, D and E for 0°, 30°, 60°, 90° wind incident angle is shown in Fig. 7. From the figure it is quite clear that at 0° wind incident angle, windward face E and F experience similar pressure. Pressure coefficient of side face A and D gradually decreases with the increment of side ratio as it increases the local pressure at the region of flow separation. B and C face experiences the suction for 0° wind incident angle. Roof face G faces negative pressure for all the side ratio.

For 30° wind incident angle, windward face E and F experience uneven pressure as the pressure distribution is no longer identical. Pressure of side face A is negative and gradually decreases where positive pressure of another side face D decreases with the increment of side ratio. Face B experiences almost equal suction where

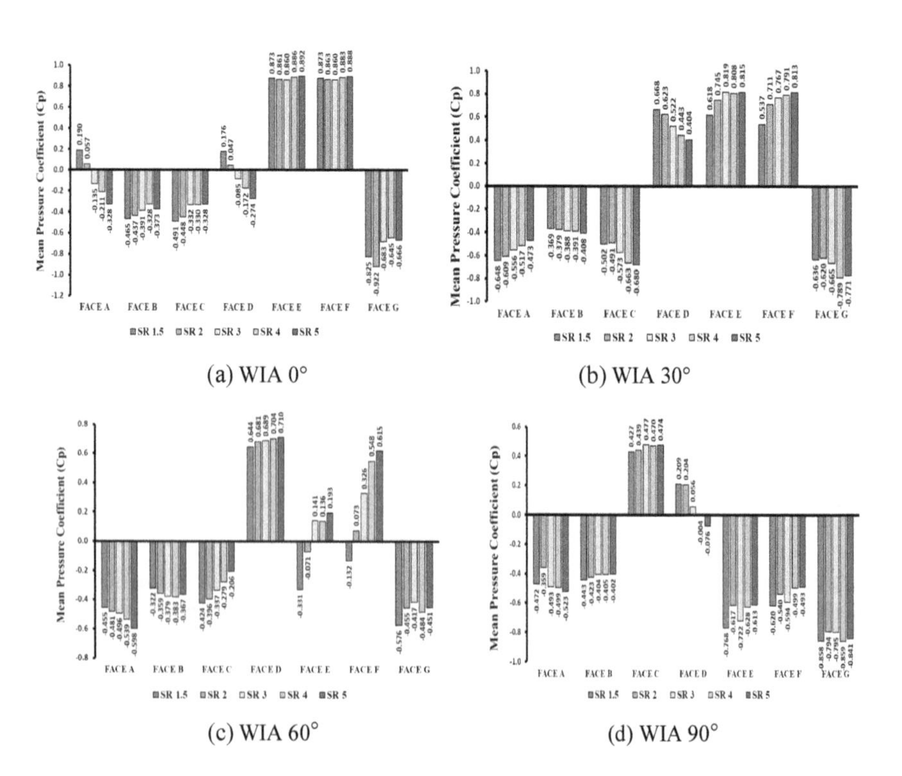

Fig. 7 Mean pressure coefficient variation of all models at different wind incident angles

suction of face C increases for 30° wind incident angle. Roof face G faces negative pressure for all the side ratio for 30° wind incident angle.

For 60° wind incident angle, windward face E and F experience pressure increment from negative to positive side where for Face F increment is more due to angle increment. Pressure of side face A is negative and gradually increases where positive pressure of another side face D increases with the increment of side ratio. Face B experiences almost equal suction where suction of face C decreases for 60° wind incident angle. Roof face G faces negative pressure for all the side ratio for 60° wind incident angle.

For 90° wind incident angle, windward face E and F experience negative pressure with a small variation with the increment in side ratio. Pressure of side face A is negative but pressure of face D decreases from positive to negative side with the increment of side ratio. Face B experiences almost equal suction where suction of face C increases for 90° wind incident angle. Roof face G faces huge negative local pressure for all the side ratio for 90° wind incident angle.

4.4 Force Coefficient Variation Study

Force coefficient along X, Y and Z direction is obtained for model A, B, C, D and E for wind angle variation from 0° to 90° with 30° regular interval which shows in Fig. 8. 0°, 30°, 60° and 90° wind angle, force coefficient along X direction is increasing with the increment of side ratio and maximum for model E. For 0° wind angle Force coefficient along Y direction is negligible at 0° wind angle for all model. For 30° wind angle force coefficient along Y direction is decreasing from negative to positive side. For 60° and 90° wind angle, suction is increasing with the increment of side ratio so maximum for model E.

Global Force coefficient along X, Y and Z direction and Global horizontal force coefficient is obtained for model A, B, C, D and E for wind angle variation from 0° to 90° with 30° regular interval which shows in Table 3. From the table, it is clearly concluded that Force coefficient along Z direction is reduced with the increment in side ratio and maximum for model A at all wind incident angle. Value of Global horizontal force coefficient is increasing with increment in side ratio and maximum for model E for each wind angle. Most critical value of Global horizontal force coefficient is found for model E at 30° wind incident angle so structural designers should be concerned about the matter.

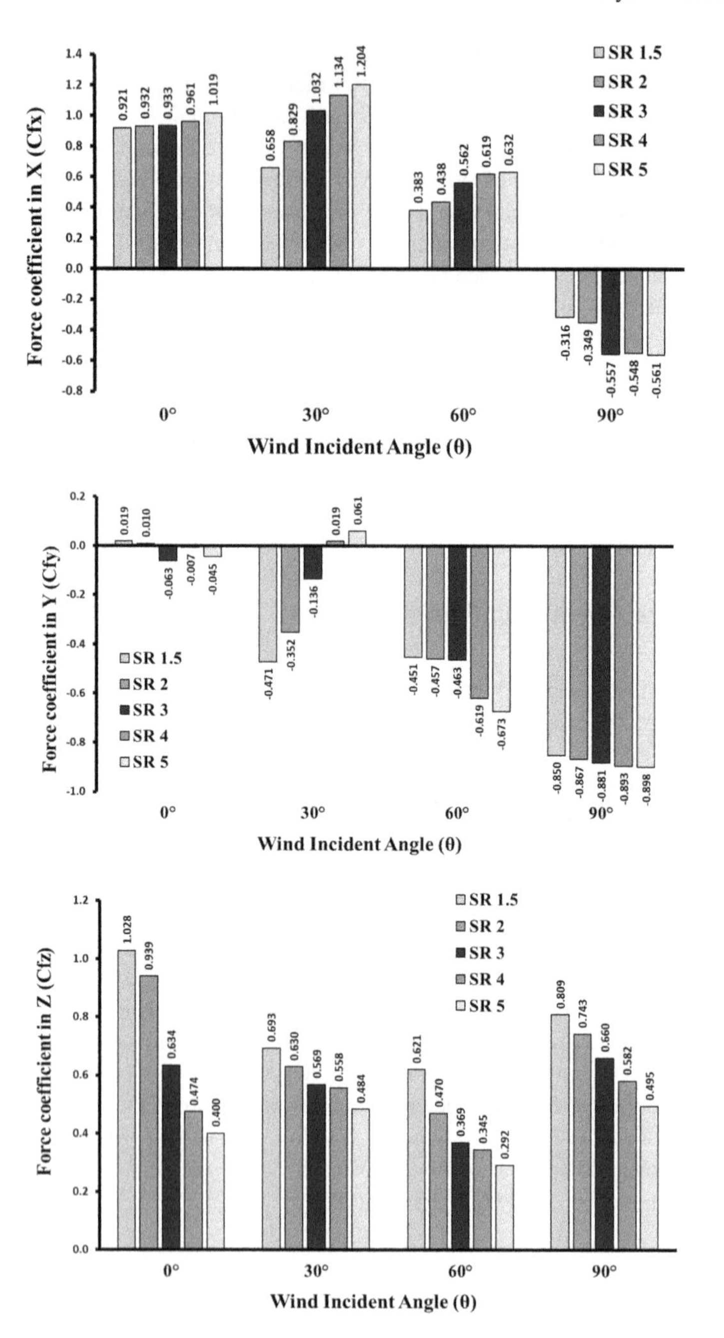

Fig. 8 Global force coefficient along X, Y and Z direction variation of all models at different wind incident angles

Table 3 Global force coefficient along X, Y and Z direction for all models

Model	SR	WIA	Lx	Ly	Cfx	Cfy	Cfz	Cfh	MAX
A	**1.5**	0°	0.335	0.279	0.921	0.019	**1.018**	0.921	1.020
B	**2**		0.366	0.274	0.932	0.010	0.939	0.932	
C	**3**		0.424	0.283	0.933	−0.063	0.634	0.936	
D	**4**		0.479	0.300	0.961	−0.007	0.474	0.961	
E	**5**		0.527	0.317	1.019	−0.045	0.400	**1.020**	
A	**1.5**	30°	0.335	0.279	0.658	−0.471	**0.693**	0.809	**1.206**
B	**2**		0.366	0.274	0.829	−0.352	0.630	0.900	
C	**3**		0.424	0.283	1.032	−0.136	0.569	1.041	
D	**4**		0.479	0.300	1.134	0.019	0.558	1.135	
E	**5**		0.527	0.317	1.204	0.061	0.484	**1.206**	
A	**1.5**	60°	0.335	0.279	0.383	−0.451	**0.621**	0.591	0.923
B	**2**		0.366	0.274	0.438	−0.457	0.470	0.633	
C	**3**		0.424	0.283	0.562	−0.463	0.369	0.728	
D	**4**		0.479	0.300	0.619	−0.619	0.345	0.875	
E	**5**		0.527	0.317	0.632	−0.673	0.292	**0.923**	
A	**1.5**	90°	0.335	0.279	−0.316	−0.850	**0.809**	0.907	1.058
B	**2**		0.366	0.274	−0.349	−0.867	0.743	0.934	
C	**3**		0.424	0.283	−0.557	−0.881	0.660	1.043	
D	**4**		0.479	0.300	−0.548	−0.893	0.582	1.048	
E	**5**		0.527	0.317	− 0.561	− 0.898	0.495	**1.058**	

5 Conclusions

The wind induced response is obtained for V plan shaped building model with different side ratio due to 0°, 30°, 60° and 90° wind incidence angle. Before proceeding to the analysis, a comparison and validation study from the previous researcher's work has been done to check the eligibility and practicability of the method.

Based on the results obtained from the analysis of the wind-induced load on V plan shaped building different side ratio due to various wind incident angle, it is observed that, the flow velocity is extremely high at edges of corner region due to side wash of flow. The size of the vortices and the wake length increases with the increment of side ratio so structural designers should be careful about local pressure at the wake region. Mean pressure coefficient is affected for side face A, D and roof G by side ratio increment. With the increment in wind incidence angle some critical local pressure increment is observed so structural designers should be concerned for cladding and roof design.

Global Force coefficient along X, Y and Z direction and Global horizontal force coefficient is obtained for model A, B, C, D and E for wind angle variation from 0° to 90° with 30° regular interval. Global Force coefficient along X and Y direction in increasing with the increment of side ratio. Global Force coefficient along Z direction reduces with side ratio increment so global tortional moment also reduces. So structural designers should be careful about considering proper side ratio to avoid too much tortional effect on V plan shaped building.

So, side ratio modification is beneficial specially for torsional moment reduction acting on a V plan shaped tall building but researchers and structural designers should give special attention for cladding and roof design due to local pressure increment.

References

1. Jain BSK, Mandal UK (1992) Dynamics of buildings with v-shaped plan. J Eng Mech 118:1093–1112
2. Stathopoulos T, Zhou YS (1994) Computation of wind pressures on L-shaped buildings. ASCE 119:1526–1541
3. Tamura T, Miyagi T (1999) The effect of turbulence on aerodynamic forces on a square cylinder with various corner shapes. J Wind Eng Ind Aerodyn 83:135–145
4. Gomes Ã, Rodrigues AM, Mendes P, Glo M (2005) Experimental and numerical study of wind pressures on irregular-plan shapes. J Wind Eng Ind Aerodyn 93:741–756. https://doi.org/10.1016/j.jweia.2005.08.008
5. Kwok BKCS, Bailey PA (2006) Aerodynamic devices for tall buildings and structures. J Eng Mech 113:349–365
6. Kim Y, You K, Ko N (2008) Across-wind responses of an aeroelastic tapered tall building. J Wind Eng Ind Aerodyn 96:1307–1319. https://doi.org/10.1016/j.jweia.2008.02.038
7. Mara TG, Case PC (2010) The effects of incremental corner modifications on a 200 m tall building 2949–2960. https://doi.org/10.1061/41130(369)266
8. Chakraborty S, Dalui SK, Ahuja AK (2014) Wind load on irregular plan shaped tall building —a case study. Wind Struct 19:59–73
9. Li Y, Li Q (2016) Across-wind dynamic loads on L-shaped tall buildings. Wind Struct 23:385–403
10. Elias S, Matsagar V, Datta TK (2017) Distributed multiple tuned mass dampers for wind response control of chimney with flexible foundation. Procedia Eng 199:1641–1646. https://doi.org/10.1016/j.proeng.2017.09.087
11. Kumar D, Dalui SK (2017) Effect of internal angles between limbs of cross plan shaped tall building under wind load. Wind Struct 24:95–118
12. Elshaer A, Bitsuamlak G, El Damatty A (2017) Enhancing wind performance of tall buildings using corner aerodynamic optimization. Eng Struct 136:133–148. https://doi.org/10.1016/j.engstruct.2017.01.019
13. Bhattacharyya B, Dalui SK (2018) Investigation of mean wind pressures on 'E' plan shaped tall building. Wind Struct 26:99–114
14. Elias S, Matsagar V (2018) Wind response control of tall buildings with a tuned mass damper. J Build Eng 15:51–60. https://doi.org/10.1016/j.jobe.2017.11.005
15. Nikose TJ, Sonparote RS (2019) Dynamic wind response of tall buildings using artificial neural network. Struct Des Tall Spec Build 28:1–18. https://doi.org/10.1002/tal.1657
16. Sanyal P, Dalui SK (2021) Effects of internal angle between limbs of "Y" plan shaped tall building under wind load. J Build Eng 33:101843. https://doi.org/10.1016/j.jobe.2020.101843
17. Sun X, Liu H, Su N, Wu Y (2017) Investigation on wind tunnel tests of the kilometer skyscraper. Eng Struct 148:340–356. https://doi.org/10.1016/j.engstruct.2017.06.052

18. Franke J, Hirsch C, Jensen AG, Krüs HW, Schatzmann M, Westbury PS, Miles SD, Wisse JA, & Wright NG (2004) Recommendations on the use of CFD in predicting pedestrian wind environment 1–12
19. Cheng CKC, Lam KM, Leung YTA, Yang K, Li Danny HW, Cheung Sherman CP (2011) Wind-induced natural ventilation of re-entrant bays in a high-rise building. J Wind Eng Ind Aerodyn 99:79–90. https://doi.org/10.1016/j.jweia.2010.11.002

Sustainable Planning of Urban Transportation Using PTV VISSIM

Leena Samuel, Muhammed Shibil, Muhuzin Nasser, Neha Shabir, and Nimson Davis

Abstract The steep increase in the number of motor vehicles on the road is primarily due to the occurrence of traffic problems such as accidents, congestion, and delays, especially in the urban areas of developing countries. In this paper, an attempt has been made to analyze the various traffic problems and to suggest sustainable traffic management measures in Kathrikadavu, Ernakulam, India. Direct field surveys were used to collect data on traffic level, land use, and pedestrian movement activities. The collected data showed that poor junction planning, shortage of traffic signals, and lack of proper traffic management are the most significant contributors to traffic congestion. Various remedial measures, based on junction enhancement, alternative service plans, and junction signalization along with sustainable measures are proposed as a feasible solution. The efficiency of the mounted signal can be evaluated using microscopic simulation models, PTV VISSIM and VISTRO. The VISSIM software shows how traffic simulation can be used to assess the efficiency and protection of an unsignalized intersection. The models are also able to produce detailed outputs such as queue length, different types of delay in various traffic inputs and time intervals, etc.

Keywords PTV VISSIM · VISTRO · Traffic · Volume · Microscopic · Simulation · Signalized

1 Introduction

Increasing demand for mobility makes decision-makers face greater challenges. This phenomenon is more present in the field of road transport. Strategies in traffic regulations show clear direction on efficiency improvement without limiting

L. Samuel · M. Shibil · M. Nasser · N. Shabir (✉) · N. Davis
Department of Civil Engineering, Federal Institute of Science and Technology, Ernakulam, India

G. C. Marano et al. (eds.), *Proceedings of SECON'21*, Lecture Notes in Civil Engineering 171, https://doi.org/10.1007/978-3-030-80312-4_76

transportation demand. This strategy applies efficiency to improve the capacity of transport infrastructure by updating traffic controls to reduce transportation problems such as congestion, pollution, and noise. The study area, Kathrikadavu Junction, is one of the most prominent junctions in Kochi. The junction, which is one of the most heavily traveled routes in Kochi, is clogged with heavy traffic. In 2010, because of increasing traffic congestion, a traffic signal was installed which was later removed due to the hindrance to free left turn. The AL Jacob Railway Overbridge, which was opened in 2014, was constructed to decongest the junction, but later it worsened the congestion at the junction. Thus, the Kathrikadavu junction is a very good example of improper planning of the junctions in Kochi.

2 Study Area

Kathrikadavu is a neighborhood in the city of Kochi, Kerala, India. It is almost in the middle of Kochi's two main intersections (junctions), Kaloor and Kadavanthra. Kathrikadavu is primarily a residential neighborhood, though more and more commercial establishments are springing up in recent years. Kathrikadavu is located on the Kaloor-Kadavanthra Road, one of Kochi's three north–south thoroughfares. One of the key advantages of Kathrikadavu is the ease with which it can be reached from other parts of Kochi city. Kaloor and Kadavanthra are at the north and south ends of the Kaloor-Kadavanthra road, while Padma Junction and MGRoad can be accessed easily by following the Pullepady road to the west side of Kathrikadavu. Thammanam and the NH47 bypass can be accessed by following the Thammanam-Kathrikadavu road towards the east (Fig. 1).

Fig. 1 Kathrikadavu junction

3 Problem Statement

A large number of vehicles are being diverted to the Thammanam-Kathrikadavu-Pullepady road after the opening of the Pullepady Railway Overbridge. The lane, on the other hand, is insufficiently large to accommodate so many vehicles. On the Thammanam-Kathrikadavu road, the Karanakudam Bridge is too narrow to accommodate more than one vehicle at a time. As a result, Kathrikadavu Junction is heavily congested. The restricted vehicle movement on the Thammanam Pullepady route also adds to the travel time.

4 Traffic Data Analysis

4.1 Road Inventory Survey

A road inventory survey was conducted on all of the study area's major roads, collecting data such as road length, cross-sections, a hierarchical pattern of roads, intersections, street vendors, and parking areas. The junction base plan was created using the Total Station Survey (Fig. 2).

4.2 Traffic Volume Study

The number of vehicles crossing a section of road per unit time at any given time is referred to as traffic volume. "classified traffic volume studies" is a term used to describe the process of determining the volume of each vehicle class separately to calculate the total volume. Each vehicle type is transformed into a passenger car (PCU). The method used to calculate traffic volume is the manual counting method and videography method. The analysis was performed on Monday, Wednesday, Thursday, and Saturday to obtain accurate traffic counts. To measure the vehicular movement, a peak hour volume count of vehicles in all directions was performed. It's also used to analyze the traffic patterns and schedule traffic operations, as well as to design pavements and signal timings. These statistics can be used to define peak traffic times, assess the impact of large vehicles or pedestrians on vehicular traffic flow, and monitor traffic volume patterns. The figure below depicts the traffic volume count (PCU) of the Kathrikadavu Junction (Fig. 3).

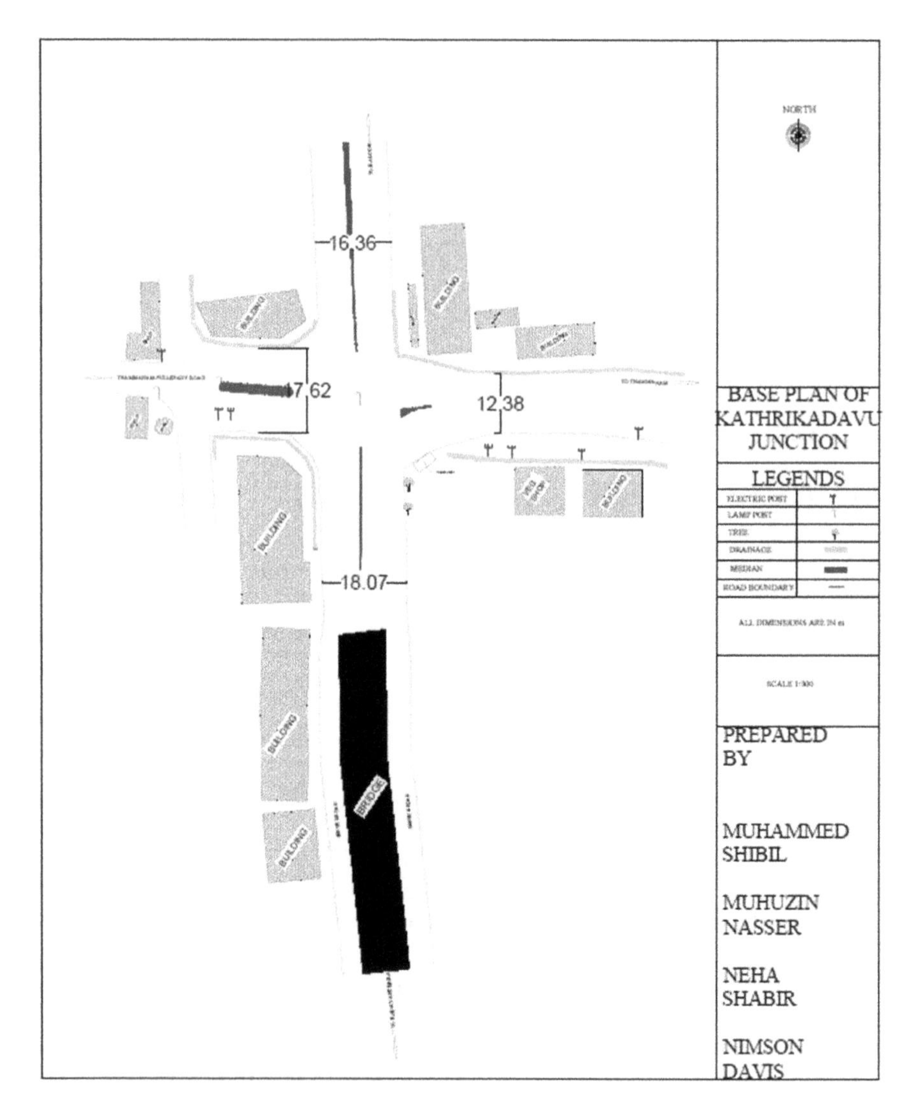

Fig. 2 The base plan of Kathrikadavu junction

4.3 Accident Data

Accident data is the most basic indicator of safety, and it is impossible to determine the scope and nature of road safety issues without it. Crash data is used to aid decision-makers in understanding the existence, causes, and consequences of crashes. This data sets the stage for developing strategies and approaches to minimize accidents and their effects. Figure 4 shows accident data obtained from the DCRB (District Crime Records Bureau) Kochi, the time periods 2016–2018.

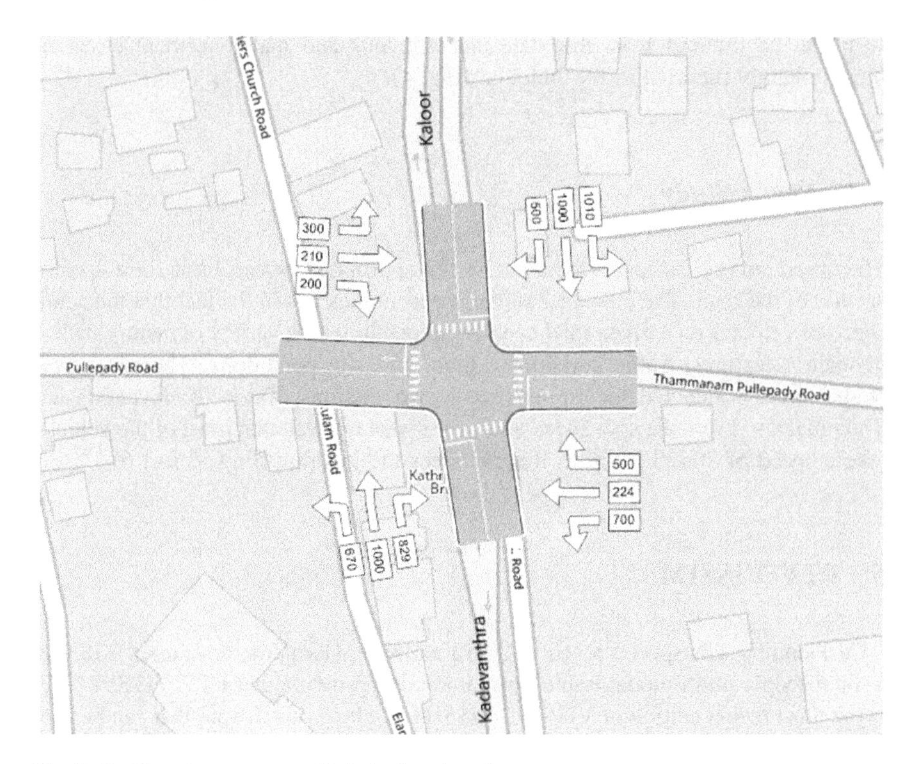

Fig. 3 Traffic volume count at Kathrikadavu junction

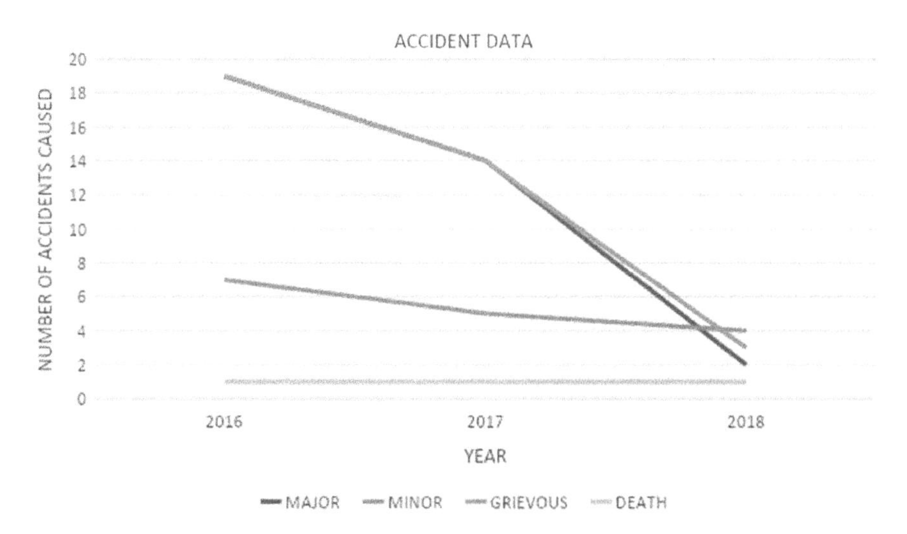

Fig. 4 Accident data

It can be inferred from this data that grievous and major accident cases are predominantly reported in the Junction (Fig. 4).

4.4 Speed Study

The speed survey can be used to assess the appropriate speed limit for a specific section of the road. The relevance of the speed studies lies in the fact that the actual speed of vehicles on a given road can vary depending on a variety of factors such as geometric features, traffic conditions, time, and site conditions. The spot speed studies were conducted for the two busy road sections, Kaloor-Kadavanthra and Thammanam. From the spot speed analysis, it was inferred that most of the vehicles take a speed of 20–30 kmph as it approaches the junction (Figs. 5 and 6).

5 PTV VISSIM

PTV Planning Transport Verkehr AG in Karlsruhe, Germany, developed VISSIM, a microscopic multi-modal traffic flow simulation program and PTV VISSIM 2021 is the most recent edition of VISSIM. VISSIM is a type of software that can be used to evaluate the private and public vehicles under a variety of conditions, including lane configuration, vehicle composition, traffic signals, and so on. It is a time and behavior-based simulation model for urban traffic and public transportation activities, as well as pedestrian flows. It allows you to simulate traffic patterns exactly, be it comparing junction geometry, analyzing public transport priority schemes, or the effects of certain signaling.

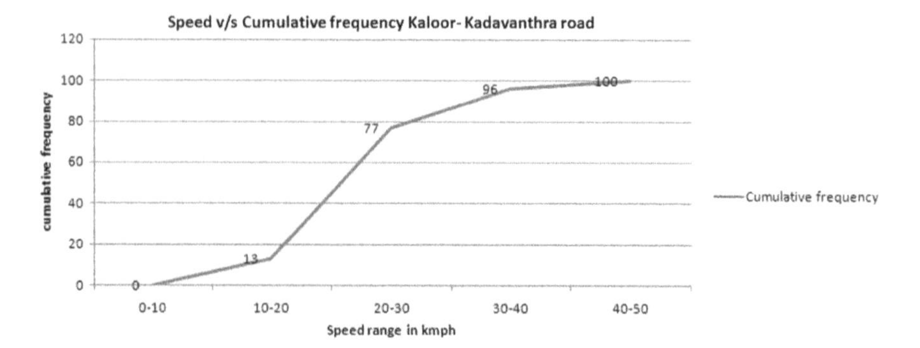

Fig. 5 Spot speed data Kaloor-Kadavanthra road

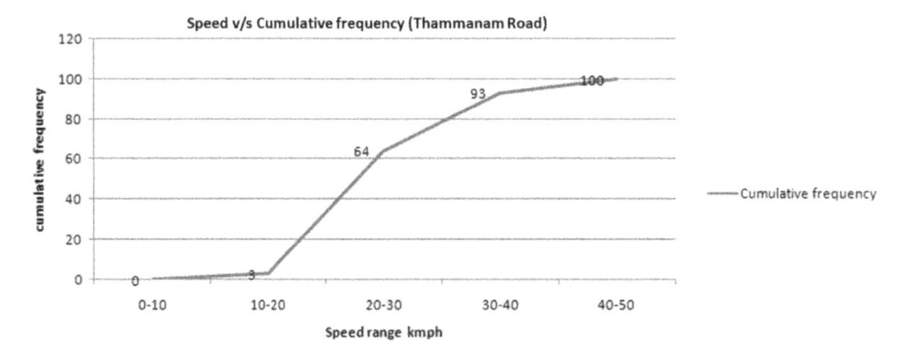

Fig. 6 Spot speed data Thammanam road

6 PTV VISTRO

PTV AG Germany developed VISTRO, which stands for Vision Traffic and Optimization. VISTRO is a transportation framework that can be used to evaluate the effect of regional growth, LOS (level of service), and intersections, as well as generate reports in the form of tables and drawings. Furthermore, this program can be used to conduct traffic effects analysis which involves trip generation distribution, assignment, traffic management scenarios, mitigation assessment and testing, comprehensive studies, and traffic forecasts for future growth.

7 Modelling of Traffic Network for a Heterogeneous System Using PTV VISSIM and PTV VISTRO

7.1 Signal Design Using Webster's Method

Determines the optimum signal cycle time corresponding to minimum total delay to all the vehicles at the approach roads of the intersection (Table 1). Optimum signal cycle,

$$C_O = (1.5L + 5)/(1 - Y)$$

Lost time per cycle (s), $L = 2n + R$.

n, No: of phases.

R, all red time or red-amber time (also provided for pedestrian crossing).

$Y = y_1 + y_2 + \cdots + y_n$, where y_1, y_2, are the ratios of flow to saturation flow

$$y_i = q_i/S_i$$

where

q, total flow in PCU/h.

S, saturation flow = 525 w.

w, width of road measured from kerb to line.

Assuming an Amber period of 2 s and all red time as 9 s No: of phases = 4.

Total lost time, L = 2n + R = 2 * 4 + 9 = 17 s.

$Y = y_1 + y_2 + y_3 + y_4 = 0.15 + 0.26 + 0.07 + 0.32 = 0.8$

Optimum cycle time = (1.5 * 17) + 5/(1 – 0.8) = 153 s

Green time for 4 phases,

$G_1 = Y_1 * (C_O - L)/Y = 0.15 * (153 - 17)/0.8 = 26$ s

$G_2 = Y_2 * (C_O - L)/Y = 0.26 * (153 - 17)/0.8 = 44$ s

$G_3 = Y_3 * (C_O - L)/Y = 0.07 * (153 - 17)/0.8 = 12$ s

$G_4 = Y_4 * (C_O - L)/Y = 0.32 * (153 - 17)/0.8 = 54$ s

Assuming all pedestrian time as 9 s and amber time as 2 s for each phase, (9 + 4 * 2 = 17 s).

Total cycle length = 26 + 44 + 12 + 54 + 17 = 153 s.

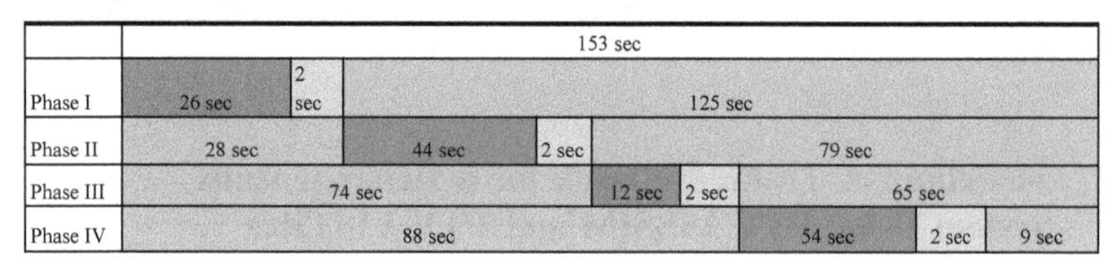

7.2 *Procedure for VISSIM*

- Input the following data:
  - Length of the road that needs to be simulated.
  - Width and other features of the road
  - Speed of the vehicle
  - Traffic volume count
  - Cycle time
- Conducted the reconnaissance survey of the study area.
- Evaluation of the current situation of the study area.
- Plotted the base plan of the study area using total station survey.
- Traffic and pedestrian studies were conducted.

Table 1 Webster method calculation

FROM	Thammanam			Kadavanthra			Pullepady			Kaloor		
TO	Kadavanthra	Pullepady	Kaloor	Pullepady	Kaloor	Thammanam	Kaloor	Thammanam	Kadavanthra	Thammanam	Kadavanthra	Pullepady
PCU/h (q)	700	224	500	670	1000	829	300	210	200	1010	1000	500
Width (w)	8.86	7.86	7.65	7.86	7.65	6	7.65	6	8.86	6	8.86	7.86
Saturation flow S = 525 w	4652	4127	4016	4127	4016	3150	4016	3150	4652	3150	4652	4127
y = q/s	0.15	0.05	0.12	0.16	0.25	0.26	0.07	0.06	0.04	0.32	0.21	0.12
y_{max}	$y_1 = 0.15$			$y_2 = 0.26$			$y_3 = 0.07$			$y_4 = 0.32$		

- The speed study was conducted to determine the average speed of the vehicles at an intersection and for the turning traffic also.
- Evaluation of the marking at the study area.
- Calculated the maximum conflict points in the intersection.
- Start the simulation.
- View results.

7.3 Procedure for VISTRO

- Create a road network (Fig. 7).
- Creation of a base scenario and includes the following.
- Setting up the intersection: Geometric data intersections such as path width, lane settings, and median existence must be completed. The input data is filled in correctly and adjusted to the condition in the field (Fig. 8).
- Volume data input: At PTV VISTRO, the volume of traffic used is the volume in the Passenger Car Unit (pcu/hour)
- Traffic control.

7.4 Output Obtained from VISTRO and VISSIM

After performing the simulation we obtained a set of results which includes:

1. Queue Length

Queue Length is the distance that vehicles queue in front of the detection line. In this study, we set 12 queue counters (4 in the surroundings of the road network and 8 inside the core area) and obtained the record of average queue length in different legs of the intersection as follows (Fig. 9).

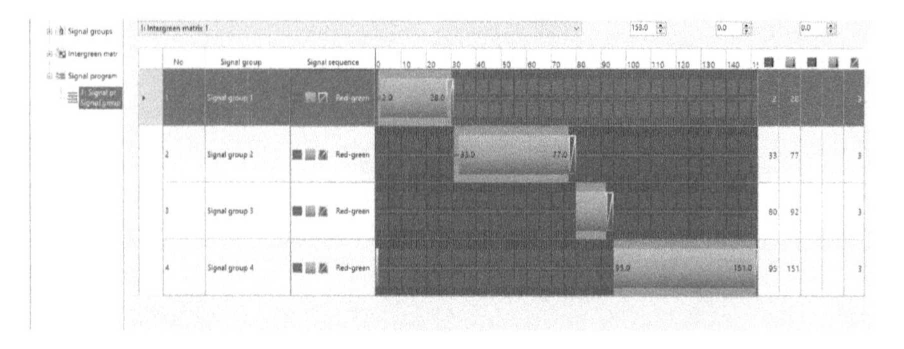

Fig. 7 Signal design

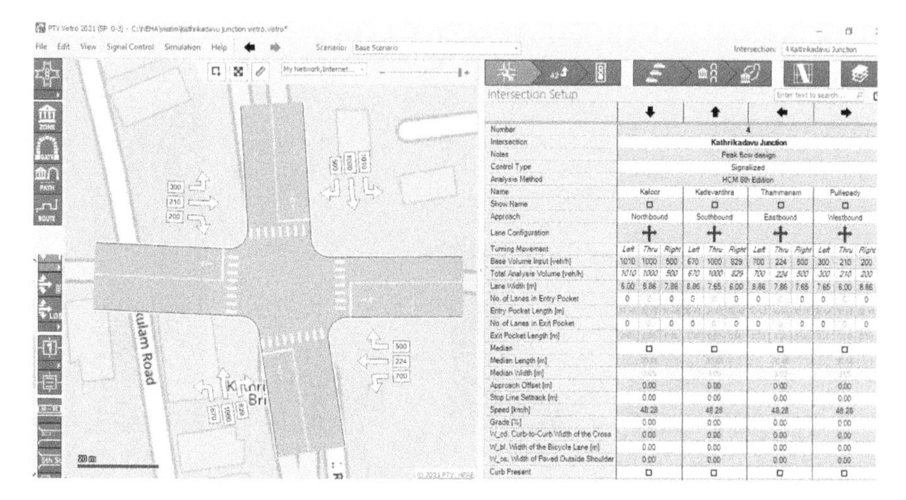

Fig. 8 Building PTV VISTRO network

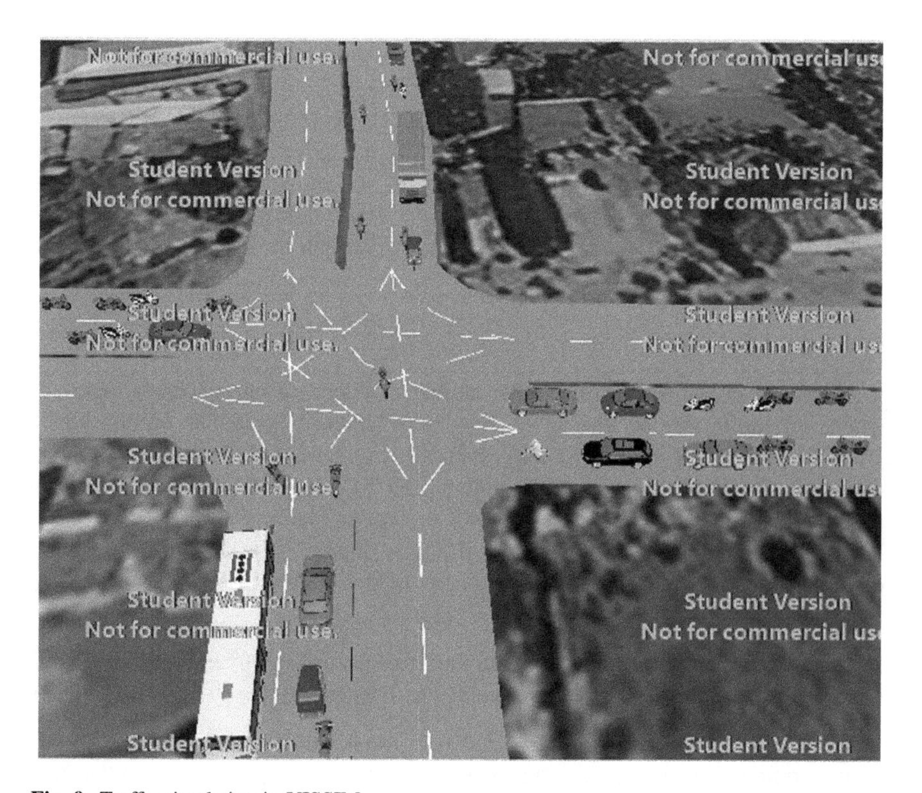

Fig. 9 Traffic simulation in VISSIM

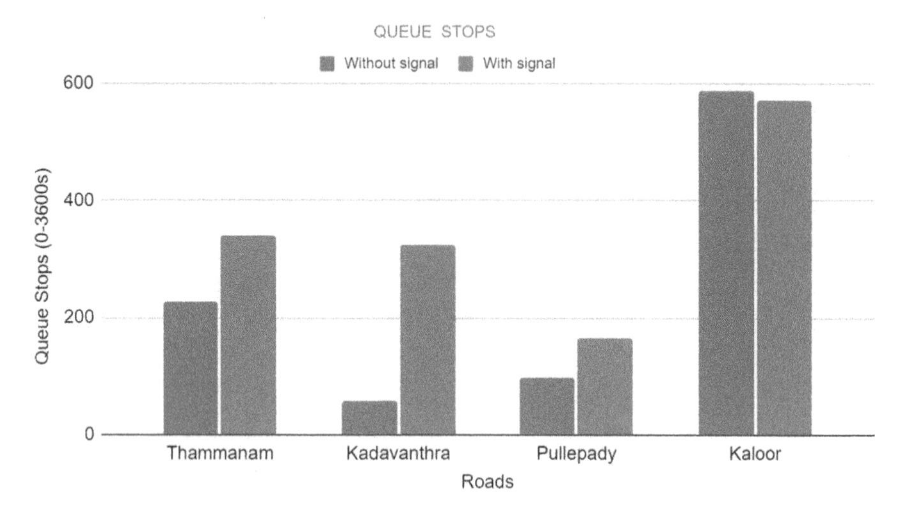

Fig. 10 Queue stops

2. Travel Time and Delay

Travel time (including parking and waiting time) is a simple and robust network performance measure that is well understood by the public, while delay means the time loss for vehicles from a starting point to an endpoint compared with free-moving traffic. In VISSIM, travel time and delay indices use the same detector called the travel time section. Thus, in this study, we put travel time and delay into consideration together.

3. Level of Service

Level-of-Service (LOS) of a traffic facility is a concept introduced to relate the quality of traffic service to a given flow rate. Level-of-Service is introduced by HCM to denote the level of quality one can derive from a local under different operation characteristics and traffic volume. HCM proposes LOS as a letter that designates a range of operating conditions on a particular type of facility. Six LOS letters are defined by HCM, namely A, B, C, D, E, and F, where A denote the best quality of service and F denote the worst. LOS obtained for the Kathrikadavu junction from VISTRO is shown below (Fig. 10).

LOS **E, Traffic at Saturation**. Low but uniform speed. Maneuverability is possible only under constraint for another vehicle. The user is frustrated (Figs. 11 and 12).

LOS **F, Congestion**. The unstable speed with the formation of waiting for lines at several points. Cycles of stop and departure with no apparent pattern because created by the behavior of other drivers. A high level of vigilance is required for the user with practically no comfort. At this level the volume to capacity ratio exceeds 1, implying that the road segment is used above design capacity.

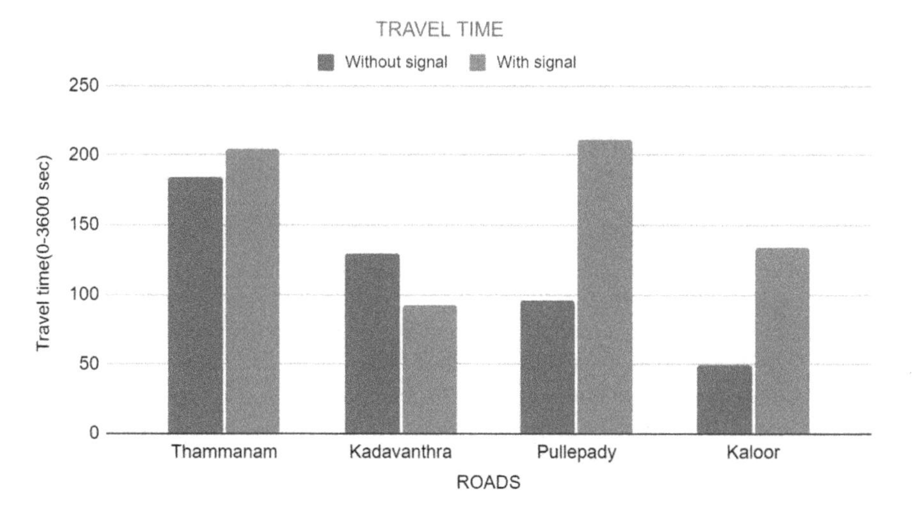

Fig. 11 Vehicle travel time data

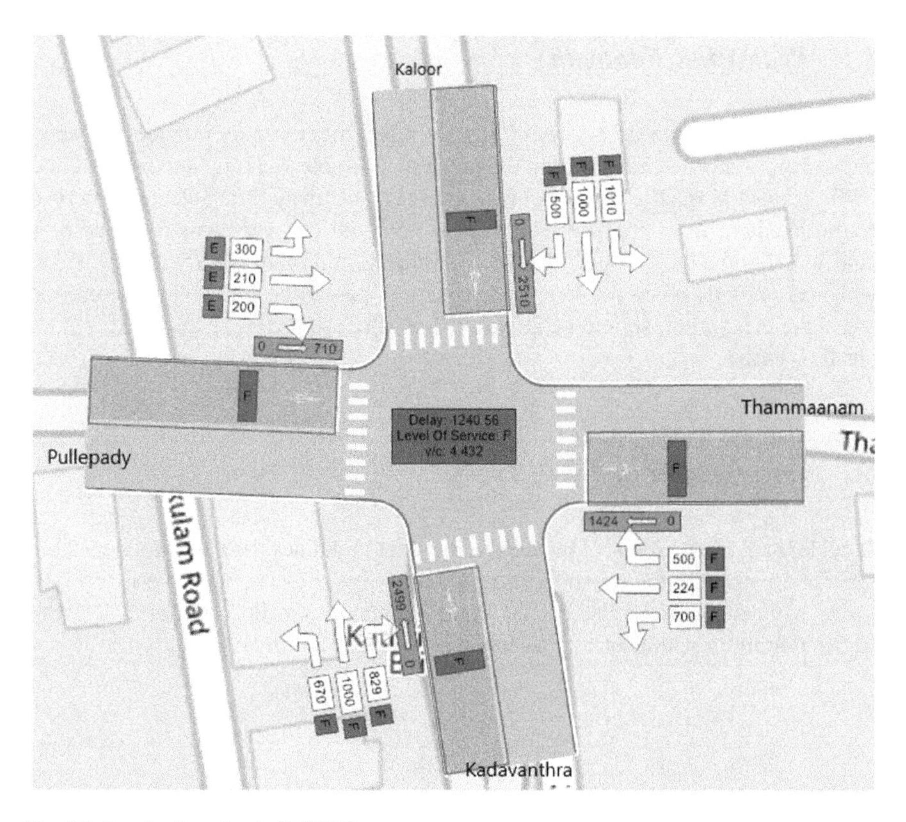

Fig. 12 Level of service in VISTRO

8 Incorporating Sustainability in Transportation Planning

Kochi city, like all other cities in India, is experiencing increased dependency on private motor vehicles for personal trips, leading to increased vehicular congestion and emissions. Pedestrians, bicyclists, and non-motorized rickshaws are the most critical elements in mixed traffic. If the infrastructure design does not meet the requirements of these elements, all modes of transport operate in sub-optimal conditions. At present, pedestrians, bicyclists, and other nonmotorized vehicle users are the captive users of these modes. Despite unsafe conditions present on the road, they continue to use these modes because their socio-economic condition does not permit them to use motorized vehicles. These users have opted for walking, cycles or public transport because it is needed for their very survival. If our future cities have to meet the demands of sustainability, then we have to ensure that these environment-friendly modes are used as a preferred choice over short motorized trips. As per the IUT Part-1 categorization of urban roads, the four roads approaching the Kathrikadavu junction are sub arterial roads.

8.1 Pedestrian Footpath

The pedestrian pathway is one of the effective measures to provide the safest movement of a pedestrian through the junction, promote walking, and hence reduce traffic congestions. The absence of a pedestrian pathway is a notable demerit of Kathirikkadavu junction. The width of the footpath can vary as per the adjacent land use. Footpaths in residential areas require a minimum clear width of 1.8 m, which is enough space for two wheelchairs to pass each other. For commercial areas, the clear width should be at least 2.5 m (IRC:103–2012, Cl 6.1 and 5.2). As per IUT Part-1.

8.2 Introduction of Cycle Tracks

A cycle track is an exclusive bicycle facility that combines the user experience of a separated path with the on-street infrastructure of a conventional bike lane. Cycle tracks have different forms, but all share common elements. As per IUT Part-1 Cl.5.2 minimum width for a two-lane cycle track is 2.0 m (Fig. 13).

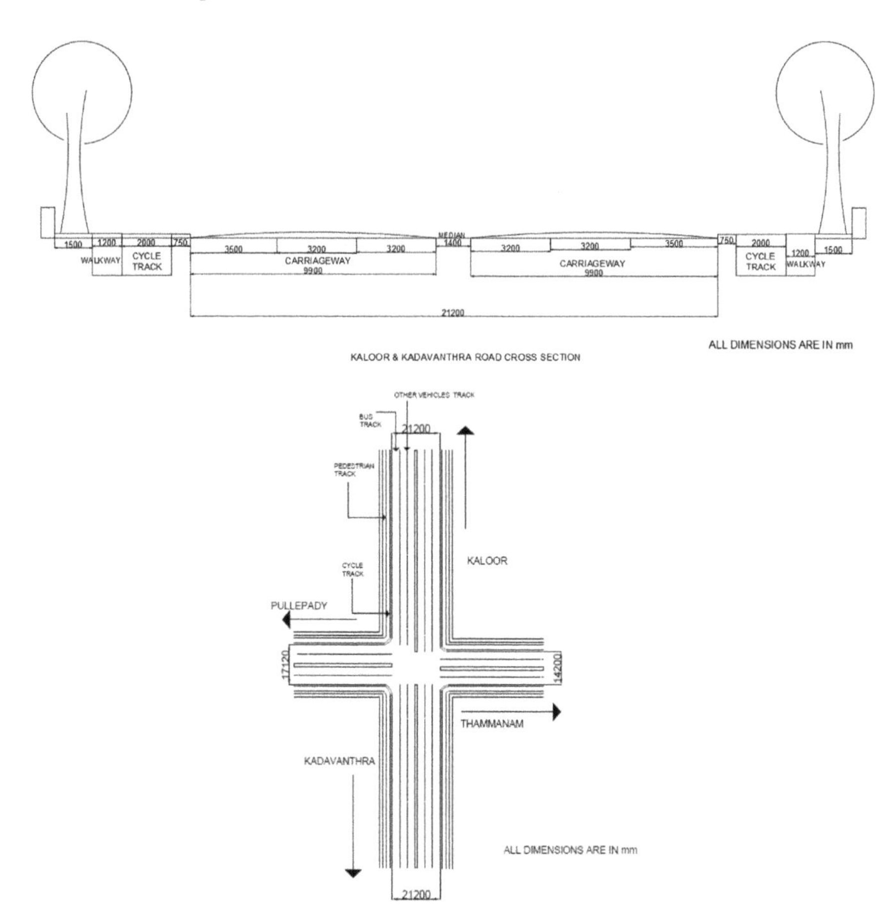

Fig. 13 Cross section and plan of the proposed road

9 Conclusion

The analysis of the current traffic demand at intersections around the project location showed that they are at saturation and sometimes facing long and intolerable delays during peak hours. The existing width of the road is not sufficient to accommodate the current traffic. Installing a signal without road widening again increases the congestion, so road widening is mandatory. It is clear from the analysis of the junction, the queue stops, vehicle delay and travel time remain the same before and after the traffic signal. Thus the junction needs to be sustainably managed to accommodate both the increasing traffic volume and should be pedestrian-friendly. Creating a sustainable transit system requires the significant tasks of planning, design, construction, operating, maintaining, and measuring performance. To make sustainable transportation a reality, we need to outline approaches that effectively contribute to our goal. Non-motorized transport strategy considers the improvement in NMT user experience by enhancing footpaths and bicycle lanes. It also addresses improvement in safety and accessibility for pedestrians and bicycles at intersections. Reducing barriers and impediments on roads to improve bicycle safety is another aspect considered under this strategy. In the current scenario, no straight traffic flow is allowed for the Thammanam-Pullepady route. Here the traffic volume for the restricted roads has been derived based on the existing traffic at the junction. VISSIM is a simulation software that has many uses. With its complete features, simulation of traffic conditions can be done to near real-world conditions. The sustainable road development plan was created in accordance with the Indian Road Congress's code provisions and Institute of Urban Transport codes. The visualization of the proposed plan was done using AutoCAD software. Thus, a future plan for the area with a capacity to accommodate the future vehicular volume was designed by making use of all the available facilities.

References

1. Kriswardhana W, Sulistyono S (2018) Intersection performance study using PTV VISTRO. Case Study, Jember
2. Friis C, Svensson L (2013) Pedestrian microsimulation, a comparative study between the software programs VISSIM and Viswalk
3. Institute of Urban Transport (IUT) guidelines Part 1 and Part 2; comprehensive mobility planning (CMP) for greater Kochin region 2016, vols I, II, III
4. Street Design Guidelines "for Equitable Distribution of Road Space"—NUTP © UTTIPEC, Delhi Development Authority, New Delhi Guidelines Adopted: Nov 2009. Revision 1: Nov 2010
5. Lal G, Divya LG, Nithin KJ, Mathew S, Kuriakose B (2016) Sustainable traffic improvement for urban road intersections of developing countries: a case study of Ettumanoor, India. Procedia Technol 25:115–121
6. Mensah S, Eshragh S, Faghri A (2011) Modern roundabouts and sustainable intersection design. In: 3rd national roundabout conference, Indiana, USA
7. Štraub D, Jaroš V (2019) Free fare policy as a tool for sustainable development of public transport services. Hum Geographies 13(1):45–59
8. Shaban A, Kourtit K, Nijkamp P (2020) India's urban system: sustainability and imbalanced growth of cities. Sustainability 12(7):2941

Rheology of Sustainable Self Compacting Concrete with Triple Blend Cementitious Materials

R. P. Divitkumar, Bharathi Ganesh, Jayesh Shrestha, Bishwombhar P. Pandit, Anish Chaulagain, and Bakesh Rawat

Abstract The concept of sustainability from conception of a project to its completion has become a priority in the civil engineering industry in which there is an acute scarcity of materials due to depletion of natural resources as ingredients of concrete. Self-Compacting Concrete (SCC), proposed in 1989 by Prof. Okamura of Japan, for the first time, had revolutionized the process of placing concrete by eliminating the compaction. SCC, a self-flowing concrete possesses flowability whilst maintaining stability/homogeneity without segregation; thus allowing filling up of congested areas in concrete between the bars in congested reinforcement. The flowability obtained from the use of high-range water-reducing admixtures helps in achieving better packing of concrete around reinforcements and superior quality surface finishes. The mineral admixtures such as Fly Ash, GGBS etc., added as powder content enhance resistance to segregation, reduce the cost of production and improve micro-structure of SCC. The experimental study focuses on the assessment of fresh properties of SCC with an equal blend of three cementitious materials; namely cement, fly ash and GGBS called Triple Blend SCC (TBSCC). The mix proportion is designed by Absolute Volume method. The conventional/control SCC mix consists of cement (70%) and fly ash (30%) with natural river sand as fine aggregates. The properties of SCC are controlled by the characteristics of materials and the mix proportions. The properties of different constituent materials used for SCC are assessed as per the standard tests

R. P. Divitkumar
SV Concrete, Bangalore, India

B. Ganesh · J. Shrestha (✉) · B. P. Pandit · A. Chaulagain · B. Rawat
Nitte Meenakshi Institute of Technology, Bangalore, India
e-mail: 1nt17cv038.jayesh@nmit.ac.in

B. Ganesh
e-mail: bharathi.ganesh@nmit.ac.in

B. P. Pandit
e-mail: 1nt17cv021.bishwombhar@nmit.ac.in

A. Chaulagain
e-mail: 1nt17cv007.anish@nmit.ac.in

B. Rawat
e-mail: 1nt17cv018.bakesh@nmit.ac.in

G. C. Marano et al. (eds.), *Proceedings of SECON'21*, Lecture Notes in Civil Engineering 171, https://doi.org/10.1007/978-3-030-80312-4_77

procedures. The acceptance criteria in the fresh state of SCC are assessed through slump flow, V-funnel and L-Box as per EFNARC 2005. The improvement in rheology of SCC mixes and enhancement of compressive strength at 28 days curing period may be attributed to usage of three mineral admixtures in equal proportion. This practice of triple blend cementitious content in SCC advocates the reduction of 67% cement content in SCC with improved fresh properties and strength of SCC, thus producing a sustainable solution for the filed applications.

Keywords Sustainability · Self-compacting concrete · Fly ash · GGBS · M-sand · Pond ash

1 Introduction

Construction is a high resource consuming process. In the process of building a simple house, one needs cement, aggregates, sand, water, wood, steel, bricks and many more. These materials are generally either non-renewable or affect the environment adversely. Sustainability deals with the methods to ensure that these resources are available for generations to come, while ensuring the reduced environmental impact. The options to adopt sustainability would be through building structures with prolonged life span and through usage of industrial byproducts replacing depleting natural resources. Hence a structured database (catalogue of sustainable material and availability) of alternative materials, source, characteristics of materials, point of contact for collection, and such key points should be prepared and awareness on the use of these materials should be provided. The discharge of CO_2 to the atmosphere is directly linked to the production of Portland cement. One solution is the use of alternative materials such as fly ash, GGBS, Silica fumes etc., for cement; pond ash, Manufactured Sand (M-Sand) to natural sand.

Coal production releases large amounts of residues which are harmful to the environment. One of the by-products, Fly Ash (FA), if disposed of in an unscientific way, affects the local ecosystems due to the heavy metal pollution through erosion and leachate generation. The generated fly ash also requires large areas for disposal. India is the third-largest generator of coal-based power after China and the USA [1]. The Indian coal is low grade and has high ash content (30–45%) as compared to imported coals (10–15%), so large quantities of fly ash are generated, 217.04 Million tons in 2018–19 [2]. In India, an area of 65,000 acres of land is being occupied by ash ponds and its generation is expected to cross 225 million tonnes by the year 2020 [3]. In order to reduce the impact of fly ash on the environment and to lower the requirement of land for its disposal, various notifications have been issued by the Ministry of Environment and Forest to achieve maximum utilization of fly ash [4]. Since 1994, under the Ministry of Science and Technology (GOI), different technologies have been adapted for safe and productive utilization of fly ash and this increased the utilization of fly ash from 6.64 million tons in 1996–97 to 168.40 Million tons in 2018–19 [5].

1.1 Alternative Materials in Concrete

Triple bend mix is a mix containing Cement and mineral admixtures (such as Fly Ash and GGBS in equal proportions as cementitious content) as binding materials. Other materials like M-Sand and Pond ash are used as fine aggregates in equal proportions as replacement to the depleting Natural River Sand. The triple bend mixes address the issues like cost, sustainability and performance simultaneously [6]. Triple bend mixtures with Silica fumes have been used extensively throughout the world, for example, in the foundation of Petronas towers, Yangshan deep water port, etc. A concrete mix of 80 MPa grade was used for superstructure with triple bend mix of 10% silica fume and 13% fly ash as cementitious content in Burj Khalifa. Thus, it can be seen that triple bend mix is quite useful if used suitably without much compromise in strength, if sufficient care is taken while designing the mix and execution.

The specification for the GGBS is given in IS 12089–1987 [7]. The chemical composition of the slag considerably varies depending on the raw materials used in the iron production. The main components of the slag are CaO (30–50%), SiO_2 (28–38%), Al_20_3 (8–24%) and MgO (1–18%). GGBS can be utilized to supplant as much as up to 80% of concrete when utilized in concrete. GGBS has better water permeability attributes and furthermore, improves protection from chloride and sulfate attack [8]. The setting time of cement is affected by numerous elements in which the two key points are temperature and water/cement ratio [9]. With GGBS as an alternate material in concrete, the setting time can be marginally varied, with the following properties:

- Better workability, placing and compaction of concrete material.
- Lower early age temperature rise which reduces the risk of thermal cracking.
- Elimination of the risk due to internal reactions in concrete constituents.
- High resistance to chloride ingress which reduces the risk of reinforcement corrosion.
- High resistance to attack by sulphate and other chemicals.

Pond ash, a coarser fly ash, a waste from Thermal Power Stations, has high amounts of silica content in it [10]. It also has pozzolanic nature when fineness is more than 320 m^2/kg and has the potential to be used as supplementary or alternative material to fine aggregates [11]. Use of pond ash in concrete will not only help the huge demand of Fine aggregates in concrete in the construction industry but will also help in the proper disposal of a huge waste that is generated from thermal power plants. 25% of the total ash produced by the thermal power plant is pond ash [6]. Pond ash consists of fused coarser particles which are porous and seem like volcanic lava. The major chemical components present in pond ash are Al_2O_3, SiO_2, and K_2O. Although these are the major compounds present in pond ash, other compounds such as CaO, TiO_2, and FeO may also be found in small quantities.

2 Present Study

The objective of the study consists of designing a mix proportion of self-compacting concrete with Cement, Fly ash and GGBS in equal proportion as cementitious materials; M-sand and pond ash in equal proportion as fine aggregates. It is followed by freezing the design mix, assessing the behaviour of SCC in its fresh state and also the compressive strength of mixes. In this system, the coarse aggregate and fine aggregate contents are fixed and self-compaction is to be achieved by adjusting the water/binder ratio. PCE based melamine admixture super plasticizer is used as admixture without the use of VMA.

3 Material Description and Methodology

3.1 Cementitious Materials

OPC 53 grade conforming to IS 12269:1987 [12] tested as per IS 4031–2013 [13] having a Fineness by Blain's permeability method of 225 m^2/kg and specific gravity of 3.15 is used as a binder.

The specific gravity of off-white colored GGBS is 2.9 with a fineness of 350 kg/m^3 as per BS:6699 [14]. The loose bulk density was found to be 1200 kg/m^3. GGBS having Fineness by air permeability of 389 kg/m^3 is finer than cement but coarser than fly ash, specific gravity of 2.89, residue of 45% of 7.10% etc. The loss of ignition is found to be 0.32% and glass content of 90%.

Class C fly ash procured from RTPS (Raichur Thermal Power Station) conforming to IS 3812–2003 [15] is assessed for its physical properties assessed as per IS 1727–1967 [16] which are summarized as follows: Fineness—Specific surface by Blaine's Air-Permeability method of 410.0 m^2/kg, Lime reactivity—Average Compressive Strength in 4.6 N/mm^2, Soundness by Autoclave Test Expansion of Specimens, of 0.025%, Residue on 45 micron sieve of 18.4%, chemical analysis of fly ash showed the main components such as Loss on Ignition (LOI) of 1.0%, 57.65% SiO_2, 26.90% of Al_2O_3, 4.26% of Fe_2O_3, Cao of 2.32% etc.

3.2 Aggregates

a. **Fine Aggregates**
   Manufactured Sand (M-Sand) and Pond Ash are used as Fine Aggregates in SCC replacing Natural River Sand in equal proportion. The physical properties assessed as per relevant codes of practice [17] conforming to IS 383:1970 [18] are presented in Tables 1 and 2.

Table 1 Typical physical properties of M-sand

S. No.	Properties	
1	Shape	Cubical
2	Particle passing 75 microns	Up to 15%
3	Grading zone	II (FM 2.6–3)
4	Specific gravity	2.5–2.9
5	Water absorption	2–4%
6	Soundness	Relatively sound (Ex. <5)
7	Alkali silica reactivity	0.001–0.008

Table 2 Physical properties of pond ash

S. No.	Properties	Result
1	Lime reactivity of pond ash	0.66
2	Specific gravity	2.16
3	Bulk density in loose state	824 kg/m^3
4	Bulk density in compacted state	990 kg/m^3
5	Atterberg's liquid limits	47.3%
6	Grain size distribution	
	Sand	72%
	Silt	28%
	Clay	Nil

b. **Coarse Aggregates**
Coarse aggregates consist of 55% of aggregates passing through 20 mm and retaining on 12.5 mm sieves and 45% of aggregates of 12.5 mm down size, conforming to All-in aggregates.

3.3 Water

Normal potable water available in the laboratory satisfying the requirements as per IS 456:2000 [19] is used for mixing SCC.

3.4 Admixture

PCE based admixture conforming to IS 9103–1999 [20] is used to obtain the workability of SCC with required rheological properties.

4 Mix Design Procedure

In the present study, the mix proportion is arrived at using "Absolute Volume Method" and the design mix is frozen by conducting extensive trials to satisfy rheological requirements as per EFNARC specifications [21].

The design mixes can be summarized as

A. Triple blend SCC mixes with cement:Fly Ash:GGBS in the Proportion of 33.33%:33.33%:33.33% by weight and Fine Aggregate content of the mix consists of M-Sand:Pond ash in the proportion of 50%:50% by weight.
B. Conventional SCC mix consists of Cement:Fly ash in the proportion by weight of 70%:30% and Fine Aggregate content of the mix consists of M-Sand:Pond ash in the proportion of 50%:50% by weight.
C. Conventional Natural River Sand SCC mix consists of Cement [11]: Fly ash in the proportion by weight of 70%:30% and Fine Aggregate content of the mix consists of only Natural River Sand.

5 Methodology, Test Results and Discussions

The results of various tests conducted on the properties of Triple Blend SCC in fresh state with alternative materials to both cement and fine aggregates are listed first, rheology of mixes assessed as per the standard procedure of EFNARC 2005 [21], next followed by the compressive strength of mixes. The rational conclusions are drawn on behavioural aspects between rheological properties of SCC mixes and compressive strength.

Results of various tests on rheological properties, namely Slump flow, T_{50} Slump, V-funnel and L-box for different mix combinations are presented in Figs. 2, 3, 5 and 7. The results of compressive strength at 28 days curing of different Triple

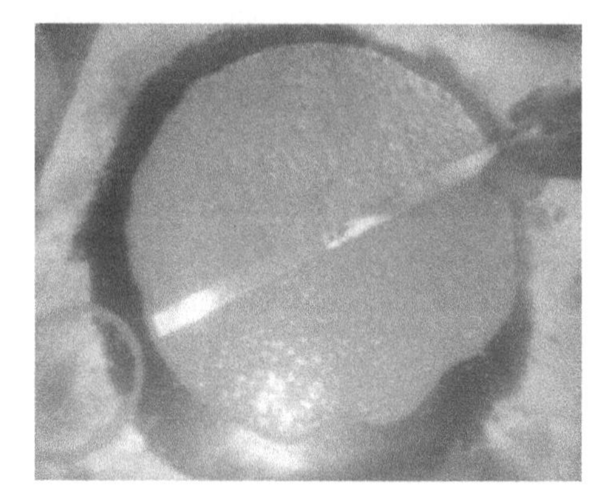

Fig. 1 Slump flow test

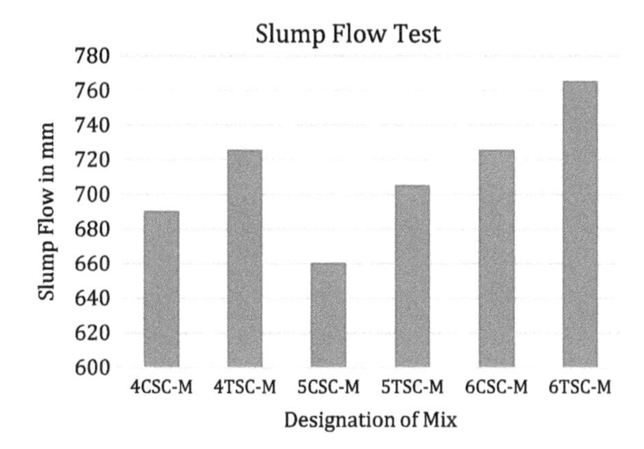

Fig. 2 Slump flow test of TSC-M and CSC-M mixes

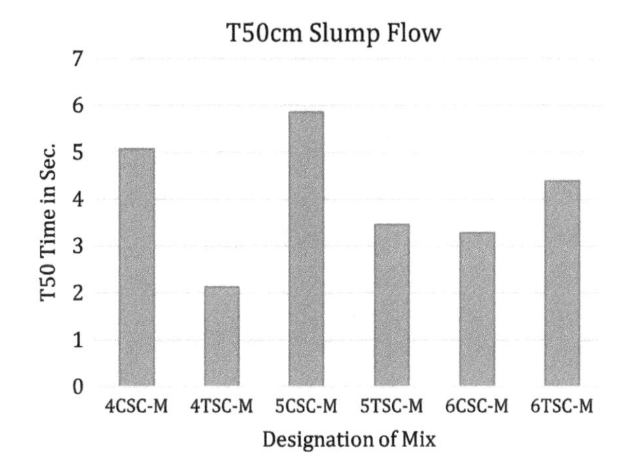

Fig. 3 T_{50}cm of TSC-M and CSC-M mixes

Blend SCC mixes (TSC mixes) and the mix (CSC mixes) with Cement (70%) and Fly ash (30%) as cementitious contents and fine aggregates consisting of M-Sand (50%) and pond ash (50%) are tabulated in Table 5. The results are discussed in comparison with that of the conventional SCC mix—the mix (CNSC-M) with Cement (70%) and Fly ash (30%) as cementitious contents and fine aggregates consisting of Natural River Sand (100%) from validated results of previous paper of the authors [11].

5.1 EFNARC Specifications for Initial Mix Composition [21]

While designing the mix, it is convenient to consider the relative proportions of the key components by volume rather than by mass.

Fig. 4 V-funnel test

Fig. 5 V-funnel test of TSC-M and CSC-M mixes

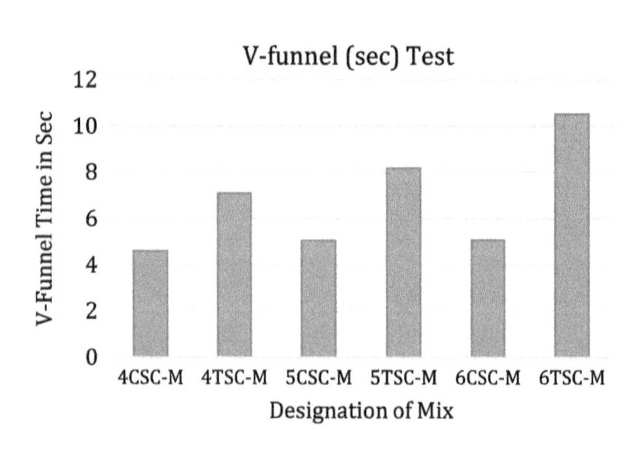

Fig. 6 L-box test

Fig. 7 L-box test of TSC-M and CSC-M mixes

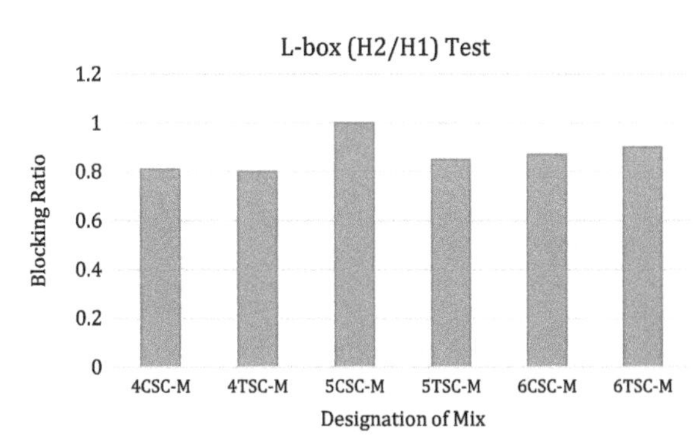

- Water/Powder ratio by volume of 0.80 to 1.10
- Total powder content—160 to 240 l (400—600 kg) per cubic meter.
- Coarse aggregate content 28–35% by volume of the mix.
- Water powder ratio selected typically ranges from 180 l/m^3 and not exceeding 200 l/m^3.
- The sand content is selected as such to balance the volume of the other constituents.

It is required to design the mix to ensure that the concrete/SCC is capable of maintaining its specified properties in fresh state/rheology, despite anticipated variations in the properties/qualities of raw materials [22]. The design mix proportion is presented in Table 3.

Table 3 Mix proportions of triple blend pond ash and M-sand SCC

SI. No. (part)	Concrete type	Designation of the mix	Mix proportion	Cementitious content (kg/m^3)	Water	Water/binder	SP dosage
					kg/m^3 (l/m^3)	Ratio	%
1	Triple blend pond ash and M-sand SCC mixes	4CSC-M	1:1.94:1.89	400	200	0.50	0.2
		4TSC-M	1:1.92:1.87	400	200	0.50	0.2
		5CSC-M	1:1:46:1.43	500	200	0.40	0.3
		5TSC-M	1:1.44:1.41	500	200	0.40	0.3
		6CSC-M	1:1.15:1.12	600	200	0.33	0.4
		6TSC-M	1:1.13:1.10	600	200	0.33	0.4

Designation of the mix	Explanation
4CSC-M	Control SCC mix with 400 cementitious content with cement and fly ash at 70%:30%, pond ash and M-sand at 50%:50%
4TSC-M	Triple blend SCC mix with 400 cementitious content with cement:fly ash:GGBS in the proportion of 33.33%:33.33%:33.33%, pond ash and M-sand at 50%:50%
5CSC-M	Control SCC mix with 500 cementitious content with cement and fly ash at 70%:30%, pond ash and M-sand at 50%:50%
5TSC-M	Triple blend SCC mix with 500 cementitious content with cement:fly ash:GGBS in the proportion of 33.33%:33.33%:33.33%, pond ash and M-sand at 50%:50%
6CSC-M	Control SCC mix with 600 cementitious content with cement and fly ash at 70%:30%, pond ash and M-sand at 50%:50%
6TSC-M	Triple blend SCC mix with 600 cementitious content with cement:fly ash:GGBS in the proportion of 33.33%:33.33%:33.33%, pond ash and M-sand at 50%:50%

5.2 *Results and Discussions—Rheology of Triple Blend SCC*

The results of fresh properties of SCC mixes assessed as per EFNARC 2005 [21] are presented in Table 4 and discussed in further paragraphs.

5.2.1 Results of Slump Flow Test and Discussion

The slump flow of 650–800 mm and slump T_{50} flow time ranging from 2 to 5 s are considered adequate for SCC mixes as per EFNARC specifications [21].

From the results of fresh properties of various mixes of TSC-M, it is observed that all the mixes of SCC have satisfied the requirement of SCC with slump flow ranging from 660 to 765 mm. The mixes are cohesive and take lesser T_{50} flow time indicating lesser frictional resistance between particles. This may be due to the presence of higher quantity of finer content of fly ash and GGBS particles and also due to the presence of higher content (nearly 20%) of particles of pond ash finer than 150 μ and finer particles of M-Sand which has influenced the flow.

Table 4 Fresh properties of mixes of triple blend SCC and CSC for different cementitious contents

Sl. No.	Designation of mix	Slump flow (dia. in mm)	T_{50}cm (s)	V-funnel (s)	L-box blocking ratio (H_2/H_1)
1	4CSC-M	690	5.07	4.6	0.81
2	4TSC-M	725	2.13	7.09	0.8
3	5CSC-M	660	5.86	5.04	1
4	5TSC-M	705	3.45	8.17	0.85
5	6CSC-M	725	3.28	5.07	0.87
6	6TSC-M	765	4.39	10.53	0.9

Hence, slump flow of fresh TSC-M mixes is better than that of the mixes of CSC-M and CNSC-M, exhibiting better achievement of slump of triple blend mixes of SCC with M-Sand and Pond Ash as fine aggregates.

5.2.2 V-Funnel Test

The time of V-funnel ranging from 6 to 12 s is considered adequate for SCC mixes. From the results of V-Funnel test, it can be noticed that V-funnel time of TSC-M mixes, CSC-M and CNSC-M mixes with time ranging from 4 to 11 s, meet the requirements of permissible limit for V-funnel test as per EFNARC 2005 for all mixes except 6TSC-M. The possible reason may be attributed to the improved flowability of SCC mixes influenced by the presence of mineral admixtures and finer particles. However, better flowability is the characteristics of lesser cementitious content.

5.2.3 L-Box Test

The requirement of allowable limit for Blocking Ratio measured from L Box test is 0.8–1.0 as per EFNARC 2005 [21]. As the blocking ratio defines the slope of the concrete at rest, the closer the ratio to 1, the better is the flowability of the mix, allowing it to pass through the congested reinforcement of RCC elements and can self-level the mix with more ease. Thus, a value closer to 1 is preferred for SCC for better packing of SCC mixes around the reinforcement of the RCC elements.

From results of blocking ratio, measured from L-Box test on fresh state of mixes of TSC-M and CSC-M; and comparing it with that of mixes of CNSC-M, it is observed that the blocking ratio is ranging from 0.8 to 0.98. It is observed that the mixes of TSC-M and CSC-M showed better workability when compared to that of conventional SCC mixes of CNSC-M, attributing it to the presence of GGBS as cementitious content in the mixes enhancing flow properties of SCC mixes.

Hence, it may be concluded that the fresh properties of TSC-M mixes assessed in terms of slump flow, T_{50} slump, V funnel and Blocking Ratio are better than that of the mixes of CSC-M and CNSC-M, showing that the better rheology can be achieved with triple blend mixes of SCC using alternative materials to cement and also to fine aggregates.

5.3 Properties of Hardened Triple Blend SCC Mixes

Hardened properties of SCC are of paramount interest to structural designers and users. In the present investigation, extensive testing is carried out to assess strength of various SCC mixes in terms of compressive strength, flexural strength and split tensile strength. However, the present paper focuses only on the 28-day compressive strength of triple blend SCC mixes.

5.4 Compressive Strength

The compressive strength of mixes is assessed as per the standard procedures of IS 516:1959 (reaff. in 2004) [23]. The results of the test on compressive strength of different mixes of SCC with fly ash, GGBS as replacement in equal proportion to cementitious content and M-Sand and Pond Ash as fine aggregates, replacing Natural River Sand are assessed for 3, 7 and 28 days curing periods is presented in Table 5. The normalized values of mixes as CSR1 and CSR2 are also presented and compared (Fig. 8).

$$CSR1 = \left(\frac{(Compressive\ Strength\ of\ TSC\ Mix)}{(Compressive\ Strength\ of\ CNSC - M\ Mix)\ at\ the\ same\ Curing\ Period}\right)$$

$$CSR2 = \left(\frac{(Compressive\ Strength\ of\ TSC\ Mix)}{(Compressive\ Strength\ of\ CSC - M\ Mix)\ at\ the\ same\ Curing\ Period}\right)$$

From the results of the compressive strength test, it is observed that the strength of TSC, CSC and CNSC are in ascending order after curing of 3, 7 and 28 days. This can be attributed to the presence of M-sand and pond ash as fine aggregates in TSC and CSC as opposed to natural sand in CNSC (Figs. 9 and 10; Table 6).

Table 5 Results of compressive strength test for different mixes of SCC

Mix designation	Compressive strength in N/mm^2 at curing period of		
	3 days	7 days	28 days
4CSC-M	7.65	12.5	19
4TSC-M	7.26	11.57	16.48
5CSC-M	10.36	18.2	25.83
5TSC-M	9.75	17.86	21.77
6CSC-M	15	23.8	39.95
6TSC-M	11.90	21.92	29.98

Fig. 8 Compressive strength test

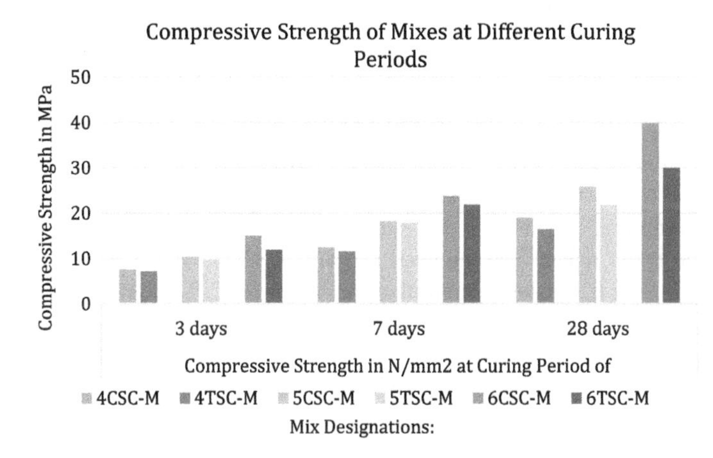

Fig. 9 Compressive strength of TSC-M and CSC-M at 3, 7 and 28 days curing

Fig. 10 Compressive strength ratio of TSC-M w.r.t CNSC-M and CSC-M

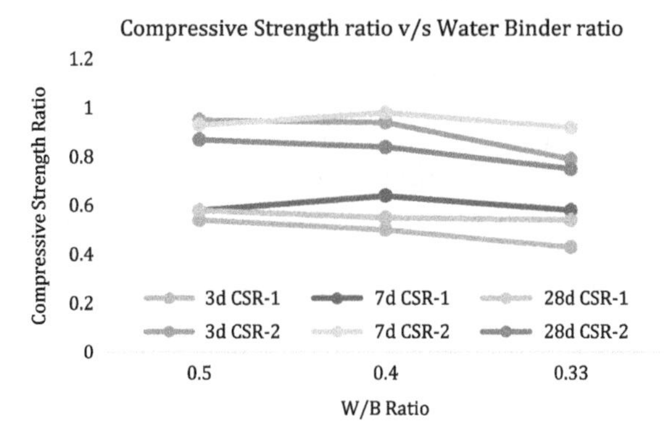

Table 6 Compressive strength ratio (CSRs) of TSC-M w.r.t. CNSC-M and CSC-M respectively

Concrete type	Designation	CSR-1			CSR-2		
		3 days	7 days	28 days	3 days	7 days	28 days
Triple blend pond ash and M-sand SCC	4TSC-M	0.54	0.58	0.58	0.95	0.93	0.87
	5TSC-M	0.5	0.64	0.55	0.94	0.98	0.84
	6TSC-M	0.43	0.58	0.54	0.79	0.92	0.75

6 Conclusion

From the discussions on the results of the various mixes, it can be found that the standard CNSC-M mixes (70% cement and 30% fly ash) achieve higher strength than CSC-M and TSC-M mixes in general. This was an expected result as the use of fine aggregates of M-sand and Pond ash as an alternative materials used in the mixes for replacing natural sand would decrease the compressive strength of the mix due to the porous nature of pond ash which absorbs water during mixing and releases water at later stage due to self-draining property of pond ash. In addition, the use of mineral admixture as a major proportion in a blend (67%) which gains strength due to secondary hydration may also be attributed for the reduced compressive strength at 3, 7 and 28 day curing periods. Further research is required to identify the exact reason and suggest alternatives to arrest the reduction in compressive strength. It is also apparent that the mixes of TSC-M and CSC-M showed better workability when compared to that of control mix CNSC-M [11]. It is observed that the same for TSC is higher due to the presence of GGBS. Hence, it is observed that the fresh properties of TSC-M, measured through tests of slump flow, T50 slump, V funnel and blocking ratio are better than that of the mixes of CSC-M and CNSC-M. Thus, it can be concluded that better flowability can be achieved with triple blend mixes in SCC by using alternative materials at the minimal expense of compressive strength.

Hence, from the discussions and conclusions of the investigation on the use of alternative materials to both cement and fine aggregates in SCC mixes encourages the use of alternative materials in SCC. However, the optimum replacement level so as to achieve the strength as that of conventional SCC needs to be assessed. This experimental study contributes to sustainable construction practices as it reduces the dependency on cement by 67% and natural fine aggregate by 100%, thus encourages the large-scale utilization of industrial wastes to facilitate sustainable human habitations, replacing fast depleting natural resources, and also helps in conserving the precious topsoil required for growing food contributing to environmental and ecological benefits [9].

References

1. Ahn SJ, Graczyk D (2012) Understanding energy challenges in India. Policies, players and issues, epublication. International Energy Agency (IEA), Climate Technology Centre and Network (CTCN) Publication
2. Kumar S, Mucsi G, Kristály F, Pekker P (2017) Mechanical activation of fly ash and its influence on micro and nano-structural behaviour of resulting geopolymers. Adv Powd Tech 28:805–813. https://doi.org/10.1016/j.apt.2016.11.027
3. Haldive SA, Kambekar AR (2013) Experimental study on combined effect of fly ash and pond ash on strength and durability of concrete. Int J Sci Eng Res 4:81–86
4. MOEF, Gazette notification for Ministry of Environment and Forests, no. 563. Ministry of Environment and Forests, New Delhi, 14 September 1999
5. CEA: Central Electricity Authority Report 2018–2019. http://cea.nic.in/reports/others/thermal/tcd/flyash_201819.pdf
6. Ganesh B, Sharada Bai H, Nagendra R (2015) Bond strength assessment of self-compacting concrete containing pond ash and manufactured sand as fine aggregate. In: 4th international engineering symposium-IES March 2015, pp 1–6
7. IS 12089–1987. Indian standard specification for granulated slag for the manufacture of Portland slag cement
8. Ganesh B, Sharada Bai H, Kowshika VR (2013) Workability retention of triple blend self-compacting concrete—proceedings of UKIERI concrete congress, innovations in concrete construction, 5–8 Mar 2013, p 38
9. Shetty MS (2018) Concrete technology. S Chand and Company Publications
10. Sharada Bai H, Ganesh B, Nagendra R (2015) Rheology and strength behavior of self-compacting concrete with ternary cementitious materials. In: 4th international engineering symposium-IES March 2015, pp C2–7–1 to 6
11. Divitkumar RP, Ganesh B, Sharada Bai H, Narendra BK, Manjunath (2020) Supplementary cementitious materials for sustainable triple blend self compacting concrete a rheological study. In: Proceedings of an international conference—second ASCE conference in India on challenges of resilient and sustainable infrastructure development in emerging economies (CRSIDE2020), 02–04 March 2020, pp 89–96
12. IS 12269:2013—Ordinary Portland Cement, 53 Grade – specification
13. IS 4031 (Part 1)-1996 (Reaffirmed 2005) Methods of physical tests for hydraulic cement
14. BS 6699:1992—Specification for ground granulated blast furnace slag for use with Portland cement
15. IS 3812:2003—Pulverized fuel ash specification. Part 1—for use as pozzolana in cement, mortar and concrete. Part 2—for use as admixture in cement mortar and concrete
16. IS 1727:1967—(Reaffirmed in 2004)—Methods of test for pozzolanic materials
17. IS 2386:1963—Methods of test for aggregates for concrete. Part I (Reaffirmed in 2002)—Particle size and shape. Part III (Reaffirmed in 2002)—Specific gravity, density, voids, absorption and bulking. Part IV (Reaffirmed in 2002)—Mechanical properties
18. IS 383:1970 (Reaffirmed 2002)—Specification for coarse and fine aggregates from natural sources for concrete
19. IS 456:2000—Code of practice for plain and reinforced concrete (fourth edition)
20. IS 9103:1999—Specification for concrete admixtures (CED 2: cement and concrete)
21. EFNARC 2005—The European guidelines for self-compacting concrete specification, production and use
22. Preetha R, Kishore GV, Pillai CS, Chetal SC (2011) Self-compacting concrete with fly ash for nuclear power plant structures. Int J Earth Sci Eng 04(06 SPL):945–948
23. IS 516:1959 (Reaffirmed in 2004)—Methods of tests for strength of concrete, test for flexural strength for moulded flexure test specimen, determination of modulus of elasticity by means of extensometer

Air-Blast Response of Free-Standing: (1) Unreinforced Brick Masonry Wall, (2) Cavity RC Wall, (3) RC Walls with (i) Bricks, (ii) Sand, in the cavity: A Macro-Modeling Approach

S. M. Anas and Mehtab Alam

Abstract In the current scenario of insecurity from different sources, the role of the compound wall is more important than ever. In addition to stability, the strength of such walls from the explosion point of view is of considerable interest to the engineers. In this paper, the finite element simulations of 6000 mm × 2500 mm 350 mm (length × height × thickness) free-standing compound unreinforced brick masonry wall subjected to experimental blast peak overpressures available in the open literature are performed. A high-fidelity FEM-based software package ABAQUS/CAE has been called for and a continuum/macro-modeling technique is adopted to optimize the computational time. Damage is simulated considering the available concrete-damaged plasticity (CDP) model for the non-linear response of the walls. Dynamic responses including maximum displacement, damage dissipation energy, and crack/damage patterns have been evaluated for the scaled distances of 2.28 and 1.81 m/kg$^{1/3}$. The height of the burst is 1250 mm from the ground. With the objective to improve the blast performance of the wall, a 230 mm thick reinforced concrete (RC) wall with a 70 mm wide cavity is considered in the study. To dissipate the explosion energy, bricks on edge, and sand are also taken as softcore materials in the cavity of the wall. The cavity is interrupted by the cross RC elements at an interval of 1.10 m. Analyses results have been discussed and a novel design of the blast-resistant free-standing wall is recommended.

Keywords Air-blast · Explosion · Reinforced Concrete (RC) wall · Cavity · Bricks · Sand · Peak overpressure · CDP model · Finite element (FE) simulation · Damage dissipation energy · Damage · Cracking

S. M. Anas is a doctoral candidate at the Jamia Millia Islamia (Central University), studying Civil Engineering, with an emphasis in Structural Engineering. His current research interests include blast dynamics, strengthening techniques, FE modeling, structural performance, concrete-steel composite compression members, and composite materials. “Scopus Author ID: 57219929545”.

S. M. Anas (✉) · M. Alam
Department of Civil Engineering, Faculty of Engineering and Technology, Jamia Millia Islamia (Central University), New Delhi, Delhi 110 025, India

G. C. Marano et al. (eds.), *Proceedings of SECON’21*, Lecture Notes in Civil Engineering 171, https://doi.org/10.1007/978-3-030-80312-4_78

1 Introduction

Multi-layered free-standing walls are usually comprised of two or three wythes (or leaves) [1]. When three in number, there are two external wythes called skin (or cladding), and one internal called core [1]. The core is usually of lower quality and may contain a significant ratio of voids. Out-of-plane response of multi-wythe free-standing walls subjected to seismic loading had been investigated employing various experimental and numerical techniques. However, their behavior against man-made disasters, especially against explosive-induced blast loading, is of considerable interest particularly in the prevailing circumstances of countries suffering from internal political rivalry and terrorism. Mazzon and Valluzzi [2] experimentally investigated the seismic response of three-leaf stone masonry walls using a shaking table. The crack patterns were compared and found closely matching with the predictions of the TNO DIANA™ 2005 program. Felice [3] numerically investigated the out-of-plane response of two-layered masonry walls with and without interlocking under quasi-static and dynamic loadings. Results showed that the interlocking between the wythes of the walls plays a significant role in enhancing the out-of-plane load-carrying capacity. Derakhshan et al. [4] investigated the out-of-plane bending effect of two-leaf (230 mm), and three-leaf (350 mm) clay brick unreinforced masonry (URM) walls through an airbag system. Ahmad et al. [5] conducted field tests on free-standing clay brick unreinforced masonry (URM) walls at different standoff distances (between 3 and 4 m); and proposed several empirical relations to estimate blast parameters. Andrea and Nigel [6] examined the failure mechanism of multi-leaf walls of Prince of Wales Fort (Canada) using a discrete element modeling approach available in LMGC90 software. Lateral bulging at mid-height of the walls was observed. Pereira et al. [7] examined the behavior of the clay brick URM wall exposed to blast pressure of 0.149 MPa using a macro-modeling approach. Large cracks were observed, mostly concentrated at the mid-height of the wall.

The present study aims to investigate the out-of-plane air-blast response of 350 mm thick un-reinforced brick masonry wall, 230 mm thick RC wall with a 70 mm wide cavity, and RC walls with bricks, and sand as soft cores in the cavity using a macro-modeling approach. A high-fidelity professional software program, ABAQUS/CAE 2017 has been employed for this purpose. The results of the study give more insight into the understanding of the importance of softcore for the monolithic response of the walls.

2 Finite Element (FE) Modeling

A high-fidelity professional software, ABAQUS/CAE 2017 has been employed in this study to simulate the air-blast response of the walls. Altogether four finite element models have been developed for this purpose. The first model (#1) is consisting of a 350 mm thick unreinforced clay brick masonry wall (Fig. 1a). The second model (#2) is consisting of a 230 mm thick reinforced concrete (RC) wall with a 70 mm wide cavity (Fig. 1b). The cavity is interrupted by the cross RC elements at an interval of 1.10 m. In the third (#3) and fourth (#4) FE models, bricks on edge, and sand are taken as softcore materials in the cavity of the wall, respectively (Fig. 1c). To optimize the running time and accuracy, brick and sand are modeled using a macro-modeling technique.

The walls are discretized with 8-node linear brick explicit elements (C3D8R) with reduced integration and hourglass control [8]. A mesh size of 20 mm has been adopted following the convergence test conducted at scaled distances of 2.28 and 1.81 m/kg$^{1/3}$. The reinforcement details of the 230 mm thick free-standing RC walls are shown in Fig. 2. The compressive strength, Young's modulus, and Poisson's ratio of the concrete are 20 MPa, 21.20 GPa, and 0.20, respectively. The ultimate tensile strength, yield strength, Young's modulus, and Poisson's ratio of the steel are 545 MPa, 500 MPa, 210 GPa, and 0.30, respectively. The thickness of the concrete cover is 25 mm. The mechanical properties of the brick, and sand have been taken from Valente and Milani [9, 10], and Payan et al. [11], respectively. The re-bars are discretized with 2-node explicit linear 3-D truss elements (T3D2) [8]. The re-bars are embedded in the RC wall using the EMBEDDED_REGION constraint command [8]. TIE_CONSTRAINT command has been utilized to model the interaction between the external wythes and the internal core (brick/sand) of the RC walls [8]. FE model #1 has 750,700 nodes and 699,960 elements. FE model #2 has 444,789 nodes and 349,882 elements. FE model #3 has 621,189 nodes and 487,382 elements. FE model #4 has 621,189 nodes and 487,382 elements. The walls are resting on an arbitrary surface which is assumed fixed. The arbitrary surface in the employed software has been modeled using the REFERENCE_POINTS command [8]. A surface-to-surface contact interaction, with "friction", "hard", and "cohesion" contact models and with penalty contact method as mechanical constraint formulation, available in the ABAQUS program has been used to model the interaction between the walls and arbitrary surface. Damage is simulated considering the

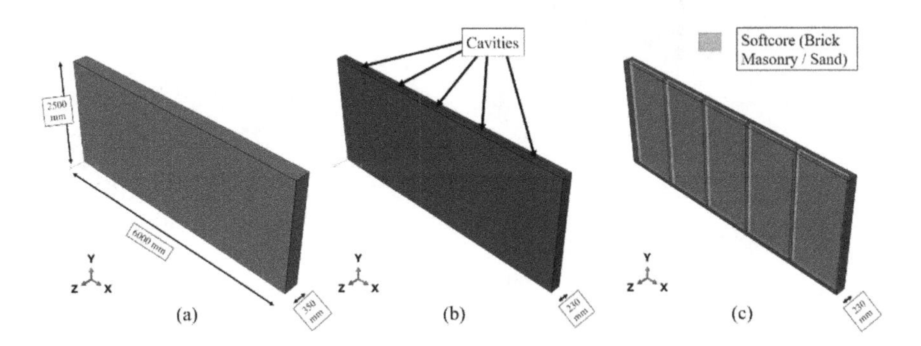

Fig. 1 FE models developed: **a** 350 mm thick URM wall (#1), **b** 230 mm thick RC wall with 70 mm wide cavity (#2), and **c** RC wall with softcore in the cavity (#3 and #4)

available concrete-damaged plasticity (CDP) model for the non-linear response of the walls [9, 10, 12]. CDP model represents the inelastic behaviour of concrete, and works on the isotropic damage elasticity concept. Previous research showed that the CDP model can simulate the cracking, crushing, and spalling of the concrete under blast loading [13–16]. The damage plasticity model considers two types of failure mechanism namely; tensile cracking and compressive crushing represented by two damage variables *dc* and *d*t [1, 17]. The automatic function in ABAQUS/CAE converts uniaxial stress-strain curve into stress versus inelastic strain curve [1]. The uniaxial compression and tensile responses of concrete are demonstrated using statistical equations under compression and tensile loading [1]. Experimental data obtained by Hafezolghorani et al. [12] is used to model the concrete.

2.1 Blast Loading

The structures may be subjected to blast loading, which comprises blast pressure, ground shock, and fragment impact, particularly in the near-range detonation; whereas, in the far-field region, structures are subjected to only blast pressure. In case of nuclear blast, the resulting shock wave produces very high blast pressure and large impulse loading, which results in destruction of structures situated at even a large distance from the source. However, the present study focuses only on explosive-induced air-blast loading rather than nuclear blasts. Air-blast generates a shock wave of very high pressure accompanying high-temperature and supersonic expansion of hot gases [18, 19]. The abrupt rise in the air pressure carried by the Mach stem causes severe damage to the buidling structures [19, 20]. Figure 3a shows the systematic time variation of air-blast pressure, $P(t)$, which is determined by the blast wave arrival time, t_A, rising time, t_1, the peak overpressure, P_{OP}, positive blast phase duration, t_d, duration of the negative blast phase, t^-, and the maximum negative

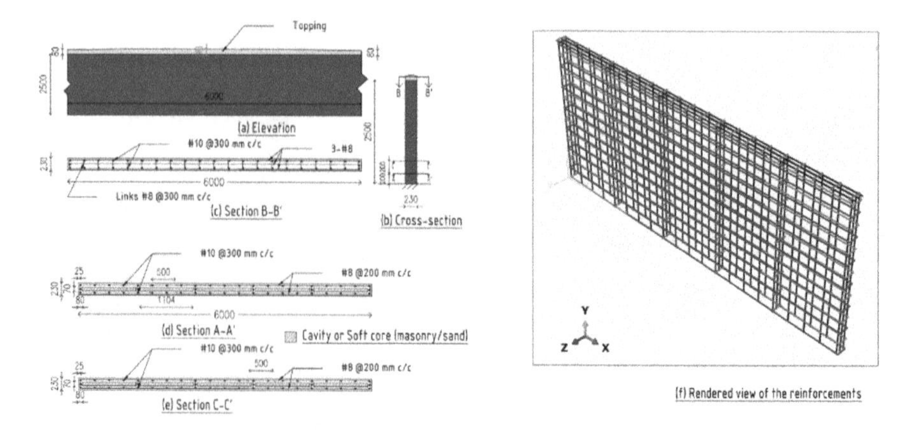

Fig. 2 Dimensions (mm) and reinforcement details of 230 mm thick free-standing RC walls

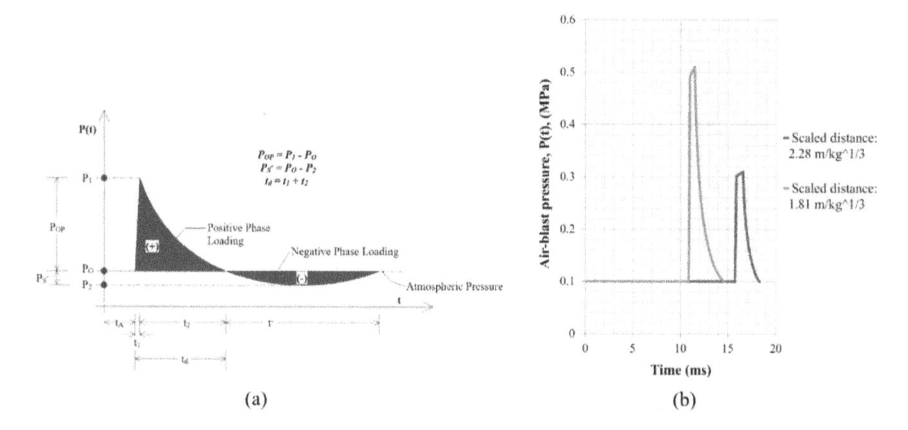

Fig. 3 **a** Time evolution of air-blast pressure due to an explosion acting on a target [21], and **b** estimated blast pressure histories

air-pressure, P_S^-. Air-blast in the ABAQUS/CAE program can be simulated using different approaches/methods. However, in the current study, empirical models based on experimental results available in the open literature have been used. Figure 3b shows the calculated air-blast profile for scaled distances of 2.28 and 1.81 m/kg$^{1/3}$ (i.e., for explosive loads of 3.60 and 7.20 kg TNT at a fixed standoff distance of 3.50 m) in free air. These profiles have been modeled following the experimental results of Ahmad et al. [5], and using the empirical blast model proposed by Wu and Hao [21]. Referring to Fig. 3b, the negative phase of the blast wave has been neglected following the guidelines of current blast design standards [18, 19, 22]. The accuracy of the employed software and used numerical method to simulate the air-blast is discussed in the previously published studies by the authors [13–17].

3 Results and Discussion

From the non-linear explicit blast analyses conducted, the following observations are worth mentioning:

- The maximum displacement, damage dissipation energy (DDE), and the maximum average horizontal crack depth at mid-height of the 350 mm thick URM wall (FE Model #1) are found to be 67.46 mm, 4456.93 J, and 260 mm, respectively, under the explosive load of 3.60 kg TNT at a detonation distance of 3.50 m in free air (Fig. 4). Increasing the yield of the explosive by a factor of 2 leads to an increase of the maximum displacement, damage dissipation energy, and average depth of horizontal cracks by factors of 1.50, 4.00, and 1.20, respectively (Table 1). The increase in the DDE indicates that the wall suffers more damage and cracking.

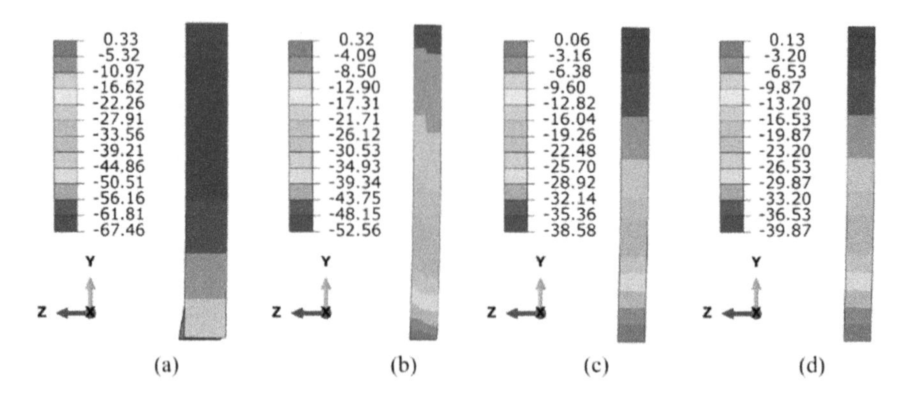

Fig. 4 Distribution of Z-displacement (mm) of different FE models for a scaled distance of 2.28 m/kg$^{1/3}$ (t = 18.30 ms): **a** 350 mm thick URM wall (#1), **b** 230 mm thick RC wall with 70 mm wide cavity (#2), **c** 230 mm thick RC wall with bricks on edge in the cavity (#3), and **d** 230 mm thick RC wall with sand in the cavity (#4)

Table 1 Summary of maximum displacement (mm) for a constant standoff distance of 3.50 m

W (kg TNT)	Z (m/kg$^{1/3}$)	Maximum transverse displacement (Z, mm) of different wall models			
		350 mm thick URM wall (#1)	230 mm thick RC wall with 70 mm wide cavity (#2)	230 mm thick RC wall with bricks in the cavity (#3)	230 mm thick RC wall with sand in the cavity (#4)
3.60	2.28	67.46	52.56 ([a]22)	38.58 ([a]43)	39.87 ([a]41)
7.20	1.81	97.44	92.99 ([a]5)	55.43 ([a]43)	61.11 ([a]37)

[a]Percentage decrease (%) in displacement with respect to 350 mm thick URM wall

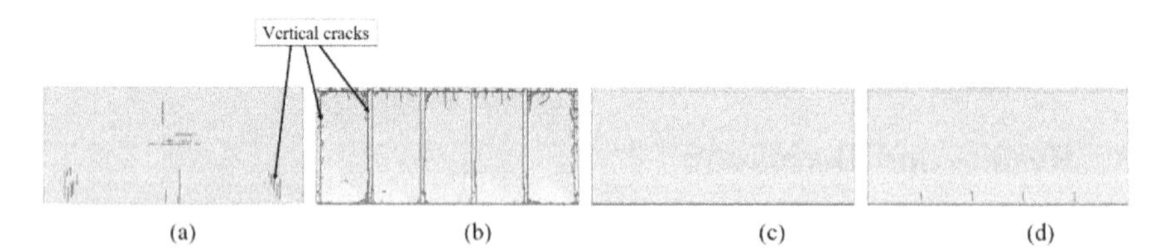

Fig. 5 Formation of cracks on the blast face (facing the explosion) of different wall models for a scaled distance of 2.28 m/kg$^{1/3}$ (t = 18.30 ms): **a** #1, **b** #2, **c** #3, and **d** #4

- The effect of the air-blast (or blast shock front) on the damage resistance of the 350 mm thick URM wall transforms the localized failure (i.e., development of closely-spaced cracks on the rear face of the wall) to globalized failure (i.e., the formation of the crater) under the considered quantities of the explosive (Figs. 5, 6 and 7). The average depth and size of the crater on the rear face of the wall are estimated to be 310 and 900 mm, respectively (Fig. 7).
- Comparing the deflection response of the 350 mm thick URM wall (FE model #1) with the 230 mm thick cavity RC wall (FE model #2) under the maximum

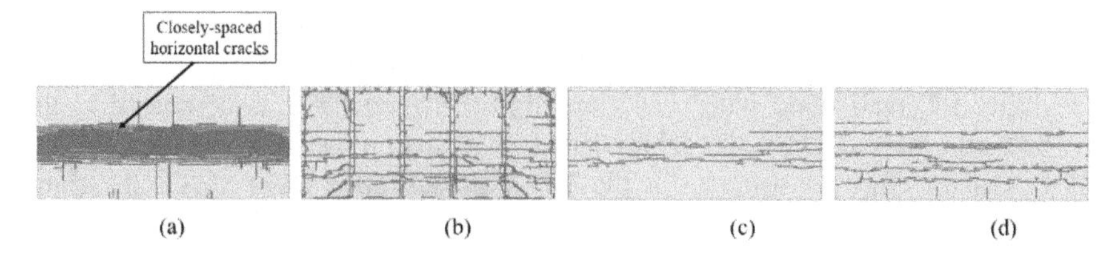

Fig. 6 Formation of cracks on the rear face of different wall models for a scaled distance of 2.28 m/kg$^{1/3}$ (t = 18.30 ms): **a** #1, **b** #2, **c** #3, and **d** #4

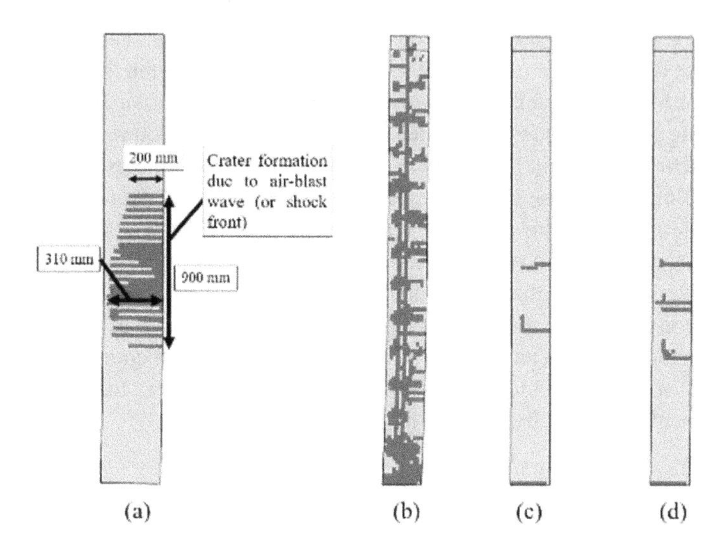

Fig. 7 Formation of cracks developed on the cross section of wall models for a scaled distance of 2.28 m/kg$^{1/3}$ (t = 18.30 ms): **a** #1, **b** #2, **c** #3, and **d** #4

Table 2 Summary of damage dissipation energy (J)

Z (m/kg$^{1/3}$)	Damage dissipation energy (J) for different wall models			
	350 mm thick URM wall (#1)	230 mm thick RC wall with 70 mm wide cavity (#2)	230 mm thick RC wall with bricks in the cavity (#3)	230 mm thick RC wall with sand in the cavity (#4)
2.28	4456.93	18,954.73 ([a]- >> 100)	2233.90 ([a]50)	3070.78 ([a]31)
1.81	17,278.72	71,467.13 ([a]->> 100)	8268.36 ([a]52)	11,576.74 ([a]33)

[a]Percentage decrease (%) in damage dissipation energy with respect to URM wall

blast load of 7.20 kg TNT, a reduction of 5% in the maximum displacement of the RC wall is found (Table 1). However, the RC wall experiences severe damage and cracking (Table 2 and Figs. 5–7).

- A significant amount of reduction in the maximum displacement (43%) as well as the damage dissipation energy (51%) has been observed in the 230 mm thick cavity RC wall filled with bricks on edge (FE model #3) as compared to the 350 mm thick URM wall (Tables 1 and 2). Hence, it can be concluded that the bricks on edge as softcore in the RC wall acts as a blast shock-absorbing material and makes the wall display a much better response than the URM wall as well as cavity RC wall against blast loading.
- A reduction of 39% in the maximum displacement and 32% in the damage dissipation energy has been observed in the 230 mm thick RC wall with sand as softcore in the cavity (FE model #4) as compared to the 350 mm thick URM wall (FE model #1) under the considered explosive loads (Tables 1 and 2).
- The use of the bricks on edge as softcore in the cavity of the 230 mm thick RC wall is found more effective with regards to damage dissipation energy and cracking under the considered yields of the explosive in comparison to the sand as softcore material in the cavity (Table 2 and Fig. 7). The reason why bricks as infilled material in the cavity is more effective to the local damage than the sand attributes to the higher value of young's modulus of bricks (approximately 18 times larger than that of sand).
- No crater has been observed in the considered RC wall models (Fig. 7).
- For a constant detonation distance (3.50 m), increased yield of the explosive by a factor of 2 leads to an increase of the damage dissipation energy of FE models #1, #2, #3, and #4 by a factor of 4 (Table 2). However, the maximum displacement increases by a factor of 1.50 (Table 1).

4 Conclusion

Out-of-plane response of the free-standing: (1) unreinforced brick masonry wall, (2) cavity RC wall, (3) RC walls with (i) bricks on edge, (ii) sand, in the cavity has been investigated using the ABAQUS/CAE, a high-fidelity professional software package. The cavity is interrupted by the cross-vertical RC elements at an interval of 1.10 m. The walls are subjected to explosive loads of 3.60 and 7.20 kg TNT at a constant standoff distance of 3.50 m in free air. The 230 mm thick RC walls with soft cores (bricks/sand) in the cavity subjected to blast loading show that most of the energy imparted by the air-blast is absorbed by the softcore materials, and a significant reduction in the maximum displacement, damage dissipation energy, and cracking has been observed with respect to the 350 mm thick URM wall, and the 230 mm thick RC wall without infilled materials. However, the RC wall with bricks on edge in the cavity is found to be the most superior to all other walls considered in the study with regards to maximum displacement, damage, and cracking under the considered explosive loads.

It is well established that increased standoff distance is the best mitigation measure against explosive-induced threats. However, in congested localities, a large standoff distance is not possible. At such places, the provision of such free-standing RC blast walls with bricks on edge as softcore provides a better option to reduce explosive-induced threats. Further, the provision of a brick-filled RC wall may serve as a sacrificial compound wall around the buildings against a higher yield of explosive detonation and protect the building from such blast.

Acknowledgements The authors wish to acknowledge Prof. T. K. Datta, Department of Civil Engineering, IIT-Delhi, India, for their precious suggestions and discussions concerning the whole numerical blast simulations presented in this study.

References

1. Szyszka M, Jasienko J, Ochsendorf AJ (2018) Out-of-plane behavior of two-layered free-standing walls: analytical solutions and small-scale tests. Eng Struct 163:11–24
2. Mazzon N, Valluzzi RM (2010) Influence of grout injection on the dynamic behavior of stone masonry buildings. Ph.D. thesis, Università degli Studi di Padova, Veneto, Italy, pp 1–272
3. Felice DG (2011) Out-of-plane seismic capacity of masonry depending on wall section morphology. Int J Archit Heritage 5:466–482
4. Derakhshan H, Griffith CM, Ingham MJ (2013) Airbag testing of multi-leaf unreinforced masonry walls subjected to one-way bending. Eng Struct 57:512–522
5. Ahmad S, Elahi A, Pervaiz H, Rahman AGA, Barbhuiya S (2014) Experimental study of masonry wall exposed to blast loading. Mater De Constr 64(313):1–11
6. Andrea I, Nigel S (2015) Discrete element modeling of stone masonry walls with varying core conditions: prince of wales fort case study. Int J Archit Heritage Conserv Anal Restor 9 (5):564–580
7. Pereira MJ, Campos J, Lourenco BP (2015) Masonry infill walls under blast loading using confined underwater blast wave generators (WBWG). Eng Struct 92(1):69–83
8. ABAQUS/CAE FEA program (2017) Concrete-damaged plasticity model, explicit solver, three dimensional solid element library. ABAQUS DS-SIMULIA User Manual
9. Valente M, Milani G (2016) Non-linear dynamic and static analyses on eight historical masonry towers in the north-east of Italy. Eng Struct 114(1):241–270
10. Valente M, Milani G (2016) Seismic assessment of historical masonry towers by means of simplified approaches and standard FEM. Constr Build Mater 108(1):74–104
11. Payan M, Khoshini M, Chenari JR (2020) Elastic dynamic young's modulus and poisson's ratio of sand-silt mixtures. J Mater Civ Eng ASCE 32(1):1–13
12. Hafezolghorani M, Hejazi F, Vaghei R, Jaafar BSM, Karimzade K (2017) Simplified damage plasticity model for concrete. Struct Eng Int 27(1):68–78
13. Anas SM, Ansari MdI, Alam M (2020) Performance of masonry heritage building under air-blast pressure without and with ground shock. Aust J Struct Eng 21(4):329–344
14. Anas SM, Alam M, Umair M (2020) Performance of one-way concrete slabs reinforced with conventional and polymer re-bars under air-blast loading. In: Chandrasekaran S, Kumar S, Madhuri S (eds) Recent Advances in Structural Engineering. Lecture Notes in Civil Engineering, vol 135. Springer, Singapore. https://doi.org/10.1007/978-981-33-6389-2_18
15. Anas SM, Alam M, Umair M (2020) Performance of one-way composite reinforced concrete slabs under explosive-induced blast loading. In: IOP Conference Series: Earth and Environmental Science, vol 614, 1st International Conference on Energetics, Civil and Agricultural Engineering 2020 14–16 October 2020, Tashkent, Uzbekistan, https://doi.org/10.1088/1755-1315/614/1/012094

16. Anas SM, Ansari Md I, Alam M (2021) A study on existing masonry heritage building to explosive-induced blast loading and its response. Int J Struct Eng (Article in press)
17. Anas SM, Alam M, Umair M (2021) Experimental and numerical investigations on performance of reinforced concrete slabs under explosive-induced air-blast loading: a state-of-the-art review. In: Structures, vol 31. Elsevier, pp 428–461. https://doi.org/10.1016/j.istruc.2021.01.102
18. Hao H, Hao Y, Li J, Chen W (2016) Review of the current practices in blast-resistant analysis and design of concrete structures. Adv Struct Eng 19(8):1193–1223
19. TM 5-1300 (1990) Structures to resist the effects of accidental explosions. Technical Manual, Joint Department of the Army, the Navy, and the Air Force, US
20. IS 4991 (1968) Criteria for blast resistant design of structures for explosions above ground. Bureau of Indian Standards, New Delhi
21. Wu C, Hao H (2005) Modeling of simultaneous ground shock and airblast pressure on nearby structures from surface explosions. Int J Impact Eng 31(6):699–717
22. Goel DM, Matsagar AV (2014) Blast-resistant design of structures. Practice periodical on structural design and construction. ASCE 19(2):040140071-9

CFD Simulation for Wind Loads on Multi–span Flat Canopy Roofs

Ajay Pratap and Neelam Rani

Abstract This paper presents the numerical simulations on multi-span flat canopy roof using Computational Fluid Dynamics (CFD) technique. Experiment has been performed for the single span canopy structure in the open circuit boundary layer wind tunnel at IIT Roorkee, India in which 1:50 reduced scaled model size of 300 mm × 150 mm × 75 mm is used. The commercially available ANSYS FLUENT software is used for the CFD simulation. Realizable k-ε turbulent model is used for the pressure and velocity distribution around the building. Atmospheric boundary layer velocity profile obtained from the numerical studies are validated with the experimental studies. The pressure coefficient obtained from the CFD simulations on the upper and lower surface of flat canopy roof at 0°, 45°, 60° and 90° wind incidence angle are compared with the experimental results. Further, pressure coefficients are measured for multi span flat canopy roofs at varying wind incidence angle. It was concluded that multi-span canopy roof surface has low-pressure coefficients and better chances of survival than the single span canopy roof surface building.

Keywords CFD · Flat canopy roof · k-ε turbulence model · Pressure coefficients

1 Introduction

Canopy structure are supported by columns only with no wall cladding having different roof shapes such as flat, mono-slope, circular or trough type. Reference [1] reviewed on the wind load on the free-standing canopy roof for mono-slope and duo pitch roof. Reference [2] determined the behavior of lift force and drag force for a range of the porous hip, gable and mono-slope canopy roof forms. The codal

A. Pratap (✉) · N. Rani
Dr. B. R. Ambedkar National Institute of Technology, Jalandhar, India
e-mail: ajayp.ce.19@nitj.ac.in

N. Rani
e-mail: ranin@nitj.ac.in

G. C. Marano et al. (eds.), *Proceedings of SECON'21*, Lecture Notes in Civil Engineering 171, https://doi.org/10.1007/978-3-030-80312-4_79

information available by various countries [3–5] on the flat canopy roofs is very limited.

Computational fluid dynamics now a days is used to evaluate the effect of wind on structures as an alternative to wind tunnel testing [6, 7]. Numerical simulation technique CFD is used to calculate the complex behaviour of fluid flow by solving the governing Navier–Stokes equations for the fluid flow. From literature it is observed that most of the numerical and experimental studied has been performed on low rise buildings with single span roof and very few studies have been performed on the multi span canopy roof [8, 9]. Numerical validation of pressure coefficient of single span flat canopy roof is performed in this paper. Further the behaviour of wind on single, double and three span flat canopy roof structure for different wind incidence angle compared with each other.

2 Experimental Setup

The experiments are carried out in open circuit boundary layer wind tunnel at the Department of Civil Engineering, Indian Institute of Technology Roorkee, India. Prototype building with flat roof have rectangular plan of size 15 m × 7.5 m and eaves height above the ground as 3.75 m. Model of flat canopy roof is made of Perspex sheet at a scale of 1:50 having plan dimension of 300 mm × 150 mm and 75 mm height (H) is used for the experiment of single span flat canopy roof [10]. Power-law velocity profile is used in the atmospheric boundary layer (ABL) to define the velocity profile inside the wind tunnel which is shown below see "Eq. 1":

$$v = v_0 \left(\frac{z}{z_0} \right)^{\alpha} \tag{1}$$

where, Z is the height above ground level; v denotes the mean wind speed at a height of Z meter above ground; v_0 denotes the mean wind speed at the reference height of Z_o m above ground and α is the power law coefficient. Wind pressure are measured at all pressure points on both top and bottom surfaces of surfaces of the model for 0°, 45°, 60° and 90° wind incidence angles see Fig. 1. From "Eq. 2", mean pressure coefficient is calculated from the measured pressure values as:

$$C_{P_{mean}} = P_{measured}/0 \cdot 6(v)^2 \tag{2}$$

where, $P_{measured}$ = wind pressure measure at pressure points in N/m^2, v = reference wind velocity at the eave height of the model 6.21 m/s.

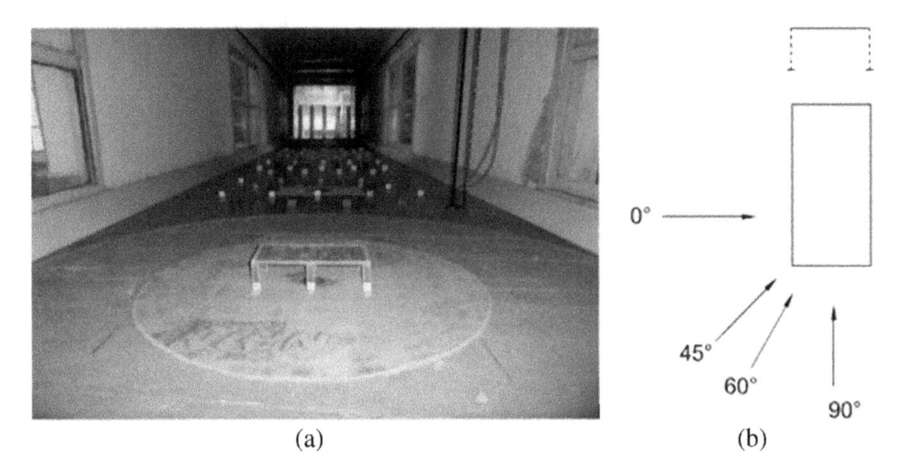

Fig. 1 **a** Single span flat canopy roof model at 0° wind incident angle inside the wind tunnel and **b** different wind incident angle representation on model

3 Numerical Simulation

Computational Fluid Dynamics technique is used for the numerical simulation of wind flow on the building models. Commercially available Ansys Fluent software is used to perform the steady Reynolds Average Navier–Stokes's computations based on the Finite volume method approach for solving flow equations.

3.1 Computational Domain and Meshing

In the pre-processing step the geometry of flat canopy roof structure is created as the solid body and the computational domain is created as the fluid body. The dimension of solid body model is selected according to experimental model dimensions. For the flow of air inside the domain an inlet surface is created and for the outflow of air outlet surface is created. References [11, 12] provided the minimum spacing of model from the inlet is 5H to satisfy the atmospheric boundary flow and to avoid backflow at the outlet spacing should be greater than 15H. The dimensions of the computational domain are i.e., $L \times B \times H = 1.6 \times 1.1 \times 0.6\ m^3$ as shown schematically in Fig. 2.

Tetrahedral element method is used for the meshing of the computational domain and for the surface of canopy. The mesh size has been kept very fine for the model in order to provide the adequate fluid flow at high gradient region. The mesh should be sufficiently fine to achieve the numerical accuracy and resolved the important flow features. The size of mesh for the model is 2 mm for the entire surfaces and 30 mm for the surface of computational domain (see Fig. 3).

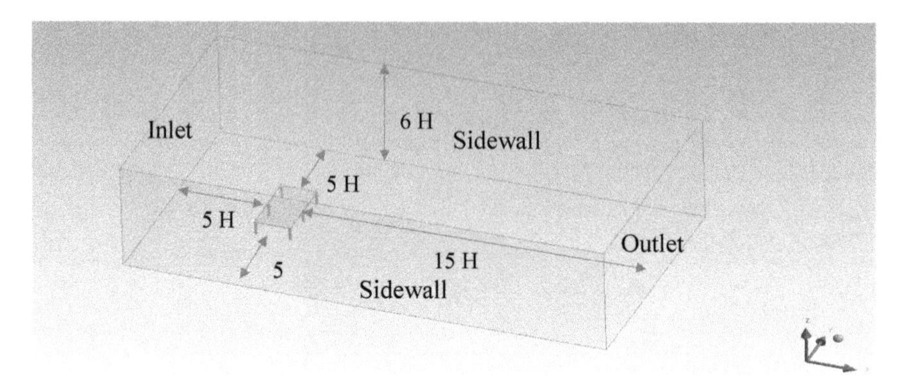

Fig. 2 Isometric view of computational domain

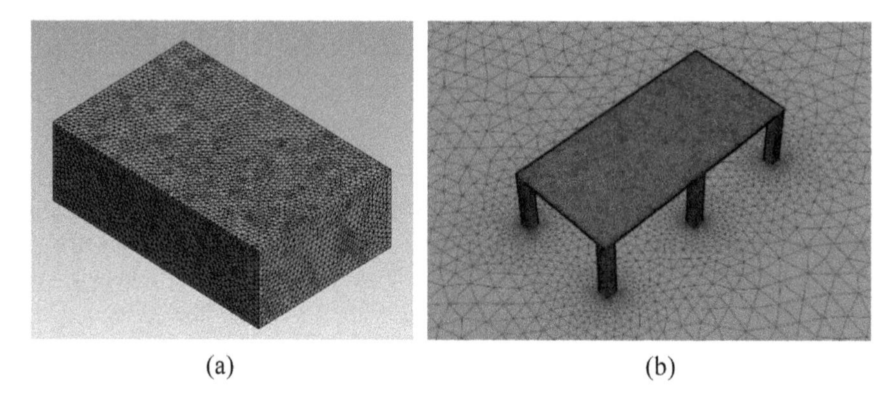

Fig. 3 Isometric view of **a** domain meshing and **b** model meshing

3.2 Turbulence Model

The fluid flowing inside domain is assumed to be turbulent and to study the effect of turbulence there are different types of model available in the ANSYS FLUENT Solver. As per the literature the most suitable model used to solve the turbulence is Realizable k-ε turbulence model suggested by [13] is very useful in estimating flow separation, reattachment and recirculation of flow around the structure. This model consists of two additional transport equation which solve the turbulent kinetic energy (k) and turbulent dissipation rate (ε).

3.3 Boundary Condition

For the actual behavior of fluid flow inside the fluid domain adequate boundary conditions should be provided. The inlet surface of the domain is used as the windward side and the velocity profile is used similar to the wind tunnel experiment. The velocity and turbulence intensity profiles found from the wind tunnel experiment has been utilized in the numerical simulation [10]. Curve fitting technique is used in excel for the experimental velocity and turbulence intensity profile which is used to define the user defined function (UDF). Logarithmic trend line for the velocity profile $V = 1.8679ln(Z) + 10.43$ and exponential trend line for the turbulence intensity (T.I) profile $I = 0.024 \times (Z) - 0.451$ is obtained from the curve fitting. The measured velocity and turbulence intensity (T.I) profiled associated with the k-ε turbulence equation model are proposed to define turbulence kinetic energy (k) and turbulence dissipation rate (ε) equation profiles as given by [14]. The defined UDF is compiled and used as the velocity-inlet in assigning boundary at inlet. On the leeward side of domain, an outlet pressure is used, in which the relative pressure is specified at 0 Pa and for to avoid backflow K and ε UDF is set similar to those used in inlet. The boundary constraint used for the side and top walls of domain as slip condition and no slip condition is used for the surfaces of the model and bottom wall of the domain.

3.4 Solver Setting

Finite volume method technique is selected for CFD simulation in Ansys Fluent to solve the governing differential equation at various nodes of model and to solve the problems associated with the applied boundary conditions for the fluid domain. Second order differencing schemes is used for solving the pressure, momentum and turbulence equations generated due to fluid flow inside the domain and "coupled" pressure–velocity coupling approach is used to solve single-phase flow problems due to its robustness for steady-state The residuals fell below the commonly applied criteria of falling to 10^{-4} of their initial values after several hundred iterations [15].

4 Numerical Validation

For the post processing unit, CFD post is used in ANSYS Fluent CFD simulation. The velocity profile obtained from the CFD simulation near the model is validated with the experimental measured profiles and the values are found close to each other as represented in Fig. 4. The wind pressure coefficient contours values obtained from the numerical studies is validated with the experimental values for 0°, 45°, 60° and 90° wind incident angles for top and bottom surfaces of single span

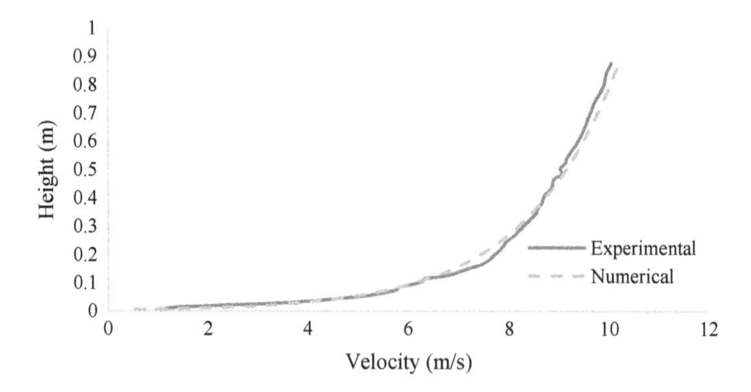

Fig. 4 Experimental and numerical validation of velocity profile inside wind tunnel

flat canopy models as shown in Figs. 5, 6, 7 and 8. From the numerical and experimental studies, it is found that maximum wind pressure coefficient generated at the windward surface and decreases as it moves towards leeward surface edge of the model. The pressure coefficient values on the top roof surface are greater as compared to lower surfaces because of the formation of strong wake region on the top surface of the roof. The wind turbulence effect is more at the edges of the windward side when wind strikes the structure and forms a wake region which produces negative wind pressure coefficient i.e., suction pressure. It is observed from numerical and experimental studies that the maximum negative pressure coefficient i.e., suction effect is more when wind flows with an angle of 45° (Cp = −2.25) and 60° (Cp = −2.31) as compared to 0° and 90° because of high turbulence is produced for oblique wind flow. The contour line of pressure coefficient formed at the windward edges of roof surfaces are denser and for oblique wind flow pressure coefficient are denser at the adjacent edges and less dense towards leeward surface side.

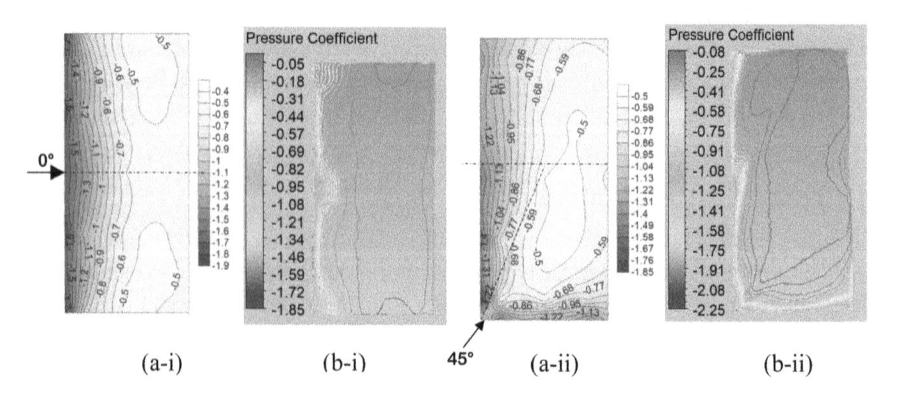

Fig. 5 Wind pressure coefficient contours obtained **a** experimentally **b** numerically on upper surface of single span canopy roof at (i) 0° and (ii) 45° wind incidence angle

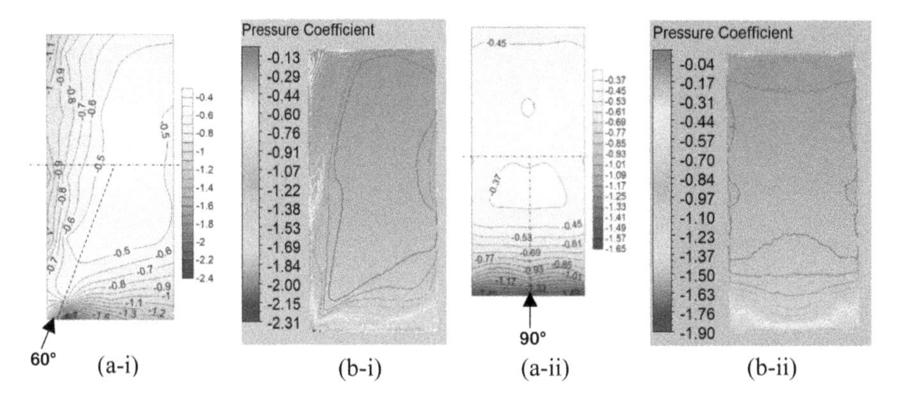

Fig. 6 Wind pressure coefficient contours obtained **a** experimentally **b** numerically on upper surface of single span canopy roof at (i) 60° and (ii) 90° wind incidence angle

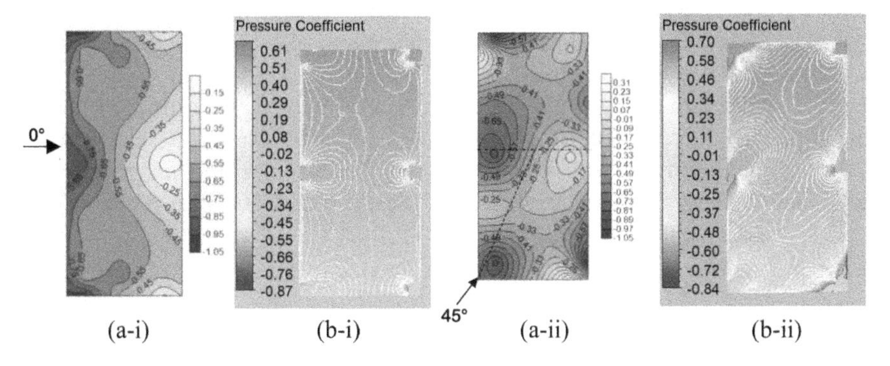

Fig. 7 Wind pressure coefficient contours obtained **a** experimentally **b** numerically on lower surface of single span canopy roof at (i) 0° and (ii) 45° wind incidence angle

5 Wind Effect Study on Multi-span Canopy Roof Surfaces

This study deals only with the effect of wind on the top surface of multi-span flat canopy roof structure for 0°, 45°, 60° and 90° wind angle and compared with each other and with the experimental observation on the single span canopy roof structure. Multi-span structure consists of two and three span canopy roof structure adjacent to each other with no spacing and the effect of wind flowing on this structure has been studied. The maximum value of negative pressure coefficient (suction) obtained on the upper surfaces of canopy structure from the numerical studies is shown in the form of graphical representation see Fig. 9. For the model with 0° wind angle, the edges of the windward surface show the highest negative pressure (suction) ($Cp = -1.85$) for single span canopy roof surface. For the 45° and 60° wind angle the negative pressure increases as compared to 0° wind angle

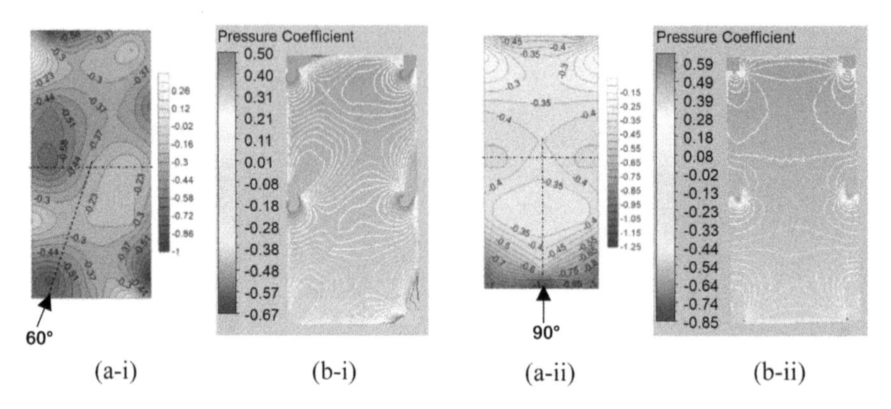

Fig. 8 Wind pressure coefficient contours obtained **a** experimentally **b** numerically on lower surface of single span canopy roof at (i) 0° and (ii) 45° wind incidence angle

for single and multi-span model. The maximum negative pressure coefficient (Cp = −2.54) value is obtained at the corner edges of two span canopy towards windward surfaces and least pressure coefficient is observed for three span model. For the model with a 90° wind angle, the edges of the windward surface show the highest negative pressure coefficient (Cp = −1.90) in case of single span while for the remaining span roof surfaces the pressure coefficient values decreases and a minimum negative pressure coefficient (Cp = −1.69) is observed for three span

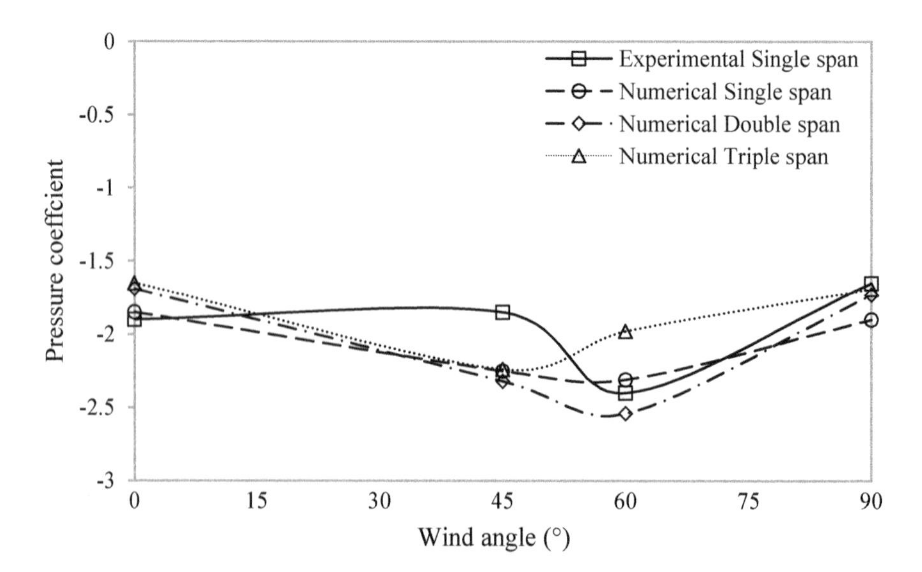

Fig. 9 Experimental and numerical result of wind pressure coefficient on upper surface for single, double, three span canopy roofs for wind incidence angle 0°, 45°, 60° and 90°

canopy roof surface model. For all wind incident angle on the surfaces of flat canopy roof model negative pressure coefficient is observed from windward surfaces to leeward surfaces and pressure coefficient values decrease respectively.

6 Conclusions

The effect of wind on the surface of single, double and three span canopy roofs for the wind incident angle of 0°, 45°, 60° and 90° have been investigated in this study to determine the wind pressure coefficient. The realizable k–ε turbulence modelling technique applied numerical simulation using CFD approach holds in good agreement with the experimental result and capable of simulating turbulence effect inside wind tunnel. It is found from numerical and experimental studies of single span flat canopy roof that the maximum negative pressure coefficient i.e., suction increases by approximately 20% when wind flows obliquely at an angle of 45° and 60° as compared to 0° and 90°. The high suction effect is developed at the windward edge of both top and bottom roof surfaces for each wind incidence angles. The pressure coefficient values decrease by more than 35% for lower surfaces when compared with the values on the top surface of flat canopy roof. The pressure coefficient contour line formed on the windward edges of roof surfaces are denser and for oblique wind flow pressure coefficient are denser at the adjacent edges and less dense towards leeward surface side. As the number of spans increases the maximum wind pressure coefficient value is observed to be decreased by approximately 18% for 60° wind incident angle in case of three span flat canopy roof and the suction pressure value on it is also low for remaining wind incidence angle. There is reduction in the wake region formation in case of multi-span canopy roof. Multi-span flat canopy roof surface has less value of pressure coefficients so, its chances of damage is very less against wind load comparing to the single span canopy roof surface building.

References

1. Uematsu Y, Stathopoulos T (2003) Wind loads on free-standing canopy roofs: a review. J Wind Eng 28(2):95_245–95_256
2. Letchford CW, Row A, Vitale A, Wolbers J (2000) Mean wind loads on porous canopy roofs. J Wind Eng Ind Aerodyn 84(2):197–213
3. Minimum Design Loads for Buildings and Other Structures (2013) ASCE Stand Am Soc Civ Eng New York 7–98
4. BSI BS 6399 (1997) Loading for buildings-Part 2. Code Practice Wind Loads, Branch of Standandard Institute
5. Bureau of Indian Standards (2015) IS 875: code of practice for design loads (other than earthquake) for building and structures—Part 3: Wind loads. BIS, New Delhi, p 51

6. Roy AK, Singh J, Sharma SK, Verma SK (2018) Wind pressure variation on pyramidal roof of rectangular and pentagonal plan low rise building through CFD simulation. Int Conf Adv Constr Mater Struct (February 1–10)
7. Singh J, Roy AK (2019) CFD simulation of the wind field around pyramidal roofed single-story buildings. SN Appl Sci 1(11):1425
8. Case PC, Isyumov N (1998) Wind loads on low buildings with 4: 12 gable roofs in open country and suburban exposures. J Wind Eng Ind Aerodyn 77–78:107–118
9. Stathopoulos T, Suresh Kumar K, Mohammadian AR (1996) Design wind pressure coefficients for monoslope roofs: a time series approach. J Wind Eng Ind Aerodyn 65(1–3):143–153
10. Rani N (2018) Wind pressure distribution on multi-span canopy roofs. Ph.D. thesis, Department of Civil Engineering, I.I.T. Roorkee, Roorkee, India
11. Franke N et al (2004) Recommendations on the use of cfd in wind engineering. In: Proceedings of the international conference on urban wind engineering and building aerodynamics, pp C.1.1–C1.11
12. Revuz J, Hargreaves DM, Owen JS (2012) On the domain size for the steady-state CFD modelling of a tall building. Wind Struct 15(4):313–329
13. Shih T-H, Liou WW, Shabbir A, Yang Z, Zhu J (1995) A new k-ϵ eddy viscosity model for high Reynolds number turbulent flows. Comput Fluids 24(3):227–238
14. Richards P, Hoxey R (1993) Appropriate boundary conditions for computational wind engineering models using the k-ϵ turbulence model. J Wind Eng Ind Aerodyn 46–47 (145–153)
15. Fouad NS, Mahmoud GH, Nasr NE (2018) Comparative study of international codes wind loads and CFD results for low rise buildings. Alexandria Eng J 57(4):3623–3639

Performance Evaluation of Masonry Infill Walls in Reinforced Concrete Frames Under Cyclic Loading Using Applied Element Method

Hemanth Kumar Karaka and Rajesh Kumar Tripathi

Abstract Open ground storey in Reinforced concrete framed buildings is considered to be the most vulnerable when it is subjected to dynamic excitation because of the formation of column sway mechanism and then leads to sudden failure of the ground storey. They must be avoided in severe earthquake zones. Many of the researcher's done experiments for the cause of failure in the ground storey and recommended increasing the stiffness and strength in the soft storey for the lateral load resistance by adding a non-structural element such as Masonry infill walls on all sides of the building. It is very complex to study RC frames' behaviour with Masonry infill walls because of different material properties. Numerical tool viz., Finite Element Method used for such complex structural behaviour requires more time for simulation and accuracy is quite acceptable but deviate from the experimental results. For this reason, a powerful numerical tool such as the Applied Element Method (AEM) has been considered. In this paper, one bay-one storey and two bay-two storey half-scale RC frames with and without infill walls under in-plane cyclic loading have been modelled using AEM. In conclusion, the results of AEM models have been compared with experimental results done by researchers. The obtained AEM results are well accepted with experimental results. The study showed that the stiffness, lateral load resistance and strength were improved for masonry infilled RC frame, compared with bare RC frame in both Single-storey and Multi-storied frames.

Keywords Applied element method · Displacement controlled cyclic loading · Lateral load resistance · Displacement ductility factor · Displacement ductility · Cracked stiffness

H. K. Karaka (✉) · R. K. Tripathi
Department of Civil Engineering, National Institute of Technology Raipur, Raipur, Chhattisgarh, India
e-mail: khkumar.phd2019.ce@nitrr.ac.in

R. K. Tripathi
e-mail: rktripathi.ce@nitrr.ac.in

G. C. Marano et al. (eds.), *Proceedings of SECON'21*, Lecture Notes in Civil Engineering 171, https://doi.org/10.1007/978-3-030-80312-4_80

1 Introduction

Analytically, the masonry structure behaviour is quite complex because of the heterogeneity in construction. Several numerical modelling techniques were developed to study the crack patterns in masonry walls. There are several limitations in numerical methods to approximate the behaviour of the infilled wall structure. Several numerical procedures were adapted to reduce approximation levels. In this context, the most popular method, i.e., the Finite Element Method (FEM), implemented for the masonry structures and constitutive laws for nonlinear behaviour in account with the material homogenisation process. Continuum idealisation, or macro-modelling, and system discontinuity, or micro-modelling, are two numerical approaches that are mainly used. In Roca et al. [1], the overview of these methods has been presented. Although, there are several extensions and hybrid methods developed for FEM, it is very difficult to achieve the element separation, rotation. The formulation for the distinct nature of masonry was started in the early seventies [2]. In the Discrete Element Method (DEM), the elements that can be modelled as rigid, partial-rigid, or deformable were considered. Unlike FEM, DEM elements remeshing is not required at the stress concentration region [3]. Cundall and Hart [4], the classical DEM have two main hypotheses, i.e. Element finite displacement and rotations, element contact or collision. In Lourenço [5], recent improvements of various methods such as MDEM, EDEM, DEM compared with FEM were depicted. The collapse behaviour of structures can be achieved in DEM. Because of this feature, many DEM related applications were developed. However, attaining real collapse behaviour of structures is difficult because inefficiency of damping factors. The examples of successful applications of DEM for earthquake analysis of masonry structures were well presented [6, 7] in their research.

Recently the Non-Smooth Contact Dynamics method adapted for the masonry structures under in-plane dynamic loading to study the behaviour [8, 9].

The rigid body spring model (RBSM) [10] partially overcome these issues in DEM. In this method, a masonry elements assembly is assumed as rigid elements connected with normal and shear springs. The structural response can be achieved by the deformation of springs. Although RBSM is well handed for static problems, dynamic analysis can also be applied; under several assumptions, adequate results can be obtained.

Meguro and Tagel-Din [11] came with a similar approach named as Applied Element Method (AEM), in which rigid elements connected by material springs. The stiffness of the springs in both normal and shear directions need to be calculated for the material behaviour. In the AEM [12], analysis can be carried out until the complete collapse of a structure. In RBSM, it is not possible to follow the complete collapse behaviour of the structure.

In the current research, Single-storey and Multi-storey RC bare frames with masonry infill walls were considered to study the behaviour of structures using the AEM.

2 Discretisation of Masonry Elements in Applied Element Method

According to Malomo [13], assembly by zero thickness springs of masonry elements as rigid units. In between the masonry elements, unit interface mortar properties are lumped. The simplified micro-modelling strategy of masonry with mortar shown in Fig. 1a.

The arrangement of interface springs and unit springs in series, stiffness k_{ni} and k_{si} shown in Eq. (1). k_{nu} and k_{su}, are unit deformation stiffnesses of springs connected between the masonry elements as shown in Eq. (2),

$$k_{nu} = \left(\frac{l_i - t_{mo}}{E_u dt_u} + \frac{t_{mo}}{E_{mo} dt_u}\right)^{-1}, \quad k_{su} = \left(\frac{l_i - t_{mo}}{G_u dt_u} + \frac{t_{mo}}{G_{mo} dt_u}\right)^{-1} \tag{1}$$

$$k_{ni} = \frac{E_u - dt_u}{l_u}, \quad k_{si} = \frac{G_u - dt_u}{l_u} \tag{2}$$

where l_i is the distance between the centroidal nodes of interface elements, l_u is the distance between the centroidal nodes of masonry elements, t_{mo} is the mortor bond thickness, t_u is the unit thickness of mortor, d is spacing of transverse springs, and E_u, E_{mo} are Young's modulus of unit mortar and actual mortar respectively, G_u and G_{mo} are Rigidity modulus of unit mortar and actual mortar, respectively.

3 Experimental Data

Ahmed Sayed [14] was conducted the experimental work on four half scaled Single-storey Reinforced Concrete frame specimens such as bare frame, hollow red brick infill wall of thickness 120 mm, hollow red brick infill wall of thickness of

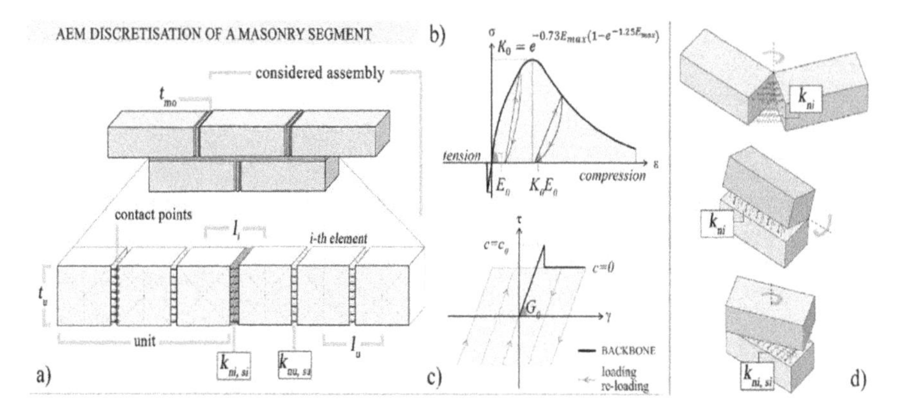

Fig. 1 Masonry assembly in AEM: **a** AEM discretisation of masonry elements, **b** σ-ε relationship of normal springs, and **c** σ-ε relationship of shear springs in cyclic loading, **d** bending and twisting of elements [13]

60 mm and cement bricks infill wall of thickness 120 mm. From his study, it was concluded that the lateral load resistance of the frame with second specimen i.e., holes red bricks masonry of thickness 120 mm was 184% greater than the first specimen i.e., bare frame, whereas for the third and fourth specimen are by 61% and 99% respectively. The initial stiffness of the second specimen is much greater than the remaining specimens.

For this reason, the current research has chosen hollow red brick infill walls of 120 mm thick to study the behaviour of two-storey two-bay RC frame with and without infill walls subjected to in-plane cyclic loading using the Applied Element Method.

4 Modelling of RC Frames

Five half scaled Reinforced Concrete models, namely SBF1, SIF1, MBF, MSF, MIF shown in Fig. 2. were considered for the nonlinear behaviour under cyclic load analysis using Applied Element Method. The first two RC models, i.e., SBF1 & SIF1 details, were taken from Ahmed Sayed's experimental work [14], namely SBF2 & SIF2, to validate numerical models. The Single-storey and Multi-storey RC bare frames were considered for infill wall models also.

5 Sectional & Reinforcement Details

The beams of cross-section 120 × 200 mm, columns of cross-section 200 × 120 mm and base of cross-section 300 × 500 mm are considered. Hollow Red Bricks of size 215 × 102.5 × 65 mm is used for masonry infill wall. High yield strength deformation bars as main reinforcement and Mild steel bars as shear reinforcement are considered. The sectional and reinforcement details of all models are shown in Table 1.

6 Material Properties

High Strength Concrete (HSC) is considered for beams and columns, and Normal Strength Concrete for Base is considered for all the models. Hollow Red Bricks is considered for Masonry infill wall (Table 2).

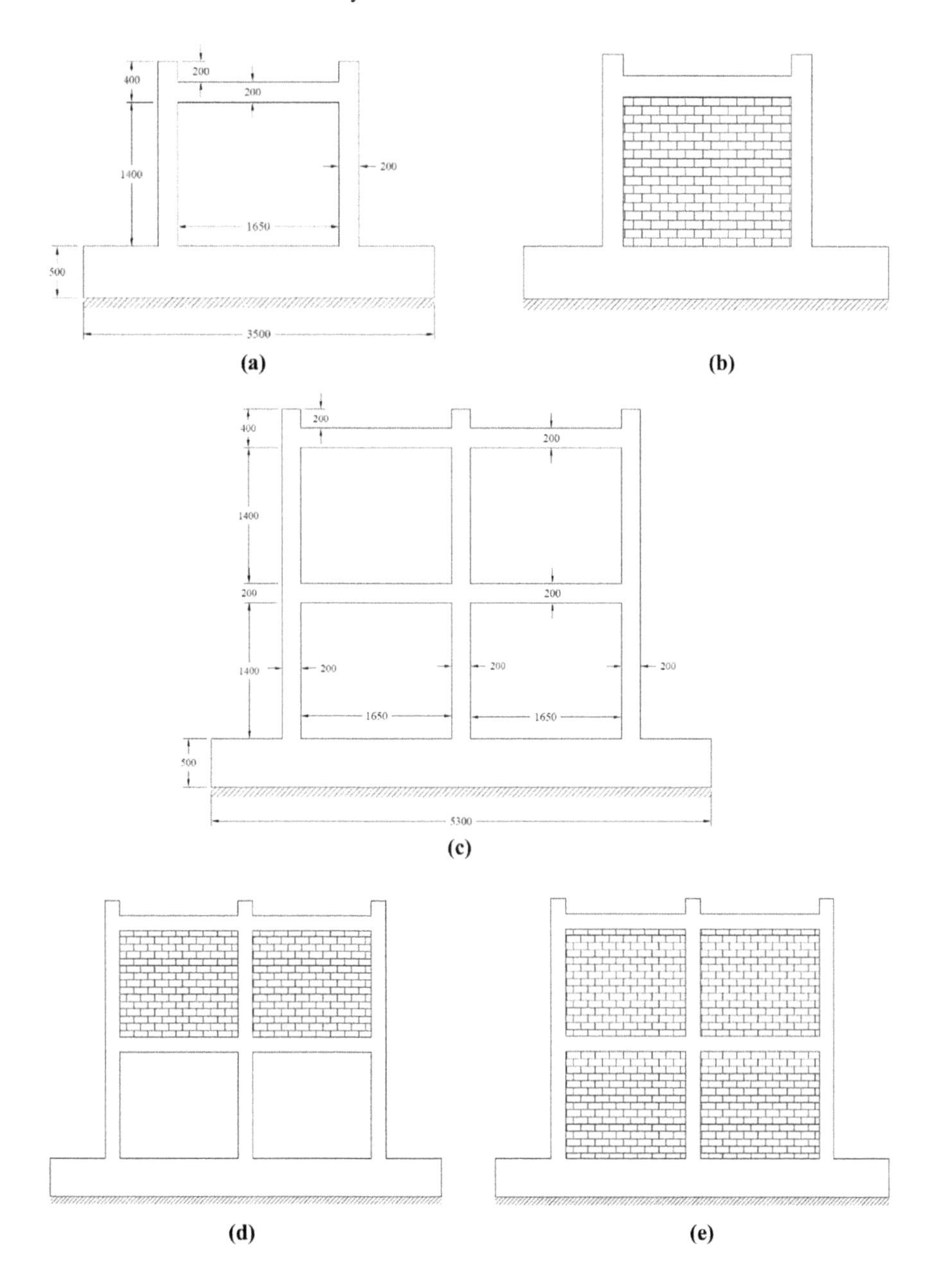

Fig. 2 Elevation of models: **a** single-storey bare frame—SBF1 & SBF2, **b** single-storey infill frame—SIF1 & SIF2; **c** multi-storey bare frame—MBF, **d** multi-storey soft-storey frame—MSF, **e** multi-storey infill frame—MIF

Table 1 Sectional and reinforcement details of all models

Name of the specimen	Member	Cross-sectional area (mm^2)	Reinforcement details
	Base	300 × 500	4–12 mmφ (top and bottom) 8 mmφ @ 180 mm c/c (stirrups)
SBF1, SIF1, MBF, MSF, MIF	Beam	120 × 200	2–12 mmφ (top and bottom) 8 mmφ @ 154 mm c/c (stirrups)
	Column	200 × 120	2–12mmφ (top and bottom) 8 mmφ @ 140 mm c/c (stirrups)

Table 2 Material properties of SBF1, SIF1, MBF, MSF, MIF [14]

Material properties	High strength concrete	Normal strength concrete	Steel	Hollow red brick
Compressive strength, GPa	0.065	0.03	0.42	0.004
Tensile strength, GPa	0.007	0.003	0.42	0.0004
Young's modulus, GPa	41.8	25	214.2	4.6
Shear modulus, GPa	17.5	11	82.6	1.8
Strain separation	0.1	0.1	0.15	0.1
Frictional coefficient	0.8	0.8	0.8	0.8
Specific weight, kg/m^3	2,750	2,500	7,850	1,635
Peak stress/yield tensile stress	–	–	1.39	–
The ratio of post-yield stiffness	–	–	0.01	–

7 Loading Configuration

All the models were subjected to cyclic displacement controlled loading condition, as shown in Fig. 3. The displacement history was the same for all tested models; Incremental Displacement considered in five stages viz., Stage 1: ±0.5 mm until 3 mm, Stage 2: ±1.0 mm until 10 mm, Stage 3: ±2.0 mm until 20 mm, Stage 4: ±4.0 mm until 40 mm, and Stage 5: ±10.0 mm up to 100 mm.

8 Results and Discussions

8.1 Mode of Failure

For frame SBF1, hair cracks observed in the columns at beam-column juncture at the end of cycle 2 mm, but for SBF2, it was observed at cycle −2 mm. The cracks were closed and opened for the increased in-plane load for each cycle. Upon application of repeated cyclic load, the flexural and shear compression cracks

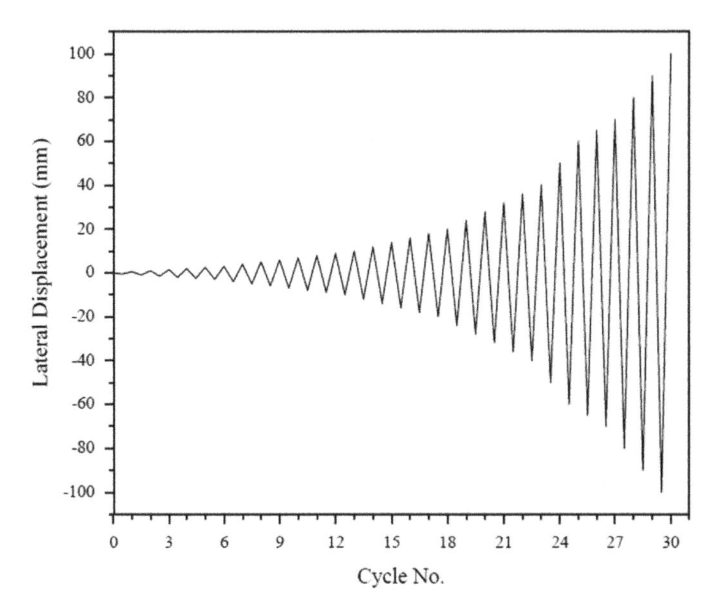

Fig. 3 Displacement history considered for all the models

observed at one-fourth of the column base and beam-column connection, respectively. At higher amplitudes of repeated cyclic load, diagonal cracks initiated and large crack openings can be seen at the beam-column junction after cycle +50 mm. The point of failure observed at +61.28 mm.

For frame SIF1, minor cracks observed in the masonry wall diagonally at the end of cycle −2 mm, but for SBF2, it was observed at cycle −1.5 mm. The cracks were closed and opened for the increased in-plane load for each cycle. Upon application of repeated cyclic load, diagonal crack further extended, and the brick element starts separating from each other at the centre at low-level loading. Lateral strength improved when compared with SBF1 but failed at cycle +12 mm. The Ultimate lateral load and Relative Displacement of SBF1, SBF2, SIF1, SIF2 shown in Table 3.

The mode of failure of SIF1 and SIF2 almost matching each other to the maximum extent.

Table 3 Numerical and experiment comparision of ultimate lateral load and relative displacement of single-storey bare frame and infill frame specimens

Specimen	SBF1	SBF2	SIF1	SIF2
	numerical	experiment	numerical	experiment
Ultimate lateral load (ton)	6.97	6.63	18.44	18.098
Relative displacement (mm)	61.58	60.32	12.195	12.06

From Table 3 and Fig. 4, it was observed that the AEM results were well accepted with experimental results. Further, the displacement controlled load condition is applied to MBF, MSF, MIF to study the behaviour.

From Table 4, the Ultimate lateral load of MSF is more compared with MIF has been observed. It is because of the weak column and strong infill wall.

From Table 4,

- For frame SIF1, the ultimate in-plane load resistance is 164% more, compared with the frame SBF1
- For frame MSF, the ultimate in-plane load resistance is 411% more, compared with the frame MBF
- For frame MIF, the ultimate in-plane load resistance is 396% more, compared with the frame MBF

The shear failure observed at the base of either side column for MSF, whereas for MIF, the flexural failure observed at the beam-column juncture of either side columns of the first floor and shear failure observed at the base of the middle column.

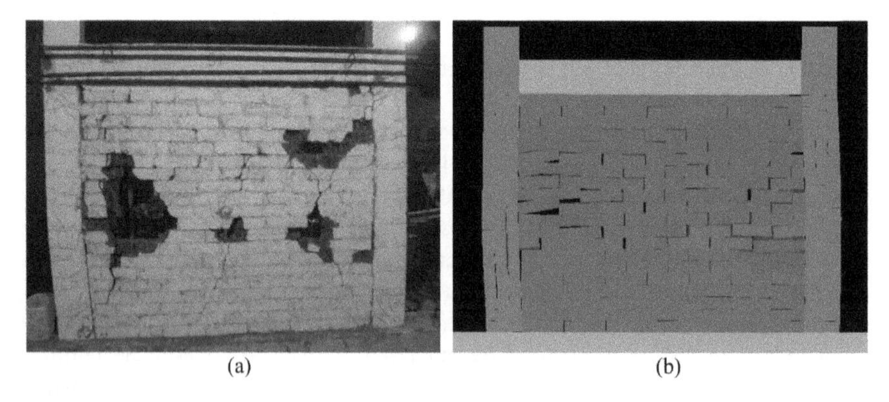

(a) (b)

Fig. 4 Mode of failure of: **a** experimental infill frame (SIF2) [14] and **b** numerical infill frame (SIF1)

Table 4 Relative displacement and ultimate in-plane load of all models

Specimen	Relative displacement (mm)	Ultimate in-plane load (ton)
SBF1	61.58	6.97
SIF1	12.195	18.44
MBF	44.90	14.70
MSF	24.06	60.44
MIF	20.23	58.23

8.2 Relationship Between In-Plane Load and Displacement

Plots of the hysteresis envelope of lateral load versus lateral displacement till the collapse of all the specimens shown in Fig. 5. The plots show the behaviour of all the specimens from the elastic stage to the collapse stage.

8.3 Ductility

According to Comite Euro-International Du Beton, 1996 [14], the Displacement ductility factor (Rμ) is the ratio between displacement at failure point (Δ_f) to displacement at the yield point of the specimen (Δ_y) shown in Eq. (3) and the Displacement Ductility is the ratio of maximum displacement in each cycle interval (Δ_i) to the displacement at the yield point (Δ_y) shown in Eq. (4) where these displacements are obtained from the plot of lateral force and lateral displacement of the specimen. The yield displacement is 80% of the ultimate load on the rising curve of lateral load versus lateral displacement plot of the specimen. The failure displacement is 80% of the ultimate load on the falling curve of load versus the displacement plot of the specimen.

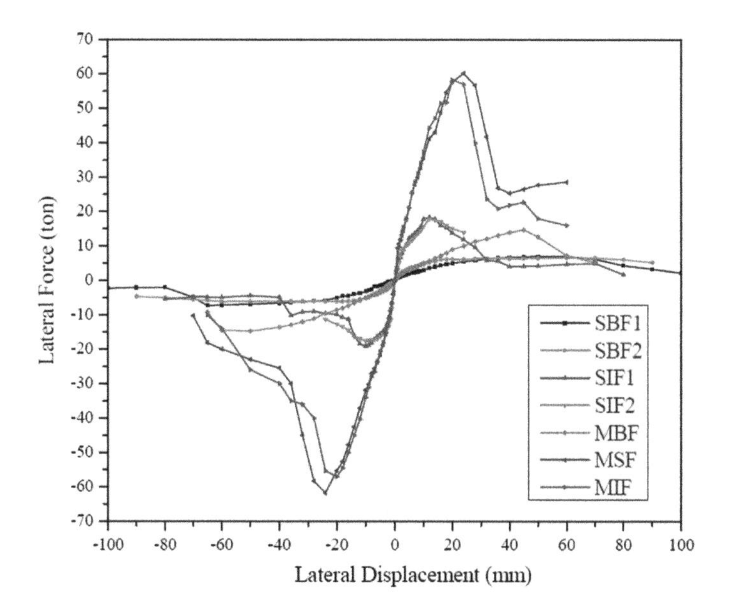

Fig. 5 Hysteresis envelope of all the specimens till the collapse

Table 5 Accumulated displacement ductility and displacement ductility factor Rμ

Specimen	Accumulated displacement ductility	Displacement ductility factor, Rμ
SBF1	23.53	2.71
SIF1	7.81	1.17
MBF	14.753	2.05
MSF	12.86	1.90
MIF	13.01	1.93

$$\text{Displacement ductility factor}, R\mu = \frac{\Delta_f}{\Delta_y} \tag{3}$$

$$\text{Displacement Ductility} = \frac{\Delta_i}{\Delta_y} \tag{4}$$

where Δ_f is the displacement at failure point, Δ_y is the displacement at the yield point, Δ_i is the maximum displacement in each cycle interval.

The Accumulated Displacement Ductility is the accumulation of the lateral displacement of first cycle until the failure load cycle and expressed by Eq. (5).

$$\text{Accumulated Displacement Ductility} = \sum\left(\frac{\Delta_i}{\Delta_y}\right) \tag{5}$$

From Table 5,

- For frame SIF1, the accumulated displacement ductility is 67% lesser than that of frame SBF1
- For frame MSF, the accumulated displacement ductility is 13% lesser than that of frame MBF and also 1.15% lesser than of frame MIF
- For frame MIF, the accumulated displacement ductility is 12% lesser than that of frame MBF but 1.16% greater than of frame MSF

8.4 *Stiffness*

According to Comite Euro-International Du Beton, 1996 [14], the stiffness of structure can be computed for the each interval of cyclic loading as the ratio of peak load at a particular cycle to the peak displacement of the same cycle.

$$\text{Cracked Stiffness}, k_i = \frac{P_i}{\Delta_i} \tag{6}$$

where, P_i is the peak load at cycle i, Δ_i is the peak displacement at cycle i (Fig. 6).

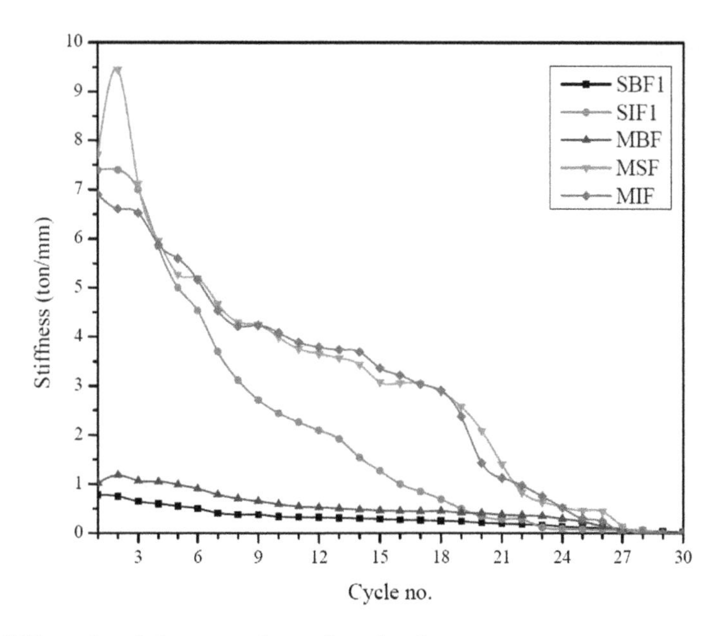

Fig. 6 Stiffness degradation versus the number of cycles

Table 6 Initial stiffness (ton/mm) of all models

Model	Initial stiffness (ton/mm)
SBF1	0.552
SIF1	2.74
MBF	0.903
MSF	3.99
MIF	4.087

The initial stiffness is the ratio of change in lateral load to change in lateral displacement for the tangent drawn at the 5th cycle. Table 6 shows the initial stiffness for all models.

From Table 6,

- For frame SIF1, the initial stiffness is 396% greater than that of frame SBF1
- For frame MSF, the initial stiffness is 341% greater than that of frame MBF
- For frame MIF, the initial stiffness is 352% greater than that of frame MBF

9 Conclusions

From this study, it can be concluded that;

1. Applied Element Method results are well accepted with experimental results.
2. The Displacement Ductility Factor of bare frames is greater than the infill frames in both Single-storey and Multi-storey frames i.e., SBF1 = 2.71, SIF1 = 1.17, MBF = 2.05, MSF = 1.90, and MIF = 1.93.
3. The accumulated displacement ductility for SIF1 with an infill wall of thickness 120 mm was less than the bare frame SBF1 by about 67%. The accumulated displacement ductility of frames MSF and MIF with infill walls is less than bare frame MBF by about 13% and 12% respectively.
4. The strength of frame SIF1 with an infill wall was 164% stronger than the bare frame SBF1. The strength of frames MSF and MIF with infill wall of 411% and 396% respectively were stronger than the bare frame MBF.
5. The ultimate in-plane load resistance for frame MSF was greater than the frame MIF by about 3.79% because of the weak column, strong infill wall behaviour.
6. The shear failure observed at the base of either side columns for MSF, whereas for MIF, the flexural failure observed at the beam-column juncture of either side columns of the first floor and shear failure observed at the base of the middle column.
7. Stiffness degradation of masonry infilled frames is very rapid compared to bare frames. Whereas the initial stiffness of SIF1 is 396% greater than that of frame SBF1 and the initial stiffness of MSF and MIF is greater than MBF by about 341% and 352%, respectively.

Acknowledgements The authors warmly acknowledge the support of Ministry of Human Resources and Development (MHRD), India and N.I.T Raipur for providing continuous support.

References

1. Roca P, Cervera M, Gariup G, Pela' L (2010) Structural analysis of masonry historical constructions. Classical and advanced approaches. Arch Comput Methods Eng 17:299–325. https://doi.org/10.1007/s11831-010-9046-1
2. Cundall AP (1971) A computer model for simulating progressive, large-scale movement in blocky rock system. Proc Int Symp Rock Mech
3. Lemos JV (2007) Discrete element modeling of masonry structures. Int J Archit Herit 1:190–213. https://doi.org/10.1080/15583050601176868
4. Cundall PA, Hart RD (1992) Numerical modelling of discontinua. Eng Comput 9:101–113. https://doi.org/10.1108/eb023851
5. Lourenço PB (2002) Computations on historic masonry structures. Prog Struct Eng Mater 4:301–319. https://doi.org/10.1002/pse.120
6. De Lorenzis L, DeJong M, Ochsendorf J (2007) Failure of masonry arches under impulse base motion. Earthq Eng Struct Dyn 36:2119–2136. https://doi.org/10.1002/eqe.719

7. Tondelli M, Beyer K, DeJong M (2016) Influence of boundary conditions on the out-of-plane response of brick masonry walls in buildings with RC slabs. Earthq Eng Struct Dyn 45:1337–1356. https://doi.org/10.1002/eqe.2710
8. Chetouane B, Dubois F, Vinches M, Bohatier C (2005) NSCD discrete element method for modelling masonry structures. Int J Numer Methods Eng 64:65–94. https://doi.org/10.1002/nme.1358
9. Meguro K, Tagel-Din H (2002) Applied element method used for large displacement structural analysis. J Nat Dis Sci 24:25–34
10. Kikuchi A, Kawai T, Suzuki N (1992) The rigid bodies-spring models and their applications to three-dimensional crack problems. Comput Struct 44:469–480. https://doi.org/10.1016/0045-7949(92)90269-6
11. Meguro K, Tagel-Din H (2000) Applied element method for structural analysis: theory and application for linear materials. Struct Eng Eng 17. https://doi.org/10.2208/jscej.2000.647_31
12. Karbassi A, Nollet MJ (2013) Performance-based seismic vulnerability evaluation of masonry buildings using applied element method in a nonlinear dynamic-based analytical procedure. Earthq Spectra 29:399–426. https://doi.org/10.1193/1.4000148
13. Malomo D, Pinho R, Penna A (2020) Numerical modelling of the out-of-plane response of full-scale brick masonry prototypes subjected to incremental dynamic shake-table tests. Eng Struct 209:110298. https://doi.org/10.1016/j.engstruct.2020.110298
14. Tawfik Essa ASA, Kotp Badr MR, El-Zanaty AH (2014) Effect of infill wall on the ductility and behavior of high strength reinforced concrete frames. HBRC J 10:258–264. https://doi.org/10.1016/j.hbrcj.2013.12.005

Numerical Studies on Seismic Behaviour of RC Structures Incorporated with Shape Memory Alloy—A Review

Geethu Elsa Thomas, P. V. Indira, and A. S. Sajith

Abstract Shape memory alloy (SMA) is an intelligent material where the essential properties emerge from its microstructural changes when exposed to outer non-mechanical stimuli like temperature. SMAs exhibit two essential properties, making them applicable for civil engineering structures, especially in earthquake-resistant designs. The first one is the superelastic effect, and the other is the shape memory effect. Superelasticity is its property to undergo large deformations and regain its original shape after the applied stress is removed. The shape memory effect is its ability to undergo distortions and then regaining its form on heating. The safety of various structures can be enhanced using SMAs. The significant expense of SMA is a demerit restricting their utilization in this area. SMAs are expected to arise as a fundamental material in the construction industry, and hence numerous researches are required in this field. Various numerical models of SMAs are studied by researchers and examined their accuracy with the experimental results. This paper presents a state of the art of the existing numerical models using nonlinear finite element analysis programs to model SMAs and examines the factors influencing their engineering properties. It also summarises the different parameters required for modeling SMAs in various finite element analysis programs.

Keywords Shape memory alloy · Smart material · Superelastic effect · Shape memory effect

1 Introduction

The national wealth of a country mainly depends on its civil infrastructure. These structures often require periodic maintenance due to deterioration. The threats from terrorist attacks like that of blast loads on buildings is a genuine matter that ought to

G. E. Thomas (✉) · P. V. Indira · A. S. Sajith
Department of Civil Engineering, National Institute of Technology Calicut, Kozhikode, India
e-mail: geethu_p180132ce@nitc.ac.in

G. C. Marano et al. (eds.), *Proceedings of SECON'21*, Lecture Notes in Civil Engineering 171, https://doi.org/10.1007/978-3-030-80312-4_81

be taken into design considerations. Also, structures are always at a risk of seismic loadings like earthquakes. Hence impact and dynamic load resistant way of designing along with infrastructure management is becoming a prior area of research. A considerable lot of these issues could be eliminated if construction is made smart, which means structures are capable of identifying their own damage, repair their condition, and can adapt according to the changes in their loading condition. SMAs are a type of smart materials with significant capacity for various civil engineering applications [1]. Two unique properties define the characteristics of this smart material, namely, the superelastic effect (PE-SMA) and the shape memory effect (SME). Superelasticity is its capacity to go through enormous distortion and to regain its structure when the given load is taken out. Shape memory effect a property through which it can take the load, deform and recover when heated. These properties could be successfully used to improve the safety of different constructions. After a severe seismic loading, SMA could offer self-centering and high damping responses, thus enhancing the overall performance.

Each SMA has a different chemical composition and atomic arrangement, which results in a variation of their mechanical behavior. Indeed, even a minor difference in the arrangement of the constitutive metals inside a similar compound may altogether influence its mechanical properties [2]. Figure 1 shows the stress–strain curve under tension/compression for SMA [3]. Similar curves are available for the characteristic behavior of SMA under torsion and shear.

The cyclic behavior of SMA is critical if they are to be utilized in seismic applications. Within the superelastic strain range, SMA can disperse a specific quantity of energy without undergoing permanent deformation when exposed to cyclic loading. Here a net release of energy is guaranteed along the stage change from austenite to martensite on loading and vice-versa on unloading [3]. A range of 5–10 cycles can be considered in the case of seismic applications. This would bring about a 40% reduction in the stress plateau in later cycles than the principle cycle [4].

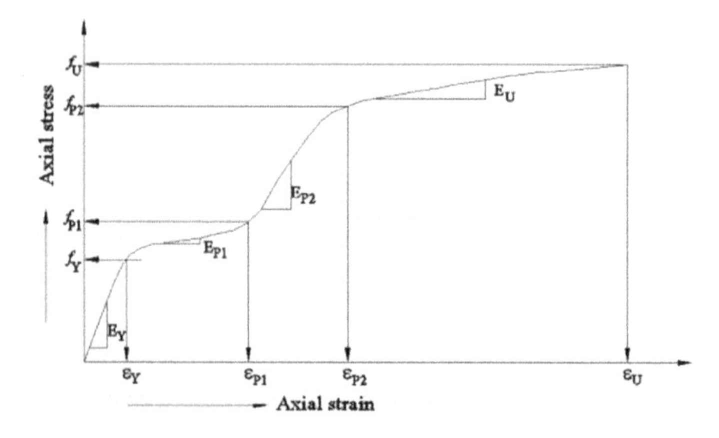

Fig. 1 Stress–strain curve under tension/compression for SMA [3]

SMAs are expected to arise as a fundamental asset in earthquake-resistant constructions, and hence numerous researches are required in this field. Various numerical models of SMAs are studied by researchers and examined its accuracy with the experimental results. This paper discusses the existing numerical models using nonlinear finite element analysis programs to model SMAs and examines the factors influencing their engineering properties.

2 Constitutive Material Modeling of SMA

The advancement of accurate and effectively reasonable constitutive models for SMAs are essential for development in this area. The unpredictable thermo-mechanical properties of SMA, with superelasticity and SME, have made SMA's constitutive modeling complex. Two very notable approaches, known as phenomenological and thermodynamics models, are utilized to demonstrate SMA.

2.1 Phenomenological Modeling (PMs)

Generally, SMA applications in structural designing are identified with using rods and wires, and hence unidimensional models (PMs) are frequently viewed as reasonable [3]. Several models are proposed by researchers Tanaka and Nagaki [5], Liang and Roger [6], Huang [7], Auricchio and Sacco [8], where the used parameters are identified experimentally. Many FE software packages like ANSYS 10.0, ABAQUS 6.4, and SeismoStruct (2004–21) include the superelastic behavior of SMA. Auricchio and Taylor [9], and Auricchio and Sacco [8] defines the material model perfectly. The mechanical features of SMA in FE software packages can be fully described using the following model-calibrating parameters: austenite to martensite finishing stress (f_{P1}), austenite to martensite starting stress (f_y), martensite to austenite finishing stress (f_{T2}), martensite to austenite starting stress (f_{T1}), maximum residual strain (ε_1), modulus of elasticity and specific weight. Figure 2 demonstrate the 1-D PE-SMA model used in the software SeismoStruct (2010).

2.2 Thermodynamics Modeling (TMs)

Laws of thermodynamics and the principles of energy consideration are the main concept behind such models. Various models have been created representing some features of SMAs. Patoor et al. [10], Goo and Lexcellent [11], Huang and Brinson [7], and others used thermodynamics laws and conducted micromechanics approaches to study crystallographic phenomena within the material. Such models are complicated and are difficult for computation than PMs since they give a

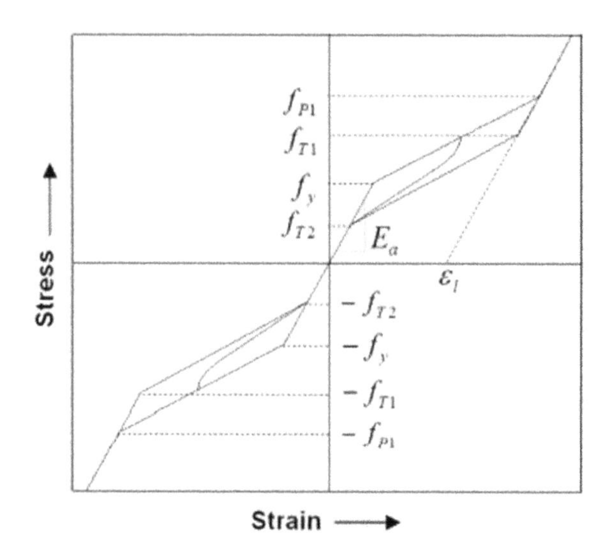

Fig. 2 1-D superelastic model of SMA (SeismoStruct 2010)

method to derive precise 3-D constitutive laws. In any case, PMs appear to be more proficient in structural designing applications.

3 Seismic Behavior of Reinforced Concrete Structures Incorporated with SMA

Seismic behavior of RC systems incorporated with SMA can be efficiently modeled using various nonlinear finite element packages. Some of these are discussed in this section showcasing the various parameters required to model SMA.

3.1 Modeling Using FE Software VecTor2

Ayoub et al. [12] tested the behavior of beams reinforced with superelastic Nickel-Titanium (Ni–Ti) SMA bars at midspan with the simply supported condition. In the finite element analysis, program VecTor2, the beams were modeled with only one homogeneous concrete layer. Transverse reinforcement was smeared in the concrete layers, whereas the embedded and external longitudinal reinforcement were modeled as discrete truss bars. Pseudoelastic SMAs are utilized as reinforcement, and the seismic behavior of a beam-column (BC) joint was evaluated by Alam et al. [3]. This study concluded that the conventional and SMA specimens had similar yield and peak load carrying capacity; however, the corresponding yield displacement was approximately 50% larger for the SMA specimen. Test results demonstrated efficient recovery of most of the post-yield displacement.

Here 7 concrete layers is taken for modeling SMA reinforced BC joint. The column was represented by the first 3 layers, while the last four layers represented the beam. Concrete and Geometric properties were similar for all the layers, whereas the steel (transverse) properties, which were smeared in the concrete, were different. Truss bars are used to model main longitudinal reinforcement. Effendy et al. [13] experimented on 3 low-rise barbell shaped shear walls under reverse cyclic loading. Superelastic SMA bars were placed externally in one of the walls in the diagonal directions, acting as structural bracing to increase ductility and energy dissipation capacity. The effect of the SMA in limiting the residual displacement was evident; however, the energy dissipation capacity was minimal in comparison to ductile walls. The geometry and material properties of the wall were modeled with four homogeneous concrete layers with smeared reinforcement. The first and the second zones were assigned to the foundation and the top beam, respectively. The third zone corresponding to the web wall, and the fourth zone to the barbell ends of the wall. Plane stress rectangular elements were used to define the mesh for the concrete layers. The diagonal SMA bars were modeled with truss bar elements. Abdulridha et al. [14] built up a primer model for understanding the hysteretic conduct of SMAs. This model was effective in anticipating the load–displacement curves of the beams subjected to cyclic loading.

3.2 *Modeling Using FE Software ABAQUS*

Abouali et al. [15] contemplated the impact of near-surface mounted (NSM) Fe-SMA strips in strengthening reinforced concrete beams, using a three-dimensional nonlinear FE model in ABAQUS. The accuracy of the model is checked using experimental data. Ghassemieh et al. [16] examined shear walls with pseudoelastic Ni–Ti reinforcement by conducting time history analysis. The outcomes demonstrated that the utilization of Ni–Ti reinforcement reduced the residual deformations. Ruiz-Pinilla et al. [17] suggested a model for Fe-SMA strip behavior using an analytical stress strain curve. Malagisi et al. [18] built up another model to deal with behavior of SMA actuators to fix cracks in concrete beams. A 1-D model was used to reenact the superelastic and SME of the SMA bars. The micro-cracking of concrete is modeled using non local damage and plasticity model options in ABAQUS. Also, the micro-cracks are designed using the transition approach from damage to fracture.

3.3 *Modeling Using FE Software OpenSees*

Tazarv [19] studied the seismic performance of bridge columns to understand the consequence of SMA on force displacement and moment curvature curves. The proposed model simulates SMA reinforced bridge column to understand its seismic behavior. This model uses a 4 node fibre section. Elastic elements are used to model footing and cap beam. Beam with hinges (force based) is used as column element, with 2 plastic hinges at the component ends. Concrete04 material was used to model concrete fibre, including a tensile stress strain behavior. ReinforcingSteel material option is available to model steel. Confined properties of concrete core are modeled utiliszing Mander's model [27]. Miller [20] studied on self-centering buckling restrained braces and its seismic behavior by subjecting to a cyclic loading condition. Here a numerical model was established for validating the experimental results.

3.4 Modeling Using FE Software Seismostruct

Alam [21] studied about implementing SMA as reinforcement and its seismic behavior on concrete frames. Two eight-storey concrete frames are subjected to 10 different ground motion records. One of the frames utilizes SMAs (plastic hinge regions) and the other with commercial steel. Results depict the resistance of SMA RC frames against strong earthquake loading. Models developed in Seismostruct successfully show the post yield deformation and recovery mechanism. Ravi [22] modeled shear walls utilizing SMA as reinforcement and studied its seismic performance. Analysis using nonlinear dynamic time history analysis and pushover analysis were conducted, and the outcomes showed an enhancement in seismic response of the system. It has been found that the Auricchio and Sacco [8] model is effective for modeling SMA in seismostruct. Nahar [23] conducted non-linear static pushover analysis on different types of SMAs utilized as reinforcement in BC joint, under reverse cyclic loading. This model depicted that SMA reinforcement has adequate energy dissipation capacity. This FE program utilizes a fibre modeling mechanism to show the distribution of material nonlinearity. 3-D, beam column inelastic displacement-based frame components are taken for demonstrating BC joint. The nonlinear uniaxial stress strain reaction of individual fibre is coordinated to acquire the sectional stress strain condition of the components.

4 Numerical Studies on Retrofitting Applications

The seismic behavior of RC frame structures retrofitted using external pseudoelastic SMA bars was studied by Elbahy [24]. The reference frame is located at a high seismic region and utilized a six-storey steel RC frame for this study. He conducted a nonlinear dynamic analysis using Seismostruct (2018) software. The retrofitted frames showed higher tolerance towards earthquake intensities with lower damage. Also, a nonlinear time-history analyses was done and identified that the adopted retrofitting technique reduces the residual drifts. Seismic behavior enhancement of RC frames using SMA bars was studied by Duran [25]. He developed a FE model of RC frames with a retrofitting system working as tension only members using SeismoStruct program. Cortés [26] modeled shear walls retrofitted with SMA. VecTor2 software is adopted in this study for simulating the model. A comparative analysis is also done with a traditional retrofit technique using steel plates. From the observations, it can be summarised that retrofitting methods with SMAs can improve the system's flexural response and energy dissipation while reducing damage to the concrete.

5 Future Scope

The studies on SMA needs future research, from literatures it's clear that Studies showcasing the shape memory effect is little addressed. Also, a numerical model simulating the utilization of SMA as fibre reinforced cement composites is not studied.

6 Concluding Remarks

This paper showcases the various numerical models utilizing the SMA in reinforced concrete structures. The different nonlinear finite element programs and the parameters to model SMA are depicted in the present study. The ability of SMA to limit residual displacements and dissipate energy leads to the possibility of using SMA bars to strengthen reinforced concrete structures designed for seismic zones. The ability of SMA to recover significant inelastic displacements permits the repairing of damaged structural elements by replacing the damaged concrete only. The high initial cost of Ni–Ti SMA is limiting its use in concrete structures to critical regions only. The efficient use of Nitinol can substantially mitigate earthquake risk and reduce the associated cost after a major earthquake. In summary, PE-SMA has the potential to be used as a replacement to steel reinforcement in seismic resistant systems. The enhanced caliber to recapture inelastic displacements, strain hardening while sustaining large displacement (ductility), energy dissipating capacity, and self-centering properties make SMAs a vital area of research.

References

1. Hardwicke C (2003) Recent developments in applying smart structural materials. JOM, ABI/INFORM Trade Ind 55:15–16
2. Strnadel B, Ohashi S, Ohtsuka H, Ishihara T, Miyazaki S (1995) Cyclic stress-strain characteristics of TiNi and TiNiCu shape memory alloys. Mater Sci Eng A 202(1–2):148–156
3. Alam MS, Youssef MA, Nehdi M (2007) Utilizing shape memory alloys to enhance the performance and safety of civil infrastructure: a review. Can J Civ Eng 34(9):1075–1086

4. Desroches R, Smith B (2004) Shape memory alloys in seismic resistant design and retrofit: a critical review of their potential and limitations. J Earthq Eng 8(3):415–429
5. Tanaka K, Nagaki S (1982) A thermomechanical description of materials with internal variables in the process of phase transitions. Ingenieur-Archiv 51:287–299
6. Liang C, Roger C (1990) One-dimensional thermomechanical constitutive relations for shape memory materials. In: Structural dynamics & materials conference 1990, pp 16–28
7. Huang M, Brinson LC (1998) Multivariant model for single crystal shape-memory-alloy behaviour. J Mech Phys Solids 46:1379–1409
8. Auricchio F, Sacco E (1997) Superelastic shape-memory-alloy beam model. J Intell Mater Syst Struct 8:489–501
9. Auricchio F, Taylor RL, Lubliner J (1996) Shape-memory-alloy superelastic behavior: 3D finite element simulations. In: SPIE, 1996, pp 487–492
10. Patoor E, Berveiller M (1997) Micromechanical modelling of the thermomechanical behavior of shape memory alloys. In Mechanics of Solids with Phase Changes (pp. 121–188). Springer, Vienna
11. Goo BC, Lexcellent C (1997) Micromechanics-based modeling of two-way memory effect of a single crystalline shape-memory alloy. Acta Mater 45(2):727–737
12. Ayoub AM, Saiidi C, Itani M (2003) A study of shape-memory-alloy reinforced beams and cubes. Reno, Nevada
13. Effendy E, Liao WI, Song G, Mo YL, Loh CH (2006) Seismic behavior of low-rise shear walls with SMA bars. In: Earth Sp. 2006—proceedings of 10th biennal international conference on engineering construction and operations in challenging environment, vol 2006, p 137
14. Abdulridha A, Palermo D, Foo S, Vecchio FJ (2013) Behavior and modeling of superelastic shape memory alloy reinforced concrete beams. Eng Struct 49:893–904
15. Abouali S, Shahverdi M, Ghassemieh M, Motavalli M (2019) Nonlinear simulation of reinforced concrete beams retrofitted by near-surface mounted iron-based shape memory alloys. Eng Struct 187:133–148
16. Ghassemieh M, Mostafazadeh M, Sadeh MS (2012) Seismic control of concrete shear wall using shape memory alloys. J Intell Mater Syst Struct 23(5):535–543
17. Ruiz-Pinilla JG, Montoya-Coronado LA, Ribas C, Cladera A (2020) Finite element modeling of RC beams externally strengthened with iron-based shape memory alloy (Fe-SMA) strips, including analytical stress-strain curves for Fe-SMA. Eng Struct 223:111–152
18. Malagisi J, Marfia S, Sacco S, Toti E (2014) Modeling of smart concrete beams with shape memory alloy actuators. Eng Struct 75:63–72
19. Tazarv M, Saiid Saiidi M (2015) Reinforcing NiTi superelastic SMA for concrete structures. J Strct Eng 141(8):04014197
20. Miller DJ, Fahnestock LA, Eatherton MR (2011) Self-centering buckling-restrained braces for advanced seismic performance. In: Structures congress 2011, pp 960–970
21. Alam MS, Youssef MA, Nehdi M (2008) Analytical prediction of the seismic behaviour of superelastic shape memory alloy reinforced concrete elements. Eng Struct 30(12):3399–3411
22. Ravi V, Krishnan PA (2019) Effect of replacing steel with shape memory alloy in shear wall systems. Mater Today Proc 11:1088–1093
23. Nahar M, Muntasir Billah AHM, Kamal HR, Islam K (2018) Numerical seismic performance evaluation of concrete beam-column joint reinforced with different super elastic shape memory alloy rebars. Eng Struct 194(2018):161–172
24. Elbahy MA, Yamen I, Youssef (2018) Seismic performance of reinforced concrete frames externally reinforced with shape memory alloy bars. In: Build. Tomorrow's Soc., vol ST118, pp 1–9
25. Duran B, Tunaboyu O, Atli KC, Avsar Ö (2019) Seismic performance upgrading of substandard RC frames using shape memory alloy bars. Smart Mater Struct 28(8)
26. Cortés-Puentes WL, Palermo D (2012) Towards design of shear walls retrofitted with shape memory alloys. In: 15th World conference on earthquake engineering, Lisbon Port., no. 2000
27. Mander JB, Priestly MJN, Park R (1988) Theoretical stress-strain model for confined concrete. J Struct Div 114(8):1804–1826

Uncertainty Quantification of Structural Response Due to Earthquake Loading

Sarit Chanda and Surendra Nadh Somala

Abstract In the performance-based structural design, the crucial component is to estimate the uncertainties in the structural responses precisely. The uncertainty may lie in many structural design parameters such as load, material properties, etc. Each parametric uncertainty has led to variation in the structural responses. Currently, structural design is performed considering the constant loading and material properties. But in reality, all these parameters are highly uncertain and can pose a wide variation in the structural responses due to earthquake loading. This study focuses on identifying the uncertainties arising from the different means and their impacts on the responses. The Monte Carlo Sampling (MCS) is employed to quantify the uncertainties in a structural deformation. The Multi-Degrees of Freedom (MDOF) structural model is constructed in the OpenSees program, and non-linear dynamic analysis is performed. The El-Centro earthquake was applied for the structural analysis. The result shows the probabilistic distribution of the earthquake response parameters. This approach is a more realistic representation of structural responses by incorporating the uncertainties in the design parameters.

Keywords Uncertainty quantification · Monte Carlo sampling · OpenSees

1 Introduction

Any civil engineering structure design deals with the uncertainty either in the material properties or the load acting on the structures. Uncertainties in the construction materials are less if the material is processed in the factory, such as in a

S. Chanda (✉) · S. N. Somala
Indian Institute of Technology, Hyderabad 502285, India
e-mail: ce16resch11006@iith.ac.in

S. N. Somala
e-mail: surendra@ce.iith.ac.in

S. Chanda
IcfaiTech, ICFAI Foundation for Higher Education, Hyderabad, India

G. C. Marano et al. (eds.), *Proceedings of SECON'21*, Lecture Notes in Civil Engineering 171, https://doi.org/10.1007/978-3-030-80312-4_82

controlled production environment with skilled laborers. Still, there will be variation in the structural properties of the material. Similarly, uncertainties will arise in the loads acting on the structures. To overcome these uncertainties, probabilistic approaches are employed in the structural design. Due to that, the loads on the structures are overestimated as well as the material properties are underestimated. Wang and Wang [7] has quantified the uncertainties from the measured data of structures. Feng et al. [3] used hybrid uncertainty model for the time depended on the response of the structure. Muscolino and Sofi [4] performed Stochastic analysis to compute bounded uncertainties. Qiu and Liu [5] has determined the safety margin of buckling columns. Risk-based cost analysis of structures is investigated by [1]. The uncertainties in the bridge-vehicle system were evaluated by [8]. Nowadays, reliability-based optimized design of the structure is in practice [6]. Beck and Gomes [2] has compared the deterministic and reliability-based structural optimization under uncertainties. In this article, we aim to quantify structural uncertainties and evaluate the variation in the structural responses. A finite element structural model is constructed in the OpenSees program, and the structure is subjected to Stochastic ground motion. The uncertainties are incorporated in the structures by assigning the variation in floor weight and the stiffness in the column. Stochastic ground motion and El-Centro ground motion are applied into the structure, and different sampling methods are employed, and the structural responses are simulated using the OpenSees program. Results show the propagation of uncertainties from input to the final responses of the structures.

2 Methodology

This research work aims to quantify the uncertainties which propagate from input to the structural response. The forward propagation method is extensively used to evaluate the probabilistic variation in the outputs by generating sample realizations and statistical parameters such as mean, standard deviation, skewness, and kurtosis. The Monte Carlo Sampling (MCS) has been used to generate the samples for the simulation.

The Monte Carlo Sampling is robust and extensively used by the researchers. The advantage of the MCS is the convergence rate is independent of the system dimension. However, the convergence rate is relatively slow compare to other sampling methods (such as Latin Hypercube Sampling and Importance Sampling).

The Latin Hypercube Sampling (LHS) is a partially random sampling approach. Here the convergence rate is better if the samples are evenly spread to the input domain.

The Importance Sampling (I.S.) is used for the rear events, such as it provides the maximum response of the structure. This sampling useful when the peak values are required.

If the structural model is large, then more computational efforts are required for the LHS or MCS methods. To overcome the computational limitation, the surrogate models are made using a fewer number of simulations. And then, using these surrogate models sample model can be generated efficiently. The Gaussian Process Regression (GPR) and Polynomial Chaos Expansion (PCE) model is the popular surrogate models.

OpenSees is used to simulate the finite element model with the dynamic ground motion applied to the structure.

3 Structural Model

The multi-degree freedom model is constructed on OpenSees. The model is a four-story single bay frame. The support of the model is considered fixed (Fig. 1). The weight of the floor and the stiffness of the columns have uncertainties. To incorporate the uncertainties in the model, we have provided the probability distribution of floor weights and stiffness of the columns. We have considered the normal distribution of in-floor weights and in the stiffness of the columns. The mean weight of the floors is 100 kip, and the standard deviation is 10 kip. Similarly, the story stiffness has a mean of 326.32 kips/in and a standard deviation of 25 kips/in. The main objective of this research work is to capture the uncertainties in the material and the loads. So, we have used a scaled structural model to investigate the uncertainties.

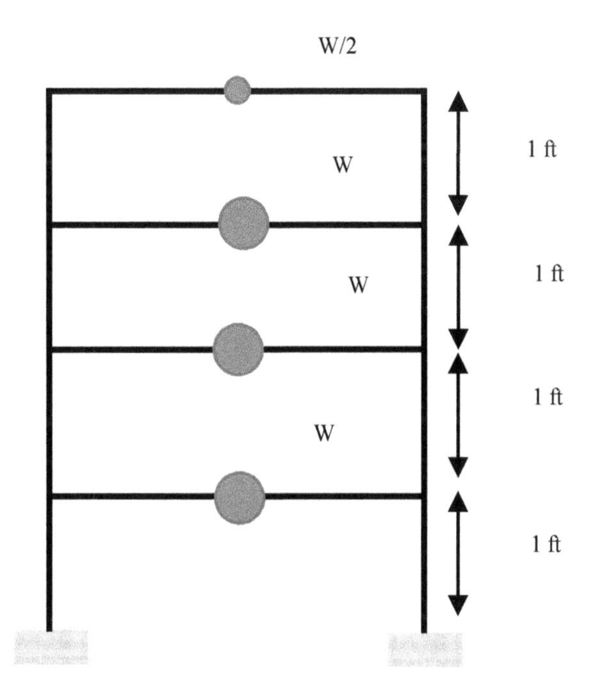

Fig. 1 Structural model

The main objective of this research work is to evaluate the mean and the standard deviation of the structural responses, such as relative displacement at the different story levels.

4 Results and Discussion

The structure is subjected to El-Centro ground motion, and finite element analysis has been carried out in the OpenSees program. Using Monte Carlo Sampling, an input sample has been prepared, and Fig. 2 displays variation in the relative displacements at the top floor. From Fig. 2, it can be observed that the relative displacement of the top floor (Vertical axis in Fig. 2) is high in a certain combination of input parameters. The uncertainty in the input has produced a wide variation in the structural responses.

Figure 3 displays the histogram of structural responses at the roof node. It can be observed that as structural responses follow a normal distribution, the uncertainties in the input are normally distributed.

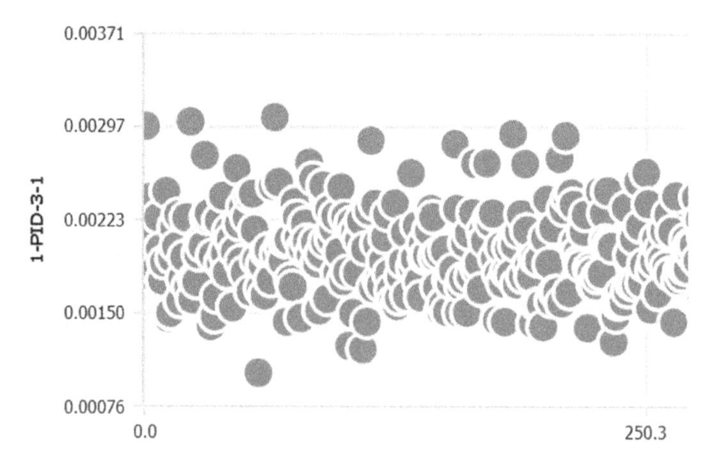

Fig. 2 Variation of relative response at the top floor

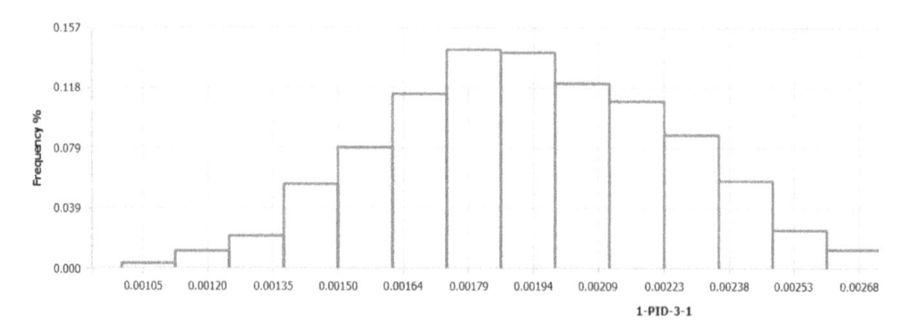

Fig. 3 Probability distribution of relative displacement at roof node

5 Conclusion

The uncertainties in the structural design produce an inaccurate estimation of structural responses due to static and dynamic loads. Uncertainty lies in the properties of construction material as well as the loads applied to the structure. The result shows the variability in the structural responses as the uncertainty in load and materials properties increases. Therefore, the conventional way of structural analysis and design did not capture the uncertainties in any form. But, to mimic the actual behavior of structures, uncertainties have to be incorporated in the design and analysis. This study has only focused on uncertainties in material properties and loads while considering a stick model. In the future, we will investigate a full-scale model with additional uncertainties included in the model.

References

1. Beck AT, da Ribeiro LR, Valdebenito M (2020) Risk-based cost-benefit analysis of frame structures considering progressive collapse under column removal scenarios. Eng Struct 225:111295. https://doi.org/10.1016/j.engstruct.2020.111295
2. Beck AT, de Gomes WJS (2012) A comparison of deterministic, reliability-based and risk-based structural optimization under uncertainty. Probab Eng Mech 28:18–29. https://doi.org/10.1016/j.probengmech.2011.08.007
3. Feng X, Wu J, Zhang Y (2018) Time response of structure with interval and random parameters using a new hybrid uncertain analysis method. Appl Math Model 64:426–452. https://doi.org/10.1016/j.apm.2018.07.043
4. Muscolino G, Sofi A (2012) Stochastic analysis of structures with uncertain-but-bounded parameters via improved interval analysis. Probab Eng Mech 28:152–163. https://doi.org/10.1016/j.probengmech.2011.08.011
5. Qiu Z, Liu D (2020) Safety margin analysis of buckling for structures with unknown but bounded uncertainties. Appl Mathe Comput 367:124759. https://doi.org/10.1016/j.amc.2019.124759
6. Wang L, Xiong C, Yang Y (2018) A novel methodology of reliability-based multidisciplinary design optimization under hybrid interval and fuzzy uncertainties. Comput Methods Appl Mech Eng 337:439–457. https://doi.org/10.1016/j.cma.2018.04.003
7. Wang X, Wang L (2011) Uncertainty quantification and propagation analysis of structures based on measurement data. Math Comput Model 54:2725–2735. https://doi.org/10.1016/j.mcm.2011.06.060
8. Wu SQ, Law SS (2012) Evaluating the response statistics of an uncertain bridge–vehicle system. Mech Syst Signal Process 27:576–589. https://doi.org/10.1016/j.ymssp.2011.07.019

Vibration Control of Bridge Suspenders Using TMD and RTLD—A Comparative Study

Nihad Mohamed Ali and A. S. Sajith

Abstract Vibration control in civil engineering structures has received much attention recently, specifically in tall buildings and slender, wind-sensitive structures. Suspender cables are fragile and sensitive elements of a suspension bridge. Extreme weather can cause significant vibrations in these cables, which can jeopardize traffic safety. As a consequence, designing effective methods to reduce such vibrations are important. In this paper, study on a relatively new technique of using a Ring Tuned Liquid Damper (R-TLD) to minimize wind induced vibrations of bridge suspenders and its comparison with a Stockbridge Damper is presented. A numerical study is carried out by modeling and analysis in MATLAB, and ideal design parameters are optimized. Also, a simulation of the vibration and forced excitation is created using SIMULINK.

Keywords Structural control · Ring-TLD · TMD · Bridge suspenders

1 Introduction

Suspender cables are the components that are most fragile in a suspension bridge. In general, the vertical loads on the bridge deck coming on to the suspender cable can be catered by adequate design; however, the dynamic loads on the suspender cable are more complicated and critical, necessitating more care. The key causes of a suspender cable's dynamic response are ambient excitation (wind loads, ground motions, etc.) and traffic loads.

Vibration control is mainly incorporated in problems related to aerospace like tracking and pointing and also in flexible space structures. The technology moved quickly into issues related to infrastructure and civil engineering, especially, in the context of protection of bridges and buildings from severe loads of wind and earthquake. Control of structural vibration caused by earthquake and wind can be done by different ways like modifying masses, rigidities, damping or shapes and

N. M. Ali (✉) · A. S. Sajith
National Institute of Technology Calicut, Calicut, India

G. C. Marano et al. (eds.), *Proceedings of SECON'21*, Lecture Notes in Civil Engineering 171, https://doi.org/10.1007/978-3-030-80312-4_83

also by providing either active or passive counter forces. A number of factors govern the selection of each type of vibration control device like efficiency, capital cost, operating cost, compactness and weight, requirements of maintenance and safety. The use of passive control devices such as Tuned Mass Dampers (TMD) [1] and Tuned Liquid Dampers (TLD) to control the response of structures has proven to be very effective in structural control. Furthermore, the implementation and monitoring are quite simple for these devices. Another very important aspect is that in an active control system, control action is affected by external power and hence is vulnerable to power failure. This is always a possibility during an earthquake and in such circumstances a passive device such as TMD or TLD is more reliable in attaining the intended purpose.

The idea of TMD was first used by Frahm in 1909 to minimise the ship's rolling motion and hull vibration. Following that, Ormondroyd and Den Hartog introduced a theory of TMD [2], which was accompanied by a thorough examination of optimal tuning and damping parameters in Den Hartog's book on mechanical vibration. An undamped SDOF device subjected to a sinusoidal excitation was the basis for the initial theory. Numerous researchers have looked into applying the principle to damped SDOF structures [3].

G. W. Housner et al. [4], established a succinct source for researchers and practitioners looking to assess the latest advances in civil engineering structure control and monitoring. They also gave a connection between the structural control and other fields of control theory, mentioning both similarities and variations along with pointing out its future scope of application.

The use of Ring-Shaped Tuned Liquid Dampers (RSTLD) for vibration mitigation was suggested by An et al. [5], who discovered that it has equally strong vibration-mitigation potential in all directions of excitation. RSTLD's inner and outer diameters, the amount of liquid, the form of liquid, and the inner liquid barriers are the different parameters on which the effect of vibration mitigation was designed.

Di et al. [6], in their paper studied the effectiveness and robustness of Stockbridge Damper (SD) in vibration control of bridge suspenders. Using a simplified design approach based on full-scale suspender experiments, SDs are designed to monitor first and second-mode cable vibrations.

Modi and Seto [7] analysed rectangular TLDs numerically when accounting for nonlinear effects. Wave dispersion, boundary layers at the walls, floating particle interactions at the free surface, and wave breaking were addressed.

2 Modelling and Analysis

Stockbridge Damper is a basic example of dynamic vibration absorber, generally used in stay cables. No significant work has been reported on the problem of lateral vibration of vertical suspenders, which is in fact a critical component of a suspension bridge considering traffic safety and serviceability. The equations governing the system is modelled along with the control force given by the dynamic vibration absorbers.

An abridged numerical model of the chosen suspenders was created in MATLAB to ease the modelling process and reduce the computational workload. The basic assumptions in the modelling ignores non-linear geometry as lateral displacements are assumed to be small and the suspender acts in the linear elastic state obeying Hooke's Law.

2.1 Cable Parameters

Cables are flexible members which can withstand tension effectively. It is helpful to imagine the cable as a piece of rope. Because of its superior flexibility, it is unsuitable for compression, bending, or shear. A cable is an important component for a cable-supported bridge's overall structural stability, such as the main suspension bridge cable, the suspension bridge hanger, and so on.

The relation between frequency and cable forces can be formulated, ignoring cable sag and accounting for bending stiffness. The cable can be treated as a beam under Axial tension. The motion equation is [8]:

$$EI\frac{\partial^4 v(x,t)}{\partial x^4} - T\frac{\partial^2 v(x,t)}{\partial x^2} + m\frac{\partial^2 v(x,t)}{\partial t^2} = 0 \quad (1)$$

where EI is the bending stiffness of the cable. The equation is solved using cable's boundary conditions. When it's simply supported at both ends:

$$\omega_n^2 = \left(\frac{n\pi}{l}\right)^2\frac{T}{m} + \left(\frac{n\pi}{l}\right)^4\frac{EI}{m} \quad (2)$$

where

ω_n Fundamental frequency of system.
n Represents mode of vibration.
T Pretension in cable.
m Cable's mass per unit length.
l Length of the cable.

2.2 Parameters of the Model Used

See Table 1 [6].

Table 1 Parameters of the pedestrian bridge model

Span (m)	Diameter (mm)	Mass/length (kg/m)	Pretension (kN)	Modulus of elasticity (GPa)	Natural frequency (rad/s)
36	55	9.6	169	210	11.579

3 Numerical Analysis

The following is the equation of motion governing the system.

$$[M]\{\ddot{x}\} + [C]\{\dot{x}\} + [K]\{x\} = F_{EXT} + F_{CTRL} \tag{3}$$

where

F_{EXT} External force on the system.
F_{CTRL} Control force given by the Damper.

3.1 State Space Formulation

In this research, the governing equation of motion is modelled and solved through the state space approach. The space whose axes are the state variables is referred to as "state space" within that space, the state of the system can be represented as a vector. The state-space representation of a linear system with u inputs, y outputs, and n state variables is as follows.

System Response = f (current state).

Set of State Variables—(Position, Velocity)

$$\dot{x}(t) = Ax(t) + Bu(t) \tag{4}$$

$$y(t) = Cx(t) + Du(t) \tag{5}$$

4 Results of Numerical Analysis

4.1 Primary System

The present study considers a generalized single degree-of-freedom system (SDOF) model for the suspender cable, the total mass is assumed to be lumped at the central node of the suspender and the structure is taken as steel with damping (ς) as 2% of critical damping. The suspender considered is the longest cable of a pedestrian bridge. The parameters are adopted from the literature [6].

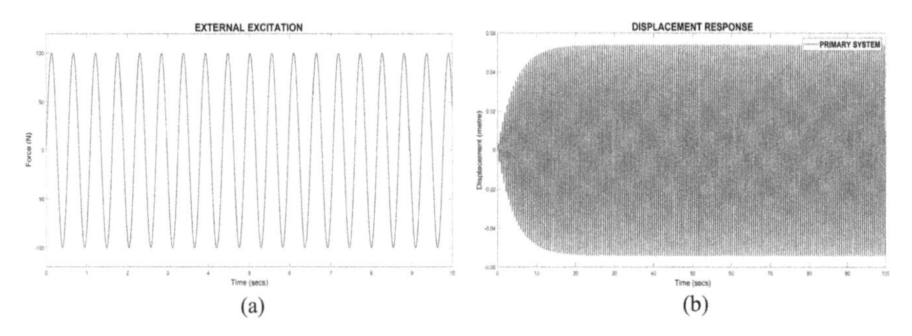

Fig. 1 **a** Sinusoidal excitation on the structure. **b** Displacement response of the suspender

Analysis is done on the structural model assuming that it is excited by a single component of force at the central node and is coded in MATLAB.

The SDOF model is analysed under sinusoidal excitation of 100 N in resonating condition and the responses without DVA shown in Fig. 1.

4.2 Control System with Stockbridge Damper (SD)

Stockbridge damper-design. As shown in Fig. 2b, A simple SD is made up of two bell- or horseshoe-shaped masses and a messenger cable. Both ends of the messenger cable are connected to the masses, and the damper is clamped to the cable. Same of masses are attached at both ends, and the clamp is located in the middle.

The MATLAB code developed for the SDOF system is amended with a Control system where a Stockbridge Damper (SD) is introduced (Fig. 2b). A two Degree of freedom model with primary system being the suspender and auxiliary system being the Stockbridge Damper attached (Fig. 2c) is analysed. Here the inherent damping of the absorber is neglected. The mass of the damper is fixed as 5% of that of the Suspender. The material of the messenger cable is an aluminium rod of diameter 12 mm. The stiffness has been fine-tuned by choosing a leg length so as to match the fundamental frequency of the suspender. A sinusoidal excitation of amplitude 100 N was applied on the controlled system at resonating condition. The response obtained is shown in Fig. 3a.

From numerical analysis the desired leg length was found to be 283 mm. The reduction in response attained was close to 99% in the ideal condition i.e., the inherent damping of the absorber being neglected. It was observed that the time to reach steady state was around 35–45 s.

Upon optimising the model parameters, a more realistic model was done having a damping ratio of around 3% for the absorber and the response was compared with the model without SD (Fig. 3b), the reduction in response was around 94%. The MATLAB code was proof checked using SIMULINK environment and the percentage error was very minimal.

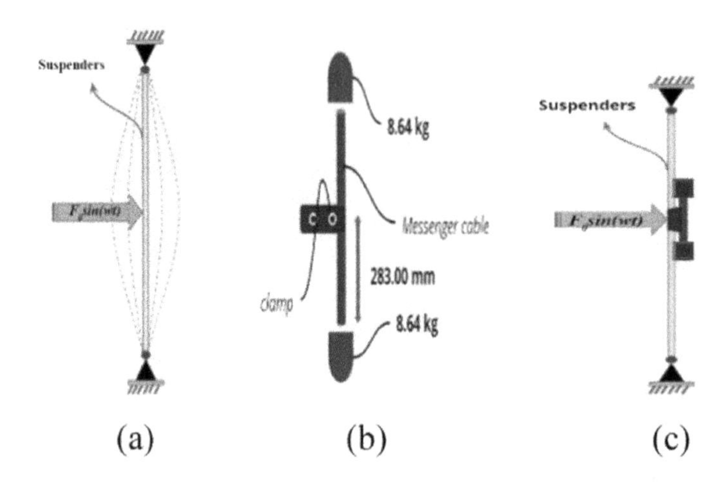

Fig. 2 **a** Primary system. **b** Stockbridge damper. **c** Suspender equipped with SD

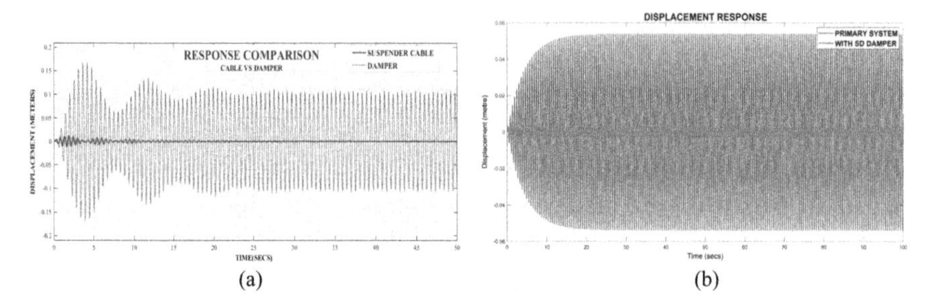

Fig. 3 **a** Ideal tuned response of a SD. **b** Realistic tuned response of a SD

4.3 Control System Annular TLD/R-TLD

RTLD Design. Annular TLD or Ring TLD are special type of Tuned Liquid Dampers which are effective in vibration mitigation of axisymmetric elements. Conventional TLDs adopted are in the form of rectangular or cylindrical sections which actually require a lot of space for installation and operation. This makes it difficult, if not impossible, for applications in tall slender sections owing to lack of space availability. Alternatively, this study presents an annular or ring-shaped TLD which can easily be installed in the cable without disturbing other mechanical devices or components of the structure. The main advantage is the ease of tuning and material cost and maintenance itself. In order to construct an efficient and secure TLD, the liquid's sloshing motion must be well understood. The key design parameters are the mass ratio, the frequency of the liquid sloshing motion, and the intrinsic damping of the TLD [9]. The strength of the sloshing motion is affected by the external excitation used, the tank's layout, the depth of the liquid layer, the properties of the confined liquid, and so on.

The natural sloshing frequency of a Ring-TLD is given by the expression, [10]

$$\omega_n{}^2 = \frac{g}{R}\varepsilon_n tanh\left(\frac{\varepsilon_n h}{R}\right) \quad (6)$$

where

ω_n = natural sloshing frequency, g = acceleration due to gravity, R = Outer radius of the tank, h = height of liquid in the tank, ε_n constant depending on value of (r/R) [11].

The parameters of the RTLD were chosen so as to meet the natural frequency of the suspender cable system. The outer diameter of the cylinder was designed to be 240 mm, height of tank was fixed as 500 mm, the liquid level required was found to be 248 mm, the mass ratio of the system was around 3%, the liquid in the system is assumed to be water. A schematic of the system is shown in Fig. 4.

A MATLAB code is developed to solve the equations of the RTLD provided, then the code was clubbed with SIMULINK to simulate the suspender cable-RTLD system,

The relevant equations used for the control force are given as follows:

$$F_{RTLD} = mM_L\left\{\ddot{q}_{(t)}F_1 d_1[1 + kR_1 k(\varepsilon_n)] + (1 + k^2)\ddot{x}_{(t)}\right\} \quad (7)$$

$$\ddot{q}_{(t)} + 2\varepsilon_l\omega_l\dot{q}_{(t)} + \omega_l{}^2 q_{(t)} = -\ddot{x}_{(t)} \quad (8)$$

$$d_1 = [2(1 - kR_1)/[(\varepsilon_n^2 - 1) - (k^2\varepsilon_n^2 - 1)R_1^2], \quad F_1 = \tanh(\varepsilon_n c)/\varepsilon_n h_0$$

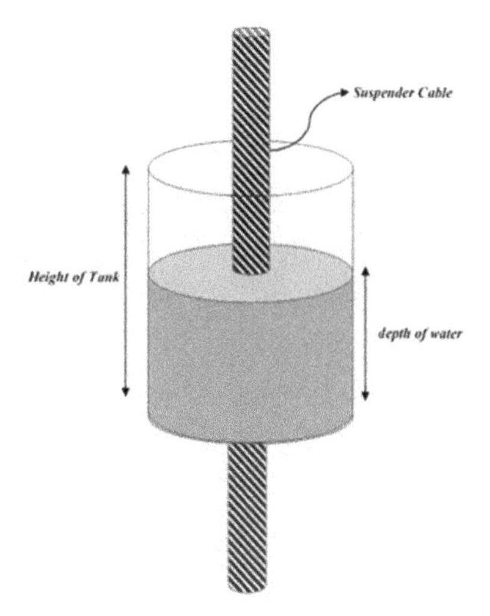

Fig. 4 Schematic of the RTLD

where,

$$R_1 = \alpha_1 J_1(\varepsilon_n k) + \beta_1 Y_1(\varepsilon_n k), \quad \alpha_1 = \frac{1}{J_1(\varepsilon_n)} + \left[\frac{J_1'(\varepsilon_n)}{Y_1'(\varepsilon_n)}\right] Y_1(\varepsilon_n)$$

$$\beta_1 = \left\{\frac{J_1'(\varepsilon_n)}{Y_1'(\varepsilon_n)}\right\}\alpha_1,$$

where

m = number of dampers, M_L = mass of liquid, $h_0 = h/R$, h being height of liquid, R being outer radius of the cylinder, k = outer radius to inner radius ratio, ε_l = liquid viscous damping ratio, ω_l = frequency of liquid sloshing, $q_{(t)}$ = generalised liquid motion in the container, $x_{(t)}$ = motion of the suspender cable. R_1, d_1 = relative parameters of the equation, $J_1(\varepsilon_n)$ = first order Bessel's function of first kind, $Y_1(\varepsilon_n)$ = first order Bessel's function of second kind.

Same set of excitation parameters that were used for the previous model are used here also and response is plotted:

The reduction in response was observed around 99%. It may be noted that the system achieved steady state in around 45–55 s. This may be attributed to the fact that SD is a discrete system where as the R-TLD a fluid continuum which require more time to achieve stability. Furthermore, it can be noted that R-TLD possess reduction capability in all directions of excitation.

A more practical model with a damping ratio of around 0.5% was given to the absorber which is water here and the response was compared with the model without control, the reduction in response was around 91% (Fig. 5b).

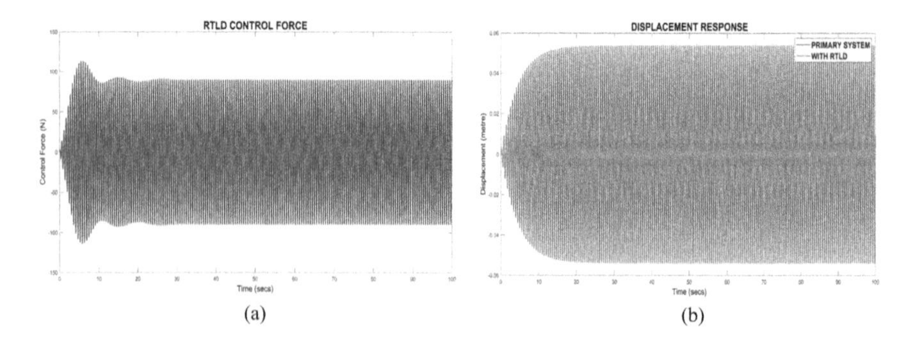

(a) (b)

Fig. 5 **a** Control force given by the RTLD. **b** Displacement response

5 Conclusions

This paper proposed a comparative study for vibration mitigation methods for slender structures such as the suspender cable in a suspension bridge. The research mainly focused on the design and comparison of a Stock bridge damper and a Ring tuned liquid damper. All the numerical procedures were coded through MATLAB and the validation of the vibration mitigation effect was done through simulation using SIMULINK. Conclusions drawn from the study are summarised below.

- Both the proposed methods dramatically reduced the vibrations in the targeted mode, the response reduction with SD and RTLD employed was comparable.
- As in the case of SD employed, better reduction in peak response was observed in the case of resonant harmonic excitations when compared with the case of R-TLD.
- In the case of Stockbridge Damper, the mitigation can be brought about adjusting the leg length of the messenger cable and cantilever tip mass attached at the ends as well as changing the material which in turn changes the inherent damping of the absorber.
- For the case of RTLD there was very good reduction in peak response but the time to achieve steady state was a bit greater compared to SD, this can be attributed to the fact SD is a discrete system where as RTLD a continuum.
- Compared to SD the designed RTLD has vibration reduction capability in all directions of excitations and thus is more suitable for axisymmetric elements such as the suspender cables. The main advantage is the ease of tuning and material cost and maintenance itself.

References

1. Manohar CS, Venkatesa S (2006) Earthquake engineering education through laboratory experiments. In: NPEEE short-term course, 16–21 Jan 2006. IISC, Bangalore
2. Den Hartog JP (2007) Mechanical vibrations. Dover Publications inc., New York
3. Li1 H-J, Hu S-LJ (2017) Tuned mass damper design for optimally minimizing fatigue damage. ASCE
4. Housner GW, Bergman LA, Caughey TK, Chassiakos AG, Claus RO, Masri SF, Skelton RE, Soong TT, Spencer BF, Yao JTP (1997) Structural control: past, present, and future. J Eng Mech 123(9):897–971
5. An Y, Ou J, Pan S, Wang Z (2019) Vibration mitigation of suspension bridge suspender cables using a ring-shaped tuned liquid damper. ASCE J Bridge Eng 24(4):04019020
6. Di F, Sun L, Qin L, Chen L, Zou Y, Jiang Y, Zhu Y (2020) Full-scale experimental study on vibration control of bridge suspenders using stockbridge damper. J Bridge Eng 25 (8):04020047
7. Modi VJ, Seto ML (1997) Suppression of flow-induced oscillations using sloshing liquid dampers: analysis and experiments. J Wind Eng Indus Aerodyn 67&68:611–625
8. Suangga M, Hidayat I, Bontan DJ (2017) The analysis of cable forces based on natural frequency. In: IOP conference series: earth and environmental science

9. Sun LM, Fujino Y, Pacheco BM, Chaiseri P (1992) Modelling of tuned liquid damper (TLD). J Wind Eng Indus Aerodyn 43(1–3):1883–1894
10. Ghaemmaghami A, Kianoush R, Yuan X-X (2012) Numerical modeling of dynamic behavior of annular tuned liquid dampers for applications in wind towers. Comput-Aided Civ Infrastruct
11. Bauer HF (1960) Theory of fluid oscillations in a circular cylindrical ring tank partially filled with liquid. NASA TND-557

Structural Performance of Cold-Formed Steel Elliptical Hollow Section Columns Under Impact Loading

K. Vyshnavi and D. C. Mitra

Abstract Due to the structural efficiency and aesthetically pleasing appearance tubular members are often used in structural applications such as buildings, roof supporting systems, offshore platforms etc. Elliptical hollow section (EHS) combines the architectural qualities of circular hollow section (CHS) and the structural advantages of rectangular hollow section (RHS). Thus, the elliptical hollow section members have been used recently as structural supporting members in different types of structures. The study consists of the numerical validation of structural performance of cold-formed steel elliptical hollow section column buckled about the minor axis. For numerical investigation finite element software ANSYS workbench is utilized. There are no other studies conducted on cold formed steel elliptical hollow section subjected to impact loading. So, the study is further extended to analyse the effect of impact on these columns. The parametric study is conducted by changing the impact velocity and by analysing the effect of impact on different eccentric height along the column length.

Keywords Elliptical hollow section · Cold formed steel · Impact

1 Introduction

Due to recent advancements in the modern architecture tubular construction is emerged and due to their aesthetic pleasing appearance and their structural advantages, these are widely used in the construction of buildings, bridges, stadia, roof supporting systems and offshore platforms etc. In addition to the familiar hollow sections such as circular, square and rectangular to meet the demands of architectural and design led to the introduction of elliptical hollow sections (EHS) due to their structural advantages and aesthetic appearance. Elliptical hollow section has advantages over other sections such as, when subjected to eccentric loading EHS can provide more efficiency than the circular hollow section, high

K. Vyshnavi (✉) · D. C. Mitra
Department of Civil Engineering, Government College of Engineering Kannur, Kannur, India

G. C. Marano et al. (eds.), *Proceedings of SECON'21*, Lecture Notes in Civil Engineering 171, https://doi.org/10.1007/978-3-030-80312-4_84

torsional stiffness, elliptical hollow section also provides good aesthetic properties, it has more bending capacity, also when compared to circular hollow section elliptical hollow section in compression have stable post buckling responses. Cold forming is one of the most commonly used steel manufacturing technologies, having several advantages. The cold forming process also result in strength enhancement and greater cross-sectional resistance than that of hot finishes sections of similar size made from the same parent material. The use of different rolling system led to the development of new innovative cross-section shape which contributes the recent developments in cold-forming technology. Elliptical hollow section is one of the most innovative sections which can be easily produced. The majority of studies have been done on the behaviour of hot-finished elliptical hollow section where as the studies on cold formed EHS is limited and this study focus on the numerical analysis of cold formed steel elliptical hollow section column subjected to impact loading since these columns are used in buildings, parking areas etc. and there are chances of vehicle impact on these columns during accidents.

Chen and Young [1] conducted study on the cold formed steel elliptical hollow section buckled about minor axis. They also investigated the material properties, residual stress and cross-sectional behaviour of these hollow sections [2]. They observed that the specimen at the curviest portion have higher proof stress and ultimate strength than its counterparts at the flattest portion. The behaviour of elliptical hollow section subjected to compression and bending [3–5], buckling behaviour [6, 7], material properties caused by different formation process [8], lateral instability of EHS subjected to bending [9], flexural buckling [10], elastic buckling response [11], shear response of EHS [12] etc. have been investigated in different studies by different researchers in the field of elliptical hollow section. However mostly the studies on elliptical hollow section are based on hot finished sections [13–16] but the studies based on cold formed steel elliptical hollow section is very limited.

This paper aims at investigating the behaviour of cold formed steel EHS columns subjected to impact and the study is thus further extended to analyse its behaviour under different impact velocity and also with different eccentric height along the column length.

2 Validation

The finite element analysis consists of mainly three stages. The very first stage is pre-processing stage, the second is analysis solver and the final stage is post-processing stage. The first stage includes defining the finite element. In the second stage the models are solving and the final stage gives the results. The finite element model is created in ANSYS workbench software. The validation of cold formed steel EHS pin ended column subjected to bending about minor axis was investigated using ANSYS workbench software. In the model, the measured cross section dimension of the test specimen was used [1]. By applying the boundary conditions and loads the model is analyzed. The validation includes the study of the performance of this column subjected to bending in minor axis direction according to the results obtained from Chen et al. (2019). In this paper the numerical result obtained from ANSYS workbench software are to be validated.

2.1 Description of the Model

The experimental model used in the paper done by Chen et al. (2019) was used to model and numerically validate in this study. In this the model taken for study have a nominal dimension of D * B * t of elliptical hollow section. The model of cold formed steel EHS pin ended column was developed with the help of geometry in ANSYS workbench software. The section was drawn and divided into four sections. That is two curved portion and two flat portions for assigning the material properties for the corresponding portions. The section is divided from the end to a distance of Dm/6 where, Dm is the largest dimension along major axis direction. The specimen taken for modelling is 140 * 85 * 3-CL440, which is having a length of 440 mm. The cold formed steel having nominal yield strength of about 388.4 MPa young's modulus of 208 GPa is used for flattest portion whereas, for curviest portion the nominal yield strength of about 401 MPa and young's modulus of about 212.7 GPa is used. The material properties of the section used is shown in Table 1. The poisons ratio of 0.3 is used for material modelling [1]. The model requires meshing in finite element analysis. Here 20 mm mesh size is applied for the model. The shell element was modelled as four nodes shell element with reduced integration that is shell element 181 in ANSYS workbench software. Figure 1 shows the finite element meshing of the model.

2.2 Loads and Boundary Conditions

The boundary conditions were replicated according to the experiments done by Chen et al. (2019). For getting a unique solution the displacement boundary conditions are required. The provided displacement and support conditions are distributed to the model as per the experimental condition. The reference point for the

Table 1 Material properties

Section	Flattest portion (TC1)			Curviest portion (TC2)		
	E (GPa)	$\sigma_{0.2}$ (MPa)	σ_u (MPa)	E (GPa)	$\sigma_{0.2}$ (MPa)	σ_u (MPa)
140 * 85 * 3CL440	208	388.4	432.9	212.7	401	457.8

Fig. 1 Finite element meshing of the model

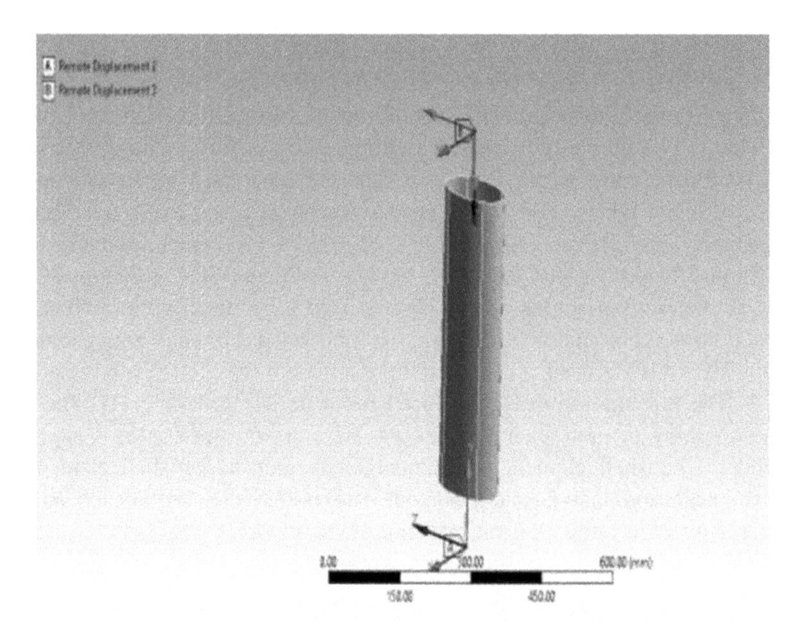

Fig. 2 Boundary conditions and loading developed using ANSYS workbench

stimulation of displacement is located at a distance of 86.8 mm away from the specimen edge which is the sum of the end plate thickness and the one wedge plate height. Also, in order to include the unintentional loading eccentricity in the model the reference points were offset from the cross-section centroid. The support condition was modelled by restraining all degrees of freedom excluding the two ends minor axis rotation and the longitudinal displacement at the loading point. At the loading end the compressive load is applied specifying the axial displacement of the reference point. Figure 2 shows the loading and boundary conditions developed using ANSYS Workbench software.

2.3 Validation of Finite Element Model

The cold formed steel elliptical hollow section column under minor axis bending was used for validation study. The model is created in ANSYS workbench and before running the solution controls were set. By inputting the number of steps for analyzing the model the load is applied as displacement. The time stepping helps to predict the number of load steps and sizes. The failure modes and ultimate load capacity predicts the accuracy of the finite element model by comparing the available experimental results with the theoretical investigation. The deformed shape of the specimen after analysis is shown in Fig. 3 The column specimen 140 × 85 × 3CL440 is validated using ANSYS Workbench software and is compared with the experimental results from the study conducted by Chen [1]. Figure 4 shows Comparison between experimental and numerical load-end shortening responses of EHS column specimen 140 × 85 × 3-CL440. The ultimate load obtained from the experiment is 394.6 kN whereas, those obtained from the finite element analysis is 396.13 kN. From this it is clear that the experimental study is validated by finite element analysis and an error of about 0.38% is observed. The test results demonstrate that the finite element model is successfully validated.

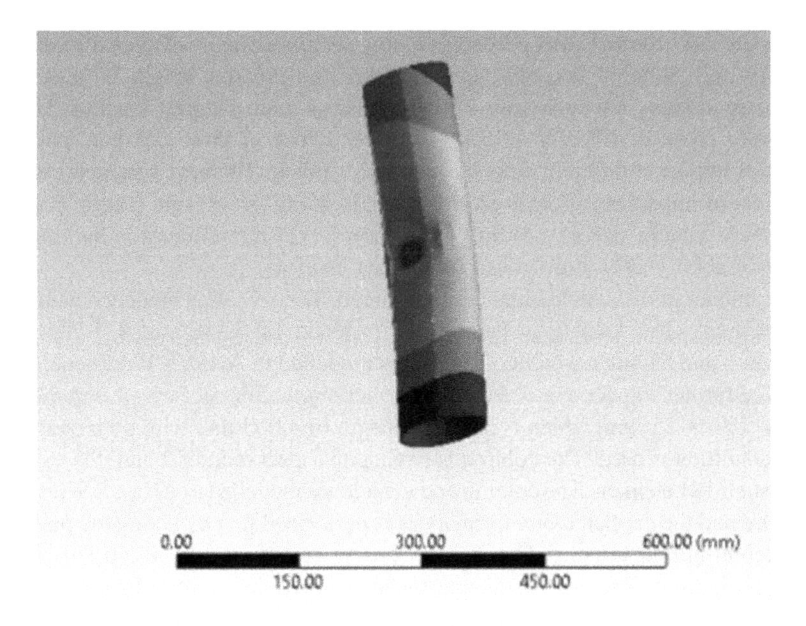

Fig. 3 Deformed shape of the specimen after analysis

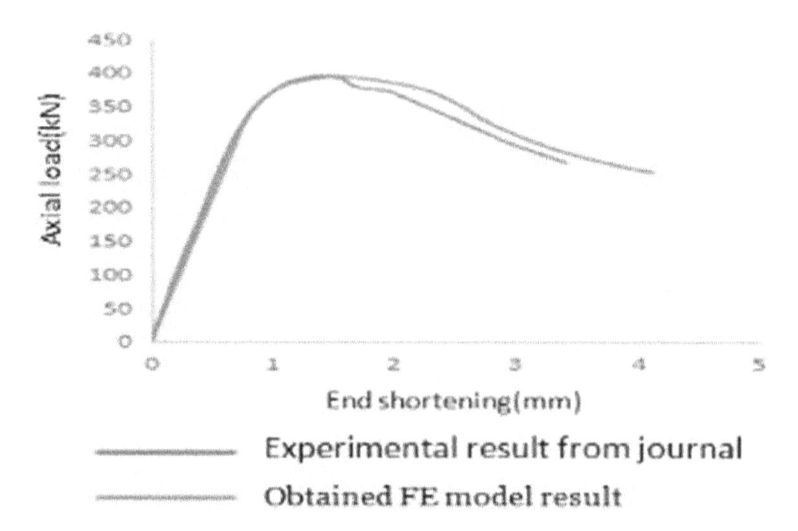

Fig. 4 Comparison between experimental and numerical load-end shortening responses of EHS column specimen 140 × 85 × 3-CL440

3 Parametric Study

In this the cold formed steel elliptical hollow section column subjected to different velocity and different eccentric height along the column length is selected as parameter to study the behaviour of these columns under impact loading. The first parameter taken is different velocity. The behaviour of these columns under the action of impact at different velocity has been studied. The next parameter taken is the effect of impact at different eccentric height along the column length. For these analyses a column size of 500 mm × 250 mm × 25 mm dimension and having a height of about 4,000 mm is used for impact analysis.

For impact study vehicle impact is considered. The vehicle is modelled using solid 186 element. The vehicle is having a dimension of 3,845 mm x 1,735 mm x 1,530 mm and having a weight of 975 kg is modelled in ANSYS Workbench and is used for further impact study. For this further study column having dimension of 500 × 250 × 25 mm, which is having a length of 4,000 mm with both ends under fixed condition is used. The column is having an aspect ratio of 2 and it is modelled using shell 181 element. The column and vehicle are modelled in ANSYS Workbench software and the explicit dynamic analysis is performed first by taking the parameter as different hitting velocity. For this analysis velocity range of 50, 60, 70, 80, and 90 kmph is used. The next parameter taken is different eccentric height along the column length and analyzing the effect of impact at those heights. For this different height of about 300, 400, 500, 600, 700 and 800 mm is taken and the effect of impact at these heights were analyzed numerically. Figure 5 indicates the modelled column and vehicle for impact study.

Fig. 5 Modelled column and vehicle for impact study

3.1 Different Impact Velocity

In this the study considered is the action of different impact velocity. A wide range of velocity varying from 50, 60, 70, 80 and 90 kmph is used for this study. In this explicit-dynamics is used for this study and the varying velocity is provided for impact in the same column having dimension of 500 × 250 × 25 mm, which is having a length of 4,000 mm. From this study it is observed that as velocity of impact in each case increases the energy also increases. Figure 6 shows the energy verses time graph under different impact velocity. From this graph it is clear that in the case of cold formed steel elliptical column as impact velocity increases the energy develop in column also increases. It is also observed that as velocity increases the deformation also increases. Figure 7 shows the deformation verses time graph under different velocity of impact. It is observed that as velocity increases deformation increases by about 10–12%.

3.2 Effect of Impact on Different Eccentric Height Along the Column Length

In this the parameter taken for analysis is the action of impact under different eccentric height along the column length. For this analysis the velocity of impact is fixed at 50kmph and the eccentricities along the column length considered for the analysis include 300, 400, 500, 600, 700 and 800 mm. Under these conditions the effect of different eccentricities along the column length and the energy developed in the column is analysed by using ANSYS workbench using explicit-dynamics. For this the same column as used in the previous case that is having dimension of 500 × 250 × 25 mm, which is having a length of 4,000 mm is used. In this analysis it is observed that as eccentricity of column along the column length increases the energy developed in the column first increases up to 500 mm height

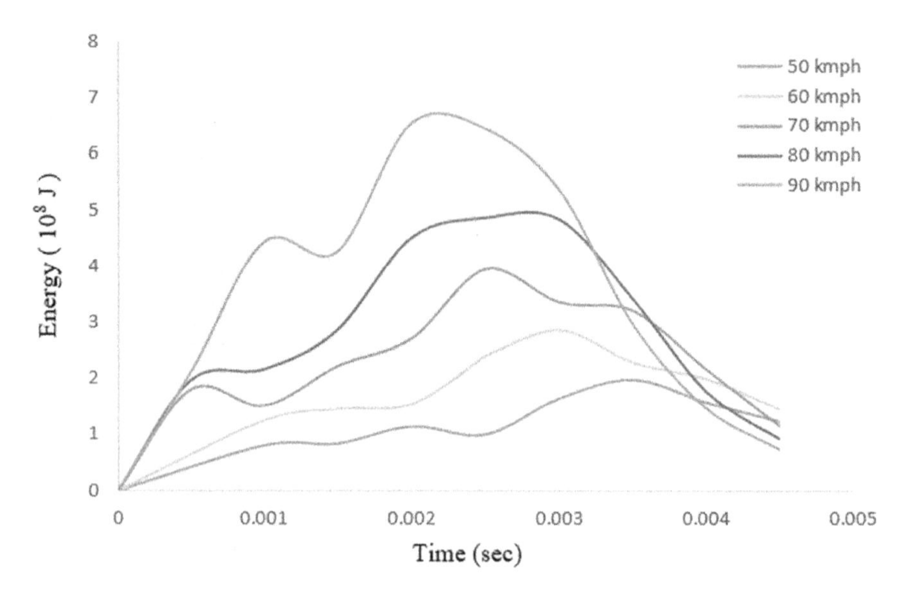

Fig. 6 Energy versus time graph under different impact velocity

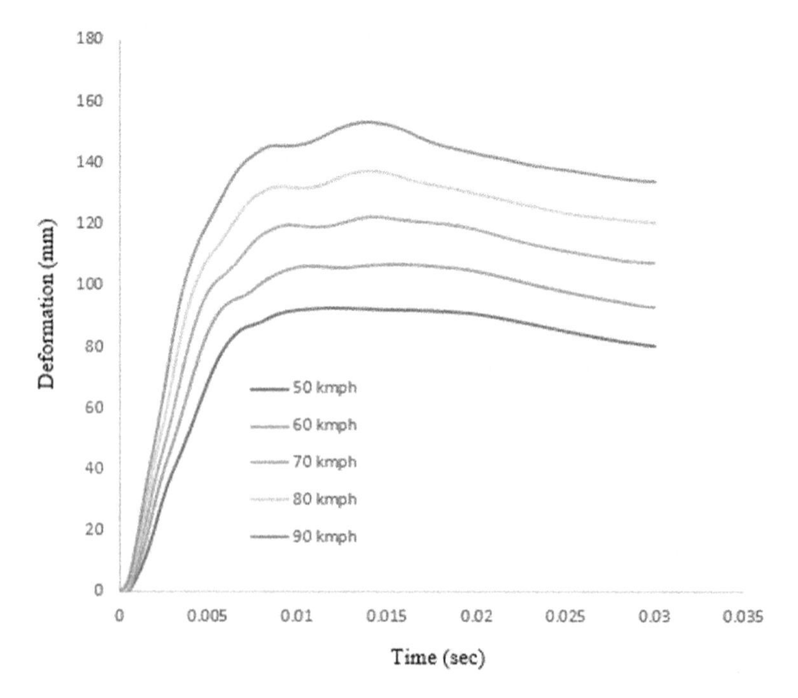

Fig. 7 Deformation versus time graph under different velocity of impact

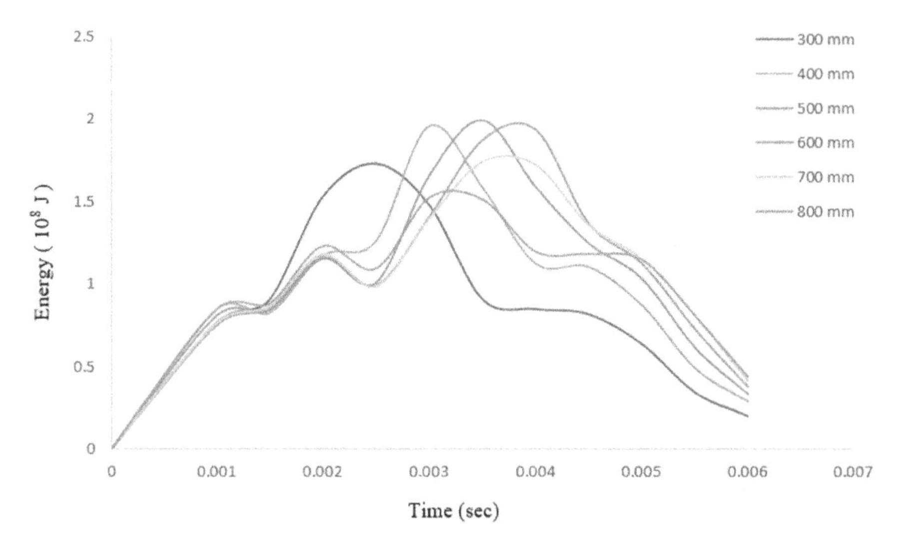

Fig. 8 Energy versus time graph under different eccentric height along the column length

and after that energy developed decreases. This is due to the dissipation of energy taking place during the impact. Figure 8 shows the energy versus time graph under different eccentric height along the column length.

4 Conclusion

The structural performance of cold formed steel elliptical hollow section column under impact loading has been studied. A numerical model was developed and that is validated against the results available from Chen et al. (2019). The parametric studies were done based on different impact velocity of 50, 60, 70, 80 and 90 kmph and also by analysing the effect of impact along different eccentric height along the column length. The energy developed in cold formed steel elliptical hollow section column by impact is found to increase with increasing the impact velocity. It is also observed that as velocity increases the deformation developed due to impact also increases by about 10–12%. Also, it is observed that the impact along the column considering different eccentric height along the column length it is found that as height increases the energy developed due to impact increased up to 500 mm height and after that the energy developed get reduced.

References

1. Chen MT, Young B (2019) Structural performance of cold-formed steel elliptical hollow section pin-ended columns. Thin-Walled Struct 136:267–279
2. Chen MT, Young B (2019) Material properties and structural behavior of cold-formed steel elliptical hollow section stub columns. Thin-Walled Struct 134:111–126
3. Chen MT, Young B (2020) Beam-column tests of cold-formed steel elliptical hollow sections. Eng Struct 210:109911
4. Law KH, Gardner L (2013) Buckling of elliptical hollow section members under combined compression and uniaxial bending. J Constr Steel Res 86:1–16
5. Law KH, Gardner L (2013) Global instability of elliptical hollow section beam-columns under compression and biaxial bending. Int J Steel Struct 13(4):745–759
6. McCann F, Fang C, Gardner L, Silvestre N (2016) Local buckling and ultimate strength of slender elliptical hollow sections in compression. Eng Struct 111:104–118
7. Silvestre N (2008) Buckling behaviour of elliptical cylindrical shells and tubes under compression. Int J Solids Struct 45(16):4427–4447
8. Quach WM, Young B (2015) Material properties of cold-formed and hot-finished elliptical hollow sections. Adv Struct Eng 18(7):1101–1114
9. Law KH, Gardner L (2012) Lateral instability of elliptical hollow section beams. Eng Struct 37:152–166
10. Chan TM, Gardner L (2009) Flexural buckling of elliptical hollow section columns. J Struct Eng 135(5):546–557
11. Abela JM, Gardner L (2012) Elastic buckling of elliptical tubes subjected to generalised linearly varying stress distributions. Thin-walled Struct 58:40–50
12. Gardner L, Chan TM, Wadee MA (2008) Shear response of elliptical hollow sections. Proc Inst Civ Eng-Struct Build 161(6):301–309
13. Chan TM, Gardner L (2008) Bending strength of hot-rolled elliptical hollow sections. J Constr Steel Res 64(9):971–986
14. Chan TM, Gardner L (2008) Compressive resistance of hot-rolled elliptical hollow sections. Eng Struct 30(2):522–532
15. Chan TM, Gardner L, Law KH (2010) Structural design of elliptical hollow sections: a review. Proc Inst Civ Eng-Struct Build 163(6):391–402
16. Cai Y, Quach WM, Chen MT, Young B (2019) Behavior and design of cold-formed and hot-finished steel elliptical tubular stub columns. J Constr Steel Res 156:252–265

Prediction of Liquefaction of Soils Using Particle Swarm Optimization (PSO)

C. Anitta Justin and N. Sankar

Abstract Prediction of liquefaction potential of soils is significant in order to mitigate risk and major damages to structures. Currently used deterministic methods have drawbacks like mismatch between the assumptions in modelling and the actual in-situ conditions, observational errors. Hence many predictive techniques are being used as an alternative solution to reach a better decision and the neural networking approaches are an ideal one. This paper presents the technique of neural network to develop an Artificial Neural Network (ANN) model optimized by Particle Swarm Optimization (PSO), based on CPT data to predict the liquefaction potential of soils. The database used in this study consists of 235 CPT-based field records from ten major earthquakes over a period of 35 years. Important parameters including normalized peak horizontal acceleration at ground surface, earthquake magnitude, total vertical stress, effective vertical stress, cone resistance and depth of penetration, are selected as the input parameters for the ANN-PSO model. PSO technique is hybridized along with Artificial Neural Network (ANN) to utilize the advantage of both the techniques.

Keywords ANN · PSO · CPT

1 Introduction

Liquefaction damages have been frequently reported after earthquakes and is commonly observed in loose saturated soil deposits. This has enhanced the increasing research into the assessment and study of liquefaction and strong contributions were given by various geotechnical engineers all over the world. Currently, many methods were developed to assess liquefaction but it is hard to choose an empirical equation since the phenomenon of liquefaction is highly complex depending on the in-situ earthquake environment and soil characteristics.

C. Anitta Justin (✉) · N. Sankar
Civil Engineering, National Institute of Technology Calicut, Calicut, India

G. C. Marano et al. (eds.), *Proceedings of SECON'21*, Lecture Notes in Civil Engineering 171, https://doi.org/10.1007/978-3-030-80312-4_85

Liquefaction assessment is generally carried out from field observations and site performance during earthquakes. The simple analytical models generated from standard penetration test (SPT) were widely used since 1960's till now. But in recent years geotechnical experts like Juang [1], Alireza [2] and Xinhua [3] achieved better results and highly recommend using cone penetration test (CPT) since it has the advantage of fast, continuous and reliable data.

Due to the inherent uncertainties associated with the above methods, it is tough to adopt a particular equation for estimating. Here comes the advancement and popularity of artificial intelligence approaches such as Artificial Neural Networks (ANN). ANN has widely explored in many engineering disciplines and has various application in the field of civil engineering. Researchers have been utilising this machine learning technique to predict liquefaction. For example, Juang [1] trained a CPT based neural network model to develop an empirical equation. Farrokhzad [4] prepared liquefaction micro zonation map for Babol City.

Although all the researchers who adopted this technique could achieve best prediction and is more efficient than statistical methods, they have pointed out several inherent drawbacks like overfitting, slow convergence, poor generalizing performance, arriving at local minimum. In order to improve the efficiency of ANN technique, Particle Swarm Optimization (PSO) is incorporated. Successful performance of PSO has been achieved by experts in diverse fields, researchers experimented with several combinations like hybrid ANN-PSO, Genetic Algorithm (GA)-PSO and are successful. For example, Alireza [2] Fuzzy Support Vector Machine (FSVM) classifier GA-PSO model gave 100% accurate results in predicting liquefaction. Xinhua [3] found that prediction from PSO-BP (Back Propagation) model is accurate than GA-BP and BP network model.

This paper discusses the development of ANN-PSO model to predict the liquefaction of soils. Six factors including including normalized peak horizontal acceleration at ground surface, earthquake magnitude, total vertical stress, effective vertical stress, cone resistance and depth of penetration are selected as the input parameters.

2 Methodology

The methodology for developing ANN-PSO model includes model input and output determination, division of data and data pre-processing, fixing suitable network architecture, optimization of connection weights (training), stopping criteria, and model validation. Artificial neural network (ANN) is a form of Machine Learning method that attempts to adopt the behaviour of the human brain and nervous system. ANN consists of three layers an input layer, an output layer and one or more hidden layers. Each layer contains a number of neurons called nodes. The input data is fed to the input layer for training and the information is propagated further according to the weights, which is adjusted by the network. On training, network utilizes a learning rule to discover a set of weights to map the input/output with least error. After the successful attainment of training phase, the performance of the trained model is verified using a testing data set.

Here the training of networks to find optimum weights is achieved through Particle Swarm Optimization (PSO) algorithm, a bird swarm simulation. In PSO each individual particle has the ability to make decision based on personal experience so far and considering the feedback of neighbor particles to reach towards an optimum goal for the entire population. A simple example is flock of birds in search of food.

The procedure begins with initialising a random group of particles. The locations of those particles are assigned randomly using representative of the ANN weights and biases. After that, the hybrid PSO–ANN is trained. The error between actual and predicted values is computed. The computed error is reduced through changing the particle positions at each iteration. This procedure is followed until the termination criteria is met.

2.1 Database

Based on thorough literature review, six parameters governing the soil liquefaction potential were taken into account. These parameters are Earthquake moment magnitude, M_W, Total vertical overburden pressure, $\boldsymbol{\sigma_V}$ (KPa), Effective vertical overburden pressure, $\boldsymbol{\sigma_V}'$ (KPa); Measured cone tip resistance q_c (KPa) value from CPT, and maximum horizontal ground acceleration, $a_{\max}$.

The liquefied case and non-liquefied case were respectively represented by 1 and 0. The proposed models were developed based on several CPT case records collected and compiled by Chern [5]. The database consists of total 235 CPT post liquefaction cases, 152 out of them are liquefied cases and other 83 are non-liquefied cases. It contains information from the CPT measurements at over 52 sites along with field performance observations of 10 different major earthquakes between 1964 and 1999 which are listed in Table 1.

Table 1 Database

Earthquake	Magnitude	Country	No of cases
1964 Niigata Earthquake	7.6	Japan	1
1971 San Fernando Valley Earthquake	6.4	USA	6
1976 Tangshan Earthquake	7.8	China	39
1977 Vrancea Earthquake	7.2	Bulgaria	3
1979 Imperial Valley Earthquake	6.6	USA	22
1981 Imperial Valley Earthquake	6.0	USA	20
1983 Nihonkai-Cho Earthquake	7.7	Japan	1
1988 Sangucnay Earthquake	5.9	Canada	2
1989 Loma Prieta Earthquake	7.1	USA	73
1999 Chi-Chi Earthquake	7.6	China	68

These dataset values are normalized using the following equation:

$$X_n = \frac{X - X_{\min}}{X_{\max} - X_{\min}} \quad (1)$$

where

X_n Normalized value.
X Actual value.
X_{max} Maximum value.
X_{min} Minimum value.

The normalized datasets are divided into two subsets: training set and testing set as 80% and 20% respectively. The datasets are divided randomly using scikit-learn's train-test-split function of Python software. It is essential that both of these subsets should represent the same population and all the patterns present in the training set should be available in testing set also. Thus, randomly different combinations are selected and then checked for its statistical consistency. This is continued until statistically consistent sets are obtained. The statistical parameters considered are mean, standard deviation, maximum and minimum are shown in Table 2.

2.2 *ANN-PSO Model*

This study is performed with MATLAB R2020b software. To obtain the optimal values of PSO parameters, the ANN-PSO model was performed with different independent tests on swarm size, coefficients of velocity and maximum number of iterations. Many of the researches suggested that more than one hidden layer would

Table 2 Statistical parameters of data

Input parameters		Mean	Standard deviation	Maximum	minimum
Ground acceleration	Training	0.273	0.153	0.08	0.8
	Testing	0.287	0.152	0.10	0.69
Earthquake magnitude	Training	7.197	0.546	5.9	7.8
	Testing	7.202	0.546	5.9	7.8
Depth	Training	6.932	3.396	1.2	13.5
	Testing	6.356	3.363	1.8	12.5
Cone tip resistance	Training	5.125	2.016	2.00	10
	Testing	4.992	2.228	2.09	9.1
Total vertical stress	Training	119.601	63.013	22.6	249.8
	Testing	119.000	63.018	33.3	231.3
Effective vertical stress	Training	76.334	33.674	20.6	182.6
	Testing	78.349	33.297	24.5	170.7

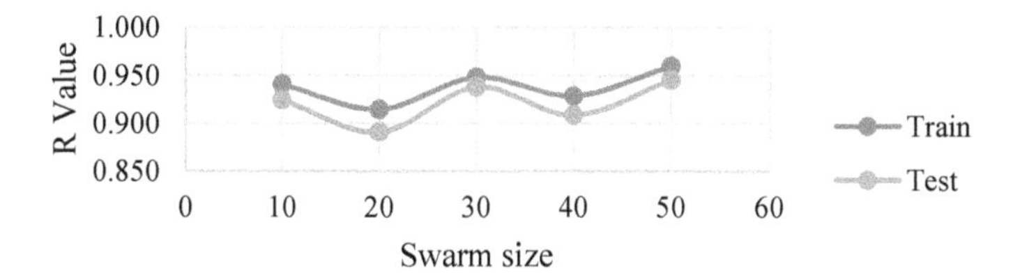

Fig. 1 Variation of R value and swarm size; C1 = C2 = 2.0, iterations = 5,000, number of hidden neurons = 6

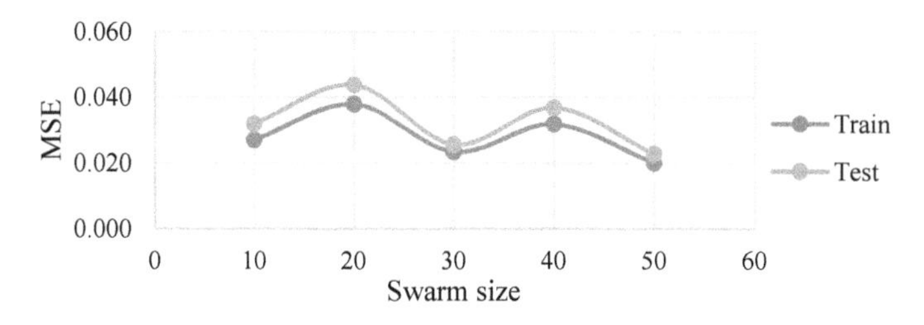

Fig. 2 Variation of MSE and swarm size; C1 = C2 = 2.0, iterations = 5,000, number of hidden neurons = 6

slow down the training process and has higher chances of getting trapped in local minima [1, 4, 7, 8]. Thus, a single hidden layer is chosen for this study. The procedure to find a best accelerating coefficient, swarm size, number of hidden neurons is purely a trial-and-error procedure that depends on the datasets taken. However, from the experience of researchers a common swarm size adopted is 10–50. Thumb rule to find the number of hidden neurons is 2i + 1 where i is the number of input parameters, here 6. Therefore, upper limit of number of hidden neurons is 13. Therefore, an interval swarm size ranging from 10 to 50 is varied fixing number of hidden neurons as 6 and C1 = C2 = 2.0 in reference with Xinhua [3]. Coefficient of correlation R and RMSE are used to assess the model performance. The plot of R versus swarm size and MSE vs swarm size is shown in Figs. 1 and 2.

As it is clear that the swarm size showed better R and reduced MSE value for both testing and training set, next trials were done fixing swarm size 50 and varying number of hidden neurons from 1 to 13 out of which 3, 5, 6, 7, 9, 10, 11, and 12 showed better results for different combinations of C1 and C2.

3 Results and Discussion

From the numerous numbers of trials performed on the datasets, it was found out that swarm size 50 and C1 = 2.0 and C2 = 2.0 showed best R and MSE values for number of hidden neurons 3, 5, 6, 7, 9, 10, 11, 12 out of which 11 showed the

Table 3 Performance of ANN-PSO model with varying no. of nodes in hidden layer

Number of hidden neurons	Train		Test	
	R	MSE	R	MSE
3	0.944	0.025	0.948	0.021
5	0.950	0.023	0.916	0.040
6	0.956	0.02	0.945	0.023
7	0.947	0.024	0.903	0.039
9	0.959	0.019	0.950	0.022
10	0.963	0.017	0.953	0.020
11	0.967	0.015	0.968	0.015
12	0.966	0.016	0.946	0.022

Table 4 Performance parameters

Performance parameters	R	MSE	R^2	RMSE
Training set	0.967	0.015	0.935	0.122
Testing set	0.968	0.015	0.937	0.122

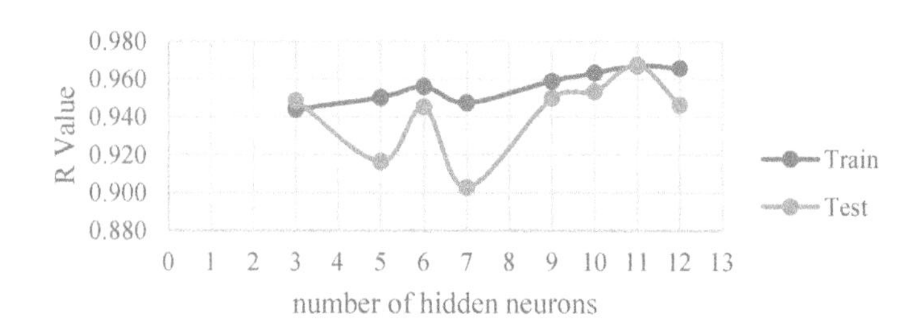

Fig. 3 Variation of R value and number of hidden neurons with swarm size 50, C1 = C2 = 2.0

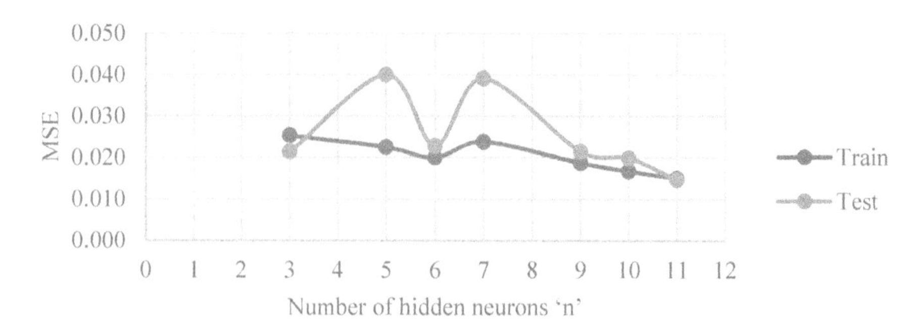

Fig. 4 Variation of MSE value and number of hidden neurons with swarm size 50, C1 = C2 = 2.0

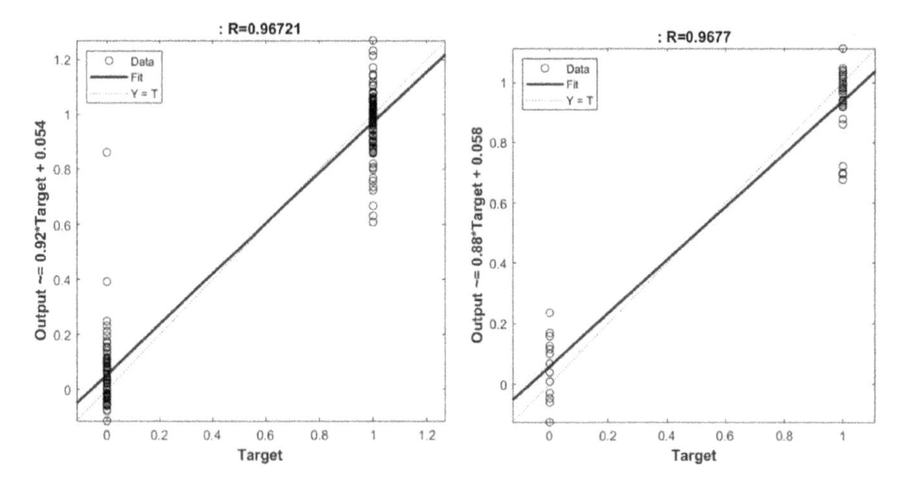

Fig. 5 Regression plot for training and testing datasets

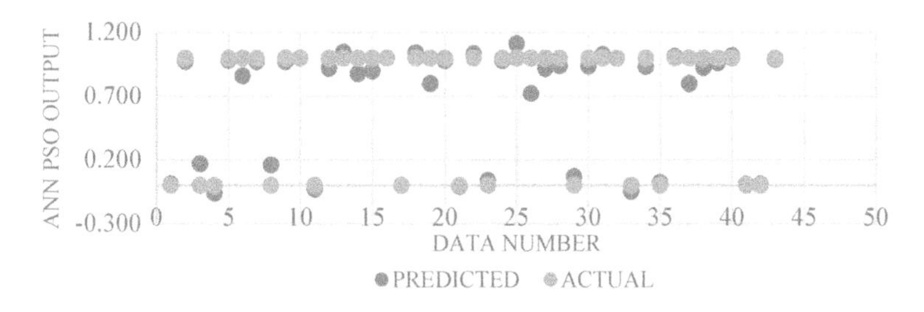

Fig. 6 Comparison between predicted and actual liquefaction for testing dataset

highest R Value for both testing and training as shown in Tables 3 and 4. R, R^2 value closer to 1 and MSE RMSE closer to 0 indicates good performance of the model. The variation of R and MSE,is shown in Figs. 3 and 4. The regression plots for training and testing datasets are shown in Fig. 5. Comparison between actual and predicted values of testing dataset is shown in Fig. 6.

4 Conclusion

In this study, an ANN-PSO model is developed to predict the liquefaction of soil. Various trails were done and the best model parameters were identified. The values of performance parameters indicated that the developed model has good predictive capability.

References

1. Juang H, Yuan H, Lee D-H, Lin P-S (2003) Simplified cone penetration test-based method for evaluating liquefaction resistance of soils. J Geotech Geoenviron Eng 129:66–80
2. Rahbarzare A, Azadi M (2019) Improving prediction of soil liquefaction using hybrid optimization algorithms and a fuzzy support vector machine. Bull Eng Geol Environ 78:4977–4987 (Springer)
3. Xue X, Liu E (2017) Seismic liquefaction potential assessed by neural networks. Environ Earth Sci 76:192
4. Farrokhzad F, Barari A, Choobbasti AJ (2010) Liquefaction micro zonation of Babol city using artificial neural network. J King Saud Univ Sci 24:89–100
5. Chern S-G, Lee C-Y, Wang C-C (2008) CPT-based liquefaction assessment by using fuzzy-neural network. J Mar Sci Technol 16(2):139–148
6. Bagheripour MH, Shooshpasha I, Afzalirad M (2012) A genetic algorithm approach for assessing soil liquefaction potential based on reliability method. J Earth Syst Sci 121:45–62
7. Rukhaiyar S, Alam MN, Samadhiya NK (2017) A PSO-ANN hybrid model for predicting factor of safety of slope. Int J Geotech Eng. ISSN: 1938–6362
8. Goharzay M, Noorzad A, Ardakani AM, Jalal M (2020) Computer-aided SPT-based reliability model for probability of liquefaction using hybrid PSO and GA. J Comput Des Eng 7(1): 107–127

Numerical Methods in the Stability Analysis of Slopes

P. Arya Babu and S. Chandrakaran

Abstract Slope failure is a complex problem that causes serious hazards throughout the world. Due to this there is large loss of life and property. So, it became very much essential to understand these slope failures, analyse and predict its vulnerability for proper mitigation of hazards. A number of methods are available for slope stability analysis and prediction. Most common among these is the convectional limit equilibrium and numerical methods. Now-a-days Artificial intelligence (AI) techniques are used widely for this purpose. This paper aims to provide a comparison of traditional methods of slope stability analysis i.e. limit equilibrium and finite element method with the recently developed artificial neural network (ANN). From the data available in the literature, a comparative study of factor of safety is carried out between the various methods through commercially available software Geostudio. The input parameters used in the study include- unit weight of soil (ϒ), cohesion (C), angle of internal friction (Ø), height of slope (H), angle of slope (α) and pore water ratio (ϒu), where as factor of safety (FOS) is the only output parameter. Also the relationship between various soil parameters and stability is established using regression analysis. The methods used in the LEM analysis are Ordinary method of slices, Bishop's method, Morgenstein Price method, Janbu's method, Spencer's method and for FEM analysis strength reduction technique is used. It is found that there is good agreement between the conventional LEM and FEM. The factor of safety value obtained with FEM is a bit higher than that obtained with LEM. Also when comparing the results with ANN, it is found to give more accurate results.

Keywords FEM · Factor of safety · LEM · Stability · ANN

Slope stability analysis is a very important area in geotechnical engineering. It has been subjected to continuous concern due to the loss of large amount of life and property annually throughout the world. Landslide is a major disaster due to the failure of slopes that causes severe damage to life and property. The analysis and

P. Arya Babu (✉) · S. Chandrakaran
Department of Civil Engineering, National Institute of Technology Calicut, Calicut, India

G. C. Marano et al. (eds.), *Proceedings of SECON'21*, Lecture Notes in Civil Engineering 171, https://doi.org/10.1007/978-3-030-80312-4_86

prediction of slope stability is necessary in identifying the critical landslide sections and mitigating damages caused by them. However, the accurate prediction of stability of slopes is very challenging as it depends on geotechnical and physical factors. The interactions between all these factors are very complex.

Slope stability has a long history started in 1920's. Many methods have been proposed to analyse and predict the slope stability. Most common among these is the limit equilibrium method (LEM) and numerical methods. Several empherical methods have also been developed. But all these methods have certain shortcomings. The limit equilibrium method is based on certain assumptions and the numerical methods are usually time consuming and their accuracy is depended on the accurate estimate of geotechnical and physical properties. LEM method makes use of number of analysis depending on the problem type whether it is circular or non-circular. All the methods requires, in general the soil or rock mass be divided into number of slices. LEM method is based on assumptions on slope geometry and the equations of equilibrium. The traditional limit equilibrium methods include the ordinary method of slices, Bishop's method, Morgenstern Price method, Janbu's method and Spencer's method. The first method developed for solving LE analysis was Fellenius or Swedich circle method (1936) also called the ordinary method of slices. This method can be applied only to circular slip surfaces and it leads to the underestimation of factor of safety (FOS). Bishop (1955) developed a more revised method for analysing circular slip which improved the accuracy of the resultant FOS. This method is still used widely in slope stability analysis software. Janbu's method is used for non circular analysis. A number of other advanced LE methods (Spencer 1967; Morgenstern-Price 1965) have been developed.

In the recent years finite element method is widely being used for slope stability analysis since assumptions need not be made in advance. This method can be used in case of complex geometries, different materials, seepage, consolidation and other coupled hydrological and mechanical behaviour. There are a number of software packages that have been developed for stability analysis which utilizes the FEM. The advancement of technology has increased the use of the FEM as it dominates a wider range of features [1]. Generally, there are two approaches to analyse the slope stability using FEM. One of these is to increase the gravity load and the second one is to reduce the strength characteristics of the soil mass. Khabbaz et al. [2] proposed that when a simple homogeneous slope is considered, the difference in the safety factors and location of critical slip surfaces are minimal and both LEM and FEM provide indistinguishable results. Both these methods have their own advantage and disadvantage. Rehman [3] carried out a comprehensive study to compare the analysis of some typical natural slopes using both limit equilibrium method and finite element method. It was found out that LEM underestimates the FOS in some cases, thus creating expensive designs. For some other cases LEM overestimates FOS thus making the structures unsafe.

Now-a-days Artificial intelligence (AI) techniques are increasingly used in the prediction of slope stability. These include soft computing methods like Artificial neural network (ANN), Support vector machines (SVM), evolutionary polynomial regression (EPR), fuzzy logic system etc. These methods doesn't require any prior knowledge of the specific model forms and have flexible non-linear modeling capabilities. It can also perform better than the traditional methods. Lu and Rosenbaum [4] used (ANN) method to analyse slope stability. Samui [5] used support vector machines to predict the slope stability as a regression problem and the stability status was modeled as a classification problem. They found out that SVM has an accuracy of 85.71%. Ahangar-Asr et al. [6] used a new approach to predict the stability of soil and rocks. They developed evolutionary polynomial regression (EPR) models and validated using results from sets of field data on the stability status of soil and rocks slopes. They found out that the proposed approach was very effective and robust in modeling the behaviour of slopes. Xue [7] proposed particle swarm optimization (PSO) to select the appropriate parameters for a SVM model, and the results showed that the combined model based on SVM and PSO is a powerful computational tool for slope stability prediction. Mohamed et al. [8] successfully applied Fuzzy logic system for slope stability prediction. Tan et al. [9] adopted Simulated annealing algorithm for determination of critical slip surface of slopes.

This paper presents a comparative study of the various methods of slope stability analysis, i.e. the traditional methods and the newly developed artificial intelligent method. Limit equilibrium and finite element methods are compared here with the artificial neural network model. The factor of safety obtained by various numerical methods is determined using commercially available software Geostudio 2021. Slope/w and Sigma/w in the Geostudio module is used in this study. Methods used for LEM analysis, include- ordinary method of slices, Bishop's method, Morgenstern Price method, Janbu's method, Spencer's method and for FEM analysis strength reduction technique is used. The main aim of this study is to establish relationship between the various soil parameters and stability. The parameters used in this study are unit weight of soil (ϒ), cohesion (C), angle of internal friction (Ø), height of slope (H), slope angle (α) and pore water pressure ratio (ϒu).

1 Methodology

In this research 192 slope cases having different geometrical and slope parameters were selected from various available literatures. Commercially available software Geostudio 2021 was used to analyse these slopes to obtain the factor of safety. SLIDE/W in Geostudio 2021 was used to find the factor of safety using Morgenstern-Price method, Bishop's method, Janbu's method, Ordinary method of slices and Spencer's method. Strength reduction technique in SIGMA/W was used to obtain the factor of safety. Along with these neural network tool box in Mathlab was also used. In this research several important parameters like unit weight of soil (ϒ), cohesion (C), angle of internal friction (Ø), slope angle (α), height of slope (H) and pore water ratio (ϒu) were considered as input parameters. Table 1 shows the various statistical parameters of the input and output value of this database in

Table 1 Statistical parameters of data

Parameter	Mean	Std.	Min	Max	Range
ϒ					
Training set	21.149	3.991	12	31.3	19.3
Testing set	20.501	3.429	16	26.89	10.89
C					
Training set	28.571	26.509	0	150.05	150.05
Testing set	64.569	59.855	9	300	291
Ø					
Training set	27.757	9.892	0	45	45
Testing set	19.972	6.324	13	38.7	25.7
α					
Training set	33.876	9.486	16	53	37
Testing set	20.057	14.934	3	59	56
H					
Training set	94.548	118.622	3.66	502.7	499.04
Testing set	48.21	90.195	3	565	562
ϒu					
Training set	0.2	0.168	0	0.5	0.5
Testing set	0.228	0.162	0	0.5	0.5

terms of maximum, minimum, mean and standard deviation. It can be seen from the table that this database consists of a wide range of values. Multiple regression analysis was carried out to establish relationship between the various input parameters and FOS.

2 Regression Analysis

Multiple regression generally establish relationship between multiple independent variables and one dependent variables. This method is now widely used in slope failures and landslides. In this study the independent variables are ϒ, C, Ø, α, H, ϒu and the dependent variable is FOS. Multiple regression analysis is carried out to correlate the factor of safety with the soil parameters (unit weight, cohesion and angle of internal friction) and slope parameters (angle of slope and height of slope). The data used for Geostudio analysis has been used for regression analysis as well. The multiple regression model revealed the following correlations between various input parameters and FOS-

For Morgenster-Price method-

$$\begin{aligned} \text{FOS} &= 4.744457 - 0.00816\Upsilon + 0.022407\text{C} + 0.07808\varnothing \\ &\quad - 0.16576\tan\alpha - 0.00294\text{H} - 0.87984\Upsilon\text{u} \end{aligned} \tag{1}$$

For Bishop's method-

$$\begin{aligned} \text{FOS} &= 4.741931 - 0.00451\Upsilon + 0.022601\text{C} + 0.079605\varnothing \\ &\quad - 0.16945\tan\alpha - 0.00302\text{H} - 0.84206\Upsilon\text{u} \end{aligned} \tag{2}$$

For Janbu's method-

$$\begin{aligned} \text{FOS} &= 55.80474 - 1.58359\Upsilon + 0.030798\text{C} - 0.34758\varnothing \\ &\quad - 0.61888\tan\alpha + 0.170221\text{H} - 18.6073\Upsilon\text{u} \end{aligned} \tag{3}$$

For Ordinary method of slices method-

$$\begin{aligned} \text{FOS} &= 4.487472 - 0.00238\Upsilon + 0.02255\text{C} + 0.077901\varnothing \\ &\quad - 0.16427\tan\alpha - 0.00289\text{H} - 0.55475\Upsilon\text{u} \end{aligned} \tag{4}$$

For Spencer's method-

$$\begin{aligned} \text{FOS} &= 49.68846 - 3.13726\Upsilon + 0.03714\text{C} + 0.411225\varnothing \\ &\quad - 0.08152\tan\alpha - 0.058838\text{H} + 32.93567\Upsilon\text{u} \end{aligned} \tag{5}$$

For Strength reduction method-

$$\begin{aligned} \text{FOS} &= 5.191455 - 0.02223\Upsilon + 0.026241\text{C} + 0.089882\varnothing \\ &\quad - 0.177728\tan\alpha - 0.00355\text{H} - 1.00519\Upsilon\text{u} \end{aligned} \tag{6}$$

For ANN-

$$\begin{aligned} \text{FOS} &= 4.092593 - 0.03058\Upsilon + 0.021246\text{C} + 0.068676\varnothing \\ &\quad - 0.16424\tan\alpha - 0.00213\text{H} - 0.13446\Upsilon\text{u} \end{aligned} \tag{7}$$

3 Results and Discussions

A comparison of factor of safety calculated from Morgenstern Price, Bishop's, Janbu's, Ordinary method of slices, Spencer's and strength reduction technique was obtained. The same was carried out ANN also. It can be noted from the table that the calculated FOS is very close to the actual FOS. Also from the figures it can be

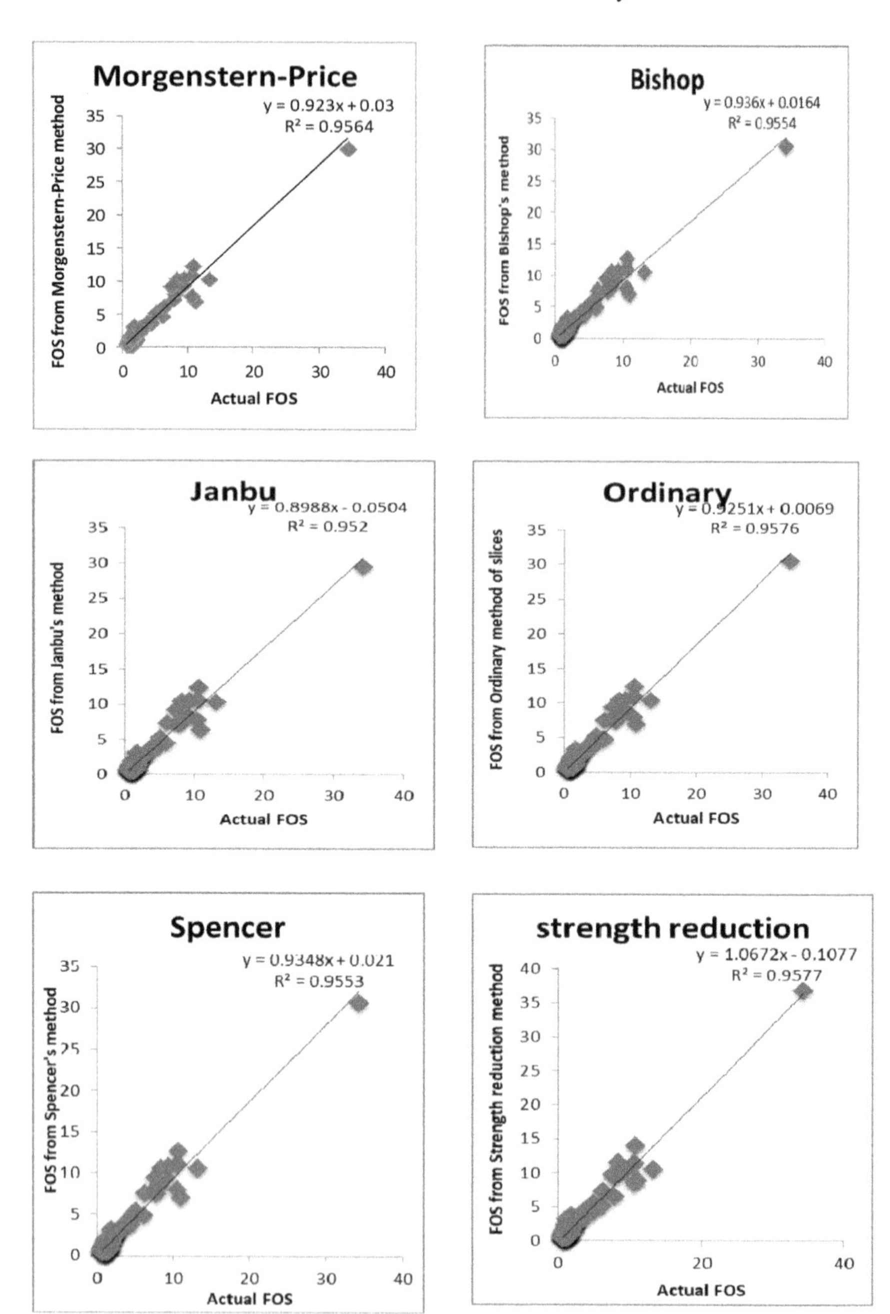

Fig. 1 FOS obtained by Morgenstern-Price method, Bishop's, Janbu's, ordinary method of slices, spencer's strength reduction method respectively

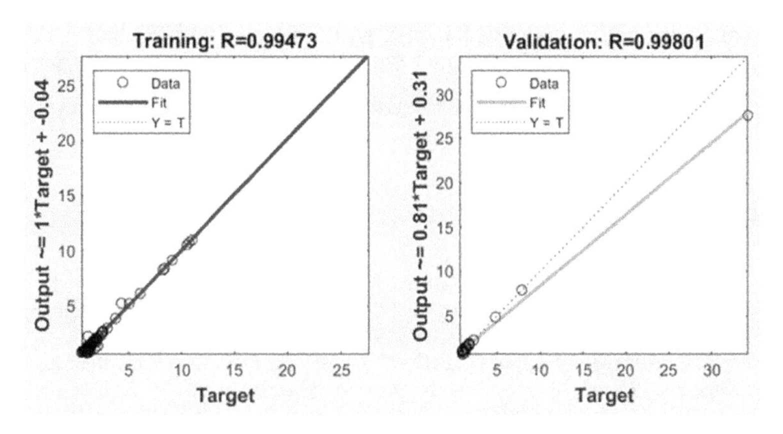

Fig. 2 Comparison of FOS predicted FOS to actual FOS

seen that the R^2 value obtained for all the methods is close to unity. From the R^2 value it is found that Spencer's method gives a more accurate result when compared to other methods (Fig. 1).

The FOS obtained from Janbu's method is always the smallest. It is also found that the strength reduction technique overestimates the FOS to a very high value when compared with other methods. All these methods sometimes underestimates FOS, resulting in expensive designs and some other times they overestimates FOS resulting in unsafe designs.

4 ANN Model

See Fig. 2.

5 Conclusion

Commercially available software Geostudio 2021 (SLIDE/W and SIGMA/W) was used to calculate the factor of safety of 192 different slopes with six input parameters (unit weight of soil, cohesion, angle of internal friction, angle of slope, height of slope and pore water ratio). It was found that all the methods gives factor of safety values almost same as the actual factor of safety. Also it can be seen very well that the FOS obtained from Janbu's method is the smallest when compared to all other methods. The R^2 value obtained from Spencer's method is very close unity. So, Spencer's method often gives more accurate result when compared to others. It is also found that the FEM method (strength reduction) often gives a bit higher value of FOS when compared to LEM. Also slope stability prediction was carried out using ANN model which gave an R^2 value of 0.99473 and 0.99801 for training and testing set respectively. Multiple regression analysis was used to establish relationship between all the input parameters and FOS.

References

1. Hammouri NA, Husein AI, Malkawi MMA, Yamin, (2007) Stability analysis of slopes using the finite element method and limiting equilibrium approach. Bull Eng Geol Environ 2008 (67):471–478. https://doi.org/10.1007/s10064-008-0156-z
2. Khabbazl H, Fatahi B, Nucifora C (2012) Finite element methods against limit equilibrium approaches for slope stability analysis. In: Proceedings of ANZ 2012 conference
3. Rehman HU (2017) Use of finite element techniques for slope stability analysis of earth dam and reservoir embankments
4. Lu P, Rosenbaum MS (2002) Artificial neural networks and grey systems for the prediction of slope stability. Nat Hazards 30:383–398, 2003
5. Samui P (2008) Slope stability analysis: a support vector machine approach. Environ Geol 2008(56):225–267. https://doi.org/10.1007/s00254-007-1161-4
6. Ahangar-Asr A, Faramarzi A, Javadi AA (2014) A new approach for prediction of the stability of soil and rock slopes. Eng Comput
7. Xue X (2016) Prediction of slope stability based on hybrid PSO and LSSVM. J Comput Civ Eng. ASCE, ISSN 0887–3801
8. Mohamed T, Kasa A, Taha MR (2014) Fuzzy logic system for slope stability prediction. Int J Adv Sci Eng Inf Technol
9. Tan W, Li J, Yu J (2015) Simulated annealing algorithm and its application for determining the critical slip surface of slopes. EJGE 20, Bund. 2

Settlement Prediction of Shallow Foundations on Cohesionless Soil Using Hybrid PSO-ANN Approach

P. Krishna Pradeep, N. Sankar, and S. Chandrakaran

Abstract Settlement estimation of shallow foundations on cohesionless soil possesses a higher level of complexity, sole reason for which can be pointed towards the uncertainties involved in factors that affect the magnitude of settlement. For a safe and perfect shallow foundation design an unerring estimation of foundation settlement is rather essential. Unlike the conventional settlement prediction techniques AI techniques have shown greater accuracy, the potential which can be exploited for the settlement prediction of shallow foundations. This study approaches the settlement prediction problem using hybrid PSO—ANN technique (Particle swarm optimization—Artificial neural network). Dataset consisting of footing dimensions, net applied pressure, depth of embedment of footing, SPT N value and depth of water table are used as input data for developing the PSO—ANN model, whereas settlement is chosen as the output data. From about more than 300 runs an optimum network of 6-13-1 was developed. The developed model obtained coefficient of correlation, R = 0.953 and mean square error, MSE = 0.119 m. For assessing effectiveness of the model developed, different performance indices such as RMSE, VAF, MAE, PI, RSR, NS etc. were chosen. All these parameters gave values corresponding to a model with good predictive capacity.

Keywords Settlement · Cohesionless soil · PSO—ANN · SPT N · Performance indices

1 Introduction

Types of settlement that are generally encountered by shallow foundations include immediate settlement, consolidation settlement and compression settlement, which solely depends on the soil type and conditions in which they are laid. Considering cohesion less soil, the major concern is the former said—immediate settlement.

P. Krishna Pradeep (✉) · N. Sankar · S. Chandrakaran
National Institute of Technology Calicut, Calicut, Kerala, India

G. C. Marano et al. (eds.), *Proceedings of SECON'21*, Lecture Notes in Civil Engineering 171, https://doi.org/10.1007/978-3-030-80312-4_87

The latter two are mostly related to cohesive soil. Higher degree of permeability exhibited by cohesion less soils lay path for instant occurrence of settlement with applied load [1], which could affect the stability of structure in short notice leaving no time to execute remedial measures. Two major criteria that control the design of shallow foundations on cohesionless soils are bearing capacity and settlement of foundation. In terms of serviceability, excessive settlement is inacceptable. Consequently settlement is considered as the controlling criterion for the foundation design rather than bearing capacity, condition being critical when least width of footing is greater than 1 m [2]. Hence settlement prediction is a major concern and is an essential criterion in the shallow foundation design. Thus accurate estimate of shallow foundation settlement is indispensable.

For the prediction of shallow foundation settlement on cohesionless soils, researchers have introduced a fair number of theoretical, empirical, semi empirical and numerical methods. Most of these available methods incorporate several assumptions related to the factors that influence settlement of shallow foundations and hence simplifies the problem. Consequently, fails to achieve consistent success during settlement prediction. Recently soft computing techniques have emerged as tool for solving such highly complex, multivariate problems. One such technique is Artificial Neural Network (ANN). Although ANN rectifies many of the shortcomings of traditional methods it still poses some drawbacks such as overfitting, getting trapped in local minima etc. If a hybrid model of ANN—PSO is used these drawbacks could be sorted out and a better predictive model could be developed [3].

In this study PSO is used as the training technique for ANN instead of the usual backpropagation method. A total of 6 input parameters and one output parameter is used, the database for which is acquired from the literature. The prediction model is developed and its performance is analyzed using various fitness parameters.

2 Model Development

The methodology considered for developing the model includes model input and output determination, division of data followed by pre-processing, determination of appropriate network architecture, model training, and performance analysis.

2.1 Database

The initial step includes determination of parameters that influences settlement estimation. For this study SPT N value, footing dimensions, depth of water table below the footing and net applied pressure are considered as the input values whereas settlement is the single output value [4]. These data were collected from literature [5]. A total of 236 data were collected, reference to which is listed in Table 1.

Table 1 Reference for database

References	No: of cases
Bazaraa (1967)	5
Briaud and Gibbens (1999)	4
Burbidge (1982)	22
Burland and Burbidge (1985)	147
Wahls (1997)	31
Maugeri et al. (1998)	2
Picornell and Del monte (1988)	1
Robert E. Kimmerling (2002)	5
Sargand et al. (1999)	19
Total	236

2.2 Data Division and Preprocessing

These dataset values are normalized using the equation given below [6].

$$X_n = \frac{X - X_{\min}}{X_{\max} - X_{\min}} \quad (1)$$

where

X_n Normalized value.
X Actual value.
X_{max} Maximum value.
X_{min} Minimum value.

After normalization the whole dataset is divided to obtain two subsets: training set and testing set. Random division is carried out for two subsets of 80 and 20% data. Data division is done in python using train_test_split function from scikit-learn library. Subset with 80% data comprises the training set whereas the second one with 20% data constitutes test set. It is essential that both of these subsets represent the same population and all the patterns present in the training set are available in testing set also [7]. Thus randomly different combinations are selected and then checked for its statistical consistency. This is continued until both the subsets are statistically consistent. The statistical parameters chosen include mean, standard deviation, maximum and minimum.

2.3 Development of Network

Network architecture can be determined by following a series of steps, starting with selection of number of hidden layers followed by fixing the number of nodes in each of these hidden layers. Studies have shown that one hidden layer is enough in a neural network to approximate any continuous function [8], thus for the present study, one hidden layer is used. Therefore a three layer architecture for neural network is finalised: input layer, one hidden layer and output layer. The number of input nodes is chosen as six which is equal to the number of input parameters and number of output neuron as one which is equal to the number of output parameters i.e. settlement. Caudell (1988) suggested that the maximum number of nodes possible in a single hidden layer model is given by (2I + 1), where I is the number of inputs [9]. Thus the upper limit for number of nodes in hidden layer is taken as 13. For the study 10 different values for number of nodes in hidden layer are considered.

In this study, optimum training of the network is performed using PSO. Here, the objective is to minimize the objective function, i.e., mean square error (MSE) of the network. The parameters associated with PSO are number of particles in the swarm (Swarm size), acceleration factor (c_1 and c_2) and inertia weight (w). The inertia weight (w) is generally assigned random value between 0 and 1. From various studies carried out, researchers have reported that usually when c_1 takes a value within the range 1–2.5 and c_2 value between 2 and 3, the network exhibits optimum performance [10]. In this study varying values of these parameters are selected, shown in Table 2, and a considerable number of runs are carried out to determine the optimum combination of network parameters with a higher R value and a lower MSE value. MATLAB R2020b is used for the network development. All the possible combinations of acceleration factors for each swarm size and number of neurons chosen are considered. Rather than adopting sensitivity analysis for determining optimum values of parameters, trial and error method is adopted. The former method has been reported ineffective.

2.4 *Stopping Criteria*

Stopping criteria considered [11]:

Maximum iteration: 20,000
If improvement in objective function for 100 successive iterations is below is 10^{-8}.

Table 2 Values of different parameters considered for network development

Acceleration factor		Swarm size	Number of neurons
c1	c2		
1	2	10	4
1.5	2.25	20	6
1.3	2.5	30	8
		50	10
		100	11
			12
			13

3 Model Validation

After successfully training the model using PSO it is required to validate the performance of trained model. The major intention behind validating the model is to ensure that the model is not simply memorizing the relationship between input and output available in the training data but has the ability to generalize within the limits of training data. Model validation is carried out using test data.

4 Fitting Examination

The result of fitting examination of PSO-ANN model is given in Table 3. The best performing acceleration factor (c_1 and c_2) corresponding to a given number of neurons in the hidden layer and swarm size is depicted in Table 3. Results for n = 6, 8, 10, 11, 12 & 13 and swarm size = 10, 20, 30, 50 & 100 are reported. Corresponding prediction evaluating indices MSE and R value for training and the testing data-set has been also presented. More the R value closer to 1 and MSE value closer to 0, better will be the performance of model. Figure 1a, b shows actual and predicted settlement values of training and testing set respectively (Fig.1).

The variation of MSE with number of iterations is shown in Fig. 2. Upto 100 iterations MSE value showed significant change. Later on the value remained constant towards 10,000. Thus maximum iteration is chosen as 20,000. All the runs performed got completed before 20,000 iterations. Iteration number for the final model and the related details are given in Table 4.The highest and lowest value of R and MSE respectively are obtained for 13 number of hidden neurons, c_1 = 1.3 and c_2 = 2.25. Thus the predictive model is developed as a 6–13-1 network. Architecture of the neural network is shown in Fig. 3.

5 Performance Analysis

Root mean square error, RMSE and coefficient of determination, R^2 are the commonly used parameters to evaluate the performance of model. In order to ensure the efficiency of prediction model, a number of fitness parameters has been considered for this study. Fitness parameters used to assess the model and their corresponding values obtained are given in Table 5.

Nash–Sutcliffe efficiency (NS) indicates the predictive power of the models. The predictive capacity increases as the NS value gets closer to 1. Here 0.909 is obtained for training set indicating good prediction. Root Mean Square Error (RMSE) closer or equal to 0 indicates marginal error in prediction. The obtained value is 0.345 which indicates marginal error. Variance Account Factor

Table 3 Best R and MSE for varying number of hidden neurons

Number of neurons	Swarm size	Acceleration factors		Train		Test	
		c1	c2	R	MSE	R	MSE
6	10	1.5	2.5	0.91	0.21	0.84	0.47
	20	1.5	2.5	0.92	0.19	0.86	0.45
	30	1.3	2.25	0.89	0.25	0.86	0.44
	50	1.3	2.5	0.92	0.19	0.92	0.267
	100	1.3	2.25	0.916	0.198	0.888	0.346
8	10	1.3	2.5	0.91	0.213	0.845	0.479
	20	1.5	2.5	0.915	0.202	0.891	0.338
	30	1.5	2.25	0.921	0.187	0.897	0.322
	50	1.3	2.25	0.921	0.187	0.869	0.404
	100	1.5	2.25	0.926	0.175	0.894	0.354
10	10	1.3	2.5	0.926	0.176	0.895	0.331
	20	2	2	0.915	0.202	0.896	0.324
	30	1	2.5	0.945	0.132	0.888	0.355
	50	2	2	0.931	0.164	0.923	0.244
	100	2	2	0.927	0.174	0.909	0.285
11	10	1.3	2.5	0.923	0.184	0.882	0.362
	20	2	2	0.919	0.192	0.904	0.3
	30	1.5	2.5	0.932	0.162	0.889	0.351
	50	1.5	2.25	0.937	0.15	0.912	0.287
	100	1.3	2.25	0.942	0.138	0.902	0.324
12	10	2	2	0.914	0.204	0.868	0.408
	20	2	2	0.912	0.207	0.882	0.362
	30	1.5	2.25	0.94	0.144	0.916	0.269
	50	1.5	2.25	0.928	0.172	0.89	0.317
	100	1.3	2.25	0.932	0.161	0.907	0.299
13	10	1.3	2.5	0.94	0.144	0.849	0.485
	20	1.5	2.25	0.929	0.168	0.88	0.357
	30	1.5	2.5	0.937	0.151	0.87	0.452
	50	1.3	2.5	0.931	0.166	0.901	0.31
	100	**1.3**	**2.25**	**0.953**	**0.119**	**0.932**	**0.217**

(VAF) equal to 100% indicates that model performance gives a reasonable result. Obtained value is 99.98% for training set which is much closer to 100%. R^2 (Coefficient of determination) and Adj.R^2 (adjusted determination coefficient) values should be closer to 1. Both closer to each other show that the model reflected most of the variability in soil parameters. Here values of R^2 and Adj.R^2 are closer to 1 and also closer to each other, 0.908 and 0.906 respectively are obtained.

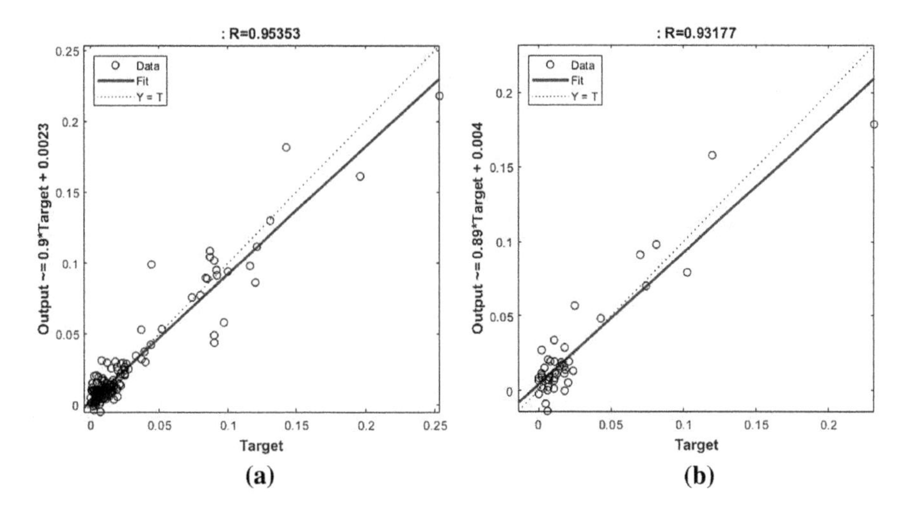

Fig. 1 Actual and predicted settlement of **a** training and **b** testing set (actual value on x-axis and predicted value on y-axis)

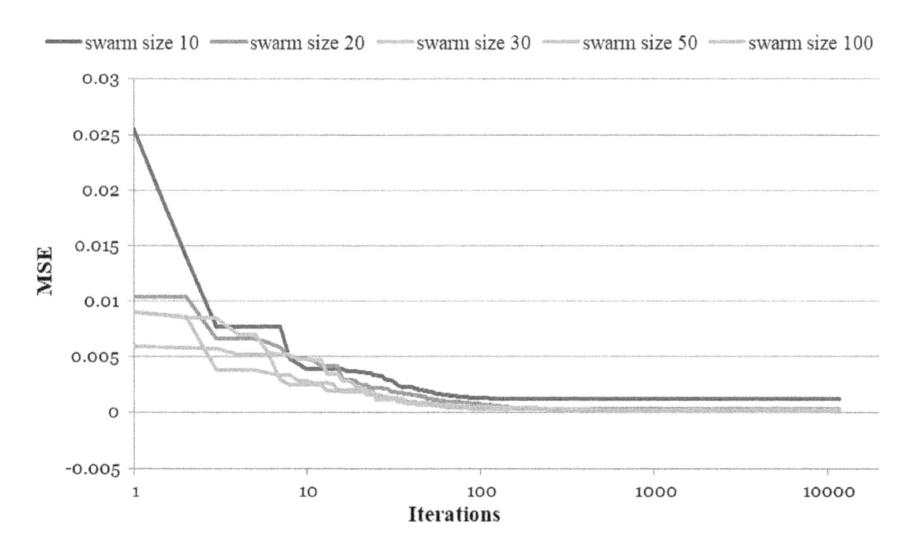

Fig. 2 MSE Variation with number of iterations

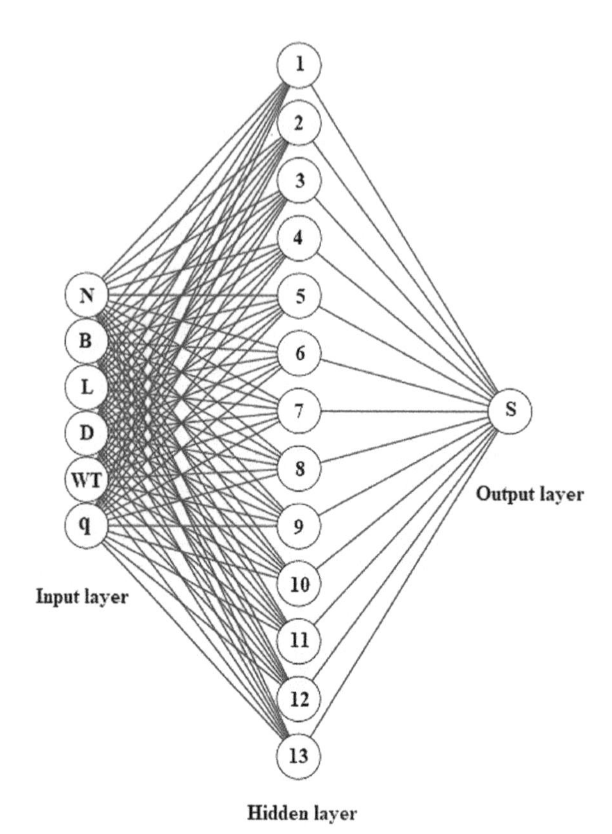

Fig. 3 Architecture of neural network developed (*N* = SPT N value, *B* = Breadth of footing, *L* = Length of footing, *D* = Depth of footing, *WT* = Depth of water table below footing, *q* = Net applied pressure, *S* = Settlement)

Table 4 Description of the hybrid ANN PSO model developed

Number of hidden neurons	13
Swarm size	100
c1	1.3
c2	2.25
Iterations	20,000[a] (13,841)
w	0.4926
Stopping criteria	Max. iteration = 20,000 or Tolerance = 10^{-8}
Space search	(−1.5, 1.5)

[a]Second criteria got satisfied and iterations got completed at 13,841

Table 5 Performance parameters

Performance parameter	Training set	Testing set
NS	0.909	0.892
RMSE	0.345	0.466
MSE	0.119	0.217
R	0.953	0.932
R^2	0.908	0.869
VAF	99.989	99.98
Adj. R^2	0.906	0.85
PI	1.561	1.384
MAPE	0.006	0.068
MAE	0.00621	0.0088

6 Conclusion

From about more than 300 runs the optimum predictive model was developed as a 6–13-1 network ie, with 6 input parameters and 13 hidden neurons. c_1 and c_2 values for the model developed are 1.3 and 2.25 respectively. An R value of 0.953 for training set and 0.932 for testing set was obtained. MSE value of 0.119 m for training and 0.217 m for testing was obtained. All the other performance parameters also gave values proving the model to be good in prediction. Thus PSO –ANN hybrid model can be used to accurately predict the settlement of shallow foundations on cohesionless soil.

References

1. Coduto DP (1994) Foundation design principles and practices. Prentice-Hall, Englewood Cliffs
2. Schmertmann JH (1970) Static cone to compute static settlement over sand. J Soil Mech Found Div Proc Am Soc Civ Eng 96:7302–1043
3. Armaghani D, Shoib R, Faizi K, Safuan A, Rashid A (2015) Developing a hybrid PSO-ANN model for estimating the ultimate bearing capacity of rock-socketed piles. Neural Comput Appl 28:391–405
4. Prasanth S, Sankar N (2015) Prediction of settlement of shallow footings on granular soils using genetic algorithm. Asian J Eng Technol 3
5. Burland JB, Burbidge MC (1985) Settlement of foundations on sand and gravel. Proc Inst Civ Eng 78:1325–1381
6. Ray R, Kumar D, Samui P, Roy L, Goh ATC, Zhang W (2021) Application of soft computing techniques for shallow foundation reliability in geotechnical engineering. Geosci Front 12:375–383
7. Shahin A, Maier R, Jaksa B (2002) Predicting settlement of shallow foundations using neural networks. J Geotech Geoenviron Eng 128(9):785–793

8. Hornik K, Stinchcombe M, White H (1989) Multilayer feedforward networks are universal approximators. Neural Netw 2:359–366
9. Caudill M (1988) Neural networks primer, Part III. AI Expert 3(6):53–59
10. Alam MN, Das B, Pant V (2015) A comparative study of metaheuristic optimization approaches for directional overcurrent relays coordination. Electr Power Syst Res 128:39–52
11. Rukhaiyar S, Alam MN, Samadhiya NK (2018) A PSO-ANN hybrid model for predicting factor of safety of slope. Int J Geotech Eng 12:556–566

Influence of Freezing–Thawing Cycles on Strength Properties of Frost-Affected Subgrade Soil Stabilised with Cement and Reinforced with Chir Pine Needles

Yamem Tamut, Ajanta Kalita, and S. K. Singh

Abstract In this investigation amelioration of the strength properties of subgrade soil found in Tawang, Arunachal Pradesh, India has been studied by adding cement as binding agent and chir pine needles (CPN) as reinforcing material. The cement percentages were varied as 2, 4 and 6% and proportions of CPN were varied as 0.5, 1, 1.5 and 2% of the dry weight of the soil. The standard proctor compaction test results show that optimum moisture content (OMC) decreases and maximum dry density (MDD) increases with the increase in cement percentage but decrease in both OMC and MDD was observed with increase CPN fibers. The California Bearing Ratio (CBR) test was performed on samples of soil mixtures under three categories namely unsoaked samples, soaked samples and samples after being subjected to 4 freeze–thaw cycles. For unsoaked and soaked conditions, an improvement in the strength was confirmed by an increase in CBR value for the test specimen containing soil, cement and CPN. However, reduction in the strength was observed after freeze–thaw cycles for all the soil mixtures. Further, to assess the strength behaviour of the frost-affected soil a series of Unconfined Compression Tests (UCS) was performed on cured samples with different soil mixtures and similar samples after subjecting them to 0, 3, 7, 14 and 28 freeze–thaw cycles. For cured samples, significant improvements in the strength were observed with the curing periods and increase in cement percentage and CPN contents. The optimum dosage of the blenders which give better performance was found to be S + 6% C + 1.5%CPN. The weakening effects of freeze–thaw cycles are confirmed by comparing the losses in strength with respect to the cured samples.

Keywords Freezing–thawing · Wood ash · Chir pine needles · Cement · CBR

Y. Tamut (✉) · A. Kalita · S. K. Singh
NERIST, Nirjuli, Arunachal Pradesh 791109, India
e-mail: ymt@nerist.ac.in

G. C. Marano et al. (eds.), *Proceedings of SECON'21*, Lecture Notes in Civil Engineering 171, https://doi.org/10.1007/978-3-030-80312-4_88

1 Introduction

Subgrade soils are constantly subjected to cyclic rolling loads due to which a firm, stable and sufficiently strong platform for the pavement from underneath is a pre-requisite for constructing an all-season road. The strength of the subgrade can be assumed by the California Bearing Ratio (CBR) values of the soil which signifies a stronger subgrade soil for a higher CBR value. Considering the design aspect, a higher CBR value indicates a lesser thickness require which in turn reduces the cost of pavement construction. In adverse climatic conditions where the pavements are subjected to seasonal freeze–thaw cycles, it affects the structure of the soil and their engineering properties [1]. When the temperature falls, entrapped water or moisture in the soil forms ice cubes. Thus, the frozen ice cubes lead to expansion in pores, and the freezing process led to aggregate separation and disintegration of soil particles [2]. The condition of the road is worsened as the temperature rises and subsequent thawing of the frozen ice cubes. Under such climatic conditions, the poor subgrade soil ideally should be avoided by replacement wherever possible. However, when complete removal of the existing weak subgrade soil may not be practically feasible; and in that case, an appropriate treatment procedure may be explored to stabilize the weak soil using cost-effective admixtures and reinforcement methods. Many researchers [3–5] have shown interest in improving the geotechnical properties of soil deteriorated by the seasonal freezing and thawing effect and have tried to find some sustainable and cost-effective solution to it. The traditionally used stabilizing agent to improve a weak subgrade soil in road construction is the cement and lime but with an ardent attempt to find sustainable and cost-effective materials is still thriving. Many researchers have contributed to finding various naturally available and industrial waste products such as wood ash [6], fly ash silica fume and fly ash [7], geofiber with synthetic fluid [8], lime and microsilica [9], lime, jaggery and gallnut powder [10], fibers-chemical additives [11], polypropylene fibres [12] as an alternative for road stabilization materials in frost affected areas.

Some studies on use of naturally available fibers such as chir pine needles as stabilisation materials have been carried out [13–15]. In developing countries like India where there are limited financial resources to invest in road construction, regular maintenance and reconstruction of roads become quite challenging. Therefore, in such a situation effective utilization of locally obtainable waste products may be more economical. From the above review, it is realized that not many studies have been done to ameliorate the strength characteristics of subgrade soil experiencing seasonal Freezing–Thawing effects by using locally available waste materials to find an economically feasible solution to this specific soil stabilization problem.

In view of the above, the authors have investigated the applicability of locally available chir pine needles as a stabilizer to mitigate this geotechnical problem of frost affected subgrade.

2 Materials Used

2.1 Soil

The soil was collected from an open pit at a mountainous pass (Sela Pass) in Tawang District, Arunachal Pradesh where the altitude is above 13,000 feet from MSL and tested for strength and other geotechnical properties. The specific gravity of the soil was found to be 2.67 and the soil was non-plastic. The grain size distribution analysis of the soil was conducted as per IS-2720 (Part-IV)-1985 and the soil was characterized as silty sand. The grain size distribution curve for the soil is shown in Fig. 1.

2.2 Chir Pine Needles

The chir pine, scientifically known as Pinus Roxburghii is a coniferous tree generally found in the higher altitudes and grows needle like leaves. Every autumn it sheds dried leaves and builds a thick carpet of needles on the ground which is highly inflammable having the potential of starting and spreading forest fires and also restricting the growth of other vegetation. The chir pine needles used in the study were collected from Dirang, West Kameng District, Arunachal Pradesh which is located along the way to Sela Pass and about 50 km away from the study area. The fibers were trimmed into small pieces with length varying from 10 to 15 mm and diameter of the fibers varied from 0.5 mm to 0.75 mm. The number of fibers per gram were approximately 250 and specific gravity of the CPN was found to be

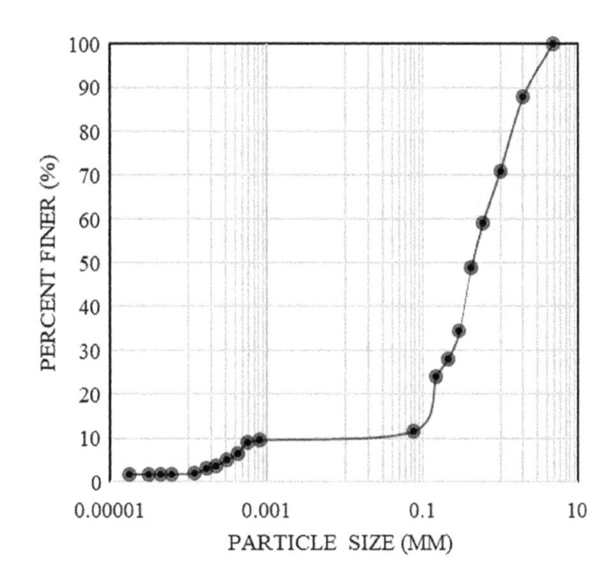

Fig. 1 Grain size distribution curve of soil

Table 1 Chemical composition of chir pine needles

Standard	Elements	Wt%	Atomic wt (%)
C $CaCO_3$	C K	70.96	77.23
O SiO_2	O K	26.94	22.02
Si SiO_2	Si K	0.45	0.21
K MAD-10 Feldspar	K K	0.44	0.15
Ca Wollastonite	Ca K	1.20	0.39

0.93. The chemical composition of the chir pine needles obtained from Energy Dispersive X-ray diffractometer (EDX) tests is presented in Table 1. The major constituents are calcium carbonate and silicon dioxide.

2.3 Cement

The cement used is fly ash-based PPC conforming to Indian Standard BIS 1489, Part-1, 2015. Normal water is used in preparation and curing of all samples. The designations used for the materials are S for the parent soil, C for cement and CPN for chir pine needles.

3 Experimental Programs

3.1 Compaction Tests

To investigate the effects of cement as stabilising agent and natural fibers like chir pine needles as reinforced materials on the strength properties of frost-affected soils, standard Proctor compaction tests as per IS 2720 (Part VII)-1980 (Reaffirmed 2011) were conducted on the untreated, treated and reinforced soils. In the treated samples, varying percentages of 2, 4 and 6% of cement and 0.5, 1, 1.5 and 2% of chir pine needles of dried soil weight were mixed carefully to obtain a homogeneous mixture. Normal tap water was used in the tests.

3.2 California Bearing Ratio (CBR) Test

The California Bearing Ratio (CBR) tests were performed according to IS 2720 (Part 16)-1987 (Reaffirmed 2002) on frost affected soil sample after treating the soil with varying percentages of cement and chir pine needles. To assess the influence of chir pine needles on CBR behaviour of cement-stabilized soil experiencing seasonal freezing–thawing effects, a series of unsoaked, soaked and 4-cycles of freezing–thawing CBR tests were conducted on numerous proportions. The soil with different mixtures was compacted at its OMC and MDD. The CBR moulds were kept in water for 96 h for the test in soaked condition and after draining the excess water CBR tests were done.

3.3 Unconfined Compression (UCS) Test

UCS tests as per IS 2720 (Part 10): 1991 (Reaffirmed 2006) were performed to investigate the strength characteristics of un-treated parent soil and treated with varying percentages of cement and chir pine needles. The samples of various proportions were prepared at their OMC and MDD, which were tested without curing and after curing for 3, 7, 14 and 28 days at room temperature using a desiccator.

3.4 Freeze–Thaw Test Methods

To investigate the potential impact of freeze–thaw cycles on CBR performance, CBR moulds compacted at OMC and MDD were alternately subjected to four cycles of freeze–thaw. The moulds were frozen in the deep freezer for 12 h at −18 °C and allowed to thaw at room temperature i.e., at +20 °C for 12 h, this constituted one cycle of freeze–thaw. On completion of 4-cycles of freeze–thaw, the CBR tests were performed on the untreated and treated samples. It has been reported [16] that when freezing at −18 °C for 8 h and thawing at 20 °C for 8 h cycles were performed, the detrimental effects on strength of compacted soils occur within the first three cycles [17]. However, four freeze–thaw cycles were maintained in the current work to represent extreme conditions.

Further, to investigate the effect of freezing thawing on UCS, the samples were subjected to 3, 7, 14 and 28 freeze–thaw cycles prior to testing.

4 Results and Discussions

After conducting standard proctor compaction test, OMC and MDD were found to be 11.42% and 19.05 kN/m^3. The variations of (MDD) and (OMC) treated with cement and reinforced with natural fibers such as chir pine needles are discussed in the following sections.

4.1 Compaction Behaviour of Soil-Cement Mixes

Figure 2 clearly shows that with an increasing percentage of cement content OMC decreases and MDD increases. The decrease in OMC with the addition of cement is owing to the cementation reaction process, as the cementation reaction process implicates an excessive consumption of water with the successive generation of heat. The losses of moisture during the evaporation process ultimately reduce the OMC. The increase in MDD could be attributed to the facts that cement being denser material than the soil sample contributes to increase in density.

4.2 Compaction Behaviour of Soil-Chir Pine Needles Mixes

It is clearly observed from Fig. 3 that OMC and MDD both decreases as the percentage of CPN increases. Decrease in MDD can be attributed to substituting the soil with chir pine needles fibers which has lower specific gravity than the soil [12]. The decrease in OMC is due to the facts that chir pine needles being fibrous, lack water holding capacity as the amount of CPN increases.

4.3 Compaction Behaviour of Soil-Cement-Chir Pine Needles Mixes

Figure 4a, b clearly show that addition of cement by 2 and 4% and varying the chir pine needles from 0.5, 1, 1.5 and 2% in the frost-affected soil increase the OMC and decrease the MDD. However, with 6% cement it shows a typical pattern of decrease in both OMC and MDD with the increase in CPN from 1.5 to 2% as shown in

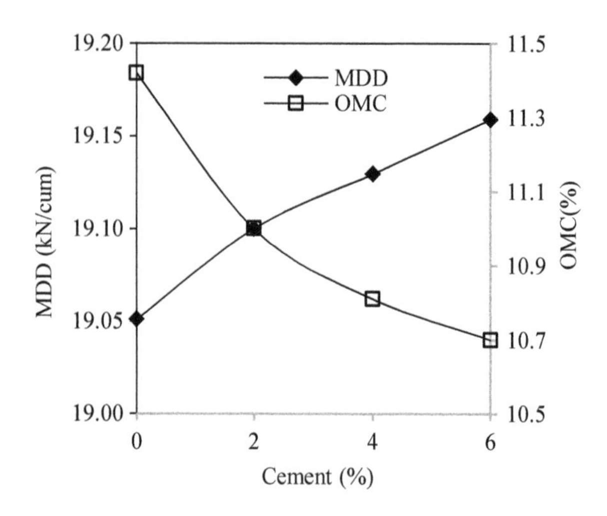

Fig. 2 MDD and OMC with cement

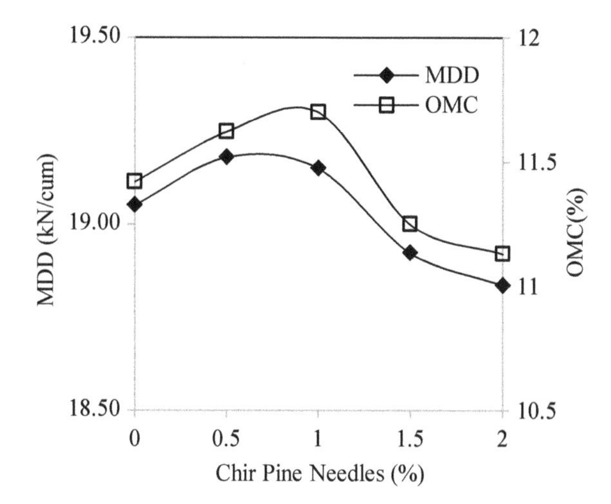

Fig. 3 MDD and OMC with CPN

Fig. 4c. The decreased MDD is attributed by replacing the soil with fibres having lower specific gravity, as the percentage of CPN the number of fibres increases resulting into decrease in the MDD.

4.4 CBR Behaviour of Soil and Cement Mixes

The CBRs of the unsoaked, soaked and freeze-thawed untreated soil were found to be 3.83%, 2.89% and 1.88% respectively. From Fig. 5 it is observed that with the increase in cement percentage, the CBR values increase in all cases of unsoaked, soaked and freeze–thaw conditions. However, the freeze-thawed CBR for the mixes is lower than the soaked and unsoaked condition. When 6% cement is added to the parent soil the CBR values of unsoaked, soaked and freeze-thawed samples increased to 16.34%, 14.47% and 13.53% respectively. The higher CBR value in unsoaked condition can be accredited to the existence of surface tension forces which extends further resistance to penetration. Further, the increase in the CBR values with the addition of cement percentage is owing to hydration processes of cement forming silicate gel and as the amount of cement increases formation of silicate gel also more which enhances the continuous increase in the CBR values [18].

4.5 CBR Behaviour of Soil and Chir Pine Needles (CPN) Mixes

To see the effect of CPN on the strength characteristics of the frost-affected soil, test specimens were prepared by reinforcing the parent soil with CPN percentage with

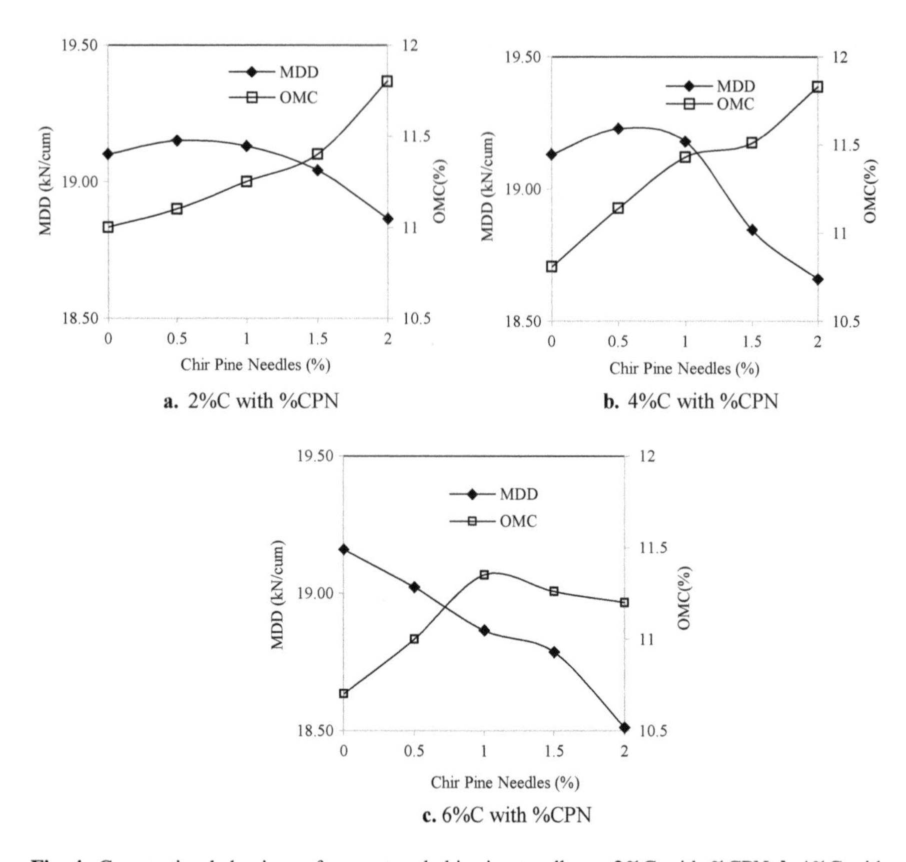

Fig. 4 Compaction behaviour of cement and chir pine needles. **a** 2%C with %CPN, **b** 4%C with %CPN, **c** 6%C with %CPN

0.5%, 1%, 1.5% and 2% respectively. Figure 5 shows that addition of CPN in the soil substantially augments the strength of CBR. It is observed that addition of CPN fibres of 1.5% improves the CBR values to a level which will be adequate for design purposes and further increase in CPN percentage may not be warranted. As the fiber content increases, the soil fiber interaction decreases and frictional resistance between the fibers and soil particles reduce due to lubrication. The CBR values further deteriorate by freeze–thaw effects.

4.6 CBR Behaviour of Soil, Cement and Chir Pine Needles Mixes

To access the impact of cement and CPN on the frost-affected soil, test samples were prepared with cement and CPN content varying from 2%C, 4%C, 6%C and 0.5%CPN, 1%CPN, 1.5%CPN and 2%CPN respectively. Figure 5 shows that

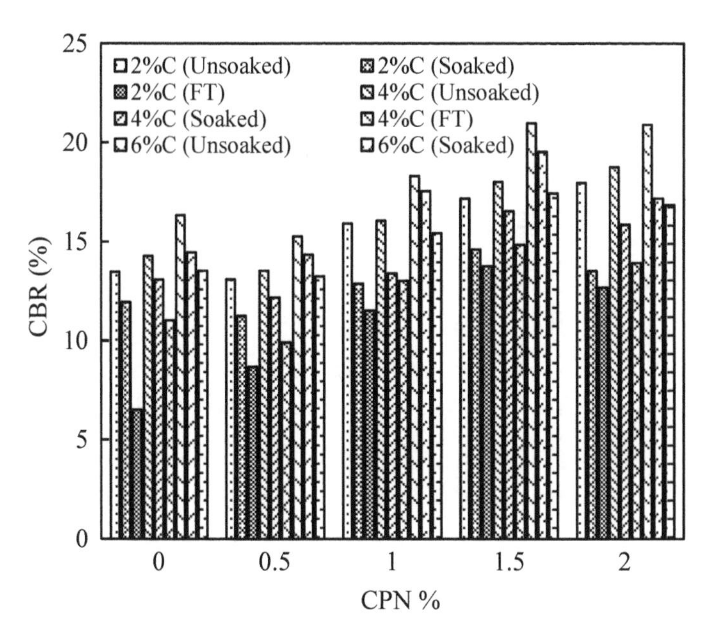

Fig. 5 Variation of unsoaked, soaked and freeze–thaw CBR with cement and chir pine needles

addition of cement and chir pine needles increases the strength of the soil as compared to the sample without treatment. However, the CBR values of soaked and freeze-thawed samples are lower than that of unsoaked samples. The mixture which gives the highest improvement in the CBR is found to be S + 6%C + 1.5%CPN with CBR value of 20.98%, 19.53% and 18.81% for unsoaked, soaked and freeze-thawed conditions. Though, repetitive freezing–thawing cycles make the soil structure loose and the mechanical strength is greatly reduced [19], agglomeration of the soil–cement particles takes place with the increase in percentage of cement. The increase in CBR could be attributed to the pozzolanic reactions and reinforcement which take place when the cement and CPN are added into the soil. Table 2 presents data on the chemical composition of untreated, treated and freeze-thawed samples obtained from EDX analysis. It is observed from Table 2 that the redistribution of contents of chemical elements takes place aided by curing and freeze–thaw cycles.

4.7 Unconfined Compressive Strength (UCS) of the Treated and Reinforced Soil

The UCS of untreated soil decreases with the curing periods. The UCS of parent soil with 28 days of curing periods and after being exposed to 28 freeze–thaw

Table 2 Chemical composition of the treated soil with 6%C + 1.5%CPN under normal condition, after 28 days curing and after 28 days freeze–thaw cycles

	S + 6%C + 1.5%CPN Normal condition			S + 6%C + 1.5%CPN After curing			S + 6%C + 1.5%CPN After freeze–thaw cycles		
Standard	Elements	Wt%	Atomic wt%	Elements	Wt%	Atomic wt%	Elements	Wt%	Atomic wt%
O SiO_2	C K	6.67	10.94	O K	54.07	69.65	O K	46.63	68.86
Na Albite	O K	50.70	62.41	Na K	1.26	1.13	Na K	1.11	1.14
Mg MgO	Na K	1.33	1.14	Mg K	1.35	1.14	Mg K	1.18	1.14
Al Al_2O_3	Mg K	1.39	1.13	Al K	8.32	6.36	Si K	19.35	16.28
Si SiO_2	Al K	7.62	5.56	Si K	20.58	15.10	KK	3.51	2.12
K MAD-10 Feldspar	Si K	17.42	12.22	KK	3.55	1.87	Ca K	4.87	2.87
Ca Wollastonite	KK	3.72	1.87	Ca K	5.10	2.62	Fe K	5.93	2.51
Fe Fe	Ca K	5.78	2.84	Fe K	5.75	2.13	BrL	14.41	4.26
Cu Cu	Fe K	5.37	1.89				Sr L	3.01	0.81

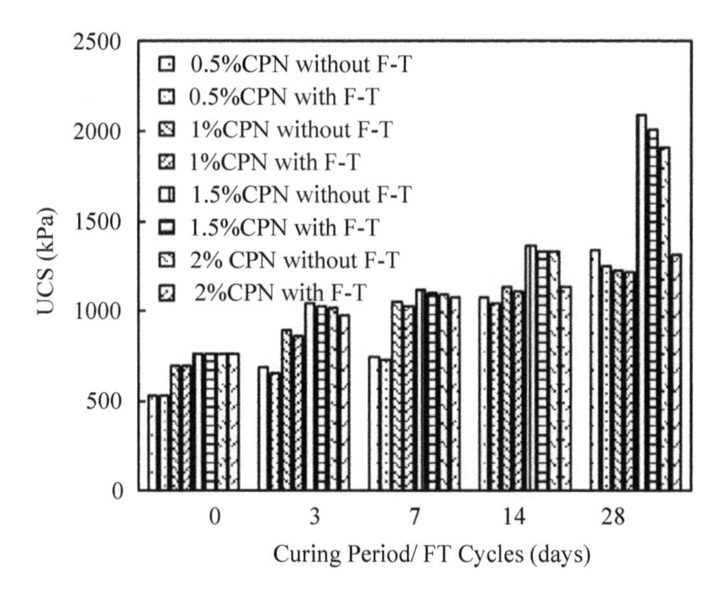

Fig. 6 Variation of UCS with 6%C and %CPN at different curing periods

cycles were found to be 258 and 237 kPa respectively. When the soil sample was treated with S + %C, S + %CPN and S + %C + %CPN, the UCS value of the mix samples increased with the increase in percentage of cement and chir pine needles as well as curing period. Further, it is observed that for S + %C and S + %CPN mixes under the same conditions, the UCS values decrease when exposed to 28 numbers of freezing–thawing cycles as compared to 28 days of curing of the test samples in desiccators. Thus, freeze–thaw cycles significantly reduce the strength, which has also been noted earlier [20]. However, for S + %C + %CPN mix samples, the UCS value is found to be better than that of S + %C and S + %CPN alone as shown in Fig. 6. The UCS value significantly increases from 258 kPa of untreated soil to 2,090 kPa of treated soil for non-freeze–thaw cycle, and 237 kPa–2,007 kPa for 28 freeze–thaw cycles. The increase in UCS can be attributed to the formation of cementitious products due to hydration of cement with the presence of water [20] and better adhesion of soil matrix with fibers [14] and [21].

4.8 Effects of Freeze–Thaw on Micro-structure Properties of the Soil

To explore the effect of curing and freezing–thawing on the micro-structural particles of the soil, scanning electron microscope (SEM) investigations were done on the treated samples which gave optimum CBR and UCS values. After proper mixing of representative samples in dry condition, the mixture was filled in a plastic

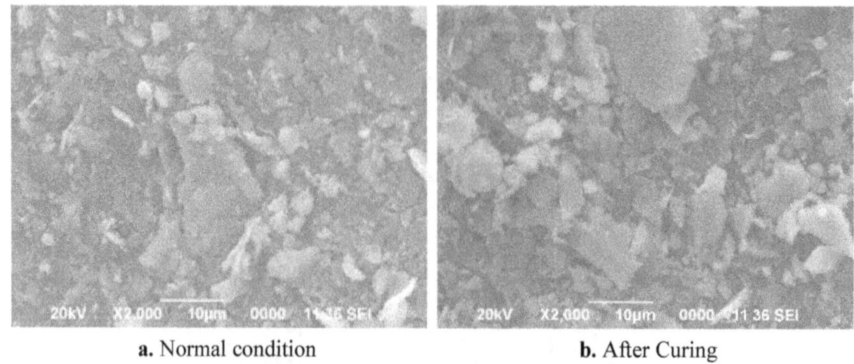

a. Normal condition **b.** After Curing

c. After Freezing-Thawing

Fig. 7 SEM of S + 6%C + 1.5%CPN on normal condition, after curing and after freeze–thaw cycles. **a** Normal condition, **b** after curing, **c** after freezing–thawing

bag and curing of the samples was done for 28 days in a desiccator. Similarly, replicate batch of samples were collected and exposed to 28 number of freeze–thaw cycles. Scanning Microscopic Electron (SEM) and Energy Dispersive X-ray diffractometer (EDX) tests were done by JEOL JSM-6390LV and Oxford Instruments model 7582 EDX at Tezpur University, Tezpur, Assam, India on the dry sample without curing and freeze–thaw cycles (Normal condition). Further, the results for sample with normal condition were compared with the samples with cured and freeze-thawed samples. Figure 7 revealed formation of bigger particle size at micro-structure after the samples had undergone for 28 days curing periods and 28 days freeze–thaw cycles. The decrease in CBR and UCS values can be attributed to the expansion in the volume of pores due to freeze–thaw cycles, which significantly weakens the cementation process and further reduces the soil-fibers interlocking and ultimately leads to decreased in cohesion and strength.

Further, comparative test results of the untreated soil with treated optimum dosage are summarized in Table 3. The best results in CBR are obtained for treated samples after curing for 4 days and after subjected to 4 cycles of freeze–thaw.

Table 3 Comparative test results of untreated soil and treated optimum dosage of S + 6% C + 1.5%CPN are summarized

			CBR (%)			UCS (kN/mm^2)		
Test	MDD (kN/m^3)	OMC (%)	Normal condition	After 4 days curing	After 4 freeze–thaw cycles	Normal condition	After 28 days curing	After 28 freeze–thaw cycles
Untreated soil	19.05	11.42	3.83	2.89	1.88	200	258	237
Treated with S + 6% C + 1.5% CPN	18.79	11.26	20.98	19.53	18.81	794	2090	2007
%age change	−1.36	−0.01	448	576	901	297	710	747

The increase reported is more than 400%. Similarly, the best results in UCS are obtained for treated samples after curing for 28 days and also after subjecting them for 28 cycles of freeze–thaw. The increase reported is more than 700%.

5 Conclusions

The present study has explored the efficiency of cement and chir pine needles as stabilizer and reinforcing material in improving the strength properties of subgrade soil under seasonal freeze–thaw effects. Based on the experimental findings following interpretations are drawn.

- OMC decreases and MDD increases with the increase in percentage of cement.
- OMC and MDD both decrease as the number of chir pine needles fiber increases.
- Increase in CBR value was observed with the increase percentage of cement. However, soaked and freeze-thawed samples show lower CBR value than the unsoaked samples.
- Similarly, CBR value increases with the increase in CPN percentage but beyond 1.5%, the increase in CBR values is not significant. Therefore, the CPN percentage beyond 1.5% of dry weight of the soil is not suggested for soil reinforcement.
- CBR value notably increased when the cement and CPN were blended together. The mixture which performed better was found to be with S + 6%C + 1.5% CPN. The unstabilized samples under unsoaked, soaked and freeze–thaw of 3.83, 2.89 and 1.88% have been significantly improved to 20.98, 19.53 and 18.81% by this mixture.
- The UCS of soil mixed with cement increases with the increase in curing period and freeze–thaw cycles. However, samples exposed to freeze–thaw cycles show lower UCS compared to those after curing.

- The mixture of Soil + Cement + CPN shows notable increase in the UCS value. Similar to CBR, the optimum UCS was obtained with S + 6%C + 1.5%CPN both in curing and freeze–thaw conditions. The UCS value increases from 258 kPa of untreated soil to 2,090 kPa of treated soil for non-freeze–thaw cycle, and 237 kPa–2,007 kPa for 28 days freeze–thaw cycles.
- Addition of CPN alone does not improve the strength properties when exposed to freeze–thaw cycles.
- SEM analysis shows that although volume of pores expands under curing and freeze–thaw cycle, the formation of cementitious materials with the addition of cement is responsible for the strength gain.
- The EDX analysis confirms the redistribution of elements with curing and freezing–thawing cycles.
- Since the CBR and UCS for the frost-affected subgrade soil considerably improved when stabilized with cement and reinforced with CPN in compared to unstabilized soil, utilization of cement-CPN is recommended for treating subgrade soil where freezing weather conditions prevail.

References

1. Konrad JM (1989) Physical processes during freeze-thaw cycles in clayey silts. Cold Reg Sci Technol 16:291–303
2. Peng W, Wang Q, Liu Y, Sun X, Chen Y, Han M (2019) The influence of freeze-thaw cycles on the mechanical properties and parameters of the Duncan-chang constitutive model of remolded saline soil in Nong'an County, Jilin Province, Northeastern China. Appl Sci MDPI 9:4941
3. Kalkan E (2009) Effects of silica fume on the geotechnical properties of fine-grained soils exposed to freeze and thaw. Cold Reg Sci Technol 58:130–135. https://doi.org/10.1016/j.coldregions.2009.03.011
4. Zaimoglu AS (2010) Freezing–thawing behavior of fine-grained soils reinforced with polypropylene fibers. Cold Reg Sci Technol 60:63–65. https://doi.org/10.1016/j.coldregions.2009.07.001
5. Ghazavi M, Roustaei M (2013) Freeze–thaw performance of clayey soil reinforced with geotextile layer. Cold Regions Sci Technol 89:22–29. https://doi.org/10.1016/j.coldregions.2013.01.002
6. Tarun RN, Rudolph NK, Rafat S (2003) Use of wood ash in cement-based materials. A CBU report, CBU-2003–19 (REP-513). http://uwm.edu/Dept/CBU/report/
7. Yarbesi N, Kalkan E, Akbulut S (2007) Modification of the geotechnical properties, as influenced by freeze–thaw, of granular soils with waste additives. Cold Reg Sci Technol 48:44–54. https://doi.org/10.1016/j.coldregions.2006.09.009
8. Hazirbaba K, Gullu H (2010) California bearing ratio improvement and freeze–thaw performance of fine-grained soils treated with geofiber and synthetic fluid. Cold Regions Sci Technol 63:50–60. https://doi.org/10.1016/j.coldregions.2010.05.006
9. Moayed RZ, Daghigh Y, Lahiji BP (2013) The influence of freeze-thaw cycles on CBR values of Silty soils Stabilized with Lime and Microsilica. In: Seventh international conference on case histories in geotechnical engineering, 29 Apr–04 May, Chicago

10. James J, Karthickeyan S, Chidambaram S, Dayanandan B, Karthick K (2018) Effect of curing conditions and freeze-thaw cycles on the strength of an expansive soil stabilized with a combination of lime, jaggery, and gallnut powder. Hindawi Adv Civ Eng (AID 1813563). https://doi.org/10.1155/2018/1813563
11. El-Maaty AEA (2016) Enhancing the CBR strength and freeze-thaw performance of silty subgrade using three reinforcement categories. Civ Eng J 2(3): 73–85. https://doi.org/10.28991/cej-2016-00000014
12. Correia, AAS, Venda Oliveira PJ, Custódio DG (2015) Effect of polypropylene fibres on the compressive and tensile strength of a soft soil, artificially stabilised with binders. Geotext Geomembranes 43(2):97–106
13. Singh V, Rani A (2018) A laboratory study on pine needle reinforced soil. Int Res J Eng Technol (IRJET) 05(05):4228–4233
14. Sharma V, Vinayak HK, Marwaha BM (2015) Enhancing compressive strength of soil using natural fibers. Constr Build Mater (Elsevier Ltd)
15. Rani P, Goyal T (2019) Improvement of engineering properties of soil using chir pine needles. Int Res J Eng Technol 06(05):4993–4997
16. Cook RD (1963) Some effects of closed system freeze-thaw cycles on a compacted, high plastic clay. M.Sc. thesis, University of Alberta
17. Hazirbaba K, Gullu H (2010) California bearing ratio improvement and freeze-thaw performance of fine-grained soils treated with geofiber and synthetic fluids. Cold Regions Sci Technol 63(2–3):50–60
18. Pandian NS, Krishna KC (2002) California bearing ratio behavior of cement-stabilized fly ash-soil mixes. J Test Eval 30(6):492–496. https://doi.org/10.1520/JTE12347J
19. Yao M, Wang Q, Ma B, Liu Y, Yu Q, Han Y (2020) Effect of freeze-thaw cycle on shear strength of lime solidified dispersion soils. Civ Eng J 6(1)
20. Xie SB, Jian-Jun Q, Yuan-Ming L, Zhi-Wei Z, Xiang-Tian X (2015) Effect of freeze-thaw cycles on soil mechanical and physical properties in the Qinghai-Tibet Plateau. J Mountain Sci 12(4):999–1099
21. Millogo Y, Morel J-C, Aubert J-E, Ghavami K (2014) Experimental analysis of pressed adobe blocks reinforced with Hibiscus cannabinus fibers. Constr Build Mater 52:71–78

Schedule Delay Modeling of Large Infrastructure Projects in India

Mohammed Abdul Wajid Siddiqui and Mir Iqbal Faheem

Abstract Construction of large infrastructure projects is one of the important endeavors in galvanizing the growth and the economic development of India. Due to various causes, many large infrastructure projects in India suffer from extensive delays in their schedule which leads to problems such as an increase in the cost of the project, claims, disputes among stakeholders, and a negative effect on the economy of the country. This study attempts to determine the main causes of these delays and to suggest mitigation strategies that can provide effective control over such delays. Based on extensive literature review 6 categories of issues contributing to schedule delays are identified and a questionnaire survey is designed in consultation with industry experts and responses from Contractors, Consultants, and Owners of the projects in the construction industry are obtained. Data collected from 135 companies include the percentage of delay attributed to the issues related to each of 6 categories by the construction personnel. Analysis of data is done and a regression model of the schedule delays in a infrastructure project is prepared to understand the main issues contributing to schedule delays. This study tries to give a relationship between the above categories of issues with the percentage of delay it may cause to the schedule of the project, which can be further used effectively to control and mitigate such delays by adopting suitable measures for construction management of future projects.

Keywords Infrastructure projects · Schedule delays · Regression analysis · Construction companies

1 Introduction

Construction industry is major sector of economic activity in India, and it accounted around 8% of its GDP prior to COVID-19 pandemic and a key indicator of economic growth [1]. The growth of construction activity assures and stimulates

M. A. W. Siddiqui (✉) · M. I. Faheem
School of Engineering, Career Point University, Kota, India

G. C. Marano et al. (eds.), *Proceedings of SECON'21*, Lecture Notes in Civil Engineering 171, https://doi.org/10.1007/978-3-030-80312-4_89

growth in other sectors which are interconnected and creates employment and investment opportunities. Some of the consequences of a delayed project can be increased cost of project, decrease in profits of contractor, loss of revenue for client on account delay in returns of investment on construction and loss in productivity.

As per government data released September 2020 [2], 30.24% of the construction projects are delayed and 25.86% of them are suffering cost overrun. Lot of effort and time is made by the construction industry and the government on controlling and minimizing construction delays. Construction projects in India face delays due to various causes such as manpower shortage, delay in material or equipment supply, disputes among stakeholders on account of payments for work done or financial matters, delay in environmental permissions/clearances, lack of project management and control over construction schedules and delays due to government shutdowns, loss of employment and migration of labor due to COVID-19 pandemic.

2 Literature Review

Many researchers have discussed the causes of construction delays in numerous ways. Some researchers found the critical delaying factors and several project categories using empirical methods, while other studies suggested ways of analyzing delays and methods minimizing it. Chan and Kumaraswamy [3] found that poor site management and supervision, unforeseen ground condition, delays in design information, slow decision making and client initiated variations were among the major causes of delay. Odeh et al. [4] ranked delay causes based on RII (Relative Importance Index) in Jordan by a survey consisting of 28 delay causes and concluded that top reason by consultants and contractors category as inadequate contractor experience and low labor productivity respectively. Venkatesh et al. [5] analyzed 45 causes of delays in India using Importance index and concluded that main causes are Shortage of labors and construction materials and rework and change orders. Ayodeji et al. [6] studied the initial and final construction time of sixty five projects in South Africa using regression analysis and found the relationship between the initial and final construction times. Pourrostam et al. [7] studied completed construction projects between 2001 and 2010 and identified 27 factors responsible for delays and subsequently ranked based on responses from contractors, clients and consultants across Iran, using SPSS and recommended use of experienced contractors and subcontractors, doing accurate initial cost estimates, as effective methods controlling delays on construction projects. Vilventhan et al. [8] studied delays related to utilities on road and bridge projects in India, using cognitive mapping techniques and revealed that top factors of delays as slow response from utility agencies, difficulty in identification of underground utilities, lack of information on underground utilities and conflict between agencies. It is observed that previous studies have focused more on finding the different causes of delay and its relative importance in construction projects, while there are few studies considering construction delay itself as an index for predicting the risk of delay in a project.

From previous studies, it identified that five major groups (i) manpower related factors (ii) Material and equipment related factors (iii) Finance related factors (iv) Environmental factors (v) Construction management and control as critical factors causing delays in construction projects. As ongoing COVID-19 pandemic has also impacted the construction sector, so it is also considered as sixth factor causing delays in the present study. It is therefore proposed to relate the delays from major group related factors in a particular project by formulating a model which can predict the delay duration for a real construction site which can minimize the delay by proper scheduling and control methods.

3 Research Methodology

The present work studied numerous factors causing delay in Indian context which were identified from literature survey and were broadly classified in five unique groups of causes based on the type and nature of factor causing delay. For example, factors like cash flow problems of contractor and delayed progress payment by client are grouped in finance related causes. A preliminary questionnaire design was made and one expert each from industry and academic field was requested to review and provide feedback and opinion on the questionnaire design. Based on this, a structured questionnaire consisting of two parts was designed to solicit response from construction personnel to know details like type of project and client, location, total cost and time of project of project, in Part–I and In Part II of the questionnaire, the construction professionals were asked to fill the percentage of delay attributed to six major group related delay factors i.e., (i) percentage of delay attributed to manpower related issues in your project and similarly next questions are for percentage of delays for (ii) material and equipment, (iii) finance, (iv) environment and (v) project management and control and (vi) COVID-19 pandemic related factors. Random sampling technique was adopted for the present study. As ongoing Covid-19 pandemic restricted personal meetings with construction professional for data collection, Microsoft Forms was used for Questionnaire survey and link was sent electronically via email through network of friends and colleagues to respondents from about 250 construction companies. 135 valid responses were received via Microsoft Forms from government and private projects in India covering road and bridges, residential, commercial and industrial projects from metro cities, other cities, towns and other areas. Collected data was categorized, checked for and analyzed by means of multiple linear regression analysis using IBM SPSS Statistics. The relationship between overall percentage of delay in the construction project and the percentage of delay by six groups mentioned above is found out by regression analysis, which is further used to test the ability of regression model.

3.1 Details of Data Collected

Type of client: Out of 135 projects 80 were from Private sector and 55 from Public sector.

Type of projects: projects studied consist of 66 residential, 36 Roads and bridges, 23 Commercial and 10 Industrial.

Cost of projects: projects studied consist of 56 projects costing under ₹100 million, 53 projects between ₹100–200 million, 18 projects costing between ₹200–500 million and 8 projects which cost more than ₹500 million (Figs. 1, 2, and 3).

4 Data Analysis

A null hypothesis was considered to find the significant relation between the each delay factor and the total percentage of delay. The descriptive statistics in Table 1 shows that Finance related issue is having a highest the mean value (23.24), then followed by COVID-19 related issue is having a mean value of 21.07, the least mean is found for Project management and control issues. It is observed that maximum delay of 63% occurs for COVID-19 pandemic issue for a particular sample project, followed by 49% for finance, 48% for manpower, 45% for environment related, 35% for poor project management and control and 33% for material and equipment.

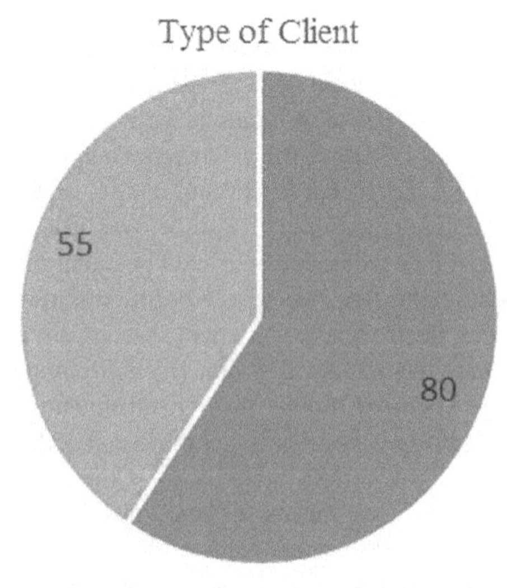

Fig. 1 Type of client for projects under study

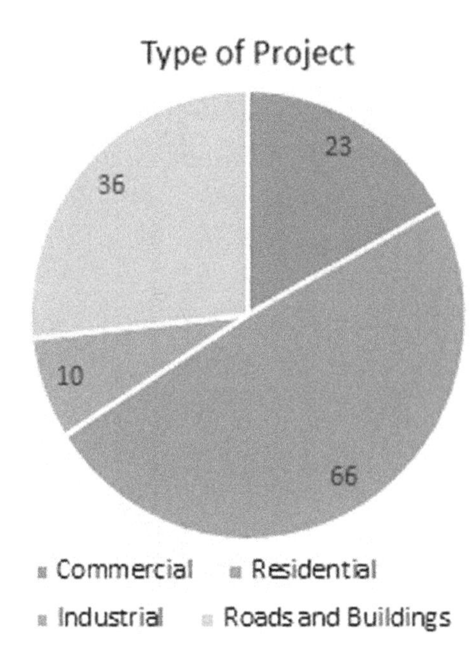

Fig. 2 Type of projects under study

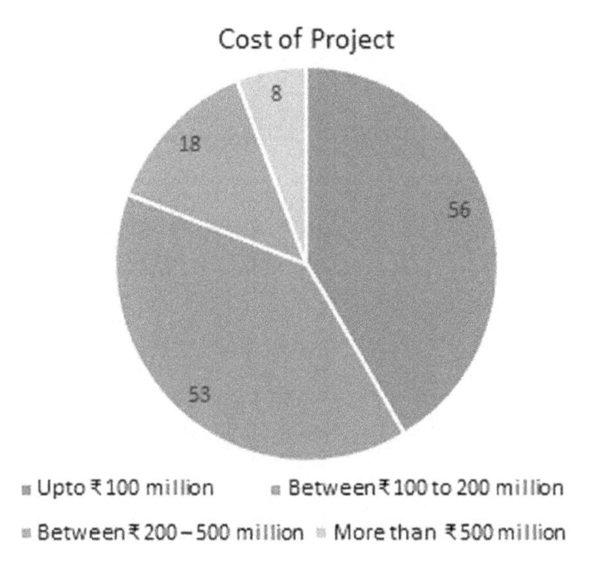

Fig. 3 Total cost of projects under study

Table 2 gives the standard ANOVA between the delay factors groups and the overall percentage of delay in a project. In this analysis results the groups like COVID-19 pandemic, material and equipment, finance and project management and control related problems are having the significance value less than 0.05. So the null hypothesis is rejected and it is concluded that there is significant relationship exists between the above factors and the overall delay percentage.

Table 1 Descriptive statistics showing overall delay percent for different types of factors

		Manpower related	Material/ equipment related	Finance related	Environmental problems	Poor project management and control	COVID-19 pandemic related	Overall delay (%)
N	Valid	135	135	135	135	135	135	135
	Missing	0	0	0	0	0	0	0
Mean		11.04	10.91	23.24	13.10	10.30	21.07	89.66
Median		8.00	8.00	25.00	10.00	10.00	21.00	89.00
Std. deviation		6.561	6.755	12.620	9.738	3.786	14.438	23.213
Range		45	32	49	45	35	63	140
Minimum		3	1	0	0	0	0	26
Maximum		48	33	49	45	35	63	166

Table 2 ANOVA: showing delay factors groups with overall delay

		Sum of squares	df	Mean square	F	Sig.
Manpower related	Between groups	2,856.565	61	46.829	1.174	0.255
	Within groups	2,912.250	73	39.894		
	Total	5,768.815	134			
Material/equipment related	Between groups	3,818.567	61	62.599	1.990	0.003
	Within groups	2,296.367	73	31.457		
	Total	6,114.933	134			
Finance related	Between groups	12,023.300	61	197.103	1.544	0.038
	Within groups	9,317.633	73	127.639		
	Total	21,340.933	134			
Environmental problems	Between groups	7,375.531	61	120.910	1.655	0.020
	Within groups	5,332.217	73	73.044		
	Total	12,707.748	134			
Poor project management and control	Between groups	1,294.581	61	21.223	2.475	0.000
	Within groups	625.967	73	8.575		
	Total	1,920.548	134			
COVID-19 pandemic related	Between groups	16,636.367	61	272.727	1.762	0.010
	Within groups	11,298.033	73	154.768		
	Total	27,934.400	134			

Higher F values and the significance value less than 0.05, implies that the calculated regression coefficient is significant.

4.1 Regression Model for Delay in Construction Project

From previous studies, it five significant groups were identified (i) manpower related factors (ii) Material and equipment related factors (iii) Finance related factors (iv) Environmental factors (v) Project management and control as critical factors causing delays in construction projects along with present COVID-19 pandemic as additional factor. A multiple linear regression model is developed with dependent variable as Overall Delay percent and six independent variables as manpower related factors, material and equipment related factors, finance related factors, environmental factors, Poor Project management and control factors and COVID-19 pandemic. There exists a casual relationship between the independent and dependent variables. The goodness of fit of the model is explained by considering the values of F test, T-test and R-squared. The above test are done using SPSS and Tables 3 and 4 shows the values of R-squared and t- test respectively.

The above regression analysis summary, the coefficient R-squared, explains the overall model fit for predicting the strength of the relationship between the model and overall percentage of delay. R coefficient was found to be 0.987 which is near to 1, it confirms that there was a linear correlation between the observed values and predicted values of the dependent variable with high predictive strength From the analysis, it was found that R-squared value is 0.975 which is also close to 1; therefore the model is able to explain 97.5% variance in the overall percentage of delay. In Table 3 above, F statistic test value was found to be 997.816 which was a higher value and the significance value was found to be less than 0.05 (0.000) which implied that the calculated regression coefficient was more significant and the variation in the independent variables contributes to variation in dependent variable.

Table 4 gives the details of coefficients of regression equation which is predicted by considering the independent variables selected.

In the present study after various analysis it was observed that poor project management and control factor do not have prediction strength as by considering only five independent variables 97.5% of variation in the dependent variable is explained by the model.

Coefficients B of the independent variables and its significant value are presented in Table 4. The independent variables having significant value, ($p < 0.05$) implies a strong relationship with the dependent variable. According to this coefficients table, the predictor variables like COVID-19 pandemic, finance related, manpower, materials and equipment, and environment related issues have a Sig value less than 0.05 ($p < 0.05$), so they were significant and it was found that there was a causal relationship between these variables and the overall percentage of delay. No multicollinearity exists between the variables by inspection of VIF (Variance influence factor) values in Table 4. Therefore the dependent variables have a direct effect on independent variables.

Table 3 Regression model summary[b]

Model	R	R square	Adjusted R square	Std. error of the estimate	Change statistics				
					R square change	F change	df1	df2	Sig. F change
1	0.987[a]	0.975	0.974	3.756	0.975	997.816	5	129	0.000

[a]Predictors: (constant), COVID-19 pandemic related, environmental problems, material/equipment related, manpower related, finance related

[b]Dependent variable: overall delay %

Table 4 Regression model equation coefficients and tolerances

Coefficients[a]								
Model		Unstandardized coefficients		Standardized coefficients	t	Sig.	Collinearity statistics	
		B	Std. Error	Beta			Tolerance	VIF
1	(Constant)	8.389	1.244		6.743	0.000		
	Manpower related	1.040	0.050	0.294	20.706	0.000	0.969	1.032
	Material/equipment related	1.043	0.049	0.303	21.214	0.000	0.955	1.047
	Finance related	0.992	0.027	0.539	36.493	0.000	0.895	1.118
	Environmental problems	1.007	0.033	0.422	30.119	0.000	0.993	1.007
	COVID-19 pandemic related	1.052	0.023	0.654	45.667	0.000	0.952	1.051

[a]Dependent variable: overall delay %

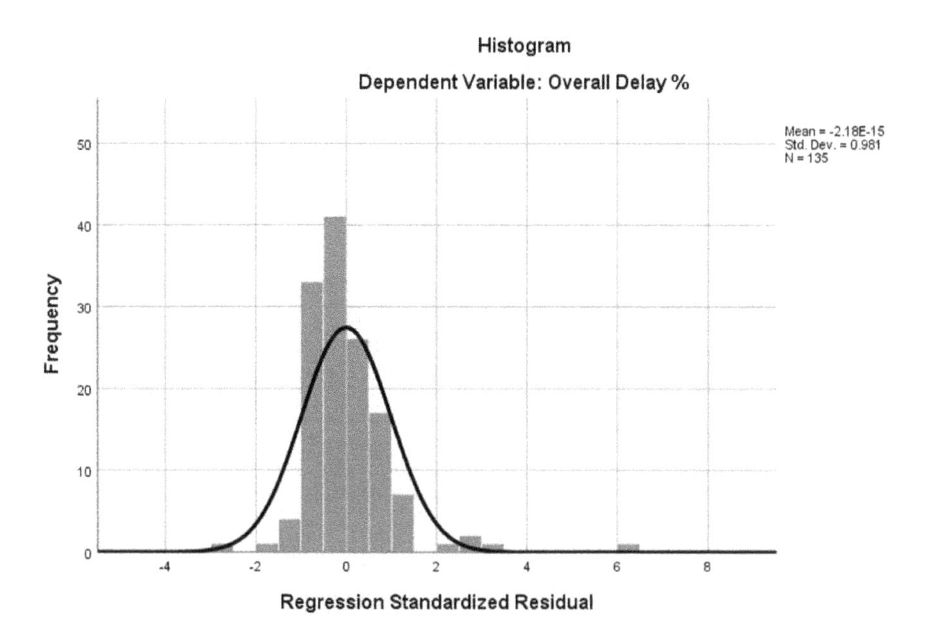

Fig. 4 Regression standard residual

Figures 4 and 5 gives the histogram and P-P plot of the regression standard residual which is distributed normally implying the goodness of fit of the model.

The regression model is represented in the form of equation as given below by using coefficients in Table 4 in Eq. (1).

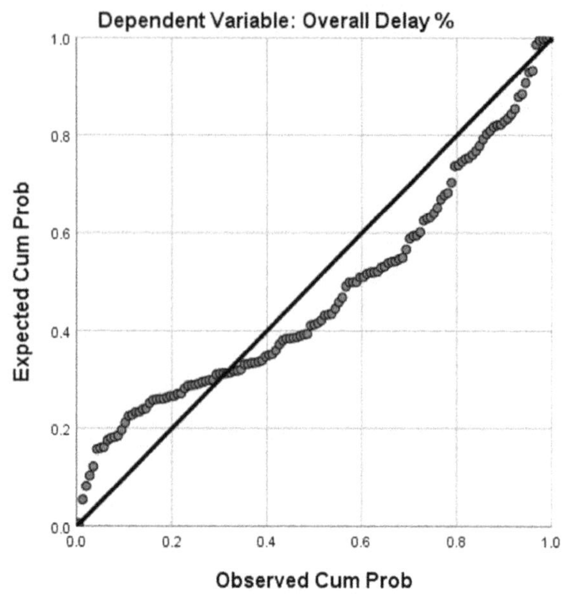

Fig. 5 Normal P-P plot of regression standard residual

$$\text{Expected overall percentage of delay} = 1.040 \times M + 1.043 \times ME + 0.992 \times F + 1.007 \times E + 1.052 \times C - 19 + 8.389 + 1.244 \quad (1)$$

where M—manpower related, ME—material Equipment related, F—finance related, E—environmental problems, C-19—COVID-19 pandemic related.

5 Conclusions

This study was done by using date from 135 construction projects. It was observed that there were very few projects which had zero delay in one of the six categories of delay factors on which the questionnaire survey was done. The regression model made was having a predictive strength of about 97.5% in relation to overall percentage of delay and the independent variables. In this study, the model is formulated by taking overall percentage of delay as y and delay related factors that cause delay as x1, x2 & x3… The planning professionals can accommodate the expected percentage of delay in preparation of the project schedule by using the above proposed equation.

The major groups such as finance related (27.2%), COVID-19 pandemic (24.7%), materials and equipment related (13%), were having the significant contribution to the overall percentage of delay, so these percentages should be taken into consideration during the planning and contracting stages to minimize the effects of construction delays. This model has been proposed as a strategic tool for real projects in forecasting the delays and improve the overall efficiency of the construction projects.

Though present study is directly applicable only to the construction projects during ongoing COVID-19 Pandemic, but can be adopted to study construction projects in critical situations involving one-time nationwide impact (for example Financial Crisis of 2008) or impacting a particular region of India (Floods in Northeastern states).

References

1. Reserve Bank of India (2020) Publications. www.rbi.org.in. Retrieved 24 Jan 2020
2. Government of India (2020) Flash report on central sector projects (Rs.150 crore and above). Infrastructure and Project Monitoring Division, Ministry of Statistics and Programme Implementation, Government of India, New Delhi, p 2
3. Chan DWM, Kumaraswamy MM (1997) A comparative study of causes of time overruns in Hong Kong construction projects. Int J Project Manage 15(1):55–63
4. Odeh AM, Battaineh HT (2002) Causes of construction delay: traditional contracts. Int J Project Manage 20(1):67–73
5. Venkatesh MP, Renuka SM, Umarani C (2012) Causes of delay in Indian construction industry. In: Applied mechanics and materials, vol 174. Trans Tech Publications Ltd., pp 2768–2773
6. Aiyetan AO, Smallwood JJ, Shakantu W (2012). A linear regression modelling of the relation-ship between initial estimated and final achieved construction time in South Africa. Acta Structilia 19(1):39–56
7. Pourrostam T, Ismail A, Mansournejad M (2011) Identification of success factors in minimizing delays on construction projects in IAU-Shoushtar Branch-Iran. In: Applied mechanics and materials, vol 94. Trans Tech Publications Ltd
8. Vilventhan A, Kalidindi SN (2016) Interrelationships of factors causing delays in the relocation of utilities. Eng Constr Archit Manage

Analysis & Design of Slender Shear Wall with and Without Openings

Pranav Shah, P. V. Indira, and Kannan C. Bhanu

Abstract Nowadays structures are going higher and higher and the effect of lateral forces by wind and earthquake on tall structures is becoming critical. Shear walls are used to resist these loads in those structures. The shear wall system is the most economic lateral load resisting system for buildings with more than 10 storeys. Shear walls have very high in-plane stiffness and strength which is very effective against lateral loads. Due to the functional requirements of structure such as doors, windows or any internal corridor, the shear wall may contain many openings, which may largely affect the overall seismic response of the structure. In this paper, results of numerical study and comparison of design codes (IS-13920:2016, ACI 318–19, NZS-3101:2006) are presented. Numerical analysis was conducted using ABAQUS software to study the effects of opening and its orientation on the slender shear wall. The Load–displacement graph was able to predict the decrease in strength and ductility of shear walls when openings were provided.

Keywords Slender shear wall · Openings · ABAQUS · Numerical analysis

1 Introduction

Shear Walls are a structural element of a building, used to resist lateral forces (parallel to the plane of the wall) through shear and bending. In other words, These are the vertical members of the horizontal force resisting system. Shear wall's behaviour generally depends on their material, length, thickness and also on their position in the building. These walls generally start at the foundation level and go continuous throughout the building height. The thickness of the wall can be as low as 150 mm or as high as 400 mm.

The shear wall system is efficient both in terms of construction cost and its effectiveness in reducing earthquake damage in structural and non-structural ele-

P. Shah (✉) · P. V. Indira · K. C. Bhanu
National Institute of Technology, Calicut, India

G. C. Marano et al. (eds.), *Proceedings of SECON'21*, Lecture Notes in Civil Engineering 171, https://doi.org/10.1007/978-3-030-80312-4_90

ments. Many modern constructions use the shear wall as the main source for lateral load resistance [1].

It is better to use a shear wall with no opening in them but sometimes due to functional requirements, it becomes unavoidable. Openings in a shear wall may largely affect the overall seismic response of the structure, therefore it is necessary to understand the effect of openings on the behaviour of the shear wall [2].

For studying the effects of different types of opening on the behaviour of shear wall, Numerical analysis has been performed by finite element based software ABAQUS and the results are presented in this paper. Also, the different codal provisions for the design of the slender shear wall in IS-13920:2016, ACI 318–19, NZS-3101:2006 are compared and presented in this paper.

2 Codal Provision for the Design of the Slender Shear Wall

In this section, The comparison between Indian (IS-13920:2016), American (ACI 318–19) and Newzealand (NZS-3101:2006) codes are presented for the design of a slender shear wall. In these three codes, many different provisions are given for the design of the shear wall but some provisions also have similarities between them. For the comparison, some of the provisions are tabulated in Table 1 and reinforcement details of the simple shear walls designed by these three codes are given in Tables 2, 3 and 4.

All three codes have given proper and sufficient information for the design of slender shear wall. But if we just compare the provisions then it can be said that provisions for providing reinforcement around the opening are different in all these codes, ACI code does not restrict maximum vertical spacing in transverse reinforcement, IS code does not have any distinct provisions for minimum reinforcement in end region and only ACI code has given the provision for providing boundary element near the opening.

A two-storey slender shear wall (in Fig. 1) is designed by these three codes and their reinforcement details are shown in Tables 2, 3 and 4.

To study the behaviour of shear walls with openings and their orientation, Numerical investigation was also carried out on the shear walls designed using IS code.

3 Numerical Analysis Using ABAQUS

3.1 Modelling and Analysis

In this study, Non-linear static analysis was done using the finite element analysis software ABAQUS. The Concrete wall was modelled using the C3D8R element.

Table 1 Some codal provisions for slender shear wall design [3–5]

Codal provisions	ACI-318:19	IS-13920:2016	NZS-3101:2006
Minimum web reinforcement (horizontal and vertical)	0.0025	0.0025	$\frac{0.7}{f_{yn}}$
Maximum allowable spacing	$\min of\left(3t, 18\,in, \frac{l_w}{3}\right)$	$\min of\left(3t, 450\,mm, \frac{l_w}{5}\right)$	$\min of\left(3t, 300\,mm, \frac{l_w}{3}\right)$
Bar diameter		Max dia $\frac{t_w}{10}$	Min dia 10 mm Max dia $\frac{t_w}{7}$
Two layer of reinforcement has to provide if	$V_u \geq 2\lambda\sqrt{f_c\prime}A_{cv}$ or $\frac{h_w}{l_w} \geq 2$	If factored shear stress $\geq 0.25\sqrt{f_{ck}}$ or $t_w > 200\,mm$	$t_w > 200\,mm$
Minimum Wall thickness		150 mm	100 mm
Relation in vertical & horizontal reinforcement	$\rho_v \geq \rho_h$	$\rho_v \geq \rho_h$	
Boundary element requirement	$f_c max > 0.2f_c'$	$f_c max > 0.2f_{ck}$	
Minimum Reinforcement in end regions	$\frac{6\sqrt{f_c\prime}}{f_y}$ For $0.15l_w$ Length		$\frac{\sqrt{f_c\prime}}{2f_y}$ For $0.15l_w$ Length
Transverse reinforcement in boundary element A_{sb}	Greater of $0.3sb_c\left(\frac{A_g}{A_c} - 1\right)\frac{f_c\prime}{f_{st}}$ $0.09sb_c\frac{f_c\prime}{f_{st}}$	$0.05s_v h\frac{f_{ck}}{f_y}$	
Maximum vertical spacing in transverse reinforcement	Min of- (1) 1/3rd of min dimension of boundary element (2) For grade 60, $6d_b$ (3) For grade 80, $5d_b$ (4)$s_o = 4 + \left(\frac{14-h_x}{3}\right)$	Min of- (1) $1/3^{th}$ of min dimension of boundary element (2) 6 times dia of smallest longitudinal bar Also, Should not greater than 100 mm	
Reinforcement around opening	Along with the minimum reinforcement, at least two no. 5 bar has to be provided in each face	Reinforcement having an area equal to the area of bars interrupted has to be provided Vertical bars will be provided throughout the length along edge & horizontal bar with development length beyond side	Reinforcement with a yield strength greater than 600 N per mm of wall thickness has to be provided around the opening

Table 2 Reinforcement details for shear wall designed by IS 13920:2016

Reinforcement details		
Web	Vertical bars	16ϕ@350 mm c/c
	Horizontal bars	10ϕ@250 mm c/c
Boundary elements	Vertical bars	6–16ϕ
	Horizontal ties	12ϕ@80mm c/c
Around openings	20 mmϕ bars in two layers around openings. The vertical bar is provided throughout the height and horizontal bars are provided development length beyond sides of the opening	

Table 3 Reinforcement details for shear wall designed by ACI 318:19

Reinforcement details		
Web	Vertical bars	20ϕ@400 mm c/c
	Horizontal bars	10ϕ@250 mm c/c
Boundary elements	Vertical bars	8–16ϕ
	Horizontal ties	12ϕ@80 mm c/c
Around openings	16 mmϕ bars in two layers around opening with 12ϕ@90 mm c/c ties for the boundary element around the opening	

Table 4 Reinforcement details for shear wall designed by NZS-3101:2006

Reinforcement details		
Web	Vertical bars	16ϕ@300 mm c/c
	Horizontal bars	10ϕ@300 mm c/c
Boundary elements	Vertical bars	6–16ϕ
	Horizontal ties	12ϕ@100 mm c/c
Around openings	16 mmϕ bars in two layers around the opening	

Reinforcement bars were modelled in the form of a two-dimensional truss element T3D2 embedded in the C3D8R element. The boundary condition at the base of the walls is simulated as a fixed end.

Figure 2 shows three meshed models, which were considered for numerical analysis. Vertical and Staggered window type openings of size 1.2 m × 1.2 m were considered for the numerical analysis.

Figure 3 shows the dimensions details and positionings of openings in different walls.

3.2 *Crack Patterns*

In solid shear wall, cracks initiated from the bottom of the wall and progressed horizontally, on further loading more flexure cracks were initiated on the upper

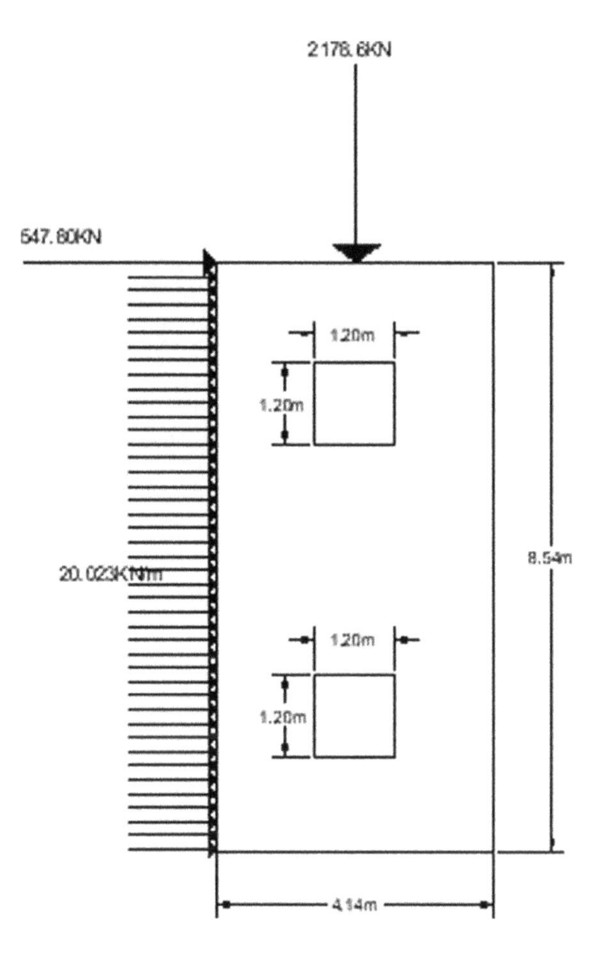

Fig. 1 Two storey slender shear wall

portion. In the wall with regular openings, cracks initiated from the level just below the bottom opening and upon further loading, cracks started from the corner of the bottom opening and moved towards the end of the wall. In the case of the staggered opening, cracks initiated from the bottom of the wall and progressed horizontally, More flexure cracks were developed on the upper portion of the specimen as the loading increased. Stress concentration on the corner of the bottom opening caused more diagonal tension cracks also (Figs. 4 and 5).

4 Results and Discussion

It is clear from the Load displacement graph that openings in shear walls decrease the load-carrying capacity (Fig. 6).

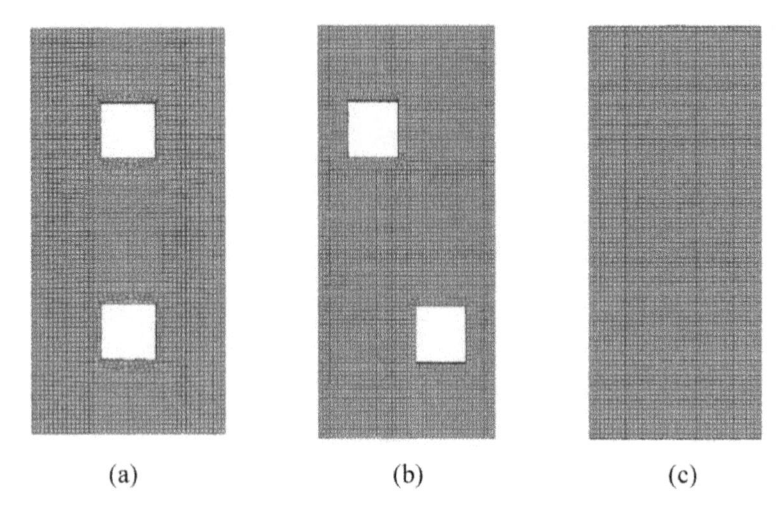

Fig. 2 Meshed models. **a** Regular window opening, **b** staggered window opening, **c** without opening

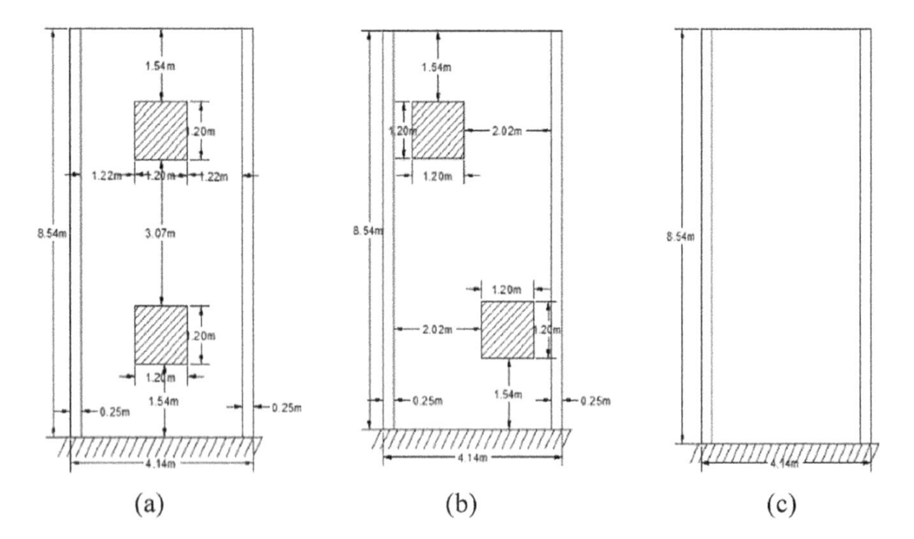

Fig. 3 Dimensions details of walls

It was also seen that openings cause disturbances in the stress flow and also in the propagation of cracks. It is affected mainly by the bottom openings as the stress value is higher at the bottom side.

In the case of the shear wall with the regular window opening, cracks do not initiate from the bottom of the wall, rather starts at a level just below the opening. This is due to the disturbance in the stress flow caused by the opening. As the opening moves away from the tension end (staggered window opening), disturbance reduced and cracks initiated from the bottom.

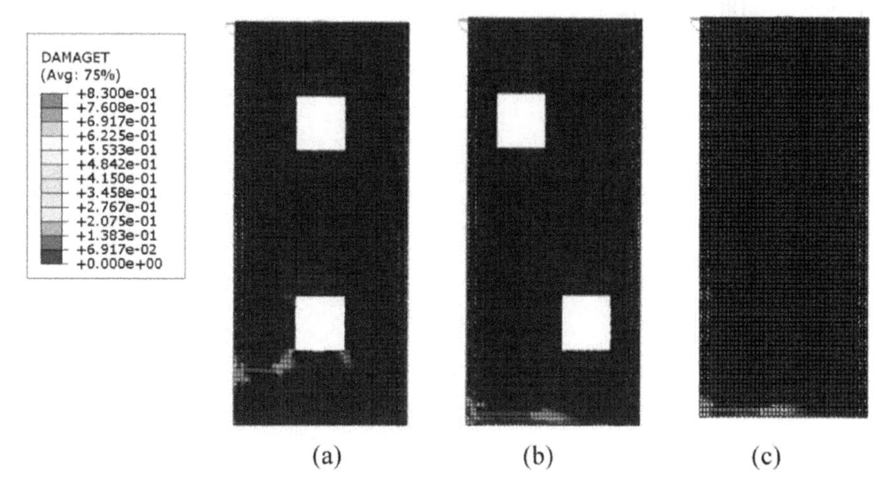

Fig. 4 Initial crack patterns [6]

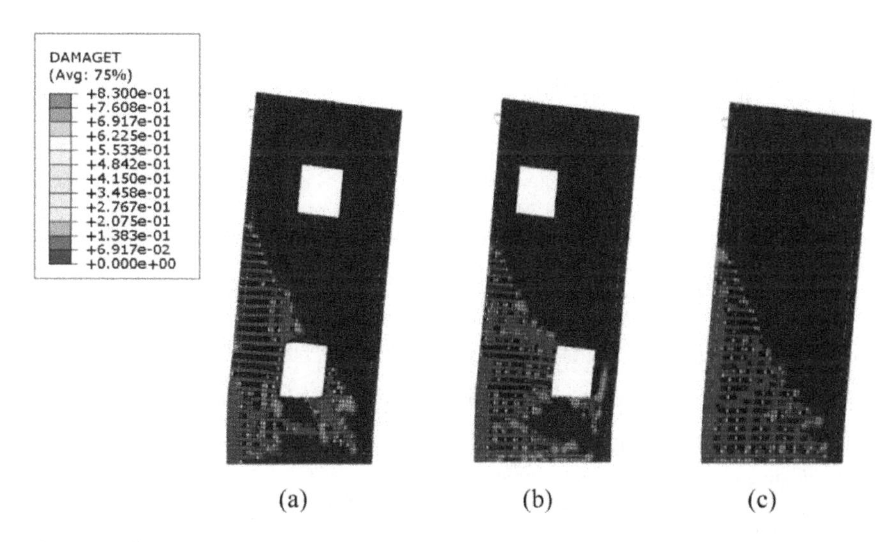

Fig. 5 Cracks pattern at 15 mm top displacement [6]

Staggered openings for this orientation shows better results than normal opening. As in this case, the bottom opening was moved away from the tension end. But from seeing crack propagation, It can be said that if the bottom opening moves toward the tension end, opposite results will come. Since Lateral load can come from any direction, regular opening equidistant from both ends will give better results.

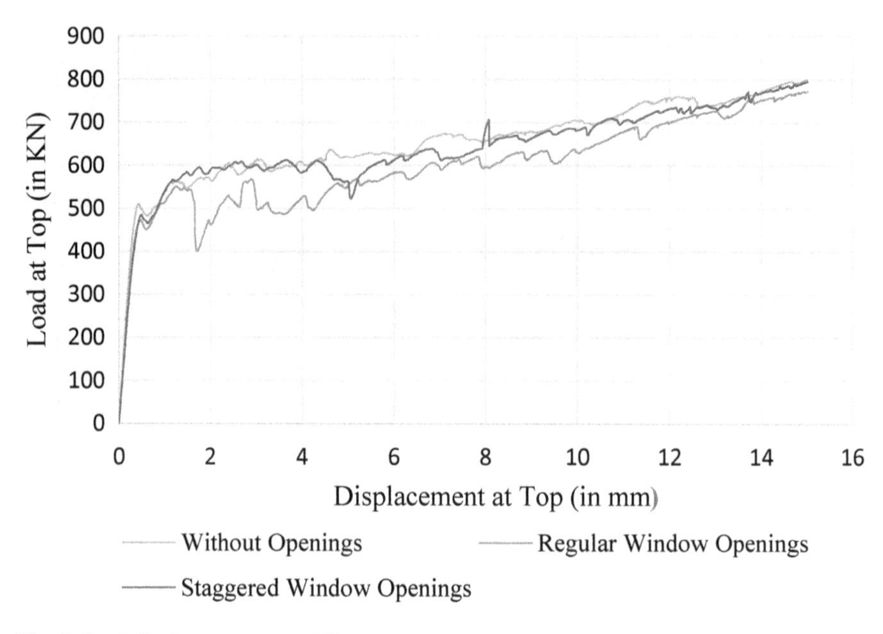

Fig. 6 Load-displacement curve [6]

References

1. Park R, Pauley T (1975) Reinforced concrete structure. Wiley, New York
2. Naresh Kumar Varma V, Praveen Kumar U (2020) Seismic response on multistoried building having shearwalls with and without openings, Material today, pp 2214–7853
3. NZS 3101:2006 (2006) Part 1: code of practice for the design of concrete structure, and part 2: commentary on the design of concrete structure. Standard association of Newzealand, Wellington
4. IS 13920:2016 (2016) Ductile design and detailing of reinforcement concete structure subjected to seismic forces
5. ACI 318-19 Building code requirements for structural concrete", ACI 318R-19, Commentary on building code requirements for structural concrete
6. Abaqus/Standard (2016) A general-purpose finite-element analyzer that employs implicit integration scheme (traditional)

Stabilization of Underground Tunnel Using Micropiles: Finite Element Approach

Irfan Ahmad Shah

Abstract For the past few decades, large scale urbanization has limited the space for the commencement of new infrastructural projects. Therefore, the modern construction industry has utilized the surface above the underground tunnel opening for construction purposes. This put a need to stabilize these underground structures due to construction above them. The present numerical study has been carried out using GeoStudio 2016 to understand the problem. The 2D Numerical model replicating the Delhi metro phase 3 tunnel projects at a standard depth of 10 m. The tunnel has a circular cross-section with a diameter of 5 m and a concrete lining of 30 cm thick. The overburden of the superstructure has been varied upto 3rd floor of the building (G + 3 building load) and the corresponding response of the tunnel has been observed at different locations (crown, spring-line and invert). Later, the ground has been stabilized by incorporating micropiles as vertical reinforcement within the footing of superstructure. It has been concluded that the micropile can stabilize the tunnel and the superstructure to a different extent. The micropiling results in the redistribution of moment in the tunnel lining with a part of the moment in the crown section getting transferred to the invert.

Keywords Underground tunnel · Micropiles · GeoStudio · Superstructure

1 Introduction

The construction of tunnels in the densely populated regions of India has become a favourable technique to divert a major part of ground traffic flow. According to the data provide by Delhi NCT (2012), the population of Delhi has increased from 9,340 to 11,297 persons per km^2 in one decade (2001–2011) and by 2020, it further increased to 15,070 persons per km^2. This increased population density puts a need to utilize the space above the underground openings for the construction of

I. A. Shah (✉)
Department of Civil Engineering, Indian Institute of Technology Bombay,
Bombay 400076, India

G. C. Marano et al. (eds.), *Proceedings of SECON'21*, Lecture Notes in Civil Engineering 171, https://doi.org/10.1007/978-3-030-80312-4_91

buildings. However, the stability of the structures is an important requirement which if compromised can results in a huge destruction of property and human lives. Therefore the demand is to stabilize these underground structures to an extent of maximum possibility. Several research studies have been done to understand the effect of construction over an underground opening [1–3]. Haiben et al. [3] studied the interaction of structures (such as buildings, piles and shallow foundation) with the tunnel. The study by Salim and Lafta [4] concluded that due to pile driving and pile loading in the adjacent areas of tunnel, the stresses in the tunnel gets increased. Further, the extent of stress increase also depends on the relative density of the surrounding medium. Analytical methods have been followed by other researchers to understand the influence of basement of the building over the existing tunnel [5, 6]. Following the similar study, Shi et al. [7] recommended that for safe excavation over the existing tunnel, the size of excavation should meet 2d $\times$ 5d plan dimension (where d is the depth of excavation).

Considering the complexities in a problem, numerical method of analysis have become a favourable tools to study the behavior of underground structure under static as well as dynamic loading conditions [8–13]. The stabilization of tunnels by different types of reinforcement has been checked over the time and analysed following numerical methods [14–16]. GeoStudio is one of the versatile software to analyse different geotechnical problems. The provision of reinforcement can be very easily incorporated in GeoStudio [17, 18]. The present study has been performed using SIGMA/W module of GeoStudio 2016 [19]. The stability of tunnel has been analysed due to construction of building (G + 3 floors). The tunnel response has been observed by increasing load of the super structure in the increment of one floor with a maximum load of the building as 80 kN/m^2. Next, the tunnel has been reinforced with micropiles and again checked for stability at different section of tunnel.

2 Finite Element Analysis

The geometry of the tunnel model is shown in Fig. 1. The tunnel is having a cross-section of 40 m $\times$ 40 m to represent the soil material and 1 m in plain element thickness. A circular tunnel of 5 m radius has been excavated at a cover depth of 10 m with a concrete lining of 30 cm thickness. The outer boundary of the model from the tunnel has been kept at a distance of 3 times the tunnel diameter which neglects the boundary effect. A footing has been provided centrally with a width of 10 m and thickness of 1 m. The response of the footing and tunnel has been observed for both the reinforced and unreinforced case. In reinforced case, the vertical micropile of length 8 m and diameter 40 cm were embedded into the ground and within the footing area only at a constant spacing of 2 m.

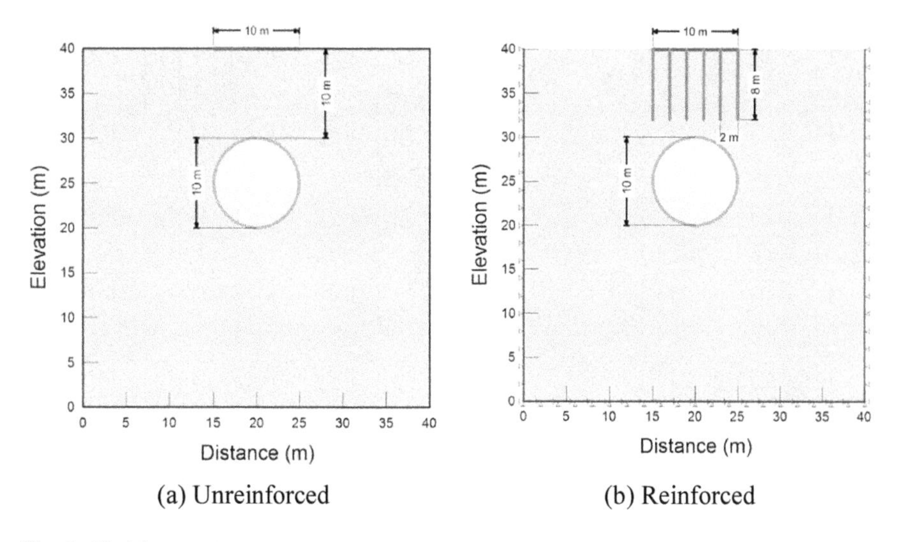

Fig. 1 Model geometry

2.1 Material Properties, Meshing and Boundary Condition

The type of soil considered in the analysis is Delhi Silty sand. Material properties of soil and the concrete lining has been taken form the study of Singh et al. [20] and are given in Table 1. The soil has been modeled as an elastoplastic material model with Mohr–Coulomb yield criteria. The footing and the vertical micropiles has been modeled as a beam element of mild steel and are capable of resisting both axial and bending forces [18, 21]. The material properties of micropiles and footing have been taken from Shah et al. [17, 18].

Figure 2 represented the meshing and boundary conditions of the numerical model for both reinforced and unreinforced case. The soil model has been discretized by taken 8 noded Quad and Triangles element having nodes at each corner and edge of the element. The size of the mesh element has been finalised by performing the mesh sensitivity analysis. Moreover, the mesh size has been refined in the area surrounding the tunnel opening in view of capturing the finer variations in details like stress and deformation etc. The model boundaries has been fixed at

Table 1 Properties of different material used in the study

Material parameter	Soil properties	Concrete lining	Reinforcement properties	
			Micropile	Footing
Dry unit weight (kN/m^3)	20	30	78.5	78.5
Modulus of elasticity (MPa)	30	3.16×10^4	2.1×10^5	2.1×10^5
Friction angle, ϕ (°)	35	–	–	–
Poisson's ratio, ν	0.25	0.15	0.3	0.3

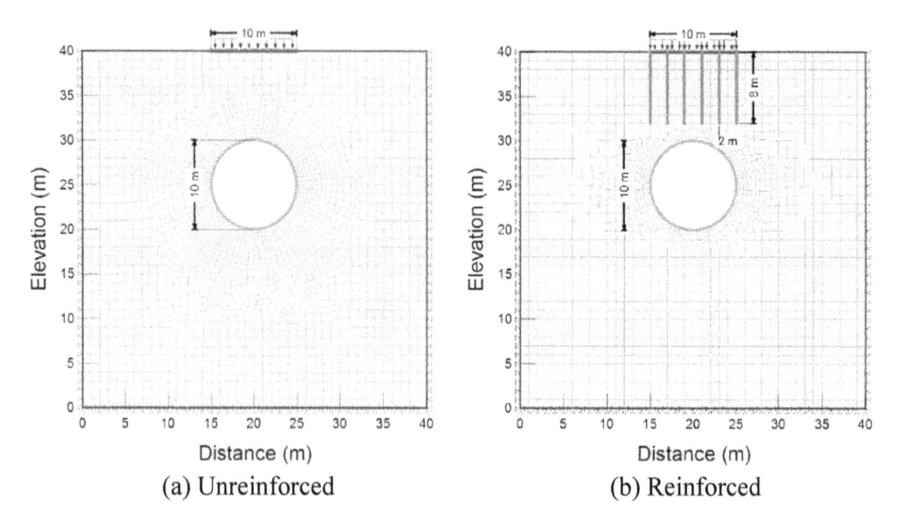

Fig. 2 Meshing and boundary condition

the bottom side and fixed in the horizontal direction only at the extreme left and right ends. The loading has been applied as an incremental stress at the footing in the incremental of 20 kPa until a maximum stress of 80 kPa.

2.2 *Validation*

The unreinforced numerical model for static loading has been validated from the experimental study of Mishra et al. [22]. The model dimensions and the boundary conditions have been considered same as given by the published paper [22]. The numerical model has a dimension of 0.3 m × 0.3 m × 0.35 m with an excavated tunnel of 0.05 diameter and an overburden depth of 0.05 m. The model has been validated for two different types of material model surrounding the tunnel opening (Geomaterial and POP). The geomaterial was prepared as a synthetic rock of composition 50% POP, 35% sand and 15% clay [22]. The material model of geomaterial has a unit weight = 14.1 kN/m^3, Modulus = 3.153 GPa, unconfined compressive strength = 4.48 MPa, cohesion of 0.79 MPa, Poisson's ratio = 0.16 and friction angle of 39°. The corresponding material parameters for POP are 11.17 kN/m^3, 4.47 MPa, 6.78 MPa, 0.79 MPa 0.22 and 51° respectively. The properties of the material surrounding are given in Table 2. Further details of the sample preparation and method of testing has been given in Mishra et al. [22, 23]. The material in the numerical validation has been modeled by elastoplastic Mohr–Coulomb constitutive material model. The loading has been applied centrally at the top surface of the model. The results are shown in Table 2.

Table 2 Model validation for static loading

Parameter	Experimental, Mishra et al. [22]		Present study		% error	
Type of rock	Geomaterial	POP	Geomaterial	POP	Geomaterial	POP
Crown deformation in (mm)	0.37	0.33	0.35	0.30	5.4%	9%

Validation of the reinforced model has been concluded from the experimental study carried out by the author elsewhere [14]. The model dimensions for the validation has been kept same as that used in the experimental setup (50 cm 50 cm × 30 cm) adopted by the author in the published paper [14]. The micropile (l/d = 100) have been embedded within the base of footing only at a spacing of 10 mm. The results for the reinforced case are given in Fig. 3. Therefore, by observing the validation results (Table 2 and Fig. 3), it can be concluded that the present numerical model can capture the material behavior to a good extent for both the reinforced and unreinforced case.

3 Results and Discussion:

The study has been carried out to understand the response of underground tunnel due to the construction of a 3-floor building. The study replicates the case of Delhi-metro phase 3 tunnel project. The load of the superstructure has been increased upto 3rd floor and the settlement of the footing has been analyzed using FE software GeoStudio 2016. The footing settlement has been analysed for the three cases:

1. Footing over unlined tunnel,
2. Footing over lined tunnel (0.3 m thick) and
3. Footing over lined tunnel stabilized by micropiles.

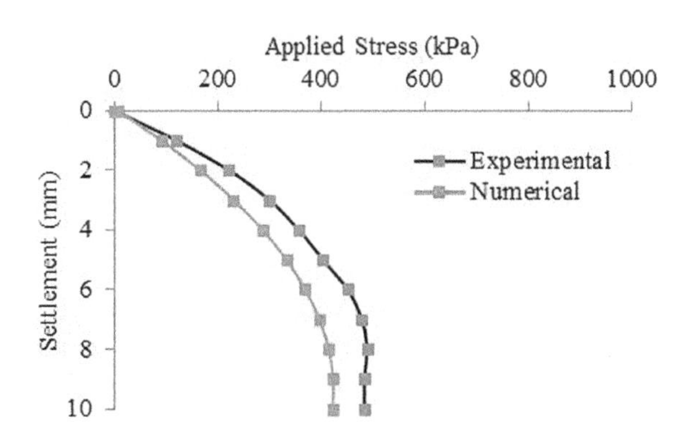

Fig. 3 Comparison of experimental and numerical results for fly-ash model reinforced with micropiles (micropile spacing = 10 mm and L/D ratio = 100)

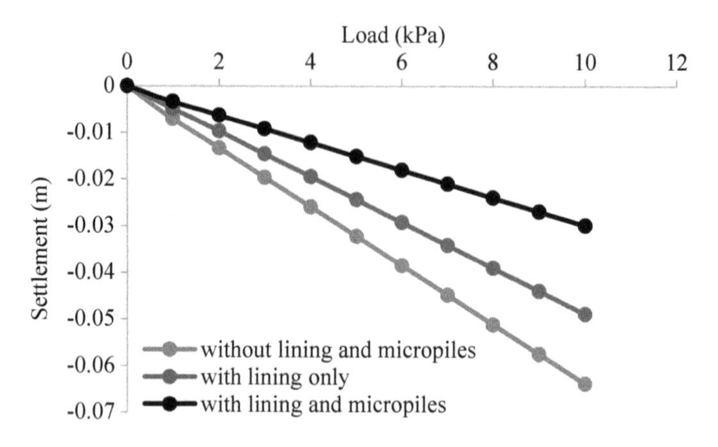

Fig. 4 Load settlement curve of foundation above underground tunnel

The results of the footing settlement have been shown in Fig. 4. The results represent a good improvement in the settlement resistance due to micropile installation. The percentage improvement in the foundation settlement for case 2 and case 3 can be observed as 23 and 53% corresponding to the superstructure load of 80 kPa. The maximum resistance can be observed at higher superstructure loading due to the better mobilization of frictional resistance of micropiles.

The study also investigated the stabilizing effect of micropiles over different sections of tunnel. The tunnel response at crown, springline and invert due to installation of micropile has been observed as shown in Fig. 5.

The results represent a reduction in the settlement at different sections of tunnel due to installation of micropiles. The maximum reduction of settlement can be observed at the crown section which represented a percentage decrease in settlement of 18.50, 18.60, 18.72 and 18.90% for the super structural load of 20, 40, 60 and 80 kPa respectively. The percentage decrease has been calculated with respect to the unreinforced case at constant loading. At spring line, the percentage decrease in settlement is 9.36, 7.54, 6.90 and 6.54% for the corresponding loading. Similarly at invert, the percentage decrease in settlement is 4.86, 3.17, 2.60 and 2.31%. It can be observed that at crown section, the settlement decreases continuously with increasing load of superstructure, however, at springline and invert, the percentage decrease is not increasing. Therefore, the influence of micropiles is primarily observed along the crown area. This can be due to the configuration of micropile embedment above the tunnel opening. Since, the micropile are located within the tunnel diameter, therefore, the resistance can be more experienced at the tunnel crown.

The bending moment of the concrete tunnel lining can be observed as shown in Fig. 6. The maximum bending moment occurs at the crown of the tunnel and decreases at the invert section. In a concrete lining without micropiles, the maximum bending moment due to loading of 80 kPa is 34.70 kN-m at the crown

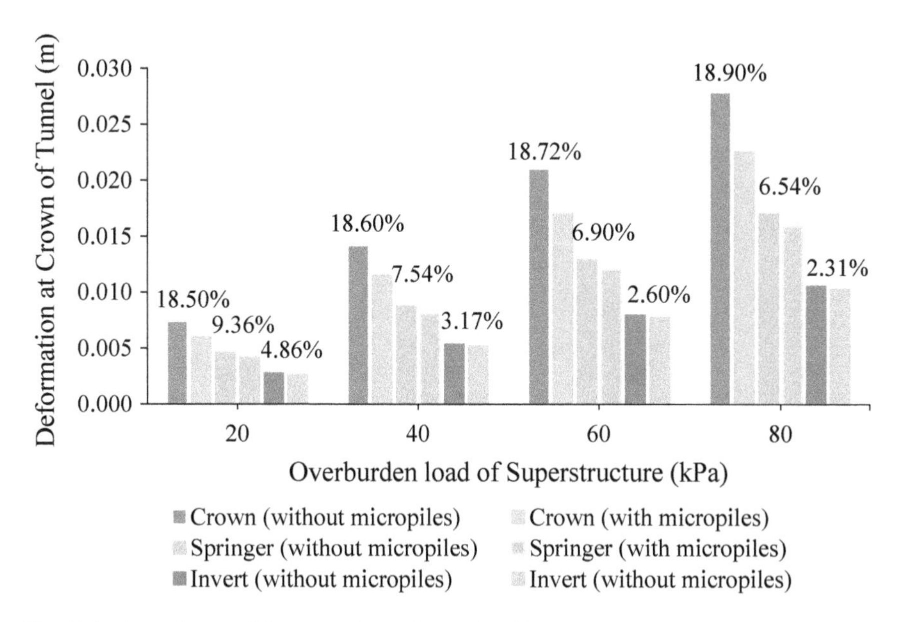

Fig. 5 Deformation at different sections of tunnel, without and with micropiles, due to super structural load

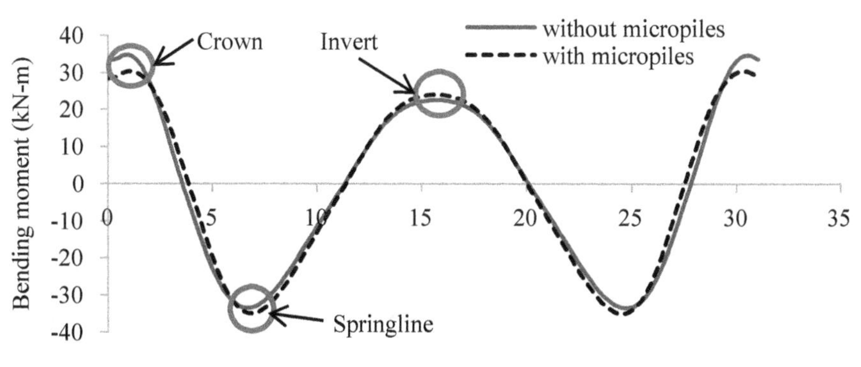

Fig. 6 Bending moment along the concrete tunnel lining

section, −33.46 kN-m at the springline and 24 kN-m at the invert. Due to micropiling system, the maximum moment comes as 30 kN-m at the crown, −34.77 kN-m at springline and 22.40 kN-m at the invert. Therefore it can be concluded that there occurs redistribution of moment in the tunnel cross-section from crown to springline due to installation of micropiles.

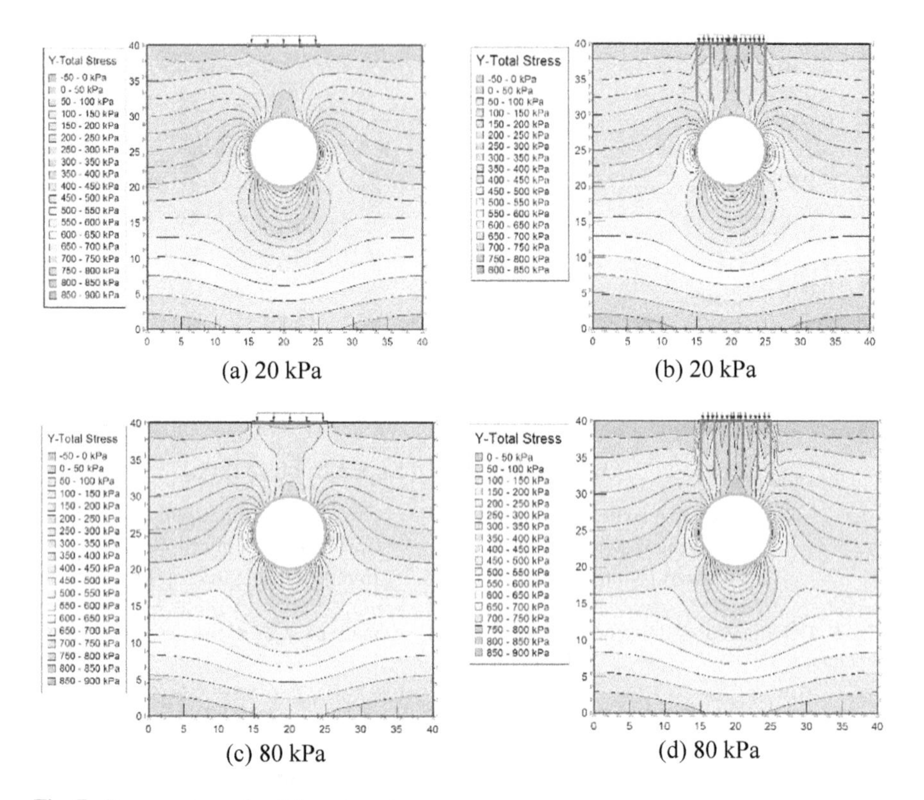

(a) 20 kPa (b) 20 kPa

(c) 80 kPa (d) 80 kPa

Fig. 7 Stress contours for different surcharge

4 Contours

The stress contours of the tunnel section for different loading are shown in Fig. 7. The stress contours of the tunnel have been represents for two extreme cases of loading (20 and 80 kPa). The contours represent that for all the cases, maximum stress occurs at the spring line of the tunnel section. However, due to installation of micropiles, there occurs stress redistribution and a part of the stress gets transferred towards the crown section of the tunnel. Due to the frictional resistance, the micropile resists a part of the coming load of the super structure and therefore stabilizes the tunnel section.

Therefore, it can be concluded that the micropiles improve foundation resistance against coming loads and hence can be concluded as a viable option for the ground improvement.

5 Conclusions

The study finds its scope in the problem of grounds improvement. The study has been performed in the view of improving the strength of the foundation and the tunnel lining against the loads of the superstructure. The ground reinforcement has been performed by the incorporation of vertical micropiles. From the study, following conclusions has been observed:

- The present study concludes that the micropiles can sufficient improve the load carrying capacity of the footing constructed over underground tunnel. By incorporating the micropiles, the settlement of the footing has been improved by 53% for the loading of 80 kPa.
- The micropile improve the tunnel performance, however, different sections of the tunnel resists the loading to a different extent. At the crown section, the percentage decrease in the settlement increases continuously with increasing load of superstructure, however, at springline and invert, the percentage decrease is not increasing.
- Due to the micropile installation, the redistribution of the bending moment occurs along the circumference of the tunnel section. A part of the bending of the crown gets transferred toward the invert section.

References

1. Liang R (2019) Simplified analytical method for evaluating the effects of overcrossing tunnelling on existing shield tunnels using the nonlinear Pasternak foundation model. Soils Found 59(6):1711–1727. https://doi.org/10.1016/j.sandf.2019.07.009
2. Zaid M, Shah IA, Farooqi M (2019) Effect of cover depth in unlined Himalayan tunnel: a finite element approach. In: The proceeding of 8th Indian rock conference, Indian International Centre, New Delhi, India, 03–04 Mar 2019. ISBN no.81-86501-27-1
3. Haibin H, Hao C, Jubo Z (2014) The Influence of foundation excavation on the existing metro tunnel in complicated environment. Electron J Geotech Eng 19:3377–3385
4. Salim NM, Lafta SJ (2017) The impact of driving and loading piles on existing tunnel. Imp J Interdiscip Res 3(7):262–270
5. Liang R, Xia T, Huang M, Lin C (2017) Simplified analytical method for evaluating the effects of adjacent excavation on shield tunnel considering the shearing effect. Comput Geotech 81:167–187. https://doi.org/10.1016/j.compgeo.2016.08.017
6. Liang R, Wu W, Yu F, Jiang G, Liu J (2018) Simplified method for evaluating shield tunnel deformation due to adjacent excavation. Tunn Undergr Space Technol 71:94–105. https://doi.org/10.1016/j.tust.2017.08.010
7. Shi J, Ng CWW, Chen Y (2015) Three-dimensional numerical parametric study of the influence of basement excavation on existing tunnel. Comput Geotech 63:146–158. https://doi.org/10.1016/j.compgeo.2014.09.002
8. Zaid M, Mishra S (2021) Numerical analysis of shallow tunnels under static loading: a finite element approach. Geotech Geol Eng 1–27. https://doi.org/10.1007/s10706-020-01647-1
9. Zaid M, Shah IA (2021) Numerical analysis of himalayan rock tunnels under static and blast loading. Geotech Geol Eng 1–21. https://doi.org/10.1007/s10706-021-01813-z

10. Shah IA, Zaid M (2020) Behavior of underground tunnel under strong ground motion. In: Proceedings of Indian geotechnical conference (IGC 2020), Vizag, India
11. Zaid M, Shah IA (2020) Dynamic stability analysis of shallow triple tunnel in rock. In Proceedings of Indian geotechnical conference (IGC 2020), Vizag, India
12. Zaid M (2021) Dynamic stability analysis of rock tunnels subjected to impact loading with varying UCS. Geomech Eng 24(6):505–518. https://doi.org/10.12989/gae.2021.24.6.505
13. Zaid M, Shah IA (2021) Blast resistant stability analysis of triple tunnel. In: Advances in geotechnical, structural and rehabilitation engineering. Springer, Berlin
14. Funatsu T, Hoshino T, Sawae H, Shimizu N (2008) Numerical analysis to better understand the mechanism of the effects of ground supports and reinforcements on the stability of tunnels using the distinct element method. Tunn Undergr Space Technol 23(5):561–573. https://doi.org/10.1016/j.tust.2007.10.003
15. Kavvadas M, Prountzopoulos G (2009) 3D analyses of tunnel face reinforcement using fibreglass nails. In: Euro, Tun 2009 conference, Bochum
16. Paternesi A, Schweiger HF, Scarpelli G (2017) Numerical analyses of stability and deformation behavior of reinforced and unreinforced tunnel faces. Comput Geotech 88:256–266. https://doi.org/10.1016/j.compgeo.2017.04.002
17. Shah IA, Zaid M, Farooqi MA, Ali K (2020) Numerical study on micropile stabilized foundation in flyash. Indian Geotech J 1–8. https://doi.org/10.1007/s40098-020-00476-6
18. Shah IA, Ali K, Farooqi MA (2021) A numerical study on the bearing capacity of fly-ash reinforced with fine micropiles. Int J Geosynthetics Ground Eng 7(1):1–14. https://doi.org/10.1007/s40891-020-00245-8
19. GEOSLOPE International Ltd (1992) SIGMA/W, for finite element stress deformation analysis user guide. Version 2, Calgary
20. Singh M, Viladkar MN, Samadhiya NK (2017) Seismic analysis of Delhi metro underground tunnels. Indian Geotech J 47(1):67–83. https://doi.org/10.1007/s40098-016-0203-9
21. Shah IA, Zaid M, Ali K (2021) Influence of micropiles on the bearing capacity of layered soil system—a numerical study. In: Advances in geotechnical, structural and rehabilitation engineering. Springer, Berlin. https://doi.org/10.1007/978-981-33-6969-6_14
22. Mishra S, Rao KS, Gupta NK, Kumar A (2017) Damage to shallow tunnels under static and dynamic loading. Proc Eng 173:1322–1329. https://doi.org/10.1016/j.proeng.2016.12.171
23. Mishra S, Rao KS, Gupta NK, Kumar A (2018) Damage to shallow tunnels in different geomaterials under static and dynamic loading. Thin-Walled Struct 126:138–149. https://doi.org/10.1016/j.tws.2017.11.051

Numerical Study of the Column Reinforced with Shape Memory Alloy

Gisha George, K. R. Bindhu, and Anagha Krishnan Nambissan

Abstract Columns are the primary load carrying members in buildings and bridges. Hence for better seismic performances the permanent drift after severe earthquake has to be reduced to make the structure functional with less maintenance. The use of the shape memory alloys in critical region can reduce the residual displacement due to its super-elastic nature. Super-elastic shape memory alloy bars have higher linear elastic range of 4–12% and can return back to its initial shape upon removal of load. In this numerical study full scale circular column with various types of nickel, iron and copper based shape memory alloy (SMA) bars in the critical region under reverse cyclic loading with constant axial load of 10%. The seismic performance of the SMA-steel reinforced column was compared with the steel reinforced normal column. The column reinforced with FeMnAlNi SMA in critical region shows better seismic performance considering load carrying capacity, displacement ductility, maximum displacement, lesser permanent deformation and energy dissipation capacity.

Keywords Shape memory alloy · Residual displacement · Super-elasticity · Ductility · Cyclic loading

1 Introduction

Columns are key structural elements participating in the load transfer in buildings and bridges. According to seismic resistant design criteria strong column and weak beam concept is followed with energy dissipated by yielding of reinforcement. Past earthquakes show major damages occur in column; this problem can rectify by using super-elastic materials like shape memory alloy which reduces the

G. George (✉) · K. R. Bindhu · A. K. Nambissan
College of Engineering Trivandrum, Thiruvananthapuram, India
e-mail: bindhukr@cet.ac.in

A. K. Nambissan
e-mail: anaghanambissan@cet.ac.in

G. C. Marano et al. (eds.), *Proceedings of SECON'21*, Lecture Notes in Civil Engineering 171, https://doi.org/10.1007/978-3-030-80312-4_92

post-earthquake maintenance. Shape memory alloys have super elastic effect, which can deform by large amount compared to steel and come back to original position with removal of load. It also has certain characteristics like hysteretic damping, corrosion resistance, fatigue resistance, strain hardening and stress plateau, providing force transmission limitations. The three basic classification of shape memory alloy are nickel based, copper based and iron based SMA. Recent studies show that the usage of NiTi SMA at plastic hinge region of the column can reduce the permanent deformation by large amount. It also shows sufficient displacement ductility and energy dissipation capacity compared to steel reinforced normal column with sufficient energy dissipation capacity and ductility.

2 Literature Review

Studies conducted during past two decades related to the influence of SMA in the behaviour of column are discussed here. Saiidi and Wang [11] investigated first experimental study on two-quarter scale down circular column with NiTi SMA in critical region and steel reinforcement was used in the remaining region. This column shows maximum displacement and less residual displacement. Billah and Alam [2] carried out numerical investigation by reinforcing NiTi SMA or stainless steel in the plastic hinge regions and with FRP or stainless-steel rebar in the remaining portion under seismic loading shows less residual displacement. Tazarv and Saiidi [13] conducted a numerical study on a full scale circular column with NiTi SMA in the critical region and compare with the normal reinforced column. The average peak displacement for the column reinforced with SMA is 13% higher than the steel-reinforced column. Billah and Alam [3] obtained numerical equation for plastic hinge length of the SMA reinforced circular concrete bridge pier by varying different parameters. Hosseini and Gencturk [6] conducted a numerical study on bridge columns with ECC and Cu–Al–Mn SMA in the critical region. There was a 90% reduction in the permanent drift of innovative column compared to conventional column specimen. Only limited studies were carried out on copper and iron based SMA as replacement of steel reinforcement in the critical regions of columns [4, 5].

Most of the studies were carried on circular column with NiTi SMA in critical region. Later new iron-based SMAs with improved super-elastic properties have been developed with recoverable strain up to 12%. Iron based SMA has low cost and good workability compared to other SMA. This numerical study investigates the performance of the full-scale circular column with different SMA in the critical region under cyclic loading with constant axial load. The numerical study compares the performance of columns in terms of maximum load, maximum displacement, ductility, energy dissipation capacity and residual displacement.

3 Geometry of the Model

A full scale circular column with a height of 3 m and 600 mm diameter with an aspect ratio of 5 to avoid shear dominated failure. Normal reinforced column (NRC) is reinforced with 12 number of 25 mm diameter steel bar with yield strength of 415 N/mm^2 resulting in longitudinal reinforcement ratio of 2%. The transverse reinforcement consists of 12 mm diameter bars at a spacing of 300 mm with special confining reinforcement for a length of 300 mm at 100 mm spacing. Five different SMA were provided in critical region or plastic hinge region of the column calculated using the equation suggested by Paulay and Priestley [10]. Mechanical properties of various SMA are represented using modulus of elasticity (E), maximum superelastic strain (ε_s), austenite to martensite starting stress (f_y), austenite to martensite finishing stress (f_{p1}), martensite to austenite starting stress (f_{T1}), martensite to austenite finishing stress (f_{T2}) are shown in Table 1 [5].

The other five column reinforced with different SMA in critical region is designated as SMA-RC-N1, SMA-RC-N2, SMA-RC-Fe1, SMA-RC-Cu, and SMA-RC-Fe2 with 1.58% steel reinforcement.

4 Numerical Modelling and Validation

For the numerical analysis, the columns are modelled using nonlinear finite element software SeismoStruct. The software is capable of predicting large displacement considering material and geometric nonlinearity under dynamic loading. The column is modelled using inelastic displacement-based frame element type, where the element deformations are interpolated from an approximate displacement field. The 2-integration section and 250 fibres were used to get results with sufficient accuracy. The confined and unconfined concrete were modelled using Mander model [8], steel using Menegotto-Pinto steel model [9] and SMA using Auricchio and Sacoo model [1].

The FE model of the normal column was validated by experimental study conducted by Kowalsky et al. [7]. The experimental value of maximum strength obtained as 167.3 kN and the numerical value was 169.4 kN with 1.25% error. The column reinforced with SMA in critical region and steel in remaining portion was

Table 1 Mechanical properties of shape memory alloy [5]

Alloy	E(GPa)	ε_s (%)	f_y (MPa)	f_{p1} (MPa)	f_{T1} (MPa)	f_{T2} (MPa)
$NiTi_{45}$	62.5	6	401	510	370	130
$NiTi_{45}$	68	8	435	535	335	170
FeNCATB	46.9	13.5	750	1,200	300	200
CuAlMn	28	9	210	275	200	150
FeMnAlNi	98.5	6.13	320	442.5	210.8	122

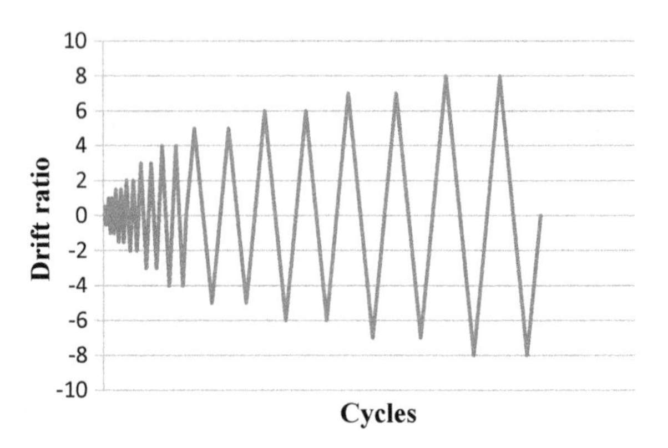

Fig. 1 Loading diagram

validated with experimental study conducted by Saiidi et al. [12]. The maximum strength obtained was 41.68 kN with displacement ductility ratio of 7.52, which closely matches with experimental results. The base of the column is fixed and the top of the column is subjected with axial load of 848.23 kN, it is taken as 10% of axial load ratio. Lateral cyclic loading is provided with a drift ratio of 0.25, 0.5, 0.75, 1, 2, etc., with an increment of 1% upto failure and each cycle is repeated two consecutive times, except 0.25% drift ratio. The graphical representation of loading protocol is shown in Fig. 1. The finite element model with cross-section showing discretisation is shown in Fig. 2.

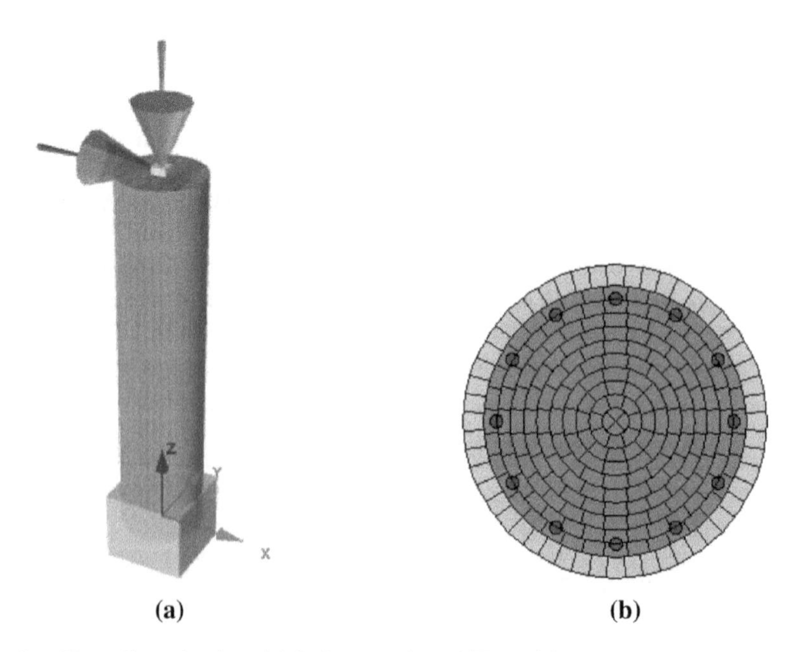

Fig. 2 **a** Three dimensional model. **b** Cross section of FE model

5 Hysteresis Curve

Hysteresis curve is obtained with respect to the cyclic load and lateral displacement. Figures 3, 4, 5, 6, 7, and 8 shows the hysteresis curve of the various specimens upto 8% drift ratio. Columns reinforced with different types of SMA in critical region shows flag shaped hysteresis curve compared to normal steel reinforced column. From the hysteresis curve the maximum strength and corresponding displacement was obtained for each specimen was given in Table 2. The column reinforced with FeNCATB (iron based) SMA in critical region shows 54.8% higher strength compared to normal column with 75% higher displacement. The normal column and column reinforced with copper based SMA in critical region reaches failure load (80% of maximum load) with in the 8% drift ratio. The column reinforced with iron based SMA in critical region with stand 10% drift ratio. The column reinforced with FeNCATB SMA in critical region shows 79.7% higher strength and FeMnAlNi shows 13.7% higher strength compared to normal column.

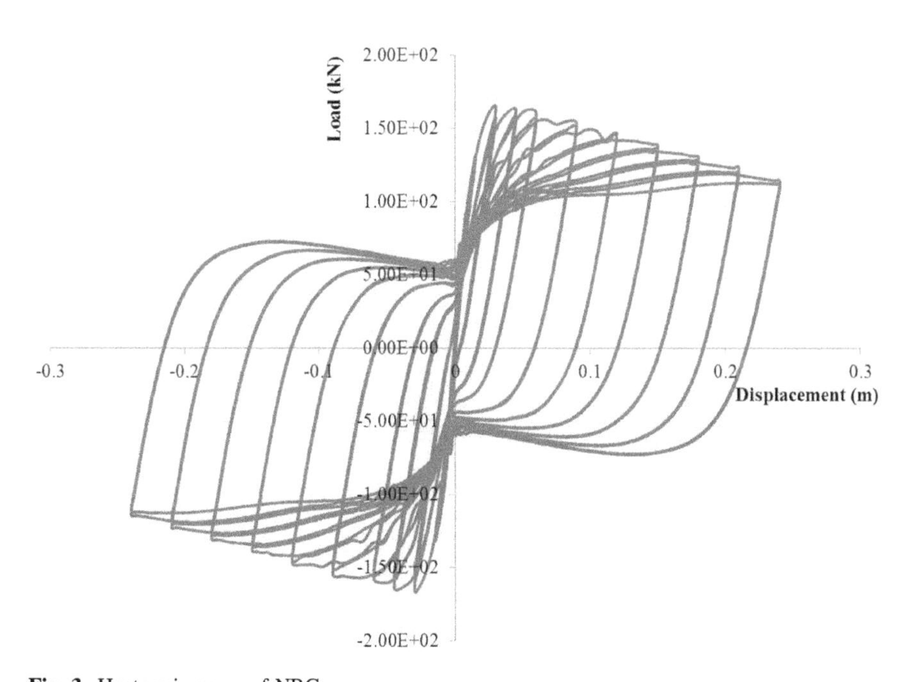

Fig. 3 Hysteresis curve of NRC

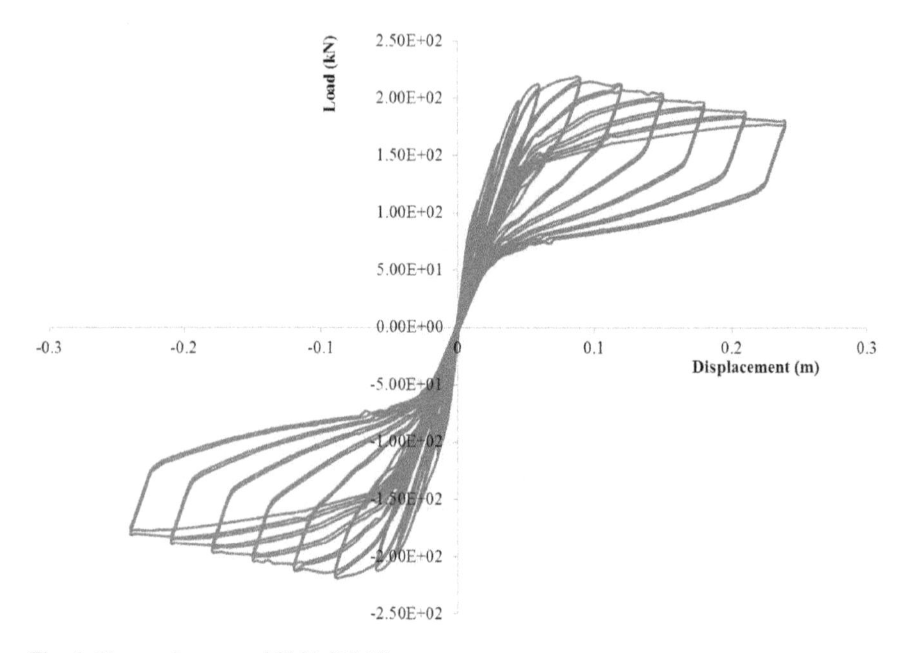

Fig. 4 Hysteresis curve of SMA-RC-N1

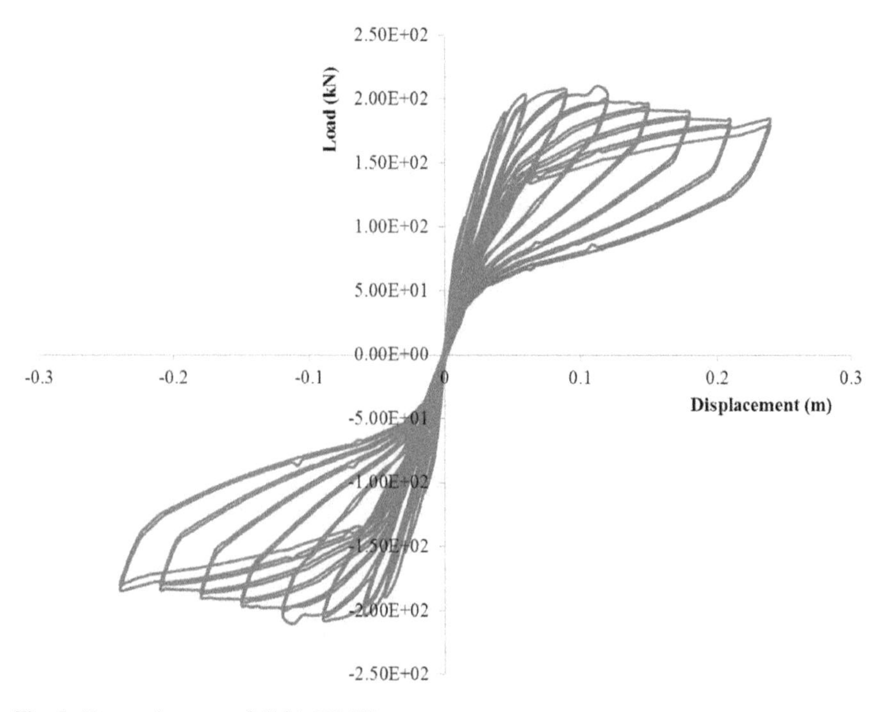

Fig. 5 Hysteresis curve of SMA-RC-N2

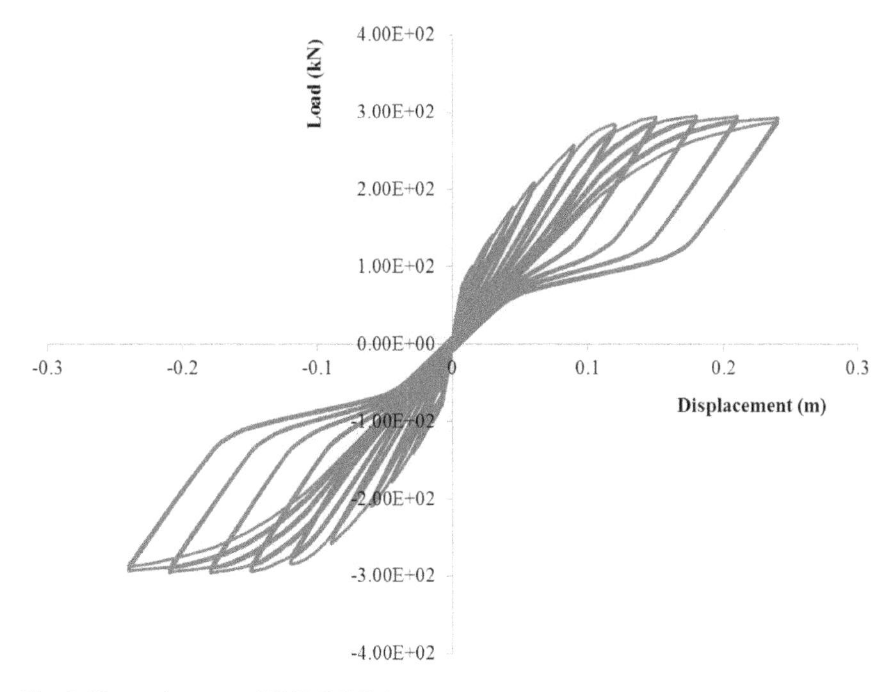

Fig. 6 Hysteresis curve of SMA-RC-Fe1

6 Displacement Ductility

Displacement ductility is the ratio of ultimate displacement to the yield displacement. Ultimate displacement was taken as the displacement corresponding to 0.8 time's maximum load of the column specimen. The yield displacement was obtained from drawing the tangent through the origin and 65% of maximum load and extent to the horizontal line through the maximum load. The displacement corresponding to the intersection of this tangent line with the horizontal gives yield displacement. The column reinforced with FeMnAlNi SMA in critical region shows 48.28% higher ductility and ultimate displacement of 0.29 m with 10% drift capacity compared to normal RC column.

7 Energy Dissipation Capacity and Residual Displacement

Energy dissipation capacity is the area under the hysteresis curve. The cumulate energy dissipation capacity of different specimen is given in Table 3. The normal columns have higher energy dissipation capacity with wide hysteresis loop, due to

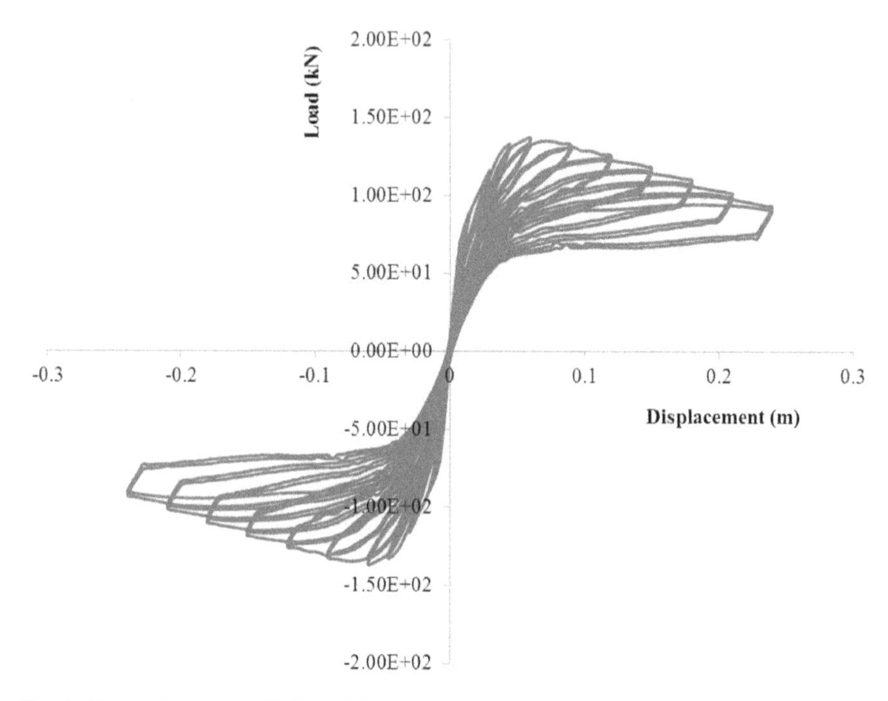

Fig. 7 Hysteresis curve of SMA-RC-Cu

yielding of reinforcement, crushing and spalling of concrete. But the column reinforced with nickel and iron based SMA in critical region does not shows 80% reduction in strength within 8% drift ratio, so it can take more cycles of load and have sufficient energy dissipation capacity. Among different SMA's FeNCATB, which is iron based SMA in the critical region shows higher energy dissipation capacity.

Residual displacement was taken as the horizontal distance at zero load of the hysteresis curve at the last load cycle provided. The residual displacement values of different specimens are shown in Table 3. The column reinforced with FeMnAlNi SMA shows zero residual displacement at 10 and 8% drift ratio but normal column shows residual displacement of 0.216 m at 8% drift.

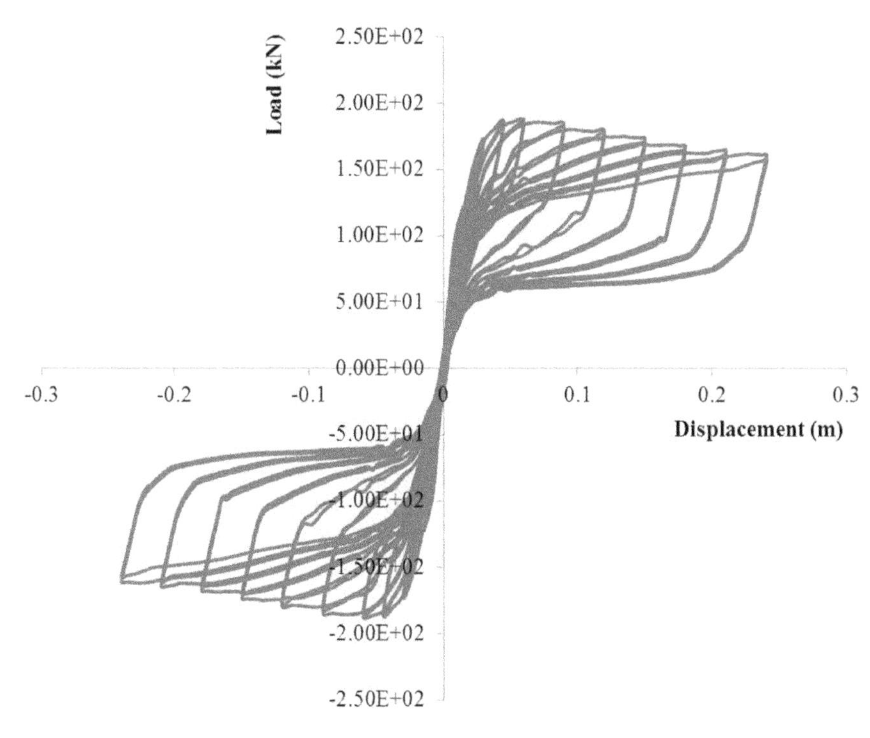

Fig. 8 Hysteresis curve of SMA-RC-Fe2

Table 2 Maximum displacement and ductility

Specimen	Maximum load (kN)	Displacement corresponding to maximum load (m)	Yield displacement (m)	Ultimate displacement (m)	Ductility
NRC	164.14	0.03	0.02	0.169	8.45
SMA-RC-N1	217.10	0.825	0.037	0.28	7.49
SMA-RC-N2	210.36	0.011	0.036	0.34	9.5
SMA-RC-Fe1	254.1	0.120	0.065	0.25	3.82
SMA-RC-Cu	136.38	0.06	0.023	0.18	7.83
SMA-RC-Fe2	186.72	0.06	0.023	0.29	12.53

Table 3 Cumulative energy dissipation capacity of different specimen

Specimen	Cumulative energy dissipation capacity (kN m)	Residual displacement (m)
NRC	610.55	0.216
SMA-RC-N1	362.33	0.004
SMA-RC-N2	484.071	0.004
SMA-RC-Fe1	1,001.464	0.289
SMA-RC-Cu	79.39	0
SMA-RC-Fe2	392.95	0

8 Conclusion

Based on the numerical study conducted on different column specimens with different SMA in critical region was compared with normal column and obtained the following conclusion:

- The column reinforced with NiTi and FeNCATB SMA in the critical region have higher load carrying capacity and maximum displacement with less ductility. It also shows permanent drift ratio higher than 0.1 %.
- The column reinforced with FeMnAlNi in the critical region shows higher load carrying capacity, ductility, maximum displacement, and zero residual drift with sufficient energy dissipation capacity.
- Considering cost, residual displacement, maximum displacement and load carrying capacity FeMnAlNi SMA is suitable to provide in the critical region of the column for better seismic performance to ensure post-earthquake functionality of the structure. This particular SMA has higher modulus of elasticity compared to other SMA's with 6% superelastic strain.

References

1. Auricchio F, Sacco E (1997) A superelastic Shape-memory alloy beam model. J Intell Mater Syst Struct 8:489–501
2. Billah, AHMM, Alam MS (2012) Seismic performance of concrete columns reinforced with hybrid shape memory alloy and FRP bars. Constr Build Mater 28:730–742
3. Billah, AHMM, Alam MS (2016) Performance based seismic design of SMA reinforced concrete bridge pier II: methodology and design example. J Struct Eng 04016140
4. Billah AHMM, Alam MS (2016) Plastic hinge length of shape memory alloy (SMA) reinforced bridge pier. Eng Struct 117:321–331
5. Billah AHMM, Alam MS (2018) Probablistic seismic risk assessment of concrete bridge piers reinforced with different types of shape memory alloys. Eng Struct 162:97–108
6. Hosseini F, Gencturk B (2018) Seismic response of bridge columns with high performance materials. In: 16th European conference on earthquake engineering, Thessaloniki, Greece

7. Kowalsky MJ, Priestly MJN, Seible F (1996) Shear and flexural behaviour of lightweight concrete bridge column in seismic region. Am Concr Inst Struct J 96(1):136–148
8. Mander JB, Priestly MJN, Park R (1988) Theoretical stress-strain model for confined concrete. J Struct Eng ASCE 114(8):1804–1826
9. Menegotto M, Pinto PE (1973) Method of analysis for cyclically loaded R.C. plane frames including changes in geometry and non-elastic behaviour of elements under combined normal force and bending. In: Symposium on the resistance and ultimate deformability of structures acted on by well-defined repeated loads, International Association for Bridge and Structural Engineering, Zurich, Switzerland, pp 15–22
10. Paulay T, Priestley MNJ (1992) Seismic design of reinforced concrete and masonry buildings. Willey, New York
11. Saiidi MS, Wang SH (2006) Exploratory study of seismic response of concrete column with SMA reinforcement. ACI Struct J 103(3):436–443
12. Saiidi MS, O'Brien M, Zadeh MS (2009) Cyclic response of concrete bridge columns using superelastic nitinol and bendable concrete. ACI Struct J 106(1):69–77
13. Tazarv M, Saiidi MS (2013) Analytical studies of seismic performance of full scale SMA reinforced bridge column. Int J Bridge Eng (IJBE) 1(1):37–50

Evaluation of Self Compacting Concrete as a Pavement Material for Construction and Retrofitting

Shashi Kant Sharma and Sandeep Singh

Abstract Pavement performance evaluation and fatigue life determination is an important part of pavement design and analysis. Laboratory testing in itself could not totally predict these parameters as it is neither possible to simulate the traffic conditions, nor costly pavement prototypes could be constructed because of lack of space and specific equipment to evaluate the prototype performance. In this study a pavement prototype was constructed, and subsequently validated for a FEM model designed on ABAQUS. Self-compacting concrete (SCC) was used as it is easier to work upon during initial construction as well retrofitting. Pavement performance was studied in terms of deflections upon loading as well as fatigue life when normal pavement quality concrete pavement was retrofitted with similar concrete, and SCC respectively. It was found that after retrofitting with SCC also, the pavement performs better as the deflections were comparable in both situations, whereas the fatigue life of the pavement improved by a magnitude of nearly 20%. Improvement in load transfer through dowel bars helped the retrofitted pavement to take a greater number of load repetitions.

Keywords Pavement quality concrete · Self-compacting concrete · Retrofitting · Dowel bars · Fatigue life · Deflections

1 Introduction

Whenever there is a distress in Pavement Quality Concrete (PQC) a decision has to be made contemplating the cost and intended life of the repair. Repairs are only intended to ensure that concrete pavements perform till design service life. Full depth repair is recommended in case a weak concrete is identified at the site of distress or the pavement have multiple type of distresses such as cracking, breaking of joint, locking of joint, etc. [1]. Repair of full depth transverse cracks requires new

S. K. Sharma · S. Singh (✉)
Department of Civil Engineering, NIT Jalandhar, Punjab 144001, India
e-mail: sandeeps.ce.19@nitj.ac.in

G. C. Marano et al. (eds.), *Proceedings of SECON'21*, Lecture Notes in Civil Engineering 171, https://doi.org/10.1007/978-3-030-80312-4_93

dowel bars to be placed and one new joint to be constructed. This necessitates that the concrete mix gets evenly distributed across the cut portion and achieve a desired level of compaction at corners and edges [2, 3]. If the concrete is not able to perform accordingly, poor bonding at the interface of old and new concrete is found. Also, it has been observed that the undulations at the interface of dowel bar and concrete causes easy cracking of new concrete which further deteriorates the condition [4, 5]. Overall, a loss in load distribution within the concrete slab, and from one slab to another has been observed. As a result, the durability as well as the fatigue life of the pavement decreases.

Self-compacting concrete (SCC) like a boon here could easily fill the entire cut portions of the slab by virtue of its high flowability and passability properties and thus imparts excellent bonding at the interface of old concrete pavement and new concrete pavement [6]. Further, because of its smooth finish there is a reduction in undulations which prevents the locking of joints [7].

SCC has a high volume of the paste which lowers its modulus of elasticity in comparison to PQC [8]. Better distribution of loads in the slab depends upon its modulus of elasticity. Hence it is very critical to say if a SCC will be beneficial for improving the fatigue life of pavement or not. Literature also suggests that compressive and flexural strength of SCC without fibers is lesser than that of PQC [9, 10]. Therefore, in this study wollastonite microfibers have been admixed to a SCC mix to check if the concrete performs better in terms of load transfer efficiency and deflections under loading. Both of these parameters indicate the expected fatigue life of the full depth repaired pavement.

2 Materials and Mixes

Ordinary Portland cement (M43) was used for casting of cubes and cylinders for all concrete mixes. The physical properties of the cement as determined from various tests confirmed to IS: 1489 (1991) [11]. The natural fine aggregates (NFA) used for the experimental program were locally procured and confirmed to IS: 383 (1970) [12]. The NFAs belonged to grading zone II. The water absorption (%) and fineness modulus of NFA is 1.2 (1-h) and 0.8 respectively. The aggregates were tested as per IS: 383 (1970) [12]. Natural Coarse aggregates (NCA) of nominal size 16 mm were used in the study as shown in Table 1. The water absorption (%) and fineness modulus of NCA is 7.4 and 2.38 respectively.

Table 1 Physical properties of concrete constituents

Physical properties	Cement	Fly Ash	Silica Fume	Wollastonite Fiber	Water	Superplasticizer (PCE based)	NCA	NFA
Specific gravity	3.14	2.2	1.75	2.92	1	1.09	2.75 (SSD)	2.6 (SSD)

Supplementary cementitious materials (SCMs) used in the investigation included FA (Class F) and SF. Amongst these, FA was procured from Ambuja Cement Plant, Ropar, India and it was ensured that the material complies with the requirements as per IS: 1727 (1967) [13]. SF was purchased from Elkem Materials Inc. Master Glenium 51 was used in different dosages by weight of binder to achieve desirable properties of SCC. It is based on modified polycarboxylic ethers (PCE), having chloride content less than 0.2% with a relative density of 1.07 and solid content 40%, pH > 6 at 25 °C, meeting the specifications of IS: 9103 (1999) [14]. The product was supplied for research purpose by M/s BASF Chemicals (India) Pvt. Ltd., Mumbai, India. Wollastonite microfiber of length 0.03 mm and aspect ratio 16.5 was used to reinforce the concrete.

2.1 *Mixes*

PQC mix was designed as per IRC: 44 [15]. The mix design has been presented in Table 2. Wollastonite microfiber reinforced SCC (FRSCC) was designed on the basis of the mix design of PQC such that the paste constituents were only altered. Excessive paste volume demand was fulfilled by putting FA, SF and wollastonite microfiber in the mix. The water to cement ratio was kept constant as 0.36 in both of the concrete mixes. FA and SF ratio was maintained at a level of 2:1 on the basis of literature review which suggests fruitful results with this ratio. Wollastonite is a weak pozzolan which contains 42% CaO, 46% SiO_2 and 8% Al_2O_3 (in this study). Therefore, the quantity of other pozzolans has been decreased in the mix.

3 Experimental Programme

3.1 *Compressive Strength and Flexural Strength*

This test is based on IS:516 (1959) [16]. The test was conducted on cubes of size 100 × 100 × 100 mm. Specimens were cured for 28 days in the curing tank. After curing, the specimens were taken out from the curing tank and left in the air to drip

Table 2 Details of design mix

Material-	Cement	Fly Ash	Silica Fume	Wollastonite Fiber	Water	Superplasticizer (PCE based)	NCA	NFA
Quantity/cum of PQC	440	0	0	0	160	–	1100	750
Quantity/cum of FRSCC	440	60	30	60	160	1.5% of Cement	1100	750

down the surface water. Then the specimens were tested on 200 tons capacity Compression Testing Machine (CTM). Similarly, the static flexural strength test was performed on standard beam specimens of size 100 × 100 × 500 mm as per the specifications laid down in IS 516 (1959) [16] after 90 days of curing.

3.2 Plate Load Test

This test is based on IS:1888 (1982) [17]. The test was conducted with the help of a loading frame attached to a 450 mm circular steel plate through which a uniformly distributed load having maximum rate of 0.1 N/mm^2 was applied. Deflections in millimeter were observed through dial gauges fixed on the edges of both pavement slabs across the joint. Test was conducted under stress control conditions at a rate of 0.032 N/mm^2/min and deflections were noted down for following load profile: 25,000, 50,000, 75,000, 100,000, 125,000, 150,000. This test was performed over a prototype having slab dimension 200 × 175 × 25 cm. Two slabs were connected via contraction joint having 7 No. 25 mm ϕ dowel bars at a spacing of 25 cm from edge as well as c/c between bars. The slab rests over 100 mm thick dry lean concrete (DLC) of 28 days compressive strength greater than 10 MPa, which itself rests over well compacted subgrade having modulus of subgrade reaction 63 MPa/m. Hence an effective modulus of subgrade reaction of 300 MPa/m was obtained over DLC.

3.3 FEM Analysis

A FEM model having same physical dimensions like the prototype was designed in the ABAQUS application and was subjected to similar load profile as that of Plate load test. Deflections under the loading were obtained from the model and their profile was subjected to chi square test with Plate load profile. This was done to assure that the model behaves in a similar fashion like the prototype.

Boundary conditions to the model were assigned on the basis of actual working of rigid pavement. The top of the slab was fixed against the movement in two horizontal conditions. The bottom face of PQC was also restricted in the same manner to allow the bending of the slab. The top and bottom face of the DLC was free to move in vertical direction and along the transverse direction but restricted in the vehicular direction. The top face of the subgrade was free to move in all directions but its bottom face was restricted in the vertical direction to simulate infinite subgrade behavior.

4 Results

4.1 Compressive Strength and Flexural Strength

PQC yielded a compressive strength of 45 MPa after 28 days curing whereas FRSCC showed an improvement with a value of 52 MPa after 28 days curing. Long term curing (90 days) complemented the strength of FRSCC much more than PQC. A value of 49.5 MPa was shown by PQC whereas FRSCC yielded 60 MPa compressive strength after 90 days curing. Hence a comparative difference of 17% and 20%, respectively was observed between both concretes after 28 and 90 days of curing.

The flexural strength values of both concretes were 4.55 MPa and 4.8 MPa after 28 days curing. After 28 days this value increased to 4.7 MPa and 5.2 MPa. Hence in this case a comparative difference of 5.5% and 10% respectively was observed between both concretes after 28 and 90 days of curing.

4.2 Plate Load Test

Table 3 shows the deflections across the joints for both cases of rehabilitated i.e., rehabilitated with PQC and rehabilitated with FRSCC respectively.

It is evident from the load deflection curves of both rehabilitated cases as shown in Figs. 1 and 2 that deflections under the load are lesser in case of FRSCC rehabilitated pavement. The maximum deflection under 15-ton load for PQC rehabilitated pavement is 0.64 mm whereas it is 0.42 mm under FRSCC rehabilitated pavement which shows that PQC rehabilitated pavement has lesser efficiency of load distribution. Also, the graphs indicate that the difference between deflections across the joints in FRSCC rehabilitated pavement is lesser than that of PQC rehabilitated pavement which proves that the load transfer efficiency improves when rehabilitated is done with FRSCC.

Table 3 Deflection (mm) across the joints

Load (N)	Deflection(mm)					
	PQC rehabilitated			FRSCC rehabilitated		
	(a)	(b)	Difference between a and b (%age)	(a)	(b)	Difference between a and b (%age)
25,000	0.22	0	100.00	0.2	0	100.00
50,000	0.45	0.08	82.22	0.36	0.09	75.00
75,000	0.7	0.21	70.00	0.45	0.14	68.89
100,000	0.93	0.34	63.44	0.67	0.27	59.70
125,000	1.17	0.53	54.70	0.85	0.4	52.94
150,000	1.5	0.64	57.33	1.1	0.51	53.64

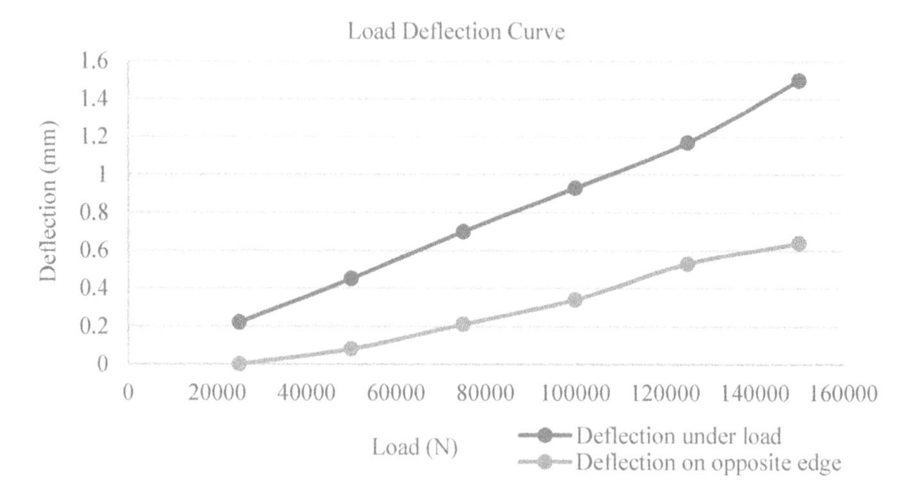

Fig. 1 Load deflection curve of PQC rehabilitated

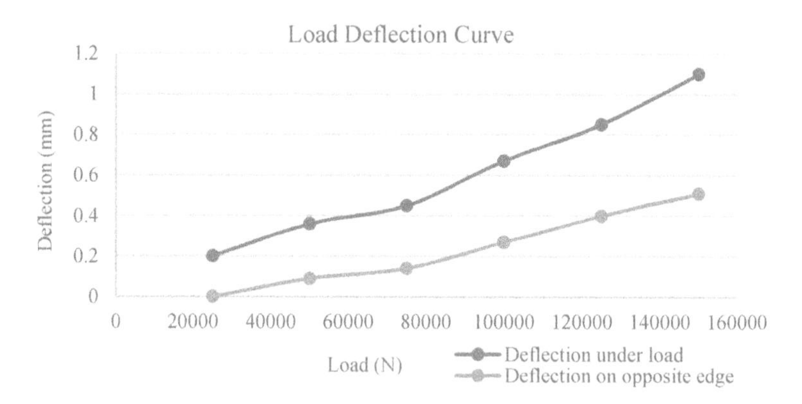

Fig. 2 Load deflection curve of FRSCC rehabilitated

4.3 *Fatigue Life*

The fatigue life was calculated from the validated models as shown in Fig. 3. The chi square values of the goodness of fit between model and prototype for FRSCC rehabilitated pavement and PQC rehabilitated pavement are 3.54 and 4.1 respectively. These are lower than chi square value (5 degree of freedom corresponding to 6 load deflection values) corresponding to 5% level of significance 4.352. From the models the fatigue life of the pavement has been determined corresponding to the point when the tensile strain in the concrete reached a value of 0.36% strain. It has been found as 2,765,763 cycles and 2,091,323 cycles for FRSCC rehabilitated pavement and PQC rehabilitated pavement. This is due to three type of

Fig. 3 Full model of PQC

improvements exhibited by FRSCC pavement: improvement in flexural strength, improvement in load transfer between old and new concrete, and improvement in load transfer efficiency across the joint.

5 Conclusion

Present study proves that the use of fibers as well as mineral admixtures is essential for a pavement to possess those properties that improves the load distribution as well as tensile strength of the concrete. Following concluding points could be earmarked in this regard:

1. The addition of mineral admixtures increases the volume of paste and reduce the stiffness of the paste, but addition of very fine micro fibers (wollastonite micro fiber in present study) compensates the loss in stiffness against deflection by improving the load transfer in the concrete. Hence, the use of very fine micro fibers having diameter in micro to nano range is very essential in SCC pavement.
2. Fine powdery material as admixtures improves the surface finishing and removes the undulations in the concrete that improves the load transfer efficiency at the joints. Fine micro fibers are also recommended to sustain this character of the concrete.
3. Fatigue life of a pavement depends upon the tensile strength and the load transfer efficiency within the concrete as well as through the concrete-dowel interphase. An improvement by a factor of nearly 33% could be achieved with respect to a pavement quality concrete if both micro fibers and admixtures are used in the pavement concrete.

References

1. Sharad SA, Gupta AK (2014) Pavement deterioration and its causes. (S. I. (SICETE), Ed.) J Mech Civil Eng (IOSRJMCE), pp 09–15
2. Ardani A, Hussain S, LaForce R (2003) Evaluation of premature PCC pavement longitudinal cracking in Colorado. In: Proceedings of the 2003 mid-continent transportation research symposium, Ames, Iowa
3. Yao JL, Weng QH (2012) Causes of longitudinal cracks on newly rehabilitated jointed concrete pavements. J Perform Constr Facil 26(1):84–94
4. Owusu-Ababio S, Schmitt R (2013) Longitudinal cracking in Widened Portland cement concrete pavements. Final Report No. 0092–12–05. Wisconsin Department of Transportation, Madison, WI
5. Jung YU, Freeman TJ, Zollinger DG (2008) Guidelines for routine maintenance of concrete pavement. Texas Transportation Institute
6. Okamura H, Ouchi M (2003) Self-compacting concrete. J Adv Conc Technol 1:5–15
7. Khayat KH, Hu C, Monty H (1999) Stability of self-consolidating concrete, advantages, and potential applications. In: Wallevik O, Nielsson I (eds) First international RILEM symposium, SCC. RILEM Publications, Stockholm, pp 134–142
8. Ozawa K, Maekawa K, Kunishima M, Okamura H (1988) Development of high-performance concrete based on the durability design of concrete structures. In: Okamura H, Shima H (eds) Second East-Asia and Pacific conference on structural engineering and construction, Kochi, Japan vol 1, pp 445–450
9. Goel S, Singh SP, Singh P (2012) Fatigue analysis of plain and fiber-reinforced self-consolidating concrete. ACI Mater J 109:573–582
10. Goel S, Singh SP, Singh P, Kaushik SK (2012) Prediction of mean and design fatigue lives of self-compacting concrete beams in flexure. J Inst Eng Ser A 93:55–61
11. IS 1489(Part 2): 1991 (2004) Portland—Pozzolana Cement—Specification, Part 2 calcined clay based, 3rd Revision. Bureau of Indian Standard, New Delhi, India
12. IS 383–1970 (1993) Specification for coarse and fine aggregates from natural sources for concrete, 2nd Revision. Bureau of Indian Standard, New Delhi, India
13. IS: 1727–1967 (1996) Methods of test for pozzolanic materials, 1st Revision. Bureau of Indian Standard, New Delhi, India
14. IS 9103: 1999 (1999) Concrete Admixtures-Specification, 1st Revision. Bureau of Indian Standard, New Delhi, India
15. IS: 44 Guide lines for cement concrete mix design for pavements, Third Revision. Bureau of Indian Standard, New Delhi, India
16. IS:516–1959 (1959) Methods of tests for strength of concrete. Bureau of Indian Standard, New Delhi, India
17. IS: 1888–1982 Method of load test on soil, Second Revision. Bureau of Indian Standard, New Delhi, India

Structural Health Monitoring of Lattice Structure Using Artificial Neural Network

Christeena Jose, Prince Thankachan, and T. M. Madhavan Pillai

Abstract Buildings need periodical health monitoring for assessing their strength and performance in the future. However, Structural Health Monitoring (SHM) plays a vital role in safeguard against failure at the member and structural level. Vibration-based structural health monitoring is the technique used to identify the structural changes using vibration measurements which causes the difference in the damage sensitive features in structures. Artificial Neural network (ANN) is an efficient Machine Learning (ML) tool widely used in many fields due to its high degree of robustness and fault tolerance. Feature selection and proper training of a network by adjusting parameters and hyperparameters are essential to get the output desired for a set of input data. The present study based on developing ANN to predict the damages in the lattice structures. Physical changes in the structure will alter the vibration parameters of the system. These vibration parameters include natural frequency, damping ratio, and mode shapes, etc. Thus, variations associated with these properties can be interpreted as damage caused to the structure. This is the idea behind the damage detection strategy of structures. Finite element analysis of lattice structure in the damaged and undamaged state is carried out to obtain various vibration parameters. Extracted vibration data is used to train neural network models to distinguish damaged conditions of the structure and thereby generalise the behaviour of the structure. The developed neural network is tested with unknown vibration data from the structure and the location of the damage is predicted.

Keywords Structural health monitoring · Artificial neural networks · Vibration parameters · Machine learning · Lattice structure

---

C. Jose (✉) · P. Thankachan · T. M. M. Pillai
National Institute of Technology Calicut, Kozhikode, India
e-mail: mpillai@nitc.ac.in

G. C. Marano et al. (eds.), *Proceedings of SECON'21*, Lecture Notes in Civil Engineering 171, https://doi.org/10.1007/978-3-030-80312-4_94

1 Introduction

Monitoring of strength and performance of civil infrastructure is essential for the comfort of the occupants. It avoids unexpected accidents and provides a safe environment for all. Damage identification is significant for preserving and sustaining the design life of civil structures. Structural Health Monitoring is the process of evaluating the changes that occurred in structures over some time. Human interference and changes in the environmental factors cause mild to severe damages in civil engineering structures. The vibration-based damage detection method measures the changes associated with the vibration parameters of the structure and determines the fault location and intensity that occurred in the structure.

Lattice structures are tower like structures similar to truss that is highly utilised in the power transmission industry, telecommunication and oil industry. They are popular because of their lightweight and fast construction. Health monitoring of these structures has more practical importance since most of them were located in inaccessible locations and a hazardous environment. Failure of these structures may interrupt power and network connection that have much more relevance in the modern world.

An artificial neural network (ANN) is a biologically inspired mathematical model made up of several layers and parallelly connected through neurons arranged in each layer. A neural network consists of three primary layers: input layer, hidden layer, and output layer (Fig. 1). It processes through a series of algorithms to make hidden relationships between input and output data.

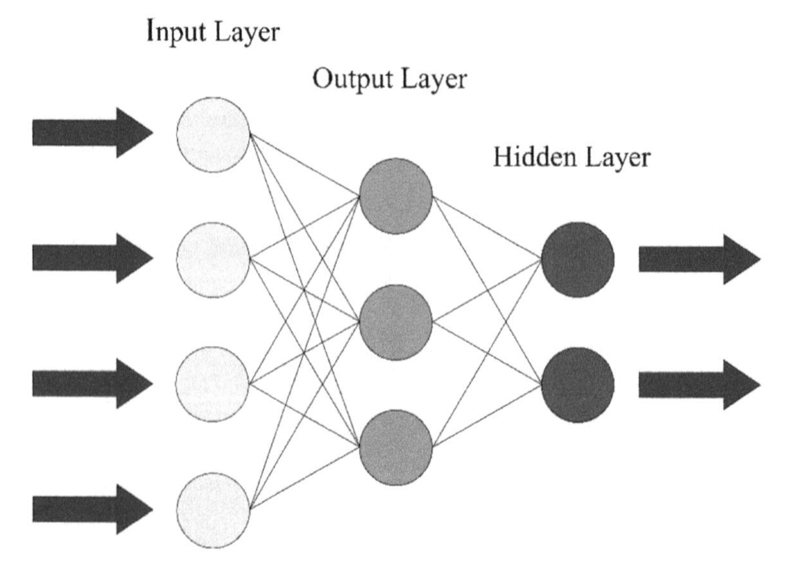

Fig. 1 The basic structure of a neural network

Studies by Idichandy et al. [1] proposed a scheme for monitoring offshore platform using natural frequency and vibration modes. An adaptive resonance theory (ART) based neural networks and backpropagation networks (BPN) were used for the damage identification of offshore platform by Mangal et al. [2]. Counter propagation neural networks were utilised instead of backpropagation to predict damages in bridges by Zhao et al. [3]. Wu et al. [4] evaluated the possibility of automated health monitoring of engineering structures with the neural network and vibration data.

Pioneering work on damage detection with the combination of neural network and cuckoo search algorithm was done by Tran-Ngoc et al. [5]. Modified modal flexibility and strain energy were precisely used by Jayasundara et al. [6] to quantify and localise damages in complex structures like deck type arch bridges. The above study showed a quick evaluation of single and multiple damages incorporated in the structure. Bao et al. [7] utilised raw strain response data from the structure to identify the damages automatically with the combination of a one-dimensional convolution neural network. The results of these analysis were highly comprising and experimentally validated by using the scale-down model of the structure.

2 Modelling of Lattice Structure

The lattice structure used in the oil industry specifically known as the offshore jacket platform is used for the present study. It consists of three types of members: main legs, inclined bracings and horizontal bracings. All sections are tubular in cross-section with different diameter. The main legs have an outer diameter of 40 mm and wall thickness of 1.5 mm. The inclined bracings have 20 mm diameter and 1.2 mm thick, while horizontal bracings are given 16 mm diameter and 1 mm thick. The details of the lattice structure model are shown in Fig. 2. The Deck slab is situated at the top of the structure and modelled using inclined and horizontal bracing. A Deck mass of 10 kg is uniformly provided over the main deck.

The structural analysis program SAP2000 was used to model the lattice structure. The members in the numerical model of lattice structure made up of three-dimensional beam elements having six degrees of freedom. Nodes were located at the intersection of longitudinal axes of each member and the model consist of 50 nodes and 124 beam elements. The support condition of the base of the model is fixed and it is provided at six unique points. Analysis of the structure was carried out by assuming it as a three-dimensional truss model.

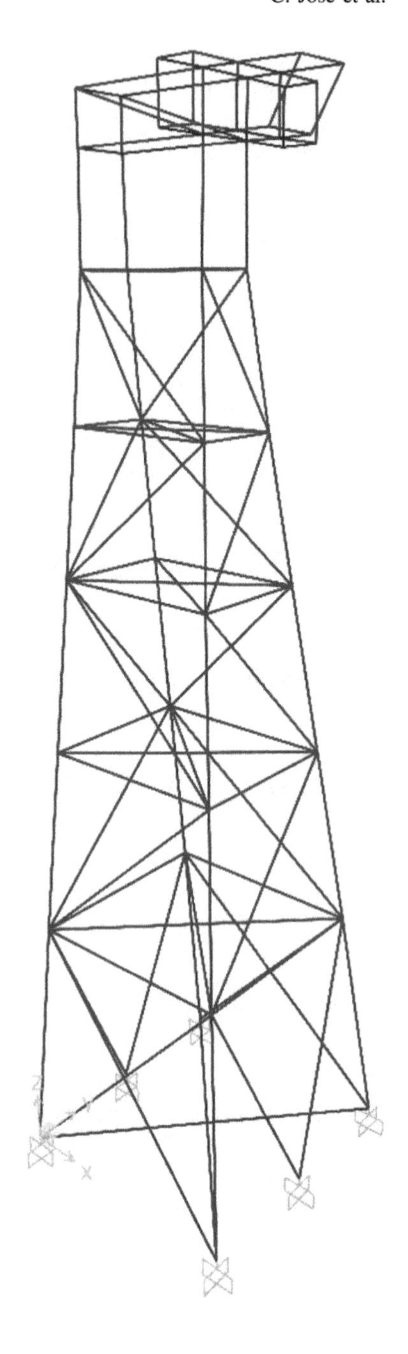

Fig. 2 Three dimensional view of the lattice structure modelled in SAP2000

3 Lattice Structure in Damaged Condition

Lattice structure in damaged condition was modelled by removing one member at a time. Members in the deck slab not considered for the damage analysis of the structure. Different damaged cases were developed similarly by removing members in chronological order. Total 78 damaged cases were modelled, out of which 6 cases used for testing and the rest were used for training the neural network.

3.1 Training of Neural Network for Damaged Cases

The natural frequency of each damaged case is extracted to create a neural network configuration to generalise the solution on the behaviour of the structure. The first five natural frequencies of translational modes along the y direction are used for damage identification of the lattice structure. Neural networks were trained using the natural frequency data as input values and centroid of removed members in Cartesian coordinates as targets. Since input and targets are already known, supervised learning algorithm is used for training the network.

Feedforward backpropagation network is employed for training and testing of data. A neural network fitting tool (nftool) from the deep learning toolbox in MATLAB was used to develop the network. The training of the neural network is a highly empirical process. The best performing network was found out by varying parameters and hyperparameters of the network. Figure 3 shows the network diagram of the best performing network in damage prediction of the lattice structure.

A neural network with two hidden layers having 25 and 15 neurons on each layer is selected based on performance. Backpropagation algorithm Levenberg Marquardt (trainlm) is applied to train the neural network. Transfer function adopted for all layers is tan sigmoid and the convergence of the network is measured using the mean square error function after each iteration. Out of 72 samples used for training, 84% is used for training and 8% is used for testing and validation. Regression coefficients (R) obtained after training are shown in Fig. 4, and overall R = 0.989, which is close to 1 and acceptable.

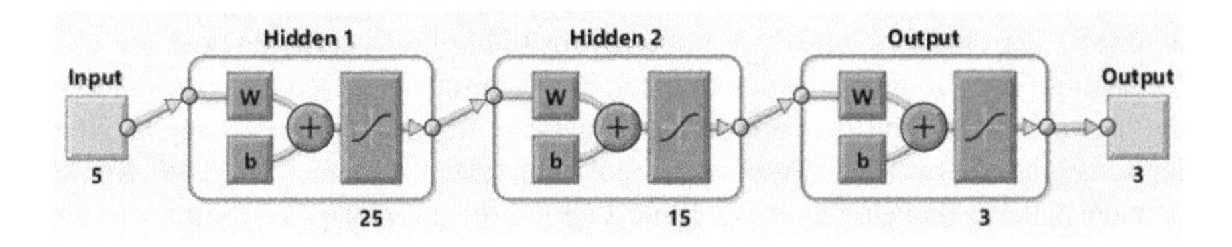

Fig. 3 Neural network configuration adopted for damage identification of the lattice structure

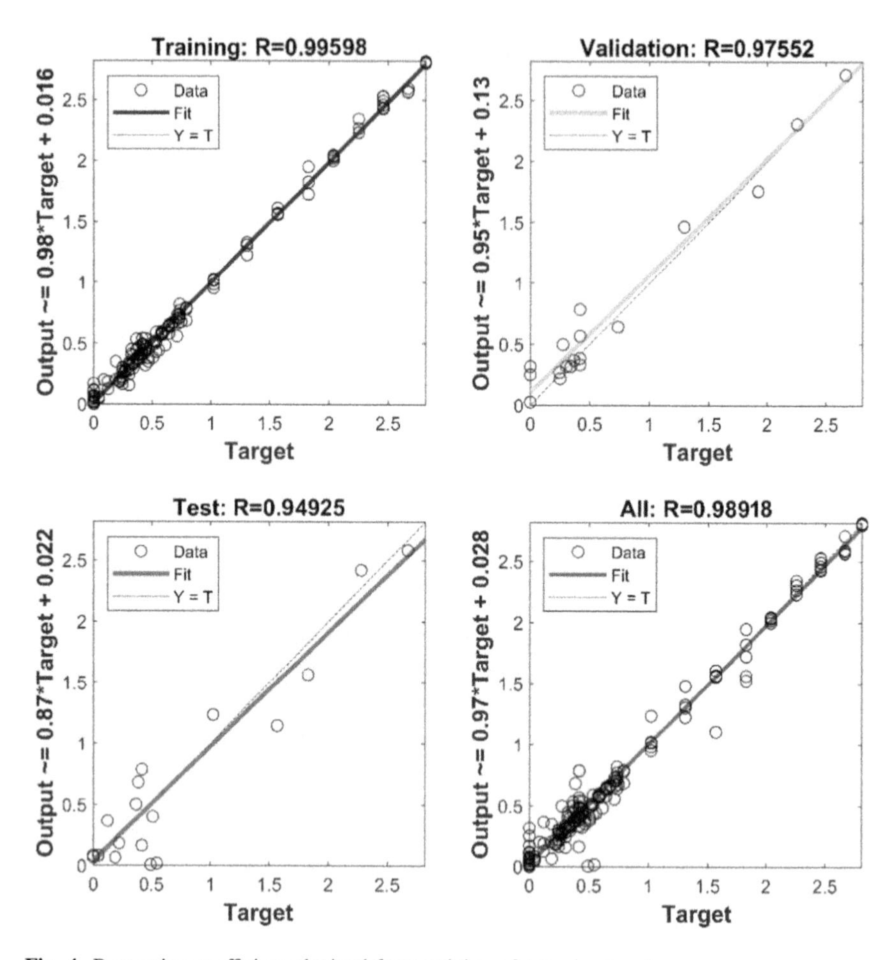

Fig. 4 Regression coefficient obtained from training of neural network

4 Results and Discussions

Testing of the neural network is carried out to analyse the accuracy of the prediction of the network in a damage scenario. In the training process network learns various damage patterns that occurred in the structure and thereby it is capable of detecting damages. Six data sets which are not used for training of neural network are used for testing. The natural frequency of the above data sets are given as input values of the network and neural network predicted damage location corresponding to given input values. Network analyses the input parameters and matches with known damage patterns that already have. Table 1 shows the unknown data used for testing of the neural network. Table 2 shows the output from the neural network.

Table 1 Data set used for testing of neural network

Member No.	*f1* (Hz)	*f2* (Hz)	*f3* (Hz)	*f4* (Hz)	*f5* (Hz)	*X*(m)	*Y*(m)	*Z*(m)
6	5.8334	31.8201	54.0737	68.5704	92.2862	0.3405	0.7922	0.368
28	5.8781	31.6375	54.0762	66.9467	83.5941	0.305	0.1570	1.309
42	5.8785	31.6481	53.9561	68.2835	85.3921	0.273	0.2215	1.823
51	5.9045	32.0808	53.9835	68.5845	91.8891	0.273	0.5868	2.0375
65	5.8593	31.7237	54.0185	68.4359	88.4789	0.2465	0.2995	2.4595
74	5.8122	31.4450	53.7213	67.7857	88.1648	0.444	0.4175	2.667

Table 2 Output from the neural network

Member No.	Actual values (y)			Neural network outcome ($\hat{y}$)			RMSE
	X(m)	*Y*(m)	*Z*(m)	*X'*(m)	*Y'*(m)	*Z'*(m)	
6	0.3405	0.7922	0.368	0.1937	0.7118	0.4382	0.1048
28	0.305	0.1570	1.309	0.3361	0.3171	1.1305	0.1395
42	0.273	0.2215	1.823	0.3085	0.5494	1.9046	0.1962
51	0.273	0.5868	2.0375	0.3359	0.3078	1.9991	0.1665
65	0.2465	0.2995	2.4595	0.2486	0.3942	2.4413	0.0557
74	0.444	0.4175	2.667	0.3682	0.4315	2.7576	0.0687

f1 to *f5* natural frequencies of damaged lattice structure from mode 1 to mode 5 along the y direction.

X, Y and *Z* are the Cartesian coordinates of centroid of the member removed.

RMSE = Root Mean Square Error.

The developed neural network configuration predicts the damage location in the member with acceptable errors. Results in Table 3 indicates the capability of the proposed neural network to detect, locate single damages with reasonable accuracy. It is observed that the output from the network gives excellent accuracy in terms of the height of location of the damaged member (z coordinate).

In most cases, the prediction results are either close to the required member or in a symmetrically opposite member. Since the structure under study was partially

Table 3 Damage prediction in lattice structure using neural network

No.	Member removed	Results from neural network
1	6	Close to member 6
2	28	Member 29 (adjacent member)
3	42	Member 45 (opposite member)
4	51	Member 50 (opposite member)
5	65	Close to member 65
6	74	Close to member 74

symmetric with respect to main and horizontal members along one direction, frequency values did not differ much for these member cases. Thus neural network predictions also seemed to be close to symmetric member. One case of prediction gave damage near a member adjacent to the damaged member (29); it is a horizontal member. For this structure, errors in the horizontal members are more than main leg members and inclined members. Stiffness of horizontal bracings are less compared to other members; hence variations in the natural frequencies are less for horizontal members and makes it more challenging to distinguish damages using neural networks.

Thus it could be stated that the height wise damage prediction gave perfect accuracy. Thus in a practical situation, the member adjacent to and symmetric to the predicted member should be checked at the same height and corrective action should be taken. The maximum root mean square error is 0.2 which shows the suitability of this method to identify damages in the lattice structure.

5 Conclusions

The study evaluated the possibility of using an artificial neural network in the damage identification of the lattice structure. Following conclusions can be derived from the results obtained.

1. Damage makes significant changes in vibration parameters of the lattice structure.
2. Natural frequency in combination with a neural network gives a good damage detection strategy.
3. Neural network precisely predicts the damage that occurred in the lattice structure having a root mean square error below 0.1. In other cases, it predicts adjacent member or opposite member at the same level.
4. Height wise prediction of damage location using neural network is highly appreciable.

Thus, it can be concluded that in lattice structures, neural network-based damage detection gives good results with the extraction of natural frequency.

References

1. Idichandy VG, Ganapathy C (1990) Modal parameters for structural integrity monitoring of fixed offshore platforms. Exp Mech 30(4):382–391
2. Mangal L, Idichandy VG, Ganapathy C (1996) ART-based multiple neural networks for monitoring offshore platforms. Appl Ocean Res 18(2–3):137–143
3. Zhao J, Ivan JN, DeWolf JT (1998) Structural damage detection using artificial neural networks. J Infrastruct Syst 4(3):93–101

4. Wu X, Ghaboussi J, Garrett JH (1992) Use of neural networks in detection of structural damage. Comput Struct 42(4):649–659
5. Tran-Ngoc H et al (2019) An efficient artificial neural network for damage detection in bridges and beam-like structures by improving training parameters using cuckoo search algorithm. Eng Struct 199(September):109637 (Elsevier)
6. Jayasundara N et al (2020) Damage detection and quantification in deck type arch bridges using vibration based methods and artificial neural networks. Eng Fail Anal 109(November 2019):104265 (Elsevier)
7. Bao X et al (2020) One-dimensional convolutional neural network for damage detection of jacket-type offshore platforms. Ocean Eng (October), 108293 (Elsevier Ltd)

Comparative Study of Predicting the Marsh Cone Flow Time of Superplasticized Cement Paste Using Machine Learning Algorithms

Kumaresh Ravi, P. Yogesh, R. Karthik, Anisha Radhakrishnan, and Dhanya Sathyan

Abstract The conventional laboratory testing of concrete for its fresh stage property is time-consuming, expensive-causing a loss of resources, materials. However, nowadays artificial intelligence-based robust systems are very well used in many engineering disciplines. As cement paste workability has a very good influence on concrete workability, a Machine-Learning model like SVM (Support Vector Machine) is calibrated, validated, and finally established to predict the Marsh Cone flow time of cement paste by giving ingredient property as the inputs. This research proposes an intelligent method for predicting the Marsh Cone flow time of cement paste based on machine learning methods. The experiment was carried out on a superplasticized cement paste made with locally available brands of Portland pozzolana cement (PPC) and superplasticizers from four different families at a water-cement ratio of 0.37. The four different families namely **PCE** (Polycarboxylate Ether), **SNF** (Sulfonated naphthalene-formaldehyde condensates), **SMF** (Sulfonated melamine–formaldehyde condensates), and (Lignosulphonates) were used. The model used water, cement, and the amount of superplasticizer (which was divided into seven different inputs based on their family and brand) as input parameters, with marsh cone flow time as the output parameter. A marsh cone flow test on more than 200 superplasticized cement paste mixes was used to obtain the model's testing and training results. The model was found to reliably predict the flow time of the Portland Pozzolana Cement paste within the experimental domain when the expected and measured values of marsh cone flow time were compared. The prediction capacity of SVM is checked with that of KNN (K-nearest neighbors algorithm) and Random forest and it is found that prediction accuracy is more for SVM compared to other models.

K. Ravi · P. Yogesh · R. Karthik · D. Sathyan (✉)
Department of Civil Engineering, Amrita School of Engineering, Coimbatore,
Amrita Vishwa Vidyapeetham, India
e-mail: s_dhanya@cb.amrita.edu

A. Radhakrishnan
Department of Computer Science, Amrita School of Engineering, Coimbatore,
Amrita Vishwa Vidyapeetham, India

G. C. Marano et al. (eds.), *Proceedings of SECON'21*, Lecture Notes in Civil Engineering 171, https://doi.org/10.1007/978-3-030-80312-4_95

Keywords Marsh cone · Superplasticizer · PPC · SVM · KNN · Random forest

1 Introduction

The Marsh cone test is used to ensure the consistency of cement pastes and to determine their specifications. In each country the standard of Marsh cone test differs, but its principles are usually the same. The time it takes for a certain volume of cement paste to flow out of the cone is measured and recorded. This calculated flow time has a relationship with the fluidity of the material that is tested. The higher the flow time, the lesser will be the fluidity. This test is a simple way to obtain information about the behaviour of cement pastes. The dosage is the optimal dose for that brand of cement and admixture (superplasticizer) and it is used in cement materials mix design to determine the congestion point.

To get better solutions on engineering problems, many studies have been done on different methods and many assumptions have been made to get the most approximate value but these techniques have sometimes shown significant errors, resulting in a large disparity between the real and expected value.

Marsh cone test is to determine the workability or consistency of cement paste. These methods give more correct description of the quality control of cement paste, but as the model is complicated to predefined as a result it complicates their uses. With these advanced machine learning methods, it was interesting to develop a new model for the Marsh cone test, By using these machine learning applications.

SVM and ANNs (artificial neural network) are used in several experiments to calculate the power of various types of concrete. By analysing the given data the machine learning model self learns from these experimental data sets and gives the functional relationships among the data, this also works when the relationships are unknown and is hard to explain the physical meaning. As the development of machine learning takes place the more accurate value are predicted, The will be no need to do the experiments once these machine learning like SVM can predict the actual value, but for these predictions, we need to give a large number of input data because the accuracy depends on the number and different types of datasets. Afterwards these models can be utilized as a tool for verifying the pre-existing modules which are used for the building codes.

2 Literature Survey

Many authors have shown the efficacy of machine learning methods such as SVM, KNN, and Random Forest in modelling the flow behaviour of concrete.

Dogra and Bhardwaj [1] analysed the behaviour of superplasticizers in cement paste with mineral admixtures including fly ash and alccofine has been studied.

According to the report, polycarboxylate ether-based superplasticizers have better compatibility than Sulphonated Naphthalene Formaldehyde-based superplasticizers.

Maroliya [2] studied, at different dosage stages, the impact of plasticizers and superplasticizers on the ingredient content of concrete-like cement and sand was studied. Furthermore, the plasticizers increased compressive strength at a lower water-cement ratio while also improving workability for a same water-cement ratio. To prevent bleeding and segregation, the cement content was decreased by increasing the sand content.

Maganti and Raju [3] studied with Sulphonated Naphthalene Formaldehyde and Lignosulphonates-based superplasticizers were tested for compatibility with Portland slag cement. Even when the water and chemical admixture family, coarse and fine aggregates, and concrete mix design system were held constant, the study found that different brands of cement behaved differently.

Sheela [4], in both hardened and fresh states, the strength behaviour and workability of superplasticized concrete and traditional concrete were studied. The use of a superplasticizer was found to improve strength and workability without raising the water-cement ratio. Non-destructive testing was also shown to be in strong alignment with the strength behaviour of superplasticized concrete. The resilience behaviour of superplasticized concrete was also discovered to be in complete sync with disruptive research values.

Roncero [5] studied the temperature effect of superplasticized cement pastes and the dosage of the superplasticizer used in the Marsh Cone test have also been observed. This research also looks at how the paste's fluidity changes over time. Polycarboxylic acid-based superplasticizers outperformed melamine-based admixtures and naphthalene, according to this research.

The paper written by Sathyan and Anand [6] explained the concrete's resiliency is improved by the combination of superplasticizers and pozzolanic compounds. Sulphate attack, chloride penetration, and acid attack have all been significant toughness concerns. The hardness of concrete was increased by lowering permeability, which was achieved by careful compaction. Superplasticizer was added to reduce porosity while increasing workability.Other researchers looked into the impact of superplasticizer on flow behaviour [7–10]

Wallevik and Wallevik [11] examined how using a rheograph and workability boxes, research examined how rheology influences concrete optimization. The rheograph detected a symmetrical shift in the rheological activity of the blend. This method was used to examine at how the behaviour of various forms of concrete shifted as the quantities of constituents changed, as well as the results of various admixtures.

Wu et al. [12], carried out a research work with the aid of the COMSOL Multiphysics programme, he created a statistical model for predicting and analysing the rheological parameters of CPB (binders, water, and tailings) under the control of temperature and hydration The predicted output values from the model on CPB were compared to data obtained from different tests to measure the software's prediction ability (rheometer, slump test, etc.) [13–18].

The compressive strength of concrete were predicted using fine-tuned Linear and nonlinear regression models [19, 20]. These models are less accurate, and they often struggle to satisfy the precision and generalisation criteria [21]. Latterly, robust systems based on artificial intelligence have been successfully implemented in this area; like neural networks [19, 20].

3 Materials Details and Sample Preparation

The cement paste was made of Portland pozzolana cement and superplasticizers from four different families. Totally Seven Superplasticizers belonging to four different families are used for this study, in that 2 superplasticizer are from Polycarboxylates Ether (**PCE1**, **PCE2**), 2 form Lignosulfonates (**LS1**, **LS2**), 2 from Sulfonated Naphthalene Formaldehyde (**SNF1**, **SNF2**), and 1 from Sulfonated Melamine Formaldehyde (**SMF1**).

The properties of the materials were assessed in accordance with IS 4031 [22–25], and IS 9103 [26] are given in the Tables 1 and 2 respectively.

4 Marsh Cone Flow Time

4.1 Experimental Setup

Previously, the Marsh cone test (Fig. 1) was used to determine the saturation superplasticizer dosage and relative fluidity of mortars and cement pastes. A metal

Table 1 Physical properties of the cement

Property	Values
Fineness	3.5
Specific gravity	2.82
Standard consistency	35.5%
Initial setting time (min)	160
Final setting time(min)	220

Table 2 Properties of superplasticizers

SP type	Solid content (%)	Density (kg/L)	pH
PCE1	36.67	1.09	6
PCE2	29.806	1.08	≥6
SNF1	37.425	1.22–1.225	7–8
SNF2	37.823	1.22	7–8
SMF1	33.056	1.226	7
LS1	30.045	1.15–1.17	≥6
LS2	31.61	1.17	≥6

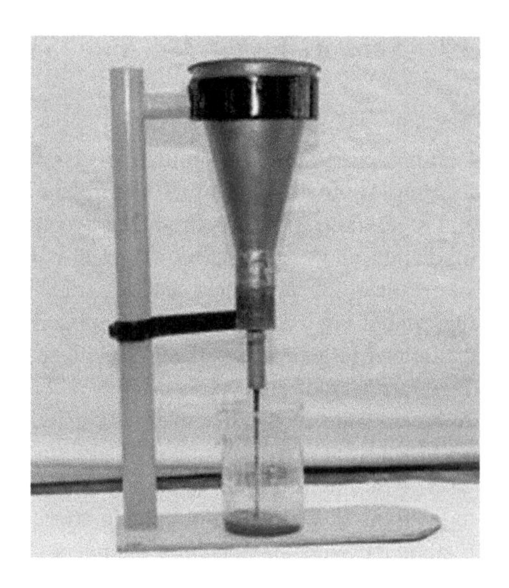

Fig. 1 Marsh cone experimental setup

cone (as per EN 445 [27]) with an 8 mm diameter nozzle was chosen as an implication to the current analysis. The time it took for 500 ml of paste to drain out of the cone was estimated after 1000 ml of paste was transferred into the cone. Gomes et al. [28] suggested a formula for determining the saturation dose based on the Marsh cone flow time curve of the pastes. The saturation dose is the Superplasticizer dosage that correlates to an internal angle of 140 10 degrees. When there are no data points that are correlative of the set of angles, interpolation is used to calculate the dose. This experiment could also be used to investigate fluidity loss over time, mineral admixture optimization in cement paste, cement mortar fluidity, and the incomparability of cement and superplasticizer. The field test is easy enough to compare the fluidity of cement pastes by adding various admixtures. Mould details are as follows: Top diameter of 160 mm and height 300 mm and nozzle diameter of 8 mm.

5 Modelling of Marsh Cone Flow Time

Software Used: XLSTAT in Microsoft Excel

XLSTAT is a powerful and flexible excel data analysis tool that allows analysing, customizing and sharing within the Excel Software. For this study, the all Prediction model (SVM, Random Forest, and KNN) were tuned and modelled in Microsoft Excel using XLSTAT software package.

5.1 *Support Vector Machine (SVM)*

A Support Vector Machine (SVM) is a powerful but versatile supervised machine learning algorithm, discriminative classifier that takes training data (supervised learning), outputs an optimal hyperplane, and can be used for both classification and regression. SVM has produced excellent results in a variety of areas, including prediction and classification. SVM has produced excellent results in a variety of areas, including prediction and classification. SVM is focused on statistical learning, which has been successfully applied to problems such as non-linear classification and large datasets. The hyperplane for classification is determined by SVM. The trajectory of all hyper-planes is determined by their precise location in space. A kernel aids in the discovery of a hyperplane in a higher-dimensional space. SVM's advantage in training is its ability to predict with multiple dimensional data. SVM is a classifier that takes labelled training data as input and uses an optimal hyperplane to predict an output. In short, SVM maximises the margin to create a hyperplane with the specified data sets. The model was fine-tuned using the following parameters (Table 3) with corresponding values like.

5.2 *KNN*

For several problems, including security techniques, K Nearest Neighbors is an efficient supervised learning process. KNN is made up of clusters of elements with identical characteristics, and it determines the class category of a test example based on its k Nearest Neighbors. The size of the dataset and the form of classification problem determine the value of k in the KNN. KNN divides the target into neighbors based on its surroundings. The following is an explanation of a k-Nearest Neighbors (KNN) classifier.

To measure the interval, we first find the elements closest to training data K from the test data using the Euclidean interval. For two elements in k dimensional space, a = [a1, a2, …, ak] and y = [b1, b2, …, bk], the Euclidean interval based on the two elements can be computed by using the Eq. (1):

$$d(a, b) = \text{sqroot of } \left(\sum \text{ki} = 1(\text{bi} - \text{ai})2\right) \tag{1}$$

Table 3 SVM tuning parameters

C	1	Gamma	0.5
Tolerance	0.001	Kernel	RBF(Radial Basic Function)

Table 4 KNN tuning parameters

Number of neighbors	5	Estimator	Median
Training set size	269	Prediction set size	14

Most of the k-Nearest Neighbors would be considered a class for the test data after having the k-Nearest Neighbors. The model was fine-tuned using the following parameters (Table 4) with corresponding values like.

5.3 Random Forest

Random forest is a collective classifier that consists of several decision trees and outputs the class, which is the mode of individual tree output of class. Individual learners are combined into decision trees. They are one of the most widely used data exploration techniques. By collecting the predictions of the many different data, the prediction is done. By Sampling X randomly, If the total cases in the training set are N, which is replaceable, from the original data, then for growing the tree, this sample will be used as the training set. For input variables of N numbers, the variable m is chosen in such a way that m«M is specified at each node, m number of variables are selected at random from the N, and the best split on n is used for separating the nodes. During the growing of the forest, n value is kept constant. Each tree is grown to the highest extent possible. No type of pruning is used. Random Forest usually exhibits a specific performance development as compared to single tree classifiers like C4.5.The model was fine tuned using the following parameters (Table 5) with corresponding values like.

5.4 Data Normalization

Normalization of data can be accomplished in a variety of ways. The Norm and Standard Deviation methods are the most often used techniques. The norm approach is used for modelling. Table 6 displays the test results for normalised input parameter values. Table 7 indicates the normalised values of test data output parameters.

Table 5 Random forest tuning parameters

Forest type	Regression	Sampling method	Random with replacement
Method	Random input	Number of trees built	500

Table 6 Input values of test data

Sample no	Water	Cement	LS1	LS2	PCE 1	PCE 2	SMF	SNF 1	SNF 2
1	0.15985	0.0589	0.0764	0	0	0	0	0	0
2	0.15701	0.0589	0.1783	0	0	0	0	0	0
3	0.16146	0.0589	0	0.0509	0	0	0	0	0
4	0.15377	0.0589	0	0.2548	0	0	0	0	0
5	0.15932	0.0589	0	0	0.0948	0	0	0	0
6	0.15991	0.0589	0	0	0.0711	0	0	0	0
7	0.15738	0.0589	0	0	0	0.0917	0	0	0
8	0.15449	0.0589	0	0	0	0.2064	0	0	0
9	0.15570	0.0589	0	0	0	0	0.0917	0	0
10	0.15074	0.0589	0	0	0	0	0.2522	0	0
11	0.16402	0.0589	0	0	0	0	0	0.0457	0
12	0.16040	0.0589	0	0	0	0	0	0.1372	0
13	0.15686	0.0589	0	0	0	0	0	0	0.0667
14	0.15686	0.0589	0	0	0	0	0	0	0.0667

Table 7 Output values of test data

Sample No:	Marsh cone flow time
1	0.1581
2	0.1461
3	0.1465
4	0.1273
5	0.1464
6	0.1521
7	0.1611
8	0.1242
9	0.1580
10	0.0786
11	0.1293
12	0.1465
13	0.1696
14	0.1696

6 Result and Discussion

6.1 Prediction Using the Model

The three different machine learning methods gave 3 different sets of predicted values. The accuracy of the training data determines the predictive performance of marsh cone flow time of cement mixtures the majority of the time. The output value

Table 8 Actual and predicted value of Marsh cone flow time

No	SVM		KNN		RF	
	Actual value	Predicted value	Actual value	Predicted value	Actual value	Predicted value
1	0.1581	0.1534	0.1581	0.1347	0.1581	0.146
2	0.1461	0.1411	0.1461	0.1358	0.1461	0.143
3	0.1465	0.1380	0.1465	0.0938	0.1465	0.144
4	0.1273	0.1261	0.1273	0.0874	0.1273	0.157
5	0.1464	0.1367	0.1464	0.1436	0.1464	0.136
6	0.1521	0.1443	0.1521	0.1981	0.1521	0.151
7	0.1611	0.1349	0.1611	0.1548	0.1611	0.153
8	0.1242	0.1198	0.1242	0.1259	0.1242	0.139
9	0.1580	0.1566	0.1580	0.1245	0.1580	0.149
10	0.0786	0.0791	0.0786	0.1237	0.0786	0.107
11	0.1293	0.1356	0.1293	0.0873	0.1293	0.147
12	0.1465	0.1206	0.1465	0.1243	0.1465	0.134
13	0.1696	0.1668	0.1696	0.1600	0.1696	0.174
14	0.1696	0.1668	0.1696	0.1600	0.1696	0.174

would be more accurate if the data point is more accurate. The experiments were carried out by varying the amount of ingredients used and the test conditions. This algorithm was fed and trained with 269 training data sets to construct this model. Quantities of cement, water, and superplasticizers are used as input parameters. The output is marsh cone flow time of cement paste.14 set of testing data was used to check the accuracy of prediction of the model. Normalised test data sets are given in

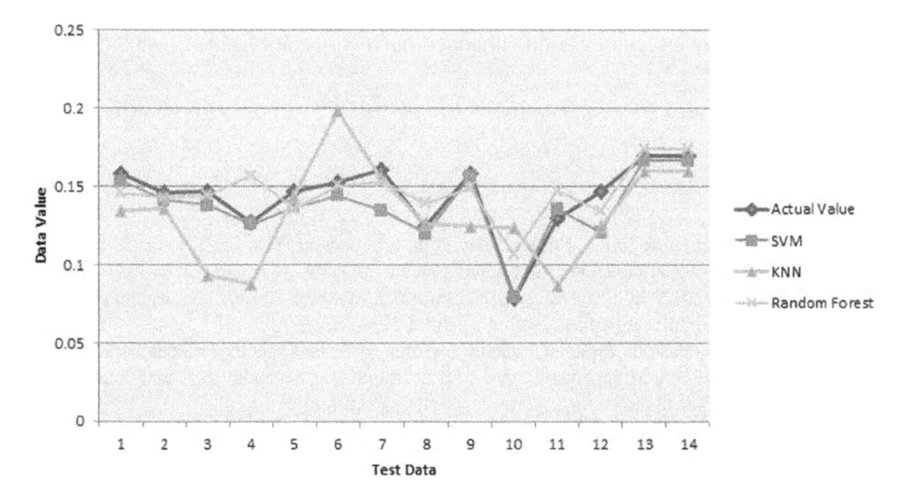

Fig. 2 Actual and predicted values

Table 9 Accuracy of prediction—Root Mean Square Error (**RMSE**) and Mean Absolute Error (MAE)

Type of machine learning	Root mean square error	Mean absolute error
Support Vector Machine (SVM)	0.01364	0.007
K-Nearest Neighbors (KNN)	0.03026	0.023
Random Forest (RF)	0.01421	0.007

Tables 6 and 7. The actual and predicted values of marsh cone flow time of test data sets using different models are given in the Table 8. To measure the accuracy of the prediction, Root Mean Square Error and Mean Absolute Error (MAE) were used.

From the predicted analysis results, it shows that the model is capable to predict with good accuracy for the flow time of the cement paste of testing data (Fig. 2). RMSE and MAE values of prediction using **KNN**, **SVM** and **RF** (Random forest) are given in the Table 9.

7 Conclusions

- The flow time measured using the Marsh cone method is a good indicator of the relative fluidity of superplasticized paste.
- RMSE values of prediction of SVM, KNN and RF are 0.01364, 0.03026, 0.01421 respectively and MAE values of prediction of SVM, KNN and RF are 0.007, 0.023, 0.007 respectively. Prediction accuracy of SVM is found to be better than KNN and RF.
- Because of the small number of training data sets, the proposed models have some drawbacks in terms of prediction. If more training datasets were available, the model's accuracy could be enhanced by incorporating them into the model.

References

1. Dogra A, Bhardwaj R (2014) Study of the admixture-cement compatibility for the economic production of special concretes. Int J Civil Eng Res 5(1):43–48. ISSN 2278-3652
2. Maroliya MK (2012) Influence of type of chemical admixtures on sand and cement content of ordinary grade concrete. Int J Adv Eng Technol 298–302
3. Maganti J, Silva Prasada Raju V Compatibility of sulphonated naphthalene formaldehyde and lignosulphonates based superplasticizer with portland slag cements. In: Third international conference on sustainable construction materials and technologies
4. Sheela S (2003) Study on workability and strength behaviour of concrete by the addition of different admixtures. In: Proceedings of the INCONTEST 2003 on 10–12, September 2003

5. Roncero J (2000) Effect of superplasticizers on the behavior of concrete in the fresh and hardened states: implications for high performance concretes
6. Sathyan D, Anand KB (2019) Influence of superplasticizer family on the durability characteristics of fly ash incorporated cement concrete. Constr Build Mater 204:864–874
7. Sindhu Menon M, Sathyan D, Anand KB (2017) Studies on rheological properties of super plasticized PPC paste. Int J Civil Eng Technol 8(10):939–947
8. Sathyan D, Anand KB, Mini KM (2016) Experimental study on Portland pozzolana cement-superplasticizer compatibility in mortar. Int J Earth Sci Eng 9(2):539–544
9. Jayasree C, Gettu R (2008) Experimental study of the flow behaviour of super plasticized cement paste. Mater Struct 41(9):1581–1593
10. Robert RC, Sathyan D, Anand KB (2018) Effect of superplasticizers on the rheological properties of fly ash incorporated cement paste. Mater Today Proc 23955–23963
11. Wallevik OH, Wallevik JE (2011) Rheology as a tool in concrete science: the use of rheographs and workability boxes. Cem Concr Res 41(12):1279–1288
12. Wu D, Fall M, Cai SJ (2013) Coupling temperature, cement hydration and rheological behavior of fresh cemented paste backfill. Miner Eng 42:76–87
13. Sathyan D, Balakrishnan AK, Mohandas SM (2018) Temperature influence on rheology of superplasticized pozzolana cement and modelling using RKS algorithm. J Mater Civ Eng 30 (9):04018221
14. Robert RC, Kuriakose NM, Gopikrishnan K, Sathyan D, Rajesh CB (2021) Modelling the rheological properties of fly ash incorporated superplasticized cement paste at different temperature using multilayer perceptrons in tensorflow. Lect Notes Civil Eng 97:635–646
15. Sathyan D, Govind D, Rajesh CB, Gopikrishnan K, Aswath Kannan G, Mahadevan J (2020) Modelling the shear flow behaviour of cement paste using machine learning-XGBoost. J Phys Conf Ser 1451(1):012026
16. Sathyan D, Anand KB, Prakash AJ, Premjith B (2018) Modeling the fresh and hardened stage properties of self-compacting concrete using random kitchen sink algorithm. Int J Concr Struct Mater 12(1):24
17. Sathyan D, Anand KB, Jose C, Aravind NR (2018) Modelling the mini slump spread of superplasticized PPC paste using RLS with the application of random kitchen sink. IOP conference series, p 310. https://doi.org/10.1088/1757-899X/310/1/012035
18. Prakash AJ, Sathyan D, Anand KB, Aravind NR (2018) Comparison of ANN and RKS approaches to model SCC strength. IOP conference series, p 310. https://doi.org/10.1088/1757-899X/310/1/012037
19. Ni Hong-Guang WJZ (2000) Prediction of compressive strength of concrete by neural networks. Cement Concr Res 30(8):1245–1250
20. Sobhani J, Najimi M, Pourkhorshidi AR, Parhizkar T (2010) Prediction of the compressive strength of no-slump concrete: a comparative study of regression, neural network and ANFIS models. Constr Build Mater 24(5):709–718
21. Zain M, Abd S (2009) Multiple regression model for compressive strength prediction of high-performance concrete. J Appl Sci 9(1):155–160
22. IS 4031, Part 5, Method for determination of initial and final setting time of cement, Bureau of Indian Standards, New Delhi, India (2005)
23. IS 4031, Part 1, Indian standard specification for method of physical tests for hydraulic cement, determination of fineness by dry sieving, Bureau of Indian standards, New Delhi (2005)
24. IS: 4031, Part 11, Indian standard specification for method of physical tests for hydraulic cement, determination of density, Bureau of Indian Standards, New Delhi (2005)
25. IS: 4031, Part 4, Indian standard specification for method of physical tests for hydraulic cement, determination of consistency of standard cement paste, Bureau of Indian Standards, New Delhi (2005)

26. IS 9103, Concrete admixture-specification, Bureau of Indian Standards, New Delhi, India (2004)
27. EN B (2007) 445, Grout for prestressing tendons-test methods. British Standards Institution
28. Gomes PCC, Gettu R, Agullo L, Bernad C (2001) Experimental optimization of high strength self-compacting concrete. In: Proceedings of second international symposium on SCC, Japan, pp 377–386

Reliability of a Timber Truss Considering Its Joints

Deborah Wadzani Dauda and Akash Malik

Abstract Timber is one of the commonly used materials for building construction and is mostly used as trusses in roofs of buildings. The joints of the truss system are critical points that need sound design and analysis to be carried out. Reliability analysis takes care of many uncertainties that arise from the design and construction of such structural systems. This paper looks into the reliability of these joints having nails as fasteners, considering variables such as the geometrical variables (cross-section of the timber truss, diameter of nails), load variables (Nodal loads of Dead, Live, and Wind loads), and resistance variables (compressive and tensile strength at right angles to the grain, shear stress and strength, splitting capacity of timber, load-carrying capacity of nails and embedment strength). The analysis was done using Monte Carlo simulations and the results showed that the probability of failure was low, indicating a high reliability of the truss. Therefore the reliability index meets the criteria of a 50-year reference period for consequence class 2.

Keywords Reliability · Timber truss · Joints · Nails · Geometrical variables · Load variables · Resistance variables · Monte Carlo simulations

1 Introduction

A wood/timber roof truss is a designed structural framework made of timber elements filling the space above a room and supporting the roof over it. Trusses typically occur at even intervals, connected together with longitudinal timber beams called purlins. The gap between the individual trusses is referred to as a bay [1].

D. W. Dauda (✉) · A. Malik
Galgotias University, Uttar Pradesh, Greater Noida, India
e-mail: deborah_dauda.socebtech@galgotiasuniversity.edu.in

A. Malik
e-mail: akash.malik@galgotiasuniversity.edu.in

D. W. Dauda
Federal Polytechnic Ede, Osun State, Ede, Nigeria

G. C. Marano et al. (eds.), *Proceedings of SECON'21*, Lecture Notes in Civil Engineering 171, https://doi.org/10.1007/978-3-030-80312-4_96

Timber trusses are found to be used widely, in Canada, in the USA, and in even Africa. Its strength-to-weight ratio is high, which allows for longer spans and also, it is known for its flexibility in its layout of floor plan [1, 2]. Failure can occur in timber structures, including timber trusses due to shear, splitting, tension perpendicular to grain, compression, instability, joint failure, amongst others; which were also observed from the case studies carried out by [3]. The crucial points of a timber truss system are its joints because the overall performance and strength of the truss system can be determined by the joints; hence, in the design of the truss, the joints are critical factors to look into. Considering not just the self-weight but also uncertainties in loading (live and wind loads) of the truss and the load duration [4], calls for reliability analysis to be carried out. Therefore, evaluating the timber truss structure using reliability analysis makes it possible to cater for those uncertainties and ensure a high reliability, which implies a low probability of failure of the system. Timber structural designs are meant to comply with the reliability standards and design requirements given in [5]. The consequences of human lives been lost, the economic, social or environmental consequences of failure determines the consequence classes, ranked as high, medium and low, that is, CC1, CC2 and CC3, although, CC2 is the consequence class for most installations where timbers are used for the structural components or systems. These consequence classes: CC1, CC2 and CC3 are linked to the reliability classes of RC1, RC2 and RC3 respectively which in turn are associated with reliability or safety indices to be obtained as a minimum required value for a particular class [6].

Although [7] considered the single response of load carrying capacity of nails for timber truss with nail joint, this paper tries to consider in addition to the load-carrying capacity of nails, the responses of shear, splitting, tension and compression at right angles to grain of timber truss with nail connections at its joints.

2 Reliability

Probabilistic approaches account for uncertainties, hence it has a lower risk/probability of failure. If treating some physical phenomena with present day scientific knowledge seem to be too complex, probabilistic methods can serve as convenient tools to describe and even model such. A given statistical data is explicitly incorporated into the design algorithms of Probabilistic design. In structural systems, the probabilistic variables to be modelled are load and resistance, and its probability or chance of failure is then calculated using this approach of reliability analysis. Reliability of structural systems is the probability that the structural system would adequately perform its intended function in its specified lifetime under specified conditions. Reliability is a method that implies limit state probabilities of a structural system under adverse environmental loadings and it always has some value of probability accuracy associated with it.

3 Limit State Function

Given, mechanical resistance properties *R*(*x*), and applied actions/load effects *S*(*x*), the structural system's reliability is *P*(*R*(*x*) > *S*(*x*)) while the chance of failure is *P*(*R* (*x*) < *S*(*x*)), and the limit state would be *P*(*R*(*x*) = *S*(*x*)).

Let *g*(*x*) indicate the limit state function (LSF), with $x = x_1, x_2$ *up to* x_n, denoting the set of random variables. Therefore

$$g(x) = R(x) - S(x) \tag{1}$$

Hence, if *g*(*x*) is less than zero, the structural system is in a failed state; if *g*(*x*) is equal to zero, the structural system is in a limit state; if *g*(*x*) is greater than zero the structural system is in a safe and reliable state.

Trying to compute the failure probability, P_F, it is written as

$$P_F = P[g(x) \leq (x)] = \int D_f \text{fx}(x)\text{dx} \tag{2}$$

With D_f referring to the subspace correlating with the failure event F, that is, the field of failure and fx(x) stands for JPDF—joint probability density function of x_1, x_2, *up to* x_n.

4 Monte Carlo Simulation Method

Simulation techniques prove their value in problems where difficulties accompany the LSF representation, especially in cases where the LSFs are not differentiable or when several design points contribute to the probability of failure [8].

In Monte Carlo simulation (MCS), the probability integral is calculated by trials, which have to be repeated many times for the results to converge. A number of random outcomes of the LSF value is calculated, also the number of outcomes where values of LSF is smaller than zero is calculated. Then that is divided by the total amount of outcomes generated and having regard to the total number of tests carried out in this manner, we can estimate the statistics of the probability of failure. Random outcomes are between zero and one.

$$\text{Probability of Failure} = (\text{No. of failures})/(\text{No. of simulations}) \tag{3}$$

There are two ways of running MCS as stated by [9]: specify the surface of failure and detect automatically any surface of failure by the program used to run the simulation. The computation procedure of MCS is also given by [10]:

(i) For the random variable, RV, select a distribution type
(ii) From the distribution, generate a sampling set

(iii) Using the distribution-generated sampling set, conduct simulations.

Table 1 shows timber properties and their statistical parameters, while Table 2 shows the types of structural timber failure for different failure modes. Depending on the kind of failure, the consequences of failure can be classified according to the following:

(i) Ductile failure having reserve strength capacity which results from strain hardening
(ii) Ductile failure having no reserve capacity
(iii) Brittle failure

Hence, there's a greater need for a high-level reliability design of structural elements/structures which are in danger of collapsing unexpectedly without warning, than for structural elements/structures in which the collapse is spearheaded by some warning signs, thus enabling courses of action to be taken in order to keep clear of severe consequences [11].

As adapted from [5], Table 3 gives the description of three classes of consequences CC1, CC2 and CC3 which the three classes of reliability RC1, RC2 and RC3 are associated with; Table 4 then shows these classes of reliability and their corresponding ultimate limit state (ULS) minimum target values of reliability/safety index, Beta. Also, the recommended minimum values of Beta for ultimate limit state and serviceability limit state (ULS & SLS) CC2:RC2 are shown in Table 5.

Table 1 Relationship between the main properties of timber (coloured-filled portion) and other timber properties, adapted from [11]

Timber Properties, [X]	Distribution type	E[X]	CoV [X]
Bending strength, fm	LN	E[fm]	0.25
Bending modulus of elasticity, Em	LN	E[Em]	0.13
Density, rm	N	E[rm]	0.10
Tensile strength parallel to the grain, $f_{t,0}$	LN	0.6 E[fm]	1.2 CoV [fm]
Tensile strength at right angles to the grain, $f_{t,90}$	Weibull	0.015 E [rm]	2.5 CoV [rm]
MOE for tension parallel to the grain, $E_{t,0}$	LN	E[Em]	CoV[Em]
MOE for tension at right angles to the grain, $E_{t,90}$	LN	E[Em]/30	CoV[Em]
Compressive strength parallel to the grain, $f_{c,0}$	LN	5 E $[fm]^{0.45}$	0.8 CoV [fm]
Compressive strength at right angles to the grain, $f_{c,90}$	N	0.008 E [rm]	CoV[rm]
Shear modulus, G_v	LN	E[Em]/16	CoV[Em]
Shear strength, f_v	LN	0.2 E $[fm]^{0.8}$	CoV[fm]

E[X] = Expected values, CoV [X] = Coefficient of Variance, LN = Lognormal

Table 2 Types of failure for different failure modes of structural timber [11]

Failure mode	Type of failure
Bending, fm	Ductile[1]
Tension parallel to the grain, $f_{t,0}$	Brittle
Tension at right angles to the grain, $f_{t,90}$	Brittle
Compression parallel to the grain, $f_{c,0}$	Ductile
Compression at right angles to the grain, $f_{c,90}$	Ductile with reserve
Shear, fv	Brittle

[1]The failure mode can be brittle for lower timber grades

Table 3 Definition of the classes of consequences

Consequences classes	Consequence class one (CC1)	Consequence class two (CC2)	Consequence class three (CC3)
Description	Low consequence if human life is lost and low or negligible economic, social or environmental consequences	Medium consequence if human life is lost, considerable economic, social or environmental consequences	High consequence to human life loss, or very large economic, social or environmental consequences
Examples of buildings and civil engineering works	Agricultural buildings where people do not normally enter, greenhouses	Residential and office buildings where consequences of failure are medium	Grandstands, public buildings where consequences of failure are high

Table 4 Minimum recommended ULS Beta values

Reliability Class (RC)	Min. values for Beta	
	1-year ref. period	50-year ref. period
RC1	4.2	3.3
RC2	4.7	3.8
RC3	5.2	4.3

Table 5 Recommended minimum values of Beta for ULS and SLS CC2:RC2 [6]

RC associated with CC2:RC2	Minimum values for Beta	
	ULS	SLS
Reference period (1-year)	4.7	2.9
Reference period (50-year)	3.8	1.5

5 Analysis Methodology

From [12], the strength class of D35 hardwood timber specie was adopted. The reliability analysis of the nail jointed timber truss was explored using a mean-centered Monte Carlo simulation (sampling analysis), and implemented with the C++ programming language in the probabilistic model library of Rt, which is a program that is object-oriented for the realization of analyses that are probabilistic in nature, with many probabilistic interactive models [13–19]. A maximum iteration of order 10^6 was chosen for the truss analysis in order for a sufficient accurate yield of probability distributions for the response of the system. The responses considered were, the timber shear, load carrying capacity of the nails, splitting capacity of the timber, tension and compression at right angles to the grain of the timber, with an Aletory type of uncertainty being assigned to the structure.

Design values are normally computed/calculated from characteristic values with partial factor and modification factor application. Hence, the design value of a strength property, F_d and the design value of a resistant property, R_d, is given by Eqs. (4) and (5). See [20].

$$F_d = K_{mod} \times (F_k/\gamma_M) \quad (4)$$

$$R_d = K_{mod} \times (R_k/\gamma_M) \quad (5)$$

where K_{mod} is the factor of modification which takes account of the effect of lasting period of the load and timber moisture content, γ_M is the partial factor for a material and resistance properties, F_k and R_k is the characteristic value of a strength property and load carrying capacity respectively. Recommended values of partial safety factors for material properties and resistances are: γ_M for solid timber is 1.3, and γ_M for connections is 1.3. The value of K_{mod} is chosen based on combined load of actions affiliated with different classes load durations which should correspond with the shortest load duration.

For safety of the nail-jointed truss considered, the following conditions need to be fulfilled [6, 20]: In case of Shear, the Design shear strength have to exceed the Design shear stress ($F_{v,d} > \tau_d$); in case of Load Carrying Capacity of Nails, the Design capacity have to exceed the Design Force ($F_{v,Rd} > F_{v,Ed}$); in case of Splitting Capacity of Timber, the Design splitting capacity of member have to exceed the Design force ($F_{90,Rd} > F_{v,Ed}$); in case of Tension, the Design tensile strength at right angles to the grain have to exceed the Design tensile stress at right angles to the grain ($F_{t,90,d} > \sigma_{t,90,d}$); in case of Compression, the Design compression strength at right angles to the grain have to exceed the Design compressive stress at right angles to the grain ($F_{c,90,d} > \sigma_{c,90,d}$). The limit state functions were formulated accordingly. The statistical parameters used in the analysis of reliability are presented in Table 6. The resistant variables were modelled as random variables, RV and partial safety factors as constants, C. The effect of the timber truss member size on strength and tension at right angles to the

Table 6 Statistical parameters used for the reliability analysis

Parameters	Description	Class	Distribution type	Mean	CoV	SD
$F_{v,k}$	Characteristic shear strength	RV	LN	4 kN	0.25	–
Gk	Dead loads	RV	N	4 kN	0.05	–
Qk	Live loads	RV	Gum	5 kN	0.2	–
Wk	Wind loads	RV	Gum	5 kN	0.4	–
γ_G	Partial FoS for dead load	C	–	1.35	–	–
γ_Q	Partial FoS for live load	C	–	1.5	–	–
γ_W	Partial FoS for wind load	C	–	1.5	–	–
γ_M	Safety factor for member and connection	C	–	1.3	–	–
K_{mod}	Factor of modification which takes account of the effect of lasting period of the load and timber moisture content	C	–	0.9	–	–
b	Breadth or width of member	DV	–	50 mm UL = 50 mm LL = 30 mm	–	1.5
h	Height of timber	DV	–	125 mm UL = 150 mm LL = 100 mm	–	2.5
ρ_k	Characteristic density	RV	N	650 kg/m^3	0.1	–
t	Thickness of members	DV	–	50 mm UL = 50 mm LL = 30 mm	–	1.5
d	Diameter of nail	DV	–	5 mm UL = 5 mm LL = 3.55 mm	–	1
$F_{v,ed}$	Design force	RV	N	3 kN	0.1	–

FoS = Factor of Safety, LN = lognormal, N = normal, Gum = gumbel

grain was taken into account hence, the geometrical variables were modelled as decision variables, DV, having upper limit (UL) and lower limit (LL) values including the current values.

6 Analysis Results and Discussion

Table 7 shows the reliability sampling (MCS) analysis results of the nail-jointed timber truss for the function "G(x)". As well, Fig. 1a depicts the Failure accumulation plot for the LSF "G(x)" of Shear and Fig. 1b depicts the Histogram accumulation plot for the LSF "G(x)" of Shear. This is respectively applicable to Fig. 2a, b for the LSF "G(x)" of Load carrying capacity of Nails, Fig. 3a, b for the LSF "G(x)" of Splitting capacity of Timber, Fig. 4a, b for the LSF "G(x)" of Tension at right angles to the grain and Fig. 5a, b for the LSF "G(x)" of Compression at right angles to the grain. The failure accumulation plots show the number of samples run (the total number of samples run for each responses is seen in Table 7) as the coefficient of variation (CoV) of the probability approached 2% which was the target CoV of probability; this implies that all the responses reached their target CoV of probability. The probability of failure was computed with CoV as low as 2% in order to have a more accurate result. The Histogram accumulation plots from the analysis of the nail-jointed timber truss show the frequencies of the PDF—probability density function, CDF—cumulative density function and the CoV—coefficient of variation against the function values. The red line is comparable to the PDF; the blue line is comparable to the CDF; the black line is comparable to the CoV of the probability. The histogram plots explains more clearly the function values in which the PDF lumps around the mean value of "G(x)" and the CoV of the probability being close to the target CoV around the mean value of "G(x)".

The probability of failure for shear is very low, which is approximately zero, hence the truss is highly safe in shear and its reliability index, 23.8066, far exceeds the criteria for a 1 & 50-year reference period for consequence classes 1, 2 & 3. The same goes for load carrying capacity for nails, tension and compression at right angles to the grain. But for splitting capacity of timber, although it did not meet the criteria for reference period of 1 year for consequence classes 1, 2 & 3, it met the criteria for reference period of 50 years for consequence class 1 & 2.

Table 7 Sampling analysis (MCS) results for the function "G(x)"

	G(x)	Total number of samples	Reliability index: Beta	Probability of failure: P_f
Shear	1.81199	74,072	23.8066	1.42804e-125
Load carrying capacity of nails	5689.14	29,097	9.99472	8.03666e-24
Splitting capacity of timber	2418.16	11,704	4.13092	1.80656e-05
Tension at right angles to the grain	0.671736	29,172	9.96759	1.05646e-23
Compression at right angles to the grain	0.064236	28,652	9.67154	1.9916e-22

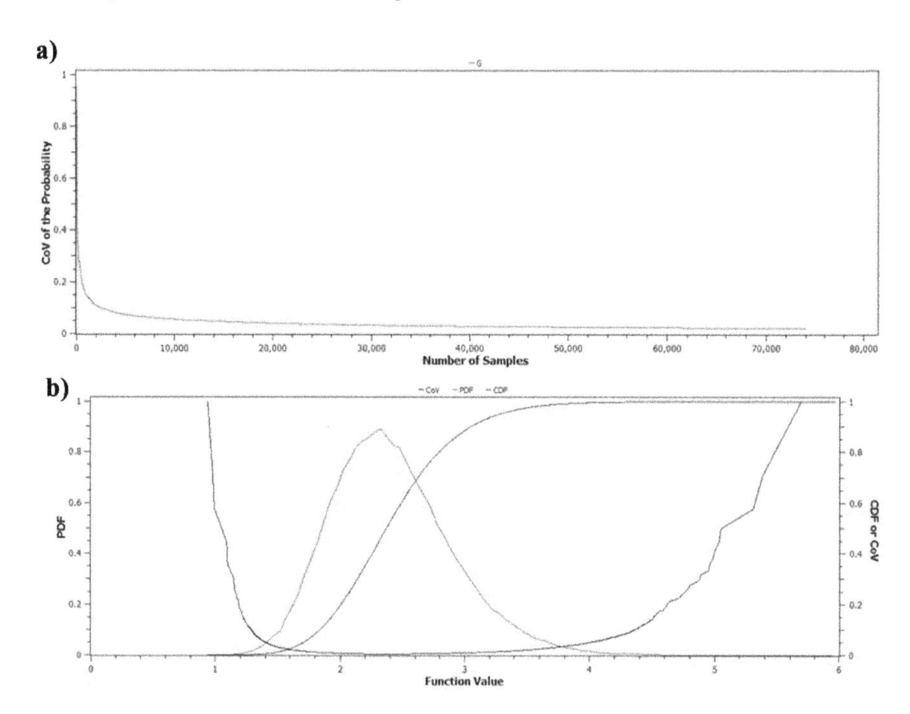

Fig. 1 **a** Failure accumulation plot for Shear. **b** Histogram accumulation plot for Shear

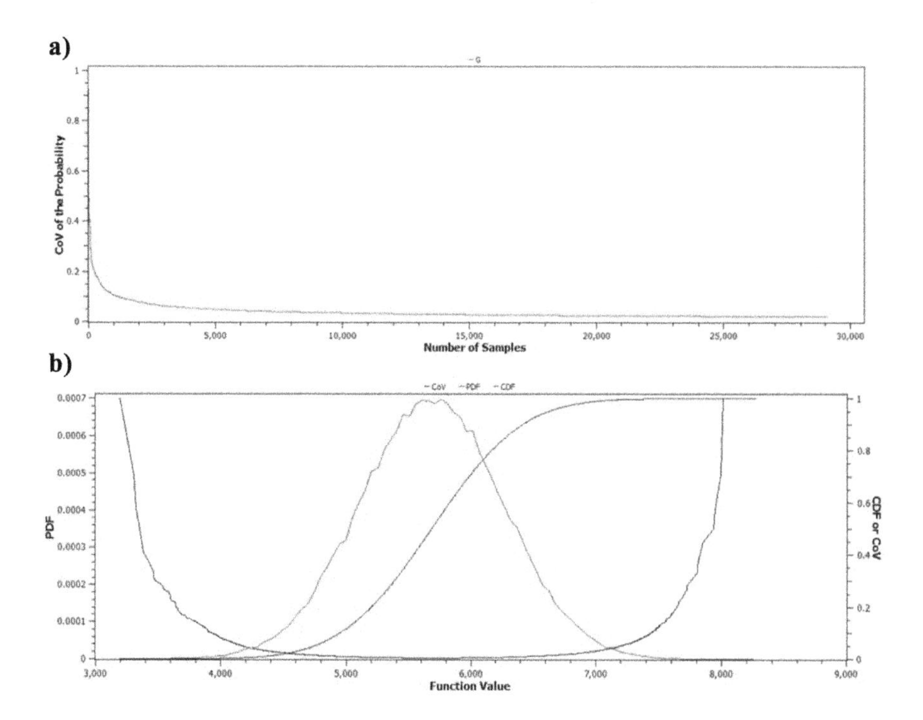

Fig. 2 **a** Failure accumulation plot for load carrying capacity of nails. **b** Histogram accumulation plot for load carrying capacity of nails

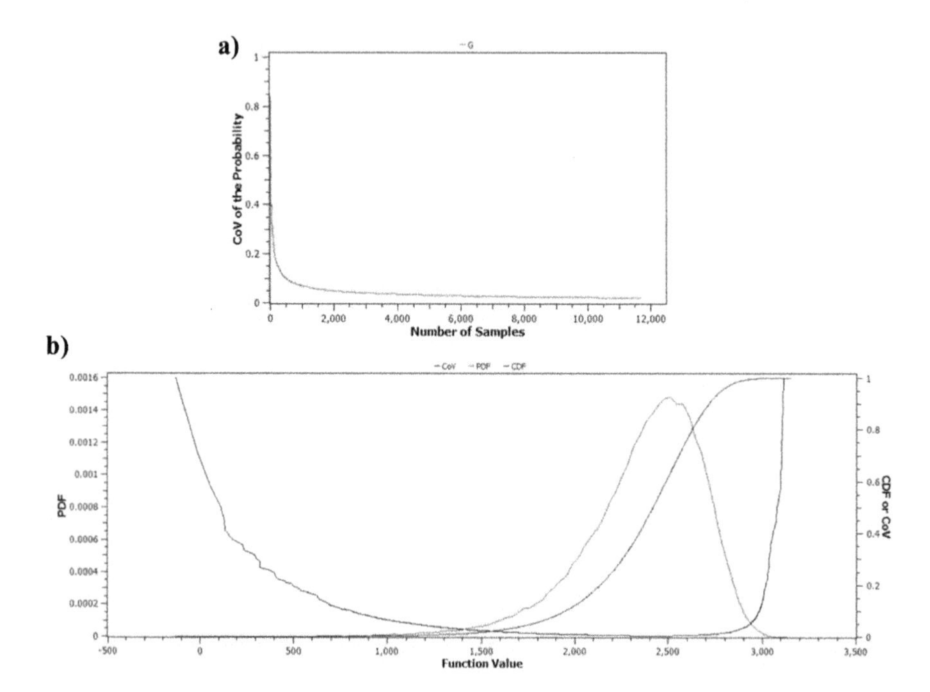

Fig. 3 **a** Failure accumulation plot for splitting capacity of timber. **b** Histogram accumulation plot for splitting capacity of timber

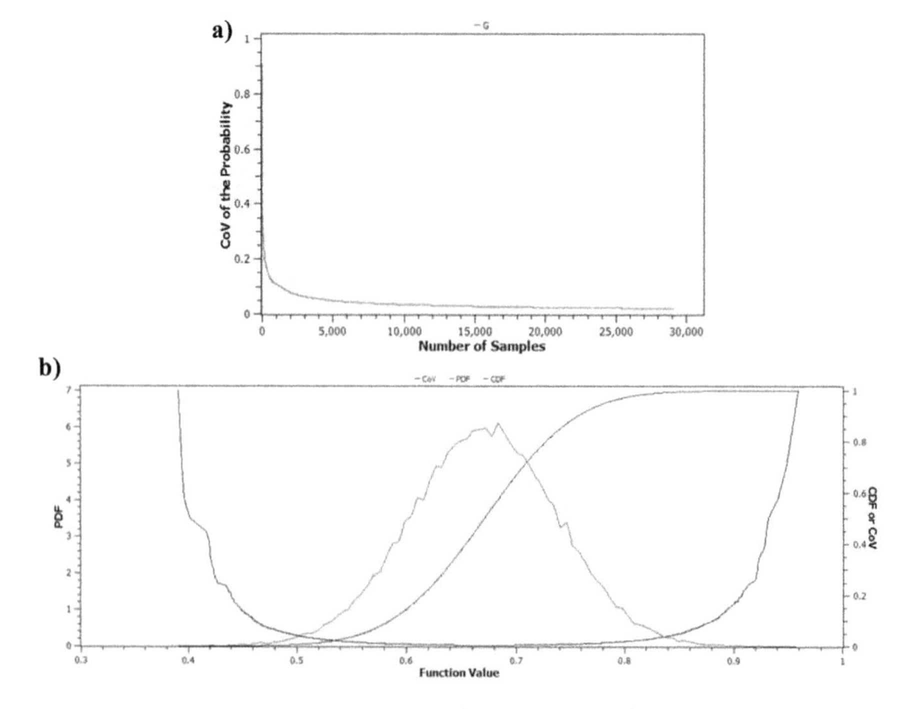

Fig. 4 **a** Failure accumulation plot for tension at right angles to the grain. **b** Histogram accumulation plot for tension at right angles to the grain

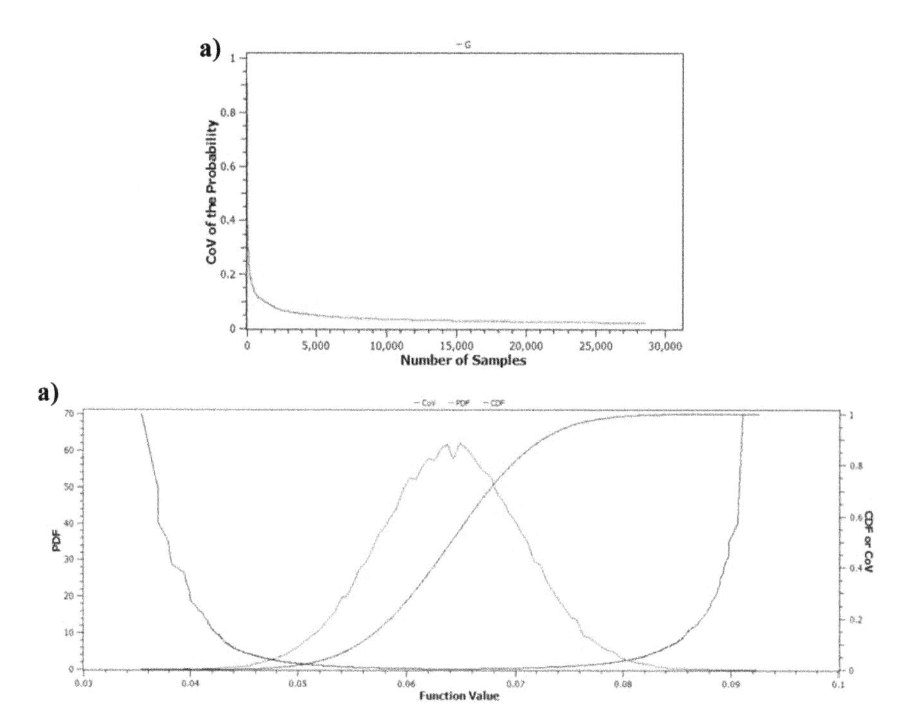

Fig. 5 **a** Failure accumulation plot for compression at right angles to the grain. **b** Histogram accumulation plot for compression at right angles to the grain

7 Conclusion

As seen in this paper, reliability analysis was conducted on a timber truss considering its joints being connected with nails. Limit state functions for shear, load carrying capacity of nails, splitting capacity of timber, tension at right angles to timber grain, and compression at right angles to timber grain were formulated and sampling (MCS) approach was used to analyze the problem. The result from the limit state functions computed shows that the timber truss considered having nailed joints, is safe in Shear, safe in the nails' load carrying capacity, safe against timber splitting, along with tension and compression at right angles to the timber grain. The respective reliability indices for each response meets the criteria of the ULS target reliability/safety index for a 50-year reference period for consequence class 2.

References

1. Curl JS (2006) "truss", A dictionary of architecture and landscape architecture (online ed.). Oxford University Press. Retrieved 1 Dec 2012
2. The constructor: https://theconstructor.org/structural-engg/timber-roof-trusses-advantages/13875/. Retrieved on 25/07/2020
3. Cabrero JM, Iraola B, Yurrita M (2018) Chapter 7—Failure of timber constructions. In: Salam A, Makhlouf H, Aliofkhazraei M (eds) Handbook of materials failure analysis. Butterworth-Heinemann
4. Li Z, Zheng X, Ni C, Tao D, He M (2020) Reliability-based investigation into the duration of load effect for the design of timber structures based on Chinese standard. Struct Safe 87: 102001–1020014
5. CEN (2002a) ENV 1990—"Eurocode 0—Basis of Structural Design". European Committee for Standardization (CEN), Brussels
6. Porteous J, Kermani A (2007) Structural timber design to Eurocode 5. Blackwell Science Ltd, 9600 Garsington Road, Oxford OX4 2DQ, UK
7. Kachalla M, Kolo ZK (2012) Strength class effect on the performance of timber nailed joint: reliability approach. Cont J Eng Sci 7(3):33–39
8. 11th Lecture: methods of structural reliability analysis
9. Tun YW, Pedroso DM, Scheuermann A, Williams DJ (2016) Probabilistic reliability analysis of multiple slopes with genetic algorithms. Elsevier Comput Geotech 77:68–76
10. Choi SK, Grandhi RV, Canfield RA (2006) Reliability-based structural design. Springer
11. Joint Committee on Structural Safety (2006) JCSS probabilistic model code, Part 3: resistance models—3.5 properties of timber. Probabilistic Model Code, Internet Publication: www.jcss.ethz.ch
12. UNI EN 338:2009. Structural timber—strength classes, UNI Ente Nazionale Italiano di Unificazione Via Sannio, 2 20137 Milano, Italia
13. Khosravikia F, Mahsuli M, Ghannad MA (2018) The effect of soil–structure interaction on the seismic risk to buildings. Bull Earthq Eng 16:3653–3673
14. Khosravikia F (2016) Seismic risk analysis considering soil–structure interaction. Master's Thesis, Department of Civil Engineering, Sharif University of Technology, Tehran, Iran
15. Mahsuli M (2012) Probabilistic models, methods, and software for evaluating risk to civil infrastructure. PhD Dissertation, Department of Civil Engineering, University of British Columbia, Vancouver, Canada
16. Mahsuli M, Haukaas T (2013) Seismic risk analysis with reliability methods, part I: models. Struct Saf 42:54–62
17. Mahsuli M, Haukaas T (2013) Computer program for multimodel reliability and optimization analysis. J Comput Civ Eng 27:87–98
18. Mahsuli M, Haukaas T (2013) Seismic risk analysis with reliability methods, part II: analysis. Struct Saf 42:63–74
19. Stroustrup B (2000) The C++ programming language. Addison-Wesley Professional, Boston
20. EN 1995-1-1 (2004) (English) Eurocode 5: Design of timber structures—Part 1-1: General—Common rules and rules for buildings

Experimental and Numerical Analysis of Axial Load Capacity of Encased Steel–concrete Composite Column

C. S. Reshma, C. P. Sudha, and P. Seena

Abstract Composite columns are usually constructed either concrete inside steel section or steel inside concrete. Composite columns can bring superior structural performance and for the same cross-sectional area these columns can carry large amount of load compared to normal columns. This paper makes a study on the axial compressive behaviour of fully encased steel–concrete composite short columns. In addition, an ordinary Reinforced Concrete (RC) short column were also tested for comparison purpose. Mix proportion of M35 conventional concrete was obtained as per IS 10262–2009. The design and detailing of composite columns were carried out as per Euro code 4 (EC4 2004). The tubular steel used for study was mild steel. Square columns having size of 100 × 100 mm with 600 mm depth are constructed for this study. The specimens were tested under uniaxial compression in a load controlled Universal Testing Machine. Linear Variable Differential Transducer was used for the measurement of axial shortening of the columns. First crack load, ultimate load carrying capacity and load-deformation behaviour of each samples were investigated. The experimentally obtained results were validated with the numerical FE software ABAQUS. Ultimate load carrying capacity was also obtained from three different codes namely Euro code 4 (2004), ACI-318 (2014) and AISC-LRFD (2010).

Keywords Composite columns · Tubular steel confinement · Uniaxial compression

1 Introduction

Composite column is widely used for the construction of multi-storey buildings to carry heavy loads with lesser cross-sectional area. It is a combined form of steel reinforced concrete and structural core steel. This core steel may be different

C. S. Reshma (✉) · C. P. Sudha · P. Seena
Department of Civil Engineering, Government Engineering College Thrissur, Thrissur, Kerala 680009, India

G. C. Marano et al. (eds.), *Proceedings of SECON'21*, Lecture Notes in Civil Engineering 171, https://doi.org/10.1007/978-3-030-80312-4_97

structural steel shapes or pipes or tubes. Since the outer concrete cover the inner structural steel, it offers higher durability and fire resistance compared to conventional steel column. In a composite column the external loading is resisted by both the concrete and the steel sections by interacting together by bond and friction. In composite structures concrete offers high compressive strength, stiffness, stability, better fire resistance and corrosion resistance while steel offers good tensile strength and reduction in construction time. Composite columns may be Fully Encased Composite (FEC) or Partially Encased Composite (PEC) or Concrete Filled Tube (CFT). Among these three types FEC columns offers better resistance to corrosion and fire. Considering strength, ductility and economy points of high-rise buildings, FEC columns makes the ideal solution for moist weather conditions.

Ellobody and Young [5] studies the behaviour of pin-ended concrete encased steel composite columns subjected to axial load. A nonlinear 3-D finite element model was developed to analyse the inelastic behaviour of concrete, structural steel, longitudinal and transverse bars. Also analyse the effect of concrete confinement of encased steel–concrete composite columns. Stud, non-slender, slender and long encased steel concrete composite columns were considered for study. Design strength of these composite columns were accurately predicted by using AISC and Eurocode-4. Lai et al. [2] studied the behaviour of axially loaded high strength concrete encased steel (CES) composite short columns experimentally. Various parameters affecting the ultimate strength of concrete ie. concrete compressive strength, transverse reinforcement bar spacing and the addition of steel fibers in the high strength concrete were studied. The experimental results shows that the design methods in Euro code EN 1994–1-1 and JGJ 138–2016 can accurately predict the maximum resistance of CES columns with normal strength steel and upto C90 grade concrete. But more than C90 grade concrete they over-estimate this value. These methods give good relation with the experimental results. To obtain the axial load carrying capacity of high strength CES columns, two analytical methods based on “reduced section” and “reduced strength” concepts are proposed. This results very well match with the experimental results. Kartheek and Venkat Das [8] presents the numerical simulation of RC and FEC column with I-section steel confinement cast with normal and high strength concrete. Due to the provision of I-section steel in composite columns ultimate load capacity has increased.

Current design guidelines for composite structures are mentioned in AISC-LRFD (2010), ACI 318 (2014), Euro code-4 (EC-4) (2004), Egyptian code (2012), Architectural Institute of Japan (AIJ 2005), and Canadian Standard Association, CSA (2009). But for the design of composite columns, ACI-318, AISC-LRFD and EC-4 are being used commonly. The main aim of this study is to experimentally and numerically evaluate the axial load capacity of encased steel–concrete composite column with tubular steel confinement. Also involves prediction of ultimate load capacity of FEC columns by using ACI-318, AISC-LRFD and EC-4.

2 Experimental Programme

The experimental stage involves casting and testing of four numbers of short columns. Out of these, two were made up of RC and the remaining two were with FEC columns with tubular steel confinement. The specimens were designed and detailed according to the Euro code 4 (EC4 2004).

2.1 Specimen Details

100 mm × 100 mm square column specimens having depth of 600 mm were constructed using normal strength concrete. 8 mm ∅ longitudinal reinforcement and 6 mm ∅ stirrups at 70 mm c/c were used. The concrete compressive strength (fck) was 43 MPa. Ys_t 240 grade steel tube was used as core steel. The outside diameter and thickness of steel tube was 33.7 and 3.2 mm respectively. The dimension of the specimen, together with the details of reinforcement is shown in Fig. 1a and the photograph of the reinforcement cage fabricated is shown in Fig. 1b. For casting of the specimens, a specially designed steel mould was used. The specimens were cast on a level floor. An electrically operated concrete mixer was used for mixing of concrete and the concrete was placed immediately after mixing. A 20 mm diameter needle vibrator was used for compaction of concrete. Figure 1c shows specimen after casting. The steel mould was removed after 24 h of

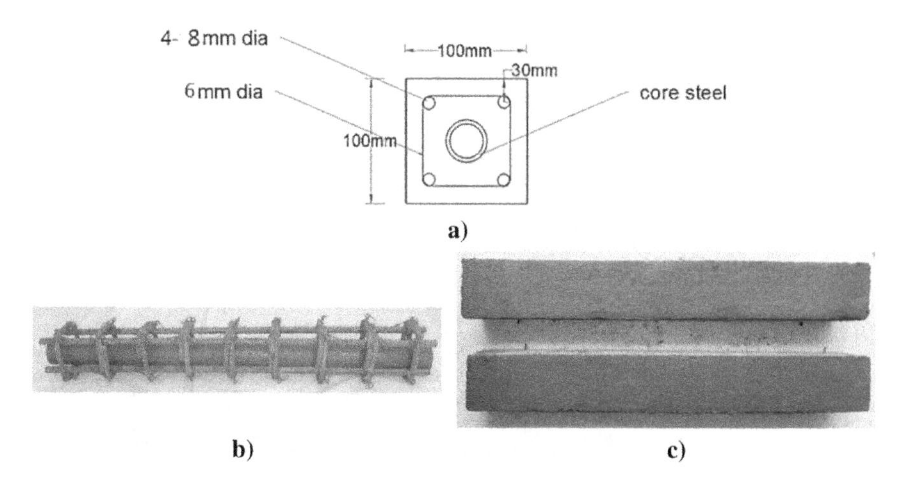

Fig. 1 **a** Typical cross section of FEC column. **b** Photograph of the reinforcement cage. **c** specimen after casting

Table 1 Reinforcement details of test specimens

S. No.	Specimen designation	Tube size (mm)	Reinforcement		Steel ratio	
			Longitudinal reinforcement	Stirrups	Steel tube % (A_s)	Rebar % (A_{sr})
1	RC	–	4-∅ 8 mm	∅ 6 mm–@70 mm c/c	–	2
2	FEC	33.2	4-∅ 8 mm	∅ 6 mm–@70 mm c/c	3	2

casting and cured for 28 days. Reinforcement details of test specimens are shown in Table 1.

2.2 Testing of Specimens

After 28 days of curing, the casted specimens were tested in a load controlled Universal Testing Machine (UTM) under uniaxial compression. The ultimate load carrying capacity of this UTM is 1000 kN. Two numbers of Linear Variable Differential Transducer (LVDT), having a range of 10 mm with a resolution of 0.01 mm was used for measuring the axial shortening of each column specimen. The specimens were subjected to monotonic compression loading until failure. The column experimental set-up is shown in Fig. 2. Two specimens each were tested for both RC and FEC short columns. The average value of two specimens were taken for analysis. The deformation at the top of the column was obtained from digital

Fig. 2 Experimental column test setup

Table 2 Experimental results of column specimens

Specimen designation	First crack load (kN)	Ultimate load (kN)	Axial shortening corresponding to ultimate load (mm)
FEC1	250	430	2.05
FEC2	220	420	2.03
RC1	120	280	1.82
RC2	150	300	1.90

indicator through LVDT. Four specimens are tested, details of these experimental results are given in Table 2 and photograph of the columns at failure under ultimate axial load are illustrated in Fig. 4. It was observed that crushing failure was occurred at top of both RC and FEC columns.

3 Finite Element Model of RC and FEC Columns

To study the behaviour and load carrying capacity of both RC and FEC columns a nonlinear 3-D finite element model was developed in the ABAQUS software. The concrete and steel tube were modelled using C3D8R element. The longitudinal and transverse bars were simulated using T3D2 three-dimensional truss elements. Loading plate was modelled using R3D4 shell element. Embedded element technique is used for steel concrete interactions. Fixed boundary conditions are applied on top and bottom of the column. Displacement control technique is used to specify the load conditions. 5 mm mesh size was selected as optimum. Modeling of concrete and reinforcement cage, meshing, loading and boundary conditions are shown in Fig. 3.

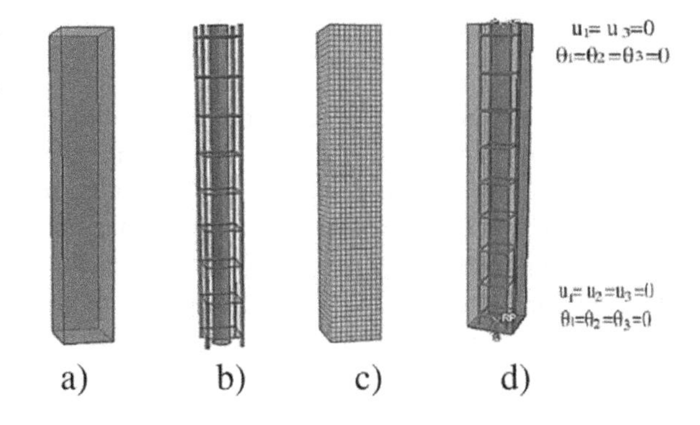

Fig. 3 A typical FEC column modeling of **a** concrete **b** reinforcement cage and tube **c** meshing of the column and **d** loading and boundary conditions

Fig. 4 Failure modes of FEC and RC column

Table 3 Material properties for concrete, reinforcement bars and steel tube

Material	Young's modulus (MPa)	Poisson's ratio	Yield strength (MPa)	Yield strain (mm/mm)	Ultimate strength (MPa)	Ultimate strain (mm/mm
Concrete	27,777.78	0.15	–	–	–	–
Bar 8 mm ∅	200,000	0.3	546	0.0032	654	0.333
Bar 6 mm ∅	200,000	0.3	513	0.0031	653	0.275
Steel tube	200,000	0.3	310	0.0038	502	0.13

Table 4 Concrete damage plasticity model input parameters

Plasticity parameter	Value used
Viscosity parameter	0.0015
Shape factor (K)	0.667
Dilation angle (ψ)	35
Eccentricity (ε)	0.1
Stress ratio (f_{b0}/f_{c0})	1.16

3.1 Material Properties for FE Model

Concrete, reinforcement bar and steel tube are the main materials incorporated in FE model for numerical simulation. Material properties of these materials used in FE model are shown in Tables 3 and 4. To simulate the concrete material behaviour of both RC and FEC columns, a concrete damage plasticity model was developed. To incorporate the compressive and tensile behaviour of concrete, a compression and tension stress–strain curve was generated by using equations of [3] and [4].

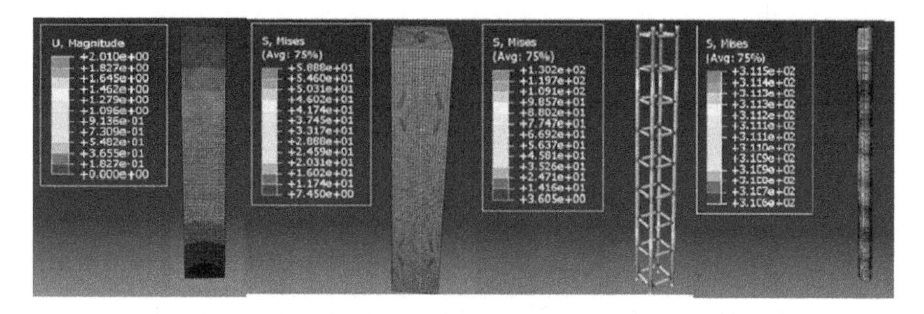

Fig. 5 Deformed shape of FEC column and Von-mises stress for concrete, reinforcement cage and steel tube at ultimate load

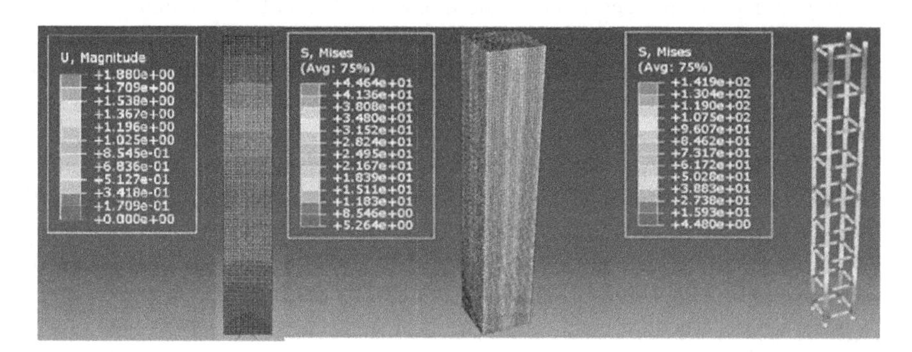

Fig. 6 Deformed shape of RC column and Von-mises stress for concrete and reinforcement cage at ultimate load

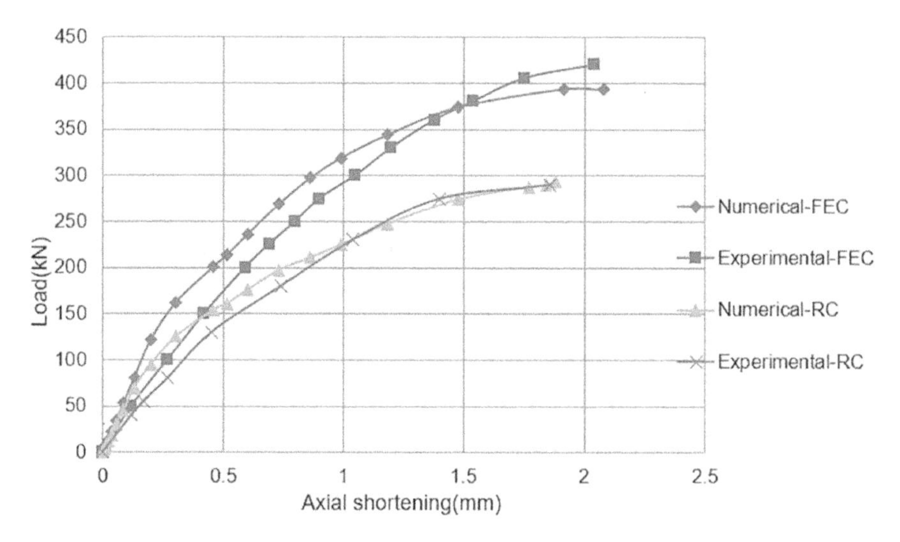

Fig. 7 Load- axial shortening response of FEC and RC columns

Table 5 Comparison between experimental versus ABAQUS results

Specimen designation	Experimental results		ABAQUS results		Ratio	
	Load (P_{test}) (kN)	Deformation (D_{test}) (mm)	Load (P_{num}) (kN)	Deformation (D_{num}) (mm)	P_{test}/P_{num}	D_{test}/D_{num}
RC	290	1.86	293.77	1.88	0.987	0.989
FEC	425	2.04	394.02	2.01	1.078	1.015

4 Comparison Between Experimental and Numerical Results

Axial load-deformation and failure pattern of each RC and FEC column specimens were observed and recorded experimentally and numerically. The experimental and numerical load-deformation behaviour of these columns were compared and are shown in Figs. 4, 5, 6 and 7. Deformed shape of both FEC and RC columns and stress distribution of each elements at ultimate load are shown in Figs. 5 and 6. Stress distribution shows that the there is some discrepancy exist between failure patterns of experimental and numerical results at ultimate load. Because local failure was occurred at the top of these column specimens under short term axial load. Stress distribution shows that failure of RC column was due to crushing of concrete and for FEC column, failure was initiated by crushing of concrete followed by yielding of structural steel. Ultimate axial load carrying capacity and corresponding deformation of both experimental and numerical results are almost similar. The mean value of experimental to numerical peak load ratio, P_{test}/P_{num} is shown in Table 5.

5 Comparisons of FEC Column Strength with Design Codes

This stage involves prediction of ultimate load carrying capacity of current composite column by using standard guidelines such as Euro Code 4 (2004) (EC-4), ACI 318 (2014), and AISC-LRFD (2010). Also obtained results were compared with experimental result and then predict which standard guidelines gives more conservative value for current study.

5.1 Euro Code 4 (2004) [6]

Plastic resistance to compression,

$$N_{pl,Rd} = A_r f_{yr} + 0.85 A_c f_{cu} + A_s f_s \tag{1}$$

$$P_{EC4} = \chi N_{pl} \tag{2}$$

$$\chi = \frac{1}{\emptyset + \left[\emptyset^2 - \lambda^2\right]^{0.5}} \leq 1 \tag{3}$$

$$\emptyset = 0.5\left[1 + \alpha(\lambda - \lambda_o) + \lambda^2\right] \tag{4}$$

$$\lambda = \frac{L_e}{\pi}\left(\frac{A_r f_{yr} + 0.85 A_c f_{cd} + A_s f_s}{E_r I_r + 0.6 E_{cm} I_c + E_s I_s}\right) \tag{5}$$

where, E_{cm} = secant modulus of elasticity of concrete, L_e = effective length of the column, E_s, E_c and E_r are the modulus of elasticity of steel, concrete and reinforcing bar respectively. I_s, I_c and I_r are the respective moment of inertia of steel, concrete and reinforcing bar. α and l_o are the factors taken as 0.49 and 0.21, respectively. Cylindrical compressive strength of concrete obtained is 33.4 MPa.

5.2 ACI 318 (2014) [1]

$$P_n = 0.8(0.85 f_{cu} A_c + F_{yr} A_r + F_y A_s) \tag{6}$$

where, f_{cu} = concrete compressive strength, F_{yr} and F_y are the respective yield strength of longitudinal reinforcement and steel shape. A_c, A_r and A_s are the area of concrete, longitudinal reinforcement and structural steel shape respectively.

5.3 AISC-LRFD (2010) [7]

$$P_{no} = \left[f_y A_s + f_{ysr} A_{sr} + 0.85 f_{cu} A_c\right] \tag{7}$$

$$P_e = \pi^2 EI_{eff}/(KL)^2 \tag{8}$$

When $\frac{P_{no}}{P_e} \leq 2.25$.

$$P_n = P_{no}\left[0.658^{\frac{P_{no}}{P_e}}\right] \tag{9}$$

where, A_c, A_{sr} and A_s are the area of concrete, longitudinal reinforcing bars and steel section respectively. EIeff = Effective moment of inertia of rigidity of composite section, f_{cu} = Specified minimum compressive strength of concrete. f_{ysr} and

Table 6 Ultimate axial load under Experimental versus analytical predictions

Specimen	P_{test} (kN)	P_{EC4} (kN)	P_{ACI} (kN)	P_{AISC} (kN)	P_{test}/P_{EC4}	P_{test}/P_{ACI}	P_{test}/P_{AISC}
FEC	425	487.45	384.38	465.24	0.872	1.106	0.913

f_y are the respective specified minimum yield stress of reinforcing bars and steel shape. 'K' is the effective length factor and 'L' is the laterally unbraced length of the column.

P_{test}, P_{EC4}, P_{ACI}, and P_{AISC} denotes the ultimate load attained from the experimental results, from Euro code 4 (2004), ACI-318 (2014) and AISC LRFD (2010) guidelines, respectively. Comparison of experimental and analytical predictions are shown in Table 6. It was observed that the overall predicted capacities using these three approaches Euro code 4, ACI 318 and AISC LRFD were shows 12.8%, 10.6% and 8.7% error than the experimental results.

6 Conclusion

This study presents the experimental and numerical analysis of RC and FEC columns subjected to uniaxial compression. Ultimate axial load carrying capacity of column has increased due to the inclusion of steel tube in reinforced concrete. The ultimate column capacity, axial shortening at failure and failure pattern of the columns were accurately predicted using ABAQUS software. The comparison between the experimental and numerical results showed that the ABAQUS models can predict the load deformation behaviour of RC and FEC columns with good accuracy. Ultimate load carrying capacity was also obtained from three different codes namely Euro code 4 (2004), ACI-318 (2014) and AISC-LRFD (2010). Among these codes, the AISC-LRFD-2010 led to the most conservative value.

References

1. American Concrete Institute (2014) Building code requirements for structural concrete (ACI 318-14) and commentary. ACI 318-14, Farmington Hills, MI
2. Lai B, Richard Liew JY, Xiong M (2019) Experimental study on high strength concrete encased steel composite short columns. J Constr Build Mater 228:116640-(1–15)
3. Carreira DJ, Chu KH (1985) Stress strain relationship for plain concrete in compression. ACI J 82(6):796–804
4. Carreira DJ, Chu KH (1985) Stress-Strain relationship for reinforced concrete in tension. ACI J 83(3):21–28
5. Ellobody E, Young B (2010) Numerical simulation of concrete encased steel composite columns. J Constr Steel Res 67:211–222

6. EN 1994-1-1 (2004) Eurocode 4: Design of composite steel and concrete structures part 1–1, General Rules and Rules for Buildings
7. Specification for Structural Steel Buildings, (AISC-LRFD), American Institute of Steel Construction, Chicago, Illinois (2010)
8. Kartheek T, Venkat Das T (2020) 3D modelling and analysis of encased steel-concrete composite column using ABAQUS. J Mater Today Proc 133:106070-(1–10)

Prediction of Bearing Capacity and Settlement from SPT Value Using PSO-ANN

R. Saranya, N. Sankar, and S. Chandrakaran

Abstract Bearing capacity and settlement are two important parameters needed for foundation design. Many of the empirical and semi-empirical methods and equations available nowadays are inconsistent because of neglecting some actual field conditions and a few of them are coping with improper input factors. During this work, a hybrid PSO (Particle swarm optimization) based ANN (Artificial neural network) model approach is utilized for predicting bearing capacity and settlement of shallow foundations on C-soil, φ-soil and C-φ soil separately based on easily detectable field conditions including SPT (Standard penetration test), which may be used for a large number of cases with high reliability and accuracy. The PSO-ANN models are developed using actual field data collected from different parts of Kerala. Six models are developed which can predict bearing capacity or settlement for C-soil, φ-soil, or C-φ soil. The predicting capability of models was checked with unseen data and located to be better. The models were compared with the performance of usually used traditional methods, separately, and were also found to be superior.

Keywords Bearing capacity · Settlement · Standard penetration test (SPT) · Artificial neural network (ANN) · Particle Swarm Optimization (PSO)

1 Introduction

Before starting any construction works on soil, bearing capacity should be calculated to reduce its risk of high settlement [1]. So every foundation design requires satisfying two major criteria, i.e. ultimate bearing capacity and limiting settlement of the foundation [2]. Different methods are there to determine bearing capacity like

R. Saranya (✉) · N. Sankar · S. Chandrakaran
Department of Civil Engineering, National Institute of Technology, Calicut, India
e-mail: sankar@nitc.ac.in

S. Chandrakaran
e-mail: chandra@nitc.ac.in

G. C. Marano et al. (eds.), *Proceedings of SECON'21*, Lecture Notes in Civil Engineering 171, https://doi.org/10.1007/978-3-030-80312-4_98

presumptive analysis, analytical method, field test like plate bearing test, centrifuge test, etc. The conventional way of predicting bearing capacity and settlement requires detailed investigations like undisturbed soil sampling and laboratory testing of samples. These activities require considerable time, manpower, and cost. So we go for simple field tests like SPT to access the bearing capacity and settlement of the foundation.

There are many empirical and semi-empirical methods developed from time to time to find out the bearing capacity of shallow foundation in cohesionless soil from SPT value. Among them, methods suggested by Terzhagi, Peck, Meyerhoff, etc. are extensively used all over the world. Some are in graphical form and others are in the form of equations. But the drawback with these traditional methods is each correlation gives different values, moreover, a variety of corrections are to be applied to the observed SPT value before using it for estimation of bearing capacity. This will create confusion among the users regarding the choice and reliability of the correlations to be applied to an unknown problem. Also, theoretical methods are based on assumptions. To cope up with these difficulties an attempt has been made to develop prediction models for predicting the bearing capacity and settlement of soil using PSO-ANN [3–7].

2 Methodology

Input parameters used for bearing capacity prediction models are Standard Penetration Test values (N), Depth of water table (d), Bulk density of soil layer (γ), Cohesion (c), Angle of friction (φ), Moisture Content (m). For the settlement prediction model, the input parameters are Standard Penetration Test values (N), Bulk density of soil layer (γ), Cohesion (c), Angle of friction (φ), bearing capacity of the soil layer (q_u). The dataset used for calibration and validation of the model was collected from bore log data and other detailed reports of various sites in different parts of Kerala. Flowchart of neural networks generated for the bearing capacity prediction model and settlement prediction model is as shown in Figs. 1 and 2 respectively. Bearing capacity and settlement models were developed separately for C-Soil, φ-Soil, and C-φ soil for the ease of comparison of results with conventional methods.

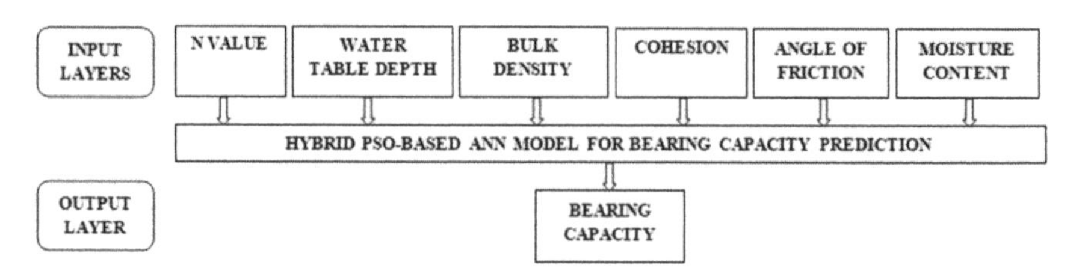

Fig. 1 Neural network for bearing capacity

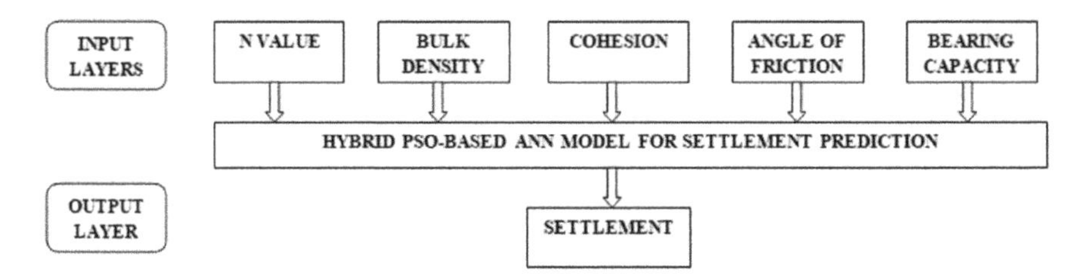

Fig. 2 Neural network for Settlement

The next important step in PSO-ANN modeling is dividing the available data into two subsets such as training and testing set. To develop a strong model, researchers suggest that all the patterns contained within the training dataset should be contained in the testing set also. To achieve this the whole data were divided into two statically consistent sets, 80% of data for training, and the remaining 20% for testing and validation such that both the subset possesses similar statistical properties including mean, standard deviation, maximum, and minimum [8]. The data division is as shown in Table 1. The PSO-ANN prediction models were developed using MATLAB R2020b software. In MATLAB there are many add-on packages to develop a deep learning MLP(Multi-Layer Perceptron) neural network using conventional algorithms like gradient descent, Levenberg- Marquardt algorithm, Bayesian Regularization, Polak-Ribiere Conjugate Gradient, etc. for training. In place of these conventional backpropagation algorithms, in this project, the PSO technique has been employed for training the neural network by integrating PSO program code with the existing neural network algorithms in the MATLAB engine.

The PSO code is iterated 20,000 times for best-fit weights and biases. The iteration is continued either 20,000 times or till the tolerance reaches the value 1×10^{-10}. After the swarm is initialized, the velocity is updated and the positions of particles are updated for the maximum number of iterations. The particle searches the search space between the values −1.5 and 1.5 for the optimum values of weights and biases. Table 2 shows the features of developed models.

Table 1 Data division

Item	Training data	Testing data	Total
C soil	262	66	328
C-φ soil	250	62	312
φ soil	108	27	135
Total			775

Table 2 Features of developed PSO-ANN model

Parameters	
Number of neurons in the hidden layer (n) Swarm size (pop)	10 100
Cognitive coefficient-(Personal learning factor, C1)	1.5
Cognitive coefficient-(Social learning factor, C2)	2.5
Inertia weight (w)	0.1 + rand*0.4
Maximum iteration	20,000
Space search (x_{min}, x_{max})	−1.5, 1.5

3 Results

There are six PSO-ANN models developed to predict bearing capacities and settlement for C-soil, φ-soil, and C-φ soil. To evaluate the performance, regression plots were generated for training and testing datasets. Based on the R-value the reliability of the model was interpreted. Regression plot of bearing capacity model of C-soil, C-φ soil, φ soil for training and the testing dataset is shown in Figs. 3a, b, 4a, b and 5a, b respectively.

The performance evaluation measures like MSE (Mean Square Error), and R (Correlation Coefficient) values were used to evaluate the prediction capability of models. MSE, R values for all the three bearing capacity prediction models are as shown in Table 3. The correlation coefficient (R) evaluates the linear correlation between the actual and predicted ultimate bearing capacity, which is good for PSO-ANN models, for training as well as testing data. The mean square error (MSE) which gives the quantitative indication of the model error in the units of the variable is low for the developed PSO-ANN model [2].

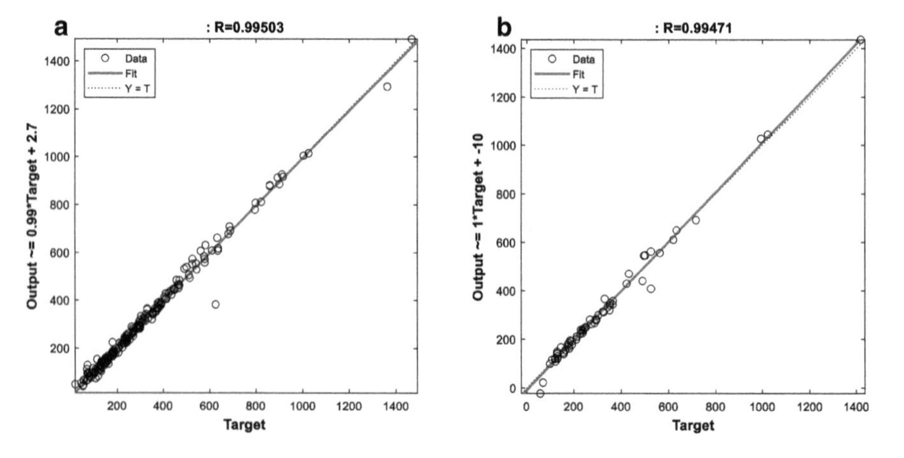

Fig. 3 Regression plot **a** C soil-Training, **b** C soil-Testing

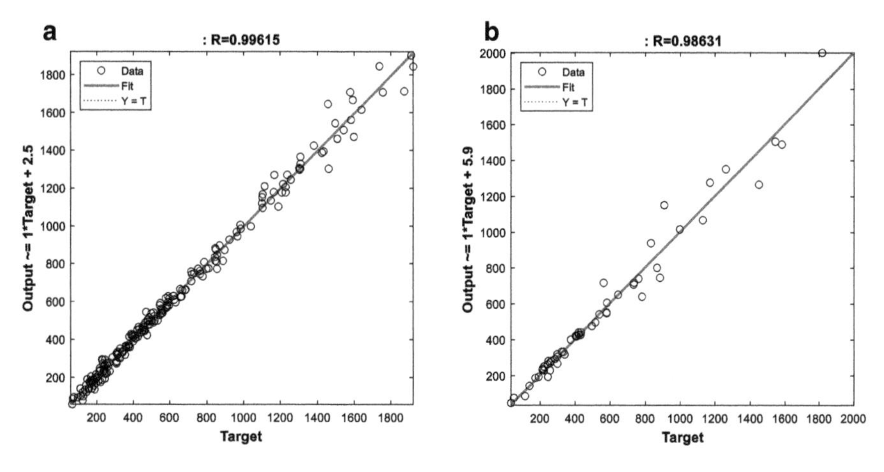

Fig. 4 Regression plot **a** C-φ soil-Training, **b** C-φ soil-Testing

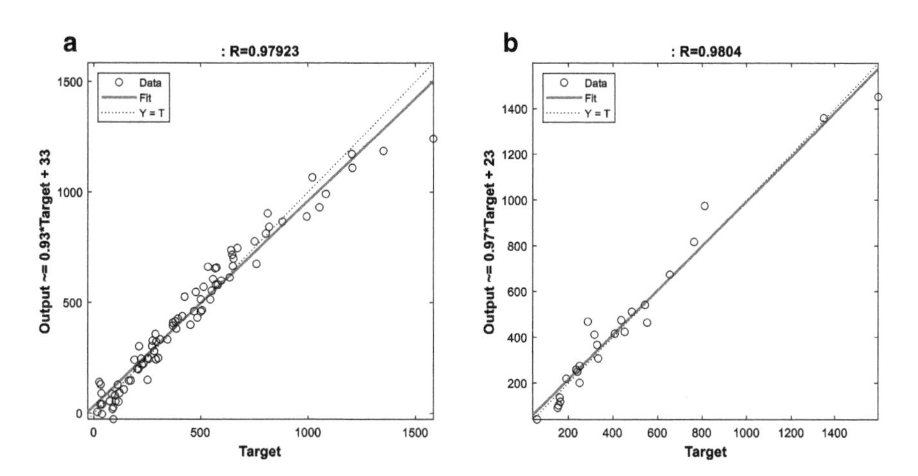

Fig. 5 Regression plot **a** φ soil-Training, **b** φ soil-Testing

Table 3 Results of developed bearing capacity prediction models

Model	Dataset	R	MSE
Bearing capacity model for C-Soil	Training	0.99503	1.8700
	Testing	0.99471	8.3500
Bearing capacity model for φ-Soil	Training	0.99615	8.3500
	Testing	0.98631	0.4700
Bearing capacity model for C-φ soil	Training	0.97923	1.0400
	Testing	0.98040	0.0035

Table 4 Results of developed settlement prediction models

Model	Dataset	R	MSE
Settlement model for C-Soil	Training	0.83348	0.1716
	Testing	0.98432	0.0788
Settlement model for φ-Soil	Training	0.83134	0.0583
	Testing	0.99979	0.0005
Settlement model for C-φ Soil	Training	0.81401	0.0682
	Testing	0.99942	0.0002

Performance of settlement model of C soil, C-φ soil, and φ soil for training and the testing dataset was also plotted. Settlement prediction model developed yield coefficient of correlation value in the range of 0.8 for the training set and 0.9 for testing and validation dataset. The performance evaluation measures like MSE, R values for settlement prediction models are as shown in Table 4.

The results obtained from these PSO-ANN models were compared with that of conventional methods to calculate bearing capacity and settlement, separately for C soil, φ soil, and C-φ soil, and the error was calculated for each data, as the difference between actual values and the values given by each of the conventional methods [9, 10]. These methods were selected for comparison as the database used for model development supports the use of these equations. From the results, it can be showed that all the six PSO-ANN models give lesser RMSE (Root Mean Square Error) value and high R-value thus proving the predictive capability and accuracy of the developed model than any other conventional method. A comparison of the performance of the PSO-ANN model with other methods in terms of the R and RMSE is shown in Table 5.

4 Conclusions

Bearing capacity and settlement calculation using conventional methods involve the use of equations and graphs which is time-consuming and tedious. To avoid such an effort, soft computing programs can be utilized. Hybrid Particle Swarm Optimization algorithm along with Multi-Layer Perceptron Neural Network can be utilized for the accurate prediction of bearing capacity and settlement. The objectives of the paper were, to check the applicability of PSO-ANN in predicting ultimate bearing capacity and settlement of the shallow foundation, and to undertake a comparative study with the commonly used conventional theories.

The results indicate that the PSO-ANN model can predict well the bearing capacity and settlement of shallow foundation and it found to significantly outperform the traditional methods. This was proved by the R and MSE values used for evaluating the model. The coefficient of correlation (R) values for the developed bearing capacity prediction model ranged from 0.99, 0.98–0.99, 0.97–0.98 for C soil, C-φ soil, φ soil respectively. This indicates that the PSO-ANN models

Table 5 Comparison with traditional methods

Model name	R	RMSE
Bearing capacity model for C-Soil		
PSO-ANN model	0.99	2.38
Prandtl's method (1921)	0.99	25.17
Ritter's method	0.99	80.16
Skempton's method (1951)	0.99	74.89
Teng's method	0.29	283.74
Bearing capacity model for φ-Soil		
PSO-ANN model	0.99	1.3900
Peck's method (1974)	0.38	4206.63
Teng's method (1962)	0.34	4100.53
Meyerhoff's method	0.38	4173.88
Bowle's method (1977)	0.35	4117.99
IS code method	0.39	4195.96
Ritter's method	0.99	2761.62
Bearing capacity model for C-φ soil		
PSO-ANN model	0.98	0.388
Hansen's theory (1961)	0.79	4549.59
Vesic's theory (1973)	0.91	2710.31
IS code method	0.88	3080
Settlement model for C-Soil		
PSO-ANN model	0.91	0.0295
Kaniraj's method (1974)	−0.10	0.16
Settlement model for φ-Soil		
PSO-ANN model	0.94	0.10
IS code method	0.92	0.13
Teng's method (1962)	0.54	0.13
Meyerhoff's method (1956, 1974)	0.88	0.13
Bowle's method (1982)	0.88	0.13
Peck's method (1974)	0.93	0.13
Settlement model for C-sφ soil		
PSO-ANN model	0.93	0.09
Bowle's method (1977, 1982)	0.93	0.13
Schultze and Sherif method (1973)	0.93	0.13
Meyerhoff's method (1974)	0.93	0.13

predicted the output near to the actual value. Also, MSE values were near equal to zero for the developed models whereas it is very high for most of the traditional methods. The developed models can be easily updated with data of soil outside Kerala and India which remain as the scope of work.

References

1. Moayedi H (2017) Comparison analysis of bearing capacity approaches for the strip footing on layered soils. Arab J Sci Eng 42:3711–3722
2. Padmini D (2008) Ultimate bearing capacity prediction of shallow foundations on cohesionless soils using neuro-fuzzy models. Comput Geotech 35:33–46
3. Harandizadeh H (2019) A new development of ANFIS–GMDH optimized by PSO to predict pile bearing capacity based on experimental datasets. Eng Comput
4. Rezaei H (2016) Bearing capacity of thin-walled shallow foundations: an experimental and artificial intelligence-based study. Appl Phys Eng 17(4):273–285
5. Momeni E (2014) Prediction of pile bearing capacity using a hybrid genetic algorithm-based ANN. Measurement 57:122–131
6. Kuo YL (2009) ANN-based model for predicting the bearing capacity of strip footing on multi-layered cohesive soil. Comput Geotech 36:503–516
7. Jin LX (2020) Ultimate bearing capacity of strip footing on sands under inclined loading based on improved radial movement optimization. Eng Optim
8. Alkroosh I (2014) Predicting pile dynamic capacity via application of an evolutionary algorithm. Soils Found 54:233–242
9. Kaniraj SR (1988) Design aids in soil mechanics and foundation engineering. Tata Mcgraw Hill, New York
10. Debnath L (2019) Pseudo-static bearing capacity analysis of shallow strip footing over two-layered soil considering punching shear failure. Geotech Geol Eng 37:3749–3770

Comparative Study of Concrete Models in OpenSEES for Performing Nonlinear Analysis

N. Sarkar and K. Dasgupta

Abstract Acknowledging the strength deterioration of Reinforced Concrete (RC) structures under nonlinear loading, along with the structural response at the ultimate load, the response throughout the analysis is of due importance for reliable nonlinear dynamic analysis. Hence, while selecting the material models for structural modelling for nonlinear analysis, it is required to strike a balance between computational efficiency and realistic representation of material response. Here, an eight storeyed RC wall-frame building, designed according to Indian design codes IS-456:2000 and IS-13920:2016, is analysed through Incremental Dynamic Analysis (IDA). Three different structural models are modelled and analysed using three different concrete models using the open source computer program OpenSEES, viz. Concrete01, Concretre02 and Concrete07. The IDA results are compared in order to study the merits and demerits of the considered concrete models. It is found that Concrete02 ensures better numerical stability than Concrete07 and also predicts the dynamic response of concrete better than Concrete01.

Keywords Concrete · Incremental Dynamic Analysis · RC structural wall · OpenSEES

1 Introduction

In the recent times, Performance-Based Earthquake Engineering (PBEE) has been widely used in the field of structural engineering. In PBEE, seismic response of a structure is used to improve the design of the structure for future seismic scenarios, rather than designing based on empirical code formulations. One of the key components of PBEE is the vulnerability assessment of the structural systems. Nonlinear dynamic analysis of the structures forms an indispensable part of the vulnerability studies. The reliability of nonlinear dynamic analysis is associated

N. Sarkar (✉) · K. Dasgupta
Department of Civil Engineering, Indian Institute of Technology, Guwahati 781039, India
e-mail: nabajit.sarkar@iitg.ac.in

G. C. Marano et al. (eds.), *Proceedings of SECON'21*, Lecture Notes in Civil Engineering 171, https://doi.org/10.1007/978-3-030-80312-4_99

with appropriate modelling of the structural models and performing the analysis in a suitable analytical platform. The material properties of concrete and steel need to be well defined for appropriate structural modelling. The aim of the present study is to investigate the modelling approach of concrete material in finite element analysis using the open source program OpenSEES [1].

In the past, several studies have been conducted to investigate the appropriate consitutive model for numerical simulation of dynamic behaviour of concrete. Sinha et al. [2] examined the stress–strain relationship for unconfined concrete when subjected to cyclic compressive loading. Karsan and Jirsa [3] have carried out an experimental research on the modelling of unconfined concrete under compressive load. Mander et al. [4] formulated the stress–strain relationship for both confined and unconfined concrete under compressive and tensile loads. Chang and Mander [5] proposed the modelling approach to include hysteretic characteristic of both confined and unconfined concrete. The stress–strain material models available in OpenSEES are largely based upon these recommendations. However, certain simplifications proposed in other studies have also been incorporated for saving computational time. Sosa and Cazia [6] compared two concrete models available in OpenSEES for simulating the dynamic behaviour of a pile-slab connection.

In the present study, Incremental Dynamic Analysis (IDA) of an eight storeyed Reinforced Concrete (RC) wall-frame building is performed in OpenSEES, modelled using three different concrete models, namely Concrete01, Concrete02 and Concrete07. The IDA results for the considered building typology are compared to investigate the advantages and disadvantages of the different concrete models.

2 Structural Modelling and Dynamic Analysis

2.1 Building Details

The structure investigated in this study is an eight storeyed RC wall-frame building with its location in Seismic Zone V as per Indian Earthquake Standard IS: 1893 (Part 1)—2016. The plan of the building is shown in Fig. 1. The dimensions of the columns and beams are 400 mm $\times$ 400 mm and 230 mm $\times$ 400 mm respectively. The thickness of slabs and RC walls are 150 mm and 200 mm respectively. The grade of steel and concrete used are Fe500 and M30 respectively. The reinforcement details of all the members are shown in Fig. 2.

2.2 Finite Element Models

In this study, an exterior 2-dimensional plane frame with RC wall, extracted from the building, is modelled using OpenSEES. The structural elements are modelled as

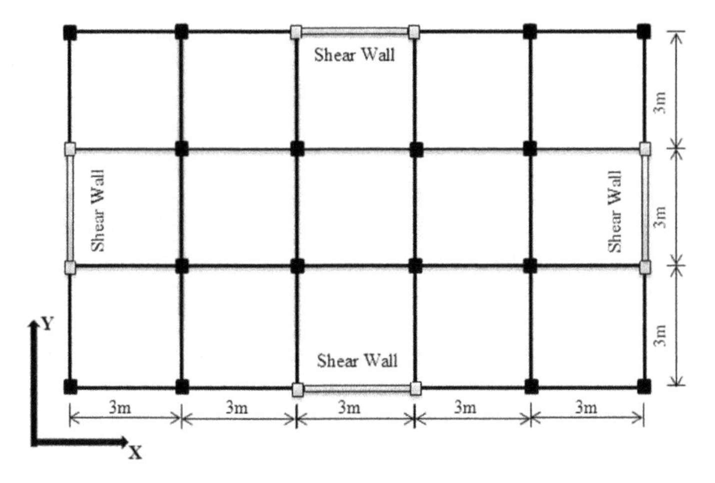

Fig. 1 Plan of the building

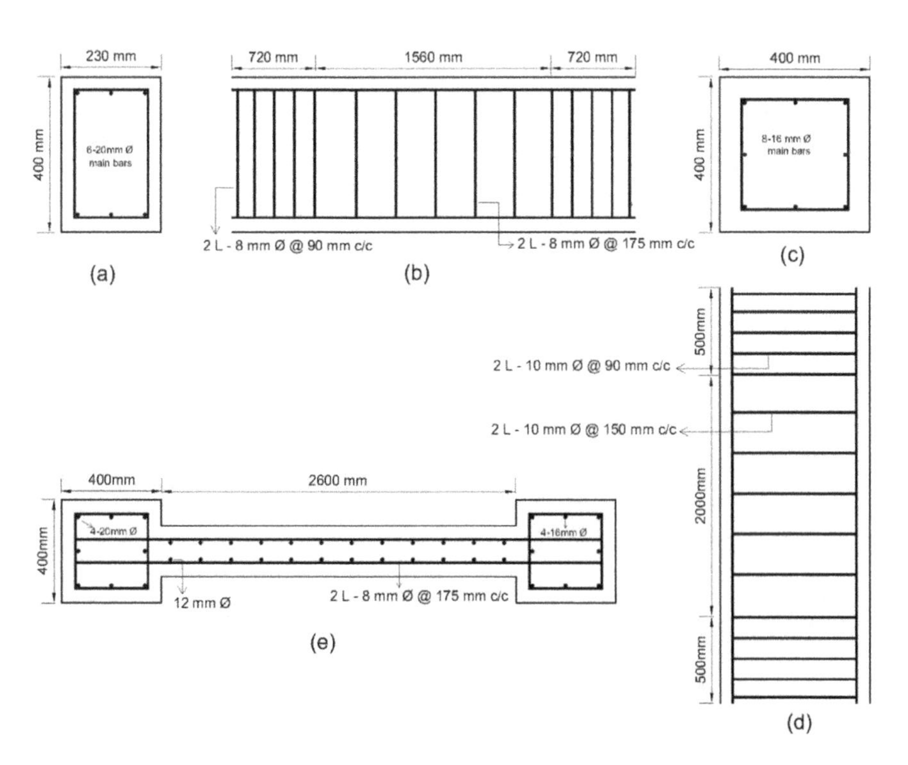

Fig. 2 Reinforcement details **a** Beam cross section; **b** Beam longitudinal section; **c** Column cross section; **d** Column longitudinal section; and **e** RC wall cross section

'NonLinear Beam—Column Elements' to capture the nonlinear response of the members, which incorporates propagation of plasticity along the element length. Fibre cross-sectional modelling approach is adopted, whereby the cross-sections are formed by the division into layers, which represent the concrete part of the section and the reinforcement. The RC structural walls are modelled employing the common method of equivalent frame element [7]. The reinforcement bars are modelled using the Steel02 material model (Fig. 3a) in OpenSEES, which utilizes the Giuffre-Menegotto-Pinto steel material model with isotropic strain hardening. For modelling the stress–strain relationship of concrete, three different cases are considered each with different concrete model, namely Concrete01, Concrete02 and Concrete07. Both Concrete01 and Concrete02 are based on the concrete model proposed by Kent-Scott-Park [8], but Concrete01 does not consider the tensile strength of concrete (Fig. 3b) and Concrete02 does consider it in the modelling approach (Fig. 3c). Concrete07 is based on the Chang and Mander concrete model [5] with modified unloading and reloading curves and without the original shift in the tension envelope for simplicity (Fig. 3d).

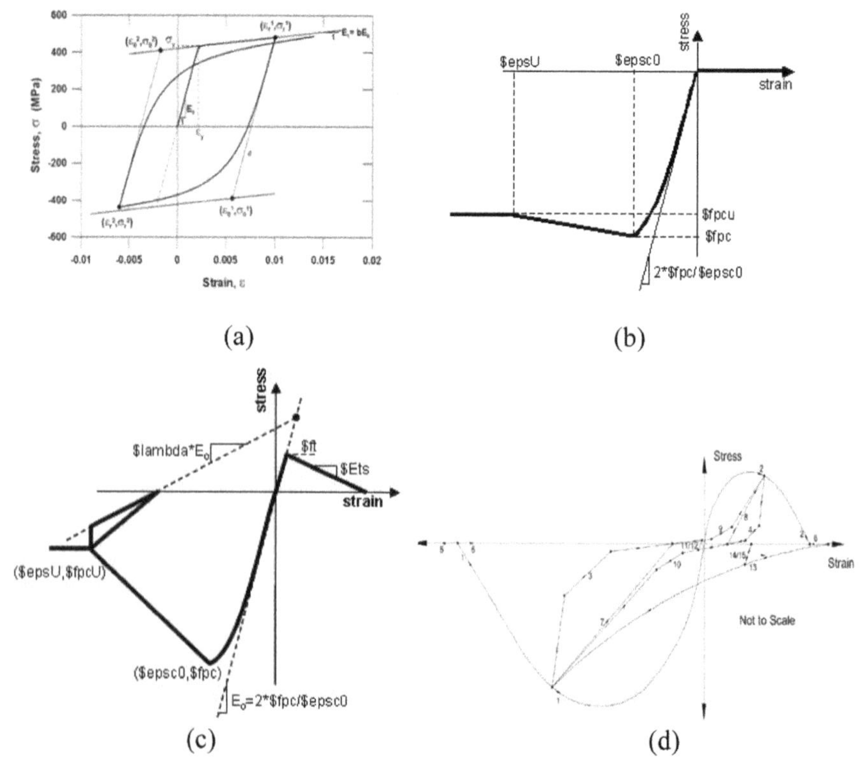

Fig. 3 Hysteretic stress–strain relation for **a** Steel02; **b** Concrete01; **c** Concrete02; and **d** Concrete07 [1]

2.3 Nonlinear Dynamic Analysis

Estimation of seismic repsonse through dynamic analysis of strucutres requires high computational effort and proper understanding of the behaviour of the structure under dynamic forces. Some of the popular techniques for performing dynamic analysis of structures are: multi-stripe analysis [9]; cloud analysis [9]; incremental dynamic analysis (IDA) [10]. IDA aids in understanding the seismic response under different levels of seismic intensity corresponding to multiple damage states in a structure. In this study, IDA is performed for the considered wall-frame structure using the hunt and fill algorithm [11] for scaling of record. IDA results are represented in terms of the relationship between Engineering Demand Parameter (EDP) and Intensity Measure (IM). EDPs quantify the amount of structural and nonstructural damages in the structures. IM quantitatively represents the principal characteristics of ground motion that affect the dynamic response of a structure. Here, the input ground motion considered for performing IDA is the record from earthquake which occurred in Christchurch, New Zealand (2011) with a magnitude of 6.2. The record is obtained from the PEER (Pacific Earthquake Engineering Research) Centre Ground Motion Database [12]. The widely used EDP and IM are employed here, namely inter storey drift (ISD) and spectral acceleration at the fundamental period (S_a (T1, 5%)) respectively. In-order to perform the IDA for the considered structure the analysis parameter adopted in this study are presented in Table 1.

3 Results and Discussion

The main objective of this study is to investigate the difference in the seismic response of the considered building typology due to modelling of the structure using three different concrete models in OpenSEES. The IDA is performed for an exterior wall-frame of the considered building in the shorter span and the results are shown with respect to the EDP and IM (Fig. 4).

Table 1 Analysis parameters for IDA in OpenSEES

Numberer type	System type	Convergence test type	Tolerance value	Max no. of iteration	Solution algorithm	Integrator object
Reverse Cuthill-Mckee	Bandgeneral Linear system of equation	Energy Increment Test	10^{-8}	10	Modified Newton	Newmark

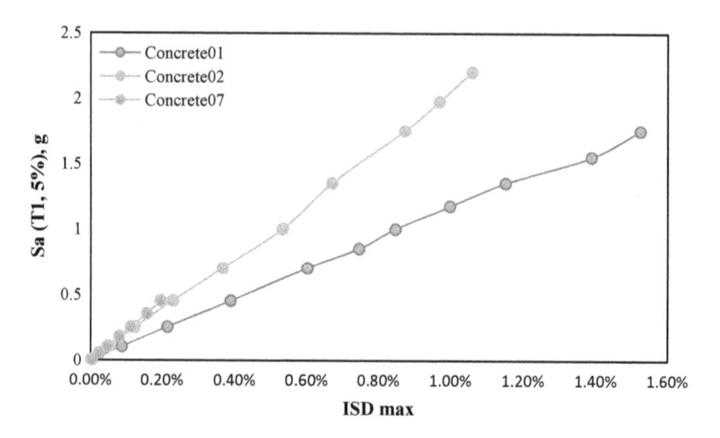

Fig. 4 IDA curve for structure modelled using three different concrete model

The Chang and Mander concrete model [5] used in Concrete07 tend to show more realistic response as compared to that of Kent-Scott-Park concrete model [8] used in both Concrete01 and Concrete02, as depicted in Fig. 3. Closer to the true response of concrete under dynamic loading, Concrete07 uses tri-linear curves for defining unloading and reloading paths. Whereas, in order to achieve computational efficiency and numerical stability, both Concrete01 and Concrete02 employ linear paths for unloading and reloading stiffness. Concrete01 is further simplified and ignores the tensile strength of concrete but is considered in modelling of Concrete02. Consequently, as seen in Fig. 4, the IDA for structure modelled with Concrete07 could be performed till a ISD of about 0.2%, but for the model using Concrete01 and Concrete02 the analysis could be performed till a much higher value of ISD. In order to understand this variation in response of the structural models due to different concrete models, the material response at the fibre cross-sectional level is analysed.

The hysteretic stress–strain output of core concrete and steel bars located at the extreme ends at the base of the RC structural wall is studied. The input ground motions records are scaled in IDA such that the structure tends to reach final stage of global dynamic instability, hence the analysis is expected to run much beyond the yielding of the steel bars. The material responses in the final run, just before the occurrence of convergence related issues, are shown in Figs. 5, 6 and 7.

The steel reinforcement stress–strain response (Fig. 7b) for the structure modelled with Concrete07 indicates that the analysis ceases to run at the onset of bar yielding, whereas IDA is designed to run till the model experiences instability due to global collapse. For the structure modelled with Concrete07, even though realistic stress–strain relation (Fig. 7a) for core concrete is exhibited but the steel stress–strain curve indicates ineffiency in performing the entire analysis. The IDA is performed accurately for the structure modelled using Concrete01 and Concrete02 (Figs. 5b and 6b), but the concrete stress–strain response (Fig. 5a) for the case of

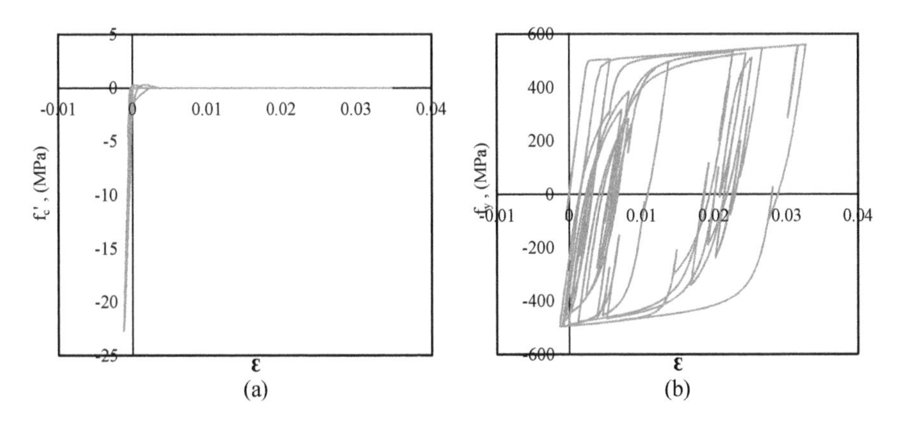

Fig. 5 Hysteretic stress-stain response of **a** concrete; **b** steel for the structure modelled with Concrete01

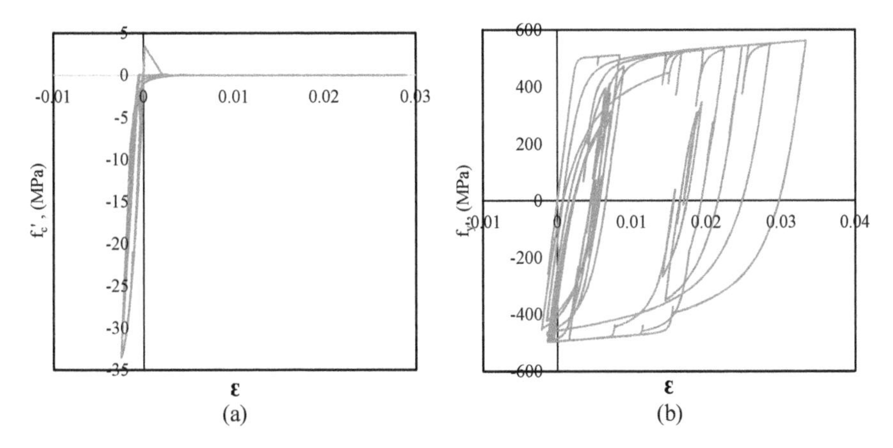

Fig. 6 Hysteretic stress-stain response of **a** concrete; **b** steel for the structure modelled with Concrete02

Concrete01 does not incorporate the tensile strength of concrete, which is included for the case of Concrete02 (Fig. 6a).

4 Conclusion

In this study, IDA is performed for an eight storeyed RC wall-frame building modelled using three different concrete models, namely Concrete01, Concrete02 and Concrete07 in OpenSEES. The IDA results are examined to understand the advantages and disadvantages for the three different concrete models. The

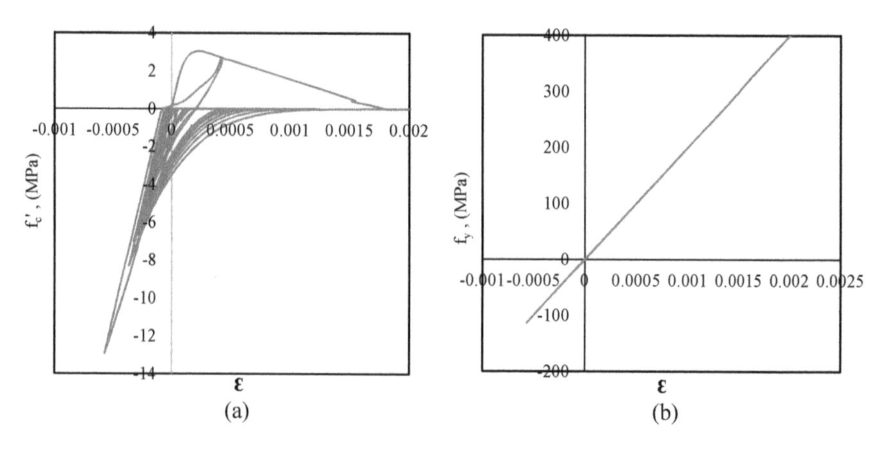

Fig. 7 Hysteretic stress-stain response of **a** concrete; **b** steel for the structure modelled with Concrete07

Concrete07 model is capable of simulating realistic behaviour of concrete under dynamic loading but due to numerical issues, the entire IDA for the structural model could not be performed. The structural models using Concrete01 and Concrete02 are able to perform the IDA more accurately but compromises in reflecting the true dynamic response of concrete, as simplifications are made to achieve computational efficiency and numerical stability. Concrete02 model proves to be advantageous as compared to Concrete01 model, due to its ability to consider tensile strength of concrete for modelling of concrete. Thus Concrete02 model serves in providing the perfect balance in between Concrete07 and Concrete01 models, in terms of computational efficiency when compared to Concrete07 model and realistic material behaviour when compared to Concrete01 model.

References

1. Mazzoni S, McKenna F, Scott MH, Fenves GL (2006) OpenSees command language manual. Pacific Earthquake Engineering Research (PEER) Center
2. Sinha BP, Gerstle KH, Tulin LG (1964) Stress-strain relations for concrete under cyclic loading. ACI J Proc 61(2):195–211
3. Karsan ID, Jirsa JO (1969) Behavior of concrete under compressive loading. J Struct Div ASCE 95(12):2543–2564
4. Mander JB, Priestley MJN, Park R (1988) Theoretical stress strain model for confined concrete. J Struct Eng ASCE 114(8):1804–1825
5. Chang G, Mander J (1994) Seismic energy based fatigue damage analysis of bridge columns: part I—evaluation of seismic capacity, NCEER technical report 94-0006. National Center for Earthquake Engineering and Research, State University of New York at Buffalo, NY, USA
6. Sosa D, Cazia D (2015) Cyclic nonlinear behavior of pile-deck connections using structural software models. Open Civil Eng J 236–248

7. Agarwal P, Shrikhande M (2006) Earthquake resistant design of structures. Prentice-Hall of India Pvt. Ltd., New Delhi, India
8. Scott BD, Park R, Priestley MJN (1982) Stress-strain behavior of concrete confined by overlapping hoops at low and high strain rates. ACI J
9. Jalayer F, Cornell CA (2009) Alternative non-linear demand estimation methods for probability-based seismic assessments. Earthq Eng Struct Dynam 38(8):951–972
10. Vamvatsikos D, Cornell CA (2002) Incremental dynamic analysis. Earthq Eng Struct Dynam 31(3):491–514
11. Vamvatsikos D, Cornell CA (2004) Applied incremental dynamic analysis. Earthq Spectra 20(2):523–553
12. PEER Homepage, http://peer.berkeley.edu

Effect of Weld Spacing on CFST Column Stiffened with Vertical Plates

T. N. Sandeep and Job Thomas

Abstract Concrete-filled steel tube (CFST) columns are increasingly used worldwide because of their various advantages over conventional RCC or steel columns. The load carrying capacity, as well as deformation characteristics of these columns, can be improved by providing supplementary confinement for the core concrete. The effectiveness of vertical stiffener plates placed internally for circular CFST columns on confinement of concrete core is determined. The influence of spacing of the intermittent welding on the column confinement and early tube buckling is studied. The weld spacing varied from 10 to 50t where 't' is the tube wall thickness. Two different tube diameters are used in the experimental study. The results showed that the weld spacing of 10t gives the highest load holding for the column and the increase in capacity is around 21% when comparing to the control test specimen.

Keywords CFST column · Vertical stiffeners · Weld in the column

1 Introduction

Concrete-filled steel tube (CFST) columns are composite columns in which steel tubes are in-filled with concrete. CFST columns have numerous advantages compared to conventional steel or RCC columns [1–3]. The two parts of a CFST column, namely steel tube and core concrete help each other to improve the overall performance of the CFST column. Improved load carrying capacity as indicated pictorially in Fig. 1, higher ductility, construction easiness (as constructed without formwork) and time saving etc. makes the CFST column an excellent choice for multistoried buildings and in structures where heavy loading is expected.

T. N. Sandeep (✉) · J. Thomas
Department of Civil Engineering, Cochin University of Science and Technology, Kochi, Kerala, India

G. C. Marano et al. (eds.), *Proceedings of SECON'21*, Lecture Notes in Civil Engineering 171, https://doi.org/10.1007/978-3-030-80312-4_100

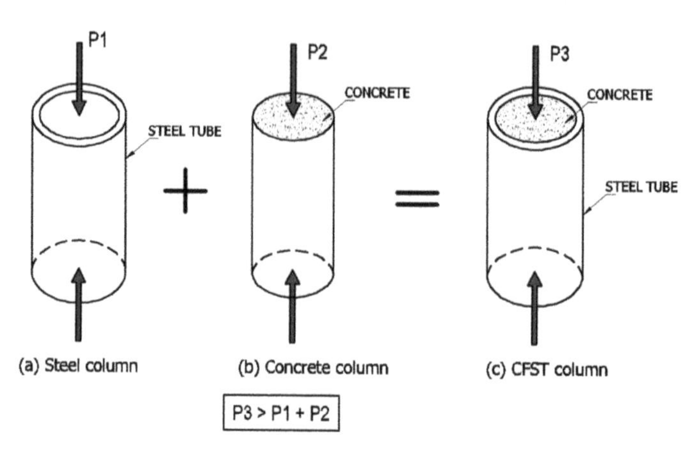

Fig. 1 Comparison of the load-carrying capacity of CFST column

The core concrete acts as a barrier against the in-ward buckling of the steel tube and thereby delaying the yielding of steel. On the other hand, the steel tube itself acts as confinement to the concrete. It is well known that confined concrete is having better mechanical properties compared to unconfined concrete and this property is well utilized in the CFST columns. The higher the confining pressure produced by the steel tube, the higher is the improvement in the concrete properties. Confining pressure depends mainly on the thickness and diameter of the steel tube, yield strength of steel and compressive strength of concrete [4]. A parameter known as 'confinement index' is commonly used to measure the degree of confinement of a CFST column.

The confinement for the CFST column can be further amplified by external confining methods suggested by researchers in the past [5–14]. The methods include the use of tie bars, spiral bars, steel jackets, through bars, FRP wrapping and use of stiffener plates. The method of use of stiffener plates is studied by Tao et al. [13] is shown in Fig. 2. Here the plates are welded continuously with the column plate either internally or externally. Fixing of the plates from inside is suitable for square CFST columns. Thus the column plates are strengthened and offer higher confinement to the concrete. This method is not suitable for circular columns due to the difficulty in the welding process from inside of the column. Thomas and Sandeep [15] proposed a new method for welding stiffener plates to circular CFST columns which are very effective from a fabrication point of view. The method of attaching steel plates to circular steel tube is shown in Fig. 4. The steel plates are intermittently welded from outer surface through predrilled holes made on the tube surface.

The present study focus on the effect of weld spacing of stiffener plates on the performance of short circular CFST columns based on experimental test results on 11 specimens.

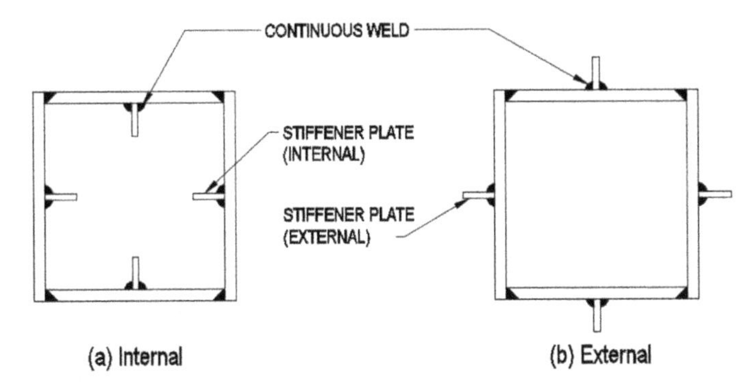

Fig. 2 Method of use of stiffener plates to strengthen square CFST column [13]

2 Materials and Testing

2.1 Concrete

Concrete is prepared using locally available materials such as crushed aggregates, manufactured sand and Portland Pozzolana Cement. Water reducing admixture is added to produce workable concrete which is essential for CFST columns. Tremie concreting method, which is commonly used for pile concreting, is suitable for the CFST columns in the site. The mix proportion of concrete used in the study is given in Table 1.

Twelve cube specimens of size 150 × 150 × 150 mm are cast using the same concrete used for column casting and kept for curing. The compressive strength of the cube specimens after 28 days is tested and the average value is found to be 35.68 MPa. A high level of workability is maintained for the mix by keeping the slump value of 100 mm. This is achieved by adjusting the super plasticizer dosage suitably.

2.2 Steel

The strength of the steel tube is tested using a tensile coupon test conducted in a Universal Testing Machine of capacity 400kN as per IS 1608 [16]. 3 numbers of test specimens of required size and shape are cut from the parent steel tube. The

Table 1 Concrete mix proportion for one cubic meter of concrete

Cement	Super plasticizer	Fine aggregate	Coarse aggregate		Water
			20 mm	10 mm	
425 kg	3.8 kg	562 kg	780 kg	320 kg	178.5 kg

average yield stress and ultimate strength are found to be 260.8 MPa and 366.0 MPa respectively.

2.3 Specimen Preparation

Two different diameter steel tubes are used in the experimental study such as 117.6 and 90 mm outer diameter. The thickness of steel is measured at different places and the average thickness is found to be 2.44 mm. The height of the column specimen is taken as 360 mm and 250 mm respectively for 117.6 mm and 90 mm diameter columns. Steel tubes are cut to a length slightly higher than column height and the height is exactly brought to the required value by grinding the extra length using a grinding machine. Thus, the edges are leveled to machine finishing thereby ensuring full contact between the specimen with the loading plates of the compression testing machine during testing. 11 specimens are prepared out of which two are control specimens without having vertical stiffener plates attached to them and the remaining are attached with stiffener plates. The number of stiffener plates kept as four numbers in all specimens to bring symmetry among specimen. Details of the test specimens are provided in Table 2. A specimen is designated using its outside diameter and spacing of weld. For example, 'C117-10' indicates a specimen with a tube outer diameter of 117.6 mm and having weld spacing of 10 mm. Stiffener plates are plug welded to the steel tube at regular intervals along the length of the tube. Holes are drilled on the steel tube to facilitate the welding process easily. A sample specimen showing holes drilled is given in Fig. 3a.

Table 2 Details of specimens

Sl. No	Specimen designation	Outer diameter of tube D (mm)	Tube wall thickness t (mm)	Height H (mm)	f_c (MPa)	f_y (MPa)	Number of stiffeners	Spacing of weld S (mm)
1	C117	117.6	2.44	360	35.37	260.8	0	–
2	C117-10	117.6	2.44	360	35.37	260.8	4	24
3	C117-15	117.6	2.44	360	35.37	260.8	4	37
4	C117-25	117.6	2.44	360	35.59	260.8	4	61
5	C117-35	117.6	2.44	360	35.59	260.8	4	85
6	C117-50	117.6	2.44	360	35.59	260.8	4	122
7	C90	90	2.44	250	35.82	260.8	0	–
8	C90-10	90	2.44	250	35.82	260.8	4	24
9	C90-15	90	2.44	250	35.82	260.8	4	37
10	C90-25	90	2.44	250	35.93	260.8	4	61
11	C90-35	90	2.44	250	35.93	260.8	4	85

Fig. 3 **a** Specimen showing drilled holes. **b** Longitudinal stiffener plates

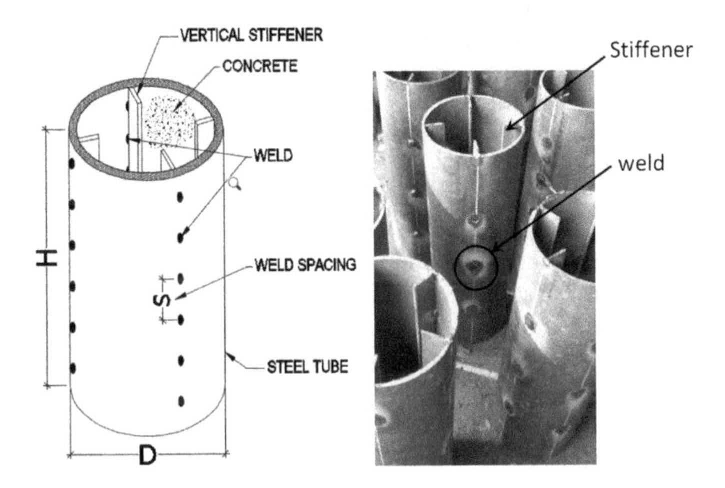

Fig. 4 Steel specimen with welded stiffener plates

The spacing of the holes varied as per the required weld spacing (S). The location of the holes is marked on the outer surface of tube wherever the welding is needed using a marker pen and a drilling machine is used to make holes of diameter approximately 6 mm. Care is taken to align the holes in the same longitudinal line so that the plate matches with all holes in that line. Welding is done through these holes to attach the stiffener plates to the inner side of the steel tube. Four stiffener plates are used with each steel tube. The stiffener plates having cross-section 30 × 2mm are cut from flat steel plate as shown in Fig. 3b. The length of the steel plate kept 2 mm less than the length of the steel tube. This tolerance ensures no projection outside tube length for the plates after fabrication thus facilitate full contact with the loading plate. The spacing of the intermittent weld used in the study (S) is 24 mm, 37 mm, 61 mm, 85 mm and 122 mm for 117 mm diameter specimens and 24 mm, 37 mm, 61 mm and 85 mm for 90 mm diameter specimens.

These values are equivalent to 10t, 15t, 25t, 40t and 50t respectively where 't' is the tube wall thickness. Steel specimens with longitudinal stiffeners welded are shown in Fig. 4.

Concrete is poured carefully to fill the steel tubes in an even manner. A needle vibrator is used to ensure complete compaction and to remove entrapped air. The top surface of the concrete is finished smoothly by keeping the level same as that of the steel tube. After 24 h all the specimens are immersed in water for curing.

3 Test Set-Up

Compression testing machine of capacity 2000 kN is used for loading the specimens. The test set-up is shown in Fig. 5. There are two loading plates each having a thickness of 50 mm to apply the compression loads uniformly in the cross-section of the column. The specimens are placed concentrically with the plates to receive loads without any eccentricity. The load is applied in a displacement-controlled manner. The rate of loading adopted is 0.3 mm/minute as followed by [17]. A seating load of 10 kN is applied to each specimen to avoid initial disturbance during loading and to have full contact with the loading plate. The deformation is noted corresponding to every 50 kN increment of axial load. Three dial gauges having least count 0.01 mm and stroke length of 25 mm placed symmetrically around the specimen to measure the axial deformation. The average of the three deformations is computed. All specimens are loaded either up to failure or up to an axial strain of 0.05 to study the post peak behavior of column specimens.

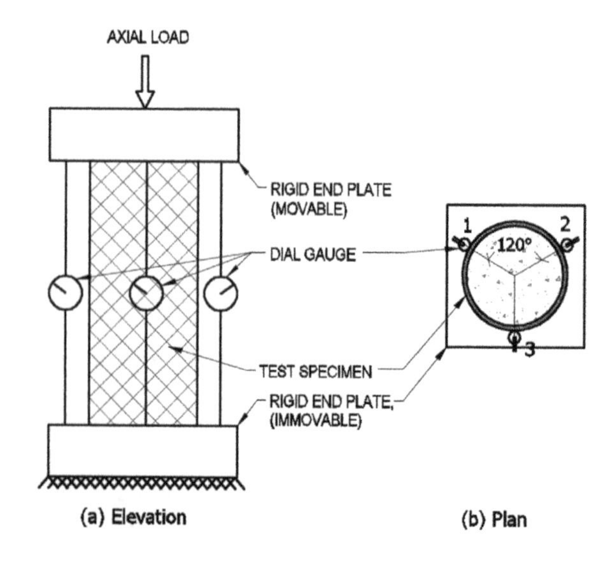

Fig. 5 Schematic of the test set-up

4 Results and Discussion

Load deformation: The specimens are tested after 28 days of curing in a Compression testing machine. The axial load-deformation curves are plotted for all specimens and are shown inFig. 6a, b. The specimen without stiffeners (control specimens C117 and C90) are having lesser load-carrying capacity. From the graph,

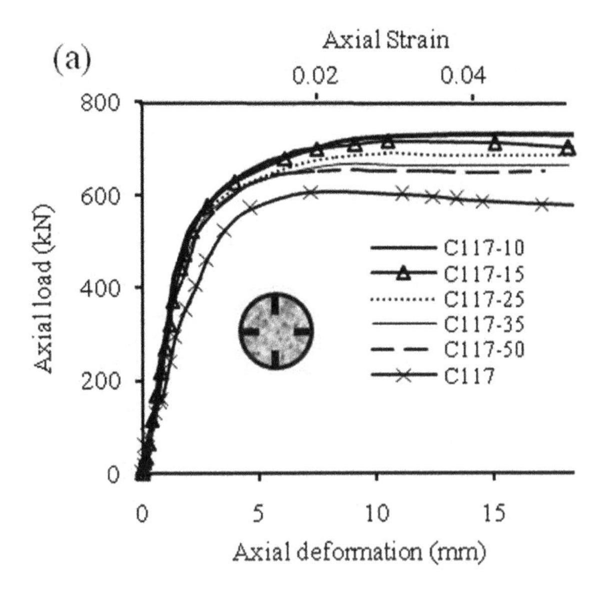

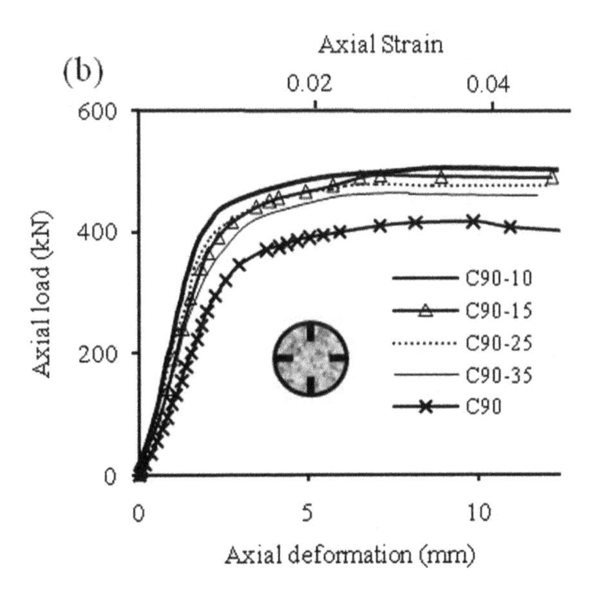

Fig. 6 Axial load—deformation of CFST columns, **a** results for 117 mm dia column, **b** results for 90 mm dia column

it is clear that the load-deformation curve remains almost the same for all specimens during the initial loading stage. The difference is predominant after reaching a certain amount of load. The stiffeners influence the load-deformation property of the CFST column. Axial load capacity or strength (N_u) is taken as the first peak load in the load-deformation curve in this study. Strength ratio is defined as the ratio between axial strength of the specimen that with the control specimen. The strength of the CFST column and the strength ratio is given in Table 3.

The initial stiffness of the CFST column is calculated using the initial slope of load-deformation curve as explained in [18]. Stiffness values are listed in Table 3. The ratio of the stiffness of a CFST specimen with that of the control specimen is termed as stiffness ratio. The presence of stiffeners increases the stiffness of the column. The minimum increase in stiffness is found to be 21% comparing to the control specimen. Stiffness increases with a decrease in weld spacing. The maximum increase is 49% and 80% for 117 mm and 90 mm diameter columns respectively.

Effect of weld spacing: The spacing between the intermittent fillet welding varies from 10 to 50t. The weld spacing is having an influence on the load resistance ability of the CFST column. As the spacing reduces the axial strength increases. The maximum increase is found to be 21% when comparing to the control specimen for both 117 mm and 90 mm diameter test specimens. Maximum load occurred for the column with stiffeners welded at a spacing of 10t. The variation of axial strength of specimens with weld spacing is given in Fig. 7. A second-order variation is observed. From the graph, it is clear that the rate of change of axial capacity is more at lesser weld spacing. This is due to the increased confinement offered by the welds when placing at nearer spacing.

Table 3 Strength and stiffness of CFST column specimens

Sl. No	Specimen Designation	N_u (kN)	Strength ratio	Stiffness (GPa)	Stiffness ratio
1	C117	605	1.00	18	1.00
2	C117-10	730	1.21	27	1.49
3	C117-15	714	1.18	26	1.39
4	C117-25	688	1.14	25	1.36
5	C117-35	666	1.10	23	1.27
6	C117-50	653	1.08	22	1.21
7	C90	416	1.00	20	1.00
8	C90-10	505	1.21	36	1.80
9	C90-15	492	1.18	28	1.40
10	C90-25	478	1.15	32	1.63
11	C90-35	462	1.11	26	1.34

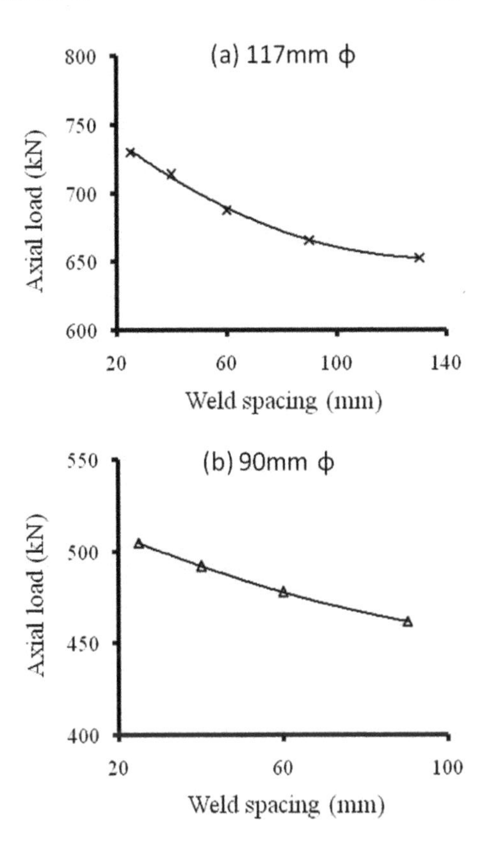

Fig. 7 Variation of Axial strength with weld spacing, **a** results for 117 mm dia column, **b** results for 90 mm dia column

5 Conclusions

The effect of intermittent welding of the stiffener plates in a steel tube on the performance of a CFST column is studied in detail in the present study. The experimental study conducted on 11 specimens having different weld spacing revealed the following conclusions.

- The intermittent welding of stiffener plates is an effective method to enhance the confinement pressure on the core concrete.
- Axial strength increases with a decrease in weld spacing. Maximum increase is found to be 21% corresponding to 10t spacing.
- The change in axial strength is 13% when the weld spacing is reduced from 50 to 10t.
- The stiffness of CFST column increases with a decrease in spacing of weld. The maximum increase in initial stiffness is found to be 49% and 80% respectively for 117 mm and 90 mm diameter columns.

It can be concluded that the increase in the axial capacity and deformation resistance is significant with the addition of vertical stiffeners in CFST columns. The spacing of weld of the vertical stiffeners is also important to increase the ultimate load carrying capacity of the CFST columns.

References

1. Uy B (1998) Ductility, strength and stability of concrete-filled fabricated steel box columns for tall buildings. Struct Des Tall Build 7(2):113–133
2. Schneider S (1998) Axially loaded concrete-filled steel tubes. J Struct Eng ASCE 125(10): 1125–1138
3. Sakino K, Nakahara H, Morino S, Nishiyama A (2004) Behavior of centrally loaded concrete-filled steel-tube short columns. J Struct Eng ASCE 130(2):180–188
4. Ho JCM, Lam JYK, Kwan AKH (2010) Effectiveness of adding confinement for ductility improvement of high-strength concrete columns. Eng Struct 32(3):714–725
5. Ho JCM, Lai MH (2013) Behaviour of uni-axially loaded CFST columns confined by tie bars. J Constr Steel Res 83:37–50
6. Lai MH, Ho JCM (2014) Confinement effect of ring confined concrete-filled steel tube columns under uni-axial load. Eng Struct 67:123–141
7. Lai MH, Ho JCM (2015) Effect of continuous spirals on uni-axial strength and ductility of CFST columns. J Constr Steel Res 104:235–249
8. Lai H, Ho JCM (2015) Axial strengthening of thin-walled concrete-filled-steeltube columns by circular steel jackets. Thin-Walled Struct 97:11–21
9. Hu YM, Yu T, Teng JG (2011) FRP-confined circular concrete-filled thin steel tubes under axial compression. J Compos Constr 15(5):850–860
10. Dabaon M, El-Khoriby S, El-Boghdadi M, Hassanein MF (2009) Confinement effect of stiffened and unstiffened concrete-filled stainless steel tubular stub columns. J Constr Steel Res 65:1846–1854
11. Dabaon M, El-Boghdadi M, Hassanein MF (2009) Experimental investigation on concrete-filled stainless steel stiffened tubular stub columns. Eng Struct 31(2):300–307
12. Petrus C, Hamida HA, Ibrahim A, Parke G (2010) Experimental behavior of concrete-filled thin-walled steel tubes with tab stiffeners. J Constr Steel Res 66(7):915–922
13. Tao Z, Han LH, Wang ZB (2005) Experimental behaviour of stiffened concrete-filled thin-walled hollow steel structural (HSS) stub columns. J Constr Steel Res 61:962–983
14. Tao Z, Han LH, Wang DY (2007) Experimental behaviour of concrete-filled stiffened thin-walled steel tubular columns. Thin-Walled Struct 45:517–527
15. Thomas J, Sandeep TN (2018) Experimental study on circular CFST short columns with intermittently welded stiffeners. Steel Compos Struct 29(5):659–667
16. IS 1608 (2005) Metallic materials—tensile testing at ambient temperature, Third Revision, Bureau of Indian Standards
17. Lai M, Li C, Ho JCM, Chen MT (2020) Experimental investigation on hollow-steel-tube columns with external confinements. J Constr Steel Res 166:105865
18. Park JW, Hong YK, Hong GS, Kim JH, Choi SM (2011) Design formulas of concrete filled circular steel tubes reinforced by carbon fiber reinforced plastic sheets. Procedia Eng 14:2916–2922

Lightning Source UK Ltd.
Milton Keynes UK
UKHW020737090922
408600UK00005B/488